D1165739

Developments in Precambrian Geology, 15

EARTH'S OLDEST ROCKS

DEVELOPMENTS IN PRECAMBRIAN GEOLOGY
Advisory Editor Kent Condie

Outcrop photograph of Earth's oldest rocks – folded migmatitic orthogneiss of the 4.0–3.6 Ga Acasta Gneiss Complex in the Slave Province, northwestern Canada. The Eo- to Paleoarchean tonalitic protoliths were affected by multiple events of migmatization and deformation from the Paleoarchean through to the Neoarchean. Scale bar in centimetres. Photo by T. Iizuka. (*Frontcover*)

View to the southwest from the northeast part of the 3.8–3.7 Ga Isua greenstone belt, of the Itsaq Gneiss Complex, towards the mountains around Godthabsfjord, West Greenland, in the far distance, over 100 km away. In the foreground, strongly deformed cherty metasedimentary rocks display strong subvertical planar fabrics and steeply-plunging lineations that formed during Neoarchean orogeny. In the middle distance on the left are black-weathering amphibolites derived from basaltic pillow lavas. 3.7 Ga tonalitic gneiss form the light coloured terrain in the centre and right middle distance. For a description of this area, see the paper by Nutman et al. in this volume. Photograph by John S. Myers. (*Backcover*)

Developments in Precambrian Geology, 15

EARTH'S OLDEST ROCKS

Edited by

MARTIN J. VAN KRANENDONK
Geological Survey of Western Australia
Perth, Australia

R. HUGH SMITHIES
Geological Survey of Western Australia
Perth, Australia

VICKIE C. BENNETT
Research School of Earth Sciences
The Australian National University
Canberra, Australia

ELSEVIER

Amsterdam - Boston - Heidelberg - London - New York - Oxford
Paris - San Diego - San Francisco - Singapore - Sydney - Tokyo

Elsevier
Radarweg 29, PO Box 211, 1000 AE Amsterdam, The Netherlands
Linacre House, Jordan Hill, Oxford OX2 8DP, UK

First edition 2007

Notice
No responsibility is assumed by the publisher for any injury and/or damage to persons or property as
a matter of products liability, negligence or otherwise, or from any use or operation of any methods,
products, instructions or ideas contained in the material herein. Because of rapid advances in the
medical sciences, in particular, independent verification of diagnoses and drug dosages should be
made

Library of Congress Cataloging-in-Publication Data
A catalog record for this book is available from the Library of Congress

British Library Cataloguing in Publication Data
A catalogue record for this book is available from the British Library

ISBN: 978-0-444-52810-0
ISSN: 0166-2635

For information on all Elsevier publications
visit our website at books.elsevier.com

Printed and bound in The Netherlands

07 08 09 10 11   10 9 8 7 6 5 4 3 2 1

Working together to grow
libraries in developing countries

www.elsevier.com | www.bookaid.org | www.sabre.org

ELSEVIER   BOOK AID
            International   Sabre Foundation

DEDICATION

M.J. Van Kranendonk would like to dedicate this book to his father, Jan, for dinnertime stories of the stars and planets that inspired him with a lifelong interest in natural science, and to his son, Damian, for continued inspiration on how to live life, and why.

CONTRIBUTING AUTHORS

D. BAKER
  Equity Engineering Ltd., 700-700 West Pender Street, Vancouver, British Columbia, Canada, V6C 1G8

R.L. BAUER
  Department of Geological Sciences, University of Missouri, Columbia, MO 65211, USA
  (bauerr@missouri.edu)

R.W. BELCHER
  Department of Geology, Geography and Environmental Science, University of Stellenbosch, Private Bag X 01,
  Matieland 7602, South Africa

V.C. BENNETT
  Research School of Earth Sciences, Australian National University, Canberra, ACT 0200, Australia (vickie.
  bennett@anu.edu.au)

A.W.R. BEVAN
  Department of Earth and Planetary Sciences, Western Australian Museum, Perth, Western Australia 6000,
  Australia (bevana@museum.wa.gov.au)

M.E. BICKFORD
  Department of Earth Sciences, Heroy Geology Laboratory, Syracuse University, Syracuse, NY 13244-1070,
  USA (mebickfo@syr.edu)

C.O. BÖHM
  Manitoba Geological Survey, Manitoba Industry, Economic Development and Mines, 360-1395 Ellice Ave.,
  Winnipeg MB, Canada, R3G 3P2 (Christian.Bohm@gov.mb.ca)

G.R. BYERLY
  Department of Geology and Geophysics, Louisiana State University, Baton Rouge, LA 70803-4101, USA
  (gbyerly@geol.lsu.edu)

A.J. CAVOSIE
  Department of Geology, University of Puerto Rico, PO Box 9017, Mayagüez, Puerto Rico 00681, USA
  (acavosie@uprm.edu)

K.R. CHAMBERLAIN
  Dept. of Geology and Geophysics, 1000 E. University, Dept. 3006, University of Wyoming, Laramie, WY 82071,
  USA (kchamber@uwyo.edu)

D.C. CHAMPION
  Geoscience Australia, GPO Box 378, Canberra, ACT 2601, Australia (David.Champion@ga.gov.au)

C. CLOQUET
  INW-UGent, Department of Analytical Chemistry, Proeftuinstraat 86, 9000 Gent, Belgium

K. CONDIE

Department of Earth & Environmental Science, New Mexico Institute of Mining & Technology, Socorro, NM 87801, USA (kcondie@nmt.edu)

B. CUMMINS

Moly Mines Pty Ltd, PO Box 8215, Subiaco East, Western Australia 6008, Australia

J.C. DANN

90 Old Stow Road, Concord, MA 01742, USA (jcdann2@yahoo.com)

J. DAVID

GÉOTOP-UQÀM-McGill, Université du Québec à Montréal, C.P. 8888, succ. centre-ville, Montreal, QC, Canada, H3C 3P8

G.F. DAVIES

Research School of Earth Science, Australian National University, Canberra, ACT 0200, Australia (Geoff.Davies@anu.edu.au)

C.Y. DONG

Institute of Geology, Chinese Academy of Geological Sciences, Beijing 100037, China; Beijing SHRIMP Centre, Beijing 100037, China

A. DZIGGEL

Institute of Mineralogy and Economic Geology, RWTH Aachen University, Wüllnerstrasse 2, 52062 Aachen, Germany (adziggel@iml.rwth-aachen.de)

C.M. FEDO

Department of Earth and Planetary Sciences, University of Tennessee, Knoxville, TN 37996, USA (cfedo@utk.edu)

D. FRANCIS

Earth & Planetary Sciences, McGill University and GÉOTOP-UQÀM-McGill, 3450 University St., Montreal, QC, Canada, H3A 2A7

C.R.L. FRIEND

45, Stanway Road, Headington, Oxford, OX3 8HU, UK

A. GLIKSON

Department of Earth and Marine Science and Planetary Science Institute, Australian National University, Canberra, ACT 0200, Australia (Andrew.glikson@anu.edu.au)

W.L. GRIFFIN

Key Centre for the Geochemical Evolution and Metallogeny of Continents (GEMOC), Department of Earth and Planetary Sciences, Macquarie University, NSW 2109, Australia (bill.griffin@mq.edu.au)

T.L. GROVE

Department of Earth, Atmospheric and Planetary Sciences, Massachusetts Institute of Technology, Cambridge, MA 02139, USA (tlgrove@mit.edu)

V.L. HANSEN

Department of Geological Sciences, University of Minnesota Duluth, 231 Heller Hall, 1114 Kirby Drive, Duluth, MN 55812, USA (vhansen@d.umn.edu)

S.L. HARLEY

Grant Institute of Earth Science, School of GeoSciences, University of Edinburgh, Kings Buildings, West Mains Road, UK (Simon.Harley@ed.ac.uk)

R.P. HARTLAUB
   Department of Mining Technology, British Columbia Institute of Technology, 3700 Willingdon Avenue, Burnaby, BC, Canada, V5G 3H2 (rhartlaub@gmail.com)

L.M. HEAMAN
   Department of Earth & Atmospheric Sciences, 4-18 Earth Science Building, University of Alberta, Edmonton, AB, Canada, T6G 2E3 (larry.heaman@ualberta.ca)

A.H. HICKMAN
   Geological Survey of Western Australia, 100 Plain St., East Perth, Western Australia 6004, Australia

H. HIDAKA
   Department of Earth and Planetary Systems Sciences, University of Hiroshima, 1-3-1 Kagamiyama, Higashi-Hiroshima 739-8526, Japan

A. HOFMANN
   School of Geological Sciences, University of KwaZulu-Natal, Private Bag X 54001, 4000 Durban, South Africa (hofmann@ukzn.ac.za)

K. HORIE
   Department of Earth and Planetary Systems Sciences, University of Hiroshima, 1-3-1 Kagamiyama, Higashi-Hiroshima 739-8526, Japan; Department of Science and Engineering, The National Science Museum, 3-23-1, Hyakunin-cho, Shinjuku-ku, Tokyo 169-0073, Japan

D.L. HUSTON
   Geoscience Australia, GPO Box 378, Canberra, ACT 2601, Australia (David.Huston@ga.gov.au)

T. IIZUKA
   Department of Earth and Planetary Sciences, Tokyo Institute of Technology, Ookayama Meguro-ku, Tokyo 152-8551, Japan (tiizuka@eri.u-tokyo.ac.jp)

B.S. KAMBER
   Department of Earth Sciences, Laurentian University, 935 Ramsey Lake Road, Sudbury, ON, Canada, P3E 2C6 (bkamber@laurentian.ca)

N.M. KELLY
   Grant Institute of Earth Science, School of GeoSciences, University of Edinburgh, Kings Buildings, West Mains Road, UK

A.F.M. KISTERS
   Department of Geology, Geography and Environmental Science, University of Stellenbosch, Private Bag X 01, Matieland 7602, South Africa (kisters@sun.ac.za)

T. KOMIYA
   Department of Earth and Planetary Sciences, Tokyo Institute of Technology, Ookayama Meguro-ku, Tokyo 152-8551, Japan

A. KRÖNER
   Institut für Geowissenschaften, Universität Mainz, 55099 Mainz, Germany (kroener@mail.uni-mainz.de)

J. LAROCQUE
   School of Earth and Oceanic Sciences, University of Victoria, P.O. Box 3055, STN CSC, Victoria, BC, Canada, V8W 3P6

D.Y. LIU
   Institute of Geology, Chinese Academy of Geological Sciences, Beijing 100037, China; Beijing SHRIMP Centre, Beijing 100037, China

D.R. LOWE
  Department of Geological and Environmental Sciences, Stanford University, Stanford, CA 94305-2115, USA (lowe@pangea.stanford.edu)

C.P. MARSHALL
  Vibrational Spectroscopy Facility, School of Chemistry, The University of Sydney, NSW 2006, Australia (c.marshall@chem.usyd.edu.au)

S. MARUYAMA
  Department of Earth and Planetary Sciences, Tokyo Institute of Technology, Ookayama Meguro-ku, Tokyo 152-8551, Japan

C. MAURICE
  Bureau de l'exploration géologique du Québec, Ministère des Ressources Naturelles et de la Faune, 400 boul. Lamaque, Val d'Or, QC, Canada, J9P 3L4

T.P. MERNAGH
  Geoscience Australia, GPO Box 378, Canberra, ACT 2601, Australia

F.M. MEYERS
  Institute of Mineralogy and Economic Geology, RWTH Aachen University, Wüllnerstrasse 2, 52062 Aachen, Germany

S.J. MOJZSIS
  Department of Geological Sciences, Center for Astrobiology, University of Colorado, 2200 Colorado Avenue, Boulder, CO 80309-0399, USA (mojzsis@colorado.edu)

P. MORANT
  Anglogold Ashanti, Level 13, St. Martin's Tower, 44 St. George's Terrace, Perth, Western Australia 6000, Australia

J.-F. MOYEN
  Department of Geology, Geography and Environmental Science, University of Stellenbosch, Private Bag X 01, Matieland 7602, South Africa (jfmoyen@wanadoo.fr)

P.A. MUELLER
  Department of Geological Sciences, Box 112120, University of Florida, Gainesville, FL 32611, USA (mueller@geology.ufl.edu)

A.P. NUTMAN
  Beijing SHRIMP Centre, Chinese Academy of Geological Sciences, 26, Baiwanzhuang Road, Beijing 100037, China (nutman@bjshrimp.cn)

J. O'NEIL
  Earth & Planetary Sciences, McGill University and GÉOTOP-UQÀM-McGill, 3450 University St. Montreal, QC, Canada, H3A 2A7 (oneil_jo@eps.mcgill.ca)

S.Y. O'REILLY
  Key Centre for the Geochemical Evolution and Metallogeny of Continents (GEMOC), Department of Earth and Planetary Sciences, Macquarie University, NSW 2109, Australia

A. OTTO
  Institute of Mineralogy and Economic Geology, RWTH Aachen University, Wüllnerstrasse 2, 52062 Aachen, Germany (otto@iml.rwth-aachen.de)

J.A. PERCIVAL
  Geological Survey of Canada, 601 Booth St., Ottawa, Ontario, Canada, K1A 0E8 (joperciv@nrcan.gc.ca)

F. PIRAJNO

Geological Survey of Western Australia, 100 Plain St., East Perth, Western Australia 6004, Australia (franco.pirajno@doir.wa.gov.au)

M. POUJOL

Géosciences Rennes, UMR 6118, Université de Rennes 1, Avenue du Général Leclerc, 35 042 Rennes Cedex, France (Marc.Poujol@univ-rennes1.fr)

O.M. ROSEN

Geological Institute of the Russian Academy of Sciences (RAS), Pyzhevsky per. 7, Moscow, 019017, Russia (roseno@ilran.ru)

M.D. SCHMITZ

Department of Geosciences, Boise State University, Boise, ID 83725, USA (markschmitz@boisestate.edu)

G.A. SHIELDS

Geologisch-Paläontologisches Institut, Westfälische-Wilhelms Universität, Correnstr. 24, 48149 Münster, Germany (gshields@uni-muenster.de)

R.H. SMITHIES

Geological Survey of Western Australia, 100 Plain St., East Perth, Western Australia 6004, Australia (hugh.smithies@doir.wa.gov.au)

C. SPAGGIARI

Geological Survey of Western Australia, 100 Plain St., East Perth, Western Australia 6004, Australia (Catherine.Spaggiari@doir.wa.gov.au)

G. STEVENS

Department of Geology, Geography and Environmental Science, University of Stellenbosch, Private Bag X 01, Matieland 7602, South Africa

R.K. STEVENSON

GÉOTOP-UQÀM-McGill and Département des Sciences de la Terre et de l'Atmosphère, Université du Québec à Montréal, C.P. 8888, succ. centre-ville, Montreal, QC, Canada, H3C 3P8

S.R. TAYLOR

Department of Earth and Marine Sciences, Australian National University, Canberra, ACT 0200, Australia (ross.taylor@anu.edu.au)

O.M. TURKINA

Institute of Geology and Mineralogy, United Institute of Geology, Geophysics and Mineralogy, Siberian Branch of RAS, UIGGM, Koptyug Avenue 3, Novosibirsk, 630090, Russia (turkina@uiggm.nsc.ru)

Y. UENO

Research Center for the Evolving Earth and Planet, Department of Environmental Science and Technology, Tokyo Institute of Technology, Midori-ku, Yokohama 226-8503, Japan (yueno@depe.titech.ac.jp)

M.J. VAN KRANENDONK

Geological Survey of Western Australia, 100 Plain St., East Perth, Western Australia 6004, Australia (martin.vankranendonk@doir.wa.gov.au)

J.W. VALLEY

Department of Geology and Geophysics, University of Wisconsin, 1215 W. Dayton, Madison, WI 53706, USA (valley@geology.wisc.edu)

Y.S. WAN
   Institute of Geology, Chinese Academy of Geological Sciences, Beijing 100037, China; Beijing SHRIMP Centre, Beijing 100037, China

M.J. WHITEHOUSE
   Swedish Museum of Natural History, Box 50007, SE-104 05 Stockholm, Sweden (martin.whitehouse@nrm.se)

S.A. WILDE
   Department of Applied Geology, Curtin University of Technology, PO Box U1987, Perth, Western Australia 6845, Australia (s.wilde@curtin.edu.au)

A.H. WILSON
   School of Geosciences, University of the Witwatersrand, Private Bag 3, 2050 Wits, South Africa (allan.wilson@wits.ac.za)

B. WINDLEY
   Department of Geology, University of Leicester, Leicester, LE1 7RH, UK (brian.windley@btinternet.com)

J.L. WOODEN
   U.S. Geological Survey, Stanford – U.S. Geological Survey Ion Microprobe Facility, Stanford University, Stanford, CA 94305-2220, USA (jwooden@usgs.gov)

J.S. WU
   Institute of Geology, Chinese Academy of Geological Sciences, Beijing 100037, China

S. WYCHE
   Geological Survey of Western Australia, 100 Plain St., East Perth, Western Australia 6004, Australia (stephen.wyche@doir.wa.gov.au)

X.Y. YIN
   Institute of Geology, Chinese Academy of Geological Sciences, Beijing 100037, China; Beijing SHRIMP Centre, Beijing 100037, China

H.Y. ZHOU
   Beijing SHRIMP Centre, Beijing 100037, China

CONTENTS

PREFACE: AIMS, SCOPE, AND OUTLINE OF THE BOOK

MARTIN J. VAN KRANENDONK, R. HUGH SMITHIES AND VICKIE C. BENNETT

The geological history of early Earth holds a certain ineluctable fascination, not just for professional Earth Scientists and geology students, but for scientists in other disciplines, as well as many in the general public. This fascination with early Earth is compelling, not least because we know so little about it, but also because – as with the search for life on ancient Earth and elsewhere in the solar system – it casts light on the fundamental issues of our existence: who are we; how are we here?

To facilitate a better understanding of these questions, we need to know how our home planet formed, what it was like in its early history, how it was able to foster the development of life, and how it evolved into the planet we live on today.

We had two main aims in mind when inviting authors to contribute papers to this book. The first aim, reflected in the main title of the book, was to compile a geological record of Earth's Oldest Rocks, with thorough descriptions of as much of the oldest continental crust as possible, and with a focus on the rocks. The second aim was to gain a better understanding of the tectonic processes that gave rise to the formation and preservation of these oldest pieces of continental crust, and when and how early tectonic processes changed to a plate tectonic Earth operating more or less as we know it today. It is the latter part of this last sentence that explains why 3.2 Ga was chosen as the upper time limit for this book, for this is the time when evidence from geological studies strongly supports the operation (at least locally) of modern-style plate tectonics on Earth (Smithies et al., 2005a). After 3.2 Ga, the geological evidence in support of some form of plate tectonics operating on Earth is compelling, although there were significant differences in how this process operated compared with plate tectonics on Proterozoic to recent Earth, but that is another story (see Van Kranendonk, 2004a, and references therein).

When considering the evolution of early Earth, it is important to keep in mind the concept of Secular Change, and to "... stretch ones' tectonic imagination with respect to non-plate tectonic processes of heat transfer ..., providing for means to test geologic histories against multiple hypotheses, aimed at understanding possible early Earth", as Vicki Hansen so nicely states in her paper on Venus towards the end of this volume. Indeed, we suggest that Secular Change be regarded as a guiding principle for studies of early Earth evolution, in the same way that Lyell's (1758) Principle of Uniformitarianism (the present is the key to the past) has guided our understanding of the more recent geological past, when Earth's primary heat loss mechanism was through plate tectonics.

Secular Change is important for early Earth studies for two main reasons. First, planetary studies have shown that Earth had a violent accretionary history starting at 4.567 Ga, and was a molten ball at 4.50 Ga due to the heat of accretion and heat from the decay of short-lived radiogenic nucleides. This contrasts dramatically with modern Earth, which is differentiated into a core, mantle, crust, hydrosphere and atmosphere, has a rapidly spinning core, a convecting and melting mantle, two types of crust, a rigid lithosphere that is divided into several plates that are moving across the planet's surface through a process we call plate tectonics, is host to a thriving biosphere, and has an oxygen-rich atmosphere. We use this contrast to directly infer secular change. What remains unanswered is the *rate* of secular change, including the rate of growth of continental crust and the time of onset of plate tectonics. Opinions in regard to these issues vary markedly. Whereas some maintain that much of this change occurred very early and that Earth has been operating in a similar fashion since 4.2 Gyr ago (e.g., Cavosie et al., this volume), others maintain that change has occurred more gradually, and that modern plate tectonics did not commence until the Neoproterozoic (Hamilton, 1998, 2003; Stern, 2005; Brown, 2006).

The second main reason why Secular Change is important when considering early Earth is based on geological evidence from Earth's oldest rocks, which shows that there are many differences between early Earth rocks and those of Proterozoic to recent Earth (e.g., Hamilton, 1993, 1998, 2005; Stern, 2005; Brown, 2006). Some of these differences include:

- Archean Earth erupted unique komatiitic magmas (Viljoen and Viljoen, 1969) from a hotter mantle (Herzberg, 1992; Nisbett et al., 1993; Arndt et al., 1998);
- Archean crust is characterised by granite-greenstone terranes, a type of crust found much less commonly in younger terrains;
- Granite-greenstone terranes are characterised by a dome-and-keel architecture that is unique to Archean and Paleoproterozoic crust (e.g., MacGregor, 1951; Hickman, 1984);
- The average composition of Archean continental crust was different (Taylor and McLennan, 1985);
- The average composition of early Archean granitic rocks is dominated by sodic (TTG) compositions, in contrast to the more potassic composition of most younger granitic rocks;
- The composition of Archean TTG is not the same as Phanerozoic adakites formed in subduction zones (Smithies, 2000);
- Archean sedimentary rocks are predominantly chert and banded iron-formation, and generally lack continental-type sedimentary rocks before ~3.2 Ga, although there are local exceptions;
- Sr-isotope data show that the chemistry of Archean seawater was essentially mantle buffered, contrasting with a riverine buffered signature after ~2.7 Ga;
- Many of the characteristic products of subduction are lacking in Archean rocks, including blueschists, ophiolites, and ultra-high pressure metamorphic terranes (Stern, 2005; Brown, 2006), and accretionary complexes with exotic blocks in zones of tectonic melange (McCall, 2003);
- Ophiolites >1 Ga are fundamentally different than younger ophiolites, according to Moores (2002).

These differences tell us that early Earth was a vastly different planet than that of to-day, primarily due to a higher mantle temperature. Follow-on effects from this include a higher geothermal gradient, which in turn resulted in greater degrees of partial melting of upwelling mantle, a thicker, but softer crust, and a softer, weaker lithosphere. A more detailed review of these differences is presented in the final paper of this book.

A note regarding terminology. For the purposes of this book, we have adopted the IUGS International Commission on Stratigraphy convention for sub-divisions of the Archean Eon into the Neoarchean Era (2.5–2.8 Ga), Mesoarchean Era (2.8–3.2 Ga), and Paleoarchean Era (3.2–3.6 Ga) (Gradstein et al., 2004). We have also used the Eoarchean Era for rocks older than 3.6 Ga as suggested by these authors, but have placed a lower limit on this sub-division at 4.0 Ga and refer to the period of time older than this as the Hadean Eon (4.0–4.567 Ga).

This book is organised into eight parts, including an Introduction, five parts describing the geology of Earth's oldest rocks, a part on early life, and a final part on the tectonics of early Earth. In Part 1, **Brian Windley** provides an overview of the history of discovery of ancient rocks on Earth, which started with publications in 1951 and was followed by seminal discoveries using advanced analytical techniques up to the present day. This is followed by **Kent Condie's** overview of the distribution of ancient rocks on Earth.

Parts 2–6 describe the geology of Earth's oldest rocks, and are divided on the basis of successive stages of early Earth evolution. Part 2 outlines the beginnings of Earth history, with a review of planetary accretion processes in the formation of the Earth and Moon by **Stuart Ross Taylor**, and an investigation of early solar system materials as represented by the meteorite record on Earth, by **Alex Bevan**. This is followed by a theoretical con-sideration of the dynamics of the mantle of early Earth by **Geoff Davies** and a review of the evidence in favour of an early terrestrial protocrust by **Balz Kamber**. Thereafter are two papers by **Cavosie and others** and **Stephen Wyche** that review the distribution and characteristics of Eoarchean zircon grains from Western Australia and their significance in terms of early Earth evolution.

Part 3 presents a series of papers that describe the geology of Eoarchean gneiss com-plexes from different cratons around the world. These include a description of the oldest rocks in the world from the Acasta Gneiss Complex in the Slave Craton in northwestern Canada by **Iizuka and others**. Other ancient, high-grade and strongly deformed rocks are described from Antarctica by **Harley and Kelly**, from the North China Craton by **Liu and others**, and from the Narryer Terrane of the Yilgarn Craton in Western Australia by **Wilde and Spaggiari**. **O'Neil and others** describe the ancient supracrustal rocks of the Nuvvuagittuq Greenstone Belt in the Superior Province, Canada. These are roughly the same age, or slightly older, than the Isua Greenstone Belt in southern West Greenland, de-scribed by **Nutman and others**, who also describe the rest of the Itsaq Gneiss Complex in which these ancient supracrustal rocks lie.

Separate chapters are devoted to the Paleoarchean development of the Pilbara (Part 4) and Kaapvaal (Part 5) Cratons, as these areas represent the best preserved, oldest rocks on Earth and are well studied. In Part 4, a review of the lithostratigraphy and structural geology of ancient rocks of the Pilbara Craton by **Van Kranendonk and others** is followed by more detailed geochemical descriptions of the geochemistry of felsic volcanic rocks by **Smithies and others**, and of granitoid rocks by **Champion and Smithies**. **Huston and others** review the genesis and tectonic environments of Paleoarchean mineral deposits in the Pilbara Craton. Combined, these papers provide several different lines of evidence for the development of early, thick continental crust and a complimentary buoyant, depleted, residual lithospheric keel through multiple mantle plume events prior to 3.2 Ga.

The papers in Part 5 review the geology of the Kaapvaal Craton, including overviews of the geochronological data throughout the craton by **Poujol**, and the geology of the well-studied Barberton Greenstone Belt by **Lowe and Byerly**. More detailed studies include those by **Kröner**, who describes the Ancient Gneiss Complex of southern Africa, **Dann and Grove** on the volcanology of flow fields in the Barberton Greenstone Belt, **Hofmann and Wilson** on the geology of chert units in the Nondweni Greenstone Belt, **Moyen and others** on the geochemistry of early granitic plutons in, and adjacent to, the Barberton Greenstone Belt, and **Stevens and Moyen** on the metamorphism of this same area. **Dziggel and others** review the tectono-metamorphic controls on gold mineralisation.

Part 6 describes Paleoarchean gneiss terranes from a number of cratons around the world and shows that formation of this type of crust was continuous through the Hadean and early Archean. **Bickford and others** describe Paleoarchean gneisses from the Minnesota River Valley and from northern Michigan in the southwestern part of the Superior Craton, in the United States of America, whereas **Böhm and others** describe Paleoarchean rocks from the Assean Lake Complex in the northwestern part of the Superior Craton, in Canada. **Chamberlain and Mueller** describe ancient rocks and zircons from the Wyoming Craton in the United States of America and **Rosen and Turkina** present an overview of ancient rocks in the Siberian Craton.

Part 7 presents several papers describing the possible evidence for life on early Earth and the geological settings in which it is found. **Whitehouse and Fedo** review the evidence for early life from the Itsaq Gneiss Complex in southern West Greenland, and **Van Kranendonk** does the same for the Pilbara Craton, Australia. **Marshall** provides a review of organic geochemical data from carbonaceous Pilbara cherts and **Ueno** provides a detailed account of the stable carbon and sulphur isotopic evidence for life from the 3490 Ma Dresser Formation in the Pilbara. At the end of this chapter, **Mojzsis** provides a comprehensive review of the sulphur system on early Earth and **Shields** reviews the marine carbonate and chert isotope records, and both authors discuss the implications of the stable isotopic data in terms of evidence for early life on our planet.

The concluding Part 8 comprises several papers on the tectonics of early Earth. **Hansen** presents a detailed description of the geology of Venus and suggests features of that planet that may be used as analogues for early Earth. **Griffin and O'Reilly** describe the composition and processes that lead to the formation of subcontinental mantle lithosphere on early Earth, and conclude that mantle plumes played a significant role in their development. The role of mantle plumes in the construction of continental crust on Earth through time is reviewed by **Pirajno**. **Glikson** reviews the evidence for, and possible geodynamic implications of, asteroid impacts on early Earth tectonics. **Percival** provides an account of the tectonic context of ancient rocks within the Superior Craton, and how they were incorporated into the craton during Neoarchean terrane accretion. The final paper by **Van Kranendonk** provides a summary of the information presented in this book together with a tectonic model for early Earth and some suggestions in regard to divisions of the early Precambrian timescale.

ACKNOWLEDGEMENTS

MVK would like to thank the Series Editor, Kent Condie, for the invitation to write/compile this book. Kath Noonan and Nell Stoyanoff of the Geological Survey of Western Australia (GSWA) are thanked for help with formatting the manuscript. Michael Prause and Suzanne Dowsett (GSWA) helped with drafting. This book is published with permission of the Executive Director of the Geological Survey of Western Australia.

PART 1

INTRODUCTION

Earth's Oldest Rocks
Edited by Martin J. Van Kranendonk, R. Hugh Smithies and Vickie C. Bennett
Developments in Precambrian Geology, Vol. 15 (K.C. Condie, Series Editor)   3
© 2007 Elsevier B.V. All rights reserved.
DOI: 10.1016/S0166-2635(07)15011-8

Chapter 1.1

OVERVIEW AND HISTORY OF INVESTIGATION OF EARLY EARTH ROCKS

BRIAN WINDLEY

Department of Geology, University of Leicester, Leicester, LE1 7RH, UK

There were many studies of Archaean rocks in the 1960s and earlier. Some described volcanic-dominated belts with important intrusive granites, such as the classic study of Rhodesian (now Zimbabwe) belts by MacGregor (1951). Others described belts of gneisses, migmatites and granites, as in NW Scotland (Sutton and Watson, 1951). The relations between such high-grade and low-grade Archaean belts were not well under-stood at the time, there being two main ideas: the gneisses formed an older basement to the greenstone belts (e.g., Bliss and Stidolph, 1969), or the gneisses were coeval roots of green-stone belts (Glikson, 1976). Windley and Bridgwater (1971) first pointed out that there are two fundamentally different types of Archaean terranes, granulite-gneiss and greenstone-granite belts, which have different protoliths and represent different erosional levels of the crust. Although there are several varieties of these two types, the general subdivision still holds and is useful for descriptive purposes (e.g., Kröner and Greiling, 1984; de Wit and Ashwal, 1997). The second major advance in understanding the broad tectonic develop-ment of Archaean rocks came much later with the realisation that there are two different types of modern orogens, namely collisional orogens like the modern Himalaya, and ac-cretionary orogens like those on the margins of the present Pacific Ocean, in particular in Japan, Alaska and Indonesia. The importance for Archaean studies is that most greenstone belts seem to be accretionary orogens (Sengör and Okurogullari, 1991), and that there are no significant Archaean continent–continent collisional orogens because continental blocks did not evolve until 2.7 Ga.

1.1-1. GRANULITE-GNEISS BELTS

Hans Ramberg (1952) published his classic textbook, The Origin of Metamorphic and Metasomatic Rocks, the field geology of which was based on his experience of the Ar-chaean high-grade gneissic rocks of west Greenland. In it, he made the prescient comment that the problem with these rocks is that they are all mutually conformable. Although he made the right observation, the answer to the problem did not come for several decades. The Scourian complex of northwest Scotland was the first Archaean granulite-gneiss belt to be analyzed in detail. Sutton and Watson (1951) demonstrated that amphibolite dykes

can be used to separate older from younger tectonic events. The dykes transect the foliation of earlier tonalitic gneisses and yet are cut by younger tonalitic gneisses. In 1968, the
oldest rock in west Greenland was a gneiss from Godthaab that had a K/Ar cooling age
of 2710 ± 130 Ma (Armstrong, 1963). Vic McGregor (1968, 1973) in his classic study of
the gneisses in west Greenland made the biggest break-through in the history of Archaean
studies, when he built upon the work of Sutton and Watson (1951) by using amphibolite
dykes to unravel the history of the gneisses in the Godthaab (Nuuk) region. His originality
was in realising that gneisses with no dykes were not ones that had not been intruded by
dykes, as most people thought at the time, but were younger tonalites that were intruded
into the dykes and were later deformed into gneisses. He called the main units the Amitsoq gneisses, the Ameralik dykes, the Malene supracrustals, and the Nuuk gneisses. In one
stroke, based solely on field work, he single-handedly broke through and pushed back one
thousand million years of Earth history, because he predicted, starting with the 2710 Ma
cooling age and estimating the ages of the many earlier events he identified, that the earliest rocks should be 3600–3660 Ma old (McGregor, 1968). Later, Stephen Moorbath dated
some samples that McGregor had sent him and produced isotopic confirmation that the
Amitsoq gneisses had a Pb–Pb whole-rock isochron age of 3620 ± 100 Ma (Black et al.,
1971).

An important development in the study of Archaean granulite-gneiss belts was the realisation by Bridgwater et al. (1974) that the gneisses in west Greenland had undergone
horizontal shortening. This paved the way for plate tectonic models to be applied to these
belts. However, this has proved difficult because of the lack of diagnostic plate tectonic
indicators in such deep crustal rocks, and because few modern orogenic belts have been
eroded or tilted to expose their deep crustal roots and so to provide a modern analogue.
The most that has been achieved so far is the recognition of different types of thrust-bound
gneissic terranes based on differences in lithological make-up, metamorphic grade and
isotopic history (Friend et al., 1987). Nevertheless, many orthogneisses do have the composition of tonalite-trondhjemite-granodiorite (TTG) that is common in Mesozoic–Cenozoic
batholiths in the Andes and elsewhere around the Pacific, and accordingly Nutman et al.
(1993b) proposed that subduction of mafic oceanic crust gave rise to continental crust
composed of TTGs that were converted to high-grade gneisses during subsequent collision
tectonics.

The Isua supracrustal belt is a strip of 3.8–3.7 Ga volcanic-sedimentary supracrustal
rocks within the orthogneisses of west Greenland. Allaart (1976) produced the first geological description and map of the belt, and made the observation that conglomerates indicate
the presence of water on Earth at that time. Shigenori Maruyama and his group made two
expeditions to Isua that resulted in two abstracts in 1991 and 1994 that led to a major paper by Komiya et al. (1999). Their main points are: there is an upwards stratigraphy from
low-K tholeiitic, pillow-bearing basalts, to cherts that contain no continent- or arc-derived
material, to mafic turbidites and conglomerates. This association is very similar to "ocean
plate stratigraphy" that is characteristic of Phanerozoic, circum-Pacific accretionary orogens, like Japan, that record the travel history of an oceanic plate from ridge to trench.
The discovery by Appel et al. (2001) of remnants of a sea-floor hydrothermal system as-

sociated with the pillow basalts at the time of lava eruption confirms the existence of an oceanic setting for these early rocks.

High-grade gneisses form the oldest Archaean rocks in the world. These are the Acasta gneisses of NW Canada that have a U/Pb age of 3.962 Ga on tonalitic gneiss (Bowring et al., 1989a); this was indeed a remarkable discovery.

1.1-2. GREENSTONE-GRANITE BELTS

Greenstone-Granite Belts

Many studies were made of gold deposits in greenstone-granite belts in the nineteenth century, which with hindsight is not surprising because they are comparable to the gold deposits in circum-Pacific accretionary orogens (like Japan) that were the source of the wealth of the Shoguns.

In the pre-mobilist (plate tectonic) era, greenstone belts were thought to be geosynclines (e.g., Glikson, 1970), to contain layer-cake stratigraphy like Mesozoic basins (e.g., Zimbabwe; Wilson, 1981), and to have the structure of synclinal basins or keels formed by gravity-induced sagduction between rising diapers of mobilised basement or domes of granitic batholiths (Anhaeusser et al., 1981; Glikson, 1972). However, several developments were to change ideas on their make-up and structure, and in turn they have transformed the world's perspective on the style and course of early Earth evolution.

1. Thrusts, recumbent nappes, structural repetitions, imbrications and inverted stratigraphic successions were discovered that indicated horizontal shortening rather than vertical movements in: Barberton (de Wit, 1982), the Yilgarn (Spray, 1985), Pilbara (Bickle et al., 1980; Boulter et al., 1987), Zimbabwe (Coward et al., 1976; Stowe, 1984), the Superior Province (Poulsen et al., 1980), and India (Drury and Holt, 1980).
2. The application of plate tectonic principles (that require horizontal translation) led to a major conceptual evaluation of all aspects of greenstone belt geology with the result that nowadays most greenstone belts are thought to contain tectonic slices of rocks derived from oceanic crust, volcanic arcs, or oceanic plateaux. Some of the first to apply such ideas to Archaean rocks were Goodwin and Ridler (1970) to the Abitibi belt in Canada, and Talbot (1973) and Rutland (1973) to Australian greenstone belts.
3. The development of the XRF spectrometer and later instruments enabled the major and trace element geochemistry of magmatic rocks to be compared with modern analogues formed in plate tectonic regimes. For example: MORB-like tholeiites occur in the Abitibi belt (Laflèche et al., 1992). The first Archaean island arc was recognized by Folinsbee et al. (1968) in Canada. Since then, many studies have demonstrated a remarkable similarity in the geochemical signature of Archaean arc rocks and their modern equivalents. Very influential at the time was the paper by Tarney et al. (1976) who compared greenstone belts and their geochemistry with the Rocas Verdes marginal basin in southern Chile. High-Mg andesites and shoshonites occur in many Archaean belts, including the Superior Province (Stern and Hansen, 1991); such rocks, like the sanukitoids in Japan, are always associated with subduction zones in the modern Earth.

In recent years there has been (rightly or wrongly!) a flood of papers on Archaean back-arc basins based almost solely on geochemical parameters.

4. The advent of precise geochronological techniques, in particular U-Pb zircon geochronology with the SHRIMP and ICP-MS, has enabled the isotopic age to be determined of minerals and rocks, which is essential for discovery of the earliest rocks, and for development of the rock record in a plate tectonic framework. Examples of isotopic-driven research today are: Hadean zircons, and >3.5 Ga rocks of the Acasta gneisses, Pilbara Craton, Isua greenstone belt, Akilia succession (North Atlantic Craton), and Barberton Greenstone Belt.

5. Geophysical data have provided a three-dimensional constraint, and indeed confirmation of, the shape and structure of many greenstone belts: for example, gravity data indicate that the Barberton Greenstone Belt (Kaapvaal Craton, southern Africa) is not a syncline, but a dipping, shallow, tabular slab (de Wit et al., 1987). Lithoprobe seismic profiles that transect the Abitibi greenstone belt (Superior Craton, North America) demonstrate that it forms a 6 to 8 km thick, shallow-dipping slab intercalated at the surface with allochthonous, thrust-imbricated rock units that are consistent with oblique subduction and imbrication (Ludden and Hynes, 2000).

1.1-3. THE HADEAN

Defined as that period of time between the formation of the planet at ~4.567 Ga and the start of the Archean at 4.0 Ga, the Hadean contained a paucity of information on the earliest Earth until Froude et al. (1983) reported that ~3.3 Ga quartzitic metasediments in the Murchison district in Western Australia contained 4 detrital zircons (out of 102 analyzed grains) with a U-Pb age of 4.1–4.2 Ga. Since then, some 70,000 zircons from these unique metasedimentary rocks have been dated (T.M. Harrison, pers. comm.), particularly from the Jack Hills supracrustal belt, in an avalanche of research that led to the discovery of 4.4 Ga detrital zircons (Wilde et al., 2001); the oldest, well-constrained isotopic age on the planet. However, further oxygen and hafnium isotopic data from these zircons has led to considerable controversy about the geological conditions on the early Earth. Current interpretations of the isotopic data suggest the presence of liquid water at the Earth's surface at 4.3 Ga (Mojzsis et al., 2001), when the early Earth was cool (Valley et al., 2002), and that continental crust formed by 4.4–4.5 Ga and was recycled into the mantle in a process akin to later plate tectonics (Harrison et al., 2005). In contrast, Nemchin et al. (2006) questioned such a model of an early Earth with cool conditions conducive to the development of oceans.

1.1-4. CONCLUSIONS AND IMPLICATIONS

The history of research on early Earth rocks reveals a secular change in ideas applied to the origin of tectonic belts, from pre-mobilist (plate tectonic) geosynclinal theory to initial, simplistic plate tectonic ideas, to increasingly sophisticated plate tectonic models. Probably the most surprising discovery has been that, although the secular decrease with time of radioactive heat production has been confirmed and further constrained, the geological record of rock associations, their geochemical signatures, and structural relationships has been found to be more similar to, rather than very different from, that of modern plate tectonic equivalents.

Abundant data accumulated over several decades indicate that Archaean rocks worldwide occur in accretionary orogens that formed by subduction–accretion processes. We have been through a recent phase when many similarities have been recorded between distinctive Archaean rocks and their modern analogues, diagnostic of certain types of plate tectonic regime. In the future, it will be more important to document dissimilarities with the modern record (e.g., different proportions of distinctive rock associations, different geochemical signatures) and from these work out the ambient plate conditions and factors responsible for them (i.e., was there a shallow or steep subduction angle, fast or slow subduction, or a more primitive, juvenile form of proto-plate tectonics than now?). An important part of this evaluation must be to define how, and to what extent, secular changes in definable parameters – such as the long-term decrease of the potential temperature of the mantle, and changes in, for example, plate rigidity, mantle composition, and sea-water chemistry – have affected, or contributed to, the underlying plate tectonic regimes. With this approach it should be possible to produce, one day, a viable paradigm for the tectonic evolution of the early Earth.

Earth's Oldest Rocks
Edited by Martin J. Van Kranendonk, R. Hugh Smithies and Vickie C. Bennett
Developments in Precambrian Geology, Vol. 15 (K.C. Condie, Series Editor)   9
© 2007 Elsevier B.V. All rights reserved.
DOI: 10.1016/S0166-2635(07)15012-X

Chapter 1.2

THE DISTRIBUTION OF PALEOARCHEAN CRUST

KENT CONDIE

Department of Earth & Environmental Science, New Mexico Institute of Mining & Technology, Socorro, NM 87801, USA

1.2-1. EARTH'S OLDEST ROCKS AND MINERALS

The oldest preserved rocks occur as small blocks tectonically incorporated within younger orogens, most commonly Neoarchean orogens (Fig. 1.2-1). These preserved blocks are generally <500 km across and are separated from surrounding crust by shear zones. Although the oldest known rocks on Earth are about 4.0 Ga, the oldest minerals are detrital zircons from the ca. 3.0 Ga Jack Hill and Mt. Narryer quartzites in Western Australia. These detrital zircons have U/Pb SHRIMP ages ranging from about 4.4 to 3.5 Ga, although only a small fraction of the zircons are >4.0 Ga (Froude et al., 1983; Amelin et al., 1999; Wilde et al., 2001; Harrison et al., 2005; Cavosie et al., this volume). One zircon, measuring 220 by 160 microns, with internal complexities or inclusions, has a concordant $^{207}Pb/^{206}Pb$ age of 4404 ± 8 Ma, which is the oldest reported mineral age from Earth and interpreted as the age of crystallization of this zircon (Wilde et al., 2001). Rare earth element (REE) distributions in this zircon show enrichment in heavy REE, a positive Ce anomaly, and a negative Eu anomaly, indicating zircon crystallization from an evolved granitic melt. Coupled with oxygen isotope results from this sample, these data (Mojzsis et al., 2001) suggest that the granitic melt was produced by partial melting of older crust, either continental crust or hydrothermally altered oceanic crust (see also Wilde et al., 2001).

The oldest isotopically dated rocks on Earth are the *Acasta gneisses* in the Slave Craton of NW Canada (Fig. 1.2-1; see Iizuka et al., this volume). These 4.03 to 3.96 Ga gneisses are a heterogeneous assemblage of highly deformed TTG (tonalite-trondhjemite-granodiorite) complexes, tectonically interleaved on a centimeter scale with amphibolite, ultramafic rocks and pink granites (Bowring et al., 1989a; Bowring, 1990; Bowring and Williams, 1999). At a few locations, metasedimentary rocks comprise part of the complex and include calc-silicates, quartzites, and biotite-sillimanite schists. Acasta amphibolites appear to represent metamorphosed basalts and gabbros, many of which are deformed dykes and sills. Some components of the complex, especially pink granites, have ages as low as 3.6 Ga, and thus it would appear that this early crustal segment evolved over about 400 My and developed a full range in composition of igneous rocks from mafic, to K-rich felsic types. Because of the severe deformation of the Acasta gneisses, the original field relations between the various lithologies are not well known. However, the chemical com-

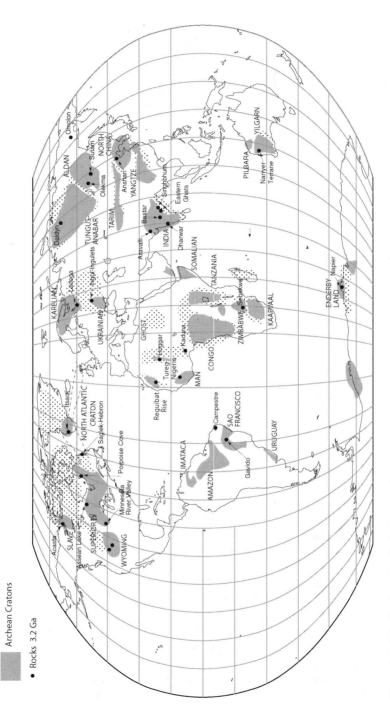

Fig. 1.2-1. Distribution of early Archean rocks.

positions of the Acasta rocks are very much like less deformed Archean greenstone-TTG assemblages elsewhere.

1.2-2. OCCURRENCES OF MAJOR PALEOARCHEAN ROCKS

The oldest rocks in Antarctica occur in the *Napier complex* in Enderby Land (Fig. 1.2-1: see Harley and Kelly, this volume). Some TTG in this complex in the Mt. Sones and Fyfe Hills areas have U-Pb zircon emplacement ages of 3.95 to 3.8 Ga, thus being comparable in age to some of the Acasta gneisses in Canada (Black et al., 1986; Sheraton et al., 1987). The Napier complex comprises chiefly TTG, some of which has sedimentary protoliths. Prior to widespread granulite-grade metamorphism at 3.1 Ga, sediments (pelites, quartzites) and volcanic rocks were deposited on the complex and mafic dykes were intruded (Sheraton et al., 1987). Thermobarometry indicates that the granulite-grade metamorphism records temperatures of 900–950 °C and pressures of 6–10 kbars. Periods of intense deformation and metamorphism are also recorded in the Napier complex by zircon growth at about 2.9 and 2.45 Ga.

The largest and best-preserved fragment of early Archean continental crust is the *Itsaq Gneiss Complex* of the North Atlantic Craton in Southwest Greenland (Nutman et al., 1996, 2002a, this volume; Friend and Nutman, 2005b) (Fig. 1.2-1). In this area, three terranes have been identified, each with its own tectonic and magmatic history, until their collision at about 2.7 Ga (Friend et al., 1988). The Akulleq terrane is dominated by the 3.9–3.8 Ga Amitsoq TTG complex, which underwent high-grade metamorphism at 3.6 Ga. The Akia terrane in the north comprises 3.2–3.0 Ga tonalitic gneisses that were deformed and metamorphosed at 3.0 Ga. The Tasiusarsuaq terrane, dominated by 2.9–2.8 Ga rocks, was deformed and metamorphosed when all of the terranes collided in the Neoarchean. Although any single terrane records <500 My of pre-collisional history, collectively the terranes record over 1 Gy of history before their amalgamation in the Neoarchean. Each of the terranes also contains remnants of highly deformed supracrustal rocks. The most extensively studied is the Isua supracrustal belt in the Isukasia area in the northern part of the Akulleq terrane (Myers, 2001; see Nutman et al., this volume). Although highly altered by submarine metasomatism, this succession comprises basalts and komatiites and intrusive ultramafic rocks interbedded with banded iron formation, intrusive sheets of tonalite and granite, and mafic volcanigenic turbidites (Rosing et al., 1996). Recent remapping of the Isua succession suggests that at least some of the schists are highly deformed tonalitic gneisses or pillow basalts (Fedo et al., 2001; Myers, 2001). Carbonate rocks are now considered to be mostly, or entirely, metasomatic in origin and some may represent the products of seafloor alteration (Myers, 2001).

When the Labrador Sea opened in the Tertiary, it split the North Atlantic Craton, and a remnant of the craton is now preserved along east coast of Labrador. The early Archean rocks in this area comprise the *Saglek–Hebron block* (Fig. 1.2-1). They are divided into the ca. 3.78 Ga Nulliak supracrustal assemblage, the widespread ca. 3.73 Ga Uivak I TTG gneisses, and the Uivak II augen gneisses (Bridgwater and Schiotte, 1990). Inherited zircons in Uivak I gneisses record ages between 3.86 and 3.73 Ga, and these were derived by rapid recycling and deposition of sediments derived largely from 3.86–3.78 Ga crustal sources. Widespread metamorphism and associated granitoid emplacement are recorded in Uivak gneisses at about 3 Ga, and extensive granulite-facies metamorphism occurred as a sequence of closely spaced events between 2.8 and 2.7 Ga.

The *Porpoise Cove supracrustal sequence* in northeastern Quebec is the oldest known greenstone in the Superior Province, at about 3.8 Ga, and is comparable in age to the Isua supracrustal belt (ca. 3.8–3.7 Ga) in southwestern Greenland (Fig. 1.2-1: see O'Neil et al., this volume). The greenstone succession consists of two assemblages (David et al., 2003). The first assemblage includes conglomerates, garnet paragneisses, banded iron formations, intermediate to felsic tuffs, and a few amphibolite horizons, whereas the second assemblage contains volcanic rocks (chiefly basalts), mafic to intermediate tuffs, ultramafic rocks, gabbros and iron formations. Rocks of the Porpoise Cove sequence are strongly deformed and transposed, and metamorphosed to upper amphibolite facies. U/Pb isotopic analyses of zircons recovered from a felsic tuff yield an age of emplacement of 3825 ± 16 Ma, and Sm-Nd isotopic analyses give a model age of 3.9 Ga and $E_{Nd}(T)$ of +2.2.

The oldest rocks of the North China Craton occur in the *Anshan* area, where some deformed granitoids have U/Pb zircon ages of 3.8 Ga (Song et al., 1996; Wan et al., 2005; Liu et al., this volume). Other trondhjemitic gneisses in this area range in age from 3.3 to 2.96 Ga, with the youngest, relatively undeformed granites emplaced at 2.47 Ga. All of the 3.8 Ga TTG in the Anshan area exhibit Nd model ages (T_{DM}) ranging from 3.9 to 2.8 Ga (Wan et al., 2005). Nd isotopic data suggest that the <3.8 Ga granitoids were produced, at least in part, by remelting of older LREE-enriched sources, perhaps similar to the exposed remnants of the 3.8 Ga rocks. The Chentaigou supracrustals, which comprise layered mafic and ultramafic rocks, banded iron formation and siliceous layered rocks of probable volcanogenic origin, have zircon ages suggesting an eruption/deposition age of 3.36 Ga. Detrital zircons from fuchsitic quartzites near Caozhuang yield ages ranging from about 3.85 to 1.89 Ga, indicating that some crustal sources in this province exceeded 3.8 Ga in age (Liu et al., 1992, this volume).

The *Narryer Terrane* in the NW part of the Yilgarn Craton, Western Australia contains widespread areas with rocks in excess of 3.0 Ga (Myers, 1988; Wilde and Spaggiari, this volume). The terrane is composed chiefly of granulite-facies gneisses (Narryer Gneiss Complex), of monzogranitic and syenogranitic compositions, that formed between 3.7 and 3.4 Ga (Muhling, 1990; Nutman et al., 1991; Pidgeon and Wilde, 1998). It also includes quartzites and other siliceous, largely detrital, metasedimentary rocks from which the oldest detrital zircons on Earth have been extracted, both at Mt. Narryer and at Jack Hills (see Cavosie et al., this volume). The entire block was deformed and metamorphosed to granulite grade at 3.3 Ga. Thermobarometry suggests metamorphic temperatures of 750–850 °C and 7–10 kbar (Muhling, 1990). However, the main deformation and metamorphism occurred at 2.7–2.6 Ga, when the Narryer and Youanmi Terranes collided.

The northern *Pilbara Craton* in Western Australia (Fig. 1.2-1) comprises five separate terranes, each with unique stratigraphy and deformational histories (Van Kranendonk et al., 2002, 2007, this volume): the East Pilbara Terrane (3.52–3.17 Ga) that represents the ancient nucleus of the craton; the West Pilbara Superterrane (3.27–3.11 Ga), which comprises three distinct terranes; and the Kuranna Terrane ($\leqslant$3.29 Ga). Each of these terranes has a distinctly different geologic history until their amalgamation at about 3.07 Ga. The oldest supracrustal rocks occur in the East Pilbara Terrane and consist of four demonstrably autochthonous groups that comprise the 3.52–3.17 Ga Pilbara Supergroup. This group is one of the best preserved early Archean greenstone successions, and chert layers within the sequence contain the oldest evidence for life in the form of macroscopic stromatolites, carbonaceous materials of probable biogenic origin, and putative microfossils (Walter et al., 1980; Schopf, 1993; Buick et al., 1995; Brasier et al., 2002; Marshall, this volume; Ueno, this volume; Van Kranendonk, this volume). The Warrawoona Group represents ~100 My of continuous volcanism from a series of mantle plume events (Hickman and Van Kranendonk, 2004; Van Kranendonk and Pirajno, 2004; Smithies et al., 2005a). Subsequent groups represent the eruptive products of discrete, younger mantle plume events that, together with the earlier events, led to the formation of thick crust on a severely depleted, buoyant, lithospheric keel (Van Kranendonk et al., 2002, 2007, this volume; Smithies et al., 2005a).

The *Kaapvaal Craton* in southern Africa assembled between 3.0 and 2.7 Ga and contains rocks as old as 3.7 Ga (Poujol et al., 2003; Schmitz et al., 2004; Poujol, this volume; Kröner, this volume) (Fig. 1.2-1). U/Pb zircon ages in the Ancient Gneiss Complex in Swaziland range from about 3.64 to 3.2 Ga and Nd model ages suggest that some crust may have formed in this region by 3.8 Ga (Kröner, this volume). Recent Nd model ages from the 3.28 Ga Sand River gneisses in the Limpopo belt in southern Africa indicate a crustal residence age for their protolith of 3.7 to 3.5 Ga (Zeh et al., 2004). New U/Pb zircon ages from the western Kaapvaal Craton reveal that two older (3.7 to 3.1 Ga) continental terranes, the Kimberley and Witwatersrand blocks, were juxtaposed by a significantly younger block between 2.93 and 2.88 Ga (Schmitz et al., 2004). Geological evidence indicates that convergence was accommodated beneath the Kimberley block, culminating in collisional suturing in the vicinity of the present-day Colesberg magnetic lineament. The timing of these convergent margin processes also correlates with the peak in Re-Os age distributions in mantle xenoliths, eclogites, and eclogite-hosted diamonds from the Kimberley block (Shirey et al., 2004). These results support the petrogenetic coupling of continental crust and lithospheric mantle throughout the Archean in southern Africa.

The Barberton Greenstone Belt in the eastern part of the Kaapvaal Craton is one of the most studied Early Archean greenstone belts (Lowe and Byerly, this volume; Dann, this volume; Moyen et al., this volume; Stevens et al., this volume; Dziggel et al., this volume). Together with coeval TTG plutons, the Barberton succession formed at 3.55 to 3.2 Ga (Kamo and Davis, 1994; Kröner et al., 1996). It is interpreted to include four tectonically juxtaposed terranes, with similar stratigraphic successions in each terrane (Lowe, 1994, 1999). Each succession begins with submarine basalts and komatiites of the Komati Formation, followed by an Archean mafic plain succession that could represent remnants of an oceanic plateau. Overlying the mafic plain succession are the Hooggenoeg and Kromberg Formations, a suite of felsic to basaltic submarine volcanics, fine grained volcaniclastic clastic sedimentary rocks, and cherts, possibly representing an oceanic arc. The terminal Moodies Group, which includes syn-orogenic sediments, may have been deposited during amalgamation of the four terranes just after 3.2 Ga.

Rocks as old as 3.5 Ga are now well documented in the central and southern parts of the Zimbabwe Craton in southern Africa (Fig. 1.2-1). Os isotope data and detrital zircon ages require continental crust at least 3.8 Ga, with a culminating event at about 3.2 Ga (Dodson et al., 1988; Nagler et al., 1997; Horstwood et al., 1999). Results suggest that the *Sebakwe protocraton*, which includes most of the central part of the Zimbabwe Craton, formed at 3.45 to 3.35 Ga. This protocraton was stabilized by widespread intrusion of granites at about 3.35 Ga.

Early Archean rocks have been reported at several localities in West Africa (Fig. 1.2-1). Although not precisely dated, blocks of early Archean rocks occur in the Leo Rise area, which is largely a region of Birimian juvenile crust. In the Archean *Man Craton*, however, detrital zircon ages suggest that formation of this craton began by 3.3–3.2 Ga (Kouamelan et al., 1997). SHRIMP zircon ages from gneisses and granites in the Man Craton in Guinea range from 3.542 to 3.53 Ga (Thieblemont et al., 2001). Evidence of early Archean rocks also comes from the *Hoggar Massif* in the Tuareg shield in southern Algeria, although dating is less precise. Nd model ages of 2.7 Ga gneisses in the Gour Oumelalen area suggest protolith ages in excess of 3.0 Ga (Peucat et al., 2003). SHRIMP zircon ages from orthogneisses of the Amsaga terrane in the *Reguibat Rise* in western Mauritania (Fig. 1.2-1) clearly indicate the presence of early Archean crust, with a range from 3.51 to 3.42 Ga, and a metamorphic overprint at about 2.7 Ga (Potrel et al., 1996).

U/Pb analyses of single zircon grains and on zircon fractions from a granodioritic or-thogneiss at *Kaduna* in northern Nigeria show early Archaean components (Fig. 1.2-1). One zircon group records an age of 3.04 Ga, which is interpreted as the age of emplace-ment and crystallization of the granodioritic magma and this age is confirmed by core-free abraded zircons yielding an age of 3.05 Ga (Bruguier et al., 1994; Dada, 1998). The oldest SHRIMP zircon age of 3571 Ma comes from the Kabala gneisses (Kröner et al., 2001). Zircon cores typically yield ages of 3.55 to 3.34 Ga and Nd model ages fall in the same range, suggesting production of juvenile continental crust at about 3.5 Ga in Nigeria (Dada, 1998; Kröner et al., 2001).

High precision U/Pb zircon ages have confirmed the antiquity of Archean TTG com-plexes in the *Minnesota River Valley* along the southwestern margin of the Superior Province (see Bickford et al., this volume), based on earlier, less precise Rb-Sr ages (Goldich et al., 1970) (Fig. 1.2-1). Zircon ages from the Montevideo Gneiss range from 3.5 to 3.48 Ga and two distinct ages from the Morton Gneiss are reported at 3.42 and 3.52 Ga (Schmitz et al., 2005; Bickford et al., 2005, this volume). In addition, high-grade metamorphism is recorded by new zircon growth at 3.35–3.3 Ma and mafic intrusions at 3.14 Ga. A second period of granite emplacement and regional metamorphism and defor-mation occurred at about 2.6 Ga, and may record accretion of the Minnesota River Valley terrane to the southern margin of the Superior Craton (Schmitz et al., 2005). Detrital zir-cons from a Neoarchean paragneiss exhibit age peaks at 3.52, 3.48, 3.38 and 3.14 Ga, reflecting input from most of the older rock units (Bickford et al., 2005).

The *Assean Lake Complex* in the northwestern part of the Superior Province (Fig. 1.2-1) contains a record of complex and prolonged crustal history spanning more than two billion years, from 3.9 to 1.8 Ga (Bohm et al., 2000, 2003, this volume). A supracrustal assemblage of amphibolite-grade graywacke, quartz arenite, arkose, mafic to intermediate volcanics, and BIF dominates the complex and probably formed in a shallow marine arc, or back-arc setting. The supracrustal rocks are intruded by a deformed and metamorphosed TTG complex. Nd model ages from Assean Lake TTG range from 4.2 to 3.5 Ga and suggest that early Archean basement may be preserved in this region. Single zircon ages indicate that tonalitic magmatism occurred at 3.2–3.1 Ga, whereas a record extending to at least 3.9 Ga is preserved in detrital zircons from paragneisses that may have a depositional age of 3.2 Ga (Bohm et al., 2003).

Local occurrences of TTG complexes with ages of 3.4–2.84 Ga are widespread in the Archean *Wyoming Craton* (Mueller et al., 1988, 1993, 1995, 1996; Chamberlain et al., 2003; Frost et al., 2006; Grace et al., 2006; Chamberlain and Mueller, this volume) (Fig. 1.2-1). The best-preserved 3.4–3.1 Ga crustal blocks are the Montana metasedimentary province of southwestern Montana (Mueller et al., 1993, 1995, 2004), the Hellroaring Plateau and Quad Creek region in eastern Beartooth Mountains (Henry et al., 1982; Mueller et al., 1985, 1992, 1995) and the Sacawee block in the northern Granite Mountains (Langstaff, 1995; Fruchey, 2002; Grace et al., 2006). In addition, U/Pb ages of detrital zircons from Archean metasedimentary rocks in the northwestern part of the craton indicate that they were derived from sources ranging in age from 3.96 to 2.7 Ga (Mueller et al., 1998). The majority of the detrital grains have ages between 3.4 and 3.2 Ga and none is younger than 2.9 Ga, suggestive of a major crust-forming event at 3.4–3.2 Ga in the Wyoming Craton.

The *Olekma Terrane* is part of the Aldan Shield and comprises a complexly deformed TTG complex with associated greenstones (Rundkvist and Sokolov, 1993; Rosen et al., 1994; Rosen, 2002; Rosen and Turkina, this volume) (Fig. 1.2-1). Metamorphic grade ranges from greenschist to granulite grade. Greenstone belts range from a few kilometers to over 100 km in length, and they are composed chiefly of basalts, with komatiites and calc-alkaline volcanics of local importance. Metasedimentary rocks include metagraywackes and minor metacarbonate. In the western part of the Olekma Terrane, late syn-tectonic to post-tectonic granitoids are widespread (Puchtel, 1992). SHRIMP zircon ages from tonalitic gneisses range from about 3.25 to 2.75 Ga (Nutman et al., 1992a). Greenstone volcanism was widespread from 3.02 to 3.0 Ga, but most of the TTG plutonism is Neoarchean in age.

Another region in the Aldan Shield that contains early Archean rocks is the *Sutam block* in the northern part of the Stanovoy Province (Fig. 1.2-1). Mafic granulites and amphibolites are widespread in the Sutam block, where they comprise up to 40% of the bedrock (Rosen et al., 1994; Rosen, 2002; Rosen and Turkina, this volume). Ultramafic massifs are also locally important in this block. Thermobarometry indicates that the Sutam block is a large slice of the middle to lower continental crust, with temperatures recorded between 820–920 °C at pressures of 8–11 kbar (Rosen et al., 1994). Some low-grade supracrustal rocks are preserved in shear zones in the central part of the Sutam block. These comprise chiefly metabasalts and meta-komatiites, with smaller amounts of metamorphosed intermediate volcanics, quartzites and carbonates. U/Pb zircon ages from this terrane are few in number. Upper intercept U/Pb discordia ages at about 3.46 Ga record the age of many of the rocks, with lower intercept ages around 2.2 Ga recording the age of widespread granulite-grade metamorphism (Bibikova et al., 1989).

Ages in excess of 3.0 Ga are also reported from the *Daldyn Terrane* in the Tungus–Anabar Shield of northern Siberia (Rosen et al., 1994; Rosen, 2002; Rosen and Turkina, this volume) (Fig. 1.2-1). This terrane comprises chiefly granulite-grade complexes of TTG and associated mafic rocks, both of which are isoclinally folded. Thermobarometry studies indicate metamorphic temperatures of 820–950 °C and pressures of 8.5–11 kbar (Rosen et al., 1994). Locally, supracrustal rocks are preserved, consisting of mafic to felsic volcanics and associated metasedimentary rocks. Unusual components in the supracrustal rocks include marble, calc-silicates, and quartzites, associated with minor BIF. Enderbites (granulite-grade tonalities) from the Daldyn Terrane yield U/Pb SHRIMP ages of about 3.16 Ga (Rosen et al., 1994; Rosen, 2002). Upper intercept discordia ages from similar rocks fall near 3.0 Ga. Metamorphic zircons yield ages of 2.76 Ga, which probably record widespread deformation and granulite-grade metamorphism in the Daldyn Terrane.

The *Omolon Terrane* in eastern Siberia (Fig. 1.2-1) comprises chiefly a TTG complex with associated mafic components. This terrane is part of the Mesozoic accretionary complex and hence its original location is unknown. U/Pb zircon ages from the Omolon Terrane range from about 3.1 Ga to possibly as much 3.65 Ga, with most ages falling in the range of 3.4 to 3.1 Ga (E. Bibikova, pers. comm.). Metamorphic zircons and zircon overgrowths record two periods of granulite-grade metamorphism at 2.75–2.65 Ga and at about 1.97 Ga.

Although poorly exposed, early Archean rocks underlie significant portions of the *Ukrainian Shield* in eastern Europe (Fig. 1.2-1). The oldest zircon ages, which range from about 3.65 to 3.4 Ga, come from tonalites and enderbites in a small region near the eastern margin of the Ukrainian shield (E. Bibikova, pers. comm.). More widespread are zircon ages of 3.1 to 3.0 Ga, and most of the shield also contains components emplaced or/and metamorphosed at 2.8 to 2.7 Ga. Less precise Pb-Pb ages from the Ingul-Ingulets region suggest major peaks in igneous activity at 3.1–3.0 Ga and 2.93–2.9 Ga in TTG complexes, and 2.8–2.7 Ga in at least one mafic complex (Yashchenko and Shekhotikin, 2000). In addition, Os isotope model ages from mafic and ultramafic bodies that are widespread in the western part of the Ukrainian Shield yield ages of approximately 3.0 Ga (Gornostayev et al., 2004).

Early Archean rocks have been recognized in the *Ladoga Terrane* of the Karelian Craton in the extreme southeastern part of the Baltic Shield (Lobach-Zhuchenko et al., 1986). Although most of the craton comprises Neoarchean granite-greenstone terrains, two older orogenic belts (3.0–2.9 Ga), one with multimodal volcanism and the other with bimodal volcanism, comprise the eastern part of Karelia. The oldest rocks are limited to a very small area northeast of Lake Ladoga, where zircons yield ages up to 3.55 Ga. Two younger belts (2.8–2.6 Ga) with the same distributions of volcanic rocks make up the western part of the Karelia and eastern Finland. U/Pb zircon ages indicate a decrease in age from east to west, with Neoarchean low-pressure granulite metamorphism recorded at 2.65 Ga in both the western and eastern domains.

Early Archean rocks with ages as old as 3.5 Ga have been reported from five localities in India (Fig. 1.2-1). Beckinsale et al. (1980) first reported a Rb-Sr whole-rock isochron age of 3358 ± 66 Ma from the *Dharwar Province* in South India. U/Pb SHRIMP zircon ages from detrital zircons in this province confirm the existence of pre-3.0 Ga rocks in this region (Nutman et al., 1992b). Goswami et al. (1995) report a 3.55 Ga zircon age from *Singhbhum Province* in eastern India. Nearby in the *Eastern Ghats granulite belt*, Bhattacharya et al. (2001) report Pb-Pb zircon ages and Nd model ages in excess of 3.0 Ga. The Eastern Ghats granulites typically have negative E_{Nd} values indicating the existence of still older crustal sources. Trondhjemitic gneisses from *Bastar Province* yield a zircon U/Pb age of 3.51 Ga, which appears to date the magmatic crystallization of the trondhjemites (Sarkar et al., 1993). Zircon ages from a TTG complex in Rajasthan in northwestern India in the *Aravalli Province* yield ages near 3.23 Ga, suggesting the existence of an old crustal block in this region (Roy and Kröner, 1996). Although not well established, the size of early Archean blocks in the Indian Craton seems to be rather small, and in some cases, could be less than a few tens of kilometers.

Early Archean rocks have been recognized in two regions of South America (Fig. 1.2-1). The *Gavido block* in the northern portion of the Sao Francisco Craton, northeast Brazil, is one of the oldest Archean fragments (Martin et al., 1997). It underwent polycyclic evolution from old juvenile components dated between 3.4 and 3.0 Ga on zircons (Nutman and Cordani, 1993; Leal et al., 2003). In the Umburanas greenstone belt, SHRIMP U/Pb isotopic analyses of detrital zircons from conglomeratic quartzites yield ages between 3.33 and 3.04 Ga. Tonalites from the Contendas–Mirante belt have intrusion ages of about 3.4 Ga (Nutman and Cordani, 1993). The Sete Voltas Massif is a composite crustal terrane built in at least three successive accretionary events: 1) generation and emplacement of magmatic precursors to the old grey gneisses at about 3.4 Ga; 2) younger grey gneisses and porphyritic granodiorites intruded the old grey gneisses between 3.17 and 3.15 Ga; and 3) granite dykes emplaced at about 2.6 Ga during a late-stage magmatic event. All units belonging to the Sete Voltas Massif yield homogeneous T_{DM} ages at 3.66 Ga, which may reflect the age of older crustal protoliths.

The oldest fragment of continental crust in South America occurs as an isolated block in the northeastern part of the Borborema province in eastern Brazil (Fig. 1.2-1). This is the *Sao Jose do Campestre Massif*, which is surrounded by 2.2–2.0 Ga gneisses of the Transamazonian Orogen. Single zircon ages from Campestre tonalities yield ages up to 3.5 Ga and T_{DM} Nd model ages of more than 3.7 Ga (Dantas et al., 2004). These old nuclei are surrounded by reworked and juvenile crustal rocks with ages of 3.22 and 3.12 Ga, and these in turn are intruded by 3.0 and 2.69 Ga granitoids.

1.2-3. MAJOR GRANITOID EVENTS IN THE PALEOARCHEAN

There are only two widespread age peaks in granitic magmatism prior to 2.8 Ga, at 3.3 and 2.86 Ga (Condie et al., 2007). The 2.86 Ga peak is recognized in the Superior, North Atlantic, Amazon and Enderby Land Cratons, and the 3.3 Ga peak occurs in the Aldan/Anabar, India, Pilbara, Yilgarn and Kaapvaal Cratons. In addition to these two age peaks, important age peaks are recognized in at least three cratons at 3.38, 3.23, 3.17, 3.02,

3.0 and 2.92 Ga. In contrast, the largest and best-defined detrital age peaks are at 3.14 Ga in Australia and Antarctica, ca. 3.16 Ga in Laurentia, and 3.38 Ga in Australia. Only the 3.38 Ga peak is recognized in both igneous and detrital zircon populations, the latter peak entirely defined by 440 single zircon ages from the Jack Hills quartzites in the Yilgarn Craton (Harrison et al., 2005).

The cratons of North America (Superior, Wyoming, Slave and North Atlantic cratons) have distinct early Archean age spectra. In contrast to the Slave and North Atlantic Cratons, all of the major granitoid age peaks in the Superior and Wyoming Cratons are less than about 3.0 Ga. The North Atlantic Craton has a distinctly bimodal distribution of age peaks around 2.8 Ga and between 3.7 and 3.6 Ga. The Slave Craton has the longest, although sporadic, early Archean history defined by igneous zircons from 4.0 to 2.9 Ga, whereas the Yilgarn Craton has the longest history defined by detrital zircons (from 4.35 to 2.8 Ga). Major granitoid age peaks older than 3.7 Ga are limited to four cratons: Yilgarn, Slave, North Atlantic and North China Cratons (Condie et al., 2007). If detrital age peaks are included, the Kaapvaal Craton can be added to the list.

The early Archean granitic age spectra do not favor Superior–Yilgarn, Slave–Yilgarn, Yilgarn–Pilbara, or Superior–Slave connections before the Neoarchean, since only one or two major early Archean granitic episodes is shared in common between any of these cratons. Unlike the Slave and Yilgarn Cratons, the Superior Craton shows no major granitoid episodes before 3.0 Ga. With our current database of zircon ages, there is no evidence to support global granitic magma events in the early Archean.

PART 2

**PLANETARY ACCRETION AND THE HADEAN TO
EOARCHEAN EARTH – BUILDING THE FOUNDATION**

Earth's Oldest Rocks
Edited by Martin J. Van Kranendonk, R. Hugh Smithies and Vickie C. Bennett
Developments in Precambrian Geology, Vol. 15 (K.C. Condie, Series Editor)
© 2007 Elsevier B.V. All rights reserved.
DOI: 10.1016/S0166-2635(07)15021-0

Chapter 2.1

THE FORMATION OF THE EARTH AND MOON

STUART ROSS TAYLOR

*Department of Earth and Marine Sciences, Australian National University,
Canberra, ACT, Australia*

The curious compositions of the Earth and the Moon have arisen as a consequence of their formation within the inner solar system. Accordingly it is necessary to place this discussion within the wider context of the formation of the planetary system itself.

2.1-1. THE SOLAR NEBULA

The Sun and planets formed from a rotating disk, the solar nebula. This contained three components, loosely "gases, "ices" and "rock". The major component was "gas" (98% H and He). The remaining two percent of the disk contained the heavier elements (the so-called "metals" of the astronomers). These had accumulated as the product of 10 billion years of nucleosynthesis, forming in previous generations of stars before being dispersed into the interstellar medium. Abundant elements such as carbon, oxygen and nitrogen formed compounds with hydrogen that included water, methane and ammonia. These compounds, constituting about 1.5%, were present in the nebula as "ices". The remaining 0.5% of the nebula was composed of dust and grains ("rock") from which the rocky planets were formed. We are well informed about its make-up because the composition of the Type 1 (CI) class of chondritic meteorites, when ratioed to a common element such as silicon, matches the composition of the solar photosphere determined from spectral analysis. As the Sun contains 99.9% of the mass of the solar system, this match informs us of the composition of the "rock" fraction of the original solar nebula.

At this stage, one might suppose that the problem of the composition of the inner rocky planets is solved. But surprisingly, both the terrestrial planets and most classes of meteorites do not match this primordial composition. To resolve this problem, we have to consider the events leading not only to the formation of the inner planets, but of the four giant planets as well (Taylor, 2001, Chapters 7, 8). It is difficult to arrive at a satisfactory definition of a planet as they are formed by stochastic processes; witness the furore over the status of Pluto or its larger colleague, 2003 UB313, which are eccentric dwarfs when placed among the planets, but are the largest bodies in the Kuiper Belt in their own right. The debate over the status of Pluto by a committee of the International Astronomical Union is an interesting example of a struggle between politics, sentiment and science. As Confucius remarked; "the beginning of wisdom is to call things by their right names" (Taylor, 2004).

2.1-2. THE FORMATION OF THE GIANT PLANETS

There is a major difference in mass between the terrestrial and the giant planets that reside beyond 5 astronomical units (AU). The latter contain a total of 440 Earth-masses of gas, ices and rock. Those nearer the Sun – Mercury, Venus, Earth, Moon and Mars – in startling contrast, contain only a trivial amount (two Earth-masses) of rock. But even the giant planets differ among themselves. Jupiter and Saturn are gas giants, bodies with massive gas envelopes surrounding cores, while Uranus and Neptune are mostly ice and rock cores with about one Earth-mass of gas. How did this difference arise?

To begin with, solar nebula material, gases, ices and rock of CI composition flowed into the Sun through the circum-solar disk. When the Sun grew large enough to initiate H to He burning, strong solar winds developed. These swept away the gas and ices from the inner nebula. As a consequence, disks around young stars survive for only a few million years. Typical disk lifetimes vary from 3 to 6 My (Haisch et al., 2001). The formation of the planets is thus a very late event in the history of the disk, beginning only after the Sun had formed and commenced the H to He burning (Taylor, 2001).

Where it was cold enough out in the disk, at about 5 AU, water condensed as ice, forming a "snow line" (Stevenson and Lunine, 1988). The resulting pile-up of ices and dust trapped at the "snow line" at 5 AU locally increased the density of the nebula. This density increase led to a rapid (10^5 year) runaway growth of large bodies (10–15 Earth-masses) of ice and dust. The ice giants Uranus (14.5 Earth-mass) and Neptune (17.2 Earth-mass) are surviving examples of these cores. At the same time, the gas (H and He) was also being dispersed by the stellar winds. These massive cores of dust and ice were able to capture variable amounts of gas by gravitational attraction. Jupiter was able to accrete about 300 Earth-masses of gas ahead of the others. It became dominant gravitationally, so that it dispersed the other cores outwards into the gas-poor regions of the nebula (Thommes et al., 2002).

The gas content of Jupiter is much less than that present in the original nebula, with the result that Jupiter does not have the composition of the Sun, but is enriched in the "ice and rock" component, or "metals" by a factor of somewhere between 3 and 13% of the solar abundances (Lunine et al., 2004; Guillot et al., 2004). Saturn, although it has a similar size core to Jupiter, managed to capture only about 80 Earth-masses of gas, whereas Uranus and Neptune managed to accrete only one or two Earth-masses of gas.

The model discussed here is referred to as the "core accretion" model for forming the giant planets. An alternative model for giant planet formation by condensation directly from the gaseous nebula is usually referred to as the disk instability model (Boss, 1997, 2003). The main attraction of this model is fast formation of the giant planets within a few thousand years, but there are two fatal flaws. First, the giant planets are predicted to be of solar composition, but Jupiter and Saturn are enriched by several times in "ices and rock" relative to the solar composition. Secondly, the interior of Jupiter is at pressures of 50–70 mbars with temperatures up to 20,000 K, so that the material is present as a plasma of protons and electrons, so-called "degenerate matter". Thus, a density contrast does not exist so that a core cannot "rain out" in the manner that the iron core formed in the Earth. Thus, the core of a giant planet has to form first, around which the gas can subsequently accrete.

2.1-3. PLANETESIMALS

In this scenario, the giant planets formed well before the terrestrial planets, while gas was still present in the nebula. The terrestrial planets accreted much later from the dry rocky refractory material (about two Earth-masses, of which the asteroids are analogues) that was leftover in the inner nebula following the dispersal of the gaseous and icy components of the nebula. Thus, following the formation of the gas and ice giants, the inner nebula was dry and free of gas. If the Earth had formed in a nebula that was gas-rich, ices would also have been present. In this case, the Earth would have accreted not only water ice, but also methane and ammonia ices. So the abundances of water, carbon and nitrogen would be orders of magnitude more than is observed, while the noble gases are highly depleted in the Earth.

But in addition to the depletion in gases and ices, the rocky component in the inner nebula is depleted in the elements that have condensation temperatures below about 1100 K. These are depleted in the entire inner nebula, in the Earth, Venus, Mars and most classes of meteorites relative to the original "rock" component of the solar nebula that is represented by the CI chondrites. (Table 2.1-1; Taylor, 2001, Chapter 5).

This depletion is illustrated by Fig. 2.1-1, in which the composition of the silicate mantle of the Earth (Table 2.1-1) is plotted relative to the composition of the CI carbonaceous chondrites. The depletion occurred in the earliest stages of the nebula, close to T_{zero}, and was not connected with the later formation of the planets. (This time is given conventionally by the ages of the oldest refractory inclusions [CAIs] in meteorites at 4567 ± 0.6 Ma and referred to as T_0 or T_{zero}: Amelin et al., 2002). Thus, the meteoritic chondrules formed 2 My after T_{zero} from material that was already depleted in the volatile elements (Amelin et al. 2002).

The causes are much debated. The nebula was cool, not hot, so that the old notion of elements condensing from a hot nebula is no longer tenable. Probably the depletion was due to early intense solar activity that swept away, along with the gases and ices, those volatile elements that were not present in grains (Yin, 2005). In the interstellar medium, elements with condensation temperatures below about 1000–1100 K are in the gas phase, while the more refractory elements are in grains. So it is plausible that the volatile elements were swept out along with the gases and ices by early intense solar winds.

It is worth noting that elements such as potassium and lead, which are much less volatile than water, are depleted, whereas the primary minerals of meteorites are anhydrous, again

Table 2.1-1. The composition of the primitive silicate mantle of the Earth (present mantle plus crust)

Element		Element		Element		Oxide	wt%
Li	2.1 ppm	Rb	0.55 ppm	Eu	131 ppb	SiO_2	46.5
Be	60 ppb	Sr	17.8 ppm	Gd	459 ppb	TiO_2	0.16
B	0.26 ppm	Y	3.4 ppm	Tb	87 ppb	Al_2O_3	3.64
Na	2500 ppm	Zr	8.3 ppm	Dy	572 ppb	FeO	8.0
Mg	23.2 wt%	Nb	0.56 ppm	Ho	128 ppb	MgO	38.45
Al	1.93 wt%	Mo	59 ppm	Er	374 ppb	CaO	2.89
Si	21.4 wt%	Ru	4.3 ppb	Tm	54 ppb	Na_2O	0.34
K	180 ppm	Rh	1.7 ppb	Yb	372 ppb	K_2O	0.02
Ca	2.07 wt%	Pd	3.9 ppb	Lu	57 ppb	Total	100.1
Sc	13 ppm	Ag	19 ppb	Hf	0.27 ppm		
Ti	960 ppm	Cd	40 ppb	Ta	0.04 ppm		
V	85 ppm	In	18 ppb	W	16 ppb		
Cr	2540 ppm	Sn	0.14 ppm	Re	0.25 ppb		
Mn	1000 ppm	Sb	5 ppb	Os	3.8 ppb		
Fe	6.22 wt%	Te	22 ppb	Ir	3.2 ppb		
Co	100 ppm	Cs	18 ppb	Pt	8.7 ppb		
Ni	2000 ppm	Ba	5.1 ppm	Au	1.3 ppb		
Cu	18 ppm	La	551 ppb	Tl	6 ppb		
Zn	50 ppm	Ce	1436 ppb	Pb	120 ppb		
Ga	4 ppm	Pr	206 ppb	Bi	10 ppb		
Ge	1.2 ppm	Nd	1067 ppb	Th	64 ppb		
As	0.10 ppm	Sm	347 ppb	U	18 ppb		
Se	41 ppb						

Source: Data from Taylor, S.R. (2001). Solar System Evolution, second ed. Cambridge University Press. Table 12.6.

indicative of a lack of water in the inner nebula both during meteorite formation and later planetary assembly.

The material in the inner nebula, initially as grains, accreted into meter-sized lumps that formed into kilometre- and eventually Moon-sized bodies. These building blocks are termed planetesimals. The best surviving analogues are the asteroids, along with Phobos and Deimos, the tiny moons of Mars. These planetesimals were dry, depleted in the volatile elements, and had wide variations in the abundance and oxidation state of iron. Some were differentiated into metallic cores and silicate mantles that are common in the bodies in the asteroid belt (Taylor and Norman, 1990; Mittlefehldt et al., 1998).

Examples of such early processes that resulted in a differentiated body are provided by the basaltic meteorites (eucrites) derived from the large asteroid 4 Vesta, 450 km in diameter. These provide evidence of the eruption of basalts on the surface of that asteroid at 4557 Ma, a date that is within a few million years of T_{zero} (Carlson and Lugmair, 2000).

The main point for the formation of the terrestrial planets is that many of the planetesimals melted and differentiated very early, within a few million years of the origin of the

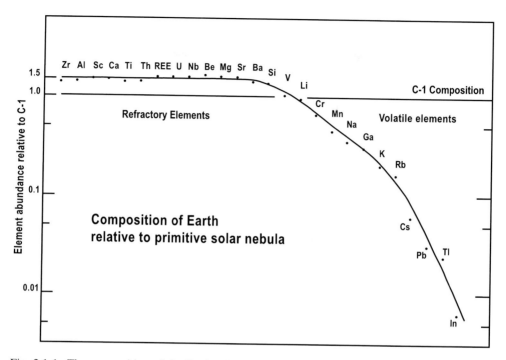

Fig. 2.1-1. The composition of the Earth relative to the primitive solar nebular abundances given by those in the Type 1 carbonaceous chondrites (CI). Data from McLennan et al. (2005) and Taylor (2001, Tables 5.2 (CI) and 12.6 (Earth primitive mantle)). The mantle data have been reduced by 0.68 to allow for the core of the Earth.

solar system at 4567 Ma (Kleine et al., 2005a). Most of the asteroids that were sunwards of 2.7 AU were melted. According to the Hf-W isotopic data, core formation on asteroids may have occurred less than 1.5 My after T_{zero} (4567 Ma) and so may predate chondrule formation, well constrained to occur more the 2 My after T_{zero} that implies separate origins for the differentiated asteroids and chondrules. The implication from what must be a very restricted sampling is that melting and differentiation were widespread. Most ordinary chondrites have also been depleted in volatile elements and display similar geochemical fractionations to those observed in the terrestrial planets.

The heat source for melting these small bodies was probably ^{26}Al ($t_{1/2} = 730{,}000$ years). ^{26}Al decays to ^{26}Mg and there is evidence of ^{26}Mg anomalies resulting from this radioactive decay in some basaltic meteorites (eucrites) from the large differentiated asteroid 4 Vesta, as well as in other meteorites.

Thus, the Earth and the other inner planets accreted from objects that had previously been melted and differentiated. Metal-sulfide-silicate equilibria were established in these bodies, under low-pressure conditions. However, following such events as the Moon-forming collision, re-equilibration of mantle and core in the Earth may have occurred under higher pressures (Halliday et al., 2000). During the accretion of the planets, further melting, perhaps as a consequence of impacts, will cause rapid and perhaps catastrophic core formation as metal segregates from silicate.

2.1-4. THE FORMATION OF THE TERRESTRIAL PLANETS

During the process of collisional accretion of the planets, the intermediary bodies grew to large sizes. Before the final sweep-up into the inner planets, computer simulations indicate there were likely over 100 objects about the mass of the Moon (1/81 Earth mass), ten with masses around that of Mercury (1/20 Earth mass), while a few exceeded the mass of Mars (1/11 Earth mass), most of which were accreted to Venus and the Earth. The stochastic nature of this process is demonstrated by the fact that Earth and Mars, the two planets on which we have most information, differ significantly in density and so in their major element composition. Venus is much closer in density, major element composition and in the abundances of the heat producing elements, K, U and Th, to the Earth, but has experienced a wildly different geological evolution (see Hansen, this volume).

Although the Earth has a general "chondritic" composition, it cannot be linked either to a specific meteorite class, or to some mixture of the many groups (see Bevan, this volume). Neither K/U ratios, volatile element compositions, nor rare gas abundances in the Earth equate to the meteoritic abundances. Oxygen isotope data show that, except for fractionated basaltic meteorites (ruled out on the other grounds), no observed class of meteorites matches the terrestrial data, except for the enstatite chondrites. Because of this and their extremely reduced nature, enstatite chondrites are often thought to be suitable building blocks for the Earth (Javoy, 1995). However their low Al/Si and Mg/Si ratios and their low volatile elements content rule them out as candidates. So, it is a coincidence that the Earth and the enstatite chondrites share the same oxygen isotopic composition. As is well known to philosophers, similarity does not imply identity.

Mercury and Mars are survivors from this final population of planetesimals that accreted to form Venus and the Earth. It took longer to form the large terrestrial planets, taking somewhere between 30 to 100 My for planetesimals to be assembled into the four terrestrial planets. This accretion of bodies into the terrestrial was hierarchical. One impacting body was at least the size of Mars and was among the last of the giant collisions with the Earth. This body, now named Theia, would have been a respectable planet in its own right had it not collided with the Earth. The consequence was the formation of the Moon as a result of a glancing collision with the Earth (Canup and Asphaug, 2001).

Some conflicting information exists on the timing of these events. One of the more useful isotopic systems is the decay of ^{182}Hf to ^{182}W that has a half-life of 9 My and so is suited to document events in the early solar system (Jacobsen, 2005). During the differentiation of planets into metallic cores and silicate mantles, separation of tungsten (into iron cores) from hafnium (retained in silicate mantles) occurred. After some initial controversy (Halliday et al., 2000; Jacobsen, 2005), it is now agreed that the presence of radiogenic ^{182}W in the terrestrial mantle indicates that the separation of core and mantle occurred within the lifetime of ^{182}Hf, and so occurred within 30–50 My of T_{zero} (Jacobsen, 2005). The oldest reliable zircon age is 4363 ± 20 Ma (Nemchin et al., 2006) that is 200 My after T_{zero}. Application of this system to the Moon, although not without difficulties, indicates crystallization of the lunar magma ocean at 4527 Ma, within 40 My of T_{zero} (Kleine et al., 2005b). This date contrasts with the younger age of 4460 ± 20 Ma (Norman et al., 2003) obtained for lunar anorthosites. Probably the best that can be said at this stage is that the accretion of the Earth and the formation of the Moon by the last giant collision occurred within 30 to 100 My after T_{zero}.

The Moon is depleted in a uniform pattern for elements that are volatile below about 1100 K relative to the Earth and other inner solar system bodies, in which these elements are depleted relative to Cl in order of volatility (Figs. 2.1-1 and 2.1-2). Refractory elements are not fractionated relative to chondritic abundances, but many investigators have suggested that the Moon is enriched in refractory elements as a group (Taylor et al., 2006). This may indicate that the Moon formed by condensation from a vapor phase following the impact. Most (85%) of the Moon is derived from Theia, the impactor. The similarity in Cr and O isotopes in both bodies may be a result of equilibration during the collision (Pahlevan and Stevenson, 2005).

The high iron/silicate ratio in Mercury was probably due to the loss of much of its silicate mantle following a collision of Proto–Mercury with an object about 20% of its mass (Benz et al., 1988).

Much radial mixing took place in the inner nebula during the final accumulation of the Earth and Venus, and the large planetesimals were widely scattered. So the material now in the Earth and Venus came from the entire inner solar system, in contrast to the accretion of the smaller planetesimals that formed from restricted radial zones.

There has been much debate over whether a "late veneer" of material was responsible for the chondritic-like patterns of the siderophile elements in the upper mantle. However, many difficulties remain, as no meteorite class seems suitable to provide this pattern (Drake and Righter, 2002). The source of water in the Earth and Mars was derived from later drift-back of icy planetesimals, or from comets from the Jupiter region. Comets, once a favourite source, are ruled out by having too high D/H ratios (Morbidelli et al., 2000).

2.1-5. THE PRE-HADEAN STATE OF THE EARTH

The consequence of such massive collisions is that these events have sufficient energy to melt the terrestrial planets (Stevenson, 1988), thus facilitating core-mantle separation. Such a collisional history also accounts for the variations in composition of the terrestrial planets, as the planets accreted from differentiated planetesimals that had already undergone many collisions. Thus, some diversity of composition can be expected. Early planetary atmospheres may also be removed or added by cataclysmic collisions, accounting for the significant differences among the atmospheres of the inner planets. Thus, "in the context of planetary formation, impact is the most fundamental process" (Grieve, 1998), while "chaos

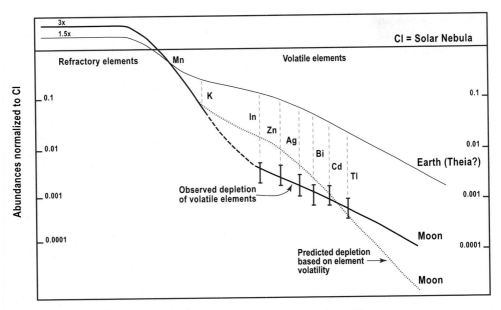

Elemental enrichment - depletion relative to CI

Fig. 2.1-2. The composition of the Moon compared with that of the Earth, both normalized to CI carbonaceous chondrites (dry basis) (Taylor et al., 2006). The abundance curve for the Earth is principally derived from McLennan et al. (2005), and other sources. The refractory elements in the Earth are 1.5 × CI, and in the Moon are 3 × CI (see text). The figure is interpreted to indicate that the material in the impactor mantle (Theia), from which the Moon was derived, was inner solar-system material already depleted in volatile elements at T_{zero} and that the abundances in the Earth provide an analogue for its composition. Mn and K provide fixed points for the Moon curve. Mn has the same abundance in the Earth and Moon. Potassium abundances are derived from K/U values (Earth 12,500; Moon 2500). The volatile-element data for the Moon are derived from Wolf and Anders (1980), who recorded a uniform depletion of 0.026 ± 0.013, for the elements listed, in lunar low-Ti basalts compared to terrestrial oceanic basalts. In the absence of more recent data for both bodies, we adopt their study as recording the Moon–Earth depletion. The significant point is that the lunar depletion is uniform and not related to volatility, which would produce a much steeper depletion pattern (lower dotted line). Thus, the lunar pattern is interpreted as resulting from a single-stage condensation from vapour (>2500 K) that effectively cut-off around 1000 K.

is a major factor in planetary growth" (Lissauer, 1999). The bizarre landscapes produced by chaotic processes are well illustrated by the Uranian satellite, Miranda (Fig. 2.1-3).

The collisions occurring during accretion are quintessential stochastic events. Of course, the probability of impacts of bodies of the right mass and at the appropriate angle and velocity to produce the Moon or remove the mantle of Mercury is low. However, other collisions involving different parameters might produce equally "anomalous" effects, such as a Moon for Venus, no Moon for the Earth, or different masses, tilts or rotation rates for

Fig. 2.1-3. Miranda, one of the icy satellites of Uranus, 242 ± 5 km in radius with a density of 1.26 ± 0.4 g cm^{-3}, showing a chaotic landscape of fault-bounded blocks, called coronae, probably the result of tidal interactions with neighbouring satellites and Uranus. The surface relief is up to 20 km. (Courtesy NASA JPL P 29505.)

the inner planets. The variations in composition and later evolution of the terrestrial planets are thus readily attributable to the random accumulation of planetesimals with varying compositions. Indeed, computer simulations have difficulty in reproducing the final stages of accretion of the inner planets, commonly producing fewer planets with large eccentricities and wider spacings, thereby emphasizing the importance of stochastic processes in planetary formation (Canup and Agnor, 2000; Levison et al., 1998). The end result of the accretion of the Earth is that it was most likely entirely molten (as was the Moon) around 4500 ± 50 Ma.

ACKNOWLEDGEMENTS

I am grateful to Dr Judith Caton for assistance and for drafting Figs. 2.1-1 and 2.1-2.

Earth's Oldest Rocks
Edited by Martin J. Van Kranendonk, R. Hugh Smithies and Vickie C. Bennett
Developments in Precambrian Geology, Vol. 15 (K.C. Condie, Series Editor)
© 2007 Elsevier B.V. All rights reserved.
DOI: 10.1016/S0166-2635(07)15022-2

Chapter 2.2

EARLY SOLAR SYSTEM MATERIALS, PROCESSES, AND CHRONOLOGY

ALEX W.R. BEVAN

Department of Earth and Planetary Sciences, Western Australian Museum, Perth, Western Australia 6000, Australia

2.2-1. INTRODUCTION

While the oldest known intact rocks on Earth, the Acasta Gneiss, are around 4.0 Ga (Bowring and Williams, 1999), and the oldest reliable age for a derived zircon in the Jack Hills metasedimentary rocks is 4363 ± 20 Ma (Nemchin et al., 2006), there is a source of much older material for study – meteorites. This simple term, however, belies the diversity and complexity of the rocks that are observed to fall, or are subsequently found on Earth, and the widely dispersed source regions in the Solar System from whence they came. Moreover, these products of early Solar System processes are a unique source of information about the mechanisms of the construction of small planetesimals and the planets, and the kinds of materials from which they might have accreted. Most importantly, the study of meteorites and their components has allowed the refinement of the time-scales for some of the earliest events in the birth and evolution of the Solar System (McKeegan and Davis, 2003).

The exact nature of the materials from which the Earth accreted is unknown. Moreover, it is not known whether any of the meteorites (or mixtures thereof) that have been recovered represent such materials (e.g., see Halliday, 2003, and references therein). A question that is fundamental to an understanding of the accretion of the Earth is constraining its age. Here meteorites have played an unique role. Since the pioneering work of Patterson (1956) it has been known that primitive meteorites (the chondrites) are ca. 4.55 Ga old, and this is used to constrain the age of the Earth. Various models for the accretion of the Earth have been proposed and, while they vary in detail, all view the Earth as 'younger' than the dust and planetesimals from which it accreted. The most widely accepted theory, which is supported by mathematical modelling, is that the Earth may have been one of the largest of a swarm of early planetesimals that accreted in the Earth's vicinity in the inner Solar System (Taylor and Norman, 1990).

2.2-2. EARLY SOLAR SYSTEM MATERIALS

The available chemical, isotopic and astronomical evidence suggests that the materials we see as meteorites originated within the Solar System, and that the great majority appear to be fragments of asteroids in solar orbits between Mars and Jupiter. Excluding meteorites recognised as of planetary origin (Lunar and Martian), some meteoritic components have extremely old formation ages (4567.1 ± 0.16 Ma), and some of these materials appear to have remained relatively unaltered since their formation (Amelin et al., 2006). As samples from minor planets that became internally, largely thermally, inactive shortly after the birth of the Solar System, meteorites record events that occurred during its earliest history.

Overall, meteoritic materials are dominated by ferro-magnesian silicates and metallic iron–nickel (Fe-Ni). Historically, meteorites were grouped into three broad categories on their contents of these two major components (Hutchison, 2004). *Iron* meteorites are composed principally of metal; *stones* (or *stony meteorites*) consist predominantly of silicates, but with varying amounts of accessory metal; and *stony irons* comprise metal and silicates in roughly equal proportions. However, the last 40 years of meteorite research has focused on meteorite classification based on their detailed mineralogy, petrology, bulk chemistry, and oxygen isotopes. One of the aims of modern meteorite research is to group together 'genetically' related meteorites, and to seek relationships between groups that may reveal their origin. The modern classification system (Tables 2.2-1 and 2.2-2) recognises groups with similar properties that reflect similar formation histories and, presumably, similar parent (asteroidal or planetary) body origins (e.g., see Krot et al., 2003 and references therein).

Essentially, only two major categories of meteorites are recognised; meteorites that contain chondrules, the *chondrites*, and the *non-chondritic meteorites* that do not. The chondrites are those meteorites characterized by small, generally mm-sized beads of predominantly silicate material, called *chondrules* (Greek *chondros* = grain), from whence their name derives, and fifteen groups are currently recognised (Table 2.2-1). The origin of chondrules remains enigmatic, but they are accepted as some of the early solids in the Solar System. Chondrites are gas-borne agglomerates, of both high- and low-temperature materials, whose individual components and whole rocks have been variably altered by retrograde (aqueous alteration) and prograde (recrystallization) metamorphism. In many chondrites, secondary (metamorphic) processes have been overprinted by tertiary (shock metamorphic) processes.

The non-chondritic meteorites (Table 2.2-2) include those meteorites that lack chondrules and have textures and chemistries that show that they formed by partial, or complete igneous differentiation of their parent bodies, or are breccias of igneous debris. They include two kinds of stony *achondritic* (silicate-rich, but lacking chondrules) meteorites; *primitive* achondrites (those that retain a chemical signature of the precursor chondritic material from which they were made), and excluding meteorites from the Moon and Mars, seven groups of highly differentiated asteroidal achondrites. Of the metal-rich meteorites, there are thirteen groups of chemically distinct iron meteorites with essentially igneous histories, and two distinct groups of igneous stony irons, *mesosiderites* and *pallasites*. In

Table 2.2-1. Meteorite classification (chondrites)

Class		Group	Petrologic type	Sub-group	Mg/Si at*	Fe/Si at*
Chondrites	Carbonaceous (C)	CI	1		1.066	8719
		CM	1–2		1.042	8177
		CO	3–4		1.053	7847
		CV	3–4	CVa, CVb, CVred	1.066	7578
		CK	3–6		1.127	7855
		CR	1–3		1.045	7875
		CH	3		1.063	15222
		CB	3	CBa, CBb		
	Ordinary (OC)	H	3–6		0.954	8177
		L	3–6		0.952	5838
		LL	3–6		0.928	4913
	Enstatite (E)	EH	3–6		0.871	8730
		EL	3–6		0.731	5934
		R (Rumuruti)	3–6		0.934	7696
		K (Kakangari)	3			

*Data from Hutchison (2004).

addition, there are a number (> 50) of meteorites (mainly irons) that do not fit into any of the recognised groups and are termed either *anomalous* (irons), or *ungrouped* (stones).

2.2-2.1. Chondritic Meteorites

While texturally, most chondrites are dominated by chondrules and the matrix in which they are set (Fig. 2.2-1(a)), mineralogically they are complex aggregates of ferro-magnesian silicates (olivine and pyroxene), Fe-Ni metal, Ca-Al-rich inclusions (often referred to as refractory inclusions, or CAIs), and rare aggregates of olivine grains (amoeboid olivine aggregates). Additionally, the mineralogy of chondrites may include magnetite, chromite (or chrome spinels), iron-sulfides (troilite, pyrrhotite and pentlandite), carbonates,

Table 2.2-2. Meteorite classification (non-chondritic meteorites)

Non-chondritic meteorites	**Primitive achondrites**	Acapulcoites Lodranites	Clan/same parent body?
		Silicates in IAB-IIICD irons Winonaites	Clan/same parent body?
	Differentiated achondrites — achondrites (asteroidal)	Angrites Aubrites Brachinites Ureilites Howardites Eucrites Diogenites	HED clan, same parent body
	achondrites (planetary)	Shergottites Nakhlites Chassignites Orthopyroxenites	Martian (SNC)
		Lunar	Moon

sulfates, and 'serpentine' group minerals (for a detailed review of meteorite mineralogy, see Rubin (1997a, 1997b)).

Ca-Al-rich inclusions contain refractory materials and range in size from sub-millimetre, to centimetre-sized, objects that occur in varying abundances in all groups of chondrites (Fig. 2.2-1(b)). The mineralogy and isotopic composition of Ca-Al-rich inclusions suggest that they are amongst the earliest solids to have formed in the Solar System, and this is confirmed by isotopic dating. Both Ca-Al-rich inclusions and chondrules are the products of very high temperature events during the early history of the Solar System, and the lat-

Table 2.2-2. (*Continued*)

Non-chondritic meteorites / **Stony irons**	Mesosiderites	Possibly related to HED clan	
	Pallasites	Main group — Possibly related to IIIAB irons Eagle Station Pyroxene	
Irons	IAB-IIICD* IC IIAB IIC IID IIE* — Possible differentiates from H-chondrite-like precursor IIF IIG IIIAB — Possibly related to Main group pallasites IIIE IIIF IVA* IVB Anomalous		

*Silicate-bearing irons.

ter probably originated from pre-existing solids in the nebula (MacPherson, 2003; Rubin, 2000; Shu et al., 2001).

Meteorites within chondritic classes and groups show the effects of varying degrees of secondary alteration, to which have been assigned petrologic labels on a numbered scale from 1–6 (Van Schmus and Wood, 1967). Petrologic types 1 and 2 refer to those chondrites containing water-bearing minerals indicative of low temperature, retrograde aqueous alteration, where type 1 chondrites have experienced greater aqueous alteration than type 2. Other chondrites show degrees of recrystallization from type 3 (least recrystallized) to type 6 (most recrystallized) that has progressively erased their chondritic textures, and is attributed to prograde metamorphism. Types 4 and 5 are intermediate between these extremes. Type 3 chondrites display prominent chondrules with abundant glasses and highly disequilibrated mineral assemblages, set in a fine-grained matrix. In type 6 chondrites, solid-state crystallization of matrix and chondrule mesostases has all but erased their chondritic textures and they contain more or less equilibrated mineral assemblages. On the basis of detailed mineral chemistry and thermoluminescence sensitivity, type 3 ordinary chondrites have been further divided into ten metamorphic sub-types (3.0–3.9) (e.g., see Sears et al. (1991) and references therein). Sub-types are also recognised in type 3 chondrites from other groups (Guimon et al., 1995; Scott and Jones, 1990).

The abundance of iron in chondrites and the distribution of this element between reduced (metal + sulfide) and oxidized (silicates + oxides) phases (Fig. 2.2-2) distinguish a number of groups of chondrites. The same groups can also be distinguished on ratios of

Fig. 2.2-1. (*Previous page.*) (a, top) A barred olivine chondrule (centre – ca. 0.4 mm across) comprising crystals of olivine in glass, together with other smaller chondrules in an ordinary chondrite. The origin of chondrules remains enigmatic, but there is general agreement that they formed from rapidly cooled molten droplets generated in the early Solar System. (b, bottom) A mass of the Allende CV3.2 chondritic meteorite, showing chondrules and large, white refractory inclusions (Ca-Al-rich inclusions) beneath black fusion crust. Bizzarro et al. (2004) have shown that whereas some chondrules in Allende formed synchronously with the earliest Ca-Al-rich inclusions, others formed over a period spanning 1.4 Ma after the earliest Ca-Al-rich inclusions. The youngest chondrules help to constrain the time of accretion (scale bar = 2 cm).

---

their refractory element (Mg, Al, Ti, Ca) contents to silicon, and oxygen isotopes (Clayton, 2003) (Fig. 2.2-3).

Thirteen groups of chondritic meteorites comprise three major classes (carbonaceous, ordinary and enstatite chondrites). Eight groups of *carbonaceous chondrites* (CI, CM, CO, CV, CR, CH, CB, and CK) are recognised (Table 2.2-1). The name *carbonaceous* is somewhat misleading since only three groups (CI, CM, and CR) contain significant amounts of carbon. Carbonaceous chondrites are characterised by Mg/Si atomic ratios >1, are generally highly oxidized, contain hydrous minerals, and can contain significant amounts of magnetite. CI chondrites lack chondrules. However, their 'chondritic' chemistries and the presence of rare high temperature mineral fragments (olivine and pyroxene) that may be chondrule remnants (McSween and Richardson, 1977; Endress and Bischoff, 1993, 1996; Leshin et al., 1997), show that they are chondrites. Compositionally the most primitive, CI chondrites provide the closest match to the photosphere of the Sun (Anders and Ebihara, 1982; Anders and Grevesse, 1989; Palme and Jones, 2003). Despite their primitive composition, however, texturally and mineralogically CI chondrites show abundant evidence of mechanical and hydrothermal processing. Only five CI chondrites are known, all observed falls, and all are microscopic breccias. Essentially, they consist of very fine-grained hydrous silicates with accessory magnetite, sulfides, and occasional veins of carbonates and sulfates as further evidence of hydrothermal processes. Gounelle and Zolensky (2001), however, have suggested that in one CI, Orgueil, sulfate veins may have been produced during post fall storage in collections, and do not represent parent body processing. Nevertheless, CI chondrites match solar abundances of the elements, so their secondary alteration must have been isochemical in an essentially closed system.

The *enstatite chondrites* comprise two groups (EH and EL, where H = high iron and L = low iron) that are distinguished on mineralogy and bulk chemistry (Fig. 2.2-2) (Sears et al., 1982). These are highly reduced materials containing abundant metal and, as their name suggests, virtually iron-free silicates. Enstatite chondrites have low Mg/Si atomic ratios (<0.9), and lithophile element bearing sulfides reflecting their highly reduced nature (e.g., oldhamite CaS) (e.g., see Brearley and Jones, 1998; Krot et al., 2003). EH and EL chondrites range in petrologic type from 3–6. There is some doubt as to whether the enstatite chondrites come from one, or two, parent bodies (e.g., see Krot et al. (2003) and references therein).

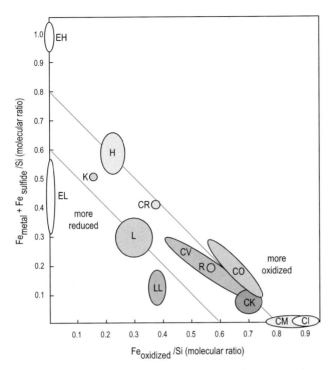

Fig. 2.2-2. The bulk molecular ratios of iron as metal + sulfide to silicon, versus iron in silicates and oxides to silicon for the chondrite groups. Not shown here are the metal-rich CH and CB chondrites (after Brearley and Jones, 1998).

Ordinary chondrites are the most abundant meteorites observed to fall and quickly recovered, accounting for more than 80% of the modern meteorite flux. Ordinary chondrites comprise three distinct groups (H, L, LL) that are depleted in refractory elements relative to CI chondrites (Mg-normalized refractory lithophile abundances of approximately $0.85 \times CI$). They contain significant amounts of both metallic and oxidized iron (Fig. 2.2-2), and have Mg/Si ratios intermediate to E and C chondrites. Their oxygen isotope compositions lie above the terrestrial fractionation line (Fig. 2.2-3). In the sequence H-L-LL, siderophile element abundances decrease and oxidation state increases. Ordinary chondrites show a wide degree in secondary metamorphic alteration from types 3–6 (severely recrystallized chondrites are sometimes labelled type 7), with some of the lowest metamorphic types ($\leqslant 3.1$) having suffered minor aqueous alteration (Hutchison et al., 1987, 1998). The ordinary chondrites appear to represent material from at least three separate parent bodies. However, a small number of chondrites that lie between the resolved groups, or apparently related chondrites extending towards more highly reduced or oxidized compositions than ordinary chondrites, may represent material from additional parent bodies.

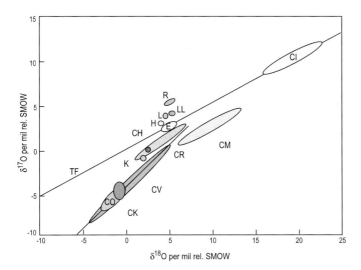

Fig. 2.2-3. Bulk oxygen isotopic compositions of the chondrite groups. TF is the terrestrial fractionation line at slope 1/2. A mixing line of slope 1 is defined by the anhydrous minerals in CO, CV and CK chondrites. Not shown are the CB chondrites (after Brearley and Jones, 1998).

The R (Rumuruti) and K (Kakangari) chondrites are distinct from each other, and from the three major chondrite classes, and may represent additional classes. R chondrites are highly oxidized materials containing nickel-bearing olivine and sulfides, magnetite, and little, or no, Fe-Ni metal. Mineralogically, Ca-Al-rich inclusions are only rarely found in R chondrites (Weisberg et al., 1991; Bischoff et al., 1994; Rubin and Kallemeyn, 1994; Schulze et al., 1994; Kallemeyn et al., 1996; Russell, 1998 and references therein), they have a high proportion of matrix to chondrules (1:1), and are commonly brecciated (Bischoff, 2000). While the R chondrites contain refractory lithophile and moderately volatile element abundances (approximately $0.95 \times CI$) close to those in ordinary chondrites, relative enrichment in some volatile elements (such as Ga, S, Se and Zn) and their oxygen isotopic compositions serve to distinguish them (Fig. 2.2-3) (e.g., see Krot et al., 2003 and references therein). R chondrites are metamorphosed (types 3.6-6), most contain solar-wind implanted gases, and they have been described as regolith breccias (Weber and Schultz, 1995; Bischoff, 2000). Some R chondrites may have undergone aqueous alteration prior to prograde metamorphism (e.g., see Greenwood et al., 2000). Some examples of unbrecciated R chondrites are also known (e.g., see Weber et al., 1997).

The K chondrites do not yet form a coherent, well established group. Instead, there are only two meteorites known with mineralogical, chemical and isotopic characteristics similar to the type meteorite Kakangari (K) (Weisberg et al., 1996). The K chondrite *grouplet* have a very high ratio by volume of matrix to chondrules (up to 3:1), metal contents varying between 6–10% by volume, overall bulk chemical compositions and oxidation states that are intermediate between H and E chondrites, and bulk oxygen isotopic compositions that plot near the CR and CH carbonaceous chondrites. All K chondrites known to date are petrologic type 3. These enigmatic meteorites with apparently diverse affinities contradict any notion of systematic variations between the chondrite groups that may be related to formation location in the Solar System, such as distance from the Sun (Hutchison, 2004).

2.2-2.2. Non-Chondritic Meteorites

Severe heating of some meteorite parent bodies in the early Solar System produced a broad range of differentiated materials from chondritic precursors. The extent of heating and melting that produced these non-chondritic meteorites varied greatly. Those with low degrees of melting include the acapulcoite, lodranite, winonaite and, perhaps, the brachinite achondrites. Imperfect separation of metal and silicate during differentiation produced silicate-bearing irons, whereas extensive melting and planetary differentiation gave rise to basaltic achondrites, pallasites and magmatic iron meteorites (Table 2.2-2).

Most achondrites are clear testimony to episodes of melting on their parent asteroids. However, while some are true igneous rocks, or accumulations of their debris (breccias), others chemically resemble chondrites. Indeed, some achondrites retain rare vestiges of chondrules as evidence of their parent materials (e.g., see Schultz et al., 1982; McCoy et al., 1996). Most of these *primitive achondrites* (Prinz et al., 1983) have bulk compositions that are approximately chondritic (Mn/Mg within chondritic ratios $3.9–9.0 \times 10^{-3}$ atomic), and textures that are either metamorphic or igneous. Primitive achondrites have been interpreted as either chondrites that have been severely recrystallized, or the results of partial melting of chondrites. In any event, they offer a 'snapshot' of an intermediate stage in the differentiation of planetesimals. The asteroidal achondrites (notably the basaltic achondrites) represent the greatest amount of igneous material available for study from planet-like bodies beyond the Earth and Moon.

Possible genetic relationships (*clans*) have been established between a number of groups of non-chondritic meteorites that may have shared the same parent body. These include: the howardites, eucrites and diogenites (HED) tentatively linked to asteroid 4 Vesta; silicate inclusions in group IAB-IIICD irons, and a small group of primitive achondrites, the winonaites; and the primitive achondrites, acapulcoites and lodranites (Table 2.2-2).

The acapulcoites and lodranites are the products of severe metamorphism, or metaigneous activity. They have granular textures and essentially chondritic chemistries, and this is reflected in their mineralogy. While their modal mineralogy superficially resembles the ordinary chondrites, there is no clear link with any known group of chondrites (Nagahara, 1992; McCoy et al., 1996; Mittlefehldt et al., 1998; Mittlefehldt, 2003).

Similar to the acapulcoites and lodranites, the winonaites and silicate inclusions with chondritic chemistries in group IAB iron meteorites mineralogically resemble ordinary chondrites, but have oxidation states that fall between H-group ordinary and E chondrites (e.g., see Bunch et al., 1970; Bild, 1977; Davis et al., 1977; Kallemeyn and Wasson, 1985; Benedix et al., 1998, 2000; Takeda et al., 2000, and references therein).

The winonaites have essentially metamorphic textures (although some contain rare relict chondrules) and are composed of olivine, orthopyroxene, clinopyroxene, plagioclase, troilite, Fe-Ni metal, chromite, daubreelite, schreibersite, graphite, alabandite, K-feldspar and apatite. Silicate inclusions in group IAB irons, however, are more heterogeneous. Benedix et al. (2000) recognised five silicate-bearing types ranging from chondritic silicates, through non-chondritic silicates, to sulfide-rich, graphite-rich, and phosphide-bearing.

Brachinites are a small, heterogeneous group of basaltic achondrites composed predominantly of olivine, but with variable subsidiary amounts of augite, plagioclase (some are plagioclase free), traces of orthopyroxene, chromite, and minor phosphates, Fe-sulfides and Fe-Ni metal. Brachinites are essentially dunitic wehrlites from a differentiated asteroid (e.g., see Mittlefehldt (2003) and references therein).

Ureilites are a large group (92) of enigmatic ultramafic achondrites containing predominantly olivine and pyroxene, but with accessory material rich in carbon (mainly graphite) interstitial to silicates. There is no consensus as the whether ureilites are cumulates, or the residues from partial melting (Goodrich, 1992). However, their igneous origin is not disputed (e.g., see Mittlefehldt (2003) and references therein).

Eight meteorites make up the angrite group, and while there is some variation in the petrology of the member meteorites, oxygen isotopes, similar and distinctive mineralogies (except the type meteorite, Angra dos Reis), and characteristic geochemistries, all suggest that they come from the same parent body. The general consensus is that angrites are mafic igneous rocks of basaltic-like composition, but significantly depleted in alkalis relative to basalts, from a differentiated parent body. However, Varela et al. (2005) have suggested a non-igneous origin for the D'Orbigny angrite. Despite their small number, there is a substantial literature on these unusual rocks (see Mittlefehldt (2003) and references therein).

The howardites, eucrites and diogenites, collectively know as the HED meteorites, represent a large amount of igneous material from the same parent body, and have been linked tentatively to asteroid 4 Vesta (e.g., McCord et al., 1970). The HED suite comprise basalts (both brecciated and unbrecciated), gabbroic cumulates, orthopyroxenites (diogenites), plus a range of brecciated mixtures (polymict and monomict eucrite breccias) and accumulated igneous debris (howardites) of various lithologies. The HED meteorites have been linked to the group IIIAB iron meteorites (see below), and other non-chondritic meteorite groups, such as the main-group pallasites and mesosiderites (for comprehensive reviews, see Mittlefehldt et al. (1998) and Mittlefehldt (2003), and references therein).

Aubrites are highly reduced, brecciated igneous rocks. Their mineralogy and O-isotopic compositions bear similarities to the enstatite chondrites to which they may be related. The dominant mineral, enstatite, is essentially FeO free, and aubrites contain variable, but subordinate, amounts of plagioclase, high-Ca pyroxene and forsterite and, like the enstatite chondrites, an accessory mineralogy of unusual sulfides. Two other meteorites, possibly related to aubrites, are Shallowater and Mount Egerton. Both are unbrecciated and show significant chemical differences from the aubrites. Mount Egerton comprises cm-sized crystals of enstatite with substantial amounts (ca. 20 wt%) of Fe-Ni metal.

Currently, thirteen chemical groups of iron meteorites are recognised and designated with roman numerals and letters (IAB-IIICD, IC, IIAB, IIC, IID, IIE, IIF, IIG, IIIAB, IIIE,

IIIF, IVA, and IVB). The accumulated chemical, structural, and mineralogical data suggest that each group represents material disrupted from a distinct parent body. Most iron meteorites show strong magmatic fractional crystallization trends in their distribution of trace elements (Ga, Ge, Ir) relative to Ni, indicating that they represent core materials from different, highly differentiated parent bodies (Scott, 1972). This conclusion is re-enforced by the determination of metallographic cooling rates that show a small variation within some of the iron groups, but that differ between groups (e.g., see Mittlefehldt et al. (1998) and references therein). Group IVA irons shows a significant variation of cooling rates indicating a complex thermal history. Group IAB-IIICD irons contain silicates with a chondritic signature and trace element compositional trends in metal that are less pronounced than the magmatic irons, indicating that they are only partial differentiates. Collectively, the iron and associated meteorites offer a unique opportunity to study the processes of metal-silicate separation, fractional crystallization, and core formation in a number of small bodies in the early Solar System.

Pallasites are essentially composed of approximately equal amounts of silicates and Fe-Ni metal. Three sub-groups are recognised: *Main-group pallasites* are composed predominantly of olivine (commonly Fa_{12}) with accessory amounts of low-Ca pyroxene, phosphates, chromite, troilite and schreibersite; *Eagle Station grouplet* pallasites are characterized by olivine that is more iron- and calcium-rich than in the main-group, and their metal also differs in composition from the main-group in having higher Ni and Ir contents; and so-called *pyroxene pallasites* contain mm-sized grains of pyroxene that make up ca. 1–3% of their volume, and different metal compositions and oxygen isotopes also serve to distinguish them from the other sub-groups.

The composition of metal in main-group pallasites is close to that in group IIIAB irons, and this has led to the suggestion that they represent mantle-core boundary materials from the same parent asteroid (Scott, 1977). A further link through oxygen isotopes to the crustal igneous HED meteorites is now considered less likely (e.g., see Drake (2001) and references therein).

Mesosiderites are complex, polymict breccias of igneous components consisting of similar proportions of silicates (clasts and matrix) and Fe-Ni metal, with accessory troilite. The clastic silicates are essentially basalts, gabbros and pyroxenites, with some dunites and anorthosites (Scott et al., 2001). The metallic component ranges from cm-sized nuggets in some mesosiderites to mm- and sub-mm-sized grains intergrown with silicates. Overall the silicate components are very similar to the HED suite of achondrites (particularly the howardites). There is general agreement that the mesosiderites represent impact mixing of asteroidal silicate crust and metallic core components. Whether mixing took place between the crust and core, respectively, of two different differentiated asteroids, or the crust and core of the same parent body has been disputed. Moreover, despite their similarity, significant differences suggest that the mesosiderites and the HED suite formed on two separate differentiated parent bodies (e.g., see Mittlefehldt et al. (1998) and Mittlefehldt (2003), and references therein).

2.2-3. EARLY SOLAR SYSTEM EVENTS AND CHRONOLOGY

From the study of meteoritic materials there are a number of broad stages of early Solar System evolution that can be identified. Radiometric ages of whole-rock samples and their individual components, using a variety of isotopic systems, have been used to construct a time-scale for significant events (e.g., see Hutchison (2004) and references therein; McKeegan and Davis (2003) and references therein). The stages include:

- nucleosynthesis and the formation interval;
- the formation of high temperature refractory solids (Ca-Al-rich inclusions) and chondrules;
- accretion of meteorite parent bodies;
- aqueous alteration and/or prograde metamorphism;
- melting, differentiation and core formation;
- impact and shock metamorphism.

Concerning the earliest history of the chondrites, the first four stages (above) are important in understanding the accretion and early evolution of their parent planetesimals. The fifth stage is important in understanding the evolution of the non-chondritic meteorites. The remaining stages (including impact and shock metamorphism) concern the break-up of the parent bodies of meteorites, the transition of the debris as small bodies in space, their fall to Earth and, in the case of ancient finds, subsequent terrestrial weathering. These stages will not be dealt with in detail here. Moreover, not all stages of early Solar System evolution are represented in all chondritic meteorites, also there is strong evidence that some stages overlapped considerably in time.

2.2-3.1. Nucleosynthesis and the Formation Interval

The earliest recognised event is the *nucleosynthesis* of elements in stars, followed by a *formation interval* marking the time between nucleosynthesis and the incorporation of material into the parent bodies of meteorites. The nucleosynthetic process was not a single event. The recognition of the former presence of short-lived, now extinct radionuclides (Table 2.2-3) with half-lives of <100 My identified from excesses of their daughter products in meteorites has helped to constrain a relative chronology of the early Solar System.

Reynolds (1960) showed that the decay of ^{129}I in the early Solar System left an excess of ^{129}Xe in a chondrite. This was the first tangible evidence of an extinct radionuclide in the early Solar System. Iodine-129 with a half-life ($T_{1/2}$) of 15.7 My had survived from a nucleosynthetic event to become incorporated in meteorites. Importantly, Jeffery and Reynolds (1961) later showed that the excess ^{129}Xe had resulted from the *in situ* decay of the now extinct ^{129}I.

Since then, the former existence of other extinct short-lived radionuclides has been detected in meteorites and utilized to constrain the relative chronology of the earliest events in the history of the Solar System. These include ^{26}Al, ^{41}Ca, ^{53}Mn, ^{107}Pd, ^{182}Hf and ^{244}Pu (Table 2.2-3). The presence of daughter products from the decay of short-lived radionuclides in meteorites shows that these early Solar System materials formed within a few

Table 2.2-3. Short-lived radionuclides detected in early Solar System materials

Radionuclide	Daughter products	Half-life (My)
^{26}Al	^{26}Mg	0.730
^{36}Cl	^{36}Ar	0.301
^{41}Ca	^{41}K	0.103
^{60}Fe	^{60}Ni	1.5
^{53}Mn	^{53}Cr	3.74
^{107}Pd	^{107}Ag	6.5
^{182}Hf	^{182}W	8.9
^{129}I	^{129}Xe	15.7
^{244}Pu	^{132,4,6}Xe $(^{238}$U)	82

Table 2.2-4. Long-lived radionuclides for absolute dating of Solar System materials

Radionuclide	Daughter products	Half-life (yrs)
^{40}K	^{40}Ar (or ^{40}Ca)	1.3×10^9
^{87}Rb	^{87}Sr	4.88×10^{10}
^{147}Sm	^{143}Nd	1.06×10^{11}
^{176}Lu	^{176}Hf	3.5×10^{10}
^{187}Re	^{187}Os	4.56×10^{10}
^{232}Th	^{208}Pb $+ 6^4$He	1.4×10^{10}
^{235}U	^{207}Pb $+ 7^4$He	7.04×10^8
^{238}U	^{206}Pb $+ 8^4$He	4.47×10^9

million years of the synthesis of their parent isotopes. The timing of the accretion of planets and their differentiation are constrained in relation to the time of nucleosynthesis, and relative chronologies based on short-lived radionuclides (Table 2.2-3) have been cross-calibrated and 'anchored' to an absolute time-scale based on long-lived radionuclides (Table 2.2-4). If high-precision Pb-Pb ages can be determined in the same sample in which the daughter products can be detected of a short-lived radionuclide that was extant at the time of crystallization, then the relative abundance of the nuclide at that time can be determined. For example, relative time-scales from ^{26}Al-^{26}Mg systematics have been tied to absolute Pb-Pb ages for Ca-Al-rich inclusions and chondrules to constrain the timing of events in the solar nebula, and ^{53}Mn-^{53}Cr systematics have been tied to Pb-Pb ages of some differentiated meteorites to constrain early planetary melting (see McKeegan and Davis (2003) and references therein).

2.2-3.2. *Ca-Al-Rich Inclusions and Chondrule Formation*

The presence of ^{26}Al ($T_{1/2}$0.730 Ma) in the early Solar System was shown independently by Lee and Papanastassiou (1974) and Gray and Compston (1974) from refractory Ca-Al-rich inclusions in the Allende (CV) carbonaceous chondrite. Lee et al. (1976) later showed that ^{26}Mg/^{24}Mg in minerals from Ca-Al-rich inclusions in Allende correlated with Al/Mg. The abundances of decay products of many short-lived radionuclides have been shown to be correlated with the abundances of their parent elements. This demonstrates that the daughter products were not inherited from some precursor material, but that radioactive decay took place within the minerals where the daughter products were detected (e.g., see MacPherson et al., 1995). Many Ca-Al-rich inclusions have similar initial ratios of ^{26}Al/^{27}Al (4–5.0×10^{-5}) and probably formed over a very short interval of $<10^5$ yrs (e.g., see Podosek and Cassen, 1994). Srinivasan et al. (1996) confirmed the short time-scale with the discovery of an excess of ^{41}K that resulted from the decay of ^{41}Ca ($T_{1/2}0.103$ Ma) in Ca-Al-rich inclusions in the Efremovka (CV3.2) carbonaceous chondrite, indicating that some Ca-Al-rich inclusions formed within $\leqslant 1$ My of the nucleosynthesis of ^{41}Ca.

The inferred initial ^{26}Al/^{27}Al ratios in Ca-Al-rich inclusions is bimodal (MacPherson et al., 1995). A peak at 4.5×10^{-5} (called the 'canonical' value) has been taken as a marker for the beginning of the Solar System, a second peak at ^{26}Al/^{27}Al $= 0$ represents an apparent absence of initial ^{26}Al in some Ca-Al-rich inclusions. Recently, Young et al. (2005) analysed Ca-Al-rich inclusions from CV chondrites and demonstrated that some had initial ^{26}Al/^{27}Al ratios at least 25% greater than the 'canonical' value, suggesting that it cannot represent the start of the Solar System. Instead, Young et al. (2005) suggest that the 'canonical' ^{26}Al/^{27}Al value represents the culmination of thousands of transient high temperature events suffered by Ca-Al-rich inclusions during residence over a period of 10^5 yrs in the solar nebula. This is consistent with correlated petrographic and isotopic evidence (e.g., see MacPherson et al., 1995) suggesting that Ca-Al-rich inclusions were disturbed, and their Al-Mg systems reset, for around 1 My or more following their formation.

Chronological studies of the carbonaceous chondrites have focused on the Ca-Al-rich inclusions as some of the oldest material in the Solar System (Fig. 2.2-1(b)). Long-lived (Rb-Sr, Pb-Pb) and short-lived (^{26}Al, ^{53}Mn, ^{129}I) radioisotope chronometry consistently give the greatest ages for Ca-Al-rich inclusions. Early work on Ca-Al-rich inclusions in the CV chondrite Allende yielded a Pb-Pb model age of 4.559 ± 0.004 Ga (e.g., Chen and Wasserburg, 1981). The oldest ages of Ca-Al-rich inclusions are taken as the oldest minimum age (sometimes referred to as T_0 or T_{zero}) of the Solar System. Evidence from the daughter products (^{26}Mg, ^{53}Cr, ^{129}Xe) of now extinct radionuclides (^{26}Al, ^{53}Mn, ^{129}I) show that Ca-Al-rich inclusions are generally older than the chondrules with which they are associated, and in some meteorites chondrules appear to have formed around 2–3 My after Ca-Al-rich inclusions (e.g., see Shu et al., 2001).

Recent work by Amelin et al. (2002) determined the Pb-Pb age of Ca-Al-rich inclusions in the CV3.2 chondrite Efremovka at 4567.2 ± 0.6 Ma extending the accepted age of the Solar System, although this has recently been revised to 4567.1 ± 0.16 Ma (Amelin et al., 2006). Amelin et al. (2002) also obtained precise Pb-Pb ages for ferro-magnesian chondrules in the unequilibrated CR3 chondrite, Acfer 059. For the most radiogenic chondrule samples, Amelin et al. (2002) obtained isochron ages ranging from 4563 Ma to almost 4565 Ma, with a preferred age of 4564.7 ± 0.6 Ma (Fig. 2.2-4). The data of Amelin et al. (2002) show that the absolute ages of Ca-Al-rich inclusions in Efremovka are older by

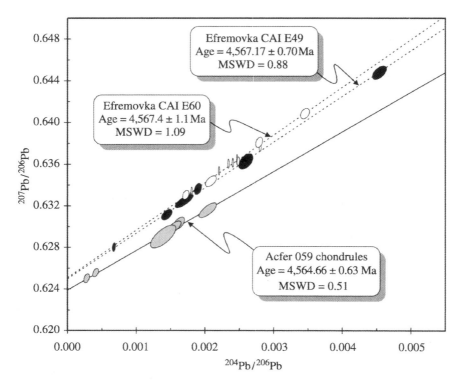

Fig. 2.2-4. Lead–lead isochrons for acid-washed fractions of two Ca-Al-rich inclusions from the CV3.2 chondrite, Efremovka, and for the six most radiogenic fractions of acid-washed chondrules from the CR3 chondrite, Acfer 059. Isochron ages for the two Ca-Al-rich inclusions overlap with a weighted mean age of 4567 ± 0.6 Ma, ca. 2.5 My older than the chondrules (after Amelin et al. (2002) – reproduced with permission of AAAS).

about 2.5 My than the dated chondrules in Acfer 059. This is consistent with the interpretation of the lower ratios of $^{26}Al/^{27}Al$ in chondrules, as opposed to Ca-Al-rich inclusions, having resulted from decay of ^{26}Al yielding an age differential of about 2 My (McPherson et al., 1995). Similar age differences between Ca-Al-rich inclusions and chondrules have been derived from ^{53}Mn-^{53}Cr and ^{129}I-^{129}Xe systematics (Swindle et al., 1996).

Most recently, Bizzarro et al. (2004) reported the presence of excess ^{26}Mg resulting from the *in situ* decay of ^{26}Al in both Ca-Al-rich inclusions and chondrules from Allende CV3.2. Individual model ages of Ca-Al-rich inclusions with uncertainties as low as ±30 ka suggest that they may have formed on a time-scale as short as 50 ka. Overall uncertainties, however, allow the oldest and youngest Ca-Al-rich inclusions to have an age differential of up to 150 ka. The chondrules from Allende analysed by Bizzarro et al. (2004) record a range of initial $^{26}Al/^{27}Al$ from 5.66 ± 0.80 × 10^{-5} to 1.36 ± 0.52 × 10^{-5}, indicating that some chondrules in that meteorite formed contemporaneously with the oldest Ca-Al-rich inclusions, and chondrule formation persisted for at least 1.4 My.

Huss et al. (2001) investigated the distribution of ^{26}Al in Ca-Al-rich inclusions and chondrules in unequilibrated ordinary chondrites. The Ca-Al-rich inclusions were found to contain radiogenic ^{26}Mg from the decay of ^{26}Al. The inferred initial ratios of ^{26}Al/^{27}Al were found to be indistinguishable from those found in Ca-Al-rich inclusions in most carbonaceous chondrites. The implication is that Ca-Al-rich inclusions in both ordinary and carbonaceous chondrites formed from the same (or similar) isotopic reservoirs, and that there was broad-scale ^{26}Al homogeneity in the nebula. This indicated that differences in the initial ratios could be interpreted in relation to formation time. The time-scale based on ^{26}Al systematics indicates that some chondrules began to form 1–2 My after Ca-Al-rich inclusions, and that metamorphism was essentially over in type 4 chondrites by 5–6 Ma after Ca-Al-rich inclusion formation. Type 6 chondrites, however, did not cool until more than 7 My after Ca-Al-rich inclusion formation (Huss et al., 2001). Huss et al. (2001) conclude that this time-scale is consistent with the decay of ^{26}Al as the principal source of heat for prograde metamorphism and differentiation in meteorite parent bodies.

2.2-3.3. Accretion, Aqueous Alteration and Prograde Metamorphism

The approximate timing of aqueous alteration and prograde metamorphism in the parent bodies of meteorites has been constrained by a variety of isotopic systems. Using Rb-Sr dating of carbonates in the Orgueil CI1 chondrite, MacDougall et al. (1984) showed that aqueous alteration occurred up to approximately 50 My after accretion of its parent body. However, ^{53}Mn-^{53}Cr systematics of carbonates in Ivuna (CI1) and Orgueil yield a much earlier time of 16.5–18.3 My after the formation of Ca-Al-rich inclusions, using an inferred initial ratio of ^{53}Mn/^{55}Mn between $1.42–1.99 \times 10^{-6}$ (Endress et al., 1996; Hutcheon and Phinney, 1996). In the Semarkona ordinary chondrite, the I-Xe chronometer suggests that aqueous alteration of its chondrules occurred approximately 10 Ma after their formation (Swindle et al., 1991a). Whereas Mn-Cr systematics of fayalite grains in chondrules in the Mokoia CV3.2 chondrite are interpreted as resulting from hydrous metamorphism on its parent body 7–16 My after the formation of Ca-Al-rich inclusions (Hutcheon et al., 1998).

Originally thought to be a nebula condensate, magnetite in the CI1 chondrite Orgueil is now considered the product of hydrous processes because the morphology of grains resemble those of magnetite formed terrestrially at low temperatures in the presence of aqueous solutions (Kerridge et al., 1979). Some apparently anomalously high I-Xe ages for components of CI chondrites, including Orgueil magnetite, have recently been determined again and found to be younger than enstatite from the Shallowater enstatite achondrite (Hohenberg et al., 2000). Although imperfectly understood, an extended aqueous alteration history for the CM carbonaceous chondrites has been inferred from the I-Xe systematics of mineral separates (e.g., see Niemeyer and Zaikowski, 1980). Using magnetite from Murchison (CM2) as a standard, closure in the Murray and Cold Bokkeveld CM2 chondrites occurred around 11 My after the formation of Ca-Al-rich inclusions. Mn-Cr systematics of carbonates in CM chondrites indicate a similar period of aqueous alteration to that inferred from

I-Xe, and initial ^{53}Mn/^{55}Mn ratios indicate that carbonates crystallized between 10–14 My after the formation of Ca-Al-rich inclusions (e.g., see Brearley and Hutcheon, 2000; Brearley et al., 2001; Brearley, 2003). Importantly, the prolonged periods of aqueous alteration indicated by the data from CI1, CM1 and CM2 chondrites are compatible with alteration while they were resident in their parent bodies. However, Mn-Cr data for carbonates in the CR chondrite, Kaidun, show an initial ^{53}Mn/^{55}Mn ratio of ca. 9.4×10^{-6}, indicating very early aqueous alteration at ca. 4569 Ma (Hutcheon et al., 1999). This age is greater than the Pb-Pb age of Ca-Al-rich inclusions in Efremovka (4567.1 ± 0.16 Ma – Amelin et al., 2006) and suggests that either 4569 Ma is only a lower limit to the age of the Solar System, or that the Pb-Pb age of Ca-Al-rich inclusions reflects a later re-equilibration event, rather than formation (Shukolyukov and Lugmair, 2006). The components of some chondrites (including CR, CO and CV), however, may have also suffered alteration prior to their incorporation into parent bodies (e.g., see Brearley (2003) and references therein).

A single, whole-rock Rb-Sr isochron for the H, LL, EH and EL chondrites (excluding the most heavily shocked) has been determined by Minister et al. (1982), yielding an Rb/Sr age of 4498 ± 15 Ma (with an initial ^{87}Sr/^{86}Sr = 0.69885). Moreover, Torigoye and Shima (1993) established an internal isochron for an EL6 chondrite (Khairpur) that is consistent with the age determined by Minister et al. (1982). However, there is no consensus as to whether these ages correspond to chondrule formation, chondrite accretion, or subsequent metamorphism. Göpel et al. (1994) studied the U-Th-Pb systematics of separated phosphates from 15 equilibrated (types 4–6) ordinary chondrites. In one H4 chondrite (Sainte Marguerite) the data suggest that phosphate had become closed to U-Pb exchange at 4563 ± 1 Ma (Göpel et al., 1994). The Pb-Pb ages of phosphates in H-group chondrites studied by Göpel et al. (1994) vary inversely with petrologic type, with the most recent closure in an H6 chondrite (Guarena) at 4505 ± 1 Ma, suggesting that this reflects closure at the end of metamorphism. However, there is no such correlation in the Pb-Pb ages of phosphates from the L- and LL-group ordinary chondrites studied by Göpel et al. (1994). More data are required to establish whether such a correlation exists in these groups, but has since been disturbed in some meteorites by shock-metamorphism.

I-Xe ages for a number of ordinary and enstatite chondrites (and their components) have been published in recent years (e.g., see Hutchison et al. (1988) and references therein; Swindle et al., 1991a, 1991b; Gilmour et al., 2000; Whitby et al., 1997, 2000; Busfield et al., 2001, 2004). There is considerable debate as to how these ages are to be interpreted (e.g., see Swindle et al., 1996). Brazzle et al. (1999) have shown that I-Xe ages of feldspars from some ordinary chondrites are concordant with Pb-Pb ages of co-existing phosphates, indicating that these ages are consistent with closure at the end of metamorphism (see also Pravdivtseva and Hohenberg, 1999). However, studies of separated chondrules from ordinary chondrites by Gilmour et al. (2000) and Gilmour and Saxton (2001) suggest that the earliest chondrule I-Xe ages may represent formation. This is supported by a general overlap in the earliest chondrule ages in the I-Xe, Mn-Cr, and Al-Mg systems (e.g., see Gilmour et al. (2000) and references therein). It is important to note that underpinning these conclusions is the assumption that isotopic homogeneity existed across the region of the early Solar System, and that there was simultaneous closure of each of the systems in the samples used for calibration (Gilmour et al., 2000).

A brecciated H3-6 ordinary chondrite (Zag) contains halite with an initial $^{129}I/^{127}I$ ratio of 1.35×10^{-4}, suggesting that the halite precipitated from saline fluids on its parent body within 2 My of the formation of Ca-Al-rich inclusions (Whitby et al., 2000). A halite crystal from another H-group ordinary chondrite breccia (Monahans, 1998) gave a model Rb-Sr age of 4.7 ± 0.2 Ga, supporting the conclusion that precipitation of halite occurred very early in the history of the H-group parent body (Zolensky et al., 1999). To constrain more accurately the timing of aqueous alteration in H-group chondrites, Busfield et al. (2004) obtained further analyses of halides from the Monahans (1998) and Zag H-group chondrites. The initial $^{129}I/^{127}I$ ratio in Monahans (1998) halide was determined to be $9.37 \pm 0.06 \times 10^{-5}$. From variation in the data, particularly in Zag, Busfield et al. (2004) concluded that it would be unreliable to interpret this in terms of a formation age. Instead, Busfield et al. (2004) propose a model whereby halides in both H-group chondrites formed at 4559 Ma, approximately 5 My after the crystallization of the Shallowater enstatite achondrite. This agrees with the estimated timing of aqueous alteration in carbonaceous chondrites and metamorphism in the ordinary chondrites.

Ar-Ar ages of H-group ordinary chondrites range from 4.45–4.53 Ga (Trieloff et al., 2003). For some unshocked ordinary chondrites, their ^{39}Ar-^{40}Ar ages are a close match to metamorphic ages determined by the Rb-Sr and Pb-Pb systems. Other ordinary chondrites have significantly lower ages indicating that the system was reset by heating, probably related to shock-metamorphism during parent body break-up (e.g., see Anders, 1964; Heymann, 1967; Turner, 1988).

To date, there are limited isotopic chronologies for the R chondrites, and none for the K chondrites. Bischoff et al. (1994) and Nagao et al. (1999) investigated the K-Ar systematics of a number of R chondrites. More recently, Dixon et al. (2003) determined the whole-rock ^{39}Ar-^{40}Ar ages of four R chondrites. The peak ^{39}Ar-^{40}Ar ages of these meteorites are $\geqslant 4.35 \pm 0.01$ Ga (Carlisle Lakes), 4.47 ± 0.02 Ga (Rumuruti), $\geqslant 4.37 \pm 0.01$ Ga (Pecora Escarpment 91002), and 4.30 ± 0.07 Ga (Acfer 217) (Dixon et al., 2003). These ages are similar to the Ar-Ar ages of undisturbed ordinary chondrites (4.52–4.38 Ga), and older than ordinary chondrites that have been shock metamorphosed ($\ll 4.2$ Ga) (Dixon et al., 2003; Turner et al., 1978; Bogard, 1995, and references therein). Of the dated meteorites, all but Carlisle Lakes are breccias, and their peak ages are disturbed to various extents by recoil, and diffusive loss of ^{40}Ar. Dixon et al. (2003) suggest that these ages probably do not record slow cooling within an undisturbed parent body, but impact and brecciation could have differentially reset their ages. Alternatively, the variation in ages may be accounted for by mixing of material with different ages.

2.2-3.4. Melting, Differentiation and Core Formation

2.2-3.4.1. Acapulcoites and lodranites

The only acapulcoite for which a formation age has been determined is the type meteorite, Acapulco. Prinzhofer et al. (1992) determined a Sm-Nd age of 4.60 ± 0.03 Ga. However, McCoy et al. (1996) questioned this age, which is somewhat older than the accepted age of the Solar System. A high precision Pb-Pb age for phosphate from Acapulco of 4.557 ± 0.002 Ga has been determined by Göpel et al. (1992), and this is taken as one of the 'anchor' ages for cross-correlation of relative chronologies based on short-lived radionuclides. The formation interval between the Bjurböle L/LL4 chondrite and Acapulco based on ^{129}I-^{129}Xe is evidently only 8 My (Nichols et al., 1994), or around 10 My using the Shallowater standard. Formation of Acapulco early in the Solar System is supported by Pellas et al. (1997), who showed that ^{244}Pu ($T_{1/2}$81.8 My) was present in Acapulco phosphates during cooling through the temperature of fission track retention. In addition, ^{40}Ar-^{39}Ar ages for a number of acapulcoites and lodranites have been measured. These are all much younger than the formation age determined for Acapulco, and range from 4.519 ± 0.017 Ga in the acapulcoite EET 84302, to 4.49 ± 0.01 Ga in the lodranite Gibson, indicating that the retention of radiogenic argon began a few tens of millions of years after their formation (McCoy et al., 1997; Mittlefehldt et al., 1996; Pellas et al., 1997). These younger ages are interpreted as representing cooling over a long time from peak temperatures (McCoy et al., 1996).

2.2-3.4.2. Winonaites and silicate inclusions in IAB-IIICD irons
The few radiometric ages available for silicate inclusions in group IAB irons suggest that the partial melting, crystallization and metal-silicate mixing in IAB irons occurred early in the Solar System. Niemeyer (1979a) showed that I-Xe ages lie within ± 3 My of the Bjurböle L/LL4 ordinary chondrite. Subsequently, Brazzle et al. (1999) recalibrated this to an absolute age of 4566 ± 0.2 Ma. Ar-Ar ages for silicate inclusions in group IAB irons range from 4.43 ± 0.03 Ga (in Landes) to 4.52 ± 0.03 Ga (in Mundrabilla, Pitts, and Woodbine), and support their early formation (Niemeyer, 1979b). Within uncertainties, an early onset of differentiation is consistent with Re-Os ages for the metallic portions of group IAB irons (see below).

2.2-3.4.3. Brachinites
Chronologically, like many other achondrites, short-lived radionuclide systematics indicate that the brachinites formed early in the Solar System (Mittlefehldt et al., 2003; Swindle, et al., 1998). A correlation between ^{129}Xe/^{132}Xe and ^{128}Xe/^{132}Xe indicates that ^{129}I was extant at the time of formation of the type meteorite, Brachina (Mittlefehldt, et al., 2003). Data for another brachinite, Eagles Nest, however, suggests that it began to retain ^{129}Xe within ca. 50 My of primitive chondrites (Swindle, et al., 1998). Wadhwa et al. (1998) showed that in Brachina, ^{53}Cr correlates with Mn/Cr indicating that ^{53}Mn was also extant at the time of its formation. By cross correlating the ^{53}Mn/^{55}Mn ratio and the Pb-Pb age of angrite LEW 86010, Wadhwa et al. (1998) calculated a formation age for Brachina of 4.5637 ± 0.0009 Ga, indicating that it is only around 5 My younger than the accepted age (from Ca-Al-rich inclusions) of the Solar System.

In terms of tertiary events, ^{39}Ar-^{40}Ar data indicate that Brachina, and another of the group, EET 99407, were outgassed ca. 4.13 Ga ago, and this has been tentatively attributed to a common impact event (Mittlefehldt, et al., 2003).

2.2-3.4.4. Ureilites

Because of their seemingly conflicting 'primitive' and 'differentiated' characteristics, the chronology of the ureilites, particularly Sm-Nd systematics, has been open to wide interpretation. However, Torigoya-Kita et al. (1995) derived a Pb-Pb age of 4.563 ± 0.021 Ga for the Antarctic ureilite MET 78008, suggesting that ureilites formed early in the Solar System. Ar-Ar data for ureilites show a history of outgassing events from approximately 4.5–4.6 Ga to 3.3 Ga which is consistent with their generally highly shocked nature (Bogard and Garrison, 1994).

2.2-3.4.5. Angrites

The importance of the angrites to the chronology of early Solar System processes far outweighs their small number and the lack of detailed knowledge of magmatic trends on their parent body. The angrite SAH 99555 is the oldest known igneous rock in the Solar System (Fig. 2.2-5). Angrites have low ^{204}Pb/^{238}U ratios which allows precise Pb-Pb ages to

Fig. 2.2-5. The Sahara 99555 angrite meteorite is the oldest known igneous rock in the Solar System, with a high precision Pb-Pb age of 4.5662 ± 0.0006 Ga. The meteorite provides evidence of melting and differentiation of at least one planetesimal very early in the history of the Solar System, prior to the formation of some chondrites (reproduced with the permission of Labenne meteorites).

be determined. Pb-Pb model ages of 4.555–4.558 Ga for the formation of LEW 86010 (4.5578 Ga), Angra dos Reis, and D'Orbigny have been determined (Lugmair and Galer, 1992; Jagoutz et al., 2003). In addition, Cr anomalies from the decay of ^{53}Mn have also been detected in mineral separates from LEW 86010 (Lugmair and Galer, 1992; Nyquist et al., 1994). The absolute Pb-Pb age of LEW 86010 has been used to calibrate the Mn-Cr chronometer (Lugmair and Shukolyukov, 1998, 2001), and these data suggest that LEW 86010 formed within 15–20 My of the most ancient materials in the Solar System. Mn-Cr studies of the D'Orbigny (Glavin et al., 2004), and recent Pb-Pb studies (Zartman et al., 2006) of the D'Orbigny and Asuka 881371 angrites, also indicate consistent Mn-Cr and Pb-Pb ages (4562.9 ± 0.7 Ma and 4563.9 ± 0.6 Ma, respectively, for D'Orbigny) and give another absolute time calibration of the Mn-Cr chronometer.

Recently, Baker et al. (2005) have re-investigated the chronometry of the angrites and determined refined Pb-Pb ages. Whole rock samples of the angrites SAH 99555, NWA 1296, and a pyroxene separate from SAH 99555 yield a high precision isochron age of 4.5662 ± 0.0006 Ga which is only ca. 1 My younger than the currently accepted minimum age of the Solar System derived from Ca-Al-rich inclusions in the CV3.2 carbonaceous chondrite Efremovka (Amelin et al., 2002, 2006).

Baker et al. (2005) found ^{26}Mg excesses from the decay of ^{26}Al in whole-rock fragments of angrites. However, separated feldspar from SAH 99555 did not yield a ^{26}Al-^{26}Mg isochron, which Baker et al. (2005) suggested was the result of re-equilibration of feldspar with low Al/Mg phases (olivine) during cooling. From united feldspar and whole-rock data for SAH 99555, however, Baker et al. (2005) obtained a ^{26}Al-^{26}Mg age 5.6 ± 0.3 My after the formation of Ca-Al-rich inclusions, in agreement with other measurements (5.7 ± 0.4 My) on purer feldspar separates from the same meteorite by Nyquist et al. (2003). Calibrated to the Pb-Pb age of Ca-Al-rich inclusions, the ^{26}Al-^{26}Mg age of feldspar of 4.5616 ± 0.0007 Ga is younger than the whole-rock Pb-Pb age of SAH 99555, which Baker et al. (2005) suggest represents thermal resetting.

By calculating ^{26}Al-^{26}Mg ages from whole rock samples of angrites, and by reference to ^{26}Al initial abundance of Ca-Al-rich inclusions, Baker et al. (2005) obtained a formation time for the angrites of 3.3–3.8 My after Ca-Al-rich inclusions, in agreement with the time difference that can be calculated from initial Sr isotopic data (Nyquist et al., 2003; Halliday and Porcelli, 2001). This age difference is greater than that (ca. 1 My) indicated by the difference between the Pb-Pb ages of angrites determined by Baker et al. (2005) and that determined for Ca-Al-rich inclusions from Efremovka by Amelin et al. (2002, 2006). When data are normalised to the standard values used by Baker et al. (2005), the Ca-Al-rich inclusion from Efremovka with the most precise Pb isotopic data defines an age of 4.5695 ± 0.0004 Ga, corresponding to an age difference between Ca-Al-rich inclusions and angrite formation of ca. 3 My. By integrating their Pb-Pb and ^{26}Al-^{26}Mg data for angrites with ^{26}Al-^{26}Mg systematics for Ca-Al-rich inclusions (Bizzarro et al., 2004; Young et al., 2005), Baker et al. (2005) suggest that some Ca-Al-rich inclusions formed at $\geqslant$ 4.5695 ± 0.0002 Ga, thus providing an older minimum age for the Solar System.

2.2-3.4.6. Howardites, eucrites and diogenites (HED)

The chronology of the HED suite of achondrites has been extensively studied. Chronometers based on long-lived isotopes Rb-Sr, Sm-Nd, and Pb-Pb and short-lived radionuclides ^{129}I, ^{244}Pu, and ^{53}Mn have been utilised to study the HED meteorites. Pb-Pb and Rb-Sr systematics yield ages for basaltic eucrite magmatism in the range 4.51 Ga (Tera et al., 1997) to 4.60 Ga (Allegre et al., 1975; Nyquist et al., 1986), respectively. From Rb-Sr systematics, Smoliar (1993) derived an age of formation of the eucrites of 4.548 ± 0.058 Ga. Following the discovery of evidence of extant ^{53}Mn and ^{60}Fe during the formation of the basaltic eucrites (e.g., see Shukolyukov and Lugmair, 1993a, 1993b), Lugmair and Shukolyukov (1997) used Mn-Cr systematics to obtain an isochron for diogenites and basaltic eucrites, and tied this to the Pb-Pb age and Mn-Cr data for the LEW 86010 angrite to obtain a time of "mantle fractionation" of the HED parent body at 4.5648 ± 0.0009 Ga.

Pb-Pb ages for three cumulate eucrites determined by Tera et al. (1997) range from 4.399 ± 0.035 Ga (Serra de Mage) to 4.484 ± 0.019 Ga (Moore County), with Moama, with larger uncertainty, intermediate at 4.426 ± 0.094 Ga. These ages are all significantly younger by about 100 My than the basaltic eucrites and, allowing for uncertainties, are less than the estimated time of differentiation of the HED body. Explanations for the discrepancy vary and may be the result of later formation, or disturbance as the result of metamorphism and annealing for which there is abundant petrological evidence in the HED suite of rocks (e.g., see Keil et al., 1997; Yamaguchi et al., 1996, 1997). The data for the diogenites suggest they are similarly younger than the basaltic eucrites. An Rb-Sr isochron for two diogenites (Tatahouine and Johnstown) derived by Takahashi and Masuda (1990) gave 4.394 ± 0.011 Ga; however, both of these meteorites show evidence of alteration by shock.

Bogard (1995) reviewed the Ar-Ar systematics of the brecciated eucrites and howardites and confirmed the extensive disturbance of the HED suite. The great majority of eucrites and howardites show significant Ar loss, and most ages indicate peak disturbance between 4.1 and 3.4 Ga ago (Bogard, 1995). However, the unbrecciated, metamorphosed basaltic eucrites yield Ar-Ar ages of approximately 4.5 Ga, older than the major period of disturbance, but younger than the estimated time of major parent body differentiation.

2.2-3.4.7. Aubrites (enstatite achondrites)

There are few modern data on the chronology of the aubrites. Compston et al. (1965) determined an Rb-Sr age for Bishopville of 3.7 Ga. In contrast, Rb-Sr and K-Ar ages for Norton County determined by Bogard et al. (1967) approach the age of the Solar System. Enstatite in Shallowater is now used as a reference standard in the I-Xe system, replacing chondrules in the Bjurböle L/LL4 chondrite that proved to have variable ^{129}I/^{127}I ratios. Shallowater has an inferred crystallization age of 4.5658 ± 0.0020 Ga which places it very early in the Solar System (Hohenberg et al., 1998; Brazzle et al., 1999). Younger ages for some aubrites may represent resetting during brecciation.

2.2-3.4.8. Iron meteorites and pallasites
Using the ^{187}Re-^{187}Os chronometer, Morgan et al. (1992, 1995) measured the ages of group IIAB and IIIAB irons and found them to be the similar to the chondrites. Subsequently, Shen et al. (1996) investigated the Re-Os systematics of some group IAB, IIAB, IIIAB, IVA and IVB irons (see also Smoliar et al., 1996). The age of the IIAB irons determined by Shen et al. (1996) is consistent with that determined by Morgan et al. (1995), whereas the group IVA irons were found to be slightly older than group IIAB. More recently, Horan et al. (1998) found that the group IAB-IIICD, IIAB, IIIAB and IVB irons have essentially the same Re-Os ages, and their data, using the same half-life of ^{187}Re, is generally in agreement with Shen et al. (1996). However, Horan et al. (1998) found that the group IVA irons were younger by approximately 80 My than the age determined by Shen et al. (1996).

Evidence for the extinct radionuclide ^{107}Pd ($T_{1/2} = 6.5$ Ma, decays to ^{107}Ag) has been found in (metal and sulfide) magmatic iron groups IIAB, IIIAB, IVA and IVB, and the metallic phases of other meteorites (Kelly and Wasserburg, 1978; Wasserburg, 1985). A correlation between excess ^{107}Ag/^{109}Ag and the ratio of Pd/Ag was shown to exist in the magmatic group IIIAB iron, Grant, by Kaiser and Wasserburg (1983). Subsequently, other irons and stony-irons have been investigated and shown to have a range of initial ^{107}Pd/^{108}Pd ratios, with most between 1.5–2.5×10^{-5} (e.g., see Chen et al. (2002) and references therein). Isochrons in the main group pallasite Brenham, and the IIIAB iron, Grant, have recently been determined by Carlson and Hauri (2002) with initial ^{107}Pd/^{108}Pd ratios of 1.6×10^{-5}, close to that determined by Kaiser and Wasserburg (1983). Apparent initial ^{107}Pd/^{108}Pd ratios for the group IAB iron Canyon Diablo and the group IVA iron, Gibeon, are the same (Chen and Wasserburg, 1990), and the data indicate that the possible magmatic pair Grant (IIIAB) and Brenham (pallasite) solidified 3.5 My after Canyon Diablo (IAB) and Gibeon (IVA). The group IVA irons have a higher radiogenic ^{107}Ag/^{108}Pd ratio than magmatic groups IIAB, IIIAB and IVB, suggesting that they are older. This is consistent with the Re-Os age of IVA irons determined by Shen et al. (1996), but at odds with that determined by Horan et al. (1998).

The ages of several main-group pallasites have been constrained by Mn-Cr systematics (Hsu et al., 1997; Hutcheon and Olsen, 1991; Shukolyukov and Lugmair, 1997). The Eagle Station pallasite has been similarly dated by Birck and Allegre (1988) and Shukolyukov and Lugmair (2006). The data suggest that ^{53}Mn ($T_{1/2} = 3.7$ My) was extant at the time that the pallasites formed. By correlation of the Mn-Cr data for the main group pallasite Omolon to the Mn-Cr and Pb-Pb age of the angrite, LEW 86010, Shukolyukov and Lugmair (1997) determined an absolute age of 4558 ± 1 Ma for Mn-Cr closure. Recent ^{53}Mn-^{53}Cr data for Eagle Station indicate that Cr isotopes equilibrated in that meteorite at 4557.5 ± 0.6 Ma, essentially at the same time as the main-group pallasite Omolon (Shukolukov and Lugmair, 2006). Moreover, Shukolyukov and Lugmair (2006) note that Cr isotope systematics in Eagle Station are similar to those in Allende, indicating that the precursor of this pallasite was a CV-like material, and this confirms previous similar suggestions made on the basis of chemical composition (Scott, 1977) and oxygen isotopes (Clayton and Mayeda, 1996). The two main-group pallasites, Springwater and Omolon, have ^{107}Pd-^{107}Ag and ^{53}Mn-^{53}Cr ages (Lugmair and Shukolyukov, 1998) that tie those determined by ^{107}Pd-^{107}Ag to the timescale derived from ^{53}Mn-^{53}Cr. However, there are discrepancies that may have resulted from the closure of systems at different times.

The evidence from irons and pallasites suggests that melting and differentiation of planetesimals, other than the angrite parent body, occurred very early in the Solar System. Re-Os and Hf-W systematics in some iron meteorites indicate that core formation occurred within 5 My of the earliest solids in the Solar System (e.g., see Horan et al., 1998). The early differentiation of planetesimals is supported by the presence of excess ^{26}Mg from the decay of now extinct ^{26}Al in the eucrites Piplia Kalan (Srinivasan et al., 1999) and Asuka 881394 (Nyquist et al., 2001b), and strengthens the case for ^{26}Al as the heat source for melting (e.g., see Keil, 2000).

The ^{182}Hf-^{182}W chronometer ($T_{1/2} = 8.9$ My) has been used to constrain rates of accretion and provide model ages of metal-silicate separation and core formation. The Hf-W system has been calibrated against Pb-Pb ages of the H4 ordinary chondrites Sainte Marguerite and Forest Vale, and cross-calibrated against the initial ratio of ^{26}Al/^{27}Al (Kleine, et al., 2002). Halliday and Lee (1999), Halliday (2000, 2003), and Jacobsen (2005) have provided detailed reviews of Hf-W systematics in meteorites and the early Earth. Halliday (2003) has cautioned that the data have to be considered carefully against uncertainties in the exact abundance of ^{182}Hf in the early Solar System, the initial ratio of ^{182}Hf/^{180}Hf, and the decay constant of ^{182}Hf. Nevertheless, there appears to be a clearly resolvable deficit in ^{182}W in iron meteorites and the metal of ordinary chondrites, relative to the atomic abundance found in the silicate Earth (Halliday, 2003). The simplest explanation for this difference is that the meteoritic metals, or silicate Earth, or both, sampled early Solar System W while ^{182}Hf was extant (Halliday, 2003, and references therein).

Recent work on the Hf/W systematics of iron meteorites has provided some constraints on their formation and evolution (e.g., Horan et al., 1998; Kleine et al., 2005a; Markowski et al., 2005, 2006). All iron meteorites have ε_W between -3.9 and -2.7, with groups II-IAB, IVB, and IC iron meteorites having the least radiogenic ^{182}W/^{184}W ratios (Kleine et al., 2005a). Provided that the ε_W values for these meteorite groups have not been altered by cosmogenic effects (see below), the implications are that the last equilibration of the Hf/W system in Ca-Al-rich inclusions apparently postdates core formation in the oldest asteroids by 2.5 ± 1.7 My. Moreover, these age constraints have led to a re-examination of models of asteroid formation and suggest that the formation of the parent asteroids of some chondrites may represent a second generation of asteroids that may be the re-accreted debris from the break-up of an earlier generation of planetesimals (Kleine et al., 2005a). However, because of the long exposure ages (several hundred Ma) of some irons, galactic cosmic ray induced effects on the isotopic composition of W cannot be excluded, and this has proved to be the case (Markowski et al., 2006). From experimental evidence of cosmogenic effects on W isotopes in the magmatic irons Carbo (IID) and Grant (IIIAB), Markowski et al. (2006) have shown that the ^{182}W/^{184}W ratio in the preatmospheric centre, compared with the preatmospheric surface of these irons, is ca. 0.5 epsilon lower. A demonstrated correlation between the ^{182}W/^{184}W ratio and ^{3}He concentration in Carbo and Grant provides a link with the flux of the relevant cosmic ray particles (Markowski et al., 2006). This allows a correction to be made for cosmic ray effects on W isotopes from ^{3}He concentrations and independent exposure ages of meteorites. When the correction is applied to some magmatic iron meteorites from groups IIAB, IID, IIAB and IVB, their W isotopic compositions are similar to initial compositions in Allende (CV3) Ca-Al-rich inclusions, and indicate segregation within 1.2 My of closure of the Hf-W system in Allende Ca-Al-rich inclusions (Markowski et al., 2006). Irons such as Carbo (IID), the oldest studied, represent the relics of the cores of an early generation of differentiated planetesimals. Overall variations in isotopic compositions in the irons studied by Markowski et al. (2006) indicate that they segregated over a time scale of ca. 4 My.

In a re-examination of the Hf-W systematics of metal in some types 3–6 ordinary chondrites, Kleine et al. (2006) tested the model proposed by Kleine et al. (2005a) that core formation in the parent bodies of magmatic irons predated the formation of chondrules in ordinary chondrites (ages constrained by Pb-Pb and Al-Mg). Metal in type 6 ordinary chondrites is enriched in W relative to metal in type 3 ordinary chondrites, while their W isotopic compositions are similar (Kleine et al., 2006). However, metal from type 6 chondrites appears slightly more radiogenic than that from the only type 3 chondrite (Julesburg L3.6) studied. The initial ^{182}Hf/^{180}Hf ratio of Julesburg metals is slightly lower than that of Allende Ca-Al-rich inclusions. This indicates that Julesburg metal formed ca. 3 My after Allende Ca-Al-rich inclusions, and is consistent with chondrule formation intervals after Ca-Al-rich inclusions, based on Pb-Pb and Al-Mg systematics (Kleine et al., 2006 and references therein). Compared with the magmatic iron Negrillos (IIAB) with a low cosmic ray exposure age (45 My), Julesburg metal has a more radiogenic W isotopic composition. This indicates that the formation of Julesburg metal postdated core formation in some magmatic irons. This is consistent with the suggestion that the accretion of parent bodies of some irons, and some achondrites, such as the angrites, predated the formation of ordinary chondrites (Kleine et al., 2006).

2.2-3.4.9. Mesosiderites
Chronological work on the mesosiderites has concentrated on their silicate components. Stewart et al. (1994) measured Sm-Nd isochron ages for three clasts from Vaca Muerta, and one from Mount Padbury, which gave 4.48–4.52 Ga for igneous clasts, and 4.42 Ga for an impact-melt clast. A U-Pb age for zircons in a basaltic clast from Vaca Muerta gave 4.563 ± 0.015 Ga (Ireland and Wlotzka, 1992), indicating crystallization very early in the Solar System. Ages from ^{39}Ar-^{40}Ar studies of a variety of clasts and whole-rock samples of mesosiderites give young ages of around 3.95 Ga, and this has been interpreted as representing either metamorphic resetting, or slow cooling (Bogard and Garrison, 1998; Bogard et al., 1990). Metal in mesosiderites yield extremely slow (<1 K My^{-1}) metallographic cooling rates (Hopfe and Goldstein, 2001, and references therein) supporting the suggestion that extended cooling may be the cause of young Ar-Ar ages. However, abundant petrological evidence of metamorphism and disruption in mesosiderites provides equal support for impact resetting as a mechanism for Ar loss (Bogard et al., 1990; Rubin and Mittlefehldt, 1992).

2.2-4. SUMMARY

For many years, the straightforward view prevailed that generally, the chondrites represent primitive Solar System materials from which the inner planets may have accreted (e.g., see Ganapathy and Anders, 1974). However, the modern view is stated by Halliday (2003): *"Though undifferentiated, chondrites, with the possible exception of CIs, are not primitive and certainly do not represent the first stages of accretion of the Earth"*. From the current, better understanding of the chronology of meteorites and their components, a much more complex picture emerges.

The parent body of the angrites evidently accreted and melted while ^{26}Al was extant, and modelling constrains accretion to $\geqslant$4.568 Ga (Baker et al., 2005). This age of accretion of the angrite parent body is older than almost all published absolute and relative ages for chondrules (Amelin et al. 2002, 2004; Bizzarro et al., 2004; Nakamoto et al., 2005). However, there are some chondrules in a CV chondrite (Allende) that appear to have formed as early as Ca-Al-rich inclusions (Bizzarro et al., 2004). Baker et al. (2005) concluded that the parent body of the angrites must have accreted before the parent bodies of most chondrites.

The existence of an early generation of differentiated planetesimals is supported by the Hf-W data for some iron meteorites (Kleine et al., 2004, 2005a, 2006) that indicate that chondrite parent bodies may have accreted after the differentiation of the parent bodies of some groups of magmatic irons. However, since some chondrules appear to be synchronous with Ca-Al-rich inclusions (Bizzarro et al., 2004), an early generation of 'chondritic' materials that were subsequently melted and are no longer represented in the current meteorite flux cannot be ruled out. Indeed, the angrite parent body appears to have formed from such a primitive source. The presence of differentiated bodies in the Solar System during, or prior to, the major period of chondrite accretion suggests that differentiated materials may have been available at the time of accretion. Inclusions of igneous rocks with ages less than a few Ma after Ca-Al-rich inclusions, which have been identified in ordinary chondrites, would seem to provide evidence to support this (e.g., see Hutchison et al., 1988).

Overall, the differential ages of Ca-Al-rich inclusions and many chondrules support models for the evolution of the early solar nebula over few million years (e.g., see Podosek and Cassen, 1994). A problem that this produces in terms of the accretion of the chondrites is the long time (ca. 1–2 My using the timescale of Amelin et al., 2002, or 2–3 My using the revised timescale of Baker et al., 2005) that Ca-Al-rich inclusions were held in the nebula before they accreted, along with chondrules, into chondrites (see MacPherson et al., 1995). No satisfactory explanation for this has been suggested, and while the age differential (based on precise absolute Pb-Pb ages of Ca-Al-rich inclusions and chondrules, respectively, from two different meteorites) is supported by Al-Mg systematics (provided that Ca-Al-rich inclusions did not contain ^{26}Al produced by spallation) it has only recently been demonstrated in a single chondrite. The data of Bizzarro et al. (2004) suggest that some chondrules in Allende not only formed contemporaneously with Ca-Al-rich inclusions, but that chondrule formation continued for a period of 1.4 My. Overall, chondrule

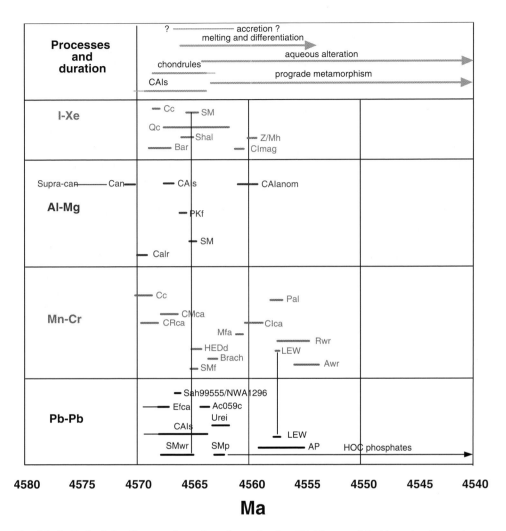

4580   4575   4570   4565   4560   4555   4550   4545   4540

Ma

Fig. 2.2-6. Early Solar System chronology from absolute Pb-Pb ages (black), and calibrated ages from Mn-Cr (red), Al-Mg (blue) and I-Xe (green) systematics. The Pb-Pb age of the angrite, Lewis Cliff 86010 (LEW), is used to calibrate the Mn-Cr system. The closure of Mn-Cr in the H4 ordinary chondrite feldspar (SMf) is correlated with Sainte Marguerite whole-rock (SMwr) and with Al-Mg (SM) and I-Xe (SM). The Pb-Pb age of Acapulco phosphate (AP) is also used as an 'anchor' age for the cross-correlation of chronologies based on short-lived radionuclides. Key to other symbols: CAIs (Ca-Al-rich inclusions), SMp (Sainte Marguerite phosphate), HOC (H-group ordinary chondrites), Urei (ureilites), Efcai (Efremovka Ca-Al-rich inclusions), Ac059c (Acfer 059 chondrules), Sah99555 (Sahara 99555 angrite), Brach (Brachina), HEDd (howardite, eucrite, diogenite differentiation), Mfa (Mokoia fayalite), Awr (Acapulco whole rock), Rwr (Richardton whole rock), CIca (CI chondrite carbonates), CRca (CR chondrite carbonates), CMca (CM chondrite carbonates), Cc (Chainpur LL3.5 ordinary chondrite chondrules), Pal (Pallasites), Calr (Al-rich chondrules),

Fig. 2.2-6. (*Continued.*) Pkf (Piplia Kalan eucrite feldspar), Can (Ca-Al-rich inclusions with the canonical ^{26}Al/^{27}Al ratio) CAIanom (Ca-Al-rich inclusions with ^{26}Al/^{27}Al = 0), CImag (CI chondrite magnetite), Z/Mh (Zag and Monahans (1998) halite), Shal (Shallowater), Qc (Quingzhen chondrules), Cc (Chainpur chondrules) (after Gilmour and Saxton, 2001; McKeegan and Davis, 2003).

formation may have continued for around 2–3 My in the early Solar System with the youngest chondrules constraining the age of accretion (Fig. 2.2-6).

The whole-rock Pb-Pb ages of ordinary chondrites may represent the end of metamorphism, and not chondrule formation or accretion. Most ages cluster around 4.555 Ga, which is 15 My less than the age of Ca-Al-rich inclusions, although U-Pb closure in phosphates from equilibrated ordinary chondrites ranges from 4.563–4.505 $\pm$ 0.001 Ga. The whole-rock Rb-Sr systematics for the ordinary chondrites and enstatite chondrites yield an age of 4.498 $\pm$ 0.015 Ga although this depends on the decay constant of ^{87}Rb used. Relative to Ca-Al-rich inclusions, the comparatively late accretion of some chondrite parent bodies after ^{26}Al and ^{60}Fe had substantially decayed, may have provided sufficient heat for metamorphism, but not enough to melt them (Baker et al., 2005). Overall, a time period of $<$100 My after T_0 appears to encompass the history of the ordinary chondrites to the end of prograde metamorphism.

Earth's Oldest Rocks
Edited by Martin J. Van Kranendonk, R. Hugh Smithies and Vickie C. Bennett
Developments in Precambrian Geology, Vol. 15 (K.C. Condie, Series Editor)
© 2007 Elsevier B.V. All rights reserved.
DOI: 10.1016/S0166-2635(07)15023-4

Chapter 2.3

DYNAMICS OF THE HADEAN AND ARCHAEAN MANTLE

GEOFFREY F. DAVIES

Research School of Earth Science, Australian National University, Canberra, ACT 0200, Australia

2.3-1. INTRODUCTION

The Earth started hot, according to current ideas about its formation. It would have cooled within a few hundred million years until its temperature was maintained by the higher radioactivity of the time, which would have maintained it at a substantially higher temperature than at present. The higher temperature of the mantle implies a lower viscosity, faster mantle convection and probably therefore a thinner lithosphere, or more accurately, a thinner thermal boundary layer. A thin boundary layer may not have had the strength to behave rigidly, so its deformation may have been distributed, rather than sharply localized like the present oceanic plate boundaries.

A higher mantle temperature also, other things being equal, implies larger degrees of melting and thicker mafic crust. Since mafic crust at the surface is less dense than the underlying mantle, its positive buoyancy might inhibit subduction (or more generally, foundering). On the other hand, mafic crust would presumably transform to eclogite at 60–80 km depths, which is denser than the mantle and would promote foundering. The interaction among these compositional buoyancies and the negative thermal buoyancy of the top thermal boundary layer might have yielded complicated dynamics. The resulting dynamics might or might not have borne much resemblance to plate tectonics, and might have included strongly episodic foundering. These dynamics have only been partially explored to date.

It had been argued that thicker oceanic crust might have inhibited or precluded subduction, and therefore plate tectonics, during roughly the first half of Earth history (Davies, 1992). However recent numerical modelling has suggested that the early mantle may have been dynamically stratified, with the upper mantle depleted of the denser eclogitic component resulting from subduction or foundering of mafic crust (Davies, 2006b). Such a relatively refractory upper mantle would yield less melt and a thinner oceanic crust, and so the possibility is reopened that subduction, and possibly plate tectonics, may have been viable, even in the Hadean. This would in turn affect the cooling history of the mantle, possibly inducing a two-stage cooling. These possibilities will be summarized here.

2.3-2. BASIC PRINCIPLES GOVERNING TECTONIC MODES

Before considering the possible dynamical and tectonic behaviour of the early Earth, two basic principles need to be appreciated. The connection between dynamics and tectonics also needs to be highlighted.

Tectonics refers to the process by which the crust is deformed, resulting in mountain building and so on. Apart from meteorite impacts, the forces producing crustal deformation are of internal origin. Thus they are the result of the dynamical behaviour of the mantle, in other words of mantle convection. Convection is driven by thermal boundary layers. In the present mantle we have evidence of two thermal boundary layers operating, each of the order of 100 km thick. A hot thermal boundary layer at the bottom of the mantle is inferred to give rise to mantle plumes and a cool thermal boundary layer at the top of the mantle essentially comprises the lithosphere. Cooling and sinking of the top thermal boundary layer at present occurs through the process of plate tectonics and associated subduction of the lithosphere (Davies, 1999).

Thus from the geologist's perspective crustal deformation and mountain building are driven by the motion of the plates. On the other hand, from the dynamicist's perspective, the cooling and foundering of the top thermal boundary layer at present takes place through a very particular mode of deformation (plate tectonics) that is evidently the result of the particular mechanical properties of the top thermal boundary layer. Loosely speaking, the top thermal boundary layer is behaving like a brittle solid: it is broken into pieces separated by large faults, and the pieces are moving.

The tectonic mode of the planet is thus, from the dynamical perspective, the mode of deformation of the top thermal boundary layer of the mantle. Other modes of deformation are quite conceivable. Indeed we ought to keep in mind that among the four terrestrial planets and a dozen or more larger satellites of the solar system only Earth is known to have plate tectonics. If the top thermal boundary is strong but unbroken the result is a so-called one-plate planet, like Mars, Mercury or Moon. The only prolonged tectonic activity among these evidently was the volcanism on Mars, plausibly due to plumes from a bottom thermal boundary layer. Europa shows evidence of distributed deformation of its icy top thermal boundary layer. Ganymede shows evidence of large rigid rafts that once moved relative to each other, but with broadly deformed intervening regions that were evidently not strong enough to be plate-like. Venus was resurfaced about 700 million years ago and appears to have been mostly tectonically inactive since (Solomon et al., 1992), so whatever the mode in which its top thermal boundary layer might move it is evidently highly episodic.

If the top thermal boundary layer is strong, then it can be called a lithosphere. Even if it is strong it may still move, if it can be broken by internal forces. If it is not strong its deformation will be more distributed, more like that of a viscous fluid, and we would not call it a lithosphere. The rheology of solids allows that deformation might be almost anything between that of a brittle solid (highly localized) and a viscous fluid (smoothly distributed). For example, a nonlinear fluid rheology can result in relatively localized mobile belts or ductile shear zones, with limited deformation elsewhere. If the thermal boundary layer is mobile, its motion might be fairly steady, like Earth's plates, or highly episodic, like Venus seems to have been. Episodes might be global, as Venus' appears to have been, or only regional.

Whatever the tectonic mode, it must satisfy two principles.

Principle 1. There must be forces capable of driving the motions of the top thermal boundary layer.

Principle 2. The tectonic mode must be capable of removing heat from the interior at a rate sufficient to explain the planet's present condition.

The first point might seem trivially obvious, but unquantified conjecture can easily fall foul of this principle. For example, it has sometimes been suggested that plates would have undergone 'flat subduction' in the past, when thick oceanic crust would have kept them buoyant (Abbott et al., 1994). This might work locally in the present Earth, if the rest of the plate is still pulled along by other subducting parts that are negatively buoyant, as may have occurred under South America for example (Gutscher et al., 2000). However it is unlikely to have worked in the past if the plate was supposed to be moved by 'ridge push' (really gravity sliding off a ridge), because ridge push is relatively weak, and only comes into play if the plate is already moving. Recall the underlying principle that convection is driven by the foundering of the thermal boundary layer, so if the thermal boundary layer is buoyant it will not founder.

Principle 2 derives from the fact that convection is a heat transport mechanism. The role of the top thermal boundary layer is to remove heat from the interior. It does this first by losing heat to the cold surface by conduction, thickening in the process. When it reaches a sufficient thickness, its accumulated negative buoyancy causes it to sink into the interior fluid, which it cools by absorbing heat. In the meantime the cool fluid is replaced at the surface by hot fluid that in turn cools and sinks. In the case of the present mantle system, heat is removed by the cycle of passive upwelling of hot (but ambient) mantle at spreading centers, conductive cooling of the near-surface material to form a thickening plate, subduction of the plate and absorption of heat from the hot interior.

To illustrate the import of Principle 2, consider the situation discussed by Davies (Davies, 1992) in which plates are rendered more buoyant by the presence of thicker oceanic crust, due to the mantle being hotter than at present. This does not preclude plate tectonics, because the plate will eventually become negatively buoyant due to conductive cooling and thickening. However subduction might be delayed long enough that the plates no longer remove heat from the mantle faster than radioactivity warms it, and the mantle would not be able to cool to its present temperature. Thus this tectonic mode is not capable, by itself, of explaining how the mantle got from a presumably hotter start to its present temperature.

2.3-3. THERMAL STATE DURING FORMATION AND THE HADEAN

Two broad modes of formation of the terrestrial planets have been debated, a relatively rapid instability of a disk of gas and dust, followed by removal of the gas, or a slower collisional accretion process, largely in the absence of gas, taking a few tens of millions of years (Wetherill, 1980). The former is no longer favored for the formation of the terrestrial planets (Boss, 1990). It has also been suggested that there might have been an early thick atmosphere, either of nebula gases or of steam, that would keep Earth's surface hot (Sasaki, 1990). Among these possibilities the most favoured seems to be collisional accretion with a thin atmosphere and a cold surface. This implies the lowest internal temperatures, but even in this case melting temperatures would have been reached near the surface when Earth was only about half of its final radius and one tenth of its final mass (Davies, 1990). As accretion proceeded, larger and larger bodies would have collided with Earth (because they would be growing too), culminating with the impact of a Mars-sized body and the consequent formation of the Moon (Wetherill, 1985). The largest impact might have melted a substantial fraction of the mantle (Melosh, 1990).

However impact melting was probably quite transient, even in the case of a magma ocean. This is because ultramafic and mafic magmas have quite low viscosities and convection driven by contact with a cold surface would be quite vigorous. Estimates based on quite general formulas characterising convective heat transport, and also on the example of Hawaiian lava lakes, indicate that even a 1000 km deep magma ocean would cool to a crystalline mush within only a few thousand years (Davies, 1990; Tonks and Melosh, 1990). Smaller impacts might generate lava seas or lava lakes lasting only centuries or decades.

Two important inferences follow from the occurrence of transient but vigorously convecting magma bodies. One is that core segregation would have been facilitated well before Earth had reached its final mass (Stevenson, 1990). On the other hand, the convection may have been vigorous enough to keep centimeter-scale silicate crystals in suspension, and thus to prevent much stratification of silicate components (Tonks and Melosh, 1990). The fact that deviations of the present mantle from compositional homogeneity have been difficult to resolve (Jackson and Rigden, 1998; van der Hilst and Kárason, 1999) also indicates that the mantle has never been so strongly stratified that subsequent mantle convection could not erase all or most of the stratification.

Segregation of the core releases enough energy to heat the interior by about 2000 °C (Flasar and Birch, 1973), so the initially cold center would have been displaced by descending core material and heated and stirred (Davies, 1980; Stevenson, 1990). Despite the high rate of energy release associated with core segregation, the mantle temperature may not have been more than about 30% greater than at present because of the lower viscosity and resulting very high efficiency of convection at high temperatures (Davies, 1980). The whole interior was probably thermally homogenized in the process of core segregation.

These inferences thus bring us to a picture in which the mantle during the final stages of accretion was considerably hotter than at present and not strongly stratified despite having been transiently molten.

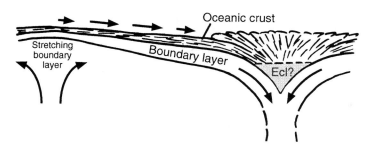

Fig. 2.3-1. Sketch of a possible alternative to plate tectonics. The thermal boundary layer is presumed to be thin and deformable, so no rigid plates form. The foundering of the thermal boundary layer is depicted as being symmetrical, since the asymmetry of modern subduction arises from the reverse faulting that characterises a strong, brittle material. A mafic (oceanic-type) crust is presumed to form and to accumulate over the downwelling zone. As the lower parts of this pile reach higher pressures and temperatures they would tend to convert to dense eclogite (Ecl). This might lead to a runaway acceleration of the foundering, resulting in the whole accumulation foundering within a short episode. After (Davies, 1992).

Although impact melting *per se* would have been transient, any upwelling in the mantle would have generated substantial pressure-release melting. Thus presumably there was early, rapid and semi-continuous formation of mafic and ultramafic crust (see Kamber, this volume). The detailed behaviour and fate of such crust is subject to the uncertainties already noted regarding the interplay between compositional and thermal buoyancies, and to the requirements of Principles 1 and 2. This kind of scenario has not been quantitatively modelled. Thus, bearing in mind the perils of unquantified conjecture, I will propose a scenario that bridges between Earth's accretion and a state in which plate tectonics might have operated and which is thus quantifiable. The usefulness of such a scenario might either be that it is justifiable in retrospect, or that it may serve as a benchmark against which other scenarios can be evaluated. In any case this approach provides a way of exploring this realm of Earth history.

To proceed, I need to assume that some form of near-surface mantle convection operated from the late stages of accretion. This requires that the top thermal boundary layer, or part of it, was able to subside. If the mantle were not melting this would be straightforward: the top surface would cool and sink according to normal thermal convection, facilitated by the relatively low viscosity of the underlying mantle. However mantle melting, and the resulting formation of mafic crust, complicates the picture and creates a chicken-and-egg question of how foundering of the surface would have begun. Presumably the continuing active bombardment of late accretion would have helped.

Pressure-release melting of fertile mantle with an excess temperature of 200–300 °C would generate a 30–40 km thickness of mafic crust (McKenzie and Bickle, 1988), which would be buoyant. Assuming concurrent degassing and the early stages of ocean formation, this material would plausibly have been hydrated after it erupted. With modest horizontal forces from convection presumed to be already active, horizontal convergence over downwelling zones might then have thickened this material to the 60 km or so required for its base to transform to eclogite, which is denser than the mantle. This would cause the mafic pile to subside, bringing more material into the eclogite range. The process thus would accelerate, quite possibly forming a runaway sinker that would founder rapidly into the mantle, as illustrated in Fig. 2.3-1.

By such a process, the mantle might be expected to have begun to differentiate very early into foundered mafic pods within a matrix of more refractory and less dense residue. The stirring action of mantle convection would tend to rehomogenize these heterogeneities, but the initial differentiation would initiate a large-scale process of vertical segregation that is summarized in the next section.

2.3-4. DYNAMICAL STRATIFICATION OF THE MANTLE

Numerical models of a convecting and differentiating hot mantle show that foundered eclogitic material can sink through the upper mantle at a sufficient rate to leave the upper mantle significantly depleted of its mafic component (Davies, 2006b). Fig. 2.3-2 shows a model run at 1650 °C. In this model the upper mantle viscosity is about 1.3×10^{18} pascal-seconds (Pa s), about 200 times less than in the present upper mantle, and the lower mantle is 30 times more viscous.

The model includes plates at the surface, simulated by imposing uniform velocities on segments of the top boundary, as indicated by the arrows in Fig. 2.3-2(a). Plate boundaries are marked by the ticks. There are two 'oceanic' plates converging towards a central 'continental' region. In the continental region the top boundary of the model represents the base of the continental lithosphere, so it is a hot, low heat flow boundary. The continental region also spreads slowly. Both of the latter features help to provide a good simulation of subduction without disturbing the main flow unduly (see Davies (2002)) for more discussion of these devices).

The model also tracks the basaltic-composition component of the mantle using four million tracers – this component exists as basalt at the surface but as other assemblages, such as eclogite, at depth. The basaltic-composition component undergoes various phase transformations with increasing depth, and is denser than average mantle at most depths below about 60 km (Hirose et al., 1999). Each tracer carries an excess mass corresponding to an excess density of $150 \, \text{kg}/\text{m}^3$. This model also includes a phase transformation in the transition zone (at about 700 km depth) whose effect tends to sharpen the contrast in tracer behaviour between the upper and lower mantles.

Melting zones are defined below the oceanic plates by an estimated depth to the solidus. Pressure-release melting is simulated by moving tracers vertically into an oceanic crust layer if they rise into a melting zone, thus also leaving the melting zone depleted of tracers. Depleted zones are less dense by virtue of having no heavy tracers within them. The thickness of the oceanic crust generated in this model can be calculated from the numbers of tracers present in the crust.

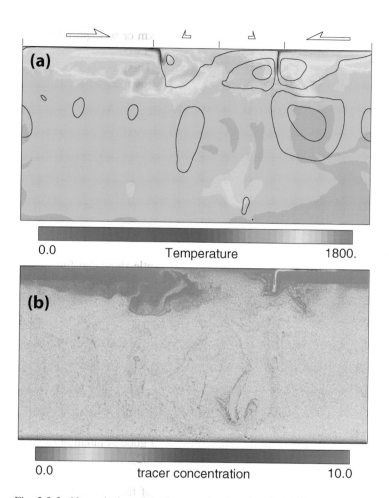

Fig. 2.3-2. Numerical model of convection in a hot (1650 °C) early mantle after 100 My: (a) temperature and streamlines (curves); (b) concentration of tracers. Plates are assumed to operate at the surface, delineated by the ticks and arrows. The central part of the top boundary is taken to be the base of continental lithosphere (see text). The tracers represent the basaltic-composition component of the mantle, and melting is simulated by transferring tracers into the oceanic crust as they enter the melting zone under the outer 'oceanic' plates. Two subduction zones, where the oceanic plates descend, carry differentiated material into the mantle. An accumulation of cooler fluid, depleted of tracers, can be seen under the left-hand plate. Details of this model are given in Davies (2006a).

Subduction injects the crustal layer and the depleted zone into the mantle. Below the 60 km eclogite transformation, denser crustal material tends to sink, and the less-dense depleted zone material tends to rise. The upper mantle viscosity is low enough that the components can separate at a significant rate, so the upper mantle becomes relatively de-

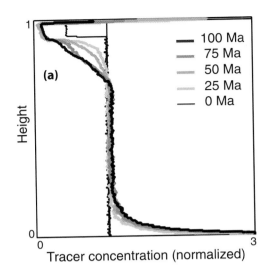

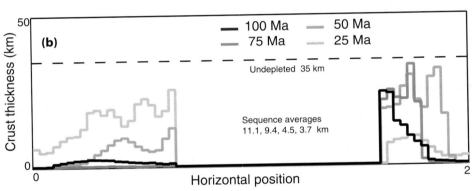

Fig. 2.3-3. (a) Horizontally averaged profiles of tracer concentration as a function of time from the model shown in Fig. 2.3-2. Note the progressive depletion of the upper mantle, and the accumulation of tracers at the base of the mantle. The viscosity of the lower mantle is 30 times higher than in the upper mantle, and is evidently sufficient to keep most of the tracers in suspension; (b) Thickness of oceanic crust calculated from the model in Fig. 2.3-2 as a function of time. A mantle undepleted of basaltic component would yield 35 km thick oceanic crust. Note the large spatial and temporal variability. Details of these models are given in Davies (2006a).

pleted of basaltic component. The lower mantle is evidently still viscous enough to keep much of the eclogite in suspension, though there is an accumulation at the bottom.

The development of this stratification is evident in Fig. 2.3-3(a), which shows profiles of the horizontally averaged tracer concentration as a function of time. The upper mantle depletion profile is established after about 75 Ma. The average thickness of oceanic crust drops from about 35 km to about 4 km within this period (Fig. 2.3-3(b)).

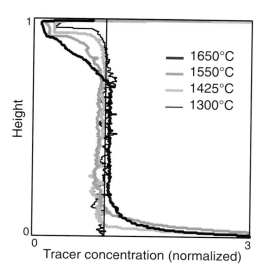

Fig. 2.3-4. Horizontally averaged profiles of tracer concentration as a function of mantle temperature. Depletion of the upper mantle is stronger at higher temperature because of lower viscosity. Details of these models are given in Davies (2006a).

At lower temperatures, this dynamical stratification is not as pronounced, because the higher mantle viscosity tends to keep more tracers in suspension. Fig. 2.3-4 shows tracer concentration profiles at several mantle temperatures. Corresponding crustal thickness are listed in Table 2.3-1. There is considerable variability in the modelled thickness with position and time, as is evident in Fig. 2.3-3(b), so these averages have uncertainties of perhaps ±3 km. Sometimes one of the two plates has essentially zero crustal thickness, as can be seen in Fig. 2.3-3(b) at 75 and 100 Ma.

These results suggest the mantle would have become dynamically stratified very early, starting during the late stages of accretion and, if mantle temperature peaked at 1600 °C or more, yielding a depleted upper mantle within 50–75 Ma. If accretion lasted for around 70 Ma, from 4.567 to around 4.5 Ga, this suggests a stratified mantle by 4.4 Ga, quite early in the Hadean.

The behaviour of the models on which these conclusions are based is obviously dependent to some degree on the assumption that plates were operating. If plates were

Table 2.3-1. Thickness of oceanic crust versus mantle temperature

Temperature (°C)	Thickness (km)	Fertile thickness (km)	Ratio
1650	4	35	0.11
1550	8	31	0.26
1425	9	20	0.45
1300	8	8	1.0

not operating, then the kind of 'drip tectonics' suggested earlier might have generated foundering basaltic bodies anyway, and these might actually have been thicker and more equidimensional than a subducted sheet of oceanic crust, which would cause them to sink faster and enhance the segregation demonstrated in the above models. Thus although the detailed veracity of the models cannot be established yet, they may nevertheless give a useful indication of the earliest dynamical behaviour of the mantle, including its possible dynamical stratification.

2.3-5. EARLY PLATE TECTONICS AND EARLY MANTLE COOLING

These results reopen the question of whether plate tectonics could have operated early in Earth history. Part of the problem has already been indicated in the earlier discussion: if the oceanic crust is 30 km thick or more then it will resist subduction because of its low density at the surface. The problem is compounded by the requirement that plates would be going faster because of the lower mantle viscosity, so they would be thinner and have less negative thermal buoyancy (remember the *plate* is defined by the thermal boundary layer). The combination of these effects is sufficient to interfere with subduction when the mantle temperature is less than 100 °C above the present temperature (Davies, 1992).

The thinner crust found in these calculations will inhibit subduction less, so a plate would not have to age as much before its net buoyancy becomes negative. In fact modern plates, with a 7 km oceanic crust, only become negatively buoyant after about 15 Ma (Davies, 1992), and a plate with a 4 km oceanic crust would become negatively buoyant within less than 5 Ma. With present plate dimensions, this permits a plate velocity of about 5000 km/5 Ma, or about 1 m/a, 20 times faster than at present. One can then calculate the heat loss that such fast plates would accomplish. With this logic, the maximum heat flow permitted by a given crustal thickness can be calculated. This is shown as the dashed line in Fig. 2.3-5(a), based on the crustal thickness variation with temperature shown in Fig. 2.3-5(c). Since the crustal thickness has a maximum (near 1500 °C), the maximum heat flow curve has a corresponding minimum

Also shown in Fig. 2.3-5(a) (solid line) is the heat flow that unimpeded plates (and associated mantle convection) would accomplish (Davies, 1993). The actual heat flow would be the lesser of these two curves. Thus at 1300 °C the crustal thickness does not inhibit the plates and the heat flow is about 35 TW (35×10^{12} W). On the other hand at 1600 °C the crust does inhibit the plates and the permitted heat flow is about 120 TW, compared with about 240 TW for unimpeded plates. Thus although heat loss from the mantle is impeded, it can still be quite high.

Given the rate of heat loss as a function of mantle temperature it is possible to calculate the thermal history of the mantle, as has been done many times before (e.g., Davies, 1993). The principle is simple enough: the rate of cooling is proportional to the rate of heat loss *minus* the rate of heat generation by radioactivity. With an initial temperature prescribed, this relationship can be integrated forward in time. A typical result of this, ignoring the effect of the oceanic crust for the moment, is shown in Fig. 2.3-5(b) by the grey curve.

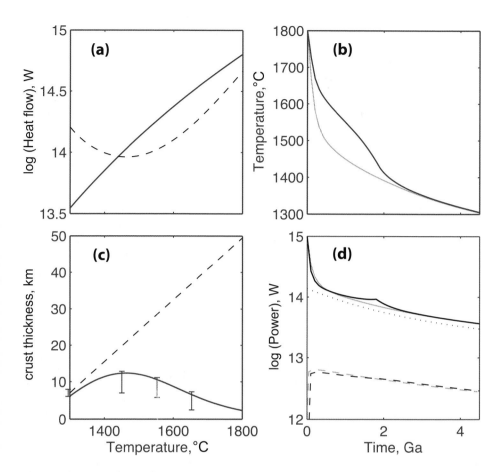

Fig. 2.3-5. Thermal evolution models and related quantities: (a) Heat flow versus temperature with no crustal hindrance (solid), and maximum heat flow allowing for crustal hindrance (dashed); (b) Mantle cooling curves in unhindered (grey) and hindered (black) cases; (c) Crustal thickness versus temperature with (solid) and without (dashed) dynamic depletion of upper mantle; (d) Heat flows versus time in unhindered (grey) and hindered (black) cases. Solid: mantle heat loss. Dotted: mantle radioactive heating. Dashed: core heat loss (input into the mantle). Unhindered case similar to those in Davies (1993). Hindered case will be described in detail in a forthcoming paper.

The early, rapid drop in temperature occurs because the rate of heat loss at high temperatures is much higher than the rate of radiogenic heating (Fig. 2.3-5(d)). As the mantle cools its rate of heat loss reduces until it approaches the rate of radiogenic heating. Thereafter the cooling history is controlled by the slow decay of the radioactive heating.

We can now look at the effect of the oceanic crust on the cooling history of the mantle. Without any crustal inhibition (Fig. 2.3-5(b), grey curve), the mantle cools through 1500 °C within about 400 Ma. With the crustal inhibition, the mantle cools more slowly at high temperatures, and does not cool through 1500 °C for about 1.5 Ga (solid curve).

The inflected shape of this cooling curve is caused by the inflections in the heat flow curves of Fig. 2.3-5(a). As the mantle cools from its high initial temperature it follows the dashed curve. It cannot cool faster than this because as it cools the mantle viscosity increases, the mantle is therefore able to entrain more mafic material, the upper mantle becomes more fertile as a result and the crustal thickness increases, slowing the rate of cooling. This self-regulation applies until the mantle temperature reaches the minimum of the dashed curve in Fig. 2.3-5(a), at which point its cooling rate is no longer constrained by crustal thickness and it goes into a transient 'free fall' until its temperature is maintained by radioactive heating.

The cooling history shown in Fig. 2.3-5 should not be taken too literally. In particular, the timing of the transition from the early hot phase to the later cooler phase is sensitive to parameters that are not very well constrained. For example some allowance has been made for the basalt-eclogite transformation assisting subduction, which raises the dashed curve in Fig. 2.3-5(a) by a factor of 1.8. Without that factor, the hot phase would only just be ending after 4.5 Ga. With a stronger effect, the hot phase would end sooner. (More details of these calculations are given in Davies (2006a)). Thus although it is tempting to associate the transition with the Archean–Proterozoic transition, this can only be conjectured at this stage.

2.3-6. DISCUSSION

The mantle cooling results just presented do not prove plate tectonics was operating during the Hadean and Archaean – that would depend on the details of the mechanical properties and the dynamical forces operating at the time. However the above results, taken at face value for the moment, do show that plate tectonics would have satisfied Principles 1 and 2 enunciated in Section 2.3-2: the negative buoyancy of the lithosphere would have sufficed to drive the plates, despite some inhibition from the buoyant oceanic crust, and the plates would have been able to operate fast enough to cool the mantle to its present state.

If plates were operating then the tectonic regime would have borne many resemblances to today's, but there would have been some differences as well. Plate velocities could have been roughly 20 times greater, so their thermal thicknesses would have been 4 or 5 times less (thickness being proportional to square root of age (Davies, 1999)), thus 20–25 km thick rather than about 100 km thick. Being thinner, they might not have been as large, but this is not a straightforward conclusion because the lower viscosity of the mantle would reduce stresses that would tend to break them. (Then, as now, there would be no need to invoke any putative 'small-scale' convection, and for the same reason – it must be relatively weak because the plate mode of convection uses most of the available negative buoyancy of the plates (Davies, 1999).)

The preceding discussion has been in terms of plate velocities, but what really counts is the areal rate of creation of plates, which is the spreading rates of mid-ocean ridges *times* the length of ridges. If plates were smaller then ridge length might be larger (again, not necessarily, it depends on the details of the geometry) and plate velocities would then have been proportionately smaller. However the analysis is unchanged by this trade-off.

The nature of the rocks produced by the scenario proposed here needs to be considered carefully. The upper mantle is more depleted in basaltic component not just because melt has previously been extracted, but because some of the previously extracted melt has been removed mechanically by gravitational settling. There is simply less of it available to melt. If that melt, from eclogitic pods, reached the surface it could be significantly enriched in incompatible elements, as at present. On the other hand, the matrix surrounding it would be more refractory than at present, because of the higher temperatures at which melting was occurring, and this would presumably increase the degree of back-reaction to which the eclogitic melt would be prone (Campbell, 1998; Yaxley and Green, 1998), but perhaps not by very much. Thus we might expect the early, thin oceanic crust to reflect some of the high temperature and reduced basaltic component of its source, but not necessarily as much as a simple two-stage melting process would imply.

ACKNOWLEDGEMENTS

The computations reported here were performed on the SGI Altix computer cluster of the Australian Partnership for Advanced Computing National Facility based at The Australian National University. The research was supported by Australian Research Council grant DP0451266.

Earth's Oldest Rocks
Edited by Martin J. Van Kranendonk, R. Hugh Smithies and Vickie C. Bennett
Developments in Precambrian Geology, Vol. 15 (K.C. Condie, Series Editor)
© 2007 Elsevier B.V. All rights reserved.
DOI: 10.1016/S0166-2635(07)15024-6

75

Chapter 2.4

THE ENIGMA OF THE TERRESTRIAL PROTOCRUST: EVIDENCE FOR ITS FORMER EXISTENCE AND THE IMPORTANCE OF ITS COMPLETE DISAPPEARANCE

BALZ S. KAMBER

Department of Earth Sciences, Laurentian University, 935 Ramsey Lake Road, Sudbury, ON, Canada, P3E 2C6

2.4-1. INTRODUCTION

The oldest known rock on Earth, the Acasta gneiss, is just over 4 billion years (Ga) old and occurs as a relatively small expanse of outcrops in the northwestern part of the Slave Craton, Canada (Bowring et al., 1990; Iizuka et al., this volume). It is the only known Hadean rock. The next oldest expanse of terrestrial rocks is the 3.9–3.6 Ga Itsaq Gneiss Complex of SW Greenland, which occupies a more substantial area (Nutman et al., this volume). But even when considering the additional few small nuclei of Palaeoarchean gneisses on other cratons, the early rock record on Earth can be termed miniscule at best.

By comparison, the Moon is vastly better endowed with ancient rocks. This is not only evident from directly dated Apollo anorthosite breccia samples (e.g., Norman et al., 2003) but also from the much higher cratering density on the lunar highlands compared to the maria. Pb-isotope data (Tera et al., 1974) and Ar-Ar dates of lunar impact glasses (Cohen et al., 2000) firmly demonstrate that the last major cratering event, the so-called late heavy meteorite bombardment (LHMB), occurred between 4.0 and 3.85 Ga, thus providing a minimum age for the heavily cratered highlands. Even on Mars, a relatively large surface area is inferred to be ancient (e.g., Solomon et al., 2005) as is indeed one of the few retrieved Martian meteorites (4.5 Ga orthopyroxenite ALH84001; e.g., Nyquist et al., 2001).

The most obvious difference between Earth, the Moon and Mars is that our planet's geology is, and for most of its history was, driven by plate tectonics. This process, whereby heat from the Earth's interior is largely lost by upwelling of hot mantle material that melts at mid-ocean ridges, and by a continuous flux of cold oceanic plates subducted back into the hot mantle, has a tendency to rework the continental fragments against which oceanic plates are moved. Hence, intuition could lead one to believe that the reason for the absence of a large volume of truly ancient rocks on Earth was that over time, they were recycled back into the mantle or reworked beyond recognition into younger continental rocks (e.g., Armstrong, 1991).

There are, however, several observations that argue for a very substantial primordial crust, as well as for relatively good chances of survival of continental crust that formed since 3.85 Ga. This evidence, which is largely indirect and often requires several layers of interpretation (and leaps of faith in the author's ability to understand data), then creates an enigma concerning the apparent, almost complete, disappearance of the primordial crust of the Earth. It also represents a formidable obstacle for researchers trying to reconstruct how the Earth might have looked prior to the LHMB.

The principal aim of this chapter is to lay out the salient arguments in favour of a voluminous and relatively long-lived early terrestrial crust and to explain the evidence for its disappearance. This discourse will hopefully allow the non-specialist reader to understand the enigma. How the enigma is solved is presently a matter of speculation only. The more plausible explanations will be discussed.

2.4-2. EVIDENCE FOR SUBSTANTIAL > 4 GA DIFFERENTIATION OF THE SILICATE EARTH

The segregation of the metallic core and loss of volatile elements that accompanied the accretion of the Earth have substantially influenced the make-up of the silicate Earth. It is widely, but not unanimously, agreed that core formation, volatile loss and the formation of the Moon by a Mars-sized impactor were completed in less than 100 Myr (e.g., Jacobsen, 2005; Taylor, this volume). Accretion of ca. 5% terrestrial mass was possible after the formation of the Moon (Canup, 2004) and was very important for the volatile budget of the Earth (Kramers, 2007).

It is necessary to mention the possibility that the silicate Earth from which Hadean crust was derived might not have been homogeneous, at least from the point of view of radiogenic isotope composition. In other words, the very earliest processes that shaped the Earth (accretion, Moon formation, magma ocean) may have already left the young mantle somewhat differentiated (e.g., Boyet and Carlson, 2005). Alternatively, these isotopic differences could actually reflect inheritance from incomplete mixing of nucleosynthetic material in the solar disc (Ranen and Jacobsen, 2006; Andreasen and Sharma, 2006). It is important to realise that all the features discussed in this chapter were superimposed on this isotopic contrast of the Earth with chondritic meteorites, regardless of its origin.

Silicate differentiation, particularly extraction and isolation of crust, leaves a traceable radiogenic isotope record, which manifests as different relative abundances of isotopes that are decay products of a different element. Next follows a discussion of the three principal lines of evidence that require the existence and temporary persistence of a voluminous Hadean crust.

2.4-2.1. Lu-Hf Isotope Systematics of Ancient Zircon

In the process of mantle melting, the trace element Hf is partitioned more strongly into the liquid than Lu on account of its more lithophile character (i.e., it plots farther to the left

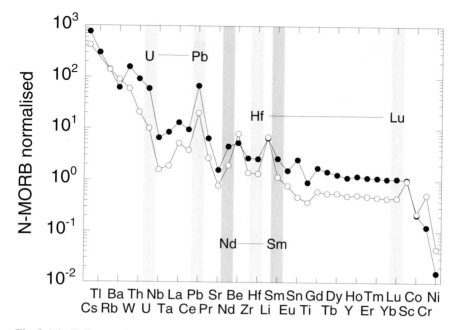

Fig. 2.4-1. Full trace element patterns of modern continental sediment composite (full circles: Kamber et al. (2005b)) and average sediment from the 3.7 Ga Isua supracrustal belt (open circles; data from Bolhar et al. (2005)). Data are normalised to MORB and elements are arranged according to relative incompatibility (Kamber et al., 2002). In mantle melting, the liquid is most enriched for elements on the left hand side of the plot and least enriched for the most compatible elements plotting at the very right. Note that for the Lu-Hf, Sm-Nd and U-Pb pairs (high-lighted), the relative incorporation into Paleoarchean and modern continental crust was comparable.

in Fig. 2.4-1). If the mantle experiences melting, its preferential depletion in Hf over Lu leads to a significant positive deviation of the Lu/Hf value relative to the original (undifferentiated Earth) value. The ingrowth of ^{176}Hf from decay of radioactive ^{176}Lu after such depletion of the mantle by long-lasting separation of the solidified melt (i.e., crust) is much stronger than in the crust. If the depleted mantle source is later tapped for remelting, the melt itself will inherit this isotopic fingerprint. Conversely, as melt has a lower Lu/Hf ratio than the undifferentiated Earth, it will develop a retarded ^{176}Hf/^{177}Hf isotopic fingerprint in the crust over time.

Zircon is the ideal time capsule to record the Lu/Hf evolution of the early Earth because it has a very low Lu/Hf ratio and hence largely preserves the Hf-isotope composition of the melt from which it formed. Zircon is also an ideal capsule because any disturbance of the system (by later geologic events) is readily monitored by U/Pb systematics (Patchett, 1983). There is now a significant database of Hf-isotope systematics of ancient (>3.7 Ga) zircons, summarised in Fig. 2.4-2. This figure shows, for the time interval 4.3 to 3.8 Ga, a relatively large spread with values both more and less radiogenic than the 'chondritic'

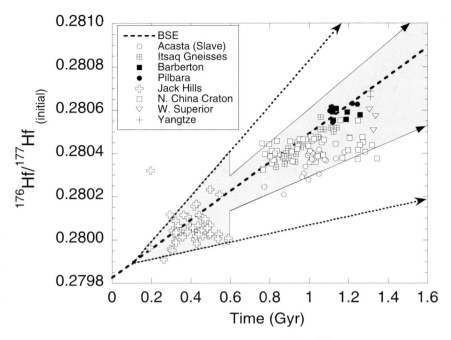

Fig. 2.4-2. Hf isotope evolution diagram, showing initial ^{176}Hf/^{177}Hf ratios recorded by Hadean and Paleoarchean zircon as a function of time, calculated from U/Pb zircon crystallisation age. Position of bulk silicate Earth (BSE) reflects ^{176}Lu decay constant of Scherer et al. (2001). Note that Hadean zircon (with time of formation < 0.57 billion years after Earth accretion) define a wide spread in Hf isotope composition. Trajectories that encompass this spread (shown as stippled arrows) extrapolate to much greater Hf-isotope variability than what is actually observed in younger zircon; namely, Paleoarchean zircons require trajectories with much less divergence between the depleted reservoir (points plotting above BSE) and the continental crust (points plotting below BSE). Combined data imply an event of rehomogenisation some time between 0.6 and 0.8 billion years after Earth accretion, here schematically outlined with a grey polygon, in which Hf isotope variability was reduced by 2/3 at 0.6 Ga. Irrespective of age and models, note that on average, zircon has unradiogenic Hf plotting well below BSE, indicating that the host melt originated by re-melting of pre-existing crust. Very few zircons reflect direct derivation from mantle. Data from: Amelin et al. (1999, 2000), Harrison et al. (2005), Wu et al. (2005), Davis et al. (2005), and Zhang et al. (2006).

reference value. Note that at present, it is unclear where exactly on this diagram the undifferentiated silicate Earth would plot, on account of the uncertainties about the Lu/Hf ratio of the Earth (Boyet and Carlson, 2005) and the ^{176}Lu decay constant (e.g., Scherer et al., 2001). Two key observations relevant for the Hadean Earth are valid, regardless. First, the Earth's oldest zircons require a substantially depleted mantle (with high time-integrated Lu/Hf ratio) and a complementary crust with low Lu/Hf ratio. Second, this large extent of silicate Earth differentiation is no longer visible in 3.8–3.5 Ga zircon from any of the cratons on which they occur (e.g., Kaapvaal, Pilbara, North Atlantic; Slave (Amelin et al.,

1999, 2000); North China (Wu et al., 2005); Western Superior (Davis et al., 2005); Yangtze (Zhang et al., 2006)).

It is of critical importance to realise that this extent of Hf-isotope spread could not have evolved if the low Lu/Hf reservoir (the crust) was dynamically recycled back into the mantle throughout the Hadean era. Rather, the spread requires the separation of a relatively long-lived crust, which was episodically internally reworked during the Hadean (Cavosie et al., 2004, this volume) and produced the scatter in unradiogenic Hf-isotope values. If this crust–mantle pair had persisted to the present day, its range in Hf-isotope values would be 300–400% that observed (Harrison et al., 2005). The lack of evidence in Mesoarchaean (and younger) zircon for such extreme Lu/Hf separation thus requires a fundamental recycling event of the Hadean crust back into the mantle some time between 3.8 and 3.5 Ga.

Hadean zircon (e.g., Froude et al., 1983; Maas et al., 1992; Wilde et al., 2001, Mojzsis et al., 2001; Iizuka et al., 2006, this volume), while itself undoubtedly the best and so far only direct evidence for the existence of a terrestrial Hadean crust, does not place any limit on the volume and bulk composition of this crust. Because zircon is the only highly resilient mineral that is amenable to radiometric dating (unlike, for example, detrital chromite that is found along with Hadean zircon in quartzites), much care must be exercised when extrapolating implications from zircon geochemical data to bulk crust as a whole. Mineral inclusions of feldspar, mica and quartz in Hadean zircon (e.g., Maas et al., 1992; Wilde et al., 2001) are clearly compatible with geochemical evidence in the zircon itself for granitoid host rocks of at least some of the zircons (e.g., Maas et al., 1992; Mojzsis et al., 2001; Crowley et al., 2005; Harrison and Watson, 2005; Iizuka et al., this volume). However, because ultramafic and mafic lithologies are devoid of igneous zircon, they are not represented in inventories. As a result, it is not surprising that the zircon geochemical evidence is in favour of granitoid host rocks, but that does not automatically imply that the bulk of the Hadean crust was not ultramafic.

In summary, the Hadean zircon evidence proves that:

(i) granitoids of diverse chemistry existed (Crowley et al., 2005);

(ii) extraction of the Hadean crust depleted portions of the mantle, in places quite strongly;

(iii) the crust and depleted mantle remained separated for a considerable period of time (200–600 Myr);

(iv) the crust was being reworked internally; and

(v) the crust and depleted mantle were largely rehomogenised by 3.5 Ga.

2.4-2.2. ^{146}Sm-^{142}Nd Isotope Systematics of Palaeoarchean Rocks

Because Nd is more lithophile than Sm (see Fig. 2.4-1), melting also leads to preferential enrichment of Sm over Nd in the mantle. Unfortunately, there is no rock forming mineral that has a truly low Sm/Nd ratio and could be used to estimate the original $^{143}Nd/^{144}Nd$-isotope composition of the rock without substantial correction for in situ ^{147}Sm decay.

However, a second Sm isotope, ^{146}Sm, with a short half-life of 103 M.yr., had essentially decayed to ^{142}Nd by the end of the Hadean. Therefore, $^{142}Nd/^{144}Nd$ data do not require correction for in situ ^{146}Sm decay since rock formation, and this system is thus ideally suited to trace Hadean Earth silicate differentiation. There is now a reliable database of ^{142}Nd/^{144}Nd measurements for Paleoarchean rocks from SE Greenland (Caro et al., 2006). The most important result from the data is that all analysed lithologies (metasedimentary rocks, orthogneisses and amphibolites) have ^{142}Nd excesses relative to modern basalts (Fig. 2.4-3). The excesses range from ca. 7 ppm (parts-per-million) in the orthogneisses,

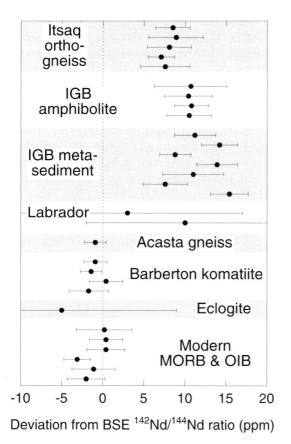

Deviation from BSE ^{142}Nd/^{144}Nd ratio (ppm)

Fig. 2.4-3. Comparison of ^{142}Nd/^{144}Nd systematics of terrestrial rocks (data from Caro et al. (2006) and Regelous and Collerson (1996)). The relative ^{142}Nd abundance is expressed as the deviation in parts-per-million from BSE. Note that the modern rocks, Barberton komatiites, and the one Acasta gneiss have no distinguishable anomalies. By contrast, all Paleoarchean rocks from SW Greenland (Itsaq gneisses, Isua supracrustal belt metasedimentary rocks and amphibolites) have clearly discernible ^{142}Nd excesses. The data obtained by Regelous and Collerson (1996) on an older generation mass spectrometer (Yakutia ecolgite and Paleoarchean Labrador samples) have larger errors. However, note the ^{142}Nd excesses in the Labrador samples, particularly in the meta-komatiite, could be real if analysed with the same technique as Caro et al. (2006).

to 15 ppm in some of the metasedimentary rocks and require Sm/Nd fractionation to have commenced before 4.35 Ga. The new data vindicate the much earlier claim for the existence of ^{142}Nd excess in a metasedimentary rock from the Isua supracrustal belt (Harper and Jacobsen, 1992), but have not confirmed the magnitude of the earlier data (33 ppm). Nevertheless, Caro et al. (2006) comment on the variability in ^{142}Nd excess shown by the clastic metasedimentary rocks and the fact that none of the currently exposed non-sedimentary lithologies in SW Greenland can provide a source for the most ^{142}Nd enriched component in the sediments. This observation is compatible with evidence from trace elements systematics of these lithologies (Bolhar et al., 2005) that require an essentially zircon-free, but strongly evolved, component that appears to be absent at the current level of exposure.

By contrast to Paleoarchean rocks from SW Greenland, the presently very limited ^{142}Nd-isotope database for up to 4.0 Ga Acasta gneisses from the Slave Craton and 3.4–3.5 Ga lithologies from the Barberton Greenstone Belt of the Kaapvaal Craton do not show resolvable ^{142}Nd excesses (Fig. 2.4-3). This may well indicate a degree of Hadean mantle heterogeneity, as the presence of $\geqslant$4.2 Ga crust at Acasta has now been demonstrated (Iizuka et al., 2006). Older vintage ^{142}Nd data were presented by Regelous and Collerson (1996) for two Paleoarchean samples from Labrador. Neither the monzodiorite gneiss nor the meta-komatiite from this area yielded anomalous ^{142}Nd within the much larger analytical errors than can be achieved with the latest generation of mass spectrometers. However, the Labrador meta-komatiite average (+10 $\pm$12 ppm) is identical to the average metabasalt value (+10.6 $\pm$ 0.4) from the Isua supracrustal belt determined by Caro et al. (2006). The possibility thus remains that the ^{142}Nd excess is a feature of the early Archaean nucleus of the entire North Atlantic Craton.

When interpreting the early Earth ^{142}Nd data, the following factors need to be considered. First, at this stage, no measurement of a terrestrial rock has yielded a relative ^{142}Nd deficit to complement the excess documented in SW Greenland. This problem is compounded further if the accessible Earth has a higher Sm/Nd than chondrites (Boyet and Carlson, 2005). The problem of the missing ^{142}Nd depleted rocks is particularly counterintuitive in the Isua supracrustal belt. Here, the lithologies with the most crustal character (i.e., metasedimentary rocks) appear to have the largest excesses, when crust as a whole has a lower Sm/Nd than the mantle and would be expected to develop a ^{142}Nd deficit. A solution to this dilemma is possibly provided by the observation that in these metasedimentary rocks the evolved component was apparently mostly juvenile (Kamber et al., 2005a). In that case, it is possible that the sediments provide a window into a depleted mantle area tapped by juvenile basaltic volcanism that fed the sediment source.

Second, regardless of this possibility, it is curious that the Hf isotopes in ancient zircon show a strong intra-crustal evolution history, while the ^{142}Nd apparently does not.

Third, in agreement with the Hf isotopes of ancient zircon, the lack of any appreciable ^{142}Nd anomaly in Archean rocks younger than 3.7 Ga is strong evidence that the reservoirs that resulted from Hadean silicate differentiation were rapidly homogenised before 3.5 Ga. In the context of the Hadean crust, this would mean effective recycling of the crust itself, as well as eradicating the depletion signature in affected mantle domains.

2.4-2.3. ^{235}U-^{207}Pb *Isotope Systematics of Palaeoarchean Rocks*

The partitioning of U and Pb is more complicated than that of Lu-Hf and Sm-Nd. Uranium is more lithophile than Pb (i.e., plotting further to the left in Fig. 2.4-1) but continental sediments are much richer in Pb than could be expected from the difference in incompatibility (Fig. 2.4-1). This is because in the process of continental crust formation, Pb is enriched due to its high solubility in fluids (e.g., Miller et al., 1994). In the trace element diagram, the enrichment manifests as a positive Pb-spike. It is not only evident in modern sediment, but was already a feature of the oldest known, 3.71 Ga, clastic metasedimentary rocks from the Isua supracrustal belt (Fig. 2.4-1). The relatively low U/Pb ratio of true continental crust (of all ages) is thus prima facie evidence for fluid induced mantle melting. The extraction of continental crust has thus left the mantle very depleted in Pb. The relatively modest U/Pb ratio of typical continental crust is also reflected in the average Pb isotope composition of average continental sediment (Fig. 2.4-4(a)) and the relatively limited Pb isotope variability of crustal rocks. In complete contrast, lunar crustal rocks have much higher and more variable U/Pb ratios and consequently, contain much more radiogenic Pb (e.g., Premo et al., 1999).

An important feature of the Paleoarchean rock record is that the spread in $^{207}Pb/^{204}Pb$ at a given $^{206}Pb/^{204}Pb$ was much greater than could be expected from extraction of Paleoarchean crust (Kamber et al., 2003). This observation is very robust because it is evident in the Pb isotopes of ores and feldspar that both contain no significant U and thus can preserve original Pb isotope composition (Frei and Rosing, 2001; Kamber et âl., 2003), as well as the positions of whole rock isochrons, whose age significance was confirmed with U/Pb zircon dates (e.g., Moorbath et al., 1973). The relatively large spread in $^{207}Pb/^{204}Pb$ (Fig. 2.4-4(b)) testifies to quite strong fractionation of U from Pb in the Hadean, rather than the Paleoarchean Earth. It is also interesting that the inferred Hadean crustal rocks

---

Fig. 2.4-4. Common Pb isotope diagrams. (A) Modelled evolution lines from 4.3 Ga to present (in 100 Myr steps) for the depleted MORB-source mantle and average sediment (after Kramers and Tolstikhin, 1997). Note the maximum divergence in Neoarchaean times and subsequent convergence caused by continental recycling into the mantle. Shown as an overlay in open symbols are initial Pb isotope compositions of continental feldspars and ores (from compilation in Kamber et al., 2003). They define a relatively modest range of compositions by comparison with the Moon (not shown). This reflects the over-enrichment of Pb in continental crust and the resulting relatively modest continental U/Pb ratio. (B) Pb isotope systematics of selected Paleoarchean lithologies from SW Greenland. Two major observations are relevant. First, the data array defined by galena from the 3.7–3.8 Ga Isua supracrustal belt plots sub-parallel to, but above, the mantle evolution line. This requires isolation of the Pb source from which the galena formed (most likely crustal in character) between 4.1 and 4.35 Ga. Second, the age regression lines of different, but largely coeval, 3.65–3.71 Ga lithologies from SW Greenland intercept the mantle evolution lines at very different 'model' ages (from 3.65 to 3.4 Ga). Shown are two examples. The juvenile type-Amîtsoq gneisses, which contrast from banded iron formation from the Isua supracrustal belt, whose Pb was derived from a high U/Pb source that separated from the mantle >4.1 Ga.

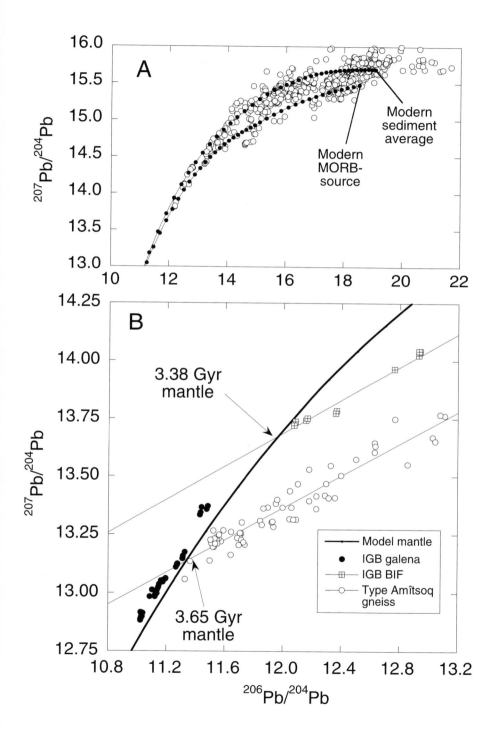

apparently had a high U/Pb, reminiscent of, if not as extreme in character as, the lunar highlands.

The best studied Paleoarchean terrain for Pb isotopes is SW Greenland, where the least radiogenic known terrestrial Pb was found in galena (Frei and Rosing, 2001). The galena data array plots significantly above the mantle evolution line (Fig. 2.4-4(b)) modelled from modern rocks by Kramers and Tolstikhin (1997). Certain other lithologies, mainly metasedimentary rocks, also have much higher $^{207}Pb/^{204}Pb$ ratios than expected in the mantle, while juvenile gneisses, such as the 3.65 Ga, type-locality Amîtsoq gneisses, have Pb isotopes in accord with a mantle origin (Fig. 2.4-4(b)).

Although the exact interpretation of these systematics depends to some extent on the validity of the mantle evolution model, which has its uncertainties, the situation is comparable to that recorded by Hf isotopes in zircon. Namely, there is considerable spread in $^{207}Pb/^{204}Pb$, which attests to U/Pb fractionation in the time period from 4.35 to 4.1 Ga (Kamber et al., 2003). The most crustal lithologies (i.e., sediments) host the Pb with the highest U/Pb source history, formally consistent with the relative incompatibility of the elements. This so-called high μ (for $\mu = {}^{238}Ub/^{204}Pb$) signature can be traced through to late Archean rocks in so-called high-μ cratons (Wyoming, Slave, Yilgarn, Zimbabwe: e.g., Mueller and Wooden, 1988). Kamber et al. (2003, 2005) noted that these are also the cratons that harbour the oldest known zircon, but whether the implication of recycled Hadean crust as an origin for the high μ source will hold up remains to be established. Regardless of the potential survival of some Hadean crustal fragments into the late Archean, the general shape of the terrestrial Pb-isotope array is not permissive of a voluminous Hadean crust that would have persisted beyond 3.5 Ga.

2.4-2.4. Summary

Isotope data from the $^{176}Lu/^{176}Hf$, $^{146}Sm/^{142}Nd$ and $^{235}U/^{207}Pb$ systems offer a coherent picture of significant silicate Earth differentiation that was operational from ca. 4.4 to 4.0 Ga. This is also the age range of the oldest known terrestrial zircon. The extent of isotopic variability requires that crust was not only formed, but that it remained isolated from the depleted mantle residue. While the isotopic data are formally consistent with crust-mantle differentiation, there are nuances whose significance are not fully understood at present. The most important of these is that the Lu-Hf and U-Pb systems contain evidence for both the crustal (low Lu/Hf, high U/Pb) and depleted mantle (high Lu/Hf and low U/Pb) complements, while ^{142}Nd apparently only records the depleted reservoir.

It has been suggested (Nutman et al., 1999; Kamber et al., 2002) that the more evolved lithologies in SW Greenland could be the melting products of garnet-amphibolites. Such rocks, albeit ultimately crustal in origin, can evolve to more radiogenic Nd-isotope compositions and the possibility of the spread in ^{142}Nd excess being partly crustal needs to be further investigated.

The most important observation is that the three systems also unanimously agree that the early crust was largely destroyed by ca. 3.8–4.0 Ga and certainly by 3.5 Ga (Bennett et al., 1993). Lu/Hf and Sm/Nd offer no evidence for the persistence of ancient crust. The very fact that some (albeit very rare) Hadean zircon has survived to the present day is more consistent with Pb isotopes that suggest accidental survival of some Hadean crustal fragments (Kamber et al., 2003, 2005). The biggest question surrounding the Hadean crust then clearly is that of its demise!

2.4-3. MODELS FOR THE DISAPPEARANCE OF THE HADEAN CRUST

2.4-3.1. *A Small Volume Depleted Mantle Portion*

Radiogenic isotope data can estimate the extent of depletion of a particular portion of mantle, but they cannot constrain the fraction of the total mantle that was depleted by crust formation. Bennett et al. (1993), working with the long-lived $^{147}Sm/^{143}Nd$ systematics of Paleoarchean rocks, which are less robust than the systems discussed above, hypothesised that the large spread in initial $^{143}Nd/^{144}Nd$ that was obvious in 3.7–3.8 Ga rocks but not in younger rocks could mean that throughout the first 700 M.yr. of Earth history, crust formation only depleted a small portion of the mantle. This part of the mantle would have become strongly depleted, but would not in itself require the formation of a volumetrically significant Hadean crust. At some time between 3.8 and 3.5 Ga, a different type of mantle convection would have initiated and would have rapidly stirred the very depleted former uppermost mantle into undepleted deeper mantle and as such erased the strongly depleted isotopic signature. Their model offered no suggestion as to what might have happened to the Hadean crust, but the idea of a volumetrically 'small' depleted Hadean mantle clearly remains attractive.

Because radioactive heat production rates in the Hadean mantle were between 3.5 and 4.5 times higher than today (e.g., Kramers et al., 2001), it is difficult (but as we shall see, not impossible) to envisage how such a small mantle domain could have remained isolated and survived convective stirring. Indeed, it is difficult to envisage how the Hadean crust itself could have withstood convective forces for hundreds of millions of years.

2.4-3.2. *A Plate Tectonic Hadean Earth*

Continental crust formation along volcanic arcs is unique in that the crustal portion of the lithosphere is equipped with a residual, refractory mantle keel (the subcontinental lithospheric mantle (SCLM); e.g., Bickle, 1986; Griffin and O'Reilly, this volume). This keel is not only depleted in fusible material but also buoyant relative to surrounding less depleted peridotite. Re-Os isotope systematics of SCLM and overlying continental crust demonstrate very convincingly that, at least in preserved Archaean cratons, the mantle and crustal portions of the lithosphere formed at the same time (e.g., Shirey and Walker, 1998; Nägler et al., 1997).

On the basis of trace element and isotope systematics of Hadean zircon, several authors (e.g., Wilde et al., 2001; Mojzsis et al., 2001; Iizuka et al., 2006), but most prominently Harrison et al. (2005), have argued that the melts from which the zircon formed were derived by subduction zone fluid-induced melting of the mantle. The ensuing picture of a 'cool', plate tectonic, Hadean Earth (see also Watson and Harrison, 2005), in which subduction operated, logically also implies the existence of the buoyant crust – SCLM keel complement. Some of the evidence for a cool early Earth has since been questioned (e.g., Whitehouse and Kamber, 2002; Glikson, 2006; Valley et al., 2006; Nemchin et al., 2006; Coogan and Hinton, 2006), but the real dilemma of the plate tectonic model for the Hadean is that it provides no mechanism by which the voluminous Hadean crust on its stable, buoyant mantle keel could have survived, isolated from the mantle, for so long and then have been destroyed between 3.8 and 3.5 Ga.

Harrison et al. (2005) appeal to the constant continental volume idea of Armstrong (1991), which is claimed to be consistent with radiogenic isotope evidence (e.g., Armstrong, 1981). However, when all the radiogenic isotope and incompatible trace element evidence available today is combined (25 years after Armstrong's calculations), it is clear that the constant continental crust volume model can no longer be supported. Briefly, in order to arrive at an average continental crust age of ca. 2 Ga (Goldstein et al., 1984), an original Hadean crust of comparable mass to that of today's would largely have been recycled. Such recycling into the mantle would have, over time, reduced the isotopic difference between the mid-ocean-ridge-basalt (MORB) source mantle and continental crust (Kramers and Tolstikhin, 1997). Many radiogenic isotope systems are indeed quite insensitive to the extent of recycling of ancient crust (e.g., Nd; Någler et al., 1998). However, this is not the case for Pb, which is overly enriched in continental crust (Fig. 2.4-1) and complementarily strongly depleted in the MORB-source mantle. A high rate of continental recycling into the mantle, as advocated by Armstrong (1981) and required to reduce the average continental crust age to 2 Ga, would lead to a MORB-source mantle Pb-isotope composition very similar to continental sediment. The actual observation, however, is that of an isotopic contrast in all isotopic systems, including in particular $^{207}Pb/^{204}Pb$ (see Fig. 2.4-4(a)). This contrast constitutes very robust evidence for a secular increase in continental crust volume.

Much additional evidence for secular increase of continental crustal mass has been accumulated since Armstrong's model, including the marine Sr-isotope record (e.g., Shields and Veizer, 2002; Shields, this volume) and the mantle Nb/Th/U evolution (Collerson and Kamber 1999), but most important for the Hadean crust is the fact that the extent of Hf-isotope variability in Hadean zircon itself requires an 'event' between 4 and 3.5 Ga that drastically converged the trajectories of Hf-isotope evolution of depleted mantle and crust (Fig. 2.4-2).

Although never explicitly addressed by the advocates of a plate tectonic Hadean Earth, it would appear that the whole-sale destruction of much of the Hadean crust in this model could only be attributed to the effects of the LHMB. As noted by Bennett et al. (1993), the appealing aspect of this proposal is the timing of the LHMB between 4.0 and 3.85 Ga (e.g., Cohen et al., 2000). Cratering of Mars' buried ancient crust and isotopic resetting in Martian meteorite ALH84001 (Turner et al., 1997) suggest that the entire inner Solar System experienced the LHMB. Neither on Mars nor on the Moon did the LHMB destroy the ancient crust. Of critical importance for the validity of the proposal of a plate tectonic

Hadean Earth is the question of whether the larger gravitational cross-section of the Earth would have attracted impacting objects of sufficient size to catastrophically destroy the crust. The single piece of evidence for the LHMB from the terrestrial rock record is isotopic in nature (Schoenberg et al., 2002), but as pointed out more recently (Frei and Rosing, 2005), additional data are required to even confirm traces of the LHMB on Earth. The effects of the LHMB on Earth are thus unconstrained.

In addition to the unresolved issue of the destruction of proposed Hadean continental crust, the model needs to address the question why the Hadean Earth was apparently so cool? The issue of mantle temperature not only compounds ideas of a subduction dominated early Earth, but presents a problem for all models that work with a relatively persistent, long-lived (ca. 500 Myr) Hadean crust.

2.4-3.3. A Long-Lived Basaltic Crustal Lid on a 'Cool' Mantle

The strongest piece of evidence for a largely basaltic composition of the early crust is the simple consideration of the amount of heat produced by the decay of K, U and Th. In the Hadean, the now virtually extinct ^{235}U was contributing almost 50% to the overall radioactive heat production, resulting in a ca. 4× higher overall heat output. Any type of crust other than low K, U, and Th, highly magnesian basalt could not have been thermally stable in the Hadean. Indeed, even basaltic crust, once it attained a thickness of > 40 km, would have melted and internally differentiated (Kamber et al., 2005), thereby producing zircon saturated crustal melts without the aid of subduction. Remelting of older Hadean crust is evident in the Hadean zircon Hf-isotope record and thermal reworking is the most elegant explanation for the complex U/Pb systematics of Hadean zircon (Nelson et al., 2000; Wyche et al., 2003).

Eruption and horizontal growth of a basaltic shell on a hot mantle causes conductive mantle cooling, which may have operated on Mars (Choblet and Sotin, 2001). On a larger planet, the relatively ineffective heat loss through the crust eventually leads to a catastrophic melting event that will destroy the primordial basaltic crust. Kamber et al. (2005) noted that this mechanism might be attractive to explain the disappearance of the Hadean crust and speculated that on Earth, the primordial crust could have persisted until 4 Ga. However, several aspects of this idea remain untested. Indeed, it is likely that the truly primordial crust of the Earth was comprehensively destroyed during the Moon-forming impact of a Mars-sized object, much earlier than envisaged by Kamber et al. (2005; see Taylor, this volume).

The problem of finding a cooling mechanism for the Hadean mantle (so as to allow stabilisation of the Hadean crust) was revisited by Kramers (2007). He argued that it was in fact the Moon-forming impact and its aftermath that could provide the solution to the problem. The heat released from the impact, the gravitational sinking of the core and the blanketing effect of the silicate vapour from the disk would have initially led to super-solidus temperatures into the very deep mantle. Once the surface had cooled, this deep magma ocean started crystallizing from the bottom up leading to a solidus-determined geotherm very different from that of a convecting solid-phase mantle. This consequence of a post Moon-forming impact is regarded by Kramers (2007) as key to the apparently cool Hadean mantle, because the resulting geotherm in the freezing magma ocean will eventually lead to mantle overturn. Mantle overturn has the capacity to establish a relatively cold lower mantle and induce immediate strong melting of the shallower mantle, which would rapidly be covered by a think basaltic crust at ca. 4.40–4.45 Ga.

The critical parameter for the likelihood of this model is the time it would have taken for a more familiar mantle geotherm to be established. Once the mantle was fully convecting, the Hadean crust would have destabilised and disappeared. Kramers (2007) provides first order calculations demonstrating that the Hadean crust could have persisted for as long as 400 Myr, in agreement with the observations outlined here, but more work is required to investigate the thermal structure and viscosity of the overturned mantle (see also Davies, this volume). It is noted that the gradual heating of a post overturn mantle, which would have resulted initially in shallow and gradually in deeper convection, could provide the mechanical rationale for the small volume of the early Hadean depleted mantle advocated by Bennett et al. (1993).

2.4-4. SUMMARY

The absence of chemically and isotopically intact Hadean terrestrial rocks obviously limits our ability to understand this crucial window of Earth history. Efforts into studying the Hadean have only gathered momentum in the last decade and it can be anticipated that more data will lead to more educated models for Hadean Earth. There are two aspects where the present data speak with some degree of confidence.

The Hadean crust started evolving between 4.45 and 4.35 Ga. The crust was, at least originally, largely basaltic in character, but internal differentiation was unavoidable, leading to formation of evolved rock types of granitoid composition from which preserved Hadean zircons presumably were sourced. The Hadean crust was not constantly recycled into the mantle, as is modern oceanic crust, but radiogenic isotopes require that a substantial portion of crust was isolated from the mantle for ca. 400 Myr. The exact mass of this crust is proportional to the portion of mantle that was depleted, which is presently unconstrained.

There is precious little evidence other than the very few recovered Hadean zircon grains for the persistence of Hadean crust beyond 3.7 Ga. Indeed, radiogenic isotope systematics of mantle derived rocks at 3.4–3.5 Ga fail to show evidence for Hadean depletion. This implies that not only the Hadean crust was destroyed, but also that the complementary depleted mantle portion disappeared, or was refertilised.

The combined evidence therefore argues against a model in which plate tectonics was established in the early Hadean and produced a crust of strictly continental character similar in volume to that on modern Earth. Indeed, the strongest evidence against a constant continental crust volume since the early Hadean is the isotopic contrast between modern MORB-source mantle and average continental crust. The only scenario in which Hadean continents are remotely feasible is if the LHMB destroyed that crust and continental formation started afresh at ca. 3.8 Ga.

The existence of zircon-bearing Hadean lithologies is not in conflict with a largely mafic/ultramafic composition of bulk Hadean crust, because radioactive heat production was so high that even such crust would have differentiated internally. It is incorrect to call such differentiates 'continental crust' because of the intimate and implicit connection of the subduction process with true continental crust. The question of the compositional character of the Hadean crust, however, is presently of only secondary nature. The true question and target for more research is whether a relatively long-lived voluminous Hadean crust required a relatively cool mantle and if so, how the superheated mantle could have cooled so effectively after the Moon-forming impact?

Earth's Oldest Rocks
Edited by Martin J. Van Kranendonk, R. Hugh Smithies and Vickie C. Bennett
Developments in Precambrian Geology, Vol. 15 (K.C. Condie, Series Editor)
© 2007 Elsevier B.V. All rights reserved.
DOI: 10.1016/S0166-2635(07)15025-8

Chapter 2.5

THE OLDEST TERRESTRIAL MINERAL RECORD: A REVIEW OF 4400 TO 4000 MA DETRITAL ZIRCONS FROM JACK HILLS, WESTERN AUSTRALIA

AARON J. CAVOSIE[a], JOHN W. VALLEY[b] AND SIMON A. WILDE[c]

[a]*Department of Geology, University of Puerto Rico, PO Box 9017, Mayagüez, Puerto Rico 00681, USA*
[b]*Department of Geology and Geophysics, University of Wisconsin, 1215 W. Dayton, Madison, WI 53706, USA*
[c]*Department of Applied Geology, Curtin University of Technology, Perth, Western Australia 6102, Australia*

2.5-1. INTRODUCTION

Little is known of the Earth's earliest history due to the near absence of a rock record for the first five hundred million years after accretion. Earth's earliest history is commonly referred to as the Hadean Eon, and comprises the time following accretion at ca. 4560 Ma, when impacts and magma oceans maintained extreme surface temperatures at or above the temperatures where oceans are vaporized to a dense steam atmosphere. The existence of buoyant crust as early as 4400 Ma is indicated by the preservation of Hadean zircons. The Earth eventually cooled, quenching the high surface temperatures of the Hadean, and gave rise to oceans. This transition to a more familiar and clement Earth ushered in the beginning of the Archean Eon (Cavosie et al., 2005a; Valley, 2006). The timing of the transition from a Hadean to an Archean Earth is inferred to pre-date the oldest known rocks, ca. 4000 to 3800 Ma orthogneisses and metasedimentary rocks that are exposed in the Slave craton of northwest Canada (Bowring and Williams, 1999) and in the North Atlantic craton of southwest Greenland (Nutman et al., 2001).

The only identified materials on Earth potentially old enough to record the Hadean–Archean transition are ancient, ⩾4000 Ma zircons found in Archean metasedimentary rocks in Australia, China, and the USA. In Western Australia, variably metamorphosed metasedimentary rocks in several localities have yielded zircons older than 4000 Ma, older than the known rock record, including the Jack Hills (Table 2.5-1), Mount Narryer (Froude et al., 1983), and Maynard Hills (Wyche et al., 2004). Rare >4000 Ma zircons have also been reported as xenocrysts in younger Archean granitoids (Nelson et al., 2000).

The Jack Hills metasedimentary rocks have received the most attention out of the above localities, primarily due to both the consistently higher concentration of ⩾4000 Ma zircon grains, as well as the presence of the oldest known detrital zircons. Given the unique win-

Table 2.5-1. Numbers of $\geqslant$3900 Ma zircons described from Jack Hills[1]

Reference	U-Pb[2]	REE	CL	δ^{18}O	ε_{Hf}	Ti	^{244}Pu	Methods
Group I: 1986 to 1992								
Compston and Pidgeon (1986)	17	–	–	–	–	–	–	SIMS
Kober et al. (1989)	5	–	–	–	–	–	–	TIMS
Maas and McCulloch (1991)	12	–	–	–	–	–	–	SIMS
Maas et al. (1992)	10	10	–	–	–	–	–	SIMS
Group II: 1998 to 2001								
Amelin (1998)	9	–	–	–	–	–	–	TIMS
Amelin et al. (1999)	(7)	–	–	–	7	–	–	TIMS
Nelson (2000)	2	–	–	–	-	–	–	SIMS
Wilde et al. (2001)	1	1	1	1	–	–	–	SIMS
Mojzsis et al. (2001)	7	–	–	7	–	–	–	SIMS
Peck et al. (2001)	4 (1)	4 (1)	(1)	4 (1)	–	–	–	SIMS
Group III: 2004 to 2006								
Cavosie et al. (2004)	48	–	24	–	–	–	–	SIMS
Turner et al. (2004)	–	–	–	–	–	–	7	LRI-MS
Cavosie et al. (2005)	–	–	24 (20)	44	–	–	–	SIMS
Dunn et al. (2005)	16	–	–	–	–	–	–	SIMS
Crowley et al. (2005)	38	36	21	–	–	–	–	LAICPMS
Watson and Harrison (2005)	–	–	1	–	–	54	–	SIMS
Harrison et al. (2005)	104	–	–	–	104	–	–	TIMS/LAICPMS
Nemchin et al. (2006)	8	–	8	8	–	–	–	SIMS
Pidgeon and Nemchin (2006)	11	–	11	–	–	–	–	SIMS
Valley et al. (2006)	–	–	(2)	–	–	36	–	SIMS
Cavosie et al. (2006)	–	42	(42)	–	–	–	–	SIMS
Fu et al. (2007)	–	–	–	–	–	(36)	–	SIMS
Total grains	292	93	90	64	111	90	7	

SIMS = secondary ionization mass spectrometry, TIMS = thermal ionization mass spectrometry. LRI-MS = Laser resonance ionization mass spectrometry. LAICPMS = laser ablation inductively coupled plasma mass spectrometry. REE = rare earth elements. CL = cathodoluminescence. δ^{18}O = oxygen isotope ratio. ε_{Hf} = epsilon hafnium.

[1] Values with no parentheses indicate the number of new grains reported. Values in parentheses indicate the number of grains in the cited study that were first reported in prior studies.

[2] Values are only listed for grains where analytical results were published.

dow these grains offer on early Earth processes, this review focuses primarily on published reports that describe the population of $\geqslant$4000 Ma zircons from the Jack Hills.

2.5-2. THE JACK HILLS

The Jack Hills, located in the Narryer Terrane of the Yilgarn Craton in Western Australia (Fig. 2.5-1), comprise a ~90 km long northeast-trending belt of folded and weakly

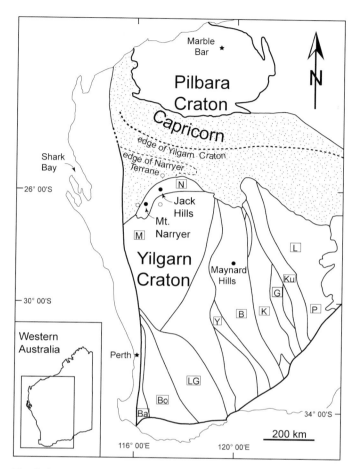

Fig. 2.5-1. Map of Archean cratons in Western Australia, after Wilde et al. (1996). Filled circles are known locations of >4000 Ma detrital zircons, open circles are locations of xenocrysts with similar ages (zircon locations referenced in text). Terranes of the Yilgarn Craton – B: Barlee, Ba: Balingup, Bo: Boddington, G: Gindalbie, K: Kalgoorlie, Ku: Kurnalpi, L: Laverton, LG: Lake Grace, M: Murchison, N: Narryer, P: Pinjin, Y: Yellowdine. Dashed lines are inferred boundaries in basement.

metamorphosed supracrustal rocks that are composed primarily of siliciclastic and chemical metasedimentary rocks, along with minor metamafic/ultramafic rocks (Fig. 2.5-2: see also Wilde and Spaggiari, this volume). Bedding strikes east-northeast and has a subvertical dip. The siliciclastic portion of the belt has been interpreted as alluvial fan-delta deposits, based on repeating fining-upward sequences consisting of basal conglomerate, medium-grained sandstone, and fine-grained sandstone (Wilde and Pidgeon, 1990). Located on Eranondoo Hill in the central part of the belt is a now famous site referred to as 'W74' (Fig. 2.5-2), the name originally assigned to a sample collected at this site by

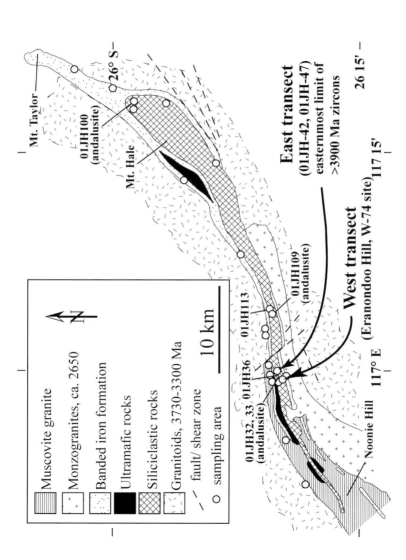

Fig. 2.5-2. Geologic map of the Jack Hills metasedimentary belt, modified from Wilde et al. (1996) and Cavosie et al. (2004). The 'West' and 'East' transects refer to the sampling transects described in Cavosie et al. (2004).

the Curtin University group. The W74 site contains a well-exposed, 2-meter thick quartz pebble metaconglomerate. This previously un-described unit was originally sampled by S. Wilde, R. Pidgeon and J. Baxter in 1984 during an ARC-funded research project, and was described by Compston and Pidgeon (1986) who reported the first ≥4000 Ma detrital zircons from the Jack Hills, including a grain with one spot as old as 4276±6 Ma. Aliquots of zircons from the original W74 zircon concentrate, and additional samples from the same outcrop, have since been the subject of many studies (see below).

2.5-2.1. Age of Deposition

The age of deposition of the Jack Hills metasediments is somewhat controversial, as it appears to vary with location in the belt. The maximum age of the W74 metaconglomerate based on the youngest detrital zircon age has long been cited as ca. 3100 Ma (e.g., Compston and Pidgeon, 1986). However, the first precise age for a concordant 'young' zircon was a 3046 ± 9 Ma grain reported by Nelson (2000). A similar age of 3047 ± 21 Ma was later reported as the youngest zircon by Crowley et al. (2005); thus, it appears that ca. 3050 is the maximum age of deposition of the metaconglomerate at the W74 site.

To explore the distribution of detrital zircon ages away from the W74 site, Cavosie et al. (2004) analyzed zircons from several samples along two transects within the conglomerate-bearing section, including a 60 m section that contains the W74 site (Fig. 2.5-3), and a 20 m conglomerate-bearing section 1 km east of W74. Both transects are dominated by chemically mature clastic metasedimentary rocks (>95 wt% SiO_2), including metaconglomerate, quartzite, and metasandstone. In the west transect, 4 out of 5 samples of quartz pebble metaconglomerate and quartzite contain detrital zircons with ages ≥3100 Ma and >4000 Ma zircons (Cavosie et al., 2004), consistent with previous results from sample W74. However, the stratigraphically highest quartzite in the west transect, sample 01JH-63, contains Proterozoic zircons with oscillatory zoning and ages as young as 1576±22 Ma (Cavosie et al., 2004), and lacks zircons older than ca. 3750 Ma (see discussion by Wilde and Spaggiari, this volume). The presence of Proterozoic zircons in this unit was confirmed by Dunn et al. (2005). Thus, independent studies have demonstrated that the youngest metasedimentary rocks in the Jack Hills are Proterozoic in age. The origin of these metasedimentary rocks remains unknown (see discussion in Cavosie et al., 2004); however, recent field investigations by the current authors have identified layer-parallel faults in the west transect between samples 01JH-63 and the W74 metaconglomerate, which suggests tectonic juxtaposition of two different-age packages of sedimentary rocks (see Wilde and Spaggiari, this volume). The minimum age of the Archean sediments in the Jack Hills is constrained by granitoid rocks which intruded the belt at ca. 2654 ± 7 Ma (Pidgeon and Wilde, 1998).

2.5-2.2. Metamorphism

The metamorphic history of the Jack Hills metasedimentary belt remains poorly documented. However, early workers described rare occurrences of andalusite, kyanite, and

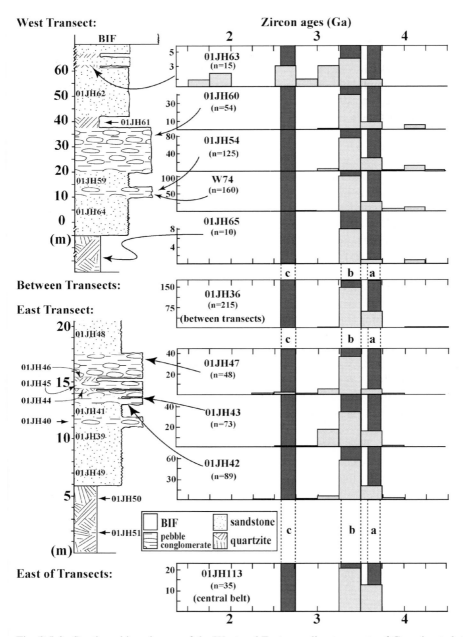

Fig. 2.5-3. Stratigraphic columns of the West and East sampling transects of Cavosie et al. (2004). Note the West transect includes two samples of the metaconglomerate at the W74 site (samples 01JH-54 and W74). Vertical bars labeled a, b, c indicate major periods of intrusive activity in the Yilgarn Craton.

chloritoid in the western part of the belt (Elias, 1982; Baxter et al., 1984). Recent petrographic studies have expanded the known occurrences of andalusite to the central and eastern parts of the belt (Cavosie et al., 2004), which suggests that the majority of the metasedimentary rocks in the Jack Hills metasedimentary belt experienced a pervasive greenschist to lower amphibolite facies metamorphism, despite the absence of index minerals in most units. The common association of metamorphic muscovite with quartz, and the absence of K-feldspar indicates that the clastic metasediments did not reach granulite facies.

2.5-2.3. Geology of Adjacent Rocks

Near Jack Hills are outcrops of the Meeberrie Gneiss, a complex layered rock that yields a range of igneous zircon ages from 3730 to 3600 Ma (Kinny and Nutman, 1996; Pidgeon and Wilde, 1998), establishing it as the oldest identified rock in Australia (Myers and Williams, 1985). Included within the Meeberrie Gneiss near both Jack Hills and Mt. Narryer are cm- to km-scale blocks of a dismembered layered mafic intrusion that together comprise the Manfred Complex (Myers, 1988b). Zircons from Manfred Complex samples yield ages as old as 3730 ± 6 Ma, suggesting it formed contemporaneously with the oldest components of the Meeberrie Gneiss (Kinny et al., 1988). Exposures of the 3490–3440 Ma Eurada Gneiss occur 20 km west of Mt. Narryer, and contain a component of younger ca. 3100 Ma zircons (Nutman et al., 1991). West of Jack Hills, the Meeberrie Gneiss was intruded by the precursor rocks of the Dugel Gneiss, which contain 3380–3350 Ma zircons (Kinny et al., 1988; Nutman et al., 1991), and, like the Meeberrie gneiss, contain enclaves of the Manfred Complex (Myers, 1988b). Younger granitoids, from 2660 ± 20 to 2646 ± 6 Ma, intrude the older granitoids in the vicinity of Jack Hills and Mt. Narryer (Kinny et al., 1990; Pidgeon, 1992; Pidgeon and Wilde, 1998). Contacts between the Jack Hills metasedimentary rocks and the older granitoids are everywhere sheared, whereas the ca. 2650 Ma granitoids appear to intrude the belt (Pidgeon and Wilde, 1998).

2.5-3. JACK HILLS ZIRCONS

Since their discovery two decades ago, compositional data and images of Jack Hills zircons have been described in more than 20 peer-reviewed articles (Table 2.5-1). In an attempt to acknowledge all those who have contributed to this research and to facilitate discussion, we have classified the published articles into three main pulses of research: articles published from 1986 to 1992 are Group I, articles published from 1998 to 2001 are Group II, and articles published from 2004 to the present are Group III. Data and conclusions from these reports are reviewed below.

2.5-3.1. Ages of Jack Hills Zircons

Many thousands of detrital grains have now been analyzed for U-Pb age using several analytical methods, including secondary ion mass spectrometry (SIMS, or ion micro-

probe), thermal ionization mass spectrometry (TIMS), and laser ablation inductively coupled plasma mass spectrometry (LA-ICP-MS). So far, analytical data of U-Pb analyses of
$\geqslant$3900 Ma zircons have been published for nearly 300 detrital grains (Table 2.5-1).

2.5-3.1.1. Group I: 1986 to 1992
The first zircon U-Pb age study in the Jack Hills was made with SHRIMP I by Compston and Pidgeon (1986), who reported 17 zircons from sample W74 with ages in excess
of 3900 Ma from a population of 140 grains, including one crystal that yielded four ages
ranging from 4211 $\pm$ 6 to 4276 $\pm$ 6 Ma, the latter constituting the oldest concordant zircon spot analysis at that time. Subsequent U-Pb studies of zircons from the W74 site by
Köber et al. (1989: TIMS), Maas and McCulloch (1991: SHRIMP), and Maas et al. (1992:
SHRIMP) confirmed that $\geqslant$4000 Ma zircons make up anywhere from 8 to 12% of the analyzed populations, and resulted in published age data for 44 $\geqslant$3900 Ma zircons. Köber et
al. (1989) used the direct Pb-evaporation TIMS method to identify >4000 Ma grains and
concluded that they originated from a granitoid rock, based on similarity of $^{208}Pb/^{206}Pb$
ratios with known rocks. Also noted at the time were the generally low U abundances
for Jack Hills zircons, including concentrations of 50–100 ppm (Compston and Pidgeon,
1986) and 60–413 ppm (Maas et al., 1992).

2.5-3.1.2. Group II: 1998 to 2001
The second wave of Jack Hills zircon research began near the end of the 1990s. New
isotopic U-Pb ages for >4000 Ma grains were published in Amelin (1998: TIMS), Nelson
(2000: SIMS), Wilde et al. (2001: SIMS), Mojzsis et al. (2001: SIMS), and Peck et al.
(2001: SIMS). Amelin (1998) demonstrated high precision $^{207}Pb/^{206}Pb$ age analyses (<1%
uncertainty) of whole grains and air abraded fragments, and again low U abundances (35–
228 ppm). Similar low U abundances were found in two grains with ages of 4080 and
4126 Ma (54–236 ppm) by Nelson (2000), and also in four grains with ages of 4039 to
4163 Ma (41–307 ppm) by Peck et al. (2001).

Perhaps one of the most significant discoveries in U-Pb studies of Jack Hills zircons
is a grain fragment that yielded a single concordant spot age of 4404 $\pm$ 8 Ma (Wilde
et al., 2001). Five additional >95% concordant spot analyses yielded a weighted mean age
of 4352 $\pm$ 10, confirming the great antiquity of the crystal (Wilde et al., 2001; Peck et
al., 2001). The assignment of 4400 Ma as the crystallization age of the zircon followed
the same methodology and rationale as that used by Compston and Pidgeon (1986) for the
4276 $\pm$ 6 Ma crystal; namely, with no analytical reason for exclusion (e.g., U-Pb concordance, ^{204}Pb, etc.), the oldest concordant spot analysis represents the minimum age of the
crystal, and the younger population of ages represent areas of the crystal affected by Pb
loss or younger overgrowths. While some authors choose to average all concordant U-Pb
analyses from a single crystal, the results of doing so generaly do not decrease the maximum age significantly. Thus, the 4400 Ma zircon extends the known age population of
zircons in Jack Hills by $\sim$125 Ma, and currently remains the oldest terrestrial zircon thus
far identified.

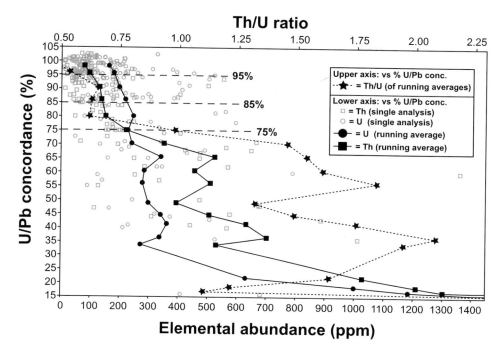

Fig. 2.5-4. Elemental Th and U abundances (lower *x*-axis) and Th/U ratio (upper *x*-axis) plotted against percentage of U-Pb concordance for 140 Jack Hills zircons (after Cavosie et al., 2004). Averages of Th, U, and Th/U ratio were calculated as 10% running averages at 5% increments of concordance beginning with 100%. Each running average includes all data points in a bin that extends over ±5% concordance from the given increment.

2.5-3.1.3. *Group III: 2004 to the present*

As of now, large numbers of ⩾4000 Ma zircon U-Pb ages are available from Jack Hills. Cavosie et al. (2004) reported ages for 42 grains ranging from 4350 to 3900 Ma, including U concentrations from 35–521 ppm. Harrison et al. (2005) reported U-Pb ages >4000 Ma for 104 Jack Hills zircons, with ages from 4371 to 4002 Ma. Additional ages and U concentrations within this range have also been reported by Dunn et al. (2005), Crowley et al. (2005), and Nemchin et al. (2006) (Table 2.5-1). In addition, Pidgeon and Nemchin (2006) identified a single, nearly concordant 4106 ± 22 Ma zircon with 21 ppm U, the lowest U concentration known for a >4000 Ma zircon from Jack Hills.

2.5-3.1.4. *Pb-loss in Jack Hills zircons*

Zircons that have experienced Pb-loss are ubiquitous in the Jack Hills metasedimentary rocks (e.g., Compston and Pidgeon, 1986; Maas et al., 1992; Cavosie et al., 2004; Nemchin et al., 2006). In an attempt to address the issue of Pb-loss, Cavosie et al. (2004) developed a method for evaluating the extent of U-Pb discordance a grain can exhibit while still yielding reliable crystallization ages, instead of picking an arbitrary cut-off value. It was shown that a correlation exists between the Th/U ratio, abundances of U and Th, and U-Pb concordance, that suggests zircons >85% concordant in U/Pb age preserve reliable isotopic ages (Cavosie et al., 2004) (Fig. 2.5-4). The cause of the observed ancient Pb-loss is unknown. However, proposed explanations include otherwise unrecognized granulite facies thermal events that might have disturbed the U-Pb systems in the grains in question (e.g., Nelson, 2002, 2004; Nemchin et al., 2006).

2.5-3.1.5. Distribution of >4000 Ma zircons in Jack Hills metasedimentary rocks
Of the studies that analyzed detrital zircons in samples away from the W74 site, all found that the percentage of $\geqslant$4000 Ma grains is highly variable, and moreover $\geqslant$4000 Ma zircons are not present in many units (Cavosie et al., 2004; Dunn et al., 2005; Crowley et al., 2005). The high percentage of >4000 Ma grains in the W74 metaconglomerate (e.g., 10–14%: Compston and Pidgeon, 1986; Maas et al., 1992; Amelin, 1998; Cavosie et al., 2004) is unique among analyzed samples, given the demonstrated heterogeneous distribution of $\geqslant$4000 Ma grains throughout the belt. The consistency of studies finding this high percentage, however, may not be surprising given that all of the studies listed in Table 2.5-1 contain analyses of zircons separated from the W74 site, and thus essentially analyzed similar populations, often from the same $\sim$2 m^3 W74 outcrop on Eranondoo Hill.

2.5-3.2. Imaging Studies of Jack Hills Zircons

Physical grain aspects of Jack Hills zircons were first described by Compston and Pidgeon (1986), who commented that grains ranged from nearly colorless to deep purplish-brown, were mostly fragments, and were rounded and exhibited pitting, suggestive of sedimentary transport. Maas et al. (1992) reported similar features, and also the occurrence of euhedral crystal terminations. The first grain images of zircons published from the Jack Hills were transmitted light images of grains analyzed by Köber et al. (1989), which showed their rounded forms and pitted surfaces. The extreme rounding of grains and pitting of surfaces was also shown in a back-scattered electron image of a rounded Jack Hills zircon mounted on carbon tape (Valley, 2005). A color image of $\sim$40 Jack Hills zircons mounted on tape prior to casting in epoxy was published by Valley (2006), and shows a population of mostly intact grains and a few grain fragments. The color image illustrates the range of deep red colors that are characteristic of the Jack Hills zircons, as well as the morphological spectrum, from essentially euhedral to completely rounded grains.
     The first cathodoluminescence (CL) image of a Jack Hills zircon was published in Wilde et al. (2001) and Peck et al. (2001), and shows a 4400 Ma zircon with oscillatory zoning (Table 2.5-1). Cavosie et al. (2004, 2005a) showed CL images and reported aspect ratios of 1.0 to 3.4 for an additional 48 zircon grains >3900 Ma from Jack Hills (Fig. 2.5-5), and interpreted that the majority of the 4400–3900 Ma population is of magmatic origin based on the common occurrence of oscillatory zoning. Crowley et al. (2005) examined 21 zircon grains >3900 Ma from Jack Hills, and also noted that oscillatory zoning was a common feature. They used the style of oscillatory and/or sector zoning to interpret that differences in CL zoning patterns between similar age zircons from Mt. Narryer implied

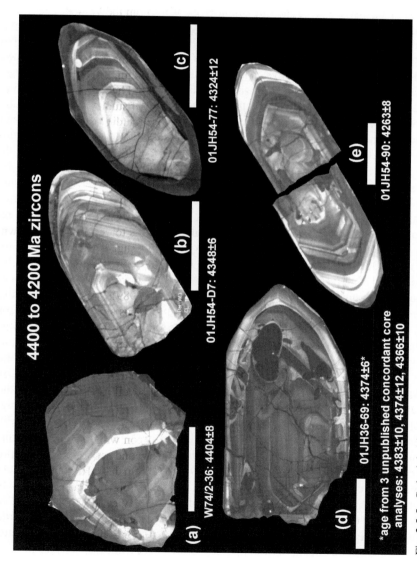

Fig. 2.5-5. Cathodoluminescence images of five 4400–4200 Ma detrital zircons from Jack Hills (additional details of these grains are presented in Cavosie et al. (2004, 2005, 2006)). Ages are in Ma. Uncertainties in Pb-Pb ages are 2 SD. Scale bars = 100 μm.

that the source rocks of the two belts were of different composition. In contrast, Nemchin et al. (2006) noted disturbed margins in CL images of oscillatory-zoned >4000 Ma zircons, and interpreted that the eight grains in their study had experienced complex histories, and that all but one zircon likely did not preserve their magmatic compositions. Pidgeon and Nemchin (2006) presented CL images for 11 additional >3900 Ma grain fragments.

2.5-3.3. Oxygen Isotope Composition of Jack Hills Zircons

Due to the slow diffusivity of oxygen in zircon (e.g., Watson and Cherniak, 1997; Peck et al., 2003; Page et al., 2006), magmatic zircon can provide a robust record of the oxygen isotope composition ($\delta^{18}O$) of host magmas during crystallization (Valley et al., 1994, 2005; Valley, 2003). Wilde et al. (2001) and Peck et al. (2001) reported $\delta^{18}O$ data, measured by SIMS for a population of five >4000 Ma Jack Hills zircons which ranged from 5.6 to 7.4‰; values elevated relative to mantle-equilibrated zircon ($\delta^{18}O = 5.3 \pm 0.6$‰, 2σ). The results were interpreted to indicate that the protolith of the host magmas to the zircons had experienced a low-temperature history of alteration prior to melting, which required the presence of liquid surface waters (Valley et al., 2002). A subsequent study by Mojzsis et al. (2001) confirmed the presence of slightly elevated $\delta^{18}O$ by reporting the same range of values (5.4 to 7.6‰) for four zircons with concordant U-Pb ages from 4282 to 4042 Ma. However, three other zircons were reported by Mojzsis et al. (2001) to have $\delta^{18}O$ from 8 to 15‰ that were interpreted to be igneous and to represent "S-type" granites. Such high values have not been reported for any other igneous zircons of Archean age (Valley et al., 2005, 2006) (Fig. 2.5-6) and the values of 8–15‰ have alternatively been interpreted as due to radiation damage or metamorphic overgrowth (Peck et al., 2001; Cavosie et al., 2005a; Valley et al., 2005, 2006). In contrast, in a study of 44 >3900 Ma zircons by Cavosie et al. (2005a), the location of *in situ* $\delta^{18}O$ analyses was correlated with the location of U-Pb analysis sites. It was found that by applying a protocol of targeting concordant U-Pb domains and discarding analyses that produced anomalous sputter pits (as viewed by SEM), the range of $\delta^{18}O$ varied from 4.6 to 7.3‰, values that overlap, or are higher than, mantle equilibrated zircon (Fig. 2.5-7). Based on results from oscillatory zoned grains with concordant U-Pb ages, Cavosie et al. (2005a, 2005b) documented that the highest $\delta^{18}O$ relative to mantle oxygen (e.g., from 6.5 to 7.5‰) only occurred in zircons with U-Pb ages younger than 4200 Ma (Fig. 2.5-7), and interpreted this to indicate that the end of the Hadean coincided with the onset of crustal weathering, which created high $\delta^{18}O$ protoliths prior to recycling and remelting that began at ca. 4200 Ma ago, or possibly even earlier. In a study of eight >4200 Ma Jack Hills zircons, Nemchin et al. (2006) also reported multiple $\delta^{18}O$ spot analyses for single grains, with grain averages ranging from 4.80 to 6.65‰, and interpreted that the $\delta^{18}O$ values represent low-temperature alteration of primary magmatic zircon. However, we note that the range of $\delta^{18}O$ values reported by Nemchin et al. (2006) lies entirely within the range of magmatic $\delta^{18}O$ values (4.6 to 7.3‰) reported by Cavosie et al. (2005a) for a larger population of >4000 Ma Jack Hills detrital zircons. No evidence for low-temperature oxygen isotope exchange has been documented thus far for any Jack Hills zircon.

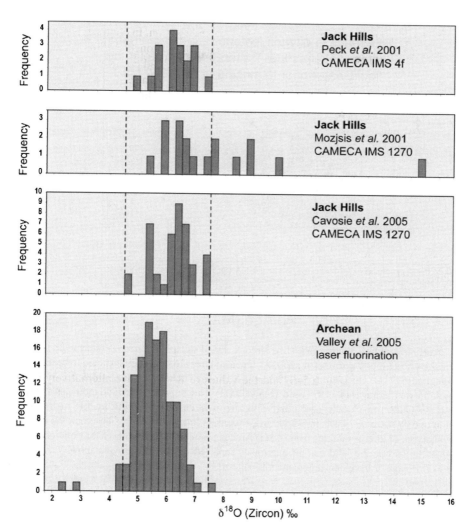

Fig. 2.5-6. Comparison of oxygen isotope ratio for Jack Hills >3900 Ma detrital zircons and Archean igneous zircons.

2.5-3.4. *Trace Element Composition of Jack Hills Zircons*

2.5-3.4.1. *Rare earth elements*

To date, four studies of rare earth elements (REE) have been conducted on >3900 Ma zircons from Jack Hills (Maas et al., 1992; Peck et al., 2001; Wilde et al., 2001; Crowley et al., 2005; Cavosie et al., 2006). All have shown that most grains have compositions that are typical of igneous zircon from crustal environments, as characterized by Hoskin and

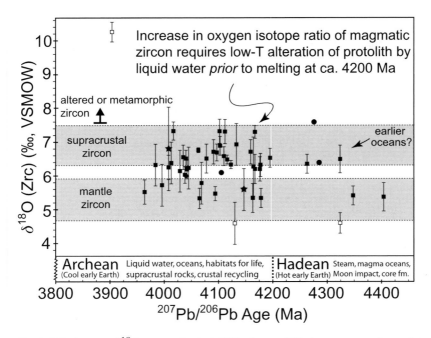

Fig. 2.5-7. Average $\delta^{18}O$ vs. age for Jack Hills zircons. Filled squares are zircons interpreted to pre-serve magmatic $\delta^{18}O$ (after Cavosie et al., 2005b). Open squares are zircons interpreted to be altered, with non-magmatic $\delta^{18}O$. Uncertainty in $\delta^{18}O = 1$ S.D. The 'mantle' zircon field is $5.3 \pm 0.6\%_o$ (2 S.D.), as defined in Valley et al. (1994) and Valley (2003). The 'supracrustal' field indicates a range in magmatic $\delta^{18}O(Zrc)$ that is elevated relative to zircon in equilibrium with mantle melts. The identification of $\delta^{18}O(Zrc) > 7.5\%_o$ in >2500 Ma zircons as 'altered zircon' is based on the obser-vation that analyses of zircon samples from >120 Archean igneous rocks by laser fluorination have not yielded $\delta^{18}O(Zrc) > 7.5$, and that all previously reported >2500 Ma zircons above 7.5‰ are discordant in U/Pb age or have non-magmatic CL patterns.

Schaltegger (2003). General enrichments in the heavy REE (HREE) over the light REE (LREE) indicate that the Jack Hills zircons crystallized in fractionated melts, suggesting the existence of differentiated rocks on the early Earth (Maas et al., 1992; Peck et al., 2001; Wilde et al., 2001; Cavosie et al., 2006). Maas et al. (1992) first demonstrated the similarity of Jack Hills zircons to crustal zircons by showing that characteristics such as positive Ce anomalies and negative Eu anomalies occurred in a population of ten >3900 Ma grains, with total REE abundances from 93 to 563 ppm. Wilde et al. (2001) and Peck et al. (2001) reported similar results, and in addition documented unusual LREE enrichments in some grains, with abundances ranging from 10 to 100 times chondritic abundance, and a much larger range of abundance for total REEs, from 414 to 2431 ppm. Crowley et al. (2005) analyzed 36 grains for REEs, and concluded that Jack Hills zircons were not similar in composition to zircons from neighboring Paleoarchean gneisses.

Hoskin (2005) noted that the unusual LREE enrichments in some of the >4000 Ma grains reported by Wilde et al. (2001) and Peck et al. (2001) were similar to the LREE enrichment measured in hydrothermal zircons from southeast Australian granites, and speculated that some of the Jack Hills zircons might have been affected by hydrothermal processes. To test the 'hydrothermal' hypothesis, Cavosie et al. (2006) analyzed REE in 42 >3900 Ma grains and correlated the location of REE analyses with locations of prior $\delta^{18}O$ and U-Pb analyses (Fig. 2.5-8). Two compositional types of REE domains were identified based on chondrite normalized abundances of La and Pr (Fig. 2.5-9). Type 1

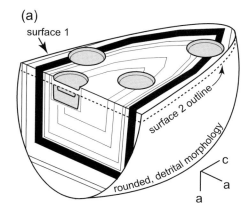

Fig. 2.5-8. Correlated microanalysis of zircon (after Cavosie et al., 2006). (a) Schematic representation of a zoned detrital zircon with correlated analyses of U-Pb, $\delta^{18}O$, and REE (the orientation of crystallographic axes are indicated to the lower-right). The 'a–c plane' [(100), approximately horizontal] represents polished surfaces 1 and 2. Surface 1 was analyzed for U-Pb (shaded ovals). The dashed line indicates the plane of surface 2, analyzed for $\delta^{18}O$ and REE. The 'a–a plane' [(001), approximately vertical] shows a hypothetical cross-section through the grain, and the volumes analyzed for U/Pb, $\delta^{18}O$ and REE. (b) Cross-section (001) of the volumes analyzed for U-Pb age, $\delta^{18}O$, and REE in (a). The dimension of the entire volume varies, but is on average 20 μm in diameter, and 10–15 μm deep.

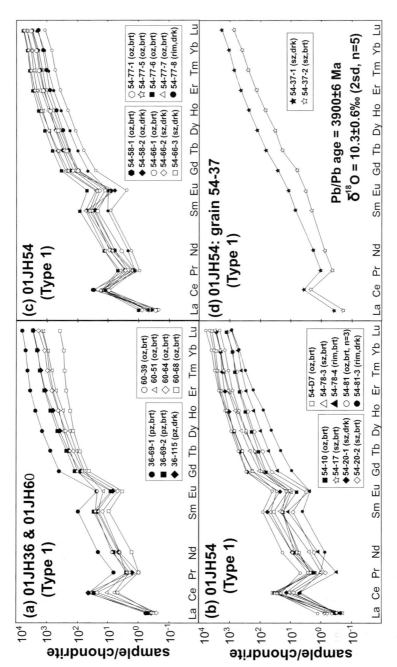

Fig. 2.5-9. Chondrite-normalized REE plots for 39 Jack Hills zircons (after Cavosie et al., 2006): (a) through (g) are 'Type 1' (magmatic); (h) is 'Type 2' (non-magmatic).

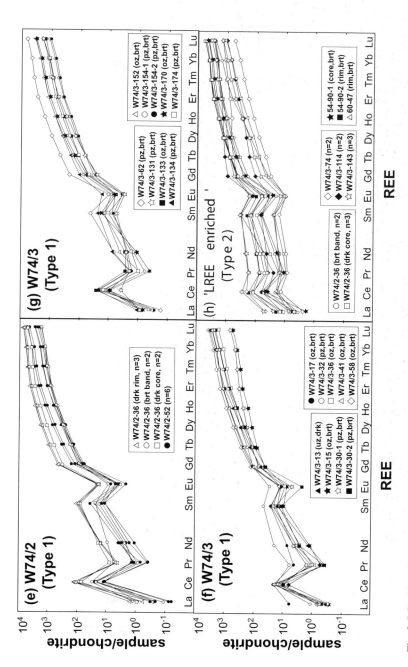

Fig. 2.5-9. (*Continued.*)

(magmatic, Fig. 2.5-9) compositions were preserved in 37 of 42 grains, and consisted of 'typical' crustal REE patterns for zircon (e.g., Hoskin and Schaltegger, 2003). Type 2 (non-magmatic, Fig. 2.5-9) compositions were found in some spots on six grains, and were defined based on the combination of $La_N > 1$ and $Pr_N > 10$. The observation that the Type 2 grain domains that yielded anomalous LREE enrichments also preserved magmatic $\delta^{18}O$ values was used to argue against a hydrothermal origin for the LREE enrichment. The Type 2 LREE enriched compositions (approx. 20% of all spots) were attributed to analysis of sub-surface mineral inclusions and/or radiation-damaged domains, and were not deemed representative of magmatic composition.

2.5-3.4.2. Ti thermometry of Jack Hills zircons

The Ti abundance in zircon (typically a few to tens of ppm) has recently been shown to be a function of melt chemistry and crystallization temperature (Watson and Harrison, 2005). Ti compositions were measured for 54 Jack Hills zircons that range in age from 4000 to 4350 Ma by Watson and Harrison (2005), who interpreted most of their Ti-in-zircon temperatures to average $696 \pm 33\,°C$ (15% of the samples yielded higher temperatures), and thus to provide evidence for minimum-melting, water-saturated granitic magmatism on Earth by 4350 Ma. Valley et al. (2006) questioned the uniqueness of this interpretation for wet-granite melting, citing new data that shows Ti-in-zircon temperatures determined for zircons from a wide range of rock types, including anorthosite, gabbro, and granitoid, overlap with compositions of Jack Hills zircons. Moreover, zirzons from peraluminous (S-type) granites and granodiorites from Sierra Nevada Batholith (USA) contain less Ti, and yield lower temperatures (avg. $= 600\,°C$) than the Jack Hills zircons (Fu et al., 2007). The measured Ti abundances for 36 >3900 Ma zircons cited by Valley et al. (2006) were reported by Fu et al. (in press), and yielded an average temperature of $715 \pm 55\,°C$, slightly higher but in general agreement with the average temperature reported by Watson and Harrison (2005). While the range of applications of this relatively new thermometer is still being explored, the similarity of Ti abundance for zircon in a wide range of felsic and mafic rock types appears to limit the usefulness of Ti composition as a specific petrogenetic indicator for detrital zircons.

2.5-3.4.3. Other trace elements

In addition to the REE, other elements have provided information about the origin of >3900 Ma zircons from Jack Hills. Maas et al. (1992) analyzed Sc by electron microprobe analysis (EMPA) and found values from <17 ppm (detection limit) to 59 ppm in 10 grains, and interpreted these low abundances as indicating an origin in felsic to intermediate rocks based on Sc partitioning behavior.

Turner et al. (2004) reported evidence that trace quantities of the now extinct isotope ^{244}Pu had been incorporated in Jack Hills zircons by measuring anomalies in Xe isotopes in 4100 to 4200 Ma grains. The detection of fissiogenic Xe allows further constraints to be placed on the original Pu/U ratio of Earth.

2.5-3.5. Hf Isotope Composition of Jack Hills Zircons

Hf isotope composition in zircon is a sensitive chronometer for crust/mantle differentiation that can be coupled to U-Pb age. Amelin et al. (1999) reported ε_{Hf} compositions ranging from −2.5 to −6.0 (ε_{Hf} values re-calculated with λ^{176} value of Scherer et al., 2001) for Jack Hills zircons with $^{207}Pb/^{206}Pb$ ages from 4140 to 3974 Ma. The paucity of positive ε_{Hf} values was cited as evidence that none of the zircons originated from depleted mantle sources, while negative ε_{Hf} compositions indicated that significant crust had formed by ca. 4150 Ma (Amelin et al., 1999). A recent Hf isotope study by Harrison et al. (2005) analyzed even older grains from the Jack Hills, including grains from 4000 to 4370 Ma. Harrison et al. (2005) reported a remarkable range of ε_{Hf} compositions, including positive values to +15 and negative values to −7, and interpreted these results to show that continental crust formation and sediment recycling was initiated by ca. 4400 Ma. This interpretation was questioned by Valley et al. (2006) who suggested that such extreme values could be caused in complexly zoned zircons by measurements of Hf in domains (LA-ICP-MS) over 100 times larger than the SHRIMP U-Pb spots, which, moreover, did not coincide. This suggestion is supported by more recent ε_{Hf} analyses reported by Harrison et al. (2006). In the newer dataset, ε_{Hf} and U-Pb were measured during the same LA-ICP-MS analysis, and by this method, no extreme compositions of ε_{Hf} were reported (Harrison et al., 2006).

2.5-4. EARLY EARTH PROCESSES RECORDED IN JACK HILLS ZIRCONS

2.5-4.1. Existence of >4300 Ma Terranes

The large number of reported concordant U-Pb ages from 4400 to 3900 Ma suggests the former existence of terranes comprised at least partially of zircon-bearing igneous rocks. The distribution of published ages for 251 detrital zircons shows that a peak of magmatic activity from 4200 to 4000 Ma dominates the detrital record of Jack Hills zircons (Fig. 2.5-10). In addition, an important population of 17 zircons older than 4300 Ma attests to an earlier period of magmatism that is not as well preserved in the detrital record. The identification of a growing number of >4300 Ma zircons suggests that pre-4300 Ma crust was present on the early Earth, and survived perhaps until the time of deposition of the Jack Hills metasedimentary rocks at ca. 3000 Ma. The location and size of this ancient crust is difficult to estimate. The fact that many of the known >4300 Ma zircons have been found in a single layer at the W74 site (Wilde et al., 2001; Cavosie et al., 2004; Harrison et al., 2005; Nemchin et al., 2006; Pidgeon and Nemchin, 2006) suggests that they were locally derived. However, one 4324 Ma grain found ∼1 km east of the W74 site (Cavosie et al., 2004), and a 4352 Ma grain described by Wyche et al. (2004) from Maynard Hills (Fig. 2.5-1), over 300 km from Jack Hills, suggests that the area of >4300 Ma crust may have been either larger or more widespread, in order to contribute to Archean sediments in multiple locations. In addition, Cavosie et al. (2004) documented younger

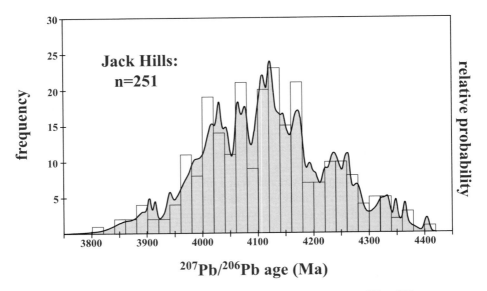

Fig. 2.5-10. Histogram and relative probability plot of 251 concordant ^{207}Pb/^{206}Pb ages for Jack Hills detrital zircons. Each datum is the oldest age assigned to a single zircon that is at least 85% concordant in the U-Pb system. Data sources are as follows: Compston and Pidgeon, 1986 ($n = 17$), Maas et al., 1992 ($n = 8$), Wilde et al., 2001 ($n = 1$), Peck et al., 2001 ($n = 3$), Mojzsis et al., 2001 ($n = 4$); Nelson, 2000 ($n = 2$); Cavosie et al. 2004 ($n = 40$), Crowley et al., 2005 ($n = 45$), Dunn et al., 2005 ($n = 16$), Harrison et al., 2005 ($= 98$), Nemchin et al., 2006 ($n = 8$), Pidgeon and Nemchin, 2006 ($n = 8$). Bin width = 20 My.

rims on two >4300 Ma zircons, interpreted as magmatic overgrowths, which yielded ages of ca. 3360 Ma and 3690 Ma. These ages match the ages of Archean granitoids in the Narryer Terrane, and suggest that ca. 4300 Ma crustal material may have been incorporated in younger Archean magmas.

2.5-4.2. The Cooling of Earth's Surface

Oxygen isotope ratios in zircon record the cooling of Earth's surface and the condensation of the first oceans before ca. 4200 Ma. The global transformation to cooler surface conditions and stable surface waters began the aqueous alteration of crustal rocks at low temperatures, a new process that would fundamentally influence the composition of igneous rocks. Magmatic recycling of this crust, altered at low-temperatures, produced igneous rocks with elevated δ^{18}O compositions, and was operative as a significant geologic process at ca. 4200 Ma, and possibly earlier. The cooling of Earth's surface, the stabilization and availability of surface water, and its influence on melting and magma generation all represent fundamental changes from a Hadean Earth. This signaled the advent of relatively cool liquid water oceans and created environments hospitable for emergence of life.

We propose that this transition at ca. 4200 Ma is a global boundary condition that can be used to define the beginning of the Archean Eon on Earth (Cavosie et al., 2005b).

ACKNOWLEDGEMENTS

The authors thank many people who contributed to our studies of Jack Hills zircons: Song Biao, Nicola Cayzar, John Craven, Scott Dhuey, Liu Dunyi, John Fournelle, Colin Graham, Matthew Grant, Brian Hess, Richard Hinton, Ian Hutcheon, Noriko Kita, William Peck, Mike Spicuzza, Mary and Matchem Walsh, Paul Weiblen. The work cited in the text by the authors was supported by grants from the NSF, DOE, and ARC.

Earth's Oldest Rocks
Edited by Martin J. Van Kranendonk, R. Hugh Smithies and Vickie C. Bennett
Developments in Precambrian Geology, Vol. 15 (K.C. Condie, Series Editor)
© 2007 Elsevier B.V. All rights reserved.
DOI: 10.1016/S0166-2635(07)15026-X

Chapter 2.6

EVIDENCE OF PRE-3100 MA CRUST IN THE YOUANMI AND SOUTH WEST TERRANES, AND EASTERN GOLDFIELDS SUPERTERRANE, OF THE YILGARN CRATON

STEPHEN WYCHE

Geological Survey of Western Australia, 100 Plain St., East Perth, Western Australia 6004, Australia

2.6-1. INTRODUCTION

The Archaean Yilgarn Craton (Fig. 2.6-1) is an extensive granite–greenstone terrain in southwestern Australia containing metavolcanic and metasedimentary rocks, granites, and granitic gneiss that formed principally between ca. 3050 and ca. 2620 Ma, with a minor older component to >3700 Ma (e.g., Kinny et al., 1990; Pidgeon and Wilde, 1990; Nelson, 1997a; Pidgeon and Hallberg, 2000; Cassidy et al., 2002; Wilde and Spaggiari, this volume). Voluminous granite intrusion between ca. 2760 and ca. 2620 Ma coincided with Neoarchaean orogeny, resulting in amalgamation and assembly of several tectonic units to form the craton (Nutman et al., 1993a; Myers, 1995; Myers and Swager, 1997). The craton is subdivided into six terranes (Fig. 2.6-1), the easternmost three of which constitute the Eastern Goldfields Superterrane (Cassidy et al., 2006). In the west, the Narryer and South West Terranes are dominated by granite and granitic gneiss, whereas the Youanmi Terrane, and the terranes of the Eastern Goldfields Superterrane, contain northeast- to northwest-trending greenstone belts separated by granite and granitic gneiss. All terranes contain rocks with pre-3000 Ma zircon xenocrysts, but they are rare (e.g., Swager et al., 1995; Cassidy et al., 2002).

There has been much discussion of the geological processes that may have produced the Eastern Goldfields Superterrane, particularly with respect to the roles of subduction (e.g., Barley et al., 1989) and mantle plumes (e.g., Campbell and Hill, 1988; Bateman et al., 2001) in its tectonic development. However, the processes that controlled deposition of greenstones in the Youanmi Terrane are not well understood. The centre of the Youanmi Terrane is marked by several large, layered mafic–ultramafic intrusions, the largest of which, the Windimurra Complex (Fig. 2.6-1), has a poorly constrained Sm-Nd age of ca. 2800 Ma (Ahmat and Ruddock, 1990). This intrusion is overlain by roof pendants of ca. 2813 Ma felsic volcanic rocks (sample 169003 in GSWA, 2006). The layered intrusions may be evidence of a major episode of plume magmatism.

Geochronological constraints on deformation suggest diachroneity of structural events across the Yilgarn Craton. However, the preserved structural grain reflects a protracted period of east–west compression between ca. 2750 Ma and ca. 2650 Ma that coincided with

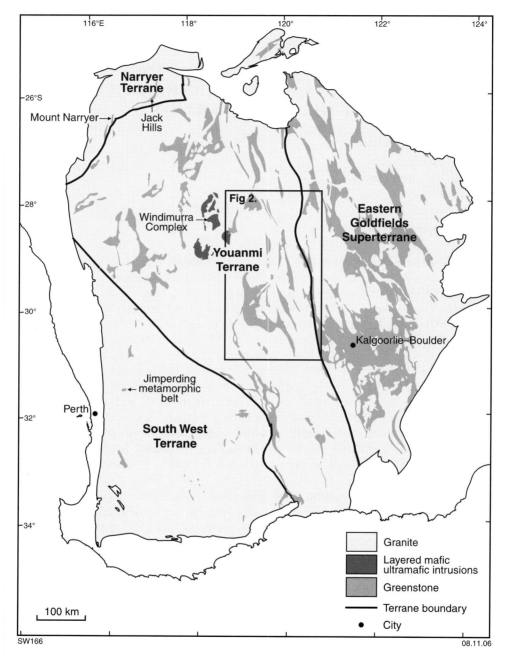

Fig. 2.6-1. Yilgarn Craton, Western Australia, showing major tectonic subdivisions and distribution of granite and greenstone.

peak metamorphism and the dominant period of granite intrusion (e.g., Chen et al., 2004). There is a distinct change in the type and style of magmatism at ca. 2655 Ma from voluminous, dominantly high-Ca granites, to less voluminous, but widespread, low-Ca granites (Cassidy et al., 2002).

Quartzite and quartz-rich, clastic metasedimentary rocks of the Narryer Terrane contain detrital zircons ranging in age from ca. 4400 Ma to ca. 1600 Ma, including the oldest terrestrial zircons yet identified (Froude et al., 1983a; Compston and Pidgeon, 1986; Wilde et al., 2001; Cavosie et al., 2004). Similar metasedimentary rocks, with detrital zircons older than ca. 3100 Ma, have been recognized in greenstone successions in the South West Terrane (Wilde and Low, 1978; Nieuwland and Compston, 1981; Kinny, 1990; Bosch et al., 1996) and the eastern part of the Youanmi Terrane (Froude et al., 1983b; Compston et al., 1984b; Wyche et al., 2004), but have not been reported from the Eastern Goldfields Superterrane.

2.6-2. EVIDENCE OF OLD (PRE-3100 MA) YILGARN CRUST

2.6-2.1. *Youanmi Terrane*

Quartzite and quartz-rich clastic metasedimentary rocks have been reported from a number of greenstone belts in the eastern Youanmi Terrane (e.g., Gee, 1982; Chin and Smith, 1983; Riganti, 2003; Wyche, 2003). Quartz sandstone and quartz–mica schist are preserved at the base of the exposed greenstone succession over a wide area in the north, with the most extensive development of these rocks in the Maynard Hills and Illaara greenstone belts (Fig. 2.6-2). These are currently the only two greenstone belts from which samples have yielded sufficient zircons for SHRIMP analysis (Wyche et al., 2004).

The quartz-rich, clastic metasedimentary rocks that were dated in the Illaara greenstone belt lie along the eastern side of the belt and represent the lowest preserved part of the succession. This unit has a maximum thickness of ∼900 m and consists of prominent, ridge-forming, fine- to coarse-grained, strongly recrystallized quartzite, and recessive intervals that range from micaceous quartzite to quartz–mica schist. Stratigraphic relationships of similar quartz-rich metasedimentary rocks in the Maynard Hills greenstone belt are less apparent due to strong deformation and recrystallization, but the rock associations in this belt suggest a stratigraphic setting similar to that of quartzites in the Illaara greenstone belt (Riganti, 2003; Wyche et al., 2004).

Primary textural and sedimentological features in the quartzites of the Illaara and Maynard Hills greenstone belts are poorly preserved due to the effects of deformation and recrystallization. A strong bedding-parallel cleavage obscures primary sedimentary structures, but they are probably thin to medium bedded and derived from mature quartz arenites. Original grainsize is difficult to determine due to extensive recrystallization, but some beds contain sub-rounded quartz pebbles. Muscovite is the dominant mica but bright green fuchsite is found locally in both the quartzite and quartz–mica schist. Very fine-grained tourmaline gives some beds a dark colouration. Their association with quartz–mica schists

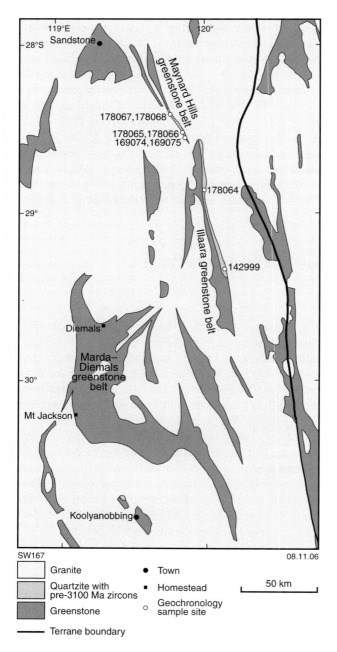

Fig. 2.6-2. Greenstone belts in the central Yilgarn Craton showing locations of quartzite samples analyzed and published by the Geological Survey of Western Australia. All data are presented in GSWA (2006).

that were probably derived from sub-mature to immature quartz arenites suggests that these rocks were deposited in a fluvial to shallow-marine environment (Wyche, 2003; Wyche et al., 2004).

In the earliest SHRIMP U-Pb zircon work in the central Yilgarn Craton, Froude et al. (1983b) identified detrital zircons in the Maynard Hills greenstone belt ranging in age between ca. 3300 Ma and ca. 3700 Ma, with a single zircon yielding an age of ca. 2900 Ma. However, no analytical data from this study have been published.

More recently, eight quartzite samples have been analyzed from the Illaara and Maynard Hills greenstone belts (Fig. 2.6-2), with a total of 293 analyses from 265 zircons (Fig. 2.6-3(a); GSWA, 2006). Of these, the youngest concordant zircon, with a $^{207}Pb/^{206}Pb$ age of 3131 ± 3 Ma (sample 169074 in GSWA, 2006), was obtained on quartzite from the Maynard Hills greenstone belt (sample 169074 on Fig. 2.6-2). This represents the maximum age of deposition of the precursor sandstone. Most zircons lie between ca. 3130 Ma and ca. 3920 Ma, but two samples yielded zircons older than 4000 Ma (Fig. 2.6-3(a)).

Of the two samples with >4000 Ma zircons, sample 169075 from the Maynard Hills greenstone belt (Fig. 2.6-2) contained one concordant zircon grain with a weighted mean $^{207}Pb/^{206}Pb$ age of 4350 ± 5 Ma (Fig. 2.6-3(b): sample 169075 in GSWA, 2006). Sample 178064 from the Illaara greenstone belt (Fig. 2.6-2) yielded one concordant zircon with a $^{207}Pb/^{206}Pb$ age of 4170 ± 6 Ma and four discordant grains older than 4000 Ma (Fig. 2.6-3(c): sample 178064 in GSWA, 2006).

2.6-2.1.1. South West Terrane

The Jimperding greenstone belt in the South West Terrane of the Yilgarn Craton consists of thin units of quartzite, pelitic schist, amphibolite, metamorphosed ultramafic rocks, and metamorphosed banded iron-formation, interleaved with a variety of gneisses, including garnetiferous gneiss. It is bounded to the east by gneiss and migmatite, and intruded to the west by granites of the Darling Range Batholith (Wilde, 2001). There are no direct ages for rocks within the greenstone belt, but the granites to the west were probably emplaced between ca. 2650 Ma and ca. 2630 Ma (Nemchin and Pidgeon, 1997). The best exposure of quartzite is in a railway cutting at Windmill Hill where flaggy and massive quartzite is commonly fuchsitic, and locally contains preserved cross-beds. The quartzite is faulted against granitic gneiss. Greenstones associated with the quartzite include amphibolite and ultramafic rock, and it is faulted against granitic gneiss (Wilde, 2001).

Following evidence of old zircons in metasedimentary rocks in the Jimperding metamorphic belt in the South West Terrane (Fig. 2.6-1; Nieuwland and Compston, 1981), Kinny (1990) analyzed 50 zircon grains from quartzite from the Windmill Hill locality, and Bosch et al. (1996) analyzed a further 12 grains from a sample of nearby pelitic schist. Zircons ranged in age mainly up to ca. 3500 Ma, with Kinny (1990) finding one zircon with an age of ca. 3735 Ma. The youngest concordant zircon obtained has a $^{207}Pb/^{206}Pb$ age of 3055 ± 6 Ma (Bosch et al., 1996), which represents the maximum age of deposition of the precursor sedimentary rock.

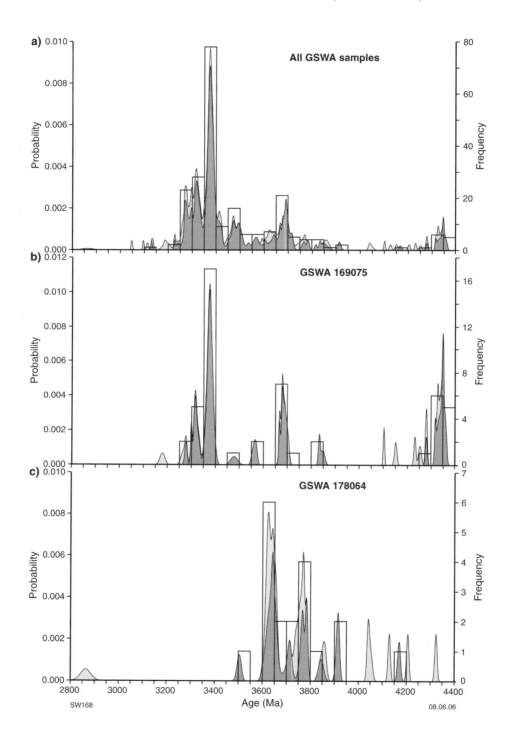

Fig. 2.6-3. (*Previous page.*) Probability density diagrams for U-Pb ages of detrital zircons in the Maynard Hills and Illaara greenstone belts, Youanmi Terrane. In each case, the dark grey area and frequency histograms (bin width 50 My) include only concordant data, defined here as analyses with f204 (i.e., fraction of common ^{206}Pb in total ^{206}Pb) <1% and discordance <10%; the light grey area includes all data: (a) all Geological Survey of Western Australia samples from the Maynard Hills and Illaara greenstone belts = 226 concordant analyses of 209 zircons (293 of 265 total); (b) Geological Survey of Western Australia sample 169075 from the Maynard Hills greenstone belt = 49 concordant analyses of 35 zircons (61 of 39 total); (c) Geological Survey of Western Australia sample 178064 from the Illaara greenstone belt = 19 concordant analyses of 19 zircons (35 of 35 total). All data are presented in GSWA (2006).

2.6-2.1.2. Eastern Goldfields Superterrane

Metasedimentary rocks like those in the Youanmi, South West, and Narryer Terranes with maximum depositional ages greater than 3000 Ma have not been recognized in the Eastern Goldfields Superterrane. Here, metasedimentary rocks with maximum depositional ages of ca. 2700 Ma are intercalated with, and in the upper part of, the volcanosedimentary greenstone succession. Detrital zircons older than 3000 Ma are rare. Campbell and Hill (1988) reported the presence of ca. 3650–3670 zircons in 'cherty' clastic sedimentary rocks of the Noganyer Formation in the south. More recently, Krapež et al. (2000) found isolated occurrences of detrital zircon grains in samples from the area around and to the south of Kalgoorlie–Boulder (Fig. 2.6-1), including a rock interpreted as turbidite from the Noganyer Formation, that range in age back to ca. 3570 Ma. They concluded that, although the sources of these older zircons could have been structural enclaves within volcanoplutonic arcs, they are more likely derived from distant cratonic basement.

2.6-2.2. Other Evidence of old Yilgarn Crust

The only known >3100 Ma rocks in the Yilgarn Craton are those in the Narryer Terrane, with metagabbroic anorthositic rocks of the ca. 3730 Ma Manfred Complex being the oldest identified component (Myers, 1988b; Kinny et al., 1988). Other gneiss components of the Narryer Terrane range in age between ca. 3730 and ca. 3300 Ma (Kinny et al., 1988; Nutman et al., 1991; Wilde and Spaggiari, this volume). Inheritance of zircons from rocks older than the Manfred Complex is rare in the Narryer Terrane. However, zircons as old as ca. 4180 Ma have been identified locally in younger (ca. 2690 Ma) gneiss (e.g., sample 105007 in GSWA, 2006).

Various authors have argued for the existence of widespread, early sialic crust, at least as old as 3000 Ma, in the eastern Yilgarn Craton. Evidence of early sialic crust includes the presence of old xenocrystic zircons, Pb-isotope studies (Oversby, 1975), Sm-Nd and Lu-Hf isotope data, granite geochemistry suggestive of extensive reworking (e.g., Champion and Sheraton, 1997), and contamination of mafic and ultramafic greenstones by crustal material (Redman and Keays, 1985; Arndt and Jenner, 1986; Barley, 1986; Compston et al., 1986b).

Pre-3100 Ma xenocrystic zircons in the Youanmi Terrane are rare, with only a few published examples. These xenocrysts are mainly found in granites, and typically range in age from ca. 3650 to ca. 3300 Ma (samples 105016 and 169076 in GSWA, 2006; Pidgeon and Hallberg, 2000; Cassidy et al., 2002). A very small number of xenocrystic zircons up to ca. 4000 Ma have been found in felsic volcanic rock (Pidgeon and Hallberg, 2000) and gneiss (sample 105018a in GSWA, 2006) from the western part of the Youanmi Terrane. However, the great majority of inherited zircons in rocks of the Youanmi Terrane are <3100 Ma (e.g., sample 142920 in GSWA, 2006; Pidgeon and Hallberg, 2000), and probably reflect ages of greenstones from within the terrane.

Old zircon xenocrysts (ca. 3500 to ca. 3000 Ma) appear to be more abundant in both greenstone and granitic rocks of the Eastern Goldfields Superterrane (Compston et al. 1986b; Campbell and Hill 1988; Claoué-Long et al. 1988; Hill et al. 1989) than in the Youanmi Terrane. This may be an artefact of the greater number of samples analyzed from the Eastern Goldfields Superterrane, or may reflect the former presence of an old protocontinent in the east of which no other trace remains.

Sm-Nd studies show that there is a fundamental difference in the isotopic nature of the source regions for granites in the Youanmi Terrane compared with the Eastern Goldfields Superterrane (e.g., Fletcher et al., 1994; Cassidy et al., 2002). Granites from the Youanmi Terrane show evidence of a long crustal prehistory, back to ca. 3000 Ma, compared with those of the Eastern Goldfields Superterrane that have more primitive and younger sources.

Griffin et al. (2004a) found detrital zircons in modern drainage systems near Sandstone in the northeastern part of the Youanmi Terrane (Fig. 2.6-2) dating back to ca. 3600 Ma. Apart from these old zircons, model ages based on ε_{Hf} values of zircons most likely derived from ca. 2600–2700 Ma granitic rocks indicate that their parent magmas were derived from continental crust dating back to ca. 3700 Ma. Griffin et al. (2004a) interpreted these data to indicate the former presence of ancient continental crust in the north-central part of the Yilgarn Craton.

2.6-3. DISCUSSION

The source of the abundant detrital zircon grains that are older than 3100 Ma in the quartzites in the Illaara and Maynard Hills greenstone belts of the Youanmi Terrane, and the Jimperding metamorphic belt of the South West Terrane, is unknown. In the Youanmi and South West Terranes, no igneous rocks older than ca. 3050 Ma have been identified, with granitic rocks in the region mainly younger than 2750 Ma. Rocks to the east in the Eastern Goldfields Superterrane are typically younger than those in the Youanmi Terrane.

Gneisses of the Narryer Terrane (Kinny et al., 1988; Nutman et al., 1991; Wilde and Spaggiari, this volume) range between ca. 3730 and ca. 3300 Ma in age. If the present-day configuration of the Yilgarn Craton reflects the situation at ca. 3100 Ma, the nearest preserved crustal remnant that pre-dates deposition of Illaara, Maynard Hills, and Jimperding quartzites is more than 300 km away.

Kinny (1990) and Bosch et al. (1996) suggested that age profiles of detrital zircons from the metasedimentary rocks in the Jimperding metamorphic belt in the South West Terrane (Fig. 2.6-1) indicated a similar provenance to those of the Mount Narryer and Jack Hills metasedimentary rocks in the Narryer Terrane (Fig. 2.6-1).

Zircon U-Th-Pb, mineral K-Ar and ^{40}Ar/^{39}Ar, and provenance studies by Kinny et al. (1990) indicated a maximum depositional age of ca. 3100 Ma for the Mount Narryer quartzite. An age of 3064 ± 17 Ma was reported for a detrital zircon from Jack Hills (sample 142986 in GSWA, 2006), thus giving a maximum age of deposition for these rocks. However, more recent studies of detrital zircons from Jack Hills have identified local populations of detrital zircons as young as ca. 1600 Ma (Cavosie et al., 2004). Whether these younger zircons indicate a more complex depositional history than has previously been recognized, or tectonic interleaving of Archaean and younger sedimentary rocks, is not yet known. No detrital zircons younger than ca. 3100 Ma have been found in quartzites in the Youanmi and South West Terranes.

Rainbird et al. (1997) have shown that large populations of zircons in quartzite can be derived from a distal source >3000 km away. However, their examples did show evidence of contributions from intervening sources and so it is likely that, if the Illaara and Maynard Hills quartzites post-dated any elements of the Youanmi Terrane, they would contain some evidence of material from greenstones or granites of the terrane. Because none of the samples contain detrital zircons formed during recorded felsic magmatism in the Yilgarn Craton, the quartzites must have been deposited before any of the presently preserved elements of the Youanmi Terrane were exposed.

Similarities in the age profiles of detrital zircons from metasedimentary rocks from the Maynard Hills and Illaara greenstone belts in the Youanmi Terrane, the Jimperding metamorphic belt in the South West Terrane, and the Mount Narryer and Jack Hills areas in the Narryer Terrane, suggest that they were derived from a similarly aged continental source (Wyche et al., 2004). Further evidence supporting a common source of detrital zircons for the Narryer and Youanmi metasedimentary rocks are the >4100 Ma zircons from the Maynard Hills and Illaara greenstone belts. The only other >4100 Ma zircons that have previously been identified are detrital zircons from Mount Narryer and the Jack Hills in the Narryer Terrane (Fig. 2.6-1; Froude et al., 1983a; Compston and Pidgeon, 1986; Wilde et al., 2001).

In the Narryer Terrane, contacts between gneisses and metasedimentary rocks that contain detrital zircons of similar ages are strongly deformed so that their relationships are unknown (Myers and Occhipinti, 2001). Kinny et al. (1990) proposed that quartzite at Mount Narryer may have been partly derived from gneissic, granitic and anorthositic rocks like those that outcrop in the Narryer Terrane, but that the presence of other zircon components suggested contributions from other source terranes. Maas and McCulloch (1991) argued that differences in REE patterns between the Mount Narryer and Jack Hills metasedimentary rocks and the nearby gneisses indicate that, although these gneisses may have contributed to the zircon populations preserved in these rocks, they were not the major source.

While it is possible to infer that the post-3700 Ma detrital zircons were at least in part derived from rocks such as the gneisses in the Narryer Terrane, or a no longer extant area of basement in the northeastern part of the Youanmi Terrane, a question remains as to the source of detrital zircons older than ca. 3730 Ma.

Maas et al. (1992), in a study of the morphological, mineralogical and geochemical characteristics of pre-3900 Ma detrital zircons from Mount Narryer and the Jack Hills, concluded that they were derived from a continental source dominated by potassic granites. Nelson et al. (2000) argued that zircon geochronology suggests the existence of an ancient (>3800 Ma) composite terrane. If the metasedimentary rocks in the Narryer Terrane were derived from a complex variety of sources, then the preserved Narryer Terrane may be a remnant of a continental mass that provided a relatively proximal source for all of the metasedimentary rocks that contain old detrital zircons in the western part of the Yilgarn Craton. Cavosie et al. (2004) said that many of the >4000 Ma zircons in the Jack Hills are igneous and may be locally derived, thus allowing the possibility of magmatic episodes in the region as far back as 4400 Ma.

According to Myers (1995), structural evidence and a gap in the zircon ages after ca. 2680 Ma suggest that the Narryer Terrane accreted to the Youanmi Terrane between 2680 and 2650 Ma. This is supported by Nutman et al. (1993a) who argued that Nd isotope compositions of Neoarchaean granites (ca. 2750 Ma and younger) that intrude the old (ca. 3730–3300 Ma) gneisses of the Narryer Terrane were derived from a younger source, similar to gneisses in adjacent terranes, and do not contain evidence of an older source. However, the rare >4100 Ma inherited zircons in some younger granitic rocks (sample 105007 in GSWA, 2006) in the Narryer Terrane indicate a greater level of complexity in the sources of at least some of these rocks than is suggested by the isotope data. Also, as discussed above, the detrital zircon data from the Youanmi Terrane suggest that pre-3100 Ma metasedimentary rocks now preserved in this terrane may have had the same provenance as similar rocks in the South West and Narryer Terranes. If true, then these regions must have had some common history before the proposed accretion of the Narryer Terrane to the Youanmi Terrane after ca. 2680 Ma.

The quartzites of the Yilgarn Craton are very similar in character to Archaean quartz arenites described from North America (Superior and Churchill Provinces: Donaldson and de Kemp (1998); Wyoming Province: Mueller et al. (1998); Slave Province: Sircombe et al. (2001)). Where detrital zircon data have been obtained from these rocks, age patterns have clear similarities to those in the Yilgarn examples, with populations in the Wyoming and Slave Provinces dating back to 4000 Ma. As in the Youanmi Terrane, there is no obvious local source for the older detrital zircons in the quartzites of the Wyoming Province, whereas gneiss and granite remnants up to ca. 4000 Ma in the Slave Province could have provided a source for detrital zircons in the Slave Province quartzites.

2.6-4. CONCLUSIONS

The quartz-rich, clastic metasedimentary rocks in the central Yilgarn Craton containing sand-sized quartz grains require a provenance with a substantial continental component, probably containing abundant granitic rocks. Rare xenocrystic zircons older than 3000 Ma in granites and greenstones; samples from modern drainage systems yielding detrital zircons that range in age back to ca. 3600 Ma, along with younger granite-derived detrital zircons that have Hf-isotope signatures suggesting the existence of crust as old as ca. 3700 Ma in the region; and granite geochemistry that requires at least a two-stage melting process, suggest the presence of older sialic crust in the region. However, no such material has been identified in the Youanmi and South West Terranes, or the Eastern Goldfields Superterrane. If pre-3100 Ma rocks in the Narryer Terrane formed part of this early crust, then it must have been adjacent to the rest of the Yilgarn Craton at an early stage in the development of the craton.

Because no potential source rocks that are older than the ca. 3055 Ma age of a detrital zircon in the Jimperding metamorphic belt in the South West Terrane have been identified in the Youanmi or South West Terranes, the early clastic sedimentary formations in the northwestern, southwestern, and eastern parts of the Yilgarn Craton may have been shallow-marine shelf deposits adjacent to a continental mass of which the Narryer Terrane is a remnant. Rifting, marked by the widespread mafic and ultramafic volcanism preserved in the Youanmi Terrane, broke up the continent some time after 3100 Ma. The large layered mafic and ultramafic intrusions in the centre of the Youanmi Terrane may also be associated with rift-related plume magmatism. Finally, a major period of voluminous granite intrusion between 2750 Ma and 2620 Ma, probably related to accretionary activity in the eastern part of the craton, completely dismembered and deformed the greenstone successions, and produced the present-day arrangement of the greenstone belts. Alternatively, widespread pre-3100 Ma continental crust that formed the proto-Yilgarn Craton has been largely reworked during the period of voluminous Neoarchaean granite intrusion associated with accretionary activity to the east, with the only preserved traces being the fragment represented by the Narryer Terrane, geochemical signatures in granitic rocks and greenstones, and old xenocrystic zircons. If this is the case, then the presence of zircons older than ca. 4100 Ma in the Youanmi quartzites indicates that there may have been a more substantial volume of pre-4000 Ma continental crust than has previously been suggested.

The similarity of the age populations of detrital zircons from metasedimentary rocks in the Youanmi Terrane, the Jimperding metamorphic belt in the South West Terrane, and the Mount Narryer and Jack Hills areas in the Narryer Terrane suggests that they are all derived from the same ancient continental source. Thus it is unlikely that the areas now occupied by the Younami, South West, and Narryer Terranes were separate entities prior to the deposition of quartz-rich, clastic sedimentary rocks at ca. 3100 Ma.

ACKNOWLEDGEMENTS

The author would like to thank Martin Van Kranendonk, Angela Riganti, and Simon Bodorkos at the Geological Survey of Western Australia for helpful suggestions and discussions. The manuscript was much improved by reviews by Peter Kinny and Bruce Groenewald. This contribution is published with the permission of the Director of the Geological Survey of Western Australia.

PART 3

EOARCHEAN GNEISS COMPLEXES

Earth's Oldest Rocks
Edited by Martin J. Van Kranendonk, R. Hugh Smithies and Vickie C. Bennett
Developments in Precambrian Geology, Vol. 15 (K.C. Condie, Series Editor)
© 2007 Elsevier B.V. All rights reserved.
DOI: 10.1016/S0166-2635(07)15031-3

127

Chapter 3.1

THE EARLY ARCHEAN ACASTA GNEISS COMPLEX: GEOLOGICAL, GEOCHRONOLOGICAL AND ISOTOPIC STUDIES AND IMPLICATIONS FOR EARLY CRUSTAL EVOLUTION

TSUYOSHI IIZUKA[1], TSUYOSHI KOMIYA AND SHIGENORI MARUYAMA

Department of Earth and Planetary Sciences, Tokyo Institute of Technology, Ookayama Meguro-ku, Tokyo 152-8551, Japan

3.1-1. INTRODUCTION

The oldest crustal rocks yet identified are 4.03–3.94 Ga magmatic protoliths of granitic-amphibolitic gneisses that outcrop within the Acasta Gneiss Complex along the Acasta River in the westernmost Slave Province (Bowring et al., 1989a; Bowring et al., 1990; Bowring and Housh, 1995; Bleeker and Stern, 1997; Stern and Bleeker, 1998; Bowring and Williams, 1999; Sano et al., 1999; Iizuka et al., 2006; Iizuka et al., 2007). In this paper, we summarize the geological, geochronological, and radiogenic isotopic data acquired from the Acasta gneisses. These data provide valuable insights into early crustal evolution by constraining the early Archean tectonothermal evolution of the complex and the relationships between the Paleoarchean crustal rocks and the Hadean (>4.03 Ga) mantle-crust system.

3.1-2. GEOLOGY

The Acasta Gneiss Complex is exposed in the westernmost Slave Province, which is in the foreland and metamorphic internal zone of the Proterozoic Wopmay Orogen (Fig. 3.1-1). The occurrence of Archean rocks in the region, which are continuous with the rocks of the Slave Province, was first indicated from regional geological mapping (St-Onge et al., 1988). The rocks yielded a minimum zircon age of 3.84 Ga and Nd model ages up to 4.1 Ga (Bowring et al., 1989b). Subsequently, several teams have conducted additional sampling and geological mapping of the Acasta Gneiss Complex and adjacent regions

---

[1]Present address: Earthquake Research Institute, The University of Tokyo, Yayoi 1-1-1, Bunkyo-ku, Tokyo 113-0032, Japan.

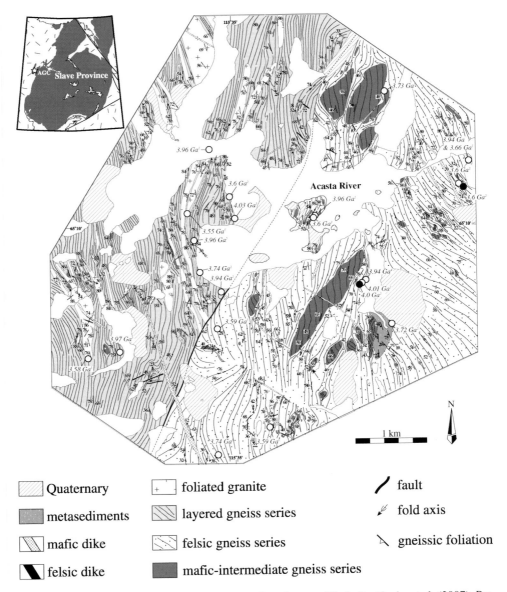

Fig. 3.1-1. Geological map of the Acasta Gneiss Complex: modified after Iizuka et al. (2007). Protolith ages of gneisses and foliated granites are shown. Solid and open circles represent the ages of amphibolitic and felsic (tonalitic-granitic) rock samples, respectively. Data sources of the ages are as follows: 1. Bowring et al. (1989a); 2. Bowring and Housh (1995); 3. Bowring and Williams (1999); 4. Iizuka et al. (2006); 5. Iizuka et al. (2007).

(Bowring et al., 1989a; Bowring et al., 1990; Bleeker et al., 1997; Stern and Bleeker, 1998; Bleeker and Davis, 1999). Iizuka et al. (2007) presented a detailed (1:5000 scale) geological map (Fig. 3.1-1) of a 6 km × 6 km area of the main part of the complex around the sample locality of the Acasta gneisses reported by Bowring et al. (1989b), and sketch maps of critical outcrops. We summarize here the geology, lithology and petrography of this area.

3.1-2.1. Geological Framework

The Acasta Gneiss Complex mainly consists of a heterogeneous assemblage of foliated to gneissic tonalite, granodiorite, trondhjemite, granodiorite and granite as well as amphibolitic, gabbroic, and dioritic gneisses (Bowring et al., 1990; Bowring and Williams, 1999; Iizuka et al., 2007). The major assemblage can be classified into four lithofacies based on the composition and texture of the gneisses (Fig. 3.1-2):

(1) a mafic-intermediate gneiss series (quartz dioritic, dioritic and gabbroic gneisses) (Fig. 3.1-2(a));
(2) a felsic gneiss series (tonalitic, trondhjemitic, granodioritic and granitic gneisses) (Fig. 3.1-2(b));
(3) a layered gneiss series of mafic-intermediate and felsic gneisses (Fig. 3.1-2(c));
(4) foliated granite, preserving an original igneous texture (Fig. 3.1-2(d)).

The main area is subdivided into two main units by a northeast-trending fault (Fig. 3.1-1). The lithology changes abruptly across this boundary, and many quartz veins, from sub-millimeters to meters thick, occur along the fault. In some places, the strike of gneissic structures also changes across the fault. The mafic-intermediate gneiss series occurs mainly as rounded to elliptical enclaves and inclusions within felsic gneisses. The felsic gneiss series occurs predominantly in the eastern area, with minor intrusions in the western area. In the eastern part of the eastern region, the felsic gneisses have northwest-trending foliations that dip 70–80° westward, but in the western part they trend north and dip 50–70° eastward. The layered gneiss series is present mainly in the western area where the gneissic foliation generally trends north-south and dips 60–80° to the west. These structures are often oblique to the boundary with the foliated granite. The foliated granite predominantly occurs as intrusions up to 200 m wide that generally trend north-south, whereas much thinner intrusions of granite and aplite are present throughout the complex. The granitic intrusions in the western region are cut by the main central fault. Northwest-trending mafic dikes are widespread and cut the main central fault.

3.1-2.2. Lithology and Field Relationships

The mafic-intermediate gneiss series (Fig. 3.1-2(a)) predominantly occurs as 3 km × 1 km to 10 cm × 10 cm enclaves within the felsic gneiss, forming blocks, boudins and bands (Fig. 3.1-3(a)). The mafic-intermediate gneiss series contains both mesocratic and melanocratic portions, and includes gabbroic, dioritic, and quartz dioritic gneisses.

Fig. 3.1-2. Four main lithofacies in the Acasta Gneiss Complex: (a) mafic-intermediate gneiss series; (b) felsic gneiss series; (c) layered gneiss series with rhythmical layering of leucocratic and melanocratic layers; (d) foliated granite, preserving original igneous texture.

Fig. 3.1-3. (a) Enclaves of quartz dioritic gneiss in granitic gneiss; (b) hornblendite inclusion in quartz dioritic gneiss; (c) hornblendite along the boundary between quartz dioritic and granodiorite gneisses; (d) folded layered gneiss and intrusion of foliated granite.

Quartz dioritic gneiss is the predominant phase, with the mineral assemblage Hbl+Pl+ Qtz+Bt±Kspar±Zrn±Ttn±Apt±Grt±opaque (mineral abbreviations after Kretz (1983)). Some gneisses, especially in the northeastern area, have abundant garnet porphyroblasts. Occasionally, massive hornblendite inclusions are present within the mafic-intermediate gneisses (Fig. 3.1-3(b)), and frequently occur along the boundary between the mafic-intermediate and felsic gneisses (Fig. 3.1-3(c)).

The felsic gneiss series (Fig. 3.1-2(b)) is widely distributed in the eastern part of the Acasta Gneiss Complex and occurs as massively or banded leucocratic gneisses including tonalitic, trondhjemitic, granodioritic and granitic gneisses. The mineral assemblage ranges from Pl+Qtz+Hbl+Bt±Kspar±Zrn±Ttn±Apt±Grt±opaque to Qtz+Kspar+Pl+Bt± Zrn±Ttn±Apt±Grt±opaque. At some localities, different types of compositions occur together, suggesting multiple generations of the protolith of the felsic gneiss series.

The layered gneiss series (Fig. 3.1-2(c)) is characterized by both continuous layering of felsic and mafic-intermediate lithological suites (gneisses), on a centimeter- to meter-scale, and a prominent preferred orientation of platy and prismatic minerals. The layered gneiss series occurs only in the western area, together with the foliated granite (Fig. 3.1-1). The mineralogy and bulk compositions of the felsic and mafic-intermediate lithological suites of the layered gneiss series are equivalent to the felsic and mafic-intermediate gneiss series in the eastern region, respectively. There are many large porphyroblasts of quartz and feldspar. In addition, some mafic-intermediate suites contain abundant garnet porphyroblasts. Thin boudins and layers of coarse-grained hornblendite are also present sporadically along the layering.

The foliated granite (Fig. 3.1-2(d)) predominantly occurs in the western region as intrusions up to 200 m wide. Original igneous textures are preserved and the unit is composed of the minerals Pl+Kspar+Qtz+Hbl+Bt±Zrn±Ttn±Apt±Grt±opaque. Some of the granites are inter-folded with the layered gneiss series (Fig. 3.1-3(d)).

Mafic dikes postdate the formation of the central fault and are generally northwest-trending. The intrusions are fine-grained and have a typical mineral assemblage of Act-Hbl+Pl+Qtz+Ep+Chl±Bt±Apt±Ttn±opaque, indicating metamorphism under epidote-amphibolite to amphibolite facies conditions. In addition, some of the gneisses contain calcite, epidote and secondary biotite, indicating that they suffered post-magmatic metasomatic alteration and infiltration of mobile elements such as Ca and K.

Field relationships between the intermediate and felsic gneisses in the eastern area are shown in Fig. 3.1-4. The outcrop consists of quartz dioritic gneiss, coarse-grained granodioritic gneiss, and granitic gneiss with pegmatites and hornblendite pods. The pegmatites mainly occur on the fringe of the granitic gneiss. The hornblendite pods are present along the boundary between the quartz dioritic and the granitic gneisses and are accompanied by relatively quartz-rich quartz dioritic gneiss. The boundary between the quartz-rich quartz dioritic gneiss and the quartz dioritic gneiss is vague. The granitic and quartz-rich quartz dioritic gneisses exhibit subparallel gneissosity to their outer margin. In contrast, the gneissosity of the coarse-grained granodioritic gneiss is obliquely cut by the granitic gneiss. These observations indicate that the protolith of the granitic gneiss intruded into the quartz dioritic and coarse-grained granodioritic gneisses and that

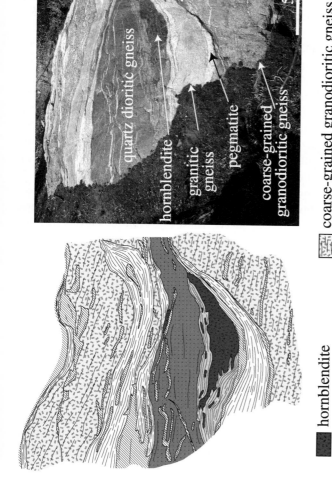

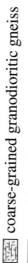

Fig. 3.1-4. Sketch map and photo of outcrop of quartz dioritic, coarse-grained granodioritic, and granitic gneisses, with minor pegmatite and hornblendite layers (after Iizuka et al., 2007). The outcrop displays that granitic gneiss occurs as intrusions into quartz dioritic and coarse-grained granodioritic gneisses. In addition, the quartz dioritic gneiss differentiated into relatively quartz-rich quartz dioritic gneiss leucosome and hornblenditic residual layers along the boundary between the quartz dioritic and granitic gneisses, indicating anatexis of the quartz dioritic blocks by intrusion of granitic magmas.

during the crystallization of the granite intrusion, fluids (highly hydrous melts) were released and formed the pegmatites on the fringe of the granite intrusion. The occurrence of hornblendite pods accompanied by quartz-rich quartz dioritic gneiss between the quartz dioritic and granitic gneisses suggests that fluid infiltration and thermal metamorphism during granite intrusion also caused partial melting (anatexis) of the quartz dioritic gneiss to form a hornblendite restite with a quartz-rich quartz dioritic gneiss leucosome.

Hence, at least five tectonothermal events in the eastern area are recognized from these fabrics: (1) and (2) emplacement of quartz dioritic magma (protolith of the quartz dioritic gneiss) and emplacement of granodioritic magma (protolith of the coarse-grained granodioritic gneiss); (3) metamorphism to produce the gneissic structures of the coarse-grained granodioritic gneiss and quartz dioritic gneiss; (4) intrusion of granitic magma (protolith of the granitic gneiss), causing anatexis and formation of hornblendites and quartz-rich quartz dioritic gneiss; (5) metamorphism and deformation to produce the gneissic structures of granitic and quartz-rich quartz dioritic gneisses.

The relationship between the quartz dioritic and granitic gneisses is also shown in Fig. 3.1-5. The outcrop comprises quartz dioritic gneiss and granitic gneiss with hornblendite pods and pegmatites. The gneissic structures of the granitic gneiss are subparallel to the direction of their outer contact. The gneissic structure of the quartz dioritic gneiss is oblique to that of the granitic gneiss at some points. Pegmatites occur along the margin with the granitic gneiss, as well as within the quartz dioritic gneiss, suggesting its derivation from fluids released during crystallization of the granite intrusion. Fourteen hornblendite pods are sporadically distributed within the quartz dioritic gneiss body and most of them are not accompanied by the quartz-rich layer, whereas the hornblendite pods in Fig. 3.1-4 occur along the boundary between quartz dioritic and granitic gneisses and are accompanied by the quartz-rich layer. In addition, the deformation structures imprinted on them are consistent with those within the quartz dioritic gneiss. These observations suggest that the hornblendite pods are remnants of older mafic material entrained by quartz dioritic magma.

Therefore, five tectonothermal events are identified from this outcrop: (1) formation of the protolith of hornblendite xenoliths; (2) emplacement of quartz dioritic magma; (3) metamorphism and deformation to produce the gneissosity of the quartz dioritic gneiss; (4) intrusion of granitic magma (granitic gneiss protolith); (5) metamorphism and deformation to produce the gneissosity of the granitic gneiss.

In the western area, field observations of the layered gneiss and foliated granite showed that the banding and gneissic structures within the layered gneiss series are obliquely cut by the foliated granite intrusion, indicating that the formation of banding structures preceded intrusion of the foliated granite. Therefore, at least five tectonomagmatic events are recognized in the western area: (1) and (2) emplacement of the mafic-intermediate gneiss protolith and emplacement of the felsic gneiss protolith; (3) metamorphism and deformation, which produced the gneissic and layering structures of the layered gneiss; (4) intrusion of granite magmas as the protolith of the foliated granite; (5) metamorphism and deformation of all lithologies.

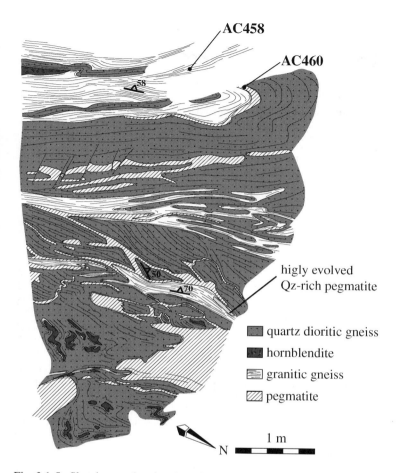

Fig. 3.1-5. Sketch map showing the relationships between quartz dioritic and granitic gneisses with minor pegmatite and hornblendite pods (from Iizuka et al., 2007). Zircon U-Pb dating, combined with a cathodoluminescence imaging study (Fig. 3.1-7), revealed the protoliths of the granitic gneiss (AC458) and pegmatite (AC460) crystallized at 3.59 Ga, and that the granitic gneiss contains zircon xenocrysts with ages up to 3.9 Ga.

3.1-3. GEOCHRONOLOGY

The first geochronological study on the Acasta Gneiss Complex was conducted by Bowring and Van Schmus (1984), using thermal ionization mass spectrometry (TIMS) zircon geochronology, for a test of the hypothesis that an early Proterozoic terrane had overthrust the western edge of the Slave Province: they obtained a zircon U-Pb age of 3.48 Ga. Bowring et al. (1989b) carried out further sampling and dating by TIMS in the region and recognized that the rocks contain zircon cores older than 3.84 Ga, with whole-rock Nd$_{CHUR}$ model ages up to 4.1 Ga. Moreover, the U-Pb data indicated that the zircons have a complex history involving Pb loss and/or zircon multi-growth events.

Subsequently, zircons from granitic to amphibolitic gneisses in the complex were analyzed in-situ by the sensitive high-resolution ion microprobe (SHRIMP), combined with imaging techniques (optical microscopy with HF etching, back scattered electron (BSE) imagery, or cathodoluminescence (CL) imagery) in order to determine crystallization ages of their igneous protoliths (Bowring et al., 1989a; Bowring and Housh, 1995; Bleeker and Stern, 1997; Stern and Bleeker, 1998; Bowring and Williams, 1999; Sano et al., 1999). Although some of the gneisses contain different U-Pb age zircon cores, these authors interpreted the oldest U-Pb ages as the crystallization ages of the protoliths of the gneisses, under the assumption that the variation of U-Pb ages was caused by Pb-loss events from one generation of magmatic zircon. These authors obtained protolith ages of 4.03–3.94 Ga, 3.74–3.72 Ga, and 3.6 Ga for the tonalitic-granitic gneisses, and of 4.0 and 3.6 Ga for amphibolitic gneisses (Fig. 3.1-1). The 4.03–3.94 Ga protoliths are now well known as the oldest terrestrial rocks. The U-Pb ages of overgrowths and altered domains in zircons from the oldest rocks indicated that they suffered metamorphic events at ca. 3.6 Ga (Bowring et al., 1989a; Bowring and Williams, 1999; Sano et al., 1999) and 3.4 Ga (Bleeker and Stern, 1997; Stern and Bleeker, 1998). Whole-rock Sm-Nd data of Acasta gneisses also yields a regression age of 3.4 Ga, suggesting that the isotope systematics had been reset during the metamorphic event at 3.4 Ga (Moorbath et al., 1997). Bleeker and Stern (1997) and Stern and Bleeker (1998) also carried out zircon U-Pb dating on felsic intrusive rocks into the Acasta gneisses, and revealed granite intrusions at ca. 3.4, 2.9 and 2.6 Ga, and syenite intrusions at 1.8 Ga.

Importantly, Bowring and Williams (1999) demonstrated that a 4.0 Ga tonalitic gneiss contains a 4.06 Ga zircon xenocryst, providing direct evidence for the existence of Hadean crust outside of the Yilgarn Craton, Western Australia. Subsequently, based on a laser ablation-inductively coupled plasma-mass-spectrometry (LA-ICPMS) and SHRIMP study, combined with CL and BSE imaging studies, Iizuka et al. (2006) reported the occurrence of a 4.2 Ga zircon xenocryst (AC012/07) within a tonalitic gneiss AC012, with a protolith age of 3.9 Ga (Fig. 3.1-6), pushing back the age of the oldest zircon outside of the Yilgarn Craton by 140 m.y.

Iizuka et al. (2007) conducted LA-ICPMS zircon U-Pb dating, combined with field observations and CL imagery, and revealed at least four tonalite-granite emplacement events in the eastern area at ca. 3.94, 3.74–3.73, 3.66 and 3.59 Ga (protoliths of felsic gneiss series), tonalite emplacement at ca. 3.97 Ga (protoliths of layered gneiss series), and granite intrusion at 3.58 Ga (protoliths of foliated granites) in the western area (Fig. 3.1-1). In addition, their comprehensive investigations demonstrated that some of the early Archean rocks contain xenocrystic zircons (Figs. 3.1-5 and 3.1-7(a)), and that some suffered anatexis/recrystallization at ca. 3.66 Ga in the western area, and at ca. 3.66 and 3.59 Ga in the eastern area, coincident with the emplacement of felsic intrusions (Fig. 3.1-7(b)).

Isotopic ages of other minerals, such as apatite, hornblende and biotite, have also been determined, in order to understand the thermal history of Acasta gneisses. Hodges et al. (1995) reported ^{40}Ar/^{39}Ar ages of 1.86 Ga for hornblende, and 1.72 Ga for biotite.

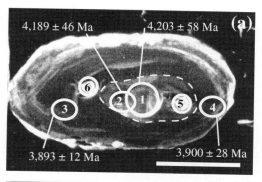

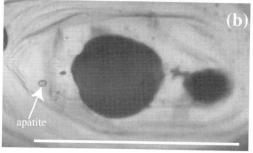

Fig. 3.1-6. (a) CL image of zircon containing 4.2 Ga xenocrystic core extracted from a 3.94 Ga tonalitic gneiss AC012; (b) transmitted light image of the xenocryst, showing location of apatite inclusion. Scale bars are 50 μm. Values record $^{207}Pb/^{206}Pb$ ages (2σ). Spot numbers correspond to those in Table 3.1-1. Modified after Iizuka et al. (2006).

Sano et al. (1999) demonstrated that apatite grains from a sample of Acasta gneiss have $^{238}U/^{204}Pb-^{206}Pb/^{204}Pb$ and $^{204}Pb/^{206}Pb-^{207}Pb/^{206}Pb$ isochron ages of 1.91 and 1.94 Ga, respectively. These ages are perhaps related to collisional and post-collisional events of the Wopmay Orogeny, and suggest an unusually slow cooling rate of $\sim2\,°C/m.y.$ following orogenesis (Hodges et al., 1995; Sano et al., 1999). Furthermore, these ages are consistent with U-Pb isotopic data from zircon and titanite, which give isotopic mixing lines between ca. 4.0 and ca. 1.9 Ga (Davidek et al., 1997; Sano et al., 1999).

3.1-4. CONSTRAINTS ON THE PROVENANCE OF THE 4.2 GA ZIRCON XENOCRYST

In order to understand the nature of Hadean crust that has been reworked into the Acasta Gneiss Complex, it is important to know the provenance of the 4.2 Ga zircon xenocryst (Fig. 3.1-6). Because zircon commonly occurs as an accessory mineral in granitoid rocks, it is reasonable to suspect that they are source rocks of the xenocryst. However, zircon occasionally occurs in other igneous rocks such as syenites, carbonatites, kimberlites, and

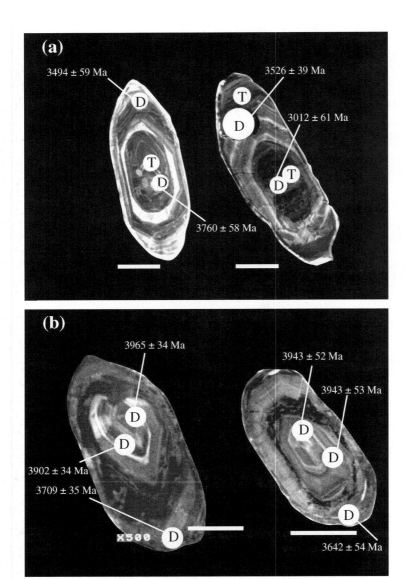

Fig. 3.1-7. CL images of zircons from: (a) granitic gneiss AC458 with a protolith age of 3.59 Ga, showing that some zircons contain xenocrystic cores; (b) granitic gneiss AC584 with a protolith age of 3.94 Ga, showing the overgrowth of oscillatory-zoned zircon during recrystallization/anatexis at 3.66 Ga, which is coincident with the protolith age of an adjacent tonalitic gneiss. Scale bars are 50 μm. Values record $^{207}Pb/^{206}Pb$ ages (2σ). "D" and "T" represent analytical spots of LA-ICPMS dating and LA-ICPMS trace element analysis, respectively. Modified after Iizuka et al. (2007).

Table 3.1-1. U-Pb-Th and trace element data for AC012/07

U-Pb-Th data

Spot	Method	U (ppm)	Th (ppm)	Th/U	$^{204}Pb/^{206}Pb$	$^{206}Pb^*/^{238}U$	$^{207}Pb^*/^{206}Pb^*$	Age (Ma)		Disc (%)
								$^{206}Pb/^{238}U$	$^{207}Pb/^{206}Pb$	
1	LA-ICPMS	N.D.[1]	N.D.[1]	N.D.[1]	0.0002	N.D.[1]	0.4874 ± 191	N.D.[1]	4203 ± 58	N.D.[1]
2	SHRIMP	699	347	0.496	0.0006	0.8611 ± 316	0.4827 ± 150	4004 ± 110	4189 ± 46	4
3	SHRIMP	621	115	0.185	0.0005	0.6900 ± 224	0.3957 ± 32	3383 ± 86	3893 ± 12	13
4	SHRIMP	649	48	0.075	0.0002	0.7567 ± 320	0.3976 ± 76	3632 ± 118	3900 ± 28	7

Trace element data (ppm)

Spot	Y	Nb	La	Ce	Pr	Nd	Sm	Eu	Gd	Tb	Dy	Ho	Er	Tm	Yb	Lu	Hf
5	1196	624	0.1	15	0.2	1.5	9.1	1.3	22	8.0	89	36	155	36	400	54	13621
6	547	663	N.D.[1]	2	0.1	0.5	2.5	0.2	6.5	2.0	37	16	94	28	295	53	16047

Note: Spot numbers correspond to those in Fig. 3.1-6. Pb^* corrected for common Pb using ^{204}Pb. All errors are quoted at 2σ.

[1] N.D. = not determined.

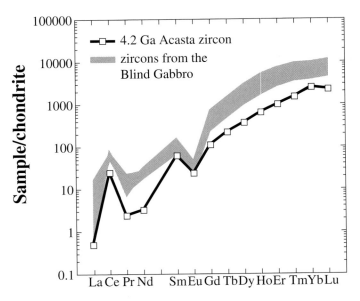

Fig. 3.1-8. Chondrite normalized REE pattern for the 4.2 Ga zircon xenocryst, compared with those of zircons from the Blind Gabbro (Hoskin and Ireland, 2000). From Iizuka et al. (2006).

mafic rocks, and it can be formed during metamorphism. The provenance of the zircon xenocryst is discussed based on its trace element compositions and mineral inclusion (Table 3.1-1 and Figs. 3.1-6, 3.1-8 and 3.1-9) (Iizuka et al., 2006).

It is widely recognized that the Th/U ratio of zircon is useful for discrimination between magmatic and metamorphic origins (Rubatto, 2002; Hoskin and Schaltegger, 2003). The 4.2 Ga zircon core has a high Th/U ratio of 0.52 (Table 3.1-1), indicating a magmatic, rather than a metamorphic, origin. Relatively high contents of incompatible elements such as Y, Nb, Hf, Th and U (Table 3.1-1) indicate that it crystallized from an evolved magma (Belousova et al., 2002; Crowley et al., 2005).

Several important points emerge from the REE pattern (Fig. 3.1-8).

(1) A prominent positive Ce anomaly. This is a typical feature of terrestrial zircons, suggesting that the 4.2 Ga zircon crystallized under oxidizing conditions (Hoskin and Schaltegger, 2003).

(2) A distinct enrichment of HREEs relative to LREEs, with prominent negative Eu anomalies, indicating that the source magma of the zircon coexisted with feldspar, not with garnet. This is consistent with a crustal, rather than a mantle, origin (e.g., from kimberlites and carbonatites) (Hoskin and Schaltegger, 2003; Belousova et al., 2002). In addition, because a prominent negative Eu anomaly is not observed in zircons from alkalic felsic rocks such as syenites (Hoskin and Schaltegger, 2003), possibly due to the high [Na_2O+K_2O/Al_2O_3] of their source magma that significantly decreases the

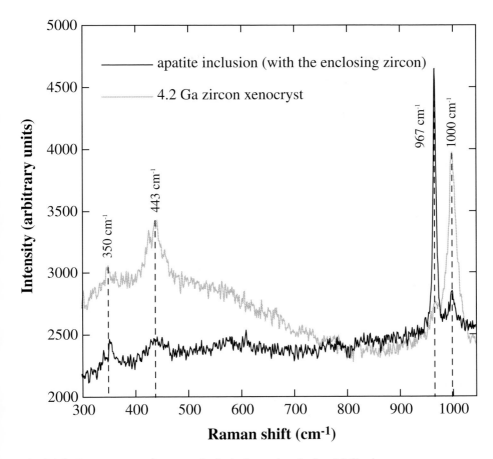

Fig. 3.1-9. Raman spectra for an apatite inclusion and enclosing 4.2 Ga zircon.

distribution of Eu in alkali-feldspars (White, 2003), it is unlikely that alkalic felsic rocks are the source of the xenocryst.

(3) Zircons from the Blind Gabbro of Australia have a concave-down curvature in the pattern of MREE to HREE ($[2Ho/(Gd+Yb)]_N = 0.84–1.01$), whereas the REE pattern of the xenocryst shows no such curvature ($[2Ho/(Gd+Yb)]_N = 0.50$); the M-HREE curvature of zircons from mafic rocks could be due to HREE depletion of melt caused by crystallization of clinopyroxene and/or orthopyroxene (Hoskin and Ireland, 2000). This suggests that the xenocryst formed in a granitic magma, rather than in a differentiated melt from a mafic parental magma.

In addition, the zircon xenocryst contains an apatite inclusion (Figs. 3.1-6 and 3.1-9). The above geochemical and mineralogical characteristics suggest that the 4.2 Ga zircon xenocryst was derived from a granitoid source.

3.1-5. RADIOGENIC ISOTOPE SYSTEMATICS

The radiogenic isotope systematics of early Archean crustal rocks potentially provides insights into the nature of their source materials. Bowring et al. (1989b) first carried out radiogenic isotopic studies on two Acasta gneiss samples and demonstrated that they have whole-rock Nd chondritic model ages of 3.9 and 4.1 Ga. Subsequently, they conducted further whole-rock Nd isotopic analyses and zircon U-Pb dating, the results of which indicated that Acasta gneisses exhibit a wide range of initial $^{143}Nd/^{144}Nd$ ratios calculated using the zircon U-Pb ages of 3.6–4.0 Ga (Bowring et al., 1990; Bowring and Housh, 1995). Based on the wide range of Nd isotopic compositions, Bowring and Housh (1995) suggested that the early Earth's mantle was highly differentiated into a geochemically enriched reservoir and depleted mantle. Additionally, the $^{142}Nd/^{144}Nd$ ratios of a few Acasta samples have been reported. They are indistinguishable from most other terrestrial rocks (McCulloch and Bennett, 1993; Caro et al., 2006), and are therefore more radiogenic relative to chondritic uniform reservoir (CHUR: Boyet and Carlson, 2005). However, because the early Archean Acasta gneisses have been subject to several periods of intense metamorphism and deformation, it is highly controversial whether the whole-rock Nd isotope systematics of these rocks have been disturbed, or represent the true isotopic composition at their time of formation, estimated from the zircon U-Pb ages (Moorbath et al., 1997; Kamber et al., 2001; Whitehouse et al., 2001).

Zircon is highly resistant and therefore potentially retains information on its crystallization age and primary isotopic signature through metamorphic events (e.g., Patchett et al., 1981; Corfu and Noble, 1992). In order to obtain more reliable isotopic information, Amelin et al. (1999, 2000) carried out Lu-Hf isotopic analysis of ca. 3.79 to 3.57 Ga-aged zircons from Acasta gneisses by solution multicollector (MC)-ICPMS. More recently, Iizuka and Hirata (2005) and Iizuka et al. (submitted) have determined in-situ Lu-Hf isotopic compositions of zircons with ages of 3.97–3.59 Ga using the LA-MC-ICPMS technique combined with CL imagery, in order to obtain more accurate isotopic information by avoiding xenocrystic/inherited cores, overgrowths, and altered domains, which are commonly observed in Acasta zircons (e.g., Fig. 3.1-7: Bowring et al., 1989a; Bleeker and Stern, 1997; Stern and Bleeker, 1998; Bowring and Williams, 1999; Rayner et al., 2005; Iizuka et al., 2006; Iizuka et al., 2007). The Lu-Hf isotopic data are graphically presented on Fig. 3.1-10 as plots of $\varepsilon_{Hf}(T)$ ratio against the U-Pb age. The figure shows that all data plot below the CHUR isotopic evolution curve, indicating that early Archean Acasta gneiss protoliths were derived from geochemically enriched crustal materials. The Lu-Hf isotopic signature is consistent with the occurrence of zircon xenocrysts within Acasta gneisses (Bleeker and Stern, 1997; Bowring and Williams, 1999; Iizuka et al., 2006; Iizuka et al., 2007) (Figs. 3.1-6 and 3.1-7(a)). It is worth noting that the Lu-Hf isotopic signature is similar to that of detrital Jack Hills zircons with ages from 3.3 to ca. 4.0 Ga (Amelin et al., 1999; Harrison et al., 2005), suggesting the similarity between the Lu-Hf isotopic evolution of the Acasta gneiss protoliths and source rocks of the Jack Hills detrital zircons.

Interestingly, early Archean Acasta rocks with depleted mantle-like whole-rock Nd isotopic signatures (Bowring et al., 1990; McCulloch and Bennett, 1993; Bowring and Housh,

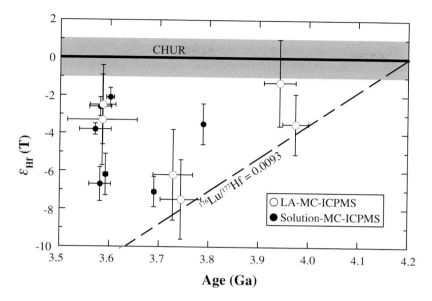

Fig. 3.1-10. Plot of age versus $\varepsilon_{Hf}(T)$ for early Archean Acasta gneiss zircons. Data sources are Amelin et al. (2000), Iizuka and Hirata (2005) and Iizuka et al. (submitted). The dotted line shows ε_{Hf} evolution of average granitoid crust with $^{176}Lu/^{177}Hf = 0.0093$ (Vervoort and Patchett, 1996) with the assumption that it originated from undifferentiated mantle (CHUR) at 4.2 Ga. The CHUR evolution line and its error limits (gray broad line) are from Blichert-Toft and Albarède (1997).

1995) have not been identified in the zircon Hf isotopic studies. Because Lu-Hf and Sm-Nd systems are generally thought to be well correlated during crustal formation processes (Vervoort and Blichert-Toft, 1999), it is likely that this decoupling was due to the redistribution of the Sm-Nd isotope system (on the whole-rock scale) during later metamorphism (Moorbath et al., 1997; Kamber et al., 2001; Whitehouse et al., 2001), or by isotope fractionation during melt segregation from a magma ocean (Albarède et al., 2000; Caro et al., 2005).

In order to estimate approximate ages of the reworked crust in the protoliths of the Acasta gneisses, we calculated Hf isotopic model ages for the protoliths following the approach of Amelin et al. (1999). Because the calculation depends on the $^{176}Lu/^{177}Hf$ ratio of the reworked crust, we obtained two alternative model ages, T_{CHUR1} and T_{CHUR2}, by using the ratio of Precambrian granitoids ($^{176}Lu/^{177}Hf = 0.0093$; based on the data of Vervoort and Patchett, 1996) and mafic crust ($^{176}Lu/^{177}Hf = 0.0193$; Taylor and McLennan, 1985), respectively. Moreover, we set the assumption for the calculation that the parental magmas of the protoliths were produced from crustal materials that were originally extracted from undifferentiated primitive mantle (CHUR). Fig. 3.1-11 illustrates histograms of T_{CHUR1} and T_{CHUR2}. These histograms clearly indicate that Hadean crust was reworked into the early Archean granitoids, consistent with the presence of zircon xenocrysts older than 4.03 Ga within the oldest known rocks (Bowring and Williams, 1999; Iizuka et al., 2006).

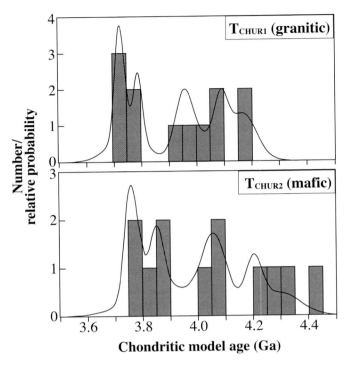

Fig. 3.1-11. Histograms and relative probabilities of Hf chondritic model ages; (a) T_{CHUR1} and (b) T_{CHUR2} for Acasta gneiss samples. Data are from Amelin et al. (2000), Iizuka and Hirata (2005), and Iizuka et al. (submitted). T_{CHUR1} and T_{CHUR2} are calculated using the ratio of Precambrian granitoid ($^{176}Lu/^{177}Hf = 0.0093$; Vervoort and Patchett, 1996) and mafic crust ($^{176}Lu/^{177}Hf = 0.0193$; Taylor and McLennan, 1985), respectively.

3.1-6. TECTONOTHERMAL EVOLUTION OF THE ACASTA GNEISS COMPLEX

The early Archean rocks in the Acasta Gneiss Complex experienced a complex geological history, as in illustrated in Fig. 3.1-12. The main area of the gneiss complex is subdivided into eastern and western regions by the central northeast-trending fault. The eastern area is dominated by mafic-intermediate gneisses and by multiple generations of felsic gneisses. In contrast, the western region is composed mainly of interlayered mafic-intermediate and felsic gneiss and foliated granites that preserve their original igneous textures.

Zircon U-Pb geochronology revealed two primary generations of granitoid emplacement in the eastern area, at 4.03–3.94 Ga and 3.74–3.72 Ga (Bowring et al., 1989a; Bowring and Housh, 1995; Bleeker and Stern, 1997; Stern and Bleeker, 1998; Bowring and Williams, 1999; Iizuka et al., 2006; Iizuka et al., 2007). These granitoids were in-

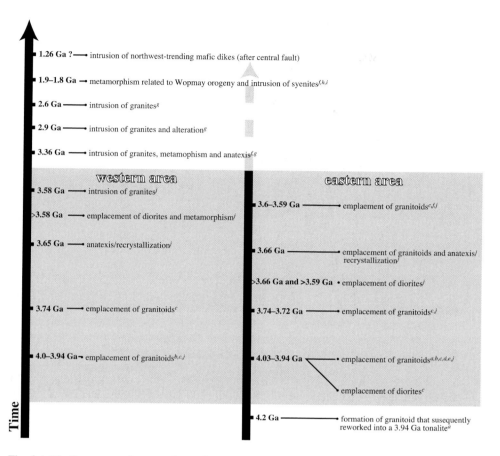

Fig. 3.1-12. Summary of tectonothermal events of the Acasta Gneiss Complex, based on geological and geochronological data obtained in this and previous studies. Data sources are: (a) Iizuka et al. (2006); (b) Bowring et al. (1989a); (c) Bowring and Housh (1995); (d) Stern and Bleeker (1998); (e) Bowring and Williams (1999); (f) Bleeker and Stern (1997); (g) Moorbath et al. (1997); (h) Hodges et al. (1995); (i) Sano et al. (1999); (j) Iizuka et al. (2007).

truded by younger granitoid magma at 3.66 and 3.59 Ga, and experienced anatexis and/or recrystallization at that time (Iizuka et al., 2007). Zircon U-Pb ages also demonstrated the emplacement age of mafic magma of 4.0 and 3.6 Ga (Bowring and Housh, 1995).

In the western area, zircon U-Pb data reveal the emplacement of felsic gneiss protoliths at 4.0–3.94 Ga and at 3.74 G.a, and the intrusion of the foliated granite protolith at 3.58 Ga (Bowring et al., 1989a; Bowring and Housh, 1995; Iizuka et al., 2007). A 3.96 Ga tonalite contained zircons with oscillatory zoned overgrowths having ages of ca. 3.66 Ga, indicating a metamorphic event at this time, consistent with field relationships that show the formation of gneissic and layered structures of the layered gneiss preceded granite intrusion. The 3.58

Ga granitic sheet is cut by the central fault, indicating that late-stage fault activity occurred after 3.58 Ga.

Zircon xenocrysts commonly occur within the early Archean Acasta gneisses (Bleeker and Stern, 1997; Stern and Bleeker, 1998; Bowring and Williams, 1999; Iizuka et al., 2006; Iizuka et al., 2007), consistent with the unradiogenic Hf isotopic compositions of zircons from these rocks (Amelin et al., 2000; Iiziuka et al., submitted). In addition, the geochemical and mineralogical signatures of the 4.2 Ga zircon xenocryst indicate that it crystallized from granitoid magma. These results, as well as the Hadean-aged Hf isotopic model ages, suggest that the early Archean granitoids were derived through reworking of Hadean continental crust.

The post early Archean history of the Acasta gneisses is also complex, and includes magmatic intrusion and metamorphic events at ca. 3.4, 2.9, 2.6 Ga (granitic sheets) and 1.8 Ga (syenites) (Bleeker and Stern, 1997). Furthermore, minerals with relatively low closure temperatures, such as apatite, biotite, titanite and hornblende demonstrate that they experienced extensive Palaeoproterozoic thermal overprinting events at ca. 1.9 Ga, related to collisional and post-collisional events of the Wopmay Orogeny (Hodges et al., 1995; Sano et al., 1999). Finally, the complex is cross-cut by northwest-tending mafic dikes, which may correspond to the ca. 1.26 Ga Mackenzie dike swarm (LeCheminant and Heaman, 1989).

3.1-7. IMPLICATIONS FOR EARLY CRUSTAL EVOLUTION

The data acquired from the Acasta Gneiss Complex provide valuable insights into early crustal evolution in this region. The occurrence of Hadean zircon xenocrysts with ages up to 4.2 Ga, together with Hadean Hf model ages obtained from magmatic zircons from Acasta gneisses, provide direct evidence for the presence of Hadean crust and for its reworking into early Archean continental crust. These results are consistent with the recent finding of pyrite-hosted inclusions with a ^{40}Ar-^{39}Ar age of ca. 4.3 Ga within banded iron formation (Smith et al., 2005), as well as a Pb-isotope model age of ca. 4.3 Ga, from parts of the Isua supracrustal belt (Kamber et al., 2003). The 4.2 Ga zircon xenocryst from the Acasta Gneiss Complex has mineralogical and geochemical signatures of derivation from a granitoid source, suggesting the existence of Hadean continental crust in two Archean cratons (see Cavosie et al., this volume). In addition, although Hf isotopic data alone rigidly constrain neither the age, composition, nor amount of the reworked crust, the geochronology, mineralogy, and geochemistry of the 4.2 Ga zircon xenocryst allows us to evaluate the amount of reworking of Hadean crust (i.e., the degree of contribution of this 4.2 Ga Hadean granitoid within each of the early Archean Acasta gneiss samples).

Assuming that the early Archean Acasta gneiss samples are a simple mix of 4.2 Ga granitoid ($^{176}Lu/^{177}Hf = 0.0093$; Vervoort and Patchett, 1996) and juvenile magma with a chondritic Hf isotopic composition, we calculated the reworking index for each sample whose Hf isotopic composition was determined. Fig. 3.1-13 shows a histogram of the calculated reworking indexes. The histogram reveals that half of the Acasta samples have a

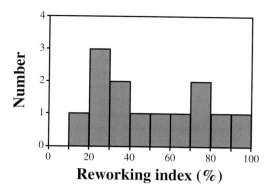

Fig. 3.1-13. Histogram of reworking indexes for the Acasta samples, assuming that they derived from a mixture of juvenile magma and 4.2 Ga granitoid crust. Data are from Amelin et al. (2000), Iizuka and Hirata (2005), and Iizuka et al. (submitted).

reworking index of greater than 50%, suggesting that extensive Hadean continental crust was reworked within the early Archean Acasta gneisses. Furthermore, on the basis of a mean 51% reworking index, as calculated here, with the approximate aerial extent of the early Archean Acasta gneisses of 20 km^2 (Nutman et al., 1996) and an average 30 km thick continental crust, we estimate the total amount of reworked Hadean continental crustal material to be $\sim$300 km^3. Although Hadean mafic crust, or proto-crust that may have formed during solidification of a magma ocean might also contribute to the formation of the early Archean continental crust (e.g., Kamber, this volume), the data presented herein are part of a growing body of evidence for the presence of voluminous continental crust on Hadean Earth, supporting the view of Ti-in-zircon thermometry and oxygen isotopic studies that granitoid formation processes in the presence of liquid water oceans might have been established by 4.2 Ga (Mojzsis et al., 2001; Peck et al., 2001; Wilde et al., 2001; Cavosie et al., 2005; Watson and Harrison, 2005).

ACKNOWLEDGEMENTS

We wish to thank Drs. T. Hirata, Y. Ueno, I. Katayama, A. Motoki, S. Rino, Y. Uehara, H. Hidaka, K. Horie, and S. P. Johnson for support of research activities leading to this paper. We are also grateful to Mr. Mark Senkiw, and Drs. Walt Humphries, Diane Baldwin and Karen Gochnauer for assistance in the geological mapping of the Acasta Gneiss Complex. Finally, B.F. Windley is thanked for encouraging us to submit a paper to this volume.

Earth's Oldest Rocks
Edited by Martin J. Van Kranendonk, R. Hugh Smithies and Vickie C. Bennett
Developments in Precambrian Geology, Vol. 15 (K.C. Condie, Series Editor)
© 2007 Elsevier B.V. All rights reserved.
DOI: 10.1016/S0166-2635(07)15032-5

149

Chapter 3.2

ANCIENT ANTARCTICA: THE ARCHAEAN OF THE EAST ANTARCTIC SHIELD

SIMON L. HARLEY AND NIGEL M. KELLY

Grant Institute of Earth Science, School of GeoSciences, University of Edinburgh, Kings Buildings, West Mains Road, UK

3.2-1. INTRODUCTION

The East Antarctic Shield comprises most of the main landmass of Antarctica, bounded by the Transantarctic Mountains and the Southern Ocean in the sector from Africa to east Australia (Fig. 3.2-1). Despite less than 0.5% of its land area consisting of exposed rock, the East Antarctic Shield preserves a remarkable record of Earth evolution that spans in time from the earliest Archaean (ca. 3850–4060 Ma) to the Cambrian. The East Antarctic Shield preserves distinct high-grade terrains that were situated adjacent to Africa, India and Australia within Gondwana and in continental reconstructions proposed for earlier time periods (e.g., Dalziel, 1991). As such, it has become a major focus for testing models of supercontinent formation and destruction (Rogers, 1996; Unrug, 1997; Fitzsimons, 2000a), as well as an important region for investigating the nature of high-grade metamorphism

Fig. 3.2-1. (*Next page.*) Map of East Antarctica in its reconstructed Gondwana context, modified from Fitzsimons (2000a, 2000b) and Harley (2003) and drawn with latitudes and longitudes referred to the present Antarctic co-ordinates. The Meso- to Neoproterozoic provinces recognised by Fitzsimons (2000a) are labelled (Wilkes, Rayner and Maud), along with a fourth province (Rauer Province) defined by Harley (2003). Latest Neoproterozoic to Cambrian aged belts (Lutzow, Prydz and Pinjarra) are also distinguished. The Maud Province is extensively reworked and reorganised in the Neoproterozoic–Cambrian, in the East Africa–Madagascar–Antarctica belt, shown here by dotted ornament. Specific areas mentioned in the text are labelled with the following abbreviations: AF (Albany-Fraser); BH (Bunger Hills); cDML (central Dronning Maud Land); DG (Denman Glacier); Dhar + Md (Dhawar and Madras); EG (Eastern Ghats); Gr (Grunehogna); KKB (Kerala Khondalite Belt); La (Lambert); LHB (Lützow-Holm Bay); M (Mawson); Madag (Madagascar); Nap (Napier Complex); nPCM (northern Prince Charles Mountains); Oy (Oygarden Islands); PZ (Prydz Bay); RI (Rauer Islands); Ru (Ruker); SL (Sri Lanka); sPCM (southern Prince Charles Mountains); Shack (Shackleton Range); SR (Sør Rondane) TA (Terre Adélie); VH (Vestfold Hills); wDML (western Dronning Maud Land); WI (Windmill Islands); YB (Yamato-Belgica). Dashed box shows the area covered by the map in Fig. 3.2-2.

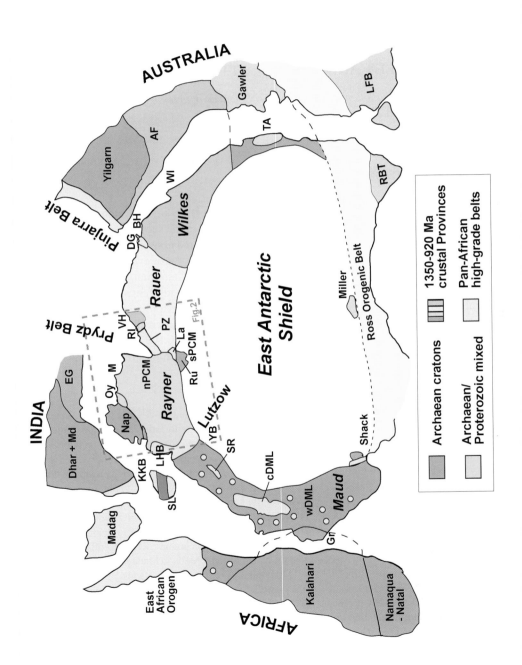

(Harley, 1998, 2003; Boger and Miller, 2004; Kelsey et al., 2003) and the character of early Archaean crust formation processes (Black et al., 1986; Choi et al., 2006). In this contribution the oldest rocks in Antarctica will be described and documented. This follows a brief overview of the Precambrian geology of the East Antarctic Shield that is included to provide a framework for understanding the geological contexts of the earliest Archaean rocks and provinces.

3.2-2. OVERVIEW OF THE GEOLOGY OF THE EAST ANTARCTIC SHIELD

Recent metamorphic, structural and geochronological studies in the East Antarctic Shield (e.g., Tingey, 1991; Fitzsimons, 2000a, 2000b; Harley, 2003; Jacobs et al., 1998, 2003), demonstrate that it consists of a variety of Archaean and Proterozoic to Cambrian high-grade terrains that have distinct crustal histories. These terrains were not finally amalgamated until the Cambrian, most likely in two separate orogenic episodes within the time interval 600–500 Ma (Boger et al., 2001; Boger and Miller, 2004; Cawood, 2005). The broader implications of this, beyond the scope of this work, are that East Gondwana was not finally assembled until the Cambrian (Hensen and Zhou, 1995; Fitzsimons, 2000a, 2000b, 2003).

The major tectonothermal events that are recorded in the East Antarctic Shield occurred in the late Archaean to earliest Proterozoic (<2840 Ma but >2480 Ma), Palaeoproterozoic (2200–1700 Ma), Meso- to Neoproterozoic (1400–910 Ma), and latest Proterozoic to Cambrian (600–500 Ma: "Pan African"). Hence, the basement terrains may be divided into four partially overlapping categories according to their age, their degree of reworking, and the nature, style and ages of the reworking events (Harley, 2003). These categories or groups are:

(1) Regions, belts or areas that preserve extensive evidence for high-grade metamorphism, magmatism and deformation at times within the broad interval 600–500 Ma, the timespan corresponding to the 'Pan-African' and to the major amalgamation events in East Gondwana;

(2) Proterozoic terranes (Provinces) dominated by Meso- to Neoproterozoic tectonic events (1400–910 Ma), affected to a minor or limited extent by overprinting at 600–500 Ma;

(3) Archaean or mixed Archaean/Palaeoproterozoic areas with strongly polyphase histories in which the early record is partially to largely obscured as a result of overprinting by younger tectonic events; and,

(4) Archaean (>2500 Ma) to Palaeoproterozoic (>1690 Ma) terranes little affected by younger overprinting effects, in which largely pristine Archaean crustal histories are preserved.

These terranes, provinces and areas are distinguished in general terms in Fig. 3.2-1. In order to provide a context for the Archaean rocks of Antarctica, the terranes and regions grouped under (1) and (2) are briefly outlined below. The main regions placed in groups (3) and (4) are divided here into those that do not preserve any extensive zircon U-Pb evidence

Table 3.2-1. Summary of protolith and event ages for Archaean terranes in the East Antarctic Shield

Age, Ma	Napier Complex	Kemp Land Coast	Mather Terrane
>3900 Ma	inherited (?) zircons Mt Sones, Gage Ridge[1,2]		
3850	Mount Sones and Gage Ridge orthogneisses[1,2,3]	Hf model ages Oygarden Group and Stillwell Hills[8]	
3650–3620		anatexis and orthogneisses, Oygarden Group[7,8]	
3550–3500	inherited zircons, Rippon Point[8]	minimum ages of older homogeneous orthogneisses on Kemp Land coast[8]	inherited zircon in ca. 2550 Ma TTG sheet; oldest composite layered gneiss component (?)[10]
3490–3420	metamorphism and anatexis, Rippon Point (ca. 3422 Ma)[8]	metamorphism (ca. 3470 Ma). Anatexis and new orthogneisses (?)[7,8]	inherited zircon components in ca. 2550 Ma TTG sheet[10]
3390–3370			
3280–3250	Riiser–Larsen orthogneiss[4]		Short Point orthogneiss[9]
3190–3160	thermal (?) resetting, Rippon Point[8]		inherited zircons in ca. 2550 Ma TTG sheet[10]
3150–3050			
3050–2990	Proclamation orthogneiss[2,3]		
2890			
2840–2800	Dallwitz orthogneiss and dominant tonalitic orthogneisses[2,3]		homogeneous tonalitic orthogneisses; layered mafic complexes[9,10]
2790–2770		granulite facies metamorphism[7,8]	
2650–2640			
2620	Tonagh orthogneiss[5]		
2550–2520	ultrahigh temperature metamorphism[3,6]		trondhjemitic TTG orthogneiss sheets[11]
2520–2460	local granitoids, waning metamorphism[2,3,6]		

(continued on next page)

for rocks older than ca. 3000–3050 My old, which are described in general terms, and those that do preserve such evidence and hence are described in more depth, and distinguished in Table 3.2-1. In the final section of this chapter those localities or areas from which the

Table 3.2-1. (*Continued*)

Age, Ma	Ruker Terrane	Vestfold Block	Denman Glacier	Terre Adélie	Grunehogna
>3900 Ma					
3850					
3650–3620					
3550–3500					
3490–3420					
3390–3370					
3390–3370	layered tonalitic orthogneisses[12]				
3280–3250					
3190–3160	dominant granitic ortho-gneisses[12,13]				
3150–3050				orthogneisses[16]	
3050–2990			granitic and tonalitic orthogneisses[15]		granitic gneiss, Annandagstop-pane[17]
2890			granulite facies metamorphism[15]		
2840–2800		inherited zircons in Grace Lake Granodiorite[14]			
2790–2770	amphibolite facies metamorphism[13]				
2650–2640	post-D pegmatite[13]		orthogneisses[15]		
2620					
2550–2520			supracrustals[16]		
2520–2460		tonalitic (ca. 2520 Ma) and dioritic (ca. 2500–2475 Ma) orthogneiss[14]	metamorphism[16]		

References:

[1]Black et al. (1986); [2]Harley and Black (1997); [3]Kelly and Harley (2005); [4]Hokada et al. (2003); [5]Carson et al. (2002); [6]Harley (2003); [7]Kelly et al. (2004); [8]Halpin et al. (2005); [9]Kinny et al. (1993); [10]Harley et al. (1998); [11]Harley, unpubl.; [12]Mikhalsky et al. (2006); [13]Boger et al. (2006); [14]Black et al. (1991); [15]Black et al. (1992); [16]Peucat et al. (1999); [17]Jacobs et al. (1996).

oldest, pre-3400 Ma Antarctic rocks have been documented are described in detail and the implications of their isotopic records for the earliest history of crust formation highlighted.

3.2-2.1. Cambrian and Proterozoic Tectonic Belts and Provinces

The recognition of intense 600–500 Ma Pan-African tectonism and high-grade metamorphism in the East Antarctic Shield has been one of the major breakthroughs in Antarctic geology in the past decade. This, coupled with the preservation of distinct earlier crustal histories in intervening terranes, has refuted the notion of a circum-East Antarctic 'Grenville' orogenic belt produced by the ca. 1100–900 Ma collision of a unified East Antarctic Craton with other parts of East Gondwana (Fitzsimons, 2000a).

High-grade Pan-African tectonism at 600–500 Ma is recognised from four potentially distinct regions (Fig. 3.2-1): Dronning Maud Land (East African – Antarctic Orogen: Jacobs et al., 1998), Lützow-Holm Bay (Lützow-Holm Belt), Prydz Bay and Southern Prince Charles Mountains (Prydz Belt) and Denman Glacier region (Pinjarra Belt) (Fitzsimons, 2000b, 2003). In all cases the high-grade metamorphism is associated with intense deformation, melting, and the emplacement of syn- to late-tectonic intrusives (e.g., Shiraishi et al., 1994; Fitzsimons et al., 1997; Jacobs et al., 1998; Boger et al., 2001). Temporal, structural and metamorphic constraints from the best-studied Pan-African areas are consistent with their formation and evolution as collisional belts, with collision terminated either by late-stage extensional collapse or the lateral escape of mid-crustal domains along high-strain zones at ca. 520–500 Ma (Fitzsimons et al., 1997; Jacobs et al., 1998, 2003; Boger et al., 2001; Harley, 2003). These high-grade belts juxtapose distinct Mesoproterozoic and Neoproterozoic crustal provinces (Maud, Rayner, Rauer and Wilkes: Fig. 3.2-1), and have reworked the margins of Archaean cratonic remnants in the Napier Complex and southern Prince Charles Mountains.

Proterozoic terranes dominated by Meso- to Neoproterozoic tectonism (1400–910 Ma) have been grouped into three broad 'provinces' that have distinct geological histories (Fitzsimons, 2000a) – the Wilkes (Wilkes Land coast), Rayner (Enderby, Kemp and Mac-Robertson Lands: Fig. 3.2-2), and Maud (Dronning Maud Land) provinces, to which Harley (2003) has added a fourth, the Rauer Province (Fig. 3.2-1).

Fig. 3.2-2. Outcrop distribution and simplified time/event information for high-grade terranes and regions in the Antarctica-Indian sector from Lützow-Holm Bay (west) to the Vestfold Hills (east). Latitudes and longitudes are referred to present Antarctic co-ordinates. Inset shows the generalised distribution of Archaean crust (Napier Complex, Kemp Land, Ruker Terrane, Rauer Terrane and Vestfold Hills), areas dominated by Proterozoic tectonothermal events (Rayner, Fisher and Lambert Provinces and Rauer–Prydz Bay basement area) and those dominated by Cambrian tectonism (Prydz and Lützow-Holm belts). Dashed boxes show the areas covered by the maps in Fig. 3.2-4 (Rauer Group) and Fig. 3.2-8 (Enderby Land).

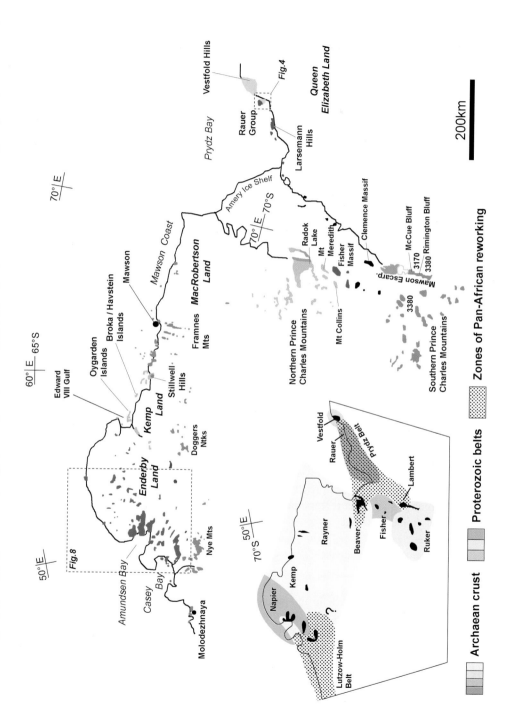

The *Wilkes Province* correlates very well with the Albany-Fraser belt of west Australia (Fig. 3.2-1). It lacks evidence for any high-grade tectonothermal events younger than ca. 1130 Ma, is dominated by low-medium pressure granulite metamorphism and the emplacement of charnockites at ca. 1210–1150 Ma, and locally preserves evidence for earlier tectonic events at ca. 1340–1280 Ma (Post et al., 1997).

In the *Rayner Province* (Fig. 3.2-1) older crust (e.g., Archaean Napier Complex, 3800–2500 Ma; 1450–1500 Ma anorthosites: Black et al., 1987; Sheraton et al., 1987) and Proterozoic sediments and younger intrusives (e.g., ca. 960 Ma charnockites: Young and Black, 1991) have undergone granulite facies deformation either at ca. 1000–980 Ma or ca. 930–920 Ma, or in both of these events. In the west and northern Rayner Province the major event is ca. 930–920 Ma in age (Kelly et al., 2000, 2002), whereas the ca. 1000–980 Ma peak metamorphic event dominates in the sub-area, known as the Beaver Terrane (Mikhalsky et al., 2001), exposed in MacRobertson Land and the Northern Prince Charles Mountains (Fig. 3.2-1; Fig. 3.2-2). This age distribution may reflect two-stage collision between the Ruker Terrane and other areas of inland Antarctica (e.g., the Mawson Continent of Fanning et al., 1999), a continent comprising much of eastern India, and a microcontinent that included the Archaean Napier Complex (Kelly et al., 2002).

The *Maud Province* (Fig. 3.2-1) is dominated by arc-related felsic plutonism at ca. 1180–1130 Ma and ca. 1080 Ma, regional high-grade metamorphism and deformation at ca. 1050–1030 Ma (Grantham et al., 1995; Jacobs et al., 1998), and further thrust-related deformation that may be as young as 980 Ma. These events are considered to reflect arc accretion followed by collision between the Kalahari Craton and an Antarctic Craton (Grantham et al., 1995; Jacobs et al., 1998).

The *Rauer Province* (Figs. 3.2-1 and 3.2-2) lies on the eastern margin of the Prydz Belt, and therefore separates the Prydz Belt from the Pinjarra Belt that occurs on the western seaboard of Australia and extends into the Denman Glacier area of Antarctica (Fig. 3.2-1). The Proterozoic tectonic history of the Rauer Province involved partial melting, high-grade tectonism and emplacement of felsic intrusives over the time interval 1080–990 Ma (Kinny et al., 1993; Harley, 2003). This event chronology bears more similarity with that of the Maud Province than with those preserved in the nearest-neighbour provinces, the Wilkes (to the E) which has an older tectonothermal record, and the Rayner (to the W) which has a major event at 930–920 Ma.

3.2-2.2. *Archaean and Archaean – Palaeoproterozoic Terranes with Little pre-3000 Ma Record*

The *Mawson Block* of Terre Adélie and other areas on the eastern fringe of the East Antarctic Shield (TA: Fig. 3.2-1; Peucat et al., 1999; Fanning et al., 1999) preserves an Archaean to Palaeoproterozoic geological record that is correlated with the Gawler Craton and adjacent areas of Australia (Fig. 3.2-1) and the Nimrod Group of the Miller Range in central Trans Antarctic Mountains (Miller: Fig. 3.2-1; Goodge and Fanning, 1995; Goodge et al., 2001). The basement geology in the Terre Adélie region includes vestiges of ca. 3050–3150 gneisses (Table 3.2-1) and ca. 2560–2450 Ma supracrustals that were metamorphosed in the earliest Proterozoic (ca. 2440 Ma; Peucat et al., 1999). These formed the basement to rift-related sediments and volcanics deposited between ca. 1775 and 1700 Ma (Fanning et al., 1999) and metamorphosed at ca. 1710–1690 Ma in the Kimban tectonothermal event (Goodge and Fanning, 1995). This event progressed under low-pressure granulite facies conditions in Terre Adélie, whereas in the Nimrod Group an event of similar age (ca. 1700–1690 Ma) progressed under eclogite facies conditions (Goodge and Fanning, 1995; Goodge et al., 2001). The correlations in events, and provenance ages in the metasediments, has stimulated the proposal that the Mawson Block may form much of the continental land-mass of East Antarctica underlying the ice, at least as far to the west as the subglacial Lake Vostok (Fanning et al., 1999; Fitzsimons, 2003; Boger and Miller, 2004).

In the *Denman Glacier/Obruchev Hills Area* (DG: Fig. 3.2-1) granitic and tonalitic orthogneiss precursors with ages between ca. 3000 and 2640 Ma have been recognised (Table 3.2-1), with the older orthogneiss affected by a granulite facies metamorphic event at ca. 2890 Ma (Black et al., 1992).

The *Vestfold Block* of eastern Prydz Bay (VH: Figs. 3.2-1 and 3.2-2) experienced its main magmatic accretion events and crust formation in the time interval ca. 2520–2480 Ma (Black et al., 1991; Snape et al., 1997). 2520 Ma tonalitic orthogneisses and pyroxene granulites, and supracrustal rocks including metavolcanics, Fe-rich semipelitic sediments and Mg-Al rich claystones (Oliver et al., 1982, Harley, 1993) were metamorphosed under granulite conditions by 2500 Ma (Table 3.2-1). The broadly syn-tectonic Crooked Lake Gneisses were then emplaced and overprinted in an upper amphibolite to granulite deformation episode at ca. 2500–2480 Ma (Black et al., 1991; Snape et al., 1997). A pre-2530 Ma geological record is preserved only in a group of migmatitic granodioritic orthogneisses, the Grace Lake Granodiorite (Table 3.2-1), which contains inherited zircons up to ca. 2800 Ma in age (Black et al., 1991) and records a $T_{DM}(Nd)$ model age of ca. 3050 Ma based on the data of Kinny et al. (1993).

The *Lambert Province* or *Terrane* (Boger et al., 2001, 2006; Mikhalsky et al., 2001, 2006a, 2006b; Fig. 3.2-2) is dominated by felsic to intermediate orthogneisses derived from granitoids emplaced at ca. 2420 Ma. These were interleaved with mafic amphibolite, calc-silicate gneiss and marble and metamorphosed to upper amphibolite facies at ca. 2065 Ma and possibly at other times in the Palaeoproterozoic (e.g., ca. 1800, ca. 1600 Ma: Boger et al., 2001; Mikhalsky et al., 2006a). Some of the orthogneisses are derived from granitoids that have elevated ε_{Nd} values (to +6.4) and T_{DM} approximating 2400 Ma and hence are considered to be juvenile crustal additions at ca. 2450 Ma. Other orthogneiss protoliths have highly negative ε_{Nd} values at 2400 Ma (to −6.4), and thus have old T_{DM} model ages (ca. 3000–3400 Ma) that imply the involvement of older Archaean crust in their genesis (Mikhalsky et al., 2006b). The Lambert Terrane is strongly overprinted by Cambrian tectonism, manifested in the intrusion and subsequent deformation of abundant granite sheets, veins and vein networks with zircon U-Pb ages between ca. 530 Ma and 495 Ma (Boger et al., 2001; Mikhalsky et al., 2006a).

The Grunehogna Craton of western Dronning Maud Land (Gr: Fig. 3.2-1) is exposed at only one nunatak (Table 3.2-1), where ca. 3000 Ma granitic basement is overlain by ca. 1100–1000 Ma shelfal sediments and volcanics of the Ritscherflya Supergroup (Grantham et al., 1995). The greater lateral extent of this basement, interpreted to be a fragment of the Kalahari Craton of southern Africa left attached to East Antarctica, is supported by geophysical surveying (Jacobs et al., 1996).

3.2-2.3. Archaean and Archaean/Proterozoic Terranes with Pre-3000 Ma Crustal Records

3.2-2.3.1. The Ruker Terrane, Southern Prince Charles Mountains
The Ruker Terrane of the Southern Prince Charles Mountains (sPCM: Figs. 3.2-1 and 3.2-2) is widely regarded as representative of a larger region of the 'inboard' EAS that collided or interacted with outboard terranes (e.g., the Archaean Napier Complex) and other shield areas (e.g., the Dharwar and Bastar cratons of India) at 990–910 Ma and 550–500 Ma (e.g., Fitzsimons, 2000b; Boger and Miller, 2004). It comprises a polyphase Archaean (ca. 3390–3155 Ma) granite gneiss basement, the Mawson Suite (Fig. 3.2-3(b)), tectonically interleaved with deformed metasedimentary and metavolcanic rocks (Fig. 3.2-3(c); Tingey, 1982, 1991; Mikhalsky et al., 2001; Boger et al., 2006; Mikhalsky et al., 2006a).

The deformed metasedimentary and metavolcanic rocks of the Ruker Terrane have been divided into three tectonostratigraphic units, the Menzies, Ruker and Sodruzhestvo Series (Mikhalsky et al., 2001). The Menzies Series shares the same Archaean deformational history as the Mawson Suite and its correlatives (Boger et al., 2006), and hence is at least late Archaean in depositional age. Quartzites, pelitic and calcareous metasediments and amphibolites of the Menzies Series exhibit medium-pressure Barrovian-style metamorphism (staurolite + kyanite ± garnet) and may preserve evidence for two distinct metamorphic events (Boger et al., 2006). The Ruker Series includes mafic to felsic metavolcanic rocks and associated metadolerite sills, metapelitic schist, slate, phyllite, and banded ironstones. The Sodruzhestvo Series consists of calcareous schist, pelite, phyllite and slate, and minor marble, quartzite, and conglomerate, and is thought to be the younger of the two metasedimentary series (Mikhalsky et al., 2001). Both the Ruker and Sodruzhestvo Series post-date the Archaean tectonics and intrusive events that have affected the Mawson Suite and Menzies Series, can be considered to represent Proterozoic cover sequences to the latter lithological groups (Boger et al., 2006), and are now metamorphosed to greenschist facies conditions.

An Archaean age for the Ruker Terrane basement orthogneiss (e.g., Mawson Suite granites) was initially established on the basis of Rb-Sr whole rock isochrons, which gave ages in the range ca. 2700–2760 Ma (Tingey, 1982), and an age of ca. 2589 Ma for a cross-cutting muscovite-bearing pegmatite. These minimum ages have been complemented by conventional multigrain zircon U-Pb data that indicate a crystallisation age of 3005±57 Ma for a granite at Mount Ruker (Mikhalsky et al., 2001), and by Sm-Nd whole rock isochron ages of 3124 ± 130 Ma and 3176 ± 140 Ma on granite gneisses from the Mawson Escarpment (Figs. 3.2-3(a,b)). Whilst indicative of the presence of older crust, these age data have large inherent errors and so do not allow detailed resolution of the age-event history of the Ruker Terrane. However, the ages of Archaean orthogneisses in the Ruker Terrane have recently been clarified by zircon U-Pb SHRIMP dating (Table 3.2-1).

Fig. 3.2-3. Geological features and relations in the Ruker Terrane, Southern Prince Charles Mountains (all photos courtesy of Dr. S. Boger, University of Melbourne). (a) View of the Lambert Glacier and surroundings, looking south along the Mawson Escarpment towards Mount Menzies, some 100 km distant. (b) View of the 1500 metre high cliff face of McCue Bluff. Typical Archaean (ca. 3160 Ma) granitic orthogneiss of the Mawson Suite is cut by several orientations of Proterozoic mafic dykes.

Fig. 3.2-3. (*Continued.*) (c) Menzies Series layered paragneisses, including mafic and fel-sic/intermediate schists interpreted as former volcanics, at Rimmington Bluff (note geologist for scale). (d) Isoclinal F_2 fold in Menzies Series layered schists and gneisses at Rimmington Bluff (note geologist for scale). (e) Boudinaged mafic unit in layered paragneiss of the Menzies Series at McCue Bluff. Syn-D3 leucosome from these boudin necks constrain an age of ca. 2772 Ma for this event (Boger et al., 2006).

Homogeneous/massive gneissic granites (Mawson Suite) that crop out in the South-ern Mawson Escarpment have yielded oscillatory zoned magmatic zircons that record crystallisation over the time interval ca. 3185–3155 Ma, with specific granitoids yield-ing population ages of 3182 ± 9 Ma (Mikhalsky et al., 2006a), 3177 ± 6 Ma, 3174 ± 9 Ma and 3160 ± 6 Ma (Boger et al., 2006). One of these granitoids preserved evidence for older crustal sources, in the form of xenocrystic zircon cores with a weighted population

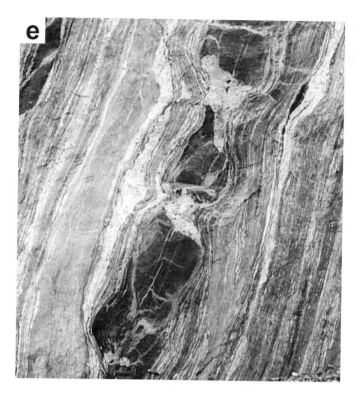

Fig. 3.2-3. (*Continued.*)

concordia age of 3370 ± 11 Ma. This age is consistent with the incorporation of crustal sources similar to a Y-depleted banded trondhjemitic gneiss and a porphyritic tonalite cobble preserved in Ruker Series metasediments and dated by Mikhalsky et al. (2006a), which yielded zircon U-Pb concordia ages of 3392 ± 6 Ma and 3377 ± 9 Ma respectively. These results demonstrate that the Ruker Terrane contains crustal components at least as old as ca. 3390 Ma, and on the basis of Sm-Nd model ages (T_{DM}) of ca. 3200–3900 Ma (Mikhalsky et al., 2006b) it is possible that even more ancient early Archaean rocks may be present.

The ages of principal Archaean regional deformation events (Fig. 3.2-3(d)) that have affected the Ruker Terrane have until recently only been constrained to pre-date cross-cutting pegmatite that has a zircon U-Pb age of ca. 2650 Ma (Boger et al., 2001, 2006). Mikhalsky et al. (2006a) inferred a ca. 3145 Ma thermal/metamorphic event based on the age of a concordant high-U zircon rim formed in one of the ca. 3380 Ma orthogneisses, but was not able to relate this to any deformation or tectonic event. However, recent zircon U-Pb dating of structurally-constrained samples (Boger et al., 2006; Fig. 3.2-3(e)) provides tight brackets on the Archaean deformation and metamorphism. Leucosomes formed between the first and second (D_1–D_2) folding events, or segregated into inter-boudin necks in more localised D_3 high strain zones, were formed over the interval ca. 2790–2770 Ma. These ages are consistent with a single, relatively short-lived, Archaean tectonothermal event and imply Rb-Sr re-equilibration on the whole rock scale during the associated upper amphibolite facies regional metamorphism (Rb-Sr ages ca. 2760–2700 Ma: Tingey, 1982).

3.2-2.3.2. The Archaean Mather Terrane, Rauer Province, Prydz Bay

The Rauer Province, which outcrops in the Rauer Group of islands on the eastern side of Prydz Bay (RI; Fig. 3.2-1; Rauer, Fig. 3.2-2), includes both Archaean and Mesoproterozoic crustal components metamorphosed and deformed at ca. 530–500 Ma during intense shearing and mylonitisation, and amphibolite to granulite facies high-strain zones associated with Pan-African tectonism in the adjacent Prydz Belt (Kinny et al., 1993; Sims et al., 1994; Harley, 2003). Here we distinguish the *Mather Terrane* as the sub-area of the Rauer Group of islands that is dominated by Archaean protoliths (Fig. 3.2-4).

The *Mather Terrane* is dominated by >3270 Ma and ca. 2820–2800 Ma tonalitic orthogneisses (Fig. 3.2-5(a); Table 3.2-1; Kinny et al., 1993), accompanied by minor supracrustal units rich in marbles and magnesian-aluminous pelites and quartzites (Mather Paragneiss: Harley et al., 1998). The oldest orthogneiss components, which contain zircons up to ca. 3500 Ma in age, will be described in depth in a later section. The ca. 2820–2800 Ma orthogneisses have a $T_{DM}(Nd)$ model age of ca. 3250–3350 Ma and are interpreted to be derived from the remelting of older protoliths (Table 3.2-1; Kinny et al., 1993). These orthogneisses intrude or enclose ca. 2840 Ma layered igneous complexes that preserve original igneous features despite the effects of later tectonothermal events (Harley et al., 1998). The extensively dyked and reworked Archaean orthogneisses and related supracrustal packages (Harley et al., 1992; Sims et al., 1994) are interleaved with Mesoproterozoic supracrustals and ca. 1030–1000 Ma felsic to mafic intrusives (Kinny et al., 1993) that experienced high-grade metamorphism and partial melting at ca. 1000 Ma and probably 510 Ma (Harley et al., 1998). Despite extensive texturally constrained geochronology, it is not clear as to whether the Archaean gneisses were reworked during the Mesoproterozoic at ca. 1000 Ma, or only affected by Archaean tectonothermal events and subsequently overprinted during interleaving with the Mesoproterozoic gneisses in the Prydz Belt event (Hensen and Zhou, 1995). This ambiguity exists because the zircon U-Pb data obtained from Archaean gneisses does not record any Mesoproterozoic disturbance (Kinny et al., 1993; Harley et al., 1998). Moreover, the polyphase reactivation of earlier gneissic fabrics in the youngest, ca. 500 Ma ductile deformation events has resulted in the Archaean and Mesoproterozoic lithological packages being rotated into parallelism (Sims et al., 1994; Harley et al., 1998).

Fig. 3.2-4. Geological map of the Mather Peninsula–Scherbinina Island area in the Rauer Group, showing the locations of the Short Point orthogneiss sample 5120 (Sheraton et al., 1984) and SH45 (Harley et al., 1998). Inset is a map of the Rauer Group as a whole, showing the general distribution of Archaean rocks ascribed to the Mather Terrane. See Fig. 3.2-2 for location of the Rauer Group and this map area within East Antarctica.

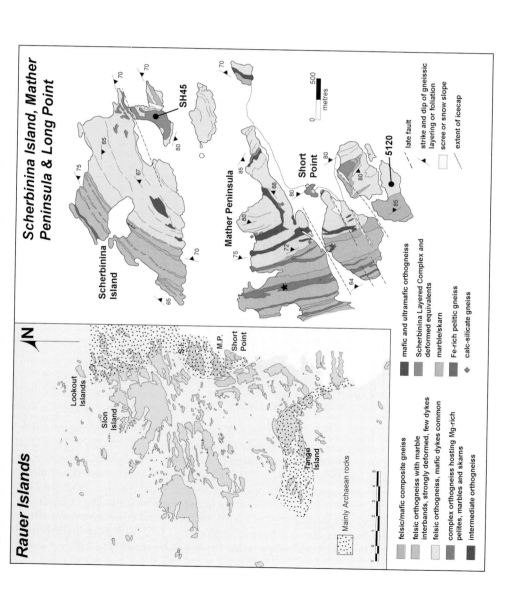

Rauer Islands

Scherbinina Island, Mather Peninsula & Long Point

Legend (Rauer Islands):
- felsic/mafic composite gneiss
- felsic orthogneiss with marble interbands, strongly deformed, few dykes
- felsic orthogneiss, mafic dykes common
- complex orthogneiss hosting Mg-rich pelites, marbles and skarns
- intermediate orthogneiss
- Mainly Archaean rocks

Legend (Scherbinina/Mather):
- mafic and ultramafic orthogneiss
- Scherbinina Layered Complex and deformed equivalents
- marble/skarn
- Fe-rich pelitic gneiss
- calc-silicate gneiss
- late fault
- strike and dip of gneissic layering or foliation
- scree or snow slope
- extent of icecap

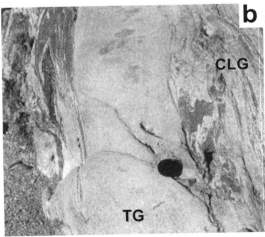

Fig. 3.2-5. Archaean geological features and relations in the Mather Terrane, Rauer Group of islands, Prydz Bay. (a) View, looking south, of the Mather Peninsula (foreground) and Short Point area. The eastern part of the area is dominated by Archaean (ca. 2820 Ma and >3200 Ma) protoliths. In the western part Archaean and Proterozoic protoliths are interleaved on tens to hundreds of metres scales. (b) 40 centimetre wide sheet of locally discordant homogeneous tonalitic orthogneiss (TG) cutting composite layered gneiss (CLG) characterised by folded mafic-felsic banding on centimetre to decimetre scales. All rocks are metamorphosed at granulite facies conditions. Scherbinina Island, Rauer Group. (c) Folded leucotonalitic–trondhjemitic orthogneiss (TrG) cutting composite layered gneiss (CLG) that preserves earlier folds in mafic-felsic banding. Scherbinina Island, Rauer Group.

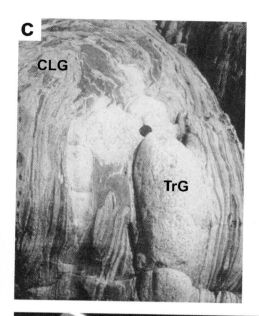

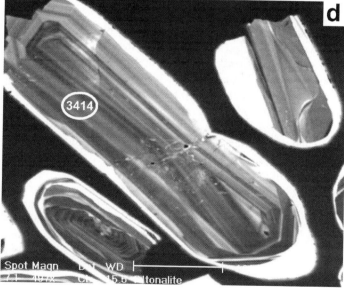

Fig. 3.2-5. (*Continued.*) (d) Cathodoluminescence (CL) image of an elongate, oscillatory zoned zircon of the type that dominates the magmatic zircon population in sample SH45 from the Rauer Group. The ca. 3414 Ma ^{207}Pb/^{206}Pb age of this concordant zircon indicates that it is inherited – much older than the ca. 2550 Ma age of the trondhjemite sheet itself. Bright CL outer rims and rounded terminations reflect zircon recrystallisation and dissolution-reprecipitation at ca. 510 Ma (Harley et al., 1998).

3.2-2.3.3. The Kemp Land Coast Region
In the Oygarden Group of Kemp Land (Oy: Figs. 3.2-1 and 3.2-2) initial deformation in the Rayner Structural Episode at ca. 930 Ma involved the deep-crustal ductile reworking and thrusting of Archaean gneisses that range in age up to ca. 3650 Ma (Kelly et al., 2002, 2004). Given the proximity of this region, and the broad first-order similarities in the age spectra of the Archaean crustal components, the Rayner Province in this area has been regarded as the reworked equivalent of the adjacent Napier Complex, described below.

Kelly et al. (2004) reported SHRIMP U-Pb zircon data for a number of structurally constrained orthogneisses from the Oygarden Group, and interpreted the quite complex isotopic results in the light of these structural constraints and detailed cathodoluminescence (CL) and back scattered electron (BSE) imaging of the separated zircons. In addition to the isotopic disturbance of zircons in the Rayner Structural Episode at ca. 930 Ma, Kelly et al. (2004) were able to identify thermal events at ca. 1600 Ma and 2400 Ma, and define a late Archaean history involving the metamorphism of early Archaean layered composite orthogneisses and intrusive homogeneous felsic orthogneisses at $\geqslant$2780 Ma (Table 3.2-1). The pre-2780 Ma events recorded in these orthogneisses and similar felsic orthogneisses from areas elsewhere in Kemp Land (Halpin et al., 2005) are considered in detail in the subsequent section on the oldest rocks.

3.2-2.3.4. The Archaean Napier Complex, Enderby Land
The Archaean Napier Complex (Nap: Figs. 3.2-1 and 3.2-2) of Enderby Land contains the oldest rocks yet recorded from Antarctica (Table 3.2-1). Whilst some granitic and tonalitic gneiss precursors, to be described in detail below, range in age up to ca. 3850 Ma (Williams et al., 1984; Black et al., 1986; Harley and Black, 1997; Kelly and Harley, 2005), the dominant granulite facies orthogneisses were derived from the metamorphism of granites, tonalites and granodiorites with mid- to late-Archaean ages of ca. 3270 Ma, 3070–2970 Ma and 2840–2800 Ma (Sheraton et al., 1987; Harley and Black, 1997; Hokada et al., 2003; Kelly and Harley, 2005). These and the rarer ancient orthogneiss precursors together with distinctive supracrustal packages, were strongly deformed and interleaved under ultrahigh temperature metamorphic conditions in the late Archaean.

The Napier Complex records the highest grades of ultrahigh temperature (UHT) crustal metamorphism worldwide (Harley, 2003). The lithological diversity of the layered paragneisses, which include mafic pyroxene granulites, metaironstones, pelites and both Fe- and Mg-rich metaquartzites, has enabled the deduction of key indicator mineral assemblages that imply temperatures of metamorphism of between 1000 and 1120 °C at depths of 20–35 kilometres. These UHT indicators, summarised by Harley (1998, 2003), include sapphirine and quartz, with orthopyroxene or garnet, aluminous orthopyroxene (9–11 wt% Al_2O_3) coexisting with sillimanite and quartz, the K-Mg-Al mineral osumilite coexisting with sapphirine or garnet, ternary mesoperthitic feldspars in pelitic gneisses, and inverted magnesian pigeonite in metamorphosed ironstones.

Following many years of controversy (Grew and Manton, 1979; Black et al., 1983b; Harley and Black, 1997; Grew, 1998), the age of the UHT metamorphism in the Napier Complex is now constrained to be younger than 2626 $\pm$ 28 Ma, which is the protolith age of a granodioritic orthogneiss from Tonagh Island (Carson et al., 2002). U-Pb ages of zircons present in syn-metamorphic pegmatitic 'sweats' and of the oldest zircon rims formed on older grains within paragneisses further constrain the age of metamorphism to between ca. 2590 Ma and 2510 Ma (Hokada and Harley, 2004; Kelly and Harley, 2005). The ubiquitous zircon age clusters at ca. 2480–2450 Ma determined in many gneisses from the Napier Complex reflect fabric formation and fluid access subsequent to the UHT event, and are interpreted to relate to final melt crystallisation and fluid expulsion that occurred as the Complex cooled, at depth in the crust, through 700–750 °C (Kelly and Harley, 2005).

Unlike many other high grade terrains, in which variations in the intensity of deformation and the presence of low-strain domains allows original rock features and relationships to be observed, any original lithological features and contact relationships in the Napier Complex have been obliterated by the intense ductile deformation and recrystallisation that was associated with the UHT metamorphism. This deformation produced pervasive flat-lying gneissic fabrics and early isoclinal folds that have been refolded and reoriented about several later phases of folds and high-strain zones. Intense down-dip linear fabrics defined by the orientations of the highest-grade minerals (e.g., sillimanite, sapphirine, orthopyroxene, quartz) are common, as is boudinage and megaboudinage of competent units during layer-parallel extension (e.g., Black and James, 1983; Black et al., 1983a, 1983b; Sheraton et al., 1987; Harley, 2003). Later deformation events are discrete, localised to pseudotachylites and metre- to tens of metres width high-strain and mylonitic zones, probably of Cambrian and Proterozoic ages (Sheraton et al., 1987).

3.2-3. THE OLDEST ROCKS: >3400 MA

Having outlined the general geological context of the Archaean of Antarctica and described the Archaean and reworked Archaean terranes themselves in more detail, we now go on to consider specific examples of the oldest rocks so far documented from the continent (Table 3.2-1). For convenience we will consider the evidence for crust, or crustal precursors, in the age range ca. 3400 Ma to 4000 Ma and further restrict the discussion to those cases where the evidence is principally provided or strongly supported by zircon U-Pb microanalysis. These restrictions limit us to consider three areas or terranes: the Mather Terrane, the Kemp Land sub-area of the Rayner Province, and the Napier Complex. Whilst crustal material as old as ca. 3390 Ma is reported from the Ruker Terrane (Table 3.2-1), as there is no zircon evidence for older crust in this terrane we will not consider it further here.

3.2-3.1. Early Archaean Orthogneisses of the Mather Terrane

The oldest gneisses and gneiss precursors in the Mather Terrane are located in the Mather Peninsula area (Fig. 3.2-4), where Kinny et al. (1993) reported an age of 3269 ± 9 Ma for oscillatory zoned magmatic zircons extracted from a charnockitic-felsic gneiss at Short

Point (Table 3.2-1). Sm-Nd data from this reasonably homogeneous orthogneiss (sample 5120: Sheraton et al., 1984) indicate that its magmatic protolith was generated from evolved crustal material ($\varepsilon_{Nd}(3270) = -2.5$) that may have been sourced from depleted mantle in the earliest Archaean ($T_{DM}(Nd)$ = ca. 3750 Ma).

The high strain associated with the latest high-grade ductile deformation episodes has led to early structures being transposed into near-parallelism with younger ones across much of the Mather Peninsula area. Archaean and Mesoproterozoic lithologies now form horizons or packages with unit-parallel boundaries, coaxial folds and refolding, and co-incident SSE-plunging mineral elongation lineations on the macro- to meso-scale (e.g., Fig. 3.2-5(a)). Within low-strain zones Harley et al. (1998) have demonstrated that the oldest gneiss components of the Rauer Terrane are represented by composite layered orthogneisses on the basis of cross-cutting relationships and relative structural chronologies (Figs. 3.2-5(b,c)). These dominantly tonalitic to granodioritic gneisses are migmatitic, containing rafts and schlieren of mafic granulite. They are cut by metre to tens of metres wide sheets of homogeneous felsic gneiss similar to those dated at ca. 2820–2800 Ma by Kinny et al. (1993) (Fig. 3.2-5(b)), or by lenses and veins of leucocratic tonalitic to trondhjemitic gneiss (Fig. 3.2-5(c)) that may also cut homogeneous tonalitic gneisses (Harley et al., 1998). The composite layered orthogneisses, homogeneous tonalitic gneisses and tonalitic-trondhjemitic gneisses are all cut by several generations of metamorphosed and deformed mafic dykes (e.g., Sims et al., 1994), including most of the suites of meta-tholeiite dykes that cut the Scherbinina Layered Complex (Harley et al., 1998).

This composite layered orthogneiss unit has not been dated directly. However, Harley et al. (1998) reported zircon U-Pb age data from a tonalitic-trondhjemitic orthogneiss sheet intruded into the ca. 2840 Ma Scherbinina Layered Complex. This sheet (SH45; Fig. 3.2-4) yielded a complex pattern of zircon U-Pb data that could be divided into a several distinct Archaean populations, each defining discordia with lower intercepts at ca. 510 Ma (the Prydz Belt event; Fig. 3.2-6(a)). With the exception of one concordant analysis at ca. 2550 Ma, all of the Archaean zircon arrays defined concordia ages older than the Scherbinina Layered Complex itself. The oldest linear array yielded a concordia intercept age of 3470 ± 30 Ma (Harley et al., 1998).

The unambiguous geological observation that this tonalitic-trondhjemitic sheet has intruded into the ca. 2840 Ma Scherbinina Layered Complex (Table 3.2-1) leads to the conclusion that the majority of the oscillatory zoned Archaean zircon in the tonalite sample is inherited from an older source rock or range of source rocks. Detailed CL and BSE imaging (e.g., Fig. 3.2-5(d)) has now been used to establish a zircon growth stratigraphy, and so texturally constrain those zircon growth zones that might be related to its crystallisation in the tonalite (Harley, unpubl.). SHRIMP analysis of these specifically targetted outer oscillatory zones yield a linear array of concordant to variably discordant zircon U-Pb analyses that define an upper intercept age of ca. 2550 Ma (Table 3.2-1), interpreted to be the crystallisation age of the tonalitic-trondhjemitic sheet itself. The 3300–3470 Ma zircon U-Pb ages are invariably obtained from elongate weakly to moderately oscillatory zoned and planar to sector zoned, elongate magmatic cores that are overgrown and trun-

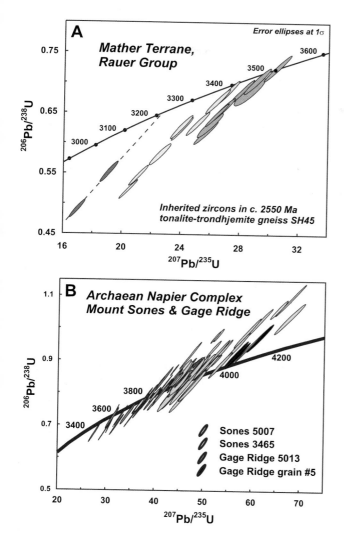

Fig. 3.2-6. (a) Concordia diagram illustrating the U-Pb systematics of inherited zircons (pre-2550 Ma) in the tonalitic-trondhjemitic orthogneiss sheet SH45 (location shown in Fig. 3.2-4). Note the presence of at least 3 sub-parallel alignments, distinguished by different shadings, controlled by Pb-loss at ca. 500 Ma from zircons with initial ages of ca. 3500–3480 Ma, ca. 3420–3400 Ma and ca. 3200 Ma. (b) Concordia diagram illustrating the collective U-Pb features of pre-3400 Ma zircon populations in the Mount Sones and Gage Ridge orthogneisses (locations shown in Fig. 3.2-8). Zircon U-Pb data (all at one sigma uncertainty) are from Williams et al. (1984), Black et al. (1986) and Harley and Black (1997). Note the presence of highly reverse discordant zircons in each of the three samples analysed, and the occurrence of distinctive near-concordant zircons that lie at older ages (>3950 Ma) than the bulk of the shared analytical population, which lies at ca. 3850 Ma. The three U-Pb analyses from grain #5 in the Gage Ridge orthogneiss are distinguished by the darkest shading.

cated either by the ca. 2550 Ma oscillatory zoned zircon or by bright CL and homogeneous rims (ca. 500 Ma).

The older zircon U-Pb analytical alignments have been re-examined in the light of the CL and BSE imaging of internal growth zones in the zircons using ISOPLOT/EX (Ludwig, 1999). Linear fits to these alignments, anchored at 500 ± 20 Ma to account for the important isotopic effects of the Prydz event, yield concordia upper intercept ages of 3488 ± 13 Ma, 3421 ± 21 Ma and ca. 3200 Ma (Fig. 3.2-6(a)). The oldest concordant zircon core, with a $^{207}Pb/^{206}Pb$ age of 3501 ± 3 Ma, provides a lower limit for the age of the oldest local crustal sources available to be melted or incorporated in the tonalitic-trondhjemitic sheet, which are most likely to be the composite layered orthogneiss that hosts such sheets in the Scherbinina area. The spectrum of inherited zircon ages from ca. 3500 Ma to 3200 Ma is consistent with the multi-phase character of the composite layered orthogneiss, the ca. 3250–3350 Ma T_{DM} estimated for the later Archaean tonalites of the Mather Terrane, and the ca. 3270 Ma age of the Short Point orthogneiss described above (Table 3.2-1). As the least CL responsive, strongly metamict and spongiform zircon cores have been avoided in SHRIMP analysis of sample SH45, it is possible that older crustal components or sources may be present in the Mather Terrane, perhaps as old as the ca. 3750 Ma Nd model age obtained for sample 5120 (Fig. 3.2-4).

3.2-3.2. Reworked Early Archaean Crust in the Rayner Province of Kemp Land

The oldest Archaean material identified in the Oygarden Group of western Kemp Land (Fig. 3.2-2) is a layered composite orthogneiss (Figs. 3.2-7(a,b)), which in domains only weakly affected by the Rayner Structural Episode preserves a banding defined by the alignment of leucocratic segregations (Fig. 3.2-7(b)). Zircon domains in OG615, an example of the layered composite orthogneiss, are of three main types: core domains with oscillatory zoning or with modified (blurred/patchy) oscillatory zoning; core domains characterised by radial, sector and firtree zoning; and rims with homogeneous luminescence. The oldest preserved cores in this case were those with radial and sector zoning, with concordia intercept ages up to 3655 ± 15 Ma (Table 3.2-1). This age is interpreted to record the partial melting of the protolith to produce the leucocratic segregations, implying that the protolith is even older (Kelly et al., 2004). These cores form an analytical array along concordia down to an age of 3469 ± 13 Ma, which is interpreted to approximate the age of a thermal/metamorphic event. A younger homogeneous felsic orthogneiss (OG525: Fig. 3.2-7(b)) intruded the layered composite orthogneiss either prior to, or synchronous with metamorphism and deformation at ca. 2840–2780 Ma (Table 3.2-1; Kelly et al., 2004).

Halpin et al. (2005) have presented zircon U-Pb age and Hf isotope data from zircons in several orthogneisses from the Kemp Land and MacRobertson Land coastal region, from the edge of the Napier Complex itself and into the Rayner Province as far as Mawson station. Four of their samples, analysed using laser ablation microprobe ICP-MS methods, yielded highly discordant zircon arrays with significant contributions from Archaean zircons that have $^{206}Pb/^{207}Pb$ ages between ca. 3000 Ma and 3600 Ma (Table 3.2-1). These gneisses included a felsic orthogneiss from Rippon Point (sample OG235), which lies

Fig. 3.2-7. Palaeoarchaean geological features and relations in the Oygarden Group of islands, Kemp Land Coast: (a) Composite layered gneisses, Shaula Island. This package includes paragneisses and banded mafic-felsic gneisses, all deformed into tight to isoclinal early D_1 folds and later intruded by homogeneous tonalitic orthogneiss similar to that depicted in Fig. 3.2-7(b) below (OG525) (the cliffs are ca. 80–100 metres high); (b) Palaeoarchaean composite orthogneiss OG615, which has yielded magmatic zircons up to ca. 3655 Ma old, intruded by ca. 2840–2780 Ma homogeneous orthogneiss OG525 (see text for details).

within the Napier Complex (Sheraton et al., 1987; Harley and Black, 1997), a duplicate sample of the Oygarden Group layered composite orthogneiss OG615 (Halpin et al., 2005, sample OG614), and orthopyroxene-bearing felsic orthogneisses from Broka-Havstein Islands and the Stillwell Islands to the east.

The zircon U-Pb data obtained on OG614 are consistent with those reported by Kelly et al. (2004). Homogeneous orthogneiss from the Broka-Havstein Islands preserves zircon cores as old as ca. 3540 Ma, a possible minimum age for its protolith, but the effects of later metamorphic events between ca. 2800 Ma and 2400 Ma are severe and hamper further interpretation. The Stillwell Hills Orthogneiss has a minimum protolith age of ca. 3460 Ma, based on the oldest surviving concordant oscillatory zoned zircon core analysis. The orthogneiss records extensive resetting to ca. 2970 Ma as well as the profound effects of recent Pb-loss on most zircons formed or reset between ca. 3460 Ma and 3000 Ma (Table 3.2-1; Halpin et al., 2005). In contrast to the Broka-Havstein orthogneiss the Stillwell Hills orthogneiss does not record any evidence in its zircon populations for resetting or regrowth at ca. 2400–2800 Ma.

The homogeneous felsic orthogneiss OG235 from Rippon Point, within the Napier Complex, contains zircons with relatively homogeneous to radial and sector zoned cores, as well as rare oscillatory zoned zircons. The latter, with concordia ages of ca. 3525 Ma, have been interpreted as inherited or xenocrystic as they are distinct in Th/U and age from the main zircon core population (Halpin et al., 2005), though it is possible that these record a minimum protolith age instead. A minimum age of 3422 ± 9 Ma for the earliest metamorphic/anatectic event (Table 3.2-1) is interpreted from the oldest zircon core preserving radial/sector zoning, whilst a second resetting event younger than 3173 ± 9 Ma has been invoked to explain the scatter of the oldest zircon analyses along concordia (Halpin et al., 2005). This sample also records a strong isotopic event affecting the zircons at 2468 ± 7 Ma, consistent with zircon U-Pb data obtained from all other localities in the Napier Complex (e.g., Black and James, 1983; Black et al., 1986; Harley and Black, 1997).

Zircon Hf isotope data from these orthogneisses is reported by Halpin et al. (2005) in terms of $T_{DM}(Hf)$, ε_{Hf}(zircon age) and using a plot of zircon $^{207}Pb/^{206}Pb$ age against $^{176}Hf/^{177}Hf$ measured on the same grain domains. For the purposes of this discussion the details of the decay constant used and the depleted mantle model applied are not critical, as the key interpretations do not depend strongly on these but rather on how younger zircon was produced (new vs recrystallised), and what Lu/Hf is applied for ^{176}Hf ingrowth in the crust following its initial formation, if that proceeded zircon crystallisation by a significant time.

The Oygarden Group layered composite orthogneiss OG615/614, and most probably the Broka-Havstein orthogneiss (despite marked outliers in the Hf isotope data) have $T_{DM}(Hf)$ similar to, or only slightly older than, their probable maximum crytallisation ages of ca. 3550–3650 Ma (Oygarden: ca. 3570 Ma; Broka: ca. 3600 Ma). These model ages are slightly older, near 3700 Ma, if the zircons are considered to have crystallised from a melt with average crustal Lu/Hf, in which some ^{176}Hf ingrowth would have occurred since initial crust formation. Older source ages are also possible if the mantle was more depleted than the DM model at ca. 3700–4300 Ma, as suggested by the data of Choi et al. (2006).

The Rippon Depot (Napier Complex) orthogneiss OG235 and Stillwell orthogneiss SW268 have, on average, lower or less evolved ^{176}Hf/^{177}Hf for their zircon ^{207}Pb/^{206}Pb age and so produce T_{DM}(Hf) model ages that are older than the most ancient U-Pb ages recorded in their zircons, and older than the samples noted above. T_{DM}(Hf) is ca. 3720–3860 Ma based on the most concordant grains (Table 3.2-1), and as old as ca. 3900–4050 Ma if ^{176}Hf in-growth is assumed in the crust prior to the melting event that produced the oldest zircons. For these orthogneisses it is entirely feasible that ancient Napier Complex (Gage Ridge type) crust (>3850 Ma depleted sources) was remelted to produce the >3460–3530 Ma orthogneiss precursors.

In detail, the Archaean event record determined from the Oygarden Group and Kemp Land is not entirely comparable to that preserved in the adjacent Napier Complex itself. Whilst it is feasible and likely, given the Hf isotope data summarised above, that some of the layered composite orthogneisses include components as old as ca. 3850 Ma, the age of the earliest Napier Complex orthogneiss precursors, the ca. 3650 Ma partial melt-ing/intrusive event and proposed metamorphism/thermal overprint at ca. 3450 Ma have not, as yet, been recorded in the Napier Complex of the Amundsen Bay region (Table 3.2-1; Fig. 3.2-8). On the other hand, the post-3200 Ma similarities are clear (Table 3.2-1). The ca. 2780 Ma age for the thermal event in the Oygarden Group is a minimum age, loosely constrained by the available data, and could potentially correspond to the ca. 2840 Ma low-pressure metamorphic episode inferred for the Napier Complex by Kelly and Harley (2005).

3.2-3.3. The Napier Complex

3.2-3.3.1. Fyfe Hills

Interest in Antarctica, and specifically the Napier Complex (Fig. 3.2-8), as a site where the most ancient rocks might be exposed was stimulated in the 1970s by the report of ca. 4000 Ma ages from the Fyfe Hills by Soviet scientists (Sobotovich et al., 1976). The Fyfe Hills (Figs. 3.2-8 and 3.2-9(a)) outcrop over an area of about 15 km^2 as two 5 km long ridges and smaller nunataks just inland from Khmara Bay in the southwest of Enderby Land (Fig. 3.2-8). Some 5–10 km west of the Fyfe Hills lie McIntyre Island and Zircon Point, sites that have also been the subjects of extensive and detailed zircon geochronology (Black et al., 1983b).

The old ages reported by Sobotovich et al. (1976) were derived from the slope of an alignment of whole rock analyses on an ^{207}Pb/^{204}Pb vs ^{206}Pb/^{204}Pb diagram, with ender-bite and gneiss yielding apparent total rock ages of 4100 ± 100 Ma and 3700 ± 200 Ma respectively. Irrespective of the validity of the actual reported ages, the high ^{207}Pb with respect to ^{206}Pb in these rocks certainly indicate a prolonged residence of the sources for the rocks in a high U/Pb environment prior to metamorphism in the late-Archaean events at ca. 2840 Ma and ca. 2550–2480 Ma that characterise the Napier Complex (Kelly and Harley, 2005).

The geology of the Fyfe Hills Soviet sample site was described in detail by Black et al. (1983a), and complemented by a more regional description and analysis of the

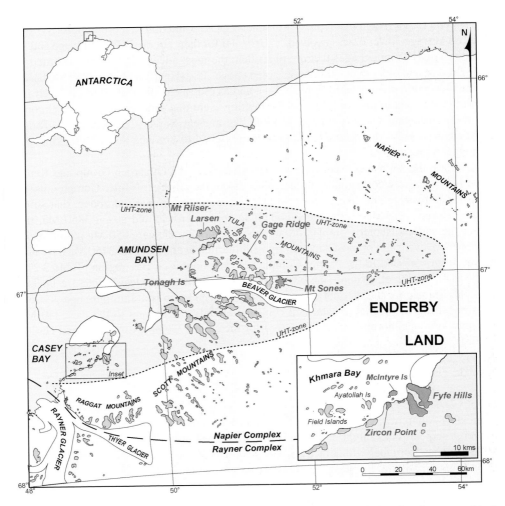

Fig. 3.2-8. Map of the western part of the Napier Complex, Enderby Land, showing the geographical positions of the localities discussed in the text (e.g., Mount Sones, Gage Ridge, Fyfe Hills, Mount Riiser-Larsen, Tonagh Island). Regional extent of ultrahigh-temperature mineral assemblages such as sapphirine + quartz is denoted by the shaded field enclosed by the dashed boundary. The Napier Complex is bound southwards by the Proterozoic Rayner Complex, along an EW trending zone that preserves Cambrian high-grade metamorphism. Area of main map is shown by the small box on the map of Antarctica, and in Fig. 3.2-2. Inset box shows the localities in the Casey Bay region (marked by a small outline box) referred to in the text.

Fyfe Hills by Sandiford (1985). The Soviet Sample site largely comprises a moderately to steeply NE-dipping (60/045) isoclinally folded charnockitic orthogneiss typified by a strong modal layering and planar fabric defined by ribbon quartz, mesoperthitic K-feldspar

Fig. 3.2-9. Archaean geological features and relations at Fyfe Hills, Enderby Land. See inset map in Fig. 3.2-8 for localities: (a) View of Fyfe Hills, looking east from Khmara Bay. The Soviet Sample Site (SSS) is denoted by the arrow; (b) F_1 isoclinal fold hinges and strong lensoidal fabric defined by trails and clusters of clinopyroxene alternating with plagioclase in a mafic granulite, Fyfe Hills Soviet Sample Site. This rock was originally interpreted by Black et al. (1983a) as a metagabbro, but subsequent geochemistry by Sandiford (1985) suggested a marl/altered basic volcanic protolith; (c) Isoclinal F_1 fold in granodioritic orthogneiss, Fyfe Hills Soviet Sample Site. A strong S_1 fabric axial planar to the fold is defined by the alignment of ribbons of quartz (dark) in the more granitic layers; (d) Lance Black contemplating the layered charnockitic gneiss sampled for Rb-Sr, Sm-Nd and zircon U-Pb from the Soviet Sample Site, and reported in Black et al. (1983a), McCulloch and Black (1984), and Black and McCulloch (1987).

Fig. 3.2-9. (*Continued.*)

and orthopyroxene (Fig. 3.2-9(b,c)). Minor interlayered quartzitic gneiss, garnet-quartz gneiss, pyroxene-magnetite-quartz metaironstone, pyroxene granulite and a clinopyroxene-plagioclase-titanite gneiss, derived from a meta-calcareous or marly precursor, are also present (Fig. 3.2-9(d)) (Sandiford, 1985). This suite of compositionally diverse gneisses, including Mn-Fe rich rocks, are considered to represent a supracrustal package deposited prior to at least ca. 3100 Ma. Leucocratic and melanocratic metre to several metre scale boudins of pyroxene-plagioclase gneisses also occur, and are interpreted by Black et al. (1983a) to be metamorphosed equivalents of layered mafic-anorthositic bodies. All rocks are intensely folded to produce isoclinal structures on decimetre to metre scales (Fig. 3.2-9(c)). Strong platey and ribbon fabrics defined by quartz, feldspars or aggregates of pyroxenes, and NE plunging lineations were produced in the first recognised UHT deformation (termed D_1; Fig. 3.2-9(c)). These flat-lying to moderately dipping structures and layering are locally refolded about close/isoclinal F_2 folds that formed also under granulite facies conditions but are distinguished by their re-orientation of the earlier fold closures and folding of the earlier-formed axial planar fabrics (e.g., Black and James, 1983; Black et al., 1983b; Sheraton et al., 1987). Overprinting both events are recrystallisation seams and deformation bands associated with a third regional deformation that domes the previously flat-lying to reclined gneiss suite on km to tens of km scales (D_3 in Black et al., 1983a, 1983b) and is responsible for the current moderate to steep orientation of the gneisses.

Black et al. (1983a) re-evaluated the earlier age data from the Soviet Sample site and presented new Rb-Sr analyses from charnockitic gneiss, leuconoritic gneiss and a variety of mafic granulites, and multi-grain zircon U-Pb TIMS analyses from a charnockitic leucogneiss. The key Rb-Sr result relevant to establishing the presence or otherwise of early Archaean crustal precursors were that large blocks of charnockitic gneiss sampled from an approximately 1 m^3 volume yielded an imprecise errorchron with an age of 3120 ($+230, -180$) Ma, and an initial ratio ($^{87}Sr/^{86}Sr$) of 0.725 ± 0.006. The high initial ratio associated with this analytical alignment yields a T_{UR} Sr model age of ca. 3900 Ma, consistent with the presence of earliest Archaean crustal precursors. The zircon U-Pb age data obtained by Black et al. (1983a) defined arrays approximating discordia between ca. 2450 Ma and ca. 3100 Ma. One clear zircon fraction, which yielded a much higher $^{207}Pb/^{235}U$ ratio than all other fractions and gave a $^{207}Pb/^{206}Pb$ upper intercept age of ca. 2900 Ma, suggested the presence of significantly older material but was also the most discordant of the analyses. Black et al. (1983a) concluded that, whilst early Archaean crust may exist at the Fyfe Hills, the effects of intense deformation and metamorphic events at ca. 3100 Ma and younger ages largely obliterated the evidence for initial crust formation.

DePaolo et al. (1982) reported Sm-Nd, Rb-Sr and Pb-Pb isotopic data on a variety of gneisses from the Soviet Sample site, provided by Soviet workers. These gneisses included the dominant quartzofeldspathic orthogneiss (charnockitic gneiss), garnet-bearing granulites, pyroxene granulite and a metaironstone. Like Black et al. (1983a) and McCulloch and Black (1984), these workers documented late Archaean disturbance of the Rb-Sr and U-Pb systems, and potentially some disturbance of the Sm-Nd whole rock system also. If we consider only charnockitic orthogneisses and the pyroxene granulites the T_{DM} of the dominant rock suites at Fyfe Hills lie in the range ca. 3200–3500 Ma, implying a significant crustal pre-history and consistent with the conclusions of Black et al. (1983a) that no ca. 3900–4000 Ma crust occurs in the Fyfe Hills, and possibly not in the entire area of the Napier Complex to the south of Amundsen Bay. This suggestion is consistent with the available zircon U-Pb data obtained from corroded zircon cores in metasedimentary paragneisses from Khmara Bay (Harley and Black, 1997; Kelly and Harley, 2005) and Mt Cronus (Harley et al., unpubl data). Inherited cores in all cases yield concordant or near-concordant ages no older than ca. 2970 Ma. If any early Archaean crust was present in this region it cannot have been available as a source of detritus for the sedimentary precursors of these paragneisses.

3.2-3.3.2. Mt Sones and Gage Ridge

Mount Sones lies in the Tula Mountains, east of Amundsen Bay (Figs. 3.2-8 and 3.2-10). The essential geology and structure of Mount Sones is typical of that observed in the Tula Mountains as a whole (Harley, 1986). Two distinctive gneiss suites or packages are present. Much of the northern and northwestern parts of Mount Sones consists of a

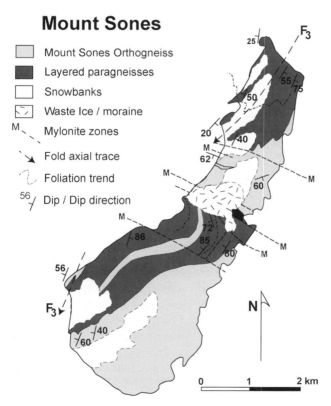

Fig. 3.2-10. Geological map of Mount Sones, modified after Harley (1986). See Fig. 3.2-8 for the location of Mount Sones within Enderby Land.

strongly deformed and compositionally diverse package of paragneisses and orthogneisses (Fig. 3.2-10), whereas the southeastern and southern sides of the mountain comprise a homogeneous, orthopyroxene-bearing, tonalitic to granodioritic orthogneiss – the Mount Sones Orthogneiss (Fig. 3.2-11(a)).

The paragneiss package (Fig. 3.2-11(b,c)) is composed of quartzo-feldspathic leuco-gneiss, pyroxene granulite, garnet quartzite, garnet-sillimanite pelitic gneiss, boudinaged ultramafic gneisses (diopsidites, bronzitites) and sapphirine-spinel-pyroxenite (Harley, unpubl.), and a rare sapphirine-cordierite-garnet-sillimanite-mesoperthite gneiss described in detail by Harley (1986). These are interleaved and layered on decimetre to metre scales, and characterised by strong SW-plunging mineral elongation lineations (Fig. 3.2-11(d)). Foliation and compositional banding in the layered gneisses and Mount Sones Orthogneiss is generally steeply SE- or NW-dipping over much of Mount Sones (Fig. 3.2-10), but SW-dipping in the hinge region of an open, kilometre-scale SW-plunging antiformal fold (F_3 of Black and James, 1983; Fig. 3.2-10) exposed on the northern and western flanks of the mountain. All gneisses and the F_3 folds in them, are cut by later ESE-trending upright mylonite zones (Fig. 3.2-11(b,c)) that may be up to 100 m wide, and which preserve amphibolite facies mineral assemblages (e.g., garnet-hornblende-plagioclase). Near the middle of Mount Sones, the layered gneiss package and Mount Sones Orthogneiss are brought into contact along a NE-trending mylonite zone that appears to be cut by the ESE-trending generation (Figs. 3.2-10 and 3.2-11(c)).

The Mount Sones Orthogneiss preserves extensive zircon U-Pb evidence for an early Archaean protolith age (Table 3.2-1; Black and James, 1983; Black et al., 1986; Harley and Black, 1997). Based on conventional multi-grain TIMS analyses of zircons in sample 78285007, a strongly HREE-depleted orthogneiss (Sheraton et al., 1987), Black and James (1983) proposed that the orthogneiss precursor may include a component as old as ca. 3700 Ma. This work was subsequently complemented by detailed SHRIMP ion microprobe analysis of zircon grain separates, with the classification of zircon type and microtexture controlled by polarised light observations (Williams et al., 1984; Black et al., 1986). These SHRIMP studies confirmed the presence of zircon overgrowths and structureless grains with ages typical of many in the Napier Complex (ca. 2950–2850 Ma and ca. 2450–2550 Ma). They also demonstrated the presence of zircon cores and moderately to weakly oscillatory-zoned zircon euhedra that produced near-concordant ages ranging between ca. 3500 and 4000 Ma. The U-Pb isotopic data and zircon features were very complex and open to interpretation in terms of the precise age significance of populations because of evidence for ancient movement of radiogenic Pb with respect to parent U and Th (Williams et al., 1984). However, Black et al. (1986) were able to infer, from the intersection with concordia of the discordia defined by the zircon U-Pb array, an age of 3927 ± 10 Ma as their best estimate of the age of the zircon cores and hence of the crustal precursor to the Mount Sones Orthogneiss. Until the discovery and dating of the Acasta Gneiss in Canada, this age gave the Mount Sones Orthogneiss sample 78285007 the status of the oldest terrestrial rock.

One of the outstanding features of the Mount Sones Orthogneiss zircon data presented by Black et al. (1986) was the large number of core analyses that were reversely discordant,

Fig. 3.2-11. Archaean geological features and relations at Mount Sones, Enderby Land: (a) View, looking SSE from its central moraine area, of the summit ridge of Mount Sones. The highest crag on the E edge of the ridge (elevation ca. 850 metres) consists of the Mount Sones Orthogneiss, which also crops out as the dark horizon/lens forming the nearer summit of the ridge. Layered paragneisses occur either side of this 100–200 metre thick horizon, and make up the exposures nearest to the tent; (b) View, looking east from near the campsite, of steeply N-dipping 10–50 metres wide mylonite and ultramylonite zones (highlighted by dashed lines and M symbols) cutting layered quartzitic, quartzofeldspathic and semi-pelitic layered paragneisses in central Mount Sones. Ridge is approximately 250 metres high; (c) Aerial view, looking west, of the eastern flank of Mount Sones, showing the extensive layered paragneiss package in contact with the Mount Sones Orthogneiss. Both are cut by mylonite zones, highlighted by dashed lines and M symbols. Cliff face in the central foreground is ca. 350 metres high; (d) SSW-plunging mineral elongation lineation in garnet-bearing quartzofeldspathic gneiss, Mount Sones. The strong lineation associated here with D_1 structures is defined by quartz rods and ribbons, and by the elongation of feldspar aggregates. The photo is taken looking onto the S_1 gneissosity plane.

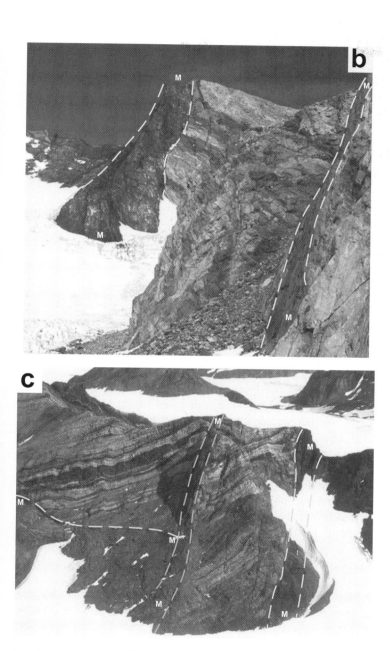

Fig. 3.2-11. (*Continued.*)

Fig. 3.2-11. (*Continued.*)

lying on a discordia that intersected the concordia curve at ca. 3930 Ma. As noted above, this feature was interpreted as resulting from ancient movement of Pb relative to its parent elements so that on the scale of an ion microprobe analysis pit (20–30 microns diameter) some zircon domains had unsupported radiogenic Pb in them (Williams et al., 1984). In the light of this, further zircon U-Pb analyses were undertaken on the Mount Sones Orthogneiss sample 78285007, another sample of orthogneiss from Mount Sones (sample 77283465), as well as of an equally ancient tonalitic orthogneiss from Gage Ridge (sample 78285013), some 10 km to the west (Harley and Black, 1997). In the case of sample 78285007, Harley and Black (1997) produced an age of 3800 (+50/−100) Ma by fitting all their new zircon analyses to a line with a lower intercept at 2950 Ma (±350 Ma). Despite the use of improved calibrational procedures their analyses also include a significant number that are reversely discordant, even though the common Pb is very low and the U contents not high (53–340 ppm). In other words, their data were similar to that reported in Black et al. (1986) and the range of U-Pb analyses must be an inherent feature of the old zircons themselves; that is, the reverse discordance is not linked to any measured chemical signal (e.g., U content, Th/U).

The Mount Sones Orthogneiss zircon datasets of Black et al. (1986) and Harley and Black (1987) have been re-examined here using ISOPLOT/EX (Ludwig, 1999). These data have been fitted by regressing analyses that are older than 3500 Ma and <5% discordant (Fig. 3.2-6(b)). A population age of 3844 ± 81 Ma is obtained based on the Black et al. (1986) data, using only the pre-3500 Ma grains, excluding the most discordant analyses, and with the lower intercept anchored at 2500 Ma. This anchored age is chosen as it approximates the age of the most ubiquitous isotopic resetting event recorded in the Napier Complex. The concordia population age is marginally older, but within error (3854 ± 90 Ma), when the lower intercept of the regression allowed to be a free parameter (1705 ± 860 Ma). By the same procedure, the Harley and Black (1997) data yields a population age of 3863 ± 53 Ma when the lower intercept is anchored at 2500 Ma. The anchored 'age' determined from all pre-3500 Ma analyses in the dataset produced by combining both studies is 3850 ± 50 Ma, identical to the free regression age of 3846 ± 48 Ma (with a highly imprecise lower intercept of 2501 ± 350 Ma) derived from the same data. The general result is that the main zircon population in the orthogneiss precursor, presumably recording the magmatic protolith age, formed at ca. 3850±50 Ma (Table 3.2-1). Older near-concordant grains some ca. 3980–4050 Ma old may be inherited, especially given the Sm-Nd systematics of sample 78285007 (Black and McCulloch, 1987), which has an average ε_{Nd} of 0 at 3840 Ma (T_{CHUR} of 3840 Ma) and hence a model age (T_{DM}) close to ca. 3960–4040 Ma (Table 3.2-1).

Limited zircon data obtained from a more layered tonalitic gneiss (sample 77283465) from the northwestern flank of Mount Sones is consistent with that described above for sample 78285007 (Fig. 3.2-6(b)), and yielded a concordia intercept age of 3773 (+13/−11) Ma in the analysis of Harley and Black (1997). Regression of this data using ISOPLOT-E/X (Ludwig, 1999) produces the same intercept age, but with a larger error (3773 ± 83 Ma). Anchoring the lower intercept at 2500 Ma following the logic applied to sample 78285007 produces a concordia upper intercept age of 3805 ± 65 Ma.

The granitic Gage Ridge Orthogneiss (sample 78285013) has been dated using SHRIMP zircon U-Pb analysis by Harley and Black (1997) and re-evaluated in the light of CL and BSE textural observations and REE analysis by Kelly and Harley (2005). The zircon U-Pb data presented by Harley and Black (1997) followed two alignments that were regressed separately, although there were no obvious differences in the morphologies or U and Th contents of the zircons placed into each alignment. The older alignment, which included reversely discordant analyses with apparent $^{206}Pb/^{238}U$ ages of >4000 Ma, yielded a concordia intercept age of 3840 (+30/−20) Ma, taken as the age of the orthogneiss protolith. Kelly and Harley (2005) distinguished elongate cores with variably preserved oscillatory zoning, traversed in some grains by networks of low luminescence healed fractures. These cores were interpreted as magmatic based on their oscillatory zoning, steep chondrite-normalised HREE patterns with Yb/Gd near 20, large negative Eu anomalies, and Th/U ratios of greater than 0.05. Undisturbed core analyses together with cores exhibiting minor to moderate textural modification (e.g., weakening or blurring of oscillations, presence of healed microfractures) define a discordant array that produces a concordia upper intercept age of 3851±62 Ma and lower intercept of 2740±110 Ma (Kelly and Harley, 2005). This same data array of 30 analytical points yields a concordia age of 3877 ± 62 Ma when a regression anchored at 2500 Ma is used, and 3889 ± 75 Ma if only the oldest grains (17 analyses at >3500 Ma) from the population are used in the anchored regression.

In detail the Gage Ridge zircon array includes four analyses (Fig. 3.2-6(b)) that are distinct in being older than ca. 3940 Ma based on both $^{206}Pb/^{238}U$ and $^{206}Pb/^{207}Pb$ ratios. The three 'oldest' and reversely discordant of these distinct analyses are all from one

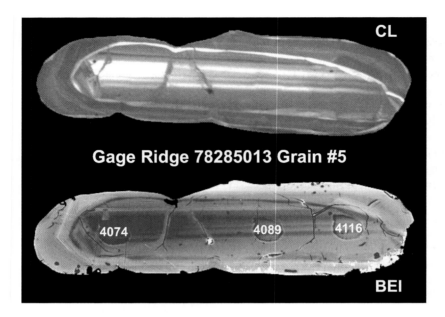

Fig. 3.2-12. Hadean zircon? The oldest zircon grain in Antarctica, Grain #5 from Gage Ridge orthogneiss sample 78285013. The averaged $^{207}Pb/^{206}Pb$ age of ca. 4090 Ma for this grain, based on the three SHRIMP U-Pb analyses (Harley and Black, 1997), when coupled with the Hf isotopic measurements of Choi et al. (2006), gives an ε_{Hf} of 8.3. This implies either that the source for the magma in which this zircon crystallised was ultra-depleted mantle, or that the U-Pb ages are too old despite their proximity to concordia.

grain (grain #5; Fig. 3.2-12), whilst a concordant zircon with an age of ca. 3940 Ma is derived from another zircon grain (#28). Exclusion of these grains from the regressed array of analyses with $^{206}Pb/^{238}U$ ratios consistent with an age >3500 Ma yields a concordia upper intercept at 3810 ± 110 Ma, compatible with the accepted ca. 3850 Ma age of the orthogneiss protolith and similar to that of the Mount Sones orthogneiss sample 78285007 (Table 3.2-1; Fig. 3.2-6(b)).

Choi et al. (2006) have analysed the Hf isotopic composition of four zircon grains from the Gage Ridge Orthogneiss, including the oldest grains #5 and #28. For a protolith age of ca. 3850 Ma the zircons yield ε_{Hf} in the range 2.5 ± 0.3 to 5.6 ± 0.4 (with $^{176}Hf/^{177}Hf$ calculated for 3850 Ma using $\lambda^{176}Lu$ of 1.865×10^{-11} yr and CHUR parameters of $^{176}Lu/^{177}Hf = 0.0332$ and $^{176}Hf/^{177}Hf = 0.282772$). These highly positive ε_{Hf} values imply juvenile sourcing of the orthogneiss protolith at ca. 3850 Ma from mantle that was highly depleted relative to CHUR, and which had been depleted from early in the Earth's history. In order to generate the high ε_{Hf} values the time integrated Lu/Hf fractionation, $f_{Lu/Hf}$ ($=[^{176}Lu/^{177}Hf]_{mantle} / [^{176}Lu/^{177}Hf]_{CHUR} - 1$) is required to be as high as 0.51, implying the presence of strongly depleted early Archaean mantle irrespective of the baseline (e.g., CHUR) parameters used (Choi et al., 2006).

Grain #5, labelled with the $^{207}Pb/^{206}Pb$ ages obtained on each analysis spot, is illustrated in Fig. 3.2-12. If this grain and grain #28 are inherited, and not related to crystallisation of the granite protolith at ca. 3850 Ma, then their ε_{Hf} would be even higher than the values given above, using the same decay constant and CHUR parameters. For an averaged $^{207}Pb/^{206}Pb$ age of ca. 4090 Ma this grain would have ε_{Hf} of 8.3, and for its $^{207}Pb/^{206}Pb$ age of ca. 3940 Ma the near-concordant grain #28 would have an initial ε_{Hf} of 7.7. These extreme ε_{Hf} values would imply the existence of an ultra-depleted early mantle source and hence the complementary existence of early, highly enriched, continental crust since the early Hadean as proposed in other studies (e.g., Bennett et al., 1993; Blichert-Toft et al., 1997; Bizzarro et al., 2003). This tantalising speculation requires further investigation through Hf isotope analysis of more of these earliest Archaean zircons from the Napier Complex, especially those that are concordant or have negligible reverse discordance.

3.2-4. CONCLUSIONS

The Archaean event histories described above are summarised on a terrane, province or area basis in Table 3.2-1. Even though the ice cap covers most of the East Antarctic Shield, limiting the total area of exposed of Archaean rocks to coastal strips and two regions of inland nunataks, it is clear from this summary that there is considerable potential for Archaean crust to form a major part of the unexplorable interior of the continent. This is particularly the case for the vast region bounded by Terre Adélie and the Miller range in the east and southern Prince Charles Mountains in the west. However, it is not likely, given the differences between the geology and age-event history of the Ruker Terrane and the Archaean preserved in the 'Mawson Block' that these represent one large craton (Table 3.2-1; Boger et al., 2006).

The oldest rocks in Antarctica, represented by early Archaean tonalites, granodiorites and granites with emplacement ages from ca. 3850 Ma to 3400 Ma, all occur in terranes that lie in the sector to the south of India in Gondwana reconstructions, bounding the Rayner Province and adjacent Prydz Belt. Although some similarities exist in the age-event chronologies recorded for each terrane, the only areas that share similar early Archaean events are the Napier Complex and the adjacent part of the Rayner Province in Kemp Land (Table 3.2-1), where an anatectic/metamorphic event at ca. 3420–3450 Ma is locally preserved. However, even in this instance there are important differences in the ages of the oldest preserved rocks: in Kemp Land these are arguably no older than ca. 3650 Ma whereas the evidence for ca. 3850 Ma orthogneiss precursors in the Napier Complex of Enderby Land is robust.

The key conclusion to emerge from this analysis is that there is no compelling reason to correlate across and between the early Archaean terranes and vestiges that now lie situated around the Rayner Province (Table 3.2-1). Based on our present state of knowledge it is likely that the Mather and Ruker terranes were always separate in the Archaean, not sharing any geological events until being reworked in either the latest Mesoproterozoic or, more certainly, in the Cambrian. The Rayner Province in Kemp Land contains some components of reworked Napier Complex, but also appears to contain Archaean crust not yet identified in the Napier Complex proper. Given this, and despite $T_{DM}(Nd)$ model ages suggesting initial extraction of crust at ca. 3700–3900 Ma, it is highly unlikely that crustal rocks of similar antiquity to the ca. 3850 Ma Mount Sones and Gage Ridge orthogneisses will be found in East Antarctica outside of the Napier Complex, where such old rocks only form a minor component of the orthogneiss suites in any case. Hence, the Napier Complex must remain the prime target of efforts in East Antarctica to find the earliest Archaean to Hadean rocks that can be used to evaluate models for Archaean tectonics, crust production and geochemical evolution of the crust-mantle system.

ACKNOWLEDGEMENTS

Discussions over the years on various aspects of Antarctic crustal evolution with numerous Antarctic petrologists and geochronologists have influenced this synthesis, and we thank all of these for their contributions to Antarctic geology. Particular thanks go to Lance Black, John Sheraton, Mark Fanning, Peter Kinny, Ian Fitzsimons, Ian Snape, Chris Carson, Steve Boger, Ed Grew, Tomokazu Hokada, Yoichi Motoyoshi, Kazuyaki Shiraishi, Joachim Jacobs, Bas Hensen and Dan Dunkley for their insights and discussions. We are indebted to Steve Boger for providing us with the photographs of Ruker Terrane geology used in Fig. 3.2-3. Microanalytical and SEM imaging work contributing to this paper (SH45, 78285013) has been supported by a NERC large grant to SLH. NMK was supported by a Royal Society of Edinburgh Fellowship for part of the work that is included in this chapter. We thank Martin Van Kranendonk for his editorial work on our behalf.

Earth's Oldest Rocks
Edited by Martin J. Van Kranendonk, R. Hugh Smithies and Vickie C. Bennett
Developments in Precambrian Geology, Vol. 15 (K.C. Condie, Series Editor)
© 2007 Published by Elsevier B.V.
DOI: 10.1016/S0166-2635(07)15033-7

Chapter 3.3

THE ITSAQ GNEISS COMPLEX OF SOUTHERN WEST GREENLAND AND THE CONSTRUCTION OF EOARCHAEAN CRUST AT CONVERGENT PLATE BOUNDARIES

ALLEN P. NUTMAN[a,b], CLARK R.L. FRIEND[c], KENJI HORIE[d,e] AND HIROSHI HIDAKA[d]

[a]*Department of Earth and Marine Sciences, Australian National University, Canberra, ACT 0200, Australia*
[b]*Beijing SHRIMP Centre, Chinese Academy of Geological Sciences, 26, Baiwanzhuang Road, Beijing, 100037, China*
[c]*45, Stanway Road, Headington, Oxford, OX3 8HU, UK*
[d]*Department of Earth and Planetary Systems Sciences, University of Hiroshima, 1-3-1 Kagamiyama, Higashi-Hiroshima 739-8526, Japan*
[e]*Department of Science and Engineering, The National Science Museum, 3-23-1, Hyakunin-cho, Shinjuku-ku, Tokyo 169-0073, Japan*

ABSTRACT

The ca. 3000 km^2 Itsaq Gneiss Complex of the Nuuk region, southern West Greenland was the first body of pre-3600 Ma crust discovered. Such ancient gneisses are now also known elsewhere, but in total form only about a millionth of the modern crust. The other 99.9999% of ancient crust was destroyed by melting and erosion over billions of years. Understanding the origin of this oldest crust is hampered by metamorphism (repeatedly) in the amphibolite or granulite facies, and most of it having been strongly deformed. The Itsaq Gneiss Complex is dominated by polyphase grey gneisses derived from several suites of tonalites, granites and subordinate quartz-diorites and ferro-gabbros that were intruded between 3870 and 3620 Ma. The hosts to these were lesser volumes of supracrustal rocks (amphibolites derived from submarine basalts with some chemical sediments and rarer felsic volcanic rocks), gabbro-anorthosite complexes and rare slivers of >3600 Ma upper mantle peridotite.

Tonalite protoliths that dominate the Itsaq Gneiss Complex resemble younger Archaean TTG (tonalite-trondhjemite-granodiorite) suites in terms of their major and trace element geochemistry. Their Sr and Nd isotopic signatures indicate that they were juvenile additions to the crust. Their bulk compositions, in comparison with other TTG suites and products from melting experiments, show that they formed by partial melting of eclogite facies hydrated mafic rocks with lesser contributions from metasomatised upper mantle. Analogous suites are now generated at convergent plate boundaries, if the subducting oceanic crust is young and thus hotter than average. In the Itsaq Gneiss Complex, younger (Neoarchaean) strong ductile deformation under amphibolite to granulite facies conditions obliterated much of the evidence of its Eoarchaean tectonic evolution. However, in the north of the Complex around the 35 km long Isua supracrustal belt, maximum metamorphic grade is lower (sub-migmatisation) and superimposed younger (Neoarchaean) deformation is less. Thus tectonic events during the $\geqslant 3600$ Ma construction of the crust can be studied. The belt contains several panels that contain amphibolites derived from submarine basalts, chemical sediments and felsic-intermediate volcanic rocks. These panels are separated by $\geqslant 3600$ Ma mylonites and then folded. Nearby smaller supracrustal belts of amphibolites derived from volcanic rocks with chemical sediments are interleaved with upper mantle peridotite and layered gabbros. It is considered that the tectonic intercalation of these unrelated rocks reflects crustal shortening, driven by compression at convergent plate boundaries. The Itsaq Gneiss Complex and all other Eoarchaean complexes, present a strong case that the oldest continental crust was built at ancient convergent plate boundaries by intrusion of tonalites into tectonically intercalated suites of supracrustal rocks (predominantly basalt), gabbros and upper mantle peridotites.

3.3-1. INTRODUCTION

Interpreting the Earth from about 3500 Ma is greatly aided by the preservation of almost undeformed and non-metamorphosed volcanic and sedimentary rocks, units of which are found in the Pilbara Craton of Western Australia and in the Barberton area of South Africa. In these exceptionally well-preserved sediments and volcanic rocks, primary textures and structures have even allowed a detailed understanding of surficial environments from 3500 Ma, including the early life record, with possibly the first microfossils and macroscopic stromatolites (see van Kranendonk, this volume). However, a major problem for interpreting the Eoarchaean (3500–4000 Ma) geological record is that it is *all* contained within gneiss complexes, where original characters of the rocks have been mostly obliterated. An example of this is the 3850–3600 Ma Itsaq Gneiss Complex of the Nuuk region, southern West Greenland (Figs. 3.3-1 and 3.3-2). The Itsaq Gneiss Complex is ideal for study of the construction of the oldest continental crust. This is because of (a) its considerable size (greater chance of sampling crustal variability), (b) good exposure, and (c) its rare areas of low deformation and alteration that are unique in the oldest terrestrial geological record (both the latter help to clarify difficult geological relationships). We compile evidence that shows that by almost 4 billion years ago, continental crust was being formed

---

Fig. 3.3-1. Summary geological map of the Itsaq Gneiss Complex, Nuuk region, West Greenland with focus on zircon age determinations on the orthogneisses. Locations of discussed $\geqslant 3850$ Ma samples G88/66 and G01/36 are shown.

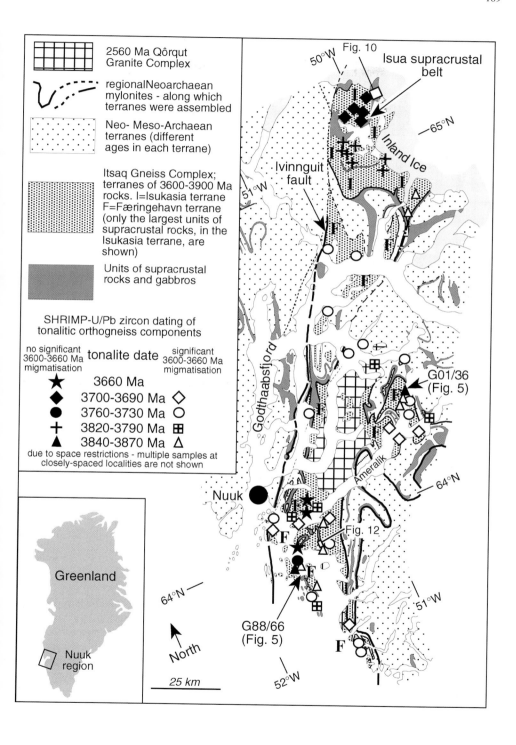

2560 Ma Qôrqut Granite Complex

regional Neoarchaean mylonites - along which terranes were assembled

Neo- Meso-Archaean terranes (different ages in each terrane)

Itsaq Gneiss Complex; terranes of 3600-3900 Ma rocks. I=Isukasia terrane F=Færingehavn terrane (only the largest units of supracrustal rocks, in the Isukasia terrane, are shown)

Units of supracrustal rocks and gabbros

SHRIMP-U/Pb zircon dating of tonalitic orthogneiss components

no significant 3600-3660 Ma migmatisation	tonalite date	significant 3600-3660 Ma migmatisation
★	3660 Ma	
◆	3700-3690 Ma	◇
●	3760-3730 Ma	○
+	3820-3790 Ma	⊞
▲	3840-3870 Ma	△

due to space restrictions - multiple samples at closely-spaced localities are not shown

Greenland

Nuuk region

64°N

North

25 km

Isua supracrustal belt

Inland Ice

65°N

Ivinnguit fault

Godthaabsfjord

Nuuk

G01/36 (Fig. 5)

Ameralik

64°N

Fig. 12

F

F

F

51°W

G88/66 (Fig. 5)

F

52°W

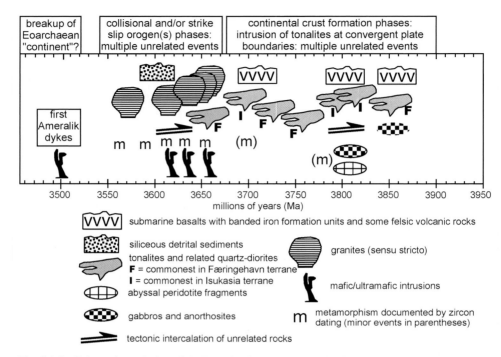

Fig. 3.3-2. Schematic evolution of the Itsaq Gneiss Complex. Each pictogram represents a generation of rocks whose age is known from zircon dating of several samples. The figure is entirely schematic and is not meant to portray a geological cross-section or the geographic distribution of units.

at convergent plate boundaries. This suggests that by that time horizontal lithospheric motions already dominated over impacts as the process controlling crustal development.

3.3-2. METHODS

The deformation of >3600 Ma rocks is polyphase, and commonly spans over a billion years. Thus in Fig. 3.3-3(a), Itsaq Gneiss Complex Eoarchaean orthogneisses are cut by

Fig. 3.3-3. (*Next page.*) (a) Polyphase orthogneisses that dominate the oldest rock record (hammer for scale). These have early tonalite and younger granite-pegmatite components in varying proportions. The granite-pegmatites can be either intrusive sheets or formed *in situ* by arrested partial melting of host tonalites during high-grade metamorphism. The Ameralik dykes were intruded at probably 3500 Ma as dolerite dykes into already complex gneisses. Later deformation under amphibolite facies conditions converted them into a tabular amphibolite bodies, with only locally preserved discordances at their margins. Thus these gneisses usually show evidence of several phases of unrelated deformation.

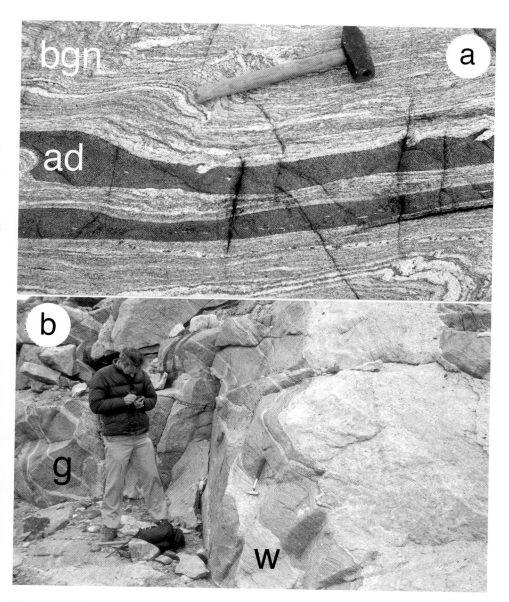

Fig. 3.3-3. (*Continued.*) (b) 3690 Ma weakly deformed tonalite (g) cut by 3650 Ma nondeformed leucogranite sheets (w) in the northern part of the Itsaq Gneiss Complex (Isukasia terrane). Such good preservation of Eoarchaean plutonic relationships is rare, and is mostly restricted on to the northern part of the Itsaq Gneiss Complex. Note that the tonalites show no evidence of *in situ* partial melting.

Mesoarchaean dolerite dykes (Ameralik dykes – McGregor, 1973), now converted into still locally discordant tabular amphibolite bodies. This demonstrates superimposed (Neoarchaean) ductile deformation under amphibolite facies conditions (McGregor, 1973). The margins of the dykes cut migmatitic structures in the orthogneiss, indicating yet earlier (Eoarchaean) high-grade metamorphism and ductile deformation.

It is fortuitous that strain during both Eo- and Neoarchaean ductile deformation events in the Itsaq Gneiss Complex was heterogeneous, and hence there are small domains that escaped severe deformation (Fig. 3.3-3(b)). It is upon these small domains that studies are best focussed. This has been the approach by several groups recently, and has yielded the most pristine samples of >3600 Ma lithologies, and thereby most concise interpretations of granitic, mafic, ultramafic and sedimentary rocks by geochemical methods (e.g., Nutman et al., 1999; Friend et al., 2002a; Polat et al., 2002; Frei and Polat, 2006).

The largest and most publicised of these low deformation areas occur in orthogneisses near the *Isua supracrustal belt* (Figs. 3.3-1 and 3.3-4) where an intrusive relationship between successive generations of tonalites and granites is displayed (Fig. 3.3-3(b); Bridgwater and McGregor, 1974; Nutman and Bridgwater, 1986; Nutman et al., 1999; Crowley et al., 2002; Crowley, 2003). Smaller low strain domains occur within the Isua supracrustal belt, such that there is very rare preservation of sedimentary and volcanic structures (Nutman et al., 1984a; Komiya et al., 1999). In the southern part of the complex there are also small domains of low total strain. These show that the banded gneisses that dominate the Complex consist of older tonalitic components in which *in situ* partial melt domains developed, and which were also veined by pegmatite. Upon strong deformation, this gives rise to the complex banded gneisses in whose components are no longer separable for individual study (Nutman et al., 2000).

Many Eoarchaean gneisses of different age are superficially similar in appearance (Nutman et al., 2000). Thus precise and accurate chronology of them is an essential starting point in order to chronicle early Earth events read from their geochemistry. This has been made possible by zircon dating using the coupled ^{235}U-^{207}Pb and ^{238}U-^{206}Pb radioactive decay systems (shortened here to U/Pb). Sensitive, high mass-resolution ion microprobes date domains typically 20 μm wide by 1 μm deep within single zircons. Guided by cathodoluminescence imaging (Fig. 3.3-5(c,d)), individual oscillatory-zoned igneous growth and recrystallised domains can be dated this way, to obtain accurate, complex geological histories for zircon populations from single rocks (Fig. 3.3-5(e,f)). U/Pb zircon analysis by thermal ionisation mass spectrometry can now date single fragments of large grains, and also provides valuable information on the early Earth (e.g., Crowley et al., 2002). There are now numerous accurate U/Pb zircon dates on the oldest rocks and minerals, with errors of only a few million years (only $\leqslant 0.1\%$ of the >3500 Ma ages). These have allowed distinction of Eoarchaean geological events at a $\leqslant 10$ million-year resolution, and have charted a 300 million year development for the Itsaq Gneiss Complex (Fig. 3.3-2; Nutman et al., 2001; Nutman, 2006).

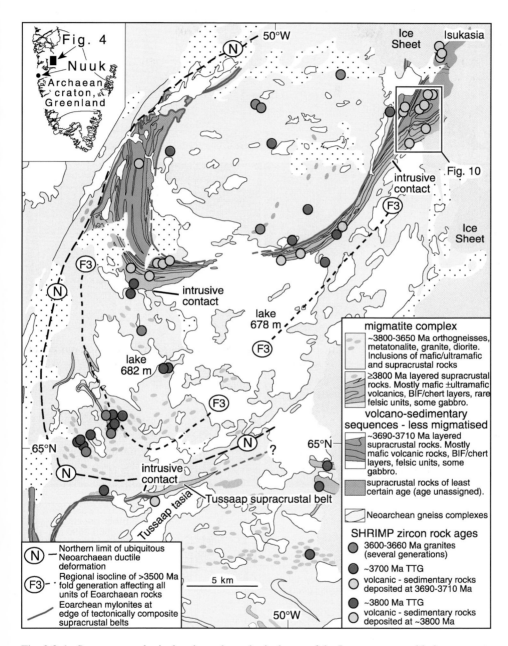

Fig. 3.3-4. Summary geological and geochronological map of the Isua supracrustal belt.

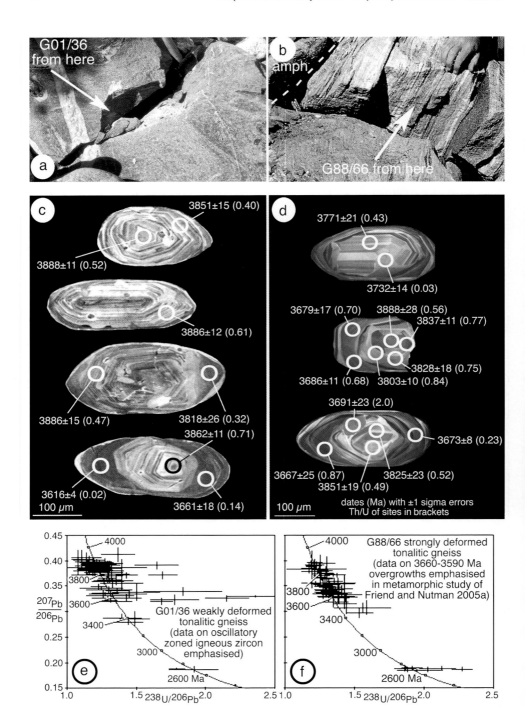

(c) and (d): dates (Ma) with ±1 sigma errors, Th/U of sites in brackets

(e) G01/36 weakly deformed tonalitic gneiss (data on oscillatory zoned igneous zircon emphasised)

(f) G88/66 strongly deformed tonalitic gneiss (data on 3660-3590 Ma overgrowths emphasised in metamorphic study of Friend and Nutman 2005a)

Fig. 3.3-5. (*Previous page.*) Oldest tonalites of the Itsaq Gneiss Complex. (a) Moderately deformed ca. 3870 Ma tonalite G01/36 from inner Ameralik (fjord). Sledgehammer head (top right) for scale. This displays *in situ* partial melt domains that infiltrate the tonalite in a too intimate fashion to physically separate them. This results in the speckled appearance of the outcrop. Thus it is impossible to obtain a tonalite sample completely free of these younger domains. (b) Strongly deformed ca. 3870 Ma tonalite G88/66 from Akilia island adjacent to amphibolites (amph) (Nutman et al., 2000, 2002b). Author's hand for scale (top right). In this rock ca. 3650 Ma granitic neosome and veins in a >3800 Ma tonalite protolith has been strongly deformed, masking evidence of their origin. (c) Typical zircons from G01/36. Dates in millions of years (Ma) are given with 1 sigma uncertainty and are followed by the analytical site Th/U ratio in parentheses. Note the kernels and whole grains of "igneous" oscillatory zoned zircons (highest Th/U) with recrystallisation domains and overgrowths of lower Th/U zircon developed in events from ca. 3810 Ma to as young as 2720 Ma. (d) Typical zircons from G88/66. Despite the different appearance of the rock, the zircons are very similar, showing that high temperature Archaean geological history was similar, apart from different degrees of deformation. (e and f) Tera-Wasserburg $^{238}U/^{206}Pb$ versus $^{207}Pb/^{206}Pb$ plot for SHRIMP analyses of G01/36 and G88/66 zircons respectively. Analytical errors are portrayed at the 1-sigma level.

3.3-3. ITSAQ GNEISS COMPLEX

The Itsaq Gneiss Complex is a term used to embrace all Eoarchaean gneisses in West Greenland (Nutman et al., 2004). The most studied and largest parts of the Complex are the *Færingehavn* and *Isukasia* terranes in the Nuuk region (Fig. 3.3-1). Two other smaller terranes occur north of the Nuuk region (Rosing et al., 2001; Nutman et al., 2004).

Nutman et al. (1993b) and Friend and Nutman (2005b) proposed that the Itsaq Gneiss Complex was constructed out of tonalites between 3850–3660 Ma, and was then affected by superimposed orogenic events between ca. 3650 and 3550 Ma, with high grade metamorphism, production of crustally-derived granites (*sensu stricto*) and polyphase deformation. These superimposed 3650–3550 Ma events were interpreted to reflect collisional and/or strike slip orogeny that affected crust formed over the previous 200 million years. This article focuses on the magmatic and tectonic processes that *built* Eoarchaean continental crust prior to the collisional and/or strike slip orogenic events.

The Eoarchaean terranes are mylonite-bounded tectonic slivers within a collage of younger terranes (each terrane has its own rock ages and internal evolution). These terranes were assembled during several Neoarchaean tectonothermal events (Friend et al., 1988; Friend and Nutman, 2005a). Thus the Eoarchaean rocks are allochthons found within a Neoarchaean orogen. Because unrelated terranes have been tectonically stacked on top of each other and some show early transient high-pressure metamorphism, terrane assembly is interpreted to have involved Neoarchaean collisional orogeny (Nutman et al., 1989; McGregor et al., 1991; Friend and Nutman, 2005a; Nutman and Friend, 2007).

3.3-3.1. Lithologies

The Itsaq Gneiss Complex contains >95% quartzo-feldspathic rocks of intrusive origin, now mostly strongly deformed into orthogneisses (Fig. 3.3-3(a)). The rare areas of relatively little deformation show that these usually formed from older tonalite and younger granite components (Fig. 3.3-3(b)). Tonalites are siliceous (typically 65–70 wt%) potassium-poor intrusive rocks largely generated by melting of hydrated mafic crust after transformation into garnet amphibolite or eclogite (summary by Martin et al., 2005). The granites were produced by partial-melting of predominantly tonalite crust during superimposed orogeny (Baadsgaard et al., 1986; Bennett et al., 1993; Friend and Nutman, 2005b). Volcanic and sedimentary rocks form <5% of the Complex. The volcanic and sedimentary rocks are scattered through the banded gneisses as enclaves and tectonic slivers, and range in size from the 35 km long Isua supracrustal belt (e.g., Moorbath et al., 1973; Allaart, 1976; Nutman et al., 1984a, 1997, 2002a; Komiya et al., 1999) of the Itsaq Gneiss Complex (Figs. 3.3-1 and 3.3-4), down to 1-metre pods.

The trace element chemistry of many Itsaq Gneiss Complex amphibolites indicates that they were derived from arc-related basalts (Polat and Hofmann, 2003; Jenner et al., 2006), rather than being plume related or being ancient mid ocean ridge basalt. Diagnostic signatures are depletions in Ti and Nb, and (variable) enrichments in Pb, Sr, Ba, Rb, and LREE. Careful sampling shows that such chemical signatures are original to the volcanic protoliths, rather than solely caused by metasomatism during later high-grade metamorphism and deformation (Polat and Hofmann, 2003). Gabbros and ultramafic rocks occur in small amounts. The ultramafic rocks are diverse in origin and chemistry (Fig. 3.3-6) and

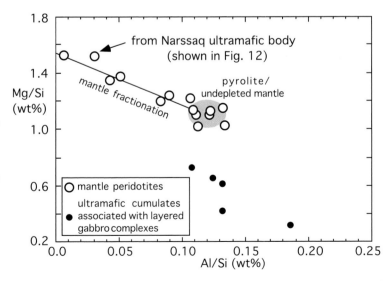

Fig. 3.3-6. Al/Si versus Mg/Si (wt%) discriminant plot for ultramafic rocks of the Itsaq Gneiss Complex. Data from Friend et al. (2002), Nutman et al. (2002b) and Nutman (unpublished).

are derived from upper mantle peridotites (e.g., Friend et al., 2002a), perhaps komatiites (McGregor and Mason, 1977) and cumulates associated with gabbros (Nutman et al., 1996; Friend et al., 2002a). Siliceous rocks derived from chemical sediments form a very small part of the Complex, and occur interlayered with amphibolites. Despite strong deformation, metamorphism and associated metasomatism, some of these display subtle trace element (rare earth element + yttrium) signatures indicating deposition from seawater (Bohlar et al., 2004; Frei and Polat, 2006). Felsic-intermediate volcanic rocks are even less common, but have been recognised in the Isua supracrustal belt (Allaart, 1976; Nutman et al., 1997a).

3.3-3.2. Isukasia Terrane

The Isukasia terrane occurs in the northern part of the Nuuk region and contains the Isua supracrustal belt (Figs. 3.3-1 and 3.3-4). At least two thirds of all international literature on Eoarchaean rocks focuses on the northern Itsaq Gneiss Complex, with most of that having been written about the Isua supracrustal belt. The southern edge of the Isukasia terrane is more deformed, and consists of banded gneisses resembling those of the Færingehavn terrane (Bridgwater and McGregor, 1974; Nutman et al., 1996). In the northern parts of the Isukasia terrane (north of line "N" in Fig. 3.3-4), Neoarchaean deformation is much less, and only weakly deformed ca. 3500 Ma Ameralik dykes cut variably deformed Itsaq Gneiss Complex lithologies (Bridgwater and McGregor, 1974; Allaart, 1976). This provides a unique opportunity to study Eoarchaean tectonics. In places the Eoarchaean rocks are almost undeformed but elsewhere, such as at the northern edge of the Isua supracrustal belt (Fig. 3.3-4), there are Eoarchaean mylonites (Nutman, 1984; Nutman et al., 1997a, 2002a; Crowley et al., 2002). The Isukasia terrane low strain domains show that in contrast to the Færingehavn terrane, *in situ* melting was very limited, and also that there is no evidence for Eoarchaean granulite facies metamorphism (Nutman et al., 1996). This lack of *in situ* migmatisation increases even more the significance of the Isukasia terrane in providing concise information on the petrogenesis of the world's oldest rocks, and hence insights into evolution of the early Earth. Geochronology of these better preserved-rocks shows that the Isukasia terrane tonalites are mostly 3810–3795 Ma and 3700–3690 Ma old, and that between 3650 and 3620 Ma they are cut by several generations of granite and pegmatite sheets (Nutman et al., 1996, 2000, 2002a; Crowley et al., 2002; Crowley, 2003). Zircon geochronology also shows that the Isua supracrustal belt comprises the remains of sequences with different ages, with volcanic rocks of ca. 3800 and 3710 Ma being present (Nutman et al., 1996, 1997a, 2002a; Crowley, 2003; Kamber et al., 2005a).

3.3-3.3. Isua Supracrustal Belt

Demonstration that the Isua supracrustal belt (*Isua greenstone belt* in some recent publications by other workers) is ⩾3700 Ma old (Moorbath et al., 1973) was a landmark in the understanding of the early Earth. This is because the Isua banded iron formations and associated metavolcanic amphibolites demonstrated that by 3700 Ma, Earth already had a hydrosphere and that by then successions were being deposited that were similar to those formed later throughout the Archaean.

Despite the Isua supracrustal belt escaping much deformation in the Neoarchaean, most of it is strongly deformed due to Eoarchaean deformation (Nutman et al., 1984a, 2002a; Myers, 2001). Thus, in most cases, primary volcanic and sedimentary structures were obliterated, and outcrop-scale compositional layering is of mostly transposed tectonic origin. It is only in rare augen of total low strain that volcanic, and even more rarely sedimentary, structures are preserved. Examples of these are illustrated in Fig. 3.3-7. The first unequivocal pillow structures in Isua supracrustal belt amphibolites were recognised at the start of the 1990s (Komiya et al., 1999; Fig. 3.3-7(a) at locality 1 on Fig. 3.3-10). Other clear examples of relict pillow structure have since been found (e.g., Solvang, 1999; Nutman et al., 2002a). These finds are important, because they demonstrate that most Isua amphibolites were derived from subaqueous volcanic rocks. This contrasts with previous ideas held in the 1970s and 1980s that amphibolites derived from altered gabbros were also an important component in the belt (e.g., Nutman et al., 1984a).

Rocks of chemical sedimentary origin such as banded iron formation (Moorbath et al., 1973; Dymek and Klien, 1988) have quartz and either calc-silicate or magnetite layering. This is mostly a transposed layering, and only locally does it represent original (albeit still deformed) sedimentary layering (Fig. 3.3-7(b) – locality 2 on Fig. 3.3-10). Felsic schist units in the belt have always elicited much debate as to their origin, with felsic volcanic protoliths (Allaart, 1976; Nutman et al. 1984a, 1997a), altered sheared tonalite sheets (Rosing et al., 1996) and radically metasomatised mafic rocks (Myers, 2001) all proposed. In the eastern part of the belt, two felsic schist units occur. The northern one has yielded several zircon dates of ca. 3710 Ma (Nutman et al., 1997a, 2002a). The southern unit crops out throughout the length of the belt, and has in several places yielded zircon dates of ca. 3800 Ma (e.g., Baadsgaard et al., 1984; Compston et al., 1986a; Nutman et al., 2002a; this paper). In the northern ca. 3710 Ma unit there are rare vestiges of graded, felsic volcanic breccia (Nutman et al., 1997a). In the southern unit there is fine-scale layering that predates the earliest folds. An example of this relict volcano-sedimentary structure is shown as Fig. 3.3-7(c) (locality 3 on Fig. 3.3-10). In other lithological units, there are structures over which debate continues as to whether they are of deformed sedimentary or entirely tectonic origin (e.g., Nutman et al., 1984a; Fedo, 2000). Typical of these is a "round pebble conglomerate" unit in a chemical sedimentary rock unit in the east of the belt (Fig. 3.3-7(d) – locality 4 on Fig. 3.3-10).

3.3-3.4. Færingehavn Terrane

The Færingehavn terrane was where >3600 Ma rocks were first identified on Earth (Black et al., 1971; McGregor, 1973). This terrane occurs in the central and southern part of the Nuuk region (Fig. 3.3-1). Overall its rocks are much more modified by strong deformation and high-grade metamorphism than the northern parts of the Isukasia terrane. This makes it harder to obtain unequivocal interpretations of its rocks. Thus it is dominated by banded, polyphase orthogneisses with tonalitic and less voluminous granitic components.

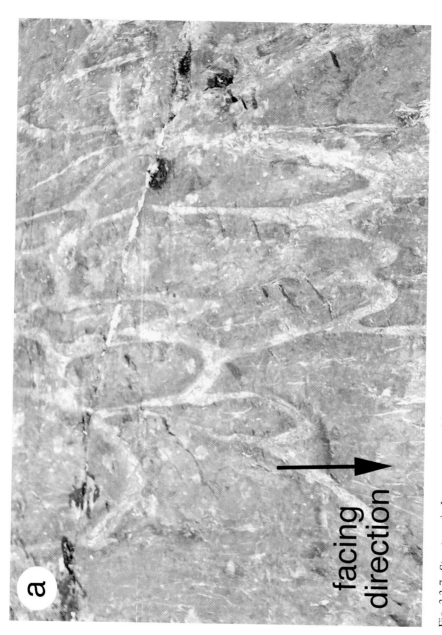

Fig. 3.3-7. Structures in Isua supracrustal belt rocks. (a) Undisputed relict pillow lava structure in amphibolites. The shape of the pillows is well enough preserved to allow determination of the top direction (arrow to foreground). Field of view is ca. 1 m.

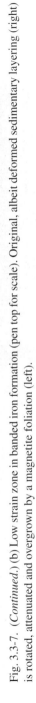

Fig. 3.3-7. (*Continued.*) (b) Low strain zone in banded iron formation (pen top for scale). Original, albeit deformed sedimentary layering (right) is rotated, attenuated and overgrown by a magnetite foliation (left).

Fig. 3.3-7. (*Continued.*) (c) Relict volcano-sedimentary layering in ca. 3800 Ma metavolcanic rocks (pen for scale). Very fine-scale depositional layering has been overprinted by coarse recrystallisation domains, very early in the history of these rocks.

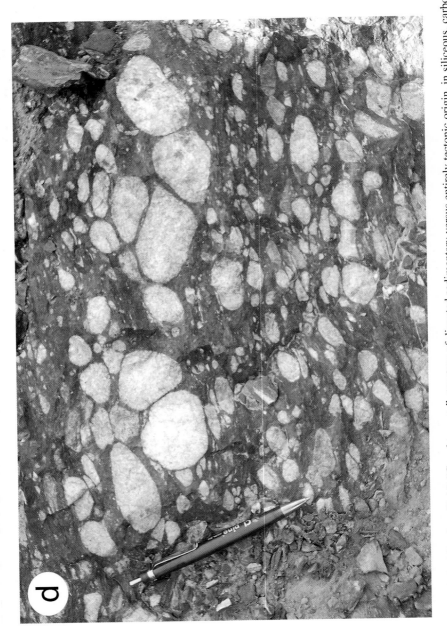

Fig. 3.3-7. (*Continued.*) (d) "Conglomerate" structure of disputed sedimentary versus entirely tectonic origin, in siliceous, carbonate-bearing rocks associated with banded iron formation (pen for scale).

Whitehouse et al. (1999) identified multiple generations of igneous zircons in samples of banded gneisses from the Færingehavn terrane. They interpreted the youngest igneous zircons as the igneous age of single protoliths, with all older zircons being xenocrysts. Nutman et al. (2000, 2002b, 2004) pointed out that as such rocks are actually complex banded gneisses, an equally permissible, if not more likely, interpretation is that several ages of igneous material are present in such rocks. Such diversity of opinion shows that working on single-component rocks from low strain zones rather than banded gneisses is important to avoid ambiguity in the interpretation of (expensive) geochronological data. By focussing geochronological and geochemical studies on these low strain domains, Nutman et al. (1993b, 1996, 2000, 2001, 2007) and Friend and Nutman (2005b) proposed that tonalites incorporated into the Færingehavn terrane are ca. 3850, 3800, 3760–3730, 3700 and 3660 Ma old. In rare domains of lower strain in the Færingehavn terrane, it is seen that the granitic components are derived from *in situ* melting of the tonalites and from granitic intrusions these formed between 3660 and 3600 Ma. These latter events have been equated with petrographic evidence for Eoarchaean granulite facies metamorphism in the Færingehavn terrane (McGregor and Mason, 1977; Griffin et al., 1980). Units of Fe-rich augen granites, monzonites and ferro-gabbros are a distinct component of the southern part of the Færingehavn terrane (McGregor, 1973). These rocks are the product of hybridisation of mantle and deep crustal magmas, and resemble A-type/within-plate-granites with high Nb, Zr, TiO_2 and P_2O_5 (Nutman et al., 1984b; 1996). These rocks have yielded zircon ages of 3640–3630 Ma (Baadsgaard, 1973; Nutman et al., 2000), and are the oldest-known rocks of this type known.

The Færingehavn terrane orthogneisses contain lenses of amphibolites, ultramafic and siliceous rocks, named the *Akilia association* by McGregor and Mason (1977). Although it is agreed that these largely represent enclaves of mafic volcanic rocks, gabbros and chemical sediments, there is lack of agreement concerning their age, because the original relationships with the surrounding orthogneisses have generally been obliterated by migmatisation and high strain (Nutman et al., 2001, 2002b). Nutman et al. (1996, 1997b, 2000) and Friend and Nutman (2005b) presented evidence that the Akilia association represents the remains of several supracrustal sequences, ranging from ⩾3850 Ma to 3650–3600 Ma. Other workers have considered there is not yet sound evidence for any of the Akilia association being ⩾3850 Ma (Whitehouse et al., 1999; Whitehouse and Kamber, 2005). Furthermore, debate continues over which Akilia association siliceous rocks are sedimentary in origin (Nutman et al., 1996; Friend et al., 2002b; Dauphas et al., 2004, versus Fedo and Whitehouse, 2002), and whether, despite having suffered granulite facies, they preserve evidence for Eoarchaean life (Mojzsis et al., 1996, versus Lepland et al., 2005; Nutman and Friend, 2006).

3.3-4. BUILDING OF ITSAQ GNEISS COMPLEX CRUST OUT OF TONALITES

The majority of Itsaq Gneiss Complex tonalites have 65–70 wt% silica, high alumina (mostly >15 wt%) and low MgO (mostly <1 wt%; Fig. 3.3-8). These are typical

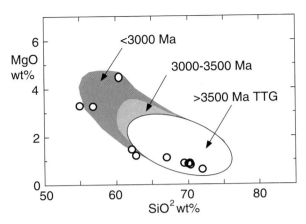

Fig. 3.3-8. SiO_2 versus MgO (wt%) for ca. 3800 Ma Itsaq Gneiss Complex tonalites from south of the Isua supracrustal belt (Nutman et al., 1999, supplemented by unpublished data). All these are homogeneous single-component samples.

high-alumina tonalites. Experimental petrology indicates that such bulk compositions are generated by partial melting of mafic rocks at high pressure in the eclogite (or high pressure granulite) stability field (summaries by Martin et al., 2005; Moyen and Stevens, 2006), with garnet + rutile as restite phases. However, coeval with the Itsaq Gneiss Complex tonalites are volumetrically minor quartz-diorites (Nutman et al., 1996, 1999) which have lower silica (55 to 60 wt%) and higher MgO (3 to 4 wt%; Fig. 3.3-8), with #Mg values approaching 50. As pointed out by Smithies and Champion (2000), the highly magnesian character of such rocks means that they cannot have been derived by melting of solely amphibolites, but must have involved some melting of mantle rocks as well. Thus it appears that the Itsaq Gneiss Complex was constructed largely by the melting of (hydrated) basaltic rocks at high pressure, with some contribution from the melting of metasomatised mantle rocks as well. In the dominant tonalites, trace element signatures such as marked depletion of the heavy over the light rare earth elements (due to garnet in the restite) and depletion of Nb and Ti (due to rutile in the restite) are also in accord with their origin by melting of high-pressure basaltic rocks (see Martin et al. (2005) for summary). Experimental petrology shows that high temperature (>900 °C) melting formed the tonalites (see Martin et al. (2005) and Moyen and Stevens (2006) for summaries).

The zircon solubility in melt relationships (Fig. 3.3-9) obtained by experimental petrology (Watson and Harrison, 1983), shows that at >900 °C, tonalite magmas with typical 120–140 ppm Zr abundances are grossly zircon undersaturated (they could carry >400 ppm Zr without starting to crystallise zircon – Fig. 3.3-9). The implication of this is that any zircons inherited from country rocks would be rapidly dissolved (Watson, 1996). Hence oscillatory-zoned igneous zircon in tonalites can be regarded as having grown out of the tonalite magma, after it had already partially crystallised and cooled substantially (Mojzsis and Harrison, 2002; Nutman, 2006). This is in keeping with the observation

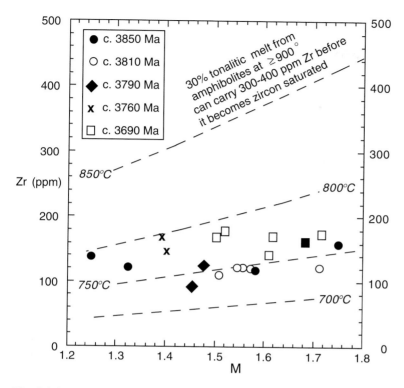

Fig. 3.3-9. Zr – versus melt composition relationships (after experimental results of Watson and Harrison (1983)). Melt composition is expressed by M = (Na + K + 2Ca/AlxSi) in cation fractions. Isotherms trace the Zr abundance at which a melt becomes zircon saturated with change of melt composition. Note that all Itsaq Gneiss Complex tonalites are grossly zircon undersaturated with respect to their likely temperature of emplacement into the crust of >900 °C.

that homogeneous tonalite samples from the Itsaq Gneiss Complex have singular ages of oscillatory-zoned "igneous" zircon (Nutman et al., 1993b, 1996, 2000). It is only when zircons are extracted from polyphase, banded gneisses (e.g., like those in Fig. 3.3-3(a)) that multiple age generations of igneous zircons are encountered in single samples (see discussion in Nutman et al. (2000, 2002b)).

Tonalites in the Itsaq Gneiss Complex with ages of ca. 3800 and 3700 Ma have been documented by several workers (e.g., Nutman et al., 1996, 2000; Crowley et al., 2002; Crowley, 2003). Tonalites and quartz-diorites in the vicinity of Amiitsoq in Ameralik fjord (McGregor, 1973) are the youngest in the Complex with zircon ages of ca. 3660 Ma (Friend and Nutman, 2005b).

However, there occur some tonalitic gneisses in different strain states that contain 3850 Ma zircons. In low strain areas these are fairly homogeneous rocks with some nebulous patches/segregations (Fig. 3.3-5(a)) in which most oscillatory zoned zircon formed

Table 3.3-1. SHRIMP U/Pb zircon analyses

Labels	Grain type	U (ppm)	Th (ppm)	Th/U	Comm. 206Pb%	^{238}U/^{206}Pb ratio	^{207}Pb/^{206}Pb ratio	^{207}Pb/^{206}Pb date (Ma)	%disc
G01/36 palaeosome-dominated part of migmatite (64°18.21'N 50°30.15'W)									
1.1	e,osc,p	181	75	0.42	0.075	1.412 ± 0.046	0.3807 ± 0.0031	3834 ± 12	−10
2.1	e,osc,p	215	147	0.69	0.067	1.322 ± 0.049	0.3853 ± 0.0026	3853 ± 10	−6
3.1	m,osc,p	135	51	0.38	0.124	1.287 ± 0.042	0.3709 ± 0.0019	3795 ± 8	−2
3.2	m,osc,p	165	74	0.45	0.103	1.361 ± 0.047	0.3427 ± 0.0037	3675 ± 17	−3
3.3	m,osc,p	133	41	0.30	0.038	1.432 ± 0.048	0.3456 ± 0.0043	3688 ± 19	−7
3.4	e,osc,p	340	236	0.69	0.020	1.291 ± 0.045	0.3794 ± 0.0021	3829 ± 8	−3
3.5	e,osc,p	312	150	0.48	0.056	1.267 ± 0.051	0.3778 ± 0.0014	3823 ± 6	−2
4.1	m,osc,p	160	77	0.48	0.005	1.248 ± 0.043	0.3874 ± 0.0023	3861 ± 9	−2
5.1	m,osc,p	121	49	0.41	0.611	1.117 ± 0.059	0.3779 ± 0.0073	3823 ± 29	8
5.2	e,osc,p	131	47	0.36	0.035	1.356 ± 0.046	0.3828 ± 0.0039	3843 ± 16	−7
6.1	m,osc,bip	176	91	0.52	0.039	1.273 ± 0.051	0.3672 ± 0.0028	3780 ± 11	−1
6.2	e,osc/rex,bip	164	73	0.44	0.023	1.381 ± 0.058	0.3353 ± 0.0047	3642 ± 22	−4
6.3	e,osc,bip	201	134	0.67	0.029	1.236 ± 0.051	0.3835 ± 0.0017	3846 ± 7	−1
7.1	e,osc,p	169	75	0.44	0.056	1.269 ± 0.041	0.3789 ± 0.0020	3827 ± 8	−2
8.1	e,osc,p	152	62	0.41	0.004	1.350 ± 0.054	0.3768 ± 0.0047	3819 ± 19	−6
9.1	e,osc,p	190	103	0.54	0.026	1.229 ± 0.047	0.3827 ± 0.0026	3842 ± 10	0
10.1	e,osc,p	208	91	0.44	0.001	1.238 ± 0.041	0.3854 ± 0.0027	3853 ± 11	−1
11.1	e,osc/rex,p,fr	207	87	0.42	0.001	1.272 ± 0.041	0.3642 ± 0.0018	3768 ± 7	−1
11.2	m,osc,p,fr	178	86	0.48	0.082	1.235 ± 0.034	0.3834 ± 0.0028	3845 ± 11	−1
11.3	e,osc/rex,p,fr	197	86	0.44	0.019	1.289 ± 0.042	0.3558 ± 0.0014	3732 ± 6	−1
11.4	e,osc,p,fr	222	102	0.46	0.027	1.232 ± 0.033	0.3842 ± 0.0024	3848 ± 10	0
12.1	e,osc,p	180	95	0.53	0.034	1.317 ± 0.039	0.3805 ± 0.0015	3834 ± 6	−5
13.1	e,osc,p	158	74	0.47	0.039	1.385 ± 0.042	0.3509 ± 0.0030	3711 ± 13	−6
13.2	m,osc,p	168	89	0.53	0.005	1.276 ± 0.086	0.3880 ± 0.0025	3863 ± 10	−3
13.3	e,osc,p	187	95	0.51	0.048	1.227 ± 0.038	0.3848 ± 0.0029	3851 ± 11	0
14.1	e,osc,p/bip	177	99	0.56	0.059	1.310 ± 0.038	0.3842 ± 0.0026	3848 ± 10	−5

Table 3.3-1. (Continued)

Labels	Grain type	U (ppm)	Th (ppm)	Th/U	Comm. $^{206}Pb\%$	$^{238}U/^{206}Pb$ ratio	$^{207}Pb/^{206}Pb$ ratio	$^{207}Pb/^{206}Pb$ date (Ma)	%disc
15.1	m,osc,p/bip	144	62	0.43	0.001	1.484 ± 0.035	0.3564 ± 0.0036	3735 ± 15	−11
15.2	m,osc,p/bip	272	202	0.74	0.067	1.204 ± 0.040	0.3585 ± 0.0016	3744 ± 7	4
15.3	e,osc,p/bip	406	272	0.67	0.007	1.378 ± 0.056	0.3454 ± 0.0036	3687 ± 16	−5
15.4	e,osc,p/bip	286	167	0.59	0.058	1.371 ± 0.074	0.3490 ± 0.0044	3703 ± 19	−5
16.1	r,h,ov	186	36	0.19	0.096	1.413 ± 0.039	0.3305 ± 0.0034	3619 ± 16	−5
16.2	r,h,ov	202	40	0.20	0.209	1.092 ± 0.078	0.3150 ± 0.0015	3545 ± 7	18
G05/15 felsic volcano-sedimentary rock (65°09.73′N 49°48.90′W)									
1.1	e,osc,p	159	129	0.81	0.154	1.257 ± 0.050	0.3700 ± 0.0028	3791 ± 11	0
2.1	m,osc,p	173	198	1.14	0.101	1.215 ± 0.048	0.3703 ± 0.0023	3793 ± 10	2
2.2	e,osc,p	225	201	0.89	0.126	1.187 ± 0.033	0.3692 ± 0.0018	3788 ± 8	4
3.1	m,osc,p,fr	118	102	0.87	0.216	1.284 ± 0.075	0.3684 ± 0.0044	3785 ± 18	−2
3.2	e,osc,p,fr	208	222	1.07	0.096	1.232 ± 0.036	0.3659 ± 0.0031	3775 ± 13	2
4.1	e,osc,p,fr	141	70	0.50	0.093	1.283 ± 0.043	0.3699 ± 0.0021	3791 ± 9	−2
5.1	e,osc,p,fr	77	35	0.46	0.109	1.122 ± 0.240	0.3610 ± 0.0247	3754 ± 108	9
5.2	e,osc,p,fr	98	54	0.55	0.481	1.273 ± 0.087	0.3601 ± 0.0039	3750 ± 16	0
6.1	m,osc,p	257	140	0.54	0.059	1.186 ± 0.030	0.3706 ± 0.0036	3794 ± 15	4
7.1	e,osc,p	466	280	0.60	0.046	1.146 ± 0.031	0.3755 ± 0.0021	3814 ± 8	6
8.1	e,osc,p	228	180	0.79	0.151	1.283 ± 0.044	0.3715 ± 0.0036	3798 ± 15	−2
9.1	e,osc,p	239	183	0.77	0.270	1.257 ± 0.043	0.3582 ± 0.0027	3742 ± 11	1
10.1	m,osc/h,bip	125	138	1.10	0.216	1.234 ± 0.051	0.3700 ± 0.0040	3791 ± 16	1
11.1	m,osc,bip	143	95	0.66	0.172	1.247 ± 0.046	0.3699 ± 0.0029	3791 ± 12	0
12.1	m,osc,bip	203	119	0.59	0.105	1.225 ± 0.043	0.3717 ± 0.0034	3799 ± 14	1
G04/85 quartz-magnetite banded iron formation (65°10.31′N 49°49.22′W)									
1.1	e,osc,bip	114	67	0.59	0.205	1.268 ± 0.033	0.3480 ± 0.0034	3698 ± 15	1
1.2	e,osc,bip	108	53	0.49	0.152	1.303 ± 0.056	0.3487 ± 0.0040	3701 ± 18	−1
2.1	e,osc,bip	157	125	0.80	0.292	1.293 ± 0.025	0.3482 ± 0.0017	3699 ± 8	0

Table 3.3-1. *(Continued)*

Labels	Grain type	U (ppm)	Th (ppm)	Th/U	Comm. 206Pb%	238U/206Pb ratio	207Pb/206Pb ratio	207Pb/206Pb date (Ma)	%disc
2.2	e,osc,bip	136	106	0.78	0.729	1.297 ± 0.039	0.3483 ± 0.0047	3699 ± 21	0
3.1	e,h/rex,bip	76	35	0.46	2.457	1.340 ± 0.041	0.3304 ± 0.0089	3619 ± 42	−1
G05/04 metadolerite dyke, Isua supracrustal belt (65°10.36′N 49°48.25′W)									
1.1	m,osc/h,p	502	70	0.14	0.034	1.391 ± 0.028	0.3071 ± 0.0007	3506 ± 4	0
1.2	m,osc/h,p	428	60	0.14	0.053	1.431 ± 0.029	0.3104 ± 0.0015	3523 ± 7	−3
1.3	m,osc/h,p	333	75	0.23	0.093	1.406 ± 0.036	0.3103 ± 0.0011	3522 ± 5	−2
1.4	e,osc/h,p	354	82	0.23	0.210	1.510 ± 0.053	0.3047 ± 0.0014	3495 ± 7	−6
2.1*	e,osc,p,fr	420	25	0.06	0.043	1.869 ± 0.034	0.1982 ± 0.0014	2812 ± 12	−2
2.2*	e,osc,p,fr	582	70	0.12	0.008	1.849 ± 0.082	0.1999 ± 0.0008	2826 ± 7	−1
G01/01 slightly altered (hydrated) metaperidotite (63°58.43′N 51°28.92′W)									
1.1	e,hd,ov	463	113	0.25	0.001	1.943 ± 0.053	0.1826 ± 0.0008	2677 ± 7	0
2.1	e,sz,fr	98	20	0.20	0.015	1.465 ± 0.049	0.2907 ± 0.0034	3421 ± 18	−2
2.2	e,sz,fr	93	19	0.20	0.001	1.493 ± 0.048	0.2764 ± 0.0040	3343 ± 23	−1
2.3	e,sz,fr	115	17	0.15	0.001	1.454 ± 0.046	0.3065 ± 0.0031	3503 ± 16	−4
2.4	e,sz,fr	63	11	0.18	0.029	1.433 ± 0.046	0.2961 ± 0.0028	3450 ± 15	−1
3.1	e,hd,p,fr	309	130	0.42	0.034	2.042 ± 0.070	0.1860 ± 0.0013	2707 ± 11	−5
4.1	e,sz,ov,fr	81	8	0.10	0.237	1.303 ± 0.041	0.3238 ± 0.0023	3588 ± 11	2
4.2	m,sz,ov,fr	296	92	0.31	0.132	1.388 ± 0.037	0.3059 ± 0.0013	3501 ± 6	0
4.3	e,sz,ov,fr	91	9	0.10	0.034	1.313 ± 0.033	0.3239 ± 0.0021	3588 ± 10	2
4.4	e,sz,ov,fr	312	109	0.35	0.022	1.417 ± 0.046	0.3062 ± 0.0019	3502 ± 9	−2
5.1	m,hd,ov	531	142	0.27	0.029	1.992 ± 0.053	0.1831 ± 0.0008	2681 ± 7	−2
6.1	e,hd,ov	674	139	0.21	0.052	1.984 ± 0.050	0.1850 ± 0.0008	2698 ± 7	−2
7.1	m,hd,ov	742	443	0.60	0.034	1.927 ± 0.049	0.1882 ± 0.0017	2726 ± 15	−1
8.1	e,hb,ov	77	13	0.17	0.284	1.340 ± 0.048	0.3211 ± 0.0037	3575 ± 18	1
8.3	e,hb,ov	59	8	0.13	0.058	1.349 ± 0.062	0.3359 ± 0.0042	3644 ± 19	−2
8.4	e,hb,ov	79	12	0.15	0.035	1.300 ± 0.063	0.3247 ± 0.0038	3592 ± 18	2

Table 3.3-1. *(Continued)*

Labels	Grain type	U (ppm)	Th (ppm)	Th/U	Comm. 206Pb%	^{238}U/^{206}Pb ratio	^{207}Pb/^{206}Pb ratio	^{207}Pb/^{206}Pb date (Ma)	%disc
10.1	e,hd,ov	340	171	0.50	0.040	2.032 ± 0.053	0.1858 ± 0.0013	2706 ± 11	−5
11.1	m,hb,ov/anh	46	4	0.09	0.001	1.700 ± 0.066	0.2681 ± 0.0066	3295 ± 39	−9
11.2	m,hb,ov/anh	49	4	0.09	0.086	1.324 ± 0.067	0.3329 ± 0.0049	3630 ± 23	0
11.3	m,hb,ov/anh	39	3	0.07	0.157	1.234 ± 0.049	0.3381 ± 0.0042	3654 ± 19	−5
12.1	m,hd,ov	336	212	0.63	0.074	1.996 ± 0.061	0.1879 ± 0.0017	2724 ± 15	−4
13.1	e,hb/sz,ov	131	25	0.19	0.265	1.672 ± 0.071	0.2147 ± 0.0017	2941 ± 13	3
14.1	m,hd,ov	418	203	0.49	0.003	1.927 ± 0.054	0.1843 ± 0.0008	2692 ± 7	0
15.1	m,sz,ov	116	75	0.65	0.097	1.963 ± 0.058	0.1841 ± 0.0014	2691 ± 13	−1
16.1	m,hd,ov	228	188	0.82	0.001	1.923 ± 0.057	0.1855 ± 0.0013	2702 ± 12	0
17.1	m,hd,ov	766	389	0.51	0.024	1.918 ± 0.054	0.1858 ± 0.0023	2705 ± 20	0
G91/06 tonalitic gneiss (64° 46.07′N 49° 59.77′W)									
A	c,sz,p	418	489	1.17	0.057	1.825 ± 0.052	0.3739 ± 0.0013	3806 ± 5	1
B	c,osc,p	132	102	0.77	0.023	1.783 ± 0.086	0.3665 ± 0.0019	3776 ± 8	−4
C	c,h,p	109	3	0.03	0.246	1.793 ± 0.043	0.3507 ± 0.0040	3704 ± 18	−5
D	c,osc,p	210	141	0.67	0.088	2.029 ± 0.052	0.3804 ± 0.0041	3832 ± 16	−3

GPS positions use WGS84 datum.

See Nutman et al. (1996, 2000) and Friend and Nutman (2005b) for G88/66 data.

p = prismatic, bip = bipyramidal – short prism, ov = oval, e = end, m = middle, r = overgrowth, c = core, rex = recrystallised, osc = oscillatory finescale zoning, sz = sector zoning, h = homogeneous, hd = homogeneous dark, fr = fragment, anh = anhedral, turb = turbid.

Corrected with model Pb of Cumming and Richards (1975) for likely rock age.

Likely laboratory contaminant grain (G05/04).

at ca. 3850 Ma, albeit has locally recrystallised and some overgrowths of younger zircon have developed on it (Fig. 3.3-5(c)). Our preferred interpretation of such rocks is that they are ca. 3850 Ma tonalites, whose zircons display the effects of superimposed polymetamorphism (Table 3.3-1; Fig. 3.3-5(e)). Other more strongly deformed rocks with ca. 3850 Ma zircon also occur, which contain variable amounts of younger zircon as well (Fig. 3.3-5(b,d,f); Nutman et al., 2000). We interpret such rocks as ca. 3850 Ma tonalites, with variable amount of younger *in situ* melt and veining that has been transformed into a banding by strong superimposed deformation. Therefore we contend that the Complex also contains ca. 3850 Ma tonalites, although others (Whitehouse et al., 1999; Whitehouse and Kamber, 2005) disagree that rocks of this age have yet been identified.

Itsaq Gneiss Complex tonalites have low initial $^{87}Sr/^{86}Sr$ ratios (Moorbath et al., 1972) and positive initial ε_{Nd} isotopic signatures (Bennett et al., 1993). Both these indicate that the tonalites are juvenile additions to the sialic crust, and not remelts of much older crust. Thus by combining the accurate and precise zircon geochronology and the Nd and Sr whole rock isotopic studies, our hypothesis is that the Itsaq Gneiss Complex was largely constructed by intrusion of several generations of tonalites (with subordinate quartz-diorites) over almost 200 million years (Fig. 3.3-2). This contrasts with an earlier hypothesis of crust-forming "super-events" in the Greenland Archaean, whereby most of the rocks in a particular gneiss complex were interpreted as being coeval (implicit, because they were placed on the same isochrons – e.g., Moorbath and Taylor (1985)). This hypothesis was based on the earlier whole rock isochron suite geochronology, rather than zircon geochronology (see Nutman et al. (2000) for discussion and early literature).

3.3-5. EOARCHAEAN TECTONIC INTERCALATION OF UNRELATED ROCKS IN THE ITSAQ GNEISS COMPLEX

3.3-5.1. Tectonic Intercalation in the Isua Supracrustal Belt, Isukasia Terrane

Over the past decade, several groups of workers have produced tectonic interpretations for the Isua supracrustal belt and the adjacent orthogneisses (e.g., Nutman et al., 1997, 2002a; Appel et al., 1998; Komiya et al., 1999; Hanmer and Greene, 2002). Most of these interpretations diverge in their detail and even in their representation of the same geology, but there is now a general consensus that the Isua supracrustal belt consists of several panels of mylonite-bounded volcano-sedimentary rocks that were juxtaposed in the Eoarchaean. Zircon geochronology demonstrates that ca. 3800 Ma and 3710 Ma volcano-sedimentary rocks are present (Fig. 3.3-4), and therefore these panels are *not* all the same age (Nutman et al., 1997a, 2002a). This is different from interpretations based on Sm/Nd whole rock

---

Fig. 3.3-10. Geological sketch map of part of the eastern end of the Isua supracrustal belt (see Fig. 3.3-1 for location). See Appel (1998) and Komiya (1999) for other maps of the same area. Zircon dates are from Nutman et al. (1995, 1997, 2002), Kamber et al. (2005a) and this paper.

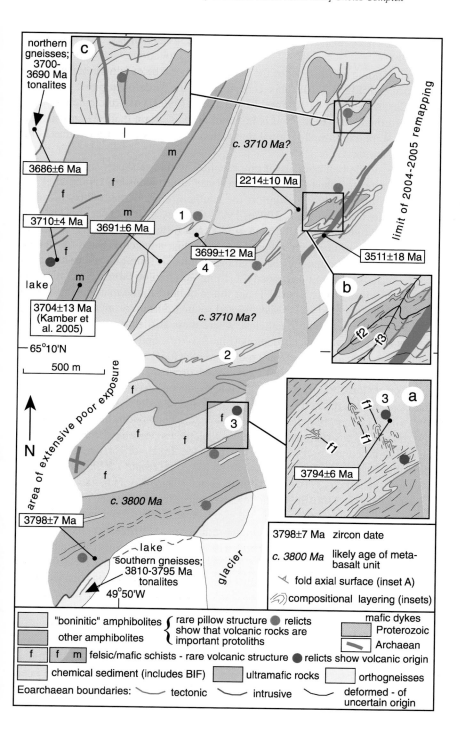

northern gneisses; 3700-3690 Ma tonalites

c

c. 3710 Ma?

2214±10 Ma

m

3686±6 Ma

f

f

m

1

3691±6 Ma

3710±4 Ma

f

3699±12 Ma

4

3511±18 Ma

lake

m

3704±13 Ma (Kamber et al. 2005)

b

f2

f3

65°10'N

c. 3710 Ma?

500 m

2

N

f

area of extensive poor exposure

f

3

f

3

a

f1

f1

f1

f

f1

3794±6 Ma

c. 3800 Ma

limit of 2004-2005 remapping

3798±7 Ma

lake

southern gneisses; 3810-3795 Ma tonalites

glacier

49°50'W

3798±7 Ma   zircon date

c. 3800 Ma   likely age of meta-basalt unit

⤇ fold axial surface (inset A)

compositional layering (insets)

"boninitic" amphibolites { rare pillow structure ● relicts show that volcanic rocks are important protoliths

mafic dykes

other amphibolites

Proterozoic

f   f   m   felsic/mafic schists - rare volcanic structure ● relicts show volcanic origin

Archaean

chemical sediment (includes BIF)

ultramafic rocks

orthogneisses

Eoarchaean boundaries: ⌢ tectonic   ⌢ intrusive   ⌢ deformed - of uncertain origin isotopic studies, which have regarded all rocks of the belt as essentially the same age (e.g., Moorbath et al., 1997).

The Eoarchaean tectonic partitioning of the Isua supracrustal belt is illustrated in a map of part of its eastern end (Fig. 3.3-10). This same area was the focus of studies by Nutman et al. (1997a, 2002a), Appel et al. (1998) and Komiya et al. (1999). All lithological boundaries are highly deformed, and for many it is impossible to ascertain their original nature. However, there are some confirmed original depositional contacts, mylonitic breaks and (deformed) intrusive contacts.

The southern side of the belt consists of amphibolites with rare preserved pillow lava structures, in which a felsic schist layer has been dated at ca. 3800 Ma (Nutman et al., 2002a). Further west, this amphibolite unit at the southern edge of the belt is intruded by ca. 3800 Ma tonalite (Nutman et al., 1997a, 1999; Crowley, 2003). Thus the southern amphibolite unit is derived substantially from $\geqslant$3800 Ma pillow lavas. To the north are ultramafic rocks with komatiitic chemistry and a felsic schist unit represented by sample G05/15 (Fig. 3.3-10(a)). The boundaries between these units are ambiguous, but are probably tectonic. The felsic schist unit has mostly a strong *LS* tectonic fabric, concordant its boundaries. However, locally early folds (F1) are found oblique to the margins of the unit (Fig. 3.3-11(a)). These early folds were sheared and rotated during, or after, assembly of different panels within the belt, but nonetheless where these earliest folds are preserved, this unit preserves relics of volcanic structures (Fig. 3.3-7(c)) and thus indicates its protolith was a volcanic or volcano-sedimentary rock. Sample G05/15 of felsic volcanic rocks from this unit has euhedral, prismatic oscillatory-zoned zircons. Excluding three younger $^{207}Pb/^{206}Pb$ ages provisionally interpreted to be due to ancient loss of radiogenic Pb, the remaining analyses have a weighted mean $^{207}Pb/^{206}Pb$ age of 3794 $\pm$ 6 Ma (Table 3.3-1; Fig. 3.3-11(a)), which is interpreted as the age of volcanic deposition. Further north is a major mylonite bounding an isoclinally folded package of chemically diverse amphibolites and chemical sedimentary rocks. Within this package, there are early mylonites that separate amphibolites of "boninitic" chemistry (high MgO, SiO_2 and LREE – Polat et al. (2002)) with a likely age of ca. 3710 Ma (Nutman et al., 1997a) from chemical sedimentary rocks containing sparse 3700–3690 Ma volcanogenic zircons (Nutman et al., 2002a). Dating of three zircons recovered from a banded iron formation (G04/85; Fig. 3.3-10) supports this timing of deposition for these chemical sediments. All three zircons are small, euhedral, have oscillatory zoning and high Th/U. They have close to concordant ages with a weighted mean $^{207}Pb/^{206}Pb$ age of 3699 $\pm$ 12 Ma (Table 3.3-1, Fig. 3.3-11(b)). This agrees with the 3691 $\pm$ 6 Ma age obtained from a few similar-looking zircons in a siliceous sample nearby in the same unit (Nutman et al., 2002a). Our preferred interpretation is that the 3690–3700 Ma zircons represent a very small volcanic component, and are a proxy for the age of deposition of the chemical sediments.

These chemical sedimentary rocks are bounded on one side by (?)3710 Ma volcanic amphibolites with "boninitic" chemistry and on their other side by compositionally different amphibolites of unknown age (Fig. 3.3-10). In places, the chemical sediments have been excised along the mylonite, so that amphibolites of different composition are in direct tectonic contact with each other (shown in detail in Fig. 3.3-10(b,c)). These earliest tectonic

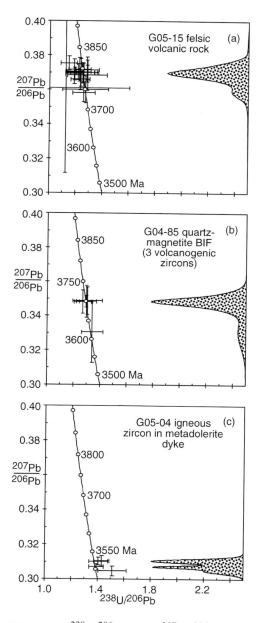

Fig. 3.3-11. $^{238}U/^{206}Pb$ versus $^{207}Pb/^{206}Pb$ Tera-Wasserburg Concordia plots for new zircon dating from the eastern end of the Isua supracrustal belt (see Fig. 3.3-10 for sample locations). Analytical errors are depicted at the 2-sigma level. (a) G05/15 felsic volcanic or volcano-sedimentary rock. (b) G04/85 banded iron formation. (c) G05/04 felsic patch (late igneous) in Ameralik dyke.

contacts were isoclinally folded (F2) and then affected by later asymmetric warps (F3). This earliest generation of isoclinally folded mylonites was recognised during supplementary mapping in 2004 and 2005. All this structural development is Eoarchaean, because an Ameralik dyke, dated by zircon at 3511 ± 18 Ma (using multiple analyses on one high Th/U prismatic zircon obtained from sample G05/04 of felsic late magmatic patch – Table 3.3-1, Fig. 3.3-11(c)), either exploited F3 fold axes or was intruded syn-F3 (Fig. 3.3-10).

To the north are panels of ca. 3710 Ma felsic to mafic volcanic schists (Nutman et al., 1997a, 2002a; Kamber et al., 2005a), and amphibolites, whose tectonic boundaries are slightly oblique to the northern margin of the belt. The northern margin of the belt is an Eoarchaean mylonite (Nutman, 1984; Nutman et al., 1997a, 2002a; Komiya et al., 1999; Crowley et al., 2002), north of which 3700–3690 Ma tonalites and ca. 3650 Ma granite sheets predominate (Nutman et al., 1996, 2002a; Crowley et al., 2002).

The eastern end of the Isua supracrustal belt is an example showing that tectonic intercalation of rocks of unrelated age and origin was an important process in building the Itsaq Gneiss Complex. Clear evidence for this has been lost from most other parts of the complex, where younger (<3600 Ma) Archaean deformation and metamorphism were more intense. However, the repetition of rocks of different age and origin throughout the complex (Nutman and Collerson, 1991; Nutman et al., 1993b; Friend et al., 2002a) suggest that similar tectonic processes occurred throughout it.

3.3-5.2. Narssaq Area, Færingehavn Terrane

On the southeast part of Narssaq (peninsula) south of Nuuk (Figs. 3.3-1 and 3.3-12), there is a ca. 1 km long body of ultramafic rocks with associated metagabbros, siliceous rocks (derived at least partly from chemical sediments) and skarned amphibolites (derived from volcanic protoliths?) embedded in polyphase orthogneisses. Most of the ultramafic rocks are highly deformed and schistose. They are also traversed by discordant veins of coarse-grained phlogopite and fibrous amphibole that formed by metasomatism, when a post-kinematic granite body was intruded nearby to the west. However, at the northern end of the body, there is a small area where the ultramafic rocks are less altered. There,

---

Fig. 3.3-12. Sketch map over a large enclave of ultramafic, mafic and siliceous rocks within polyphase orthogneisses, southeastern Narssaq peninsular, Færingehavn terrane (see Fig. 3.3-1 for location). (a) Denotes area of low strain where dunite veins traverse harzburgites. (b-b') Is where ultramafic rocks and associated metagabbros are truncated by a package of siliceous rocks of probable chemical sedimentary parentage and skarned amphibolites probably derived from basalts. On the right hand side of the figure are Tera-Wasserburg ($^{238}U/^{206}Pb$ versus $^{207}Pb/^{206}Pb$) plots for SHRIMP analyses of rare metamorphic zircons from ultramafic sample G01/01 and reconnaissance data for zircons from orthogneiss G91/06 at the margin of the ultramafic body. In G01/01 the oldest detected metamorphic zircons are ca. 3650 Ma, giving the *minimum* age of the ultramafic rocks. In G91/06, >3800 Ma determinations on oscillatory zoned zircons give the likely age of the tonalitic protolith.

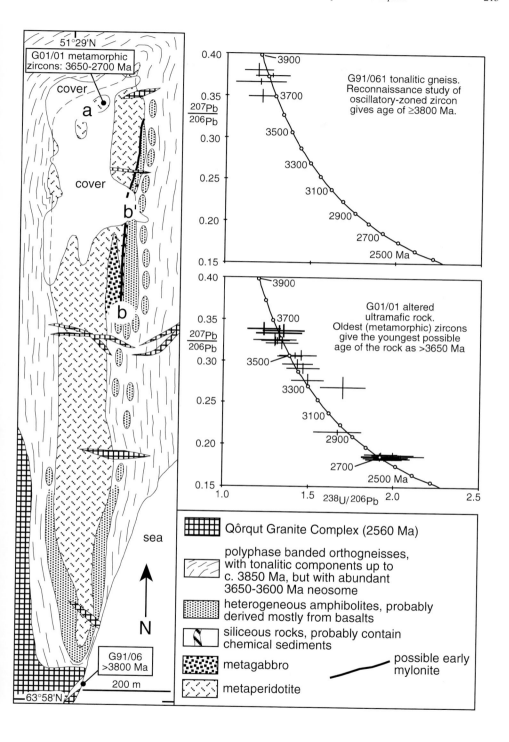

G91/061 tonalitic gneiss. Reconnaissance study of oscillatory-zoned zircon gives age of ≥3800 Ma.

G01/01 altered ultramafic rock. Oldest (metamorphic) zircons give the youngest possible age of the rock as >3650 Ma

Qôrqut Granite Complex (2560 Ma)

polyphase banded orthogneisses, with tonalitic components up to c. 3850 Ma, but with abundant 3650-3600 Ma neosome

heterogeneous amphibolites, probably derived mostly from basalts

siliceous rocks, probably contain chemical sediments

metagabbro

metaperidotite

possible early mylonite

dunite dykes cut at a high angle weak compositional layering in their harzburgite host. The dunites have chemical characteristics of strongly depleted upper mantle peridotites (Fig. 3.3-6). The zircons are all equant to anhedral grains devoid of oscillatory zoning, and are interpreted to have grown *in situ* during metamorphism. They have yielded ages between 3650 and 2700 Ma (Table 3.3-1; Fig. 3.3-12). In which case, the oldest of these zircons show that these mantle-derived ultramafic rocks must be at least 3650 Ma old. On the eastern margin of the ultramafic body, there is a lens of layered gabbroic amphibolite. This is truncated obliquely by a ca. 1 m thick layer of siliceous rocks, that were probably derived at least in part from chemical sediments. East of these are skarned, heterogeneous amphibolites that are likely to be derived from volcanic protoliths. Thus, upper mantle and deep crustal rocks might be in contact with upper crustal rocks at this locality. A likely explanation for how this occurred is that they were tectonically juxtaposed prior to being engulfed in tonalitic protoliths of the surrounding orthogneisses. Similar relationships have been observed south of the Isua supracrustal belt, where ⩾3800 Ma upper mantle peridotites, ultramafic cumulates, layered gabbros and upper crustal rocks were tectonically juxtaposed prior to being engulfed in ca. 3800 Ma tonalite (Nutman et al., 1996; Friend et al., 2002a). For the Narssaq body, the banding of the gneisses is concordant to its margins, and neosome emanating from the orthogneisses has broken-up parts of its margin (Fig. 3.3-12). Reconnaissance zircon dating on a tonalitic orthogneiss near the body (G91/06) yielded ⩾3800 Ma oscillatory zoned, high U/Th zircon (Table 3.3-1; Fig. 3.3-12), indicating the presence of very old rocks in the vicinity. Due to the superimposed deformation, the relationship between the ⩾3800 Ma tonalites and the body dominated by ultramafic rocks is unclear. An age of >3800 Ma (i.e., pre-tonalite) is possible, whereas ca. 3650 Ma metamorphic zircons in the ultramafic rocks give the absolute minimum age.

3.3-6. DISCUSSION

3.3-6.1. Construction of Eoarchaean Continental Crust

The Itsaq Gneiss Complex provides robust evidence for the formation of 3900–3600 Ma continental crust by intrusion of tonalites of several ages into suites of older rocks that are dominated by metabasalts with lesser amounts of chemical sediments, felsic volcanic rocks, gabbros and upper mantle peridotites. In rare cases, field relationships are preserved that demonstrate that these were being tectonically intercalated in the same period that tonalites were being intruded (Friend et al., 2002a).

The amphibolites of basaltic origin have compositional affinities with arc basalts, and are neither plume-related nor mid ocean ridge basalt (Polat and Hofmann, 2003; Jenner et al., 2006). Partial melting of hydrated basalt converted into eclogite produced abundant tonalites (Nutman et al., 1999). However, also associated with the tonalites are more magnesian quartz-diorites (Fig. 3.3-8), whose genesis must have involved melting of (metasomatised) mantle. In modern settings this occurs when young, hot oceanic crust is being

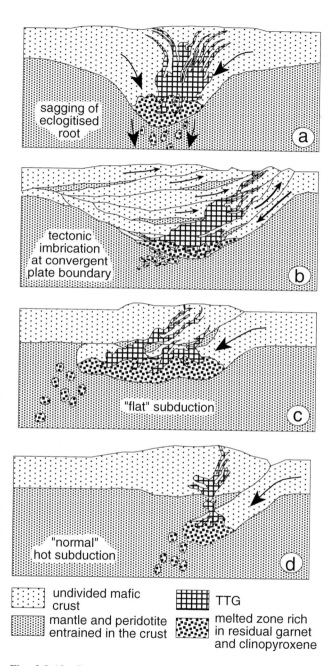

Fig. 3.3-13. Cartoon cross-sections of choices for architecture at Eoarchaean convergent plate boundaries and TTG producing environments (see also Smithies and Champion (2000) and references therein).

subducted, and although most melt is generated from the down-going slab, some is produced by melting and reaction between a mantle wedge and melts ascending from the slab (Smithies and Champion, 2000; Martin et al., 2005). The key geochemical characteristics of the amphibolites, tonalites and quartz-diorites make it likely that >3600 Ma continental crust in the Itsaq Gneiss Complex (and in all other ancient gneiss complexes) was constructed by magmatism at ancient convergent plate boundaries. It is stressed however, that the internal architecture of these ancient plate boundaries has not yet been resolved. The presence of the quartz-diorites with their mantle origins probably excludes models of ductile down-welling, where tonalite melts were produced in an eclogitised thickened crustal root (Fig. 3.3-13(a)). Instead, either imbrication of mafic crust in which some slices of upper mantle became trapped (Fig. 3.3-13(b)) or various forms of more uniformitarian shallow subduction (Fig. 3.3-13(c,d)) are more likely, because they provide a setting suitable for incorporation of some upper mantle in the melting processes. The latter three geometries have all been discussed as likely later in the Archaean (e.g., Smithies and Champion, 2000), but we contend that models involving mantle melting are permissible for continental crust formation in the Eoarchaean as well. Regardless of the exact geometries, the chemistry of igneous rocks from the Itsaq Gneiss Complex indicates the existence of convergent plate boundary magmatism in the Eoarchaean. The assemblage of the intercalated metabasalts ± chemical sediments, gabbros and upper mantle peridotites then intruded by tonalites ± quartz-diorites is evocative of supra-subduction zone ophiolites. However, because "ophiolite" is such a catch-phrase, we prefer not to apply it to the Eoarchean rocks we describe here. Instead, we prefer to conclude they formed at convergent plate boundaries, as do supra-subduction zone ophiolites in more recent times.

3.3-6.2. How Representative is the Preserved Oldest Geological Record?

The presently known area of Eoarchaean rocks underlies about 10,000 km^2 of Earth's crust. About a third of that is in the Narryer Gneiss Complex of Western Australia, and is very poorly exposed – this reduces even further the amount to study. Although new occurrences will be revealed in the future (see Stevenson et al. (2006) for the most recent discovery from Hudsons Bay, Canada), the total amount of Eoarchaean crust preserved on Earth is definitely small. Thus it should always be asked – is this small sample of crust representative of geological processes at that time?

ACKNOWLEDGEMENTS

Work was undertaken with support of NERC grant NER/A/S/1999/00024 and ARC grant DP0342798. The late Vic McGregor is thanked for many discussions of Nuuk region geology over the years, prior to his death in 2000. APN thanks Richard Arculus for an Honorary Fellowship at the Department of Earth and Marine Sciences, ANU. Kensuke Tanigaki and Yoshiki Goto are thanked for assistance in the field and zircon separation.

Earth's Oldest Rocks
Edited by Martin J. Van Kranendonk, R. Hugh Smithies and Vickie C. Bennett
Developments in Precambrian Geology, Vol. 15 (K.C. Condie, Series Editor)
© 2007 Elsevier B.V. All rights reserved.
DOI: 10.1016/S0166-2635(07)15034-9

Chapter 3.4

THE GEOLOGY OF THE 3.8 GA NUVVUAGITTUQ (PORPOISE COVE) GREENSTONE BELT, NORTHEASTERN SUPERIOR PROVINCE, CANADA

JONATHAN O'NEIL[a], CHARLES MAURICE[b], ROSS K. STEVENSON[c], JEFF LAROCQUE[d], CHRISTOPHE CLOQUET[e], JEAN DAVID[f] AND DON FRANCIS[a]

[a]*Earth & Planetary Sciences, McGill University and GÉOTOP-UQÀM-McGill, 3450 University St. Montreal, QC, Canada, H3A 2A7*
[b]*Bureau de l'exploration géologique du Québec, Ministère des Ressources naturelles et de la Faune, 400 boul. Lamaque, Val d'Or, QC, Canada, J9P 3L4*
[c]*GÉOTOP-UQÀM-McGill and Département des Sciences de la Terre et de l'Atmosphère, Université du Québec à Montréal, C.P. 8888, succ. centre-ville, Montreal, QC, Canada, H3C 3P8*
[d]*School of Earth and Oceanic Sciences, University of Victoria, P.O. Box 3055, STN CSC, Victoria, BC, Canada, V8W 3P6*
[e]*INW-UGent, Department of Analytical Chemistry, Proeftuinstraat 86, 9000 Gent, Belgium*
[f]*GÉOTOP-UQÀM-McGill, Université du Québec à Montréal, C.P. 8888, succ. centre-ville, Montreal, QC, Canada, H3C 3P8*

3.4-1. INTRODUCTION

Our knowledge of the first billion years of the Earth's evolution is limited and early magmatic processes, such as mantle differentiation and crustal formation, remain poorly understood. Zircons from the Jack Hills conglomerates (Wilde et al., 2001) suggest the existence of continental crust as old as 4.4 Ga. These early ages, however, are obtained on detrital zircons from much younger rocks, whose protolith has long since been destroyed or reworked. Other than timing, these ancient zircons provide little information about the chemistry of the Earth's early mantle. Rather, rare preserved relicts of Eoarchean mantle-derived crust provide the best compositional and isotopic constraints on early crust-mantle differentiation of the Earth. The 3.6–3.85 Ga Itsaq Gneiss Complex (West Greenland), which includes the Isua Greenstone Belt, is the most extensive early Archean terrain preserved. The Nd isotopic compositions for these mantle-derived rocks indicate that their mantle source was already strongly depleted at 3.8 Ga (Bennett et al., 1993; Blichert-Toft et al., 1999; Frei et al., 2004), implying that significant volumes of continental crust had already formed during the Hadean (see Kamber, this volume). Such remnants of Eoarchean mantle-derived rocks are, however, rare, and models for the evolution of the mantle are still poorly constrained for the first billion years of Earth's history (see Davies, this volume).

In this paper, we report the first detailed description of the Nuvvuagittuq (originally named Porpoise Cove) greenstone belt, dated at 3.8 Ga. As one of the world's oldest known mantle-derived suite of rocks, the Nuvvuagittuq greenstone belt offers an extraordinary opportunity to further our understanding of the early Earth. Preliminary results for this newly discovered Eoarchean supracrustal assemblage indicate that both aluminum-depleted (ADK) and aluminum-undepleted (AUK) komatiitic magmas existed at 3.8 Ga and that the mantle had already experienced a long-term depletion at that time. Furthermore, a prominent banded iron formation, which serves as a stratigraphic marker horizon within the belt, displays Fe isotopic compositions that are systematically heavier than their enclosing igneous rocks, similar to results obtained at Isua. Although it has yet to be demonstrated that such isotopic fractionation requires an organic origin, the possibility that the formation of such Archean Algoma-type banded iron formations involves biological activity has major implications for the timing of the appearance of life on Earth.

3.4-2. GEOLOGICAL FRAMEWORK

The Nuvvuagittuq greenstone belt is located on the eastern coast of Hudson Bay, in the Northeastern Superior Province (NESP) of Canada (Fig. 3.4-1: see also Percival, this volume). Early work on this portion of the Superior Province suggested that it was composed mostly of granulite-grade granitoids (Stevenson, 1968; Herd, 1978; Card and Ciesielski, 1986; Percival et al., 1992). More recent work has shown, however, that it is dominantly composed of Neoarchean plutonic suites in which amphibolite- to granulite-grade greenstone belts occur as relatively thin keels (1–10 km) that can be traced continuously for up to 150 km along strike (Percival and Card, 1994; Percival et al., 1995, 1996, 1997a; Leclair, 2005). The magmatic and metamorphic evolution of the NESP spans nearly 2 billion years of Earth history (3.8–1.9 Ga), as determined by $\sim$220 U-Pb zircon ages acquired by governmental surveys (Leclair et al., 2006, and references therein). On a regional scale, distinct lithological assemblages appear as large linear positive and negative aeromagnetic anomalies, which have led to the partitioning of the NESP into lithotectonic domains (Percival et al., 1992, 1997b). These domains have subsequently been modified following further field mapping and the acquisition of more isotopic data, with the result that the NESP is now separated into two isotopically distinct terranes (Leclair et al., 2006; Boily et al., 2006a). To the east, the Arnaud River Terrane includes rocks that are younger than ca. 2.88 Ga and characterized by juvenile isotopic signatures and Nd depleted-mantle model ages (T_{DM}) < 3.0 Ga. To the west, rocks of the Hudson Bay Terrane, which includes the Nuvvuagittuq greenstone belt, represent a reworked Meso- to Eoarchean craton, with zircon inheritance ages and T_{DM} as old as 3.8 Ga (Stevenson et al., 2006).

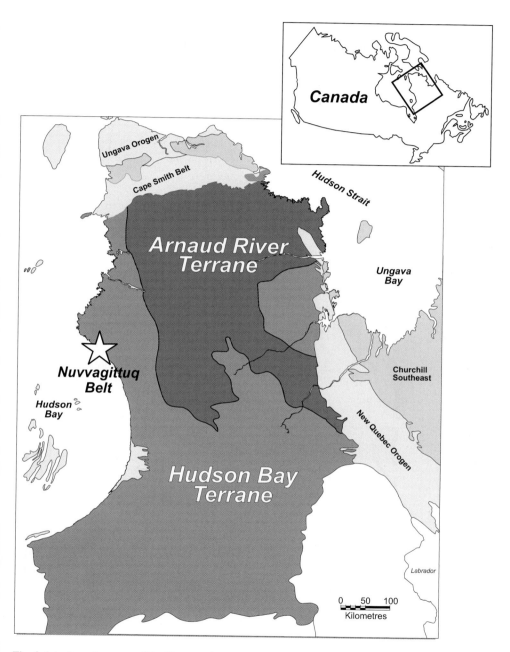

Fig. 3.4-1. Location map of the Nuvvuagittuq greenstone belt in the northeastern Superior Province.
Isotopic terranes from Boily et al. (2006), Leclair (2005) and Leclair et al. (2006).

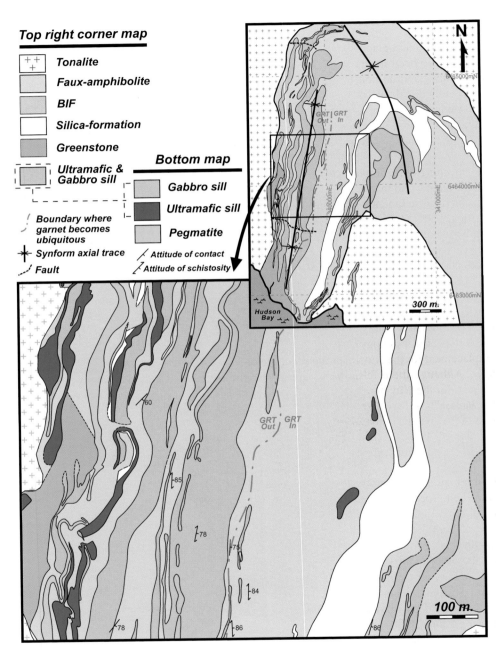

Fig. 3.4-2. Geological map of the Nuvvuagittuq greenstone belt. Coordinates: UTM zone 18 NAD 27.

3.4-3. GEOLOGY OF THE NUVVUAGITTUQ BELT

Lee (1965) first mapped the Nuvvuagittuq greenstone belt and small portions of it have subsequently been mapped in more detail by Nadeau (2003). We have now mapped the entire Nuvvuagittuq belt at the scale of 1:20 000, and the western limb of the belt at a more detailed scale of 1:2000 (Fig. 3.4-2). The Nuvvuagittuq belt is a volcano-sedimentary succession that occurs as a tight to isoclinal synform refolded into a more open south-plunging synform, with bedding largely parallel to the main, steeply-dipping schistosity. The supracrustal assemblage of the belt is essentially composed of three major lithological units:

(1) cummingtonite-amphibolite that is the predominant lithology of the belt;
(2) ultramafic and mafic sills that intrude the amphibolites; and
(3) chemical sedimentary rocks that comprise a banded iron formation and a silica formation.

The Nuvvuagittuq belt is surrounded by a 3.6 Ga tonalite, itself surrounded by a younger 2.75 Ga tonalite (Stevenson and Bizzarro, 2006; David et al., 2002, 2004; Simard et al., 2003).

The Nuvvuagittuq belt contains rare felsic bands, 15 to 50 cm in width (Fig. 3.4-3(a)), which have been interpreted by Simard et al. (2003) to be a felsic tuff. U-Pb ages obtained on zircons from one of these felsic bands, a plagioclase-quartz-biotite schist suggest an age of emplacement possibly as old as 3825 ± 18 Ma (David et al., 2002). Subsequent high-resolution geochronology done by Cates and Mojzsis (2007a) confirmed a minimum age of emplacement for the Nuvvuagittuq sequence of 3751 ± 10 Ma based on $^{207}Pb/^{206}Pb$ zircon ages. Although these felsic bands are geochemically similar to the surrounding tonalite and rare in the Nuvvuagittuq belt, a U-Pb age of ca. 3.66 Ga on zircons (Stevenson and Bizzarro, 2006; David et al., 2004) from the surrounding tonalites makes it unlikely that these felsic bands represent remobilized tonalite.

3.4-3.1. Cummingtonite-Amphibolite

Cummingtonite-amphibolites are the predominant lithologies of the Nuvvuagittuq green-stone belt. These peculiar amphibolites are dominated by cummingtonite, which gives this lithology a light grey to beige color, rather than the dark green to black color character-istic of hornblende-dominated amphibolites typical of the Superior Province. Because of the unusual color of the amphibolites in this region, they have been referred to as "faux-amphibolite" in the field, a term which we will use for the rest of this contribution.

The faux-amphibolite is a heterogeneous gneiss consisting of cummingtonite + quartz + biotite + plagioclase ± anthophyllite ± garnet, with the majority of the biotite having been replaced by retrograde chlorite. It is generally characterised by a finely laminated texture defined by the alternation of biotite-rich and cummingtonite-rich laminations en-hanced by ubiquitous mm- to cm-scale quartz ribboning that generally follows the main schistosity (Fig. 3.4-3(b,c)). Variations in the proportion of cummingtonite and biotite also occur on a meter scale with large bands dominated by cummingtonite + quartz + plagio-clase, within more biotite-rich faux-amphibolite.

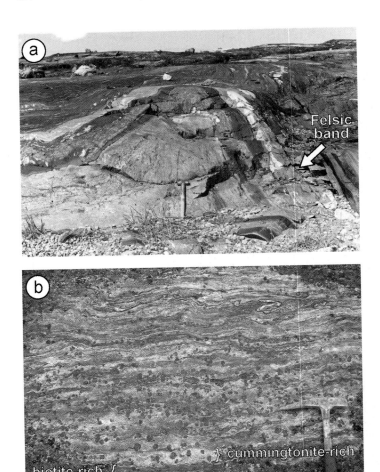

Fig. 3.4-3. Photographs of Nuvvuagittuq rocks. (a) Felsic band on which the 3.825 Ga U-Pb zircon age has been obtained (David et al., 2002). (b) Garnet-bearing faux-amphibolite.

One of the striking features of the faux-amphibolite is the variation in garnet content. The faux-amphibolite in the western limb of the belt rarely contains garnet, whereas cm-sized garnets are ubiquitous in the eastern limb (Fig. 3.4-2), although its proportion varies substantially, with alternating garnet-poor and garnet-rich layers (Fig. 3.4-3(c)). In addition, there is a gradational transition in the southwestern corner of the belt from faux-amphibolite gneiss to massive aphanitic greenstones that are interpreted to be massive volcanic flows (Fig. 3.4-2). In contrast to the typical faux-amphibolite, these rocks are

Fig. 3.4-3. (*Continued*). (c) Garnet-poor and garnet-rich layers within the faux-amphibolite. (d) Ultramafic sill with gabbroic top.

characterized by a greenschist-facies mineral assemblage of chlorite + epidote + quartz + plagioclase ± actinolite ± carbonate.

The abundance of garnet in the faux-amphibolites of the eastern limb of the belt, along with the compositional layering, lead to their first being mapped as paragneiss (Simard et al., 2003). However, although the faux-amphibolites are compositionally somewhat variable, they are generally basaltic in composition and are significantly more mafic than

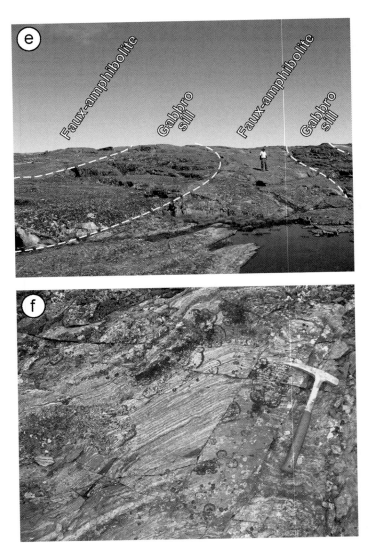

Fig. 3.4-3. (*Continued*). (e) Gabbro sills intruding the faux-amphibolite. (f) Banded iron formation with alternating quartz-rich and magnetite-rich laminations.

Archean shales, with higher MgO and lower SiO_2 contents (40–56 wt% SiO_2, 4–16 wt% MgO) (Table 3.4-1), similar to those of gabbro sills and greenstones in the belt. The garnet-bearing faux-amphibolites are compositionally similar to the biotite and cummingtonite-rich units, but have lower Mg# and higher Al_2O_3 contents. Both the garnet-bearing and garnet-free faux-amphibolite are, however, Ca-poor relative to the gabbro sills that intrude them. Although similar in composition, the greenstones on the southwestern limb

Table 3.4-1. Major (wt%) and trace (ppm) element data for Nuvvuagittuq rocks

Major elements

Sample	SiO$_2$	TiO$_2$	Al$_2$O$_3$	MgO	FeO	MnO	CaO	Na$_2$O	K$_2$O	P$_2$O$_5$	LOI	UTM Easting	UTM Northing
Cummingtonite-rich faux-amphibolite													
PC-54	54.27	0.41	18.60	6.88	7.22	0.16	7.20	1.40	1.86	0.06	1.54	339865	6464451
PC-58	54.53	0.79	14.57	6.98	9.81	0.19	8.78	1.28	0.62	0.10	1.40	339959	6464353
PC-131	49.95	0.34	16.34	11.44	8.78	0.18	5.14	2.08	1.92	0.03	2.91	339574	6464197
PC-149	52.67	0.51	16.07	9.58	8.46	0.17	6.77	1.25	1.47	0.04	2.18	339686	6463935
PC-151	50.12	0.37	17.36	9.34	9.20	0.20	7.40	1.17	1.76	0.03	2.02	339727	6463946
PC-159	54.47	0.47	16.09	6.81	10.04	0.26	5.10	2.05	1.38	0.04	2.22	339920	6463917
PC-162	52.07	0.45	15.42	9.43	9.02	0.21	7.55	1.23	1.02	0.04	2.46	339961	6463911
PC-171	48.86	1.08	14.70	8.31	14.54	0.24	4.53	1.12	3.07	0.05	1.46	340127	6463551
PC-173A	45.59	0.73	15.22	15.61	12.17	0.18	1.02	1.07	2.78	0.17	4.47	340111	6463566
PC-173B	47.34	0.66	14.20	15.42	12.40	0.19	2.04	1.25	1.56	0.08	3.81	340111	6463566
Biotite-rich faux-amphibolite													
PC-129	52.01	0.76	14.12	10.70	10.70	0.09	3.89	1.96	0.88	0.07	3.91	339550	6464197
PC-132	51.19	0.89	14.42	8.68	10.54	0.21	8.20	1.92	1.08	0.08	1.83	339582	6464222
PC-135	48.22	0.35	16.38	13.62	11.65	0.13	0.45	0.06	1.80	0.03	6.37	339632	6464229
PC-150	54.53	0.51	16.13	10.28	8.23	0.18	4.94	1.80	0.10	0.06	2.57	339689	6463919
PC-152	48.52	0.61	16.18	15.79	9.90	0.19	0.28	0.36	0.43	0.04	6.79	339740	6463925
Garnet-bearing faux-amphibolite													
PC-157	52.67	0.60	17.39	6.33	10.11	0.37	2.85	1.98	2.72	0.07	3.50	339912	6463958
PC-160	55.07	0.61	17.53	6.77	12.38	0.10	0.15	0.19	1.79	0.06	4.20	339933	6463921
PC-161	53.83	0.62	17.42	3.99	9.31	0.24	7.28	1.12	2.56	0.08	2.68	339955	6463924
PC-163	52.61	0.42	19.16	6.96	9.17	0.28	2.28	2.32	1.73	0.03	4.00	339977	6463823
PC-164	55.30	0.38	18.64	4.95	7.42	0.16	4.92	1.88	1.72	0.05	3.63	339989	6463855
PC-176	49.76	0.92	13.87	8.71	21.16	0.12	0.51	0.35	0.86	0.11	1.74	340079	6463584

Table 3.4-1. (*Continued*)

Major elements

Sample	SiO$_2$	TiO$_2$	Al$_2$O$_3$	MgO	FeO	MnO	CaO	Na$_2$O	K$_2$O	P$_2$O$_5$	LOI	UTM Easting	UTM Northing
Greenstone													
PC-177	64.05	0.58	11.00	3.76	6.94	0.13	2.96	2.77	2.49	0.07	4.16	339436	6463037
PC-178	56.01	0.91	13.61	4.87	12.20	0.10	1.83	0.13	3.95	0.08	4.89	339464	6463078
PC-179	39.62	0.70	16.32	7.16	26.63	0.10	0.02	0.07	1.41	0.01	5.52	339472	6463067
PC-180	44.52	0.77	13.88	8.60	10.61	0.18	6.29	1.16	4.12	0.06	8.21	339489	6463045
PC-181	44.69	1.03	17.89	10.07	10.29	0.14	2.73	2.47	2.54	0.10	6.96	339514	6463050
PC-182	51.00	0.74	13.80	9.05	11.33	0.12	2.74	1.69	2.18	0.07	6.04	339526	6463050
PC-183	40.65	1.39	15.79	11.39	11.33	0.17	5.21	1.52	1.55	0.12	9.53	339540	6463086
PC-184	43.26	0.40	19.25	7.34	7.22	0.14	5.98	1.78	4.36	0.05	9.25	339555	6463096
PC-185	43.13	0.45	15.36	11.06	9.67	0.18	5.97	2.01	1.09	0.05	9.96	339575	6463114
PC-186	51.45	0.56	15.57	6.43	8.48	0.17	4.99	0.45	2.83	0.06	7.84	339589	6463114
PC-189	47.15	0.66	15.20	9.06	8.92	0.15	5.56	1.73	1.44	0.06	9.04	339694	6462839
Gabbro													
PC-81	49.01	0.74	15.80	8.74	9.90	0.21	10.82	2.27	0.70	0.06	1.08	339669	6464278
PC-82	49.20	0.76	15.64	7.91	9.93	0.19	11.37	1.58	0.68	0.06	1.68	339669	6464278
PC-83	48.26	0.75	15.57	8.62	10.37	0.20	11.24	1.83	0.75	0.06	1.41	339669	6464278
PC-85	49.03	1.07	14.89	7.94	12.68	0.19	9.72	3.09	0.34	0.08	0.50	339669	6464278
PC-86	49.40	1.20	14.67	7.55	12.88	0.24	9.41	2.98	0.36	0.09	0.58	339669	6464278
PC-87	50.62	1.16	13.09	9.08	13.52	0.28	7.96	2.54	0.33	0.07	0.48	339669	6464278
PC-118	48.03	0.97	15.25	7.62	12.01	0.22	9.87	1.55	1.89	0.07	1.17	339534	6464230
PC-119	48.80	1.03	15.24	7.17	12.00	0.23	8.53	2.00	2.22	0.09	1.20	339537	6464234
PC-121	48.09	1.12	14.91	7.43	12.53	0.24	8.10	1.54	2.99	0.09	1.40	339536	6464220
PC-147	52.26	0.84	9.48	10.36	13.02	0.25	10.08	0.95	0.47	0.08	0.79	339649	6463952
PC-187	50.02	0.74	11.49	8.95	12.98	0.20	10.73	1.53	0.91	0.05	0.74	339830	6462843
PC-188	47.54	1.00	13.89	5.92	17.91	0.22	8.78	1.69	0.40	0.09	0.69	339772	6462769
PC-191	47.09	1.23	8.57	5.84	22.29	0.17	8.30	0.67	0.17	0.13	2.81	339789	6462840

Table 3.4-1. (Continued)

Major elements

Sample	SiO_2	TiO_2	Al_2O_3	MgO	FeO	MnO	CaO	Na_2O	K_2O	P_2O_5	LOI	UTM Easting	UTM Northing
Ultramafic Type-1													
PC-25	43.90	0.46	3.71	26.55	11.99	0.18	5.79	0.13	0.05	0.02	5.99	339633	6464857
PC-26	41.65	0.32	2.32	31.36	11.52	0.19	2.64	0.09	0.04	0.02	8.73	339633	6464857
PC-27	45.42	0.54	4.59	24.05	11.56	0.18	8.33	0.17	0.05	0.02	4.21	339633	6464857
PC-28	45.42	0.52	4.45	25.12	10.75	0.18	7.33	0.18	0.05	0.02	4.98	339633	6464857
PC-29	43.90	0.46	3.71	26.55	11.99	0.18	5.79	0.13	0.05	0.02	5.99	339633	6464857
PC-113	42.64	0.42	3.53	27.98	12.05	0.19	4.43	0.20	0.05	0.02	6.99	339513	6464278
PC-114	39.96	0.15	1.15	34.67	10.50	0.19	1.30	0.01	0.02	0.02	11.09	339524	6464261
PC-115	38.62	0.27	2.18	33.96	9.27	0.19	2.38	0.06	0.03	0.02	11.62	339519	6464268
PC-116	39.88	0.25	2.02	35.27	9.34	0.19	0.54	0.00	0.40	0.02	11.13	339529	6464245
PC-125	40.52	0.27	2.22	30.86	12.08	0.19	2.73	0.05	0.02	0.02	9.89	339520	6464113
27525A	40.28	0.41	3.19	31.03	11.40	0.18	3.94	0.19	0.04	0.06	7.80		
27521A	38.55	0.35	2.58	31.62	11.57	0.19	3.61	0.21	0.39	0.09	9.30		
19095A	42.70	0.24	4.60	25.70	13.77	0.09	4.44	0.29	0.06	0.01	7.00		
Type-1 amphibolitic chilled margin[M] and layer[L]													
PC-15[M]	46.77	0.64	6.15	18.14	14.58	0.26	10.69	0.25	0.06	0.03	1.18	339633	6464857
PC-33[M]	49.44	0.65	5.51	17.79	11.8	0.29	11.34	0.48	0.13	0.03	1.45	339608	6464731
PC-110[M]	50.87	0.5	4.41	19.87	9.02	0.17	12.16	0.44	0.10	0.02	1.36	339505	6464264
PC-20[L]	43.64	0.59	5.65	22.69	13.47	0.18	8.02	0.22	0.05	0.02	3.82	339633	6464857
PC-21[L]	46.1	0.61	5.8	20.79	13.57	0.18	9.4	0.24	0.06	0.03	2.21	339633	6464857
Ultramafic Type-2													
PC-72	42.66	0.18	5.04	29.88	7.81	0.13	4.38	0.23	0.06	0.02	8.51	339738	6464332
PC-73	40.56	0.19	5.28	31.59	8.08	0.13	2.90	0.11	0.04	0.02	9.95	339738	6464332
PC-74	40.73	0.16	4.27	31.71	8.37	0.12	3.16	0.11	0.04	0.02	9.98	339738	6464332
PC-75	38.40	0.16	4.32	31.99	9.00	0.14	2.99	0.06	0.04	0.02	11.33	339738	6464332
PC-76	40.26	0.20	5.12	31.11	8.36	0.13	3.59	0.10	0.04	0.02	9.96	339738	6464332

Table 3.4-1. (*Continued*)

Major elements

Sample	SiO$_2$	TiO$_2$	Al$_2$O$_3$	MgO	FeO	MnO	CaO	Na$_2$O	K$_2$O	P$_2$O$_5$	LOI	UTM Easting	UTM Northing
PC-91	42.27	0.11	4.71	29.48	7.49	0.15	5.18	0.23	0.05	0.01	8.94	339669	6464278
PC-92	40.82	0.08	3.85	33.03	8.19	0.13	2.13	0.07	0.03	0.01	10.67	339669	6464278
PC-93	39.68	0.08	3.86	32.59	8.40	0.15	2.81	0.13	0.03	0.01	10.81	339669	6464278
PC-94	40.37	0.11	5.21	31.67	8.06	0.14	3.05	0.14	0.04	0.01	9.85	339669	6464278
PC-138	42.17	0.24	6.27	28.22	8.58	0.13	4.80	0.10	0.03	0.03	8.05	339623	6463969
PC-141	40.28	0.17	4.56	31.23	8.87	0.14	3.16	0.05	0.03	0.02	9.98	339614	6463947
Type-2 amphibolitic chilled margin^M and layer^L													
PC-71^M	46.51	0.39	11.28	20.89	10.51	0.17	6.56	0.66	0.12	0.04	2.42	339738	6464332
PC-142^M	44.69	0.46	11.73	22.27	9.99	0.12	5.49	0.30	0.12	0.05	3.49	339619	6463967
PC-88^L	47.34	1.14	14.21	16.48	12.41	0.29	1.77	0.64	1.91	0.09	2.8	339669	6464278
BIF													
PC-192	51.88	0.02	0.29	2.93	39.65	0.38	1.39	0.04	0.02	0.04	0.00	339782	6462834
PC-193	80.57	0.03	0.75	0.75	15.77	0.19	0.39	0.04	0.03	0.02	0.00	339777	6462849
PC-197	46.32	0.03	0.28	3.18	45.98	0.20	0.12	0.03	0.11	0.06	0.08	339506	6463764
PC-198B	34.04	0.02	0.25	1.90	59.06	0.15	0.37	0.02	0.05	0.08	0.00	339512	6463744
PC-199	59.88	0.03	12.31	1.51	20.34	0.58	2.67	0.09	0.37	0.02	0.25	339515	6463764
PC-200	32.20	0.02	0.27	3.49	59.14	0.22	0.09	0.02	0.15	0.06	0.00	339517	6463770
Si-formation													
PC-194	75.69	0.07	3.14	2.43	9.11	0.78	2.33	0.03	0.11	0.02	5.08	339701	6462834
Grt Si-rich unit													
PC-165	65.53	0.40	15.92	2.18	9.12	0.16	0.21	0.34	2.86	0.05	2.29	339977	6463773

Table 3.4-1. (Continued)

Tace elements

Sample	Rb	Sr	Zr	Nb	Y	Ni	Cr	V	Co	La	Ce	Pr	Nd	Sm	Eu	Gd	Tb	Dy	Ho	Er	Tm	Yb	Lu
Cummingtonite-rich faux-amphibolite																							
PC-54	59	62	39	1.6	15	96	121	152	35														
PC-58	22	51	161	6.7	27	61	88	208	35														
PC-131	65	123	23	1.0	13	188	410	205	50														
PC-149	34	44	40	1.7	13	87	233	219	51														
PC-151	44	53	25	0.7	14	126	193	228	50														
PC-159	38	58	63	3.1	21	90	127	199	40														
PC-162	29	55	50	2.5	15	137	349	174	47														
PC-171	66	50	91	2.6	44	115	266	274	48														
PC-173A	81	60	39	1.7	20	172	314	221	55														
PC-173B	47	42	37	1.5	24	176	326	245	53														
Biotite-rich faux-amphibolite																							
PC-129	29	47	56	2.2	17	92	171	249	41														
PC-132	32	74	57	2.3	18	53	130	267	38														
PC-135	58	8	23	1.0	9	186	416	203	42														
PC-150	2	24	40	1.6	13	80	233	201	41														
PC-152	10	4	67	3.1	16	93	142	213	54														
Garnet-bearing faux-amphibolite																							
PC-157	84	37	70	3.6	18	90	155	218	40														
PC-160	60	16	69	3.9	15	140	177	224	50														
PC-161	85	52	67	3.5	20	71	159	230	54														
PC-163	53	51	28	1.1	16	104	247	245	41														
PC-164	54	53	31	0.9	14	99	200	195	48														
PC-176	37	5	52	2.2	18	415	1321	252	84														

Table 3.4-1. (*Continued*)

Trace elements

Sample	Rb	Sr	Zr	Nb	Y	Ni	Cr	V	Co	La	Ce	Pr	Nd	Sm	Eu	Gd	Tb	Dy	Ho	Er	Tm	Yb	Lu
Greenstone																							
PC-177	62	98	42	1.7	10	107	323	87	29														
PC-178	131	63	47	1.8	15	86	242	77	32														
PC-179	58	7	36	1.2	6	266	367	210	68														
PC-180	293	124	38	1.4	17	138	333	252	40														
PC-181	73	71	106	5.3	13	96	63	171	53														
PC-182	49	59	76	4.2	11	88	51	95	52														
PC-183	60	87	73	2.8	12	109	255	314	50														
PC-184	174	97	28	0.9	11	123	289	237	42														
PC-185	43	96	51	2.6	15	130	171	150	48														
PC-186	74	66	61	3.1	15	85	131	188	40														
PC-189	49	93	49	2.4	18	92	279	201	41														
Gabbro																							
PC-81	16	94	39	1.4	18	159	350	236	43	2.2	0.8	5.7	4.7	1.6	0.7	2.2	0.4	2.7	0.6	1.9	0.3	1.8	0.3
PC-82	21	94	43	1.5	17	120	324	232	41														
PC-83	18	92	40	1.7	16	166	324	227	43	2.3	0.9	5.8	4.6	1.6	0.6	2.2	0.4	2.6	0.6	1.8	0.3	1.8	0.3
PC-85	4	78	57	2.1	21	132	180	257	52	3.1	1.2	8.1	6.5	2.2	0.9	2.9	0.5	3.4	0.7	2.3	0.3	2.1	0.3
PC-86	10	96	62	2.4	26	75	214	301	42														
PC-87	5	77	62	2.4	25	82	192	291	49	3.2	1.5	9.5	8.1	2.7	0.8	3.5	0.6	4.0	0.9	2.7	0.4	2.6	0.4
PC-118	73	138	53	2.1	21	122	294	260	53														
PC-119	76	112	58	2.2	23	116	228	252	56														
PC-121	114	62	60	2.2	23	113	224	281	53														
PC-147	8	46	80	4.4	18	119	547	192	63														
PC-187	26	104	49	1.9	18	199	870	218	60														
PC-188	5	65	66	2.6	27	56	155	298	64														
PC-191	1	3	51	3.1	15	501	2782	214	157	1.9	0.9	5.5	5.5	2.1	0.9	3.0	0.5	2.9	0.6	1.7	0.3	1.6	0.2

Table 3.4-1. (*Continued*)

Tace elements

Sample	Rb	Sr	Zr	Nb	Y	Ni	Cr	V	Co	La	Ce	Pr	Nd	Sm	Eu	Gd	Tb	Dy	Ho	Er	Tm	Yb	Lu
Type-1 ultramafic																							
PC-25	0	12	22	1.1	8	1237	2422	116	92														
PC-26	7	10	17	0.8	4	1854	1769	64	124														
PC-27	0	14	30	1.2	10	900	2181	129	83														
PC-28	6	15	29	1.2	9	1096	1886	113	85														
PC-29	6	12	22	1.1	8	1237	2424	115	92														
PC-113	0	7	21	0.9	7	1295	2023	106	96														
PC-114	0	5	6	0.4	4	1921	959	50	126														
PC-115	0	9	14	1.0	4	2046	1891	54	110														
PC-116	28	4	11	1.1	2	2053	1616	51	117														
PC-125	2	14	14	0.8	3	1757	1717	66	128														
27525A	0	0	0	0	0	0	0	0	0														
27521A	0	0	0	0	0	0	0	0	0														
19095A	0	0	0	0	0	0	2400	0	0														
Type-1 amphibolitic chilled margin[M] and layer[L]																							
PC-15[M]	0	8	31.3	1.5	11.7	284	1596	166	80	3.65	9.8	1.33	6.66	1.87	1.0	2.12	0.37	2.22	0.44	1.34	0.19	1.16	0.18
PC-33[M]	0	15.4	33.8	1.6	10	193	2051	175	60	2.28	5.32	0.73	4.0	1.46	0.85	1.8	0.32	2.0	0.41	1.22	0.18	1.13	0.16
PC-110[M]	0	10.3	26.9	1.4	6.8	712	2049	114	75	2.77	6.07	0.74	3.53	1.0	0.49	1.23	0.22	1.32	0.27	0.8	0.12	0.77	0.11
PC-20[L]	0	11	30.9	1.1	9.4	647	2520	154	66	1.0	3.05	0.49	3.14	1.2	0.43	1.48	0.26	1.67	0.35	1.06	0.15	0.94	0.14
PC-21[L]	0	13.7	32.7	1.3	11.1	457	1832	161	82	2.6	6.2	0.83	4.48	1.54	0.61	1.88	0.33	2.0	0.42	1.21	0.17	1.06	0.15
Type-2 ultramafic																							
PC-72	4	12	18	0.6	7	1687	5851	76	85														
PC-73	3	14	19	0.8	6	1935	5722	77	99														
PC-74	4	16	17	0.8	6	2141	5257	62	121														
PC-75	4	15	18	0.9	6	2270	5580	63	118														
PC-76	4	19	22	1.0	7	1941	4551	67	113														

Table 3.4-1. (*Continued*)

Trace elements

Sample	Rb	Sr	Zr	Nb	Y	Ni	Cr	V	Co	La	Ce	Pr	Nd	Sm	Eu	Gd	Tb	Dy	Ho	Er	Tm	Yb	Lu
PC-91	0	12	6	0.3	3	1559	5175	76	72														
PC-92	3	8	4	0	3	1621	4672	55	79														
PC-93	0	9	3	0	2	1848	6905	58	78														
PC-94	4	7	5	0	3	1795	5790	76	109														
PC-138	0	16	27	1.2	10	1611	4190	90	108														
PC-141	0	15	19	0.9	7	1935	4855	63	103														

Type-2 amphibolitic chilled margin[M] and layer[L]

Sample	Rb	Sr	Zr	Nb	Y	Ni	Cr	V	Co	La	Ce	Pr	Nd	Sm	Eu	Gd	Tb	Dy	Ho	Er	Tm	Yb	Lu
PC-71[M]	0	43.7	44.7	2.1	18.1	616	1892	150	81	4.7	11.2	1.39	6.4	1.8	0.72	2.24	0.43	2.84	0.62	1.96	0.3	1.93	0.29
PC-142[M]	2	34.6	52.5	2.4	17.5	649	2137	161	80	5.34	11.8	1.53	7.0	2.0	1.1	2.4	0.4	2.9	0.6	2.0	0.3	2.0	0.3
PC-88[L]	34.3	82.5	62.8	2.3	18.8	102	241.5	277	63	2.67	6.24	0.91	5.0	1.9	0.4	2.5	0.5	3.3	0.7	2.3	0.4	2.4	0.4

BIF

Sample	Rb	Sr	Zr	Nb	Y	Ni	Cr	V	Co	La	Ce	Pr	Nd	Sm	Eu	Gd	Tb	Dy	Ho	Er	Tm	Yb	Lu
PC-192	0	3	3	0.5	12	56	8	11	0	1.9	3.7	0.5	2.1	0.6	0.3	1.0	0.2	1.2	0.3	0.9	0.1	0.9	0.1
PC-193	0	1	7	0.0	3	44	6	9	8	1.9	3.3	0.4	1.4	0.3	0.2	0.3	0.1	0.4	0.1	0.3	0.0	0.2	0.0
PC-197	6	2	1	0.4	8	78	23	11	4	2.2	4.4	0.5	2.1	0.5	0.3	0.8	0.2	1.0	0.2	0.8	0.1	0.8	0.1
PC-198B	3	2	3	0.5	9	67	8	9	0	3.2	5.7	0.6	2.4	0.5	0.3	0.7	0.1	0.9	0.2	0.8	0.1	0.8	0.1
PC-199	18	6	48	0.5	3	5	1	1	0	1.8	3.1	0.4	3.2	1.9	4.6	1.4	0.2	0.6	0.1	0.2	0.0	0.2	0.0
PC-200	9	2	1	0.3	13	77	9	11	0	2.3	4.6	0.6	2.6	0.8	0.4	1.2	0.2	1.5	0.4	1.2	0.2	1.2	0.2

Si-formation

Sample	Rb	Sr	Zr	Nb	Y	Ni	Cr	V	Co	La	Ce	Pr	Nd	Sm	Eu	Gd	Tb	Dy	Ho	Er	Tm	Yb	Lu
PC-194	8	10	16	1.1	18	46	55	23	8	20.4	33.0	3.4	11.2	1.6	0.5	1.6	0.3	2.2	0.5	1.7	0.3	1.5	0.2

Grt Si-rich unit

Sample	Rb	Sr	Zr	Nb	Y	Ni	Cr	V	Co	La	Ce	Pr	Nd	Sm	Eu	Gd	Tb	Dy	Ho	Er	Tm	Yb	Lu
PC-165	99	17	36	1.6	13	114	110	206	59	3.8	7.8	0.9	4.1	1.1	0.3	1.5	0.3	2.6	0.6	2.1	0.3	2.1	0.3

Alteration-free samples were crushed in a steel jaw crusher and ground in an alumina shatter box. Major and trace elements were analyzed by X-ray fluorescence (XRF) by the McGill Geochemical Laboratories, using a Philips PW2400 4 kW automated XRF spectrometer system. Major elements, Ba, Co, Cr, Cu and V were analyzed using 32 mm diameter fused beads, while Rb, Sr, Zr, Nb and Y were analyzed using 40 mm diameter pressed pellets. The accuracy for silica is within 0.5% and within 1% for other major and trace elements. REE concentrations were determined by Activation Laboratories, using a Perkin Elmer SCIEX ELAN 6000 coupled-plasma mass-spectrometer (ICP-MS) using a lithium metaborate/tetraborate fusion technique for digestion. Coordinates: UTM zone 18 NAD 27.

of the belt tend to have slightly lower SiO_2 contents than the faux-amphibolites at similar MgO contents. The relatively high loss on ignition (LOI; 4–10 wt%) and high K_2O contents (up to 4 wt%) of the greenstones suggest that they may have been extensively altered.

3.4-3.2. Gabbro and Ultramafic Sills

The striking feature of the western limb of the Nuvvuagittuq belt is the presence of numerous ultramafic and gabbroic conformable bodies within the faux-amphibolite (Fig. 3.4-3(d,e)). These bodies are interpreted to be sills because of the absence of any volcanic features, as well as the lack of asymmetry of the upper and lower margins typical of lava flows. The ultramafic sills range from 5 to 30 meters in width and consist of brown weathering serpentine-rich interiors with thin grey to dark green amphibole-rich margins. The ultramafic interiors of the sills consist mainly of serpentine and talc, with lesser tremolite, hornblende and chromite, but also contain amphibole-rich layers 10 to 20 cm in thickness. The amphibolitic margins of the ultramafic sills are composed dominantly of hornblende and talc and are interpreted to be chilled margins, while the amphibole-rich layers within the sill interiors are thought to have been pyroxene cumulate horizons. Locally, the presence of gabbroic tops suggests the separation of a residual liquid.

Two types of ultramafic sills can be recognized in the western limb of the Nuvvuagittuq belt, with those on the western side of the BIF being compositionally distinct from those on the eastern side of the BIF. The sills on the western side of the BIF (Type-1) are relatively poor in Al and rich in Fe, whereas those on the eastern side on the BIF (Type-2) are relatively rich in Al and poorer in Fe (Fig. 3.4-4(a c)). The serpentine-rich rocks of these two types of ultramafic sills fall along distinct olivine control lines in a Pearce-type plot (Fig. 3.4-5), suggesting that they are both olivine cumulates. Most strikingly, Cr increases with decreasing MgO within the olivine cumulates of Type-1 sills, but decreases with MgO in the Type-2 sills (Fig. 3.4-4(b)). The calculated CIPW-normative mineralogy of the amphibolite layers and margins also support the existence of two types of ultramafic sills in that normative clinopyroxene is abundant in Type-1 sills, whereas orthopyroxene predominates in Type-2 sills (Fig. 3.4-6). Moreover, metamorphic orthopyroxene is observed in the amphibolites of Type-2 sills, but not in Type-1 sills. The amphibolitic margin of both sill types exhibit slightly fractionated light rare earth element (LREE) profiles, but have different heavy rare earth element (HREE) profiles, with the chill margin of Type-2 sills displaying relatively flat HREE, whereas those of the Type-1 sills have slightly fractionated HREE profiles (Fig. 3.4-7(a)).

Type-2 ultramafic sills disappear approximately 75 meters to the east of the BIF in the western limb of the Nuvvuagittuq belt, but are replaced by numerous gabbro sills. The gabbro sills are typically meters to tens of meters in width, are distinctly darker than their cummingtonite-amphibolite host, and also lack the ubiquitous quartz ribboning of the latter. The gabbros consist of coarse- to medium-grained hornblende + plagioclase + quartz ± orthopyroxene ± cummingtonite. They are commonly characterized by a fine-

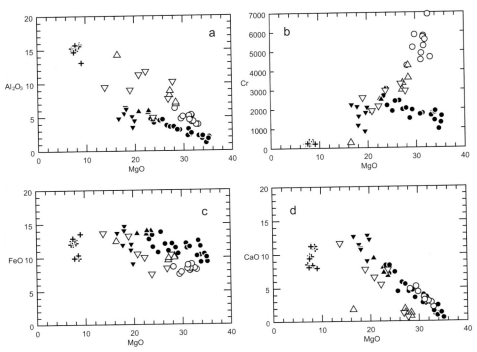

Fig. 3.4-4. Selected major (wt%) and trace (ppm) elements vs MgO for the gabbro and ultramafic sills. Symbols: black circles = Type-1 ultramafic; black triangles = Type-1 amphibolitic layer; black inverted triangles = Type-1 amphibolitic chill margin; open circles = Type-2 ultramafic; open triangles = Type-2 amphibolitic layer; open inverted triangles = Type-2 amphibolitic chill margin; black plusses = gabbro.

scale gneissosity defined by plagioclase-rich and amphibole-rich bands such that they were referred to as the "black and white" gneiss in the field. Despite their metamorphic mineral assemblage, the igneous term "gabbro" will be used to distinguish these dark hornblende-rich units from the dominant lighter coloured, cummingtonite-rich faux-amphibolites of the belt. The gabbros are relatively uniform in terms of major and trace elements, with SiO_2 ranging from 46–52 wt% and MgO from 5–10 wt% (Fig. 3.4-4 and Table 3.4-1). The gabbros have CaO contents (7.5–11.5 wt%) that are systematically higher than the faux-amphibolite at equivalent MgO contents. All of the gabbroic sills have flat to slightly depleted LREE profiles (Fig. 3.4-7(b)).

The $^{147}Sm/^{144}Nd$ ratios in the gabbros and peridotites of one of the Type-2 ultramafic sills range from 0.1519 to 0.2175, with calculated ε_{Nd} values at 3.8 Ga ranging from −1.8 to +3.4 (Table 3.4-2), with most of them being positive. The gabbros and ultramafics define a coherent array that is dispersed along a calculated 3.8 Ga isochron in a plot of $^{147}Sm/^{144}Nd$ versus $^{143}Nd/^{144}Nd$ (Fig. 3.4-8).

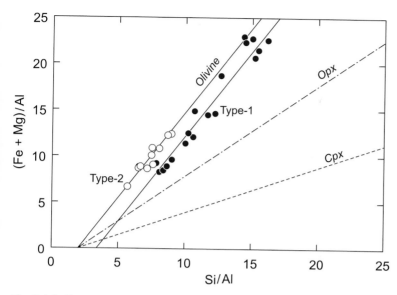

Fig. 3.4-5. Pearce type plot of (Mg + Fe)/Al vs Si/Al. Symbols as in Fig. 3.4-4.

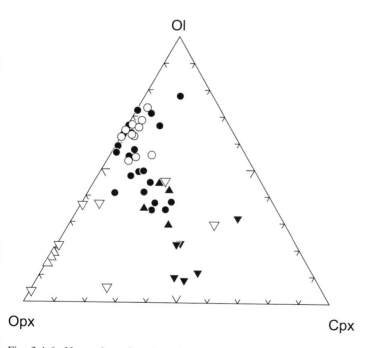

Fig. 3.4-6. Normative mineralogy for the ultramafic sills, the amphibolitic layers and the amphibolitic chill margins. Symbols as in Fig. 3.4-4. Ol: olivine, Opx: orthopyroxene, Cpx: clinopyroxene.

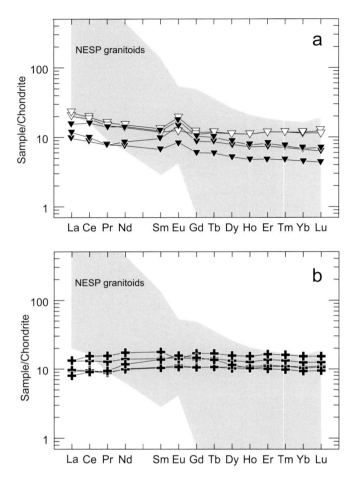

Fig. 3.4-7. Chondrite-normalized REE profiles. (a) Amphibolitic chill margin of the ultramafic sills. (b) Gabbro sills. Symbols: black inverted triangles = Type-1 amphibolitic chill margin; open inverted triangles = Type-2 amphibolitic chill margin; black plusses = gabbro. Data for NESP granitoids recovered from the SIGEOM database (available at www.mrnf.gouv.qc.ca/english/products-services/mines.jsp). Chondrite composition is from Sun and McDonough (1989).

3.4-3.3. *Banded iron formation and silica-formation*

A 5–30 m wide banded iron formation (BIF) can be traced continuously along the western limb of the belt and discontinuously along the eastern limb. The BIF is essentially a finely laminated quartz + magnetite + grunerite rock, with thin alternating quartz-rich and magnetite-rich laminations 0.1–1 cm wide (Fig. 3.4-3(f)). Grunerite is preferentially associated with the oxide-rich laminae, although it also commonly occurs as disseminated grains in the quartz-rich laminae. Actinolite is also found and locally both amphiboles have

Table 3.4-2. Sm-Nd isotopic data for Nuvvuagittuq gabbro and ultramafic sills

Sample	Rock type	Nd (ppm)	Sm (ppm)	$^{147}Sm/^{144}Nd$	$^{143}Nd/^{144}Nd$	2σ error	$\varepsilon Nd_{(3.8\ Ga)}$
por 21	Ultramafic	2.5	0.8	0.1963	0.512716	0.000009	1.75
por 25	Ultramafic	3.3	1.1	0.2069	0.512886	0.000010	−0.18
WP78	Gabbro	5.1	1.4	0.1637	0.511759	0.000010	−0.98
WP62b	Gabbro	5.8	2.0	0.2125	0.513012	0.000010	−0.48
WP43	Gabbro	7.2	2.4	0.2020	0.512792	0.000009	0.43
WP42a	Gabbro	7.5	2.5	0.2011	0.512807	0.000010	1.12
WP42c	Gabbro	7.0	1.8	0.1519	0.511682	0.000010	3.38
WP47a	Gabbro	5.4	1.8	0.2046	0.512838	0.000001	0.00
WP47b	Gabbro	4.7	1.7	0.2175	0.513070	0.000001	−1.77
PC-81	Gabbro	5.2	1.8	0.2047	0.512872	0.000001	0.63
PC-83	Gabbro	5.0	1.7	0.2006	0.512797	0.000001	1.22
PC-85	Gabbro	7.2	2.3	0.1956	0.512722	0.000001	2.21
PC-89	Chill margin	0.9	0.3	0.2068	0.512997	0.000017	2.05
PC89-D	Chill margin	0.8	0.3	0.2065	0.512985	0.000010	1.98
PC-93	Ultramafic	0.3	0.1	0.2154	0.513138	0.000031	0.57
PC-94	Ultramafic	0.4	0.2	0.2058	0.512899	0.000019	0.62

Samples for Nd isotope analysis were crushed to powder form and dissolved with a HF-HNO$_3$ mixture in high-pressure Teflon containers. A ^{149}Sm-^{150}Nd tracer was added to determine the Nd and Sm concentrations. The REE were concentrated by cation exchange chromatography and the Sm and Nd were extracted using an orthophosphoric acid-coated Teflon powder after Richard et al. (1976). Sm and Nd isotopic ratios were measured on a VG SECTOR-54 mass spectrometer using a triple filament assembly, in the GEOTOP laboratories at the Université du Québec à Montréal. Repeated measurements of LaJolla Nd standard yielded a value of $^{143}Nd/^{144}Nd = 0.511849 \pm 12$ ($n = 21$). The total combined blank for Sm and Nd is less than 150 pg. The reported Sm and Nd concentrations and the $^{147}Sm/^{144}Nd$ ratios have accuracies of 0.5%, corresponding to an average error of $0.5\varepsilon Nd$ unit for the initial Nd isotopic composition. $^{146}Nd/^{144}Nd$ was normalized to 0.7219 for mass fractionation corrections. The reference value for $^{143}Nd/^{144}Nd_{CHUR}$ was taken to be 0.512638, while that for $^{147}Sm/^{144}Nd_{CHUR}$ was taken to be 0.1967, and the decay constant for ^{147}Sm was assumed to be 6.54×10^{-12} a^{-1}.

been replaced by minnesotaite. Garnetiferous quartz-rich horizons occur locally within and adjacent to the BIF and the silica formation. In the southwestern corner of the belt, the iron formation transgresses from the lower grade greenstones, where the quartz-rich laminae have a distinct jasper-like colour, into the cummingtonite-bearing faux-amphibolites.

A 100 m wide silica formation in the eastern limb of the belt is composed almost entirely of massive recrystallized quartz with minor disseminated pyrite. At the southern-most edge of this limb, the silica-formation grades into BIF, suggesting that it may be a silica-rich facies of the BIF. Such an interpretation is supported by the local presence of a thinner silica-rich unit (1–15 m wide) adjacent to the BIF in the western limb of the belt.

The Nuvvuagittuq BIFs display concave-up depleted LREE profiles with relatively flat HREE profiles when normalized to post-Archean shales (PAAS) in a REE + Y plot. Their trace element profiles are characterized by strong positive Eu anomalies and weak positive Y anomalies (Fig. 3.4-9), which are thought to be common features of Archean BIFs precipitated from seawater (Bolhar et al., 2004). Although the HREE profile of the silica

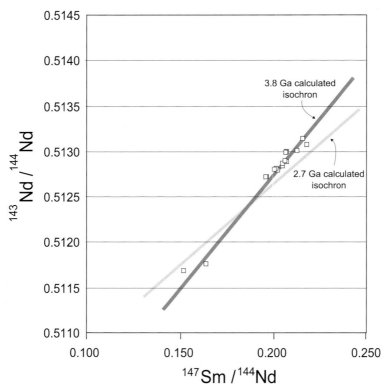

Fig. 3.4-8. $^{147}Sm/^{144}Nd$ vs $^{143}Nd/^{144}Nd$ for gabbro and ultramafic sills. Also shown are 3.8 and 2.7 Ga reference isochrons.

formation (sample PC-194) closely parallels those of the BIF samples, exhibiting similar positive Eu and Y anomalies, their LREE are relatively unfractionated relative to PAAS.

Fe isotopic compositions were determined for both BIF samples and samples of the enclosing faux-amphibolite and gabbro (Table 3.4-3). All BIFs have heavier Fe isotopic compositions (F_{Fe} = 0.25–0.48‰/amu; notation defined in Table 3.4-3 notes) than the adjacent magmatic lithologies, whose Fe isotopic compositions range from 0–0.2‰/amu (Fig. 3.4-10) (O'Neil et al., 2006). Sample PC-133, an amphibolite at the contact with the BIF, gave a F_{Fe} value intermediate between those of the BIF and the enclosing mafic lithologies. These values are similar to those measured by Dauphas et al. (2007).

3.4-4. DISCUSSION

3.4-4.1. Protolith of the Cummingtonite-Amphibolites

The dominance of amphiboles like hornblende and cummingtonite in most lithologies, along with the occurrence of metamorphic orthopyroxene and the abundance of garnet,

Table 3.4-3. Fe isotopic data for BIFs and igneous rocks from Nuvvuagittuq

Sample	δ^{56}Fe (‰)	2sd	δ^{57}Fe (‰)	2sd	F_{Fe} (‰/amu)	2sd
PC-101	0.00	0.02	0.03	0.02	0.00	0.01
PC-103	0.07	0.06	0.10	0.07	0.03	0.03
PC-110	0.07	0.03	0.09	0.02	0.03	0.01
PC-111	0.25	0.10	0.32	0.10	0.12	0.04
PC-114	0.09	0.05	0.12	0.05	0.04	0.02
PC-119	0.10	0.12	0.19	0.19	0.06	0.06
PC-119[a]	0.02	0.03	0.07	0.07	0.02	0.02
PC-123	0.03	0.01	0.06	0.01	0.02	0.01
PC-133	0.48	0.16	0.72	0.24	0.24	0.08
PC-133[a]	0.56	0.17	0.96	0.06	0.30	0.05
PC-140	0.36	0.21	0.58	0.34	0.18	0.11
PC-144	0.36	0.22	0.52	0.22	0.18	0.09
PC-146	0.19	0.06	0.33	0.06	0.10	0.03
PC-146[a]	0.38	0.10	0.54	0.21	0.18	0.06
PC-147	0.30	0.16	0.45	0.25	0.15	0.08
PC-149	−0.04	0.11	0.10	0.06	0.01	0.04
PC-150	0.37	0.09	0.67	0.13	0.20	0.05
PC-187A	0.10	0.14	0.14	0.10	0.05	0.05
PC-187B	0.12	0.07	0.19	0.05	0.06	0.03
PC-189	0.06	0.05	0.09	0.11	0.03	0.03
PC-190	0.22	0.10	0.32	0.09	0.11	0.04
PC-194	0.17	0.15	0.29	0.25	0.09	0.08
PC-195	0.26	0.30	0.35	0.54	0.12	0.17
PC-191	0.59	0.18	0.86	0.38	0.29	0.11
PC-192	0.71	0.10	0.97	0.16	0.34	0.05
PC-193	0.67	0.19	1.00	0.40	0.34	0.12
PC-197	0.97	0.10	1.42	0.16	0.48	0.05
PC-198B	0.57	0.16	0.83	0.21	0.28	0.07
PC-199	0.49	0.12	0.74	0.24	0.25	0.07
PC-200	0.62	0.13	0.90	0.17	0.30	0.06

Samples for Fe isotope analysis were digested using HNO_3-HF and HNO_3-$HClO_4$ mixtures as well as HCl in Teflon containers. Fe was extracted using AGMP-1 anion exchange chemistry in HCl media following a procedure similar to that of Dauphas et al. (2004a). Fe isotopes were measured using a VG Multicollector-ICPMS Isoprobe at the GEOTOP laboratories. A standard-sample-standard analytical protocol was used to correct for mass bias using the IRMM014 international isotopic standard as reference. The reference materials used for precision and accuracy were BCR-1 (basalt), AC-E (granite), and IF-G (iron formation from Isua, Greenland). F_{Fe} for reference materials gave: BCR-1 = 0.07 ± 0.02‰/amu ($N = 2$ measurements), AC-E = 0.17 ± 0.09‰/amu ($N = 2$ measurements) and IF-G = 0.31 ± 0.04‰/amu ($N = 3$ measurements). Standard values are similar to those previously published (Beard et al., 2003; Butler et al., 2005; Dauphas and Rouxel, 2006; Poitrasson et al., 2004; Rouxel et al., 2005). $F_{Fe} = \delta ij/(i - j)$, and $j = {}^{54}$Fe, i equals either ^{56}Fe, ^{57}Fe, or ^{58}Fe, and $\delta ij = ((i\text{Fe}/j\text{Fe})/(i\text{Fe}/j\text{Fe})\text{standard} - 1) \times 1000$, with IRMM-014 as the reference standard.
[a]Duplicate.

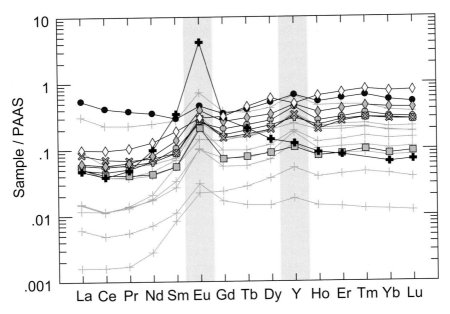

Fig. 3.4-9. Post-Archean Australian Shale (PAAS)-normalized REE + Y profiles for Nuvvuagittuq BIFs. Symbols: open diamonds = PC-165; grey circles = PC-192; grey squares = PC-193; black circles = PC-194; grey plus = PC-197; grey crosses = PC-198b; black plusses = PC-199; grey diamonds = PC-200; small light grey plusses = BIF from SW Greenland (Bolhar et al., 2004).

suggest that the metamorphic conditions in the Nuvvuagittuq belt reached at least upper amphibolite facies. Most biotite is, however, altered to chlorite, indicating the existence of extensive retrograde metamorphic effects. The progression from the chlorite-epidote greenstones to garnet-free amphibolites and then to garnet-bearing amphibolites (Fig. 3.4-2) suggests the presence of a map-scale metamorphic gradient from upper greenschist in the west to upper amphibolite facies in the east. This interpretation is supported by the observation that BIF in the western limb of the Nuvvuagittuq belt cuts across this gradient. The highest temperatures obtained using the garnet-biotite geothermometer (Ferry and Spear, 1979) on the least altered biotite ranges from 550 to 600 °C.

The compositions of the faux-amphibolites are similar to the gabbros and cluster along the gabbroic cotectic, as defined by MORB glasses (Fig. 3.4-11). Moreover, the faux-amphibolites are compositionally different from Archean shales and thus they are interpreted to be meta-igneous rocks rather than metamorphosed shales. Although the faux-amphibolites are compositionally similar to the gabbro sills, the dominance of cummingtonite over hornblende in the faux-amphibolite appears to reflect their lower CaO contents (Fig. 3.4-12). The abundance of garnet in the faux-amphibolites of the eastern limb of the Nuvvuagittuq belt may reflect their systematically lower Mg# and higher Al content compared to the garnet-free faux-amphibolite. The abundance of quartz ribboning within the faux-amphibolite suggests that Si and probably other elements such as

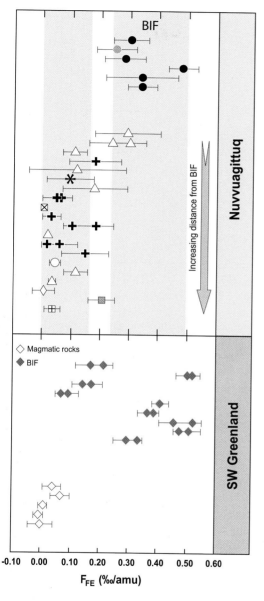

Fig. 3.4-10. Fe isotopic compositions of the BIFs and surrounding lithologies expressed as F_{Fe} (‰/amu). Symbols: black circles = BIF; grey circle = garnet-bearing BIF; black plusses = gabbro; open circle = ultramafic; black asterisk = silica-formation; open triangles = amphibolitic sill margin and layer; open diamond = cummingtonite-rich faux-amphibolite; grey square = biotite-rich faux-amphibolite; open square with black plus = tonalite; open square with black cross = felsic band. SW Greenland data are from Dauphas et al. (2004b).

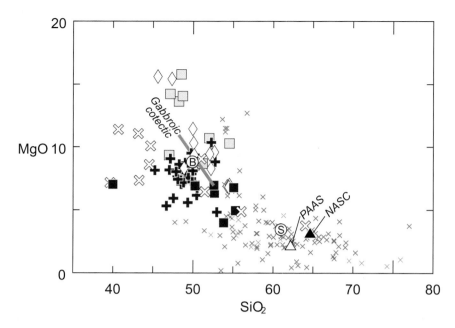

Fig. 3.4-11. SiO_2 vs MgO (wt%) for Nuvvuagittuq faux-amphibolites and gabbros. Symbols: black plusses = gabbro; open diamonds = cummingtonite-rich faux-amphibolite; grey squares = biotite-rich faux-amphibolite; black squares = garnet-bearing faux-amphibolite; open crosses = greenstone; small grey crosses = Archean shales from literature; B = average Archean basalt; S = average Archean shale. Data for shales are from Feng and Kerrich (1990), Hofmann et al. (2003), Bolhar et al. (2005), Fedo et al. (1996) and Wronkiewicz and Condie (1987). Data for average Archean shales and basalts are from Condie (1993). The gabbroic cotectic is from MORB 22-25 North glasses (data from Bryan, 1981). PAAS: Post-Archean Australian Shale, NASC: North American Shale Composite.

Ca were relatively mobile in the faux-amphibolite during metamorphism. These features, combined with the fine compositional layering of the faux-amphibolite, suggest that the faux-amphibolites may represent mafic pyroclastic deposits that were more susceptible to alteration and metasomatism than the more massive gabbroic sills that intrude them. According to this interpretation, loss of Ca during alteration and/or metamorphism favoured the formation of cummingtonite in the faux-amphibolites over hornblende in the gabbros, with the more magmatically evolved faux-amphibolite compositions developing garnet because of their higher Al and Fe contents.

3.4-4.2. Significance of the Ultramafic and Gabbro Sills for the Chemistry of the 3.8 Ga Mantle

A comparison of the REE profiles of gabbros from the Nuvvuagittuq belt with the LREE-enriched profiles of over 500 NESP granitoids argues against any significant interaction of the gabbros with surrounding felsic crust (Fig. 3.4-7). The slightly fractionated LREE

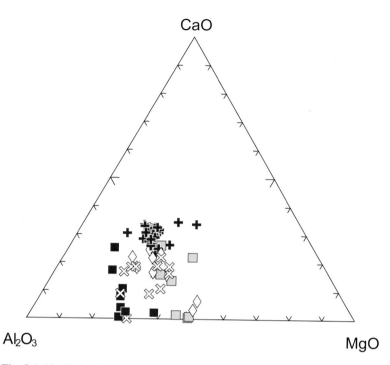

Fig. 3.4-12. CaO-Al$_2$O$_3$-MgO ternary diagram for the faux-amphibolites and gabbros. Symbols as in Fig. 3.4-11.

profiles of the chilled margins of the ultramafic sills indicate possible interaction with the host faux-amphibolites. The similarity between the HREE profiles of the gabbros and the chilled margins of the Type-2 sills suggests that they are cogenetic, an interpretation supported by field evidence that gabbroic tops are best developed on Type-2 sills.

Although it is possible that the ultramafic and gabbro sills are somewhat younger than the faux-amphibolite, a number of observations argue that they are comagmatic feeders to the Nuvvuagittuq volcanic succession, despite their chemical differences due to alteration. First, there is a systematic progression in the western limb of the Nuvvuagittuq belt from west to east: tonalite, Type-1 ultramafic sills, BIF, Type-2 ultramafic sills and gabbro sills (Fig. 3.4-2). This sequence is mirrored in the eastern limb of the belt, although with many fewer sills. The consistency of this sequence suggests that, despite the complexities of deformation, the original volcanic stratigraphy of the belt, with its comagmatic sills, is preserved. Second, the felsic unit that yielded an age of ca. 3.8 Ga occurs as a structurally conformable layer within a gabbro sill that can be traced around outcrop-scale folds for many meters. This felsic unit is either coeval with its gabbroic host, or intrudes it.

The different compositions and mineralogy of the Type-1 and Type-2 ultramafic sills suggest that they were derived from distinct magmas. Since the cumulate rocks of both types of ultramafic sills define olivine control lines, it is possible to estimate the composi-

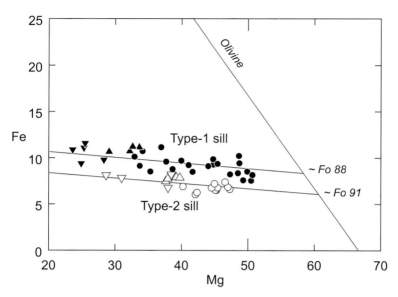

Fig. 3.4-13. Fe vs Mg in cation units for Type-1 and Type-2 ultramafic sills. Symbols as in Fig. 3.4-4.

Table 3.4-4. Major (wt%) and trace (ppm) elements for the average calculated liquids for Nuvvuagittuq ultramafic sills

	Type-1 sill average calculated liquid	Type-2 sill average calculated liquid
SiO$_2$	49.85	50.55
TiO$_2$	0.67	0.33
Al$_2$O$_3$	5.92	11.45
MgO	19.22	18.22
FeO	15.04	9.82
MnO	0.24	0.17
CaO	8.13	6.85
Na$_2$O	0.24	0.27
K$_2$O	0.07	0.09
P$_2$O$_5$	0.04	0.04
Rb	8.6	8.7
Sr	19	38
Zr	19	29
Nb	7.4	6.2
Y	21	12
Ni	672	1236
Cr	3070	13330
V	168	164
Co	155	248

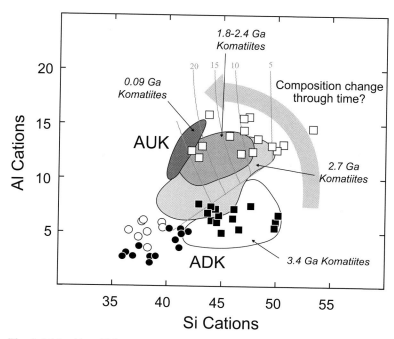

Fig. 3.4-14. Al vs Si in cation units for Type-1 and Type-2 ultramafics and for Type-1 and Type-2 calculated liquids. Symbols: black circles = Type-1 ultramafic; black squares = Type-1 calculated liquid; open circles = Type-2 ultramafic; open squares = Type-2 calculated liquid. Grey grid indicates the pressures in kbar of experimental partial melts of primitive mantle. Data are from Francis (2003) and Casey (1997).

tion of the parental magmas of both sill types. The parental liquids for each sill type were calculated by mathematically extracting the olivine defined by their olivine control lines ($\sim$Fo$_{88}$ for Type-1 sill and $\sim$Fo$_{91}$ for Type-2 sill) (Fig. 3.4-13) until the remaining composition would be in equilibrium with the extracted olivine if it were a liquid, assuming an Fe/Mg K_D ([Fe/Mg]$_{ol}$/[Fe/Mg]$_{liq}$) of 0.3.

The calculated liquids range from komatiitic basalt to komatiite in composition, with the average calculated liquid being komatiite for both sill types (Table 3.4-4). Although the compositions of the calculated liquids for both sill types have similar MgO contents (18–20 wt%), they have distinctly different Al contents (Table 3.4-4, Fig. 3.4-14). The estimated composition of the parental magma for the Type-1 sills is similar to that of aluminum-depleted komatiite (ADK; $\sim$6 wt% Al$_2$O$_3$), whereas that of the Type-2 sills resembles that of aluminum-undepleted komatiite (AUK; $>$10 wt% Al$_2$O$_3$). The presence of both ADK and AUK in the same 3.8 Ga volcanic sequence contradicts the view that there has been a secular evolution from ADK to AUK during the Archean (Fig. 3.4-14; Francis, 2003).

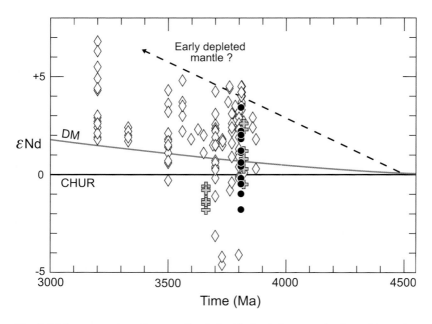

Fig. 3.4-15. ε_{Nd} vs time diagram. Symbols: black circles = rocks from this study; grey plusses = Nuvvuagittuq's rocks from Stevenson and Bizzarro (2005) and David et al. (2002); open diamonds = data from Jahn et al. (1982), Cloué-Long et al. (1984), Grau et al. (1987), Hamilton et al. (1987), Wilson and Carlson (1989), Collerson et al. (1991), Bennett et al. (1993), Vervoot et al. (1996) and Blichert-Toft et al. (1999). DM (depleted mantle) curve is modeled after DePaolo (1981). Chondritic ratios used for ε_{Nd} are $^{143}Nd/^{144}Nd = 0.512638$ and $^{147}Sm/^{144}Nd = 0.1966$.

A number of early Archean rocks from southern West Greenland, Labrador and South Africa have yielded positive initial ε_{Nd} values (up to $+4$ at 3.8 Ga) requiring an early and rapid depletion of the Earth's mantle (Collerson et al., 1991; Bennett et al., 1993; McCulloch and Bennett, 1994; Brandl and de Wit, 1997; McCulloch and Bennett, 1998; Blichert-Toft et al., 1999; Kamber, this volume). The Sm-Nd isotopic compositions of the Nuvvuagittuq rocks also support an early depletion of the mantle, based on data from gabbro and ultramafic sills yielding ε_{Nd} values as high as $+3$ (Fig. 3.4-15). Although the preliminary results indicate that the Nuvvuagittuq rocks are not as depleted as the 3.8 Ga rocks of SW Greenland (Bennett et al., 1993; Blichert-Toft et al., 1999), ε_{Nd} values greater than $+1.9$ in a mantle source at 3.8 Ga require a significant earlier depletion equivalent to that seen in the present-day MORB source. The positive ε_{Nd} values commonly obtained in mantle-derived rocks of Eoarchean supracrustal sequences imply that the Earth's mantle had already experienced an extensive trace element depletion well before 3.8 Ga. For example, an ε_{Nd} of $+4$ at 3.8 Ga would require a trace element depletion factor more than twice the average value for the present-day depleted mantle. Such depletion is not supported by the flat to slightly depleted REE profiles of the Nuvvuagittuq's gabbros (Fig. 3.4-7(b)) and the general lack of evidence for extensive trace element depletion in Eoarchean mantle-derived rocks is a continuing puzzle.

3.4-4.3. *Significance of Banded Iron Formation*

Although iron formations are generally thought to represent chemical precipitates produced by marine exhalations (Graf, 1978; Appel, 1983; Gross, 1983; Jacobsen and Pimentel-Klose, 1988; Olivarez and Owen, 1991), the causes of Fe precipitation throughout geologic time are less well understood. The Proterozoic Superior-type banded Fe-formations are believed to be formed by oxidation on shallow continental shelves of Fe^{2+} rising from deep ocean basins. This scenario cannot, however, explain the Algoma-type Fe-formations that typify earlier Archean greenstone belts, because they would have been precipitated under anoxic conditions (Holland, 1994, 2005). It has recently been discovered that Fe^{2+} can be directly oxidized by microbial activity under anoxic conditions (Lovley et al., 1987; Widdel et al., 1993). Indeed, Konhauser et al. (2002) have proposed that Algoma-type banded Fe-formations were precipitated by the activity of anaerobic bacteria in the Archean. The origin of life on Earth and its evolution over time has been highly debated in past years (e.g., Mojzsis et al., 1996; Fedo and Whitehouse, 2002a, 2002b; Mojzsis and Harrison, 2002b; Lepland et al., 2005; Moorbath, 2005), and establishing the chemical sedimentary origin for a rock is a prerequisite to demonstrating potential biological activity and addressing the controversial question of the timing of the origin of life on Earth. The concave-up depleted LREE profile of the Nuvvuagittuq BIF, combined with its flat HREE profile and positive Eu and Y anomalies with respect to PAAS, are characteristic features of Archean Algoma-type banded Fe-formations of marine exhalite origin, with a seawater signature (Fryer, 1977; Graf, Jr., 1978; Fryer et al., 1979; Jacobsen and Pimentel-Klose, 1988; Fig. 3.4-9).

BIF from the Nuvvuagittuq belt has heavier Fe isotopic compositions than the surrounding igneous lithologies (Fig. 3.4-10), a feature that is also consistent with an origin as a chemical precipitate. The Fe isotopic compositions of the BIF from the Nuvvuagittuq belt are similar to those in the Akilia BIF (SW Greenland), which are also enriched in heavy Fe isotopes (0.1 to 0.5‰/amu) relative to their surrounding igneous lithologies (Dauphas et al., 2004b), suggesting a common depositional process for both these examples of Eoarchean BIF. Although the Fe isotopic enrichment observed in the amphibolites directly adjacent to the BIF suggests some degree of local exchange, the fact that such diverse meta-igneous lithologies, including the faux-amphibolites that are interpreted to have been altered, the amphibolitic sill margins, the gabbro and ultramafic sills, as well as the tonalite and the felsic bands, all share the same Fe isotopic composition suggests that the heavy Fe isotopic enrichment displayed by the Nuvvuagittuq BIF is not a metasomatic or alteration feature. The REE+Y profiles and the Fe isotopic compositions of Nuvvuagittuq's BIF confirm their origin as marine exhalites and, although the mechanism(s) responsible are not well understood, the Fe isotopic fractionation observed in the BIF of Nuvvuagittuq and Akilia raises the possibility that life was already established on the Earth at 3.8 Ga.

3.4-5. CONCLUSIONS

The newly discovered 3.8 Ga Nuvvuagittuq greenstone belt represents one of the Earth's oldest mafic mantle-derived supracrustal suites and constitutes an important constraint for modelling the evolution of the early Earth. The Nuvvuagittuq greenstone belt differs from other greenstone belts in the Northeastern Superior Province in that it is dominated by amphibolites composed mainly of cummingtonite, unlike typical Archean amphibolites that are dominated by hornblende. This feature is thought to reflect the loss of Ca in altered and metamorphosed mafic pyroclastic rocks. The chemical compositions of the calculated parental liquids for the ultramafic sills that intrude these rocks indicate that both aluminium-depleted komatiite (ADK) and aluminium-undepleted komatiite (AUK) magmas were present at 3.8 Ga, arguing against a temporal evolution from ADK to AUK during the Archean. Despite their relatively undepleted trace element profiles, most Nuvvuagittuq's mafic and ultramafic rocks display positive ε_{Nd} values, implying derivation from a depleted mantle source and supporting evidence from other >3.6 Ga supracrustal suites for early depletion of the Earth's mantle. The concave-up LREE profiles, positive Eu and Y anomalies, and the heavy Fe isotopic enrichment in the Nuvvuagittuq BIF confirms its origin as a marine chemical precipitate and, along with similar rocks in SW Greenland, may indicate that life was already established on the Earth at 3.8 Ga.

ACKNOWLEDGEMENTS

This research was supported by Natural Science and Engineering Research Council of Canada (NSERC) Discovery Grants to Francis (RGPIN 7977-00). The whole rock XRF analyses were performed by Glenna Keating and Tariq Ahmedali, the thin sections were made by George Panagiotidis, and the Microprobe work for the garnet-biotite geothermometer was performed by Lang Shi at McGill University. We would like to thank the municipality of Inukjuak and the Pituvik Landholding Corporation for permission to work on their territory. We also thank the Inukjuak community and especially Mike Carroll, Valerie Inukpuk Morkill, Rebecca Kasudluak, and Johnny Williams for their hospitality and support. We thank Witold Ciolkiewicz and Alexandre Jean for assistance in the field and Shoshana Goldstein for late night discussions in the office. GEOTOP publication No. 2007-0006. Ministère des Resources naturelles et de la Faune contribution #84999-20087-01.

Earth's Oldest Rocks
Edited by Martin J. Van Kranendonk, R. Hugh Smithies and Vickie C. Bennett
Developments in Precambrian Geology, Vol. 15 (K.C. Condie, Series Editor)   251
© 2007 Elsevier B.V. All rights reserved.
DOI: 10.1016/S0166-2635(07)15035-0

Chapter 3.5

EOARCHEAN ROCKS AND ZIRCONS IN THE NORTH CHINA CRATON

DUNYI Y. LIU[a,b], Y.S. WAN[a,b], J.S. WU[a], S.A. WILDE[c], H.Y. ZHOU[b], C.Y. DONG[a,b] AND X.Y. YIN[a,b]

[a]*Institute of Geology, Chinese Academy of Geological Sciences, Beijing 100037, China*
[b]*Beijing SHRIMP Centre, Beijing 100037, China*
[c]*Department of Applied Geology, Curtin University of Technology, Perth, Western Australia 6845, Australia*

3.5-1. INTRODUCTION

There are about 35 Archean crustal blocks around the globe and along the margins of most of them Paleoproterozoic rifts developed, suggesting that these cratons and fragments are the result of break-up of still larger supercratons (Bleeker, 2003). A wealth of zircon dating indicates that $\sim$2.7 Ga (mainly in the range 2.65–2.75 Ga) tectonothermal events occurred extensively in these cratons (Condie, 1998). In contrast, a $\sim$2.5 Ga tectonothermal event is most significant in the North China Craton (NCC), which makes its Neoarchean history markedly different from most other cratons. Extensive Paleoproterozoic tectonothermal events also occurred in the NCC. Distinctively, the NCC is one of the four rare regions in the world where $\geqslant$3.8 Ga continental rocks are found (Liu et al., 1992; Song et al., 1996; Wan et al., 2005a). In the Anshan area, large volumes of >3.6 Ga rocks have been identified, including three different types of rocks older than 3.8 Ga. Abundant >2.8 Ga rocks and zircons are also present in many other areas of the NCC. Following an introduction to the general geology of the NCC, this chapter focuses on the advances made in recent years in the study of its Eoarchean rocks.

3.5-2. GENERAL GEOLOGY

Located in the eastern Eurasian continent, the NCC is broadly confined to longitudes 100–130° E and latitudes 32–42° N, with a total area of $\sim$1 million km^2 (Fig. 3.5-1). Since the 1980s, many specialised studies and regional summaries have been conducted, which laid a good foundation for more intensive studies (Sun et al., 1984; Qian et al., 1985; Bai et al., 1986, 1996; Lan et al., 1990; Wu et al., 1991, 1998; Zhao, 1993; Li et al., 1994; Shen et al., 1994; Lu et al., 1996; Shen et al., 2000; Chen et al., 2005). Zhao et al. (2000, 2001a, 2001b, 2003) identified a nearly N-S-trending Paleoproterozoic collisional orogenic

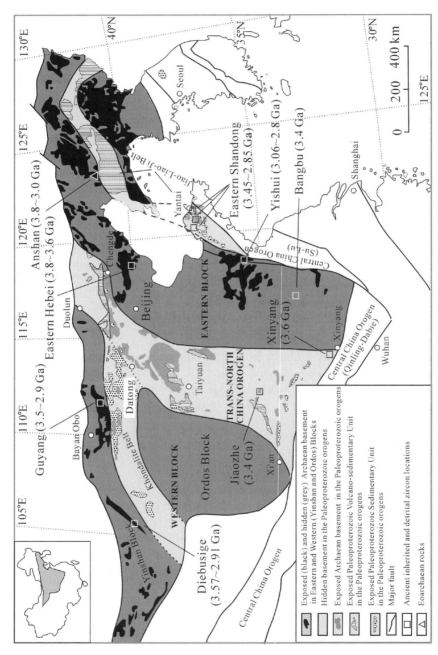

Fig. 3.5-1. Map of the North China Craton (modified after Zhao et al. (2005) and Wan et al. (2006)). Also shown are locations where rocks and zircons older than 2.8 Ga are present (data from Liu et al., 1992 and in review; Ji, 1993; Song et al., 1996; Wang et al., 1998; Li et al., 2001; Lu et al., 2002b; Jin et al., 2003; Zheng et al., 2004; Shen et al., 2004; Xu et al., 2004; Wan et al., 2005; and Gao et al., 2006).

belt extending through Wutai and Hengshan in Shanxi province to Kaifeng and Lushan in Henan province, called the "Trans-North China Orogen", which separates the NCC into the Eastern and Western blocks (Fig. 3.5-1). Recently Zhao et al. (2005) delineated a Khondalite (metasedimentary) Belt in the Western Block. They considered that the time of collision of the Yinshan and Ordos blocks along the Khondalite Belt might be slightly earlier than the collision along the Trans-North China Orogen. In addition, they also delineated a Paleoproterozoic Jiao–Liao–Ji tectonic belt in the Eastern Block. The identification of the Paleoproterozoic Trans-North China Orogen was an important step in understanding Precambrian events in the NCC. However, some problems merit further study. For example, from the formation of the arcs represented by the Hengshan–Wutai–Fuping complexes at the end of the Archean, to collision of the Eastern and Western blocks at the end of the Paleoproterozoic, spanned 0.6–0.7 billion years. Such lengthy evolution is anomalous on a global scale. Furthermore, there are several early Paleoproterozoic tectonothermal events in the Trans-North China Orogen (Wan et al., 2006), so the Paleoproterozoic evolution is more complex than previously thought.

Shen et al. (2005) presented a histogram of the early Precambrian zircon age data from the NCC, which shows a maximum at 2.55–2.5 Ga. In addition, minor peaks appear at ~1.85, 2.15 and ~2.7 Ga (Fig. 3.5-2(a)). Recent zircon geochronological studies further support this result (Geng et al., 2006; Tian et al., 2006; Kröner et al., 2006; Wan et al., 2006). Wan et al. (2006) applied SHRIMP U-Pb zircon dating to metasedimentary rocks of various metamorphic grade across the NCC. The ages of metamorphic zircons, as identified by their textures and compositions, cluster at ~1.85 Ga (Fig. 3.5-2(b)), representing the time of the late Paleoproterozoic collisional orogeny and unification of the NCC. Besides >2.4 Ga detrital zircons, large numbers of 2.35–2.0 Ga detrital zircons are also present in the sedimentary rocks (Fig. 3.5-2(b)), which not only supports the view that the early Paleoproterozoic tectonothermal event was prevalent in the NCC, but also indicates that metasedimentary rocks in the NCC, including the khondalite series, formed in the Paleoproterozoic rather than the Archean (as previously thought by many researchers).

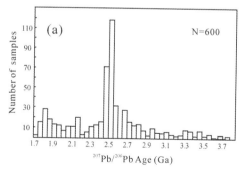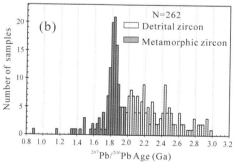

Fig. 3.5-2. Age histograms of: (a) zircon data for precambrian rocks in the North China Craton (after Shen et al., 2005); (b) metamorphic and detrital zircon data from Proterozoic metasedimentary rocks (after Wan et al., 2006) for Precambrian rocks in the North China Craton.

Paleoproterozoic rocks are mainly distributed in the above-mentioned three tectonic belts defined by Zhao et al. (2005) (Fig. 3.5-1). The Paleoproterozoic high pressure granulite assemblages preserved in the Trans-North China Orogen are interpreted to be the result of Paleoproterozoic continental collision (Guo et al., 2005; Kröner et al., 2006). According to the rock associations and their ages, two associations have been distinguished: an older "volcano-sedimentary unit" that formed between 2.37 and 2.0 Ga and a younger "sedimentary unit" that formed mainly between 2.0 and 1.88 Ga (Wan et al., 2006).

After unification at ∼1.85 Ga, the NCC entered an extensional phase, resulting in characteristic anothosite, rapikivi granite and mafic dyke emplacement (Sun et al., 1991; Song, 1992; Li et al., 1995, 1997; Lu, 2002; Lu et al., 2002; Zhao et al., 2004; Peng et al., 2005). The Mesoproterozoic Changcheng System unconformably overlies the early Precambrian metamorphic basement. The youngest detrital zircon age obtained from the Changzhougou Formation in the lowest part of the Changcheng System is ∼1.8 Ga. Therefore, it is evident that the Changcheng rocks formed after 1.8 Ga (Wan et al., 2003a) and represent a cover sequence over blocks of old continental crust amalgamated by ∼1.85 Ga.

Whereas the zircon age histogram indicates that the period 2.55–2.50 Ga is the most important tectonothermal event in the NCC (Fig. 3.5-2(a)), whole-rock Nd isotope data (Wu et al., 2005a) suggest that the Nd depleted mantle model ages of both mafic and felsic rocks in the NCC mainly cluster at 3.0–2.6 Ga (Fig. 3.5-3), i.e., 0.4–0.2 Ga older than the zircon U-Pb age of the rocks. Undoubtedly the Neoarchean was the main period of formation and accretion of continental crust in the NCC. The areas where zircon chronological studies have been most intensive include Fushun, Anshan, Dashiqiao and Jianping in Liaoning Province (Kröner et al., 1998; Song et al., 1996; Wu et al., 1998; Wan, 1993; Wan et al., 2003b, 2005b, 2005c), eastern Hebei Province (Liu et al., 1990; Wan et al.,

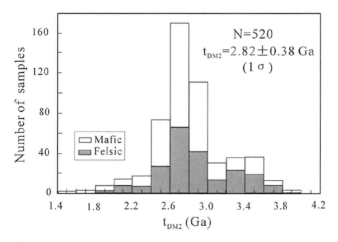

Fig. 3.5-3. Whole rock t_{DM} data from mafic and felsic rocks in the North China Craton (after Wu et al., 2005).

2003b), Hengshan, Wutai and Fuping in Shanxi Province (Liu et al., 1985; Wilde et al., 1997, 2005; Guan et al., 2002; Zhao et al., 2002; Chen et al., 2005; Kröner et al., 2005), Jiaozuo and Dengfeng in Henan Province (Li et al., 1987; Wang et al., 2004; Gao et al., 2006), western Shandong Province (Jahn et al., 1988; Cao et al., 1996; Shen et al., 2004) and at Jining and Guyang in Inner Mongolia (Shen et al., 1987; Jian et al., 2005).

There are two main views about the nature of the Neoarchean tectonothermal event in the NCC: one is arc magmatism culminating in collisional orogeny. Zhao et al. (2000, 2001a, 2001b, 2002, 2003) considered that the Hengshan–Wutai–Fuping complexes in the Trans-North China Orogen are the product of arc magmatism, but that the Eastern and Western blocks did not collide until late in the Paleoproterozoic. Alternatively, Li et al. (2000, 2002) and Kusky et al. (2001) proposed collision in the Neoarchean. The south-central segment in this model is similar in spatial distribution to that of the Paleo-proterozoic collisional orogenic belt proposed by Zhao et al. (2000, 2001a, 2001b, 2003) although the northern portion of the belt turns eastward into western Liaoning – southern Jilin provinces. Li et al. (2000, 2002) and Kusky et al. (2001) proposed that the Western and Eastern blocks collided and amalgamated as early as the end of the Archean, forming the unified NCC. Evidence proposed for this includes a $\sim$2.5 Ga linear granite-greenstone belt from Fushun-Qingyuan in western Liaoning province, through Hengshan–Wutai–Fuping in western Hebei – northern Shanxi provinces, to Kaifeng–Lushan in Henan province. They suggest the rock associations and geochemistry show island arc features and there are pos-tulated ophiolites of similar age (Kusky et al., 2001; Li et al., 2002). Likewise, the studies of Wan et al. (2005a, 2005b) in the northern Liaoning area show that island arc magmatism has great significance for Neoarchean continental crustal accretion of the NCC. This means that plate tectonics played a role in the NCC as early as the end of the Archean and was the main cause for continental crustal accretion of the NCC. However, this view is ques-tioned, since: (1) no reliable $\sim$2.5 Ga metamorphic ages have been obtained up to now in the central and southern parts of the Trans-North China Orogen; (2) the age of high-pressure granulite is not Neoarchean but Paleoproterozoic (Kröner et al., 2006); and (3) many geologists now consider that the so-called "Archean ophiolite" is neither Archean, nor an ophiolite (Zhao et al., 2007).

The other possible mode of magma generation is underplating from mantle plumes. The main features supporting this view are as follows: (1) different types of magmatic rocks formed in a very short time period at $\sim$2.5 Ga; (2) in some places, 2.5 supracrustal rocks and trondhjemite-tonalite-granodiorite (TTG) granitic rocks were metamorphosed soon after their formation (Kröner et al., 1998; Zheng et al., 2004b; Wan et al., 2005a); (3) 2.5 Ga rocks are widespread in the NCC rather than only distributed in the so-called "Trans-North China Orogen". For example, $\sim$2.5 Ga rocks and zircons are also found in the Guyang area of Inner Mongolia in the Western Block and the Xuzhou-Huaihe and Bangbu areas in the Eastern Block (Jin et al., 2003; Xu et al., 2004a, 2004b; Jian et al., 2005). There is thus no strong linear control on the distribution of 2.5 Ga magmatic rocks.

As stated above, tectonothermal events in many areas of the NCC took place at $\sim$2.5 Ga. In these areas, $\sim$2.7 Ga rocks are rare, but do occur (Guan et al., 2002). Large amounts of $\sim$2.7 Ga rocks have only been recognised in eastern and western Shandong province

(Jahn et al., 1988; Jhan et al., unpublished data; Wan et al., unpublished data), including TTG granitic rocks and volcano-sedimentary supracrustal rocks. Their ages are notably different from those in other areas of the NCC, but very similar to those of other cratons around the world. It is notable that the granite-greenstone belt in western Shandong province contains typical komatiites (Cao et al., 1996; Polat et al., 2006), a feature of ~2.7 Ga granite-greenstone belts around the world (Nelson, 1998). However, this granite-greenstone belt is invaded by abundant ~2.5 Ga crustally derived granite, so it still shows significant differences from ~2.7 Ga greenstone belts in other cratons.

3.5-3. OLDEST ROCKS AND ZIRCONS IN THE NCC

3.5-3.1. Anshan Area

Anshan, located in the northeastern part of the Eastern Block (Fig. 3.5-1), is the area where the oldest rocks in the NCC outcrop (Fig. 3.5-4). In this area, three different types of 3.8 Ga rocks have been discovered: the Baijiafen mylonitized trondhjemitic rocks, the Dongshan banded trondhjemitic rocks and the Dongshan meta-quartz diorite. The ages and compositions of associated younger Archean rocks include, the 3.3 Ga Chentaigou supracrustal rocks, 3.3 Ga Chentaigou granite, 3.1 Ga Lishan trondhjemite, 3.0 Ga Donganshan granite, 3.0 Tiejiashan granite, 2.5 Ga Anshan Group supracrustal rocks, and 2.5 Ga Qidashan granite (Liu et al., 1992; Song et al., 1996; Wu et al., 1998; Wan et al., 1997, 1998, 1999, 2001, 2002, 2005a). Eoarchean rocks, including those found in the past five years, are mainly distributed at three sites (Fig. 3.5-4). These are described below.

3.5-3.1.1. Baijiafen complex
The site of the first discovery of 3.8 Ga rocks in the Anshan area is the Baijiafen quarry (Fig. 3.5-4). At the entrance of the quarry, there crops out a suite of strongly mylonitized rocks dominated by trondhjemite and called the Baijiafen granite (Song et al., 1996) or Baijiafen gneiss (Wan et al., 2005a). Here we refer to it as the Baijiafen complex, because it is actually composed of rocks of different ages and origins. The complex, ~700 m long and ~50 m wide, is elongated in a NW-SE direction and is in tectonic contact with the 3.3 Ga Chentaigou granite on the SW side. SHRIMP zircon dating of samples AB87-7 and A9011 (Liu et al., 1992) gave rise to mostly discordant data, showing strong lead loss (Fig. 3.5-5). Three data points close to concordia and distributed together gave a weighted mean ^{207}Pb/^{206}Pb age of 3804 ± 5 Ma, which was considered to be the formation age of the Baijiafen granite (Liu et al., 1992). The samples also contain ~3.3 Ga and other zircon ages, these being interpreted as resulting from later granitoid intrusion and metamorphism.

Recently, we have undertaken more intensive geochronology on the Baijiafen complex at the quarry site (Liu et al., in review). Sample A0518 was taken from a ~10 cm wide mylonitized trondhjemitic layer in the Baijiafen complex, ~2 m away from the boundary between the 3.3 Ga Chentaigou granite and the Baijiafen complex. It is distinguished from the surrounding trondhjemitic rocks by its lower biotite content. The zircons are of

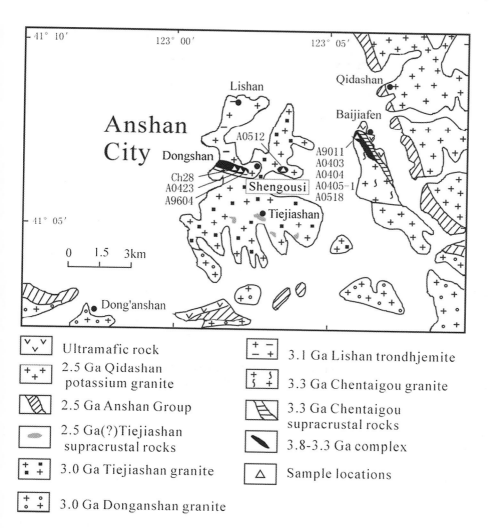

Fig. 3.5-4. Geological sketch map of the Anshan area (after Wan et al., 2005), showing distribution of sample locations. Baijiafen: A9011, A0403, A0404, A0405-1 and A0518; Dongshan: Ch28, A0423 and A9604; and Shengousi: A0512 are from Liu et al. (in review).

magmatic origin and most SHRIMP data points lie on, or near, concordia, but some show loss of radiogenic lead (Fig. 3.5-6(a)). The weighted mean $^{207}Pb/^{206}Pb$ age of ten near-concordant analyses is 3784 ± 16 Ma, but with a very high MSWD (11.7). Four analyses (2.1, 3.1, 9.1 and 9.2) recording the maximum $^{207}Pb/^{206}Pb$ ages give a weighted mean age of 3800 ± 5 Ma (MSWD = 0.53), being indistinguishable from the zircon ages determined on samples AB87-7 and A9011 (Liu et al., 1992). Another three analyses (1.1, 6.1 and 14.1) with similar $^{207}Pb/^{206}Pb$ ratios give a slightly younger weighted mean age

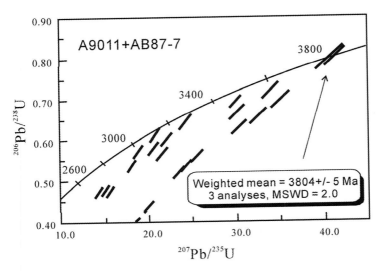

Fig. 3.5-5. SHRIMP U-Pb concordia diagram for zircons from trondhjemitic gneiss (samples A87-7 and A9011) from the Baijiafen Complex in the Anshan area (after Liu et al., 1992).

of 3752 ± 8 Ma (MSWD = 0.27). Zircons with ages of ~3.5 Ga and ~3.3 Ga are also present. We consider that the 3.8 Ga age represents the formation age of the trondhjemite, while other ages record later tectonothermal events (Liu et al., in review).

Sample A0405 is a grey, mylonitized trondhjemitic rock collected northeast of sample A0518. Zircons are prismatic in shape and show magmatic zoning, and some zircons enclose relict cores (Liu et al., in review). Magmatic zircons have undergone strong lead loss, but many data are distributed on a single discordia line (Fig. 3.5-6(b)). The upper intercept age controlled by three data points (3.1, 3.2 and 5.1) that are closest to concordia is $3585 + 25/-10$ Ma (MSWD = 1.7) and their weighted mean ^{207}Pb/^{206}Pb age is 3573 ± 21 Ma (MSWD = 4.8). This age is interpreted as the formation time of the trondhjemite. Data point 7.1, analyzed on a core enclosed in magmatic zircon (as indicated by the cathodoluminescence (CL) image), is closest to concordia, with a ^{207}Pb/^{206}Pb age of ~3790 Ma (Fig. 3.5-6(b)).

In the Baijiafen complex there are some volumetrically minor biotite schists, which form lenses a few centimeters to half a meter wide and alternate with the trondhjemitic rocks. The schists were originally considered to be altered basic dikes (Song et al., 1996), but the intercalation of thin siliceous bands interpreted as chert (Fig. 3.5-7) indicates that they are probably of supracrustal origin (Liu et al., in review). Zircons in the biotite schist sample A0403 are columnar in shape with magmatic zoning. Four data points (6.1, 9.1, 10.1 and 12.1) lie on the same discordia (Fig. 3.5-6(c)), with an upper intercept age of 3725 ± 65 Ma (MSWD = 6.1) and a weighted mean ^{207}Pb/^{206}Pb age of 3723 ± 17 Ma (MSWD = 4.1) (Liu et al., in review). If the protolith of the biotite schist was of volcanic origin, this age should represent the formation age of the supracrustal rocks, but if the

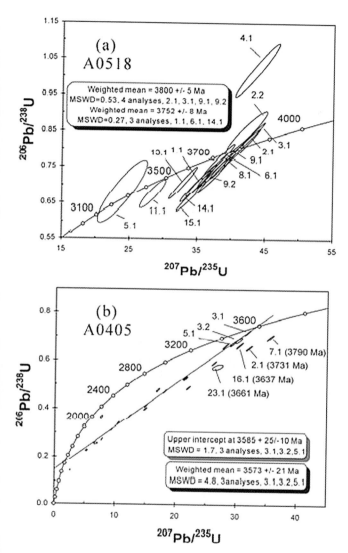

Fig. 3.5-6. SHRIMP U-Pb concordia diagrams for zircons from the Baijiafen Complex in the Anshan area (a) fine-grained trondhjemitic rock (A0518); (b) grey trondhjemitic rock (A0503); (c) biotite schist (A0403); (d) trondhjemitic dyke (A0404) (from Liu et al., in review).

protolith was of sedimentary origin, the age should limit the maximum formation age of the supracrustal rocks. Either way, because the rock is clearly cut by 3.62 Ga trondhjemitic dykes (Fig. 3.5-7), the formation age of the protolith of the biotite schist must be older than 3.62 Ga. This sample also contains zircons with younger ages, of which some are broadly distributed along a discordia with an upper intercept age of ~3.6 Ga.

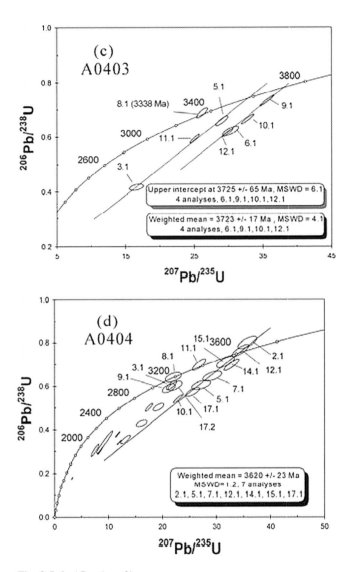

Fig. 3.5-6. (*Continued.*)

From the field relations, it is clear that the biotite schist represented by sample A0403 is cut by trondhjemite dykes (Fig. 3.5-7). The dykes vary in width from several centimeters to more than ten centimeters and show a gneissic structure, suggesting that the dykes have undergone metamorphism and deformation after their formation (Liu et al., in review). Zircons from a trondhjemite dyke (sample A0404) have well-developed magmatic zoning and eight data points define a discordia (Fig. 3.5-6(d)). The upper intercept age defined

Fig. 3.5-7. Field photograph showing relationships of chert and biotite schist, cut by 3.6 Ga trondhjemitic dyke, in the Baijiafen Quarry. The hammer is 40 cm long.

by seven of these data points, and excluding point 10.1 with the strongest discordance, is 3622 ± 33 Ma (MSWD = 1.6); the weighted mean ^{207}Pb/^{206}Pb age is 3620 ± 23 Ma (MSWD = 1.2). It is considered that 3.62 Ga represents the intrusion age of the dyke. Some ~3.3 Ga zircons also occur in sample A0404 and probably reflect emplacement of the nearby 3.3 Ga Chentaigou trondhjemite (Liu et al., in review).

These studies demonstrate that the diverse components of the Baijiafen complex show a range of ages. The following main conclusions can be drawn:

(1) the new 3.8 Ga zircon age determinations further confirm that there is Eoarchean crustal material in the Baijiafen complex (Liu et al., 1992; Liu et al., in review);

(2) there are also rocks or zircons with ages of 3.72, 3.62 and 3.57 Ga;

(3) it is very likely that >3.6 Ga supracrustal rocks occur in the complex, and if this is the case, these are the first Eoarchean supracrustal rocks found in the Anshan area and they are the only recognized occurrence in Asia;

(4) the zircons in the old rocks record ~3.3 Ga ages, which are the result of superimposition of a ~3.3 Ga tectonothermal event that included the emplacement of the nearby 3.3 Ga Chentaigou trondhjemite (the Baijiafen complex itself contains ~3.3 Ga mylonitized trondhjemitic rocks (Liu et al., in review)); and

(5) hence the Baijiafen complex is composed of several rock types of diverse origin and age.

3.5-3.1.2. Dongshan complex

Situated in the Dongshan park in Anshan, the Dongshan complex extends in a nearly E-W direction with a strike length of >1000 m and is commonly less than 10 m wide, between the 3.1 Ga Lishan trondhjemite and 3.0 Ga Tiejiashan granite (Fig. 3.5-4). Banded migmatite of overall trondhjemitic composition, with ages of 3.81 and 3.32 Ga (Song et al., 1996), constitute the main part of the complex, but there are also some 3.7 Ga banded trondhjemitic rocks (see below), which show no appreciable difference in appearance. The migmatitic banding generally strikes 310–340°, parallel to the trend of the complex. Apart from 3.8–3.7 Ga banded trondhjemitic rocks, there are also other 3.8–2.5 Ga rocks, including phlogolite-tremolite rocks, amphibolite, mica schist, quartzite, dioritic gneiss, meta-quartz diorite, fine-grained trondhjemite, monzogranite and pegmatite veins. Below we only consider the >3.6 Ga rocks.

The sample (Ch28) of 3.8 Ga banded trondhjemitic rock selected for SHRIMP zircon dating was collected near a pavilion beside a narrow road in Dongshan park (Fig. 3.5-8) and the rock consists of fine-grained grey bands and coarse-grained leucocratic bands, a few millimeters to 1–2 cm wide (Song et al., 1996). The minerals in the leucocratic bands are coarser in grain size, mainly consisting of microcline, plagioclase and quartz; whereas the minerals in the grey bands are finer in grain size, mainly consisting of plagioclase, quartz and biotite, with minor microcline. On the concordia diagram, zircon data points clearly define two discordia lines (Fig. 3.5-9(a)), giving upper intercept ^{207}Pb/^{206}Pb ages of 3811 ± 4 Ma (MSWD = 1.2) and 3322 ± 12 Ma (MSWD = 3.9), respectively. They are interpreted as the ages of formation of the trondhjemite and the time of late-stage anatexis, respectively (see also Song et al., 1996; Liu et al., in review), as indicated by the banded appearance of the rock.

Fig. 3.5-8. Field photograph of 3.8 Ga banded trondhjemitic rock in Dongshan. The knife is ~13 cm long.

SHRIMP zircon dating of another trondhjemitic rock sample (A0423), collected several tens of metres away, gives different ages (Liu et al., in review). This sample is the grey, fine-grained part of a banded trondhjemite, with only a few thin coarse-grained bands. The zircons commonly show strong lead loss but the data are mainly located on the same discordia (Fig. 3.5-9(b)), suggesting lead loss from zircons at a common time. Three data points (4.1, 17.1 and 18.1) with the least lead loss give a weighted mean $^{207}Pb/^{206}Pb$ age of 3680 ± 19 Ma (MSWD = 4.0), apparently representing the age of the trondhjemite. Data point 12.1 is the core in magmatic zircon and, although discordant, its $^{207}Pb/^{206}Pb$ age is 3795 ± 25 Ma, being in agreement with the 3.8 Ga zircon age obtained from nearby trondhjemites in the complex. There is no evidence of 3.3 Ga zircon in this sample.

A 3.8 Ga meta-quartz diorite in the Dongshan complex occurs as small "enclaves" in banded trondhjemitic rocks (Wan et al., 2005a). However, their original relationships are unclear, due to strong deformation. The rock has a massive or weakly gneissic structure and consists predominantly of plagioclase, biotite and quartz, with minor K-feldspar, and secondary chlorite and epidote. Some plagioclase grains show polysynthetic twinning and

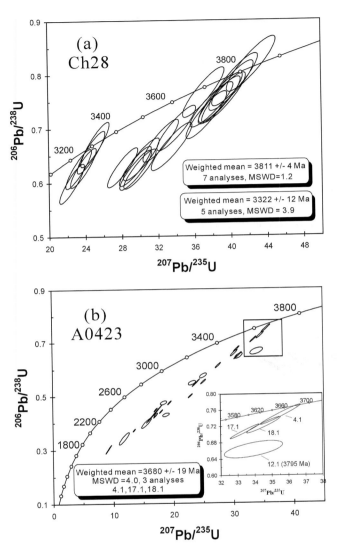

Fig. 3.5-9. SHRIMP U-Pb concordia diagrams for zircons from the Dongshan and Shengousi Complexes in the Anshan area. (a) Banded trondhjemitic rock (Ch28) (after Song et al., 1996). (b) Grey portion of banded trondhjemitic rock (A0423). (c) Meta-quartz diorite (A9604) (after Wan et al., 2005). (d) Banded trondhjemitic rock (A0512). Samples Ch28, A0423 and A9604 are from the Dongshan Complex and sample A0512 is from the Shengousi Complex (Liu et al., in review).

growth zoning of magmatic origin, suggesting that its protolith is igneous. Magmatic zoning is also distinct in CL images of the zircons, although some show recrystallization and metamorphic overgrowths. The upper intercept age of a discordia defined by eight data

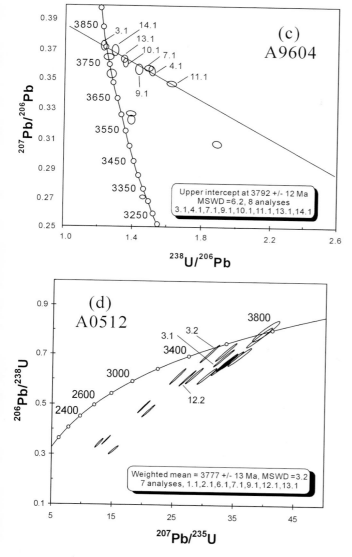

Fig. 3.5-9. *(Continued.)*

points is 3792 ± 12 Ma (MSWD = 6.2) (Fig. 3.5-9(c)). Data point 3.1 is concordant and gives the oldest $^{207}Pb/^{206}Pb$ age of 3792 ± 4 Ma; this is therefore considered the formation age of the quartz diorite (Liu et al., in review). There are also variably discordant zircons with ~3.7, ~3.6 and ~3.3 Ga ages in the sample (Fig. 3.5-9(c)), indicating that the rock has undergone repeated modification by later geological processes.

3.5-3.1.3. Shengousi complex
The Shengousi complex is another area where 3.8 Ga old rocks have been found recently
(Liu et al., in review). It occurs east of the Dongshan Park (Fig. 3.5-4) and the complex is
composed of rocks of different types and ages from 3.8 to 3.0 Ga. Here we only introduce
the results of age dating of the oldest rock, a 3.8 Ga banded trondhjemitic gneiss. This
unit (represented by sample A0512), with an outcrop width of several meters, resembles
in appearance the banded trondhjemites in the Dongshan complex (Fig. 3.5-10). Seven
data points from magmatic zircon are clustered near concordia (Fig. 3.5-9(d)) and give a
weighted mean ^{207}Pb/^{206}Pb age of 3777 ± 13 Ma (MSWD = 3.2), which is interpreted as
the formation age. The rock also contains younger zircons. Of these, data point 3.2, which
was measured at the edge of a zircon grain, falls near concordia (Fig. 3.5-9(d)) and it has a
Th/U ratio of 0.02, suggesting that it is metamorphic in origin (Williams, 2001). Therefore,
its ^{207}Pb/^{206}Pb age of 3555 ± 6 Ma is interpreted to reflect metamorphism at ~3.56 Ga
(Liu et al., in review).

It is noteworthy that 3.3 and 3.0 Ga granitic rocks and 3.5–3.0 Ga detrital zircons are
present in the Gongchangling area, 40 km east of Anshan (Wan, 1993) and that detrital

Fig. 3.5-10. Field photograph of 3.8 Ga banded trondhjemitic rock in the Shengousi Complex. The
pen is 14 cm long.

and inherited zircons of 3.61, 3.3, 3.1 and 2.8 Ga also occur in Neoarchean supracrustal and granitic rocks in the Haicheng-Dashiqiao area, ~40 km south of Anshan (Wan et al., 2003b). This suggests that Eoarchean components occur more widely outside the Anshan area.

3.5-3.2. Other Areas Containing Ancient Rocks and Zircons in the NCC

3.5-3.2.1. Eastern Hebei Province

The Caozhuang rock series is distributed in the vicinity of Huangbaiyu in the southern part of the Qi'an gneiss dome in eastern Hebei Province (Fig. 3.5-1). It consists of amphibolite, biotite-plagioclase gneiss, banded iron formation (BIF), fuchsite quartzite, impure marble, biotite schist and barium-adularia-bearing gneiss. The units are interleaved with, and also occur as small enclaves in, ~2.5 Ga granitic rocks. According to a whole-rock Sm-Nd isochron age from amphibolite, the Caozhuang rock series formed at ~3.5 Ga (Jahn et al., 1987). However, as the measured rock samples included different types of amphibolite, the reliability of the Sm-Nd age is uncertain. Liu et al. (1992) performed SHRIMP dating of detrital zircons from fuchsite quartzite in the area. A total of 81 spots were analyzed on 61 zircon grains. The data are variably discordant (Fig. 3.5-11(a)) and the $^{207}Pb/^{206}Pb$ ages vary from 3.85 to 3.55 Ga. The 3.85 Ga zircons are the oldest materials so far obtained from the NCC. Most detrital zircons show well-developed oscillatory zoning and have Th/U ratios markedly higher than 0.1, indicating their derivation from a magmatic source (Williams, 2001). However, one grain with an age of 3.55 Ga has a very low Th/U ratio (0.02) and therefore may be of metamorphic origin. This could indicate that a tectonothermal event occurred at this time in the source region from which the detritus of the quartzite was derived, restricting its deposition time to after ~3.55 Ga (note that a zircon of similar age was obtained from the Shengousi Complex, as discussed above). Some detrital zircons have overgrowths with metamorphic ages ranging from 1.95 to 1.84 Ga (Fig. 3.5-11(a)) (Liu et al., 1992). These are considered to have formed in situ, post-deposition, and to result from a younger metamorphic event.

Wu et al. (2005) carried out LA-ICP-MS U-Pb dating and laser LA-MC-ICP-MS measurements of Hf isotopic compositions on detrital zircons from a separate sample of the fuchsite quartzite. The results of zircon dating are similar to those of Liu et al. (1992), but the zircons commonly show stronger lead loss. With the exception of sample 04QA04-02, the $^{176}Hf/^{177}Hf$ values of analyzed zircons vary from 0.28065 to 0.28034, and the Hf depleted mantle model ages vary between 3.96 and 3.63 Ga. In the $\varepsilon_{Hf}(t)$ vs $^{207}Pb/^{206}Pb$ age diagram, data points have a linear distribution (Fig. 3.5-11(b)). Four zircon grains with a $^{207}Pb/^{206}Pb$ age of ~3.8 Ga have $\varepsilon_{Hf}(t) = 1.8 \pm 3.7$ and show a Hf isotopic composition similar to that of chondrite. Wu et al. (2005b) interpreted these results to show that the ~3.8 Ga crust beneath eastern Hebei province was derived from a mantle source region where no pronounced crust–mantle differentiation occurred, implying that no continental material older than 3.8 Ga exists in the area.

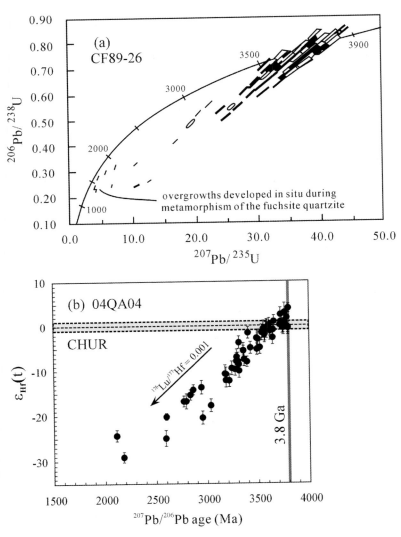

Fig. 3.5-11. (a) SHRIMP U-Pb concordia diagram (after Liu et al., 1992) and (b) $\varepsilon_{Hf}(t)$ versus $^{207}Pb/^{206}Pb$ age diagram (after Wu et al., 2005) for detrital zircons from fuchsite quartzite of the Caozhuang series (CF89-26 and 04QA04) in Eastern Hebei.

3.5-3.2.2. Xinyang area, Henan province

The Precambrian basement exposed in the Dengfeng–Lushan–Xinyang area in the southern segment of the Trans-North China Orogen is Paleoproterozoic to Neoarchean in age, and the oldest components are ~2.8 Ga TTG rocks (Kröner et al., 1988). Zheng et al. (2004a) have reported U-Pb ages and Hf isotopic compositions of zircons from felsic granulite xenoliths in Mesozoic volcanic rocks in the Xinyang area. LA-ICP-MS and LA-MC-ICP-MS were used for zircon analyses, and TIMS U-Pb dating was also performed on a few individual zircon grains. The characteristics, ages and Hf isotopic compositions of zircons in samples XY9951 and XY9928 are rather similar. The zircons show core-rim structure or lack internal structure in CL images. The cores of zircon grains with core-rim structure show oscillatory or irregular zoning, and strong lead loss is common (Fig. 3.5-12(a)). The upper and lower intercept ages of the discordia fitted for the data points of samples XY9951 and XY9928 are 3670 ± 120 and 1981 ± 260 Ma and 3655 ± 100 and 1767 ± 210 Ma, respectively. By combining the data of the two samples and excluding three discordant data points, we obtain 3659 ± 59 Ma and 1786 ± 140 Ma for the upper and lower intercept ages of the discordia, respectively (Fig. 3.5-12(a)). In the $\varepsilon_{Hf}(t)$ vs ^{207}Pb/^{206}Pb age diagram, the data points generally show a linear distribution, with negative $\varepsilon_{Hf}(t)$ values (Fig. 3.5-12(b)). According to these data, Zheng et al. (2004) concluded that the protolith of felsic granulite samples XY9951 and XY9928 had already been separated from the mantle before 4.0–3.9 Ga, underwent anatexis at 3.7–3.6 Ga, and experienced granulite-facies metamorphism at ~1.9 Ga.

3.5-3.2.3. Jiaozuo area, Henan province

3.4 Ga detrital zircons have been found in the Precambrian metamorphic basement in the Jiaozuo area, ~400 km NNW of Xinyang in the Trans-North China Orogen (Fig. 3.5-1; Gao et al., 2006). According to regional correlation and zircon isotope chronology, the Precambrian basement in the area formed in the Neoarchean and is unconformably overlain by Mesoproterozoic-Paleozoic strata (BGMRHP, 1989). In outcrops covering an area of up to 100 m^2, where sample 04516-1 was taken for zircon dating, the metamorphic basement consists principally of felsic gneiss and small amounts of amphibolite. The felsic gneiss is recumbently folded and has undergone potassic alteration. Despite strong metamorphism and deformation, Gao et al. (2006) suggested that lamination might still represent the original sedimentary bedding. Zircons in potassic-altered biotite-plagioclase gneiss (sample 04516-1) commonly have a core-rim structure and magmatic zoning is present in the inherited cores. The Th/U ratios of the cores range from 0.24 to 0.81 (16 data points), indicating that they were probably derived from magmatic source rocks (Williams, 2001). Detrital zircons generally show strong lead loss, but most data are distributed along a discordia (Fig. 3.5-13) of which five points closest to the upper intercept give a weighted mean ^{207}Pb/^{206}Pb age of 3399 ± 8 Ma and the upper intercept age of the discordia line is 3404 ± 6 Ma. Data point 17.1 is very close to concordia with a ^{207}Pb/^{206}Pb age of 3302 ± 13 Ma. These data indicate that the source region of the detrital material is mainly made up of 3.40 Ga rocks and small amounts of 3.30 Ga rocks. The metamorphic overgrowths along zircon rims are usually narrow with no internal structure. Two analyses (16.1 and 16.2 in Fig. 3.5-13) have Th/U ratios of 0.05–0.09. However, it is difficult to accurately determine the metamorphic rim age due to the strong lead loss.

There are reports of old zircon ages in several other areas in the NCC. Neoarchean hornblende granite in the Guyang area, Inner Mongolia, on the northwestern margin of the

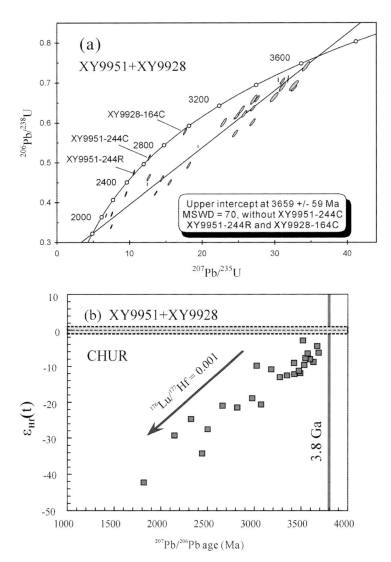

Fig. 3.5-12. (a) SHRIMP U-Pb concordia diagram and (b) $\delta_{Hf}(t)$ versus $^{207}Pb/^{206}Pb$ age diagram for zircons from felsic granulite xenoliths (XY9951 and XY9928) in the Xinyang area, Southern Henan Province (after Zheng et al., 2004).

NCC (Fig. 3.5-1) contains 3.48–2.91 Ga inherited zircons (Jian et al., 2005). The Diebusige Group in the Diebusige area, Alxa (Fig. 3.5-1), contains 3.57 Ga and 3.02–2.91 Ga zircons (Shen et al., 2005). There are 2.9 Ga supracrustal rocks and TTG rocks in eastern Shandong province, (Liu, unpublished data), 3.45–3.10 Ga inherited zircons in the Yanshanian granite

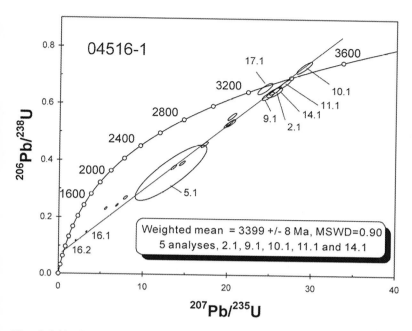

Fig. 3.5-13. Concordia diagram of SHRIMP U-Pb data for detrital zircons from paragneiss (04516-1) in the Jiaozuo area, Northern Henan Province (after Gao et al., 2006).

(Wang et al., 1998) and 3.34–2.85 Ga detrital zircons in the Paleoproterozoic Jinshan and Fenzishan groups (Ji, 1993; Wan et al., 2003b, 2006). Some 3.15–2.8 Ga inherited zircons have been found in 2.5 Ga garnet-bearing charnockite in the Yishui area, western Shandong province (Shen et al., 2004) and 3.40 Ga inherited zircons are also present in Mesozoic granodiorite in the Bangbu area (Fig. 3.5-1) on the southeastern margin of the NCC (Jin et al., 2003).

3.5-4. DISCUSSION

The Neoarchean was the main period of crustal growth in the NCC, with Neoarchean continental rocks accounting for over 85% by volume. However, the NCC has a very long history of continental evolution and Eoarchean rocks or zircons have been found in many areas of the craton. Of these areas, Anshan is the most important because many rocks older than 3.6 Ga have been identified here. Although the geological implications of some zircon ages remain to be refined, the Eoarchean tectonothermal events in the Anshan area are summarized in Table 3.5-1.

The spread of zircon ages from the Anshan area indicate several granitoid forming events between 3.8–3.6 Ga. Moreover, the Baijiafen complex contains Eoarchean supracrustal rocks, the first such record from China. The Th/U values of some zircons

Table 3.5-1. Eoarchean tectonothermal records in the Anshan area

Sample No.	Rock type	Zircon age
Ch28	Banded trondhjemitic rock	3811 ± 4 Ma
A9011 + A87-7	Mylonitized trondhjemitic rock	3804 ± 5 Ma
A0518	Fine-grained trondhjemitic rock	3752 ± 8 Ma
		3800 ± 5 Ma
A9604	Meta-quartz diorite	3794 ± 4 Ma
A0512	Banded trondhjemitic rock	3777 ± 13 Ma
A0403	Biotite schist	3723 ± 17 Ma
A0423	Banded trondhjemitic rock	3680 ± 19 Ma
A0404	Trondhjemitic dike	3620 ± 23 Ma
A0405	Grey trondhjemitic rock	3573 ± 21 Ma

are noticeably <0.1, a compositional feature common to metamorphic zircon (Williams, 2001), which might indicate that metamorphism also occurred during this period. According to their composition, these early-formed rocks were derived not only from a mantle source region, but also continental crustal sources (Wan et al., 2005a).

Similar to Anshan, a spread of Eoarchean rock formation ages ($\geqslant 3.6$ Ga) has also been identified in other Eoarchean gneiss complexes around the world, including the Itsaq Gneiss Complex of Greenland (Nutman et al., 2000; Nutman, 2006), the Uivak gneisses of Labrador (Schiotte et al., 1989), rocks in the Mt Sones area, Antarctica (Harley and Black, 1997), the Acasta Gneisses of northwestern Canada (Bowring and Williams, 1999) and the Narryer Gneiss Complex of Western Australia (Nutman et al., 1991, 1993). Therefore, magmatic events from 4.0 to 3.6 Ga are preserved in crustal fragments on a global scale. It is quite possible that plate tectonics started to play a role, at least in some way, as early as the Eoarchean (Nutman, 2006 and references therein). However, we consider that the Anshan area might have been at an elevated temperature during that period. This would explain the weak light versus heavy REE fractionation of all rocks older than 3.6 Ga found in the area, which indicates that the elevated geotherm of the subduction zone led to melting at low pressures, below the garnet stability field (Wan et al., 2005a).

Anshan and adjacent areas thus form an old continental nucleus, whose importance not only lies in the discovery of diverse rock types >3.6 Ga, but also in their record of almost continuous Archean crustal evolution from $\geqslant 3.8$ to 2.5 Ga. In many respects, Anshan represents in miniature the early crustal evolution of the NCC. Intensive study of this one area may therefore provide the possibility of establishing the Archean tectonochronological framework of the whole craton. Moreover, compared with three of the other areas of >3.8 Ga crust – Greenland, Antarctica and the Acasta area in Canada – Anshan is more easily accessible, which makes it the first choice for Archean research in China.

However, early Archean rocks and zircons also exist in many other areas of the NCC. The Caozhuang rock series in eastern Hebei Province contains fuchsite quartzite with detrital zircons having ages between 3.85 and 3.55 Ga (Liu et al., 1992) and their Hf depleted mantle model ages vary between 3.96 and 3.63 (Wu et al., 2005b). Rocks as old as this have not yet been identified in the surrounding orthogneisses. The ages of detrital zircons in paragneiss in the Jiaozuo area, Henan province, mainly cluster at 3.4 Ga (Gao et al., 2006). The felsic granulite xenoliths in Mesozoic volcanic rocks in the Xinyang area, Henan province, have a U-Pb age of ∼3.65 Ga and Hf depleted mantle model ages of 4.0–3.9 Ga, indicating that very old continental crustal rocks may occur at depth (Zheng et al., 2004a). In addition, rocks and zircons older than the Mesoarchean occur in many areas of the NCC, including the Eastern and Western blocks and the Trans-North China Orogen, but interestingly, most have been found in the Eastern Block (Fig. 3.5-1). One of the reasons for the apparent variable distribution of Paleo-Eoarchean rocks may be the difference in the intensity of research. With on-going study, it is likely that more old rocks and zircons will be found in other areas of the NCC. It is also likely that the NCC may contain several Paleo- to Eoarchean nucleii.

ACKNOWLEDGEMENTS

The authors are grateful to F.Y. Wu, J.P. Zheng and L.Z. Gao for providing relevant figures and original data. Many thanks also to Q.H. Shen, Z.Q. Zhang, Y.S. Geng, C.H. Yang, L.L. Du, Y.X. Zhuang, J.H. Guo, M.G. Zhai, S. Gao, X.H. Li, Y.F. Zheng, S.N. Lu, H.M. Li, P.S. Miao, F.Q. Zhao, G.C. Zhao, Z.X. Li, W. Compston, A.P. Nutman, I.S. Williams, B.M. Jahn, A. Kröner, A. Nemchin and M. Cho for their kind help, discussion and advice during our ongoing study of the Eoarchean of China. We specifically thank A. Nutman and M.J. Van Kranendonk for their valuable comments on the manuscript. The study was supported financially by the Key Program of the Land and Resource Ministry of China (DKD2001020-3 and 1212010711815).

Earth's Oldest Rocks
Edited by Martin J. Van Kranendonk, R. Hugh Smithies and Vickie C. Bennett
Developments in Precambrian Geology, Vol. 15 (K.C. Condie, Series Editor)
© 2007 Elsevier B.V. All rights reserved.
DOI: 10.1016/S0166-2635(07)15036-2

Chapter 3.6

THE NARRYER TERRANE, WESTERN AUSTRALIA: A REVIEW

SIMON A. WILDE[a] AND CATHERINE SPAGGIARI[b]

[a]*Department of Applied Geology, Curtin University of Technology, PO Box U1987, Perth, Western Australia 6845, Australia*
[b]*Geological Survey of Western Australia, 100 Plain Street, East Perth, Western Australia 6004, Australia*

3.6-1. INTRODUCTION

The Narryer Terrane occupies an area of $\sim$30,000 km^2 in the northwestern corner of the Archean Yilgarn Craton of Western Australia (Fig. 3.6-1). It is one of the earliest crustal terranes on Earth, containing rocks with U-Pb zircon ages ranging up to 3730 Ma, the oldest known rocks in Australia, and detrital zircons up to 4404 Ma, the oldest terrestrial material on Earth. In this review, we will document the history of investigations in the region, outline the geology of the two areas in which most studies have been undertaken – Mt. Narryer and Jack Hills – and comment on several key aspects that have emerged in recent years. A more detailed account of studies into the ancient detrital zircon crystals from Jack Hills is presented in Cavosie et al. (this volume).

3.6-2. HISTORICAL

The first systematic geological investigations of the area commenced with the 1:250,000 mapping program of the Geological Survey of Western Australia in the late 1970s and early 1980s on the Belele (Elias, 1982) and Byro (Williams et al., 1983) sheets. Of particular interest was a sequence of granulite facies quartzite, conglomerate and other metasedimentary components, surrounded by granitic gneisses, identified at Mt. Narryer (Fig. 3.6-1) by I.R. Williams during preliminary investigations in 1979. The similarity of the quartzites to rocks of equivalent metamorphic grade in the Toodyay-York area near Perth (Wilde, 1980), some 800 km to the south (Fig. 3.6-2), suggested a possible extension of the high-grade gneisses already identified in the southwestern Yilgarn Craton (Arriens, 1971) northward along the western margin of the craton. This led to the concept of the 'Western Gneiss Terrain' of Gee et al. (1981); one of the major crustal components identified in the first subdivision of the Yilgarn Craton into terrains and provinces (Fig. 3.6-2).

Initial geochronological investigations on granitic gneisses in the vicinity of Mt. Narryer, using Rb-Sr whole-rock techniques (de Laeter et al., 1981a), established an age of

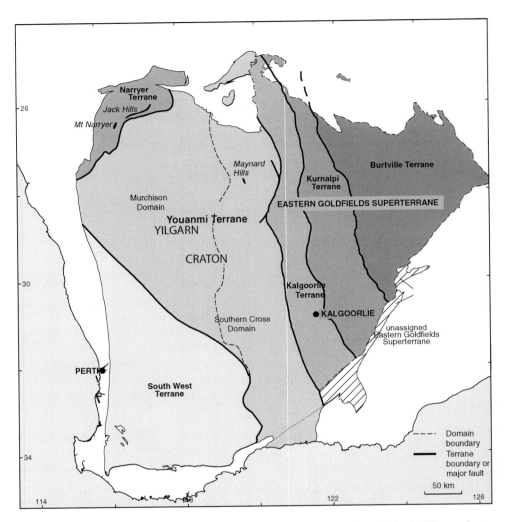

Fig. 3.6-1. Map showing Superterranes, Terranes and the Murchison Domain in the Yilgarn Craton of Western Australia (based on Cassidy et al., 2006).

3348 ± 43 Ma using multicomponent samples. This age was comparable to those obtained by Arriens (1971) from the southwestern Yilgarn Craton and appeared to substantiate the

Fig. 3.6-2. (*Next page.*) Map of the Western Gneiss Terrane as originally defined by Gee et al. (1981) and modified from Myers (1990), illustrating the earlier interpretation of distribution of gneissic rocks along the western margin of the Yilgarn Craton.

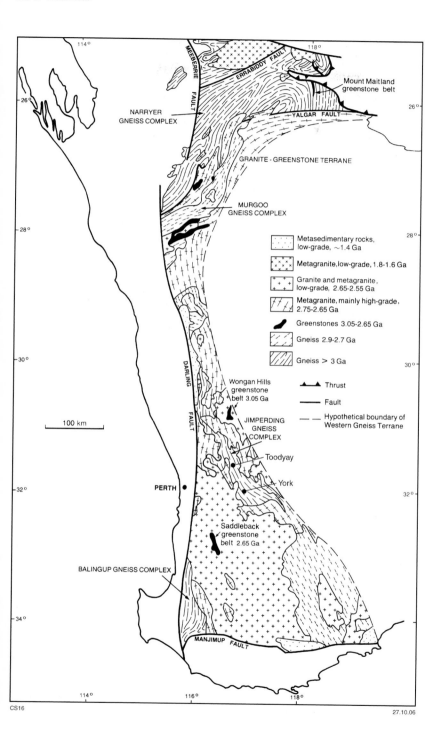

link between the southern and northern parts of the 'Western Gneiss Terrain'. Subsequent dating (de Laeter at al., 1981b), using the Sm-Nd whole-rock technique, of two of these samples yielded model ages (T_{CHUR}) of 3630 ± 40 Ma and 3510 ± 50 Ma, considerably older than any ages previously obtained from the Yilgarn Craton. The oldest model age was obtained from layered monzogranitic gneiss, whereas the younger age was obtained from more leucocratic granitic gneiss that was intrusive into the former.

A detailed map of the Mt. Narryer area was prepared by I.R. Williams in 1981–1982 and J.S. Myers in 1983, but not published until 1987 (Fig. 3.6-3) (Williams and Myers, 1987). As a consequence, the subdivision of the gneisses proposed therein was published first in Myers and Williams (1985). The rocks at Mt. Narryer were later referred to as the 'Narryer Gneiss' (Myers, 1988a).

Investigations of the area coincided with the development of the SHRIMP ion microprobe at the Australian National University (ANU) in Canberra (Compston et al., 1984). This instrument allowed, for the first time, precise *in situ* Th-U-Pb analysis of individual zircon grains at a ~25 micron spot size, with the possibility of placing more than one analytical site per grain. Following a detailed investigation of the ancient gneisses and quartzites in the Toodyay area by Nieuwland and Compston (1981) using conventional multigrain zircon U-Pb methods, it was a natural progression for the Canberra group to undertake a detailed study of the recently-identified rocks of apparently similar age near Mt. Narryer.

An initial investigation of the Mt. Narryer area commenced using the SHRIMP ion microprobe, including not only a study of the ancient gneisses, but also of the detrital zircon population in the granulite facies quartzites and conglomerates. This led to the exciting discovery of four detrital zircon cores with ages ranging from 4110 to 4190 Ma, the oldest crustal remnants identified on Earth at that time (Froude et al., 1983). The four grains were interpreted to have the same age (~4150 Ma; no uncertainty quoted) and were obtained from a quartzite (GSWA sample 71932) collected ~2.5 km NNE of Mt. Narryer. This finding led to additional studies and ultimately to a major field and analytical program by the Australian National University (the 'First Billion Years' project) aimed at further characterising the area.

In 1983, as part of a program to examine the five greenstone belts that had been identified within the Western Gneiss Terrain by Gee et al. (1981) (Fig. 3.6-2), the first detailed mapping of the Jacks Hills at a scale of 1:10,000 was completed by final year undergraduate students of the Western Australian Institute of Technology (now Curtin University of Technology) under the guidance of John Baxter, Robert Pidgeon and Simon Wilde. The Jack Hills was initially considered to be a greenstone belt (Elias, 1982), but was found to consist predominantly of metasedimentary rocks, including both chemical and clastic varieties, enclosed by granitic gneisses (Baxter et al., 1984). However, unlike Mt. Narryer, the metamorphic grade was greenschist to amphibolite facies (Baxter et al., 1984). A suite of samples, including quartzite and conglomerate, was collected for U-Pb zircon

---

Fig. 3.6-3. (*Next page.*) Detailed map of Mt. Narryer (modified from Williams and Myers (1987)).

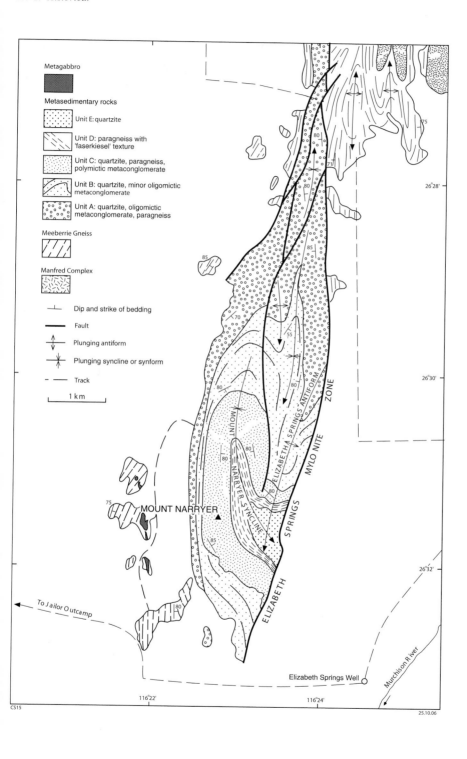

Metagabbro

Metasedimentary rocks

Unit E: quartzite

Unit D: paragneiss with 'faserkiesel' texture

Unit C: quartzite, paragneiss, polymictic metaconglomerate

Unit B: quartzite, minor oligomictic metaconglomerate

Unit A: quartzite, oligomictic metaconglomerate, paragneiss

Meeberrie Gneiss

Manfred Complex

Dip and strike of bedding

Fault

Plunging antiform

Plunging syncline or synform

Track

1 km

To Jailor Outcamp

MOUNT NARRYER

MOUNT NARRYER SYNCLINE

ELIZABETH SPRINGS ANTIFORM

ELIZABETH SPRINGS MYLONITE ZONE

Elizabeth Springs Well

Murchison River

26°28'

26°30'

26°32'

116°22'

116°24'

CS15

25.10.06

geochronology in both 1983 and 1984, when the mapping exercise was repeated. In order to test if there was any similarity with the Mt. Narryer metasedimentary rocks, zircons obtained from a greenschist facies conglomerate exposed on Eranondoo Hill (Fig. 3.6-4) were run on the ANU SHRIMP I ion microprobe. Two zircons with ages of 4276 ± 12 Ma were identified, making these the oldest known crustal components on Earth at that time (Compston and Pidgeon, 1986).

In subsequent years, further detailed mapping and geochronological investigations in the northern Yilgarn Craton led to the better characterisation of the Narryer Terrane (Williams et al., 1983; Williams and Myers, 1987; Myers, 1988a, 1988b; Kinny et al., 1988; Kinny et al., 1990; Wilde and Pidgeon, 1990; Myers, 1990a; Nutman et al., 1991; Maas and McCulloch, 1991; Maas et al., 1992; Myers, 1993; Kinny and Nutman, 1996; Pidgeon and Wilde, 1998). However, the level of activity had started to wane by the mid-1990s.

Following Compston and Pidgeon (1986), some further work was undertaken on the same suite of zircons extracted from the metaconglomerate at Eranondoo Hill. Köber et al. (1989) published the results of stepwise single zircon evaporation measurements, whilst Amelin (1998) analysed crystal fragments by standard U-Pb isotope dilution using thermal ionisation mass spectrometry; both studies essentially reproducing the earlier SHRIMP results.

In 1998, a re-investigation of the ancient zircon population from Eranondoo Hill at Jack Hills was undertaken by a team from Curtin University and the University of Wisconsin, incorporating both SHRIMP U-Pb and CAMECA IMS 4f oxygen data on a new suite of zircon grains extracted from the same sample (W74) analysed by Compston and Pidgeon (1986). This led to the identification of a zircon that contained a portion with an age of 4404 ± 8 Ma (Wilde et al., 2001); by far the oldest age ever obtained from crustal material on Earth. A companion investigation began in 1999 with a team from UCLA and Curtin University collecting new material from the area, including from the same sample site where W74 was obtained. Importantly, both studies obtained evidence from the W74 site for elevated δ^{18} oxygen values in the ancient zircon population, implying zircon growth in rocks previously subjected to interaction with surface waters (Wilde et al., 2001; Mojzsis et al., 2001; Peck et al, 2002). This has led to the hypothesis of a cool early Earth (Valley et al., 2002; Valley, 2005), implying the early development of oceans on the planet. As with the initial studies in the 1980s, these recent investigations coincided with new technological advances, this time involving the ability to analyse oxygen isotopes of single zircon grains *in situ* with great precision using the CAMECA IMS 1270 (and now the 1280) ion microprobe (Cavosie et al., 2005) and this has opened up an exciting new time for studying these earliest remnants of Earth's continental crust. The ability to measure precise hafnium ratios in single zircon crystals has also provided an important new advance (Harrison et al., 2005). These aspects are more fully described in a companion paper (Cavosie et al., this volume).

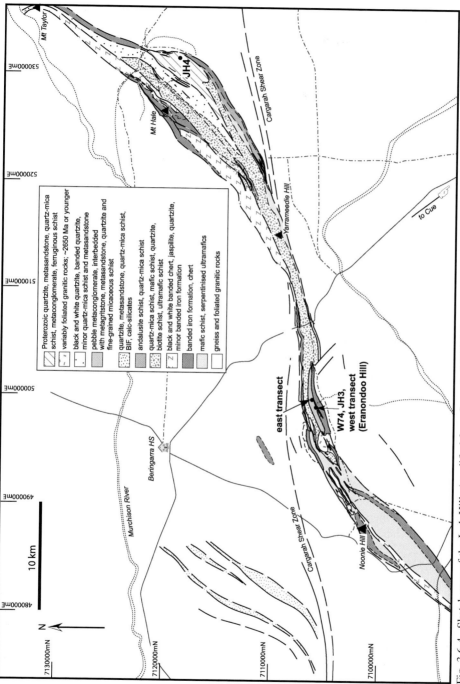

Fig. 3.6-4. Sketch map of the Jack Hills (modified from Spaggiari (2007a)).

3.6-3. CHARACTERISTICS OF THE NARRYER GNEISS COMPLEX

3.6-3.1. Overview

Myers (1990a) presented the first attempt at subdividing the Yilgarn Craton into a number of tectonostratigraphic terranes, and further developed this idea in subsequent years (Myers, 1993, 1995). He proposed the term 'Narryer Terrane' to cover the area previously referred to as the 'Narryer Gneiss Complex', at the same time recognising that the latter is merely a subcomponent of the terrane, since extensive areas of Late Archean granites had subsequently been recognised throughout the area. In the most recent revision of terranes in the Yilgarn Craton, Cassidy et al. (2006) have retained this terminology and we have adopted this in the present paper. However, we retain the term "Narryer Gneiss Complex" when referring to the >3 Ga rocks present within the area.

The Narryer Gneiss Complex is one of the largest intact pieces of ancient continental crust remaining on Earth, occupying a significant portion of the Narryer Terrane, which exceeds 30,000 km^2 in area (Myers, 1988a). The western margin of the Narryer Terrane (Fig. 3.6-5) is defined by the Darling Fault, and the northern margin by the Errabiddy Shear Zone, along which the Glenburgh Terrane was accreted to the Yilgarn Craton during the 2005 to 1960 Ma Glenburgh Orogeny (Occhipinti et al., 2004; Occhipinti and Reddy, 2004). The southern margin marks the boundary with the Younami Terrane, a major granite-greenstone association in the Yilgarn Craton (Cassidy et al., 2006), and is defined by the Balbalinga and Yalgar Faults; the latter interpreted as a major dextral strike-slip fault (Myers, 1990b, 1993). The structural history of this fault is poorly understood, and the boundary is likely to have been reworked and cryptic (Fig. 3.6-5) (Nutman et al., 1993a; Spaggiari, 2007a; Spaggiari et al., in review).

The Narryer Gneiss Complex (Myers, 1988a) is composed of granitoids and granitic gneisses ranging in age from Paleo- through to Neoarchean (Kinny et al., 1990; Nutman et al., 1991; Pidgeon and Wilde, 1998). These rocks are locally interlayered with deformed and metamorphosed banded iron formation (BIF), mafic and ultramafic intrusive rocks, and metasedimentary rocks (Williams and Myers, 1987; Myers, 1988a; Kinny et al., 1990). More extensive sequences of amphibolite to granulite facies metasedimentary rocks occur at Mt. Narryer, whereas greenschist to amphibolite facies BIF, mafic and ultramafic rocks, and both Archean and Paleoproterozoic clastic metasedimentary rocks occur at Jack Hills (Elias, 1982; Williams et al., 1983; Wilde and Pidgeon, 1990; Cavosie et al., 2004; Dunn et al., 2005; Spaggiari, 2007a).

The early Archean gneisses of the Narryer Terrane have been interpreted as an allochthon that was thrust over ca. 3000 to 2920 Ma granitic crust of the Youanmi Terrane, prior to, or during, the period of late Archean granitic magmatism that stitched the two terranes (Nutman et al., 1993a). Although there is evidence of deformation prior to the intrusion of these granites, the deformation that produced the main tectonic grain is believed to have occurred at amphibolite facies between ca. 2750 and 2620 Ma, and to have affected both terranes (Myers, 1990b). Three phases of folding are recognised; recumbent folding

Fig. 3.6-5. Reduced to pole, total magnetic intensity image of the northwestern Yilgarn Craton showing the Narryer Terrane and part of the Murchison Domain of the Youanmi Terrane. Terrane boundaries and major faults modified from Myers and Hocking (1998), and after Occhipinti et al. (2004) and Spaggiari (2006).

associated with thrusting (D$_1$), followed by two phases of upright folding with generally northeast or east-northeast trending axes (D$_2$ and D$_3$; Myers, 1990b).

The Paleoproterozoic Capricorn Orogeny (1830–1780 Ma) produced intracratonic, predominantly greenschist facies, dextral transpressional reworking of both the northern and southern margins of the Narryer Terrane. These effects are evident in the Errabiddy Shear Zone (Occhipinti and Reddy, 2004), the Jack Hills metasedimentary belt (Spaggiari, 2007a; Spaggiari, 2007b; Spaggiari et al., 2007) and the Yarlarweelor Gneiss Complex in the northern part of the Narryer Terrane. The latter was deformed, metamorphosed and in-truded by granites and dykes between approximately 1820 and 1795 Ma (Sheppard et al., 2003). This involved a two stage process of intrusion during compression, followed by intrusion during dextral strike slip deformation. Deformation related to the Capricorn Orogeny has been interpreted to extend as far south as the northern end of the Mingah Range greenstone belt within the Murchison Domain of the Youanmi Terrane (Fig. 3.6-5; Spaggiari, 2006; Spaggiari et al., in review). There is evidence that the Yalgar Fault was probably active during the Capricorn Orogeny, but whether the Yalgar Fault is a true terrane boundary is unclear.

3.6-3.2. Mt. Narryer Region

Myers and Williams (1985) recognised that most rocks at Mt. Narryer have been metamor-phosed to granulite facies and undergone multiple deformation episodes. They identified two major suites of granitic gneisses: the older Meeberrie gneiss and the younger Dugel gneiss, which yielded Sm-Nd (T_{CHUR}) model ages of 3630 ± 40 Ma and 3510 ± 50 Ma, respectively (de Laeter et al., 1981b). Within the Dugel gneiss, fragments of a layered basic intrusion were recognised and referred to as the Manfred Complex (Myers and Williams, 1985; Williams and Myers, 1987). In addition, quartzites and metaconglomerates make up the core of Mt. Narryer itself.

3.6-3.2.1. Meeberrie gneiss
The Meeberrie gneiss is composed of alternating bands of quartzo-feldspathic and biotite-rich units that define a prominent layering (Fig. 3.6-6(a,b)). Some components were originally porphyritic, now occurring as deformed augen porphyroclasts, and the whole sequence is dominantly composed of monzogranite. However, later work has shown that many rocks are really part of the tonalite-trondhjemite-granodiorite (TTG) suite (Nutman et al., 1991; Pidgeon and Wilde, 1998). The early biotite granitoid was intruded by peg-matites and the whole complex brought into layer-parallelism by subsequent deformation. Sm-Nd (T_{CHUR}) model ages of 3630 ± 40 Ma (de Laeter et al., 1981b), and 3710 ± 30 Ma and 3620 ± 40 Ma (de Laeter et al., 1985) were obtained from near Mt. Narryer, whereas zircon cores from the sample giving the 3630 ± 40 Ma Sm-Nd age, obtained 7.5 km NNE of Mt. Narryer, gave SHRIMP U-Pb ages of 3678 ± 6 Ma (Kinny et al., 1988). The oldest known component of the Meeberrie gneiss is a tonalite collected 3 km south of the Jack Hills, with a SHRIMP U-Pb age of 3731 ± 4 Ma (Nutman et al., 1991): this is the old-est known rock in Australia. Another Meeberrie gneiss sample, collected 1 km east of the above, has a distinctly younger age of 3597 ± 5 Ma, indicating the composite nature of the Meeberrie gneiss. Biotite from two samples of Meeberrie gneiss recorded K-Ar ages of 1887 ± 34 Ma and 1778 ± 32 Ma (Kinny et al., 1990), clearly establishing that younger events have affected the Mt. Narryer area.

A later SHRIMP ion microprobe zircon U-Pb study by Kinny and Nutman (1996), cen-tred around Mt. Narryer, identified three main age populations in the Meeberrie gneiss: at $\sim$3670, 3620 and 3600 Ma, with minor older components at $\sim$3730 Ma and younger

Fig. 3.6-6. Field photographs of gneisses and granitoid rocks in the Narryer Terrane. (a) Meeberrie gneiss at site C of de Laeter et al. (1981a) that formed part of the data-set recording a Rb-Sr isochron age of 3348 ± 43 Ma; hammer is ∼32 cm long. (b) Meeberrie gneiss at site Y21, 3 km south of Jack Hills, where sample 88–173 was collected. It yielded an age of 3731 ± 4 Ma (Nutman et al., 1991), making it the oldest rock in Australia; length of photo ∼1 m. (c) Dugel gneiss at type locality, 10 km NNE of Mt. Narryer. The gneiss is cut by a Neoarchean aplite dyke; hammer is ∼30 cm long. (d) Single plagioclase crystals and anorthosite pods in Dugel gneiss at same locality as (c); hammer is 32 cm long. (e) Eurada gneiss at type locality near Eurada Bore, where a ∼3480 Ma zircon age was obtained from sample MN 10 by Kinny (1987); length of photo ∼60 cm. (f) Deformed porphyritic granite, ∼1 km south of Meeberrie gneiss at site C of de Laeter et al. (1981a); pencil is 15 cm long. All photographs courtesy of Pete Kinny.

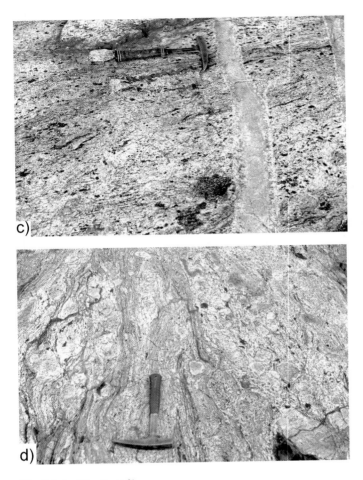

Fig. 3.6-6. (*Continued.*)

components at ~3300 Ma. Kinny and Nutman (1996) showed that the gneiss consisted of multiple components down to centimetre scales and they emphasized the importance of working in low strain zones, where original intrusive relationships may still be preserved: elsewhere, the rocks approach migmatite. A simple age population of ~3620 Ma was also obtained from a gneiss at Mt. Murchison (Fig. 3.6-5) by Kinny and Nutman (1996), who emphasised that it was impossible to quote a single age for the Meeberrie gneiss in the Mt. Narryer region; it is a polyphase migmatite with ages ranging from 3730–3300 Ma.

3.6-3.2.2. Dugel gneiss
The Dugel gneiss occupies large areas of the Narryer Gneiss Complex and is dominantly syenogranite, although it ranges to monzogranite in composition (Fig. 3.6-6(c)). Veins of

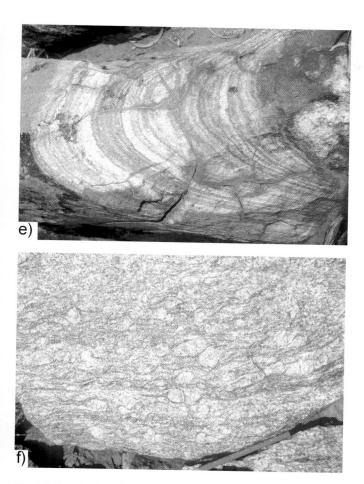

Fig. 3.6-6. (*Continued.*)

leucocratic syenogranite cut the Meeberrie gneiss and show a less prominent foliation than the latter, although there are diffuse variations in grain size and some pegmatite veins. Whereas areas of one gneiss type may dominate over the other, they are intimately interdigitated over large areas, with the Dugel gneiss originally intrusive into the monzogranite, and both units now strongly deformed together. Where intrusive into the Meeberrie gneiss, the Dugel gneiss is commonly pegmatitic. Some deformed and metamorphosed mafic lenses have been interpreted as synplutonic dykes intruded contemporaneously with the granitic protolith of the Dugel gneiss at ∼3.4 Ga (Myers, 1988a).The gneiss has Sm-Nd (T_{CHUR}) model ages of 3510 ± 50 Ma (de Laeter et al, 1981b), 3520 ± 30 Ma (Fletcher et al., 1983), and 3540 ± 30 Ma (de Laeter et al., 1985). Zircon cores gave a SHRIMP U-Pb upper intercept age of 3381 ± 22 Ma (Kinny et al., 1988); a revised age for this sample (Nutman et al., 1991) is 3375 ± 26 Ma, with rims defining an age of 3284 ± 9 Ma. Although this age is similar to the $\sim$3300 Ma component of the Meeberrie gneiss, the two appear quite distinct in that there is no evidence for anatexis/migmatisation of the Dugel gneisses at $\sim$3300 Ma (Kinny and Nutman, 1996).

3.6-3.2.3. Manfred Complex

Dismembered fragments of gabbro, anorthosite and ultramafic rocks form prominent trails within the Dugel gneiss northeast of Mt. Narryer (Myers and Williams, 1985), but are also recorded from the Meeberrie gneiss (Myers, 1988a, 1988b). They range in size from a few centimetres (Fig. 3.6-6(d)) to one kilometre long. Relict igneous textures are locally preserved, especially plagioclase crystals, although most rocks have been recrystallised to an equigranular metamorphic texture. The most characteristic rock type is leucogabbro, with much of the material converted to amphibolite during subsequent metamorphism. Igneous layering is evident in rocks of all compositions. The variable deformation of adjacent fragments suggests they were deformed prior to incorporation in the Dugel gneiss (Myers and Williams, 1985). Samples of the Manfred Complex gave a Sm-Nd whole-rock isochron age of 3680 ± 70 Ma and a Pb-Pb isochron age of 3689 ± 146 Ma (Fletcher et al., 1988). The Pb-Pb data suggest incorporation of pre-existing felsic material in either the mantle source region or the mafic magma (perhaps from >4 Ga crust). More precise U-Pb zircon ages of 3730 ± 6 Ma were obtained for the Manfred Complex anorthosite by Kinny et al. (1988). Hornblendes from leucogabbro and anorthosite units recorded K-Ar ages of 2787 ± 50 Ma and 2667 ± 34 Ma (Kinny et al., 1990), whereas plagioclase from the same samples gave ages of 1954 ± 38 Ma and 2066 ± 38 Ma, respectively. An ^{40}Ar/^{39}Ar age of 2704 ± 14 Ma on hornblende from the anorthosite sample confirms the K-Ar age and indicates either Neoarchean to Paleoproterozoic resetting, or slow cooling of the terrane.

3.6-3.2.4. Eurada gneiss

Kinny (1987) was the first to recognise that a younger series of gneisses are also present in the Narryer Gneiss Complex when he identified $\sim$3300 Ma granitoid near Mt. Narryer and $\sim$3480 Ma layered gneisses at Eurada Bore, 20 km NW of Mt. Narryer (Fig. 3.6-6(e)). The rocks are commonly tonalitic and show no evidence of partial melting or of reaching granulite facies. Additional samples collected $\sim$20 km W of Mt. Narryer (Nutman et al., 1991) yield weighted mean ^{207}Pb/^{206}Pb ages of 3489 ± 5 Ma and 3490 ± 6 Ma; the latter also containing a zircon population with a ^{207}Pb/^{206}Pb age of 3439 ± 3 Ma, taken to be the true age of this sample, with the older grains interpreted as being inherited from the protolith or adjacent gneiss. A further sample of Eurada gneiss, taken 5 km W of the above samples, gave ^{207}Pb/^{206}Pb ages of 3466 ± 3 Ma, taken to be the age of the igneous protolith. A population of distinctly younger grains gave a ^{207}Pb/^{206}Pb age of 3055 ± 3 Ma, which was interpreted to reflect pegmatite intrusion into the gneiss or else to record a partial melting event (Nutman et al., 1991).

3.6-3.2.5. *Porphyritic granitoids*

Although most of the >3 Ga rocks in the Mt. Narryer region are gneissic in character, Kinny et al. (1990) identified a deformed porphyritic granitoid considered to have been emplaced between the D_1 and D_2 phases of deformation. Importantly, this rock crops out immediately east of the main metasedimentary sequence at Mt. Narryer and only 1 km south of the type locality of the Meeberrie gneiss (Fig. 3.6-6(f)). Zircons from this sample have core $^{207}Pb/^{206}Pb$ ages of 3302 ± 6 Ma and rims that are zoned but with much higher uranium contents. The age of the rims is less precise and slightly younger, but they have suffered ancient Pb loss and were interpreted to be of igneous origin (Kinny et al., 1990).

3.6-3.2.6. *Younger gneisses*

Nutman et al. (1991) analysed a garnetiferous trondhjemitic gneiss collected 10 km S of Byro Homestead (and ∼40 km NNW from Mt. Narryer). The rock appeared similar to Dugel gneiss but the most concordant zircon gave a weighted mean $^{207}Pb/^{206}Pb$ age of 2620 ± 12 Ma, taken to be the age of the igneous protolith. That rocks of such age appear superficially similar to much older gneisses argues that great caution is required when extending the gneissic classification away from well-dated areas.

3.6-3.2.7. *Metasedimentary rocks*

These make up approximately 10% of the gneiss complex and are dominantly quartzite and BIF, with minor pelitic and semi-pelitic rocks and calc-silicate gneisses. Mt. Narryer itself is dominated by quartzites (Fig. 3.6-7(a)), with some lenticular bodies of polymictic and quartz-pebble metaconglomerates. Metamorphic minerals include sillimanite, cordierite, clinopyroxene, garnet, amphibole and biotite, consistent with granulite facies metamorphism, with local amphibolite facies retrogression. The rocks are continuously exposed over a distance of 21 km in a north-south direction and are up to 2.5 km wide. They dip steeply to the west and are folded at Mt. Narryer into a synform that plunges steeply to the southeast (Fig. 3.6-3). Williams and Myers (1987) distinguished the more highly-deformed metasedimentary rocks in the north of the Mt. Narryer range near Mt. Dugel, in which all sedimentary structures have been obliterated and bedding is transposed into tectonic layering, from their less deformed equivalents in the south at Mt. Narryer. On Mt. Narryer, the quartzites show cross- and graded-bedding that allow the way up to be determined (Williams and Myers, 1987). Williams and Myers (1987) identified five lithostratigraphic units that preserve a primary stratigraphy (units A to E). Kinny et al. (1990) interpreted the quartzite sequence to represent a fluvio-deltaic regime, with the nature of the sediments, coupled with the absence of volcanic rocks, favouring a shallow continental shelf environment. Contacts with the gneisses are everywhere strongly sheared, obscuring the original relationships. However, the abundance of detrital zircons with ages similar to those obtained from the gneisses (Froude et al., 1983; Schärer and Allègre, 1985; Compston and Pidgeon, 1986; Kinny et al., 1988; Maas et al., 1992; Pidgeon and Nemchin, 2006) suggests that the gneisses may have formed basement to the clastic sedimentary rocks. Trails of metasedimentary rocks also occur in the gneisses to the north of Mt. Narryer. These are dominated by sillimanite- or pyroxene-bearing quartzite and granulite facies

Fig. 3.6-7. Field photographs of supracrustal rocks in the Narryer Terrane. (a) Garnet-sillimanite quartzite on the eastern flank of Mt. Narryer; hammer is ∼32 cm long. (b) Looking east along the Jack Hills range from the central part of the belt. Note the sigmoidal curvature of the belt: ultramafic rocks defining the southern limit of the belt are in the foreground. (c) Brecciated banded iron formation horizon near Mt. Taylor, near the eastern limit of the Jack Hills belt; hammer head is ∼16 cm long. (d) Quartzite just east of Noonie Hill, showing strong lineation; hammer is ∼32 cm long. (e) Interlayered metaconglomerate and meta-sandstone in the west-central part of the Jack Hills belt, ∼1 km north of the W74 site; scale in cm. (f) Granite pavement formed from Neoarchean granite at the 'blob', ∼2.5 km WSW of W74 site.

iron formations composed of quartz-magnetite-pyroxene. The latter also occur as pebbles in metaconglomerates at Mt. Narryer and indicate a sequence that pre-dates the deposition of the Mt. Narryer clastic sedimentary rocks.

The initial dating of Mt. Narryer quartzite (sample GSWA 71932) by Froude et al. (1983) identified a spectrum of zircon ages from 3100–4190 Ma. The zircons with ages

Fig. 3.6-7. (*Continued.*)

>3500 Ma were interpreted as detrital magmatic grains, whereas those with ages between 3300 and 3100 Ma were considered to be the result of metamorphism, based on the belief that the 3348 ± 43 Ma Rb-Sr age of de Laeter et al. (1981a) recorded the time of this event (Kinny et al., 1990), rather than on any inherent property of the zircons. An additional study on the same aliquot of zircons used by Froude et al. (1983) was undertaken by Schärer and Allègre (1985), who analysed 27 individual grains, and 12 fragments obtained from 5 additional grains, using thermal ionisation mass spectrometry. They failed to identify any zircons >4 Ga, instead obtaining $^{207}Pb/^{206}Pb$ ages ranging from 2707 to 3597 Ma. This naturally caused concern at the time and so additional work was undertaken by Kinny et al. (1990) who provided further details on a total of 275 individual zircon grains, based partly on unpublished work of Froude et al. (1983). These were obtained from a total of three samples: the initial plagioclase-biotite bearing quartzite sample of Froude et al.

Fig. 3.6-7. (*Continued.*)

(1983) containing the >4.1 Ga zircons (GSWA 71932) and two samples of sillimanite-cordierite-biotite-garnet quartzite (GSWA 71921 and 71924). A total of 7 grains >4 Ga were identified, including those originally published in Froude et al. (1983). Kinny et al. (1990) concluded that the youngest grains were ~3280 Ma and magmatic in origin and that the 3100 apparent ages of Froude et al. (1983) were artefacts of ancient Pb disturbance of older grains with high uranium contents. They interpreted the depositional age of the Narryer quartzites to be ⩽3100 Ma, partly based on correlations with Jack Hills (Compston and Pidgeon, 1986) and the Toodyay area in the southwest Yilgarn Craton (Nieuwland and Compston, 1981). The metasediments are thus younger than the adjacent Dugel gneiss and post-date the possible metamorphic event at ~3350 Ma. Kinny et al. (1990) suggested that there was a later amphibolite facies metamorphic event at ~2700 Ma. Samarium-neodymium T_{DM} model ages from Mt. Narryer quartzites and pelites gave a range of values from 3.97–3.72 Ga (Maas and McCulloch, 1991), reflecting the large contribution made by ancient components.

K-Ar dating of two samples of hornblende collected from quartz-bearing amphibolites within the metasedimentary sequence on Mt. Narryer (Kinny et al., 1990) yielded ages of 2610 ± 34 Ma and 3664 ± 78 Ma, whereas biotite from a pelitic unit gave an age of 2022 ± 44 Ma; indicating thermal disturbances in the Neoarchean and Paleoproterozoic.

Separate metasedimentary sequences at Mindle Well (the Mindle metasedimentary rocks) and Mt. Murchison (Mt. Murchison metasedimentary rocks) were identified and mapped (Fig. 3.6-5) by Williams and Myers (1987). The Mindle rocks are separated from the main Narryer sequence by a zone of highly sheared Meeberrie gneiss and are composed of clinopyroxene-bearing quartzite and hypersthene-bearing banded iron formation, with minor units of feldspathic quartzite, biotite gneiss and calc-silicate gneiss. Locally, they are intruded by components of the Dugel gneiss. An attempt to date these rocks failed due to a lack of zircon (Nutman et al., 1991). However, two samples analysed by Maas and McCulloch (1991) yielded Sm-Nd T_{DM} model ages of 3.90 and 2.83 Ga.

The rocks at Mt. Murchison are steeply-dipping quartzites enclosed in strongly deformed gneiss of Meeberrie type. There are also numerous streaks and lenses of banded iron formation, cordierite-bearing quartzite and gneiss, and calc-silicate rocks in the adjacent gneisses. A fuchsite-bearing quartzite, collected 15 km SE of Mt. Murchison was analysed by Nutman et al. (1991). The main population yielded $^{207}Pb/^{206}Pb$ ages of ∼3600 Ma, but other grains recorded a variety of ages at ∼3725–3700 Ma, ∼3510–3420 Ma and ∼3300 Ma, indicating a mixed provenance.

Other metasedimentary sequences were dated by Kinny et al. (1990) and Nutman et al. (1991). A sequence of BIF, quartzite and mica schist (Nutman et al., 1991) occurs ∼35 km SSW of Mt. Narryer. The mica schist has detrital zircon cores ranging in $^{207}Pb/^{206}Pb$ age from 3675 to 2760 Ma, with two analysed rims giving $^{207}Pb/^{206}Pb$ ages of 2711 ± 9 Ma and 2650 ± 9 Ma, respectively. The quartzite zircons record detrital cores with $^{207}Pb/^{206}Pb$ ages ranging from 3900 to 2990 Ma, with the majority of apparent ages between 3370 and 3220 Ma, similar to the population at Mt. Narryer and Jack Hills (Compston and Pidgeon, 1986). The zircon rim ages are 2687 ± 10 Ma, similar to those in the mica schist and interpreted as a period of metamorphic overgrowth (Nutman et al., 1991). A metasedimentary rock from Mt. Dugel composed of hypersthene and garnet porphyroblasts in a cordierite-biotite-quartz matrix was analysed by Kinny et al. (1990). This yielded zircon core ages of 3500–3100 Ma, considered to be detrital grains of magmatic origin, surrounded by rims with ages of ∼2680 Ma, interpreted as metamorphic overgrowths and thus similar in age to the metamorphic zircon rims identified by Nutman et al. (1991) from the site ∼35 km SSW of Mt. Narryer.

3.6-3.2.8. Neoarchean intrusions

The above sequence of rocks is cut by a variety of mafic and felsic intrusives that range in metamorphic grade from granulite facies to being unmetamorphosed. The mafic rocks largely represent variably deformed and metamorphosed mafic dykes whereas the granitoid rocks include both minor intrusives and small plutons that are likewise variably deformed and metamorphosed. The youngest felsic rocks are largely unmetamorphosed, although locally sheared (cf. Pidgeon and Wilde, 1998), and are related to the main phase of ~2660 Ma Yilgarn granitoids that occur throughout the craton. Kinny et al. (1990) dated a fine-grained monzogranite obtained 2.5 km SW of Mt. Narryer. Approximately one third of the zircons show distinct cores and rims which record weighted mean ^{207}Pb/^{206}Pb ages of 2919±8 Ma and 2646 ± 6 Ma, respectively; the latter age obtained by also incorporating data from zircons devoid of cores and together interpreted as the crystallisation age of the rock.

3.6-3.2.9. Timing of deformation and metamorphism

Three major deformation events are recognised in the Mt. Narryer region. D_1 represents the earliest layering in the Meeberrie gneiss (also possibly in the Dugel gneiss) (Myers and Williams, 1985), but is not recorded in the metasedimentary rocks. It is present as a weak layering defined by grain-size and compositional variations and as rootless isoclines, but no major folds related to this event have so far been recognised.

The main deformation episode is D_2, defining the gneissic fabric and major isoclinal folds in both the Meeberrie and Dugel gneisses and also recorded in the metasedimentary rocks at Mt. Narryer (Williams and Myers, 1987). This event was considered to coincide with the peak of granulite facies metamorphism and has been variously dated at ~3350 Ma by Rb-Sr methods (de Laeter et al., 1981a), 3280 ± 140 Ma by Sm-Nd (T_{CHUR}) model age (Fletcher et al., 1988) and 3296 ± 4 Ma by SHRIMP U-Pb analysis of zircon rims from Meeberrie gneiss at Mt. Narryer (Kinny et al, 1988). However, Kinny and Nutman (1996) later re-interpreted the latter age as a subsequent igneous event. The sample contained a few igneous cores with an age of 3678±6 Ma, enclosed in magmatic rims giving the age of 3296 ± 4 Ma, but with a main population of igneous zircons giving an age of 3620 Ma. An imprecise age of ~3284 Ma was obtained from two zircon rim analyses of Dugel gneiss, 10 km NNE of Mt. Narryer (Kinny et al., 1988), and was initially believed to be consistent with the Meeberrie gneiss data. A minimum age of 3298 ± 8 Ma was also obtained from zircon rims from Manfred Complex leucogabbro (Kinny et al., 1988) and was believed to be consistent with the gneissic data. However, following the re-interpretation, Kinny and Nutman (1996) concluded that the only identifiable metamorphic event occurred in the Neoarchean at ~2700 Ma. Nonetheless, they considered that upper amphibolite to granulite facies metamorphism with anatexis may have occurred several times between 2730 and 3600 Ma and also at ~3000 Ma, although the evidence has been destroyed (Kinny and Nutman, 1996).

The D_3 episode consists of tight to more open folds that form large-scale interference patterns where they intersect D_2. This event is generally considered to equate to the time of retrograde amphibolite facies metamorphism (Myers and Williams, 1985). The timing of this event was initially considered to be dated at ~2900 Ma using Pb-Pb data from the Manfred Complex (Fletcher et al., 1988). However, Myers (1988a) also considered that the major deformation episode (D_2?) occurred between 2700–2600 Ma during collision of the Narryer Terrane with the granite-greenstone terrane of the Murchison Domain to the south (Fig. 3.6-1). Data presented by Kinny et al. (1990) also supported metamorphism at ~2700 Ma, as did metamorphic rims on zircons from metasedimentary samples presented in Nutman et al. (1991). This Neoarchean event obliterated almost all evidence of any earlier metamorphic episodes, of which the scant remaining evidence suggests that these occurred at ~3055 Ma and ~3300 Ma (Nutman et al., 1991).

Several mylonitic shear zones are recorded in the Narryer area, including the Elizabeth Springs zone that bounds the eastern margin of the Mt. Narryer sedimentary package (Fig. 3.6-3; Myers and Williams, 1985). However, no recent studies have been undertaken on these zones and so their timing with respect to Archean and Proterozoic events in the Narryer Terrane is unknown.

3.6-3.3. Jack Hills Metasedimentary Belt and Environs

3.6-3.3.1. Setting

The Jack Hills metasedimentary belt lies ~60 km NE of Mt. Narryer and is approximately 90 km long with a pronounced sigmoidal curvature, typical of a dextral shear zone (Fig. 3.6-7(b), Spaggiari, 2007b). The northeastern and southwestern parts of the belt are within fault splays off a major east-trending shear zone named the Cargarah Shear Zone (Fig. 3.6-4; Williams et al. 1983; Spaggiari, 2007b). The Cargarah Shear Zone cuts through gneiss and granite just north of the belt on the western side, through metasedimentary rocks and mafic and ultramafic schists at its center near the Cue-Berringarra Road, and through granitic rocks and gneisses south of the belt on the eastern side. Lenses of quartzite, mafic schist, BIF, calc-silicate and chert within gneiss and granite are enclosed by fault splays off the western end of the shear zone. The continuation of the shear zone east of the belt is evident in aeromagnetic images (Fig. 3.6-5), with fault splays containing predominantly BIF, calc-silicate and ultramafic rocks to the east. The central region of the belt is dominated by high strain, including pronounced stretching and boudinage, whereas the fault splays contain areas of lower strain with more localised high strain zones. The majority of kinematic indicators within the Cargarah Shear Zone indicate dextral, strike-slip movement, that most likely developed in a transpressional regime (Williams, 1986; Spaggiari, 2007a, 2007b).

The margins of the belt are sheared and have steep dips, indicating that the belt itself is also steeply dipping. Across strike, the belt is interpreted to be wedge-like, pinching out within the gneiss and granitoid at depth, and consisting of a series of internally folded fault slices (Spaggiari, 2007a, 2007b). This interpretation differs from Elias (1982), who interpreted the belt as a synformal structure. The belt has undergone a long and complex structural history that has produced the present-day geometry. ^{40}Ar-^{39}Ar data from the Jack Hills belt and surrounding areas indicate that the main phase of movement along the Cargarah Shear Zone took place during the Capricorn Orogeny (Spaggiari et al., 2004; Spaggiari, 2007a; Spaggiari et al., 2007). Previous studies have also speculated that certain structures in the southern Narryer Terrane are Proterozoic in age and related to the Capricorn Orogeny (Williams et al., 1983; Williams, 1986). Williams (1986) inferred that what is now defined as the Cargarah Shear Zone formed during the Proterozoic because it trends parallel to major shears in the Capricorn Orogen. Reactivation of the Cargarah Shear Zone and associated structures, and formation of new, semi-brittle or brittle structures, took place after the main phase of shearing, but the timing of these is poorly constrained. The ^{40}Ar-^{39}Ar data, and Rb-Sr data of Libby et al. (1999), indicate that a deformation event may have occurred at approximately 1400–1350 Ma. Faulting at around 1200 Ma, indicated by the ^{40}Ar-^{39}Ar data, may have been related to the intrusion of mafic dykes at around 1210 Ma (Wingate et al., 2005). Intrusion of cross-cutting mafic dykes at 1075 Ma marks the final stage of substantial tectonic activity in the Jack Hills region (Wingate et al., 2004; Spaggiari, 2007a).

3.6-3.3.2. *Metasedimentary sequences*

The dominant lithologies in the Jack Hills belt are BIF, chert, mafic and ultramafic rocks, and siliciclastic rocks that include quartz-mica schist, andalusite schist, quartzite, meta-sandstone, and oligomictic metaconglomerate (Fig. 3.6-7(e)) (Elias, 1982; Williams et al., 1983; Wilde and Pidgeon, 1990). Wilde and Pidgeon (1990) divided the supracrustal rocks of the Jack Hills belt into three informal associations:

 (i) chemical sediment association consisting of BIF, chert, mafic schist (amphibolite) and minor ultramafic intrusions (Fig. 3.6-7(b,c)), developed along the margins of the belt;

 (ii) pelite-semipelite association characterised by quartz-biotite and quartz-chlorite schists, associated with local mafic and ultramafic schists, that were possibly part of a turbidite sequence, and is located in the central areas of the belt, and

(iii) clastic sedimentary association that occurs as a more restricted succession of mature clastic sedimentary rocks comprising conglomerate, sandstone, quartzite (Fig. 3.6-7(d,e)), and siltstone, developed away from the belt margins.

The latter were interpreted by Baxter et al. (1986) and Wilde and Pidgeon (1990) as having been deposited in the medial to distal part of an alluvial fan-delta, and are where the majority of detrital zircons $\geqslant$4.0 Ga have been found. Based on work undertaken in the 1980s, it was not clear whether these associations were all part of the same sedimentary event, or whether they were unrelated successions juxtaposed by deformation, though the latter was favoured (Wilde and Pidgeon, 1990). Spaggiari et al. (2007) have divided the belt into four informal associations, not unlike those above, but including known Proterozoic rocks. These are:

(1) an association of banded iron formation, chert, quartzite, mafic and ultramafic rocks;

(2) an association of pelitic and semi-pelitic schist, quartzite, and mafic schist;

(3) an association of mature clastic rocks including pebble metaconglomerate; and

(4) an association of Proterozoic metasedimentary rocks.

Some of the second association may be part of the same succession as association 1, but relationships are obscured. Given that the present-day margins of the belt are largely a consequence of Proterozoic shearing, the original extent of each association is unknown.

Chert and BIF, together with quartzite, mafic and ultramafic schist, and some pelitic and semi-pelitic rocks (associations 1 and 2), have been intruded by monzogranite, muscovite granite and pegmatite. However, there are no clear intrusive relationships with the mature clastic sedimentary rocks that host the ancient detrital zircons. The monzogranites have SHRIMP U-Pb zircon ^{207}Pb/^{206}Pb ages of 2654 ± 7 Ma and 2643 ± 7 Ma (Pidgeon and Wilde, 1998), but have undergone late shearing, especially adjacent to the Jack Hills belt.

Jack Hills metaconglomerate from Eranondoo Hill, together with pelite-semipelite from nearby and south of Mt. Hale (Fig. 3.6-4), yielded T_{DM} Sm-Nd model ages of 4.11–3.30 Ga (Maas and McCulloch, 1991), whereas the quartzites showed two groups of ages at 3.75–3.53 Ga and 3.11–2.82 Ga: BIF T_{DM} model ages fell into the latter group. An ultramafic shale from Mt. Taylor, in the far northeast of the Jack Hills belt (Fig. 3.6-4), gave a T_{DM} model age of 3.61 Ga (Dobros et al., 1986). A rare felsic volcanic unit that lies south of the main belt, ∼5 km SE of Eranondoo Hill has a T_{chur} Sm-Nd model age of 3.68 Ga (Baxter et al., 1984).

Although a significant number of detrital zircons have been analyzed from Jack Hills, most of the work has focussed on zircons that are ⩾4.0 Ga because they yield significant information about early Earth processes (e.g., Valley et al., 2002; Harrison et al., 2006; Cavosie et al., this volume). Most zircon analyses have come from the same pebble metaconglomerate outcrop on Eranondoo Hill (W74) (Fig. 3.6-4) since it has been shown in several studies to yield the greatest percentage of ancient zircons – about 12% (Compston and Pidgeon, 1986). The mature clastic rocks have yielded detrital grains with domains up to 4.4 Ga (Wilde et al., 2001). However, from over 70,000 grains analyzed it has been shown that there is a complete spread of ages between 4.4 to 4.0 Ga, with the greatest percentage between 4.0 and 4.2 Ga (Cavosie et al., 2004; Harrison et al., 2005). The mature clastic rocks also contain abundant detrital zircons with ages between 3.7 and 3.1 Ga, indicating deposition after ∼3 Ga (Compston and Pidgeon, 1986; Nutman et al., 1991).

Until 2002, all metasedimentary rocks in the Jack Hills belt were thought to be of early Neoarchean age, since no zircon younger than ∼3100 Ma had been recorded from the samples. However, Cavosie et al. (2004) and Dunn et al. (2005) both found concordant Late Archean and Paleoproterozoic zircons in some of the clastic metasedimentary units. The eastern transect (Fig. 3.6-4) of Cavosie et al. (2004) included metaconglomerate sample 01JH47 which yielded two magmatic grains with ages of 2724 ± 7 Ma (88% concordance) and 2504 ± 6 Ma (94% concordance). This was the first evidence that some of the sediments contained late Neoarchean zircons. Metaconglomerate sample 01JH42 from the same transect also contained a single grain with an age of 2.3 Ga, but was only analysed with a one-cycle SHRIMP run and the data were not published.

Along the western transect of Cavosie et al. (2004) on Eranondoo Hill, which included the W74 site (Fig. 3.6-4), quartzite sample 01JH63 contained concordant late Archean grains (2736 ± 6, 2620 ± 10, and 2590 ± 30 Ma), two concordant Paleoproterozoic grains (1973 ± 11 and 1752 ± 22 Ma), and a single concordant grain at 1576 ± 22 Ma. This quartzite is from the southern end of the transect, close to the contact with BIF, and is interpreted to form part of a Proterozoic succession of predominantly quartz-mica schist. In addition, Dunn et al. (2005) analyzed a fine-grained quartz-mica schist from what appears to be the same succession (sample JH3, Fig. 3.6-4). The sample contained concordant Paleoproterozoic grains mostly ranging in age from ∼1981–1944 Ma, with the youngest grain having a $^{207}Pb/^{206}Pb$ SHRIMP age of 1797 ± 21 Ma. The sample also contained a high proportion of concordant late Neoarchean grains.

Dunn et al. (2005) also dated a metaconglomerate from the northeastern part of the belt (sample JH4, Fig. 3.6-4) that contained a similar range of ages to JH3, with the youngest concordant grain being 1884 ± 32 Ma and older grains ranging between 3725–3500 Ma. This metaconglomerate is part of a thick sequence of predominantly quartzite and metasandstone, but it is not clear whether it is part of the same sequence as at Eranondoo Hill.

At present, the full extent of Proterozoic rocks within the belt is unclear, and it is not possible to distinguish them from Archean metasedimentary rocks, due to lithological similarity, Proterozoic deformation effects, and recrystallization. However, taken together, these data from independent studies indicate that part of the Jack Hills succession was deposited during or following the Capricorn Orogeny, with a detrital contribution from rocks formed during the Glenburgh Orogeny (Dunn et al., 2005) and also possibly derived from weathering and erosion of the Neoarchean granitoids. The single grain with an age of 1576 ± 22 Ma (Cavosie et al., 2004) is problematical and does not correspond with any known sources of detritus so far recorded from the region. Although more data are needed, they indicate that at least some of the sedimentary rocks were deposited after intrusion of the Neoarchean granites. Importantly, sample JH3 also contained a single concordant grain with an age of 4113 ± 3 Ma, plus a few grains of 3500–3300 Ma zircon (Dunn et al., 2005). This has implications for understanding the source of the $\geqslant 4.0$ Ga zircons and implies that it may have been extant during the Proterozoic. Alternatively, the $\geqslant 4.0$ Ga zircons in Proterozoic rocks may have been recycled from older sedimentary sources.

In summary, it is evident that at least three (but possibly four) metasedimentary successions are present in the Jack Hills belt; an older succession of BIF, chert, quartzite, quartz-mica schist, amphibolite and ultramafic rocks (intruded by Neoarchean granites); a turbiditic sequence of pelite and semi-pelite, for which there are no precise age data on the time of deposition; a siliciclastic succession, including the mature clastic units that host the majority of $\geqslant 4.0$ Ga detrital zircons; and a Paleoproterozoic succession of quartz-mica schist, quartzite and metasandstone, with local metaconglomerate.

The depositional age of the mature clastic rocks devoid of Paleoproterozoic zircons is uncertain. The chemical sediment association in the southwestern part of the belt must be older than ~ 2.6 Ga (the maximum Rb-Sr age obtained on micas from an intrusive muscovite-bearing granite at Noonie Hill: Fletcher et al. (1997) – see below). However, the clastic rocks could potentially be Paleoproterozoic in age, since the zircon population merely reflects the source that is being eroded and it is possible that either the Neoarchean granites were not exposed and eroding at the time of deposition, or that the zircons sampled are of restricted, local derivation and do not fully reflect the total exposure at the time of formation. The absence of late Archean detrital zircons from most samples, however, favours an early Neoarchean age of deposition.

3.6-3.3.3. Granitoids

A study of the granitoids adjacent to the Jack Hills Metasedimentary Belt (Pidgeon and Wilde, 1998) revealed several differences from the features described from Mt. Narryer and its immediate environs. Firstly, many of the rocks are less deformed and retain some of their igneous features, including original phenocrysts. Whilst all rocks show some degree of deformation, revealed as granulation in thin section, many are best classified as deformed granitoids rather than as orthogneisses. In general, it appears that the amount of deformation increases toward discrete shear zones. Furthermore, pluton-scale bodies of TTG rocks are present, ranging in composition from tonalite and trondhjemite to granodiorite.

Overall, the ages obtained from the granitoids at Jack Hills closely match those obtained from Mt. Narryer, ~60 km to the southwest, with a spread of SHRIMP U-Pb ages from 3.7–3.3 Ga (Pidgeon and Wilde, 1998). Within this range, discrete peaks are evident at 3.75–3.65 Ga, ~3.50 Ga and at 3.3 Ga, interpreted to define major isotopic disturbances associated with magmatic activity. The oldest group of ages was considered by Pidgeon and Wilde (1998) to record the time of extraction of the earliest tonalites from a garnet-amphibolite source. The presence of 3.3 Ga zircon rims around older cores in several rocks suggests that metamorphism also occurred at this time. Importantly, evidence for 3.3 Ga activity appears to be restricted to the southern side of the Jack Hills belt, suggesting that the belt may lie along a major suture within the Narryer Terrane.

Geochemical data suggest an evolutionary trend within the TTG suite (Pidgeon and Wilde, 1998), commencing with the earliest tonalites and trondhjemites with $^{207}Pb/^{206}Pb$ ages of 3608 ± 22 Ma, but containing zircons as old as 3756 ± 7 Ma – the oldest material so far identified from the TTG rocks in the Narryer Gneiss Complex. Some ~3.5 Ga zircons occur in these rocks, suggesting a disturbance in the isotopic system at this time. Furthermore, a granodiorite immediately north of the belt has an age of 3516 ± 32 Ma, indicating that this was a significant period of activity in the region. In rocks south of the belt, there are also rims of ~3.3 Ga zircon around 3.6–3.5 Ga cores. Younger granodiorite and porphyritic granite with $^{207}Pb/^{206}Pb$ ages of ~3300 Ma have more prominent negative Eu anomalies and appear to be the result of remelting of the older tonalite-trondhjemite, since cores as old as 3707 ± 12 Ma are present. These rocks also appear to have undergone a late Archean disturbance at ~2700 Ma, similar in age to the event recognised near Mt. Narryer (Kinny et al., 1990).

A suite of monzogranites is ubiquitous in the area (Fig. 3.6-7(f)), locally cross-cutting the TTG rocks, though commonly brought into subparallelism in high strain zones. Two monzogranites gave SHRIMP U-Pb ages of 2643 ± 7 Ma and 2654 ± 7 Ma (Wilde and Pidgeon, 1990), similar within error to the age obtained for monzogranite at Mt. Narryer by Kinny et al. (1990). This indicates that much of the Narryer Terrane was assembled by this time; such ages being common throughout the Yilgarn Craton. Zircon cores with significantly older ages are not common in monzogranite at Jack Hills, but two such examples, with ages of ~3.6 and 3.5 Ga, were recorded by Pidgeon and Wilde (1998). Monzogranites have been affected by Proterozoic shearing adjacent to the Jack Hills (Spaggiari, 2007a, 2007b) and this appears to be reflected in the non-zero disturbance of the zircon isotopic systems (Pidgeon and Wilde, 1998).

In the southwest near Noonie Hill (Fig. 3.6-4), the Jack Hills belt is intruded pervasively by muscovite-bearing granites and pegmatites. These granitoids dismember the belt and enclose rafts of metasedimentary rocks (especially BIF), with the latter commonly showing no rotation or deviation from the trend observed in the adjacent belt. The muscovite-bearing granitoids are devoid of major deformation and therefore post-date the monzogranites. No U-Pb ages are available for these rocks; the only data being Rb-Sr and K-Ca ages determined on white micas (Fletcher et al., 1997). The Rb-Sr ages have large errors (~50 Ma) and range from 2556–2530 Ma for the granite and from 2595–2569 Ma for the pegmatite, whereas the K-Ca ages are younger, ranging from 2117–2099 Ma in the granite and 2435–2383 Ma in the pegmatite. These data are difficult to interpret, since the ages are older in the pegmatites, yet these clearly cross-cut the muscovite granite in the field.

3.6-3.3.4. Metamorphism and deformation

The supracrustal rocks at Jack Hills are at upper greenschist to lower amphibolite facies, and are therefore of lower metamorphic grade than the metasedimentary rocks at Mt. Narryer, and most of the granitic gneisses of the Narryer Gneiss Complex. Amphibolite facies metamorphism at Jack Hills is denoted by the presence of grunerite in BIF and the association of calcic plagioclase and hornblende in mafic schist and amphibolite (Wilde and Pidgeon, 1990). Hornblende in the mafic schists is commonly overprinted by actinolite, suggesting some retrogression to greenschist facies. The majority of pelitic rocks are semi-pelites that lack diagnostic mineral assemblages, other than andalusite. Local andalusite-staurolite-chloritoid-chlorite assemblages in siliciclastic rocks near Noonie Hill (Fig. 3.6-4) indicate greenschist-amphibolite transition facies metamorphism (Wilde and Pidgeon, 1990).

Baxter et al. (1986) and Wilde and Pidgeon (1990) recognised five phases of deformation in the Jack Hills belt (D_1–D_5); these had the following characteristics. The earliest phase of deformation (D_1) is represented by rootless isoclinal folds with a strong mineral stretching lineation in BIF and chert. The D_2 folds are open to tight minor folds with a strong axial planar fabric, although a major D_2 anticline was considered to control the geometry of the belt. Folds associated with this event are likewise found mainly in BIF and chert. The D_3 folds are gentle sub-horizontal open folds. Importantly, this is the earliest phase of folding identified in the clastic sedimentary sequence that hosts the ancient detrital zircons (Wilde and Pidgeon, 1990). All these events appear to have developed in the Neoarchean. The D_4 phase is a set of minor to major conjugate shears that cut both the belt and the surrounding granitoids; there is local folding associated with this event, which probably took place in the Proterozoic. Structures associated with this deformation phase commonly mark the boundaries between the various lithological associations, with the implication that the belt was not finally assembled until the Proterozoic. The final deformation phase (D_5) is a set of local kinks and buckles. Wilde and Pidgeon (1990) suggested that D_1 at Jack Hills might be equivalent to the regional scale D_2 identified in and around Mt. Narryer (Myers and Williams, 1985); that D_2 and D_3 form part of the regional D_3 event; whereas D_4 and D_5 were local and Proterozoic in age.

In contrast to Baxter et al. (1986) and Wilde and Pidgeon (1990), Spaggiari (2007a, 2007b) has not attempted to classify structures within the belt according to "D_1, D_2", etc. This is because the belt is dominated by overprinting coaxial structures, many of which formed due to major shearing involving both flattening and rotation, and are therefore impossible to distinguish chronologically. These include tight to isoclinal folds at various scales, well-developed mineral lineations, predominantly dextral shear bands, and low-angle shear bands (Spaggiari, 2007a, 2007b). The deformation history has been interpreted to include an early phase of folding, including recumbent folding that was possibly related to thrusting, followed by upright folding and major transpressive shearing. Most shears are reworked and overprinted by widespread brittle and semi-brittle faulting, and localised folding including kink folds (Spaggiari, 2007a, 2007b). The early recumbent folding and thrusting may have taken place during the regional deformation event associated with intrusion of Neoarchean granites (~2700 to 2620 Ma). This deformation is evident in associations 1 and 2 of Spaggiari (2007a) and Spaggiari et al. (2007), but not in the mature clastic rocks that include the outcrop of W74. That association is dominated by moderately southwest plunging, tight folds inclined to the southeast, cut by shears (Spaggiari, 2007a, 2007b). Based on ^{40}Ar-^{39}Ar data and regional relationships, major shearing and formation of the Cargarah Shear Zone is interpreted to have taken place during, and/or following, the Capricorn Orogeny between 1830 to 1780 Ma (Spaggiari, 2007a; Spaggiari et al., 2007 and in review). Reworking of the belt, mainly in the form of semi-brittle faults and kink folding, took place after major shearing, possibly during several discrete deformation episodes prior to the intrusion of cross-cutting mafic dykes at 1075 Ma (Spaggiari, 2007a; Wingate et al., 2004).

3.6-4. SOME OUTSTANDING ISSUES

As will be evident from the above review, work in the Narryer Terrane has so far been largely concentrated in two main areas; at Mt. Narryer and in the Jack Hills. In recent years, the renewed interest in the area has been driven very much by the discovery of even older zircons at Jack Hills (Wilde et al., 2001), the elevated δO^{18} of many of these from at least 4.2 Ga onwards (Cavosie et al., 2005) and the implications that these results have for the nature of the early Earth (Valley et al., 2002; Harrison et al., 2005; Valley et al., 2006).

There are a number of issues that currently remain unresolved, the most significant of which are discussed below.

3.6-4.1. Regional Correlations

The highly complex nature of the gneisses, commonly interlayered on a centimetre scale, means that unravelling the geological history is a difficult process. Since the original discovery of >4 Ga zircons was at Mt. Narryer, it was natural that this area was the first to be studied in detail. Hence, subsequent work meant that ideas and interpretations at Mt. Narryer were extended, with various levels of success, to adjacent areas. Nutman et al. (1991) attempted a re-evaluation of the Narryer Terrane, based on the available geochronology at that time. They considered that early Archean rocks are largely restricted to the eastern part of the Narryer Terrane, with younger gneisses dominating in the western and northern parts. They introduced the idea of classifying the early Archean gneisses into the Nookawarra and Eurada Gneiss Associations, with the Narryer Supracrustal Association representing a unit of similar status (see below). However, it should be noted that this suggested subdivision has not been widely used in subsequent publications, including those by the authors in question.

Nookawarra Gneiss Association: this includes the Meeberrie and Dugel gneisses and the Manfred Complex. Further work on the Meeberrie gneiss (Kinny and Nutman, 1990) showed that the protolith ages vary considerably and include components with ages of ~3730, 3670, 3620 and 3600 Ma, commonly invaded by pegmatite and granite veins with ages of 3300–3280 Ma. There is also a significant component of 3300–3280 Ma granitoids and gneisses at both Mt. Narryer (Kinny et al., 1990) and Jack Hills (Wilde and Pidgeon, 1990), which are not found in the Eurada Gneiss Association (see below), leading to the idea that the rocks may not have experienced a common history until after 3280–3300 Ma. However, the presence of ~3300 Ma rims on zircons in the Eurada gneisses (Kinny, 1987) and migmatitic components of this age near Jillawarra Bore (Wiedenbeck, 1990) suggest that the Nookawarra and Eurada associations started to have a common history from this time. Nutman et al. (1991) argue that the ~3380 Ma Dugel gneiss is much more restricted in occurrence than originally proposed by Myers and Williams (1985), being present only in the Mt. Narryer area.

Eurada Gneiss Association: this was first identified in the southern part of the Narryer Terrane and is a group of gneisses with protolith ages of 3490–3440 Ma (Kinny, 1987). However, rocks of this age are also present in the northeast part of the terrane (Wiedenbeck, 1990).

Narryer Supracrustal Association: Nutman et al. (1991) considered that the metasedimentary rocks were all the same age (something we now know to be incorrect) and thus constituted a single metasedimentary package. Support for this view came from the Sm-Nd data of Maas and McCulloch (1991), who determined that the maximum depositional age, based on Sm-Nd systematics, was ~3 Ga for both Mindle and Jack Hills quartzite and BIF. The detrital zircon populations show peaks at 3750–3600, 3500–3300 and 3100–3050 Ma, corresponding closely to the age of orthogneisses in the Eurada and Nookawarra Gneiss Associations, implying that these were the dominant source materials (Nutman et al., 1991; Pidgeon and Nemchin, 2006). However, the missing component is the source that provided the >4 Ga zircon components, estimated by Nutman et al. (1991) as making up ~1% of the area.

Whilst an attempt to make such a regional correlation is to be commended, the reality is that it requires extensive geochronology. Indeed, because the Meeberrie gneiss itself is a complex of components ranging in age from ~3750 to ~3300 Ma, it is currently impossible to separate out these components without the aid of U-Pb zircon geochronology. This means that almost each individual outcrop would need to be dated in order to get a realistic idea of the distribution of the various components; not a viable task.

3.6-4.2. Sedimentary Environment

Baxter et al. (1984) and Pidgeon et al. (1986) interpreted the clastic sedimentary sequence at Jack Hills as a fan-delta sequence, probably deposited in an intracratonic fault-bounded basin. Kinny et al. (1990) considered that the metasedimentary rocks at Mt. Narryer formed at a continental margin, whilst Maas and McCulloch (1991) argued that the association of clastic, near-shore fluviatile sequences with chemical sedimentary rocks and turbiditic sediments at both localities suggested a possible rifted cratonic margin. This was supported by distinctly different rare earth element (REE) patterns in the metasediments compared to the adjacent orthogneisses; the former showing negative Eu anomalies indicating derivation from K-rich granitoids. The presence of low ε_{Nd} values in some of these sedimentary rocks was interpreted to indicate a component of detritus ($\geqslant$10%) from 4.2–4.1 Ga LREE-enriched continental crust (Maas and McCulloch, 1991).

However, the recent realization that not all metasediments at Jack Hills are $\gtrsim$ 3100 Ma in age (Cavosie et al., 2004; Dunn et al., 2005), together with the recognition that there may be up to four discrete packages of sediments (Wilde and Pidgeon, 1990; Spaggiari et al., 2007), means that alternative models need to be considered. The similarity between the clastic sediments containing no zircons younger than ~3 Ga with those that contain zircons as young as 1576 $\pm$ 22 Ma (Cavosie et al., 2004) makes this task difficult. The least studied component is the pelite-semipelite association, characterised by quartz-mica schists, with local mafic and ultramafic schists. These possibly form part of a turbidite sequence (Wilde and Pidgeon, 1990) and may not have formed in the same sedimentary environment as either the chemical or clastic sedimentary associations. Furthermore, no rocks from the pelite-semipelite association have been dated using U-Pb zircon techniques, so currently it is unclear if this represents an Archean or a Paleoproterozoic sequence.

3.6-4.3. The Source of the Detrital Zircons, Including those >4 Ga

Many of the early papers have speculated as to the likelihood of finding samples of >4 Ga rocks in the currently-exposed Narryer Terrane. It is known from the zircon data (Wilde et al., 2001; Mojzsis et al., 2001; Peck et al., 2002; Cavosie et al., 2004; Harrison et al., 2005) that events prior to the oldest known rocks in the area (~3730 Ma) were complex and that a single source is unlikely. Furthermore, the recent studies by Crowley et al. (2005) and Pidgeon and Nemchin (2006) have suggested that there may be differences in the source regions between the Mt. Narryer and Jack Hills metasediments. In a LAM-ICPMS study, Crowley et al. (2005) observed differences in morphology, zoning, inclusion type and some trace element concentrations in zircons from the two areas, especially between the western Jack Hills (including the W74 site) and the other sample locations, which included the eastern part of the Jack Hills belt. They also noted the low proportion of grains >3900 Ma (3%) from Mt. Narryer, similar to earlier studies. However, the recent study at Mt. Narryer by Pidgeon and Nemchin (2006) has identified a population of 25 grains that were older than 3900 Ma, making up ~12% of the total zircon population and therefore similar to the concentration at the W74 site in Jack Hills as determined by Compston and Pidgeon (1986). They also identified zircons older than those recognised by Froude et al. (1983) and Kinny et al (1990) at Mt. Narryer, with the oldest grain having a $^{207}Pb/^{206}Pb$ age of 4281 $\pm$ 11 Ma.

Pidgeon and Nemchin (2006) showed that the zircon population from the W74 meta-conglomerate site at Jack Hills lacks the strong age peak at 3750–3600 Ma found in the Mt. Narryer quartzite, whereas the 3500–3350 Ma peak, so prominent at Jack Hills, is extremely weak at Mt. Narryer. They thus imply a difference in provenance between the two metasedimentary belts. However, this may be more an artefact of their limited sampling, since sample JH4 from the northern part of the Jack Hills belt near Mt. Hale (Fig. 3.6-4) does show a significant peak at 3750–3600 Ma (Dunn et al., 2005) and may share the same provenance as the Narryer metasedimentary rocks (Crowley et al., 2005).

Although precise rock types yielding the detrital zircon populations at Jack Hills and Mt. Narryer are difficult to quantify, Crowley et al. (2005) suggest that zircons in metasedimentary rocks from Mt. Narryer and the eastern Jack Hills were from a granitic source that closely matches the adjacent gneisses in terms of age, oscillatory zoning and U-content. On the contrary, those from the western Jack Hills (including the W74 site) do not match those found in either the high-grade gneisses of the Narryer Terrane or other known granitoids of similar age.

With respect to the source of >4 Ga zircons, as pointed out in many recent studies (Maas et al., 1992; Wilde et al., 2001; Mojzsis et al., 2001; Peck et al., 2001; Harrison et al., 2005; Crowley et al., 2006), at least some of the ~4200 Ma grains were derived from an evolved granitic source. Since >4.2 Ga zircons are now known from both Jack Hills and Mt. Narryer (Pidgeon and Nemchin, 2006) and that some differences exist between these two populations, it may be that diverse rocks of this age still exist as remnants caught up in the high-grade gneisses that enclose the metasedimentary belts. Ancient zircons have also been recorded as rare components in Neoarchean granitoids near Mt. Narryer (Nelson et al., 2000) and further afield as detrital grains in quartzite in the Maynard Hills (Wyche et al., 2004). These discoveries suggest the original widespread distribution of >4 Ga granitoid crust.

Although many attempts have been made to locate clasts of ancient granitoid in the conglomerates at Jack Hills and Mt. Narryer, these have so far proved unsuccessful. It is important to note that the presence of quartzite, sandstone, chert and BIF (in addition to vein quartz) in the Jack Hills metaconglomerates, indicates that much of the detritus is the result of re-working of earlier sedimentary rocks. Significantly, quartzite clasts in the metaconglomerate at Eranondoo Hill (sample W74) contain detrital zircons with a similar age range to those of the whole rock (Pidgeon et al., 1990), implying that the zircon signature is inherited from an earlier sedimentary source. This being the case, it is quite likely that all remnants of >4 Ga granitic rocks have been destroyed during the long interval between formation and final deposition after ~3100 Ma.

ACKNOWLEDGEMENTS

We thank Michael Prause of the Geological Survey of Western Australia for drafting Figs. 3.6-2 and 3.6-3. Pete Kinny kindly provided most of the field photographs of the gneissic rocks. Simon Wilde acknowledges on-going support from the Australian Research Council for his work at Jack Hills. Catherine Spaggiari publishes with permission of the Director of the Geological Survey of Western Australia. This is The Institute for Geoscience Research (TIGER) publication number 59.

PART 4

THE PALEOARCHEAN PILBARA CRATON, WESTERN AUSTRALIA

Earth's Oldest Rocks
Edited by Martin J. Van Kranendonk, R. Hugh Smithies and Vickie C. Bennett
Developments in Precambrian Geology, Vol. 15 (K.C. Condie, Series Editor)
© 2007 Elsevier B.V.   All rights reserved.
DOI: 10.1016/S0166-2635(07)15041-6

Chapter 4.1

PALEOARCHEAN DEVELOPMENT OF A CONTINENTAL NUCLEUS: THE EAST PILBARA TERRANE OF THE PILBARA CRATON, WESTERN AUSTRALIA

MARTIN J. VAN KRANENDONK[a], R. HUGH SMITHIES[a], ARTHUR H. HICKMAN[a] AND DAVID C. CHAMPION[b]

[a]*Geological Survey of Western Australia, 100 Plain St., East Perth, Western Australia 6004, Australia*
[b]*Geoscience Australia, GPO Box 378, Canberra, ACT 2601, Australia*

4.1-1. INTRODUCTION

The 3.53–2.83 Ga Pilbara Craton of Western Australia is one of only two areas on Earth that contain large, well-exposed areas of little deformed, low-grade Paleoarchean rocks – the other being the Kaapvaal Craton in southern Africa – and as such is important for understanding early Earth and the processes involved in the formation of continental crust.

The Pilbara Craton is famous for its well preserved Paleoarchean volcano-sedimentary succession that includes evidence of the oldest life on Earth, and for its classic dome-and-keel geometry of ovoid, domical granitic complexes and flanking arcuate, synclinal greenstone belts. As with many things, however, both the evidence for early life and the origin of the dome-and-keel structural map pattern have been strongly debated.

In this paper, we describe the lithostratigraphy, geochemistry, and structural and meta-morphic geology of the ancient, eastern nucleus of the Pilbara Craton – known as the East Pilbara Terrane – and then discuss the possible mechanisms that led to its formation and preservation. Separate papers describe the geochemistry of felsic volcanic rocks (Smithies et al., this volume) and granitic rocks (Champion and Smithies, this volume) in more detail, while Huston et al. (this volume) describe the mineral deposits of this terrane and the in-ferences that can be drawn from them for geodynamic models. Three other papers describe the evidence for ancient life in these rocks (see Van Kranendonk, this volume; Marshall, this volume; Ueno, this volume).

4.1-2. GEOLOGY OF THE PILBARA CRATON

The 3.53–2.83 Ga Pilbara Craton is a nearly circular piece of crust in the northwest-ern part of Western Australia, whose boundaries are defined by aeromagnetic and gravity anomalies, and by orogenic belts (Fig. 4.1-1). The northern part of the craton is well ex-posed over an area of 530 km × 230 km, but much of the rest is unconformably overlain

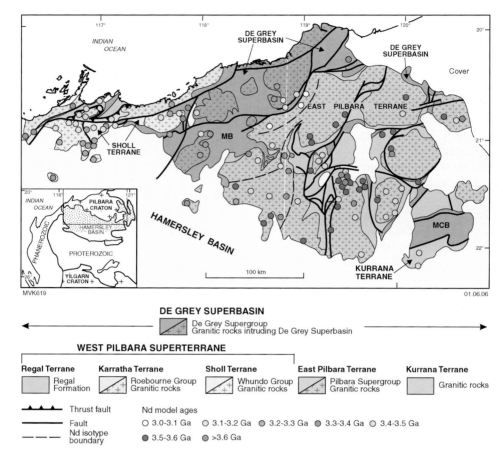

Fig. 4.1-1. Simplified geological map of the Pilbara Craton, showing terranes, late tectonic basins of the De Grey Supergroup, and Nd model age data. Inset shows craton in Western Australia.

by volcanic and sedimentary rocks of the ca. 2.78–2.45 Ga Mount Bruce Supergroup deposited in the Hamersley Basin (inset of Fig. 4.4-1: Arndt et al., 1991; Blake et al., 2004; Trendall et al., 2004).

Hickman (1983, 1984) provided a regional lithostratigraphic interpretation of the 'Pilbara Block', based on reconnaissance geological mapping, and concluded there was a relatively uniform stratigraphy across the craton and that the major tectonic structures resulted from essentially solid-state granitoid diapirism. Subsequent geochronology and mapping, however, showed that the greenstones are much more laterally variable in age across the craton than previously thought (Horwitz and Pidgeon, 1993; Hickman, 1997; Smith et al., 1998; Buick et al., 2002), and include exotic terranes accreted during the later stages of tectonic development of the craton (Tyler et al., 1992; Barley, 1997; Smith et al., 1998; Van Kranendonk et al., 2002, 2006a, 2007a; Hickman, 2004).

Recent compilations of stratigraphic, geochronological, structural, and isotopic data collected during a 12-year geoscience program by the Geological Survey of Western Australia, Geoscience Australia and a variety of university-based researchers have been used to divide the granite-greenstone rocks of the northern Pilbara Craton into five terranes and five late tectonic, dominantly clastic sedimentary basins (Fig. 4.1-1: Van Kranendonk et al., 2002, 2006a, 2007a). Terranes have unique lithostratigraphy, granitic supersuites, structural map patterns, geochemistry and tectonic histories, as outlined in more detail elsewhere (Hickman, 1997, 2004; Van Kranendonk et al., 2006a, 2007a; Smithies et al., 2007b).

Terranes include: the 3.53–3.165 Ga East Pilbara Terrane (EPT), representing the ancient nucleus of the craton; the ⩾3.18 Ga Kurrana Terrane in the southeastern part of the craton, which has EPT-type Sm-Nd model ages and may represent a rifted fragment of this crust; and the 3.27–3.11 Ga West Pilbara Superterrane (WPS). The WPS is a collage of three distinct terranes that include: the 3.27 Ga Karratha Terrane (KT), which may represent a rifted fragment of the EPT (Hickman, 2004) and which is juxtaposed against; ca. 3.20 Ga, MORB-type basaltic rocks of the Regal Terrane across a major thrust; the 3.13–3.11 Ga Sholl Terrane in the southern part of the WPS, which is juxtaposed against the other two terranes of the WPS to the north across the crustal-scale, strike-slip Sholl Shear Zone, which has a long history of formation and re-activation from ca. 3.07 to 2.92 Ga (Hickman, 2004).

The rocks of the WPS, EPT and Kurrana Terrane are separated by, and unconformably overlain by, the dominantly clastic sedimentary rocks of the De Grey Supergroup that was deposited across the whole of the craton between 3.02–2.93 Ga (Van Kranendonk et al., 2006a, 2007a). The supergroup comprises the basal Gorge Creek Group (ca. 3.02 Ga), the Whim Creek Group (3.01 Ga), the Croydon Group (2.99 or 2.97 Ga, through to 2.94 Ga) and the Nullagine Group (⩾2.93 Ga). Major depositional basins include the Gorge Creek Basin, which formed across the whole of the craton, the Mallina Basin between the WPS and EPT, and the Mosquito Creek Basin between the EPT and the Kurrana Terrane (Fig. 4.1-1). Recent models suggest deposition of the De Grey Supergroup (and intrusion of contemporaneous granitic supersuites) mainly as a result of lithospheric extension following accretion of the WPS at 3.07 Ga (Van Kranendonk et al., 2007a). Extension was viewed by these authors as the result of orogenic relaxation and breakoff of the slab of oceanic crust that was subducted during the formation of the Whundo Group (Sholl Terrane). The latter stages of deposition accompanied compressional deformation that affected the whole of the WPS and western half of the EPT at ca. 2.94–2.92 Ga, and the southeastern part of the northern Pilbara Craton at 2.905 Ga, the latter during accretion of the Kurrana Terrane with the EPT across the Mosquito Creek Basin (Van Kranendonk et al., 2007a).

Rocks of the WPS show lithostratigraphic, geochemical and structural features that are distinctly different from the EPT and thus have been used to argue for a change in tectonic style, from crust formation processes dominated by mantle plumes prior to 3.2 Ga, to processes dominated by plate tectonics after ca. 3.2 Ga (Smith et al., 1998; Hickman, 2004; Van Kranendonk et al., 2006a, 2007a). A key element in this model is geologic, isotopic, and geochemical data relating to the 3.13–3.11 Ga Whundo Group of the Sholl Terrane, which is a fault-bounded, >10 km thick succession of juvenile (ε_{Nd} values of $\sim +2$ to

+3 on volcanic rocks: Sun and Hickman, 1998), bimodal basaltic to felsic volcanic and volcaniclastic rocks that show affinities with an oceanic island arc (Smithies et al., 2005a, 2007a). Geochemical data indicate that the Whundo Group has a complex, but compelling, arc-like assemblage of rock types, including boninites (that have all of the compositional hallmarks of modern boninites), interlayered tholeiitic and calc-alkaline volcanics, Nb-enriched basalts, adakites, and rhyolites (Smithies et al., 2004b, 2005a, 2007b). Unusual trends of increasing trace element enrichment (i.e., higher La/Sm), which measures increasing degrees of partial melting (e.g., increasing Cr, decreasing HFSE, HREE, etc.) in the calc-alkaline rocks, can only be explained if the degree of melting is sympathetically tied to metasomatic enrichment (i.e., flux melting) of a mantle source. Trends to higher Th/Ba upsection in the Whundo Group stratigraphy reflect an increasing slab-melt component, and the calc-alkaline basalts and andesites eventually give way to adakites and Nb-enriched basalts, which require a major slab-melt component in the petrogenesis. Thus, the Whundo Group provides strong geochemical evidence for a subduction-enriched mantle source that strongly supports existing geological relationships (e.g., exotic terrain, no continental sediments, arc-like volcanic assemblage) for modern-style subduction-accretion processes at 3.12 Ga (Smithies et al., 2005a, 2007a, 2007b; Van Kranendonk et al., 2007a).

4.1-3. GEOLOGY OF THE EAST PILBARA TERRANE

The East Pilbara Terrane (EPT) is composed of rocks that formed between ca. 3.53 Ga and 3.165 Ga, prior to the time of onset of common geological events across the craton at 3.07 Ga (Van Kranendonk et al., 2006a, 2007). Volcano-sedimentary rocks (greenstones) of the EPT are assigned to the Pilbara Supergroup, which consists of four demonstrably autochthonous groups deposited from ca. 3.53–3.19 Ga. These are distributed in arcuate greenstone belts that wrap around domical, multiphase granitic complexes (Fig. 4.1-2). Granitic rocks of the EPT include five granitic supersuites that were emplaced at discrete intervals from ca. 3.50 to 3.165 Ga. Granitic rocks are exposed in domical granitic complexes and in syn-volcanic laccoliths. Periods of deformation and metamorphism accompanied each of the groups and emplacement of granitic rocks into the EPT prior to deposition of unconformably overlying supracrustal rocks (De Grey Supergroup) and intrusion by younger granitic rocks, both of which relate to common geological events across the craton between 3.07–2.83 Ga.

4.1-3.1. Lithostratigraphy in Greenstone Belts

The EPT is underlain by the generally well-preserved Pilbara Supergroup, which consists of four demonstrably autochthonous volcano-sedimentary groups deposited between ca. 3.53–3.19 Ga (Fig. 4.1-3: Van Kranendonk et al., 2006a, 2007a). Detailed geological mapping and extensive SHRIMP geochronology has revealed no stratigraphic repetitions in any of the greenstone belts in the EPT (Van Kranendonk et al., 2004, 2006a).

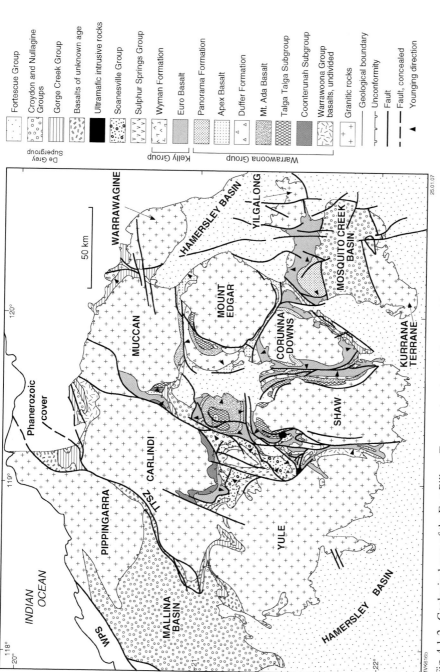

Fig. 4.1-2. Geological map of the East Pilbara Terrane, showing the dome-and-keel architecture and the distribution of major units of the Pilbara Supergroup.

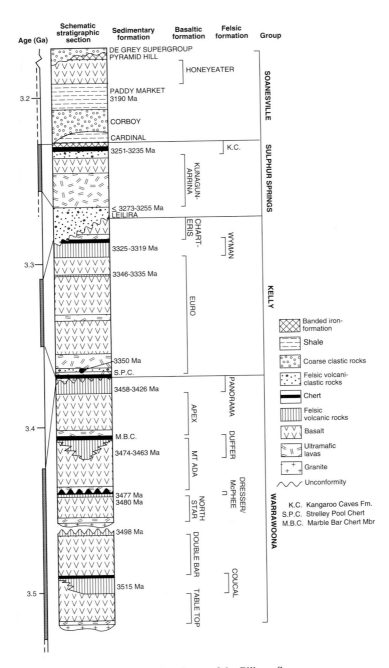

Fig. 4.1-3. Lithostratigraphic column of the Pilbara Supergroup.

The maximum *preserved* thickness of the Pilbara Supergroup is ~20 km. However, contacts of the lower part of the supergroup are everywhere with intrusive granitic rocks and the upper parts of each of the groups has been eroded beneath unconformably overlying groups or formations, such that the original thickness of any of the component groups, and the supergroup as a whole, is unknown. Furthermore, there is good evidence from changes in the dip of bedding between groups, and from the nature of the unconformities between groups, that the supergroup was deposited as a laterally accreting succession within developing synclinal basins on the flanks of amplifying granitic domes, during periods of extension (Hickman, 1984; Van Kranendonk et al., 2004, 2007).

4.1-3.1.1. Warrawoona Group

The stratigraphically lowest Warrawoona Group is at least 12 km thick and was deposited over 100 My as a result of continuous volcanism from ca. 3.53–3.43 Ga (Fig. 4.1-3: Van Kranendonk et al., 2002, 2006a, 2007a). The bulk of this succession comprises well-preserved pillow basalt, komatiitic basalt, and komatiite (Fig. 4.1-4(a,b)). Lesser felsic volcanic intervals are typically restricted to thin tuffaceous horizons a few 10's of metres thick, but dacite dominated deposits locally reach a few kilometers thick in the Coucal, Duffer and Panorama Formations, where coarse volcaniclastic breccias are a common lithofacies (Fig. 4.1-4(c): see Smithies et al., this volume).

Sedimentary units in the Warrawoona Group include a variety of lithology, such as volcaniclastic rocks, quartz-rich sandstone, and carbonates, but mainly consist of grey, white, blue-black, and red layered cherts derived from silicified carbonates (Fig. 4.1-4(d): Buick and Barnes, 1984; Van Kranendonk, 2006). An unusual sedimentary unit is the 3470±2 Ma Antarctic Creek Member of the Mount Ada Basalt, which contains thin beds with sand-size grains of altered quench-textured impact spherules (Lowe and Byerly, 1986; Byerly et al., 2002). Many chert units in the Warrawoona Group, including the jaspilitic cherts of the ca. 3.47 Ga Marble Bar Chert Member of the Duffer Formation, are well layered at a millimetre to centimetre scale, indicating deposition under quiet water, deep marine conditions, consistent with evidence from underlying and overlying pillow basalts. In contrast, the chert-barite unit of the ca. 3.49 Ga Dresser Formation in the North Pole Dome shows evidence of deposition under tectonically active conditions, as indicated by rapid lateral facies variations, beds of olistostrome breccia, and internal erosional unconformities (Nijman et al., 1998; Van Kranendonk, 2006; Van Kranendonk et al., 2006c, 2007c). The local preservation of desiccation cracks and the more widespread occurrence of rippled carbonate sedimentary rocks and stromatolites in this unit indicate periods of shallow water deposition, including subaerial exposure (Lambert et al., 1978; Walter et al., 1980; Groves et al., 1981; Buick and Dunlop, 1990; Van Kranendonk, 2006). Stromatolites are well developed in this unit and vary in morphology, suggesting diverse assemblages of micro-organisms at this early stage in Earth history (see Van Kranendonk, 2006, this volume, Chapter 7.2). Whereas original studies suggested deposition of the Dresser Formation in a restricted shallow marine basin, more recent studies suggest deposition within a felsic volcanic caldera affected by syn-depositional growth faults and voluminous hydrothermal

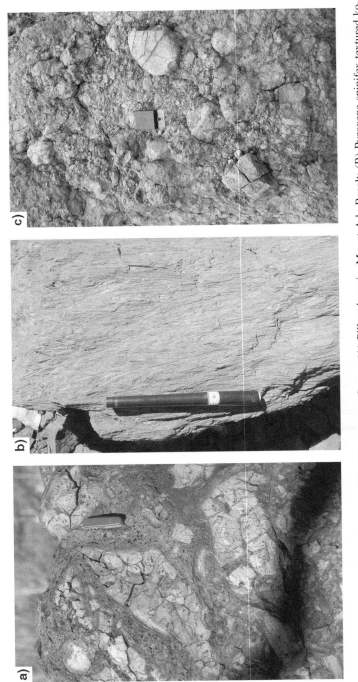

Fig. 4.1-4. Representative rock types of the Warrawoona Group: (A) Pillow breccia, Mount Ada Basalt; (B) Pyroxene spinifex-textured komatiitic basalt, Mount Ada Basalt; (C) Coarse volcaniclastic breccia, Duffer Formation; (D) Typical laminated grey and white chert, Dresser Formation; (E) Carbonate-altered, cross-bedded volcaniclastic sandstone (basal surge deposit?), Panorama Formation.

MVK728

05.12.06

Fig. 4.1-4. (*Continued.*)

circulation (Nijman et al., 1998; Van Kranendonk, 2006; Van Kranendonk et al., 2006c, 2007c).

The stratigraphy of the Warrawoona Group varies across the EPT, but dating confirms stratigraphic correlations across most of the terrane. The Warrawoona Group is uplifted and eroded away in the west, were the oldest components of the group (the Coonterunah Subgroup) are exposed and unconformably overlain by the Kelly Group (see Fig. 17 in Van Kranendonk and Pirajno, 2004). The thickest (12 km) and most complete section through the group is exposed in the Marble Bar greenstone belt (Hickman, 1983), although the lowermost part of the stratigraphy is excised by intrusive granitic rocks. A distinctive feature of the group in this belt is the great thickness (4 km) of volcaniclastic rocks of the Duffer Formation (Hickman, 1983) and the presence of a voluminous mafic dyke swarm that feeds the Apex Basalt (Van Kranendonk et al., 2006b). To the west and south of the Marble Bar area, most of the group is preserved, except that the Apex Basalt is missing and the Panorama Formation lies directly on the Duffer Formation across an erosional unconformity (Di Marco and Lowe, 1989b: see Fig. 11 in Van Kranendonk et al., 2004). The youngest Panorama Formation is present in all greenstone belts except in the far western part of the EPT, where it has been eroded away.

Detailed mapping and extensive SHRIMP U-Pb zircon geochronology suggest that the volcanic rocks were erupted as eight (ultra)mafic through felsic volcanic cycles of ca. 15 My duration (Hickman and Van Kranendonk, 2004), each of which was capped by thin sedimentary rock units silicified by syn-depositional hydrothermal fluids (Van Kranendonk, 2006). Warrawoona Group volcanism closed with Panorama Formation andesitic to rhyolitic volcanism that was developed in several stratigraphically and compositionally distinct centres across the terrane (Smithies et al., 2007b, this volume) from 3.458–3.426 Ga, deposited under shallow water to subaerial conditions (Fig. 4.1-4(e): Di Marco and Lowe, 1989a).

4.1-3.1.2. Kelly Group

Deposition of the Warrawoona Group was followed by a 75 My hiatus in volcanism, during which time the terrane was uplifted and eroded under at least locally subaerial conditions (Buick et al., 1995). Sedimentary rocks of the 30–1000 m thick Strelley Pool Chert at the base of the Kelly Group were deposited on the Warrawoona Group during this time interval, across a regional unconformity (Fig. 4.1-3: Van Kranendonk et al., 2006a, 2007a). This formation comprises a lower unit of fluviatile to shallow marine conglomerates and quartzite, a middle unit of stromatolitic marine carbonates, and an upper unit of coarse clastic rocks (Lowe, 1983; Hofmann et al., 1999; Van Kranendonk et al., 2003; Van Kranendonk, 2006; Allwood et al., 2006a), and was deposited on a carbonate platform that extended across the EPT (Van Kranendonk, this volume, Chapter 7.2).

The conformably overlying Euro Basalt consists of a ⩽1.5 km thick basal unit of komatiite and up to 5 km of overlying, interbedded komatiitic basalt and tholeiitic basalt that was erupted in ~25 My, from 3.35–3.32 Ga (Fig. 4.1-5(a,b)). This was followed by eruption of ca. 3.325 Ga high-K rhyolites of the Wyman Formation (Fig. 4.1-5(c)) that was accompanied by the emplacement of genetically related, voluminous, monzogranitic plutons of the Emu Pool Supersuite (see 3.2 Granitic Rocks, below). Basaltic volcanism continued with eruption of the conformably overlying, but undated, Charteris Basalt, which is locally 1000 m thick (Hickman, 1984).

Fig. 4.1-5. Representative volcanic rocks of the Kelly Group: (A) Olivine spinifex-textured komatiite, Euro Basalt: top of flow unit to right (width of view is 30 cm across); (B) Large, lobate pillows, Euro Basalt; (C) Columnar rhyolite, Wyman Formation.

MVK729   05.12.06

4.1-3.1.3. Sulphur Springs Group
The ≤3.27–3.23 Ga Sulphur Springs Group was deposited across an unconformity on older greenstones in the western part of the EPT (Figs. 4.1-2, 4.1-3: Van Kranendonk, 2000; Buick et al., 2002). In the type area in the Soanesville greenstone belt, this group is up to 4000 m thick and consists of basal sandstone and felsic volcaniclastic rocks of the 3.27–3.25 Ga Leilira Formation (Buick et al., 2002; sample 178045 in GSWA, 2006), up to 2000 m thick of komatiite to komatiitic basalt of the ca. 3.25 Ga Kunagunarrina Formation (geochronology from sample 160957 in GSWA, 2006), and up to 1500 m thick of andesite-basalt through to rhyolite of the 3.245–3.235 Ga Kangaroo Caves Formation, which is capped by <30 m of silicified epiclastic and siliciclastic rocks (Van Kranendonk, 2000; Buick et al., 2002). Felsic volcanic rocks of the formation thin laterally away from the thickest part of the group and pass into ≤500 m of banded iron-formation (Pincunah Member of the Kangaroo Caves Formation: Van Kranendonk et al., 2006a). The banded iron-formation, along with panels of silicified epiclastic sediments and large blocks of rhyolite, are incorporated within a unit of coarse olistostrome breccia at the top of the formation, over the apex of the Strelley Monzogranite (Fig. 4.1-6).

The group is intruded by the shallow level, syn-volcanic Strelley Monzogranite, a K_2O-rich subvolcanic laccolith with rapakivi textures (Brauhart, 1999; Van Kranendonk, 2000, 2006). Heat from this intrusion drove hydrothermal circulation that precipitated volcanogenic massive sulphide (Cu-Zn) deposits (Fig. 4.1-6: Morant, 1998; Vearncombe et al., 1998; Huston et al., 2001; Van Kranendonk, 2006). Eruption of the Sulphur Springs Group was coeval with widespread monzogranite plutonism of the 3.275–3.225 Ga Cleland Supersuite across the northern and western parts of the EPT.

4.1-3.1.4. Soanesville Group
Disconformably overlying the Sulphur Springs Group is the 3.235–3.19 Ga succession of clastic sedimentary rocks, basalt, and banded iron-formation of the Soanesville Group (Figs. 4.1-2, 4.1-3: Buick et al., 2002; Van Kranendonk et al., 2006a: Rasmussen et al., 2007). This group includes, from base to top: shale of the Cardinal Formation; the Corboy Formation of dominantly sandstone; the Paddy Market Formation of shale and siltstone; a 1150 m thick succession of interbedded high-Mg and tholeiitic pillow basalts known as the Honeyeater Basalt; and the Pyramid Hill Formation of banded iron-formation and ferruginous shale (Van Kranendonk et al., 2006a). The Honeyeater Basalt is associated with the Daltons Suite of layered mafic-ultramafic intrusions that range in composition from dunite, through pyroxenite, to dolerite (with local granophyre: Van Kranendonk, 2000).

These rocks lie conformably, or with an onlapping lower contact, on silicified epiclastic rocks of the Sulphur Springs Group (Van Kranendonk, 2000). The lower sedimentary formations of the group are linked to waning activity of the Sulphur Springs Group on

---

Fig. 4.1-6. Schematic evolution of deposition of the Sulphur Springs Group during emplacement of the Strelley Monzogranite, in a felsic volcanic lacco-caldera setting.

a) Deposition of lower Sulphur Springs Group

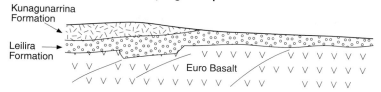

Kunagunarrina Formation

Leilira Formation

Euro Basalt

b) Onset of Kangaroo Caves Formation magmatism

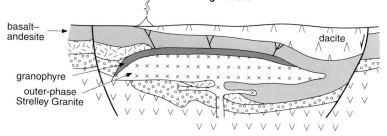

basalt–andesite

dacite

granophyre

outer-phase Strelley Granite

c) Upper Kangaroo Caves Formation and VHMS mineralization

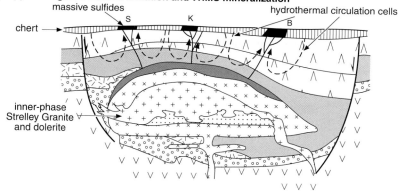

massive sulfides

hydrothermal circulation cells

chert

inner-phase Strelley Granite and dolerite

d) Final caldera collapse

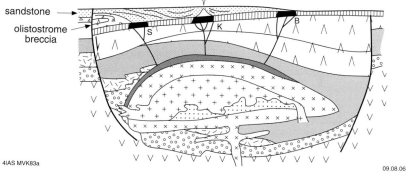

sandstone

olistostrome breccia

4IAS MVK83a

09.08.06

the basis of: (1) hydrothermal alteration associated with Zn-Cu mineralization at Sulphur Springs extends into the overlying shales (Hill, 1997); (2) the lowermost formations were deposited on relict topography defined by the underlying volcanic rocks, indicating close contemporaneity of deposition. For these reasons, it is considered likely that the Soanesville Group was deposited soon after, and following on from, Sulphur Springs Group volcanism, from ca. 3.235 to 3.19 Ga (see Tectonic Evolution, below). Deposition of the lower, clastic components of the group occurred during horst-and-graben style block faulting of the basement, indicating extension (Wilhelmij and Dunlop, 1984).

4.1-3.2. Granitic Rocks

Eruption of the Pilbara Supergroup was accompanied by the emplacement of five igneous supersuites, each of which was derived from melting of yet older crust (Smithies et al., 2003). The Warrawoona Group was accompanied by emplacement of the 3.50–3.46 Ga Callina Supersuite and the 3.45–3.42 Ga Tambina Supersuite (Fig. 4.1-7: Van Kranendonk et al., 2006). The bulk of each of the supersuites was emplaced contemporaneously with the felsic volcanic Duffer and Panorama Formations, respectively. The granitic rocks in these supersuites are predominantly tonalite-trondhjemite-granodiorite (TTG) and include the calc-alkaline rocks of the North Shaw Suite described by Bickle et al. (1993) (Fig. 4.1-8(a)). The two supersuites are differentiated on the basis of cross-cutting relationships, age and state of deformation, with the younger Tambina Supersuite being markedly less affected by migmatization (Fig. 4.1-8(b): Van Kranendonk, 2003b). Widespread leucogranite of the Tambina Supersuite forms a significant component in the Shaw Granitic Complex, and is interpreted to represent melt derived from Callina Supersuite protoliths (Figs. 4.1-7, 4.1-8(c): Van Kranendonk, 2003b; Pawley et al., 2004). Callina and Tambina Supersuite magmas were predominantly emplaced as a mid- to upper-crustal, sheeted sill complex that did not erupt to the surface (Van Kranendonk et al., 2002; Pawley et al., 2004; Smithies et al., this volume), although local, high-level subvolcanic porphyries are near-surface feeders for the Duffer Formation in the Marble Bar greenstone belt (Van Kranendonk et al., 2006b). The thermal effects of this initial magmatic episode continued to as young as 3.40 Ga, as indicated by leucosomes and zircon overgrowths in the Shaw Granitic Complex (Zegers, 1996; M. Van Kranendonk, unpublished data).

Callina Supersuite TTG have low Mg#, and Cr and Ni concentrations when compared with modern-style subduction zone-related adakite, and provide no evidence for evolution in a subduction zone environment (Smithies, 2000). Significantly, despite contemporaneity between the Callina Supersuite and felsic volcanic rocks of the Duffer Formation, the vast majority of these rocks have different trace element and Nd-isotopic compositions and were derived from different sources (Smithies et al., 2007b, this volume): whereas the bulk of the Duffer Formation was likely a result of fractionation of a crustally (TTG) contaminated tholeiitic parent magma, Nd-isotopic and Pb-Pb data from the North Shaw Suite indicate that Callina Supersuite TTG were derived through melting of basaltic crust and earlier TTG rocks, some of which were >3.7 Ga (Bickle et al., 1993).

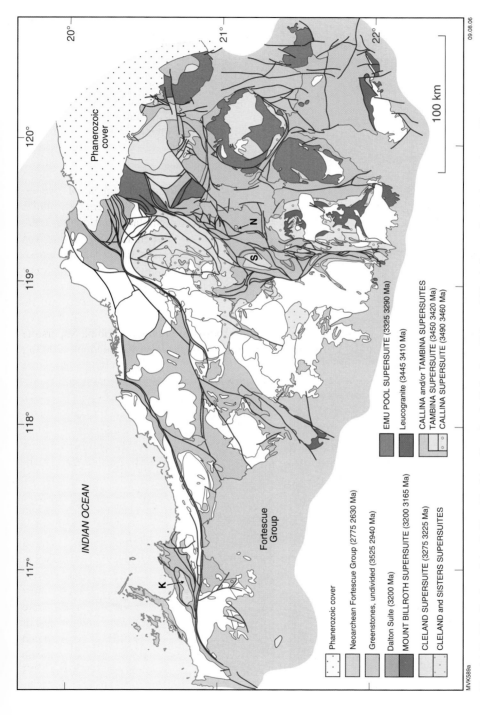

Fig. 4.1-7. Distribution of the Callina, Tambina, Emu Pool, Cleland, and Mount Billroth Supersuites in the northern Pilbara Craton.

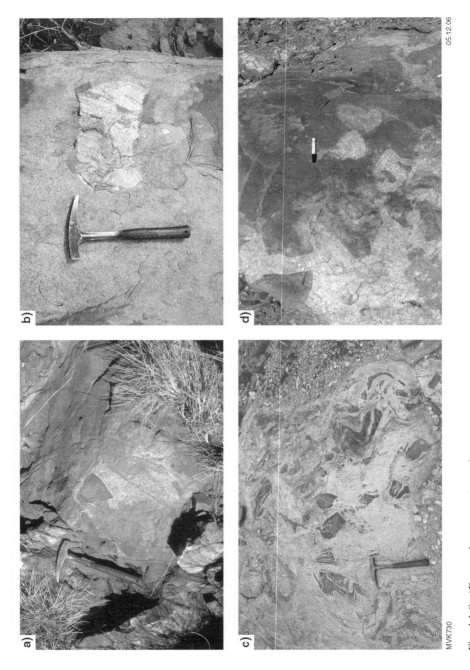

05.12.06

MVK730

Fig. 4.1-8. *(For caption, see next page.)*

Fig. 4.1-8. (*Previous page.*) Representative granitic rocks of the East Pilbara Terrane: (A) Intrusive contact between the ca. 3469 Ma Coolyia Creek Granodiorite, Callina Supersuite of the Shaw Granitic Complex and contact metamorphosed Warrawoona Group basalt; (B) Xenolith of Callina Supersuite tonalitic orthogneiss in low-strain granodiorite of the Tambina Supersuite, northwest Shaw Granitic Complex; (C) Leucogranite diatexite of the ca. 3440 Ma Pilga Leucogranite, Tambina Supersuite, with xenoliths of amphibolite and Callina Supersuite orthogneiss (above, left of hammer), Shaw Granitic Complex; (D) Outcrop evidence of mingling between dolerite and monzogranitic magmas in the inner phase of the ca. 3.24 Ga Strelley Monzogranite, Cleland Supersuite.

The Kelly and Sulphur Springs Groups were each accompanied by the intrusion of widespread and voluminous granitic supersuites, emplaced into the cores of the progressively evolving, domical granitic complexes (e.g., Van Kranendonk et al., 2002, 2004). It is interesting to note that whereas the older Emu Pool Supersuite (3.32–3.29 Ga) is developed in the eastern half of the EPT where it overlaps with the contemporaneous Wyman Formation, the younger Cleland Supersuite (3.27–3.22 Ga) is developed across the EPT, *except* in the southeast where the Emu Pool Supersuite was most voluminously developed (Fig. 4.1-7). Granitic rocks of the Emu Pool and Cleland Supersuites are dominantly monzogranitic in composition and were derived from crustal melting of older granitic supersuites (Barley and Pickard, 1999; Smithies et al., 2003), although compositions range from trondhjemite to diorite to syenite (see Table 2 in Van Kranendonk et al., 2006a). The ca. 3.24 Ga Strelley Monzogranite (S on Fig. 4.1-7) is atypical in that it has rapakivi textures, evidence for both mingling and mixing between basaltic and felsic magmas (Fig. 4.1-8(d)), and a within-plate geochemical signature (Brauhart, 1999; Van Kranendonk, 2000).

Granitic rocks of the Mount Billroth Supersuite include the ca. 3.17 Ga Flat Rocks Tonalite that was emplaced into the western margin of the EPT at broadly the same time as protoliths to the 3.20–3.18 Ga gneissic components of the Kurrana Terrane were being emplaced in the southeast (Fig. 4.1-7). Dated samples of granitic rocks of the Mount Billroth Supersuite in the EPT contain xenocrystic zircons of the Emu Pool Supersuite/Kelly Group and Callina Supersuite/Warrawoona Group. As with older granitic supersuites of the EPT, these data indicate derivation of the Mount Billroth and Elizabeth Hill Supersuites through recycling of mafic to felsic crust (e.g., Smithies et al., 2003).

4.1-3.3. Evidence for Sialic Basement to the Greenstones

A significant result of the extensive SHRIMP geochronology of volcanic and sedimentary rocks from the Pilbara Supergroup is the discovery of abundant inherited zircons that are older than the oldest dated supracrustal rocks (i.e., >3.5 Ga: Van Kranendonk et al., 2002, 2007a). The inherited zircon component includes xenocrystic zircons in volcanic rocks (e.g., ca. 3.72 Ga xenocrystic zircon in ca. 3.46 Ga Panorama Formation rhyolite: Thorpe et al., 1992a), and detrital zircons in quartz-rich sandstones in the Warrawoona Group (e.g., Van Kranendonk, 2004b).

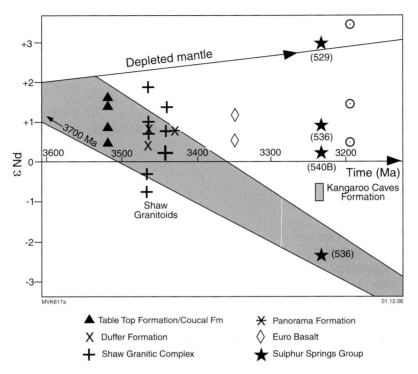

Fig. 4.1-9. ε_{Nd}–time diagram for rocks of the East Pilbara Terrane, showing evidence for contamination of even the oldest intrusive and volcanic rocks by still older sialic crust to 3.7 Ga.

The presence of older basement to the Pilbara Supergroup is also supported by Nd-isotopic and Pb-isotopic data. A single result from the Table Top Formation gives an ε_{Nd} value of $\sim +1.4$, which is slightly below the depleted mantle value of $+2.4$ at 3.5 Ga (Fig. 4.1-9). Results from the ca. 3.49 Ga North Star Basalt and ca. 3.47 Ga Mount Ada Basalt from the Marble Bar greenstone belt give model ages of ca. 3.56 Ga and ca. 3.71 Ga, and an ε_{Nd} value of $+1.6$ (Fig. 4.1-9). Syngenetic Pb in hydrothermal barite from the ca. 3.49 Ga Dresser Formation has anomalously high-μ values ($^{238}U/^{204}Pb$) of 9.45, indicating derivation of the Pb from highly evolved (felsic) crust, at least 3.70 Ga in age (Thorpe et al., 1992b).

More direct evidence for the presence of older rocks beneath the greenstones is present in the form of xenoliths of ca. 3.58 Ga anorthosite in the Shaw Granitic Complex (McNaughton et al., 1988), and ca. 3.65 Ga xenoliths of folded trondhjemitic gneiss in younger granitic rocks in the Warrawagine Granitic Complex (Williams, 1999). Nd model age data from granitic rocks of the Callina, Tambina, Emu Pool, and Cleland Supersuites yield Nd model ages of between 3.4–4.06 Ga, from which the widespread presence of older crust can also be inferred (see Fig. 4 in Smithies et al., 2003).

4.1-3.4. Geochemistry

4.1-3.4.1. Ultramafic rocks

Ultramafic to mafic volcanic rocks from the various formations of the Warrawoona Group have remarkably similar geochemical characteristics (Green et al., 2000; Van Kranendonk and Pirajno, 2004; Smithies et al., 2005b, 2007b). Thin (1–5 m scale), vesicular and spinifex-textured komatiite flow units at the base of the Warrawoona Group (Table Top Formation) are Al-undepleted komatiites with flat normalised trace-element patterns and Al_2O_3/TiO_2 ratios of 20–24. Komatiites and komatiitic basalts in the stratigraphically higher formations of the Warrawoona Group are also predominantly Al-undepleted. Komatiites and komatiitic basalts of the Euro Basalt range from Al-depleted to Al-undepleted, the latter being indistinguishable from those of the Al-undepleted komatiites in the Coonterunah Subgroup (Smithies et al., 2007b). Al-depleted rocks ($Al_2O_3/TiO_2 \sim 6$) are restricted to the stratigraphically highest parts of the Euro Basalt and are LREE-enriched compared to all of the other komatiites and komatiitic basalts in the EPT.

Komatiites and komatiitic basalts from the Sulphur Springs Group (Kunagunarrina Formation) can be divided into two units. A stratigraphically lower unit consists of Al-undepleted komatiitic basalts with LREE depleted REE patterns that likely reflect melting of a refractory mantle source. ε_{Nd} values for these rocks decrease from +3.2, which falls on the depleted mantle curve at 3.24 Ga, to +1.1 with increasing stratigraphic height, indicating increasing degrees of crustal contamination (samples 529 and 536, respectively, on Fig. 4.1-9). The stratigraphically upper, conformably overlying, unit consists of Al-depleted komatiites and komatiitic basalts, with relatively low Al_2O_3/TiO_2 ratios (8.5–16.6), similar to Barberton-type Al-depleted komatiites. Normalised trace element patterns for these rocks vary from slightly LREE depleted patterns to those with significantly LREE enriched patterns that, together with high La/Sm ratios, and low Sm/Nd, La/Nb, and Nb/Th ratios, suggest that some of these magmas have been contaminated by continental crust, a feature which is confirmed by $\varepsilon_{Nd(3250\ Ma)}$ values of between −2.3 to 0.2 (samples 536 and 540B on Fig. 4.1-9).

4.1-3.4.2. Mafic rocks

Basaltic rocks (strictly basalts and basaltic andesites) of the entire Pilbara Supergroup follow tholeiitic trends that Smithies et al. (2005b) showed could be subdivided into a high-Ti group ($TiO_2 > 0.8$ wt%: ~65% of analysed basaltic rocks) and a low-Ti group ($TiO_2 < 0.8$ wt%: ~35% of analysed basaltic rocks) that are interbedded and cannot be distinguished in the field. REE patterns for basaltic rocks show very weakly fractionated N-MORB-like trace-element patterns, which typically vary from 4 to 10 times primitive mantle values. Several studies have shown that Warrawoona Group basalts were contaminated by a component of older felsic crust (Green et al., 2000; Bolhar et al., 2002; Van Kranendonk and Pirajno, 2004; Smithies et al., 2005b, 2007b).

The high-Ti basalts have relatively high concentrations of HFSE and REE compared to the low-Ti basalts, are generally more Fe-rich, have very low Al_2O_3/TiO_2 (8.9–18.7) and high Gd/Yb ratios (1.12–2.23). The concentration ranges of incompatible trace elements and the primitive mantle normalised patterns for high-Ti tholeiites from the Kelly Group (Euro and Charteris Basalts), Sulphur Springs Group, and Soanesville Group (Honeyeater Basalt) are almost identical to those of the lowermost Warrawoona Group (Smithies et al., 2005b).

Basaltic rocks of the low-Ti group from the Euro Basalt, Sulphur Springs Group and Soanesville Group, however, extend to progressively more incompatible trace-element depleted compositions compared with Warrawoona Group basalts, reflecting an increasingly refractory mantle source (Smithies et al., 2005b, this volume). Trace element geochemical analyses of the Honeyeater Basalt indicate they are typically low-Ti basalts with very little evidence for contamination by crustal components (Smithies et al., 2007b). REE data show unusual, slightly LREE depleted patterns that resemble in most respects the low-Ti basalts of the underlying Sulphur Springs Group and indicate derivation from a depleted mantle source, consistent with the pattern of depletion of the subcontinental mantle lithosphere described for the lower parts of the Pilbara Supergroup.

4.1-3.4.3. Felsic volcanic rocks

Felsic volcanic rocks of the Warrawoona Group can be divided into a number of distinct geochemical series, as discussed in more detail in Smithies et al. (2007b, this volume). Andesite to dacite rocks of the lowermost Coucal Formation of the Warrawoona Group are typically sodic, with K_2O/Na_2O between 0.05 and 0.45. A more enriched series show features clearly more typical of fractionating tholeiitic series than of calc-alkaline series or of the Archean TTG series, being low in K_2O (<1.0 wt%), rich in Fe, and with HREE and Y concentrations positively correlated with SiO_2 content and La/Yb ratios. The less enriched series have low Yb and high La/Yb, which are consistent with small amounts of contamination of basaltic magmas with TTG-like material.

The lower part of the 3.47 Ga Duffer Formation is principally composed of two basalt to dacite series that overlap extensively in silica range and are sodic ($K_2O/Na_2O < 0.5$). These rocks show some features more typical of modern calc-alkaline series or the Archean TTG series, including similarly high ranges in La/Nb, Th/Nb and La/Yb. However, they contrast with Andean arc rocks in having a lower proportion of rocks in the basaltic silica range and typically more sodic compositions, and in these respects are more like the Archean TTG series. The upper part of the formation ($\sim$1 km thick north of Marble Bar) contains rhyolites with $K_2O/Na_2O = 1.5$–2.6, high Fe, HREE, Zr and Nb (Glikson and Hickman, 1981), which share features with felsic volcanic rocks that have elsewhere been interpreted as strongly fractionated tholeiites (e.g., Hollings and Kerrich, 2000), although moderately high K_2O ($\sim$2.6 wt%) for the bulk of the formation, and notable Nb and Ti anomalies, likely also require a degree of crustal assimilation (e.g., Smithies et al., 2005a).

Rocks of the 3.43 Ga Panorama Formation show the widest compositional range, with compositionally distinct geographical (volcanic) centres. Some of these volcanic centres are similar in composition to the main components of the Duffer Formation, but typically range to higher K_2O/Na_2O. Others have low Yb concentrations (0.6 ppm) and high La/Yb, and represent the oldest preserved major outcrops of TTG-like volcanic rocks.

Rhyolite of the Wyman Formation (Kelly Group) is highly potassic and thought to derive from melting of pre-existing continental crust, as are contemporaneous high-K_2O granitic rocks of the Emu Pool Supersuite (Glikson et al., 1987; Collins, 1993; Barley and Pickard, 1999).

Brauhart (1999) recognised two felsic volcanic series in the Kangaroo Caves Formation (Sulphur Springs Group) that belong to a high-K magma series, as indicated by high K_2O (up to 5 wt%) and high Th/Nb ratios. He concluded that these rocks derived from a common partial melt source and were tholeiitic, but were not comagmatic. Flat HREE profiles of felsic volcanic rocks indicate minimal influence of garnet during partial melting or fractionation and thus shallow levels of partial melting. Negative ε_{Nd} model ages of -1 for felsic rocks of the formation indicate contamination by older continental crust (Fig. 4.1-9).

4.1-3.4.4. Granitic rocks

Geochemical characteristics of early granitic supersuites (Callina, Tambina, Emu Pool, and Cleland) from the East Pilbara Terrane, and the resulting conclusions, are discussed by Champion and Smithies (this volume). These authors found that TTGs of the Callina and Tambina Supersuites have low Mg#, and Cr and Ni concentrations when compared with modern-style subduction zone-related adakite, and provide no evidence for evolution in a subduction zone environment (Smithies, 2000). Although a slab melting origin for the high-Al TTGs of the EPT can not be unequivocally ruled out, Champion and Smithies (this volume) suggest that the geochemical data more strongly favours derivation via partial melting of thickened mafic crust. This interpretation is consistent with both the plume-like geochemistry (Van Kranendonk and Pirajno, 2004; Smithies et al., 2005b) and the cyclic nature of the volcanism within the thick Warrawoona Group, which has been interpreted as a reflecting a series of mantle plume events (Hickman and Van Kranendonk, 2004; Van Kranendonk and Pirajno, 2004; Van Kranendonk et al., 2007a). With such a multi-plume hypothesis, it can be speculated that granite magmatism resulted from conductive heat distribution from the plume head and derived mafic magmas (Fig. 4.1-10: cf. Campbell and Hill, 1988).

Significantly, despite contemporaneity, the TTG of the Callina Supersuite and the felsic volcanic rocks of the Duffer Formation have different trace element and Nd-isotopic compositions and were derived from different sources (Smithies et al., 2007b, this volume). Whereas the Duffer Formation was likely a result of fractionation of a crustally (TTG) contaminated tholeiitic parent magma, Nd-isotopic and Pb-Pb data from the North Shaw Suite indicate that Callina Supersuite TTG were derived through melting of basaltic crust and earlier TTG rocks, some of which were >3.7 Ga (cf. Bickle et al., 1993).

4.1-3.5. Deformation and Metamorphism

The EPT is characterised by a classical dome-and-keel geometry defined by broad (av. 60 km wide), ovoid granitic domes and intervening synclinal greenstone keels (Hickman, 1984; see Fig. 2.6-1 in Hickman and Van Kranendonk, 2004; Van Kranendonk et al., 2004).

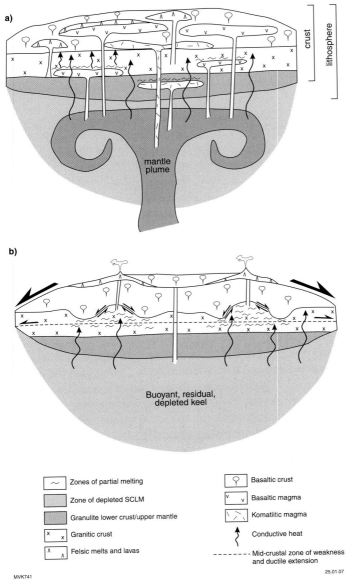

MVK741   25.01.07

Fig. 4.1-10. Schematic tectonic model of crust formation in the East Pilbara Terrane: (A) Voluminous melting from mantle plumes causes crustal thickening and eruption of the Pilbara Supergroup.
Basalts and komatiitic magmas derive from the plume, whereas contemporaneous felsic melts derive
from differentiation of mafic magmas and from partial melting of pre-existing sialic crust and basaltic
crust. (B) Magmatic overthickening and uplift above the mantle plumes lead to radial extension of
the upper and middle crust across mid-crustal detachment zones and facilitates partial convective
overturn of the upper and middle crust.

Geological relationships and age data from across the EPT indicate no structural duplication of the stratigraphy, negating previous suggestions of thrusting (Van Kranendonk et al., 2002, 2004, 2007a).

Four main sets of structures have been identified in the EPT, prior to accretion with the WPS at ca. 3.07 Ga (Van Kranendonk et al., 2007a). D1 structures in the EPT include a set of tight folds in the Coonterunah Subgroup and curved bedding that wraps around the southeastern edge of the Carlindi Granitic Complex (see Fig. 24 in Van Kranendonk, 2006). These rocks are unconformably overlain by the ca. 3.40 Ga Strelley Pool Chert, which provides a minimum age of D1 deformation in this area. Further east, D1 deformation resulted in gentle tilting of rocks older than ca. 3.469 Ga and gave rise to the non-deposition, or erosion, of the 4 km thick Apex Basalt in the western part of the EPT (cf. DiMarco and Lowe, 1989b) and the erosional unconformity at the base of the Panorama Formation in the Coongan greenstone belt (cf. Van Kranendonk et al., 2004).

North of Marble Bar, D1 structures include a set of recumbent isoclinal folds of bedding in the ca. 3.477 Ga McPhee Formation, which verge to the northwest, away from the Mount Edgar Granitic Complex (Collins, 1989). Although undated, these structures are interpreted to have formed during an early period of doming of the Mount Edgar Granitic Complex, probably at ca. 3.47 Ga (Collins, 1989; Collins et al., 1998). D1 structures are also preserved in the core of the Shaw Granitic Complex, where folded leucosomes in migmatitic TTG gneisses of the ca. 3.47 Ga Callina Supersuite are cut by non-migmatitic components of the ca. 3.43 Ga Tambina Supersuite (Fig. 4.1-8(b)). Similar relationships have been observed in the Mount Edgar, Muccan and Warrawagine Granitic Complexes.

D2 deformation was synchronous with, or post-dated, deposition of the Panorama Formation and intrusion of the Tambina Supersuite (i.e., 3.46–3.42 Ga), but predated deposition of the unconformably overlying Kelly Group ($\leqslant$3.35 Ga). This event resulted in tilting of bedding and erosion or non-deposition of the Panorama Formation in the west and south, and produced a strong foliation in pyrophyllite-altered metabasaltic schists of the Mount Ada Basalt in the North Pole Dome (e.g., Fig. 10d in Van Kranendonk and Pirajno, 2004). D2 deformation also resulted in the onset of doming of the Shaw Granitic Complex (Van Kranendonk et al., 2004). This was accompanied by partial melting and generation of leucogranite, which continued to as young as 3.40 Ma (Fig. 4.1-8(c): Zegers, 1996; Van Kranendonk et al., 2004; M. Van Kranendonk, unpublished data).

D3 deformation resulted in the formation of the main amplitude of the dome-and-keel geometry in the eastern part of the EPT (Collins et al., 1998; Van Kranendonk et al., 2002, 2004). This event commenced at between 3.324–3.310 Ga and was largely completed by 3.304 Ga (Williams and Collins, 1990; Collins et al., 1998; also data in Zegers et al., 1999), contemporaneous with the emplacement of the main component of the Emu Pool Supersuite: younger components of this supersuite (ca. 3.29 Ga) cut D3 structures in the far southeast of the EPT (Farrell, 2006). D3 structures include a set of radial metamorphic mineral elongation lineations and stretching lineations that plunge away from the cores of the granitic complexes, forming a zone of pure constrictional strain characterised by vertical L-tectonites in the highest strain area at the triple point junction between the Shaw, Mount Edgar and Corunna Downs Granitic Complexes (Fig. 4.1-11: Collins, 1989;

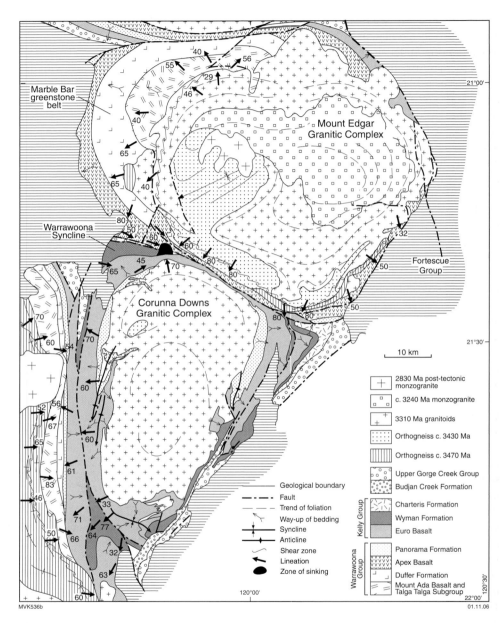

Fig. 4.1-11. Geological map of part of the East Pilbara Terrane, showing the radial distribution of metamorphic mineral elongation and stretching lineations around domical granitic complexes, lending support to models of partial convective overturn of upper and middle crust, and precluding an origin of the domes as metamorphic core complexes (from Van Kranendonk et al., 2004).

Collins et al., 1998; see also Fig. 12 in Van Kranendonk et al., 2004). D3 structures also include broadly circular, or arcuate ductile shear zones along some granite-greenstone contacts, and ring faults along the axial surfaces of greenstone synclines, along the interface between structural domes (cf. Van Kranendonk, 1998). Whereas the former shear zones display greenstone-side-down shear kinematics and emplacement of syn-kinematic granites at ca. 3.32 Ga (Collins, 1989; Collins et al., 1998; Van Kranendonk et al., 2004), some shear zones within greenstone belts display complex local deformational histories and local reverse shear sense indicators (Kloppenburg et al., 2001), consistent with the patterns of diapiric synclines (e.g., Ramberg, 1967: Van Kranendonk et al., 2004). Steeply-plunging, to overturned, hook-shaped D3 folds of greenstones occur on either side of the northern Corunna Downs and Shaw Granitic Complexes (Fig. 4.1-11), and indicate a tilted, mushroom-shaped, cross-sectional geometry for these complexes, identical to those developed around salt diapirs (Van Kranendonk et al., 2004). These folds are bound outside by ring faults, and inside by undeformed intrusive contacts with ca. 3.32 Ga granitic rocks, indicating contemporaneous formation of folds with granite intrusion (see Collins et al., 1998; Pawley et al., 2004; Van Kranendonk et al., 2004). D3 foliations are developed in greenstones and older granitic supersuites and are oriented subparallel to the margins of granitic domes and to bedding in the greenstones (Hickman, 1983, 1984).

Heat from intrusion resulted in amphibolite-facies contact aureoles in greenstones around the margins of granitic complexes (Hickman, 1983). Kyanite-bearing schists from the high-strain shear zone bounding the Mount Edgar Granitic Complex indicate maximum pressure-temperature (P-T) conditions of 5.5-6 kbars, 500–600 °C at this time (Delor et al., 1991) and have distinctive P-T paths diagnostic of granitic diapirism (Collins and Van Kranendonk, 1999; Van Kranendonk et al., 2002).

D4 deformation at ca. 3.24 Ga was synchronous with emplacement of the Cleland Supersuite, and was completed by ca. 3.20 Ga, the time of closure of the Ar-Ar and Rb-Sr systems across much of the EPT (Wijbrans and McDougall, 1987; Collins and Gray, 1990; Zegers et al., 1999). D4 structures include growth faults around the Strelley Monzogranite (Van Kranendonk, 2006), a component of uplift of the high-pressure rocks around the southern margin of the Mount Edgar Granitic Complex previously mentioned (Van Kranendonk et al., 2004), the development of dome-and-syncline structures at amphibolite facies in the northeastern part of the Yule Granitic Complex, and tilting of older greenstones throughout the EPT. Previously formed shear zones between the Corunna Downs and Shaw Granitic Complexes and along the eastern margin of the Shaw Granitic Complex were also re-activated at this time, as indicated by Ar-Ar data (Davids et al., 1997; Zegers et al., 1999). This technique also indicates thermal resetting (or closure) of metamorphic mineral systems at ca. 3.18 Ga in the western part of the EPT (Wijbrans and McDougall, 1987; Zegers et al., 1999).

4.1-3.6. Later Events

The EPT was affected by several tectono-magmatic events after a primary formation, as described in detail elsewhere (Van Kranendonk et al., 2002, 2007a). The main event was regional compressional deformation at ca. 2.93–2.88 Ga (Zegers et al., 1998; Rasmussen et al., 2007), which caused shearing and folding of rocks across the western part of the EPT, and extensive crustal melting that resulted in emplacement of granitic rocks of the Sisters Supersuite. Domes in the western part of the EPT were re-activated and amplified at this time, and the transpressional Lalla Rookh – Western Shaw structural corridor was formed (Fig. 4.1-2: Van Kranendonk and Collins, 1998; Zegers et al., 1998). The dome-and-keel architecture was re-amplified again during emplacement of ca. 2.85 Ga post-tectonic granites (Split Rock Supersuite), and during deposition of the 2.775–2.63 Ga Fortescue Group (Van Kranendonk, 2003a).

4.1-4. TECTONIC MODELS, OLD AND NEW

4.1-4.1. Tectonic Setting of Greenstone Belt Development

Previous models for the tectonic setting of Pilbara Supergroup volcanic rocks include formation of lower Pilbara Supergroup basaltic rocks as either a continuous succession of greenstones erupted onto older continental basement (Glikson et al., 1981; Hickman, 1983, 1984), as an oceanic plateau (Arndt et al., 2001), or as mid-ocean crust (Ueno et al., 2001; Kato et al., 2003). The formation of felsic volcanic rocks of the Warrawoona Group was considered to be in a volcanic arc above a subduction zone (Bickle et al., 1983, 1993; Barley et al., 1984; Barley, 1993) during westward continental growth (Barley, 1997).

However, more recent geochronological, geochemical, and isotopic data show that an oceanic plate and subduction zone environment does not apply to the Pilbara Supergroup, but rather lend support to an origin through a succession of discrete mantle plume events, with eruptive products deposited on a basement of older sialic crust in a tectonic setting analogous to the Kerguelen Plateau (Van Kranendonk et al., 2002, 2007; Hickman and Van Kranendonk, 2004; Van Kranendonk and Pirajno, 2004).

Most significant in regard to the tectonic setting of Pilbara Supergroup volcanic rocks are:

(1) Stratigraphic evidence, confirmed by extensive geochronology and geochemistry, that the Pilbara Supergroup is everywhere a right way up and upward-younging succession, with no major tectonic duplications, with upwards changes in geochemistry (Glikson et al., 1981; Smithies et al., 2005b, 2007b), and with demonstrably autochthonous contacts between groups, as well as between formations within groups (Van Kranendonk, 2000; Van Kranendonk et al., 2007).

(2) Inherited and detrital zircon, geochemical, and isotopic evidence that the Warrawoona Group was extensively contaminated by older felsic crust, and thus not erupted in mid-ocean ridge, or intra-oceanic arc settings, but rather was deposited on a basement of older felsic crust (Glikson et al., 1981; Green et al., 2000; Van Kranendonk and Pirajno, 2004; Smithies et al., 2007b; Van Kranendonk et al., 2002, 2007).

(3) Geochemical evidence for derivation of the ultramafic and mafic volcanic rocks of the supergroup from mantle plume sources (Arndt et al., 2001; Van Kranendonk and Pirajno, 2004), including a component derived from a subcontinental mantle lithosphere

that became progressively more depleted through time as a result of repeated and vo-luminous melt extraction events (Smithies et al., 2005b).

(4) Geochemical evidence that most felsic volcanic rocks in the Warrawoona Group are the result of fractionation of tholeiitic parental magmas and are *not* comparable to subduction-generated calc-alkaline volcanic rocks, as previously suggested. Contem-poraneous TTGs of the Callina and Tambina Supersuites, and the minor volume of TTG-like felsic volcanic rocks, are thought to derive from melting of basalt and older felsic crust at the base of the volcanic pile, rather than from any form of subduction (see Smithies et al., this volume).

4.1-4.2. Tectonic Models of Deformation

There are three competing tectonic models of deformation of the EPT. One is a diapiric model of formation, first advocated by Hickman (1975, 1983, 1984) and Collins (1989; Collins and Gray, 1990; Williams and Collins, 1990), which has been refined into a model of partial convective overturn of the upper and middle crust at several discrete periods from 3.46–2.75 Ga, caused by a variety of tectonic events including mantle plume events, regional sinistral transpression, and extension (Collins et al., 1998; Van Kranendonk and Collins, 1998; Collins and Van Kranendonk, 1999; Van Kranendonk et al., 2002, 2004, 2007; Hickman and Van Kranendonk, 2004; Pawley et al., 2004; Sandiford et al., 2004).

The second tectonic model of deformation of the EPT, which invokes Alpine-type thrusting and core complex formation during multiple episodes of orogenesis, is based on interpretations of structural duplication of the greenstones due to horizontal tectonics (Bickle et al., 1980, 1985; Boulter et al., 1987; Zegers et al., 1996, 2001; van Haaften and White, 1998; Kloppenburg et al., 2001; Blewett, 2002) and a calc-alkaline, volcanic arc interpretation for the origin of ca. 3.47 Ga felsic volcanic rocks and tonalites (Barley et al., 1984, 1998; Bickle et al., 1983, 1993). Horizontal tectonic models invoke episodes of Alpine-style thrusting at ca. 3.47 Ga and ca. 3.3 Ga, followed by periods of extension, during which the granitic domes formed as metamorphic core complexes, analogous with Basin-and-Range style extension (cf. Zegers et al., 1996, 2001; van Haaften and White, 1998; Barley and Pickard, 1999; Kloppenburg et al., 2001).

A third model suggested that the dome-and-keel map pattern of the EPT was the product of cross-folding (Noldart and Wyatt, 1962; Blewett, 2002).

The evidence presented above that indicates the Pilbara Supergroup is everywhere a right way up and upward-younging succession that is unaffected by tectonic dupli-cation precludes Alpine-style thrusting in the geological evolution of the East Pilbara Terrane (Van Kranendonk et al., 2001, 2002). This is supported by metamorphic studies that indicate contact-style metamorphism due to granite intrusions, rather than regional metamorphism as a result of crustal thickening in Alpine-style orogens (Collins and Van Kranendonk, 1999; Van Kranendonk et al., 2002; see also Bickle et al., 1985). The area of high-grade rocks and tight, overturned folds in the northwest Shaw area, which was widely cited as the first, best evidence of recumbent isoclinal folding due to thrusting in the Archean (Bickle et al., 1980, 1985), and that was interpreted to have formed at 3.47 Ga

(Zegers et al., 1996, 2001), has since been shown to be the result of ca. 2.93 Ga transpressional deformation within a restraining bend of a sinistral strike-slip shear system known as the Lalla Rookh – Western Shaw structural corridor (Van Kranendonk and Collins, 1998; Van Kranendonk et al., 2004).

A core complex origin for the granitic domes is inconsistent with numerous features of the regional geology, as outlined in detail in Hickman and Van Kranendonk (2004) and Van Kranendonk et al. (2004). The main points against a core complex model of development of the domes include:

- no linear chain of domical core complexes [domes and greenstone belts are randomly oriented];
- no unidirectional orientation of lineations across domes [lineations are radial around domes];
- no evidence of unidirectional vergence of greenstones across domes [ubiquitous kinematics of greenstone-side-down displacements];
- no extensional detachments at many granite margins [many granitic complexes have well preserved intrusive contacts with greenstones];
- no extensional juxtaposition of *older*, high-pressure granitic basement against low-grade, syn-kinematic sediments [granites are *younger* than greenstones, and flanking rocks are greenstones affected by contact metamorphism];
- no shallow granite-greenstone contacts [all contacts are steep to overturned].

An origin of the dome-and-keel map pattern through cross-folding is precluded by an internally chaotic structural pattern within the domes despite their relatively simple ovoid outlines, and by the presence of shallow-dipping foliations in the crests of domes, where vertical foliations should occur according to the cross-folding model (Van Kranendonk et al., 2004). Given the 60 km diameter average diameter of the domes, an origin through cross-folding would require buckling of the double the thickness of the crust, which is clearly impossible.

Rather, the structural geometry, metamorphic features, and plutonic history of the dome-and-keel map pattern is fully consistent with the diapiric rise of granitic domes during punctuated episodes of partial convective overturn of the upper to middle crust. Major features in support of vertical tectonics in the development of the dome-and-keel pattern are:

- random distribution of large, ovoid granitic domes and narrow, linear, synclinal greenstone belts;
- a radial pattern of metamorphic mineral elongation and stretching lineations around domes, culminating in zones of pure L-tectonite deformation in triple points between domes;
- the emplacement of successive granitic phases into the cores of progressively developing domes;
- the internally chaotic structural pattern within domes that have simple ovoid outlines;
- granite-side-up, greenstone-side-down kinematics around domes;
- lateral accumulation of greenstones in developing synclinal basins between domes;

- the contact style of metamorphism and the style of variation in P-T-t paths away from dome margins (Hickman, 1983, 1984; Collins, 1989; Collins et al., 1998; Collins and Van Kranendonk, 1999; Van Kranendonk et al., 2002).

4.1-4.3. Current Model of Crust Formation

The Paleo- to Mesoarchean evolution of the East Pilbara Terrane is considered to arise from three successive mantle plume events, resulting in voluminous outpouring of dominantly basaltic lavas and crustal thickening through magmatic intraplating and underplating (Fig. 4.1-10(a)). During plume events, conductive heat from the mantle and from intracrustal magmas chambers combined with heat from radioactive elements in older granitic rocks, which were buried into the mid crust by the thick eruptive volcanic lid, to result in widespread melting of the granitic lower to mid crust (Fig. 4.1-10(a): cf. Sandiford et al., 2004; Bodorkos and Sandiford, 2006). This melting resulted in weakening of the mid crust, which facilitated sinking of newly erupted greenstones, which in turn drove the partial melts in the mid crust upwards, into the cores of rising domes where it crystallised to form new granitic intrusions (Figs. 4.1-10(a), 4.1-11, 4.1-12).

This process was repeated three times, during eruption of the Warrawoona, Kelly, and Sulphur Springs Groups. Each event served to amplify the pre-existing dome-and-keel structure. Evidence form volcanic geochemistry shows that melting events occurred progressively deeper in the lithosphere and mantle through time, and that successive plume events caused ever greater degrees of depletion of the subcontinental lithospheric mantle (Smithies et al., 2005b), resulted in it becoming more buoyant than surrounding mantle and forming an unsubductable lithospheric keel (Fig. 4.1-10(b): cf. Griffin and O'Reilly, this volume). This is the principal reason for the preservation of the East Pilbara Terrane and led to the subsequent growth of the craton through subduction-accretion of younger material onto this ancient nucleus (Van Kranendonk et al., 2002, 2007).

Plume magmatism during Warrawoona Group time passed through, and erupted onto an older sialic basement, forming a sort of hybrid oceanic-continental plateau, in many ways analogous with the Phanerozoic Kerguelen Plateau (cf. Van Kranendonk and Pirajno, 2004; Van Kranendonk et al., 2007). Periods of shallow water deposition (Dresser Formation, Panorama Formation) suggest periods of magmatic inflation and crustal uplift, as is common during plume events (cf. Sengor, 2001). However, the presence of thick pillow basalt successions interbedded with the shallow water episodes suggests intervening periods of basin deepening. Evidence in the form of the Salgash Dyke Swarm, and from caldera formation at different times in the Warrawoona Group, suggests that basin deepening was achieved through crustal extension (cf. Nijman and DeVries, 2004; Van Kranendonk, 2006; Van Kranendonk et al., 2006b). The combination of uplift and extension is interpreted to represent a linked duality (Van Kranendonk et al., 2007a), whereby crustal extension resulted from magmatic overthickening and was achieved by the development of horizontal shear zones in the ductile middle crust (Fig. 4.1-10(b): cf. Moser et al., 1996). Crustal extension facilitated the rise of granitic domes. The radial pattern of lineations around the domes indicates that extension was through pure shear and equal in all horizontal direc-

Stage 1: Early crust formation (3530–3430 Ma)

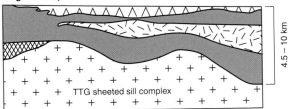

Stage 2: Plume volcanism (3350–3325 Ma)

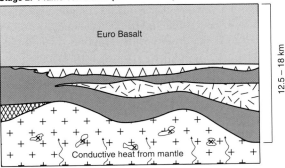

Stage 3: Overturn (3325–3308 Ma)

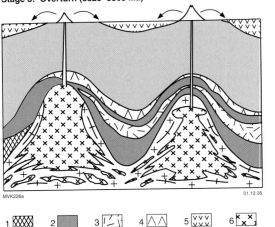

Fig. 4.1-12. Schematic model of the onset of partial convective overturn of the upper and middle crust: (A) Eruption of the Warrawoona Group and emplacement of TTG as a mid-to upper-crustal sheeted sill complex is followed by uplift, local folding, and erosion (A). Burial of the Warrawoona Group by the newly erupted, 8-km-thick Euro Basalt (B), in combination with conductive heat from the associated plume head, and heat from radioactive decay within earlier TTG, causes partial melting of the earlier TTG (cf. Sandiford et al., 2004). This results in a weak middle crust into which the overlying greenstones sink (Collins et al., 1998). Sinking causes migration of granitic melt into rising domical granitic complexes (C).

tions (σ_1 vertical, $\sigma_2 = \sigma_3$). In Sulphur Springs Group time, extension continued after the end of volcanism and led to the formation of deep basins on the EPT filled by a thick clastic sedimentary succession (lower part of the Soanesville Group), and to whole-lithosphere rifting of the margins of the EPT (see Van Kranendonk et al., 2007a).

Geochemical results from felsic igneous rocks show that, in addition to mantle melting, heat from plume events caused melting of basalt and granitic lower crust during Warrawoona Group time, and melting of granitic mid crust during Kelly and Sulphur Springs Groups time (Fig. 4.1-12(c)). This process resulted in large-scale recycling and progressive differentiation of the crust over the course of evolution of the craton. Mass transfer of the more differentiated products to the surface was achieved through the cores of domes, which can thus be viewed as long-lived, crustal scale boils.

ACKNOWLEDGEMENTS

Discussions with David Huston, Bill Collins and Mark Pawley are gratefully acknowledged. Suzanne Dowsett is thanked for expert drafting of figures, often at short notice. This paper is published with permission of the Executive Director of the Geological Survey of Western Australia.

Earth's Oldest Rocks
Edited by Martin J. Van Kranendonk, R. Hugh Smithies and Vickie C. Bennett
Developments in Precambrian Geology, Vol. 15 (K.C. Condie, Series Editor)
© 2007 Elsevier B.V. All rights reserved.
DOI: 10.1016/S0166-2635(07)15042-8

339

Chapter 4.2

THE OLDEST WELL-PRESERVED FELSIC VOLCANIC ROCKS ON EARTH: GEOCHEMICAL CLUES TO THE EARLY EVOLUTION OF THE PILBARA SUPERGROUP AND IMPLICATIONS FOR THE GROWTH OF A PALEOARCHEAN PROTOCONTINENT

R. HUGH SMITHIES[a], DAVID C. CHAMPION[b] AND MARTIN J. VAN KRANENDONK[a]

[a]*Geological Survey of Western Australia, 100 Plain St., East Perth, Western Australia 6004, Australia*
[b]*Geoscience Australia, GPO Box 378, Canberra, ACT 2601, Australia*

4.2-1. INTRODUCTION

Geochemical studies of mafic rocks can potentially provide direct clues on mantle evolution, mantle conditions and juvenile crust formation, while geochemical studies of felsic rocks can provide clues as to how the crust differentiated. Such data are readily available for the Palaeozoic Earth, but significantly less common for the Archean Earth. The Archean rock record is overwhelmingly dominated by Neoarchean (2.4–3.0 Ga) rocks, and rapidly becomes more poorly represented with increasing age such that we have few remaining sequences that provide any direct clues to Paleoarchean geological process, and only cryptic glimpses of process that may have operated before then.

The Warrawoona Group, in the Pilbara Craton of northwestern Australia (see Fig. 4.1-2), is a basalt-dominated volcano-sedimentary sequence formed through essentially continuous volcanism from 3.515 to 3.426 Ga. The group represents the world's oldest well-preserved volcano-sedimentary succession and, as such, is a crucial link in our understanding of early crustal evolution. Felsic volcanic units locally form a significant part of this succession, forming the 3.515 Ga Coucal Formation, the 3.47 Ga Duffer Formation and the 3.43 Ga Panorama Formation (Fig. 4.1-3).

This paper describes the geology and geochemistry of felsic and mafic volcanic units from the Coonterunah Subgroup (Table Top and Coucal Formations) and Duffer Formation of the Warrawoona Group. Building on earlier geological and geochemical studies in the area (see discussions and references in Van Kranendonk et al., this volume), it provides some new insights into intracrustal processes involved in the development of an ancient protocontinent. A companion paper (Champion and Smithies, this volume) looks at the evolution of contemporaneous felsic intrusive rocks of the Pilbara Craton.

4.2-2. REGIONAL GEOLOGICAL SUMMARY

The Pilbara Craton is divided into the 3.53–3.17 Ga East Pilbara Terrane, the 3.27–3.11 Ga West Pilbara Superterrane, and ~3.2 Ga Kurrana Terrane, distinguished by unique lithostratigraphy, structural map patterns, geochemistry and tectonic histories. The early volcano-sedimentary history of greenstones within the East Pilbara Terrane is described by the Pilbara Supergroup, which is composed of four demonstrably autochthonous groups, of which the 3.53 to 3.426 Ga Warrawoona Group is the stratigraphically lowest group. The felsic volcanic units described here are interlayered with a greater volume of typically pillowed tholeiitic basalts at the preserved base of the Warrawoona Group (i.e., the Coucal Formation of the Coonterunah Subgroup), or at a higher stratigraphic level (Duffer Formation) within the lower part of the Warrawoona Group. A detailed account of the regional geology of these regions is provided elsewhere (Van Kranendonk et al., 2002, 2005, 2006a, 2007a, this volume).

4.2-3. ANALYTICAL METHODS

All rocks sampled have undergone at least lower greenschist-facies metamorphism and samples in the lower part of the Coonterunah Subgroup have an amphibolite-facies thermal overprint close to contacts with rocks of the Carlindi Granite Complex. Such processes can result in mobility of some major elements (particularly Si, Na, K and Ca, but also Fe) and trace elements (particularly the large ion lithophile elements (LILE)). The rare earth elements (REE) and high field strength elements (HFSE), however, are generally immobile under these conditions (e.g., Arndt et al., 2001) and so description of the trace elements characteristics of the volcanic rocks presented here concentrates mainly on these trace elements.

Major elements were determined by wavelength-dispersive XRF on fused disks using methods similar to those of Norrish and Hutton (1969). Precision is better than $\pm 1\%$ of the reported values. Loss on Ignition (LOI) was determined by gravimetry after combustion at 1100 °C. FeO abundances were determined by digestion and electrochemical titration using a modified methodology based on Shapiro and Brannock (1962). The trace elements Ba, Cr, Cu, Ni, Sc, V, Zn and Zr were determined by wavelength-dispersive XRF on a pressed pellet using methods similar to those of Norrish and Chappell (1977), while Cs, Ga, Nb, Pb, Rb, Sr, Ta, Th, U, Y and the REE were analysed by ICP-MS (Perkin Elmer ELAN 6000) using methods similar to those of Eggins et al. (1997), but on solutions obtained by dissolution of fused glass disks (Pyke, 2000).

4.2-4. COONTERUNAH SUBGROUP

Buick et al. (1995), Green et al. (2000), Van Kranendonk (2000) and Arndt et al. (2001) have described the geology of the Coonterunah Subgroup of the Warrawoona Group from

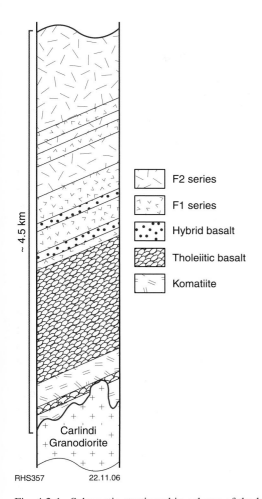

~ 4.5 km

F2 series

F1 series

Hybrid basalt

Tholeiitic basalt

Komatiite

Carlindi
+ Granodiorite

RHS357   22.11.06

Fig. 4.2-1. Schematic stratigraphic column of the lower part of the Coonterunah Subgroup (Coucal Formation) in the Pilgangoora area.

along the southern margin of the Carlindi Granitic Complex, where it is best exposed (Fig. 4.1-2). Although typically dominated by basaltic and gabbroic rocks, andesitic to dacitic volcanic and volcaniclastic deposits are locally abundant and volumetrically dominate some sections (Fig. 4.2-1). Sampling of these felsic sequences was restricted to the more massive, coherent, feldspar porphyritic parts of units and to thick units of feldspar-phyric vitric tuff. Banded, sorted, graded, brecciated and agglomeratic portions of depositional units were avoided. Sampling traverses were across strike, broadly perpendicular to bedding, which dips at a moderate to steep angle to the south and is right-way-up. There was no indication from mapping, or from geochemical trends, of any large-scale structural repetition of the sequence.

The exposed base of the Coonterunah Subgroup includes a unit of interleaved vesicular tholeiitic basalts and thin (1–5 m scale) vesicular and spinifex-textured komatiite flow units (Fig. 4.2-1). These units are overlain by a thick sequence of locally vesicular and pillowed tholeiitic flows (Fig. 4.2-2(a)) with rare interflow accumulations of carbonate-rich sandstone and siltstones, chert and quartzite. This lower komatiite-basalt unit comprises the Table Top Formation and is overlain by the Coucal Formation – a lithologically diverse sequence of basalt and andesitic to dacitic rocks. The boundary between the Table Top and Coucal Formations is locally marked by a transition zone in which there occurs geochemical and textural evidence for limited magma mingling and mixing between the tholeiites and overlying magmas, producing hybrid compositions (Fig. 4.2-2(b)). Tholeiitic basalt does not reoccur within the Coonterunah Subgroup above this transition zone.

The sequence overlying the transition zone is dominated by andesitic to dacitic volcanic and volcaniclastic units. These include massive plagioclase porphyry, feldspar-phyric vitric tuff (Fig. 4.2-2(c)) and, near the top of the unit, multiple $\sim$5 m thick bed sets of graded fine- to coarse-grained volcaniclastic sandstones (Fig. 4.2-2(d)) each capped by a 0.5–1 m thick layer of silicified ash (banded chert). Two geochemical series of felsic rocks are recognised; Coonterunah F1 volcanics range from andesite to dacite and dominate the lower half of the sequence. Coonterunah F2 volcanics range from basalt to andesite and dominate the upper half.

4.2-4.1. Geochemistry of the Coonterunah Subgroup

4.2-4.1.1. Table Top Formation – basalts and komatiites

The presence of Mg-rich basaltic rocks (MgO up to $\sim$12 wt%) at the exposed base of the Coonterunah Subgroup (Table Top Formation) was described by Green et al. (2000). Our data include a basal unit of interleaved vesicular tholeiitic basalts and thin (1–5 m scale) vesicular and spinifex-textured komatiite flow units. These are the oldest known komatiites of the Pilbara Craton. They have SiO_2 from 45.5 to 49.4 wt%, MgO from 22 to 30 wt% and $Mg^\#$ from 79 to 85 wt% (Table 4.2-1), and have flat normalised trace-element patterns with values typically between 0.9 to 2.4 times primitive mantle values ($\sim$3 to 6 $\times$ chondritic) and $[Ce/Yb]_{PM} \sim 0.9$ (Fig. 4.2-3). With near-chondritic Gd/Yb and Al_2O_3/TiO_2 ratios of 1.1–1.3 and 20–24, respectively, these rocks are Al-undepleted komatiites.

Basaltic rocks in the Table Top Formation are high-Ti tholeiites ($TiO_2 > 0.8$ wt% – Smithies et al., 2005b) (Table 4.2-1). They show very weakly fractionated mid-oceanic ridge basalt (MORB)-like trace element patterns which typically vary from 4 to 10 times primitive mantle values, with $[La/Yb]_{PM}$ from 0.8 to 2.0 (Fig. 4.2-3). Most incompatible trace-element ratios are correspondingly close to primitive mantle values, including $[La/Nb]_{PM}$ values which range from 0.89 to 1.66. $Mg^\#$ varies from 69 to values as low as 32 with very little change in SiO_2 values, which lie around 50.5 wt%. The most primitive rocks have Th/Nb and Nb/La ratios slightly lower than primitive mantle values, reflecting a slightly depleted source, but these ratios, and particularly Th/Nb, remain higher than MORB values. Th/La ratios, which are reliable indicators of crustal input (e.g., Plank, 2005), are low (0.08 to 0.15) and close to primitive mantle values ($\sim$0.12). A single Nd-

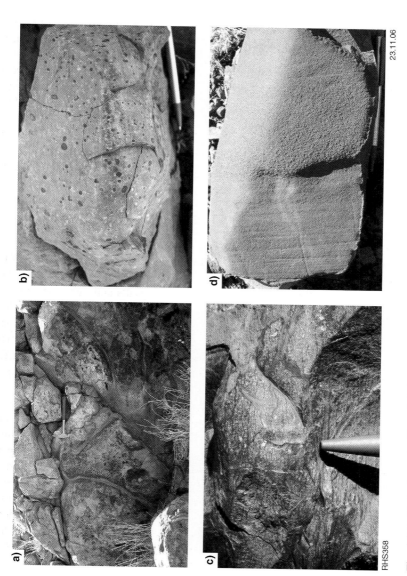

Fig. 4.2-2. Vesicular and pillowed tholeiitic flows (a); mafic globules within basalts at the 'transition zone' between the tholeiitic basalts and the F1 series (b); plagioclase porphyry, feldspar-phyric vitric tuff (c); fine- to coarse-grained graded beds within the F2 series (d).

Table 4.2-1. Representative analyses of komatiites and basalts from the Table Top Formation (Coonterunah Subgroup)

Series	Komatiites			Basalts				Hybrids		
Sample No	179757	179756	179755	179763	179766	179769	179770	179811	179746	179776
SiO_2	41.49	43.67	46.64	48.20	48.80	49.34	51.87	46.88	51.49	53.88
$aSiO_2$	45.52	47.00	49.38	51.17	49.67	50.38	52.48	52.55	53.14	55.04
TiO_2	0.22	0.25	0.27	1.33	0.94	0.94	1.19	0.64	1.27	1.09
Al_2O_3	5.03	5.74	6.48	15.95	13.91	14.64	15.63	12.67	15.50	15.67
MnO	0.15	0.14	0.17	0.14	0.26	0.20	0.26	0.20	0.23	0.20
MgO	29.95	26.92	22.89	6.74	6.32	8.65	3.77	4.20	4.22	4.38
$Fe_2O_3{}^T$	10.18	10.03	9.76	18.66	14.81	12.88	12.99	10.83	11.50	10.31
CaO	2.95	5.24	7.82	1.64	11.72	8.41	10.92	9.16	9.81	9.24
K_2O	0.01	0.01	0.02	0.22	0.20	0.21	0.38	0.42	0.61	0.58
Na_2O	0.02	0.08	0.17	2.31	2.31	3.46	2.63	3.67	2.76	3.11
P_2O_5	0.02	0.02	0.02	0.12	0.06	0.08	0.11	0.05	0.19	0.14
LOI	9.71	7.64	5.89	6.17	1.78	2.12	1.19	12.08	3.21	2.15
Total	100.10	100.13	100.10	100.01	100.05	100.06	100.07	100.08	100.03	100.06
Cr	5034	4699	3636	309	188	424	128	314	47	171
Ni	1589	1286	1005	110	94	156	65	93	35	80
Sc	23	25	27	53	46	44	42	37	34	26
V	100	99	112	339	298	298	323	294	272	175
Rb	1.7	1.5	2.3	10.6	4.6	5.0	4.4	12.6	26.6	28.7
Ba	25	18	26	24	41	50	93	105	147	78
Sr	12.0	44.9	22.6	44.8	83.8	88.7	66.8	84.0	178.5	185.9
Th	<0.1	<0.1	<0.1	0.5	0.3	0.3	0.9	2.7	2.4	3.9
U	<0.1	<0.1	<0.1	0.1	<0.1	<0.1	0.2	0.6	0.5	1.0
Nb	0.4	0.6	0.7	4.3	2.2	2.5	3.9	3.6	8.4	7.5
Hf	0.3	0.4	0.5	2.6	1.5	1.7	3.1	2.1	3.3	3.0
Zr	10.6	13.8	15.4	86.6	43.7	52.5	97.7	64.8	135.5	125.5
Y	3.7	4.8	5.6	14.8	17.0	19.5	28.0	22.2	29.7	30.7
La	0.12	0.22	0.60	3.58	2.64	2.78	6.18	8.84	17.09	13.97
Ce	1.51	2.20	2.44	11.56	7.62	8.50	14.94	17.64	33.77	26.33
Pr	0.26	0.34	0.40	1.82	1.19	1.42	2.22	2.25	4.18	3.27
Nd	1.27	1.71	2.06	8.92	6.26	7.01	10.99	9.27	19.20	15.22
Sm	0.33	0.63	0.65	2.52	1.89	2.31	3.19	2.77	4.47	3.78
Eu	0.10	0.14	0.28	0.55	0.82	0.90	1.03	0.84	1.35	1.10
Gd	0.55	0.70	0.87	2.75	2.65	3.14	4.05	3.12	5.20	4.46
Tb	0.10	0.13	0.15	0.53	0.48	0.58	0.76	0.59	0.81	0.70
Dy	0.67	0.84	0.98	3.31	2.98	3.55	5.04	3.76	5.07	4.26
Ho	0.16	0.20	0.24	0.75	0.68	0.83	1.11	0.86	1.09	0.92
Er	0.46	0.61	0.65	2.02	1.91	2.30	3.24	2.45	3.13	2.65
Yb	0.48	0.61	0.71	2.17	1.87	2.40	3.14	2.44	2.85	2.62
Lu	0.08	0.10	0.11	0.34	0.30	0.39	0.56	0.39	0.49	0.46

Table 4.2-1. (*Continued*)

Series	Komatiites			Basalts				Hybrids		
Sample No	179757	179756	179755	179763	179766	179769	179770	179811	179746	179776
K_2O/Na_2O	0.21	0.12	0.09	0.10	0.08	0.06	0.14	0.11	0.22	0.19
$Mg^{\#}$	85	84	82	42	46	57	37	43	42	46
Sr/Y	3.24	9.35	4.04	3.03	4.93	4.55	2.39	3.78	6.01	6.06
La/Yb	0.25	0.36	0.85	1.65	1.41	1.16	1.97	3.62	6.00	5.33

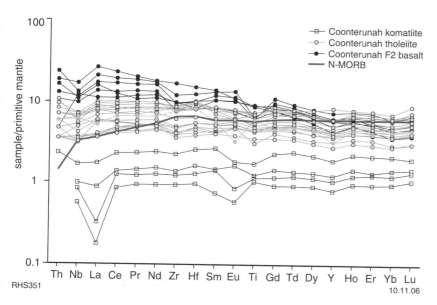

RHS351

10.11.06

Fig. 4.2-3. Primitive mantle normalised trace element diagram comparing Coonterunah tholeiitic basalts, komatiites and basaltic members of the F2 series. N-MORB also shown. (N-MORB and normalisation factors from Sun and McDonough (1989)).

isotopic determination gives an ε_{Nd} of $+1.39$, is only slightly below the depleted mantle value of $+2.37$ at 3.51 Ga, and is consistent with La/Yb and La/Nb values that bracket primitive mantle values. These data indicate that the more primitive basalts in the lower part of the tholeiitic pile underwent very little interaction with felsic crust.

4.2-4.1.2. Coucal Formation – basalts to rhyolites

The transition zone between the tholeiitic basalts and the overlying felsic-dominated sequence includes basalts with higher concentrations of the more highly incompatible trace elements (Th, U, Nb, Zr and LREE) than the underlying tholeiites (Table 4.2-1). In terms of all major and trace elements, and trace element ratios, these basalts invariably plot between the compositional ranges for tholeiites and the interbedded and directly overlying Coon-

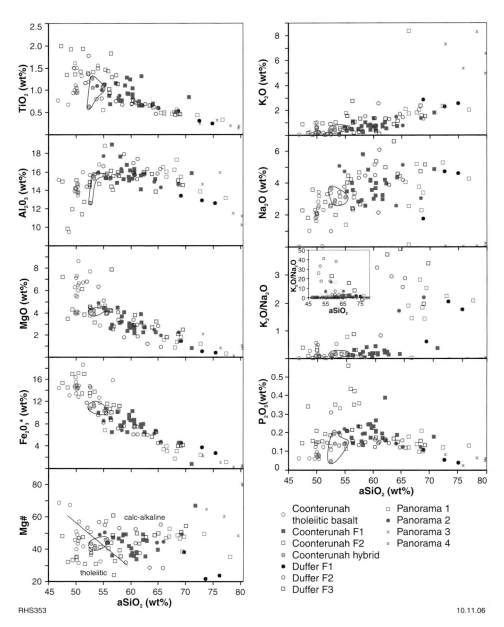

Fig. 4.2-4. Major element variation diagrams for volcanic rocks of the Coonterunah Subgroup, Duffer Formation and Panorama Formation. Panorama 1–4 refers to four geographically separated volcanic centres of Panorama Formation rocks (see Smithies et al., 2007). Outlined field is for the hybrid rocks found near the transition between the Table Top and Coucal Formation of the Coonterunah Subgroup. Note that SiO_2 is calculated on an anhydrous basis (a SiO_2).

Table 4.2-2. Representative analyses of volcanic rocks from the Coucal Formation (Coonterunah Subgroup)

Series	Coonterunah F1				Coonterunah F2				Adakite
Sample No	179742	179787	179865	179740	179789	179792	179794	179791	179741
SiO_2	51.26	58.28	59.47	67.28	46.47	53.44	54.79	56.60	66.06
$aSiO_2$	56.18	60.44	60.08	70.74	47.49	56.02	55.55	57.53	69.02
TiO_2	0.88	0.84	1.24	0.85	2.00	1.43	1.46	1.34	0.48
Al_2O_3	15.34	15.70	15.26	15.87	14.97	15.24	15.80	15.95	15.04
MnO	0.14	0.12	0.13	0.01	0.24	0.19	0.24	0.19	0.05
MgO	2.61	2.49	2.93	0.82	7.22	4.27	4.02	3.41	1.14
$Fe_2O_3^T$	9.04	8.17	10.28	0.80	15.35	10.74	11.64	10.37	3.58
CaO	7.19	6.59	5.46	2.32	9.24	5.38	5.79	5.60	3.28
K_2O	0.48	0.70	1.37	1.83	0.91	0.54	0.63	0.53	0.90
Na_2O	4.00	3.69	3.01	4.88	2.30	4.23	4.46	4.67	5.06
P_2O_5	0.17	0.27	0.24	0.18	0.14	0.44	0.56	0.36	0.12
LOI	9.59	3.71	1.03	5.15	2.20	4.82	1.40	1.64	4.48
Total	100.07	100.06	100.05	100.08	100.01	100.06	100.06	100.04	100.05
Cr	10	43	6	28	145	80	61	38	12
Ni	35	36	15	18	172	57	40	45	18
Sc	21	17	19	25	27	20	19	19	7
V	199	107	157	174	174	114	102	113	42
Rb	18.8	33.6	42.6	46.5	24.8	13.1	13.9	11.2	17.3
Ba	153	185	659	271	94	94	118	102	334
Sr	129.5	242.9	194.3	119.5	135.1	138.3	176.7	156.3	174.3
Th	3.8	4.8	5.1	3.8	0.4	2.7	3.9	3.1	5.7
U	0.8	1.0	1.2	0.9	<0.1	0.6	0.8	0.7	1.3
Nb	9.2	12.9	11.2	9.3	8.6	20.9	21.2	18.0	7.9
Hf	4.2	5.6	6.2	4.1	3.2	7.7	8.0	7.3	4.1
Zr	170.4	240.3	214.0	177.0	108.7	308.5	309.7	290.7	190.0
Y	29.8	36.0	33.1	26.0	27.4	47.7	48.9	43.6	11.5
La	19.24	33.90	26.96	20.90	7.62	31.46	51.29	29.87	26.22
Ce	37.33	66.15	57.38	38.95	19.91	71.09	109.50	63.51	44.45
Pr	4.46	7.84	7.06	4.55	3.29	9.59	14.51	8.48	4.78
Nd	19.56	32.89	28.05	20.03	16.45	40.09	57.07	35.26	19.16
Sm	4.76	6.80	6.26	4.65	4.69	9.02	10.99	7.61	3.28
Eu	1.18	1.70	1.53	1.22	1.71	2.49	2.73	2.31	0.90
Gd	5.11	6.41	6.10	4.84	4.98	8.68	9.57	7.84	2.68
Tb	0.82	0.95	0.97	0.71	0.87	1.48	1.55	1.30	0.35
Dy	5.05	5.74	5.86	4.32	5.09	8.48	8.51	7.71	1.89
Ho	1.09	1.22	1.28	0.89	1.04	1.84	1.90	1.65	0.36
Er	3.06	3.38	3.75	2.40	2.67	5.11	5.17	4.60	0.98
Yb	2.86	3.17	3.46	2.14	2.40	5.04	4.98	4.37	0.98
Lu	0.51	0.57	0.56	0.37	0.40	0.84	0.82	0.71	0.16

Table 4.2-2. (*Continued*)

Series	Coonterunah F1				Coonterunah F2				Adakite
Sample No	179742	179787	179865	179740	179789	179792	179794	179791	179741
K_2O/Na_2O	0.12	0.19	0.45	0.38	0.39	0.13	0.14	0.11	0.18
$Mg^{\#}$	36	38	36	67	48	44	41	39	39
Sr/Y	4.35	6.75	5.87	4.60	4.93	2.90	3.61	3.58	15.16
La/Yb	6.73	10.69	7.79	9.77	3.18	6.24	10.30	6.84	26.76

terunah F1 volcanic series (Fig. 4.2-4). They are almost certainly binary hybrid magmas. The stratigraphic transition zone in which they occur provides common evidence favouring this interpretation, in the form of globules of basalt in andesite and of andesite in basalt (e.g., Fig. 4.2-2(b)).

Rocks belonging to the two volcanic series in the upper sequences (Coonterunah F1 and Coonterunah F2 – Table 4.2-2) are not readily distinguishable from each other in the field, but are compositionally distinct. The Coonterunah F1 rocks are dominantly andesites with silica values typically between 55 and 65 wt% (one sample at 71 wt%), whereas Coonterunah F2 rocks range from basalt to andesite, with a silica range of 48 to 58 wt% (Fig. 4.2-4). The two groups overlap extensively in terms of Al_2O_3, MgO, K_2O and Na_2O. The Coonterunah F1 rocks straddle the calc-alkaline-tholeiite transition, whereas the Coonterunah F2 rocks appear to be part of a tholeiitic series (Fig. 4.2-4). Both groups are low- to medium-K, with a K_2O/Na_2O range between 0.05 and 0.45, and with no evolutionary trend to higher K_2O. The Coonterunah F1 rocks, however, differ from the Coonterunah F2 rocks in typically having lower concentrations of TiO_2, ΣFe (as Fe_2O_3) and P_2O_5.

A single dacite (69 wt% SiO_2), not belonging to either group, has compositions typical of Archean TTG, including low Yb (0.98 ppm) and very high La/Yb (26.7) (Fig. 4.2-5), reflecting a source with residual garnet.

Basaltic rocks within the Coonterunah F2 series show considerable overlap in major element compositions with tholeiites in the lower part of the Coonterunah Subgroup, but range to slightly more evolved members with MgO as low as 2.5 wt% ($Mg^{\#} \sim 24$) and SiO_2 as high as 58 wt%, and with significantly lower ranges in Cr and Ni. Concentration ranges for Yb also overlap extensively, but the more incompatible trace elements are significantly enriched in the Coonterunah F2 basalts (Fig. 4.2-3), with very little overlap in La and Th concentrations. Values of La/Yb for the Coonterunah F2 basalts range from 3.2 to 6.5 (cf. 1.08 to 2.9 for the lower tholeiites), but La/Nb ratios are low (0.89 to 2.06) and similar to those of the lower tholeiitic basalts (Fig. 4.2-6).

Fig. 4.2-5. Primitive mantle normalised trace element diagram comparing felsic volcanic rocks of the Coucal, Duffer and Panorama Formations. The average composition of TTG (Martin, 1999) is also shown for comparison (normalisation factors from Sun and McDonough, 1989).

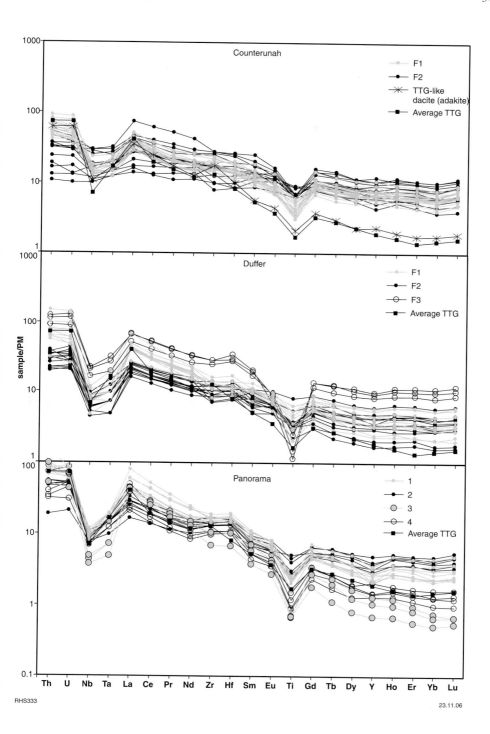

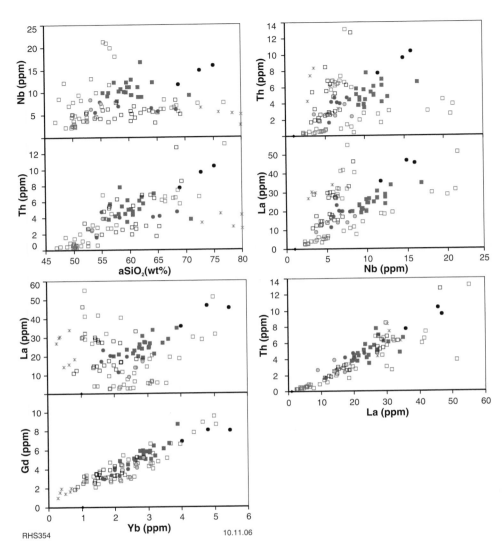

RHS354   10.11.06

Fig. 4.2-6. Trace element variation diagrams for volcanic rocks of the Coonterunah Subgroup, Duffer Formation and Panorama Formation. Symbols as for Fig. 4.2-4.

The Coonterunah F1 andesites and dacites show wide ranges in La/Nb (1.95–2.72) and Th/Nb (0.34–0.76) ratios over a narrow range of Nb concentrations (8–13 ppm) (Fig. 4.2-6). In contrast, Coonterunah F2 basalts and andesites have narrower ranges of La/Nb (1.35–1.66) and Th/La (0.04–0.19, one value at 0.28), at significantly lower Nb values. They also have generally lower La/Yb ratios (3.2–6.8 cf. 6.6–10.7) but similar Gd/Yb ratio (1.6–2.5) (Fig. 4.2-6).

Primitive mantle normalised trace element patterns form a tight array for the Coonterunah F1 series, whereas the Coonterunah F2 series shows a wider range, but with more or less parallel patterns (Fig. 4.2-5). The trace element patterns for Coonterunah F1 and F2 rocks overlap extensively, except for Th and U, which are generally higher in the Coonterunah F1 series.

4.2-5. DUFFER FORMATION

The geology and geochemistry of the Duffer Formation has previously been described by Jahn et al. (1980), Glikson and Hickman (1981), Barley et al. (1984), Glikson et al. (1987) and Barley (1993). Samples for the present study were taken along two traverses in the Talga Talga anticline of the Marble Bar greenstone belt. The basal part of the Duffer Formation (Fig. 4.1-3) comprises interbedded massive, pillowed to tuffaceous basalt and massive to graded, coarse- to fine-grained plagioclase porphyritic andesite and dacite. Two compositional types of basalts are recognised. Duffer B1 basalts are tholeiitic and compositionally similar to the Mount Ada Basalt; they are typically plagioclase-phyric and form massive flow units or, less commonly, tuffaceous volcaniclastic deposits. Duffer B2 basalts are strongly enriched in incompatible trace elements and range from plagioclase porphyritic pillowed flow units to, more commonly, graded and strongly feldspar porphyritic fine- to medium-grained volcaniclastic deposits.

The andesitic to dacitic rocks interbedded with B1 and B2 basalts dominate the lower half of the Duffer Formation. They include a high proportion of fine- to medium-grained volcaniclastic material (typically graded tuff), although volcaniclastic conglomerate as well as 1–5 m thick ungraded units of medium- to coarse-grained plagioclase porphyry, welded plagioclase porphyritic tuff, and feldspar porphyry sills also occur. These rocks can be divided geochemically into a minor group showing characteristics of fractionated tholeiites (Duffer F1) and a larger group having similarities to the associated Duffer B2 basalts (Duffer F2).

Much of the upper half of the Duffer Formation is typically coarser-grained, with coherent crystal tuffs and massive plagioclase porphyritic rocks. These Duffer F3 rocks are compositionally distinct (more enriched) from the underlying Duffer F2 rocks. Pillowed units of Duffer B2 basalts are interbedded with the Duffer F3 sequence. It must be noted that the uppermost part of the Duffer Formation crops out to the southwest of the traverse sampled during this study. This apical section includes rocks that range into the rhyolitic field, have higher K_2O and K_2O/Na_2O (Glikson and Hickman, 1981) than rocks from lower in the section, and in these respects are much more like units from the slightly younger (3.458–3.426 Ga) Panorama Formation (see below). They are likely to represent extensive fractionation of the Duffer F2 and F3 series.

4.2-5.1. Geochemistry of the Duffer Formation

Tholeiitic basalts that are interleaved with felsic rocks of the Duffer Formation, but that do not form part of either recognised geochemical basalt–dacite series (i.e. Duffer F2 and F3;

Table 4.2-3. Representative analyses of volcanic rocks from the Duffer and Panorama Formations

Series	Duffer F1	Duffer F2		Duffer F3		Panorama 1 Sandy Creek		Panorama 2 McPhee Dome		Panorama 3 Panorama Ridge	Panorama 4 Kitty's Creek
Sample No	179726	176800	179705	179714	176770	179895	179873	180208	180209	179898	180223
SiO_2	70.77	50.38	61.89	52.41	67.73	64.25	71.25	51.63	62.54	82.65	74.64
$aSiO_2$	72.57	53.95	64.68	55.36	68.30	65.67	72.29	54.88	65.20	86.22	75.86
TiO_2	0.31	0.81	0.48	1.13	0.45	0.49	0.44	1.00	0.60	0.15	0.33
Al_2O_3	12.90	15.84	15.78	15.30	15.07	16.50	13.83	17.70	14.30	12.41	15.93
MnO	0.06	0.14	0.07	0.17	0.05	0.08	0.09	0.13	0.10	0.00	0.00
MgO	0.53	6.34	2.83	3.94	2.09	1.54	1.19	4.40	1.94	0.02	0.83
$Fe_2O_3{}^T$	3.73	9.55	4.51	9.87	4.01	3.43	2.39	8.54	4.77	0.14	1.08
CaO	2.12	6.77	3.85	6.13	2.55	5.24	1.57	5.19	5.71	0.01	0.04
K_2O	2.31	0.08	1.73	1.57	2.39	2.14	2.35	0.69	0.77	0.24	5.38
Na_2O	4.74	3.43	4.44	4.04	4.72	4.08	5.28	4.73	5.14	0.02	0.04
P_2O_5	0.05	0.13	0.12	0.21	0.15	0.12	0.12	0.22	0.12	0.03	0.02
LOI	2.54	7.09	4.51	5.63	0.85	2.22	1.46	6.30	4.26	4.33	1.64
Total	100.06	100.56	100.20	100.39	100.04	100.08	99.97	100.53	100.24	99.98	99.92
Cr	12	225	24	77	69	50	52	168	21	8	13
Ni	9	81	53	48	36	36	22	78	28	9	15
Sc	8	30	8	22	7	9	9	21	12	4	5
V	<5	190	83	202	47	78	54	186	92	17	25
Rb	51.6	<1.0	62.5	26.8	62.5	79.8	38.7	31.9	36.2	4.2	148.1
Ba	802	102	456	1294	811	455	1073	143	186	58	371
Sr	126.6	273.6	256.7	383.8	275.2	194.1	80.8	202.7	148.9	39.9	23.4
Th	9.6	1.8	1.7	4.8	12.7	6.2	6.6	4.7	4.2	8.4	6.3
U	2.4	0.4	0.5	1.0	2.9	1.4	1.8	1.1	1.1	1.5	1.8
Nb	14.9	4.3	3.2	6.1	8.3	6.3	7.5	7.8	5.8	3.5	5.6
Hf	9.0	2.4	2.7	3.5	5.2	4.8	5.2	4.9	4.2	3.2	4.7
Zr	315.0	80.0	95.0	119.0	180.0	174.0	185.0	189.0	150.0	119.0	155.0

Table 4.2-3. (Continued)

Series	Duffer F1	Duffer F2		Duffer F3		Panorama 1 Sandy Creek		Panorama 2 McPhee Dome		Panorama 3 Panorama Ridge	Panorama 4 Kitty's Creek
Sample No	179726	176800	179705	179714	176770	179895	179873	180208	180209	179898	180223
Y	42.4	16.4	7.7	20.1	11.1	11.7	15.1	22.5	13.6	4.7	6.7
La	46.76	12.87	11.03	26.95	46.37	32.21	31.91	19.69	17.23	30.06	34.06
Ce	91.58	26.98	22.76	52.52	78.90	60.27	60.47	40.93	33.69	49.16	47.02
Pr	11.10	3.51	2.85	6.55	8.74	6.50	6.59	5.09	3.89	5.26	5.78
Nd	45.19	14.75	11.94	26.94	30.50	23.43	23.88	21.11	15.53	17.91	20.74
Sm	8.65	3.57	2.21	4.99	4.87	3.82	4.24	4.26	3.17	2.61	3.05
Eu	1.49	1.08	0.84	1.39	1.22	1.10	1.11	1.37	0.95	0.66	0.86
Gd	8.05	3.36	1.84	4.45	3.25	2.70	3.16	3.97	3.01	1.63	2.06
Tb	1.36	0.56	0.24	0.66	0.44	0.39	0.50	0.66	0.52	0.21	0.28
Dy	7.55	3.41	1.41	3.84	2.26	2.24	2.83	3.94	2.74	0.91	1.37
Ho	1.69	0.76	0.28	0.81	0.43	0.46	0.62	0.86	0.62	0.16	0.27
Er	4.82	2.21	0.85	2.35	1.08	1.22	1.71	2.37	1.75	0.39	0.72
Yb	4.79	2.05	0.78	2.19	1.06	1.14	1.78	2.38	1.61	0.32	0.64
Lu	0.77	0.31	0.12	0.35	0.15	0.19	0.28	0.40	0.25	0.05	0.10
K_2O/Na_2O	0.49	0.02	0.39	0.39	0.51	0.52	0.44	0.15	0.15	13.44	138.05
$Mg^#$	22	57	55	44	51	47	50	50	45	21	60
Sr/Y	2.99	16.68	33.34	19.09	24.79	16.59	5.35	9.01	10.95	8.49	3.49
La/Yb	9.76	6.28	14.14	12.31	43.75	28.25	17.93	8.27	10.70	93.94	53.22

see below), are high-Ti tholeiites (Table 4.2-3) identical in trace element concentration and normalized patterns to the high-Ti tholeiites of the underlying Mount Ada Basalt (Smithies et al., 2005b). Normalized trace-element patterns overlap extensively with those of the high-Ti basalts from the Coonterunah Subgroup, with a similar range in $[La/Yb]_{PM}$ (1.21–1.74) and $[La/Nb]_{PM}$ (1.20–1.65), but extend to higher concentrations (up to $\sim$13 times primitive mantle values).

The Duffer F2 and Duffer F3 basalt–dacite series (felsic series) have a similar silica range (Duffer F2 = 51.8–65: Duffer F3 = 52.4–68.8), and show extensive overlap in most other major elements. Exceptions are Al_2O_3, which, for samples with >55 wt% SiO_2, is generally higher in the Duffer F2 series (>15.7) than in the Duffer F3 series (<15.7) and P_2O_5, which is slightly higher in the Duffer F3 series (Table 4.2-3; Fig. 4.2-4). Both series are classified as calc-alkaline in terms of their silica and $Mg^{\#}$ range, and medium-K, with K_2O/Na_2O generally <0.5. Compared to rocks of the Duffer F2 and Duffer F3 series, the Duffer F1 rhyolites are silicic (SiO_2 68.8–75 wt%) and have relatively low Al_2O_3 (<14 wt%), high K_2O/Na_2O ($\sim$0.5 and higher, although they are still sodic) and have high ΣFe (as Fe_2O_3; 2.7–4.6 wt%) and correspondingly low $Mg^{\#}$ (22–38) at a given silica value. The Duffer F3 series is characterised by elevated incompatible elements relative to Duffer F2 (Fig. 4.2-5). For example, Th and LREE concentrations are 2–3 times higher in Duffer F3, as are Th/Nb and LREE/Nb ratios (Fig. 4.2-6). Incompatible element ratios (e.g., Th/La) vary from similar to significantly higher, in the Duffer F3 series. Importantly, incompatible elements in the Duffer F3 series increase strongly with increasing SiO_2 (e.g., Th from <3 to >8 ppm), unlike the Duffer F2 series. The Duffer F1 series have LILE and Th contents similar to Duffer F3, but have higher LREE, HFSE (including Nb – 12–16 ppm), and HREE, and less fractionated trace element patterns with lower Th/La, Th/Nb, La/Nb, La/Yb, Gd/Yb (Fig. 4.2-6).

The Duffer F1 rocks have the highest concentrations of highly incompatible trace elements of all the Coonterunah and Duffer felsic units, but have MREE and HREE concentrations similar to those of the more enriched members of the Coonterunah F2 series and to tholeiitic basalts interlayed with the felsic rocks of the Duffer Formation.

Compared to the Coonterunah felsic rocks, the Duffer F2 and Duffer F3 series range to significantly higher silica values and generally have lower ΣFe for a given silica value (Fig. 4.2-4). The Duffer felsic series have higher normalised La/Nb ($\sim$3.2 for Duffer F2 and Duffer F1, and 5.3 for Duffer F3) (Fig. 4.2-6). They are also notably depleted in HREE, and have more fractionated normalised trace element patterns (average $[La/Yb]_{PM}$ $\sim$6.7 for Duffer F2 and Duffer F1 and $\sim$13.3 for Duffer F3).

4.2-6. FELSIC VOLCANIC UNITS AT HIGHER STRATIGRAPHIC LEVELS

The proportion of basaltic lava directly interbedded with felsic volcanic and volcani-clastic layers of the Warrawoona Group decreases with decreasing age. However, the geo-chemical diversity of the felsic rocks appears to increase with decreasing age, reflecting an increasing diversity of crustal source regions. The felsic volcanic sequences sampled from the stratigraphically higher 3.458–3.426 Ga Panorama Formation, at the top of the Warra-woona Group (Fig. 4.1-3), contain very few interbedded mafic rocks and occur as discrete felsic volcanic centres, each with a unique compositional range (Table 4.2-3; Fig. 4.2-5).

Felsic volcanic rocks of the Panorama Formation range to higher silica values than those of the older formations (Fig. 4.2-4). Rocks from two of the four centres sampled have undergone extensive alteration, having lost nearly all Na_2O, coupled with either re-moval of K_2O or enrichment of K_2O (up to $\sim$8.2 wt%). Nevertheless, even the least altered rocks from the remaining two centres show slightly higher K_2O/Na_2O (0.35–0.7) than fel-sic rocks from the Coucal Formation and Duffer Formation. Felsic volcanic rocks of the Panorama Formation from the four different felsic volcanic centres (Smithies et al., 2007a) each show very different normalized trace element patterns (Fig. 4.2-5) and, in particular, have very different HREE concentrations (with overlapping LREE contents). Apart from the rocks from McPhee Dome, which have $[La/Yb]_{PM} \sim 8$, felsic volcanic rocks from the Panorama Formation have the lowest Yb concentrations and highest normalized La/Yb ra-tios (average $\sim$19 for Sandy Creek; $\sim$24 for Kitty's Gap; $\sim$67 for Panorama Ridge) of all of the sampled felsic volcanic rocks of the Warrawoona Group; like the Duffer F3 series, there is a strong increase in $(La/Yb)_n$ with increasing SiO_2. Felsic rocks from the Panorama For-mation, along with the Duffer F3 series, also have amongst the highest normalized La/Nb ratios.

Stratigraphically higher felsic volcanic units of the Pilbara Supergroup (including the 3.32 Ga Wyman Formation of the Kelly Group and the 3.24 Ga Kangaroo Caves Formation of the Sulphur Springs Group), along with their subvolcanic equivalents, are strongly potas-sic crustal melts (Jahn et al., 1981; Glikson et al., 1987; Collins, 1993; Cullers et al., 1993; Brauhart, 1999; Champion and Smithies, 2000, 2001, this volume). The Wyman Forma-tion, and associated granitic rocks of the Emu Ponds Supersuite (Corunna Downes Granitic Complex), for example, are silicic ($SiO_2 = \sim$64–78 wt%) and K-rich ($K_2O = \sim$4–6 wt%) rocks with high K_2O/Na_2O (0.8–1.8) and very high La/Yb (up to 80).

4.2-7. PETROGENESIS OF MAFIC VOLCANIC ROCKS

The petrogenesis of the basalts that dominate the Pilbara Supergroup and that bracket the felsic stratigraphic units is not the primary focus here. However, the major felsic vol-canic units of the lower Pilbara Supergroup clearly show a compositional range that extends back to basaltic compositions and so some insight into the petrogenesis of the basalts, and particularly on the extent to which they have interacted with felsic crust, is warranted.

The geochemistry of various mafic rocks within the Pilbara Supergroup has previously been described by Glikson and Hickman (1981), Brauhart (1999), Green et al. (2000), Arndt et al. (2001), Bolhar et al. (2002), Van Kranendonk and Pirajno (2004) and Smithies et al. (2005b). The majority of the basaltic rocks (strictly basalts and basaltic andesites) follow tholeiitic trends and Smithies et al. (2005b) showed that they could be subdivided, irrespective of age, into a high-Ti group ($TiO_2 > 0.8$ wt%) and a low-Ti group ($TiO_2 < 0.8$ wt%). High- and low-Ti basalts are interbedded and cannot be distinguished in the

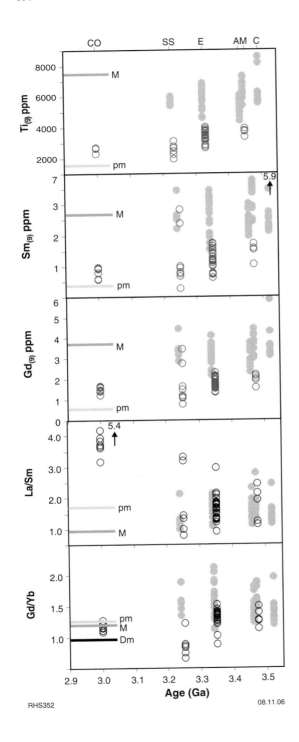

Fig. 4.2-7. (*Previous page.*) Trace element variations with age for the high-Ti (solid symbols) and low-Ti (hollow symbols) basalts of the Warrawoona Supergroup. Trace element concentrations have been extrapolated to a MgO content of 9 wt% (see Smithies et al., 2005b).

---

field. Compared to the low-Ti basalts, the high-Ti basalts also have relatively high concentrations of HFSE and REE (Fig. 4.2-7), are generally more Fe-rich, have very low Al_2O_3/TiO_2 (18.7–8.9) and high Gd/Yb ratios (1.12–2.23). The high-Ti group accounts for 65% of basaltic rocks within the Pilbara Supergroup. Glikson and Hickman (1981) showed that some of the lower basalt formations in the Pilbara Supergroup showed a progressive upward decrease in HFSE and REE. Smithies et al. (2005b) confirmed that this was also the case for the tholeiitic basalts in the lower Table Top Formation of the Coonterunah Subgroup.

Green et al. (2000) suggested that basalts in the Warrawoona Group assimilated up to 25% felsic crustal material and on that basis argued for interaction with continental crust. Arndt et al. (2001), however, argued that the lower Warrawoona Group represents a Paleoarchean analogue of oceanic plateau crust, erupted subaqueously and essentially free of any felsic crustal contamination, although earlier Nd-isotopic data clearly require at least some contamination from older crustal sources (e.g., Hamilton et al., 1983; Gruau et al., 1987). Others have suggested that the Warrawoona Group was a volcanic plateau that was erupted onto continental crust (Van Kranendonk et al., 2002, this volume; Van Kranendonk and Pirajno, 2004).

Bolhar et al. (2002), suggested that the great majority of the samples used by Green et al. (2000) showed very little evidence for contamination at all, and that a minority of 'enriched' samples, on which the high estimate of contamination was based, represent an unrelated population. In agreement with Bolhar et al. (2002), many of the tholeiitic samples we analysed show only small, or no, primitive-mantle normalised depletions in Nb with respect to Th and La, do not show evidence for significant enrichment of Th relative to the LREEs (cf. Plank, 2005), and provide very little evidence for significant crustal contamination (Fig. 4.2-3). Our data for Coonterunah Subgroup basalts do include rocks compositionally very similar to the Th- and LREE-enriched samples studied by Green et al. (2000), but these are the hybrid basalts (Fig. 4.2-4), found mainly in the transition zone between the tholeiites and the F1 andesite to dacite series (Fig. 4.2-1), where rock textures also provide field evidence for magma mingling (Fig. 4.2-2(b)). While these hybrids clearly have incorporated considerable felsic material, contamination has occurred more or less in situ, and the composition of these rocks says little about the abundance of pre-existing felsic crust.

Low minimum La/Nb, La/Sm and La/Yb ratios (N-MORB-like or lower) indicate that the most primitive high- and low-Ti basalt magmas from throughout the Pilbara Supergroup erupted essentially free of significant input from felsic crust (Smithies et al., 2005b). Nevertheless, individual basalt units (high- and low-Ti) do show a range in Th/La, La/Sm and La/Yb ratios and in LREE concentrations that likely reflects at least some interaction with felsic crust. Using the same mixing models as Green et al. (2000), the maximum amount of contamination is typically <10% for the majority of rocks. However, neither the degree of contamination nor the proportion of contaminated rocks appears to increase significantly with decreasing age despite regional geological and geochronological evidence that large amounts of granitic material evolved throughout the depositional period of the group. The evidence from the Warrawoona and Kelly Groups cannot be used to support a setting similar to typical modern-day oceanic plateau crust or to continental crust, but rather a hybrid setting between these two – possibly with similarities to the modern Kerguelen Plateau (e.g., Van Kranendonk and Pirajno, 2004; and see Van Kranendonk et al., this volume).

4.2-8. PETROGENESIS OF FELSIC VOLCANIC ROCKS

It is relevant to compare Warrawoona Group felsic volcanic rocks with rocks of the tonalite-trondhjemite-granodiorite (TTG) series, which are widely believed to have dominated Archean felsic crust (e.g., Glikson, 1979; Jahn et al., 1981; Martin, 1987; Martin et al., 2005). Rocks of the TTG series have compositions expected from <20% melting of mafic crust at high pressure (e.g., Rapp et al., 1991; Rapp and Watson, 1995; Sen and Dunn, 1994). These rocks typically have a high silica content averaging ~68 wt%, are sodic (Na$_2$O ~ 4.65; K$_2$O/Na$_2$O ~ 0.4), with fractionation trends to only slight, or no, increase in K$_2$O/Na$_2$O. They have high La/Yb (~48) and low Yb (~0.6 ppm) (Martin et al., 2005) reflecting a garnet and/or amphibole residue in the source. Possible environments where such magmas may have formed include a subducted basaltic slab or in overthickened mafic crust (Barker and Arth, 1976; Martin, 1986, 1994; Atherton and Petford, 1993; Drummond and Defant, 1990). Granites similar in age to Warrawoona Group felsic volcanic rocks, and with compositions consistent with the TTG series, are known to form a voluminous component of the 3.49–3.45 Ga Callina Supersuite in the East Pilbara Terrane, with the North Shaw Suite being the best documented example (Bickle et al., 1989, 1993; Champion and Smithies, this volume).

In contrast to rocks of the TTG series, the Coucal Formation (Coonterunah Subgroup) and Duffer Formation volcanic units contain a high proportion of low-silica rocks in the basaltic to andesitic range, and the felsic rocks of the Duffer Formation typically have slightly higher K$_2$O/Na$_2$O (>0.5) than is typical of TTG, particularly in the stratigraphically highest part of the section (e.g., Glikson and Hickman, 1981). Normalising the Coonterunah and Duffer felsic volcanic units against average TTG (Fig. 4.2-8) further identifies significant enrichments in MREE and HREE, and lower La/Yb ratios in the felsic volcanic rocks. Only a single felsic volcanic layer with TTG compositions was found in the Coonterunah Subgroup and none was found in the Duffer Formation – despite the similarity in age with TTG-like granites of the Callina Supersuite. Two of the four distinct centres of the stratigraphically overlying Panorama Formation have trace element compositions similar to TTG.

Barley et al. (1984) and Cullers et al. (1993) suggested the compositional spectrum of volcanic rocks in the Duffer Formation is a result of fractional crystallisation from a

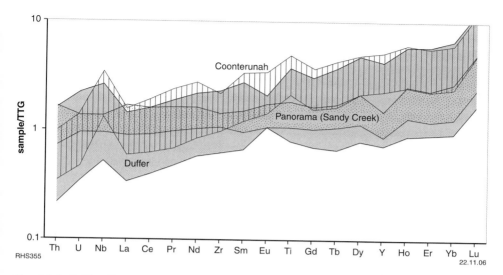

Fig. 4.2-8. Spider diagram showing the trace element compositions of felsic volcanic rocks of the Coucal, Duffer and Panorama Formations normalised against the average TTG composition of Martin (1999).

basaltic parental magma. A continental arc parental basalt was preferred by Barley et al. (1984). However, low La/Nb ratios in the Coonterunah F2 series (Fig. 4.2-6), tholeiitic major element trends for the primitive members of the Coonterunah F1 and Coonterunah F2 series (Fig. 4.2-4), and flat normalised HREE patterns for the Duffer F1 series (Fig. 4.2-5), better reflect a tholeiitic parental magma for the Pilbara felsic volcanic rocks.

In Fig. 4.2-9 we compare the felsic volcanic series of the Warrawoona Group with Phanerozoic rock series that reflect magmas formed in some of the tectonic environments that have previously been proposed for the East Pilbara Terrane. To investigate the possibility that at least some of the Warrawoona Group felsic volcanics have tholeiitic affinities, we have included tholeiitic series rocks from Iceland. These represent a thick, and possibly plume-related, series of basalts and intermediate rocks derived from extensive fractionation of tholeiitic basaltic magmas, and include rhyolites, representing extremely evolved compositions. To investigate suggestions that the felsic volcanic rocks of the Warrawoona Group share close compositional affinity with modern calc-alkaline rocks, and particularly those in continental settings (Barley et al., 1984), we compare these rocks with basaltic to rhyolitic calc-alkaline rocks from the Andean arc.

The Coonterunah F1 and Coonterunah F2 series have low K_2O (<1.0 wt%) and are Fe-rich. The Coonterunah F2 series, in particular, has similarly low La/Nb and Th/Nb ratios to the lower Coonterunah basalts and, despite a restricted silica range (up to 58 wt% anhydrous), show no significant change in these ratios through up to four-fold increases in La and Th concentrations (Figs. 4.2-4 and 4.2-6). These lavas, along with Duffer F2, and possibly Duffer F1 lavas, show positive correlations between Th and Yb and on this

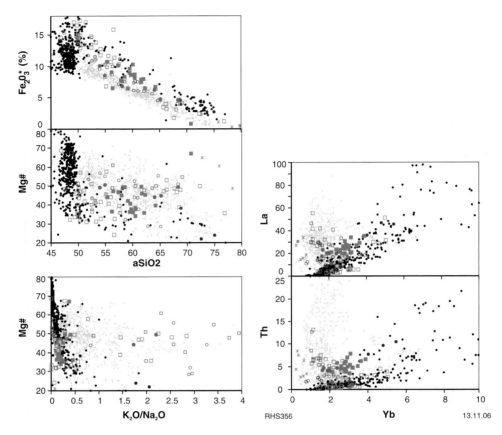

Fig. 4.2-9. Major and trace element variation diagrams comparing the volcanic rocks of the Coonterunah Subgroup, Duffer Formation and Panorama Formation (symbols as for Fig. 4.2-4) against primitive and evolved tholeiitic rocks from Iceland (black dots) and calc-alkaline basaltic to rhyolitic rocks from the Andean arc (gray dots). Data for the Iceland tholeiitic rocks and Andean arc rocks are from the GEOROC database (http://georoc.mpch-mainz.gwdg.de/georoc/).

basis can be distinguished from the remaining, and stratigraphically higher lavas of the Duffer and Panorama Formations, which either show a negative trend or extend to high concentrations of Th with minimal or no change in Yb (Fig. 4.2-9).

Many of the geochemical features of the Coonterunah and lower Duffer felsic volcanic series are indeed reminiscent of trends in fractionated tholeiitic series (Fig. 4.2-9), where minerals like pyroxenes, with low and relatively similar partition coefficients for a range of incompatible trace elements, control compositional trends. This is particularly the case for the rare Duffer F1 rhyolites (68.8–75 wt% SiO_2), which have high Fe (for their silica contents), HREE, Zr and Nb, and share these features with other Archean felsic volcanic rocks that have elsewhere been interpreted as strongly fractionated tholeiites (e.g., Hollings and Kerrich, 2000; Smithies et al., 2005b). Moderately high K_2O ($\sim$2.6 wt%) and notable Nb and Ti depletions in the Duffer F1 rhyolites likely require a degree of crustal assimilation, but similar features are found in modern Iceland rhyolites, which form in an oceanic setting. Likewise, elevated La/Sm ratios in the Coonterunah F2 series likely reflect assimilation of felsic (high La/Sm) crust, prior to fractional crystallisation.

Compared to the Coonterunah F2 series, slightly higher La/Sm, La/Nb and La/Yb in the Coonterunah F1 series (Fig. 4.2-6) possibly reflect an even greater degree of interaction between mafic magmas (or their source) and crustal material. In support of this, ratios of similarly highly incompatible trace elements (e.g., La/Th, Th/Nb, La/Zr – Fig. 4.2-6) in the Coonterunah F1 rocks match those of the single sample of TTG-like dacite. Low Yb and high La/Yb in the Coonterunah F1 rocks are consistent with small amounts of contamination of basaltic magmas with this TTG-like material. Very modest changes in La/Nb and Th/Nb ratios over significant La and Th ranges could suggest that such contamination occurred before closed system fractionation that did not involve hornblende – presumably in a lower crustal magma chamber.

The Duffer F3 series, rhyolitic units from the stratigraphically highest parts of the Duffer Formation (Glikson and Hickman, 1981), and all geochemical populations within the Panorama Formation, in contrast, show some features more typical of modern calc-alkaline series or the Archean TTG series. They show strong overlap with the compositional range for calc-alkaline rocks related to the continental Andean arc system, including similarly high ranges in La/Nb, Th/Nb and La/Yb (Fig. 4.2-9), although, as noted by Barley et al. (1984) they typically have higher Ni and Cr (a feature also noted by Glikson (1980) for a significant proportion of Archean felsic rocks). They contrast, however, with Andean arc rocks in having a lower proportion of rocks in the basaltic silica range. They typically also have more sodic compositions, falling in the low K_2O/Na_2O part of the range for the Andean arc rocks (Fig. 4.2-9), and show no clear trends to K-enrichment. In these respects, the Duffer F3 series and the Panorama Formation are more like the Archean TTG series.

Thus, in a general sense, the felsic volcanic rocks of the Warrawoona Group show a systematic change with time, as exemplified in Figs. 4.2-5 and 4.2-6. Along with a general decrease in the amount of basalt directly associated with felsic magmatism, and with a general trend to magmas with higher silica contents with decreasing age, felsic magma compositions within the silica range of andesite to dacite change from those with trace and major element attributes more consistent with fractionated members of a tholeiitic magma series to those with a TTG affinity. Apart from a single TTG-like unit, the felsic volcanic rocks of the Coonterunah Subgroup are andesites and dacites with Yb from $\sim$2 to 5 ppm and La/Yb from $\sim$3 to 10. In contrast, two of the four volcanic centres within Panorama Formation have trace-element compositions compatible with the TTG series (Yb <0.8 ppm; La/Yb = 23–108), while the remaining two centres include andesites and dacites with Yb from $\sim$1 to 2 ppm and La/Yb from $\sim$10 to 50. Andesites and dacites of the Duffer Formation overlap these two compositional ranges.

Barley et al. (1984) recognised two source components for the Duffer Formation and for some of the units within the Panorama Formation. These components are basalt and a HREE-depleted component equivalent to TTG. These authors suggested that the magmas result from complex supra-subduction zone processes involving mixing between mantle wedge-derived basalt and slab-derived felsic melts. The hybridisation processes required for such interactions have been the subject of several studies (cf. Martin et al., 2005) with two broad end-members reflecting the ultimate fate of the slab melt. High-silica adakites (including some TTG) are produced from slab-melts variably contaminated by mantle in the wedge as the melt ascends to the crust, whereas and low-silica adakites are produced by remelting of the subduction mantle following metasomatism by slab-melt (Martin et al., 2005, and references therein). On a major element basis, Duffer Formation felsic rocks would be analogous with low-silica adakites, but are, in general, significantly enriched in HREE compared to low-silica adakites. The Coonterunah Subgroup felsic volcanic series have Y concentration (20–55 ppm) more than double the average low-silica adakite concentrations (~13 ppm – Martin et al., 2005). Clearly, the proportion of any high La/Yb 'TTG' component in the Coonterunah and Duffer felsic volcanic rocks is extremely small. Even if a TTG component is required, it is not clear whether this was added into the mantle source or simply reflects later assimilation. The observation that the proportion of the TTG component in the felsic rock systematically increases, with decreasing age, over the ~100 m.y. time span, and the lack of evidence that a subduction-like enriched component was added to the source of the thick basaltic formations that separate the felsic units, suggests that the TTG component was incorporated into the basaltic magmas, rather than into the mantle source of those magmas, and that the basaltic magmas had increasing access to this TTG-like crust with decreasing age.

We suggest that geochemical variations within the felsic volcanic series that comprise the Coucal Formation and the Duffer Formation are essentially the result of fractionation of tholeiitic basaltic parental magmas. These magmas were variably contaminated, prior to the bulk of fractionation, through assimilation of felsic crust or contemporaneous felsic magmas to produce compositional trends that exhibit some hallmarks of TTG-like magmatism (or magma). They are not, however, part of the classic Archean TTG series. Nor are they typical calc-alkaline arc-related rocks. Although they show some compositional overlap with continental arc calc-alkaline rocks, they also show significant differences and there is no evidence that they acquired their enriched compositional features directly from an enriched mantle source, as do arc-related calc-alkaline magmas.

The amount of TTG-like crust assimilated into the typically tholeiitic magmas was less for the Coucal Formation magmas than it was for the Duffer F2 and Duffer F3 magmas. It was greatest for magmas of the Panorama Formation and two of the four volcanic centres from the Panorama Formation possibly have compositions consistent with belonging to true TTG series. If so, they represent the oldest significant volume of TTG volcanism in the Pilbara Craton.

4.2-9. IMPLICATIONS FOR EARLY PILBARA CRUSTAL EVOLUTION

4.2-9.1. Early Felsic Crust – TTG or Tholeiitic Andesite?

The felsic volcanic history of the Pilbara Craton did not involve significant TTG magmas. Such magmas were certainly produced and may have been locally voluminous (e.g., in the 3.49–3.45 Ga Callina Supersuite), but they did not erupt. Condie (1993) was perhaps one of the first to recognise the compositional dichotomy between Archean TTGs and Archean felsic volcanic rocks. He noted that the latter are enriched in HFSE, HREE and depleted in Sr, Ba, Th, U, LREE relative to TTGs. This, however, is not always the case, as felsic volcanics in the Kalgoorlie Terrane of the Yilgarn Craton are clearly the extrusive equivalents of TTGs (e.g., Brown et al., 2001).

The abundance of TTG-like crust (of intrusive or extrusive origin) relative to other felsic crustal components is difficult to constrain, and the relative proportions of volcanic products may not directly reflect the proportions of intrusive counterparts. Van Kranendonk et al. (2002) proposed that the 3.49–3.45 Ga Callina Supersuite, including TTG-like granites, was emplaced into mid- to upper-crust as a sheeted sill complex. However, based on the compositional trends to increasing contamination of tholeiitic felsic components by TTG-like components with decreasing age, we would suggest that TTG-like material was a less common component of crust older than ca. 3.45 Ga. Felsic magmatism in the Pilbara Craton also shows a progressive trend to more K-rich compositions for rocks younger than ca. 3.45 Ga (e.g., Champion and Smithies, this volume). Hence, it appears that the peak of magmatism with typical TTG-like characteristics occurred in the Pilbara Craton somewhere between around 3.49 and 3.45 Ga.

Conversely, the relative proportion of felsic tholeiitic material appears to have decreased (relative to TTG) with decreasing age up to at least ca. 3.45 Ga. Moreover, geochronological studies of inherited zircon populations in granites from the Pilbara Craton indicate at least 200 m.y. of crustal pre-history before deposition of the Coucal Formation. If the observed trend in relative abundance of felsic components is extended back in time, it seems likely that pre-3.5 Ga Pilbara crust might have been a heterogeneous sequence comprising mafic lavas interlayered with voluminous felsic (dominantly andesitic to dacitic) lavas which were derived almost entirely through fractionation of the associated tholeiitic magmas, perhaps in the way that the intermediate to felsic component of Icelandic crust has formed.

These findings challenge, at least for the Pilbara Craton, a widely held belief that rocks of the TTG series overwhelmingly dominated early Archean felsic crust, but they may also have interesting and wider implications in terms of petrogenetic models for the TTG series.

Numerous experimental studies and theoretical modelling of trace element partitioning (see Moyen and Stevens, 2006) have established that rocks of the high-Al TTG series can be derived from a basaltic source that was melted at pressures high enough to stabilize HREE-sequestering garnet. Arguments persist as to the origins of significant normalised negative Nb-anomalies (i.e., high La/Nb and Th/Nb) and sub-chondritic Nb/Ta ratios. High-pressure experiments demonstrate stabilisation of Nb-sequestering rutile in residues of tholeiitic basalt melting (e.g., Green and Pearson, 1986), while the observed range in Nb/Ta and Zr/Sm ratios in TTG appears to require either residual (source) hornblende (Foley et al., 2000), or inheritance of these specific compositional features from their sources (e.g., Rapp et al., 2003).

Compositional similarities between the TTG series and high-silica adakites (sodic dacites with low-HREE and high La/Yb that form predominantly through melting of sub-ducted hydrated slab-basalt) have led to suggestions that TTGs result from subduction in the Archean (Martin, 1999; Drummond et al., 1996). TTGs, however, show a secular decrease in such compositional features as $Mg^\#$, Ni and Cr concentrations from the Pale-oarchean to the Neoarchean (Smithies and Champion, 2000; Smithies, 2000; Martin and Moyen, 2002; Martin et al., 2005). Modern adakites are more like Neoarchean TTG, with the elevated $Mg^\#$, Cr and Ni likely reflecting interaction between the high-pressure felsic melts of hydrated basaltic-slab and the overlying mantle wedge (i.e., a steep subduction origin). Typically low $Mg^\#$, Cr and Ni in Paleoarchean TTG, however, can be interpreted to reflect very little, or no, interaction with mantle peridotite, either due to production at slightly higher structural levels in a subduction zone, to very low-angle subduction, or to melting of thickened basaltic lower crust in a non-subduction environment. Delamination of eclogitic lower crust (possibly itself partially a result of previous TTG-forming events) and subsequent mantle upwelling is thought to provide the heat to form TTG (and some modern adakites) through lower crustal melting in a non-subduction environment (Davies, 1995; Zegers and van Keken, 2001).

The geochemistry of igneous rocks (felsic to mafic) in the Pilbara Craton shows a re-markable spatial and temporal switch. Post-3.2 Ga rocks, mainly in the west and north of the craton, show strong evidence for the development of a subduction-enriched mantle source region (Smithies et al., 2005a, 2007a). This signature, however, is absent in pre-3.2 Ga magmas, and there are no structural or stratigraphic features in the Pilbara Super-group that require a modern-style convergent margin setting. These observations suggest that modern-style steep-subduction models, involving development of a significant mantle wedge, are not appropriate to the evolution of the pre-3.2 Ga Pilbara Craton.

Because the Pilbara TTG's are typically low $Mg^\#$, Cr and Ni rocks, Smithies et al. (2003) invoked a process of 'Archean flat-subduction', whereby an oceanic basaltic slab (source) was subducted at an angle low enough to affectively exclude a mantle wedge con-taminant. In modern subduction zones, however, it is the thermal input from the mantle wedge that induces slab-melting, and the Archean flat-subduction model fails to identify an alternative source of heat, unless shear heating played a considerably greater contri-bution to the thermal budget of subduction zones in the Archean than it has in modern subduction zones. In the modern context, flat subduction is typically associated with very little magmatism (e.g., Gutscher et al., 2000a).

4.2-9.2. Early Crust Formation through Internal Differentiation

Based on the new geochemical data presented here, we now suggest that TTG not only form a less volumetrically significant component of early Pilbara crust than previously thought, but also that the Pilbara TTG series rocks that did form were the result of lower crustal melting in a non-subduction environment. Crustal melting is consistent with the contemporaneous production of both low- and high-Al TTGs in the Pilbara Craton (Champion and Smithies, this volume). A non-subduction environment also accords well with the plume-like geochemistry (Van Kranendonk and Pirajno, 2004; Condie, 2005; Smithies et al., 2005b) and the thick, cyclic, continuously upward-younging and unconformity-bound lithological architecture of the Pilbara Supergroup, which is most satisfactorily interpreted as a result of a series of mantle plume events (Van Kranendonk et al., 2002; Van Kranendonk and Pirajno, 2004; Smithies et al., 2007a).

Furthermore, rather than direct melting of purely basaltic lower crust, the source of the Pilbara TTG-series rocks may have been, or at least included, a Coonterunah-like andesitic supracrustal sequence, and their intrusive equivalents, which we infer to have been buried at middle to lower crustal levels. A trace element enriched low-K_2O (tholeiitic) andesitic source would potentially have been more hydrated (through igneous fractionation alone) than typical lower-crustal mafic crust and would have yielded larger volumes of high-silica melt at lower melting temperatures. The Coonterunah F1 and Coonterunah F2 rocks have low K_2O/Na_2O (up to 0.45 but mostly <0.2) and so the resulting melts will also be sodic. An obvious requisite, imposed by the high La/Nb and La/Yb ratios of the 3.41–3.47 Ga high-Al TTGs (Champion and Smithies, this volume), is that at least some portions of this heterogeneously interlayered crust (or the intrusive equivalents) initially melted at pressures high enough to leave residual garnet and rutile.

This model explains the rather high, but limited, silica range of the TTG series, and because the felsic source components already have elevated LILE concentrations (and yield high-silica melts at lower melt fractions than do basaltic sources), it would also explain why many TTG rocks (in the Pilbara Craton and in other Archean terrains) have higher LILE and Th concentrations than would be expected from melting of typical Archean basalt alone (see Champion and Smithies, this volume). Furthermore, residual (source) assemblages directly resulting from TTG formed in this way would be considerably less voluminous, for a given amount of TTG, than TTG produced through basalt melting. Earlier residual assemblages accumulated during fractionation of basalt to andesite remained in very deeply buried magma chambers, perhaps as seismically invisible eclogite lenses within the crustal lithosphere.

4.2-9.3. Continued Crustal Cycling

Felsic magmatism occurred throughout the Pilbara Craton during >7 separate events up to ca. 2.8 Ga (Van Kranendonk et al., 2002, 2006a). Nd-isotopic data show that the source of these granites, irrespective of granite intrusive age, was >3.4 Ga, with most being >3.5 Ga (Smithies et al., 2003). Geochronological (U-Pb SHRIMP) studies of zircons from Pilbara Craton granites (Van Kranendonk et al., 2002; Barley and Pickard, 1999) typically identify very complex zircon inheritance patterns that reflect a protracted history of crustal recycling events. Hence, the bulk of the source for Pilbara felsic crust was produced very early and since has simply been continuously recycled.

The upper felsic volcanic units of the Pilbara Supergroup (e.g., the 3.32 Ga Wyman Formation and 3.24 Ga Kangaroo Caves Formation), along with their subvolcanic equivalents, are strongly potassic crustal melts (Jahn et al., 1981; Glikson et al., 1987; Collins, 1993; Cullers et al., 1993; Brauhart, 1999; Champion and Smithies, 2000, 2001). In this respect, the Pilbara Supergroup shows a secular evolution in felsic volcanic units from tholeiitic, to TTG-like, to intra-crustal high-K melts. While no plutonic equivalents of the Coucal Formation andesites or dacites are known, granites of the Pilbara Craton also show a secular compositional evolutionary trend from TTG-series rocks in the 3.49–3.45 Ga Callina Supersuite, to 'transitional' (higher LILE) TTG-like rocks to high-K granites. Champion and Smithies (2001) explained the trend for the granites in terms of progressive recycling of an initially basaltic crust. We now suggest that the first stage in this secular trend is voluminous mafic magmatism, likely in a plume-related plateau setting, with fractionation of a significant volume of this magma to andesitic or dacitic compositions. This mafic + andesitic to dacitic crust subsequently melted to produce TTG-like felsic magmas during continued (or renewed) plume magmatism and mafic intraplating at ~3.5–3.45 Ga. Champion and Smithies (this volume) note that the compositions of these early felsic magmas requires melting over a range of pressures from mid to lowest (<50 km) crust. The variable LILE contents of the TTG-like magmas produced during this period reflects the relative contributions made to their source by the evolved (LILE-rich) andesitic to dacitic crustal component. While rare granites have low-LILE contents consistent with a predominantly basaltic source, the typically LILE-enriched compositions of the majority of granites (Champion and Smithies, this volume) produced at this time suggest that evolved andesitic to dacitic crustal component, like Coonterunah Subgroup and Duffer Formation felsic volcanic rocks, typically formed a major source component. Felsic magmas formed during a later major crustal melting, between 3.3 and 3.24 Ga, have a compositional range that overlaps that of the earlier granites but extends to more evolved (SiO_2-, K_2O-rich) compositions. The source for these magmas likely included a variable contribution (over a variable crustal depth) from basaltic crust, the andesitic to dacitic component, as well as TTG formed during the earlier crustal melting events. This process can be compared, in some respects, to the process of zone refining, resulting in progressive transfer of more incompatible components towards the surface.

4.2-10. CONCLUSIONS

The Coonterunah Subgroup (Table Top and Coucal Formations) and Duffer Formation of the Warrawoona Group, Pilbara Craton, contain the world's oldest well-preserved felsic and mafic volcanic units. The geology and geochemistry of these volcanic rocks provide insights into intracrustal processes involved in the Paleoarchean development the Pilbara protocontinent.

Significantly, the early felsic volcanic history of the Pilbara Craton did not involve eruption of voluminous TTG magmas. Such magmas were produced and may have been locally voluminous (e.g., in the 3.49–3.45 Ga Callina Supersuite), but rather than erupting, they formed a mid- to upper-crustal sill complex. Instead, early felsic volcanism was dominated by andesitic to dacitic lavas that fractionated from tholeiitic parental magmas variably contaminated by crustal components. While basalt dominates the Warrawoona Group, the proportions of these evolved (andesite + dacite) units to basalt increases with increasing age. It is likely that the evolved units formed a major component of crust formed prior to the 3.15 Ga Coucal Formation.

Furthermore, rather than direct melting of purely basaltic lower crust, the source of TTG-series rocks that did form in the early Pilbara Craton may have been, or at least included, an evolved Coonterunah-like supracrustal sequence, and their intrusive equivalents, which we infer to have been buried at middle to lower crustal levels. Apart from satisfying all of the major element features of TTG (also see Champion and Smithies, this volume), such a source would explain some of the discrepancies between the trace element composition of typical Archean TTG and melts expected through high-pressure melting of typical Archean basalt, and in particular, the higher LILE and Th concentrations in the former. Residual (source) assemblages directly resulting from TTG formed in this way would be considerably less voluminous, for a given amount of TTG, than TTG produced through basalt melting. Earlier residual assemblages accumulated during fractionation of basalt to andesite remained in very deeply buried magma chambers, perhaps as seismically invisible eclogite lenses within the crustal lithosphere.

Consequently, we suggest that TTG not only form a less volumetrically significant component of early Pilbara crust than previously thought, but also that the Pilbara TTG series rocks that did form were the result of lower crustal melting in a non-subduction environment. A non-subduction environment also accords well with the plume-like geochemistry (Van Kranendonk and Pirajno, 2004; Condie, 2005; Smithies et al., 2005b) and the thick, cyclic, continuously upward-younging and unconformity-bound lithological architecture of the Pilbara Supergroup, which is most satisfactorily interpreted as a result of a series of mantle plume events (Van Kranendonk et al., 2002; Van Kranendonk and Pirajno, 2004; Smithies et al., 2007a). While a non-subduction environment adequately accounts for the evolution of Paleoarchean Pilbara crust, modern-style plate tectonic processes involving steep subduction of oceanic crust are strongly implicated in the evolution of post-3.2 Ga Pilbara crust (Smithies and Champion, 2000; Smithies et al., 2003, 2004b, 2005b; Van Kranendonk et al., 2002, this volume).

ACKNOWLEDGEMENTS

This study has benefited greatly through numerous discussions with, and suggestions from, Arthur Hickman. Reviews by Andrew Gilkson and Mark Barley helped to greatly improve the manuscript. Published with the permission of the Director, Geological Survey of Western Australia and the Chief Executive Officer, Geoscience Australia.

Earth's Oldest Rocks
Edited by Martin J. Van Kranendonk, R. Hugh Smithies and Vickie C. Bennett
Developments in Precambrian Geology, Vol. 15 (K.C. Condie, Series Editor)   369
© 2007 Published by Elsevier B.V.
DOI: 10.1016/S0166-2635(07)15043-X

Chapter 4.3

GEOCHEMISTRY OF PALEOARCHEAN GRANITES OF THE EAST PILBARA TERRANE, PILBARA CRATON, WESTERN AUSTRALIA: IMPLICATIONS FOR EARLY ARCHEAN CRUSTAL GROWTH

DAVID C. CHAMPION[a] AND R. HUGH SMITHIES[b]

[a]*Geoscience Australia, GPO Box 378, Canberra, ACT 2601, Australia*
[b]*Geological Survey of Western Australia, Department of Industry and Resources, 100 Plain St., East Perth, Western Australia 6004, Australia*

4.3-1. INTRODUCTION

The world's pre-3.2 Ga rock record is poorly preserved, typically strongly structurally modified and, in spite of the significant progress that has been achieved through advances in geochronology over the last 20 years, remains difficult to decipher. Geological studies of these early Earth fragments not only provide constraints on the nature and style of the prevailing tectonic processes and of early crustal growth, but – paradoxically – require an understanding of the range of tectonic processes and crustal growth mechanisms that could have operated during that period.

A variety of philosophical approaches have been applied to this conundrum. These range from theoretical studies based on physical and thermal properties (Sleep and Windley, 1982; Abbot and Hoffman, 1984; Davies, 1995) or geochemistry (Martin, 1986), to comparative studies relating Archean rock types (or typically their geochemistry) to modern counterparts and, by corollary, modern tectonic environments (e.g., Drummond and Defant, 1990; Martin, 1993, 1999; Hollings and Kerrich, 2000; Smithies and Champion, 2000; Polat et al., 2002; Smithies et al., 2004a, 2004b; Martin et al., 2005), to more data driven studies (e.g., Smithies, 2000; Martin and Moyen, 2002; Smithies and Champion, 2002; Smithies et al., 2003). All these approaches eventually invoke some *a priori* assumptions of particular tectonic environments (e.g., Martin and Moyen, 2002; Smithies, 2000).

One result of this is that while the petrogenesis of pre-3.2 Ga granites[1], and especially the tonalite-trondhjemite-granodiorite suite (TTGs), which volumetrically dominate crust

---

[1]The term 'granites' when used non-generically, e.g., 3.2 Ga granites, is used *sensu lato*, and includes all intrusive rock types from quartz diorites to true granites (*sensu stricto*), including granodiorite, trondhjemite and tonalite.

of this age, seems reasonably well understood, many crucial questions remain unanswered. Two of the more controversial aspects are the actual sites of TTG magma formation (i.e., thickened mafic crust, subducting slab, or both), and the roles of intracrustal differentiation and evolved crustal contributions. The timing and extent of intracrustal differentiation has implications for the onset of potassic granite magmatism that typically dominates the late stages of many Archean crustal evolution cycles, and for understanding the influences on the more expanded evolutionary sequence from fractionated tholeiitic rocks to TTGs to potassic magmatism (e.g., Champion and Smithies, 2001).

Perhaps the best exposed and best understood Paleoarchean rocks are those of the East Pilbara Terrane (EPT), in the Pilbara Craton of north Western Australia (Van Kranendonk et al., 2006a, 2007a, this volume). The Paleoarchean rock record of the East Pilbara Terrane comprises quasi-continuous greenstone development in conjunction with similar aged, though more episodic, felsic magmatism (extrusive and intrusive), all in an autochthonous geological environment. The structural integrity of the terrane, combined with the well preserved geology of the sequences, provide strong constraints both on tectonic models and on petrogenetic models of granite formation. Particularly helpful are the concurrent records of felsic and mafic magmatism and, through these, the links that can be established between the mantle and the crust during Paleoarchean times (e.g., Smithies et al., 2005b, 2007a).

This paper describes the geology and geochemistry of the pre-3.2 Ga TTGs, TTG-like granites and more potassic granites of the East Pilbara Terrane and the constraints on their petrogenesis. A companion paper by Smithies et al. (this volume) looks at the evolution of contemporaneous felsic and mafic extrusive rocks of the East Pilbara Terrane. Insights from that study are included here to establish more robust petrogenetic and tectonic models for the pre-3.2 Ga felsic magmatism, and to provide constraints on crustal growth models for the early Archean. In particular, these models emphasise the role of compositionally evolved crust in TTG formation, and may possibly have wider application to early Archean granite genesis, in general.

4.3-2. REGIONAL GEOLOGICAL SUMMARY

The Pilbara Craton is divided into: the 3.53–3.07 Ga East Pilbara Terrane (EPT), which includes the Pilbara Supergroup; the 3.27–3.11 Ga West Pilbara Superterrane (WPS); and the ⩾3.2 Ga Kurrana Terrane (KT). Each of these are distinguished by unique lithostratigraphy, structural map patterns, geochemistry and tectonic histories and they are separated by late tectonic, dominantly clastic sedimentary rocks of the De Grey Supergroup deposited in the Mallina and Mosquito Creek basins (see Fig. 4.1-1). A detailed account of the geology of these regions is provided by Van Kranendonk et al. (2002, 2006a, 2007a, this volume) and will not be repeated here.

4.3-3. GRANITE GEOCHEMISTRY AND PETROLOGY

4.3-3.1. Introduction

The exposure of preserved granite totals an area of ~24,000 km². Most of the known periods of granite intrusion in the EPT either broadly correspond to periods of greenstone development (i.e., are contemporaneous with felsic volcanism in the Pilbara Supergroup), or postdate greenstone formation. These magmatic periods include ca. pre-3.5 Ga, 3.50–3.42 Ga, ca. 3.32 Ga, ca. 3.24 Ga, 2.95–2.93 Ga, and ca. 2.85 Ga, with most magmatism relating to the 3.50–3.42, ca. 3.32, ca. 3.24 and 2.95–2.93 Ga events (see Fig. 4.1-7: Bickle et al., 1983, 1989, 1993; Buick et al., 1995, 2002; Nelson, 1996–2002; Barley et al., 1998; Barley and Pickard, 1999; Van Kranendonk et al., 2002, 2006a, 2007a). The younger, post-3.1 Ga events are concentrated within the western part of the EPT (Champion and Smithies, 2000), as well as the adjoining Mallina Basin (Smithies and Champion, 2000; Smithies et al., 2004a; Van Kranendonk et al., 2006a, 2007a, this volume). Sm-Nd isotope data (see Smithies et al., 2007b) show that there is a marked change in average model crustal age (T_{2DM}) between the dominantly 3.6–3.2 Ga EPT (T_{2DM} ages of >3.6–3.3) and the 3.02–2.94 Ga Mallina Basin (T_{2DM} ages of 3.2 to 2.96 Ga) (Figs. 4.1-1, 4.1-7). This Nd-isotopic discontinuity represents largely juvenile crustal growth and accretion to the west of the EPT, related to the tectonic evolution of the WPS and Mallina Basin (Smith et al., 1998; Champion and Smithies 2000; Smithies and Champion 2000; Smithies et al., 2004a). The geological and Sm-Nd data further indicate that post-3.2 Ga crustal growth extended to the western part of the EPT (Champion and Smithies, 2000), and the post 3.2 Ga granites in that region are, accordingly, petrogenetically grouped with those found in the Mallina Basin. Broadly coinciding with the Nd isotopic break in the western EPT is the change to dominantly younger granites, emplaced at 2.94 and ca. 2.85 Ga (e.g., Champion and Smithies, 2000). Magmatism of this age occurs elsewhere within the EPT, but is not as voluminous.

4.3-3.2. Ca. 3.5–3.42 Ga granites

The oldest, widely preserved granites range in age from ca. 3.50–3.42 Ga (Fig. 4.1-7). These are best documented from the Shaw Granitic Complex (Bickle et al., 1983, 1993; Van Kranendonk et al., 2002, 2006a), but are also present within the Carlindi, Yule, Warrawagine, Mount Edgar, Corunna Downs and Muccan Granitic Complexes (Fig. 4.1-7: Buick et al., 1995; Nelson, 1998, 1999; Williams and Collins, 1990; Bagas et al., 2003; Van Kranendonk et al., 2006a, this volume). Units dominantly comprise hornblende-biotite and biotite tonalites, trondhjemites and granodiorites (e.g., Jahn et al., 1981; Bickle et al., 1983, 1989). They vary from discrete plutons of variably deformed granite, such as in the northern part of the Shaw Granitic Complex (Bickle et al., 1993), to strongly deformed, locally migmatitic and variably-banded gneisses. The latter range from large zones of probably mixed ages (e.g., Shaw Granitic Complex; Bickle et al., 1993; Van Kranendonk, 2003: Mt Edgar Granitic Complex; Collins, 1993), to marginal facies in granitic complexes, commonly extensively intruded by younger granites and locally forming abundant xeno-liths/enclaves within them (e.g., Yule Granitic Complex; Blewett and Champion, 2005).

Van Kranendonk et al. (2006a) subdivided the pre-3.42 Ga granites into the 3.5–3.46 Ga Callina Supersuite and the 3.45–3.42 Ga Tambina Supersuite (Fig. 4.1-7, Table 4.3-1). These authors also showed that the intrusive rocks were emplaced contemporaneously with development of the Pilbara Supergroup and, more specifically, with felsic extrusive units within the supergroup (e.g., the Duffer Formation/Callina Supersuite and Panorama Formation/Tambina Supersuite extrusive/intrusive pairs).

Although there is an apparent age difference between the two supersuites and also some evidence for structural differences (i.e., the Tambina Supersuite is locally significantly less migmatised; Van Kranendonk, 2003), distinguishing between members of each supersuite is difficult, such that granites of both supersuites are discussed together in this paper (see Table 4.3-1). Older granites, preserved as remnants in gneiss, are also included within this group (e.g., 3.655 Ga remnants within >3.42 Ga gneiss of Warrawagine Granitic Complex: Nelson, 1999; Williams, 2001). Similarly, composite units (such as the Nandingarra Granodiorite in the Corunna Downs Granitic Complex; Bagas et al., 2003), which comprise a mixture of ca. 3.45 and 3.315 Ga ages (Nelson, 2002), are here treated with the ca. 3.5–3.42 Ga granites. The gneissic rocks of Collins (1993), which may be composite in part, are treated similarly. A full listing of granite units included in this group is given in Table 4.3-1.

4.3-3.3. Ca. 3.32–3.24 Ga granites

The ca. 3.32–3.24 Ga granites are best documented in the Mount Edgar (Collins, 1983, 1993; Williams and Collins, 1990) and Corunna Downs Granitic Complexes (Davy, 1988; Barley and Pickard, 1999; Bagas et al., 2003), where they comprise the bulk of the complexes (e.g., Van Kranendonk et al., 2006a: Fig. 4.1-7). Recent data suggests that the older granites (ca. 3.325 to 3.290 Ga) of this group also dominate the Warrawagine Granitic Complex (Williams, 2002) and are present in the Muccan Granitic Complex (Williams, 1999; Fig. 4.1-7). They appear to be largely absent from the Shaw, Yule and Carlindi Granitic Complexes (Fig. 4.1-7), although the presence of inherited zircons of this age may suggest that such granites once formed a component of these complexes as well.

Rock types range from hornblende-biotite to biotite, tonalite, granodiorite, and trond-hjemite to monzogranite (e.g., Davy, 1988; Collins, 1993; Barley and Pickard, 1999; Williams, 1999, 2001; Bagas et al., 2003). Van Kranendonk et al. (2006a) grouped granites of this age (ca. 3.325 to 3.290 Ga) into the Emu Pool Supersuite. Younger granites of this group (ca. 3.27-3.24 Ga granites) are widespread and occur as minor components within the Yule, Mount Edgar, Muccan and Warrawagine Granitic Complexes (Nelson, 1998, 1999; Williams, 1999, 2001; Van Kranendonk et al., 2002, 2006a), as well as forming discrete intrusions, such as the Strelley Monzogranite (Buick et al., 2002) (Fig. 4.1-7). Rock types comprise hornblende-biotite and biotite tonalite, granodiorite and granite (Collins, 1993; Williams, 2001; Buick et al., 2002). Van Kranendonk et al. (2006a) grouped granites

Table 4.3-1. Geochemical subdivisions for granite batholiths, as used in this paper, showing current high-Al low-Al groupings and previous classifications, and data sources

Groups	Granite complex	Classification	Data sources	Previous classifications
ca. 3.33–3.25 Ga group				
ca. 3.3 to 3.25 Ga				
				Emu & Cleland Supersuites (Van Kranendonk et al., 2006)
Munganbrina	Mount Edgar	high-Al?	Collins (1983); GSWA-GA unpublished	Munganbrina Suite (Collins, 1983, 1993)
Muccan	Muccan	high-Al, low-Al	GSWA-GA unpublished	
ca. 3.3 Ga				Emu Supersuite (Van Kranendonk et al., 2006)
Boobina	Corunna Downs	low-Al	Davy (1988), Bagas et al. (2003), GSWA-GA unpublished	Boobina – Group 2 (Bagas et al., 2003)
Boodallana	Mount Edgar	high-Al	Collins (1983); GSWA-GA unpublished	Boodallana Suite (Collins, 1983, 1993)
Carbana	Corunna Downs	low-Al	Davy (1988), Bagas et al. (2003), GSWA-GA unpublished	Carbana – Group 2 (Bagas et al., 2003)
Carbana low-Th	Corunna Downs	low-Al	Davy (1988), Bagas et al. (2003), GSWA-GA unpublished	Carbana – Group 2 (Bagas et al., 2003)
Chimingadgi	Mount Edgar	high-Al	Collins (1983); GSWA-GA unpublished	Chimingadgi Suite (Collins, 1983, 1993)
Coppin Gap	Mount Edgar	high-Al	Collins (1983); GSWA-GA unpublished	Coppins Gap Suite (Collins, 1983, 1993)
Corunna Unit 2	Corunna Downs	low-Al	Davy (1988), Bagas et al. (2003), GSWA-GA unpublished	Unnamed unit – Group 2 (Bagas et al., 2003)
Corunna Unit 1	Corunna Downs	low-Al	Davy (1988), Bagas et al. (2003), GSWA-GA unpublished	Unnamed unit – Group 2 (Bagas et al., 2003)
Mondana	Corunna Downs	low-Al	Davy (1988), Bagas et al. (2003), GSWA-GA unpublished	Mondana – Group 3 (Bagas et al., 2003)

Table 4.3-1. (Continued)

Groups	Granite complex	Classification	Data sources	Previous classifications
Triberton	Corunna Downs	low-Al	Davy (1988), Bagas et al. (2003), GSWA-GA unpublished	Triberton – Group 1 (Bagas et al., 2003)
Warrulinya	Mount Edgar	high-Al	Collins (1983)	Warrulinya Suite (Collins, 1983, 1993)
Yandicoogina	Mount Edgar	high-Al	Collins (1983); GSWA-GA unpublished	Yandicoogina Suite (Collins, 1983, 1993)
ca. 3.5–3.42 Ga group				
ca. 3.5 to 3.3 Ga granites				Emu & Callina & Tambina Supersuites (Van Kranendonk et al., 2006)
Nandingarra	Corunna Downs	low-Al	Davy (1988), Bagas et al. (2003), GSWA-GA unpublished	Nandingarra – Group 1 (Bagas et al., 2003)
ca. 3.5 to 3.42 Ga granites				Tambina & Callina Supersuites (Van Kranendonk et al., 2006)
Carlindi	Carlindi	high-Al	GSWA-GA unpublished	
Mt Edgar high-Al	Mount Edgar	high-Al	Collins (1983); GSWA-GA unpublished	
Mt Edgar low-Al	Mount Edgar	low-Al	Collins (1983); GSWA-GA unpublished	
Mt Edgar high-Al, low-Th	Mount Edgar	high-Al	Collins (1983); GSWA-GA unpublished	
Muccan	Muccan	high-Al	GSWA-GA unpublished	
Shaw high-Al	Shaw	high-Al	Bickle et al. (1989, 1993)	layered unit (Bickle et al., 1989, 1993)
Shaw layered	Shaw	high-Al	Bickle et al. (1989, 1993)	
Yule	Yule	low-Al mostly	GSWA-GA unpublished	Yule (Champion & Smithies, 2000)

Age subdivisions follows Van Kranendonk et al. (2006). Refer to Fig. 4.1-7 for distribution of granite complexes and granite age groups of this age into the Cleland Supersuite. A full listing of granite units included in the ca. 3.32–3.24 Ga granite group is given in Table 4.3-1.

4.3-4. GEOCHEMISTRY

4.3-4.1. Analytical Methods

Geochemical data (major elements, trace elements and Sm-Nd isotopes) used in this paper include both previously collected published (Bickle et al., 1983, 1993; Collins, 1993; McCulloch, 1987; Davy, 1988) and unpublished data (Collins, 1983), and new data for either newly collected rocks (GA and GSWA unpublished data) or reanalysis of previously collected samples (from Davy, 1988). All new major and trace element geochemical data were analysed at Geoscience Australia, Canberra. Major elements were determined by wavelength-dispersive XRF on fused disks using methods similar to those of Norrish and Hutton (1969). Precision is better than $\pm 1\%$ of the reported values. Loss on Ignition (LOI) was determined by gravimetry after combustion at $1100\,°C$. FeO abundances were determined by digestion and electrochemical titration using a modified method based on Shapiro and Brannock (1962). The trace elements Ba, Cr, Cu, Ni, Sc, V, Zn and Zr were determined by wavelength-dispersive XRF on a pressed pellet using methods similar to those of Norrish and Chappell (1977), while Cs, Ga, Nb, Pb, Rb, Sr, Ta, Th, U, Y and the REE were analysed by ICP-MS (Perkin Elmer ELAN 6000) using methods similar to those of Eggins et al. (1997), but on solutions obtained by dissolution of fused glass disks (Pyke, 2000). Sm-Nd isotopic analyses were determined by isotope dilution at VIEPS Radiogenic Isotope Laboratory, Department of Earth Sciences, La Trobe University, Victoria. Analytical techniques follow the method reported in Waight et al. (2000).

4.3-4.2. Ca. 3.5–3.42 Ga Granites

The geochemistry of 3.5–3.42 Ga granites was previously discussed by Bickle et al. (1983, 1993) for the Shaw Granitic Complex, by Collins (1993) and Davy and Lewis (1986) for the Mt Edgar Granitic Complex, by Davy (1988) and Bagas et al. (2003) for the Corunna Downs Granitic Complex, and by Champion and Smithies (2000) for the Yule Granitic Complex. Granites of this age have an expanded silica range (62–76% SiO_2). TiO_2, Al_2O_3, FeO^*, MgO, CaO and P_2O_5 are all negatively correlated with SiO_2, generally forming broad but well-defined trends (Fig. 4.3-1). In detail, chemical differences between and within granitic complexes are evident for most of the major elements; for example, Nandingarra granites in the Corunna Downs Granitic Complex have elevated MgO (Fig. 4.3-1) and low P_2O_5 relative to other granites with the same SiO_2 contents. The most consistent differences are for Al_2O_3, FeO^* and Na_2O, with low Al_2O_3, Na_2O, high FeO^* and high Al_2O_3, Na_2O, low FeO^* subgroups present (Fig. 4.3-1). These will be referred to here as low- and high-Al subgroups, respectively. The majority of the ca. 3.5–3.42 Ga granites belong to the high-Al group, with low-Al granites occurring in the Yule, Corunna Downs

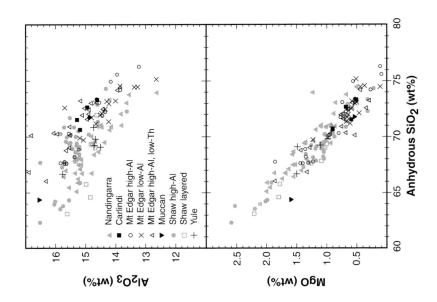

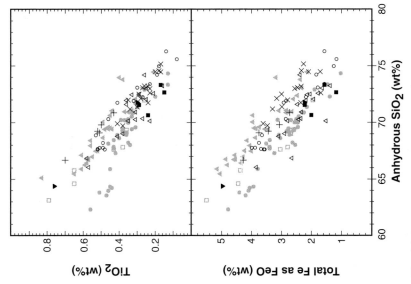

Fig. 4.3-1.

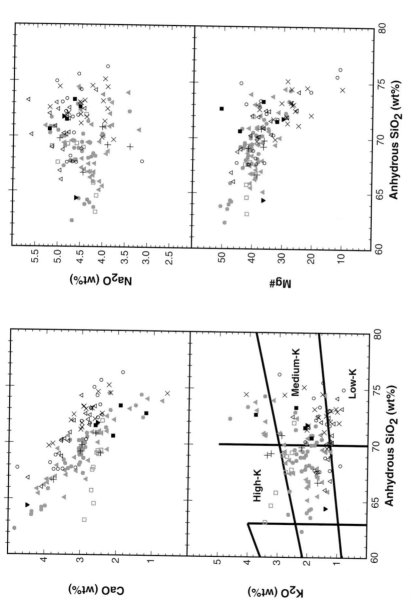

Fig. 4.3-1. (*Continued*). Major element Harker diagrams for the ca. 3.5–3.42 Ga granites; SiO₂ (anhydrous) versus Mg#, TiO₂, Al₂O₃, FeOtot, MgO, CaO, Na₂O and K₂O. Mg# = 100 molecular (Mg/(Mg + total Fe)). Low-Al units and granite complexes shown as '+', '×', and upright grey triangles. Data from Davy (1988), Bickle et al. (1989, 1993), Collins (1983) and GA-GSWA unpublished. Unit divisions as given in Table 4.3-1. Low-, medium- and high-K subdivisions after Gill (1981).

and Mount Edgar Granitic Complexes (Table 4.3-1). Mg# (mostly 45-10) are moderate to low (Fig. 4.3-1), as are Cr (mostly <50 ppm) and Ni (<40 ppm) contents, consistent with other early Archean TTGs (e.g., Smithies, 2000). Mg# is largely independent of subgroup, reflecting the variable MgO contents of these rocks.

Despite the variation in Na_2O, all granites are sodic with Na_2O/K_2O varying from >1 to >4. K_2O shows no strong correlation with Na_2O, or with SiO_2, MgO, Mg#, etc. (Fig. 4.3-1; see also Fig. 15 of Bickle et al., 1993). Granites vary from medium- to high-K (for given SiO_2) for both the low-Al and high-Al subgroups (Fig. 4.3-1). The majority of the granites straddle the Ca-enriched boundary of the 'trondhjemite' field on the Ca-Na-K plot (Fig. 4.3-2) and there is evidence for limited K-enrichment within all granitic complexes and both subgroups (Fig. 4.3-1, 4.3-2: see also Bickle et al., 1993). The behaviour of Rb mirrors that of K_2O (Fig. 4.3-3), varying over a considerable range of values (from 40 to 150 ppm for both subgroups), but with no obvious correlation with SiO_2 or Mg#. Ba appears to behave differently to other LILEs, with a pronounced geographic distribution in abundance. This is best shown by the almost universally elevated values (500 to >1500 ppm) in the Shaw and Yule Granitic Complexes relative to the nearby Mount Edgar and Corunna Downs Granitic Complexes (mostly <500 ppm; Fig. 4.3-3). Sr broadly correlates with Na_2O and amplifies the subgroups evident in Al_2O_3, FeO^* and Na_2O (Fig. 4.3-3), with low values (mostly <350 ppm) in the low-Al group, and high values (to >800 ppm) in the high-Al group (Fig. 4.3-4(b)). Sr is negatively correlated with SiO_2 for both groups (Fig. 4.3-3).

The ca. 3.5–3.42 Ga granites are slightly to strongly LREE-enriched when normalised to primitive-mantle (e.g., $(La/Sm)_N$[2] varies from 1 to 10; $(La/Yb)_N$ from 7 to >100), but vary from HREE- and Y-depleted, to Y-undepleted (e.g., $(Gd/Yb)_N$ ranges from 1.3–4.5 (Fig. 4.3-5(a–c)), with Yb from 0.2 to 2.5 ppm. HREE and Y decrease with increasing SiO_2, following two distinct trends: a low HREE and Y (Y < 12 ppm, Yb < 1.2 ppm at 65% SiO_2) trend is best seen in the granites of the Shaw Granitic Complex (see Bickle et al., 1993); and an elevated HREE and Y (Y < 25 ppm, Yb < 2.5 ppm at 65% SiO_2) trend that is most common in the Corunna Downs and Mount Edgar Granitic Complexes (Fig. 4.3-3). Samples with elevated $(Gd/Yb)_N$ or $(Tb/Yb)_N$ are restricted to low Yb (<1.2 ppm) rocks (Fig. 4.3-5(a–c)). The Y and HREE trends largely reflect the high- and low-Al subgroups (low and high-Y, respectively; Fig. 4.3-4), though a small subset of samples from a number of high-Al units are characterised by elevated Y (up to 25 ppm; Fig. 4.3-4(a)).

LREE broadly correlate with the HREE (Fig. 4.3-5(a–c)) and generally lie along either a high or low LREE/HREE trend, for low- or high-Al subgroups, respectively. LREE/LREE ratios (e.g., La/Sm) are variable but generally greater for the high-Al subgroup. Both positive and negative Eu anomalies are present. Small to moderate negative Eu anomalies (Eu/Eu^* from ~1.0 to <0.4) characterise the low-Al subgroup, consistent with the low Sr contents (<500 ppm). These are typical of low-Al TTGs, which are dominantly Y-undepleted and Sr-depleted (e.g., Barker and Arth, 1976; Barker, 1979; Fig. 4.3-5(a–c)).

---

[2] All normalised elements and ratios are normalised to Primitive Mantle using the values of Sun and McDonough (1989).

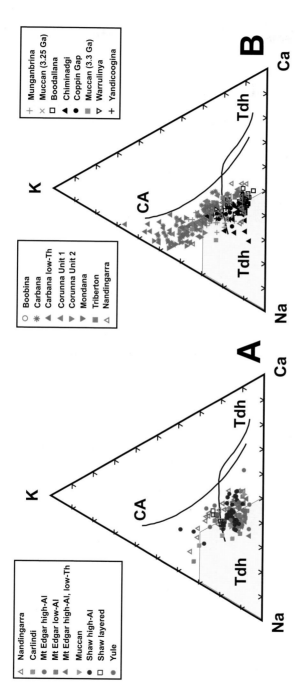

Fig. 4.3-2. Ca-Na-K ternary diagrams for (A): ca. 3.5–3.42 Ga; and (B): ca. 3.32–3.25 Ga Pilbara granites. High-Al units and granite complexes shown in blue, green and black, low-Al units and granite complexes in red and magenta. Data sources as for Figs. 4.3-1 and 4.3-7. Abbreviations: TDH – trondhjemite, CA – calc-alkaline and trondhjemite trends, and shaded trondhjemite field, from Martin (1994).

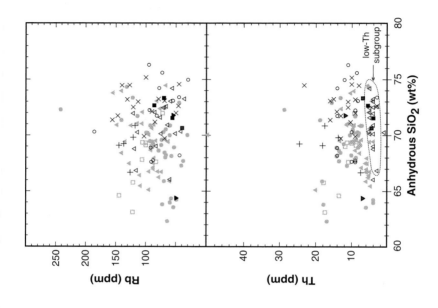

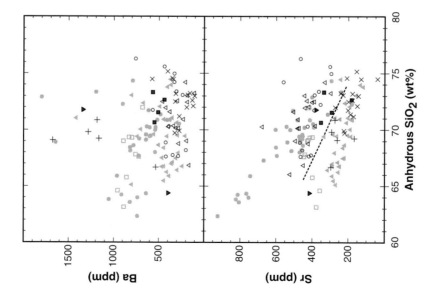

Fig. 4.3-3.

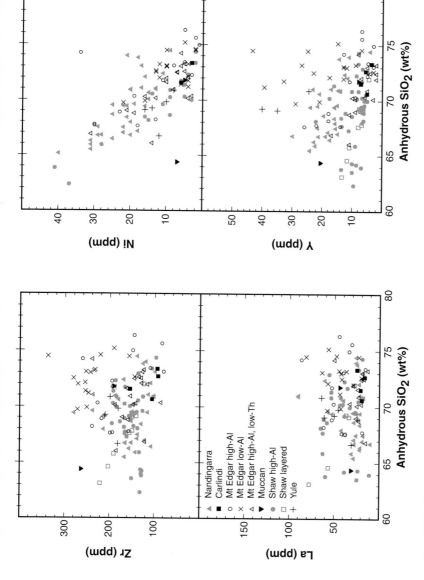

Fig. 4.3-3. (*Continued*). Selected trace element Harker diagrams for the ca. 3.5–3.42 Ga granites: SiO₂ (anhydrous) versus Ba, Rb, Sr, Th, Zr, Ni, La and Y. Low-Al units and granite complexes shown as '+', '×', and upright grey triangles. Low-Th Mount Edgar subgroup shown circled on Th-SiO₂ plot. Dotted line on Sr-SiO₂ plot shows high and low Sr contents of the high-Al and low-Al groups, respectively. Data sources as for Fig. 4.3-1.

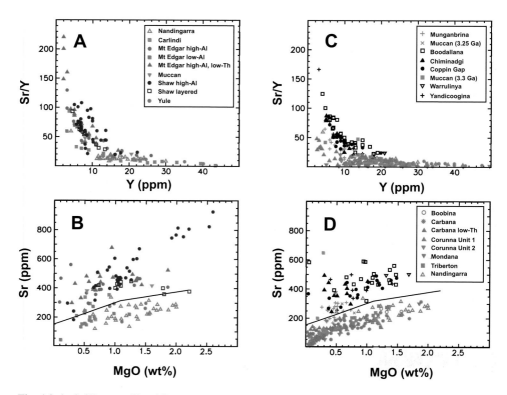

Fig. 4.3-4. Sr/Y versus Y and Sr versus MgO for (A,B): ca. 3.5–3.42 Ga; and (C,D): ca. 3.32–3.25 Ga Pilbara granites. High-Al units and granite complexes shown in blue, green and black, low-Al units and granite complexes in red and magenta. Data sources as for Figs. 4.3-1 and 4.3-7. Note legend for ca. 3.32–3.25 Ga Pilbara granites split into two. Line, in C and D, illustrates the common split into 2 subgroups for granites of both age groups.

Notably for this subgroup, Eu/Eu* shows a good negative correlation with Yb, though has little correlation with indicators of differentiation such as SiO_2, or MgO. Granites of the high-Al subgroup have small positive to small negative Eu anomalies (Eu/Eu* mostly 1.1 to 0.8).

Importantly, there appears to be no correlation between LREE and LILE (K_2O, Rb, Ba, Pb) abundances. The HREE show a similar relationship, although the lowest LILE values are related to lowest HREE abundances (but not vice versa); e.g., the high-Al subgroup granites extend to lower Y values than the low-Al subgroup, but have similar elevated Rb contents.

Unlike the case for LILE, the concentrations of Th, Nb, and possibly Zr correlate with LREE abundances. Thorium contents (to 20 ppm) in most of the granites are elevated above crustal averages (e.g., 3.5 ppm in average continental crust; Taylor and McLennan, 1985), and are 10 to >20 times higher than in similar-aged mafic extrusive rocks of the

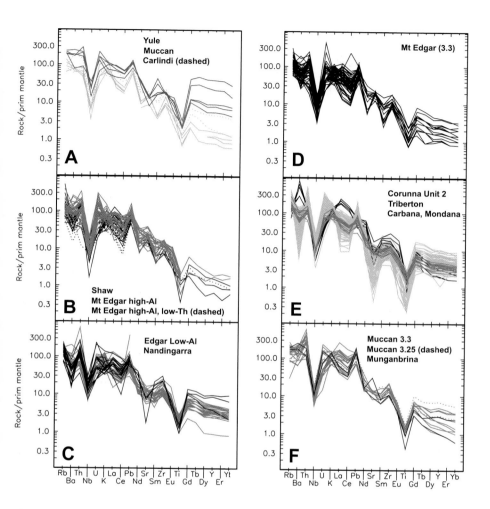

Fig. 4.3-5. Primitive-mantle normalised multi-element plots for selected granite units. A–C = ca. 3.5–3.42 Ga granites; D–F = ca. 3.32–3.25 Ga granites. Data sources as for Figs. 4.3-1 and 4.3-7. Mt Edgar (3.3) is for all Mt Edgar 3.3 Ga units. Primitive mantle values from Sun and McDonough (1989). Element list and order follows Hoffman (2005).

EPT (mostly <1 ppm). There is a pronounced low Th subgroup (~3–5 ppm) that is most evident in the Mount Edgar Granitic Complex (Fig. 4.3-3), but which appears to include samples from other granitic complexes. These rocks also have low K_2O contents (even at high SiO_2) and high Na_2O and Na_2O/K_2O (Fig. 4.3-1, 4.3-2(a), 4.3-3).

A variety of trends (with increasing SiO_2) are evident in the Th data (Fig. 4.3-3). Firstly, the low-Th Mount Edgar (and others) subgroup show approximately constant Th with increasing SiO_2. Secondly, the Nandingarra granites mostly exhibit strongly increasing Th from low values similar to those of the low-Th Mount Edgar subgroup. The remaining granites have elevated Th at all SiO_2 levels, and perhaps even decreasing Th for the Shaw Granitic Complex (Fig. 4.3-3).

Ce/Th ratios ($\sim$7) and Th/La ratios (cluster around 0.2 to 0.3) are relatively constant, similar to average values for TTGs reported by Martin (1994) and Condie (1993) and slightly lower than average upper continental crust (0.36; Taylor and McLennan, 1985). The lowest Th/La ratios are from the low-Th Mount Edgar subgroup (0.1 to 0.15) and are similar to, or higher than, primitive mantle values (0.12; Sun and McDonough, 1989). There is no correlation between Th and Y contents, with both the high- and low-Al subgroups having similar Th abundances.

Available isotope data for the 3.5–3.42 Ga granite group show a moderate range of ε_{Nd} (+2.1 to −1.5), with most +1.8 to −0.5 (Fig. 4.3-6: McCulloch, 1987; Bickle et al. 1993; Smithies et al., 2007b). This range requires at least a minor contribution from older (pre-existing) felsic crust, and the reasonable negative correlation between ε_{Nd} and K_2O (or versus Th; Fig. 4.3-6) suggests some form of efficient mixing process (e.g., mixed sources or assimilation). One sample falls below the ε_{Nd}–K_2O trend. This is from a unit of the Warrawagine Granitic Complex that is known to contain interlayered components as old as ca. 3.65 Ga (Nelson, 1999; Williams, 2001). One sample lies above the main ε_{Nd}–K_2O trend. Although interpreted as a 3.45 Ga felsic band (e.g., McCulloch, 1987), the sample actually falls within a unit that Bickle et al. (1989) mapped as a ca. 2.93 Ga granite and has a composition more consistent with this latter interpretation. Also shown on Fig. 4.3-6 are data for three samples from the ca. 3.47 Ga Coolyia Creek Granodiorite (formerly the Chocolate Hill suite of Bickle et al., 1993). The data are further complicated by the lack of major-element analysis (i.e., no K_2O for the two least evolved samples). The Th-ε_{Nd} plots show similar broad negative correlations, although with poorer correlation and steeper negative slopes (Fig. 4.3-6).

Champion and Smithies (2000) proposed that the low-Al subgroup was largely confined to the western part of the EPT (although age constraints are poor) and to younger members of the 3.5–3.42 Ga granites. Certainly, those granites with elevated HREEs tend to occur as enclaves and pendants within the western part of the Yule Granitic Complex (Fig. 4.1-7). However, it is clear that low-Al granites are also present to the east, in the Mount Edgar and Corunna Downs Granitic Complexes (Table 4.3-1). In addition, gneissic xenoliths of high-Al granites occur in the eastern Yule Granitic Complex (Pawley and Collins, 2002).

4.3-4.3. Ca. 3.32–3.24 Ga Granites

The 3.32–3.24 Ga granites have an expanded silica range (65–76%), but include much more silica-rich members than the older granites (Fig. 4.1-7, 4.3-7). Like the latter, the oxides (except Na_2O and K_2O) are all negatively correlated with SiO_2, generally forming broad but well-defined trends (Fig. 4.3-7). Also, like the older granites, two clearly distinctive geochemical types are evident within the 3.32–3.24 Ga granites, again with high-Al and low-Al signatures. The 3.32–3.24 Ga high-Al subgroup is characterised by higher Al_2O_3, CaO, Na_2O, P_2O_5, Sr, and Na/K, and lower FeO^*, K_2O, Rb, Y, and the HREE

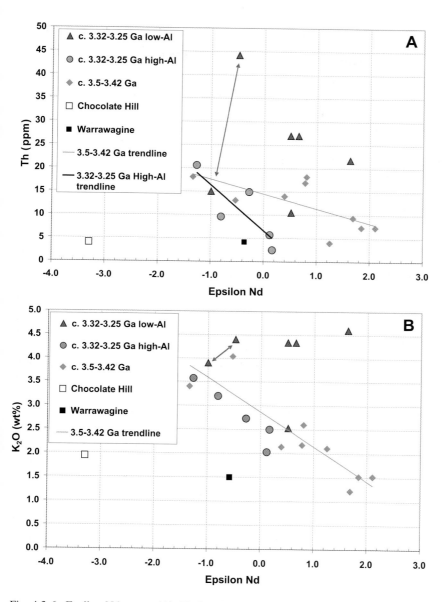

Fig. 4.3-6. Epsilon Nd versus (A): Th (ppm); and (B): K_2O (wt%), for the ca. 3.5–3.42 Ga and ca. 3.32–3.25 Ga Pilbara granites. The Chocolate Hill unit is part of the ca. 3.5–3.42 Ga high-Al group; K_2O and Th contents are only available for one Chocolate Hill sample; the other 2 samples (ε_{Nd} of −2.6 and −0.77) are not shown. The Warrawagine sample may be part of the ca. 3.5–3.42 Ga granites, the age is, however, uncertain (see text). The arrowed line (in A and B) points to 2 samples from the one ca. 3.32–3.25 Ga unit. Data from McCulloch (1987), Bickle et al. (1993), Brauhart et al. (2000) and Smithies et al. (2007).

compared with the low-Al subgroup (Figs. 4.3-4, 4.3-7, 4.3-8). MgO, Mg# and TiO_2 contents are similar in both subgroups (Fig. 4.3-7). The 3.32–3.24 Ga high-Al subgroup is the dominant and most widespread subgroup and is best documented in the Mount Edgar Granitic Complex (Collins, 1993; Davy and Lewis, 1986). This subgroup has compositions broadly similar to the high-Al subgroup of the older granites, particularly at <70% SiO_2. They are felsic (65–78% SiO_2), mostly high-Al (>15% at 70% SiO_2), sodic (4-5% Na_2O, Na_2O/K_2O = 1–4), and mostly medium-K, with moderate to low Mg# (45 to less than 15; Figs. 4.3-1, 4.3-7). The majority of these rocks are Sr-undepleted, Y- and HREE-depleted, and characterised by low HREE, Y (<20 ppm), and elevated Sr (>550 ppm at 66% SiO_2; Figs. 4.3-8, 4.3-4(d)). They are moderately LREE enriched (($La/Sm)_N$ = 3–10; $(La/Yb)_n$ = 5 to >40), with unfractionated HREE (($Gd/Yb)_N$ = 1.5 to >3.5). HREE patterns become flatter with increasing SiO_2 (Fig. 4.3-5(d–f)). Europium anomalies vary from slightly positive to commonly slightly negative (Eu/Eu* largely ∼1.1–0.7). Like the ca. 3.5–3.42 Ga granites, there is a negative correlation between Eu/Eu* and Yb, but no apparent correlation between Eu/Eu* and SiO_2 or MgO.

Differences between the younger and the older subgroups of high-Al granites are minor. The younger subgroup lacks the less silicic end-members (being mostly >66% SiO_2), but has more silica-rich members (>72% SiO_2) and correspondingly more potassic compositions. Hence, there is clear evidence for increasing K_2O with increasing SiO_2 (e.g., Collins, 1993; Figs. 4.3-1, 4.3-7). This is especially true for the Munganbrina Suite (Collins, 1993) that, based on current work (see Van Kranendonk et al., 2006a), is, at least in part, younger (∼3.24 Ga) than the rest of the Mount Edgar granites. LILE contents and ratios (e.g., K/Rb, K/Ba) show wide overlap for high-Al granites of both age groups (Figs. 4.3-3, 4.3-8). Th contents are similar with similar Ce/Th and La/Th ratios, as is Eu/Eu*.

The second and less abundant, low-Al subgroup of 3.32–3.24 Ga granite (Table 4.3-1) is best seen in the ca. 3.3 Ga Corunna Downs Granitic Complex (Davy, 1988; Barley and Pickard, 1999; Bagas et al., 2003), but also includes some 3.24 Ga granites (e.g., Strelley Monzogranite; Vearncombe and Kerrich, 1999). Granites of this type extend to very siliceous compositions (>65 to 78 wt% SiO_2), markedly more so than all other 3.3 Ga, and older, Pilbara granites (Fig. 4.3-7). They are K_2O-rich (mostly high-K, with 2.5 to >5 wt% K_2O) and exhibit pronounced K-enrichment, with Na_2O/K_2O anti-correlated with SiO_2 (from ∼2 to <0.5; Figs. 4.3-2, 4.3-4). The combination of low Al_2O_3, Na_2O (<4.5%) and Sr (300 ppm), and high FeO* with high K_2O, Rb (70–250 ppm) and HREE is diagnostic (Figs. 4.3-2, 4.3-4, 4.3-7, 4.3-8). K_2O, Rb, Rb/Sr (up to 8), Pb (up to 30 ppm), Th (up to 30 ppm), Th/La (to >2), and U (up to 7 ppm) are highest at the highest SiO_2 contents. K_2O and Th/La are strongly positively correlated with increasing SiO_2. Ba and Sr contents decrease to levels as low as <50 ppm and <20 ppm, respectively (Fig. 4.3-8). Both correlate positively with each other and with Eu/Eu*. K/Rb (∼250–160) and Ca/Sr (∼110–70) ratios also decrease with increasing SiO_2. The granites are typically moderately LREE-enriched (($La/Yb)_N$ = 3.5–8, becoming smaller for the more siliceous end-members), with flat to slightly fractionated HREE (($Gd/Yb)_N$ ranges from 2.3 to <1.0 in the more siliceous end-members). They have pronounced negative Eu anomalies (Eu/Eu* mostly 0.8 to <0.2) and are Y- and HREE-undepleted (20–40 ppm Y at 68% SiO_2; Figs. 4.3-5, 4.3-7, 4.3-8).

There is no discernible correlation between Eu/Eu* and the HREE, although the highest Yb values occur in rocks with the lowest Eu/Eu*. The major difference between the older and younger low-Al groups is that the latter extends to higher SiO_2 contents (77 wt% vs 75 wt%), with correspondingly greater concentrations of K_2O, Rb, Pb and Th (at $\sim$77 wt% SiO_2), and stronger evidence for fractional crystallisation. For example, Ba (to 40 ppm), Zr (to <60 ppm), LREE (Ce to <20 ppm) and compatible elements such as Sr (to <20 ppm), together with such ratios as Eu/Eu* (to <0.1), decrease to low values with increasing silica in the ca. 3.32–3.24 Ga low-Al group (Figs. 4.3-5, 4.3-8).

A number of chemical differences also exist within the ca. 3.32–3.24 Ga low-Al sub-group. This is evident, for example, in trace elements such as Zr (Fig. 4.3-8) and Nb, which exhibit significant inter-unit variation independent of major element variation. The most distinctive of the ca. 3.32–3.24 Ga low-Al subgroup is 'unit 2' of Bagas et al. (2003), which is characterised by very high LREE (Ce >200 ppm), very high Eu, strongly frac-tionated REE (with high (La/Yb)$_N$ = 75–130 and (Gd/Yb)$_N$ = 3–4), high Zr, Zr/REE (e.g., Zr/Sm), Sr, extreme Ba (1500–5000 ppm) and low Rb, Pb and variably low Th (Figs. 4.3-5, 4.3-8). In addition to all of these subgroups, there is also a low Th (Th < 14 ppm at high SiO_2) subgroup, which is most evident in the Carbana Monzogranite (e.g., Fig. 4.3-8). Members of this subgroup, which tend to have compositions intermediate between 'unit 2' and the bulk of the ca. 3.32–3.24 Ga low-Al subgroup, are characterised by typically higher Sr, Ba and Eu, very high Eu/Eu* (0.5 to >4.0), lower Pb, Rb, Th, HREE and Y, and weakly to strongly fractionated REE ((La/Yb)$_N$ = 4–55 and (Gd/Yb)$_N$ = 0.6–3.4). Notably, there is a reasonably strong positive correlation between Ba contents and the de-gree of REE fractionation in this low-Th subgroup. Again there are no strong differences in major elements between this subgroup and the rest of the low-Al subgroup.

The ca. 3.32–3.24 Ga granites have ε_{Nd} values between +1.6 and −1.3 (Fig. 4.3-6). The high-Al granites show a reasonable negative correlation between ε_{Nd} and K_2O or Th (Fig. 4.3-6), similar to that observed for the ca 3.45 Ga granites (although displaced to-wards higher K_2O; Th values are similar). The ca. 3.32–3.24 Ga low-Al granites, however, exhibit no such relationship. This, in part, reflects unit variability, although the highest Th values are associated with more evolved (more negative) ε_{Nd} signatures. The Strelley Monzogranite (Brauhart and Morant, 2000), in part, lies on the trend shown by the high-Al granites (Fig. 4.3-6).

4.3-5. GRANITE PETROGENESIS

4.3-5.1. Introduction

Petrogenetic models for both the ca. 3.5–3.42 Ga and ca. 3.32–3.24 Ga granites must account for a variety of competing chemical signals. Both age groups contain low- and high-Al subgroups, with corresponding differences in CaO, Na_2O, Sr, HREE, and Y. Con-versely, all subgroups show a significant variation in the LILEs for given SiO_2 contents (also mirrored in the trondhjemitic versus calc-alkaline trends), and a range in ε_{Nd} which

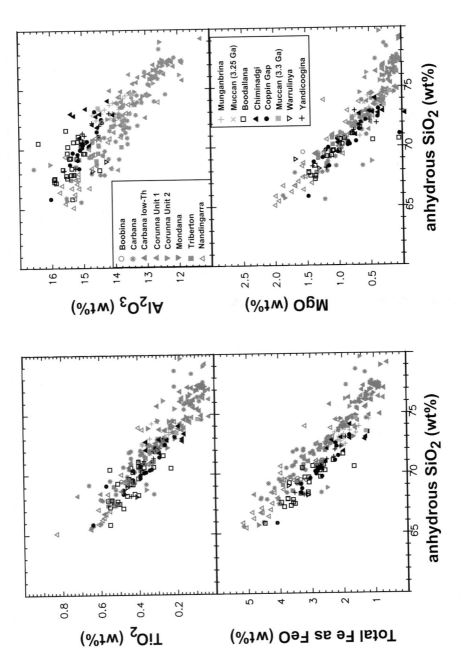

Fig. 4.3-7.

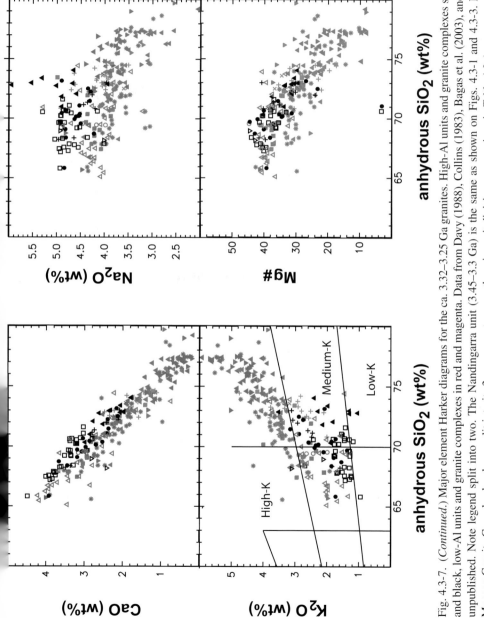

Fig. 4.3-7. (*Continued.*) Major element Harker diagrams for the ca. 3.32–3.25 Ga granites. High-Al units and granite complexes shown in blue and black, low-Al units and granite complexes in red and magenta. Data from Davy (1988), Collins (1983), Bagas et al. (2003), and GA-GSWA unpublished. Note legend split into two. The Nandingarra unit (3.45–3.3 Ga) is the same as shown on Figs. 4.3-1 and 4.3-3. Note that the Muccan Granite Complex has been split into its 2 component ages, otherwise unit divisions are as given in Table 4.3-1.

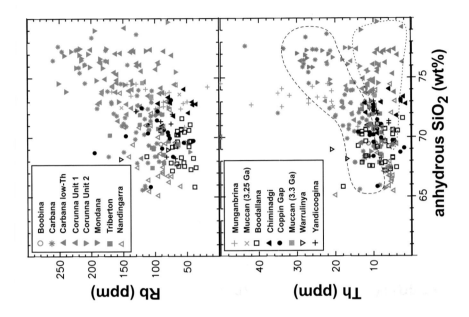

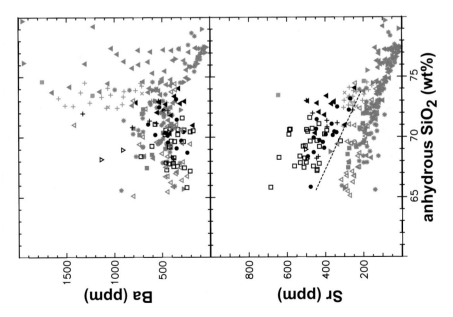

Fig. 4.3-8.

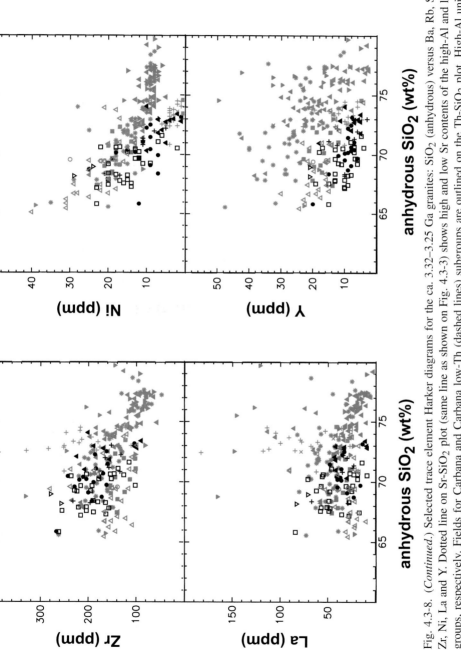

Fig. 4.3-8. (*Continued.*) Selected trace element Harker diagrams for the ca. 3.32–3.25 Ga granites: SiO₂ (anhydrous) versus Ba, Rb, Sr, Th, Zr, Ni, La and Y. Dotted line on Sr-SiO₂ plot (same line as shown on Fig. 4.3-3) shows high and low Sr contents of the high-Al and low-Al groups, respectively. Fields for Carbana and Carbana low-Th (dashed lines) subgroups are outlined on the Th-SiO₂ plot. High-Al units and granite complexes shown in blue and black, low-Al units and granite complexes in red and magenta. Data sources as for Fig. 4.3-7.

correlates with the LILEs. In addition, there is chemical variability evident between, and within, individual units (within each subgroup).

4.3-5.2. Partial Melting Versus Fractional Crystallisation

The lack of significant Eu anomalies in the majority of the granites (Fig. 4.3-5), especially the high-Al subgroup, coupled with the obvious lack of correlation between Eu/Eu* with MgO, or SiO$_2$, suggests that fractional crystallisation was not a major process for most of the East Pilbara granites, at least at pressures where plagioclase was stable. The evidence is less definitive for the low-Al subgroup, where clearly there has been some role for plagioclase, either residual after partial melting and/or via fractional crystallisation. The moderate (to high) Mg# for many of the East Pilbara granites, however, argues against any significant fractional crystallisation for either the low-Al or high-Al subgroups. Similar conclusions have been reached for Archean TTGs in general (e.g., Martin, 1994; Clemens et al., 2006; Moyen et al., this volume), and for the East Pilbara granites (e.g., Bickle et al., 1993; Collins, 1993), although some of these studies demonstrated that fractionation and cumulate processes were locally important.

Only in the silica-rich end-members of the ca. 3.32–3.24 Ga low-Al subgroup is there compositional variation that is consistent with moderate to extensive fractional crystallisation (e.g., Bagas et al., 2003). Here, elements such as Sr, Ba, Eu, and Eu/Eu* (Fig. 4.3-5) decrease to very low levels with correspondingly high to very high Rb/Sr and Rb/Ba ratios. These trends are matched by compatible behaviour of Zr and the LREE (i.e., decreases with increasing SiO$_2$; Fig. 4.3-8), suggesting accessories such as zircon, and perhaps allanite, are also involved. In this regard, the low-Th Carbana subgroup and 'unit 2' (Corunna Downs Granitic Complex) of the ca. 3.32–3.24 Ga low-Al subgroup – with higher Sr, Ba, Eu, very high Eu/Eu* (0.5 to >4.0), lower Pb, Rb, Th, HREE and Y (Fig. 4.3-8), and weakly to strongly fractionated REE (Fig. 4.3-5) – may reflect concentration of such phases (i.e., plagioclase, zircon, allanite) and could be interpreted as partly cumulate in origin. Significantly, it is the ca. 3.32–3.24 Ga low-Al granites that show the least correlation of the LILE with ε_{Nd} (Fig. 4.3-6), also consistent with the effects of fractional crystallisation.

4.3-5.3. Partial Melting

The geochemical differences between the high-Al and low-Al subgroups of the EPT mirror variations evident in sodic granites in general, such as TTGs (e.g., Barker and Arth, 1976; Barker, 1979; Martin, 1987). These variations are commonly ascribed to varying residual mineralogy after partial melting of mafic precursors. For example, variations in the HREE contents and corresponding changes in LREE/HREE ratios (e.g., (La/Yb)$_N$ and (Gd/Yb)$_N$ ratios) of TTGs are typically attributed to the presence or absence of residual garnet (versus amphibole) during partial melting of a mafic source (e.g., Barker and Arth, 1976; Arth et al., 1978; Barker, 1979; Martin, 1986; Drummond and Defant, 1990; Rapp et al., 1991, 2003; Rapp and Watson, 1995; Rushmer, 1991; Winther and Newton, 1991; Sen and Dun, 1994; Klein et al., 2000; Willie et al., 1997; Barth et al., 2002; Clemens et al., 2006; Moyen et al., this volume). Similarly, the correlations observed for the Pilbara granites between HREE concentrations and Al_2O_3, Na_2O, Sr, Eu/Eu^*, and Sr/Y ratios are consistent with such models and suggest an antithetic role for garnet and plagioclase; i.e., residual garnet but not plagioclase for the high-Al granites, and residual plagioclase but not garnet for the low-Al granites (Fig. 4.3-4).

High-Al granites – in particular the low-LILE end-members – have compositions very similar to the typical Archean high-Al TTGs of Martin (1994; see Bickle et al., 1993) (Fig. 4.3-9). Accordingly, simple trace-element modelling was undertaken to test whether the Pilbara high-Al granites, especially the closely TTG-like low-LILE end-members, have compositions consistent with derivation by partial melting at high pressures (garnet stable) from a mafic (basaltic) source. Bulk partition coefficients were calculated assuming 5%, 10% and 20% partial melting (e.g., see arguments in Barth et al., 2002). Also, as a measure of the suitability of the modelled source compositions, the degree of partial melting was calculated for highly incompatible elements assuming bulk partition coefficients of zero (i.e., perfectly incompatible). Batch partial melting equations (e.g., Shaw, 1970) were used. Averages of the Coonterunah and Apex basalts, part of the sequence of basalts from the base of the Pilbara Supergroup (Smithies et al., 2005b; Van Kranendonk et al., this volume), were used as analogues for modelled lower crustal source compositions (Table 4.3-2). Incompatible element abundances for these rocks, however, were modified to overcome element mobility caused by greenschist-facies metamorphism. This was achieved by increasing contents of Rb and Ba to create smooth primitive mantle-normalised multi-element curves assuming Th and LREE immobility. Source compositions (Table 4.3-2) are of slightly-enriched low-K tholeiites similar in composition to many Archean basalts (e.g., Condie, 1989). The LILE-poorer TTGs (the ca. 3.5–3.42 Ga, low-

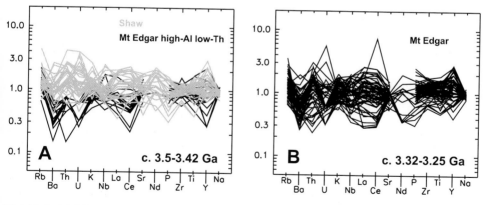

Fig. 4.3-9. Multi-element plots for selected high-Al granite units, normalised to average TTG composition of Martin (1994). Data sources as for Figs. 4.3-1 and 4.3-7. A = ca. 3.5–3.42 Ga granites; B = ca. 3.32–3.25 Ga granites (Mt Edgar is for all Mt Edgar 3.3 Ga units). Element list and order follows Hoffman (2005).

Table 4.3-2. Geochemical averages for selected high-Al subgroups (ca. 3.45 Ga Edgar low-Th, ca. 3.3 Ga Yandicoogina), and for mafic rocks from Warrawoona Group (Apex Basalt, Coonterunah Subgroup). Minimums and maximums (top and bottom 10% trimmed) also shown for granites

	Apex basalt		Coonterunah subgroup basalt			Edgar high-Al low-Th			Yandicoogina high-Al		
Group:	all	mod.	fract	mafic	fract						
Category	M	R	M	M	M	M	max	min	M	max	min
Majors											
SiO_2	48.83		49.79	49.22	46.47	70.56	72.48	65.98	70.35	72.37	67.45
TiO_2	0.94		0.96	1.05	1.92	0.31	0.57	0.21	0.35	0.47	0.27
Al_2O_3	13.03		12.64	14.19	12.77	15.35	16.81	14.42	14.83	15.59	14.18
FeOtot	11.25		11.03	13.04	13.81	2.35	3.10	1.52	2.40	3.52	1.91
MgO	6.41		6.24	6.47	4.31	0.73	1.43	0.34	0.77	1.41	0.40
MnO	0.21		0.19	0.25	0.25	0.04	0.07	0.03	0.05	0.06	0.01
CaO	9.56		9.99	10.41	9.24	3.26	4.09	2.62	2.76	3.49	2.08
Na_2O	2.07		3.36	2.31	2.30	4.92	5.60	4.42	4.35	4.96	3.96
K_2O	0.09	0.21	0.09	0.26	0.62	1.31	1.99	1.04	2.54	3.12	1.59
P_2O_5	0.09		0.09	0.10	0.16	0.11	0.16	0.07	0.11	0.16	0.09
LOItot	5.39		5.12	1.95	8.79	1.10	1.55	0.80	1.15	2.34	0.76
mg#	52		52	46	35	38	46	25	38	42	24
Traces											
Ba	31	52	41	49	51	280	422	162	548	1330	381
Rb	1.0	4.7	<1.0	5.6	15.8	70	119	42	80	130	68
Sr	92.6	138	97.2	99.7	136.7	442	524	331	414	499	256
Pb	2.0	1.5	0.7	4.0	8.0	11.5	15	9.0	14.3	19	12
Th	0.6	0.6	0.6	0.4	1.1	4.0	5.0	3.0	8.5	11	6
U	0.1	0.2	0.2	0.1	0.3	1.0	1.0	<1.0	1.0	2.2	1.0
Nb	3.4	4.5	3.8	3.5	8.3	5.0	11.0	4.0	6.3	9.7	3.0
Zr	67	67	77	64	112	144	188	116	159	232	132
Y	21	21	23	21	27	6	11	2	10	15	3
Ga	15.0	15.0	14.9	16.0	18.5	19	21	18	18.5	21	15
La	4.4	4.4	5.3	4.0	10.9	23.0	32.8	15.0	32.1	67.0	23.0
Ce	11.6	11.6	13.7	10.9	23.6	29.5	56.7	14.0	60.4	115.0	40.0
Nd	8.9	8.9	10.4	8.9	18.1	19.6	19.6	19.6	22.8	23.8	21.7
Sm	2.6	2.6	2.9	2.5	4.7	2.8	2.8	2.8	3.5	3.8	3.2
Eu	0.95	0.95	1.01	0.90	1.74	0.90	0.90	0.90	0.85	0.90	0.80
Gd	3.29	3.29	3.63	3.20	4.98	2.0	2.0	2.0	2.7	2.9	2.4
Tb	0.59	0.59	0.64	0.61	0.87	0.2	0.2	0.2	0.4	0.4	0.3
Dy	3.70	3.70	3.99	3.79	4.65	1.1	1.1	1.1	1.8	1.9	1.7
Ho	0.88	0.88	0.89	0.86	0.97	0.2	0.2	0.2	0.4	0.4	0.3
Er	2.43	2.43	2.52	2.44	2.67	0.5	0.5	0.5	1.1	1.1	1.1
Yb	2.32	2.32	2.44	2.43	2.40	0.4	0.4	0.4	1.1	1.2	1.0
Lu	0.38	0.38	0.38	0.40	0.40	<0.1	<0.1	<0.1	0.2	0.2	0.1

Table 4.3-2. (*Continued*)

	Apex basalt		Coonterunah subgroup basalt			Edgar high-Al low-Th			Yandicoogina high-Al		
Group:	all	mod.	fract	mafic	fract						
Category	M	R	M	M	M	M	max	min	M	max	min
Sc	38	38	39	46	27	7	7	7	4	4	4
Ni	104	104	89	94	90	8	21	4	11	21	5
V	320	320	293	305	282	27	42	11	35	54	18
Cr	345		379	197	145	13	30	6	16	26	6
Cu	95		82	20	154	9	20	3	13	32	7

Modified Apex Basalt refers to trace element compositions (Rb, Ba, Pb, Sr, U) recalculated by smoothing multi-element curves using assumed immobile Th and La. Data sources for granites listed in Table 4.3-1; for basalts from Smithies et al. (2007). Category abbreviations: M = median, R = recalculated

Th granites and the Yandicoogina Suite from the Mount Edgar Granitic Complex; Collins, 1993) were used as melt compositions (Table 4.3-2).

From the results, summarised in Table 4.3-3, it is apparent that calculated bulk partitions for the HREEs fall within experimentally determined brackets (e.g., Klein et al., 1997, 2000; Barth et al., 2002; Klemme et al., 2002; compilation in Rollinson, 1993) for 5 to 20% partial melting, assuming significant garnet in the residue (i.e., a garnet-clinopyroxene ± amphibole residue consistent with high pressure experimental data; e.g., Rapp et al., 1991; Wyllie et al., 1997). The general agreement between the modelled and experimental data is to be expected given that the behaviour of the HREEs is very sensitive to the amount of garnet in the residue, but largely insensitive to source enrichment (e.g., N-MORB versus E-MORB). Consideration of the more incompatible elements, especially Th and the LREE (and the less robust Rb), show that even with the slightly enriched source composition used (Coonterunah basalt), only relatively small degrees (<20%) of partial melting are permissible (Table 4.3-3). For larger degrees of partial melting, and for the more LILE-rich granites that typify the EPT, more enriched compositions are required. Finally, it is evident that the modelling requires either a residual phase sequestering Nb (Table 4.3-3) and Ti and Ta (most likely rutile, as has been recognised previously by e.g., Green and Pearson, 1986; Barth et al., 2002; Rapp et al., 1991; Xiong et al., 2005), but also possibly amphibole (e.g., Foley et al., 2002), or a source with a Nb- and Ti-depleted signature. The latter option appears less likely, given the lack of evidence for such rocks in the East Pilbara (e.g., Smithies et al., 2005b) and in the Paleoarchean in general.

As detailed by Martin (1994), Archean TTGs are dominantly of the high-Al type, to the extent that the TTG terminology has become synonymous with high-Al compositions. It is evident, however, that Archean low-Al TTGs not only exist, but are locally voluminous (e.g., Barker and Arth, 1976; Champion and Smithies, 2000, this study; Bagas et al., 2003). Accordingly, it is important to establish whether the Pilbara low-Al TTG groups can also be derived from basaltic protoliths, and particularly from compositions similar to those that may have produced the high-Al TTGs.

Table 4.3-3. Calculated bulk partition coefficients (Kd) with 5%, 10% and 20% partial melting for selected high-Al subgroups (Mt Edgar high-Al low-Th and Yandicoogina)

Mt Edgar high-Al, low-Th (3.5–3.42 Ga)

Source	Apex basalt source				mafic Coonterunah source				fract Coonterunah source			
	f	f	f	Kd	f	f	f	Kd	f	f	f	Kd
Value	0.05	0.1	0.2	0	0.05	0.1	0.2	0	0.05	0.1	0.2	0f
	Kd	Kd	Kd	f	Kd	Kd	Kd	f	Kd	Kd	Kd	f
Ba	0.14	0.09	−0.02	0.19	0.13	0.08	−0.03	0.17	0.13	0.09	−0.02	0.18
Rb	0.02	−0.04	−0.17	0.07	0.03	−0.02	−0.15	0.08	0.18	0.14	0.03	0.23
Sr	0.28	0.24	0.14	0.31	0.18	0.14	0.03	0.23	0.26	0.23	0.14	0.31
Pb	0.08	0.03	−0.09	0.13	0.30	0.28	0.18	0.35	0.65	0.66	0.62	
Th	0.11	0.06	−0.06	0.15	0.05	0.00	−0.13	0.10	0.23	0.19	0.09	0.28
Nb	0.90	0.90	0.88		0.65	0.67	0.63		1.62	1.73	1.83	
Zr	0.44	0.41	0.33	0.47	0.40	0.39	0.31	0.45	0.73	0.76	0.73	
Y	3.40	3.53	3.85		3.29	3.59	3.91		4.22	4.61	5.06	
La	0.15	0.10	−0.01	0.19	0.12	0.08	−0.04	0.17	0.42	0.41	0.34	0.47
Ce	0.36	0.33	0.24	0.39	0.32	0.30	0.21	0.37	0.75	0.78	0.75	
Nd	0.43	0.40	0.32	0.46	0.41	0.39	0.32	0.46	0.88	0.91	0.90	
Sm	0.93	0.93	0.92		0.85	0.89	0.88		1.63	1.75	1.84	
Eu	1.1	1.1	1.1		1.0	1.0	1.0		1.9	2.0	2.2	
Gd	1.7	1.7	1.8		1.6	1.7	1.8		2.5	2.7	2.9	
Tb	3.0	3.1	3.4		3.0	3.3	3.6		4.3	4.7	5.2	
Dy	3.5	3.6	4.0		3.4	3.7	4.1		4.2	4.6	5.0	
Ho	4.6	4.8	5.2		4.3	4.7	5.1		4.8	5.3	5.8	
Yb	6.1	6.3	7.0		6.1	6.6	7.3		6.0	6.6	7.3	
Ni	14.5	15.3	17.1		12.5	13.8	15.4		12.0	13.2	14.8	
V	12.4	13.1	14.6		11.3	12.4	13.9		10.4	11.5	12.8	

Yandicoogina – high-Al (3.32–3.25 Ga)

Source		mafic Coonterunah source				fract Coonterunah source			
		f	f	f	Kd	f	f	f	Kd
Value		0.05	0.1	0.2	0	0.05	0.1	0.2	0
		Kd	Kd	Kd	f	Kd	Kd	Kd	f
Ba		0.04	−0.01	−0.14	0.09	0.05	−0.01	−0.13	0.09
Rb		0.02	−0.03	−0.16	0.07	0.16	0.11	0.00	0.20
Sr		0.20	0.16	0.05	0.24	0.29	0.26	0.16	0.33
Pb		0.24	0.20	0.10	0.28	0.54	0.51	0.45	0.56
Th		0.00	−0.06	−0.19	0.05	0.08	0.03	−0.09	0.13
Nb		0.53	0.51	0.44	0.56	1.33	1.35	1.40	
Zr		0.37	0.34	0.26	0.41	0.69	0.67	0.63	0.71
Y		2.2	2.3	2.4		2.8	2.9	3.2	
La		0.08	0.03	−0.10	0.12	0.30	0.26	0.17	0.34
Ce		0.14	0.09	−0.02	0.18	0.36	0.32	0.24	0.39

Table 4.3-3. (*Continued*)

Yandicoogina – high-Al (3.32–3.25 Ga)

Source	mafic Coonterunah source				fract Coonterunah source			
	f	f	f	Kd	f	f	f	Kd
Value	0.05 Kd	0.1 Kd	0.2 Kd	0 f	0.05 Kd	0.1 Kd	0.2 Kd	0 f
Nd	0.36	0.32	0.24	0.39	0.78	0.77	0.74	0.79
Sm	0.71	0.69	0.65	0.72	1.36	1.38	1.43	
Eu	1.1	1.1	1.1		2.1	2.2	2.3	
Gd	1.2	1.2	1.3		1.9	2.0	2.1	
Tb	1.8	1.8	1.9		2.6	2.7	2.9	
Dy	2.2	2.2	2.4		2.7	2.8	3.0	
Ho	2.5	2.6	2.8		2.9	3.0	3.2	
Yb	2.3	2.3	2.5		2.2	2.3	2.5	
Lu	2.8	2.9	3.1		2.8	2.9	3.1	
Ni	8.9	9.4	10.4		8.6	9.0	10.0	
V	9.3	9.7	10.8		8.6	9.0	10.0	

Also shown are f values (percentage of melting) assuming Kd = 0 for incompatible elements – these values only shown for moderately incompatible elements. Source rock and granite compositions listed in Table 4.3-2. Calculations using Apex basalt (the more LILE-poor of the 3 modelled sources) only given for the low-LILE Mt Edgar high-Al low-Th subgroup

Modelling results, using the parameters and methods outlined earlier, are listed in Table 4.3-5, for the ca. 3.5–3.42 Ga Mount Edgar low-Sr, the ca. 3.45–3.3 Ga Nandingarra, and ca. 3.3 Ga Triberton subgroups (Tables 4.3-4, 4.3-5), the most LILE-poor of the low-Al granites. The results of partial melting models (Table 4.3-5) are consistent with the postulated largely garnet-free residue and ~20 to <40% residual plagioclase (using partition coefficients from Klein et al., 1997, 2000; compilation in Rollinson, 1993), consistent with low-pressure partial melting experiments (<50 km; e.g., Wyllie et al., 1997). As shown for the high-Al granites, elevated modelled concentrations of the LILEs and Th again suggest a slightly more enriched source than those modelled (or some fractional crystallisation); e.g., Rb in the postulated source is not high enough (Table 4.3-5).

4.3-5.4. LILE, Th and ε_{Nd} Variability

As discussed by Bickle et al. (1993) for the Shaw Granitic Complex, the ca. 3.5–3.42 Ga granites include members with elevated LILEs (e.g., 2–3 wt% K_2O at 60–65 wt% SiO_2, compared with 1.76 wt% K_2O at 69.8 wt% SiO_2 for the average TTG of Martin, 1994) and K_2O enrichment trends. This K-enrichment is evident in all granitic complexes and in both the high-Al and low-Al subgroups, although there is no systematic correlation with SiO_2, or MgO. Given that even the most LILE-poor high-Al and low-Al granites require an enriched basaltic precursor, the LILE variability evident in the East Pilbara granites

Table 4.3-4. Geochemical averages for selected low-Al subgroups (ca. 3.5–3.42 Ga Mt Edgar low-Sr, ca. 3.45–3.3 Ga Nandingarra, ca. 3.32–3.25 Ga Triberton)

	Mt Edgar low-Al ca. 3.5–3.42 Ga			Nandingarra low-Al ca. 3.45–3.3 Ga			Triberton low-Al ca. 3.32–3.25 Ga		
	M	max	min	M	max	min	M	max	min
Majors									
SiO_2	71.86	73.92	69.23	68.74	72.6	65.2	69.22	71.48	67.16
TiO_2	0.26	0.38	0.17	0.48	0.61	0.31	0.39	0.54	0.32
Al_2O_3	14.35	15.2	13.27	14.74	15.55	13.8	14.52	15.27	13.79
FeOtot	2.47	3.47	1.71	3.91	4.86	2.48	3.26	4	2.66
MgO	0.58	0.98	0.37	1.35	1.87	0.85	1.1	1.35	0.78
MnO	0.05	0.07	0.02	0.06	0.08	0.03	0.05	0.06	0.04
CaO	2.93	3.79	2.19	2.99	3.95	2.08	2.83	3.19	2.22
Na_2O	4.37	4.98	3.91	4.2	4.56	3.81	4.21	4.36	3.87
K_2O	1.47	2.81	1.04	2.05	3.46	1.45	2.73	3.55	2.34
P_2O_5	0.1	0.13	0.05	0.09	0.13	0.07	0.09	0.11	0.07
LOItot	1.05	1.9	0.75	0.96	1.6	0.27	0.75	1.16	0.27
mg#	27	38	21	39	41	33	37	38	34
Traces									
Ba	272	401	116	409	737	291	518	737	404
Rb	100	147	46	90	141	67	97	124	86
Sr	188	242	129	243	289	177	203	280	171
Pb	14	21	11.5	11.6	20	7.6	12.5	16.4	10.5
Th	11	16	8	6.2	12.9	4.8	8.5	10.9	7.9
U	1	4.4	<1	1.1	1.9	<1	1.4	1.8	1.1
Nb	11	17.1	7	7	8.5	5.3	7.5	8	6.5
Zr	247	281	157	143	214	108	139.5	154	102
Y	21	43.5	7	16	23.2	12.6	18.2	21.2	13.8
Ga	18	20.5	17	19.2	20.5	16	18.7	19.6	17.3
La	44	64	23	26.1	57.2	18	40.3	45.8	30.3
Ce	70	108	44	47.5	101	36.4	74.3	80.8	53.5
Nd	42.7	55.7	31.8	20.4	43.3	14.5	26.9	29.3	20.3
Sm	7.4	9.4	6.2	3.8	6.2	2.6	4.4	4.8	3.2
Eu	1	1.2	0.8	1	1.5	0.7	0.9	1.1	0.7
Gd	6.2	8.1	5.8	3.4	4.6	2.6	3.95	4.5	2.8
Tb	1.1	1.3	0.8	0.5	0.8	0.4	0.6	0.7	0.4
Dy	5.9	6.8	3.3	2.6	3.9	2	3.1	3.4	2.2
Ho	1.2	1.4	0.5	0.5	0.8	0.4	0.6	0.7	0.5
Er	3.6	4.3	1.4	1.5	2.2	1.2	1.75	2	1.3
Yb	3.2	4.6	1.1	1.5	2	1.1	1.65	1.9	1.4
Lu	0.4	0.7	0.2	0.2	0.3	0.2	0.25	0.3	0.2
Sc	4	7	2	6	8	3	6	6	4
Ni	10	16	2	20	31	13	19	21	14
V	17	29	6	47	66	26	40	47	26
Cr	12	21	5						
Cu	9	22	1	19	58	10	16	20	11

Minimums and maximums (top and bottom 10% trimmed) also shown for granites. Data sources for granites listed in Table 4.3-1. Abbreviations: M = median.

Table 4.3-5. Calculated bulk partition coefficients (Kd) with 5%, 10% and 20% partial melting, of Apex Basalt, for selected low-Al subgroups (ca. 3.5–3.42 Ga Mt Edgar low-Al, ca. 3.45–3.3 Ga Nandingarra, ca. 3.3 Ga Triberton)

Apex Basalt source												
Granite Mt Edgar low-Al					Nandingarra				Triberton			
f	f	f	Kd		f	f	f	Kd	f	f	f	Kd
Value 0.05	0.1	0.2	0		0.05	0.1	0.2	0	0.05	0.1	0.2	0f
Kd	Kd	Kd	f		Kd	Kd	Kd	f	Kd	Kd	Kd	f
Ba	0.15	0.10	−0.01	0.19	0.08	0.03	−0.09	0.13	0.05	0.00	−0.12	0.10
Rb	0.00	−0.06	−0.19	0.05	0.00	−0.05	−0.18	0.05	0.00	−0.06	−0.19	0.05
Sr	0.72	0.70	0.67		0.54	0.52	0.46		0.66	0.64	0.60	0.68
Pb	0.06	0.01	−0.12	0.11	0.08	0.03	−0.09	0.13	0.07	0.02	−0.10	0.12
Th	0.00	−0.05	−0.18	0.05	0.05	0.00	−0.13	0.10	0.02	−0.03	−0.16	0.07
U	0.11	0.06	−0.06	0.15	0.09	0.04	−0.08	0.14	0.06	0.01	−0.12	0.11
Nb	0.38	0.35	0.26	0.41	0.63	0.61	0.56		0.58	0.56	0.51	0.60
Zr	0.23	0.19	0.09	0.27	0.44	0.41	0.34	0.47	0.45	0.42	0.35	0.48
Y	1.0	1.0	1.0		1.3	1.3	1.4		1.2	1.2	1.2	
Ga	0.82	0.81	0.79		0.77	0.75	0.72		0.79	0.77	0.75	0.80
La	0.05	0.00	−0.13	0.10	0.12	0.07	−0.04	0.17	0.06	0.01	−0.11	0.11
Ce	0.12	0.07	−0.04	0.17	0.20	0.16	0.06	0.24	0.11	0.06	−0.05	0.16
Nd	0.17	0.12	0.01	0.21	0.41	0.38	0.30	0.44	0.30	0.26	0.16	0.33
Sm	0.32	0.28	0.19	0.35	0.67	0.65	0.61		0.57	0.55	0.49	0.60
Eu	0.95	0.94	0.94		0.95	0.94	0.94		1.06	1.06	1.07	
Gd	0.51	0.48	0.41		0.97	0.96	0.96		0.82	0.81	0.79	0.83
Tb	0.51	0.48	0.41		1.18	1.19	1.21		0.97	0.97	0.97	0.98
Dy	0.61	0.59	0.53		1.45	1.47	1.53		1.20	1.22	1.24	
Ho	0.71	0.70	0.66		1.79	1.83	1.94		1.48	1.51	1.57	
Er	0.66	0.64	0.59		1.65	1.69	1.78		1.41	1.43	1.49	
Yb	0.71	0.69	0.66		1.58	1.61	1.68		1.43	1.45	1.51	
Lu	0.95	0.94	0.94		1.95	2.00	2.13		1.55	1.58	1.65	
Sc	9.9	10.4	11.6		6.6	6.9	7.7		6.6	6.9	7.7	
Ni	10.9	11.4	12.8		5.4	5.7	6.3		5.7	6.0	6.6	
V	19.8	20.8	23.3		7.1	7.5	8.3		8.4	8.8	9.8	

Also shown are f values (percentage of melting) assuming Kd = 0 for incompatible elements – these values are only shown for moderately incompatible elements. Source rock and granite compositions listed in Table 4.3-4.

requires, if produced by partial melting alone, a source that was more-enriched than the slightly enriched basalt used for modelling. This conclusion is supported by generalised modelling considerations. For example, Th, which is not mobile during metamorphic or hydrothermal alteration processes (as shown, for example, by the relatively constant Ce/Th ratios in the East Pilbara granites), is clearly enriched in many of the ca. 3.5–3.24 Ga granites (Fig. 4.3-8), where it can reach concentrations as high as 15–20 ppm at ~65 wt%

SiO_2. If these rocks were the result of 10–20% partial melting, their source would require at least 1.5 to >3 ppm higher Th contents than are in the pre 3.42 Ga basalts from the East Pilbara (typically between 0.1 and 1 ppm; Smithies et al., this volume) and higher than in NMORB and EMORB in general.

Another possible explanation for the elevated LILEs and Th is via fractional crystallisation, with or without assimilation. However, there is no geochemical evidence for significant fractional crystallisation in the East Pilbara granites, particularly in the high-Al rocks, and in Archean high-Al TTG in general (e.g., Martin, 1994; Clemens et al., 2006; Moyen at al., this volume). More complex models involving magma mixing, assimilation or mingling are not only inconsistent with field evidence that shows that microgranular (or other) enclaves are very rare (e.g., Collins, 1993), but do not actually solve the problem, since there is still a requirement for LILE-enrichment.

Given that the LILE variability largely reflects source compositions, then apparent broad correlations between ε_{Nd} and the LILEs and Th should provide further constraints. The (limited) Nd isotope data show a less-radiogenic ε_{Nd} signature associated with higher K_2O and Th contents. This indicates that the LILE-enrichment represents a crustal component with significant time-integrated crustal residence time (needed to develop the ε_{Nd} signature) as high as 200 My. Similarly, Bickle et al. (1993) interpreted the isotope data for the Shaw Granitic Complex as a mixture of depleted mantle-derived material and older crust. There is, however, no reason that the radiogenic end-member itself could not have some crustal residence, with the amount of pre-history largely dependent on Sm/Nd ratios that vary from sub- to supra-chondritic (e.g., $(La/Yb)_{PM}$ ranges from 0.8 to 2.0; Smithies et al., 2005b) for the early Pilbara basalts. Together, the data indicate that additional and/or different components (i.e., some form of intermediate to felsic crustal input) is required to augment a purely basaltic protolith for the genesis of both the high- and low-Al East Pilbara granites.

Potential analogues for possible enriched sources can be found within the well understood Paleoarchean volcanic stratigraphy of the Warrawoona Group at the base of the Pilbara Supergroup (see Fig. 4.1-3: Glikson and Hickman, 1981; Green et al, 2000; Arndt et al., 2001; Smithies et al., 2005b, this volume; Van Kranendonk et al., 2006a, this volume). Although largely a mafic to ultramafic sequence dominated by tholeiitic basaltic rocks (basalts to basaltic andesites; Smithies et al., 2005b), the sequence also includes intermediate and more felsic rocks (Hickman, 1983). For example, Smithies et al. (this volume) document voluminous andesitic to dacitic (and rare rhyolitic) volcanic rocks in the ca. 3.515 Ga Coucal and ca. 3.47 Ga Duffer Formations of the Warrawoona Group. These rocks do not have TTG compositions and are more compatible with derivation by fractionation (± assimilation of local crust) from tholeiitic precursors (Smithies et al., this volume). They have low K_2O (<1.0 wt%), slightly to moderately fractionated REE (La/Yb from ~3 to 10), elevated HREE (Yb from ~2 to 5 ppm) and La/Nb and Th/Nb ratios similar to the associated tholeiitic basalts, despite significant variation in LREE and Th concentrations (up to four-fold increases; Smithies et al., this volume).

Trace-element modelling of East Pilbara granites was undertaken using a source protolith with composition similar to the andesitic-dacitic rocks of the Coucal Formation of

Table 4.3-6. Geochemical averages for selected ca. 3.32–3.25 Ga high-Al (Chiminadgi) and low-Al (Carbana) subgroups, and for intermediate to felsic rocks of the Coonterunah Subgroup (see Smithies et al., this volume)

	Coonterunah Felsic-2		Carbana low-Al ca. 3.32–3.25 Ga			Chiminadgi high-Al ca. 3.32–3.25 Ga		
	M	R	M	max	min	M	max	min
Majors								
SiO_2	53.44		71.40	77.23	67.19	71.03	72.14	69.53
TiO_2	1.42		0.34	0.55	0.08	0.29	0.37	0.17
Al_2O_3	15.08		13.66	14.70	12.36	14.70	15.28	14.22
FeOtot	12.09		2.86	4.35	1.11	1.98	2.54	1.28
MgO	3.85		0.63	1.22	0.11	0.51	0.73	0.30
MnO	0.24		0.04	0.06	0.02	0.04	0.05	0.03
CaO	6.25		2.00	2.82	0.56	2.36	2.86	1.97
Na_2O	3.10		3.75	4.06	3.15	4.90	5.53	4.52
K_2O	0.56		3.65	5.30	2.77	1.88	2.41	1.30
P_2O_5	0.33		0.06	0.12	0.01	0.10	0.12	0.05
LOITOT	3.64		0.78	1.92	0.25	1.70	2.67	1.13
Mg#	36		29	35	15	31	38	30
Traces								
Ba	76	234	485	774	146	407	551	267
Rb	11.5	21.2	130	219.4	99.5	63	89	41
Sr	157	180	140	220	37	393	453	297
Pb	5	6.75	18.6	28.5	14.3	15	21	11
Th	2.7	2.7	15.9	28.2	10	8	12	2.2
U	0.62	0.62	2.5	5.6	1.3	1	1.9	0.8
Nb	11.2	11.2	10	13.5	5.7	7	7	2.6
Zr	198	198	156.5	206	92	169	208	99
Y	33.1	33.1	25.3	35.8	15.8	7.0	11.4	5.0
Ga	19.6	19.6	17.2	19.5	15.6	19.0	21.4	18.0
La	19.5	19.5	46.4	67.9	23.4	33.0	42.5	9.3
Ce	42.7	42.7	88.0	118.0	43.9	45.0	69.0	17.9
Nd	25.2	25.2	30.6	42.3	16.4	8.2	26.4	7.4
Sm	6.3	6.3	5.3	7.3	3.4	1.5	4.5	1.4
Eu	1.9	1.9	0.9	1.3	0.3	0.5	0.9	0.4
Gd	6.1	6.1	4.5	6.9	2.6	1.3	3.9	1.2
Tb	1.0	1.0	0.7	1.1	0.4	0.2	0.6	0.1
Dy	5.9	5.9	3.8	5.7	2.1	0.9	2.9	0.8
Ho	1.3	1.3	0.8	1.1	0.5	0.2	0.5	0.1
Er	3.6	3.6	2.3	3.3	1.4	0.5	1.6	0.4
Yb	3.1	3.1	2.4	3.1	1.3	0.5	1.6	0.4
Lu	0.5	0.5	0.4	0.5	0.2	<0.1	0.2	<0.1
Sc	20	20	3	7	<2	3	3	2
Ni	45	45	13	23	9	5	10	2

Table 4.3-6. (*Continued*)

	Coonterunah Felsic-2		Carbana low-Al ca. 3.32–3.25 Ga			Chiminadgi high-Al ca. 3.32–3.25 Ga		
	M	R	M	max	min	M	max	min
V	174	174	20	42	4	22	40	14
Cr	38	38	9	9	9	11	16	<1

Minimums and maximums (top and bottom 10% trimmed) also shown for granites. Modified Coonterunah refers to mobile trace element compositions (Rb, Ba, Pb, Sr) recalculated by smoothing multi-element curves using assumed immobile Th and La. Data sources for granites listed in Table 4.3-1; for intermediate and felsic volcanics from Smithies et al. (2007). Category abbreviations: M = median, R = recalculated.

the Warrawoona Group (Tables 4.3-6, 4.3-7). This modelling shows that 10 to 20% partial melting of such an enriched source can, at least theoretically, produce magmas with compositions similar to the LILE-enriched members of the high-Al and low-Al granites, and even the more evolved compositions such as the Carbana and Mondana Monzogranites. The only discrepancies are perhaps higher than expected bulk partition coefficients for the LILEs (Table 4.3-7), especially for Sr in the low-Al Carbana subgroup. The results for the LILE, however, are consistent with the evidence for fractionation within the more potassic granites (as discussed earlier). It is notable that calculated bulk partition coefficients for Nb are larger (i.e., more compatible) for the high-Al than for the low-Al granites (Table 4.3-7), consistent with interpreted variations in depth of melting and experimental evidence for the presence of residual rutile at higher pressures (e.g., Xiong et al., 2005).

Although the trace element modelling shows that compositions such as the Coucal Formation-type intermediate rocks are possible source protoliths for the Pilbara granites, there are other potential models for explaining the LILE enrichment in the granites. Given the commonly assumed slab-melting origin for many Archean TTG (e.g., Martin, 1987, 1994; Drummond and Defant, 1990; Martin et al., 2005), perhaps the most viable alternative model would be one involving melting of subducted sediments along with the slab, as has been proposed for parts of some modern arcs (e.g., Elliot et al., 1997; Class et al., 2000; Pearce et al., 2005). Although these and other models, such as simple or complex assimilation (e.g., MASH; Hildreth and Moorbath, 1988), may in part be valid for the East Pilbara granites and could produce the LILE variability, they cannot adequately account for the coexistence of both the high-Al and low-Al subgroups (and corresponding chemical differences). This requires melting at different pressures (see earlier). Simple plots, like Sr vs MgO (Fig. 4.3-4) show that the East Pilbara granites almost exclusively fall into two distinct groups, with Sr levels ~2–3 times higher in the high-Al subgroup than the low-Al subgroup for the same MgO contents, and, more importantly, similar Mg#, Ni and Cr.

In summary, it would appear that the compositional control on the East Pilbara granites is largely a function of two, at least partly independent, processes. Firstly, the effects of

Table 4.3-7. Calculated bulk partition coefficients (Kd) with 5%, 10% and 20% partial melting, of average intermediate-felsic Coonterunah volcanics, for selected ca. 3.32–3.25 Ga high-Al (Chiminadgi) and low-Al (Carbana) subgroups

Intermediate-felsic Coonterunah source							
Granite	Carbana low-Al				Chiminadgi high-Al		
	f	f	f	Kd	f	f	f	Kd
Value	0.05	0.1	0.2	0	0.05	0.1	0.2	0
	Kd	Kd	Kd	f	Kd	Kd	Kd	f
Ba	0.46	0.42	0.35	0.48	0.55	0.53	0.47	0.57
Rb	0.12	0.07	−0.05	0.16	0.30	0.26	0.17	0.34
Sr	1.30	1.32	1.36		0.43	0.40	0.32	0.46
Pb	0.33	0.29	0.20	0.36	0.42	0.39	0.31	0.45
Th	0.13	0.08	−0.04	0.17	0.30	0.26	0.17	0.34
U	0.21	0.16	0.06	0.25	0.60	0.58	0.53	0.62
Nb	1.1	1.1	1.2		1.6	1.7	1.8	
Zr	1.3	1.3	1.3		1.2	1.2	1.2	
Y	1.3	1.3	1.4		4.9	5.1	5.7	
Ga	1.1	1.2	1.2		1.0	1.0	1.0	
La	0.39	0.35	0.27	0.42	0.57	0.54	0.49	0.59
Ce	0.46	0.43	0.36	0.49	0.95	0.94	0.94	0.95
Nd	0.81	0.80	0.78		3.18	3.30	3.58	
Sm	1.2	1.2	1.2		4.3	4.5	5.0	
Eu	2.1	2.2	2.3		3.9	4.0	4.4	
Gd	1.4	1.4	1.4		4.9	5.1	5.6	
Tb	1.4	1.4	1.5		5.1	5.3	5.8	
Dy	1.6	1.6	1.7		6.8	7.1	7.9	
Ho	1.6	1.7	1.7		6.6	6.9	7.7	
Yb	1.3	1.3	1.4		6.6	6.9	7.6	
Sc	7.0	7.3	8.1		7.0	7.3	8.1	
Ni	3.6	3.7	4.1		9.4	9.9	11.0	
V	9.1	9.6	10.6		8.3	8.7	9.6	
Cr	4.4	4.6	5.0		3.6	3.7	4.1	

Also shown are f values (percentage of melting) assuming Kd = 0 for incompatible elements – these values are only shown for moderately incompatible elements. Source rock and granite compositions listed in Table 4.3-6.

partial melting at different pressures primarily control the prevailing residual mineralogy (particularly garnet and feldspar), exerting a strong control on the behaviour of the HREE, Al_2O_3, Sr, and Y. This has been long recognised for TTGs in general. Secondly, and less well documented previously and largely hidden in discussions of TTG petrogenesis, is that typically high LILEs, Th and U require either a source more enriched than typical Archean basalts, or an evolved source component in addition to such basalts.

4.3-5.5. Towards a Petrogenetic Model

The simplest model to explain both the extrusive (e.g., Smithies et al., this volume) and intrusive magmatic record of the East Pilbara Terrane is that of a variably thick(?), initially mafic, crust, similar to oceanic plateaux (e.g., Arndt et al., 2001; Van Kranendonk and Pirajno, 2004; Smithies et al., 2005b). In this setting, the mafic magmas may undergo fractional crystallisation ($\pm$ assimilation), as seen in the volcanic record of the terrane (Smithies et al., this volume), but there is also subsequent partial melting at a range of crustal levels to produce high-Al and low-Al granites (Fig. 4.3-10). Likely spots for TTG generation in such a scenario would be where mantle-derived magmas are ponding, especially underplating at the base of the crust (e.g., Gromet and Silver, 1987; Atherton and Petford, 1993), though perhaps also in hot zones within the crust (Annen et al., 2006). There are granitic complexes within the East Pilbara that are dominated by either low-Al granites (e.g., Corunna Downs Granitic Complex and most of the Yule Granitic Complex; Fig. 4.1-7, Table 4.3-1) or by high-Al granites (e.g., Shaw Granitic Complex). So, for these, the type of TTG present (high- or low-Al) may simply reflect a range of crustal thickness, as has been proposed for sodic intrusive rocks in the Peninsular Ranges Batholith (e.g., Gromet and Silver, 1987). The presence of either high-Al or low-Al TTG type could then potentially be used as an indicator of paleo-crustal thickness, as speculated by Champion and Smithies (2000) for the western part of the Yule Granitic Complex. It is apparent, however, that granitic complexes such as the Mount Edgar Granitic Complex contain both spatially and temporally associated high- and low-Al TTGs, requiring partial melting at different depths within the same broad crustal column. These may correspond to crustal hot zones as advocated by Annen et al. (2006).

Seismic data (e.g., Drummond, 1988) show that the present-day crust in the Pilbara Craton is not only relatively thin ($\sim$30 km), but also dominantly felsic. Even if we assume that this crust may have once been thick enough to generate the high-Al granites at the pressures required, it is still probably too felsic to contain the required mafic-ultramafic, or probably eclogitic, residuum from the proposed partial melting events responsible for generation of the ca. 3.5–3.24 Ga EPT granites. Potential solutions to the lack of Archean mafic lower crust commonly invoke some form of recycling of these (eclogitic) rocks to the mantle, such as delamination (e.g., Smithies and Champion, 1999; Zegers and van Keken, 2001), although there is little evidence for this in the EPT. Perhaps a more realistic potential solution lies in the interpretation of what the seismic Moho actually represents (e.g., Griffin and O'Reilly, 1987); i.e., does it represent the boundary between felsic crust and the eclogitic residues required by the plateau model proposed here?

One important feature of the above plateau model is that it does not require a subduction environment to produce the TTGs. Although a slab melting origin for the East Pilbara high-Al TTGs can not be unequivocally ruled out, we suggest that the spatial and temporal association of the high-Al TTGs with low-Al TTGs, the preference of some granitic complexes for either high- or low-Al TTGs, the similar range in ε_{Nd} and LILEs in both types, and the lack of elevated Mg#, Cr and Ni in the high-Al TTGs that may indicate interaction with mantle wedge (e.g., Rapp et al., 1999), more strongly favours derivation through par-

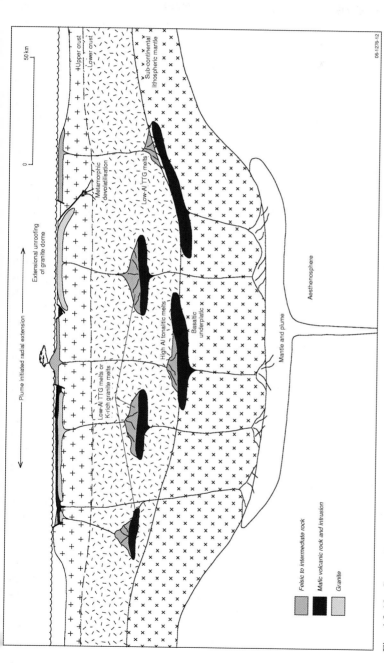

Fig. 4.3-10. Generalised petrogenetic model for the Paleoarchean granites of the East Pilbara Craton. Crustal melting is suggested to occur largely at the base of the crust, and within hot zones in the middle to lower crust, in response to uprising mantle melts related to mantle plumes (e.g., Annen et al., 2006), in an oceanic plateau-like environment (cf. Van Kranendonk et al., 2006). Melting of the base of crust produces either high-Al or low-Al granites, dependent on crustal thickness, whereas melting in crustal hot zones largely produces low-Al granites. In the model above, the Yule Granite Complex of the East Pilbara (see Fig. 4.1-7), with its common low-Al granites, would lie above thin crust, whereas the Mount Edgar Granite Complex (dominated by high-Al granites but with some low-Al granites) would lie above thicker crust. Note vertical scale = horizontal scale.

tial melting of thickened mafic crust, such as proposed for younger terranes (e.g., Gromet and Silver, 1987; Atherton and Petford, 1993). This interpretation is certainly consistent with both the plume-like geochemistry (e.g., Arndt et al., 2001; Smithies et al., 2005b) and the cyclic nature of the dominantly basaltic volcanism within the thick Warrawoona Group, which has been interpreted as reflecting a series of mantle plume events (Van Kranendonk et al., 2002; Hickman and Van Kranendonk, 2004; Smithies et al., 2005b). With such a multi-plume hypothesis, it can be speculated that the episodic granite magmatism in the East Pilbara would be related to the arrival of successive plumes, with the relative time scales of melting related to mechanisms of conductive or convective heat distribution (e.g., Campbell and Hill, 1988). Gromet and Silver (1987) have shown that derivation of high- and low-Al TTGs in this manner does not rule out an overall subduction environment, although there is no strong evidence or requirement for such an environment in the Paleoarchean development of the East Pilbara Terane (Van Kranendonk et al., 2007a). Similarly, it is obvious that oceanic plateaux, once formed, may eventually end up within subduction zones, or have subduction zones initiate around their margins, and generate adakitic (TTG-like) magmatism, such as now occurring for the Carnegie Ridge in Ecuador (e.g., Bourdon et al., 2003). Although this possibility can not be ruled out for the Pilbara at some stage in its history, it is noted that the adakitic magmatism in Ecuador is accompanied by a variety of associated magmatic products (e.g., magnesian andesites, Nb-enriched basalts; Bourdon et al., 2003) that are not observed in the EPT. Such rocks are present in the younger (Mesoarchean) West Pilbara Superterrane of the Pilbara Craton, where good evidence for subduction-type processes are evident (e.g., Smithies et al., 2005a, 2007a).

Importantly, Smithies et al. (2005b, this volume) noted a variety of secular changes in the volcanic rocks of the East Pilbara Terrane, including a general decrease in the amount of basalt directly associated with felsic magmatism, a trend to magmas with higher silica contents, and a change in the geochemistry of the intermediate to felsic volcanic rocks from compositions more consistent with fractionated tholeiites to compositions more akin to TTGs. They argued that these secular changes, coupled with the lack of evidence for a subduction-like enriched component in the mafic rocks, indicated that the basaltic magmas had increasing access to TTG-like crust with decreasing age (i.e., pointing to increasing crustal differentiation). As discussed by Champion and Smithies (2001), similar secular trends are evident in Archean granites which range from true TTGs, through LILE- and Th-enriched TTGs (called 'transitional' TTGs by those authors), to more potassic compositions, recording a change from basaltic through to tonalitic and then more felsic (more heterogeneous) protoliths.

Finally, as shown by Huston (this volume), the oceanic plateau model for the East Pilbara Terrane can also explain the majority of mineralisation observed therein, and in Paleoarchean terranes in general. Furthermore, the oceanic plateau model, with its inferred large vertical transfer of heat into and through the crust, may also help explain the map pattern of ovoid, domical granitic complexes surrounded by generally synformal greenstone packages (Fig. 4.1-7; see also Van Kranendonk et al., 2002, 2004, 2007a, this volume). This pattern, commonly, though somewhat controversially, ascribed to partial convective overturn (e.g., Collins et al., 1998), may largely be driven by episodic thermal effects related to mantle plumes and contemporaneous extension of upper to middle crust as a result of magmatic overthickening and inflation of the crust (see Fig. 4.1-10 in Van Kranendonk et al., this volume).

4.3-6. DISCUSSION

Both Smithies (2000) and Martin and Moyen (2002) investigated secular changes in Archean TTGs (dominantly high-Al) and documented clear chemical distinctions between early and late Archean TTGs. Specifically, the data show that late Archean TTGs have higher maximum Mg#, Ni, Cr, Sr and Eu/Eu* relative to early Archean TTGs. These authors, however, differed in their interpretation of these trends. Martin and Moyen (2002) interpreted the changes as gradual, reflecting increasing depth of melting, whereas Smithies (2000) suggested a significant change in chemistry around ~3.2 Ga, marking the onset of true slab melting (from either flat subduction and/or intracrustal melting of thickened crust). Regardless, both studies suggested that the elevated Mg#, Ni, and Cr in the late Archean TTGs reflects interaction of slab melt with mantle wedge, as shown experimentally by Rapp et al. (1999). Mg#, and Ni (and Cr) contents for the Pilbara high-Al granites are low (e.g., Mg# up to 50, but mostly <45, Ni < 45 ppm; Figs. 4.3-1, 4.3-3, 4.3-7, 4.3-8), consistent with values documented by Smithies (2000) and Martin and Moyen (2002). More importantly, the high-Al data overlap with values for the Pilbara low-Al groups, consistent with the interpretation of minimal mantle wedge input in the high-Al magmas.

4.3-6.1. Transitional TTGs

Champion and Smithies (2001) introduced the term 'transitional' TTGs for granites, such as the ca. 3.5–3.42 Ga East Pilbara granites, which have TTG-like compositions coupled with elevated LILE, Th and U contents. Champion and Smithies (2001) further pointed out that such granites are a relatively common feature of most Archean terranes and, where present, are either temporally associated with, or postdate, true TTGs. As shown here, this LILE-enrichment does not reflect fractional crystallisation and cannot be produced by melting of purely basaltic precursors, thus an additional component must be involved. Potential candidates for this enrichment are many and include: crustal contamination, especially MASH-type processes (Hildreth and Moorbath, 1988); subducted sediment input (e.g., Plank, 2005); restite-rich magmas (i.e., essentially remobilising pre-existing TTG crust; Champion and Sheraton, 1997); and enriched crustal sources (this work).

Perhaps some of the best examples of transitional TTGs are found in the voluminous and dominant Neoarchean granites of the Yilgarn Craton (Champion and Sheraton, 1997; Champion and Smithies, 2001). Champion and Sheraton (1997) discussed the origin of transitional TTGs (called High-Ca granite in the Yilgarn) in some detail and, based on zircon inheritance, Sm-Nd data and geochemical considerations, suggested that a significant component of older continental crust was involved in their genesis. This would also appear to be the case for the LILE-enriched high-Al and low-Al granites of the East Pilbara, and possibly Archean LILE-enriched TTGs in general. Regardless of the ultimate origin of such rocks (crustal or slab melts), their presence in Archean terranes can be used as an indicator of the presence of pre-existing felsic crust and to provide temporal constraints on the crustal evolution of a craton (e.g., Champion and Smithies, 2001).

Champion and Smithies (2001) pointed out, however, that many compilations of TTGs include such rocks (e.g., Kamber et al., 2002). Notably, even estimates for average Archean TTGs (such as Martin, 1994 and Condie, 1993), which are based on sodic end-members, show some elevated concentrations of incompatible elements. In particular, Th contents, and Th/La ratios are high for the transitional TTGs as shown by the Pilbara granites (Figs. 4.3-3, 4.3-8), and more generally, by values for average Archean TTGs (e.g., Martin, 1994 and Condie, 1993). Plank (2005) discussed at length the Th/La ratios in various rock types and showed not only that crustal values are significantly higher than mantle values, but that Th/La ratios in arc rocks appear to be strongly controlled by (i.e., inherited from) continental crust. Plank (2005) showed that, for modern arcs, this most probably occurs via the subducted sediment component, which, however, still requires some mechanism to produce pre-existing crust with elevated Th/La. The crustal melting model for TTGs, especially transitional TTGs, advocated here, provides one potential avenue for producing continental crust with these elevated average Th/La ratios.

4.3-6.2. Conclusions

Paleoarchean granites form a major component of the East Pilbara Terrane of the Pilbara Craton where they occur largely within circular to ovoid granite complexes surrounded by Archean greenstones. Granite formation was episodic, with ages of 3.5–3.42, ca. 3.32, and ca. 3.24 Ga. The majority of these granites are felsic (62–76% SiO_2), and sodic (Na_2O/K_2O from >1 to >4), and have low Mg#, Ni and Cr contents, as is typical of Early Archean TTGs. They vary from medium- to high-K, however, and have chemistry that varies from trondhjemitic to calc-alkaline (e.g., Bickle et al., 1993). Two distinctive geochemical groups are evident in both the ca. 3.5–3.42, and ca. 3.32–3.24 Ga granites: a high-Al group with elevated Al_2O_3, Na_2O, Sr, LREE/HREE, and low FeO^*, HREE, Y, and $(Gd/Yb)_N$; and a low-Al group with low Al_2O_3, Na_2O, Sr, LREE/HREE, and high FeO^*, HREE, Y, and $(Gd/Yb)_N$. Sr contents, in particular, are diagnostic (Fig. 4.3-4). Within each group (and largely regardless of SiO_2 content), LILE and Th contents range from relatively LILE- and Th-poor, to LILE- and Th-enriched. This range is apparent in both the ca. 3.5–3.42 and ca. 3.32–3.24 Ga granites, although LILE-enriched compositions are more strongly developed in the younger group. ε_{Nd} is reasonably well correlated with LILE and Th compositions, with LILE-enriched granites having the least radiogenic signatures. This correlation is evident for all except the 3.32–3.24 Ga low-Al group, which is also the only group to have undergone significant fractional crystallisation.

Geochemical differences between the high- and low-Al groups correspond to those identified for sodic intrusive rocks in general (e.g., Barker and Arth, 1976), where they have been convincingly demonstrated to reflect the relative roles of residual garnet, amphibole and plagioclase at differing pressures of partial melting of basaltic precursors (e.g., Barker and Arth, 1976; Martin, 1986, 1994; Rapp et al., 1991). Trace element modelling using average compositions of local, marginally older, Pilbara basalts demonstrate that the LILE-poor end-members of both the high-Al and low-Al Pilbara granites can also be derived in such a manner. LILE-rich end-members, however, require a more enriched source component, which is consistent with the Nd isotope data. Trace-element modelling suggests compositions similar to exposed intermediate volcanic rocks – themselves differentiates from basaltic precursors (Smithies et al., this volume) – are suitable source protoliths.

The spatial and temporal juxtaposition in the EPT of low-Al TTGs that are unlikely to be slab melts, with high-Al TTGs that are commonly proposed to reflect slab melting (e.g., Martin, 1986; Drummond and Defant, 1991), is notable. Although a slab melting origin for the high-Al TTGs can not be ruled out, we suggest that they were derived via partial melting of thickened mafic crust, such as proposed for younger terranes (e.g., Gromet and Silver, 1987; Atherton and Petford, 1992). Features more in favour with the crustal model include: the association of high-Al and low-Al TTGs; the preference of some granitic complexes for either high- or low-Al TTGs; the similar range in ε_{Nd} and LILEs in both types; the lack of elevated Mg#, Cr and Ni in the high-Al TTGs that may indicate interaction with mantle wedge (e.g., Rapp et al., 1999); the presence of demonstrably suitable source compositions in the associated volcanics; and the contemporaneous volcanic record in the EPT which favours successive plume environments and not arc magmatism (Hickman and Van Kranendonk, 2004; Smithies et al., 2005b, 2007a). As shown by Gromet and Silver (1987), deriving high- and low-Al TTGs in this manner does not constitute an argument against a subduction environment, although as argued above there is no strong evidence, at least in the Paleoarchean record of the East Pilbara, for such an environment.

Finally, we emphasise the importance of identifying LILE- and Th-enriched TTGs (transitional TTGs of Champion and Smithies, 2001). These rocks, which are often not recognised and are commonly grouped (in many Archean granite studies) with more primitive (LILE-poor) TTGs, are a good indicator of the presence of pre-existing felsic (non-basaltic) crust and can provide temporal constraints on the crustal evolution and maturity of Archean cratons.

ACKNOWLEDGEMENTS

This contribution is a product of the North Pilbara National Geoscience Mapping Agreement between the Geoscience Australia and the Geological Survey of Western Australia. It has benefited from discussions with many colleagues, including Leon Bagas, Richard Blewett, Carl Brauhart, Kevin Cassidy, Bill Collins, David Huston, Arthur Hickman, Jean-Francois Moyen, Shen-Su Sun (deceased) and Martin Van Kranendonk. David Huston and Anthony Budd are thanked for reviews at Geoscience Australia. John Clemens, Hervé Martin, and an anonymous reviewer are thanked for their concise and helpful reviews. Angie Jaentsch and Chris Fitzgerald are thanked for the timely drafting of figures. This manuscript is published with permission of the Chief Executive Officer of Geoscience Australia and the Director, Geological Survey of Western Australia.

Earth's Oldest Rocks
Edited by Martin J. Van Kranendonk, R. Hugh Smithies and Vickie C. Bennett
Developments in Precambrian Geology, Vol. 15 (K.C. Condie, Series Editor)
© 2007 Published by Elsevier B.V.
DOI: 10.1016/S0166-2635(07)15044-1

Chapter 4.4

PALEOARCHEAN MINERAL DEPOSITS OF THE PILBARA CRATON: GENESIS, TECTONIC ENVIRONMENT AND COMPARISONS WITH YOUNGER DEPOSITS

DAVID L. HUSTON[a], PETER MORANT[b], FRANCO PIRAJNO[c], BRENDAN CUMMINS[d], DARCY BAKER[e] AND TERRENCE P. MERNAGH[a]

[a]*Geoscience Australia, GPO Box 378, Canberra, ACT 2601, Australia*
[b]*Sipa Resources Limited, PO Box 1163, West Perth, Western Australia 6872, Australia*
[c]*Geological Survey of Western Australia, 100 Plain St., East Perth, Western Australia 6004, Australia*
[d]*Moly Mines Pty Ltd, PO Box 8215, Subiaco East, Western Australia 6008, Australia*
[e]*Equity Engineering Ltd., 700-700 West Pender Street, Vancouver, British Columbia, Canada, V6C 1G8*

4.4-1. INTRODUCTION

The Pilbara Craton of Western Australia contains the oldest known ore deposits, *sensu stricto*; the North Pole (or Dresser) barite deposits. These syngenetic deposits, which are hosted by the ca. 3490 Ma Dresser Formation in the East Pilbara Terrane (Van Kranendonk et al., 2007a, this volume), were mined in the 1970s, producing a total of 0.129 Mt of barite. The fact that barite-bearing deposits are common before 3200 Ma, but uncommon from 3200 to 2400 Ma (Huston and Logan, 2004), is but one of several important differences in the metallogeny of the Paleoarchean relative to the Meso- and Neoarchaean. Other important differences include the presence of significant porphyry and epithermal deposits in the Paleoarchaean, important differences in the metallogeny of volcanic-hosted massive sulphide (VHMS) deposits, and the abundance and setting of lode-gold deposits.

The purpose of this contribution is to describe mineral deposits formed prior to 3200 Ma in the Pilbara Craton, assess the tectonic setting of these deposits, and then compare them with other Paleoarchean terrains, analogous deposits of Mesoarchaean to Neoarchaean age – including the richly mineralized Abitibi-Wawa Subprovince of the Superior Craton, Canada and the Eastern Goldfields Superterrane of the Yilgarn Craton, Western Australia – and Phanerozoic deposits. This comparison assists not only in determining what makes highly endowed mineral provinces, but provides constraints on the evolution of the atmosphere, hydrosphere and tectonics of the early Earth.

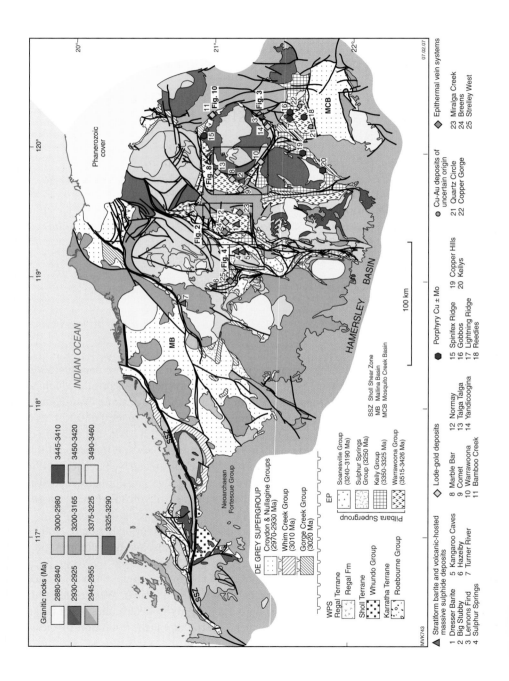

Fig. 4.4-1. Generalised geological map of the Pilbara Craton showing the location of mineral deposits formed before 3200 Ma. MB = Mallina Basin. MCB = Mosquito Creek Basin. SSZ = Sholl Shear Zone.

4.4-2. PALEOARCHEAN GEOLOGICAL EVOLUTION OF THE PILBARA CRATON

The Pilbara Craton (Fig. 4.4-1) has a geological history that spans over 800 million years, from ca. 3655 Ma, the age of a tonalitic gneiss in the Warrawagine Granitic Complex, to ca. 2830 Ma, the youngest age of granites of the Split Rock Supersuite (Van Kranendonk et al., 2006a, 2007a, this volume). Supracrustal rocks of the craton were deposited in two supergroups, from at least 3515 through to 3165 Ma (Pilbara Supergroup), and from 3020 Ma to 2930 Ma (De Grey Supergroup). The Pilbara Craton is unconformably overlain by the volcano-sedimentary Mount Bruce Supergroup, with an age of 2775–2450 Ma (Nelson et al., 1999), giving a total geologic history in the Pilbara region of over one billion years. For a detailed geologic history of the Pilbara Craton, see Van Kranendonk et al. (this volume).

4.4-3. PALEOARCHEAN MINERAL DEPOSITS OF THE PILBARA CRATON

Of all Paleoarchean terrains, the Pilbara Craton is perhaps the best mineralized. In addition to barite deposits, Paleoarchean rocks in the Pilbara Craton contain lode-gold, VHMS, porphrry Cu-Mo and epithermal deposits (Fig. 4.4-1). Of these, lode-gold deposits have been mined historically and, at this writing, feasibility studies are being undertaken on VHMS and porphyry Cu-Mo deposits. Although not economically important, BIFs are also present and provide important constraints on the composition of Paleoarchean seawater.

4.4-3.1. Volcanic-Hosted Massive Sulphide Deposits and Stratiform Barite Deposits

The Warrawoona Group is host to the oldest VHMS deposits. Although this group is dominated by basaltic rocks, significant andesitic to rhyolitic intervals are present, including the ca. 3471–3463 Ma Duffer Formation and ca. 3459–3428 Ma Panorama Formation (Thorpe et al., 1992a; Nelson et al., 1999; Smithies et al., this volume). The ca. 3490 Ma Dresser Formation, which contains minor felsic volcaniclastic rocks, hosts the VHMS-related North Pole barite deposits. The Duffer Formation contains the world's oldest significant base metal accumulations at the Big Stubby deposit and the Lennons Find district (Fig. 4.4-1).

4.4-3.1.1. The North Pole barite deposits
The North Pole Dome contains massive barite lenses hosted by the Dresser Formation. This unit comprises pillowed basalt interbedded with up to five stromatolitic, carbonate-

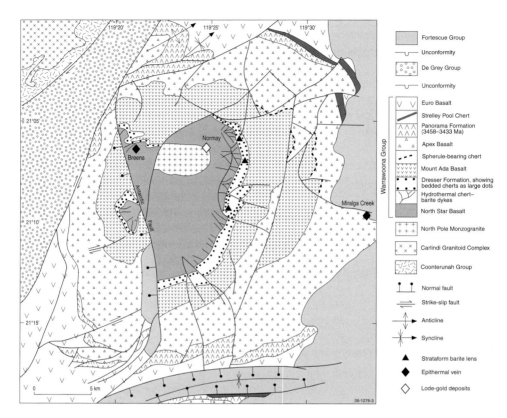

Fig. 4.4-2. Geology of the North Pole Dome, showing the locations of the Dresser barite deposits, the Normay lode-gold deposit, and the Miralga Creek epithermal Au-Ag-base metal prospect. Modified from Van Kranendonk et al. (2001).

and barite-bearing chert units (Van Kranendonk, 2006). The basal North Pole Chert hosts barite lenses over an 8 km strike length (Fig. 4.4-2), with the largest barite accumulation in the southern part of the dome, at the Dresser mine, which was active between 1970 and 1990, with total production of nearly 0.130 Mt of barite (Abeysinghe and Fetherston, 1997). Lead isotope analyses of galena indicate a model age of around 3490 Ma (Thorpe et al., 1992b).

As the Warrawoona Group in the North Pole Dome is one of the best preserved Paleoarchean supracrustal belts in the world, and as the Dresser Formation within it contains some of the earliest known stromatolites, its sedimentological and structural history is well described (e.g., Barley et al., 1979; Walter et al., 1980; Buick and Dunlop, 1990; Nijman et al., 1998; Ueno et al., 2001; Van Kranendonk et al., 2003; Van Kranendonk, 2006). The North Pole Chert and barite lenses formed in shallow water to emergent conditions (i.e., <200 m: Lambert et al., 1978; Van Kranendonk, 2006). Nijman et al. (1998) indicated that deposition of the North Pole Chert was controlled by syn-volcanic, listric growth faults, with the barite mounds forming in slightly deeper water adjacent to these faults.

Individual barite lenses are commonly more than 20 m thick, with strike lengths in excess of 50 m. The barite mounds contain 10–20 cm thick layers of coarsely crystalline barite. Barite also forms thin layers within the chert, locally with habits suggestive of replacement of gypsum, and in chert veins in the underlying basalts (Lambert et al., 1978; Buick and Dunlop, 1990; Abeysinghe and Fetherston, 1997; Nijman et al., 1998). Minor base metal sulfides (to 3.7% Zn) have been intersected in some barite lenses (Fergusson, 1999). These deposits differ from other VHMS deposits by the lack of significant base-metal mineralization and the extreme abundance of barite, characteristics that may result from low temperatures of hydrothermal deposition and are consistent with formation at shallow water depths (cf. Van Kranendonk and Pirajno, 2004; Van Kranendonk, 2006).

Van Kranendonk and Pirajno (2004) indicate that the barite lenses are associated with a zone of intensely altered rocks that extend up to 1.5 km stratigraphically below the baritic horizon. This contrasts with rocks above the baritic horizon, which are not altered. Within this alteration zone, massive kaolinite at the area of most massive barite mineralization passes downwards into a white-mica-rutile assemblage. This alteration zone is associated with a swarm of silica ± barite veins that are interpreted to have fed the syngenetic barite lenses (Nijman et al., 1998; Ueno et al., 2001; Van Kranendonk, 2006). Quartz veins with epithermal-like textures, consistent with the emplacement of the veins and the barite lenses under a low confining pressure, are also present (Van Kranendonk and Pirajno, 2004).

4.4-3.1.2. *The Big Stubby deposit*

The first direct indication of the antiquity of mineral deposits in the East Pilbara Terrane came from the small (0.1–0.2 Mt), but high grade (13.8% Zn, 4.5% Pb, 350 g/t Ag and 20% Ba; Reynolds et al., 1975) Big Stubby VHMS deposit, which is hosted by the Duffer Formation. Sangster and Brook (1977) reported a Pb-Pb model age of ca. 3500 Ma that was later recalculated at ca. 3472 Ma (Thorpe et al., 1992b), an age consistent with the independently determined age of the host rocks using the U-Pb on zircon method. Descriptions of the Big Stubby deposit have been presented by Reynolds et al. (1975) and, more recently, by Samieyani (1993), upon which the following summary is based.

The deposit consists of six stacked lenses hosted by felsic to intermediate volcanic rocks. The basal Big Stubby tuff-breccia contains felsic clasts in an intermediate matrix. This unit is intruded by rhyolite domes, which are overlain by the massive sulfide lenses, jaspilite, and intermediate volcaniclastic rocks. This sequence is intruded by dikes that disrupt both the host sequence and the massive sulfide lenses (Samieyani, 1993).

The ores are simple, consisting of low-Fe sphalerite (0.68–1.7 mole% FeS), pyrite, and galena, with minor chalcopyrite and trace tetrahedrite and acanthite. Gangue minerals include tourmaline, barite, quartz, sericite, chlorite and calcite. The associated alteration zone is characterized by a quartz-sericite-pyrite-carbonate assemblage (Reynolds et al., 1975; Marston, 1979; Samieyani, 1993). Samieyani (1993) reported fluid inclusion homogenization temperatures of 230–390 °C, with most data between 250–350 °C.

4.4-3.1.3. The Lennon's Find deposits

Of the VHMS districts of the northern Pilbara Craton, the Lennon's Find district is one of the more poorly described, despite containing the earliest (ca. 3470 Ma; Thorpe et al., 1992b), significant VHMS deposit in the world. Production in the district was 50.8 tonnes and a non-JORC-compliant 1.2 Mt resource grading 0.43% Cu, 7.76% Zn, 1.94% Pb, 100 g/t Ag and 0.3 g/t Au has been outlined for the Hammerhead zone (Ferguson, 1999), one of five mineralized zones that occur over a strike length of 5 km (Fig. 4.4-3).

VHMS deposits in the Lennon's Find district are located just below the contact between mafic schist of the Apex Basalt and underlying, mainly felsic, schist of the Duffer Forma-

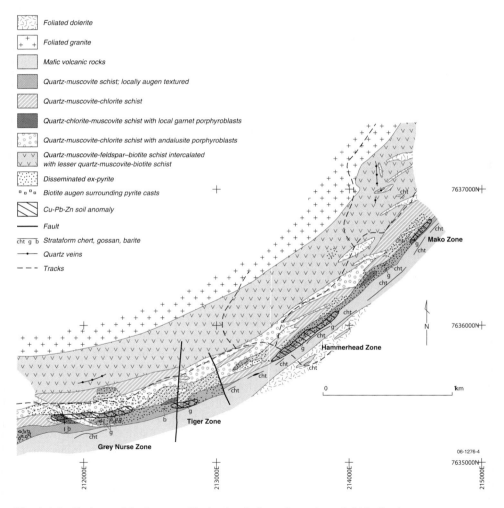

Fig. 4.4-3. Geology of the Lennons Find volcanic-hosted massive sulphide district.

tion (Fig. 4.4-3: Ingram, 1977). Thorpe et al. (1992a) presented a conventional zircon U-Pb date for the Duffer Formation along strike from the Lennon's Find district of 3465 ± 3 Ma, which is consistent with a galena Pb-Pb model age of ca. 3473 Ma (Thorpe et al., 1992b; see also Huston et al., 2002).

The Duffer Formation comprises three laterally persistent units in the Lennon's Find area (Fig. 4.4-3). The basal quartzo-feldspathic schist contains mainly fine-grained quartz-muscovite-biotite $\pm$ feldspar schist and lesser, knotty quartz-feldspar-muscovite-biotite schist with quartz microphenocrysts. This unit, interpreted as metamorphosed felsic volcanic rocks, has a true thickness of 250–300 m. These schists are overlain by a 80–150 m thick metasedimentary unit containing fine-grained quartz-muscovite-chlorite schist with up to 20%, 5–20 mm andalusite porphyroblasts (metapelite), and fine-grained quartz-muscovite-chlorite schist (metapsammite). Bedding is defined in pelites by andalusite-rich layers (Fig. 4.4-4(a)). The uppermost unit of the Duffer Formation consists of 0–75 m thick quartz-muscovite schist that is locally augen-textured near the top. The unit contains all significant gossans and massive sulfide lenses and thins in unmineralized zones (Fig. 4.4-3). It is interpreted as hydrothermally altered felsic volcanic rock. The Apex Basalt consists of strongly foliated mafic schist, with lesser talc $\pm$ carbonate schist and chert units near the base.

Foliated leucogranite intrudes quartzo-feldspathic rocks at the base of the Duffer Formation. Minor, apparently late, pegmatite dykes intrude the foliated granite. The granite and the volcano-sedimentary sequence are intruded by foliated 0.5–20 m dolerite sills and dykes. The last intrusive phase is narrow (0.5–1 m) felsite sills and dykes most likely associated with the nearby ca. 3324 Ma Wilina Pluton (Collins et al., 1998).

The rocks hosting the Lennon's Find district form a broadly arcuate belt that trends 075° in the west and 045° in the east (Fig. 4.4-3). These rocks dip moderately (40–70°) to the south. Although no map-scale folds are present in the area other than this arching, folds with amplitudes up to 2 m are present in outcrop.

Five mineralized zones have been defined at Lennon's Find based on the location of old workings and on surficial geochemistry (Fig. 4.4-3). These zones are localized at two stratigraphic horizons: an upper horizon within the quartz-muscovite schist, and a lower horizon near the upper contact of the underlying psammo-pelitic rocks. The upper horizon is more intensely mineralized and contains the Hammerhead and three other mineralized zones, two of which (Tiger and Bronze Whaler) have significant (>1% Zn) drill-hole intersections. The lower horizon contains the Grey Nurse zone, which also has significant drill-hole intersection. A third horizon, defined by geochemical anomalies and by weak pyritic gossan, occurs below the Grey Nurse position, near the contact between the psammo-pelitic unit and the underlying quartz-feldspathic schist (Fergusson, 1999; not shown in Fig. 4.4-3).

The Hammerhead zone consists of a thin (<5 m true thickness) lens of baritic semi-massive to massive sulfide (gossan at surface: Fig. 4.4-4(b)) that dips moderately to the SE. This lens has a moderate to steep rake to the east, which may be controlled by regional structural elements. Mineralogical studies by Marston (1979) indicated that this mineralized zone consists of sphalerite (1.46–3.56 mole% FeS), chalcopyrite and galena

Fig. 4.4-4. Photographs showing rock relationships at the Lennons Find volcanic-hosted massive sulphide district: (a) Bedding in pelitic rocks defined by andalusite-rich layers. (b) Surface gossan expression of Hammerhead zone massive sulphide-barite lens. (c) Iron-oxide (after pyrite) stringers in chlorite-sericite schist in the stratigraphic footwall to the Hammerhead zone.

in association with barite and pyrite. Three textural styles are present in the Hammerhead zone: (a) massive banded, which contains 25–80% sulfide with quartz, barite, chlorite, carbonate, tourmaline, muscovite and biotite gangue; (b) thinly-banded and disseminated,

Fig. 4.4-4. (*Continued.*)

which contains 5–20% sulfide, mainly pyrite, in 1–20 mm bands and as disseminated grains; and (c) disseminated, which contains 1–2% disseminated sulfide, again mainly pyrite (Sjerp, 1983). The massive-banded sulfide is located at the top of the mineralized sequence, overlying thinly-banded and disseminated sulfide. The thinly-banded zone described by Sjerp (1983) is probably a stringer zone (Fig. 4.4-4(c)). The Hammerhead zone has a Pb-Ag-rich top and a Zn-Cu-rich base.

Alteration facies in the Lennon's Find area are restricted to disseminated pyrite in the upper part of the Duffer Formation and quartz-sericite schist at the top of the Duffer Formation (Fig. 4.4-3). No hydrothermal alteration assemblages were noted in the Apex Basalt.

4.4.3.1.4. Deposits hosted by the Sulphur Springs Group

The Panorama (or Strelley) Zn-Cu-Pb deposits are located 120 km southeast of Port Hedland in the East Pilbara Terrane (Figs. 4.4-1 and 4.4-5). The history of mineral exploration at Panorama is discussed in Morant (1998) and Buick and Doepel (1999).

Base metal mineralization is hosted by the Sulphur Springs Group in the Soanesville, Pincunah and East Strelley greenstone belts of the East Pilbara Terrane (Fig. 4.4-5(a)): Van Kranendonk, 1998; Van Kranendonk and Morant, 1998). Volcanic rocks of the Sulphur Springs Group and the consanguineous Strelley Monzogranite have been dated at 3235–3238 Ma (Buick et al., 2002). Model Pb-Pb ages for galena from VHMS and vein prospects are mostly 3220–3260 Ma (data from Vearncombe (1995), Brauhart (1999), and Sipa Resources Ltd, using the Pb evolution model of Thorpe et al. (1992b)). Metamorphic grade and strain are mostly low, such that primary textures and structures are well preserved in the deposits (Vearncombe et al., 1995), and in volcanic and sedimentary rocks (Brauhart et al., 1998).

The Sulphur Springs Group includes three volcano-sedimentary formations and the Strelley Monzogranite (Fig. 4.4-5(a): Van Kranendonk et al., 2006a). The lowermost Lelira Formation comprises a ~3900 m succession of turbidites and chert, which are locally intruded by subvolcanic rhyolite. The Kunagunarrina Formation comprises up to 3000 m of mafic and komatiitic volcanics, with thin intercalated volcaniclastic deposits and chert. The uppermost Kangaroo Caves Formation comprises up to 1700 m of tholeiitic volcanics and subvolcanic sills ranging from basalt to rhyolite (Morant, 1998; Brauhart, 1999; Vearncombe and Kerrich, 1999). A regionally extensive marker chert at the top of this formation comprises silicified sedimentary rock with locally abundant black kerogenous chert veining. The marker chert, which defines a transition from volcaniclastic to epiclastic sedimentation, is about 5 m thick regionally, but is up to 100 m thick at Sulphur Springs. It includes unusual local rocks in the hangingwall of the Sulphur Springs (oistostrome and tholeiitic rhyodacite sills) and Kangaroo Caves (calc-alkaline rhyodacite) deposits.

The Strelley Monzogranite is a subvolcanic intrusion, which geochemical and geochronological studies have shown to be consanguineous and coeval with the tholeiitic volcanics of the Kangaroo Caves Formation (Brauhart, 1999; Vearncombe and Kerrich, 1999; Buick et al., 2002). It is a magnetite series hornblende-biotite bearing granite that is interpreted to have driven hydrothermal circulation in the Kangaroo Caves Formation that resulted in the formation of the VHMS deposits (Brauhart et al., 1998).

Three major (Sulphur Springs, Kangaroo Caves and Bernts) and seven small VHMS deposits and prospects are hosted by the Sulphur Springs Group in the Panorama district. Of these, eight are hosted by volcanic rocks of the Kangaroo Caves Formation, including the Bernts deposit, which is in a structurally detached block (Fig. 4.4-5(a)). Other styles of mineralization hosted by the Sulphur Springs Group include:

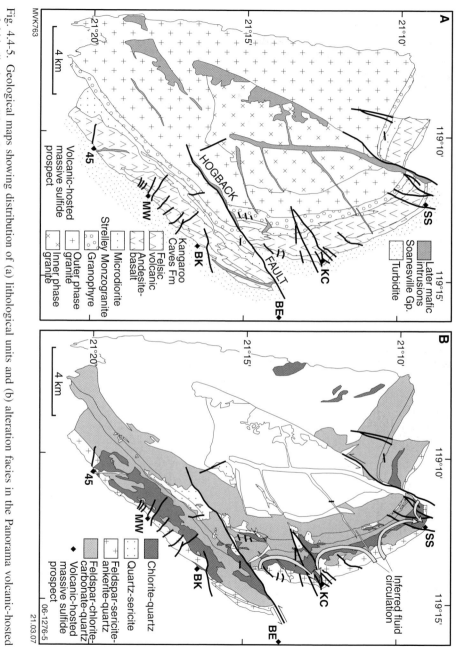

Fig. 4.4-5. Geological maps showing distribution of (a) lithological units and (b) alteration facies in the Panorama volcanic-hosted massive sulphide district (modified after Brauhart et al., 1999).

(1) magmatic Cu-Zn-Pb-Sn veins near the top of the Strelley Monzogranite (Wheal of Fortune prospect: Brauhart et al., 1998);

(2) molybdenite, fluorite and topaz associated with greisen zones in the Strelley Monzo-granite (Drieberg, 2004);

(3) Cu-Au mineralization in quartz stockworks hosted by spherulitic dacite of the Kanga-roo Caves Formation (Electra-Riviera prospects):

(4) Ni mineralization near the base of a peridotitic intrusion in the Kunagunarrina Forma-tion (Daltons prospect); and

(5) Au mineralization in the Leilira Formation (Obelix prospect) and in granite-hosted quartz veins.

The Panorama VHMS deposits and prospects are mainly of the Zn-Cu type (Large, 1992) and sulfide mineralization is typically zoned downwards from Zn ($\pm$Pb)-rich to Cu-rich. Bernts and Mad Hatters are of the Zn-Pb-Cu type, and are comparatively rich in barite, but poor in Cu. The deposits are low in gold (generally $\leq$0.2 g/t) and silver (generally <40 ppm) relative to other Australian VHMS deposits (Large, 1992).

At Sulphur Springs (10 Mt @ 1.4% Cu, 3.5% Zn and 17 g/t Ag: Fig. 4.4-6(a)) and Kangaroo Caves (1.7 Mt @ 0.6% Cu, 9.8% Zn, 0.6% Pb, 18 g/t Ag and 0.1 g/t Au: Fig. 4.4-6(b)), massive sulfide mineralization is hosted at the top of a laterally extensive, 200 m thick tholeiitic dacite sill intruded near the top of a sequence of pillowed and hyalo-clastic andesite-basalt (McPhie and Goto, 1996; Morant, 1998). The Bernts deposit (0.6 Mt @ 0.3% Cu, 7.8% Zn and 1.7% Pb) and the Breakers, Man O'War, Anomaly 45 and Jamesons prospects, are hosted by a comagmatic suite of tholeiitic rhyolite and rhyolitic volcaniclastic rocks. The Cardinal prospect is associated with felsic volcaniclastic rocks towards the base of the formation.

Zinc-Cu zones at Sulphur Springs are immediately beneath the marker chert, within the marker chert, and, rarely, in the hangingwall rhyodacite. Mineralization in the marker chert is generally Zn-rich, and the larger accumulation beneath the marker chert has a massive Zn-rich cap, a massive Cu-rich middle, and a stringer-style, Cu-bearing base. The Zn and Cu zones may be separated by up to 10 m of sub-economic pyrite mineralization (Morant, 1998).

Much of the high-grade Zn mineralization at Kangaroo Caves is hosted by the marker chert in a shallowly northeast-plunging shoot that has been intersected by drilling up to 1.5 km down plunge from the gossan outcrop (Fig. 4.4-6(b)). Copper-rich mineralization, although not well developed, is present in strongly altered volcanic rock beneath the marker chert.

The ore assemblage at the three main deposits includes pyrite, low-Fe sphalerite, chalcopyrite and galena. Secondary Cu sulfides, arsenopyrite, tennantite-tetrahedrite, pyrrhotite, rutile, bournonite, famatinite, bismuthinite and Se-bearing weibullite are present at Sulphur Springs (Outokumpu Research Oy, unpublished data, 1994; Vearncombe, 1995). Quartz, chlorite, sericite, ankerite and barite are the main gangue minerals at both deposits. The very low strain and metamorphic grade have preserved sulfide textures (Fig. 4.4-7), from which textural zoning has been recognized at the Sulphur Springs and Kanga-roo Caves deposits (Fig. 4.4-7; Vearncombe, 1995; Vearncombe et al., 1995). Dendritic

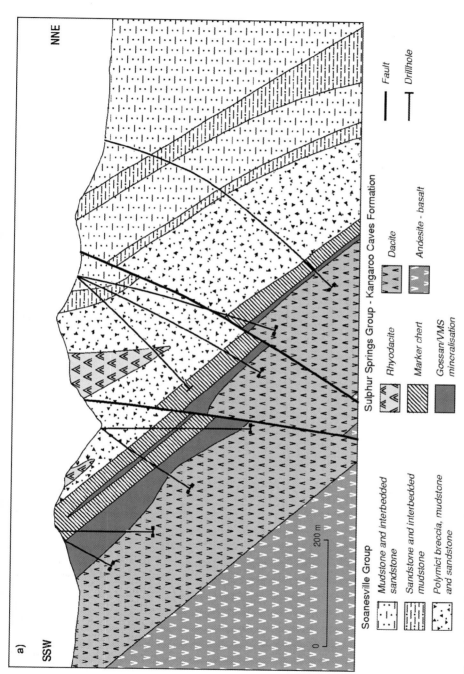

Fig. 4.4-6. Geological cross sections of the (a) Sulphur Springs and (b) Kangaroo Caves deposits.

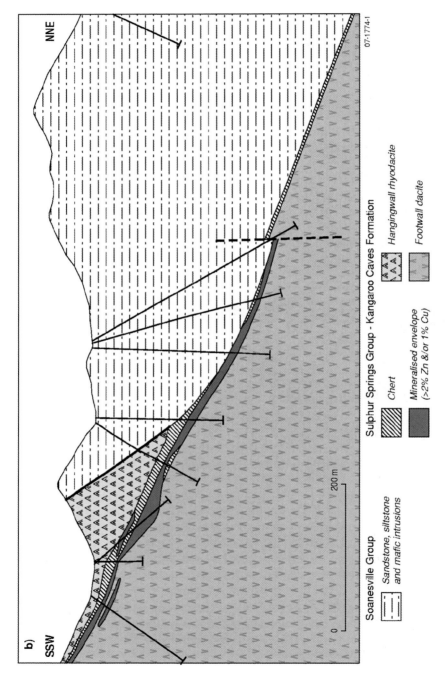

Fig. 4.4-6. (*Continued.*)

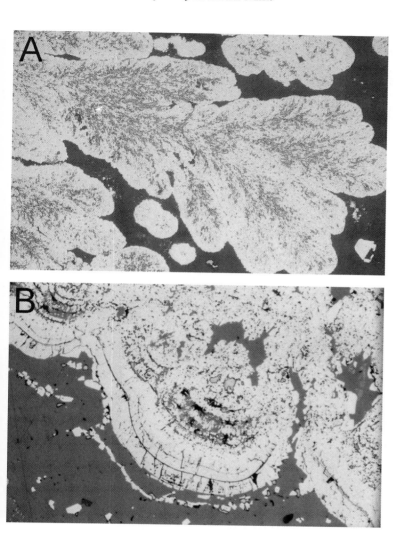

Fig. 4.4-7. Ore textures from volcanic-hosted massive sulphide deposits in the Panorama district. (a) Photomicrograph of dendritic pyrite from Sulphur Springs deposit (reflected light). (b) Photomicrograph of colloform pyrite from the Sulphur Springs deposit (reflected light). (c) Photomicrograph of spherical pyrite pellets from the Sulphur Springs deposit (reflected light). (d) Photography of bladed barite (recessively weathering) from marker chert south of the Kangaroo Caves deposit.

(Fig. 4.4-7(a)), colloform (Fig. 4.4-7(b)) and botryoidal textures that characterise the upper parts of the mineralization are interpreted to have formed by open space precipitation of sulfides (Morant, 1998). Vearncombe et al. (1995) interpreted some delicate sulfide textures, such as spherical pellets (Fig. 4.4-7(c)) and globular and stromatolitic types, as analogues of sulfide chimneys at present-day submarine hydrothermal vents. Barite is

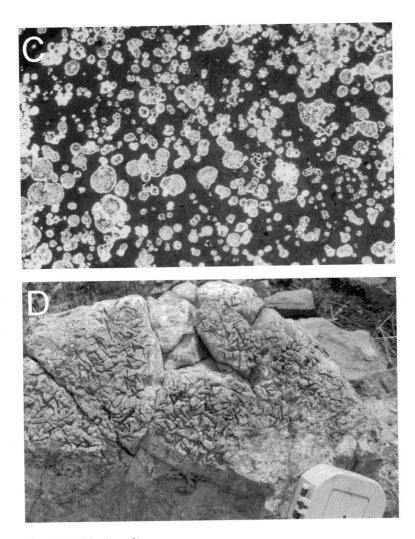

Fig. 4.4-7. *(Continued.)*

also preserved in the upper parts, as tabular blades, rosettes (Fig. 4.4-7(d)) and colloform masses. The lower parts of the massive to semi-massive sulfide lenses contain massive, granular, honeycomb and filigree sulfide textures, as well as veining (Vearncombe et al., 1995). Stringer zones comprise volcanic rocks that are strongly altered to chlorite, and sericite-rich, feldspar-destructive facies, with veins that contain pyrite, chalcopyrite and some sphalerite (Morant, 1998). Textural evidence for a replacement origin for the mineralization includes the preservation of volcanic textures such as perlite, pumice and shards in massive to semi-massive sulfides. Lithogeochemical evidence includes the preservation of immobile element ratios in texturally similar volcanic rocks (Morant, 1998). At the Bernts deposit, orange sphalerite progressively replaces coarse volcaniclastic units.

The close spatial relationship at all the main VHMS deposits between mineralization and synvolcanic faults indicates a primary structural control on the distribution of mineralization (Fig. 4.4-5: Morant, 1998; Vearncombe et al., 1998).

The VHMS deposits are associated with regional hydrothermal alteration assemblages that affect the upper Sulphur Springs Group, including the Strelley Monzogranite. Brauhart et al. (1998) identified a feldspar-destructive, semi-conformable, chloritic alteration zone located 1.5–2.0 km beneath the VHMS-bearing position. Beneath VHMS deposits, this alteration assemblage transgresses the stratigraphy, connecting VHMS deposits to the semi-conformable zone (Fig. 4.4-5(b)). Brauhart et al. (2001) and Huston et al. (2001a) showed that this semi-conformable feldspar-destructive zone lost most (>80%) ore re-lated elements, including Zn, Pb, Cu, Mo, and S, which is used to infer that this zone was the source of metals in the mineral deposits. Brauhart et al. (1998) interpreted semi-conformable alteration zones to represent seawater recharge, and transgressive alteration zones to represent discharge zones in a seawater-dominant hydrothermal system driven by the Strelley Monzogranite.

Footwall alteration at the Sulphur Springs deposit is localized within a broad mushroom-shaped pipe between synvolcanic faults (Young, 1997). Feldspar-destructive alteration comprises proximal chlorite-dominant alteration surrounded by strong sericite-ankerite alteration. Hangingwall alteration extends 450 m into the Soanesville Group sedimentary rocks above the Sulphur Springs deposit (Hill, 1997). The alteration comprises an inner semi-conformable silica-sericite-chlorite ± pyrite zone and an extensive outer carbonate zone that envelops a silica-carbonate zone.

In addition to the deposits hosted near the top of the Kangaroo Caves Formation, strati-form Zn-Cu and Zn-Pb prospects are present in chert and volcaniclastic rocks at the top of the Leilira Formation (Roadmaster prospect) and in underlying metamafic volcanic rocks of the Euro Basalt (Kelly Group).

4.4-3.1.5. Other possible volcanic-hosted massive sulphide prospects
In addition to deposits in greenstones surrounding the Mount Edgar Granitic Complex, a VHMS origin is possible for Cu-Zn prospects at Hazelby and in the McPhee Dome. The Hazelby deposit (Fig. 4.4-1) is hosted at the top of a silicified black shale lens in mafic to ultramafic schists (Marston, 1979) assigned to the ca. 3498 Ma Double Bar Formation of the Warrawoona Group (Blewett and Champion, 2005; Van Kranendonk et al., 2006a). If this interpretation is correct, the Hazelby prospect is the oldest mineral occurrence in the Pilbara Craton. Marston (1979) described the prospect at surface as a limonitic bed a few metres thick carrying minor malachite and azurite. At depth, mineralization is character-ized by up to 30% pyrite, with minor to accessory chalcopyrite, sphalerite, pyrrhotite and arsenopyrite.

A number of Cu-Au and Zn-Cu-Pb-Ag prospects hosted by the Duffer Formation in the McPhee Dome (Fig. 4.4-1) have been interpreted as VHMS deposits by the current explorers (Graynic Metal Ltd Prospectus, 2005). However, other origins are possible and these prospects are described later in this contribution.

In late 2005, DeGrey Mining Ltd discovered a series of VHMS prospects to the south of the Tabba Tabba Shear Zone in a steeply dipping, ~200 m-thick package of felsic and intermediate volcanic rocks that are variably altered to quartz-sericite schist. Drilling indicates that the prospects are Pb- and Au-rich (DeGrey Mining Ltd, 2006 Annual Report). Although the host rocks have not been directly dated, Pb-isotope model ages from galena in one of the prospects were ca. 3294 Ma (D. Huston, unpubl. data). If this age is correct, it indicates a fourth period of Paleoarchean VHMS mineralization in the Pilbara Craton.

4.4-3.2. Lode-Gold Deposits

As noted by Zegers et al. (2002), the Pilbara Craton contains the oldest known gold deposits with ages of constrained to between 3450 and 3300 Ma. This gold mineralizing event was the first of two that affected the Pilbara Craton (Huston et al., 2002). It produced global resources of about 22 tonnes, about a third of the Pilbara Craton total. The largest global resources are from the Marble Bar, Comet, Warrawoona, Bamboo Creek and North Pole districts, with minor production from the Talga Talga and Yandicoogina areas. With the exception of North Pole, these districts are related to ring faults in greenstone belts that wrap around the Mount Edgar Granitic Complex (Fig. 4.4-1: see Van Kranendonk et al., 2004).

4.4-3.2.1. The Marble Bar, Comet and Warrawoona districts
The Marble Bar greenstone belt along the southwestern margin of the Mount Edgar Granitic Complex (Fig. 4.4-8) produced in excess of 6.0 tonnes of gold, about 44% of the total for the older lode-gold deposits. This area contains a further 11.6 tonnes or resources, for a total global resource of 17.6 tonnes. This area, which contains the Marble Bar, Comet and Warrawoona districts, comprises mafic and felsic volcanic rocks of the Talga Talga and overlying Coongan Subgroups of the Warrawoona Group. The Duffer Formation, the basal unit of the Coongan Subgroup, has an age of ca. 3470 Ma (Van Kranendonk et al., 2002).

The Marble Bar district produced 1.30 tonnes of gold, with most production before 1950. Gold was produced from 0.15–2 m thick quartz reefs that dip shallowly (10–20°) to the west (Finucane, 1936: Fig. 4.4-9(a)). The veins are conformable with a north-south trending unit of carbonated chloritic schist within the Duffer Formation. Finucane (1936) indicates that in most, if not all, deposits, the auriferous veins are located several metres above, and are sub-parallel to, the contact between the chloritic schist and ca. 3470 Ma Homeward Bound Granite (Van Kranendonk et al., 2006b) of the Mt Edgar Granitic Complex. In detail, the veins are associated with dolerite sills, and gold is restricted to the margins of the veins (Finucane, 1936). Unoxidized ores contained pyrite, chalcopyrite and minor galena.

At the Halley's Comet deposit, located 9 km south-southwest of Marble Bar (Fig. 4.4-8), gold is hosted by talc-chlorite-carbonate schist, which Finucane (1937) interpreted as sheared and carbonated mafic lava. Unlike the deposits at Marble Bar, gold is hosted by

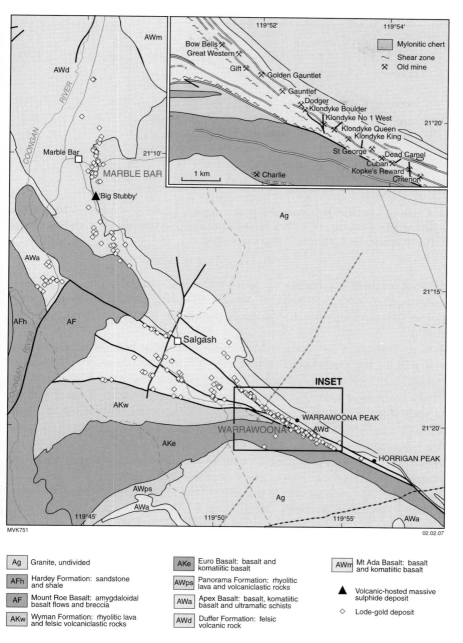

Fig. 4.4-8. Geology of the Marble Bar greenstone belt showing the locations of lode-gold deposits (modified after Fergusson and Ruddock, 2001). The inset shows the details of the Warrawoona district (modified after Jupiter Mines Ltd [www.jupitermines.com]).

a)

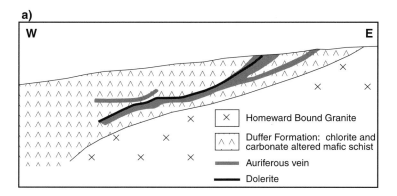

Fig. 4.4-9. Cross sections showing geological relationships at (a) the Jo-Jo deposit, Marble Bar District (modified after Finucane, 1936), and (b) the Klondyke Queen deposit, Warrawoona district (modified after Jupiter Mines Ltd, [www.jupitermines.com]).

pyritized chlorite-fuchsite-carbonate rock with small quartz-carbonate veins. The mineralized zone was 40 m long and dipped 70° to 165° (Finucane, 1937). The Comet group of deposits was the most significant producer in the Marble Bar district, producing 3.82 tonnes of gold.

Of the three districts in the southern Marble Bar greenstone belt, the most information is available for the Warrawoona district, which historically produced 0.94 tonnes (Fergusson and Ruddock, 2001) and has JORC-compliant resources of 9.10 tonnes Au (Jupiter Mines Ltd website [www.jupitermines.com]). The deposits are associated with the Klondyke Shear Zone, which cuts an attenuated synclinal structure between the Mt Edgar and Corunna Downs Granitic Complexes. The Klondyke Shear Zone is part of the fault system that rings the Mount Edgar Granitic Complex and is cut by ca. 3324 Ma Wilina Pluton (Huston et al., 2001b), providing a minimum age for shearing and mineralization. In the Warrawoona district, the Klondyke Shear Zone and associated (S_3) foliation are sub-vertical and strike ~100°, sub-parallel to the strike of the Mount Edgar Granitic Complex contact (J. Vearncombe, written comm., 1995).

Most deposits are localized 50 m north of, and parallel to, the Klondyke Shear Zone, in altered mafic and ultramafic volcanic rocks of the Apex Basalt (Fig. 4.4-8 inset: Fergusson and Ruddock, 2001). The largest deposits are located within sericitic schist to the south of a narrow (<2 m), sub-vertical mylonitized chert known locally as the "Kopcke's Leader" (Fig. 4.4-9(b)). Vearncombe (written comm., 1995) identified two alteration phases; an early carbonate-chlorite±talc±quartz assemblage, followed by an ore-stage sericite-pyrite-fuchsite assemblage. The gold is hosted by 10–30 mm wide, boudinaged veins with sericite-pyrite selvages. Other vein minerals include pyrite, chalcopyrite, and minor sphalerite and galena. Huston et al. (2001b) indicate that these veins show chocolate-tablet boudinage, with horizontal and vertical extension. Fluid inclusion data indicate the veins formed from low salinity (2–7 eq. wt.% NaCl), CO_2-rich fluids that homogenized be-

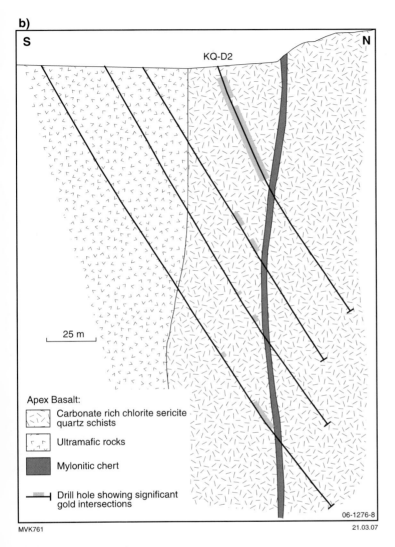

Fig. 4.4-9. (*Continued.*)

tween 160–380 °C (average; Huston et al., 2001b), characteristics typical of lode-gold ore-forming fluids (Goldfarb et al., 2005).

4.4-3.2.2. Bamboo Creek district

The Bamboo Creek district, which is located along the northeastern margin of the Mount Edgar Granitic Complex (Fig. 4.4-1), yielded 6.5 tonnes of gold (Fergusson and Ruddock, 2001). This district consists of over 30 occurrences over a strike length of 5 km that are

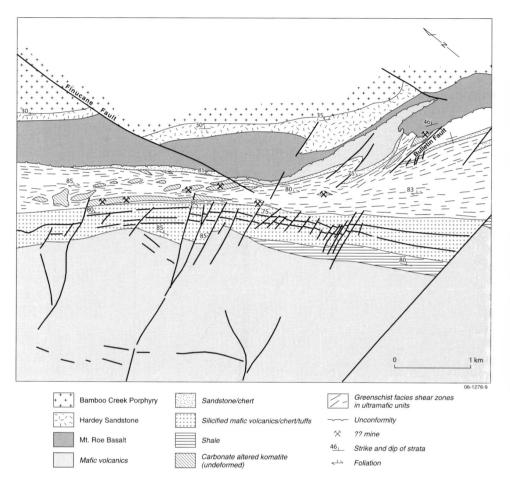

Fig. 4.4-10. Geology of the Bamboo Creek lode-gold district (modified after Zegers et al., 2002).

localized in large-scale boudins within the Bamboo Creek Shear Zone (Fig. 4.4-10: Fergusson and Ruddock, 2001; Zegers et al., 2002), a west-northwest striking shear zone that is part of the shear system that rings the Mount Edgar Granitic Complex. This shear zone is cut to the southeast by an unnamed, ca. 3310 Ma phase of the Mount Edgar Granitic Complex (Zegers et al., 2002; Van Kranendonk et al., 2002).

The Bamboo Creek Shear Zone is a 0.5–1 km wide zone of talc-chlorite mylonitic schist that has deformed ultramafic rocks, including komatiites from the ca. 3.35 Ga Euro Basalt. The gold lodes are generally hosted within less-deformed, 100 m-scale boudins where they are associated with extensional R ductile shears related to a sinistral, northeast-side up kinematic framework (Zegers et al., 2002).

The gold is associated with up to 1-m-thick laminated quartz-carbonate veins that are mostly restricted to boudins. Free gold is associated with tourmaline, pyrite, arsenopyrite, chalcopyrite, gersdorffite, galena and tetrahedrite in the veins (Zegers et al., 2002; Hickman, 1983). The gold lodes are associated with chlorite-quartz and fuchsite-carbonate alteration zones that pass outward to an outermost chlorite-carbonate zone (Zegers et al., 2002).

4.4-3.2.3. North Pole district
Some 1.16 tonnes of gold was produced in the North Pole district, located 45 km west-southwest of Marble Bar (Fergusson and Ruddock, 2001). Virtually all the gold was extracted from the Normay deposit, which is hosted by a 1.3 m wide, east-west trending quartz vein that dips 75° to the north and has a strike length over 1 km (Fig. 4.4-2: Hickman, 1983). This vein is hosted by sheared metabasalt of ca. 3470 Ma Mt Ada Basalt near the northern contact of the ca. 3460 Ma North Pole Monzogranite (Thorpe et al., 1992b). It is cut by a gabbro dyke interpreted to be related to the ca. 2700 Ma Fortescue Group. In addition to gold, the veins contain minor pyrite and Cu-Zn-Pb sulphide minerals.

4.4-3.2.4. Minor lode-gold districts
In addition to the above districts, we have assigned the Talga Talga and Yandicoogina groups of deposits to the older period of gold mineralization. These groups of deposits are associated with the ring fault around the Mount Edgar Granitic Complex, but have produced only 0.06 and 0.58 tonnes of hard-rock gold, respectively (Fergusson and Ruddock, 2001). Auriferous quartz veins in the Talga Talga group are hosted by quartz-carbonate-chlorite schist of the McPhee Formation. The Yandicoogina group differs from other groups in that it is hosted by shears in felsic volcanic rocks of the Duffer Formation and in sheared ca. 3470 Ma granite (Fergusson and Ruddock, 2001; Van Kranendonk et al., this volume).

4.4-3.2.5. The age of mineralization
Although absolute age data do not exist for the deposits described above, geological relationships constrain the age to between ca. 3470 Ma – the age of the oldest host rocks, and ca. 3320 Ma – the age of intrusions that transect the ring faults with which many of these deposits are associated. With the exception of the Bamboo Creek deposits, all districts described above are hosted by ca. 3470 Ma rocks of the Coongan Subgroup or similar aged granites. The Bamboo Creek deposits are localized within a shear zone that cuts undated mafic volcanic rocks that are correlated with the Euro Basalt. However, the only local age constraint is the ca. 3455 Ma age of the Panorama Formation, which stratigraphically underlies the mafic rocks. Based on these local geological relationships, the maximum age of the Bamboo Creek deposits are ca. 3455 Ma (see also Zegers et al., 2002). However, based on the age of an interbedded felsic tuff from the base of the group in the East Strelley greenstone belt, 100 km to the west-southwest, Van Kranendonk et al. (2006a, 2007a, this volume) suggest that the Euro Basalt has an age of ca. 3350 Ma. Some uncertainty must be attached to this interpretation due to the distance of correlation.

Independent age constraints are available for deposits in the Warrawoona, Bamboo Creek and North Pole districts using lead isotope data from ore-related galena and the Pb-evolution model of Thorpe et al. (1992b). Model ages calculated from least radiogenic analyses are 3406 Ma, 3431 Ma and 3385 Ma for these districts (Thorpe et al., 1992b; Huston et al., 2002), implying an overall age of around 3400 Ma (see also Zegers et al., 2002).

Alternatively, Van Kranendonk et al. (2007b) have suggested that gold mineralization took place at ca. 3315 Ma, during their initial period of partial convective crustal overturn and development of the ring faults that encircle the granitic complexes. However, Huston et al. (2002) found that, in most cases, the Pb model ages accurately reflected independently established ages of mineralization in the Pilbara Craton. Hence, we prefer an age of ca. 3400 Ma for lode-gold mineralization in the ring faults associated with the Mount Edgar Granitic Complex, although a younger age of ca. 3315 Ma is plausible.

4.4-3.3. *Deposits of the Porphyry-Epithermal Environment*

The East Pilbara Terrane contains several deposits that formed at high crustal levels in the porphyry-epithermal environment before 3200 Ma, the most important of which is the Spinifex Ridge (also known as Coppin Gap) porphyry Mo-Cu(Ag-W) deposit (B. Cummins in Van Kranendonk et al., 2006b). Other porphyry Mo-Cu(W) deposits include the Gobbos, McPhee Creek East, and Lightning Ridge deposits, which are located in the McPhee Dome (Barley, 1982: Fig. 4.4-1) and in the Copper Hills district in the Kelly greenstone belt (Marston, 1979).

In addition to these porphyry deposits, the East Pilbara Terrane contains a number of localities in which veins with epithermal-like textures are present, the most significant of which is the Miralga Creek Au deposit (Geollnicht et al., 1989; Groves, 1987). Of potential significance to the evolution of life, several of the stromatolite occurrences described in the East Pilbara Terrane are associated with epithermal-textured quartz veins (see below).

4.4-3.3.1. *The Spinifex Ridge Mo-Cu porphyry deposit*
The Spinifex Ridge Mo-Cu deposit, the oldest known porphyry-related mineral deposit, is associated with the Coppin Gap Granodiorite along the northern margin of the Mount Edgar Granitic Complex (Figs. 4.4-1, 4.4-11). This granodiorite has an age of 3314 ± 13 Ma (Williams and Collins, 1990) and intrudes a volcanic sequence that dips $70°$ to the north and comprises the ca. 3460 Ma Apex Basalt, felsic volcanic rocks of the Panorama Formation, and komatiitic and high-Mg basaltic rocks of the Euro Basalt (Fig. 4.4-11: de Laeter and Martyn, 1986; B. Cummins in Van Kranendonk et al., 2006b). Unaltered Coppin Gap Granodiorite consists of porphyritic microgranodiorite with phenocrysts of oligoclase, quartz and biotite in a microcrystalline groundmass of oligoclase, quartz, biotite and orthoclase (de Laeter and Martyn, 1986). It belongs to the high Al tonalite-trondhjemite-granodiorite (TTG) suite of Champion and Smithies (this volume). Contacts of the granodiorite are commonly brecciated and the ore-related porphyry is interpreted to have been emplaced along fault zones.

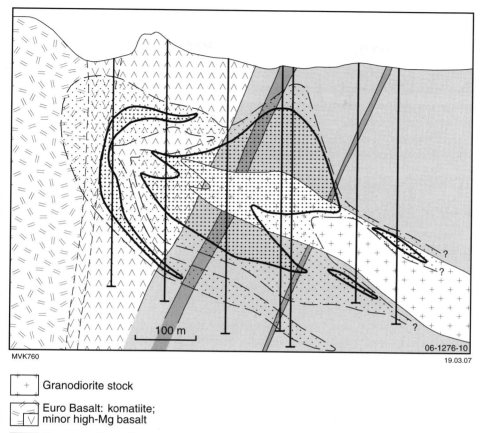

MVK760

06-1276-10

19.03.07

Granodiorite stock

Euro Basalt: komatiite;
minor high-Mg basalt

Panorama Formation:
felsic volcaniclastic rocks

Apex Basalt: basalt;
minor felsic volcanic rocks

——— 0.1% Mo

— – 0.04% Mo

Fig. 4.4-11. Cross section showing geological relationships at the Spinifex Ridge porphryry Mo-Cu deposit (modified after Molymines Ltd website [www1.molymines.com]).

The mineralized zones are developed about a sub-horizontal, 200 m-thick by 50–80 m-wide phallic stock at the apex of the Coppin Gap Granodiorite (Fig. 4.4-11). A total JORC-compliant resource of 500 Mt grading 0.06% Mo, 0.09% Cu and 1.7 g/t Ag has been established at Spinifex Ridge. Although all rocks in the host sequence are mineralized, the most intense mineralized zone is in the granodiorite, and high Mo grades are restricted in the host sequence to within 100 m of the granodiorite contact.

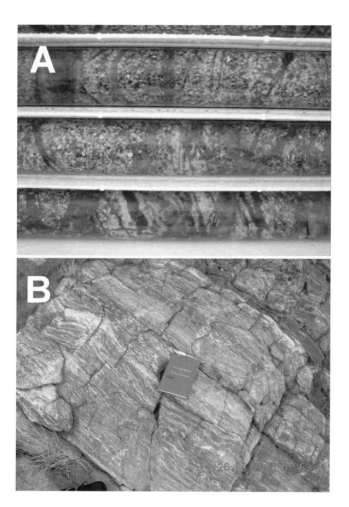

Fig. 4.4-12. Photographs of textures of >3200 Ma veins and related rocks from the Pilbara Craton: (a) Stockwork quartz-molybdenite veins cutting altered granodiorite, Spinifex Ridge deposit; (b) Siliceous laminates that resemble sinter deposits at the west Strelley stromatolite occurrence; (c) Crustiform banding in quartz veins associated with barite occurrences in the Dresser Formation.

Although the mineralized stock currently plunges to the southeast at 35° (B. Cummins in Van Kranendonk et al., 2006b: Fig. 4.4-11), at the time of mineralization, the stock was likely sub-vertical and the host rocks sub-horizontal. The mineralized zone comprises stockwork quartz and quartz-carbonate veins (Fig. 4.4-12(a)), with varying amounts of molybdenite, chalcopyrite, pyrrhotite, pyrite, and scheelite, and minor to trace sphalerite, galena, arsenopyrite, tetrahedrite, stibnite and native copper (de Laeter and Martyn, 1986; Jones, 1990; B. Cummins in Van Kranendonk et al., 2006b). Potassium feldspar is a com-

Fig. 4.4-12. (*Continued.*)

mon accessory phase of the veins. Although Cu and Mo are broadly correlated, the orebody is zoned outwards from the granodiorite stock as follows: Cu → Cu-Mo → W, with Ag associated with Cu. In the more distal parts of the mineralized zone, molybdenite is present in veins that are virtually devoid of other minerals (B. Cummins in Van Kranendonk et al., 2006b). Galena from a vein south of the Spinifex Ridge deposit (Richards et al., 1981) yielded a Pb isotope model age of ca. 3326 Ma using the Pb-evolution model of Thorpe et al. (1992b).

At Spinifex Ridge, the potassic alteration assemblage is characterized by the presence of potassium feldspar in molybdenite and chalcopyrite-bearing veins, the replacement of oligoclase in the granodiorite by potassium feldspar, and pervasive biotite development peripheral to the high-grade mineralized zone. This potassic assemblage is surrounded by and, in places, overprinted by a phyllic zone characterized by intense sericite alteration that has replaced all minerals except quartz (B. Cummins in Van Kranendonk et al., 2006b). The phyllic zone at Spinifex Ridge differs from that at younger deposits in that pyrrhotite, rather than pyrite, is the iron sulphide mineral accompanying the phyllic assemblage.

4.4-3.3.2. *Cu-Mo deposits in the McPhee Dome*
The McPhee Dome contains a number of prospects hosted by the Warrawoona Group, the most significant of which is the Gobbos Cu-Mo prospect (Bagas, 2005). Other small Cu-Mo prospects include Lightning Ridge, and a number of Cu-Au and Zn-Pb-Cu-Ag-Au prospects are present in the Quartz Circle area (Fig. 4.4-1).

The Cu-Mo prospects are associated with sodic biotite granodiorite stocks that intruded felsic volcanic rocks of the Duffer Formation and the Apex Basalt (Barley and Pickard, 1998). The stock at the Gobbos deposit was emplaced at 3313 ± 4 Ma and has a calc-alkaline character with fractionated rare earth element patterns (Barley and Pickard, 1998). These rocks belong to the low-Al TTG group of Champion and Smithies (this volume)

Chalcopyrite and molybdenite, along with variable scheelite and pyrite and rare galena are present in stockwork quartz-carbonate veins with variable amounts of potassium feldspar, biotite and chlorite (Marston, 1979; Barley, 1982). At the Gobbos deposit, the stockwork zone is associated with a small silicified and sericitized quartz-plagioclase porphyry stock.

4.4-3.3.3. *Other possible porphyry-related systems*
Other possible porphyry-related systems have been identified in the Kelly greenstone belt to the west of the McPhee Dome (Fig. 4.4-1). Like deposits in the McPhee Dome, these deposits are associated with 3321–3315 Ma quartz-feldspar porphyritic stocks that intrude, or are temporally associated with, the ca. 3320 Ma Wyman Formation (Barley and Pickard, 1999). Unlike the porphyry deposits at Spinifex Ridge and in the McPhee Dome, the deposits in the Kelly greenstone belt lack Mo and are generally not associated with quartz stockwork. At the Copper Hills deposit, minor to accessory pyrite is disseminated through the porphyritic host along with chalcopyrite and magnetite. Trace quantities of sphalerite and pyrrhotite are also present (Marston, 1979). In the vicinity of the Copper Hills mine and at the Kellys prospect 15 km to the southwest (Fig. 4.4-1), prospects were developed in quartz veins along shears that cut feldspar porphyry. Where described, primary mineralization comprises chalcopyrite and pyrite in quartz veinlets or disseminated through the host chlorite ± sericite altered quartz-feldspar porphyry (Marston, 1979).

4.4-3.3.4. *Cu-Au and Zn-Pb-Ag-Au deposits in the McPhee Dome*
About 15 km to the west-southwest of the Gobbos prospect (Fig. 4.4-1), several small Cu-Au and Zn-Pb-Cu-Ag-Au prospects are localized in the Quartz Circle area near a small tonalitic stock (Graynic Metals Pty Ltd prospectus, 2005). The prospects are zoned from Cu-Au to Zn-Pb-Ag-Au over a distance of 1.5 km laterally away from the stock. The most significant of these prospects is the Igloo (also known as Quartz Circle) deposit, for which a non-JORC-compliant resource of 0.127 Mt grading 4.1% Cu has been established. This prospect is largely a secondary zone of malachite, chalcocite and native copper, with primary chalcopyrite, galena and sphalerite at depth. The primary zone is characterized by chalcopyrite-quartz breccia and contains significant gold grades (Graynic Metals Ltd June 2006 quarterly report). A copper-gold vein has also been identified at the Millers prospect, 300 m to the south of Igloo, and disseminated to semi-massive sphalerite-galena-rich zones have been intersected at the Emporer (also known as Lead-Zinc) prospect 800 m to the southeast.

About 2 km to the north of the Igloo prospect, the Gold Show Hill Au prospect consists of quartz veins hosted mostly by fragmental felsic volcanic rocks that are altered to a quartz-sericite-pyrite ± carbonate assemblage. Within the veins, pyrite and arsenopyrite are the dominant ore minerals, with pale sphalerite, tetrahedrite, chalcopyrite, acanthite and galena present in varying amounts. Preliminary fluid inclusion data (T. Mernagh, unpubl. data) from this prospect indicate that the ore fluids were CO_2-rich and low salinity (0–6 eq. wt% NaCl), with homogenization temperatures for the CO_2-rich inclusions between 230-300 °C.

The Copper Gorge prospect, 15 km to the east-northeast of the Igloo prospect (Fig. 4.4-1), is localized within felsic volcanic rocks of the Duffer Formation. Pyrite, sphalerite and chalcopyrite are present as disseminated grains or in stockwork quartz veins. The surface expression is malachite-hematite impregnations with minor cuprite (Marston, 1979).

The origin of Cu-Au and Zn-Pb-Ag-Au prospects in the McPhee Dome is unclear, with contradictory evidence. The current explorers in the Quartz Hill area (Graynic Metals Ltd) favour a VHMS origin, although previous explorers (Pancontinental Mining Pty Ltd) preferred an epigenetic origin, with mineralization occurring 85 million years after deposition of the Duffer Formation, possibly in association with emplacement of the tonalitic stock (cf. Graynic Metals Prospectus, 2005). An epigenetic origin is supported by Pb isotope model ages of ca. 3350 Ma (Huston et al., 2002) for galena from both the Gold Show Hill (sample RC61) and Emperor (sample 110012) prospects, about 120 million years younger than the age of the Duffer Formation. As fluid inclusion data from the Gold Show Hill prospect are carbonic and unlike VHMS fluids, an epigenetic, possibly epithermal, origin is proposed for the deposits, although the data are not conclusive. The spatial association of these hydrothermal veins with granitic stocks, together with the metal zoning and fluid inclusion data, suggest that they are intrusion-related vein deposits (cf. Sillitoe and Thompson, 1998).

4.4-3.3.5. Miralga Creek epithermal gold deposit

A more definitive epithermal deposit is present in the North Pole Dome at the Miralga Creek prospect (Fig. 4.4-2). This prospect, which is associated with high level porphyritic felsic dykes and stocks that are geochemically similar to the Panorama Formation, is hosted by the Apex Basalt and characterized by cockade and comb structures, as well as crustiform banding. Goellneicht et al. (1989) indicate that the prospect is characterized by disseminated, stringer, vein and hydrothermal breccia mineralization with an overall Zn-Pb-Cu-As-Ag-Sb-Au $\pm$ Hg $\pm$ Bi $\pm$ Mo $\pm$ Tl $\pm$ Se $\pm$ Pd metal assemblage. These authors recognized two stages of mineralization; an early stage characterized by pyrite and chalcopyrite, followed by a later stage characterized by sphalerite, galena, tetrahedrite, silver minerals and gold. These stages are associated with an alteration assemblage of quartz-muscovite-pyrophyllite $\pm$ pyrite $\pm$ rutile $\pm$ phlogopite. Although high-temperature (>600 °C), high-salinity (>70 eq. wt% NaCl) fluids were present in the porphyry stock, mineralization formed from lower temperature (<250 °C) and salinity (<13 eq. wt% NaCl) fluids, leading Goellnicht et al. (1989) to present a model in which a weakly developed magmatic system was overprinted by epithermal mineralization.

4.4-3.3.6. Other possible porphyry-epithermal mineral deposits and alteration assemblages

Marshall (2000) identified the Breens Cu prospect in the North Pole Dome (Fig. 4.4-2) as having potential epithermal characteristics. This prospect, which is hosted by the carbonate altered basalt of the ca. 3470 Ma Mount Ada Basalt, comprises two moderately dipping breccia zones that are present near the base of a belt of bleached (kaolinitized), silicified and brecciated volcanic rocks. At depth, the breccia zones contain local zones of massive pyrite-chalcopyrite with chalcocite and other secondary minerals (Marston, 1979).

Van Kranendonk and Pirajno (2004) described pyrophyllite-quartz-sericite-leucoxene alteration assemblages in basalt underlying the ca. 3400 Ma Strelley Pool Chert in the North Pole Dome (Fig. 4.4-1). As the overlying basalts are unaltered, this assemblage, which is similar to advanced argillic assemblages developed in epithermal environments, formed during the hiatus associated with the unconformity at the base of the Strelley Pool Chert (cf. Buick et al., 1995). Approximately 40 km to the west, also along this unconformity, veins with epithermal textures and siliceous laminates that resemble sinter deposits (Fig. 4.4-12(b)) are present at the west Strelley stromatolite occurrence (Fig. 4.4-1). Crustiform veins are also developed in the centres of chert veins near the Dresser barite deposits (Fig. 4.4-12(c); Van Kranendonk and Pirajno, 2004).

4.4-3.4. Banded Iron Formations

In contrast with the Mount Bruce Supergroup of the Hamersley Basin (Nelson et al., 1999), one of the largest iron ore provinces in the world, the Pilbara Craton is relatively poor in iron ore. The only significant deposits (Shay Gap-Sunrise Hill, Nimingarra and Mount Goldsworthy: Podmore, 1990) in the Pilbara Craton are hosted by the ca. 3020 Ma Cleaverville Formation (Gorge Creek Group, De Grey Supergroup: Van Kranendonk et al., 2006a). Banded-iron formations (BIFs) in Paleoarchean units are restricted to the ca. 3515 Ma (Buick et al., 1995) Coucal Formation of the Coonterunah Subgroup and the Pyramid Hill Formation of the Soanesville Group (Van Kranendonk, 2000). In the Coucal Formation, several 2–10-m-thick units of cherty BIF are present at the top and bottom of the formation, which contains mostly mafic and lesser felsic volcanic rocks. In the Pyramid Hill Formation, BIF is interbedded with slate (Van Kranendonk, 2000) and overlies the Honeyeater Basalt. Although neither the Pyramid Hill Formation nor the Honeyeater Basalt have been dated, these units are constrained between ages of 3235 and 3020 Ma (Van Kranendonk et al., this volume), with the likely age around 3200 Ma. The close association of BIFs in the Coucal and Pyramid Hill Formations with volcanic rocks suggests that they are Algoma-type BIFs.

4.4-4. METALLOGENESIS IN OTHER PALEOARCHEAN TERRAINS

Although the Pilbara Craton is the most mineralized Paleoarchean terrain, other terrains, including the Barberton Greenstone Belt in southern Africa, and the Dharwar and Singhbhum–Orissa Cratons in India, also contain significant mineralization. Of these terrains, the best described is the Barberton Greenstone Belt (Ward, 1999). Although the main economic interest in this area is ca. 3084 Ma (de Ronde et al., 1991) lode-gold deposits that produced about 320 tonnes of gold (Ward, 1999), the Barberton Greenstone

Belt also contains sub-economic mineral deposits of Paleoarchean age, including strat-iform barite deposits that range in age from 3550 to 3260 Ma (Table 1 of Huston and Logan, 2004), VHMS deposit with an age ca. 3260 Ma (Bien Venue: Ward, 1999; Nelson et al., 1999), epithermal Hg $\pm$ Au mineralization of unknown ("late tectonic") age (Ward, 1999), and iron ore deposits associated with BIFs deposited at ca. 3450 and ca. 3260 Ma (Ward, 1999; Table 2 of Huston and Logan, 2004). In the Indian subcontinent, the Dhar-war Craton contains both stratiform barite and BIFs with an age of ca. 3300 Ma, and the Singhbum–Orissa Craton contains BIFs with an age of ca. 3210 Ma (Huston and Logan, 2004). Other Paleoarchean terrains, including the Isua supracrustal belt in Greenland, and the Amazonia Craton in South America, contain Paleoarchean BIFs, but no other signifi-cant mineralization, and the Aldan Shield in Russia contains stratiform barite (Huston and Logan, 2004).

4.4-5. A COMPARISON OF PALEOARCHEAN, NEOARCHAEAN AND PHANEROZOIC METALLOGENY

Paleoarchean terrains around the world are characterized by stratiform barite deposits and BIFs as unifying metallogenic elements, with some terrains also containing, in decreas-ing abundance, VHMS deposits, epithermal vein, porphyry Cu-Mo, and lode-gold deposits. This metallogenic assemblage is distinct, with important differences, yet surprising similar-ities, to other time slices through geologic history. To establish how metallogenic evolution is controlled by mantle dynamics (e.g., Pirajno, this volume), by the evolution of geologic processes such as tectonics, and by secular changes in the atmosphere and hydrosphere, the metallogeny of the Pilbara Craton and other Paleoarchean terrains is compared below with younger periods, particularly the highly endowed Neoarchaean and Phanerozoic.

4.4-5.1. *Volcanic-Hosted Massive Sulphide Deposits and Stratiform Barite Deposits*

Because of their abundance through nearly all geologic time, VHMS and strataform barite deposits provide a window into secular changes in metallogeny. As these deposits form by the interaction of hydrothermal fluids with ambient seawater at, or near, the seafloor, they record changes not only in hydrothermal processes, but also in seawater chemistry.

Volcanic-hosted massive sulphide deposits are generally zoned, grading from a Cu-rich base to a Zn-rich top. Although there are variations in the details, this general pattern, which is interpreted to be the consequence of temperature variations and "zone refining" (Eldridge et al., 1983), appears to have been consistent through geologic time. Moreover, as ore textures from undeformed Paleoarchean deposits are similar to those in "black smok-ers", the modern analogues of VHMS deposits (Vearncombe et al., 1995), depositional processes involving the quenching of hydrothermal fluids by seawater also appear to have been relatively uniform through time.

In detail, secular changes in VHMS metallogeny have implications to changes in tec-tonic style, changes in seawater composition, and the availability of seawater sulphate for incorporation into hydrothermal systems. Most Archaean and Proterozoic VHMS deposits are depleted in Pb relative to their Phanerozoic equivalents. This trend, which is well expressed by variations in the Zn ratio (100Zn/[Zn+Pb]: Fig. 4.4-13) has been interpreted

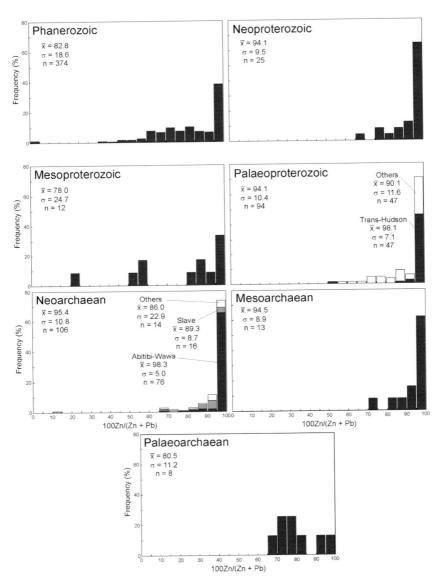

Fig. 4.4-13. Variations in the zinc ratio (100Zn/[Zn+Pb]) for volcanic-hosted massive sulphide deposits through geologic time.

to be controlled in part by the higher abundance of Pb-poor mafic volcanic rocks in Precambrian VHMS districts (cf. Huston and Large, 1987). However, Paleoarchean VHMS deposits have Zn ratios significantly lower than Neoarchaean deposits, particularly those from the Abitibi-Wawa Subprovince of the Superior Craton (Canada), and more similar to Phanerozoic deposits. The apparent Pb-poor character of Archean deposits is due to the dominance of VHMS deposits from the Abitibi-Wawa Subprovince during this period. Unlike the Pilbara Craton, the Abitibi-Wawa Subprovince lacked pre-existing sialic crust, and VHMS deposits formed over a short time period (2740–2698 Ma) within juvenile arc and back-arc systems (Corfu, 1993; Ayer et al., 2002). The Pilbara Craton and the Slave Province (Canada), which are characterized by Pb-rich VHMS deposits (Fig. 4.4-13), both had pre-existing sialic crust (Davis et al., 1996; Van Kranendonk et al., 2002; Champion and Smithies, this volume). Based on these observations and the association of major VHMS deposits with primitive crust in the Murchison domain of the Youanmi Terrane and Eastern Goldfield Superterrane in the Yilgarn Craton, Huston et al. (2005) suggested that more primitive volcanic terranes are better endowed in VHMS deposits than more evolved terrains, at least for the Archaean.

Another "atypical" characteristic of Paleoarchean VHMS deposits relative to Neoarchean examples is the presence of barite. All four Paleoarchean VHMS districts contain barite, and stratiform barite is present in fourteen other districts (Huston and Logan, 2004). In contrast, only one of 38 Neo- to Mesoarchaean VHMS districts contains barite, and only one other Neoarchean occurrence of stratiform sulphate minerals is known (Golding and Walter, 1979). In the Pilbara Craton, barite is present in all VHMS districts older than 3200 Ma, but lacking in VHMS deposits younger than 3.2 Ga. Barite is also uncommon in Proterozoic deposits, but is common once more in Phanerozoic deposits.

Huston and Logan (2004) argued that the presence of syngenetic barite prior to 3200 Ma indicated that sulphate was a significant component of the upper part of a layered Paleoarchean ocean, whereas the virtual lack of syngenetic or diagenetic barite during the Meso- and Neoarchaean indicated a relatively homogeneous, virtually sulphate-free ocean. Paleoarchean barite is characterized by negative mass independent ^{33}S fractionation, which implies that the sulphate formed by photochemical dissociation of SO_2 in the atmosphere (Farquhar et al., 2000; Mozjsis, this volume, and references therein). Although photochemical dissociation of SO_2 continued until at least 2450 Ma, either changes in tectonic style (see below), the flourishing of sulphate-reducing bacteria, or hydrothermal sulphate reduction prevented the establishment of a surface sulphate-rich layer and the formation of sulphate deposits between 3200 Ma and 2300 Ma, after which syngenetic and diagenetic sulphate became increasingly common (Huston and Logan, 2004).

Data from the Panorama district indicate that, unlike barite, sulphide minerals in the VHMS deposits are not characterized by mass independent ^{33}S fractionation (Golding and Young, 2005), suggesting that seawater sulphur was not incorporated into the ore-forming fluid. This interpretation is supported by a restricted range in δ^{34}S (-4 to $3\permil$) for Paleoarchean (and Neo- to Mesoarchaean) deposits, which is unlike the large range for Phanerozoic deposits (-20 to $+27\permil$) and most consistent with derivation either from direct magmatic emanations or leaching of volcanic rocks (Huston, 1999). Huston et al.

(2001a) demonstrated leaching of volcanic rocks for the source of sulphur in the Panorama district. In contrast, the covariance between VHMS sulphide $\delta^{34}S$ and seawater sulphate $\delta^{34}S$ (Sangster, 1968) during the Phanerozoic suggests incorporation of a large component of seawater sulphate into younger VHMS ore fluids (Solomon et al., 1989). This difference may relate to much lower overall sulphur abundances in Archaean oceans (2 ppm: Huston, 1999) relative to Phanerozoic oceans (modern value: 900 ppm [or 2700 ppm SO_4^{2-}]).

4.4-5.2. Lode-Gold Deposits

Although Neoarchaean provinces such as the Eastern Goldfields Superterrane of the Yilgarn Craton and the Abitibi-Wawa Subprovince of the Superior Craton are among the most richly gold-mineralized provinces in the world, Meso- and, in particular, Paleoarchean provinces are weakly mineralized. The ~3400 Ma deposits associated with the Mount Edgar Granitic Complex are the only Paleoarchean deposits of any consequence. Although many aspects of these deposits, including the style and textures of the ores, the alteration assemblages and the ore fluids are similar to many Neoarchaean and younger deposits, the structural and tectonic settings, both at the deposit and regional scale, appear to be different.

The ~3400 Ma lode-gold event was the first of two gold events in the Pilbara Craton. The second event, which at ~2900 Ma (Huston et al., 2002) accounts for two-thirds of the Pilbara's total gold resources, differs from the first event in being associated with linear shear zones in the Mallina and Mosquito Creek basins that can be traced over 100 km. Similarly, ca. 3084 Ma lode-gold deposits in the Barberton Greenstone Belt, and Neoarchaean deposits in the Eastern Goldfields Superterrane and the Abitibi-Wawa Subprovince are localized along second and third order structures associated with laterally extensive, broadly linear shear zones (Vearncombe et al., 1989; Goldfarb et al., 2005).

Lode gold deposits in these Neo- and Paleoarchean provinces and in Phanerozoic provinces tend to be late-orogenic and commonly associated with transpressional and/or compressional shears that may mark accreted terrane margins (Goldfarb et al., 2005 and references therein). In contrast, ~3400 Ma lode gold deposits in the Pilbara Craton are associated with faults that ring granite complexes and, at least at the Bamboo Creek district, were extensional. This difference may relate to a different tectonic regime prevailing before 3200 Ma (see below and Hickman, 2004 and Van Kranendonk et al., 2007a, this volume).

4.4-5.3. Porphyry Mo-Cu and Epithermal Au Deposits

The porphyry Mo-Cu deposits at Spinifex Ridge and in the McPhee Dome are most similar to the quartz monzonitic-granitic porphyry Mo-Cu subclass of Seedorff et al. (2005). As the composition of the stocks associated with the Pilbara deposits is more mafic than this subclass, a more direct analogy may be with the monzodiorite-granite porphyry Cu-(Mo) subclass, which is transitional into the quartz monzonitic-granite porphyry Mo-Cu subclass (Seedorff et al., 2005). Young deposits of both of these subclasses formed in continental arcs, in contrast to island arcs inferred for Cu-(Au) porphyry deposits (Barton, 1996; See-dorff et al., 2005). The association of Mo-rich porphyry Cu deposits with continental crust is consistent with the environment of the Pibara Craton at ~3310 Ma, where granites of this age yield Nd depleted mantle model ages in excess of 3500 Ma, indicating derivation from old continental crust (Smithies et al., 2003; Champion and Smithies, this volume).

The association of porphyry Cu-Mo deposits with both Al-rich (Spinifex Ridge) and Al-poor (Gobbos) granites of the TTG suite suggest that the progenitor magmas were derived from both shallow (<45 km; garnet unstable) and deeper (>50 km; garnet stable) melting (Champion and Smithies, this volume). Although the geochemical data of the TTG suites are compatible with derivation from slab and mantle wedge melting and with the arc setting inferred by Barley and Pickard (1999), the geochemistry of the associated volcanic rocks is inconsistent with derivation from subduction-related processes (Smithies et al., 2005b, this volume). The first definitive evidence of subduction, at ca. 3120 Ma in the West Pilbara Superterrane (Smithies et al., 2005a), post-dates formation of the Pilbara porphyry Cu-Mo deposits. Hence, these deposits must either be associated with "cryptic" subduction, or formed in a setting other than that of a convergent margin.

There are many porphyry systems that are not associated with convergent margins. The 10–18 Ma Gangdese porphyry Cu belt and the 33–52 Ma Yulong porphyry Cu, both in Tibet, are not linked to subduction zones (Hou et al., 2003, 2004). Instead, these deposits are related to post-collisional extensional tectonism and N-S trending rift zones. More-over, porphyry Mo deposits in the 300–280 Ma Oslo Rift (Neumann, 1994) are closely associated with a ring-type volcano-plutonic complex (Schonwandt and Petersen, 1983). Silicic plutonic ring complexes associated with mantle plume magmatism are known from the Kerguelen archipelago (Bonin et al., 2004). These form caldera structures, which show evidence of intense magma degassing and emplacement of ring complexes. At depth, these plume-related volcano-plutonic systems may well host porphyry-epithermal systems.

Vein systems with "epithermal" textures appear to have been present through much of the early history of the Pilbara Craton. The oldest, well described "epithermal" veins are associated with the ca. 3490 Ma Dresser barite deposits (Van Kranendonk and Pi-rajno, 2004), with possible older veins at the Breens Cu deposit. The most significant "epithermal" veins are at the Miralga Creek deposit and have an age of ca. 3450 Ma. Un-mineralized epithermal-like veins are also associated with the unconformity at the base of the ca. 3400 Ma Strelley Pool Chert. The apparent abundance of epithermal veins in these old rocks is in contrast with the perception that epithermal veins are largely a Phanerozoic phenomenon. Rather, processes that produce these veins appear to have been present at least at ca. 3490 Ma, suggesting that the relative lack of epithermal veins in old rocks is a consequence of preservation and not age.

The association of "epithermal" veins and syngenetic barite deposits with stromatolites in the Dresser Formation and the Strelley Pool chert suggests that low temperature, shallow water to locally emergent hydrothermal systems may have been crucibles for early life (Van Kranendonk, 2006b, this volume). As argued by Russell et al. (2005), hydrothermal environments may have provided the heat and nutrients necessary for the emergence of life.

4.4-5.4. Banded Iron Formations

Paleoarchean BIFs, both in the Pilbara Craton and elsewhere, are exclusively of the Algoma-type, which are small in size and associated with contemporaneous volcanic suites. In contrast, laterally extensive deposits of Superior-type (Hamersley-type in Australia) formed in a shelf environment without a volcanic association (Gross, 1965), and began to form at ca. 3000 Ma, and with major peaks in development at ca. 2500 and 1900 Ma (Isley and Abbott, 1999; Huston and Logan, 2004). Initial deposition of Superior-type BIFs appears to coincide with the first development of a continental platform/passive margin sequence at ca. 2970 Ma in the Kaapvaal Craton (De Wit et al., 1992; Skulski et al., 2004).

In contrast, Algoma-type BIFs are known from ca. 3810 to ca. 1800 Ma, at which time BIF deposition virtually ceased (Isley and Abbott, 1999; Huston and Logan, 2004). Algoma-type BIF are relatively thin, discontinuous, and typically exhibit facies changes from oxide to carbonate, silicate and sulphide. The origin of Algoma-type BIFs is probably volcanic-exhalative, and Isley and Abbott (1999) correlated major peaks in BIF deposition to periods of plume-related mafic magmatism, with the major peak at ca. 2700 Ma corresponding to a major period of Neoarchaean crustal growth.

Differences in the temporal distribution of Algoma- and Superior-type BIFs in the Archaean may point to tectonic controls on their distribution. The presence of Algoma-type BIFs through all of the Archaean is consistent with models tying crustal growth with mantle plumes during this period (Isley and Abbott, 1999; Smithies et al., 2005a). In contrast, the lack of Superior-type BIFs in the Paleoarchean is consistent with the lack of passive margins, which may point to a different style of tectonics during this period (see below).

4.4-6. IMPLICATIONS FOR TECTONIC PROCESSES DURING THE PALEOARCHEAN

The metallogeny of the Paleoarchean, which is best exemplified by the Pilbara Craton, has important differences with the metallogenically more prolific Neoarchaean, but surprising similarities to the Phanerozoic. Although some of these differences can be attributed to the composition of the hydrosphere, many are more likely related to differences in mechanisms of crustal growth and tectonic style.

The Paleoarchean is characterized by a suite of mineral deposits that are more typical of Phanerozoic, rather than Neoarchaean, terrains. For instance, porphyry Cu-Mo and epithermal deposits appear to be more common in the Paleoarchean than in the Neoarchaean or Proterozoic. Moreover, some mineral deposit types in the Paleoarchean differ in apparent structural and tectonic setting to younger analogues. For instance, despite similarities in fluid composition and depositional mechanisms, Paleoarchean lode-gold deposits in the Pilbara appear to have formed in a significantly different structural, and possibly different tectonic, environment to the orogenic environment characteristic of younger lode-gold systems. These observations may be consistent with differences in tectonic style between the

Paleoarchean and younger geologic periods inferred by previous workers (Hickman, 2004; Smithies et al., 2005a, 2005b; Van Kranendonk et al., 2007a).

Several papers have shown that there is little geological evidence for modern-style plate tectonics during early crustal growth in the Pilbara Craton from 3515 to 3240 Ma (Van Kranendonk et al., 2002, 2004, this volume; Van Kranendonk and Pirajno, 2004; Smithies et al., 2005b, this volume). Rather, these authors inferred that crustal growth occurred as a series of mantle plumes impinged on subcontinental lithospheric mantle, causing melting and producing a series of voluminous mafic basalts that formed an oceanic plateau. In contrast, Smithies et al. (2005a) demonstrated strong geochemical evidence for modern-style convergent plate tectonics in the formation of the ca. 3120 Whundo intra-oceanic arc in the western Pilbara Craton.

If this model is correct, differences in the mechanism of crustal growth of the Pilbara Craton and, possibly other Paleoarchean terrains, has significant implications for metallogeny (Fig. 4.4-14). Plume initiated crustal growth would favour equidimensional geological features, rather than belt-like geologic features (e.g., recent plume-related geological features, such as the Ontong Java plateau, do not tend to form linear belts). In addition, impingement of successive plumes on evolving sub-continental mantle lithosphere would result in successive periods of isotropic extension, leading to the potential development of extensional dome complexes (see also Van Kranendonk et al., this volume) and compartmentalized, volcanic-dominated, largely sub-aqueous basins. Importantly, extensional dome complexes may have been driven, in part, by density differences between granites and greenstones (e.g., Hickman, 1983; Collins et al., 1998; Sandiford et al., 2004).

The periodic development of volcanic-dominated, extensional basins upon an oceanic plateau as a consequence of plume impingement would create an ideal environment for the generation of VHMS deposits. However, as these basins are likely to be relatively small and compartmentalized, the potential for the development of giant deposits is possibly limited. The repeated nature of the inferred plume activity would produce repeated pulses of VHMS activity that correspond to pulses of plume-related magmatism.

Perhaps the most significant metallogenic difference is the environment in which lode-gold deposits develop. Lode-gold deposits through most of geologic time are associated with linear, through-going crustal-scale shear zones in orogenic belts associated with convergent plate margins (Goldfarb et al., 2005). In contrast, lode-gold deposits formed before 3200 Ma are associated with ring faults around granitic complexes and, in at least the case of the Bamboo Creek deposits (Zegers et al., 2002), formed in an extensional environment. These contrasts are consistent with the change in tectonic style advocated by Smithies et al. (2005a, 2005b) and Van Kranendonk et al. (2007a). As lode-gold deposits in both tectonic settings appear to share similar ore fluids and depositional mechanisms, it may be that lode-gold deposits formed in extensional environments at other times in Earth's history.

Another possible consequence of plume-related crustal growth prior to 3200 Ma is that thick, passive margin sequences did not develop. Deposition of the earliest known passive margin sequence, the ca. 2970 Ma Pongola Supergroup on the Kaapvaal Craton (Skulski et al., 2004), coincides with development of the first Superior-type BIF. As some models for the genesis of Superior-type BIFs infer iron deposition when deep, Fe^{2+}-rich oceanic wa-

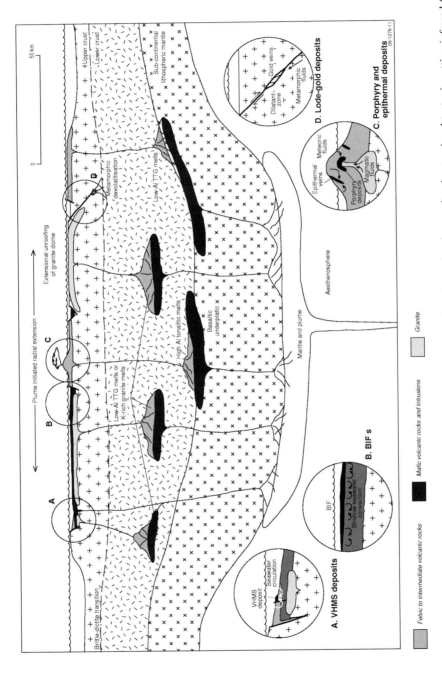

Fig. 4.4-14. Schematic diagram showing the relationship of mineral deposits to the plume-related extensional tectonic setting inferred for the Pilbara Craton prior to 3200 Ma.

ters upwell onto passive margins, the lack of significant passive margins prior to 3200 Ma would preclude these deposits. In contrast, as Algoma-type BIFs are generally inferred to have formed as a consequence of submarine volcanic activity, these deposits should be present through geologic history, including the Paleoarchean, until the final oxygenation of the Earth's hydrosphere at ~1800 Ma (Barley et al., 2005).

Modern oceanic plateaux currently crest at depths ranging from 1300 to 500 m, with some evidence of local emergence (e.g., Kerguelen and Ontong Java plateaux: Richardson et al., 2000; Taylor, 2006). This environment is conducive to the formation of shallow water to emergent mineral deposits, including epithermal veins, such as are observed in the Pilbara Craton, where there is evidence of shallow water deposition (Buick and Dunlop, 1990; Van Kranendonk, 2006). It may also be important to the understanding of early life as, at least in some cases, there appears to be a spatial association of early stromatolites with epithermal-like veins and barite deposits (Van Kranendonk, 2006). Moreover, if, as suggested by Huston and Logan (2004), the Paleoarchean seas had a surface layer rich in sulphate, syngenetic mineral deposits formed on the inferred plateau would be prone to contain barite.

Finally, periodic remelting of sub-continental lithospheric mantle during plume events would produce progressively more evolved melts, a trend noted by Smithies et al. (2005b), that would be progressively less prospective for komatiite-associated Ni-Cu deposits, but more prospective for large-ion lithophile element deposits (Sn, Ta, etc.). Although no significant komatiite-associated Ni-Cu deposits are known in the East Pilbara Terrane, major Ta and Sn pegmatite deposits are associated with Mesoarchaean granites in the Pilbara Craton (Sweetapple et al., 2002).

The most important deposits that do not apparently fit into a plume-related tectonic model are the porphyry Cu-Mo deposits. The closest Phanerozoic analogues to these deposits are quartz monzonitic-granitic porphyry Mo-Cu deposits in the Basin-and-Range Province of western North America. At the time that these deposits are interpreted to have formed, western North America was interpreted to have been a convergent margin, with a predominately compressional tectonic style (Barton, 1996). Candela and Picoli (2005) and Cooke et al. (2005) indicate that in young porphyry systems, volatiles released as subducting slabs descend into the mantle cause flux melting of the mantle wedge, producing H_2O-rich magmas that generate porphyry style deposits as they ascend to high crustal levels. This contrasts with the plume-related extensional tectonic style inferred for the Pilbara Craton during the Paleoarchean. However, H_2O-rich, sodic melts can also form by melting of an amphibole-bearing mafic source in thickened crust (Beard and Lofgren, 1991; Rushmer, 1991). This source is consistent with Nd isotope data suggestive of a crustal source and the enrichment of large-ion lithophile elements that characterise magmatic suits associated the Spinifex Ridge and other porphyry-related mineral deposits in the Paleoarchean Pilbara Craton (Champion and Smithies, this volume).

Data from Paleoarchean lode-gold and porphyry Cu-Mo deposits suggest that these deposits can form in tectonic environments other that those typically inferred in younger terrains. In particular, extensional environments may have potential for lode gold deposits, and plume-related crustal melting may produce porphyry deposits. Moreover, experience from the Pilbara Craton suggests that in regions that do not have significant erosion, epithermal deposits of both the high-sulphidation and low-sulphidation type, can be preserved, irrespective of age.

4.4-7. CONCLUSIONS

The Pilbara Craton, which is the best mineralized of all Paleoarchean terrains around the world, is characterized by a large variety of mineral deposits. The most common mineral deposits in the Pilbara Craton, and most other Paleoarchean terrains, are stratiform barite deposits and Algoma-type BIFs. Other Paleoarchean deposits in the Pilbara Craton include barite- and Cu-Zn-rich VHMS deposits, porphyry Cu-Mo deposits and epithermal vein systems, all of which are more similar to deposits formed in Phanerozoic rather than Meso- or Neoarchaean terrains. Although lode-gold deposits are also present, the extensional structural setting differentiates these from the syn- to late-orogenic setting of many younger deposits. These characteristics of Paleoarchean mineral deposits can be related to possible secular changes in tectonic processes and in the composition of the atmosphere and hydrosphere.

The presence of abundant syngenetic barite is likely to be a consequence of the availability of photolytic sulphate in the upper layer of the Paleoarchean ocean. The presence of epithermal vein deposits and, to a lesser extent porphyry Cu-Mo deposits, is the consequence of high levels of preservation in the Pilbara Craton. The association of lode-gold deposits with extensional ring faults around major granite complexes and the development of porphyry Cu-Mo deposits in a nonconvergent environment may be related to plume-related crustal growth prior to about 3200 Ma.

ACKNOWLEDGEMENTS

This contribution is a product of the North Pilbara National Geoscience Mapping Agreement between the Australian Geological Survey Organization (now Geoscience Australia) and the Geological Survey of Western Australia. It has benefited from discussions with many colleagues, including Leon Bagas, Richard Blewett, Carl Brauhart, David Champion, Arthur Hickman, Hugh Smithies, Shen-Su Sun and Martin Van Kranendonk. Leon Bagas, David Champion and Hugh Smithies are thanked for comments on earlier drafts. Angie Jaentsch and Chris Fergusson are thanked for the timely drafting of figures. This manuscript is published with permission of the Chief Executive Officer of Geoscience Australia and the Executive Director of the Geological Survey of Western Australia.

PART 5

THE PALEOARCHEAN KAAPVAAL CRATON, SOUTHERN AFRICA

Earth's Oldest Rocks
Edited by Martin J. Van Kranendonk, R. Hugh Smithies and Vickie C. Bennett
Developments in Precambrian Geology, Vol. 15 (K.C. Condie, Series Editor)
© 2007 Elsevier B.V. All rights reserved.
DOI: 10.1016/S0166-2635(07)15051-9

Chapter 5.1

AN OVERVIEW OF THE PRE-MESOARCHEAN ROCKS OF THE KAAPVAAL CRATON, SOUTH AFRICA

MARC POUJOL

Géosciences Rennes, UMR 6118, Université de Rennes 1, Avenue du Général Leclerc, 35 042 Rennes Cedex, France

5.1-1. INTRODUCTION

The Kaapvaal Craton of Southern Africa (Fig. 5.1-1) is one of the oldest and best-preserved Archean continental fragments on Earth. Its assembly during the Archean took place episodically over a 1000 million year period (3500 to 2500 Ma) and involved magmatic arc formation and accretion as well as tectonic amalgamation of numerous discrete terranes or blocks (de Wit et al., 1992; Lowe, 1994; Poujol and Robb, 1999). Delineation of these domains, and constraining the timing and duration of the principal magmatic events within them, is best achieved by the craton-wide application of accurate and precise U-Pb dating (Eglington and Armstrong, 2004; Poujol et al., 2003; Schmitz et al., 2004), together with appropriate tectonic and metamorphic studies. This paper summarizes some of the recent age data that identify the main tectono-magmatic events that have contributed to the assembly of the craton prior to 3200 Ma.

Based on geological, structural or aeromagnetic lineaments, faults and thrusts, the Kaapvaal Craton has been divided into three main blocks separated by major, craton-wide, lineaments (Fig. 5.1-1; Schmitz et al., 2004): the Witwatersrand Block in the east, the Kimberley Block in the west and the Pietersburg Block in the north. To the north, the Thabazimbi–Murchison lineament (TML) separates the Witwatersrand Block from the Pietersburg Block. To the west, the Colesberg lineament separates the Witwatersrand and Kimberley blocks. Eglington and Armstrong (2004), however, identified the Kimberley terrain, Witwatersrand terrain and Pietersburg terrain and describe two additional terrains, the Swaziland terrain to the south-east and an unnamed terrain to the north-west, separated from the Kimberley and Pietersburg terrains by the Palaba Shear Zone (PSZ). For the purposes of this paper, however, the Kaapvaal Craton has been subdivided (Fig. 5.1-2) into eastern, central, northern and western domains as in Poujol et al. (2003). These four domains are not intended to represent geological terranes, although each does have its own identifiable geological and chronological characteristics.

Much of the Kaapvaal Craton is covered by Neoarchean-to-Palaeoproterozoic volcano-sedimentary sequences and good exposures of basement exist in only a few areas. Of these, the Barberton Mountain Land is a region of superb three-dimensional exposure and repre-

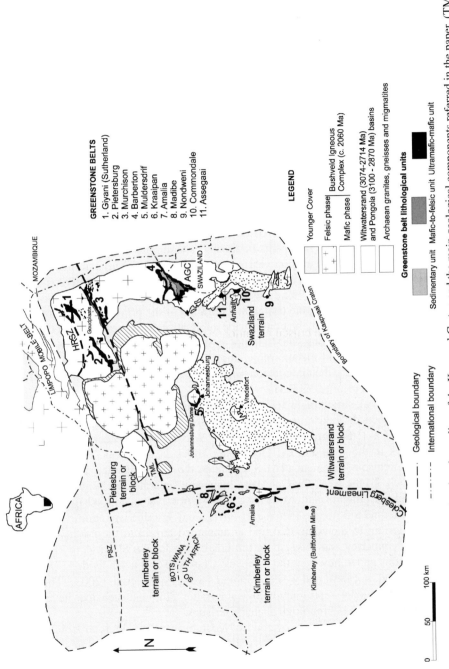

Fig. 5.1-1. Simplified diagram showing the outline of the Kaapvaal Craton and the main geological components referred in the paper. (TML): Thabazimbi–Murchison lineament, (HRSZ): Hout River Shear Zone and (PSZ): the Palaba shear zone.

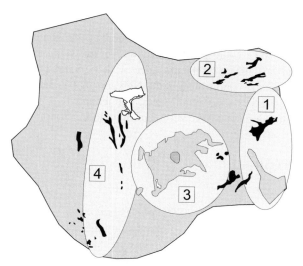

1. Eastern domain   3. Northern domain
2. Central domain   4. Western domain

Fig. 5.1-2. Simplified diagram showing the locations of the eastern, central, northern and western domains of the Kaapvaal Craton as used in this paper.

sents the type area for Archean crustal evolution in the craton. The majority of all studies and age dating of the crust has also been carried out in the Barberton region and, accordingly, crust-forming processes are best described from this portion of the craton. As emphasized by Schmitz et al. (2004), much of the geochronological data to the west of the craton concerns mantle xenoliths. Consequently, the direct comparison between the crustal evolution of the eastern, central and northern domains and the mantle evolution of the western domain is equivocal. This may, at least in part, be responsible for the discrepancies between the geochronological records.

In this paper, we synthesize the accurate geochronological data applicable to the Barberton Mountain Land as well as three other geographic domains of the craton where a useful body of data has now been generated.

5.1-2. OVERVIEW OF THE PRE-MESOARCHEAN EVOLUTION OF THE KAAPVAAL CRATON

5.1-2.1. Eoarchean (>3600 Ma)

5.1-2.1.1. Barberton Mountain land
The oldest rock yet dated on the Kaapvaal Craton is a sample of banded tonalite gneiss from part of the Ancient Gneiss Complex (AGC) in northwest Swaziland (Fig. 5.1-1). This sample provided a population of zircons which yielded an U-Pb isotope age of 3644 ±

4 Ma (Compston and Kröner, 1988). Crystallization and recrystallization of other zircons in the same sample indicate that this rock has been affected by subsequent events, the most notable of which are at 3504 ± 6 Ma and 3433 ± 8 Ma. Other evidence for the existence of Eoarchean crust in the Kaapvaal Craton is provided by the presence of xenocrystic zircons in the Vlakplaats weakly foliated to massive granodiorite and the Ngwane gneiss, dated at 3702 ± 1 Ma and 3683 ± 10 Ma, respectively (Kröner et al., 1996; Kröner and Tegtmeyer, 1994). There are no other Eoarchean intrusions that have yet been found in the craton as a whole.

5.1-2.2. Palaeoarchean (3600 to 3200 Ma)

5.1-2.2.1. Eastern Domain
(i) 3600 to 3400 Ma tonalite-trondhjemite-granodiorite (TTG) gneiss plutons and volcanism   Tonalite and trondhjemite gneisses occur as discrete plutons along the southwestern part of the Barberton Greenstone Belt (BGB). They contain greenstone xenoliths of variable size, which are thought to be mainly remnants, but could also partly represent infolded material. Some of the plutons are unequivocally intrusive into the adjacent greenstones, whereas others display sheared margins indicating that they were structurally emplaced into their present position. A granitoid gneiss clast found in the Moodies Formation was dated at 3570 ± 6 Ma (Kröner and Compston, 1988). A zircon xenocryst found in the Hooggenoeg Formation was dated at 3559 ± 27 Ma while an age of 3563 ± 3 Ma was also determined for a well foliated tonalitic gneiss (Kröner et al., 1989) in the Ancient Gneiss Complex (AGC). A weakly foliated porphyritic granodiorite (Vlakplaats) yielded a Pb-Pb zircon age of 3540 ± 3 Ma (Kröner et al., 1996). A tonalitic gneiss wedge in the Theespruit Formation, described as a possible basement for the BGB, was dated at $3538 +4/-2$ Ma and 3538 ± 6 Ma respectively (Armstrong et al., 1990; Kamo and Davis, 1994). This age is identical within error to the age of 3531 ± 4 Ma of a granitoid gneiss clast found in the Moodies Group (Kröner and Compston, 1988). In the AGC, the Ngwane tonalic gneiss was dated at 3521 ± 23 Ma and 3490 ± 3 Ma (Kröner and Tegtmeyer, 1994), while a granite boulder from the Moodies conglomerate in the BGB yielded an age at 3518 ± 11 Ma.

The Steynsdorp Pluton, a banded trondhjemitic gneiss, yielded ages between of 3510 ± 4 Ma and 3505 ± 5 Ma (Kamo and Davis, 1994; Kruger, 1996). A granodioritic phase of the pluton has been dated at 3502 ± 2 Ma (Kröner et al., 1996). An identical age of 3504 ± 24 Ma was found for a trondhjemitic gneiss from the AGC (Kröner et al., 1989). Younger ages at 3490 Ma have also been reported for both the trondhjemitic and tonalitic phases of the pluton (Kröner et al., 1991a, 1992). Three zircon xenocryst found in the gneiss were dated at 3553 ± 4 Ma, 3538 ± 9 Ma and 3531 ± 3 Ma, respectively (Kröner et al., 1991, 1996).

The Stolzburg Pluton, which comprises a biotite-bearing trondhjemitic gneiss which is intrusive into the lower part (Theespruit Formation) of the BGB, was dated at $3460 +5/-4$ Ma by Kamo and Davis (1994) and 3445 ± 3 Ma (Kröner et al., 1991). An age of 3431 ± 11 Ma (Dziggel et al., 2002) has also been reported, a range interpreted to reflect the age of crystallization of the gneisses. A pebble found in the Moodies Group was dated at $3474 + 35/-31$ Ma (Tegtmeyer and Kröner, 1987). A quartz-feldspar porphyry dyke cutting the Komati Formation in the BGB was dated at $3470 + 39/-9$ Ma on zircons and 3458 ± 1.6 Ma on titanites, while metagabbros from the Komati Formation were dated at 3482 ± 5 Ma (Armstrong et al., 1990). Within the AGC, Kroner et al. (1989) reported an age of 3458 ± 3 Ma for a tonalitic gneiss, while the Tsawela Gneiss, a foliated tonalite, was dated between at 3455 and 3436 Ma (Kröner and Tegtmeyer, 1994).

The trondhjemitic Theespruit Pluton unequivocally intrudes and cross-cuts the adjacent lower portion (Sandspruit and Theespruit Formations) of the BGB. The reported single-zircon U-Pb ages vary only slightly, between 3443 and 3437 Ma (Armstrong et al., 1990; Kamo and Davis, 1994; Kamo et al., 1990; Kröner et al., 1991, 1992), and are indistinguishable from the age for the Stolzburg Pluton, indicating that both bodies belong to the same igneous event.

The Doornhoek Pluton is the smallest (4×2 km) of the intrusive trondhjemite gneiss bodies and is entirely surrounded by greenstones of the Theespruit Formation of the BGB. U-Pb isotopic determinations on single zircons yielded an age of $3448 + 4/-3$ Ma (Kamo and Davis, 1994). A titanite U-Pb age ($\sim$3215 Ma; Kamo and Davis, 1994) has also been obtained from the Doornhoek Pluton, as well as imprecise Rb-Sr whole-rock ages (Barton et al., 1983) of 3176 ± 203 Ma (coarse-grained phase) and 3191 ± 46 Ma (fine-grained variety). Both the titanite and Rb-Sr ages are interpreted as thermal re-setting, possibly manifest as numerous quartz veins especially in the western portions of the body.

The Theeboom Pluton comprises two varieties of gneiss, an early well-foliated trondhjemite and a younger cross-cutting, more homogeneous variety. For the older phase a single zircon U-Pb age of $3460 + 5/-4$ Ma, and for the younger variety titanite and zircon ages of 3237–3201 Ma were reported by Kamo and Davis (1994). These workers regarded the Theeboom Pluton as the southern extension of the Stolzburg Pluton.

The TTG gneisses southwest of the BGB are virtually unmetamorphosed and are now interpreted to be the plutonic equivalents of, and hence comagmatic with, felsic volcanics and shallow subvolcanic intrusions occurring in the stratigraphically lower portions of the adjacent greenstone belt (de Wit et al., 1987a). The age correlation between plutonic and extrusive components of this magmatic suite have been confirmed by Kröner and Todt, (1988), Armstrong et al. (1990) and Kroner et al. (1991, 1992), who showed that felsic rocks of the Theespruit and Hooggenoeg Formations, with zircon U-Pb ages of 3453–3438 Ma, overlap with the 3460–3430 Ma ages of the Doornhoek, Theespruit and Stolzburg plutons. The one exception to this is the ca. 3510 Ma old Steynsdorp pluton which is clearly older than most of the metavolcanic sequences in the southwestern part of the BGB, but is broadly coeval with an adjacent suite of felsic volcanics previously correlated with the Theespruit Formation and dated between 3548 and 3530 Ma (Kröner et al., 1992, 1996). These observations are central to the hypothesis by Lowe (1994) that the BGB comprises a series of discrete tectono-stratigraphic blocks that were progressively accreted to one another between 3550 and 3220 Ma to form a composite granite-greenstone terrane that represents the core of an ancient continental fragment now referred to as the Kaapvaal Craton.

(ii) 3400–3250 Ma rocks   This period experienced very little plutonic or volcanic activity within this part of the craton. In south-central Swaziland, the granodiorite of the Usuthu suite in the south central Swaziland was dated at 3306 ± 4 Ma (Maphalala and Kröner, 1993). Activity was also recorded in the BGB where pegmatitic to medium grained gabbros and metagabbros present in the Komati Formation were dated at around 3350 Ma (Armstrong et al., 1990; Kamo and Davis, 1994). Within the Medon Formation in the BGB, a volcanic unit has been dated at 3298 ± 3 Ma (Byerly et al., 1996). A coarse-grained granodiorite described as the older part of the Stentor Pluton yielded a U-Pb age of 3347 + 67/−60 Ma (Tegtmeyer and Kröner, 1987), whereas zircon xenocrysts found mostly in volcanic rocks from the Fig Tree Group gave ages between 3334 and 3310 Ma (Byerly et al., 1996; Kröner et al., 1991). Finally, a pebble found in the Moodies Group yielded a poorly constrained age of 3306 + 65/−57 Ma (Tegtmeyer and Kröner, 1987).

(iii) Circa 3250 Ma tonalite-trondhjemite-granodiorite (TTG) gneiss plutons and associated volcanism   During this period, volcanic activity was recorded with the deposition of the Fig Tree Group in the central part of the BGB, consisting mostly of dacitic tuffs and agglomerates interbedded with ferruginous cherts. Several publications (Armstrong et al., 1990; Byerly et al., 1996; Kamo and Davis, 1994; Kohler et al., 1993; Kröner et al., 1991, 1992) indicate that the deposition of the Fig Tree Group took place between 3259 and 3225 Ma. A feldspar-quartz porphyry that intrudes the Fig Tree Group metasediments dated at 3227 ± 3 Ma provides a minimum age of the Fig Tree Group (de Ronde, 1991).

   Gneisses flank the western and northwestern margin of the Barberton Greenstone Belt and include the Nelshoogte, Kaap Valley and Stentor plutons. They are generally larger bodies than the older intrusions that are found along the southern and southwestern flank of the belt.

   The Kaap Valley Pluton is the largest of the three bodies. Towards the centre, the pluton displays only a weakly developed fabric. It contains a number of mafic and ultramafic xenoliths. The Kaap Valley Pluton differs significantly from most of the other gneiss plutons in the region in that it is entirely tonalitic and is one of the best-dated granitoids in the Barberton region. Results of precise single-grain U-Pb zircon dating by various workers differ only slightly, ranging from 3229 to 3223 Ma (Armstrong et al., 1990; Kamo and Davis, 1994; Layer et al., 1992; Tegtmeyer and Kröner, 1987). $^{40}Ar/^{39}Ar$ hornblende and biotite ages are slightly lower, being 3214 and 3142 Ma, respectively, which can be attributed to variable closure temperature effects in the slowly cooling pluton (Layer et al., 1992).

   The Nelshoogte Pluton is an oval-shaped body of trondhjemitic gneiss, situated between the Badplaas domain and Kaap Valley pluton. A Pb-Pb zircon age of 3212 ± 2 Ma for the intrusion has been obtained by York et al. (1989). Studies by these authors on muscovite and biotite separates gave $^{40}Ar/^{39}Ar$ ages of 3080–2860 Ma, suggesting that the pluton experienced a prolonged cooling history. An imprecise Rb-Sr whole-rock age of 3180 ± 75 Ma was reported by Barton et al. (1983) while an U-Pb age of 3236 ± 1 Ma was found for a strongly foliated sample (Kamo and Davis, 1994).

   The Stentor Pluton occurs as an elongate intrusion of trondhjemitic to granodioritic gneisses that intrudes between the BGB and the Nelspruit batholith. Conventional zircon dating yielded U-Pb ages of 3250 ± 30 Ma (Tegtmeyer and Kröner, 1987). A more recent U-Pb zircon age of 3107 ± 5 Ma obtained by Kamo and Davis (1994) from the eastern portion of the pluton suggests that certain parts of the body should be attributed to the Nelspruit Suite.

The Dalmein Pluton is granodioritic in composition and is situated at the southeastern end of the BGB, where it cuts into both the older (ca. 3450 Ma) TTG gneiss plutons as well as the various lithologies of the BGB. Single zircon U-Pb dating has yielded an age of 3216 + 2/−1 Ma (Kamo and Davis, 1994).

The poorly documented Badplaas domain, located to the south of the Barberton Greenstone Belt, is made up of a variety of compositionally and texturally distinct, variably gneissose trondhjemites. Geochronological work on the five main TTG phases (Kisters et al., 2006) reveals that the Badplaas domain was assembled over a period of ca. 50 Ma between ca. 3290 and 3240 Ma. This is interpreted as a 50 Ma record of convergence-related TTG plutonism.

Other intrusive bodies that fall into the ca. 3230 Ma category are; the tonalitic to granodioritic Wyldsdale pluton that occurs along the Mgudugudu thrust in the Piggs Peak area of northern Swaziland, that yielded a single zircon U-Pb isotope age of 3234 + 17/−4 Ma (Fletcher, 2003); a granodiorite from the Usuthu Suite in North-Central Swaziland (Maphalala and Kröner, 1993) dated at 3231 ± 4 Ma and 3224 ± 4 Ma, suggesting that this body is not equivalent to the Usuthu Suite dated farther south; a K-feldspar-rich granitic gneiss from the AGC dated at 3227 ± 21 Ma (Kröner et al., 1989); and, a tonalitic intrusion in the Weltevreden area of the BGB, as well as a tonalitic dyke dated at ca. 3229 Ma (de Ronde and Kamo, 2000). In the Tjakastad schist belt (TSB), a 10 km long N–S trending extremity of the BGB that occurs along the southern margin of the belt, titanites have been dated at 3229 ± 25 Ma in a garnet-bearing metabasite (Diener et al., 2005). This age has been interpreted as dating the peak of the metamorphism in the TSB. Similar results have been found by Dziggel et al. (2005) for a greenstone remnant located 10 km to the west of the TSB. Titanites extracted from a clastic metasedimentary unit yield an upper intercept age of 3229 ± 9 Ma, providing a minimum age for the peak of metamorphism (~650–700 °C) while zircons separated from the same unit record a range of $^{207}Pb/^{206}Pb$ dates between ~3560 and 3230 Ma, the youngest group yielding a weighted mean age of 3227 ± 7 Ma. They also defined a minimum age for the timing of deformation with the emplacement age of 3229 ± 5 Ma found for a late-kinematic trondhjemite.

Finally, a series of granitoid samples were dated around 3.2 Ga. A pebble found in the Moodies Group yielded an age of 3224 ± 6 Ma (Tegtmeyer and Kröner, 1987), and a porphyry intrusion from the Stolzburg Syncline was dated at 3222 + 10/−4 Ma (Kamo and Davis, 1994). In the AGC, a tonalitic gneiss was dated at 3214 ± 20 Ma (Kröner et al., 1989) and an unfoliated granodiorite intrusive into the Dwazile greenstone belt yielded an age of 3213 ± 10 Ma (Kröner and Tegtmeyer, 1994).

In the southeastern part of the Kaapvaal Craton, pre-3200 Ma granitoids occur in two areas: one in the vicinity of Piet Retief, and the other adjacent to the Nondweni greenstone belt (Fig. 5.1-1). In the Piet Retief-Paulpietersburg region, Hunter et al. (1992) reported the presence of layered granitoid rocks consisting of alternating light and dark TTG gneisses and amphibolites preserved as inliers in younger granitoids. The gneisses are described as similar to the ca. 3640 to 3460 Ma layered gneisses of the bimodal suite of the Ancient Gneiss Complex in Swaziland. South and west of the Assegaai and Commondale Archean greenstone remnants (Fig. 5.1-1) are exposures of Luneburg gneisses (Hunter et al., 1992), which consist of medium-grained tonalitic to trondhjemitic rocks described as chemically and mineralogically indistinguishable from the ca. 3458–3362 Ma Tsawela gneisses of southwestern Swaziland dated by Kröner et al. (1991). A third variety of gneiss (referred to as the De Kraalen gneiss), consists of layered tonalitic rocks and occurs southeast of the De Kraalen greenstone remnant in the valley of the Assegaai River. No precise radiometric ages are available for the various gneisses previously described, but they have been intruded by the ca. 3250 Ma Anhalt Granitoid Suite described below.

The Anhalt Granitoid Suite (Fig. 5.1-1) comprises the most extensive development of granitoid rocks in the southeastern part of the Kaapvaal Craton. The Anhalt granitoids can be subdivided into a number of different phases, including trondhjemites, granodiorites and quartz monzonites (Hunter et al., 1992). A Rb-Sr whole-rock age of 3250 ± 39 Ma for the Anhalt Suite rocks has been documented by (Farrow et al., 1990). In the area adjacent to the Nondweni greenstone belt, a homogeneous, fine- to medium-grained rock of batholithic proportions, known as the Mvunyana Granodiorite, is the most widespread granitoid (Matthews et al., 1989). Rb-Sr, Pb-Pb and limited U-Pb zircon isotopic data obtained for the Mvunyana granodiorite yielded an age of ca. 3290 Ma (Matthews et al., 1989). Further south, in the Natal Spa, a granite sensus-stricto was dated at 3210 ± 25 Ma (Reimold et al., 1993).

In the southeastern region of the Kaapvaal Craton, only the Commondale and Nondweni greenstone belts have been dated. The Commondale greenstone belt yielded a precise Sm-Nd age of 3334 ± 18 Ma on an exceptionally well preserved peridotite suite of komatiitic affinity (Wilson and Carlson, 1989). An age of 3406 ± 6 Ma was obtained on zircons from a rhyolite flow in the uppermost Witkop Formation of the Nondweni Group (Versfeld and Wilson, 1992) while a Re-Os isochron regression age of 3321 ± 62 Ma was obtained for komatiite and komatiitic basalt flows collected from the same formation (Shirey et al., 1998).

5.1-2.2.2. Northern part of the craton

Very little pre-Mesoarchean geochronological data are available for this part of the craton. This area is characterized by the widespread occurrence of the Goudplaats-Hout River Gneiss Suite, which comprises a wide spectrum of granitoid gneisses, types and compositions. These gneissic bodies range from homogeneous to strongly layered, from leucocratic to dark grey, and from fine-grained to pegmatoidal varieties. They underlie both the high-grade ("Southern Marginal Zone") and low-grade terranes of the northern part of the Kaapvaal Craton, mainly to the north of the Pietersburg and Giyani greenstone belts (Fig. 5.1-1). The oldest known age in this region comes from a 3364 ± 18 Ma zircon xenocryst found in the Mac Kop conglomerate from the Murchison greenstone belt (Poujol et al., 1996).

In most of the area around the Giyani greenstone belt (GGB, Fig. 5.1-1), medium-grained, whitish, or locally pinkish, tonalitic or trondhjemitic gneisses are the dominant phase, with the exception of the southern boundary of the GCB where massive granites are developed. Ages of ca. 3282.6 ± 0.4 and 3274 + 56/−45 Ma from dark-grey tonalitic gneisses located to the north of the GGB have been obtained (Kröner et al., 2000). Within the GGB, a meta-andesite was dated at 3203.3 ± 0.2 Ma (Kröner et al., 2000) while a felsic metavolcanic yielded an age of 3203 ± 4 Ma (Brandl and Kröner, 1993).

At the Goudplaats locality (Fig. 5.1-1), light-to dark-grey gneisses, together with minor leucocratic gneisses and hornblende amphibolite and hornblende-biotite tonalitic gneisses, are prominently exposed in a river section. A dark-grey migmatitic tonalitic gneiss from this locality has been dated at 3333.3 ± 5 Ma (Brandl and Kröner, 1993).

In the Lowveld, south of the Murchison greenstone belt, two principal types of gneisses are present; a layered composite variety termed the Makhutswi Gneiss and a "homogeneous gneiss" which has not yet been named. The Makhutswi Gneiss extends from the Murchison greenstone belt southwards for some 50 km, and then forms another smaller occurrence still further to the south. Limited geochemical data suggest that the Makhutswi Gneiss has a tonalitic to granodioritic composition. A single zircon U-Pb age of 3228 ± 12 Ma, has been obtained for a gneiss sampled close to the contact with the Murchison greenstone belt (Poujol et al., 1996). An imprecise Rb-Sr whole-rock age of 3268 ± 113 Ma was also reported from the Phalaborwa area (Barton, 1984).

5.1-2.2.3. Central domain

Very few Paleoarchean ages have been documented in this part of the Craton. The oldest granitoid phase recognised so far is a Palaeoarchean trondhjemite gneiss from the Johannesburg Dome (Fig. 5.1-1) that yielded an age of 3340 ± 3 Ma (Poujol and Anhaeusser, 2001). Still older xenocrystic zircon ages, of ca. 3425 Ma (Hart et al., 1999) and 3310 Ma (Kamo et al., 1996), have, however, been reported from a paragneiss and a pseudotachylitic breccia, respectively, sampled on the Vredefort Dome, and from the Neoarchean Makwassie Quartz Porphyry of the Ventersdorp Supergroup, with a zircon xenocryst age of ca. 3480 Ma (Armstrong et al., 1991). Two additional zircon xenocrysts found in the highly amygdaloidal volcanic horizon of the Crown Lava from the Witwatersrand Supergroup have been dated at 3259 ± 9 and 3230 ± 8 Ma, respectively (Armstrong et al., 1991). Finally a tonalitic gneiss from the Johannesburg Dome has been dated at 3199.9 ± 2 Ma (Poujol and Anhaeusser, 2001).

5.1-2.2.4. Western domain

The Archean crystalline basement of this part of the Kaapvaal Craton is only poorly exposed as it is covered by Neoarchean to Phanerozoic sedimentary and volcanic strata. Consequently, it has produced very little geochronology. Indication of old crust is provided by the presence of several zircon xenocrysts. Two xenocrysts from a felsic schist collected in the Western succession of the Madibe greenstone belt (Fig. 5.1-1) provided ages of 3428 ± 11 and 3201 ± 4 Ma (Hirner, 2001). Tonalitic and trondhjemitic gneisses and migmatites occur as windows in Karoo cover rocks in the Kimberley–Boshoff–

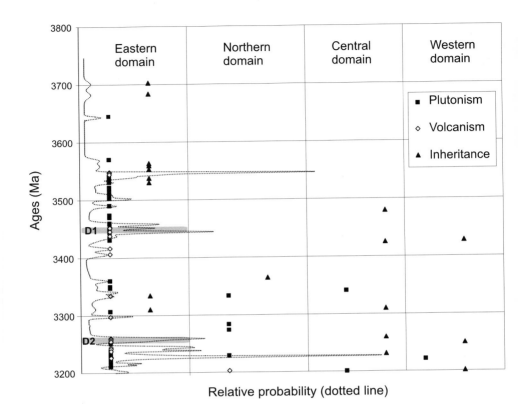

Fig. 5.1-3. Compilation of the geochronological constraints on the evolution of the Kaapvaal craton (see text for source references). Dotted line compiles the geochronological data relative probability for the entire craton ($N = 122$). D1 (development of an active margin) and D2 (main "orogenic" stage) refer to the main deformation events in the Barberton granitoid-greenstone terrain.

Koffiefontein area of the Northern Cape and Free State Provinces. Zircon cores from samples of these rocks from the Bultfontein diamond mine in the Kimberley area (Fig. 5.1-1) defined a mean $^{207}Pb/^{206}Pb$ age of circa 3250 Ma (Drennan et al., 1990). A peraluminous tonalitic gneiss obtained from the Bultfontein diamond mine dump yielded abundant prismatic zircons with $^{207}Pb/^{206}Pb$ dates ranging from 3246 to 3069 Ma (Schmitz et al., 2004). The proposed crystallization age for this tonalite is ca. 3.22 Ga.

5.1-3. CONCLUSIONS

As illustrated in this paper, the Kaapvaal Craton of Southern Africa experienced a protracted evolution with different terrains exhibiting distinct histories (Eglington and Armstrong, 2004). The magmatic record for the four Kaapvaal domains has been plotted in Fig. 5.1-3. It emphasises the earlier development of crust in the south east, followed by crust formation in the central, eastern and northern sectors. Three main peaks of magmatic activity, which contributed to the assembly of the craton, can be defined at ca. 3550, 3450 and 3250–3220 Ma respectively (Fig. 5.1-3). The oldest rocks so far recognized on the Craton are located to the east, in the Ancient Gneiss Terrane and the Barberton Mountain Land (Fig. 5.1-3), providing evidence of the early development of crust in this region. This region, often referred as the Kaapvaal Shield (de Wit et al., 1992), is therefore the most likely candidate for the initial development of what was to become the Kaapvaal Craton. Evidence of early Mesoarchean xenocrysts exists, however, in other part of the craton, as far west as the Madibe greenstone belt (Figs. 5.1-1 and 5.1-3), indicating that the Kaapvaal Shield might have been bigger than its present-day known limit. Development of new crust occurred in the eastern, south-eastern and northern part of the craton before being eventually amalgamated into a coherent entity by granitoid intrusions and convergent margin tectonics by ca. 3.2 Ga. Details on the tectonic and magmatic evolution of the different parts of the craton are provided in the following chapters.

Earth's Oldest Rocks
Edited by Martin J. Van Kranendonk, R. Hugh Smithies and Vickie C. Bennett
Developments in Precambrian Geology, Vol. 15 (K.C. Condie, Series Editor)
© 2007 Elsevier B.V. All rights reserved.
DOI: 10.1016/S0166-2635(07)15052-0

Chapter 5.2

THE ANCIENT GNEISS COMPLEX OF SWAZILAND AND ENVIRONS: RECORD OF EARLY ARCHEAN CRUSTAL EVOLUTION IN SOUTHERN AFRICA

ALFRED KRÖNER

Institut für Geowissenschaften, Universität Mainz, 55099 Mainz, Germany

5.2-1. INTRODUCTION

The origin and evolution of the early Archean tonalite-trondhjemite-granodiorite (TTG) suite in the oldest shield areas of the world has been the subject of considerable debate (Moorbath, 1977; Barker, 1979; Jahn et al., 1981; Condie, 2005) and is still one of the central issues in the discussion on gneiss–granite–greenstone belt relationships. Proponents advocating an intraoceanic origin for the greenstones consider the TTG gneisses to result from anatectic melting of mafic–ultramafic greenstone lithologies, culminating in the diapiric emplacement of large plutons (Anhaeusser, 1973; Glikson, 1979). The banded nature of the gneisses and evidence for high-strain deformation is ascribed to later deformation that has affected both the plutonic and metavolcanic rocks (e.g., Anhaeusser and Robb, 1981).

The alternative interpretation has been that many of the TTG gneisses originated through melting of mafic granulites, preferably quartz eclogite or garnet amphibolite, in the lower crust and are genetically unrelated to the adjacent greenstones (Barker, 1979; Hunter et al., 1979). Supporters of this concept consider that at least part of the gneisses predate the greenstones (Hunter, 1974; Collerson and Bridgwater, 1979; Kröner, 1985; Myers and Williams, 1985; Compston and Kröner, 1987) and constitute some of the oldest preserved remnants of continental crust on Earth.

Southern Africa preserves one of the most complete and detailed records of early Precambrian crustal evolution (McCarthy and Rubridge, 2005), and the Ancient Gneiss Complex (AGC) of Swaziland and related rocks along the southern margin of the 3.5–3.2 Ga Barberton Greenstone Belt (BGB, Lowe and Byerly, this volume) have played a prominent role in models for the early evolution of continental crust (e.g., Kröner, 1985; Hunter and Wilson, 1988). The controversy on the gneiss–greenstone relationships centers around the age relationships between the Barberton Greenstone Belt (BGB) of eastern Transvaal and NW Swaziland, and the Ancient Gneiss Complex (AGC) of central, NE and NW Swaziland (Fig. 5.2-1).

The term Ancient Gneiss Complex was coined originally to accommodate a number of lithologically distinct Palaeoarchean units occurring mainly in central Swaziland

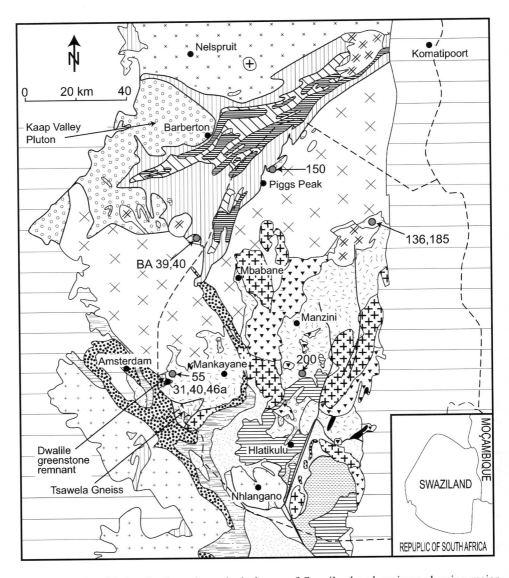

Fig. 5.2-1. Simplified and schematic geological map of Swaziland and environs showing major Precambrian rock units and locations of dated samples. Based on 1:250,000 Geological Map of Swaziland (1982). Locations of dated samples mentioned in the text are shown, but sample prefixes such as AGC and BA left off. Modified from Kröner et al. (1989). (*For legend see next page.*)

Age in Ga   **Lithologic unit**

2.5-0.2   Proterozoic and younger cover sequences
            (Transvaal and Karoo Supergroups)

2.7-2.8   Mswati Granites

2.72   Kwetta and Mtombe Granites

2.72   Hlatikulu Granite

2.86   Usushwana Complex

            Undifferentiated granitoids and gneisses,
            e.g. Nhlangano gneiss

            Mozaan Group

2.96   Insuzi Group

-3.1   Mpuluzi (Lochiel) Granite Batholith

2.8-3.2   Granodiorite and syenite plutons

-3.23   Usutu Intrusive Suite

3.2-3.5   TTG diapiric granitoids (Barberton Mountain Land)

-3.2   Moodies Group
-3.25   Fig Tree Group   Swaziland Supergroup
-3.45   Onwerwacht Group   (greenstone belt association)

3.40-3.45   Tsawela Gneiss

-3.45   Dwalile Supracrustal Suite (greenstone belt association)

3.2-3.64   Ngwane Gneiss (Bimodal Suite), Tsawela and Mhlatuzane
            Gneisses (TTG gneisses and amphibolites), Mkhondo
            Metamorphic Suite suite and other orthogneisses.

            Sample locality for zircon geochronology

            International border

Fig. 5.2-1. (*Continued.*)

(Fig. 5.2-1; see Hunter, 1970) and which are summarized below. The AGC is a typical early to mid-Archean terrain comprising granitoids and multiply deformed granitoid gneisses of the tonalite-trondhjemite-granodiorite (TTG) suite (Hunter et al., 1978) and interlayered amphibolites, most of which are probably derived from gabbroic dykes (Hunter et al., 1984; see also Fig. 5 in Jackson, 1984). This tectonic sequence of siliceous and mafic gneisses is also known as the Bimodal Suite (Hunter, 1970), now renamed as Ngwane Gneiss (NG, Wilson, 1982), and is the oldest part of the AGC (see Fig. 5.2-2(a)). The complex also includes remnants of greenstone belt supracrustal assemblages that vary in size from xeno-liths a few centimetres long to inliers several kilometres across. The largest of these is the Dwalile greenstone remnant in southwestern Swaziland (Fig. 5.2-1).

The AGC is separated from the BGB by a large granitoid sheet-like pluton some 3 Ga in age and known as the Mpuluzi Batholith (Hunter, 1974; Barton et al., 1983; see Fig. 5.2-1). In northwest Swaziland, however, small inliers of AGC gneisses occur in faulted and sheared contact with BGB rocks (e.g., near Piggs Peak, see Fig. 5.2-1), and this, together with the presence of a tectonic wedge of tonalitic gneiss in the lower Onwerwacht Group of the BGB (De Wit et al., 1987) suggests that the two units were in direct contact prior to about 3.0 Ga.

Fig. 5.2-2. Field photographs of rocks from the Ancient Gneiss Complex in Swaziland. (A) Banded tonalitic-trondhjemitic Ngwane Gneiss with layer-parallel leucocratic veins and cross-cutting pegmatites. Sample AGC 150 with a maximim age of 3644 ± 4 Ma was taken from the darkest and most homogeneous portion of this roadcut along the asphalt road immediately east of 'The Falls' and some 7 km northeast of Piggs Peak (see Fig. 5.2-1 for location). (B) Mylonitized tonalitic gneiss (tight isoclinal folds in leucogranite veins still preserved) from contact zone of Ngwane Gneiss with ultramafic rocks of the Barberton Greenstone Belt. Small waterfall at 'The Falls' below homestead and due west of sample AGC 150. (C) Ductile shear zone in Ngwana Gneiss E of Bloemendal, central Swaziland. (D) Contact between isoclinally folded Ngwane Gneiss (bottom) and foliated but non-layered Tsawela Gneiss (top). Small tributary of Ngwempesi River, central Swaziland. (E) Tight fold in metagreywacke of Dwalile Supracrustal Suite. This is location of sample AGC 40 shown in Fig. 5.2-1. (F) Garnet–sillimanite gneiss (khondalite) from metapelite sequence of sedimentary inlier in Ngwane Gneiss at Shiselweni, central Swaziland. This is near location of sample AGC 155 shown in Fig. 5.2-1.

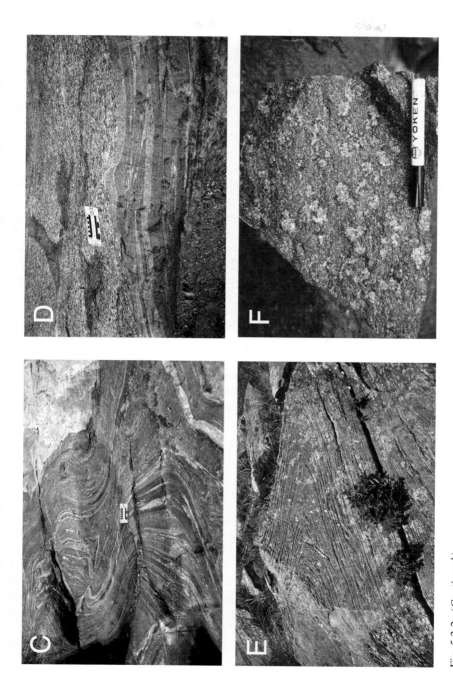

Fig. 5.2-2. (*Continued.*)

The AGC has its widest distribution in a broad belt through Swaziland (Fig. 5.2-1), and the area so far studied in most detail centers around the town of Mankayane (Hunter, 1970; Hunter et al., 1978; Jackson, 1984). The NG represents the oldest unit of the AGC as suggested by its structural evolution (Jackson, 1984; Jackson et al., 1987) and geochronology, and its preserved metamorphic grade is upper amphibolite-facies. However, at least part of the suite has been to granulite-facies as shown by inliers of >3.3 Ga metasedimentary rocks with high-grade mineral assemblages indicating temperatures of 700–900 °C and pressures of 6–7.5 kbar (Kröner et al., 1993; Condie et al., 1996). The NG gneisses and greenstone remnants are intruded by the Tsawela Gneiss (TG, see Fig. 5.2-1), a weakly to well foliated tonalitic to trondhjemitic pluton which was emplaced after the oldest NG gneisses and greenstones had already been deformed at least once (Jackson, 1984; see Fig. 5.2-2(d)). Age relationships are discussed below.

Scattered remnants of mafic–ultramafic metavolcanic rocks associated with BIF and clastic metasedimentary rocks occur throughout the NG and are infolded with the granitoid gneisses (Fig. 5.2-2(e)). They are collectively referred to as the Dwalile Supracrustal Suite (DSS, Wilson, 1982; Jackson, 1984). The Dwalile greenstone remnant is the largest of these. Jackson (1984) considered these rocks, some of which are lithologically somewhat similar to the Onverwacht Group greenstones (BGB), but of higher metamorphic grade, to be younger than the NG on structural grounds. However, direct field evidence for this interpretation was not provided, since all contacts are tectonic.

Supracrustal rocks of dominantly sedimentary origin are common in the Mkondo Valley of central Swaziland and in the southern part of the country and were previously included in the AGC. The precise correlation of these rock-types is still debated, but there is a possibility that they represent metamorphic equivalents of the post-3.0 Ga Pongola Supergroup (Wilson and Jackson, 1988), and this is perpetuated in the Geological Map of Swaziland (1982). Other granitoid gneisses, locally garnetiferous and containing metasedimentary relicts, have been given local names (e.g., Mahamba and Nhlangano gneisses, see Fig. 5.2-1 and Geological Map of Swaziland, 1982) and are most likely late Palaeo- to Mesoarchean in age. Although originally included in the AGC, they are not further considered here.

5.2-2. FIELD RELATIONSHIPS AND ORIGIN OF COMPONENTS OF THE AGC

Layered gneisses of the Ngwane Gneiss are characterized by the alternation of medium- to fine-grained light and dark coloured layers ranging in thickness from a few mm to 50 cm, as a consequence of variations in the amount of hornblende and biotite. Planar foliation is commonly defined by aligned hornblende and/or biotite laths. Quartzofeldspathic veins and pegmatites of several generations are common, and the earliest are thin ($\sim$2–5 mm wide) veinlets arranged parallel to the dominant foliation and occasionally displaying intrafolial folds. Subsequent generations cross-cut the foliation but may be locally attenuated and/or folded (Fig. 5.2-2(a); see also Hunter et al., 1992).

Three groups of chemically distinct quartzofeldspathic gneiss have been recognized in the NG (Hunter et al., 1984; Hunter, 1991). The most common has Rb and (Nb + Y) contents similar to Phanerozoic subduction-related granites (Pearce et al., 1984) but is distinguished from them by heavy rare earth element (HREE) and high field strength element (HFSE) depletion. The second group comprises high-SiO_2 gneisses with large negative Eu-anomalies, high Th/Ba ratios, enriched contents of HFSE and flat HREE slopes. The third type is characterized by strongly fractionated REE patterns, small to large positive Eu-anomalies, and high contents of Ba and Sr (Hunter et al., 1984).

There have been few detailed studies on the structural evolution of the AGC, but descriptions of various field relationships are summarized on the published 1:50,000 sheets of the Geological Survey of Swaziland. A more detailed analysis of the Ngwane and Tsawela Gneisses and the Dawlile greenstone remnant in southwestern Swaziland was provided by Jackson (1984), who demonstrated that these rocks reflect evolution from early homogeneous ductile strain at upper amphibolite to granulite grade, to late inhomogeneous brittle strain, indicating deformation at successively higher crustal levels. This uplift probably occurred along large-scale N- and NW-directed shear zones some of which have spectacular exposures (Fig. 5.2-2(b,c)) and resulted in a vertical displacement estimated at ~20 km (Jackson, 1984).

Geochronological data discussed below demonstrate that most of the NG has a complex history and that gneisses at outcrop scale may range in age between 3.64 and 3.2 Ga (Kröner et al., 1989) and are intimately interlayered without any obvious structural break. This is similar to the situation in the early gneiss terrain of West Greenland (see Nutman et al., this volume) and is best explained through the mechanism of banded gneiss formation as proposed by Myers (1978).

Jackson (1984) mapped the NG, the Dwalile greenstone remnant and the Tsawela gneiss in the Mankayane area (Fig. 5.2-1) and recognized a large overturned antiformal structure in these rocks. The cumulative strain pattern in these repeatedly and ductilely deformed assemblages indicates flexural flow, preceded and followed by flattening. High strains during the early phases of deformation resulted in highly attenuated folds (e.g., Fig. 5.2-2(d,e)) boudins, dyke contacts and other primary igneous structures. Such gently dipping, high-strain structures result from lateral translation of soft, hot, middle to lower continental crust as documented by Myers (1978) in West Greenland. Jackson (1984) also concluded that this style of deformation differs from that induced by diapirism around the margins of the BGB.

The boot-shaped, amphibolite-facies Dwalile greenstone remnant occurs in the extreme SW of Swaziland (Fig. 5.2-1), south of the small village of Dwalile. It was mapped by Jackson (1984) who concluded from a structural study that contacts between rocks of the multiply deformed Dwalile and Ngwane rocks are everywhere conformable, but that the Dwalile suite structurally overlies the Ngwane gneisses and may therefore be younger. The structurally lowest rocks in the foliated suite are magnesian schists and serpentinites, interlayered with, and followed by, komatiitic to andesitic metavolcanic rocks and tuffs, now largely preserved as amphibolite, amphibole schist and actinolite schist. Structurally overlying this mafic–ultramafic suite are metasedimentary rocks comprising metaquartzite, metagreywacke (Fig. 5.2-2(e)), metapelite, calc-silicate gneiss and BIF. The metasedimentary rocks are locally interlayered with the metavolcanic rocks, often with sharp contacts.

Although this could be the result of early thrusting, similar to what has been inferred in the lower parts of the BGB (De Wit et al., 1983), Jackson (1984) and Kröner and Tegtmeyer (1994) interpreted many of these contacts as non-tectonic, suggesting a primary sequence of interbedded komatiitic–basaltic lavas and predominantly clastic sediments.

Tegtmeyer (1989) and Kröner and Tegtmeyer (1994) chemically analysed and dated several units of the Dwalile sequence, and these and additional data are summarized in Kröner et al. (1993); the age data are reviewed below. Peak metamorphic conditions at 550–600 °C and ~4 kbar were estimated from mineral compositions, and the coexistence of quartz, staurolite, andalusite, muscovite, biotite and chlorite in one semi-pelitic sample supports this estimate. The Dwalile mafic and ultramafic rocks display remarkable similarities with those of the BGB, and Tegtmeyer (1989) suggested on the basis of REE data that some Dwalile meta-komatiites and metabasalts reflect up to 10% contamination with older continental crust, a feature which is supported by the Sm-Nd isotopic data.

The Tsawela gneisses occur in the Mankayana area of central Swaziland (Fig. 5.2-1) and comprise tonalites with 63–69% SiO_2, and MgO and CaO contents reflecting the relative abundances of hornblende and biotite (Hunter, 1993). Chondrite-normalized REE display steep light REE and flat to gently sloping heavy REE patterns that are reminiscent of those of the Kaap Valley Pluton at Barberton (Hunter et al., 1978; Hunter 1993).

5.2-3. GEOCHRONOLOGY AND IMPLICATIONS FOR GNEISS–GREENSTONE RELATIONSHIPS

There is a large number of published age data on the AGC. However, with few exceptions, these do not resolve age differences within the AGC or the chronological relationship with the BGB as discussed below. Barton et al. (1980) reported a 10-point Rb-Sr whole-rock isochron age of 3555 ± 111 Ma with a $^{87}Sr/^{86}Sr$ initial ratio of 0.6999 ± 16 for banded Ngwane gneisses from a quarry near the Njoli Dam in northeast Swaziland (AGC 136, see Fig. 5.2-1). Carlson et al. (1983) analysed samples from widely scattered rock types including Ngwane gneisses, Dwalile greenstones and younger granitoid intrusives by the Sm-Nd method and obtained a best-fit line fulfilling isochron criteria with an age of 3417 ± 34 Ma and $\varepsilon_{Nd(t)} = 1.1 \pm 0.6$. These authors argued that the excellent collinearity of their data points supports derivation of all analysed rock types from an isotopically homogeneous source and that the calculated age reflects the time of igneous emplacement.

However, Kröner et al. (1989) demonstrated with U-Pb single zircon SHRIMP ages that igneous emplacement of the NG precursors occurred over a time span of at least 400 My. The above Sm-Nd 'isochron' age of 3417 ± 34 Ma is therefore suspect and does not provide evidence for the AGC as a whole to be time-equivalent or to be significantly younger than the BGB, the more so since Carlson et al. (1985) revised their age to 'approaching 3550 Ma' (no analytical data and errors given). The 3555 Ma Rb-Sr age of Barton et al. (1980) is also suspect, since Kröner et al. (1989) dated zircons from one of the gneisses at the Njoli Dam quarry (AGC 136, see Fig. 5.2-1) and obtained a magmatic emplacement age of 3214 ± 20 Ma.

Direct evidence for pre-3500 Ma ages in the AGC was provided by Compston and Kröner (1988), who SHRIMP-dated single zircons from a tonalitic Ngwane Gneiss north of Piggs Peak (AGC 150 in Fig. 5.2-1) that reflect four distinct age groups, interpreted as igneous and metamorphic episodes, and whose oldest group yielded a precise age of 3644 ± 4 Ma (Fig. 5.2-3). The other groups are defined by mean ages of 3580, 3504, and 3433 Ma (see Compston and Kröner, 1988, for detailed discussion).

There is no doubt that the 17 oldest zircon grains of this sample crystallized at 3644 ± 4 Ma, on the evidence of the excellent replication of age within and between grains. The crucial question is what geological event do these grains represent? Compston and Kröner (1988) suggest two alternative possibilities. (1) Magmatic crystallization in the original gneiss protolith, the zircon age would reflect emplacement and freezing of the tonalite pluton prior to its deformation, and all younger zircon ages would represent later growth or alteration due to metamorphism. Points consistent with this interpretation

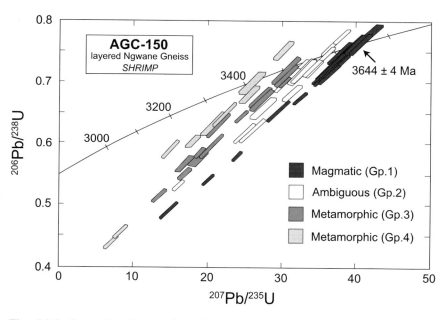

Fig. 5.2-3. Concordia diagram for all analysed zircons from banded tonalitic gneiss sample AGC 150. Error boxes are 1-sigma. The oldest magmatic episode at 3644 ± 4 Ma (2σ) produced the dominant type of zircon which probably precipitated from the original magma. Recrystallization accompanied (and obscured) by early Pb loss occurred within the oldest grains at 3504 and ~3433 Ma. Whole new grains also grew at these times. The post-3644 Ma growth is interpreted as due to episodic deformational and metamorphic events that transformed the original tonalite pluton into a foliated banded gneiss. In addition, many grains are visibly overgrown by two layers of younger zircon of different colour and texture, dated at 2986 and 2867 Ma. Euhedral, finely-zoned whole grains having the 2986 Ma age are also present, evidently contributed by thin felsic veins associated with the nearby Mpuluzi granite (Compston and Kröner, 1988).

are the uniformity of the 3644 Ma ages, the relic igneous characteristics seen in some grains (subhedral morphology and internal euhedral lamellae), and the large proportion of grains that have this age. (2) Xenocrysts older than the tonalite. In this case, zircons characterizing emplacement of the tonalite magma would be one of the other groups that generate the continuum of $^{207}Pb/^{206}Pb$ ages down to ca. 3400 Ma, as shown in Fig. 5.2-3. For example, the magmatic grains could be the group at 3504 ± 3 Ma, whereas the younger group at 3433 ± 4 Ma could be interpreted as new growth or recrystallization of zircon that accompanied the development of gneissic foliation at that later time. There is no objective or conclusive way of choosing between these alternatives, but Compston and Kröner (1988), using estimates for zircon solubility in the original tonalite melt, favoured 3644 Ma as the magmatic emplacement age. If this interpretation is correct, this is the oldest emplacement age yet reported for a crustal rock in the African continent.

Further indications for the antiquity of the NG gneisses come from zircon ages summarized below, and for a 3538 ± 3 Ma old tonalitic gneiss wedge within the lower Onverwacht Group (Theespruit Formation) in the southern part of the Barberton Mountain Land (Armstrong et al., 1988). Lastly, there is a zircon age of 3570 ± 6 Ma for a granite clast in the Moodies conglomerate of the BGB (Kröner and Compston, 1988), believed to be derived from erosion of the AGC terrain during greenstone basin evolution (Eriksson, 1980; Jackson et al., 1987).

The oldest date so far reported from the supracrustal assemblage of the BGB is a $^{207}Pb/^{206}Pb$ mean zircon age of 3547 ± 3 Ma for strongly deformed felsic metavolcanic samples of the Theespruit Formation, exposed in the Steynsdorp greenstone remnant (BA 39 and 40 in Fig. 5.2-1) and in tectonic contact with the rest of the BGB (Kröner et al., 1996). These rocks are ∼100 My older than those of the Komati Formation in the main belt and constitute the oldest components of the BGB, named Steynsdorp suite by Lowe and Byerly (this volume) and onto which successively younger units were tectonically and magmatically accreted (Kröner et al., 1996).

The available isotopic data therefore indicate that at least parts of the NG are older than the oldest parts of the BGB, but direct evidence for a basement-cover relationship is lacking. Age relationships and chemical similarities between felsic volcanic rocks of the BGB and TTG gneiss domes surrounding the BGB suggest a genetic relationship, and the presence of zircon xenocrysts up to 3.7 Ga in age in samples of both the felsic volcanics and the granitoids suggests that older crust was involved in their formation (Kröner et al., 1996).

5.2-4. NGWANE GNEISS

Apart from the 3.64 Ga zircon age for the tonalitic gneiss in northern Swaziland, there are several other pre 3.5 Ga ages for similar rocks of occurring in the NE and central parts of the country. Sample AGC 55 is a tonalitic gneiss collected close to the contact with the Dwalile greenstone remnant (Fig. 5.2-1). The igneous zircons provided nearly concordant ages with a mean of 3521 ± 23 Ma, and one zircon xenocryst has a slightly discordant

^{207}Pb/^{206}Pb age of 3683 ± 10 Ma (Kröner and Tegtmeyer, 1994). This suggests that material of similar age as exposed in northern Swaziland was involved in the generation of 3.5 Ga gneisses and that at least some of the Ngwane gneisses are not juvenile but represent, at least in part, remelts of earlier crustal material. This is also suggested by the whole-rock Nd isotopic systematics for many Ngwane gneiss samples which provide depleted mantle Nd mean crustal residence ages of 3.6–3.7 Ga (Kröner et al., 1993).

Further indications for the antiquity of the Ngwane gneisses come from a near-concordant SHRIMP zircon age of 3563 ± 3 Ma (Fig. 5.2-4(a)) for a tonalitic gneiss in central Swaziland (AGC 200 in Fig. 5.2-1; Kröner et al., 1989) and for a concordant SHRIMP age of 3505 ± 24 Ma (Fig. 5.2-4(b)) for a trondhjemitic gneiss from the Njoli Dam area of NE Swaziland (AGC 185 in Fig. 5.2-1; Kröner et al., 1989). Similar to the 3.64 Ga gneiss from NW Swaziland, the zircon population of this sample shows a complex pattern of Pb-loss, even within single grains, which gives rise to exceptionally large scatter in the Concordia diagram (Fig. 5.2-4(b)). A small, clear, euhedral zircon population in this sample is slightly discordant (symbols with broken lines in Fig. 5.2-4(b)) and defines a distinctly younger age group with a Concordia intercept at 3166 ± 4 Ma. These grains represent thin leucocratic lit-par-lit veins that cut the banded gneiss and reflect an event of leucogranite magmatism that is also widespread elsewhere in the AGC.

Many TTG gneisses mapped as Ngwane Gneiss in the field, and resembling those mentioned above, yielded a surprising variety of igneous emplacement ages between 3455 and 2745 Ma, as summarized by Kröner et al. (1993). This either implies that the Ngwane Gneiss is an extremely heterogeneous assemblage in which the various members are not all genetically related, or many gneisses so far labelled as Ngwane Gneiss on account of their field appearance do not belong to this unit. Gneisses as young as 3200 Ma have acquired a fabric that is macroscopically indistinguishable from that of much older varieties. This attests to the fact that intense ductile deformation has affected many of the Palaeoarchean rocks of the AGC at various times and has often obliterated earlier structures (Jackson, 1984).

All samples of Ngwane Gneiss with zircon ages between 3.66 and 3.2 Ga have Sm-Nd isotopic compositions that can be fitted to a common regression line on a ^{143}Nd/^{144}Nd versus ^{147}Sm/^{144}Nd diagram, defining an "age" of 3760 ± 210 Ma and $\varepsilon_{Nd(t)} = +2.7$ (Kröner et al., 1993). The most plausible interpretation of this linear relationship is that all these rocks were derived from a crustal protolith which separated from its mantle source some 3.7–3.8 Ga ago. The consistency of the age defined by the above reference line with the individual Nd model ages led Kröner et al. (1993) to suggest that the Ngwane gneisses probably inherited their Nd isotopic composition from their (common?) protolith(s).

5.2-5. DWALILE SUPRACRUSTAL SUITE

Detrital zircons from Dwalile metagreywacke samples display typical rounding ascribed to sedimentary transport, but preserve several features such as oscillatory zoning that suggest a primary magmatic origin. Kröner and Tegtmeyer (1994) reported detrital

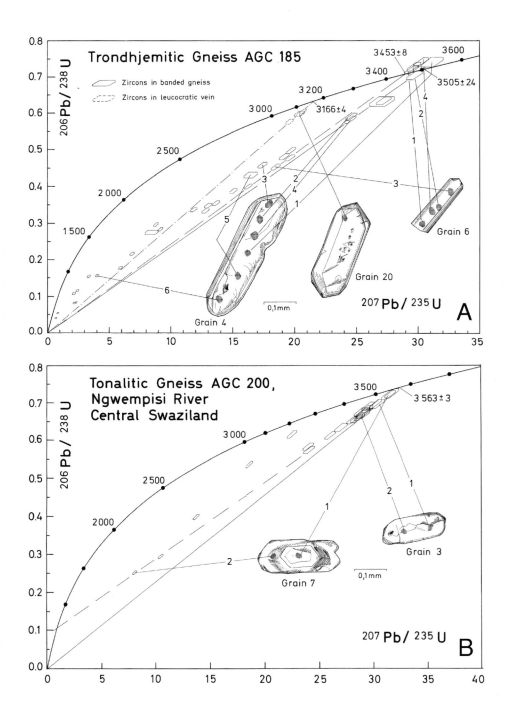

Fig. 5.2-4. Concordia diagrams showing isotopic data for SHRIMP zircon analyses of Ngwane TTG gneiss samples from Swaziland (from Kröner et al., 1989). For sample locations see Fig. 5.2-1: (a) Tonalitic gneiss AGC 200 showing well grouped, slightly discordant zircon analyses with mean $^{207}Pb/^{206}Pb$ age of 3563 ± 3 Ma; (b) Trondhjemitic gneiss AGC 185 showing complex Pb-loss patterns and maximum age of 3505 ± 24 Ma.

U-Pb zircon SHRIMP ages from three such samples (AGC 31, 40, 46a in Fig. 5.2-1) with an array of data points in the Corcordia diagram, some of which are grossly discordant and are not considered further here. The least discordant data (Fig. 5.2-5(a,b)) represent individual $^{207}Pb/^{206}Pb$ ages between 3423 ± 30 Ma and 3566 ± 33 Ma, which probably reflects the original chronological heterogeneity of the source terrain, presumably the old components of the Ngwane Gneiss. The youngest near-concordant detrital zircon age of 3423 ± 30 Ma defines the upper age for Dwalile Supracrustal Suite deposition (Kröner and Tegtmeyer, 1994), and this is compatible with the age of the upper part of the Komati Formation in the BGB (see Poujol, this volume).

Tegtmeyer (1989) investigated the Sm-Nd isotopic systematics in nine small slabs cut from one sample of banded, impure calc-silicate rock of the DSS. The results display considerable scatter but can be fitted to a regression line corresponding to an age of 3220 ± 220 Ma ($\varepsilon_{Nd(t)} = -4.2 \pm 2.4$; see Fig. 15 in Kröner et al., 1993). This scatter probably represents incomplete re-equilibration during the tectono-thermal event that led to intense deformation and amphibolite-facies metamorphism in the Dwalile Supracrustal Suite and Ngwane Gneiss rocks.

5.2-6. TSAWELA GNEISS

This coarse-grained and generally well foliated, but not layered, hornblende–biotite tonalite can locally be observed to intrude already deformed Ngwane Gneiss (see Fig. 5.2-2(d)), but at most localities contacts with the Ngwane Gneiss are concordant and thus precludes the establishment of relative age relationships in the field. Kröner et al. (1989) determined precise single zircon evaporation $^{207}Pb/^{206}Pb$ ages for three Tsawela samples which vary between 3436 ± 6 Ma and 3458 ± 3 Ma. Zircons from a further sample were analysed conventionally and yielded a Concordia intercept age of 3402 ± 9 Ma (Wendt, 1993).

These variable zircon ages suggest that the Tsawela Gneiss, although appearing lithologically and chemically distinct (Hunter, 1993), may consist of several igneous phases which were emplaced over a period of about 60 My, between ~3460 and ~3400 Ma. Thus, the Tsawela Gneiss is older than several members of the Ngwane Gneiss and only intrudes the oldest phases of the Ngwane Gneiss. Also, the Tsawela Gneiss is time-equivalent to several of the TTG plutons intruding the southern margin of the Barberton Greenstone Belt (Kröner et al., 1991; Kamo and Davis, 1993).

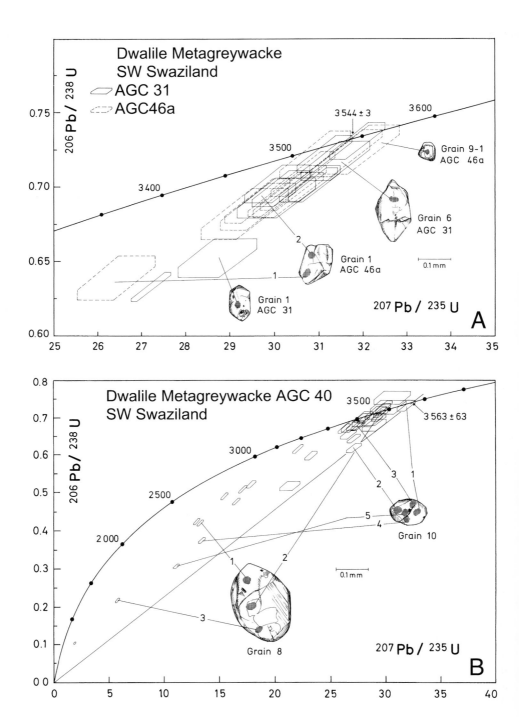

Fig. 5.2-5. Concordia diagrams showing isotopic data for SHRIMP detrital zircon analyses of Dwalile metagreywacke samples from Swaziland (from Kröner and Tegtmeyer, 1994). For sample locations see Fig. 5.2-1: (a) Samples AGC 31 and 46a show virtually identical isotopic patterns and grouping and suggest a very homogeneous source terrain with a mean $^{207}Pb/^{206}Pb$ age of 3544 ± 3 Ma. Sample 40 shows slightly more scatter but similar grouping and suggests a mean $^{207}Pb/^{206}Pb$ source age of 3563 ± 63 Ma.

5.2-7. DISCUSSION AND CONCLUSIONS

The Ancient Gneiss Complex of Swaziland is a chronologically complex and heterogeneous Palaeoarchean crustal unit containing a variety of similar looking TTG gneisses that vary in age between ~3640 and ~3200 Ma. The last major tectono-metamorphic event responsible for intense flattening of virtually all AGC lithologies must have occurred around, or shortly after, 3200 Ma ago, and this event may also have been responsible for much of the deformation in the neighbouring Barberton Greenstone Belt (Kröner et al., 1989; Kamo and Davies, 1993; Dziggel et al., 2002). Many rocks of the AGC show evidence of intracrustal melting and/or derivation from older continental material, and pronounced negative Eu-anomalies, as well as negative $\varepsilon_{Nd(t)}$-values for some gneisses and metasedimentary rocks, suggest the presence of evolved crust as early as 3.5 Ga ago.

The Dwalile Supracrustal Suite constitutes the remnant of an early Archean greenstone belt and may be correlated with parts of the Onverwacht Group of the Barberton Greenstone Belt. It contains mature clastic sedimentary rocks interlayered with komatiite and was probably deposited on an early phase of the Ngwane Gneiss.

The Sm-Nd isotopic ratios for samples of the Ngwane Gneiss, Dwalile metasedimentary rocks and the Tsawela Gneiss show a surprisingly good linear correlation and define a reference line in a $^{143}Nd/^{144}Nd$ versus $^{147}Sm/^{144}Nd$ diagram defining an age of 3625 ± 94 Ma and $\varepsilon_{Nd(t)} = +1.0$ (Fig. 5.2-6). It is therefore likely that all these rocks were ultimately derived from protoliths of broadly the same age. The implication of this interpretation is that most, if not all, of the granitoid gneisses and Dwalile supracrustal rocks in the AGC originated from crustal sources of considerable antiquity.

Much of the present exposure of the AGC is probably underlain by crust that first formed ~3.6–3.8 Ga ago and was subsequently modified, particularly through remelting and ductile deformation, to give rise to the various compositional gneiss varieties that now occur in the AGC. It is likely, following Johnson and Wyllie (1988), Jérbak and Harnois (1991) and Van der Laan and Wyllie (1992), that the gneiss protoliths were derived from still older tonalite, and not from a mafic source as has been suggested from geochemical modelling (Hunter, 1979).

A possible explanation for the existence of geochemically and chronologically distinct groups of layered gneisses that make up the Ngwane Suite may lie in a multi-stage genetic model such as proposed by Jahn and Zhang (1984), which would be consistent with the above experimental studies. In such a model, two generations of TTG magma are produced, first by partial melting of mafic rocks, and second by partial melting of the first generation

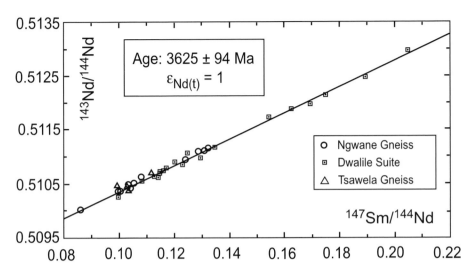

Fig. 5.2-6. ^{143}Nd/^{144}Nd versus ^{147}Sm/^{144}Nd diagram for whole-rock samples of NG, TG and DSS. Good linear correlation of all data is interpreted to approximate age of protolith from which all three suites are derived (from Kröner et al., 1993).

tonalites. The spectrum of single zircon ages reported for AGC gneisses may be a reflection of the multi-stage origin of the Ngwane Gneisses. Continued ductility of the evolving crust would be maintained by repetitive magmatism as a consequence of underplating by new pulses of magma, partial melting, and metamorphism. The early phases of repetitive magmatism and associated tectonic thickening of the crust contributed to increasing rigidity of the AGC, as has been demonstrated from structural work in Swaziland (Jackson, 1984).

ACKNOWLEDGEMENTS

This review includes partly unpublished data from the MSc and PhD theses of Claudio Milisenda, Axel Tegtmeyer and Immo Jobst Wendt whom I thank for their collaboration in studying the Archean rocks of Swaziland. The advice and collaboration of Bill Compston and Ian Williams of the SHRIMP-Centre at the Australian National University in Canberra in interpreting the zircon age data is greatly appreciated. Richard Maphalala and the Geological Survey and Mines Department in Swaziland provided substantial logistic support during field studies, and the German Science Foundation (DFG) supported this research through several grants. Kent Condie, Simon Johnson, Martin Van Kranendonk and Brian Windley suggested improvements of the manuscript. Petra Koppenhöfer prepared the illustrations. This is contribution no. 41 of the Geocycles Cluster of Mainz University, funded by the State of Rhineland-Palatinate.

Earth's Oldest Rocks
Edited by Martin J. Van Kranendonk, R. Hugh Smithies and Vickie C. Bennett
Developments in Precambrian Geology, Vol. 15 (K.C. Condie, Series Editor)
© 2007 Published by Elsevier B.V.
DOI: 10.1016/S0166-2635(07)15053-2

Chapter 5.3

AN OVERVIEW OF THE GEOLOGY OF THE BARBERTON GREENSTONE BELT AND VICINITY: IMPLICATIONS FOR EARLY CRUSTAL DEVELOPMENT

DONALD R. LOWE[a] AND GARY R. BYERLY[b]

[a]*Department of Geological and Environmental Sciences, Stanford University, Stanford, CA 94305-2115, USA*
[b]*Department of Geology and Geophysics, Louisiana State University, Baton Rouge, LA 70803-4101, USA*

5.3-1. INTRODUCTION

Rocks in the 3.55 to 3.22 Ga Barberton Granite Greenstone Terrain (BGGT), South Africa and Swaziland (Fig. 5.3-1), represent one of the oldest, well-preserved pieces of continental crust on Earth. Together with similar rocks of nearly identical ages in the Pilbara Craton of Western Australia, rocks of the BGGT have provided most of the direct geologic evidence on the nature and evolution of the pre-3.0 Ga Earth, its crust, surface environment, ocean, atmosphere, and biota. The BGGT includes two main components: the supracrustal succession of the Barberton Greenstone Belt (BGB), which will be the primary focus of this discussion, and associated generally deeper-level intrusive units, orthogneisses, and metamorphosed supracrustal xenoliths. Supracrustal units of the BGB form a mountainous area, the Barberton Mountain Land, which affords many excellent outcrops accessible by vehicle and foot. Detailed geological studies have focused in the southern half of the belt (Fig. 5.3-1) and in several mining districts in the northern part of the belt.

Intrusive units surrounding the BGB represent two major families of igneous rocks: the TTG group (trondhjemite-tonalite-granodiorite), notable for the predominance of plagioclase over alkali feldspar, even in rocks of high silica content, and the GMS group (granite-monzonite-syenite), in which plagioclase is subordinate to alkali feldspar. The TTGs make up broad, low-relief valleys around the BGB, whereas the GMS group tends to form more distant high, flat plateaus. The TTGs are coeval with the BGB, whereas the great bulk of the GMS plutons were intruded long after sedimentation and peak deformation of the supracrustal sequence had ceased.

We here review the geology and evolution of the BGB based on our work over the last 30 years and the published work of others and consider the implications of the results toward the nature of early crust and tectonics before 3.2 Ga.

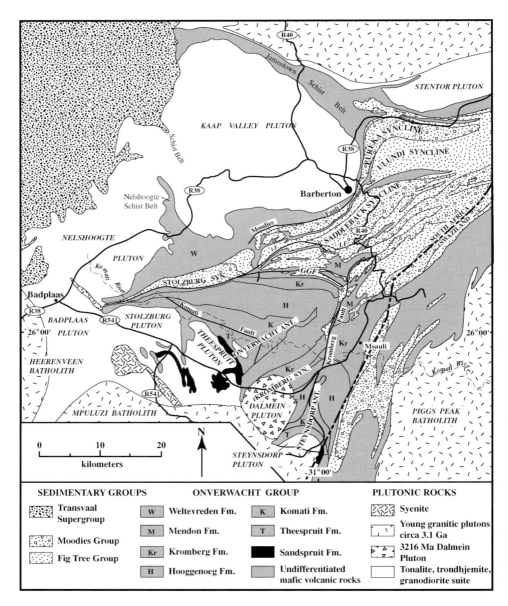

Fig. 5.3-1. General geology of the Barberton Greenstone Belt and vicinity. Abbreviations: IF: Inyoka Fault; GGF: Granville Grove Fault.

5.3-2. GENERAL GEOLOGY OF THE BGB

5.3-2.1. Stratigraphy

The BGB includes volcanic, sedimentary, and shallow intrusive rocks ranging in age from >3547 to <3225 Ma (Figs. 5.3-1 and 5.3-2). The rock have traditionally been divided into three main lithostratigraphic units: from base to top, the Onverwacht, Fig Tree, and Moodies Groups (Viljoen and Viljoen, 1969a, 1969b; Lowe and Byerly, 1999). The Onverwacht Group is composed largely of mafic and ultramafic volcanic rocks with subordinate felsic volcanic flow units and tuffs. Mostly thin, interbedded sedimentary units marking breaks in eruptive activity have been widely silicified to form impure cherts. The Onverwacht Group is diachronous in age across the belt. Rocks in the southern part of the belt range from >3547 to ~3260 Ma and exceed 10 km in stratigraphic thickness. Ultramafic and mafic rocks fringing the western and northern edge of the belt, which may be as little as 1000 m thick, appear to range from ~3330 Ma to perhaps as young as 3240 Ma (summarized in Lowe, 1999b). The Fig Tree group is a transitional unit up to ~1800 m thick composed of interlayered volcaniclastic strata, marking the final stages of greenstone belt volcanism, and terrigenous clastic units eroded from uplifted portions of the underlying greenstone succession. It was deposited between ~3260 and 3225 Ma. The post-3225 Ma Moodies Group is composed of up to 3000 m of coarse, quartzose and feldspathic sandstone and chert-clast conglomerate derived by erosion of underlying greenstone units and uplifted plutonic rocks.

The present discussion is focused on the southwestern half of the contiguous greenstone belt, mostly in South Africa and west of road R40 from Barberton to Swaziland, and the surrounding deeper-level (TTG) plutons, gneisses, migmatites, and supracrustal xenoliths (Fig. 5.3-1), which together comprise the Barberton Granite-Greenstone Terrain. Rocks in the northeastern part of the belt, beyond Fig. 5.3-1, have not been well studied and many show greater strain than those in the study area. BGB rocks in Swaziland (Fig. 5.3-1) are as yet poorly known. The Jamestown and Nelshoogte Schist Belts (Fig. 5.3-1) have been studied locally (e.g., Anhaeusser, 1972, 1985, 2001) but stratigraphic and age relationships are poorly resolved. These more highly strained belts are not considered here.

5.3-2.1.1. Structure

Although well preserved in many respects, rocks of the BGB have been subject to multiple episodes of folding, faulting, and metamorphism (e.g., de Wit, 1982; de Ronde and de Wit, 1994; Lowe et al., 1999). Today, the supracrustal sequence is broken up into tectonic blocks by both large and small faults: the major faults discussed here are shown in Fig. 5.3-3(A). Both bedding and fault planes throughout the belt have generally been rotated to vertical or subvertical dips. Over wide areas, small-scale strain effects such as cleavage, foliation, and lineation, are absent or have been partitioned into more ductile units, resulting in the wide preservation of textural features down to a few microns across. A more detailed summary of the structures and structural development of the BGB is presented later in this paper.

5.3-2.1.2. Alteration

Virtually all BGB rocks have been altered at temperatures in excess of 300 °C (Xie et al., 1997; Cloete, 1999; Tice et al., 2004). However, many units display original textures and sedimentary structures down to a few microns across, generally as a consequence of early silicification and the local partitioning of strain into adjacent more ductile units, such as serpentinized ultramafic lavas. However, widespread metasomatic alteration has obliterated all but the most refractory primary minerals in most rocks. Preserved detrital and early diagenetic minerals in sedimentary units include chromites, zircon, apatite, rutile, coarse quartz, carbonates, iron oxides, and barite. The bulk of the original sediments have been replaced by microquartz (chert) or carbonate or have been recrystallized into fine mosaics of quartz, carbonate, sericite, chlorite, and other alteration products. Igneous units have been similarly altered, although, again, textures and structures are locally preserved with remarkable clarity. Peridotitic komatiites are commonly altered to mosaics of serpentine, magnetite, tremolite, and chlorite. Primary chromite is commonly preserved and rare enclaves show primary orthopyroxene, augite, fresh olivine (Smith and Erlank, 1982), and even optically glassy melt inclusions (Kareem and Byerly, 2002, 2003). Extreme metasomatic alteration of komatiitic and basaltic flow rocks marks the tops of most volcanic sequences. These altered rocks commonly underlie regionally developed silicified sedimentary units (cherts) and appear as brownish carbonate-rich rocks or greenish cherts. The remaining mineralogy is dominated by Cr-rich sericite. They are generally cut by complexly anastomozing quartz veins. These zones have variously been interpreted as sea-floor flow-top alteration zones (Lowe and Byerly, 1986a; Lowe et al., 1999), hydrothermal alteration zones (Duchac and Hanor, 1987; Hanor and Duchac, 1990), or shear zones marking major thrust faults (de Wit, 1982, 1983).

Basaltic volcanic and volcaniclastic units are typically altered to assemblages of albite, tremolite, chlorite, and occasionally epidote and sphene. Primary pyroxenes and plagioclase are preserved locally. Coarser-grained volcaniclastic units may be substantially replaced by iron-rich dolomite and magnesite. Felsic volcanic and volcaniclastic units, where best preserved, contain primary zircon, apatite, quartz, and less commonly plagioclase, hornblende, and sphene. More typically, these rocks are altered to lithologies dominated by microcrystalline quartz and sericite.

Fig. 5.3-2. (*Next page.*) Generalized stratigraphies of the principal tectono-stratigraphic suites in the BGGT (modified from Lowe, 1999c, Fig. 2). Within the study area, the Kromberg suite is an overlap assemblage deposited on the Songimvelo suite. The outcrop areas of the Steynsdorp, Songimvelo, and Kromberg suites include rocks of the individual suites as well as overlap assemblages deposited during formation of adjacent, younger suites. Ages with asterisks (∗) below stratigraphic columns are from xenocrysts within the magmatic rocks on that block. Ages with daggers (†) are from detrital zircons and rock fragments within sedimentary units. Age with number symbol (#) is from a gneiss block within a shear zone along the southern edge of the Songimvelo Block. Ages with double asterisks (∗∗) are from the Theespruit Formation of the Steynsdorp suite south of the Komati Fault on the west limb of the Onverwacht Anticline.

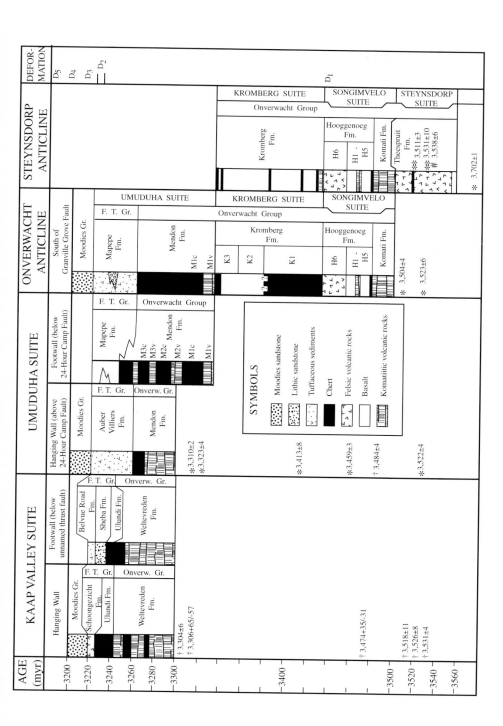

Major and trace element compositions of BGB rocks have been used with moderate success in petrogenetic studies (e.g., de Wit et al., 1987a; Vennemann and Smith, 1999; Byerly, 1999). In the most highly altered lithologies, immobile elements (Al, Ti, Zr and Cr) and REE have proven useful in the identification of the protoliths (e.g., Duchac and Hanor, 1987; Lowe, 1999a, 1999b).

5.3-3. TECTONO-STRATIGRAPHIC SUITES

While the classical Onverwacht, Fig Tree, and Moodies Groups provide a useful lithostratigraphic framework for the BGB, regional usage of these names conveys the impression of a layer-cake type stratigraphy that obscures the complex and diachronous evolution of the belt. In order to better characterize the development of the BGB and surrounding granitoid intrusive units, we here divide rocks of the Kaapvaal Craton in the BGB area into five major architectural elements here termed tectono-stratigraphic suites (Figs. 5.3-1 and 5.3-2). These suites correspond generally to the tectono-stratigraphic blocks of Lowe (1999c) and the "magmato-tectonic" events of Poujol et al. (2003) and Moyen et al. (this volume). Each suite consists of a sequence of volcanic and sedimentary rocks and related deeper-level TTG plutons, gneisses, and migmatites. These form the basic architectural building blocks of the BGGT. They are not simply fault-bounded structural blocks or exotic terranes that have been assembled tectonically. While some suites are in fault contact, in many instances, volcanic or sedimentary rocks of one suite lie at least locally with apparent conformity on strata of earlier suites.

The BGGT tectono-stratigraphic suites include (Figs. 5.3-1 and 5.3-2):
(1) the Steynsdorp suite, which includes pre-3.5 Ga rocks of the BGB, exposed in the southernmost and southeasternmost parts of the belt, and the 3509 Ma Steynsdorp Pluton;
(2) the Songimvelo suite, which includes much of the south-central part of the BGB and the adjacent ~3445 Ma TTG plutons and metamorphic units;
(3) the Kromberg suite, which includes volcanic and sedimentary rocks of the Kromberg Formation;
(4) the Umuduha suite, occupying the central part of the BGB; and,
(5) the Kaap Valley suite, which includes the BGB north of the Inyoka Fault and the 3236 to ~3225 Ma Kaap Valley, Nelshoogte, and Badplaas TTG Plutons.
The general relationships among these tectono-stratigraphic suites are shown schematically in Fig. 5.3-4. We have not included the Steyndorp and Songimvelo suites in a single suite, block, or domain as do Kisters et al. (2003) and Moyen et al. (2006), which they term the Stolzburg domain, because, although they may have been metamorphosed and structurally affected together by later events, they clearly represent geochronologically and stratigraphically separate eruptive and intrusive cycles.

5.3-3.1. Steynsdorp Suite

The Steynsdorp suite includes rocks older than 3500 Ma, including the Theespruit Formation and the 3509 Ma Steynsdorp Pluton in the Steynsdorp Anticline and the Theespruit

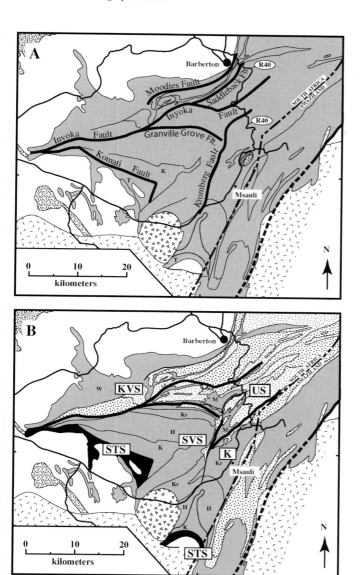

Fig. 5.3-3. (A) Map showing the major faults in the BGB discussed in the text. (B–F) Outcropping greenstone-belt rocks of the Barberton tectono-stratigraphic suites (black). (B) Steynsdorp suite (STS). (C) Songimvelo suite (SVS). (D) Kromberg suite (K). (E) Umuduha suite (US). (F) Kaap Valley suite (KVS).

Formation in the Onverwacht Anticline south of the Komati Fault (Fig. 5.3-3(B)). Locally, komatiitic and basaltic rocks of the Sandspruit Formation may represent older parts of this suite. In the Onverwacht Anticline, the Theespruit Formation includes a variety

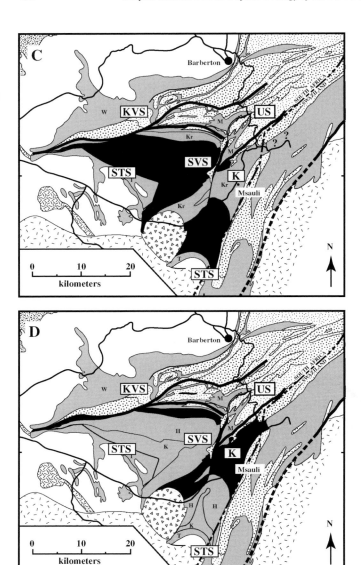

Fig. 5.3-3. (*Continued.*)

of metamorphosed and faulted felsic volcanic rocks, felsic breccias, banded cherts, and metamorphosed mafic and komatiitic volcanic rocks (Fig. 5.3-5). These are separated from rocks of the less altered Komati Formation by the Komati Fault (Figs. 5.3-1, 5.3-3(A)). Felsic volcanic rocks have yielded ages of 3531 ± 10 Ma (Armstrong et al., 1990) and 3511 ± 3 Ma (Kröner et al., 1992). Gneissic TTG blocks with crystallization ages of

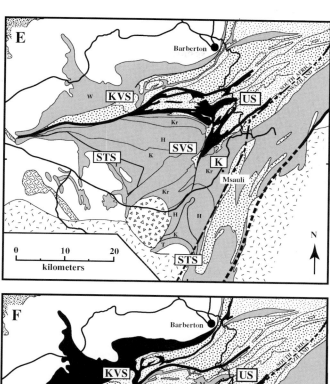

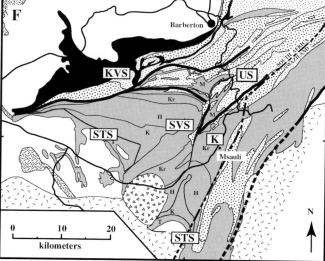

Fig. 5.3-3. (*Continued.*)

3538 ± 6 Ma (Armstrong et al., 1990) and 3538 + 4/−2 Ma (Kamo and Davis, 1994) have also been brought up along faults.

In the Steynsdorp Anticline, the Steynsdorp suite includes a thick sequence of mafic and felsic schists representing altered volcanic units and interlayered metaquartzites representing metamorphosed cherts assigned to the Sandspruit and Theespruit Formations

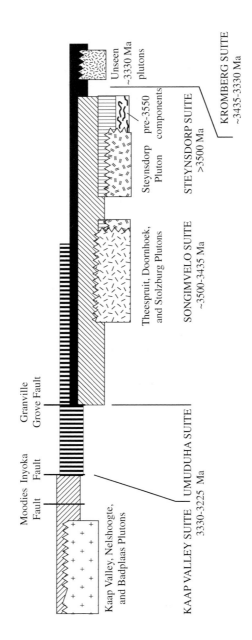

Fig. 5.3–4. Schematic diagram showing inferred pre-deformation relationships of the tectono-stratigraphic suites making up the BGGT. Modified from Lowe (1999c; Fig. 4). The magmatic center for the Kromberg suite is shown schematically on the right side of the diagram.

Fig. 5.3-5. Rocks of the Theespruit Formation, Steynsdorp suite, west limb of the Onverwacht anticline. (A) Layered tuffaceous aluminous metasediments. The coarser layers are composed of large andalusite porphyroblasts, not primary conglomeratic material. (B) Aluminous metasediments showing layering that may represent primary cross laminations. Pen in both figures is 15 cm long.

(Viljoen and Viljoen, 1969a) and the associated Steynsdorp Pluton. Theespruit schists and cherts show tight isoclinal folds with strong axial plane cleavage (Kisters and Anhaeusser, 1995b). Theespruit felsic schists have been dated at 3548 ± 3 to 3544 ± 3 Ma (Kröner et al., 1996).

The composite Steynsdorp Pluton includes older tonalitic-trondhjemitic gneisses and younger foliated granodiorite (Kisters and Anhaeusser, 1995b; Kröner, et al., 1996), both dated between 3502 ± 2 and 3511 ± 4 Ma (Kamo and Davis, 1994; Kröner et al., 1996). The late-stage granodiorites are clearly intrusive into the oldest portions of the BGB (Kröner, et al., 1996). These rocks also contain zircon xenocrysts 3553 ± 5 to 3531 ± 3 Ma, the same age as felsic tuffs of the Theespruit Formation (Kröner et al., 1996). The somewhat younger, weakly foliated Vlakplaats granodiorite, 3450 ± 3 Ma, intrudes the Komati Formation in the Steynsdorp Anticline. It is notable for zircons 3702 ± 1 Ma, the oldest xenocryts reported from the BGGT or Kaapvaal Craton, indicating that the source region was not simply mafic crust but crust that was compositionally and geochronologically complex (Kröner et al., 1996).

The contact between the Theespruit and Komati Formations, marking the contact between the Steynsdorp and Songimvelo suites, is poorly exposed but was termed a "plane of décollement" by Viljoen et al. (1969) and a tectonic contact by Kröner et al. (1996). Kisters and Anhaeusser (1995b) noted the contrast in style and intensity of deformation across this contact.

The northern extent of the pre-3.5 Ga Steynsdorp suite beneath the younger rocks is unknown, although similar-aged intrusive units are recorded from the Ancient Gneiss Complex in northeastern Swaziland (Kröner et al., 1991b) and xenocrystic zircons older than 3.5 Ga are present in younger felsic volcanic units in the Songimvelo, Umuduha, and Kaap Valley suites (Fig. 5.3-4).

5.3-3.2. Songimvelo Suite

Classic sections of the Komati and Hooggenoeg Formations (Viljoen and Viljoen, 1969a, 1969b) totaling ~6 to 7.5 km thick in the Onverwacht Anticline, Kromberg Syncline, and Steynsdorp Anticline (Fig. 5.3-3(C)) form the Songimvelo suite (Fig. 5.3-6). In these sections, Songimvelo suite rocks are overlain with apparent conformity by rocks of the Kromberg suite.

5.3-3.2.1. Komati Formation

The Komati Formation (Viljoen and Viljoen, 1969a; Viljoen et al., 1983) includes peridotitic and basaltic rocks between the Komati Fault and the Middle Marker, a distinctive chert marking the base of the Hooggenoeg Formation. The type section on the west limb of the Onverwacht Anticline is about 3500 m thick and composed almost exclusively of komatiitic and subordinate basaltic volcanic rocks (Viljoen and Viljoen, 1969a; Dann, 2000). Although de Wit et al. (1987b) suggested that the layered komatiites are vertical dikes within a sheeted dike complex and Grove et al. (1997) interpreted them as horizontal sills intruded into the deep Archean crust, most workers have supported their origin as ultramafic, high-magnesian lava flows (e.g., Cloete, 1999; Dann, 2000).

Komatiites are composed of high magnesian olivines and pyroxenes, and chromites, set in a fine-grained matrix that was probably glass. Olivine and pyroxene in komatiitic flow units commonly display spinifex textures formed through magma quenching (Fig. 5.3-7(A) and (B)). Pillows are usually poorly developed but do occur in komatiitic basalts (Fig. 5.3-7(C)). Chromites are commonly preserved, both morphologically and compositionally, in even the most altered rocks. However, primary olivine and pyroxene minerals are rarely preserved. Olivine is largely pseudomorphed by serpentine and magnetite, and pyroxene and glass by tremolite and chlorite, respectively.

The Komati Formation contains only one known sedimentary layer, a 5–10 cm thick layer of felsic tuff (Fig. 5.3-7(D)), and, except at the top, lacks alteration zones that might represent breaks in the eruptive sequence. Their absence probably reflects the short time required for eruption of the 3.5 km of komatiitic lavas. Dann (2000) reports an age on the felsic tuff in the middle part of the formation of 3481 ± 2 Ma.

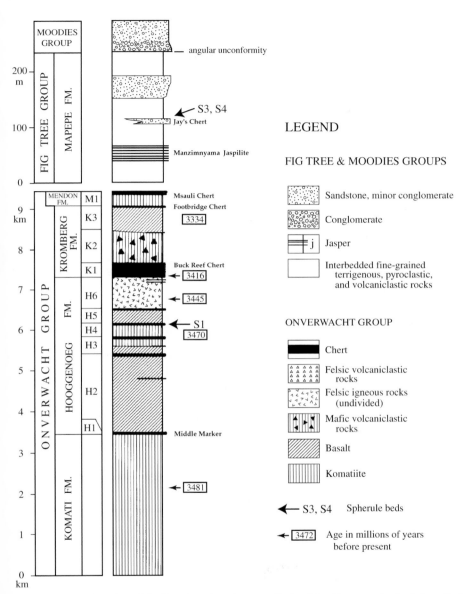

Fig. 5.3-6. Stratigraphy of the Songimvelo suite, west limb of the Onverwacht Anticline, Barberton Greenstone Belt. The greenstone portion of the Songimvelo suite includes the Komati and Hooggenoeg Formations, which are capped by H6, a thick unit of felsic volcanic and volcaniclastic rocks and shallow hypabyssal intrusions. These are comagmatic with the 3445 Ma TTG suite that includes the Theespruit, Stolzburg, and Doornhoek Plutons (Fig. 5.3-1). The overlying Kromberg Formation is an overlap assemblage from a younger suite. The Mendon Formation here includes only the lowest volcanic-sedimentary cycle, M1, and is an overlap assemblage from the Umuduha suite.

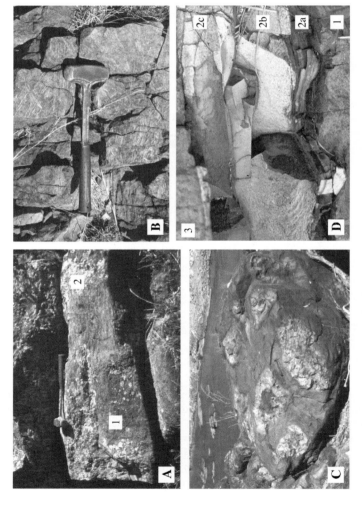

Fig. 5.3-7. Rocks of the Komati Formation. (A) Thin flow unit that includes a lower layer of randomly oriented spinifex textured komatiite (1) and an upper layer of vertically oriented pyroxene spinifex (2). (B) Olivine spinifex in a komatiitic flow. (C) Pillowed komatiitic basalt. (D) Layer of felsic tuff (2), 13 cm thick, between underlying (1) and overlying (3) flow units in the Komati Formation. This tuff has been dated at 3481 ± 2 Ma by Dann (2000). The tuff includes three individual tuff beds, 2a, 2b, and 2c from base to top. The middle bed shows normal grain-size grading.

5.3-3.2.2. Hooggenoeg Formation

The Hooggenoeg Formation (Viljoen and Viljoen, 1969b) comprises a thick sequence of tholeiitic basalts, komatiitic basalts and rarer peridotitic komatiites; dacitic igneous rocks; and thin cherty units. The thickness of the formation on the central part of the west limb of the Onverwacht Anticline is about 3900 m. The upper parts of all sections of the Hooggenoeg Formation in this area have been disturbed by shearing and intrusion (Lowe and Byerly, 1999; Lowe et al., 1999). The stratigraphy is more coherent along most of the east limb, where a complete section of the formation is 2900 m thick, and on the far western part of the west limb. The upper, felsic part of the formation is well exposed along the Komati River (Viljoen and Viljoen, 1969b; Lowe and Knauth, 1977). The formation also occurs to the southeast on the east limb of the Kromberg Syncline and in the Steynsdorp Anticline but has not been investigated in any detail in these areas.

The Hooggenoeg Formation has been divided into six informal members (Fig. 5.3-6) that are traceable around the Onverwacht Anticline and Kromberg Syncline (Lowe and Byerly, 1999): (H1) the Middle Marker; (H2) massive and pillowed tholeiitic basalt capped by a thin unit of gray chert; (H3 and H4) units of komatiitic basalt and basalt, with at least two zones of spinifex-bearing komatiite at the top of H3v and in the lower third of H4v, capped by thin units composed largely of silicified volcaniclastic debris (H3c and H4c); (H5) basalt and variolitic basalt capped by a thin unit of gray chert and silicified volcaniclastic sediment; and (H6) felsic volcanic and volcaniclastic rocks. Within each member, the volcanic units and capping cherts are designated with "v" and "c", respectively (e.g., H2v and H2c). Each member and the component volcanic and cherty divisions are traceable around the entire Onverwacht Anticline. Evidently the rapid eruption of mafic to komatiitic volcanic rocks formed extensive, low-relief subaqueous volcanic plains that were mantled by thin veneers of volcaniclastic and carbonaceous sediments during regional breaks in eruptive activity.

H1: The Middle Marker

The Middle Marker (H1) is the oldest, relatively unmetamorphosed sedimentary unit in the BGB. It has been dated at 3472 ± 5 Ma (Armstrong et al., 1990) and its regional stratigraphy and sedimentology have been studied by Lanier and Lowe (1982). The Middle Marker is everywhere a thin unit, 1 to about 5 m thick, that caps the 3.5+ km thick sequence of komatiitic volcanic rocks of the Komati Formation. It is underlain by a zone about 50 m thick formed by the alteration and silicification of the uppermost komatiites of the Komati Formation. The Middle Marker is composed of silicified komatiitic ash and carbonaceous materials. The komatiitic ash is now represented by pale green micaceous chert. Chemically, these and similar ash layers throughout the Hooggenoeg Formation are composed largely of SiO_2, Al_2O_3, and K_2O, and were widely regarded as silicified felsic ashes. However, examination of immobile element ratios, especially Al_2O_3, Zr, TiO_2, and Cr, indicates unambiguously that these units represent altered komatiitic ashes (Lowe, 1999a, 1999b). Several normally graded beds of accretionary lapilli and ash are present toward the top of the Middle Marker in most sections. Carbonaceous material occurs as a few thin layers of relatively pure carbonaceous chert but is more common as particles mixed with komatiitic ash in detrital units. There are also a few coarser conglomeratic units composed of coarse ash particles, rip-up ash clasts, rounded carbonaceous clasts, and accretionary lapilli.

In most sections, the Middle Marker shows evidence of deposition by waves or currents, including detrital units showing large-scale cross-stratification. The wide presence of current structured intervals suggests that the sea floor was widely swept by currents, perhaps alternating with intervals of quieter water when only air-fall komatiitic ash accumulated. However, the regional extent, continuity, and consistency of the Middle Marker in the Onverwacht Anticline suggests that the depositional environment did not vary greatly laterally and that deposition may have taken place on a flat, open, tide- or wave-influenced shelf at depths of a few 10's to perhaps as much as 100 m.

H2: Tholeiitic basalt

The Middle Marker is overlain directly and conformably by a thick sequence of tholeiitic basalt (H2v) capped by a thin unit of black chert (H2c). This subdivision is about 1800 m thick on the west limb of the Onverwacht Anticline and 1200 to 1400 m thick along the northern part of the east limb. It consists largely of alternating units of pillowed and massive tholeiitic basalt (Fig. 5.3-8(A) and (B)). Individual pillowed and massive units range from several meters to nearly 500 m in thickness, although most are between 10 and 50 m thick, and show complex lateral and vertical interstratification and interfingering (Fig. 4 of Viljoen and Viljoen, 1969b; Williams and Furnell, 1979). Pillows, which average between 0.5 and 1.5 m across, commonly show radial pipe vesicles around the outer edges and downward projections between underlying rounded pillow tops.

H3: Komatiitic and tholeiitic basalt and komatiite

Member H3 of the Hooggenoeg Formation, about 220–380 m thick, is made up largely of pillowed tholeiitic basalt, variolitic pillow basalt (Fig. 5.3-8(C) and (D)), and, at the top, massive spinifex-bearing komatiitic basalt and komatiite (H3v). At the top, H3v widely includes a layered to massive ultramafic unit termed the Rosentuin Ultramafic Body that locally reaches 150 m thick (Viljoen and Viljoen, 1969b) and is directly overlain by H3c. H3c is a distinctive layer of silicified volcaniclastic and carbonaceous sediment (Figs. 5.3-8(C) and 5.3-9(A)). Along the west limb of the Onverwacht Anticline, this chert includes a basal zone up to 15 m thick of massive, intensely silicified and metasomatically altered ash, overlain by 1 to 7 m of silicified airfall dust, ash, and accretionary lapilli, current-deposited volcaniclastic debris and carbonaceous black chert (Fig. 5.3-9(B)). The silicified ash in H3c is high in Cr and shows trace element and REE compositions indicating that the original ash was komatiitic basalt (Lowe, 1999a). H3c is about 20 m thick along the east limb but lacks the massive basal ash. H3c is underlain by a regionally traceable unit of altered, spinifex-bearing komatiite to komatiitic basalt containing Cr-bearing micas and stratiform, fibrous veinlets of silica and carbonate (Lowe and Byerly, 1986a).

H4: Komatiitic basalt, basalt, and komatiite

H3c is overlain by 250 to 350 m of volcanic rocks of H4v. The lower part of this member is dominated by komatiitic basalt and most sections include a zone, 10 to 20 m thick,

Fig. 5.3-8. Mafic volcanic rocks of the Hooggenoeg Formation. (A) Outcrop of tholeiitic pillow basalts and lava tubes from member H2v. Hammer at lower right for scale. (B) Variolitic pillows of H2v. (C) Chert unit H3c (left) about 20 meters thick, underlain by 40–50 meters of pillow basalts. (D) Variolitic pillow basalt of H3v. Large flattened cavity is probably a drain-out cavity.

showing coarse, vertical to subvertical pyroxene spinifex up to 2 m high. In the middle and upper parts of H4v, thick massive flow rock containing fine, randomly oriented spinifex is interstratified with pillowed spinifex-free basalt. The uppermost units generally consist of pillowed spinifex-free high-Mg basalt.

H4c is 0 to 6 m thick and is composed of silicified airfall and current-worked volcaniclastic debris and carbonaceous chert. It inclues spherule bed S1, containing sand-sized

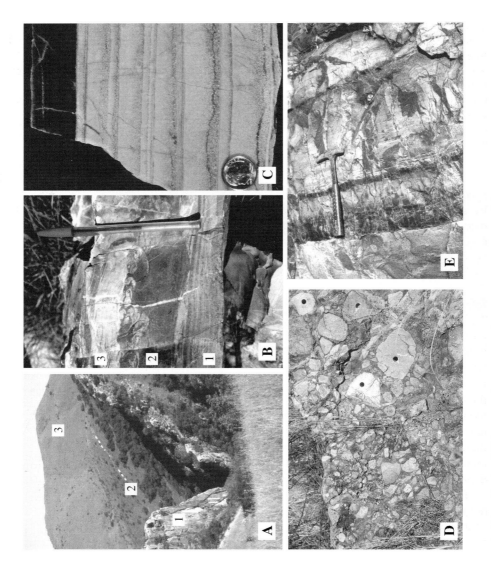

Fig. 5.3-9. (*Previous page.*) Sedimentary units in the Hooggenoeg Formation. (A) Outcrops of the resistant chert unit H3c (1) and the next higher chert unit H4c (2). The upper flank of the ridge is made up of felsic intrusive and volcaniclastic rocks of H6 (3) with the Buck Reef Chert of the Kromberg Formation at the skyline. (B) Chert from H3c showing silicified flat laminated current-deposited komatiitic ash and particulate carbonaceous matter (1) overlain by massive carbonaceous chert (2), and a layer of cross-laminated komatiitic ash (3), the lowest part of which has foundered into the soft carbonaceous sediment. This association reflects the interplay of the fall of pyroclastic debris, reworking of the pyroclastic sediments by currents, and the local production of carbonaceous matter by organisms in the formation of chert units like H1c and H3c in the Hooggenoeg Formation. (C) Slab from H5c showing 5–6 thin graded layers of accretionary lapilli and ash. The lack of current structures implies that these were fall-deposited layers that were not reworked by currents. (D) Coarse conglomerate composed exclusively of clasts of felsic volcanic rock in H6 along the Komati River in the Komati Gorge section, west limb of the Kromberg Syncline. These conglomerates are interbedded with angular felsic breccias and grade upward into thick-bedded turbidites and then thinner-bedded turbidites, as shown in (E). Holes that deface outcrop are where samples for paleomagnetic studies were drilled out. They are approximately 2.5 cm across. (E) Medium-bedded, normally graded turbidites near the top of H6 on the west limb of the Kromberg Syncline along the Komati River. Stratigraphic top is toward the left. The dark bed tops are composed of chert formed by the silicification of fine-grained cross-laminated felsic volcaniclastic sand and silt (Lowe and Knauth, 1977).

---

spherules formed as quenched liquid silicate droplets from a rock vapor plume produced by a large meteorite impact (Lowe et al., 2003). This is the only impact layer found to date in the Onverwacht Group. Others are present in the Fig Tree Group (Lowe and Byerly, 1986b; Lowe et al., 1989, 2003). S1 has been identified and correlated between the BGB and Pilbara Craton, Western Australia. Detrital zircons from S1 have been dated at 3470 ± 2 Ma (Byerly et al., 2002).

H5: Basalt

The uppermost part of the mafic volcanic sequence of the Hooggenoeg Formation is made up largely of pillowed and massive high-Mg and tholeiitic basalt (H5v). This member is poorly preserved on the west limb of the Onverwacht Anticline because of disruption within the Geluk Fault zone, which separates H4 or H5 from overlying dacitic rocks of H6. It is well developed, however, along the eastern limb, where it averages about 390 m thick. Most of the rocks are pillow basalts with pillows 0.5 to 1.5 m across, commonly with radial pipe vesicles and chilled selvages. Near the top of the section is a regionally developed unit of variolitic pillow basalt. The varioles are separated around the pillow margins but coalesce toward the interiors (Viljoen and Viljoen, 1969b). This unit includes some of the largest varioles we have seen in the greenstone belt, locally reaching 3 cm in diameter.

Thin chert units are locally interbedded with the volcanic rocks. The volcanic sequence is capped by a layer of chert (H5c) from less than 1 to 2 m thick. This chert consists of a basal zone of massive black chert overlain by interbedded volcaniclastic sediments and black chert. Several thin layers of accretionary lapilli occur regionally at the top of

H5c (Fig. 5.3-9(C)). The chert is underlain by a zone up to 10 m thick of silicified basalt, locally cut by black chert dikes.

H6: Felsic igneous and volcaniclastic rocks

The uppermost member of the Hooggenoeg Formation, H6, is a complex association of felsic igneous and volcaniclastic rocks. On the central part of the west limb of the Onverwacht Anticline, H6 consists largely of massive quartz-bearing dacitic intrusive rock overlain by a thin cover of volcaniclastic breccia, conglomerate, and sandstone. This intrusion (Viljoen and Viljoen, 1969b; Smith, 1981) has been dated at 3445 ± 4 Ma by de Wit et al. (1987a). It crops out for 9 to 10 km along strike, ranges from 1 to 2.5 km thick, and contains large, detached, rotated masses of basaltic volcanic rock of underlying Hooggenoeg members H4 and H5. Immediately underlying and peripheral to the intrusion is a zone of shearing and block rotation from a few tens to several hundreds of meters thick, the Geluk Fault zone (Lowe and Byerly, 1999; Lowe et al., 1999). It is made up largely of overturned blocks of Hooggenoeg members H2v to H4v and is sharply bounded below by undeformed rocks of members H4 and H5.

The intrusion and zone of overturned blocks are overlain and flanked by thick sequences of massive, coarse, felsic volcaniclastic breccia and conglomerate containing sparse clasts of mafic and komatiitic volcanic rock. These clastic units thin and fine away from the intrusion in both directions. On the far west limb of the Onverwacht Anticline, H6 is represented by a sequence of fine-grained dacitic tuff less than 100 m thick. Along the east limb of the anticline, massive fan-delta and alluvial conglomerates over 750 m thick grade southward into a section in Komati Gorge along the Komati River that includes about 10 m of dacite-clast conglomerate (Fig. 5.3-9(D)) and breccia overlain by 150 m of dacitic debris-flow deposits, coarse-grained thick-bedded turbidites, and fine-grained, thin bedded, silicified turbidites (Fig. 5.3-9(E)) (Viljoen and Viljoen, 1969b; Lowe and Knauth, 1977). Paleocurrent indictors in the Komati Gorge section suggest flow from north to south. The felsic turbidites at the top of the Hooggenoeg Formation have been correlated with the Buck Reef Chert on the west limb of the anticline (Viljoen and Viljoen, 1969b), but regional mapping indicates that these turbidites are a local facies of the felsic volcanic sequence in the type section of the Hooggenoeg Formation (Lowe and Byerly, 1999).

The dacitic intrusion appears to have been emplaced beneath a roof of mafic volcanic rocks of H4 and H5 and conglomerates of H6 that was no more than two hundred meters thick and which fractured and foundered into the silicic magma. We have identified no evidence, such as foundered blocks of Buck Reef Chert, that overlying sedimentary units of the Kromberg Formation existed at the time of intrusion. Ages of 3445 ± 6 Ma and 3438 ± 12 Ma on volcaniclastic units laterally equivalent to the intrusion, of 3445 ± 4 Ma on the intrusion itself (de Wit et al., 1987a), and of 3416 ± 7 Ma on rocks in the lowest Kromberg Formation overlying the intrusion (Kröner et al., 1991a) indicate that the intrusion is Hooggenoeg in age and was emplaced before deposition of overlying sedimentary units of H6 and the Buck Reef Chert. The intrusion probably formed a shallow dacitic dome mantled by roof rock and its own chilled and fragmented debris that was subsequently eroded to form the flanking volcaniclastic aprons. The intrusion may represent only the uppermost hypabyssal portion of a much larger body that was connected at depth to the 3445 Ma-old tonalite-trondhjemite-granodiorite (TTG) suite bounding the southern margin of the BGB and upward to vents through which felsic extrusives of the Hooggenoeg Formation were erupted (de Wit et al., 1987a). Felsic volcanic rocks roughly equivalent to H6 in other areas have been dated at 3457 ± 15 to 3438 ± 6 Ma (Kröner and Todt, 1988).

Knauth and Lowe (2003) have shown that below H6, oxygen isotopic compositions of cherts of the Onverwacht Group have partially equilibrated with surrounding volcanic rocks whereas above H6, they have not been homogenized and retain original systematic differences and a non-igneous range of compositions. They attribute resetting below H6 to high water/rock ratios during metasomatism associated with the emplacement of the 3445 Ma TTG suite south of the Komati Fault and emplacement and eruption of comagmatic rocks of H6.

5.3-3.2.3. Songimvelo intrusive and related rocks

Supracrustal rocks of the Songimvelo suite are nowhere in contact with the corresponding deep-level TTG plutons. However, south of the Komati Fault (Fig. 5.3-1), rocks mapped as Theespruit and Sandspruit Formations are widely intruded by TTG bodies, the largest of which are the ~3445 Ma Theespruit, Stolzburg, and Doornhoek Plutons. These rocks are mostly trondhjemites displaying moderate gneissosity parallel to intrusive margins with the BGB (Robb and Anhauesser, 1983; Kisters and Anhaeusser, 1995a) and dated between ~3460 and 3443 Ma (Kamo and Davis, 1994; de Wit et al., 1987a).

Unlike most supracrustal rocks, many TTGs retain much of their original mineralogy. The trondhjemites are dominated by plagioclase and quartz, with minor biotite and microcline, and the tonalities by plagioclase, quartz, biotite, and hornblende with minor microcline. These are all typically coarse-grained rocks. Alteration, when present, is mostly confined to sericite after feldspar and epidote-chlorite after the ferromagnesian phases.

Within the intrusive and gneiss complex south of the Komati Fault, many metamorphosed supracrustal blocks are preserved as xenoliths within the intrusive units. These include komatiitic rocks mapped as the Sandspruit Formation (Fig. 5.3-1) by Viljoen and Viljoen (1969a). To date, the stratigraphy, structure, and ages of these rocks remain poorly resolved. The Sandspruit Formation (Fig. 5.3-1) was considered by Viljoen and Viljoen (1969a) to be the lowest stratigraphic unit in the Onverwacht Group. More recent studies (Dziggel et al., 2002, 2006a) have shown that the Sandspruit Formation locally includes thin clastic sedimentary units containing detrital zircons with ages ranging from 3540 to 3521 Ma. These represent source rocks equivalent in age to the Theespruit Formation and suggest that the associated volcanic units may also be age equivalents of rocks within the main body of the BGB. Other greenstone remnants, such as the Schapenburg Schist Belt (Anhaeusser, 1983), include detrital zircons with ages of 3250 and 3231 Ma, equivalent in age to the Fig Tree Group in the BGB (Stevens et al., 2002). These rocks and similar greenstone outliers appear to represent portions of an originally more extensive greenstone terrain represented by the BGB that have been isolated through a combination of structural, intrusive, and erosive events.

5.3-3.3. Kromberg Suite

The Kromberg suite includes rocks of the Kromberg Formation (Figs. 5.3-1, 5.3-2, and 5.3-3(D)). In the Komati Gorge section, silicified volcaniclastic turbidites and debris-flow deposits of H6 at the top of the Hooggenoeg Formation are overlain by 100–200 m of massive serpentinized ultramafic rock that mark the base of the type section of the Kromberg Formation (Viljoen and Viljoen, 1969b). The Kromberg Formation here includes about 1700 m of volcanic and sedimentary rocks representing three principal lithofacies: (1) massive and pillowed basalt and komatiite (Fig. 5.3-10(A)), (2) mafic lapilli tuff and lapillistone (Fig. 5.3-10(B) and (C)), and (3) black and banded chert (Fig. 5.3-10(D)). The upper contact of the Kromberg Formation and suite is placed at the top of the Footbridge Chert, a regionally traceable unit of black and banded chert cropping out at the footbridge across the Komati River (5700 ft. in stratigraphic section of the Kromberg Formation in Fig. 9 of Viljoen and Viljoen (1969b)). Viljoen and Viljoen (1969b) included in the Kromberg Formation an additional 300 m of basaltic and serpentinized peridotitic komatiite above the Footbridge Chert. Based on regional stratigraphic relationships, Lowe and Byerly (1999) included these in the Mendon Formation.

On the west limb of the Onverwacht Anticline, felsic volcaniclastic conglomerates and sandstones of H6 are overlain with apparent conformity by cherts of the Kromberg Formation. A well-exposed section 1500 to 1800 m thick is present on farm Granville Grove 720 JT. This section can be divided into three members: (K1) the Buck Reef Chert, (K2) mafic lapilli tuff and lapillistone, and (K3) tholeiitic basalt.

5.3-3.3.1. Buck Reef Chert (K1)
On the west limb of the Onverwacht Anticline and on the east limb north of the Msauli River, the basal member of the Kromberg Formation is a 150 to 350 m-thick unit of chert named the Buck Reef Chert by Hall (1918) in reference to the fact that it was a "reef" that lacked gold. This unit was mislabeled the Bucks Ridge Chert by Heinrichs (1980). The contact between dacitic volcaniclastic sandstone at the top of H6 and the base of K1 is transitional, with thin beds of black chert, silicified wave-rippled carbonaceous sediment, and silicified evaporite interbedded in the upper 5 to 50 m of the volcaniclastic section (Lowe and Fisher Worrell, 1999).

The Buck Reef Chert includes three and, in some sections, four subdivisions: (1) a basal division of silicified evaporite (Fig. 5.3-11(A) and Lowe and Fisher Worrell, 1999) that is widespread along the west limb of the Onverwacht Anticline but replaced by non-evaporitic strata on the east limb and the far western part of the west limb, (2) a lower division of black-and-white banded chert (Fig. 5.3-11(B)), (3) a division of banded ferruginous chert, and (4) a upper division of black-and-white banded chert. The basal evaporite includes from 5 to 40 m of silicified evaporites, silicified laminated and wave-rippled shallow-water sediments, and silicified evaporite-solution collapse layers on the central part of the west limb (Fisher Worrell, 1985; Lowe and Fisher Worrell, 1999). Sediments of the Buck Reef Chert were deposited on a subsiding volcanic platform representing the eroded felsic volcanic edifice of H6. The black-and-white banded cherts contain a wide variety of detrital

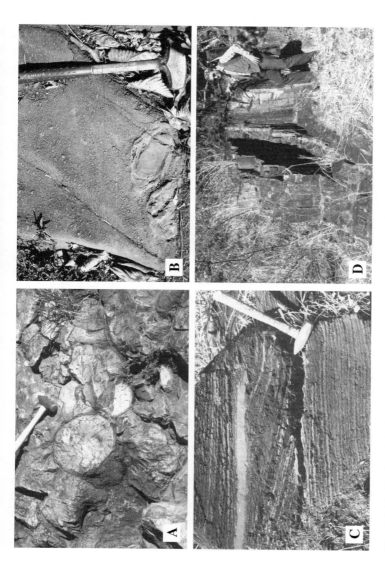

Fig. 5.3-10. Rocks of the Kromberg Formation. (A) Pillow basalt in the lower part of the Kromberg Formation in Komati Gorge, west limb of the Kromberg Syncline. The pillows show well-developed radial pipe vesicles. This pillowed sequence is laterally equivalent to part of the Buck Reef Chert on the west limb of the Onverwacht Anticline. (B) Member K2v of mafic lapillistone showing cobble-sized coated bombs. (C) Large-scale cross-stratification in mafic lapillistone of K2v. (D) Unit of black and banded chert in the lower part of the Kromberg Formation, west limb of the Kromberg Syncline, along the Komati River. This unit, K1c2 of Lowe and Byerly (1999), is the lateral equivalent of the middle part of the Buck Reef Chert on the west limb of the Onverwacht Anticline.

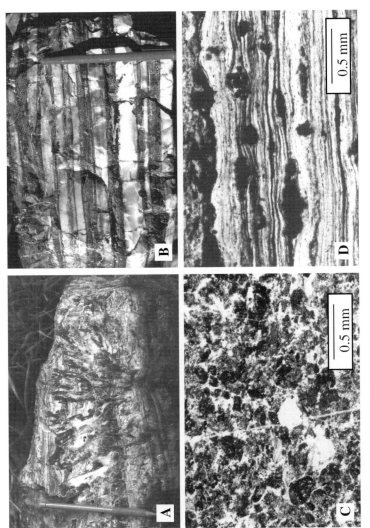

Fig. 5.3-11. Photographs of the Buck Reef Chert member (K1) of the Kromberg Formation. (A) Silicified evaporates from the basal part of the Buck Reef Chert (Lowe and Fisher Worrell, 1999). Upward radiating evaporate crystals, 5–10 cm long, have been dissolved and the cavities infilled by an initial lining of fine silica (dark) and an inner fill of coarse quartz (white). The crystals were growing upward through wave-rippled sediments, also now silicified. (B) Banded black-and-white chert of the Buck Reef Chert. The dark bands contain abundant organic carbon, both as particles and more continuous mat-like laminations. (C) Photomicrograph of detrital carbonaceous particles now cemented by silica in the BRC. (D) Fine, mat-like carbonaceous laminations in the Buck Reef Chert. Larger clots of carbonaceous matter are detrital grains that were rolled by currents and deposited on the in situ mats.

(Fig. 5.3-11(C)) and in situ mat-like (Fig. 5.3-11(D)) carbonaceous matter that has been shown to represent early photosynthetic microbial communities (Tice and Lowe, 2004, 2006a, 2006b).

In the hinge region of the Onverwacht Anticline, the evaporitic and lower black-and-white banded chert divisions can be identified but the upper divisions are represented by a relatively uniform sequence of thinly layered banded ferruginous chert. Volcanic units are interstratified in the thinning chert sequence on the northern part of the east limb of the anticline and, along the Komati River, the equivalent section consists of three chert layers interbedded with basaltic volcanic units (cherts at 750, 1250, and 2450 ft. in stratigraphic section of the Kromberg Formation in Fig. 9 of Viljoen and Viljoen (1969b)). For several kilometers north and south of the Komati River, the top of H6 and the lowest of these three cherts are separated by a ~100–200 m thick unit of serpentinized peridotitic komatiite overlain by ~50 m of komatiitic lapillistone with thin accretionary lapilli beds at the top. The evaporitic member of the Buck Reef Chert is absent in this section and these komatiitic units are thought to correlate with thin layers of komatiitic ash and accretionary lapilli near the top of the evaporitic member on the west limb of the Onverwacht Anticline (Lowe and Fisher Worrell, 1999).

In all sections on the west limb, the Buck Reef Chert is cut by anastomosing dikes and sills of mafic and komatiitic igneous rock.

5.3-3.3.2. Mafic lapilli tuff and lapillistone (K2)
The middle part of the Kromberg Formation on the west limb is made up largely of coarse, altered, mafic lapillistone and tuff of K2v (Ransom, 1987; Ransom et al., 1999). The basal 10 to 100 m consists of fine-grained mafic to komatiitic tuff and fissile, non-silicified, tuffaceous, carbonaceous shale containing thin komatiitic flows. The overlying unit of coarse mafic to ultramafic lapillistone ranges from 300 to 1000 m thick. The lower third to half consists mainly of massive lapillistone, with lapilli 0.5 to 4 cm in diameter, containing admixed accidental clasts of Buck Reef Chert, fine-grained mafic flow rock, and rarely coarse-grained pyroxenite. Mafic flow units are widely interstratified within the fragmental sequence. The upper part of K2v is composed of stratified, generally finer-grained lapilli and includes thin layers of altered silicified ash. The topmost 50 to 100 m consist of lapilli generally less than 1 cm in diameter showing abundant current structures, including large-scale cross-stratification. The relatively high Cr content of the lapillistone (1450, 1492, 2060 ppm for three samples of carbonated lapillistone) suggests that the original debris was komatiitic in composition.

Near the central part of the west limb of the Onverwacht Anticline, K2v reaches its maximum thickness of over 1000 m, and the lower half includes abundant large angular blocks of Buck Reef Chert. In this area, the upper 200 to 300 m of the Buck Reef Chert are missing over an outcrop distance of 1.5 to 2 km (Lowe and Byerly, 1999). The vertical-sided depression left by removal of the chert is filled with mafic pyroclastic debris and mafic lavas of K2v, and the adjacent and underlying chert is cut by dikes of similar intrusive rock. This area probably represents the site of a major phreatomagmatic explosion that marked the initiation of K2v pyroclastic volcanism (Ransom et al., 1999).

The pyroclastic sequence is capped locally along the west limb by a thin unit of silicified ash and dust and black chert (K2c) showing abundant large-scale cross-stratification and other evidence of deposition in shallow water (Ransom et al., 1999).

K2 is traceable around the hinge region of the Onverwacht Anticline but is extensively faulted within the Kromberg Fault zone along the east limb north of the Komati River. In the type section along the Komati River, the member includes only 75 m of mafic to komatiitic pyroclastic debris interbedded with basalt flows (Fig. 9 of Viljoen and Viljoen, 1969b). Near the top, the volcaniclastic section shows well-developed current layering and large-scale cross-stratification (Plate XIa of Viljoen and Viljoen, 1969b). Most of the lapillistone and other coarse-grained fragmental units have been extensively replaced by chlorite, tremolite, iron-rich dolomite, and ankerite.

5.3-3.3.3. Basalt (K3)

On the west limb of the Onverwacht Anticline, K2 is overlain by silicified pillow basalt (K3v) 500 to 600 m thick (Ransom, 1987). The upper 350 m of the formation is composed of thick units of tholeiitic basalt, thin layers of komatiitic basalt, tuffaceous layers, and pillow breccia. The formation is capped by the Footbridge Chert (K3c), which, on the west limb, consists of 15 to 25 m of black and black-and-white banded chert. A second unit of black chert, poorly silicified carbonaceous shale, and carbonate 14 m thick is present 60 m below the Footbridge Chert in the type section.

The base of the Kromberg suite is $>3416 \pm 7$ Ma, based on the age of zircons collected from the top of the evaporitic member at the base of the Buck Reef Chert (Kröner et al., 1991a), $<3445 \pm 4$ Ma, based on the age of shallow intrusive rocks in H6 (de Wit et al., 1987a), and probably $<3438 \pm 6$ Ma, based on zircon ages of H6 volcanic rocks elsewhere (Kröner and Todt, 1988). For calculation purposes, we have taken the age as 3435 Ma. The top of the Kromberg suite is $<3334 \pm 3$ Ma, based on zircons from a thin felsic tuff in the Footbridge Chert in its type section (Byerly et al., 1996).

5.3-3.3.4. Intrusive rocks

The Kromberg suite lacks an obvious felsic volcanic stage, but like the Mendon Formation south of the Granville Grove Fault, available exposures of the Kromberg suite are thought to represent the distal parts of a suite that was more fully developed in other areas (Fig. 5.3-4). Widespread xenocrystic zircons 3352 to 3310 Ma in younger BGB felsic tuffs and lavas (Fig. 5.3-2), $3347 + 67/-60$ Ma granitoid phases in the Stentor Pluton north of the BGB (Tegtmeyer and Kröner, 1987), and 3303 ± 6.4 Ma tonalitic xenoliths (Kamo and Davis, 1994) in the 3.1 Ga Nelspruit Batholith north of the BGB all point to the presence of a felsic magmatic event of this age in the BGGT area that may have closed the Kromberg cycle. Intrusive units 3360–3300 Ma are also widely developed in the Ancient Gneiss Complex of Swaziland (Kröner et al., 1991b). Finally, Van Kranendonk et al. (2004) indicate that the second major tectonic-magmatic event in the Pilbara Craton occurred during the 3340–3315 Ma interval. It included the intrusion of major potassic granitoid rocks, such as the Mount Edgar Batholith and Corunna Downs Complex. Rhyolites of the Wyman Formation are the volcanic correlatives of these intrusive phases. Nelson et al. (1999) refer to this period as the Wyman Event and suggest that it ranged from 3325–3290 Ma. The specific geologic and geographic relationships of Archean rocks in the Kaapvaal and Pilbara Cratons is not fully resolved, but there are many stratigraphic, tectonic, and magmatic features and events suggesting that they developed in close proximity to one another (Zegers et al., 1998b; Nelson et al., 1999).

5.3-3.4. Umuduha Suite

5.3-3.4.1. Mendon Formation

On the west limb of the Onverwacht Anticline, the Granville Grove Fault marks the northern limit of the exposed parts of the Songimvelo and Kromberg suites (Fig. 5.3-3). North of the Granville Grove Fault and south of the Inyoka Fault (Figs. 5.3-1 and 5.3-3(E)), komatiitic volcanic rocks of the Onverwacht Group have been assigned to the Mendon Formation by Lowe and Byerly (1999). Throughout its outcrop, the Mendon Formation is made up of volcanic cycles (Fig. 5.3-12), each consisting of an ultramafic flow unit capped by a chert layer (Byerly, 1999). The volcanic units are termed M1v, M2v, etc. and the capping cherts M1c, M2c, etc. (Fig. 5.3-12).

South of the Granville Grove Fault, M1v is the only volcanic unit present in the formation (Fig. 5.3-2). It is about 300 m thick and appears to represent a single thick flow of peridotitic komatiite (Byerly, 1999). It is capped by M1c, the Msauli Chert, a 20–30 m-thick unit of silicified komatiitic ash and carbonaceous matter. The Msauli Chert is noteworthy because it includes numerous graded layers of komatiitic ash and accretionary lapilli (Lowe and Knauth, 1978), many of which can be traced throughout the entire outcrop of the unit (Lowe, 1999b). The stacked graded ash and lapilli layers, many of which are massive at the base and show cross-laminations at the top, have been interpreted as turbidites deposited in deep water (Stanistreet et al., 1981; Heinrichs, 1984). However, more careful analysis of the sedimentary structures and textures show that the principal sediment types include current-deposited komatiitic ash (Fig. 5.3-13(A)), graded layers of fall-deposited komatiitic accretionary lapilli and ash (Fig. 5.3-13(B)), and interlayered units of current-deposited ash and carbonaceous matter deposited under alternating current-active and quiet-water conditions (Fig. 5.3-13(C) and (D)). Deposition involved the fall of pyroclastic debris from komatiitic eruption clouds into a marine setting in which quiet-water intervals alternated with periods of current activity, probably a tide-influenced shallow-water to shelf environment (Lowe, 1999b). South of the Granville Grove Fault, M1c is overlain by 20–50 m of black, banded, and ferruginous chert succeeded by rocks of the Fig Tree Group.

North of the Granville Grove Fault on the west limb of the Onverwacht Anticline exposed parts of the Mendon Formation young northward across a series of fault-bounded outcrop belts. This northward younging involves both the disappearance of older cycles at the base of the exposed sections and the progressive appearance of higher, younger cycles at the tops (Figs. 5.3-2 and 5.3-12). The northernmost belt includes rocks of M3, M4, and M5 capped by black cherts overlain by Fig Tree strata.

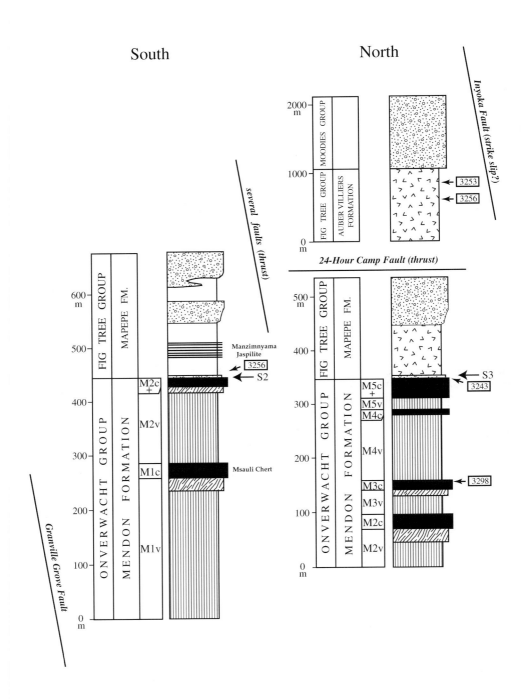

Fig. 5.3-12. (*Previous page.*) Generalized stratigraphy of the Umuduha suite. The mafic to ultramafic portion of this suite includes rocks of the Mendon Formation. It is made up of volcanic-sedimentary cycles, each consisting of a lower komatiitic volcanic unit and a capping chert. South of the Granville Grove Fault (Fig. 5.3-1), M1v is the only volcanic unit present, overlying rocks of the Kromberg Formation and overlain by M1c and succeeding black cherts that may include the equivalents of higher volcanic cycles to the north. North of the Granville Grove fault, rocks of the Kromberg Formation are not exposed and may not have been present. Traced to the north across a series of fault-bounded structural blocks, rocks of older Mendon volcanic-sedimentary cycles disappear and younger volcanic units appear at the top of the formation below the overlying Fig Tree Group.

The Footbridge Chert at the top of the underlying Kromberg Formation on the Songimvelo block has been dated at 3334 ± 3 Ma (Byerly et al., 1996) and a felsic tuff in M2c or M3c at 3298 ± 3 Ma (Byerly et al., 1996). The top of the formation youngs to the north. A felsic tuff near the base of the Fig Tree Group immediately north of the Granville Grove Fault has an age of 3258±3 Ma (Byerly et al., 1996). A second felsic tuff at the base of the Fig Tree Group just south of the Inyoka Fault has yielded an age of 3243 ± 4 Ma (Kröner et al., 1991a). The Mendon Formation therefore ranges from ~3330 to ~3245 Ma.

5.3-3.4.2. Fig Tree Group
The Fig Tree Group consists of interstratified terrigenous clastic units and dacitic to rhyo-dacitic volcaniclastic and volcanic rocks. The base of the group is widely marked by layers of spherical particles (Fig. 5.3-14) produced by the condensation of rock vapor clouds resulting from the impacts of large meteorites on the early Earth (Lowe and Byerly, 1986b; Lowe et al., 1989, 2003; Shukolyukov et al., 2000; Kyte et al., 2003; Krull-Davatzes et al., 2006). South of the Inyoka Fault, the Fig Tree includes two formation-level units, the Mapepe and Auber Villiers Formations, separated by a regional thrust fault, the 24-hour Camp Fault (Fig. 5.3-12). The Mapepe Formation, on the lower plate of the fault, includes up to 700 m of predominantly fine-grained felsic tuffaceous strata (Fig. 5.3-15(A)), shale, chert-grit sandstone (Fig. 5.3-15(B)), and chert-clast conglomerate (Fig. 5.3-15(C) and (D)). Spherule bed S2 (Lowe et al., 2003) widely marks the base of the Mapepe Formation in southern sections (Fig. 5.3-12). It is overlain by iron-rich sediments that in southern sections include the Manzimnyama Jaspilite (Fig. 5.3-12). This unit of hematitic iron formation thickens to the south onto the Songimvelo block and thins and becomes more tuffaceous to the north toward the Inyoka Fault. Immediately south of the Inyoka Fault, the lowest 100 m of Mapepe strata consists mainly of fine-grained felsic tuff (Fig. 5.3-12). The bulk of the Mapepe Formation consists of a complex association of fan delta terrigenous clastic units, felsic pyroclastic rocks, and minor chert, jasper, and barite. The thickest and best studied barite deposits (Fig. 5.3-15(E)) are those in the Barite Syncline (Heinrichs and Reimer, 1977). Fine-grained dacitic tuffs in the Mapepe Formation immediately north of the Granville Grove Fault have yielded ages of 3258 ± 3 Ma, immediately above the Manzimnyama Jaspilite, and 3227 ± 4 Ma, at the top of the formation, and a tuff near the

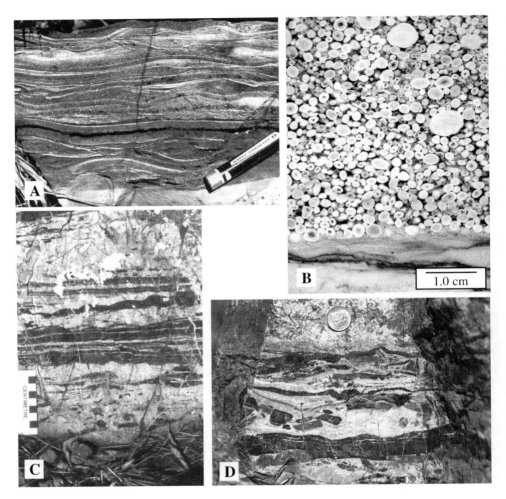

Fig. 5.3-13. Rocks of the M1c, the Msauli Chert member of the Mendon Formation (Lowe, 1999b). (A) A cross-stratified, current-deposited ash unit. (B) Graded, fall-deposited layer of volcanic accretionary lapilli overlying cross-laminated, current-deposited ash. Trace element and REE analyses indicate that the original ash was komatiitic in composition (Lowe, 1999a, 1999b). Many lapilli show nuclei. (C and D) Interlayered dark gray carbonaceous sediment and cross-laminated, current-deposited komatiitic ash, some with rip-up clasts of carbonaceous matter, both now silicified to form impure chert. The regular alternation of current-deposited materials and fine carbonaceous matter suggest regularly alternating periods of current activity and quiet water as characterize many tide-influenced settings.

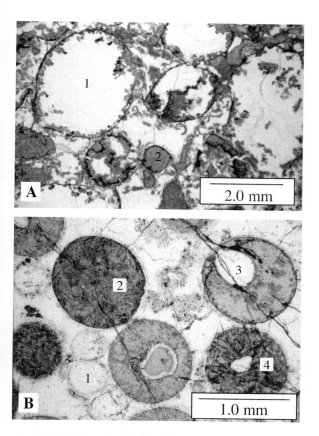

Fig. 5.3-14. Spherules from spherule bed S3. These compositionally complex particles formed through the condensation and solidification of a rock vapor plume produced by the impact of a large chondritic meteorite (Lowe et al., 2003; Shukolyukov et al., 2000). (A) bimodal spherules including large, translucent, fine-grained quartz (chert) spherules (1) and small, mica-rich spherules (2). The quartz spherules show rims and internal chunks of mica-rich material. The original compositions have been severely altered by pervasive metasomatism. The large quartz spherules were probably originally composed of low-alumina glass. (B) Complex spherules including small, quartz spherules (1), larger mica-rich spherules (2–4), some of which show internal cavities that are now filled by coarser cavity-filling quartz (3) and some ghosts of fine altered crystallites, possibly feldspar (4).

base immediately south of the Inyoka Fault has been dated at 3243 ± 4 Ma (Kröner et al., 1991a; Byerly et al., 1996),

The Auber Villiers Formation includes 1500 to 2000 m of dacitic tuff; coarse volcani-clastic sandstone, conglomerate, and breccia; and terrigenous chert-clast conglomerate, chert-grit sandstone, and shale. This sequence and locally overlying Moodies strata form the upper plate succession above the 24-hour Camp Fault. Volcanic breccia and tuff in

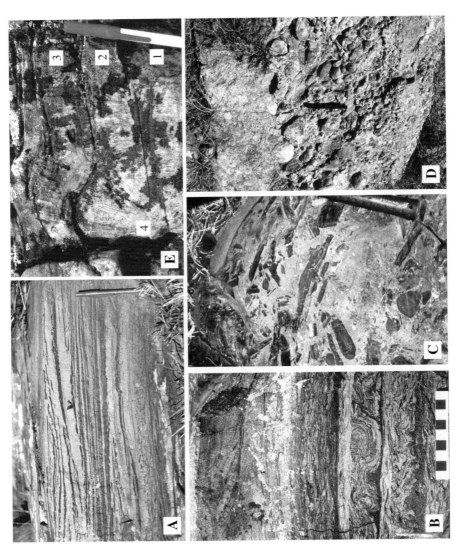

Fig. 5.3-15. Rocks of the Mapepe Formation of the Fig Tree Group. (A) Cross-stratified, current-deposited felsic tuff. Cross-sets are defined by laminations rich in darker iron oxides. (B) Fine-grained terrigenous sandstone, siltstone, and mudstone representing the distal part of a fan delta developed during the earliest stages of uplift and orogenesis (Lowe, 1999a). (C) Chert-clast conglomerate representing locally derived sedimentary materials eroded during the initial stages of uplift. (D) Conglomerate composed of a variety of chert and silicified volcanic materials representing more extensive uplift and deeper erosion into the lower parts of the greenstone sequence, upper Mapepe Formation. (E) Barite. The unit shows three flat layers (1–3), each composed of barite blades, and a rounded mound (4) composed of upward-radiating barite blades.

the Auber Villiers Formation have yielded ages of 3253 ± 2 Ma (Byerly et al., 1996) and 3256 ± 4 Ma (Kröner et al., 1991a).

5.3-3.4.3. Intrusive rocks

There are no reported TTG equivalents of the 3260–3250 Ma Auber Villiers Formation or the oldest parts of the Mapepe Formation. Along the northwestern and western margins of the BGB, the Nelshoogte Pluton has been dated at 3236 Ma (de Ronde and Kamo, 2000) and the Kaap Valley and Badplaas Plutons at 3230–3225 Ma. These plutons appear to be partial age equivalents of the Mapepe Formation in the Umuduha suite and are generally regarded as comagmatic with felsic units in the Schoongezicht and Belvue Road Formations of the Fig Tree Group in the Kaap Valley suite. Many small felsic igneous bodies ranging from 3241–3229 Ma (Kroner et al., 1991a; de Ronde et al., 1991b; Kamo and Davis, 1994) are intrusive into rocks of the Umuduha and Kromberg suites on the west limb of the Onverwacht Anticline.

5.3-3.5. Kaap Valley Suite

North of the Inyoka Fault (Figs. 5.3-1, 5.3-3(A)), the Onverwacht Group includes a succession of komatiitic and basaltic volcanic rocks and tuffs, layered ultramafic complexes, and thin cherty units (Fig. 5.3-16) that are tightly folded and extensively faulted. These form the Kaap Valley suite. This suite is cut by a number of large faults, especially the Moodies Fault (Fig. 5.3-3(A)), which may mark major boundaries between different crustal blocks (de Ronde and Kamo, 2000). Thickness estimates are difficult because of the lack of marker units or datable zircon-bearing tuffs, but probably > 1000 m of rocks are present, assigned to the Weltevreden Formation by Lowe and Byerly (1999). Although these rocks are generally heavily altered, they locally include some of the freshest peridotitic komatiites in the world, including flow units with fresh olivine, pyroxene, and optically isotropic glass (Kareem and Byerly, 2002, 2003). The Weltevreden Formation also includes a number of large layered ultramafic bodies. Along the northern margin of the belt, these include the Sawmill, Pioneer, and Emmenes "intrusions" (Wuth, 1980; Anhaeusser et al., 1981; Anhaeusser, 1985). Similar bodies occur in the Jamestown Schist Belt (Anhaeusser, 1985), the Nelshoogte Schist Belt (Anhaeusser, 2001), the Mendon Formation, and units H3 and H4 of the Hooggenoeg Formation (Viljoen and Viljoen, 1969b; Anhaeusser, 1985), but are absent in the Komati Formation. They are made up of alternating units of serpentinized dunite, peridotite, orthopyroxenite, clinopyroxenite, and gabbro (Wuth, 1980; Anhaeusser, 1985). The presence of both cross-laminated tuffs and spinifex zones (Fig. 5.3-17(A)) within these bodies in the Kaap Valley suite suggests that all may represent thick, layered komatiitic flows and interstratified komatiitic tuffs. Coarse komatiitic sands and conglomerates are present locally (Fig. 5.3-17(B)). A Nd/Sm age date of 3286 ± 29 Ma (Lahaye et al., 1995) suggests that the Weltevreden Formation is equivalent to at least the upper part of the Mendon Formation. The wide presence of spherule bed S3 at the base of the Fig Tree Group in the Kaap Valley suite suggests that the topmost cherts of the Weltevreden Formation are about 3243 Ma.

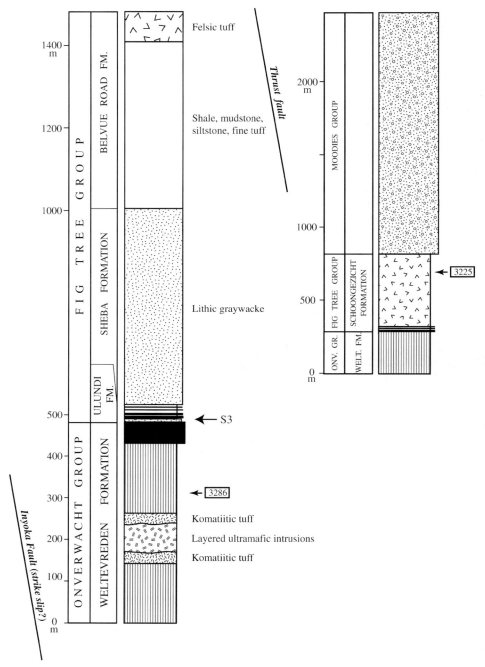

Fig. 5.3-16. Stratigraphy of the Kaap Valley suite north of the Inyoka Fault (Fig. 5.3-3(F)).

Fig. 5.3-17. Rocks from the Weltevreden Formation. (A) Silicified olivine spinifex. (B) Sedimentary breccia composed of clasts of komatiitic volcanic rock.

The Fig Tree Group in the Kaap Valley suite (Fig. 5.3-16) includes the Ulundi, Sheba, Belvue Road, and Schoongezicht Formations (Condie et al., 1970). The first three form a continuous 1500-m-thick Fig Tree succession on the lower plate of an unnamed thrust fault. The Schoongezicht Formation occurs on the upper plate. The Ulundi Formation is a 20 to 50 m thick unit of carbonaceous shale, ferruginous chert, and iron-rich sediments at the base of the Fig Tree Group. Spherule bed S3 occurs at its base. The overlying Sheba Formation includes between 500 and 1000 m of predominantly fine- to medium-grained lithic turbiditic graywacke. The Belvue Road Formation consists of 200–500 m of mudstone and thin, fine-grained turbiditic sandstone. Toward the top, it includes an increasing proportion of dacitic volcaniclastic rocks. Along the northeast end of the Stolzburg Syncline, sedimentary rocks of the Belvue Road Formation are succeeded by serpentinized komatiitic volcanic rocks. Extensive shearing and brecciation along the contact suggest that it is a fault. This komatiitic unit and 10–30 m of overlying chert are interpreted to be the upper part of the Weltevreden Formation (Lowe and Byerly, 1999). They are overlain, apparently conformably, by nearly 450 m of turbiditic, plagioclase-rich sandstone and mudstone of the Schoongezicht Formation that is succeeded by >1000 m of quartzose sandstones of the Moodies Group. This sequence is interpreted to have been thrust over the structurally underlying Weltevreden through Belvue Road sequence (Lowe and Byerly, 1999; Lowe et al., 1999). Similar stratigraphic and structural relationships characterize upper and lower plate sections throughout western BGB north of the Inyoka Fault. Juvenile dacite-clast conglomerate near the top of the Schoongezicht Formation has yielded maximum single-crystal zircon ages of 3225 ± 3 Ma (Kröner et al., 1991a). The overall age of the Kaap Valley suite is probably ~3300 to 3225 Ma.

5.3-3.6. Moodies Group – All Suites

The youngest rocks in the BGB are lithic, feldspathic, and quartzose sandstone, conglomerate, and siltstone of the Moodies Group (Fig. 5.3-18) deposited in a wide range of fluvial, tidal, and shallow marine settings (Eriksson, 1978, 1979, 1980). These strata reach about 3500 m thick along the northern edge of the BGB. The wide development of conglomerate at the base of the Moodies and, in southern areas, of conglomerates resting with angular unconformity on rocks of the Onverwacht Group suggest that the base of the Moodies is an unconformity in much of the Songimvelo block. In the Umuduha and Kaap Valley blocks, the Moodies Group is underlain with apparent conformity by the Auber Villiers or Schoongezicht Formations, respectively. Regionally, this sequence has been thrust over rocks that include the Mendon and Mapepe Formations in the Umuduha block and the Weltevreden and overlying Ulundi, Sheba, and Belvue Road formations in the Kaap Valley block.

The Moodies Group appears to include two and possibly more distinct facies. Rocks north of the Inyoka Fault comprise sections commonly exceeding 2000 m thick that include microcline and clasts of potassic plutonic rock. South of the Inyoka Fault, Moodies sections are generally less than 1000 m thick and individual northeast-trending outcrop belts are characterized by distinctive conglomerate-clast and sandstone compositions. These facies contrasts suggest that Moodies sediments in each belt were derived from different sources and either deposited in separate parts of a large basin, with incomplete mixing of detritus from different sources, or in several small basins. The Moodies everywhere marks orogenic uplift, erosion down through the greenstone sequence to deep-seated plutonic rocks, and deposition in adjacent foreland and intermontane basins (Heubeck and Lowe, 1999).

The age of the Moodies Group is poorly constrained and may be different in different parts of the BGB. On the east limb of the Onverwacht Anticline, tightly folded Moodies strata lie with angular unconformity on folded Onverwacht and Fig Tree rocks. Folded Onverwacht rocks in the southernmost Kromberg Syncline are truncated by the post-tectonic $3216 + 2/-1$ Ma Dalmein Pluton (Kamo and Davis, 1994). In the central BGB, a felsic dike intrusive into Moodies strata has been dated at 3207 ± 2 Ma, providing a minimum age

Fig. 5.3-18. Rocks of the Moodies Group. (A) Cross-stratified quartzose sandstone showing what appears to be reversing directions of flow, suggesting deposition under tidal conditions. (B) Cross-stratified sandstone interlayered with thin units of mudstone and containing mudstone rip-up clasts. This has also been interpreted to have been deposited under tidal conditions (Eriksson, 1979).

of sedimentation (Heubeck and Lowe, 1994a). Toward the northeastern end of the BGB, the late-stage Salisbury Kop Pluton, dated at $3109 + 10/-8$ Ma (Kamo and Davis, 1994), truncates tightly folded Moodies strata.

5.3-4. STRUCTURAL DEVELOPMENT OF THE BGGT

Many previous studies have addressed the structural development of the BGB (Ramsay, 1963; Gay, 1969; Anhaeusser, 1976, 1983; de Wit, 1982, 1983; de Wit et al., 1983, 1987a; Lamb, 1984; Lamb and Paris, 1988; de Ronde and de Wit, 1994; de Ronde et al., 1991a, 1991b; Kamo and Davis, 1994; Lowe et al., 1985, 1999; Lowe, 1999c; de Ronde and Kamo, 2000). These have been complemented by studies in deeper-level portions of the

BGGT, including structural features of the TTG plutons and gneisses, schist belts, and meta-greenstone remnants along the southern and western margins of the BGB (Kisters and Anhaeusser, 1995a, 1995b; Stevens et al., 2002; Kisters et al., 2003; Diener et al., 2005, 2006) and the 3.1 Ga GMS batholiths (Jackson and Robertson, 1983; Westraat et al., 2005; Belcher and Kisters, 2006a, 2006b).

5.3-4.1. Pre-D_1 Deformation

The earliest well-documented deformational event in the BGB, D_1, coincided with ~3445 Ma felsic magmatic activity. However, some data suggest an older pre-Komati Formation episode of deformation involved faulting, folding, and structural repetition of rocks of the Theespruit Formation in both the Onverwacht and Steynsdorp Anticlines and the formation of foliation and gneissic textures in the Steynsdorp Pluton (Kisters and Anhaeusser, 1995a, 1995b; Kröner et al., 1996; Kisters et al., 2003). This event is poorly constrained at present and is not numbered separately.

5.3-4.2. D_1

D_1 coincided with magmatic activity represented by H6, the felsic igneous member of the Hooggenoeg Formation, and the comagmatic ~3445 Ma TTG plutons in the southern BGGT (de Wit 1982, 1983; de Wit et al., 1987a; de Ronde and de Wit, 1994; Lowe et al., 1999). Based on the age of felsic igneous rocks of H6 affected by faulting and the age of strata overlying H6 that are not, de Ronde and de Wit (1994) estimated D_1 to have occurred between 3445 ± 4 and 3416 ± 7 Ma. However, the effects and extent of D_1 are controversial. De Wit (1982), de Wit et al. (1983, 1987b), and de Ronde and de Wit (1994) interpreted D_1 as forming "recumbent nappes, downward facing sequences, and the emplacement of ophiolite allochthons" during severe horizontal shortening (de Ronde and de Wit, 1994). The classic sections of the Onverwacht and Fig Tree Groups in the Onverwacht Anticline and Kromberg Syncline were interpreted as a tectonic complex composed largely of allochthonous sheets assembled by structural stacking and repetition of a stratigraphic section estimated to range from as little as 500 m (Paris, 1985) to 3 or 4 km thick (de Wit, 1982). The thrust faults were interpreted to be zones of alteration and gneissic fabrics that underlie many cherts in the Onverwacht Group (de Wit, 1982, 1983). However, high-precision zircon age dating (Armstrong et al., 1990; Kröner et al., 1991a, 1996; Byerly et al., 1996, 2002) and regional geologic mapping (Lowe et al., 1985, 1999; Lowe and Byerly, 1999) have shown that the basic stratigraphy of Viljoen and Viljoen (1969a, 1969b) from the Komati Formation through the lower Fig Tree Group around most of the Onverwacht Anticline and Kromberg Syncline is largely intact and that units young continuously from the bottom to the top of the 9–11 km thick stratigraphic sequence (Lowe and Byerly, 1999). A zone of D_1 faulting involving the local overturning of large blocks of underlying portions of the Hooggenoeg Formation is present on the west limb of the Onverwacht Anticline (Lowe and Byerly, 1999; Lowe et al. 1999) but not on the east limb and does not interrupt or repeat the overall stratigraphic sequence. The Kromberg

Formation, described as the Kromberg allochthon by de Ronde and de Wit (1994), is an intact, autochthonous stratigraphic unit relative to adjacent units and ranges from about 3416 ± 7 Ma (Kröner et al., 1991a) at the base to 3334 ± 3 Ma (Byerly et al., 1996) at the top. Studies of the alteration zones interpreted as shear zones marking D_1 thrust faults have shown that they formed largely through sea floor and/or hydrothermal alteration unaccompanied by shearing (Lowe and Byerly, 1986a; Duchac and Hanor, 1987; Hanor and Duchac, 1990; Lowe et al., 1999). De Ronde and Kamo (2000) correlated formation of "flaser-banded green cherts" in what Lowe and Byerly (1999) named the Weltevreden Formation along the northern edge of the BGB with D_1 shearing, but geochronological studies suggest that no rocks older than about 3330 Ma are exposed north of the Granville Grove Fault (Lahaye et al., 1995; Byerly et al., 1996, 2002). The known effects of D_1 in the BGB, therefore, are limited to the west limb of the Onverwacht Anticline where deformation accompanied the shallow intrusion of felsic magmas (de Wit et al., 1987a; Lowe et al., 1999).

The high-level felsic magmatic activity was followed by an interval of cooling and extension, forming fault-bounded extensional basins in which evaporitic sediments and banded cherts of the basal part of the Buck Reef Chert member of the Kromberg Formation accumulated (Lowe and Fisher-Worrell, 1999). More recently, de Vries et al. (2006) have suggested that this extension was associated with the large-scale collapse of the H6 felsic volcanic platform, in part during deposition of the Buck Reef Chert.

5.3-4.3. ~3330 Ma Deformation

Komatiitic volcanic rocks of the Mendon Formation, ~3330 to ~3245 Ma, are repeated across a number of fault-bounded belts between the Granville Grove and Inyoka Faults (Figs. 5.3-1, 5.3-3(A)) in the western BGB (Lowe and Byerly, 1999). Northward across these faults, age relationships suggest deposition of Mendon Formation lavas in a progressively widening rift (Lowe et al., 1999), implying a regional episode of crustal extension. North of the Granville Grove Fault, no rocks are known older than the Mendon Formation. The grossly similar stratigraphies and ages of supracrustal rocks in the Umuhuha and Kaap Valley suites (Figs. 5.3-2, 5.3-12, 5.3-16) suggest that rifting at about 3330 Ma initiated formation an extensive tract of new mafic crust, represented by rocks of the Mendon and Weltevreden Formations, adjacent to blocks of the older Kromberg, Songimvelo, and possibly Steynsdorp suites. No specific structures can be identified as having been formed during this extension.

5.3-4.4. D_2

D_2 of de Ronde and de Wit (1994) and Lowe et al. (1999) was the first regional and perhaps the main deformational event of BGB history. It was associated with deposition and subsequent deformation of rocks of the Fig Tree Group and probably extended from about 3230 to 3225 Ma (de Ronde and de Wit, 1994; Lowe et al., 1999; Kisters et al., 2003). D_2 was characterized in the BGB by the formation of inward-verging fold and thrust complexes in both the northwestern and southeastern parts of the BGB (Lowe et al., 1999). Many of the faults that repeat rocks of the Mendon Formation and overlying Fig Tree Group, including the Granville Grove Fault (Figs. 5.3-1 and 5.3-3(A)) and a related series of thrust faults now exposed in the western part of the belt, formed at this time (Lowe et al., 1999). Thrust faulting was probably coincident with deposition of thick sequences of Fig Tree sandstone and conglomerate in footwall foreland basins in both the northwestern and southeastern parts of the BGB.

In the deeper-level terrain south of the BGB, 3230–3225 Ma was the time of peak metamorphism (Dziggel et al., 2002; Stevens et al., 2002; Diener et al., 2005). Clemens et al. (2006) postulated that the magmatic event represented by the 3236–3225 Ma Kaap Valley, Nelshoogte, and Badplaas Plutons along the northern and western margins of the BGB was associated with "significant and rapid crustal thickening". (Kisters et al., 2003) have suggested that this interval of deformation and crustal thickening was followed closely by extensional orogenic collapse and core complex formation that juxtaposed lower crustal TTGs in the south against less metamorphosed upper crustal greenstone-belt rocks to the north along the Komati Fault and related shear zones. They proposed that collapse may have been reflected in the formation of extensional basins in which sediments of the Moodies Group accumulated.

This event is widely regarded as having involved the amalgamation and suturing of an older, ~3445 Ma terrane in the SE, represented by the Steynsdorp, Songimvelo, Kromberg, and Umuduha suites, with the younger post-3330 Ma terrane to the NW, represented here by the Kaap Valley suite (de Ronde and Kamo, 2000; Anhaeusser, 2006). The suture is widely regarded as the Inyoka-Saddleback Fault zone (Fig. 5.3-3(A)). However, the stratigraphic sequences on both sides of the Inyoka and Saddleback Faults are essentially identical. Both the Weltevreden Formation north of the faults and the Mendon Formation to the south show no components older than about 3334 Ma. Moreover, terrains NW and SE of the fault are both characterized by structurally repeated sequences, with upper-plate sections including Moodies rocks underlain by proximal felsic volcanic units of the Fig Tree Group and Onverwacht komatiitic volcanic rocks, and lower-plate sections that lack Moodies sediments and consist of more distal felsic volcaniclastic and epiclastic facies of the Fig Tree Group underlain by komatiitic volcanic units. These relationships make it apparent that (1) the boundary between the young NW and older SE terrains lies along the Granville Grove Fault, not the Inyoka-Saddleback faults, and (2) the Inyoka-Saddleback Faults did not form at 3230–3225 as suggested by de Ronde and Kamo (2000) but after sedimentation and thrusting of the Moodies Group and other upper plate sequences over the structurally underlying lower-plate successions. Moreover, the overlap of the basal member of the Mendon Formation, M1, onto the older Songimvelo/Kromberg block indicates that the contact between the younger and older crustal units formed at about ~3330 Ma, not 3230–3225 Ma, that this event appears to have involved rifting of the older suites, not suturing, and that the younger crustal suites formed close to and upon the edge of the older suite and the two are not amalgamated exotic blocks or "terranes".

5.3-4.5. D_3

Post-Fig Tree deformational events are less well resolved in the BGB, in part because magmatism largely ceased in Fig Tree time. It is clear that after initial deformation of rocks of the Fig Tree Group, later, but perhaps nearly continuous deformation accompanied deposition of rocks of the Moodies Group. The thick Moodies sequences appear to have accumulated in a number of separate basins. These basins may have formed in part during thrusting (Jackson et al., 1987; Heubeck and Lowe, 1994a, 1994b, 1999) and in part in extensional settings (Lowe, 1994, 1999c; Kisters et al., 2003). This deformation has been termed early D_3 by de Ronde and de Wit (1994) and D_3 by Lowe et al. (1999b).

5.3-4.6. D_4

Throughout the northern part of the BGB north of the Granville Grove Fault, rocks of the Moodies Group underlain with apparent conformity by Fig Tree felsic volcanic and volcaniclastic rocks and, locally, cherts and komatiites of the Onverwacht Group, are thrust over previously deformed sections of Fig Tree and upper Onverwacht rocks (Figs. 5.3-12 and 5.3-16). To date, we know of no stratigraphic section north of the Granville Grove Fault where the Moodies Group and underlying felsic Fig Tree units are not in fault contact with "underlying" more heavily deformed Fig Tree and Onverwacht rocks. This stage of deformation has not been recognized by other investigators and was termed D_4 by Lowe et al. (1999).

The more distal character of felsic volcaniclastic suites in the lower plate assemblages and their interlayering with thick clastic sequences derived by weathering and erosion of older greenstone units, the more proximal character of the Schoongezicht and Auber Villers Formations on the upper plates, and the probable comagmatic relationships of the Schoongezicht volcanic rocks with 3229–3225 Ma TTG plutons along the northern margin of the belt suggest that D_4 thrusting was generally from northwest to southeast.

5.3-4.7. D_5

D_5 of Lowe et al. (1999) and late D_3 and D_4 of de Ronde and de Wit (1994) represent still younger post-Moodies deformation, but age constraints are poor. De Ronde and de Wit (1994) suggest that late D_3 and D_4 occurred between 3216 + 2/−1 and 3084 ± 54 Ma, but this deformation could have been initiated as early as ∼3220 Ma if deposition and thrusting of the Moodies Group occurred rapidly following ∼3225 Ma Fig Tree felsic magmatic activity volcanism. Later deformation included tightening of previous folds and tight folding and faulting of the upper-plate sequences of the thrust faults formed during D_4. Subsequent events included rotation of bedding, previous structures, and fabrics to the vertical; formation large more-or-less linear, generally NE-SW trending faults including the Moodies, Inyoka, and Saddleback Faults (Fig. 5.3-3(A)); and deformation during high-level structural emplacement of surrounding granitoid plutons. Kisters et al. (2003) have suggested that the steepening of fabrics to their present vertical orientation may reflect "late-stage isostatic instability caused by the horizontal stretching and vertical thinning of crust during which less dense and hot basement gneisses rose ... as solid-state diapirs through the overlying denser, mafic to ultramafic supracrustal sequence". More recently, Belcher and Kisters (2006) have suggested that the Heerenveen and related large 3.1 Ga granitoid batholiths south and east of the BGGT (Fig. 5.3-1) were emplaced during crustal shortening represented by late D$_3$ of de Ronde and de Wit (2004). Overall, post-Moodies events within the BGB and in surrounding areas are too poorly resolved and dated to establish specific correlations and histories at this time. We also suspect that many additional structural events during the 3216 to ∼3100 Ma interval, some probably local, others of more regional extent, will be identified and add to the complexity.

5.3-5. EVOLUTION OF THE BGGT

The formation of the continental crust in the BGGT and vicinity can be described in terms of three general evolutionary periods:
(1) a period from >3550 Ma, and probably >3700 Ma, to 3230 Ma in which the main tectono-stratigraphic suites formed through magmatic accretion (Lowe, 1994);
(2) a period of deformation, shortening, and amalgamation of the assemblage of tectono-stratigraphic suites through tectonic processes from about ∼3230–3216 Ma; and
(3) thickening and stabilization of the crust through continued deformation from ∼3216 to <3100 Ma and intrusion of large granitic batholiths at 3100 Ma.
The initial magmatic period involved at least four and possibly more magmatic cycles over an interval of 320+ myr, each cycle represented by one or more of the tectono-stratigraphic suites:
(1) >3547 to ∼3500 Ma, the Steynsdorp cycle,
(2) ∼3500 to ∼3435 Ma, the Songimvelo cycle,
(3) ∼3435–3330 Ma, the Kromberg cycle,
(4) ∼3330 to 3225 Ma, the Umuduha and Kaap Valley cycle.
The presence of pre-3700 Ma zircons as xenocrysts in the Steynsdorp Anticline suggests that even older tectono-magmatic cycles may have preceded these. The contrasting Onverwacht and Fig Tree stratigraphies in the Kaap Valley and Umuduha suites suggest that they represent approximately coeval suites that are now in tectonic contact along the late-formed Inyoka Fault (Fig. 5.3-4). An additional structural repetition of a similar suite is represented by komatiites, felsic volcaniclastic and quartzose epiclastic rocks of the Onverwacht, Fig Tree and Moodies Groups, respectively, that make up the upper plates of thrust faults in the Umuhuha and Kaap Valley suites (Figs. 5.3-12 and 5.3-16).

The formation of each tectono-stratigraphic suite involved (1) an initial ultramafic and mafic volcanic stage characterized by the eruption of a thick komatiitic and basaltic volcanic sequence, and (2) a felsic magmatic stage during which the mafic and ultramafic volcanic sequence was intruded by TTG plutons and covered by comagmatic dacitic to rhyolitic volcanic rocks (Fig. 5.3-4).

The product of the magmatic period of BGGT evolution was a complex assemblage of tectono-stratigraphic suites. Each suite probably included a nucleus consisting of an early mafic to ultramafic volcanic sequence that was melted at depth, intruded at higher levels by TTG magmas, and overplated at the surface by comagmatic felsic volcanic rocks. Flanking this nucleus away from intrusive centers were areas where the greenstone sequence remained essentially unmetamorphosed and undeformed, such as the Umuhuha suite. The only intrusive rocks in these areas are small hypabyssal felsic intrusive bodies, satellites to the main nuclear TTG plutons. The Steynsdorp suite may have suffered early deformation and shortening and its contact with overlying rocks of the Songimvelo suite may be a fault, analogous to the Komati Fault, or a sheared unconformity, where Songimvelo lavas lapped onto the older suite. The mélange of Umuhuha and Kaap Valley suites and upper and lower plate sequences north of the Granville Grove Fault probably represents a separate younger crustal unit formed mainly <3330 Ma, repeated by thrust stacking during D_4, and offset along the late strike-slip Inyoka and Saddleback Faults during D_5.

The widespread overlap of younger volcanic suites onto older suites and the presence in younger felsic magmatic units of xenocrystic zircons derived from older suites (Fig. 5.3-2) indicate that most BGGT suites formed adjacent to or on the margins of older suites, a process termed magmatic accretion. In addition, fragmentation of older blocks during younger magmatic and rifting events may have resulted in isolated rafts of older suites that were buried beneath the younger volcanic suites, as is suggested schematically by the isolated blocks of the Steynsdorp suite beneath younger Songimvelo lavas in the southwestern part of the BGB (Fig. 5.3-4). The Ancient Gneiss Complex of Swaziland probably formed as a separate block or assemblage of blocks that was amalgamated through tectonic accretion with the BGGT during D_2 or later shortening events.

Some deformation probably accompanied or closely followed felsic magmatism in each suite. In the Steynsdorp suite, deformation appears to have followed magmatism at 3509 Ma (Kisters and Anhaeusser, 1995a). In the Songimvelo suite, deformation that accompanied volcanism and intrusion at ~3445 Ma has been termed D_1. In northern areas, D_2 accompanied late-stage felsic volcanism at 3230–3225 Ma. However, D_2 was more widespread and crustal thickening more severe than during preceding felsic magmatic events and involved the formation of fold-and-thrust belts, orogenesis, and sedimentary basins (Lowe et al., 1999). Regional shortening and amalgamation of these nuclei, marking the final stage of BGB evolution and the initial stage of the formation of the Kaapvaal Craton, is represented by sediments of the upper Fig Tree and Moodies Group and accompanying D_2, D_3, and D_4 deformation.

The tectonic implications of the individual stages of these cycles and of the tectonostratigraphic cycles themselves remain controversial. The length of individual cycles ranges from >50 myr for the Steynsdorp suite, the base of which is not exposed, to about 100 myr for the Kromberg suite, with an overall average of about 80 myr/suite. Their duration and style suggest that they represent basic crustal cycles. Similar mafic to felsic cycles have been widely described from greenstone belts in other Archean cratons (e.g., Thurston et al., 1985).

Stratigraphic relationships of the Mendon Formation of the Umuduha suite suggest that it formed in a progressively widening rift, but the tectonic setting(s) of komatiitic volcanism in general remain controversial. Until recently, the prevailing view has been that komatiites reflect relatively deep, plume-related melting of hot, dry mantle to produce high-temperature ($\sim$1700 °C) anhydrous magmas (Campbell et al., 1989; Arndt et al., 1998). A few workers have compared the BGB volcanic sequence to modern ophiolites and argued that the Komati Formation is a rifting-related sheeted dike complex (de Wit et al., 1987b; Parman et al., 1997). More recent studies have suggested that many komatiites resemble modern subduction-related boninites and formed as hydrous (4–6% water by weight), lower temperature ($<$1500 °C) magmas by shallower melting along subduction zones (e.g., Allegrè, 1982; Parmen et al., 2001). However, boninite-like chemical signatures can also originate through crustal contamination (Smithies et al., 2003) and the magmatic accretionary evolution of the BGGT provides a fertile ground for such contamination. Evidence for the involvement of older components in the evolution of younger tectono-stratigraphic suites is provided by the abundance of xenocrystic zircons in all felsic and TTG units (Fig. 5.3-2). Our observations suggest that the early mafic stages of the BGGT tectono-magmatic cycles involved rifting and the formation of tracts of new crust, perhaps in back arc (Lowe, 1994) or plume settings.

Late TTG intrusion and felsic volcanism in the BGB and in Archean greenstone belts in general have been regarded a signature of subduction-related magmatism (e.g., Drummond and Defant, 1990; Martin, 1999; Martin et al., 2005). More recently, a number of workers have argued that some TTG magmas have originated through the melting of deep greenstone-belt roots in an intraplate setting (e.g., Smithies, 2000). Moyen et al. (this volume) suggest that both varieties are represented by the TTG plutons in the BGGT. Smithies et al. (2003) argue that TTG generation may result from flat-slab subduction beneath 40–70 km thick mafic to ultramafic volcanic, arc-related crust, with emplacement of the TTG plutons into the upper levels of this mafic crustal section. In the long complex history of the BGGT, tectono-stratigraphic cycles may have terminated both through episodes of melting of the roots of a thickened mafic crust and, at other times, through subduction-driven melting, as suggested by Moyen et al. (this volume). Moyen et al. (2006) have also argued that high-pressure, low-temperature mineral assemblages in meta-supracrustal rocks in the BGGT imply that subduction was active during their formation 3230–3225 Ma.

A remarkable aspect of BGGT history is the 320+ myr-long interval of cyclic magmatic accretion from at least $\sim$3550 to 3230 Ma that apparently never involved tectonic events that formed major uplifts or sedimentary basins at the Earth's surface. Sedimentary layers in the Onverwacht Group are composed mainly of silicified and altered chemical, biogenic, and pyroclastic sediments (Lowe and Knauth, 1977; Lowe, 1980, 1982, 1999a). Detrital sand-sized quartz is limited to sparse grains derived from quartz-phyric felsic volcanic units and in the BGB microcline is absent in sedimentary units below the Fig Tree Group. During the felsic magmatic stages, local topographic relief was produced by both volcanism and local deformation, and relatively thin, predominantly volcaniclastic units eroded from these high-standing edifices were deposited adjacent to the eruptive centers, such as the turbiditic and debris-flow facies of H6 along the Komati River and felsic breccia and conglomerate in H6 on the far west limb of the Onverwacht Anticline. An isolated, 8-m-thick unit of microcline-bearing "arkose" has been reported from a high-grade supracrustal xenolith 10–15 km south of the BGB (Dziggel et al., 2006), but the regional implications of this occurrence and the possible role of metamorphism in forming the mineral assemblage are not resolved. Neither of these accumulations represents a significant tectonics-related sedimentary basin. In contrast, post-3230 Ma deformation saw the formation of mountains and associated sedimentary basins, exposure of the intrusive and metamorphic roots of the BGGT, and the erosion and deposition of thick sequences of basinal, often quartz- and microcline-rich clastic sediments.

Recent studies of pre-4.0 Ga detrital zircons have suggested that the inventory of large blocks of continental crust was perhaps as great in the Hadean as it is today (Harrison et al., 2005). While referring specifically to the Earth prior to 4.0 Ga, this picture does not appear to describe the Earth's crust during formation of the BGGT. The development of the BGGT involved over 300 myr of history and was dominated by the cyclic formation and evolution of ultramafic and mafic volcanic platforms that were subsequently thickened by TTG magmatism. These events may reflect a variety of tectonic settings, including rifts, plumes, and convergent plate boundaries. However, throughout this history, here is no evidence for the existence, appearance, or influence of identifiable blocks of thick, substantially older, continental crust as basement to the developing blocks, as buoyant plates or exotic terranes that became involved in BGB evolution through collision or mountain making, or as sources of clastic sediment. On a hot young earth with elevated rates of heat loss and a plate-tectonic style of geodynamics, large continental blocks, if they existed, should have played a significant role in sedimentary systems and crustal tectonics, as they do today. These observations suggest either that a geodynamic system involving large-scale horizontal movements of crustal blocks did not operate during the Paleoarchean or that large cratons and continent-sized blocks of fractionated crust did not exist then, or both. The broader geologic record points to a progressive increase in the influence of continental blocks after about 3.1 Ga (Lowe, 1992), and the geologic record and history preserved in the BGB are consistent with the interpretation that the 3.5–3.2 Ga Earth lacked large blocks of buoyant, stable continental crust. Prior to 3.2 Ga, collections of tectono-stratigraphic suites like those in the BGGT may have served as some of the earliest buoyant, protocontinental domains. They would have included large amounts of differentiated and fractionated "continental" components, such as TTG plutons, felsic volcanic rocks, and even more highly fractionated igneous rocks locally, but until 3.2–3.1 Ga did not evolve into thickened, stable cratons.

5.3-6. CONCLUSIONS

The formation of the continental crust in the eastern Kaapvaal Craton can be described in terms of three generalized evolutionary stages. (1) The first stage involved the formation of a complex of tectono-stratigraphic suites. Each suite formed through an initial stage of ultramafic and mafic volcanism, perhaps initiated through rifting or plume activity, followed by thickening of the ultramafic and mafic volcanic sequence by the intrusion of TTG magmas at depth and eruption of comagmatic felsic volcanic units at the surface. Although including differentiated and fractionated "continental" TTG and felsic volcanic components that may have been older 3700 Ma, these suites were not of continental proportions or stability. The long sequential magmatic development of these blocks, from >3550 to ~3230 Ma, was followed by (2) tectonic amalgamation and suturing of the suites to form a larger quasi-continental block. This stage of development is represented by sediments of the Fig Tree and Moodies Groups and the tectonic activity that they reflect. The final stage of continent formation involved (3) late-stage deformation, intracrustal melting, and potassic GMS plutonism at about 3.1 Ga forming a large, thick, buoyant continental block represented today by the Kaapvaal Craton.

The evolution of the BGB can be traced in some detail from >3547 to <3225 Ma. Throughout the first 320 myr of this history, until about 3230 Ma, the BGB developed largely through magmatic, not tectonic accretion and lacks evidence for associated orogenesis and major uplift, either of the underlying parts of the greenstone sequence itself or of flanking areas that are no longer preserved (Lowe, 1980, 1982). At no time during its history does the BGB show evidence, such as deformation, the development of sedimentary basins, or thick sequences of quartz- or microcline-bearing clastic sediments, for the existence of or interaction with large blocks of substantially older continental crust. The development of small protocontinental nuclei that included differentiated plutonic rocks was a integral part of the evolution of the BGB but these blocks were not fully amalgamated, thickened, and stabilized to form a true craton or continental block until about 3.1 Ga. On a young, dynamic Earth, these features argue strongly against suggestions that there was a significant inventory of large, stable, blocks of old continental crust (Armstrong, 1968, 1981; Harrison et al., 2005). Finally, unlike post-3.0 Ga greenstones (e.g., Cawood et al., 2006), the BGB did not develop through magmatism, sedimentation, and accretion within an oceanic island arc system. The differences between pre-3.0 and post-3.0 Archean greenstone belts are substantial (Lowe, 1980, 1982) and point to fundamental changes in the tectonic regime on the early Earth at about 3.1–3.0 Ga, perhaps associated with or in response to the appearance of the first cratons and/or modern plate tectonics.

ACKNOWLEDGEMENTS

This research was supported by NASA Exobiology Program grants NCC2-721, NAG5-9842, NAG5-13442, and NNG04GM43G to DRL, by NSF grant EAR-9909684 to GRB, and by grants to DRL and GRB from the UCLA Center for Astrobiology. The authors are grateful to the Mpumalanga Parks Board and especially Louis Loock (Regional Manager) and Property Mokoena for allowing access to the Songimvelo Game Reserve. We would also like to thank Sappi Limited and J.M.L. van Rensburg, Forestry Manager, for permission to access private forest roads and many key areas during this study and Mr. Collin Willie for permission to access outcrops in Taurus forests. We are also grateful to the Barberton Mines Limited, especially Roelf le Roux, Chris Rippon, and Charles Robus, and to Cluff Mining SA for access to properties, cores, and logistical support throughout this study. We would also like to thank Carl Anhaeusser and John Percival for reviewing the manuscript and their many valuable comments and suggestions.

Earth's Oldest Rocks
Edited by Martin J. Van Kranendonk, R. Hugh Smithies and Vickie C. Bennett
Developments in Precambrian Geology, Vol. 15 (K.C. Condie, Series Editor)
© 2007 Elsevier B.V. All rights reserved.
DOI: 10.1016/S0166-2635(07)15054-4

Chapter 5.4

VOLCANOLOGY OF THE BARBERTON GREENSTONE BELT, SOUTH AFRICA: INFLATION AND EVOLUTION OF FLOW FIELDS

JESSE C. DANN[a,b,c] AND TIMOTHY L. GROVE[b]

[a]*Department of Geological Sciences, University of Cape Town, Rondebosch 7701, South Africa*
[b]*Department of Earth, Atmospheric and Planetary Sciences, Massachusetts Institute of Technology, Cambridge, MA 02139, USA*
[c]*90 Old Stow Road, Concord, MA 01742, USA*

5.4-1. INTRODUCTION

In greenstone belts, most volcanic sequences were mantle melts that erupted in submarine tectonic settings before they were deformed during assembly of continental lithosphere. Thus, we have three challenges: (1) to understand the original volcanic setting, (2) to reconstruct its path from ocean to continent, and (3) to explain the role of the mantle, not only in generating magmas but also in becoming the thick and buoyant lithospheric mantle that underlies and stabilizes Archean cratons (Wyman and Kerrich, 2002; Parman et al., 2004). Since subduction zones create continental lithosphere, both by magmatic and tectonic processes, these three challenges are closely linked, and in fact, many greenstone belt volcanics have the geochemical signatures of modern magmatic arcs. For example, supra-subduction zone ophiolites represent submarine crust that formed by seafloor spreading above a subduction zone before they were deformed during continental assembly (e.g., Dann, 2004). However, the origin of komatiite-bearing volcanic sections remains controversial (Arndt et al., 1999): do these high-temperature magmas require a mantle plume or can more moderate but hydrous conditions above a subduction zone suffice? Komatiites represent a large spectrum of compositions with as many geochemical shades as basalts, yet in the controversy over their origin (plume v. arc), local evidence has not been placed within this broader context. In addition, their stratigraphic associations and basic volcanology have been poorly integrated into the debate.

The Barberton Greenstone Belt (BGB) offers a unique opportunity to research Earth's oldest remnants of seafloor crust – komatiite-bearing volcanic sections with well preserved textures and minerals (igneous augite and olivine). The volcanic rocks are locally hydrothermally altered on the seafloor beyond recognition to cherts and carbonates. Some of these silicified horizons preserve fossil bacteria (Westall et al., 2001), a discovery that supports the hypothesis that hydrothermal waters vented on the seafloor and supported early life. Furthermore, traces of bacterial alteration of submarine glass have recently been discovered in these ancient rocks (Staudigal et al., 2006, for review). Understanding this Early Archean ecosystem, a volcanic environment, is a crucial step to understanding the conditions of life's origin. On a more practical note, reconstructing the architecture of submarine flow fields is an important guide in exploration for volcanogenic ore deposits, for example, komatiite-hosted nickel deposits (e.g., Hill, 2001; Beresford et al., 2002). Finally, the ultramafic volcanics have no modern analogues, so we have the opportunity to extend the principles of modern flow fields and reconstruct these ancient volcanic settings.

Volcanologists who describe the evolution of basaltic flow fields have provided us with compelling analogues for komatiites (Walker, 1991; Hon et al., 1994; Rossi and Gudmundsson, 1996; Self et al., 1998; Keszthelyi and Self, 1998; Thordarson and Self, 1998). In subaerial basaltic flow fields, lava spreads as a series of thin *flow lobes*, which under some conditions form thick *compound flows*. In others, the lobes coalesce into sheets that inflate to thick flows. Flows thicken or inflate when the crust holds incoming lava that lifts the roof or upper crust. Thick *sheet flows* record episodes of inflation in textural and vesicle layering in the upper crust. When flow is focused in channels, inflation rotates the upper crust of lava channels, forming *tumuli*, craggy domes and ridges on flow surfaces. When one area of a flow inflates more than another, the crust breaks and rotates, forming the edge of a *lava rise*. When the crust breaks, lava can escape as a *break out*, a lobe that may flow over the original flow surface before the flow's interior has solidified. In addition to inflation by lava influx or *endogenous intrusion*, flows can also deflate if more lava drains out of an area of a flow than comes in, forming collapse structures and *withdrawal cavities*. In the inflation model, the lava flowing under a thickening crust remains thin and within the laminar flow regime (versus turbulent) and is increasingly insulated and slowly cooled.

The question here is whether these observations of subaerial flow fields can explain features of submarine komatiites exposed as cross sections in greenstone belts. Basalt erupted underwater cools rapidly as pillows, and the added pressure inhibits vesiculation, so pillowed flows and sparse vesicles distinguish submarine flows. Beyond these superficial effects of water, the submarine environment plays a diminishing role because as the upper crusts of sheet flows thicken, lava is increasingly insulated and free to travel across submarine lava plains the great distances that characterize komatiite flow fields.

5.4-2. TECTONO-VOLCANIC HISTORY OF BARBERTON GREENSTONE BELT (BGB)

Volcano-sedimentary sequences of the BGB record over 400 million years of crustal evolution, a history punctuated by three TTG volcano-plutonic events and three phases of deformation (de Ronde and de Wit, 1994; Lowe and Byerly, 1999). The volcano-sedimentary component has three stages: (1) an early stage dominated by mafic and ultramafic, submarine volcanism of the Onverwatch Group, (2) the basin-filling sequence of the Fig Tree Group, and (3) the unconformably overlying Moodies Group, molasse deposits associated with fold-and-thrust deformation (Fig. 5.4-1). When the 3.445 Ga TTG

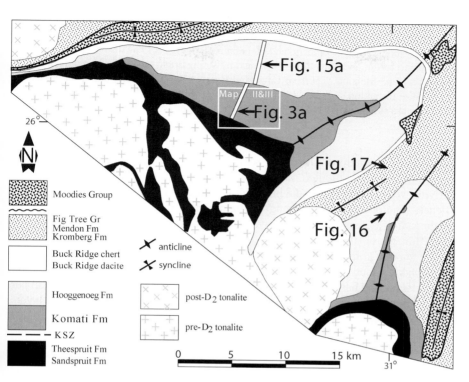

Fig. 5.4-1. Geologic map of the southern Barberton Greenstone Belt, showing location of stratigraphic columns and maps.

suite intruded and erupted, the crustal section was thickened, deformed, and eroded near sea level, a pivotal event that separates the Onverwatch Group volcanics into three distinct groups: (1) the Sandspruit-Theespruit section that hosts 3.445 Ga batholiths, (2) the Komati–Hooggenoeg section, basement for the 3.445 Ga felsic volcanics and host to numerous subvolcanic sills and dikes, and (3) the Kromberg–Mendon volcanics that started about 30 million years after the TTG event. This gap in volcanism is filled by the Buck Ridge Chert, a 150–350 m thick sequence of silicified sediments that overlies the volcanic expression of 3.445 Ga TTG intrusions, the Buck Ridge Dacite. This paper describes the mafic and ultramafic, submarine lavas of the Komati–Hooggenoeg and Kromberg–Mendon sections. Our work focuses primarily on the Komati Formation (e.g., Dann, 2000, 2001; Grove et al., 1997; Parman et al., 1997, 2001, 2004; Parman and Grove, 2004), with lesser work in the Hooggenoeg and Kromberg Formations.

The volcanic stratigraphy of submarine lavas divides into three magma types: basalt, komatiitic basalt, and komatiite (Fig. 5.4-2), each with distinct flow morphologies and geochemical and petrological characteristics. Komatiite volcanism repeats in three formations in association with komatiitic basalt at ca. 3.48, 3.46, and 3.33 Ga. The dominant stratigraphic trend within the Onverwatch Group is an increasing frequency of chert horizons

Fig. 5.4-2. (*Next page.*) Ages, thickness, rates, and composition of volcanism in the Onverwatch Group's Komati, Hooggenoeg, Kromberg, and Mendon formations: (A) The histogram of lava types – basalt, komatiitic basalt, and komatiite – is plotted by stratigraphic thickness of members or formation. Kromberg and Mendon data from Lowe and Byerly (1999); (B) The rate of volcanism is average over entire formation and does not include tectonic, stratigraphic, or analytic uncertainties. U/Pb zircon ages are tabulated in de Ronde and de Wit (1994) and Lowe and Byerly (1999).

---

within the volcanic formations. The cherty sediments represent gaps in effusive volcanism when the sediments were silicified on the ocean floor. The alteration process penetrated the underlying volcanic rocks, producing silicified pillows and even converting komatiites into cherts and/or carbonates. Along with more chert and alteration, the duration of the formations also increases. The oldest volcanic section – the Komati Formation – has no cherty sediments or alteration beyond greenschist metamorphism: the komatiites are serpentinites, and the komatiitic basalts are greenschists, both without penetrative fabrics. Based on available zircon ages for a 2.3 km section, the Komati Formation erupted during a 10 million year period between 3481 ± 2 Ma and 3472 ± 5 Ma for a rate of 0.25 km per million years. In contrast the 2.7 km thick Hooggenoeg Formation with 4–5 chert layers and seafloor alteration zones may record up to 27 million years of volcanism between 3.47 and 3.445 Ga – 0.1 km per million years. The Mendon Formation with the most cherts and seafloor alteration may represent only 0.03 km per million years. In summary, the younger formations record longer periods with less volcanism, presumably due to longer gaps between volcanic events (as opposed to actual rates of eruption). Although more work is needed to rigorously constrain the timing of volcanism, the ultramafic formations erupted at higher rates within each pair.

5.4-3. VOLCANOLOGY OF THE BARBERTON GREENSTONE BELT

The volcanological component of research in the BGB has had three hurdles: (1) distinguishing the volcanic stratigraphy from intrusions, especially mafic and ultramafic sills, (2) defining flow fields in vertical dipping sequences despite deformation and locally extreme alteration, and (3) seeing the dynamics of flowing lava recorded by discontinuous outcrops. With interesting variations, the same problems occur in each volcanic formation, and our research focuses on the Komati Formation, primarily because of its exquisite exposure of the oldest, well-preserved komatiites on Earth. Their composition extends the field of volcanology into ultramafic realms with no modern analogues.

Submarine flows of spinifex-textured komatiite (>23% MgO) erupted at least three times over 180 millions years from the ca. 3.48 Ga Komati Formation and overlying Hooggenoeg Formation (<3.445 Ga) to the 3.3 Ga Mendon Formation (Fig. 5.4-2). This repeated volcanic history within one crustal section implies that we might find equivalents of the Mendon komatiites intruding the underlying formations and raises the question: Can we distinguish massive komatiite flows in the Komati Formation from komatiite sills that

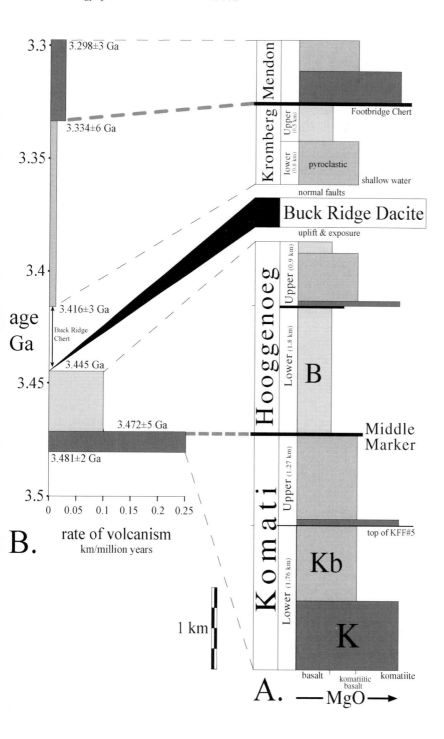

may be 130 million years younger and equally serpentinized? During this interval, the crustal section evolved tectonically, especially during intrusion of ca. 3.445 Ga tonalite batholiths, so we might expect the komatiite magmatic suites to reflect those changes. All three magma types – basalts, komatiitic basalts, and komatiites – have the same possibilities for creating a complex, evolving magmatic architecture that we can unravel with detailed mapping, coupled with geochemical and petrologic analysis. Confusion in both the Komati and Kromberg Formations has centered on ultramafic sills – are they flows, subvolcanic sills, or much younger intrusions unrelated to the volcanic stratigraphy?

In the Komati Formation the volcanic stratigraphy is intruded by a suite of wehrlite dikes and sills, black serpentinites that locally blend with massive komatiite flows. Previous researchers considered these wehrlites to be feeder dikes for the komatiite flows (Viljoen et al., 1983; de Wit et al., 1987b), but the following features distinguish this intrusive suite from the komatiites: (1) cross-cutting contacts, (2) co-intrusive gabbroic and diabasic phases, (3) coarse-grained, poikolitic pyroxene and igneous hornblende, and (4) higher TiO_2. This suite intrudes the overlying Hooggenoeg Formation but with less outcrop-level ambiguity because the Hooggenoeg has no black, serpentinized komatiite lava flows. This suite also intrudes some tonalite sills, a fact that indicates this intrusive suite is at least 30 million years younger than the komatiites in the Komati Formation. This widespread suite includes several layered intrusions, and a comprehensive study including geochronology is needed to understand its role in the evolution of the crustal section and to test the possibility that a single magmatic event links structural blocks or stitches terrains.

In the second challenge, we need to divide the volcanic stratigraphy into distinct units or flow fields. A flow field is an eruptive episode of geochemically coherent lava, a batch of mantle melt that has undergone the same AFC processes, possibly to varying degrees. Therefore, when reconstructing flow fields, we must exclude sill complexes, divide the stratigraphy into compositional groups, and recognize any volcanic hiatus marked by sediments or horizons of seafloor alteration or weathering. In the Komati Formation, alternating layers of komatiite and komatiitic basalt define distinct flow fields. In addition to compositional grouping in the Hooggenoeg Formation, distinct episodes of volcanic activity are bound by gaps marked by chert beds and severe alteration of underlying volcanic rocks. In the Kromberg Formation, a pyroclastic deposit of komatiitic basalt – possibly erupted subaerially – is overlain by submarine basalt flows. In this case, both the composition and style of eruption define distinct volcanic episodes. Since modern volcanism varies with its tectonic environment, we assume that the geochemical evolution of flow fields of the Onverwatch Group records the early tectonic evolution of the crustal section.

Besides the outcrop scale differences – komatiites are black, basalts are green – geochemistry is essential for defining flow fields. The volcanic rocks of Barberton divide into three main groups – komatiites, komatiitic basalts, and basalts, listed in order of the number of available analyses. Komatiites and komatiitic basalts are strongly fractionated within flows: olivine phenocrysts settled to form basal cumulates. As a result komatiitic basalts with erupted liquid compositions around 15% MgO fractionated to make rock compositions with 6 to 22% MgO, a compositional range that suggests a continuum of compositions between the basalt and komatiites. However, the gap in liquid compositions between komatiitic basalts (15% MgO) and komatiites (25% MgO) is real, making their association truly bimodal in Barberton. In the larger context of all greenstone volcanics, basalts and komatiites may be end members in a compositional continuum of mantle melts, but in many greenstone belts, komatiites and komatiitic basalt occur together repeatedly with a broad gap in parental compositions. Besides the bimodality, the komatiite-komatiitic basalt association shows overwhelming evidence for fractionation within flows and very little for pre-eruptive fractionation in magma chambers.

The third challenge – seeing the dynamics of flowing lava – attempts to move modern volcanology into the Archean. The flow field concept provides an important perspective while we interpret the cross sections of steeply dipping volcanic stratigraphy, the map view in most orogenic belts. First, a flow starts at the vent and ends at its distal reaches, so lava has passed through almost every cross section – our map views – whether we see evidence of flow or not. In the old view, spinifex flows were viewed as closed systems of ponded lava (e.g., Silva et al., 1997), but now they are understood to have been lava conduits that record the flux of lava and inflation/deflation in their textures and zoning. Similarly, sheet flows of komatiitic basalt have pyroxene-spinifex layers and mingling features, both recording inflation of lavas in that compositional range. The contacts between pillowed and massive basalt flows reveal complex systems of lava channelized in tubes with levees of pillows, pillowed flows with prograding foresets, flows with in situ flow-top brecciation, and other structures that freeze and record lava flow dynamics. What is interesting about flow fields in submarine lava plains is how flows create topography that determines the course of successive flows, a self-organizing system that we just start to appreciate in mapping some of the best outcrops.

5.4-4. KOMATI FORMATION

The Komati Formation has two members: (1) a lower member of alternating layers of komatiite and komatiitic basalt, and (2) an upper member dominated by pillowed and massive flows of komatiitic basalt (Fig. 5.4-3(A)). In the Lower Komati, five layers of komatiite share the same zoning – spinifex komatiite overlying massive komatiite – yet each layer is geochemically distinct (Fig. 5.4-3(B)). The repeated pattern – spinifex over massive – along with the changing geochemistry defines komatiite flow fields within the Komati Formation (Fig. 5.4-3(C)). Similar to komatiites, the flow fields of komatiitic basalt are geochemically distinct and have two flow types – pillowed and massive – but they do not show a repeated pattern of pillowed over massive flows (or visa versa). Thus, we consider all the major layers in the Lower Komati to be distinct flow fields, defined by the stratigraphy, flow morphology, petrology, and geochemistry. These flow fields represent distinct batches of mantle melt because low-P, olivine fractionation can not account for the difference in Al_2O_3/TiO_2. The Upper Komati is not as continuously exposed as the Lower Komati, it lacks stratigraphic markers, and geochemical coverage is too sparse, so we cannot divide this section of komatiitic basalt into flow fields without further work.

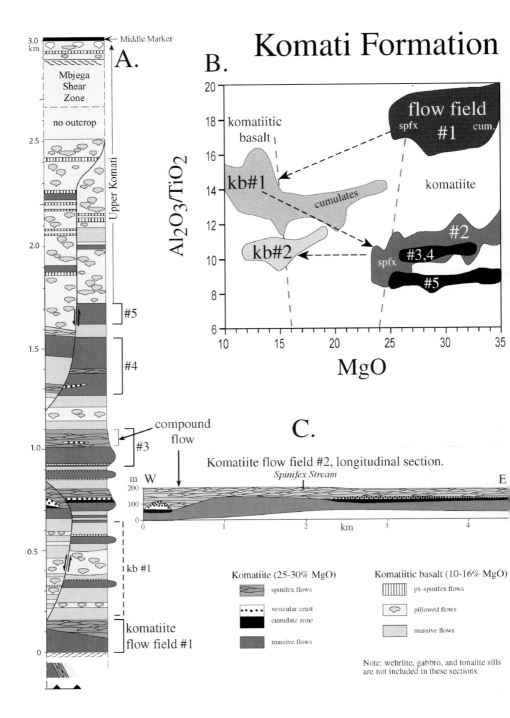

Komati Formation

C. Komatiite flow field #2, longitudinal section.

Komatiite (25-30% MgO)
- spinifex flows
- vesicular crust
- cumulate zone
- massive flows

Komatiitic basalt (10-16% MgO)
- px-spinifex flows
- pillowed flows
- massive flows

Note: wehrlite, gabbro, and tonalite sills are not included in these sections.

Fig. 5.4-3. The Komati Formation: (A) Stratigraphic column of the Upper and Lower Komati Formation with komatiite flow fields labeled #1–5, solid brackets; (B) Plot of Al_2O_3/TiO_2 vs MgO distinguishes flow fields for both magma types and reveals shared trend; (C) Longitudinal section of komatiite flow field #2 with massive, vesicular, and spinifex komatiites.

The Komati Formation only contains volcanogenic sediments produced directly from endemic volcanic processes with one exception: the Komati Tuff, a dacitic airfall tuff preserved between lava flows (Fig. 5.4-4). Several thin beds grade from greywacke to porcelainite, and the greywacke contains volcanic fragments and crystals of quartz and plagioclase. A single zircon age of 3481 ± 2 Ma (S. Bowring, pers. comm.) records a nearby felsic volcanic event and provides the best age for Komati volcanism.

5.4-4.1. Lower Komati Formation: Komatiite Flow Fields

With the Lower Komati neatly divided into flow fields, we can focus on the architecture of komatiite flow fields by examining the morphology and relationship between the three flow types: massive, vesicular, and spinifex. The most important volcanological discovery of our mapping is probably the thick vesicular tumulus (Fig. 5.4-5) because its textural zones illustrate how channelized komatiite lava inflated to 50 m thick, creating relief on the seafloor and potentially transporting lava long distances from a vent under water (Dann,

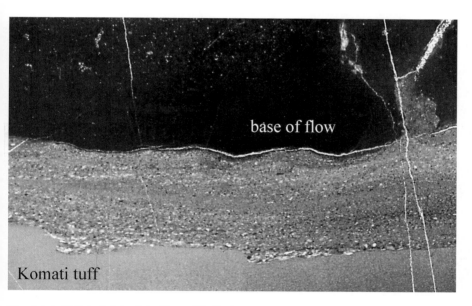

Fig. 5.4-4. Polished slab of the Komati Tuff, showing graded beds of dacitic airfall tuff, overlain by flow of komatiitic basalt (black).

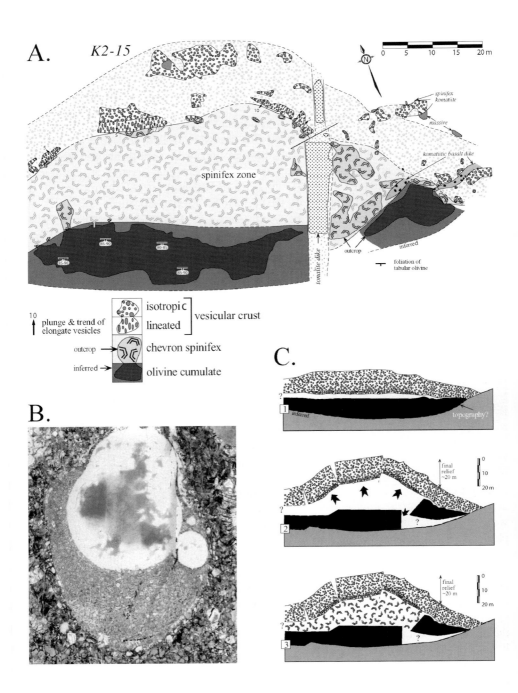

Fig. 5.4-5. Thick vesicular tumulus or lava tube: (A) Outcrop map showing cumulate, spinifex, and vesicular zones of 50 m thick, lenticular komatiite; (B) Thin section of a segregation vesicle, showing a serpentine-filled compound vesicle (top) and segregation (bottom) of augite and glass; (C) Model based on restoring rotated contacts to horizontal (1). Influx of lava rotated thick crust (2) and spinifex zone crystallized (3).

2001). Vesicular komatiite occurs discontinuously within flow field #2 (Fig. 5.4-3(C)) between overlying spinifex flows and underlying massive flows and, in total, within three of the five komatiite flow fields. Although vesicular komatiites are volumetrically rare, they record stages in the emplacement of komatiite flows and link the massive and spinifex flows within single komatiite flow fields.

5.4-4.1.1. Vesicular flows

The thick vesicular tumulus has three extraordinary textural zones: a lower olivine-cumulate zone, a central olivine-spinifex zone, and an upper vesicular zone (Fig. 5.4-5(A)). The cumulate, an orthocumulate dunite, consists of tabular olivine crystals (1 cm), tightly packed to form an originally horizontal magmatic foliation. The interior spinifex zone crystallized fist-sized, skeletal olivine with interstitial augite and glass. The vesicular zone consists of large serpentine-filled segregation vesicles (Fig. 5.4-5(B)) that are distinctly elongate near the lower contact, parallel to a cryptic vertical fabric of skeletal olivine. Forming the tumulus' lenticular shape, the vesicular carapace broke into blocks and rotated 35–45° outward, rotating also the originally vertical, linear fabric of vesicles. The lenticular cavity is filled by the spinifex zone, indicating that it crystallized after the rotation of the upper carapace. As shown in the model (Fig. 5.4-5(C)), the flow thickened in two stages: (1) vesicles accumulated beneath a downward crystallizing roof, and (2) the roof broke and rotated with the influx of lava, forming the domed cross section or tumulus.

The tumulus is probably the cross section of a lava channel or tube that focused flow like a river along a topographic low (the lower contact is not exposed). Long-lived flow within a conduit explains the exceptional features of this tumulus. First, an extensive lava-tube system was needed to transmit enough hydraulic pressure downstream to inflate the 20 m thick crust and form the 50 m thick tumulus. Second, the long-lived flux of lava within a tube facilitated the growth and accumulation of both exceptionally large olivine phenocrysts and vesicles in, respectively, the cumulate and upper crust. The 20 m thick vesicular crust effectively insulated lava from seawater, potentially allowing a lava tube of this size to rapidly deliver large volumes of lava long distances from the vent, prior to crystallization of the spinifex zone. Similar lenticular units, thick and vesicular, occur in the younger Kambalda komatiites of Australia, which Stolz and Nesbitt (1981) interpreted as lava channels, key arteries within komatiite flow fields. What makes the tumulus exceptional is how rotation of the upper crust is recorded in textures and contacts, well exposed in outcrops of the Komati Formation (Fig. 5.4-5).

Along with the thick tumulus in flow field #2, a vesicular sheet flow has inflation structures exposed in every outcrop for several km along strike (Fig. 5.4-3(C)). Locally, the

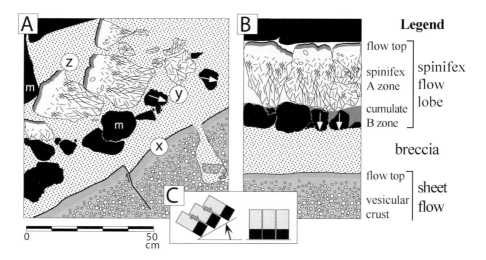

Fig. 5.4-6. Vesicular sheet flow: (A) Outcrop map of tilted flow top (x) and breccia with blocks of spinifex flow with glassy flow base (y) and top (z); (B) Flow top and blocks restored to horizontal; (C) Model of domino-style rotation with tilting; (D) Outcrop map showing vesicle-spinifex-layered crust intruded by cumulate B-zone lava, linking chilled, cross-cutting contacts with cryptic, layer-parallel, intrusive contacts; (E) Possible drill core view of upper crust in (D) with cryptic intrusive contacts; (F) Model of sheet-flow inflation: initial crust-forming stage (1), intermediate stage of differential inflation and breakout of flow lobes (2), and final stage that rotates flow lobes in thick breccia (3).

upper crust is intruded by dikes from the flow's interior, tilted 35°, and overlain by a thick breccia (Fig. 5.4-6(A)). The breccia fills fissures that cut through the flow top into the vesicular crust. The breccia contains blocks of a spinifex flow with its glassy flow top, spinifex zone, and lower cumulate zone, blocks that can be reassembled (Fig. 5.4-6(B)), indicating that the spinifex flow fractured and rotated like dominoes as the crust tilted (Fig. 5.4-6(C)). The lower part of the crust is intruded by a network of dikes emerging from the flow's lower cumulate zone (not shown here, see Dann (2000), Fig. 5), indicating that massive komatiite remained mobile beneath the insulating crust, behavior not usually associated with a cumulate.

Where cumulate komatiite intrudes the upper crust (Fig. 5.4-6(D)), the chilled margins of dikes are well formed near the flow top (x) and progressively disappear downward (y), a feature that illustrates (1) how cryptic intrusive contacts pose as layering and (2) how dikes might fractionate lava. Without chilled margins or cross cutting contacts (e.g., 'z'), layer-parallel intrusions of this crystal-rich lava can not be distinguished from crustal layers or zones, a problem that plagues research limited to drill core (Fig. 5.4-6(E)) and may explain vesicular and massive zones, repeatedly layered in thick Kambalda vesicular komatiites described by Beresford et al. (2000). Even thin sections of these cryptic contacts are ambiguous. Furthermore, with massive komatiite making up 62% of the volcanic stratig-

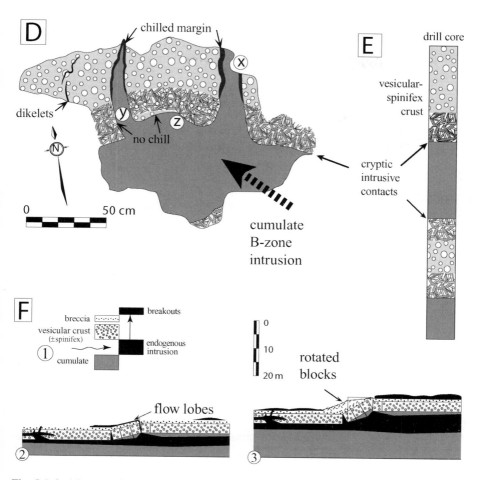

Fig. 5.4-6. (*Continued.*)

raphy, we are missing important information on the volcanology of komatiites because so many features that must be there are not distinguishable: flow tops, contacts between endogenous intrusions and upper crust, and layers that record episodes of crystallization or accumulation. Finally, any discernible chilled margins solidify from phenocrysts-free liquid, indicating that phenocrysts are filtered by the flow through the dike, fractionating the lava. Therefore, as the lava flows through a network of dikes in the crust and erupts, it is likely to form a spinifex flow (Fig. 5.4-6(B)) with a proportionately thin B zone, a minimal load of phenocrysts. This fractionation of olivine may mimic on a small scale what occurred from proximal to distal facies within each flow field.

In a model based on surface morphology of modern flows, we bring together the features of this vesicular sheet flow – intrusion, crustal tilting, breakout, and brecciation (Fig. 5.4-

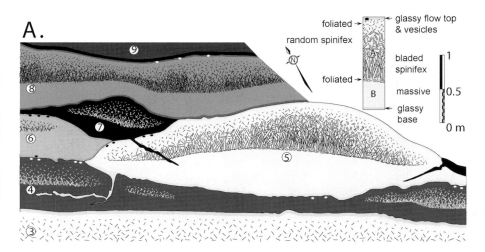

Fig. 5.4-7. Spinifex komatiite flow lobes: (A) Outcrop map showing chilled contacts with vesicles and spinifex and cumulate zones; (B) Top of spinifex zone from glassy flow top to top of bladed spinifex zone. Note that a subhorizontal flow fabric of olivine laths (serpentine pseudomorphs) dominates the top 3 cm.

6(F)). Vesicular and spinifex zones alternate in the upper crust, a pattern that indicates episodes of vesiculation during influx and inflation of this sheet flow (#1). When this upper crust solidified, the lower cumulate zone was still hot fluid lava, flowing and lifting crust preferentially in some areas more than others (i.e., a lava rise). The solid crust responded by breaking, tilting, and accumulating a thick breccia, features that mark the edge of a lava rise (#2). While the crust was tilting, lava intruded fractures and erupted locally as thin flows or breakouts. On the edge of the lava rise, one of these flows broke up and rotated like dominoes in the breccia as the underlying crust tilted (#3).

This vesicular sheet flow and overlying breccia have two implications for flow field development. First, flow lobes started to cover the sheet flow before its interior had solidified. This physical overlap means that flows can insulate underlying flows, delaying their cooling and crystallization, so we must be aware that cooling units are larger than individual flows. Second, these two flow types overlap in time, supporting our contention that komatiite flow fields evolve from massive to spinifex flows and do not consist of two separate effusive events. In this way, the vesicular flows provide a missing link between the overlying spinifex flows and underlying massive flows.

5.4-4.1.2. Massive flows

Massive komatiite forms ridges and most of the outcrop, but not finding any of the textural features of spinifex flows, one may question whether they really are flows. Weathered surfaces and thin sections reveal a framework of equant olivine (1–2 mm) in a matrix of augite laths and glass. Could this orthocumulate really flow for at least 6–10 km across the

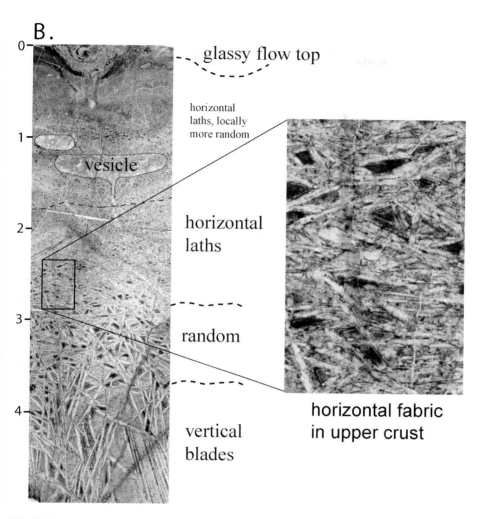

Fig. 5.4-7. (*Continued.*)

ocean floor? Despite these questions, we are certain they are flows. First, they are equally fine-grained throughout even the thickest units (i.e., equally rapidly cooled), and they are texturally indistinguishable from the B zone, the lower olivine cumulate of spinifex flows (see Fig. 5.4-7). Second, they are locally interlayered with the spinifex flows as the end member of a continuum: the B zone makes up 10 to nearly 100% of spinifex flows. Third, massive flows have a few unique transitional features including internal lobate chills (Dann, 2000, Fig. 6), vestiges of earlier surfaces, and spinifex veins or vein networks (Dann, 2000, Fig. 11). Finally, the mobility of massive komatiite and how it forms such thick flows is illustrated by how it intrudes, erodes, engulfs, and covers the upper crust of the vesicular sheet flow (Dann, 2001, Figs. 5–9). Since phenocrysts make a more viscous lava, massive komatiite must have been less fluid than aphyric komatiite, a difference that may explain variable zoning in spinifex flows.

5.4-4.1.3. *Spinifex flows*

Spinifex flows, about one-third of the komatiites, are characterized by two distinct textural zones: an upper olivine spinifex or A zone in sharp contact with a lower olivine cumulate or B zone (Fig. 5.4-7(A)). The A zone records the downward crystallization of an olivine framework that culminates in paper-thin blades of olivine up to 50 cm in length. In sharp contrast, the B zone records the accumulation of olivine phenocrysts (equant, 1–2 mm). In weathered outcrops the contact between the A and B zones stands out because of the huge contrast in grain size. The proportion, A or B zone to total thickness, ranges from 10 to 90% due to a large variation in accumulated olivine. Some A and B zones have distinct horizontal layering that record episodes of lava influx and inflation. Flows vary from 20 cm thick lobes to 8 m thick sheets, both of which are rare compared to 1–2 m thick flow lobes (Fig. 5.4-7(A)). The size and shape of spinifex flows compares well with flow lobes and thin sheets that make up compound flows in *pahoehoe* flow fields. Based on this comparison and coherent stratigraphic trends in thicknesses and A–B zoning, we think that these spinifex flows make up compound flows, the upper part of komatiite flow fields. Any emplacement model must explain the variety of spinifex flows, in particular, their shapes, thicknesses, and A–B zoning.

In response to cooling from the flow top, the A zone records downward growth of the upper crust in four stages marked by distinct zones of increasing grain size (Fig. 5.4-7(B)): (1) the glassy flow top and breccia (with vesicles), (2) microlites of olivine aligned in the plane parallel to the flow top (with vesicles and cryptic layering), (3) randomly oriented olivine laths, and (4) vertically oriented olivine blades. The depth of this last transition, 5 cm below the flow top here, is greater in some flows. At the base of the thickest A zones, dozens of mm-thin blades of olivine up to 50 cm in length grow interconnected in optically continuous books (Fig. 5.4-8(A)). Books are cut off by adjacent books, forming a continuous framework around interstitial space (∼50%) filled with augite laths and glass. In some flows, this framework is in cone-shaped growths radiating downwards. In about half of the thickest flows, a layer of random olivine laths interrupts the normal size-gradient of the bladed spinifex zone, recording the influx of lava with different internal properties (i.e., higher ratio of crystal-nucleation density to crystal-growth rate).

The B zone is commonly featureless and so unlike the spinifex zone that if they were not bound between glassy flow top and base, they might appear to be two separate lavas. The glassy flow base is choked with olivine phenocrysts, so it is much thinner than the glassy flow top. At the top of the B zone (B1), the olivine is more

---

Fig. 5.4-8. Spinifex komatiite flow lobes: (A) Base of spinifex zone showing interlocking books of bladed olivine; (B) B-zone graded layering from smooth olivine-rich to rough olivine-poor (augite grew interstitial to olivine and was converted to amphibole, which weathers in relief).

bladed-
olivine
spinifex

A zone
B

elongate and aligned in a horizontal plane. More rarely, the B zone is layered with modally graded beds (Fig. 5.4-8(B)). The B zone often contains a subhorizontal spinifex vein, which locally connects to the overlying A zone. This filter-pressed vein indicates that the B zone is the last part of the flow to solidify (i.e., the liquid interstitial to the framework of olivine phenocrysts is last to solidify). Unlike the much thicker vesicular sheet flow, the B zone never intrudes the upper crust of spinifex flows except in very rare cases where the upper crust is disrupted by differential inflation/deflation.

5.4-4.1.4. Inflation and evolution of komatiite flow fields

Any model of komatiite volcanology must explain (1) the huge variety of flow types from 20 cm thick lobes to the 50 m thick vesicular tumulus, (2) within spinifex flows, the variable layering and proportions of A and B zones, and (3) the evolution of flow fields from massive to vesicular to spinifex flows. The best model to explain the range of sizes and shapes of flows is the inflationary model: a thin flow front (20 cm high) fed by laminar flow inflates as the outer visco-elastic skin holds incoming lava. The lenticular shapes taken by such fluid lava reflects this ballooning effect (Fig. 5.4-7(A)). Without this effect, highly fluid lava would always have a flat flow top even in small lobes. The thickest flows have proportionately thicker A zones with layers, two features that record lava flowing under the upper crust in episodes, preferentially feeding spinifex growth of the A zone. Only the thickest flows exhibit rotation of thick crusts by (1) endogenous intrusion of mobile B-zone lava (Fig. 5.4-6) and (2) by influx of A-zone lava under the thick vesicular crust of the tumulus (Fig. 5.4-5). So, the inflation model explains the variation in flow thickness and the shapes and structures, but what about the variable zoning in the spinifex flows?

The B-zone, a framework of tiny equant phenocrysts, makes up 100% of massive flows and as little as 10% of spinifex flows, with a continuum of A–B zoning between these extremes. In addition, stratigraphic variation of flows along Spinifex Stream reveals an orderly pattern: a series of thin flows oscillates between 0 and 20% A zone, followed by a trend toward proportionately thicker A zones, up to 70% (see Dann, 2000, Fig. 8), a trend that mimics the overall zoning of the flow fields – massive to spinifex. To explain these patterns, we compare two flow models, turbulent and laminar.

If spinifex flows were emplaced by turbulent flow that kept olivine phenocrysts suspended until they settled within each flow, then the proportion of phenocrysts would be maintained from flow to flow. However, this model does not fit the observations. First, phenocrysts frozen in the glassy flow top are rare (less than 1%) whereas phenocrysts in the B zones compose 5–45% of the flow. The lack of phenocrysts in the first part of the flow to freeze indicates that the phenocrysts were not evenly distributed but must have been concentrated in the lower part of the flow during emplacement, primarily in the laminar flow regime. Second, laminar flow beneath a thickening upper crust is indicated by the ubiquitous flow fabric in the upper A zone (Fig. 5.4-7(B)) and by other A and B zone layering (Fig. 5.4-8(B)). In the laminar flow model, the phenocrysts settle, (1) enriching the base, where most of the phenocrysts are carried as a suspended bed load, and (2) depleting the upper part, where spinifex textures can grow in supercooled conditions without crystal nuclei. As a result, a single lobe or conduit had two lavas with different properties that allowed them to flow and respond independently to changing hydrologic and thermal conditions, accounting for the variable proportions of A–B zoning in spinifex sheets and lobes.

More dense and viscous, the lava that became the B zone tended to fill topographic lows and be left behind as flow surged from lobe to lobe. As a result, succeeding lobes filled with more aphyric lava, rendering proportionately thicker spinifex zones. The tumulus (Fig. 5.4-5) best illustrates the late influx of aphyric lava forming the interior spinifex zone. Inflation/deflation would also play a role in fractionating the lava. As lobes coalesced and inflated into thick sheet flows, incoming lava filled space rather than rushing through, and these conditions – slowly cooled and relatively static – facilitated the growth of a thick spinifex crust. When flow through the sheet resumed, deflating it, only phenocryst-rich lava was mobile, leaving the sheet with the relatively thick A zone typical of the thickest spinifex flows. Another mechanism would be whether lava breaks out through (1) a basal toe, tapping B-zone lava, or (2) the flow top, tapping aphyric lava. Although we cannot see in the outcrop exactly why A–B zoning varies from lobe to lobe, these mechanisms contributed to variable A–B zoning and the stratigraphic trend to flows with proportionately thicker A zones.

The five thickest flow fields share the same stratigraphy: a compound flow of multiple spinifex flows overlies a sequence of massive flows. These fields represent a single eruptive episode, as indicated by two types of transition: (1) interlayered spinifex and massive flows, and (2) vesicular flows that link underlying massive lava to overlying spinifex flows. The difference between the two flow types is primarily the phenocryst load – less olivine phenocrysts leaves more liquid space, free of nuclei, for the growth of spinifex olivine. This transition in flow types reflects several mechanisms: (1) superposition of distal and proximal facies and (2) maturation of the vent. We think that the spinifex flows have a bulk composition closer to the parental magma (25–28% MgO) than the massive flows (33–37% MgO). Therefore, massive flows trapped olivine as komatiite lava flowed away from the vent, and with the decrease in phenocryst load, distal facies were marked by spinifex flows. Distal refers to the extent that lava flows from the vent or away from a lava tube, so if eruption rates decreased, distal facies may overlie medial or proximal facies of earlier flows, a possible explanation for the stratigraphy of komatiite flow fields.

On the other hand, this sequence may reflect thermal maturation of the vent. Initially, magma cooled and degassed in the conduit and fissure, and erupted as olivine-phyric lava. As magma passed through an increasingly heated conduit, it erupted hotter, closer to its liquidus, thus less degassed and more aphyric. Both the absence of nuclei and degassing during emplacement contributed to supercooling and rapid growth of skeletal olivine. The vesicular komatiites record degassing of lava under thickening crust, while most vesicles escaped the thinner spinifex flows. The large skeletal crystals at the base of A zones reflect high crystal growth rates (supercooling), combined with a low cooling rate, due to an insulating upper crust.

5.4-4.2. Lower Komati Formation: Flow Fields of Komatiitic Basalt

The flow fields of komatiitic basalt in the Lower Komati are 80% sheet flows with interlay-ered pillows (Fig. 5.4-9(A)). What distinguish pillows of komatiitic basalt from those of basalt are the large number of withdrawal cavities, a darker color with contrasting varioles, pale selvages, and the lack of vesicles. The sheet flows display some of the most interesting features: ocelli, internal compositional layering, and pyroxene-spinifex zones. These flows are highly fractionated internally, and most of these internal zones and structures record episodes of inflation, which mingled fractionated and fresh lava.

Many pillows have quartz-filled cavities, commonly stacked in the upper third of the pillow (Fig. 5.4-9(B–C)). Note the arcuate upper surface and flat lower surface of the up-permost cavities. Above the arc is the thickness of the pillow wall when lava first partially drained. The flat lower surface is the level of lava within the pillow when water quenched it. This liquid surface was originally horizontal and makes an easily measured proxy for bedding, roughly parallel to the flow field. In contrast to the smooth shapes of the outer cav-ities, the inner ones are rough and irregular with lava drips or stalactites that were frozen on their upper surfaces (Fig. 5.4-9(C)). The drainage cavities record the pulse of lava as it flowed through a pillow (tube). When a pillow forms (Fig. 5.4-9(D)), it is full and begins to solidify from the outer selvage (#1). When it buds and feeds a new pillow, lava drains out faster than it flows in, so the lava level falls. Drainage creates a vacuum that pulls water in through cracks, water that chills the lava surface (#2). Each level of cavities represents an event in the progression of the flow (#3), and some pillows have eight levels. Withdrawal cavities are more common in these pillows because komatiitic basalt was more fluid than basalt.

Sheet flows in the Lower Komati are of three types: layered, mingled, or spinifex. All flows are strongly fractionated with lower cumulate zones (22% MgO) and upper zones en-riched in incompatible elements (Viljoen et al., 1983; Dann and Grove, unpublished data). Layered flows are sheets with a cryptic layering (Fig. 5.4-10(A)) preserved by varying proportions of metamorphic plagioclase and actinolite (Fig. 5.4-10(B)). Augite is rarely preserved. All the dark layers have a sharp upper contact that locally appears to truncate layering in the overlying leucocratic phase, layering partly defined by vesicles (traced in Fig. 5.4-10(C)). All the dark layers grade into the underlying leucocratic phase. This com-positional gradation below a sharp contact is typical of all the layering in these sheet flows.

Since all thick sheets must inflate by lava flowing beneath a thickening upper crust that is rapidly cooling against seawater, each layer (six in this outcrop) records an episode of lava influx (Fig. 5.4-10(D)). In this model, fresh lava solidifies quickly at the flow top, while olivine and pyroxene settle and fractionate the interior. As the upper crust solidifies, some vesicle layers accumulate beneath this roof. When a pulse of lava inflates the flow, fresh hot lava locally truncates the roof and chills against the cooler, fractionated lava (or crust), forming the sharp upper contact. The interior of the inflating lava cools more slowly and fractionates more, further from the chills, creating the compositional gradation. The influx of lava occurs episodically, inflating the flow and juxtaposing fractionated and fresh lava.

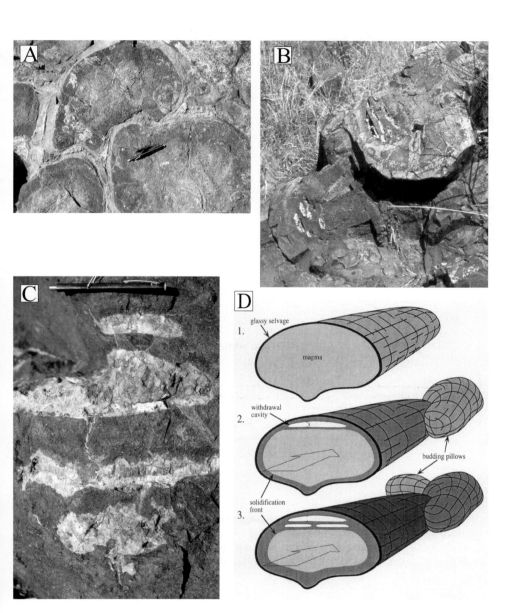

Fig. 5.4-9. Pillowed flows of komatiitic basalt: (A) Pillows; (B) Pillows with withdrawal cavities filled with quartz (white) and carbonate (hollow). These pillows are overturned; (C) Stacked withdrawal cavities; (D) Model linking drainage cavities to budding of pillows.

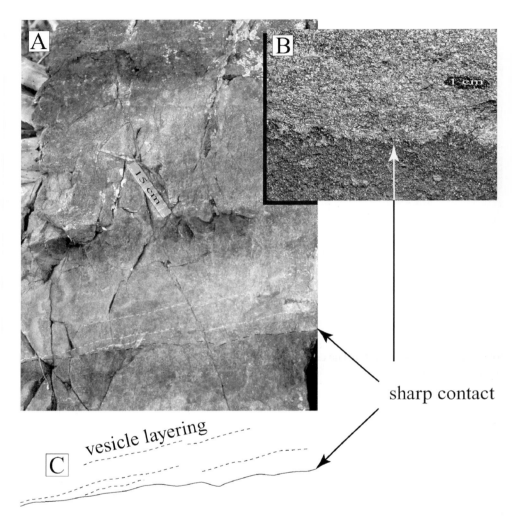

Fig. 5.4-10. Massive komatiitic basalt: (A) Compositional layering that grades downward from mafic to intermediate composition; (B) Close-up of sharp upper contact (1 cm mark) in meta-igneous texture; (C) Trace of vesicle layering truncated by sharp contact; (D) Model of episodic inflation and fractionation to explain compositional layering and internal truncations.

Some massive sheet flows have, in their upper interiors, dark green basaltic lava in sharp contact with pale green, coalescing andesitic lobes (ocelli) that form both horizontal layers and wild spiral-like shapes (Fig. 5.4-11; also see de Wit (1987), Cloete (1999), and Dann et al. (1998)). Cloete (1999) proposed that these shapes reflect flow within a lava tube, and we think that inflation and mingling provide the best explanation for their origin. In this model, flow tops – aphyric, aphanitic with chilled lobes like the overlying pillows –

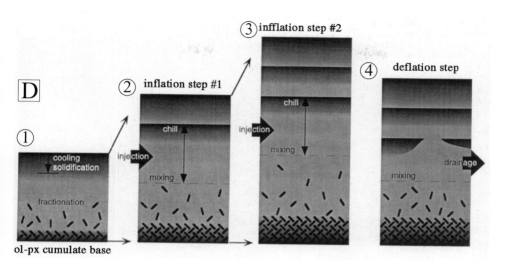

Fig. 5.4-10. (*Continued.*)

preserve unfractionated komatiitic basalt (14% MgO). This composition fractionates by crystal settling to form (1) a lower cumulate (22% MgO) and (2) the andesitic layers and lobes (6% MgO, 60% SiO_2). The cusp-and-lobe contacts represent mingling of two silicate liquids: fresh basaltic komatiite – unfractionated, hot, and highly fluid – and the andesite – cooler and more viscous. These compositions are not immiscible pairs, but within many sheet flows, the temperature differences and rapid cooling prevent these compositions from mixing. These mingling structures (ocelli) are unique in Barberton to komatiitic basalts because of the efficiency of fractionation within flows to create a contrast in lava viscosity.

Ocelli are ubiquitous within Archean volcanic sequences (de Wit and Ashwal, 1997), inspiring a variety of explanations from sediment xenomelts to magma mingling or immiscibility, and they are not always easily distinguishable from varioles, usually smaller spheres, which appear to be devitrification structures enhanced by metamorphism. Despite this ambiguity we think that the occelli and related layering within sheet flows of komatiitic basalt record the mingling of fresh and fractionated lava within inflating lava flows.

5.4-4.3. Upper Komati Formation

The Upper Komati is 85% pillowed flows of komatiitic basalt, interlayered with massive units that form ridges with poorly exposed contacts. Without clear contact relationships, we cannot distinguish sheet flows from subvolcanic sills. To address this problem, we mapped three sheet-like units interlayered with pillows in nearly continuous outcrop with well exposed contacts (Fig. 5.4-12; for location, see Dann et al. (1998), Dann (2000, Map#1, K15-7)). Two sheets (~10 m thick) cross the Mlala Stream outcrop and are representative of the two types that form prominent outcrops and ridges throughout the Upper Komati

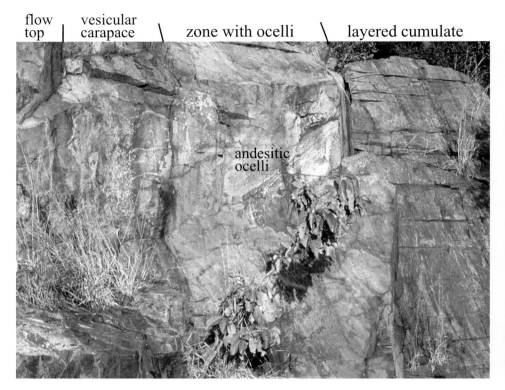

Fig. 5.4-11. Sheet flow of komatiitic basalt with inflation-related mingling structures. The lower cumulate zone is finely layered while the andesitic differentiate (pale) is mingled with undifferentiated lava (dark), an indicator of a later episode of lava influx.

– pyroxene-spinifex units and massive sheets without textural zones. The third unit (1 m thick) is closely associated with pillows and a layer of pillow breccia intruded by a vertical dike, a peculiar combination. The breccia weathers recessively, so this outcrop provides a rare glimpse of the lithologies poorly represented in hillside outcrops elsewhere.

The lowest sheet has three zones: a lower cumulate, a pyroxene-spinifex layer (Fig. 5.4-12, #1), and an upper, fine-grained zone with a flat, featureless flow top. This zoning is typical of many sheets of komatiitic basalt in the Upper Komati, but here the stream erodes a vertical dike, rendering an excellent section (#2). The flow top is covered by pillows (Fig. 5.4-12, #3), one of which protrudes downward into the underlying flow as a ropey and sinuous dikelet (Fig. 5.4-13(A)). This connection between pillows and the underlying unit proves that this spinifex sheet is a lava flow – not a sill. Quartz-filled cavities between the fissure walls and dikelet indicate that the lava poured into an open fissure, in contrast to magma injected into a fracture. Fissures commonly open modern flow tops (Fig. 5.4-13(B)), but they are rarely exposed in vertical sequences. What other features characterize submarine sheet flows? The spinifex zone is not diagnostic, but pyroxene-spinifex units

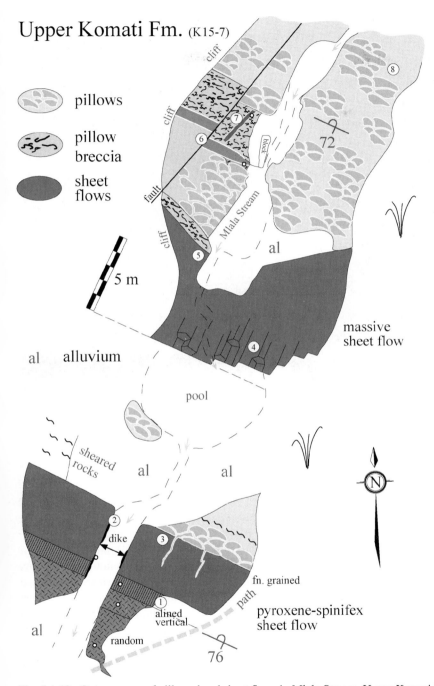

Fig. 5.4-12. Outcrop map of pillowed and sheet flows in Mlala Stream, Upper Komati Formation.

Fig. 5.4-13. Komatiitic basalt of Upper Komati Formation: (A) Ropey lava (10 cm wide) continuously exposed between pillow and lava-filled fissure in underlying spinifex flow (#3, Fig. 5.4-12); (B) Modern example of pahoehoe lobe flowing into open fissure in underlying sheet flow (field book for scale; Canary Islands, courtesy of Stillman); (C) Columnar joints in lower half of sheet flow; (D) Pyroxene spinifex zone in thick sheet flow; linear igneous fabric has kink band (diagonal above compass for scale).

with clear intrusive contacts have symmetrical zoning (see Fig. 14 in Dann (2000)) and do not have the thick upper zone. This thick, fine-grained crust was rapidly cooled from a submerged flow top and is diagnostic of sheet flows.

The next sheet (6–11 m thick) lacks a spinifex zone (Fig. 5.4-12, #4), but its lower half is columnar jointed (Fig. 5.4-13(C)), cooling fractures found in both lava flows and sills. Its flow top is locally convoluted with quartz infillings (#5), indicative of open spaces created by viscous deformation at the top of a sheet flow. The overlying breccia may be a flow-top breccia, but it accumulated only on the low side of the V-shaped flow top, the side inclined 20° to the presumed paleohorizontal (defined within the outcrop by 6 contacts and layers and many withdrawal cavities) compared to 35° for the other side. This inflexion represents about 5 m of flow-top topography filled by pillows of the succeeding flow. Features indicative of a sheet flow include (1) the basal columnar jointed zone and (2) the variable upper surface correlative with overlying stratigraphy.

Fig. 5.4-13. (*Continued.*)

The third sheet tapers with multiple upper and lateral chills over a surface swell in the underlying pillowed flow (Fig. 5.4-12, #6). The base of this sheet contours over underlying pillows as expected for a flow. The flow top has three layers that successively end in the breccia as the sheet thins (Fig. 5.4-14(A)), indicating that the flow tops formed and fragmented in episodes, contributing to the breccia. The sheet's tapered end consists of four thin "tentacles" and one thicker "trunk" (Fig. 5.4-14(B)), recording episodes of lava influx, drainage, and collapse. The overlying breccia consists of pillow fragments in a hyaloclastite matrix. The fragments are isolated pillows and broken tubular shapes (Fig. 5.4-14(C)). The breccia is intruded by a vertical, stratabound dike (Fig. 5.4-12, #7), but how the dike connects to the underlying and overlying units is not clear. Neither the breccia nor its contacts are sheared, so the layer-parallel shear found between the two thick sheets (Fig. 5.4-12) cannot explain why this dike appears stratabound.

The simplest explanation for the similarity of the sheet and dike is for lava to flow from the sheet into the dike, a connection that implies that lava continued to flow through the sheet when it was covered by the breccia. This timing is supported by how the breccia truncates the succession of flow-top layers (Fig. 5.4-14(A)) and overlies the thin chilled layers

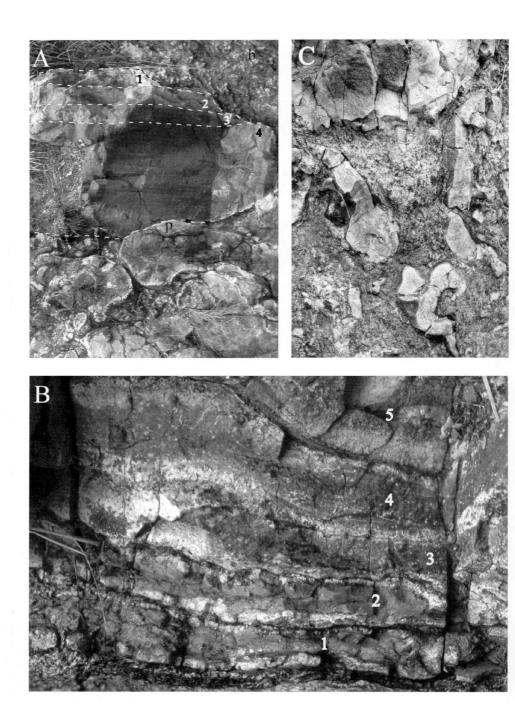

Fig. 5.4-14. (*Previous page.*) Sheet flow (1 m thick, #6, Fig. 5.4-12) of komatiitic basalt tapers in stages (1–4) between pillows (p) and overlying breccia (b): (A) The outermost crust #1 breaks up into breccia and is replaced by new crust #2 that also breaks up and is replaced by crust #3 and finally #4 from the interior; (B) Five glassy layers at terminus of sheet (just to right of (A)) with distinct chilled margins (white rims), also record episodes of emplacement; (C) Pillow breccia with isolated pillow (top) and fragments of outer crusts of drained pillow tubes.

(Fig. 5.4-14(B)), suggesting that the breccia formed by successive inflation and collapse of this sheet. Similar thin dikes occur throughout the Upper Komati and appear to be related to the volcanic stratigraphy because they intrude neither the Lower Komati nor the Middle Marker.

This outcrop exposes the contacts of two types of sheet flows: the massive columnar jointed sheet and the pyroxene-spinifex sheet, the two major types of ridge-forming units interlayered with pillows throughout the Upper Komati. Based on the contacts exposed in this outcrop, we feel confident that most of these ridge-forming units are sheet flows and not subvolcanic sills. The spinifex sheet has a flat upper surface, and the fissures in the upper zone may have opened when lava flowing in the interior fractured and extended the rapidly cooled upper crust. The columnar jointed sheet has variable thickness creating local topography. These slopes and variable thickness of the flow indicate that this sheet is not ponded lava with a flat top, but the relief reflects differential inflation and/or deflation.

The top of the Komati Formation is marked by the Middle Marker, a chert bed or silicified tuff of komatiitic basalt (Lowe, 1999b). Graded beds of lapilli are overlain by crossbedded tuff. The volcanic rocks beneath the Middle Marker are altered by a seafloor hydrothermal process that occurred during a hiatus in effusive volcanism.

5.4-5. HOOGGENOEG FORMATION

The Hooggenoeg Formation contains all three magma types – basalt, komatiitic basalt, and komatiite – in distinct flow fields separated by gaps in volcanism marked by cherts and underlying zones of alteration (Fig. 5.4-15). During the gap represented by the Middle Marker, komatiitic volcanism of the Komati Formation changed to Hooggenoeg basalt, nearly 2 km of basalt or 60–70% of the formation. Komatiitic volcanism returned in the Upper Hooggenoeg (Fig. 5.4-15), marked by interlayered pillows and pyroxene-spinifex sheet flows. Although geochemical coverage is sparse, we are fairly confident that the basalt and komatiitic basalt are not interlayered because of their distinct field characteristics. The only komatiite flow field (Fig. 5.4-15(A)) consists of olivine–spinifex flows that are entirely altered (Fig. 5.4-15(B)) beneath a chert horizon that separates two flow fields of komatiitic basalt. The komatiite appears to fill a broad channel in the underlying surface about 80 m thick. Although komatiite flows are not interbedded with flows of komatiitic basalt, the two magma types are closely associated – not with basalt.

Hooggenoeg Formation

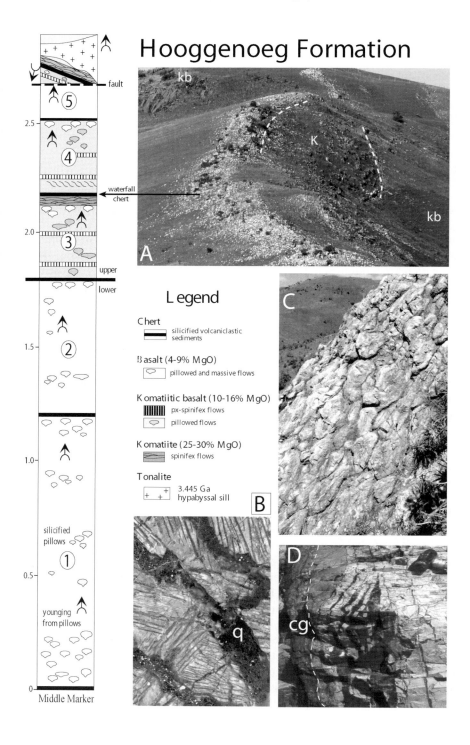

Legend

Chert
silicified volcaniclastic sediments

Basalt (4-9% MgO)
pillowed and massive flows

Komatiitic basalt (10-16% MgO)
px-spinifex flows
pillowed flows

Komatiite (25-30% MgO)
spinifex flows

Tonalite
3.445 Ga
hypabyssal sill

Fig. 5.4-15. (*Previous page.*) Hooggenoeg Formation. Stratigraphic column shows lower member of basalt and upper member with komatiitic basalt and komatiite: (A) The only komatiite flow field is locally 80 m thick and highly altered beneath the Waterfall Chert; (B) Silicified olivine-spinifex komatiite cut by network of quartz veins (q) peppered with pyrite (white); (C) Highly altered pillows of komatiitic basalt beneath the Waterfall Chert; (D) Silicified sediment is locally dilated by black chert, eroded forming local uniformity (dashed line), and covered by cherty conglomerate (cg).

The chert horizons are not really cherty sediments but rather volcaniclastic deposits, probably komatiitic basalt, reworked by currents, then silicified on the ocean floor (Lowe, 1999b). Beneath the sediments, the lavas are intensely altered; green pillows become white with 80% SiO_2 (Fig. 5.4-15(C)), and komatiite becomes a complex assemblage of three alteration products: silica, carbonate, and talc. The alteration is quite a dynamic process. Over-pressured fluids opened bedding-parallel, carbonate extension veins. Black chert locally dilated silicified sediments (Fig. 5.4-15(D)), creating an angular unconformity that may be more apparent than real due to extensive dissolution along bedding-parallel stylolites. The extent of alteration and the complexity of recorded processes attest to the long periods of seafloor exposure and gaps in volcanism that, for our purposes here, divide the volcanic stratigraphy into distinct flow fields.

To better understand the relationship between massive and pillowed flows, we mapped the most continuous outcrop of the Hooggenoeg Formation in the Mlondozi Stream on the northeast-facing limb of the Steynesdorp Anticline (Fig. 5.4-1). Geochemistry confirms that these pillowed and massive flows are the low-Ti Hooggenoeg basalts as opposed to the high-Ti Kromberg basalts. These two basalts look similar and could be confused in structurally complex areas. In this stream, continuous 3-D outcrop is cut by many concordant intrusions or sills that are greenschists similar to the lavas. They are characterized by (1) planar contacts that are locally cross-cutting, (2) coarser textures, (3) more rarely xenoliths or multiple chilled margins, and (4) more Fe and Ti. In contrast, the massive basalt flows are very closely associated with pillows and may show irregular contacts at angles to, but not cross cutting, the stratigraphy.

One aphyric, massive flow has a triangular shape bound by pillows (Fig. 5.4-16). The lower chilled contact drapes over the underlying pillows (Fig. 5.4-16(A), #1) and extends 0.5 m upward in the flow (#2), features that are characteristic of lava flows. The pillows have outer rims of vesicles (calcite amygdules), and some of the larger pillows are elongate (#3) parallel to the base of the flow and other paleohorizontal markers. One such marker is a plane of vesicle accumulation about 2 m from the flow base (#4). Linear vesicles trains occur only beneath and perpendicular to this plane (#5). Above this plane the upper two-thirds of the flow is featureless except for dispersed large vesicles near the top (#6). The base of a nearby massive flow has pipe-vesicles, which may represent an earlier stage in the development of the vesicles trains and plane of vesicle accumulation. These features are unique to the basalts.

The shape of this flow defies the commonly held belief that massive flows ponded in topographic lows formed by underlying pillowed flows. Is it a coincidence that the apex

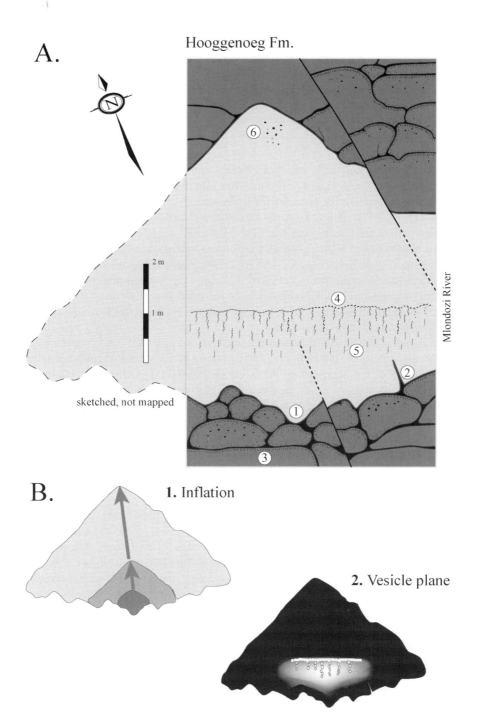

A. Hooggenoeg Fm.

2 m

1 m

sketched, not mapped

Mlondozi River

B. **1.** Inflation

2. Vesicle plane

Fig. 5.4-16. (*Previous page.*) Hooggeneog Formation in the Mlondozi River: (A) Outcrop map of vesicular pillow basalts and triangular massive flow with plane of vesicle accumulation; (B) Inflation model for lava tube.

---

mirrors the lowest point of the flow base? Lava flows into topographic lows that serve as (1) a channel, focusing flow, or (2) as a pond, and most flow involves a combination of the two. Where lava flows, it also feeds a cooling roof that grows thicker, so the first part of the flow to solidify may have been the apex, possibly when the flow was only a meter thick and resembled a large pillow or tube located by the topographic low, a channel. In this model (Fig. 5.4-16(B), #1), the flow top inflates, and the margins extend, with continued influx of lava. With water rapidly cooling the flow top, the upper crust solidifies more rapidly. When the vesicles accumulated along the plane, the upper crust must have been about 4 m thick (Fig. 5.4-16(B), #2). The linear vesicle trains are the frozen path of vesicles rising from the base of the flow after flow within the tube had ended.

5.4-6. KROMBERG FORMATION

The Kromberg Formation consists of two volcanic members: a lower pyroclastic deposit of komatiitic basalt and an upper sequence of pillow basalts (Lowe and Byerly, 1999). The section is dilated by ultramafic sills, locally to such an extent that early workers described the section as interlayered basaltic and ultramafic lavas (Viljeon and Viljeon, 1969; Vennemann and Smith, 1999). The ultramafic rocks are well exposed in the river, and de Wit et al. (1987b) described a sheeted intrusion to support an ophiolite model. To better understand these relationships, we mapped two outcrops in the Komati River Gorge: (1) a 75 m outcrop of pillowed and massive basalt flows (Fig. 5.4-17), and (2) an 80 m thick, ultramafic intrusion with multiple chilled margins (Fig. 5.4-19). Despite the narrow outcrop along the river, the maps reveals surprisingly angular contacts between pillowed and massive flows, a lava tube, foresets of elongate pillows, and a flow-top breccia. The sheeted sill complex includes screens of chert and is offset by faults parallel to the river, explaining why the stratigraphy does not correlate across the river.

In addition to more vesicles, the Kromberg Formation basalts have a unique signature of volcanic features. They also are geochemically distinct from earlier Hooggenoeg basalts in having much higher TiO_2. Following TTG magmatism and unroofing of the underlying mafic-ultramafic sequence and preceding deposition of the Fig Tree Group, the Kromberg and Mendon Formations may representing rifting of the crustal section, as proposed by Lowe and Byerly (1999). Their detailed stratigraphic work on the underlying Buck Ridge Chert reveals normal growth faults, and ideally we would like to test the rift hypothesis against the volcanic stratigraphy of the Kromberg Formation, as well.

The lower pyroclastic unit of the Kromberg Formation is unique in the BGB for its well sorted mafic lapilli, commonly crossbedded, but not silicified. Ransom et al. (1999)

Kromberg Formation

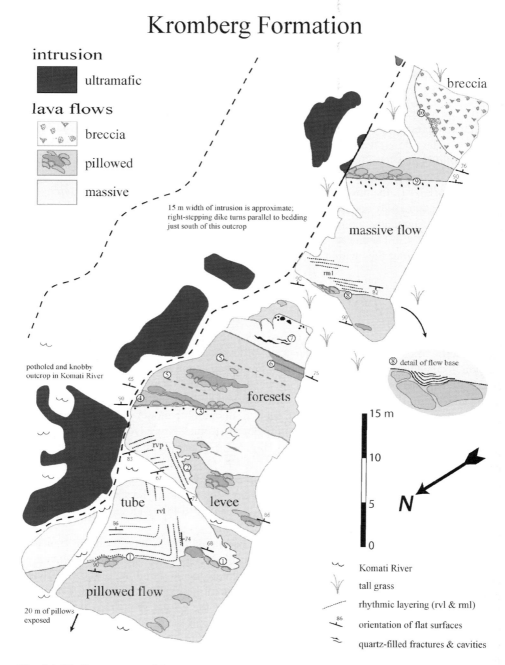

Fig. 5.4-17. Outcrop map of the Kromberg Formation, the type locality along the Komati River. The submarine lavas are intruded by an ultramafic dike.

mapped a proposed explosion crater marked by (1) excavation of the underlying Buck Ridge Chert, (2) the thickest deposit (1000 m), (3) the most abundant and largest bombs, and (4) abundant thin, olivine-phyric dikes. They concluded that subaerial to subaqueous phreatomagmatic eruptions created the crater and pyroclastic deposits. Their analyses of dikes (15% MgO) indicate komatiitic basalt. This volcanic deposit is overlain by a chert horizon, marking a time gap before the onset basaltic volcanism.

The mapped basalt stratigraphy (Fig. 5.4-17) is cross cut by a 15 m thick, ultramafic dike that turns parallel to bedding just to the southeast where contact relations are poorly exposed and non-diagnostic. The outcrop exposes a series of aphyric massive and pillowed flows, but what is unusual is that even within this narrow section, only one massive unit is through-going. The others have inclined or lateral contacts with pillows. Five massive-pillow contacts have the same orientation (N40E/~90), a good proxy for the paleo-horizontal. We rotated the southeast-facing section to horizontal on the page, a view that emphasizes the contacts inclined to horizontal. Standing on the outcrop, we cannot see these lateral contacts and inclined flow surfaces that reveal the flow dynamics; they only emerge from detailed mapping of the outcrop.

The lowest massive unit has a sharp lower contact that drapes over the underlying pillows (Fig. 5.4-17, #1). This contact has abundant elongate vesicles like the rinds of pillows. This well defined flow base contrasts with the lateral contacts that are diffuse over 1–2 meters (#2) and not marked by vesicles. Parallel to the lower contact are rhythmic vesicles planes (RVPs) that bend around orthogonal to the flow base and sub-parallel to the lateral contact with pillows and bend again forming a layered roof. We interpret these RVPs to represent successive pulses of lava solidified within a lava tube. The intersection of these steeply dipping RVPs defines a steeply plunging flow vector for the flow in the lava tube, an orientation that makes the lateral transition to pillows a pillowed levee. This lateral levee was overridden, forming a predominantly flat flow top marked by large vesicles (#3).

The next pillowed flow is particularly noteworthy because the pillows are elongate and inclined. The wedge-shaped pillow (Fig. 5.4-17, #4) has a flow top inclined 20° to the underlying surface, and successive large pillows are inclined and elongate, forming foreset beds inclined up to 30° (#5). Note that the most inclined pillows are so elongate that they resemble thin sheet flows more than pillows (#6). These flows are over-ridden by a more massive flow with lots of quartz-filled cavities or gashes (#7), evidence that the flow was structurally unstable and gashes opened as it solidified on the paleo-slope.

The succeeding pillowed flow reestablished the horizontal surface (Fig. 5.4-17, #8) for the next massive sheet, 14 m thick, with roughly parallel flow top and base. The flow top (#9) is marked by large vesicles. Lacking vesicles, the base of the flow fills the topography of the underlying pillows (#8). In one 30–40 cm deep trough, the flow divides into multiple lower chills or thin layers (Fig. 5.4-18(A)). Each layer thins laterally and merges with the base of the flow. Similarly, the lower 1–4 m of this massive flow is rhythmically layered (Fig. 5.4-18(B)), not by vesicles, but as modal layering (RML). Both types of layering may record successive pulses of lava influx. In the first case, the layers record successive pulses of lava that progressively filled the trough as the flow inflated. Laminar flow along the base of the flow would tend to isolate the trough from the full velocity of the flow,

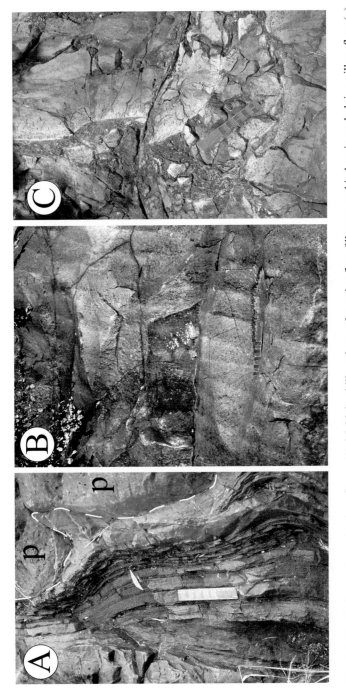

Fig. 5.4-18. Features of massive lava flows: (A) Multiple chills at base of massive flow fill topographic low in underlying pillow flow (p); (B) Rhythmic modal layering in massive flow; (C) Flow-top breccia.

allowing chills to grow thicker during a pulse. The rhythmic layering may also occur when crystallization oscillates between plagioclase and pyroxene.

A thin pillowed unit separates this massive flow from the next that has a flow top breccia at an oddly inclined angle. The flow top is partly intact and marked by vesicles and partly broken into fragments that can be refit (Fig. 5.4-18(C)). The inclined contact between the flowtop and breccia is probably a slope, coming out of the plane of the map. Whatever the explanation, the number of inclined contacts or sloped surfaces within this narrow window is a remarkable characteristic of this volcanic section.

The ultramafic unit best exposed along the Komati River is 80 m thick and intruded both chert and itself, forming a sheeted sill complex (de Wit et al., 1987b). It outcrops along the bank and within the river (Fig. 5.4-19(A)). Detailed maps illustrate intrusive contacts against (1) screens of course-grained ultramafic rock (Fig. 5.4-19(B)) and (2) screens of chert (Fig. 5.4-19(C) and 5.4-20(B)). The pale green chilled margins finely protrude and fragment the wall rock and grade to the dark green/black of the coarser-grained, slowly cooled, interior portions of the sill (Fig. 5.4-20(A)).

Along with the ultramafic dike cross cutting the volcanic section (Fig. 5.4-17), this 80 m thick sill belongs to an intrusive suite that intrudes this part of the Kromberg Formation but does not belong to the Kromberg Formation. It is not genetically related to the volcanic stratigraphy, and no komatiite lavas were found within the Kromberg Formation. The ultramafic sill complex may be related to similar sills around the Barberton Greenstone Belt and/or be related to later komatiite volcanism of the Mendon Formation.

5.4-7. MENDON FORMATION

The Mendon Formation, as described by Lowe and Byerly (1999), conformably overlies the Kromberg Formation and consists of lava flows of komatiite and komatiitic basalt interlayered with cherts or silicified volcaniclastic sediments. Byerly (1999) illustrates one sediment-bound volcanic sequence, 200 m thick, with a lower 20 m of komatiite and an upper 180 m of komatiitic basalt. The sediments contain some zircon-bearing ash with ages around 3.3 Ga, 150 million years younger than the Komati Formation. Lowe and Byerly (1999) correlated distinctive marker beds across fault-bound blocks to construct an aggregate section of the Mendon Formation about 1 km thick, a section that includes some thick coarse-grained, layered ultramafic rocks that may be intrusions. The sediments divide the volcanic rocks into four units, the top of which are extensively altered with bedding-parallel veins, zones extending from 10 to 50 m down into the ultramafic rocks. Like the Hooggenoeg Formation, these alteration zones mark volcanic quiescence and exposure of the top of the flow fields to corrosive fluids. Byerly (1999) describes komatiites and komatiitic basalts with spinifex textures and zones similar to flows in the Komati Formation (see Byerly, 1999, Fig. 5). Whether the cumulates are flows or sills remains problematic as the fine-grained massive komatiite flows interlayered with spinifex flows are difficult to distinguish from courser-grained intrusions. What is remarkable about the Mendon Formation is that spinifex komatiites locally have well-preserved igneous olivine, despite the

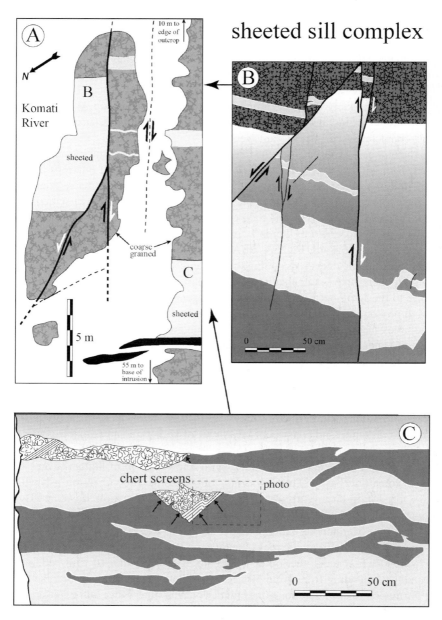

Fig. 5.4-19. Outcrop maps of sheeted ultramafic intrusion, type locality of Kromberg Formation: (A) Outcrop map of bank and island outcrops along the Komati River, showing sheeted sill, chert screens, and faults parallel to the course of the river; (B) Detail of sheeted intrusion showing one-way chill against coarse-grained screen and multiple dikes; (C) Sheeted intrusions chilled against screens of chert.

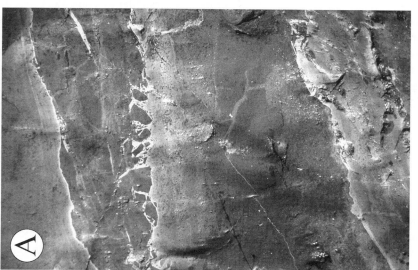

Fig. 5.4-20. Outcrop photos of sheeted ultramafic sill: (A) Pale-green chilled margins of sheeted intrusion; (B) Screens of chert with bedding-parallel chilled margin, cross cut by later sill.

widespread seafloor alteration, and some komatiite flows are interlayered with sediments (Byerly, 1999), a feature absent from any other formation.

5.4-8. COMPARISON OF FLOW MORPHOLOGIES

All three lava compositions – basalt, basaltic komatiite, and komatiite – share the same bifurcation into two fundamental, submarine flow types: thin lobes or pillows and thick sheets, a pattern similar to subaerial flow fields. However, composition partly determines the shapes and sizes of flows, their extents, and their relationships. In the Lower Komati, olivine-phyric massive sheets make up 62% of the komatiites and extend for 11 km across the entire map area (Fig. 5.4-1). Where the thickness of massive sheets is well defined by interlayered spinifex flows, they are up to 23 m thick, whereas spinifex flows are rarely 8 m thick and mostly are 1–2 m thick lobes. In the komatiitic basalts of the Upper Komati, pillows dominate, and sheet flows extend for a kilometer or less. In contrast the basaltic massive flows are locally not even sheets (Fig. 5.4-16), or they seem to be lava tubes within pillowed flows (Fig. 5.4-17) and probably are not very extensive on a map scale. This morphological trend from extensive sheets of komatiite to a predominance of pillowed basalts may reflect a decreasing eruption rate or greater viscosity that prevents the basalts from forming underwater sheets with thick upper crusts and thick lava tubes.

The lobes of spinifex komatiite, similar in size to basaltic pillows, have different shapes. Although basaltic pillows conform to underlying topography and have convex upper surfaces, they do not fit tightly and leave open spaces that are filled with quartz. In contrast, the komatiites are so fluid that they leave no empty spaces and even have pointed lateral margins, a feature never formed by more viscous basaltic lava. Rapid cooling underwater forces the more viscous basaltic lava into more rounded shapes, but subaerial pahoehoe is not so confined and takes on lenticular, tightly fitting shapes similar to komatiite, hence the resemblance between komatiite and pahoehoe flow lobes (e.g., Dann, 2000, Fig. 16). Another feature related to fluidity is the stack of drainage cavities that are much more common in pillows of komatiitic basalts than in the basalts. The komatiites in Barberton do not form pillows with the convexity needed to support a cavity, so any drainage probably collapses the crust and leaves no telltale sign.

5.4-9. RECONSTRUCTING FLOW FIELDS

Ideally we would like to see in our maps a cross section from the vent to distal reaches of the flow field. However, komatiite vents have not been convincingly documented anywhere, and the chances are very small that an erosional surface will intersect the vent in vertical sequences. The flow fields of the Komati Formation do not show lateral facies transitions or thickening that point toward the vent. But, we have other lines of evidence: (1) the vertical transitions – massive-vesicular-spinifex, (2) the thick tumuli, a large lava tube, and (3) spinifex flow lobes, the major component of the thinnest komatiite flow fields, bound by sediments, in the Hooggenoeg and Mendon Formations. Also, Hill et al. (1995), Lesher and Arndt (1995), Pering et al. (1995), and Hill (2001) provide volcanological models of younger Australian komatiites, based largely on drill core, but they add a perspective from a larger scale, showing proximal, medial, and distal facies and representative cross sections of model flow fields.

The lenticular shape and low aspect ratio indicates that the tumulus is a cross section of a lava tube (as argued above), flowing at nearly right angles to the map view. If the tube flows away from a fissure vent, then this direction implies that the fissure may be roughly parallel to the map view. Perhaps all the flow fields of the Komati Formation flow at a high angle to the map section and away from the vent, a claim that could be tested by plotting a large number of flow lineations. What direction the komatiitic basalts came from is even more poorly constrained, but the lack of compositional mixing suggests that they vented separately from the komatiites.

The tube developed in two stages: an early stage of olivine-phyric lava that fractionated and degassed during emplacement, creating the cumulate and the vesicular crust, and a late stage of aphyric lava that rotated the crust and crystallized as the interior spinifex zone. This evolution of komatiite lava reflects both fractionation within the plumbing system and vent maturation and explains the recurring flow-field stratigraphy: massive flows overlain by spinifex flows. Compound flows of spinifex flow lobes may mark distal facies of komatiite flow fields (Cas et al., 1999), an indicator that is supported by features in Hooggenoeg and Mendon Formations. First, the flow fields are thin, 20–80 m, compared to 100–270 m in the Komati. Second, the flows are primarily thin spinifex flow lobes, closely associated with sediments marking gaps in volcanism, compared to the Komati's dominant massive flows and lack of sediments. Thus, the distal facies of the Komati flow fields may have looked like those of the Mendon and Hooggenoeg Formations.

5.4-10. FLOW FIELDS: PETROGENESIS AND PLATE TECTONIC SETTING

In the Barberton Greenstone Belt, komatiite (K) and komatiitic basalt (KB) erupted together at three different times in the ca. 3.48 Ga Komati, 3.47 Ga Hooggenoeg, and 3.3 Ga Mendon Formations. Where cherts, representing gaps in volcanism, bound 200–400 m thick volcanic sections, K and KB erupt in succession, but the order of eruption is variable – K before KB in some sections and KB before K in others. Interlayered in the Lower Komati, Ks and KBs show the same pattern of increasing trace element contents with time, a trend of decreasing Al_2O_3/TiO_2 (Fig. 5.4-3) that reflects a common petrogenesis for the two magma types. In addition, this bimodal K–KB volcanism is widespread in Archean greenstone belts around the world – clearly a long lasting association that characterizes an Archean tectonic environment. However, Archean K–KB volcanism has no exact analogue with submarine volcanism in modern plate tectonic settings. Therefore, we can only find the best fit, acknowledging that the Earth has cooled and evolved over last 3.5 billion years and that researchers choose different criterion for the best fit. Besides acknowledging their shared petrogenesis, we support a supra-subduction zone model for K–KB petrogenesis because (1) KBs are geochemically similar to boninites, and (2) experimental petrology reveals that Barberton komatiite was a hydrous magma.

Most geologists agree that continental crust has been assembled throughout Earth history by convergent tectonism at subduction zones (Kusky, 2004), such that continental crust inherits its average composition from arc magmatism. However, convergence may sample and accrete crustal sections with submarine lavas erupted in a variety of oceanic settings: mid-ocean ridges; oceanic plateaus and islands; and fore-, intra-, and back-arc basins, all of which have served as modern analogues to Archean greenstone belts. In Phanerozoic orogenic belts, convergence preferentially sampled arc-type ophiolites and submarine crustal sections because they form adjacent to thick arc crust that is destined to be accreted, in contrast to thin oceanic crust that is continuously subducted. Based on this mechanics, we might expect that submarine sequences in Archean greenstone belts represent a collage of primarily arc environments.

Petrogenetic models for the K–KB association are split between three camps. Some believe that the high-MgO content of the lavas requires the high temperatures of mantle plumes to feed oceanic plateaus, which accrete with arc rocks to form greenstone belts. Others believe that hydrous melting of depleted mantle above a subduction zone would better explain the geochemistry and variety of associated magma types. The third camp recognizes the arc signature of KBs (especially boninites), holds to the plume origin of K, and explains their widespread association with plume-arc interaction. Our research on the Komati Formation supports the supra-subduction zone model, but does not refute the plume model for komatiites with different stratigraphic associations and different compositions and mineral assemblages (e.g., Belingwe, Kambalda, etc.).

Cameron et al. (1979) note the similarity between KB and boninites and suggests a similar origin by hydrous melting of a depleted mantle source, possibly in a forearc. Kerrich et al. (1998) find KB interlayered with K, which is geochemically identical to modern boninites, and resort to a hybrid plume-arc model to explain bimodal K–KB volcanism. In the Komati Formation, KB not only has high MgO (11–14%) but also high SiO_2 (54%), even in areas devoid of any evidence for silicification. So, the KB is really high-Mg basaltic andesite and belongs to the vast compositional spectrum of subduction-related volcanic rocks, now recognized in Precambrian orogenic belts (Polat and Kerrich, 2004). The KB has the low TiO_2 (0.6%) typical of low-TiO_2 arc tholeiites that tend toward boninites. Hooggenoeg KB has even lower TiO_2 (0.30%) with lower MgO ($\sim$11–12%), indicating melting of a more depleted source than in the Upper Komati (Lower degree melts are enriched in moderately compatible elements like Ti, so its depletion must be in the mantle source). Parman and Grove (2004) compare the KB to modern boninites, conclude that the geochemical systematics are remarkably similar, and suggest that KB also formed by hydrous melting above a subduction zone but from a less depleted source. If KB and K share the same petrogenesis, and KB erupted in arcs, then what about K?

Parman et al. (1997) crystallized K magma under hydrous and dry conditions and compared the experimental pyroxene compositions to the natural pyroxenes in the Komati Formation. They discovered that the K flows must have contained water when they crystallized, results that do not necessarily apply to all komatiites (Arndt et al., 1998). We think that the water was inherited during melting (high P) because the lavas could not have absorbed sea water before pyroxene crystallized (low P/high T). In modern settings, water is released from the down-going slab by metamorphic reactions, melts the overlying mantle, and partitions into the melt that feeds arc volcanism. Similarly, we think K formed by hydrous melting of depleted mantle in a hotter Archean subduction zone (Parman et al., 2001; Parman and Grove, 2004) and that the residue of komatiite melting formed the buoyant rigid tectosphere beneath the Kaapvaal Craton (Parman et al., 2004), which has been instrumental in stabilizing and preserving this ancient crustal section, as well as serving as the source of diamonds in southern Africa.

The supra-subduction zone model explains bimodal K–KB volcanism. The plume model does not explain why K is interlayered with arc rocks. The arc-plume model implies that vents for plume and arc eruptions were close together, a coupling widespread in the Archean, which is completely absent from modern plate tectonics, a discrepancy difficult to explain since the hypothesis relies on modern analogues. On the other hand, the arc model can explain the variety of high-Mg, bimodal volcanism as a product of hydrous melting of variably depleted mantle at different depths, melting triggered by a sequence of metamorphic dehydration reactions (i.e., amphibole vs. serpentine) at distinct P–T conditions along the subducting slab. In addition, subduction zones are sites of episodic, evolving, but sustained magmatism, including TTGs, and convergent and strike-slip tectonism, all the forces that created Early Archean orogenic belts.

Besides the high-Mg lavas, the Hooggenoeg and Kromberg formations contain tholeiitic basalts. Arndt et al. (1997) compare examples of Archean tholeiites to modern basalts, note affinities with island arc basalts (IAB), but conclude that they 'are matched by no common type of modern basalt' (p. 243), having less Ti than MORB but more Ni than IAB. As a result, assigning modern tectonic settings based on chemistry is not straightforward. Based on Ti, we think the Hooggenoeg tholeiites ($TiO_2 < 1\%$) are IATs that represent melting under different conditions than the K and KB. In contrast, the Kromberg basalts with 1–2% TiO_2 are more analogous to back-arc basin basalts that tend toward MORB compositions. The Kromberg suite also includes a prominent high-Ti basaltic dike in the Komati Formation, an intrusion that may attest to coherent stratigraphic succession (Fig. 5.4-2). The KB in the Kromberg and overlying Mendon Formation may also indicate an arc environment. The Kromberg basalts erupted after 3.45 Ga TTG magmatism thickened the crust, creating shallow water to subaerial conditions. So, these back-arc basalts may represent rifting and formation of a basin that accumulated the overlying Fig Tree Group. If these interpretations are correct, the entire pre-Moodies history of the Barberton Greenstone Belt took place in an evolving arc.

5.4-11. KOMATI–HOOGGENOEG SECTION: A FORE-ARC OPHIOLITE?

The Komati–Hooggenoeg section (KHS) was seafloor crust that may have formed in fore-arc environment, but whether we decide to call it an ophiolite is debatable, especially since the sheeted-dike complex indicative of seafloor spreading is absent. De Wit (2004) suggests that the KHS is the upper part of a supra-subduction complex of ophiolitic nature and contends that Archean ophiolites are rarely recognized because they do not fit the expected model. Based on the KB–boninite connection, Parman and Grove (2004) consider the KHS to be an ophiolite. The KHS is 6 km thick, about the thickness of modern oceanic crust and many ophiolites. However, the KHS structurally overlies partly younger arc rocks along the Komati Shear Zone (Armstrong et al., 1990). As a result the base of this thrust sheet is not exposed and the preserved thickness is structurally controlled. For example, the lower half of the Komati Formation is entirely missing on the southeast-facing limb of the Onverwacht fold (Fig. 5.4-1). Unlike an ophiolite, the 6 km thick KHS is entirely volcanic – almost an order of magnitude thicker than the volcanic sections (0.5–1 km) of ophiolites or oceanic crust, which overlie much thicker intrusive sections. On the other hand, thrusting may have sampled the upper 6 km of much thicker crust and left behind the intrusive layer, a possibility that raises the question: What features of a volcanic layer point to a missing intrusive layer and/or an origin by seafloor spreading?

The gabbroic layer of ophiolites records magma storage and commonly fractionation and mixing expressed as the compositional trends in the overlying dikes and flows. The flows of K and KB erupted directly from the mantle without storage and fractionation within magma chambers, so they have no complementary missing intrusive section. The volcanic rocks of ca. 2 Ga Jormua ophiolite also were not fractionated within magma chambers, and Peltonen and Kontinen (2004) see that they erupted directly on mantle tectonites intruded by sheeted dikes. Based on this example, the KHS may represent 6 km thick volcanic crust, which delaminated along the boundary between volcanic and mantle rocks during thrusting.

The volcanic layers of ophiolites and oceanic crust are so thin because extension pulls the crust apart, focusing magma into axial zones where most of it solidifies as dikes and subvolcanic plutons. Besides thin volcanic layers, this rift setting limits the aerial extent of flows and commonly produces lots of pillow breccia. In contrast the Lower Komati has two distinct lavas, K and KB, interlayered in a lava plain setting but erupted from separate vent areas, preventing the K and KB from mixing. Although these magmas were not focused into an axial zone of seafloor spreading, the crust may have been extending. Normal faults, offsetting the volcanic stratigraphy several 100 m, are cut by dikes of komatiitic basalt (Dann, 2000, Map I & II), indicating extension on-going during ultramafic magmatism. However, the same sense of normal faulting also accompanied the eruption and intrusion of tonalitic magma, 30 million years later (de Vries, 2004), so the regional tectonics of the over-riding plate may have been extensional without seafloor spreading and without creating submarine crust with the pseudostratigraphy characteristic of ophiolites and oceanic crust.

ACKNOWLEDGEMENTS

We thank Maarten de Wit for facilitating our research in South Africa. Support from both the US National Science Foundation and the National Research Foundation of South Africa is gratefully acknowledged. We also thank Steve Barnes for his helpful review.

Earth's Oldest Rocks
Edited by Martin J. Van Kranendonk, R. Hugh Smithies and Vickie C. Bennett
Developments in Precambrian Geology, Vol. 15 (K.C. Condie, Series Editor)
© 2007 Elsevier B.V. All rights reserved.
DOI: 10.1016/S0166-2635(07)15055-6

Chapter 5.5

SILICIFIED BASALTS, BEDDED CHERTS AND OTHER SEA FLOOR ALTERATION PHENOMENA OF THE 3.4 GA NONDWENI GREENSTONE BELT, SOUTH AFRICA

AXEL HOFMANN[a] AND ALLAN H. WILSON[b]

[a]*School of Geological Sciences, University of KwaZulu-Natal, Private Bag X 54001, 4000 Durban, South Africa*
[b]*School of Geosciences, University of the Witwatersrand, Private Bag 3, 2050 Wits, South Africa*

5.5-1. INTRODUCTION

Silicification of volcanic rocks immediately below cherts representing equally silicified sedimentary horizons is a common phenomenon in Palaeoarchaean supracrustal successions, including the Onverwacht Group of the Barberton Greenstone Belt, Kaapvaal Craton, in southern Africa (Paris et al., 1985; Lowe and Byerly, 1986a; Duchac and Hanor, 1987; Hofmann, 2005a) and the Warrawoona Group of the Pilbara Craton in Western Australia (e.g., Kitajima et al., 2001; Terabayashi et al., 2003; Van Kranendonk and Pirajno, 2004; Van Kranendonk, 2006). The origin of the silicification in the Barberton Greenstone Belt was initially a matter of controversy. Interpretations cited weathering of volcanic flow tops (Lowe and Byerly, 1986a) or hydrothermal alteration (de Wit et al., 1982; Duchac and Hanor, 1987; Hanor and Duchac, 1990) as the principal causes. Hofmann (2005a) and Hofmann and Bolhar (2007) recently proposed a detailed model of low-temperature (100–150 °C) seafloor hydrothermal processes for the origin of the silicification in the Barberton Greenstone Belt. Hydrothermal activity is attributed by these authors to high heat flow in pre-Mesoarchaean times, resulting in the establishment of shallow subseafloor convection cells and a diffuse upflow of hydrothermal fluids over broad areas. Because silica alteration zones underlie silicified sedimentary horizons that contain some of the oldest morphological evidence for life on Earth in the form of microfossils (Pflug, 1965; Walsh, 1992; Westall et al., 2001), it is crucial to learn more about the hydrothermal activity and to investigate any link between hydrothermal processes and the habitat of early life.

The Nondweni greenstone belt is situated ca. 300 km south of Barberton in northern KwaZulu-Natal and contains an exceptionally well preserved volcano-sedimentary succession (Wilson and Versfeld, 1994a, 1994b). The Nondweni and Onverwacht Groups formed at roughly the same time and share many similar lithological features. These include silicification of lava flow tops beneath chert horizons and the presence of barite, carbonaceous matter, and stromatolite-like structures in chert horizons. Silicification and the presence of barite indicate seafloor hydrothermal activity, whilst the remainder may point to the establishment of early life on the seafloor at that time. In this paper, we focus on a single aspect of the geology of the Nondweni greenstone belt, namely the origin of silica alteration in this belt, in order to emphasize that this phenomenon is characteristic for pre-Mesoarchaean greenstone successions and that important inferences can be drawn from the effect of low-temperature hydrothermal processes on the geological and biological processes that were operating on the early Earth.

5.5-2. GEOLOGICAL SETTING

5.5-2.1. Regional Geology of the Southeastern Part of the Kaapvaal Craton

The Nondweni greenstone belt is one of a number of Palaeoarchaean supracrustal fragments preserved in the southeastern part of the Kaapvaal Craton, south of the Barberton Greenstone Belt (Fig. 5.5-1). These supracrustal fragments are associated with migmatites and granitoid basement rocks and are unconformably overlain by the cratonic cover succession of the ca. 3.0 Ga Pongola Supergroup. Contacts of the greenstone fragments with older basement granitoid rocks are not observed, and the Archaean rocks occur as inliers within extensive Mesozoic cover.

The preserved Archaean supracrustal fragments exhibit a wide variety of rock types and include clastic, chemical and biological sedimentary rocks, pyroclastic rocks, and volcanic rocks ranging from komatiites and basalts to andesites. This range of rock types precludes simple lithostratigraphic correlations; however, they may be approximately coeval. The proportion of the rock types between the various occurrences varies greatly. These successions are generally steeply dipping and have experienced varying degrees of deformation. In contrast, the overlying Pongola Supergroup is relatively undeformed and generally shallowly dipping and represents a major transition in the Archaean history of the Kaapvaal Craton, although regions of intense deformation also occur.

The supracrustal greenstone fragments include, from north to south, the Dwalile, Assegai, de Kraalen, Commondale and Nondweni suites (Fig. 5.5-1). The Dwalile suite (Jackson, 1984) comprises talc schists, amphibolites, serpentinites and hornblende schists with minor intercalations of metasedimentary rocks. The Assegai suite consists predominantly of clastic and chemical metasedimentary rocks, including quartzites, quartz-sericite schists, garnet-bearing mica schists and banded iron formation. Metavolcanic rocks are represented by amphibolites and ultramafic schists (Hunter et al., 1983). The Commondale ultramafic suite is a succession of siliceous komatiites containing thick cumulate layers and spinifex zones and is characterized by the presence of orthopyroxene (both as cumulates and spinifex) and exceptionally Mg-rich olivine (up to Fo_{97}; Wilson, 2003; Wilson et al., 2003). Other lithologies in the Commondale greenstone fragment include ferruginous quartzites, talc schists and amphibolites. The Nondweni greenstone belt is dominated by volcanic rocks, which include komatiites, basalts, komatiitic andesites and dacites.

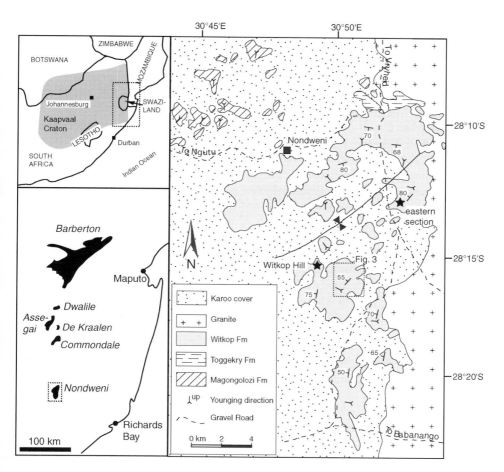

Fig. 5.5-1. Geological map of the Nondweni greenstone belt, showing the distribution of the Magongolozi, Toggekry and Witkop Formations (after Versfeld, 1988). Also shown is the location of other supracrustal greenstone belts at the southeastern edge of the Kaapvaal Craton.

Metasedimentary rocks are restricted to chert bands and to a succession of clastic, biogenic and chemical sedimentary rocks at the top of the succession (Hunter and Wilson, 1988).

5.5-2.2. *Geology of the Nondweni Greenstone Belt*

Rocks of the Nondweni Group (Wilson and Versfeld, 1992a) are exposed over an area of approximately 100 km^2 as a series of inliers within Mesozoic sedimentary cover (Fig. 5.5-1). The belt has been intruded on the east side by a granodiorite dated at ca. 3.29 Ga (Rb-Sr; Matthews et al., 1989) thereby providing a minimum age for the suite.

A SHRIMP age of 3406 ± 3 Ma on zircons from a rhyolite flow in the middle part of the succession (Armstrong, 1989) is considered the most reliable absolute age for the belt. A younger age of 3.2 Ga was indicated by Sm-Nd data of the individual mafic lithologies (Wilson and Carlson, 1989), but the combined isochron for all lithologies gave an age of ca. 3.4 Ga. The basement rocks of the supracrustal succession are not exposed, but are likely to be granitoid gneisses.

The Nondweni Group comprises three formations: the lower Magongolozi Formation, intermediate Toggekry Formation, and the uppermost Witkop Formation (Wilson and Versfeld, 1992a, 1992b; Versfeld and Wilson, 1992a, 1992b). The volcano-sedimentary sequence is arranged in a broad, steeply dipping synclinal structure with the fold axis trending NE-SW. At least one earlier deformational event is indicated by low-angle thrust zones approximately parallel to the bedding planes, and now also steeply dipping. A later minor cross-folding event is also indicated.

Many greenstone belts have a major sedimentary unit developed at the top of the succession that was deposited following cessation of major volcanic activity and during, or prior to, compressional deformation (e.g., Barberton Greenstone Belt). Such sedimentary rocks are absent from the Nondweni greenstone belt, possibly due to limited exposure, but sedimentary sequences in other supracrustal fragments of the southeastern part of the Kaapvaal Craton may be lateral equivalents of such successions.

5.5-2.2.1. Magongolozi Formation

This basal formation (Fig. 5.5-1), with an exposed thickness of ca. 7000 m, is vertically dipping, strikes E–W, and comprises mainly pillowed and spinifex-textured mafic volcanic rocks. It is also intruded by mafic sill complexes. Sedimentary rocks comprise less than 5% and are mainly cherts interlayered with the volcanic rocks, while clastic sedimentary rocks are absent (Wilson and Versfeld, 1994a).

5.5-2.2.2. Toggekry Formation

This formation is dominated by felsic tuffs, rhyolite flows, quartz-sericite-feldspar schists and intercalated basalt flows. It is located in the northeastern extremity of the greenstone belt (Fig. 5.5-1) and may represent outpourings from a single volcanic centre. Estimates of the thickness are highly variable, but the thickest section in the east reaches approximately 2000 m. The rocks tend to be more highly deformed and metamorphosed than other parts of the belt, partly because they are located close to an intrusive granite and to the hinge of the synclinal fold closure. A small deposit of massive Pb-Zn sulphide is associated with quartz-sericite schists and has been interpreted as resulting from seafloor exhalative discharge.

5.5-2.2.3. Witkop Formation

The Witkop Formation is the uppermost formation and areally the most extensive. It is approximately 7000 m thick and occurs on both limbs of the syncline. Well developed younging indicators include features such as pillow lavas, graded bedding in tuffs, and cross-lamination in chert layers and epiclastic deposits. The dominant rock types are basalts and intermediate volcanic rocks, as well as komatiite units. A well developed sequence of clastic sedimentary rocks, which is mainly derived from felsic volcanic rocks and includes stromatolitic structures and silicified evaporites or hydrothermal precipitates (Wilson and Versfeld, 1994a), is an important but localized occurrence. It marks the only continuous sedimentary succession exposed in the greenstone belt.

5.5-2.3. Volcanic Rocks of the Nondweni Group

Volcanic rocks dominate all three formations in the Nondweni Group (Wilson and Versfeld, 1994b). In the Magongolozi Formation these are marked by basalts, basaltic andesites and a group of rocks with relatively high SiO_2 and MgO contents, termed komatiitic andesites, which have similarities to arc-related volcanic rocks, such as boninites, or crustally contaminated mantle melts (Riganti, 1996). Pyroxene spinifex is common in many of these rocks and several occurrences are regarded as ancient lava lakes (Wilson and Riganti, 1998).

The Witkop Formation is made up mainly of basalts and komatiitic basalts, occurring as pillowed or sheet flows (Fig. 5.5-2(a)). Komatiite occurs in several distinct stratigraphic zones and comprises about 17% of the total exposure (Wilson and Versfeld, 1994a). Fractionated intrusions are present as diorite and gabbro. The komatiites are distinct from those of the Barberton Greenstone Belt in that the MgO content does not exceed 22%, but they have a similar CaO/Al_2O_3 ratio >1 (Riganti and Wilson, 1995a). A unique feature is the development of pyroxene (now amphibole) bladed spinifex with crystals up to 5 m in length, commonly formed in cone-type structures (Fig. 5.5-2(b); Wilson et al., 1989). These cones are always developed at the tops of flow units and fan downwards. Their consistency allows them to be used as primary way-up indicators.

The volcanic rocks of the Toggekry Formation contrast with those of the other two formations in that basalts are only a minor component ($<10\%$) and the sequence is made up of massive rhyolite and rhyodacite flows, and felsic tuffs which include graded air-fall tuffs. An AFM plot indicates these to be of calc-alkaline affinity and have compositions similar to those of the calc-alkaline occurrence at Welcome Well (Yilgarn Craton, Western Australia; Giles and Hallberg, 1982). The felsic rocks can be divided into two groups on the basis of Cr ($<$ or $>$ than 30 ppm), Ni and Sc contents. It is uncertain whether these compositions reflect mixed ash deposits derived from multiple vents, or by interaction of felsic and komatiite magmas.

5.5-2.4. Sedimentary Rocks of the Nondweni Group

A highly localized, but well preserved, sedimentary succession, 500 m thick, occurs at the top of the Witkop Formation and has been interpreted as a shallow subaqueous, or intermittently subaerial, succession deposited in a rifted continental margin setting (Wilson and Versfeld, 1994a). The basal sequence is characterized by banded and fragmented cherts grading upwards into coarse-grained felsic lapilli tuffs, which are com-

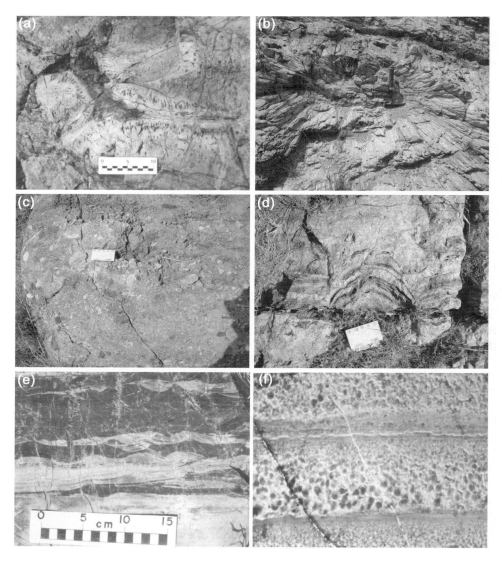

Fig. 5.5-2. Features of volcanic and sedimentary rocks of the Witkop Formation. (a) Pillow basalt showing well developed radial pipe vesicles. (b) Large-scale spinifex texture forming a radiating, cone-shaped structure in a komatiite. (c) Polymict conglomerate; clasts are mainly chert and felsic volcanic rocks. (d) Silicified stromatolite. (e) Wavy bedding in banded black-and-white chert. (f) Polished rock slice of accretionary lapilli tuff showing normal grading. The bed in the centre is ca. 2.5 cm thick.

monly cross-bedded by fluvial or coastal reworking. The upper conglomerate sequence is made up of normally graded massflow to sheetflood deposits that contain a wide variety of both rounded and angular rock fragments (chert, rhyolite, silicified basalt; Fig. 5.5-2(c)). The uppermost part of the succession reflects a marked decrease of depositional energy and is made up of a stromatolite zone (Fig. 5.5-2(d)) overlain by silicified chemical precipitates, possibly evaporites, and rhythmically layered accretionary lapilli tuffs.

Other rocks of sedimentary origin in the Nondweni greenstone belt include variably recrystallized chert bands, typically a few metres thick, that are intercalated with the volcanic rocks. These include black-and-white banded cherts and more massive green cherts. The banded cherts locally exhibit sedimentary structures (Fig. 5.5-2(e)), such as flaser bedding, cross-lamination, ripples and slump structures, suggesting deposition in a shallow-water environment. In many cases, white chert layers are fragmented into tabular blocks enclosed within black chert. Green cherts locally contain normally graded beds of accretionary lapilli (Fig. 5.5-2(f)), which suggest that they represent silicified pyroclastic deposits. An unusual association of carbon-bearing quartzites and subordinate cherts occurs in the western limb of the Witkop Formation as relatively thick bodies (10–100 m in size). These may have originated by chemical precipitation of silica emitted by fumaroles in a subaqueous environment followed by later recrystallisation (Riganti and Wilson, 1995b; van den Kerkhof et al., 2004), or in the more unlikely case, to have resulted from structural dismemberment of laterally continuous layers. The association of chert layers in contact with basaltic or komatiitic volcanic sequences of the Witkop Formation is the main thrust of this study.

5.5-3. ANALYTICAL METHODS

A total of 42 samples were analysed for major and trace elements. They include silicified volcanic rocks and silicified sedimentary rocks, now consisting of chert, as well as cross-cutting chert veins. All samples are from surface exposures, but most samples do not show visible evidence for weathering due to high silica contents.

Fresh rock samples were reduced by jaw crusher into small chips. After cleaning in distilled water, the chips were pulverized using a carbon steel mill. Bulk samples were processed, which include chert-filled vesicles and veins in some highly silicified volcanic rocks. Major elements were analysed by XRF using flux fusion disks against primary standards and certified standard reference materials. Selected trace elements were determined using pressed powdered disks. XRF correction procedures were carried out using in-house computer programmes. Trace elements were determined using the Elan 6100 ICP-MS against primary standard solutions and validated against certified standard rock materials (Wilson, 2003). Rock dissolutions were carried out by dissolving 50 mg of each sample using the Anton-Paar Multiwave high pressure and temperature microwave digester with 40 minute digestion times. Solutions were diluted with internal standard and made up to 50 ml for analysis.

5.5-4. FIELD RELATIONSHIPS AND GEOCHEMISTRY

Several successions of basalt and komatiitic basalt intercalated with bedded carbonaceous cherts of the Witkop Formation were sampled in order to investigate geochemical changes with depth. Two sections, herein referred to as the main and southern sections, were investigated along the Vumankula River (Fig. 5.5-3). Additional samples were also obtained from Witkop Hill and an eastern section (Fig. 5.5-1), as described below.

The Witkop Formation of the Nondweni greenstone belt was studied predominantly by Versfeld (1988), who also conducted a study on the alteration of the Nondweni Group. Silicification of a variety of volcanic rocks was observed, while carbonatization and minor talcification affected komatiitic lithologies only. These alteration processes were attributed to sea-floor alteration which took place during periods of volcanic quiescence. Both silicification and carbonatization are associated with an increase in SiO_2, and a decrease in MgO and CaO. Fe_2O_3 was strongly depleted during silicification. K_2O (and Rb, Ba) were enriched during silicification, but depleted during carbonatization. A more detailed study

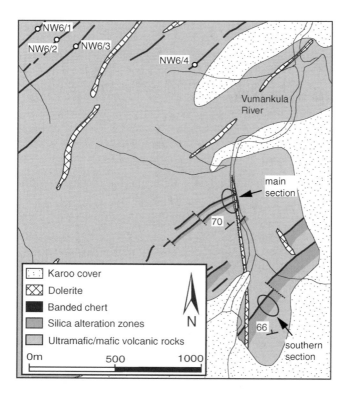

Fig. 5.5-3. Geological map of a portion of the Witkop Formation (modified from Versfeld (1988)). See Fig. 5.5-1 for locality. Strata young and dip steeply towards the northwest.

of element changes during alteration is presented in this paper, with the aid of detailed major and trace element geochemistry of several alteration zones.

5.5-4.1. *Vumankula River, Main Section*

A well exposed section of massive and pillowed basalts, which are interlayered with two chert horizons, was sampled in detail (Fig. 5.5-4). The section starts with a unit of massive basalt that is capped by a continuous horizon of bedded chert. The basalt appears to become more altered upsection, and chert veins up to 5 cm wide cross-cut the basalt, especially in the uppermost three metres. The chert horizon is overlain by a sequence of pillow basalts that are capped by a second horizon of black-and-white banded chert (Fig. 5.5-5(a,b)). The upper half of the sequence of pillow lavas is silicified and transected by veins of massive carbonaceous chert up to 50 cm wide. The veins are mainly oriented perpendicular to the stratification and contain angular fragments of the immediate host rock (Fig. 5.5-5(c)). The veins become more abundant upsection, where they preferentially occupy the domains between individual pillows (Fig. 5.5-5(d)). Thin chert veins up to 2 cm wide also cut across the banding in the chert bed (Fig. 5.5-5(b)). Chert is overlain by massive, unaltered komatiitic basalt, but poor exposure and faulting prevented any further studies.

A total of 12 samples of volcanic rocks were taken from this stratigraphic interval, covering a range of relatively unaltered to silicified rocks (Fig. 5.5-4). In addition, 5 samples of carbonaceous chert derived from bedded horizons and cross-cutting veins were also analysed. The least altered basalt sample (NW6/24) has a SiO_2 content of 57%, which defines it together with MgO and CaO/Al_2O_3 values as a basaltic andesite according to the scheme proposed by Wilson and Versfeld (1994b) and Riganti and Wilson (1995a). One sample of the uppermost, unaltered volcanic unit straddles the boundary between komatiitic basalt and komatiitic andesite. Immobile element ratios, such as Al_2O_3/TiO_2, are essentially constant for the basalts (Fig. 5.5-4), suggesting that they were derived from compositionally similar magmas, while the unit of komatiitic basalt above the second chert bed shows different values. With respect to major elements, there is a general increase in SiO_2 content towards the top of each lava unit, reaching a maximum value of 73%. Contents of K_2O are generally higher in more altered samples. Most other major elements show decreasing concentrations, both in absolute terms and relative to immobile elements, although P_2O_5 and Na_2O contents remain constant or show an erratic distribution. The cherts have very high SiO_2 contents, generally >95%, and, as expected, low concentrations of most other elements.

Most trace elements were immobile during alteration, except for some minor depletion of Cu and Zn in the most altered rocks (Fig. 5.5-4). Interestingly, the cherts have relatively high values of Cu, Zn, and Ni, despite high silica contents (e.g., 76 ppm Zn in average basalt vs 58 ppm Zn in average chert of this section), and also have much higher ratios of these metals relative to immobile elements (e.g., Hf) compared to the basalts (Fig. 5.5-4), although they probably contain a detrital component compositionally similar to the basalts. Cr/Th ratios (113–858) and Th/Sc ratios (0.05–0.31), which are very useful indicators for provenance (Condie and Wronkiewicz, 1990), are slightly elevated relative to the basalts.

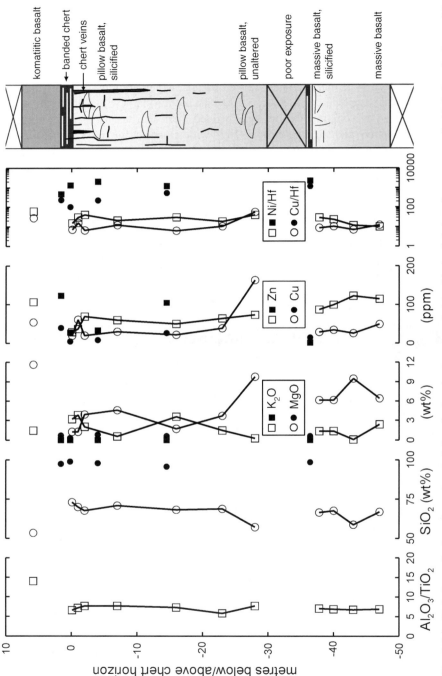

Fig. 5.5-4. Variations of selected geochemical data for basalts (open symbols) with depth below the uppermost bedded chert horizon of the Vumankula River main section. Analytical data of bedded cherts and chert veins (filled symbols) are also shown.

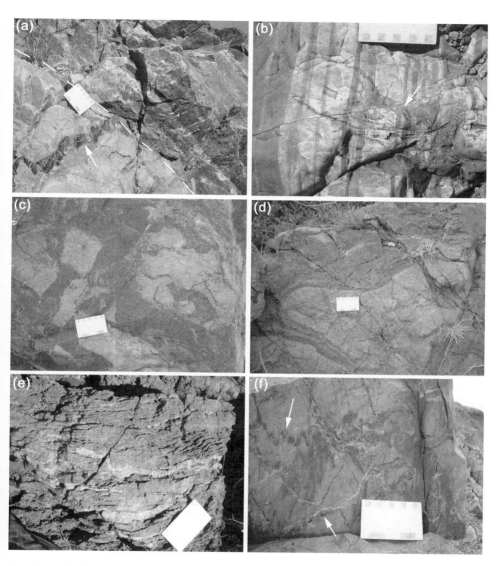

Fig. 5.5-5. Field relationships of bedded cherts, chert veins and silicified volcanic rocks. (a) Contact between silicified basalt and overlying banded chert (marked with stippled line), Vumankula River main section. Note black chert vein cutting across basalt (arrow). (b) Black-and-white banded chert, Vumankula River main section. Note cross-cutting chert vein (arrow). (c) Silicified basalt intensely veined by black carbonaceous chert, Vumankula River main section. (d) Silicified pillow basalt surrounded and transected by carbonaceous chert, Vumankula River main section. (e) Highly altered komatiitic rock, Vumankula River south section. Note abundant stratiform veins of clear chert. (f) Massive greyish green chert transected by ptygmatically folded veins of carbonaceous chert. Bedding is vertical.

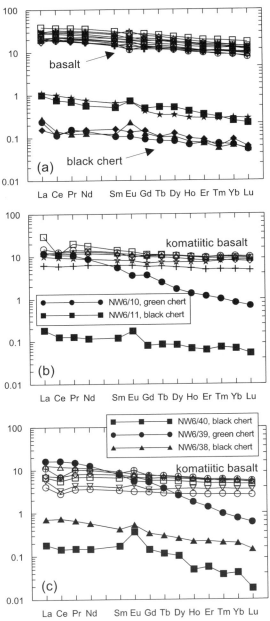

Fig. 5.5-6. Chondrite-normalized (Boynton, 1984) REE plots of unaltered and altered volcanic rocks (open symbols) and cherts (filled symbols). (a) Vumankula River main section. (b) Vumankula River south section. (c) Eastern section. (d) Carbonaceous cherts from the Wikop Formation northwest of the Vumankula River main section. (e) Witkop Hill.

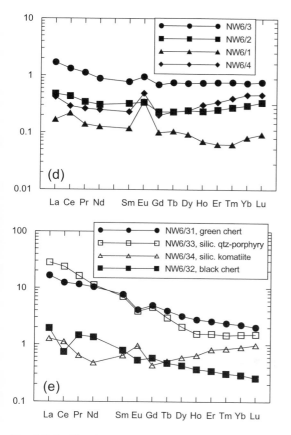

Fig. 5.5-6. (*Continued.*)

The basalts have smooth, LREE-enriched and HREE-depleted patterns (Fig. 5.5-6(a); $La_N/Yb_N = 1.4$–2.37). The REE patterns are similar to those of the cherts, which are slightly steeper ($La_N/Yb_N = 1.7$–4.12), although the patterns are shifted towards lower values due to lower ΣREE contents in these rocks. The basalts do not show a consistent Eu anomaly, while most cherts have a positive Eu anomaly (average $Eu/Eu^* = 1.33$).

In order to evaluate relative gains and losses of elements during alteration of the volcanic rocks, an isocon analysis (Grant, 1986, 2005) was carried out (Fig. 5.5-7(a)). Averages of the two most altered samples were compared with the averages of the two least altered samples in the sample suite. If there was no change in volume during alteration, elements that show no change in concentration will fall on a line with a slope of 1, passing through the origin. Elements that are enriched relative to the least altered sample will plot above this line, and those that are depleted will plot below this line (Grant, 1986, 2005). Al_2O_3, V, Nb, Hf, Ta were found to lie on an isocon with a slope of 1, indicating immobility of these elements and isovolumetric alteration. This is in accord with field observations, as

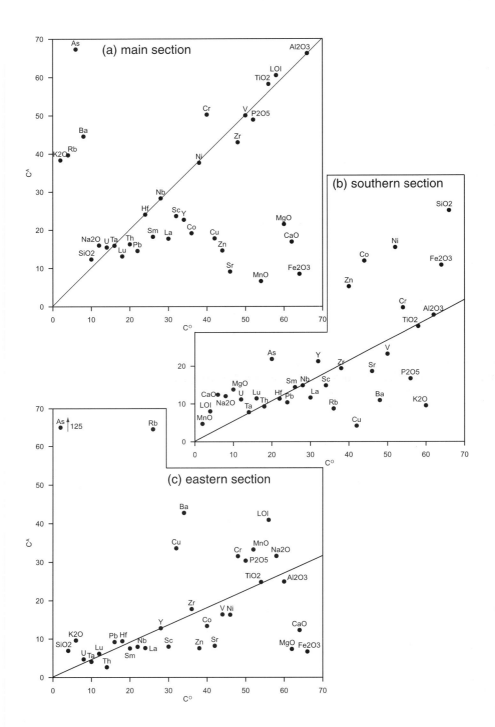

Fig. 5.5-7. (*Previous page.*) Isocon diagrams (Grant, 1986, 2005) for least-altered (C^O) and most altered (C^A) volcanic rocks from the Nondweni greenstone belt. Elements and oxides above the isocon line have been added during alteration, while those below have been depleted. (a) Vumankula River main section (C^O, average of samples NW6/20, NW6/24; C^A, average of NW6/14, NW6/29). (b) Vumankula River southern section (C^O, average of samples NW6/6, NW6/12; C^A, average of NW6/8, NW6/9). (c) Eastern section (C^O, NW6/35; C^A, NW6/43).

---

the silicified rocks sampled for this study do not contain secondary chert veins. Significant gains (>100%) for K_2O, Rb, Ba, and As are noted, while Na_2O, SiO_2 and Cr increased in the order of 20–30%. Cu, MgO, Zn, CaO, Sr, Fe_2O_3, and MnO were all depleted and lost more than 50% (in order of increasing loss). The REE were slightly depleted, and the LREE were generally more depleted than the HREE.

5.5-4.2. Vumankula River, Southern Section

This section crops out approximately 600 m southeast of the main section (Fig. 5.5-3) and is separated from it by a poorly exposed and partly faulted sequence of basalts and cherts. The section consists of a sequence of altered volcanic rocks at least 20 m thick that are overlain by bedded chert (Fig. 5.5-8). Alteration is mainly in the form of silicification and carbonatization. Local spinifex textures and weathering/alteration patterns indicate a komatiitic parentage. The altered rocks are transected by irregular veins of translucent chert that are typically 10 cm in length and oriented sub-parallel to the stratification (Fig. 5.5-5(e)). These veins are particularly common in the uppermost 3 metres of the section. The overlying chert horizon is 1.8 m thick and consists of thinly bedded, possibly ripple-laminated to massive black carbonaceous chert and green chert. Cross-cutting black chert veins are common.

Six samples of volcanic rocks and two chert beds were analysed. All volcanic rocks sampled are altered, although the least altered samples (NW6/6, NW6/12) have major element characteristics close to komatiitic andesites (Riganti and Wilson, 1995a). The Al_2O_3/TiO_2 ratio is constant for the section, but Al_2O_3 values do differ, suggesting either mobility of Al or volume changes. An isocon analysis (Fig. 5.5-7(b)) was carried out using the averages of the two most altered samples compared with the averages of the two least altered samples in the sample suite. An isocon could be fitted with an average slope of 0.535, using Al_2O_3, TiO_2, Nb, Hf, and Ta. The slope of the isocon being <1 indicates an increase in the rock volume, if the elements were immobile. Element gain and loss patterns are somewhat different to those observed for the main section. This is partly related to the absence of a relatively unaltered rock in the sample suite, as, for example, one of the samples used for the analysis (NW6/6) is actually enriched in K_2O (3%), resulting in apparent losses of this element (and Rb, Ba) in more altered rocks. Nevertheless, there is a clear increase in CaO, both in absolute and relative terms (Fig. 5.5-8). There is also an increase in LOI, which can be attributed to the presence of carbonates in the samples, as CaO and LOI show a positive linear relationship ($r^2 = 0.82$). Alteration of this unit was in the form of both silicification

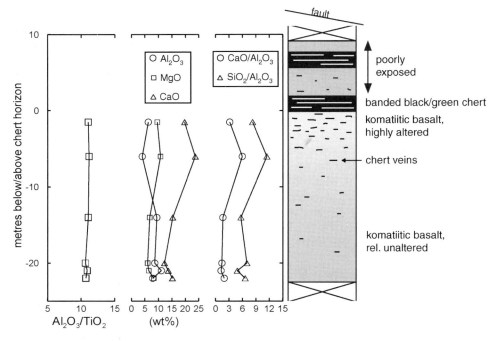

Fig. 5.5-8. Variations of selected geochemical data of komatiitic basalt with depth below the upper-most bedded chert horizon of the Vumankula River southern section.

and carbonatization, which resulted in element depletion and enrichment patterns slightly different to those derived from silicification alone.

The REE patterns of the volcanic rocks are rather flat (Fig. 5.5-6(b)), with La_N/Yb_N ratios typically in the order of 1.3. While the REE pattern of the carbonaceous chert is similar to those of the volcanic rocks, with the exception of a positive Eu anomaly ($Eu/Eu^* = 1.77$), the REE pattern for green chert is steep ($La_N/Yb_N = 14.7$), indicating that its composition is unrelated to both the volcanic rocks and the carbonaceous chert, and probably includes a more felsic component.

5.5-4.3. Eastern Section

A third section situated along a small stream east of the Babanango-Vryheid road, herein termed eastern section, was investigated (Fig. 5.5-1). It consists of massive komatiitic basalt capped by a massive greyish green chert (Fig. 5.5-9). The volcanic rocks are highly silicified, and the extent of silicification increases upsection. Numerous veins of translucent chert oriented sub-parallel to the stratification are present immediately below the chert, resulting in a lenticular pattern (zebraic texture of Duchac and Hanor, 1987). Veins up to 5 cm wide and filled with carbonaceous chert are oriented both sub-parallel and perpendicular to the stratification. They are present in the uppermost 1.5 metres of the volcanic

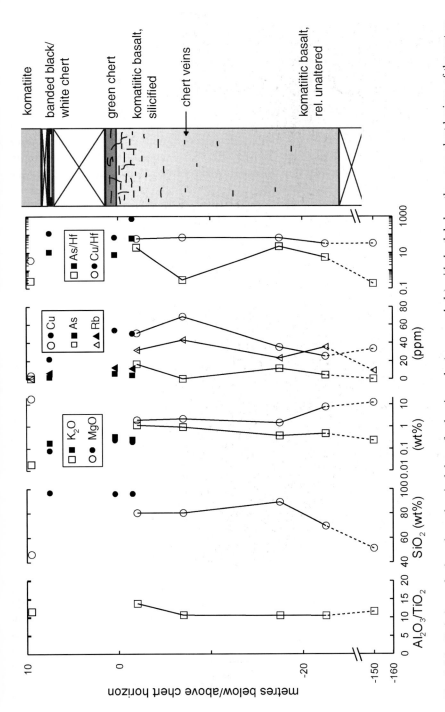

Fig. 5.5-9. Variations of selected geochemical data of volcanic rocks (open symbols) with depth below the green chert horizon of the eastern section. Analytical data of bedded cherts and chert veins (filled symbols) are also shown.

rocks as well as in the overlying green chert. Cross-cutting veins in the chert appear to have been ptygmatically folded (Fig. 5.5-5(f)). The chert is overlain by an interval of very poor exposure, probably occupied by highly altered komatiitic rock and possibly some black carbonaceous chert at the base. This interval is overlain by a bed of banded black-and-white chert, followed by poorly exposed, weathered, unsilicified massive komatiitic rock.

All rocks exposed immediately below the green chert bed are altered to various degrees. One sample of fresh komatiitic basalt (NW6/35) was obtained from an outcrop approximately 150 metres downsection. This sample shows very similar immobile element ratios to the rest of the sample suite (e.g., Al_2O_3/TiO_2; Fig. 5.5-9) and, although it is most probably derived from a different volcanic flow unit, was regarded to most closely reflect the composition of the lavas prior to alteration.

With respect to major elements, there is an increase in SiO_2 contents to values of >80% towards the top of the section. K_2O equally increases upsection, while MgO, CaO, and Fe_2O_3 contents decrease. Overlying and cross-cutting cherts have very high SiO_2 contents (>95%) and low concentrations of most other elements. Komatiite above the chert sequence, despite its high LOI value (probably due to recent weathering), does not show evidence for silicification, carbonatization and K-metasomatism (Table 5.5-1).

Most trace elements were immobile during silicification, except for the addition of As, Rb, Ba, and Cu (Fig. 5.5-9). The concentrations of these elements are also high in the cherts, at least relative to immobile elements (Fig. 5.5-9). The REE patterns of the volcanic rocks are similar to those of the southern section, with slight LREE-enrichment and HREE-depletion (Fig. 5.5-6(c)), and with La_N/Yb_N ratios in the order of 1.5. The REE patterns of the carbonaceous cherts are steeper ($La_N/Yb_N = 3.5$–4.4), and both samples have a positive Eu anomaly ($Eu/Eu^* = 1.4$–2.35). The REE pattern for green chert is, again, much steeper ($La_N/Yb_N = 21.7$), and the ΣREE much higher.

In an isocon analysis of the least altered vs the most altered (silicified) volcanic rock sample (Fig. 5.5-7(c)), an isocon with a mean slope of 0.45 was constructed using the relative concentrations of Al_2O_3, TiO_2, Hf, and Ta. The abundance of chert veins in the altered rocks explains the slope of the isocon <1, which indicates an increase in the rock volume during alteration. As, Rb, SiO_2 and K_2O are the elements that are enriched the most (>100%), while MgO and Fe_2O_3 were depleted by more than 30%.

5.5-4.4. Other Sites

In order to obtain more geochemical information on the variety of silicified rocks in the Witkop Formation, a number of additional samples were studied. Four samples of carbonaceous chert from different horizons (1–2 metres thick) of banded black-and-white chert that crop out several hundreds of metres upsection from the Vumankula River main section (Fig. 5.5-3) were investigated. The cherts are geochemically very similar to carbonaceous cherts from the other sections; they have very high SiO_2 contents (generally >95%) with all other major elements having concentrations below 1.3%. Al_2O_3 contents are 0 to 1.17% (average 0.46%), suggesting a low amount of primary detrital material. The concentrations of most trace elements are equally low, except for several metals, such as Cu, Zn, and

Table 5.5-1. Major element (in wt% oxide) and trace element (in ppm) concentrations for samples from the Witkop Formation

Locality	Witkop Fm (see Fig. 5.5-3)				Vumankula River (southern section)				
Sample	NW6/1	NW6/2	NW6/3	NW6/4	NW6/12	NW6/6	NW6/5	NW6/7	NW6/8
Lithology	CBC	CBC	CBC	CBC	K	K	K	K	K
m below chert					−22	−21	−20	−14	−6
SiO_2	98.00	97.54	95.64	96.99	51.99	53.63	60.37	53.04	45.72
Al_2O_3	0.08	0.24	1.17	0.36	7.88	11.09	8.65	9.40	3.99
Fe_2O_3	0.58	0.76	1.20	0.98	12.30	8.73	8.74	11.14	7.91
MnO	0.00	0.01	0.01	0.01	0.51	0.38	0.34	0.39	1.14
MgO	0.06	0.05	0.24	0.08	8.31	6.43	5.99	6.77	10.77
CaO	0.03	0.01	0.02	0.01	15.21	13.66	12.25	15.32	23.74
Na_2O	0.00	0.00	0.00	0.00	0.18	0.33	0.13	0.12	0.22
K_2O	0.02	0.06	0.27	0.05	0.22	3.00	0.69	0.68	0.30
TiO_2	0.00	0.01	0.04	0.02	0.74	1.02	0.81	0.85	0.36
P_2O_5	0.00	0.04	0.02	0.00	0.06	0.07	0.06	0.07	0.03
LOI	0.45	0.64	0.85	0.82	2.52	1.70	1.63	2.29	5.93
Total	99.22	99.37	99.46	99.31	99.92	100.03	99.66	100.05	100.10
Sc	0.28	0.62	1.81	0.71	33.89	55.37	40.38	34.65	15.12
V	1.05	4.10	22.87	15.46	216.27	316.29	218.66	233.74	96.10
Cr	8.16	15.92	71.11	61.26	1128.01	1979.28	1830.50	1225.08	796.57
Co	1.02	5.88	6.99	5.26	69.59	67.22	43.86	52.19	55.61
Ni	10.10	43.14	68.49	44.05	296.95	272.61	204.68	305.73	225.78
Cu	17.58	53.33	73.17	19.81	113.05	23.20	24.49	94.05	6.40
Zn	37.78	342.21	58.10	34.21	72.37	59.26	49.54	82.75	63.74
As	0.08	0.46	0.92	1.12	0.27	1.11	0.22	0.23	0.49
Rb	0.42	1.89	10.91	2.11	4.72	100.34	18.68	18.20	15.98
Sr	6.20	7.49	8.37	6.20	122.52	66.80	53.92	71.60	27.48
Y	0.15	0.50	1.91	0.68	19.97	20.62	19.76	17.50	11.42
Zr	1.01	2.19	16.43	3.99	46.09	67.59	50.28	55.41	23.05
Nb	0.04	0.10	0.89	0.17	2.56	3.66	2.68	3.05	1.31
Ba	5.44	28.68	30.25	14.37	371.51	90.70	40.90	20.64	31.06
La	0.05	0.15	0.53	0.13	9.19	3.76	4.73	3.52	1.92
Ce	0.18	0.35	1.07	0.23	8.89	10.14	10.33	9.56	4.79

(continued on next page)

Pb (e.g., 342 ppm Zn in one sample). HFSE ratios are variable (Nb/Ta = 10.3–35; Zr/Hf = 30.8–67.6; Th/U = 0.1–1.04), which may reflect interaction with hydrothermal fluids (Bau, 1996). Cr/Th ratios (182–1332) and Th/Sc ratios (0.06–0.22) are similar to material of basaltic composition (Hofmann, 2005b), although with possible contributions of minor ultramafic and felsic material. Y/Ho ratios are slightly elevated relative to the chondritic value (29.5–35.4). ΣREE contents are very low (0.5–3.4 ppm) and the REE patterns are relatively flat to slightly concave with some LREE enrichment (Fig. 5.5-6(d); La_N/Yb_N = 0.9–2.3). All samples are characterized by a positive Eu anomaly (Eu/Eu* = 1.24–3.25).

Table 5.5-1. (*Continued*)

Locality	Witkop Fm (see Fig. 5.5-3)				Vumankula River (southern section)				
Sample	NW6/1	NW6/2	NW6/3	NW6/4	NW6/12	NW6/6	NW6/5	NW6/7	NW6/8
Lithology	CBC	CBC	CBC	CBC	K	K	K	K	K
m below chert					-22	-21	-20	-14	-6
Pr	0.02	0.04	0.14	0.03	2.47	1.57	1.76	1.43	0.74
Nd	0.08	0.19	0.53	0.15	10.70	7.79	8.33	7.12	3.78
Sm	0.02	0.06	0.15	0.05	2.73	2.31	2.43	2.15	1.21
Eu	0.03	0.03	0.07	0.04	0.95	0.83	0.71	0.77	0.45
Gd	0.03	0.06	0.18	0.05	2.89	2.67	2.74	2.52	1.50
Tb	0.01	0.01	0.04	0.01	0.52	0.50	0.52	0.46	0.28
Dy	0.03	0.08	0.24	0.08	3.33	3.33	3.40	2.94	1.80
Ho	0.01	0.02	0.05	0.02	0.70	0.72	0.72	0.61	0.38
Er	0.01	0.05	0.16	0.07	1.92	2.00	2.02	1.67	1.03
Tm	0.00	0.01	0.03	0.01	0.29	0.32	0.33	0.26	0.16
Yb	0.02	0.06	0.16	0.10	1.92	2.04	2.12	1.64	1.03
Lu	0.00	0.01	0.03	0.02	0.29	0.30	0.31	0.24	0.15
Hf	0.02	0.06	0.53	0.10	1.24	1.73	1.35	1.52	0.63
Ta	0.00	0.01	0.09	0.01	0.16	0.22	0.18	0.19	0.08
W	0.05	0.04	0.15	0.08	0.10	0.59	0.16	0.12	0.21
Pb	1.71	9.80	22.38	2.31	1.63	2.03	1.77	1.75	0.61
Th	0.02	0.07	0.39	0.05	0.26	0.35	0.28	0.29	0.12
U	0.21	0.19	0.38	0.28	0.07	0.09	0.06	0.08	0.04
ΣREE	0.47	1.12	3.36	0.99	46.77	38.26	40.45	34.88	19.24
Eu/Eu*	3.25	1.24	1.29	2.28	1.04	1.02	0.84	1.00	1.03
La$_N$/Sm$_N$	1.42	1.48	2.17	1.82	2.12	1.02	1.23	1.03	1.00
Gd$_N$/Yb$_N$	1.23	0.77	0.93	0.43	1.22	1.06	1.04	1.24	1.18
La$_N$/Yb$_N$	2.06	1.58	2.27	0.90	3.23	1.25	1.50	1.45	1.26
Al$_2$O$_3$/TiO$_2$	22.22	20.17	27.19	23.68	10.70	10.90	10.64	11.06	11.17
CaO/Al$_2$O$_3$	0.38	0.04	0.02	0.03	1.93	1.23	1.42	1.63	5.94
Nb/Ta	35.00	14.71	10.26	14.08	16.18	16.69	15.30	16.19	16.10
Zr/Hf	67.60	39.18	30.82	41.54	37.14	39.16	37.28	36.47	36.60
Y/Ho	30.60	29.53	35.35	30.77	28.65	28.60	27.41	28.79	29.88
Cr/Th	371	227	182	1332	4273	5639	6487	4225	6445
Th/Sc	0.079	0.113	0.216	0.065	0.008	0.006	0.007	0.008	0.008

(*continued on next page*)

Samples were also obtained from Witkop Hill, where a characteristic sequence of rocks is complexly folded (Fig. 5.5-1; mapping in progress). The sequence starts with a highly silicified komatiite or komatiitic basalt, which locally shows pseudomorphed spinifex textures. Reliable younging indicators are absent, but the silicified komatiite is likely overlain by banded black-and-white chert, a greenish grey massive chert consisting entirely of silicified lapilli, and a highly silicified quartz-phyric rock, probably of originally rhyodacitic to rhyolitic composition and representing a felsic lava or pyroclastic flow.

Table 5.5-1. (Continued)

Locality				Vumankula River (main section)					
Sample	NW6/9	NW6/10	NW6/11	NW6/19	NW6/20	MW6/21	NW6/22	NW6/23	NW6/24
Lithology	K	GC	CBC	mB	mB	mB	mB	CBC	pB
m below chert	−1.5	1.4	1.6	−47	−43	−40	−37.8	−36.4	−28
SiO_2	51.52	96.40	97.98	66.85	58.67	67.56	66.43	98.41	57.15
Al_2O_3	6.25	1.00	0.30	13.54	13.33	12.00	12.43	0.16	13.82
Fe_2O_3	7.36	0.49	0.63	6.40	9.46	6.15	6.14	0.61	9.75
MnO	0.97	0.01	0.01	0.22	0.33	0.19	0.19	0.01	0.31
MgO	9.60	0.19	0.09	2.14	3.61	3.23	3.54	0.07	3.16
CaO	19.73	0.09	0.05	3.00	8.08	3.36	3.84	0.09	9.29
Na_2O	0.85	0.01	0.06	0.64	0.38	0.80	1.08	0.07	0.32
K_2O	0.21	0.27	0.08	2.37	0.09	1.36	1.35	0.03	0.28
TiO_2	0.57	0.50	0.01	1.97	1.99	1.75	1.77	0.00	1.80
P_2O_5	0.01	0.02	0.00	0.22	0.22	0.18	0.18	0.00	0.22
LOI	2.58	0.30	0.24	2.63	3.20	3.01	3.03	0.31	3.29
Total	99.65	99.27	99.44	99.98	99.36	99.59	99.97	99.77	99.39
Sc	24.02	4.40	0.36	20.12	20.90	16.72	14.83	0.03	24.57
V	150.49	32.70	6.63	296.25	330.14	305.68	279.33	1.27	311.41
Cr	1240.78	313.83	60.86	26.76	24.52	32.84	60.28	3.93	137.66
Co	92.41	17.32	11.07	30.10	26.98	50.59	24.63	3.22	69.79
Ni	334.09	75.96	47.88	39.67	41.89	76.44	95.15	27.28	115.72
Cu	7.25	59.07	72.56	48.70	25.11	33.87	28.19	14.14	163.02
Zn	70.88	32.18	8.48	115.08	122.75	99.59	86.61	1.02	72.65
As	1.01	5.44	0.95	3.39	2.33	16.37	11.67	1.56	0.43
Rb	9.57	11.50	2.67	101.09	3.45	62.34	32.54	0.37	13.40
Sr	49.16	5.05	1.36	55.85	129.35	39.30	62.66	0.60	119.73
Y	15.52	2.01	0.14	30.34	43.94	24.09	21.90	0.17	31.83
Zr	34.91	29.85	2.13	139.34	146.17	114.90	126.44	0.47	129.68
Nb	2.00	5.38	0.11	7.33	7.23	6.35	6.82	0.03	6.95
Ba	74.09	25.98	10.93	119.57	25.33	51.74	54.03	1.66	8.63
La	3.10	3.73	0.06	8.86	12.03	6.72	5.36	0.07	9.23
Ce	7.58	8.79	0.10	24.30	29.77	17.91	17.33	0.09	23.27

(continued on next page)

The silicified komatiite is enriched in SiO_2 (88.7%) and K_2O (1.8%), while all other major element are strongly depleted. High Cr contents (2712 ppm), but Ni concentrations too low for a komatiite (78 ppm) suggest depletion of Ni during alteration, a feature characteristic for highly silicified komatiites of the Barberton Greenstone Belt (Hofmann, 2005a). The REE pattern is irregular (Fig. 5.5-6(e)), possibly as a result of REE mobility during alteration. The silicified rhyolite shows high SiO_2 (82.8%) and K_2O (3.1%) values, a depletion of most other elements, and relative steep REE patterns, with LREE enrichment and HREE depletion ($La_N/Yb_N = 18.7$). The green chert differs from black chert by higher

Table 5.5-1. (*Continued*)

Locality				Vumankula River (main section)					
Sample	NW6/9	NW6/10	NW6/11	NW6/19	NW6/20	MW6/21	NW6/22	NW6/23	NW6/24
Lithology	K	GC	CBC	mB	mB	mB	mB	CBC	pB
m below chert	−1.5	1.4	1.6	−47	−43	−40	−37.8	−36.4	−28
Pr	1.11	1.30	0.02	3.62	4.44	2.61	2.16	0.02	3.43
Nd	5.27	5.25	0.07	17.65	21.34	12.52	10.05	0.09	16.04
Sm	1.59	1.06	0.02	5.05	6.00	3.49	3.02	0.02	4.38
Eu	0.15	0.59	0.26	0.01	1.56	1.95	1.18	0.98	0.01
Gd	1.80	0.94	0.02	5.57	6.98	3.96	3.47	0.02	5.01
Tb	0.35	0.12	0.00	0.94	1.20	0.69	0.62	0.00	0.87
Dy	2.36	0.55	0.03	5.83	7.40	4.25	4.00	0.03	5.37
Ho	0.53	0.10	0.00	1.16	1.53	0.86	0.83	0.00	1.10
Er	1.53	0.25	0.01	3.07	4.06	2.25	2.25	0.02	2.91
Tm	0.25	0.03	0.00	0.46	0.61	0.33	0.35	0.00	0.43
Yb	1.72	0.17	0.01	2.74	3.78	2.03	2.18	0.01	2.67
Lu	0.26	0.02	0.00	0.40	0.55	0.29	0.32	0.00	0.39
Hf	0.90	0.85	0.05	3.97	3.59	3.27	3.26	0.01	2.90
Ta	0.13	0.40	0.01	0.48	0.45	0.40	0.43	0.00	0.41
W	0.33	1.30	0.04	0.47	1.26	1.25	4.97	0.06	0.96
Pb	0.97	0.56	1.25	1.66	1.90	1.75	2.41	0.53	3.04
Th	0.20	0.41	0.01	0.78	0.77	0.62	0.55	0.01	0.62
U	0.10	0.16	0.17	0.18	0.21	0.18	0.16	0.01	0.15
ΣREE	28.04	22.57	0.37	81.22	101.65	59.10	52.93	0.39	76.76
Eu/Eu*	1.06	0.81	1.77	0.90	0.92	0.97	0.93	1.08	1.08
La_N/Sm_N	1.23	2.21	1.52	1.10	1.26	1.21	1.11	2.22	1.32
Gd_N/Yb_N	0.85	4.45	1.18	1.64	1.49	1.57	1.28	1.36	1.51
La_N/Yb_N	1.22	14.71	2.67	2.18	2.15	2.23	1.65	3.49	2.33
Al_2O_3/TiO_2	11.06	2.01	24.39	6.86	6.71	6.85	7.02	32.00	7.67
CaO/Al_2O_3	3.16	0.09	0.17	0.22	0.61	0.28	0.31	0.56	0.67
Nb/Ta	15.51	13.46	19.20	15.19	16.09	15.87	15.98	27.37	
Zr/Hf	38.72	35.02	45.94	35.14	40.70	35.11	38.83	38.32	44.65
Y/Ho	29.54	20.69	29.18	26.14	28.79	27.96	26.24	34.53	28.91
Cr/Th	6361	760	4541	34	32	53	110	584	223
Th/Sc	0.008	0.094	0.038	0.039	0.037	0.037	0.037	0.198	0.025

(*continued on next page*)

element and oxide contents other than SiO_2 (e.g., 28.4 ppm vs 2.9 ppm ΣREE) and high transition metal contents (1510 ppm vs 11 ppm Cr), resulting in a Cr/Th ratio for the green chert similar to komatiitic basalt, while the Cr/Th ratio for black chert is typical for basaltic rocks. However, the REE patterns of both chert types are similar: both show moderately steep LREE-enriched and HREE depleted patterns (La_N/Yb_N = 6.7 and 7.4), suggestive of a more evolved, possibly felsic detrital/tuffaceous component similar in composition to the overlying rhyolitic rocks. The presence of a mafic signature in terms of immobile el-

Table 5.5-1. (*Continued*)

Locality	Vumankula River (main section)								
Sample	MW6/25	NW6/26	NW6/28	NW6/16	NW6/29	NW6/14	NW6/27	NW6/17	NW6/30
Lithology	pB	pB	pB	pB	pB	pB	CVC	CVC	CBC
m below chert	−23	−16	−7	−2	−1	−0.1	−14.5	−4	0.2
SiO_2	68.73	68.18	70.82	67.65	69.95	73.01	95.59	97.80	98.84
Al_2O_3	12.50	15.04	10.72	13.70	15.43	11.81	1.17	0.17	0.18
Fe_2O_3	3.67	1.72	4.56	3.89	1.28	1.27	0.62	0.83	0.32
MnO	0.15	0.07	0.18	0.15	0.04	0.04	0.02	0.01	0.00
MgO	3.07	1.26	2.76	2.32	1.33	1.10	0.22	0.14	0.06
CaO	3.64	3.26	5.51	5.39	2.43	2.30	0.85	0.05	0.07
Na_2O	1.45	1.25	0.48	0.30	0.72	0.21	0.03	0.00	0.04
K_2O	1.48	3.54	0.57	2.03	3.80	3.18	0.03	0.02	0.03
TiO_2	2.15	2.06	1.40	1.79	2.15	1.79	0.01	0.01	0.01
P_2O_5	0.28	0.20	0.15	0.21	0.19	0.22	0.00	0.00	0.00
LOI	2.50	2.86	2.60	2.46	2.43	4.34	0.70	0.63	0.16
Total	99.63	99.44	99.75	99.89	99.74	99.27	99.24	99.66	99.72
Sc	24.17	18.55	17.82	22.70	18.61	14.98	0.64	0.16	0.19
V	365.41	307.82	225.95	301.90	374.63	267.11	6.25	4.04	2.55
Cr	102.12	133.38	102.18	107.79	124.93	78.73	16.70	12.12	26.77
Co	32.90	64.87	19.67	43.36	41.91	9.53	10.48	8.96	7.82
Ni	67.57	103.50	50.51	121.49	115.49	40.62	58.83	72.71	56.56
Cu	38.04	21.30	29.30	19.44	60.42	19.20	25.87	8.08	4.44
Zn	63.34	49.53	59.18	68.51	36.52	28.41	104.92	32.79	27.26
As	16.31	42.53	1.53	7.58	29.60	1.45	5.93	1.88	7.52
Rb	54.34	36.20	34.78	75.58	64.40	102.75	0.96	0.29	0.60
Sr	76.57	35.18	106.04	71.94	31.01	18.33	8.36	0.84	0.84
Y	37.24	21.14	21.91	27.53	29.77	20.79	0.90	0.31	0.17
Zr	141.18	135.20	98.90	117.08	146.87	100.07	2.01	1.35	1.57
Nb	8.47	7.80	5.33	6.79	8.32	6.02	0.07	0.05	0.08
Ba	53.72	68.18	31.83	141.94	48.84	140.35	3.48	12.37	1.63
La	9.24	5.96	6.33	8.58	6.21	6.35	0.31	0.05	0.09
Ce	26.57	18.37	15.57	21.90	18.49	15.47	0.61	0.10	0.12

(*continued on next page*)

ement ratios, but steep REE patterns unlike Nondweni mafic volcanic rocks, may suggest contemporaneous felsic and mafic/ultramafic magmatism in the source area of these rocks.

5.5-5. PETROGRAPHY

Unaltered or little altered mafic to ultramafic volcanic rocks of the Witkop Formation exhibit greenschist facies mineral assemblages (Fig. 5.5-10(a)). Pyroxene is generally altered to tremolite-actinolite, and plagioclase is partially sassuritized. Chlorite and talc

Table 5.5-1. (*Continued*)

Locality	Vumankula River (main section)								
Sample	MW6/25	NW6/26	NW6/28	NW6/16	NW6/29	NW6/14	NW6/27	NW6/17	NW6/30
Lithology	pB	pB	pB	pB	pB	pB	CVC	CVC	CBC
m below chert	−23	−16	−7	−2	−1	−0.1	−14.5	−4	0.2
Pr	3.82	2.50	2.27	3.16	2.74	2.41	0.08	0.02	0.01
Nd	18.70	12.01	10.31	14.83	13.56	11.58	0.34	0.09	0.08
Sm	5.36	3.41	2.81	4.04	4.11	3.16	0.10	0.03	0.02
Eu	2.26	1.34	1.14	1.42	1.44	0.84	0.05	0.01	0.02
Gd	6.27	3.62	3.27	4.37	4.76	3.46	0.13	0.04	0.03
Tb	1.09	0.62	0.59	0.74	0.89	0.57	0.02	0.01	0.00
Dy	6.62	3.88	3.67	4.79	5.66	3.67	0.16	0.04	0.03
Ho	1.34	0.81	0.75	0.98	1.18	0.73	0.03	0.01	0.01
Er	3.53	2.22	2.04	2.73	3.20	2.03	0.07	0.02	0.02
Tm	0.50	0.34	0.30	0.40	0.49	0.29	0.01	0.00	0.00
Yb	2.98	2.09	1.79	2.65	2.98	1.82	0.05	0.02	0.01
Lu	0.43	0.30	0.26	0.37	0.44	0.25	0.01	0.00	0.00
Hf	3.67	3.43	2.43	2.97	3.81	2.70	0.05	0.04	0.04
Ta	0.51	0.47	0.32	0.40	0.50	0.35	0.00	0.00	0.01
W	7.70	5.62	1.22	0.26	1.51	0.77	0.04	0.00	0.02
Pb	1.61	2.26	1.67	3.05	1.99	1.27	1.84	0.69	0.43
Th	0.72	0.53	0.51	0.58	0.64	0.49	0.03	0.03	0.03
U	0.22	0.12	0.18	0.14	0.27	0.12	0.02	0.02	0.06
ΣREE	88.70	57.49	51.09	70.95	66.17	52.62	1.98	0.43	0.44
Eu/Eu*	1.19	1.17	1.15	1.03	1.00	0.78	1.45	0.89	1.84
La_N/Sm_N	1.08	1.10	1.42	1.33	0.95	1.26	1.96	0.97	2.29
Gd_N/Yb_N	1.70	1.40	1.47	1.33	1.29	1.54	1.98	1.61	1.88
La_N/Yb_N	2.09	1.92	2.38	2.18	1.40	2.35	4.04	1.70	4.12
Al_2O_3/TiO_2	5.81	7.29	7.67	7.66	7.19	6.60	98.33	15.89	16.22
CaO/Al_2O_3	0.29	0.22	0.51	0.39	0.16	0.19	0.73	0.29	0.39
Nb/Ta	17.01	16.60	16.62	16.65	16.66	16.99	14.22	16.67	13.30
Zr/Hf	38.48	39.47	40.71	39.42	38.55	37.04	41.24	37.53	36.36
Y/Ho	27.88	25.95	29.11	28.20	25.21	28.47	29.93	43.71	29.92
Cr/Th	142	250	200	185	196	161	485	404	858
Th/Sc	0.030	0.029	0.029	0.026	0.034	0.033	0.054	0.183	0.168

(*continued on next page*)

are commonly present in the more Mg-rich lithologies. Silicified volcanic rocks contain a large proportion of relatively coarse-grained (i.e., recrystallized), granular microquartz (Fig. 5.5-10(b)), which occurs as matrix material. With increasing silicification, quartz replaces phenocrysts and forms veins. Anastomosing, stratiform quartz veins are particularly common in silicified komatiites where they impart a lenticular fabric to the rocks. Carbonate-altered rocks are characterized by the presence of irregular aggregates or euhedral crystals of calcite and possibly dolomite/ankerite. The relative timing of silicification

Table 5.5-1. (*Continued*)

Locality	Vumankula River (main section)		Witkop Hill				Eastern section	
Sample	NW6/15	NW6/13	NW6/34	NW6/32	NW6/31	NW6/33	NW6/35	NW6/44
Lithology	CBC	KB	K	CBC	GC	Porphyry	KB	KB
m below chert	1.6	5.8					−150	−22.5
SiO_2	97.37	53.50	88.70	98.61	90.99	82.80	51.48	69.61
Al_2O_3	0.27	10.57	6.21	0.08	3.82	11.32	6.99	4.96
Fe_2O_3	0.71	11.64	1.05	0.53	0.88	0.40	13.78	7.05
MnO	0.01	0.22	0.04	0.00	0.01	0.00	0.24	0.17
MgO	0.14	9.76	0.56	0.03	0.53	0.14	12.31	7.75
CaO	0.09	8.87	0.01	0.01	0.03	0.02	12.60	7.17
Na_2O	0.03	0.73	0.01	0.00	0.05	0.23	1.30	1.07
K_2O	0.03	1.45	1.82	0.01	1.08	3.10	0.24	0.47
TiO_2	0.02	0.75	0.10	0.01	1.29	0.19	0.60	0.47
P_2O_5	0.00	0.09	0.01	0.01	0.03	0.02	0.05	0.04
LOI	0.79	2.19	1.03	0.07	0.56	1.40	0.78	0.93
Total	99.46	99.77	99.53	99.36	99.27	99.63	100.36	99.70
Sc	0.52	31.42	4.68	0.36	3.16	1.63	30.84	24.82
V	5.71	219.57	72.26	1.92	111.17	24.60	170.93	129.99
Cr	18.11	637.98	2711.56	10.71	1509.81	17.67	1512.77	930.78
Co	8.97	53.11	27.48	2.42	204.65	1.26	82.85	43.71
Ni	77.85	117.65	77.64	21.01	604.48	7.53	456.67	172.27
Cu	39.22	53.37	9.65	6.94	219.89	7.99	33.56	25.36
Zn	123.52	106.49	64.24	1.47	45.09	11.30	72.52	48.96
As	2.15	0.52	0.32	3.23	60.63	0.10	0.19	4.41
Rb	1.05	75.99	52.87	0.11	31.78	85.81	9.40	35.46
Sr	10.69	119.96	3.63	0.58	8.33	12.04	116.82	49.46
Y	0.69	22.61	1.14	0.67	5.18	2.77	13.52	10.52
Zr	4.88	72.95	6.36	2.13	89.31	99.24	39.02	29.11
Nb	0.46	4.76	0.18	0.10	5.41	6.94	2.56	1.76
Ba	70.22	238.54	122.20	2.37	67.30	86.83	33.62	69.73
La	0.34	6.12	0.39	0.60	5.12	8.69	3.96	2.11
Ce	0.80	15.14	0.90	0.59	10.00	19.06	9.58	4.47

(*continued on next page*)

and carbonatization cannot be determined with absolute certainty, although intergrowth of carbonate and quartz and the lack of unequivocal overprinting relationships strongly suggest that both alteration processes took place at the same time.

Black-and-white banded chert horizons mainly consist of granular microcrystalline quartz to megaquartz, depending on the degree of recrystallization. White chert layers consist of pure granular quartz, whereas black chert bands contain numerous equant to ovoid grains/aggregates of quartz and some carbonaceous matter up to 500 μm in diameter within

Table 5.5-1. (*Continued*)

Locality	Vumankula River (main section)		Witkop Hill				Eastern section	
Sample	NW6/15	NW6/13	NW6/34	NW6/32	NW6/31	NW6/33	NW6/35	NW6/44
Lithology	CBC	KB	K	CBC	GC	Porphyry	KB	KB
m below chert	1.6	5.8					−150	−22.5
Pr	0.11	2.21	0.08	0.18	1.41	1.98	1.40	0.82
Nd	0.45	10.08	0.28	0.81	6.23	6.76	6.60	4.13
Sm	0.13	2.77	0.12	0.15	1.50	1.38	1.73	1.27
Eu	0.05	0.86	0.07	0.04	0.31	0.29	0.59	0.50
Gd	0.11	3.06	0.11	0.15	1.29	1.19	1.99	1.48
Tb	0.02	0.56	0.02	0.02	0.19	0.14	0.34	0.27
Dy	0.11	3.62	0.19	0.14	1.02	0.66	2.16	1.76
Ho	0.02	0.78	0.05	0.03	0.20	0.11	0.45	0.37
Er	0.06	2.17	0.17	0.07	0.54	0.33	1.23	1.02
Tm	0.01	0.34	0.03	0.01	0.08	0.05	0.18	0.16
Yb	0.06	2.15	0.19	0.06	0.47	0.31	1.16	0.99
Lu	0.01	0.33	0.03	0.01	0.07	0.05	0.17	0.15
Hf	0.17	1.93	0.19	0.05	2.15	2.65	1.01	0.79
Ta	0.04	0.30	0.01	0.01	0.33	0.66	0.15	0.10
W	0.05	0.51	0.32	0.09	6.32	13.14	0.09	0.11
Pb	2.52	2.14	0.76	0.48	2.44	1.46	0.96	1.75
Th	0.16	0.71	0.03	0.04	0.38	2.32	0.49	0.16
U	0.24	0.18	0.14	0.10	0.20	0.82	0.12	0.07
ΣREE	2.27	50.17	2.64	2.86	28.41	41.00	31.53	19.49
Eu/Eu*	1.39	0.90	1.82	0.78	0.67	0.68	0.97	1.12
La_N/Sm_N	1.72	1.39	1.98	2.46	2.14	3.95	1.44	1.05
Gd_N/Yb_N	1.43	1.15	0.47	1.99	2.22	3.06	1.39	1.21
La_N/Yb_N	3.78	1.92	1.41	6.74	7.38	18.67	2.30	1.44
Al_2O_3/TiO_2	11.64	14.02	61.09	7.55	2.96	58.16	11.64	10.49
CaO/Al_2O_3	0.33	0.84	0.00	0.13	0.01	0.00	1.80	1.45
Nb/Ta	12.67	16.08	22.46	13.55	16.50	10.50	16.93	17.20
Zr/Hf	29.04	37.76	32.94	45.31	41.50	37.44	38.75	37.08
Y/Ho	32.62	29.10	24.69	25.80	26.13	24.86	30.12	28.23
Cr/Th	113	894	90317	242	3978	8	3087	5875
Th/Sc	0.307	0.023	0.006	0.124	0.120	1.427	0.016	0.006

(*continued on next page*)

the quartz matrix (Fig. 5.5-10(c)). Carbonaceous chert that fills veins below bedded black chert horizons is petrographically very similar to the bedded chert (Fig. 5.5-10(d)). Accretionary lapilli and glass shards have been observed in some black chert horizons (Versfeld, 1988). Green chert mainly consists of granular quartz, but green, probably Cr-rich sericitic mica, and zircon are also common. Green cherts are either homogeneous or comprise silicified accretionary lapilli and ash particles, indicating a volcaniclastic origin of these rocks.

Table 5.5-1. (*Continued*)

Locality	Eastern section						
Sample	NW6/43	NW6/42	NW6/41	NW6/40	NW6/39	NW6/38	NW6/37
Lithology	KB	KB	KB	CVC	GC	CBC	K, weathered
m below chert	−17.5	−7	−2	−1.5	0.4	7.5	9.5
SiO_2	89.27	80.25	80.22	96.12	96.06	96.95	46.63
Al_2O_3	2.90	6.52	6.54	1.01	0.95	0.54	7.03
Fe_2O_3	1.40	2.54	2.74	0.83	0.87	0.71	14.17
MnO	0.15	0.20	0.29	0.01	0.02	0.01	0.21
MgO	1.46	2.24	1.94	0.20	0.24	0.08	17.66
CaO	2.42	4.24	5.59	0.43	0.35	0.21	8.28
Na_2O	0.71	0.53	0.06	0.00	0.04	0.00	0.27
K_2O	0.38	0.93	1.13	0.26	0.34	0.18	0.02
TiO_2	0.28	0.61	0.47	0.02	0.48	0.05	0.61
P_2O_5	0.03	0.05	0.01	0.00	0.02	0.00	0.07
LOI	0.57	1.18	0.86	0.62	0.18	0.45	5.05
Total	99.57	99.30	99.87	99.50	99.54	99.17	99.99
Sc	8.21	28.26	20.58	3.03	6.04	0.61	32.11
V	63.66	156.88	120.42	11.54	36.71	6.30	167.76
Cr	993.80	1778.89	1734.67	149.44	386.30	37.86	1960.61
Co	27.67	74.15	112.36	37.54	43.11	18.00	87.47
Ni	161.91	269.76	276.45	133.89	120.96	119.81	727.68
Cu	35.23	69.14	50.54	50.17	53.78	22.14	3.94
Zn	14.46	28.42	34.78	189.65	27.20	103.83	74.92
As	11.87	0.29	16.65	4.29	6.00	2.02	0.26
Rb	23.31	43.21	31.97	11.70	12.76	6.65	0.31
Sr	22.82	38.96	54.04	2.55	4.03	2.16	17.52
Y	6.17	11.22	8.07	0.19	2.78	0.45	14.40
Zr	19.29	38.60	29.53	3.24	30.21	7.88	42.06
Nb	0.92	2.14	1.60	0.12	5.53	0.48	2.44
Ba	42.27	38.89	98.85	11.12	10.46	9.64	2.07
La	1.26	2.28	1.78	0.06	5.04	0.22	3.45
Ce	2.27	5.60	2.69	0.12	13.14	0.59	5.56

(*continued on next page*)

The silicified rhyolite of Witkop Hill consists of rounded and corroded quartz phenocrysts and sericite-replaced K-feldspar phenocrysts in a matrix of granular quartz and sericite.

5.5-6. DISCUSSION

5.5-6.1. *Compositional Changes in Silica Alteration Zones*

The silicification of volcanic rocks in zones immediately below chert beds that show an increase in the intensity of silicification stratigraphically upward towards the contact with

Table 5.5-1. (*Continued*)

Locality	Eastern section						
Sample	NW6/43	NW6/42	NW6/41	NW6/40	NW6/39	NW6/38	NW6/37
Lithology	KB	KB	KB	CVC	GC	CBC	K, weathered
m below chert	−17.5	−7	−2	−1.5	0.4	7.5	9.5
Pr	0.43	0.97	0.53	0.02	1.83	0.08	1.25
Nd	2.17	4.84	2.59	0.09	7.56	0.34	5.99
Sm	0.65	1.49	0.76	0.03	1.52	0.08	1.64
Eu	0.23	0.51	0.33	0.03	0.41	0.04	0.72
Gd	0.80	1.65	0.90	0.04	1.35	0.09	1.90
Tb	0.14	0.31	0.17	0.01	0.18	0.01	0.35
Dy	0.89	2.01	1.17	0.03	0.82	0.08	2.23
Ho	0.20	0.44	0.27	0.00	0.13	0.02	0.47
Er	0.56	1.18	0.75	0.01	0.29	0.05	1.27
Tm	0.09	0.19	0.12	0.00	0.03	0.01	0.20
Yb	0.55	1.15	0.74	0.01	0.16	0.04	1.21
Lu	0.08	0.17	0.11	0.00	0.02	0.00	0.18
Hf	0.53	1.00	0.86	0.07	0.77	0.19	1.05
Ta	0.06	0.14	0.11	0.01	0.37	0.04	0.14
W	0.17	0.36	0.37	0.12	0.47	0.20	0.11
Pb	0.55	1.23	2.11	0.46	0.83	1.02	0.39
Th	0.09	0.23	0.16	0.05	0.41	0.13	0.23
U	0.07	0.05	0.08	0.09	0.16	0.13	0.08
ΣREE	10.32	22.77	12.91	0.44	32.47	1.65	26.40
Eu/Eu*	0.98	0.99	1.22	2.35	0.87	1.40	1.24
La_N/Sm_N	1.21	0.96	1.48	1.06	2.09	1.68	1.33
Gd_N/Yb_N	1.16	1.16	0.99	3.45	6.98	1.69	1.27
La_N/Yb_N	1.53	1.34	1.63	4.40	21.72	3.52	1.92
Al_2O_3/TiO_2	10.52	10.63	13.80	44.93	1.98	10.98	11.59
CaO/Al_2O_3	0.83	0.65	0.86	0.42	0.37	0.39	1.18
Nb/Ta	15.08	15.16	15.17	12.11	14.93	13.67	17.11
Zr/Hf	36.59	38.77	34.47	47.19	39.14	41.80	40.09
Y/Ho	31.08	25.72	30.32	54.85	21.57	28.77	30.48
Cr/Th	10725	7774	10537	3138	943	297	8359
Th/Sc	0.011	0.008	0.008	0.016	0.068	0.210	0.007

K, komatiite; KB, komatiitic basalt; pB, pillow basalt; mB, massive basalt; GC, green chert; CBC, carbonaceous banded chert; CVC, carbonaceous vein chert.

the chert, indicates a link between the silicification process and the formation of the bedded chert horizons. Silica alteration zones are found below cherts through the entire sequence of the Witkop Formation and are also common in the Magongolozi Formation.

The silica alteration zones are characterized by a gradual increase in SiO_2 content from the original igneous values ($\sim$50–57%) up to 89%, in zones tens of metres thick. There

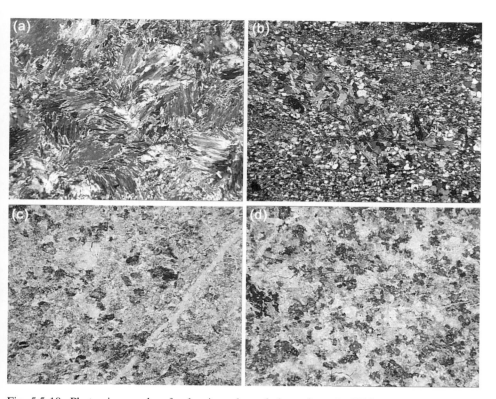

Fig. 5.5-10. Photomicrographs of volcanic rocks and cherts from the Witkop Formation. (a) Komatiitic basalt unaffected by silicification (sample NW/6/35), consisting mainly of actinolite and exhibiting a variolitic texture. (b) Highly silicified komatiitic basalt (sample NW/6/43). (c) Layer of carbonaceous chert in black-and-white banded chert (sample NW/6/30). Note disseminated aggregates containing carbonaceous matter (100–200 μm in diameter) in a groundmass of recrystallized microquartz (30–60 μm in diameter). (d) Carbonaceous vein chert located 14.7 metres below the bedded chert horizon depicted in (c). Note that both cherts are petrographically indistinguishable. In all photomicrographs the field of view has a width of 3 mm; (a,b) cross-polarized and (c,d) plane-polarized light.

are no significant changes in immobile element ratios in genetically related sample suites. Enrichment of SiO_2 is associated with a depletion of Fe_2O_3, MgO, and CaO. K_2O, together with Rb and Ba, are strongly enriched in these zones, whereas Na_2O is slightly enriched. Transition metal contents were mainly unaffected by the alteration, although Cr shows a tendency for enrichment, while Co was depleted. Zn was also lost during silicification. However, the true nature of relative changes of these elements is uncertain, as they show considerable variation in concentration in igneous rocks. The HFSE and REE were mostly immobile, although LREE may have been slightly depleted relative to the HREE. Other elements that were affected during silica alteration include Sr, which was depleted, and As, which was the element most enriched during alteration.

Carbonatization is mainly associated with alteration of ultramafic lithologies. Our data are insufficient to fully explore element changes and redistributions during this process, although compositional changes in carbonatized samples are very similar to those observed for silicification, as the two alteration processes were closely related. Enrichments of CaO was observed in this study, in contrast to Versfeld (1988), who observed depletion of CaO. This discrepancy can be accounted for if the release of Ca from some portions of volcanic rocks during silicification was precipitated in carbonates elsewhere in the volcanic pile, such as in void spaces of the altered volcanic rocks. Redistribution of Ca is in line with the findings of Nakamura and Kato (2004), who did not identify a net loss of Ca during seafloor carbonatization of 3.46 Ga basalts of the Pilbara Craton. The carbonate may be a sink for other elements as well (e.g., Fe, Mg). Therefore, the regional (3-D) extent of carbonatization versus silicification needs to be established before element fluxes between seawater and oceanic crust can be fully established.

5.5-6.2. *Origin of Bedded Cherts*

Black-and-white banded chert horizons originated as sediments that were deposited on the seafloor and later silicified by circulating hydrothermal fluids. Therefore, these cherts do not represent pure chemical sedimentary rocks, but consist of clastic, tuffaceous and, possibly, biogenic material, which is surrounded and partially replaced by chemically precipitated chert. Versfeld (1988) locally observed ripple cross-lamination and flaser bedding (Fig. 5.5-2(e)), which were regarded to indicate deposition of the chert precursor in a shallow-water environment. Black-and-white banded cherts show REE patterns and immobile trace element ratios similar to interbedded silicified mafic to ultramafic volcanic rocks, suggesting that they contain a detrital and/or tuffaceous protolith component similar in composition to such rocks. Slight differences from unaltered mafic/ultramafic protoliths do occur however, commonly in the form of elevated Cr/Th and Th/Sc ratios, which may indicate minor contributions of ultramafic and felsic tuffaceous material. In addition, these rocks contain abundant carbonaceous matter that is identical in appearance to that found in cherts of a similar age, such as in the Barberton Greenstone Belt, and is most likely of biogenic origin (e.g., Walsh and Lowe, 1999; Hofmann and Bolhar, 2007). Most carbonaceous cherts also show positive Eu anomalies and elevated Y/Ho ratios, which is typical for Archaean marine precipitates (e.g., Bolhar et al. 2004; Hofmann, 2005b). The seawater signal is, however, obscured by the presence of clastic/tuffaceous detritus in the cherts.

The common presence of accretionary lapilli in green cherts indicates their pyroclastic derivation. The distinct REE patterns for green cherts relative to the carbonaceous cherts indicate their derivation from intermediate volcanic material (possibly andesitic in composition), whereas high Cr contents also point to an ultramafic component. More samples of this chert type would need to be analysed in order to obtain a more firm understanding of its origin.

Both carbonaceous and green cherts also have relatively high alkali contents and are enriched in Cu and Zn. These metals were probably derived from the underlying volcanic rocks, as they generally were depleted of these metals during silicification.

5.5-6.3. Origin of Chert Veins

Most chert veins cut across volcanic rocks stratigraphically beneath chert beds. They can also be found in bedded chert horizons, but are absent in overlying rocks. Chert veins are mainly of two types: stratiform veins of pure, translucent chert, and cross-cutting veins of carbonaceous chert. The veins provide direct evidence for the formation of the silica alteration zones immediately below the seafloor and at very shallow burial depths. Carbonaceous vein chert contains sand-sized carbonaceous grains that are petrographically and chemically similar to carbonaceous material in the overlying chert beds (Figs. 5.5-6, 5.5-10(c,d)). The geometry of the veins, and evidence of hydraulic fracturing and wall rock replacement, suggests that they formed as a result of hydrothermal activity from hydrothermal fluids flowing upwards through the footwall. Syndepositional silicification of seafloor sediments resulted in the formation of impermeable barriers for ascending hydrothermal fluids. Buildup of hydrothermal fluid pressure eventually resulted in breaching and brecciation of these cap rocks, as indicated by the local presence of chert veins within the sedimentary units. This resulted in the formation of fractures in the alteration zones that were then infilled by pure chert that precipitated out of the hydrothermal fluid or by not yet silicified carbonaceous sedimentary material flowing downward (Lowe and Byerly, 1986a; Hofmann and Bolhar, 2007). Interaction between hot hydrothermal fluids and colder, sediment-ladden seawater probably resulted in further hydrothermal brecciation of vein material and host rocks. The capping cherts were not, however, entirely solidified, as indicated by local ptygmatic folding of carbonaceous chert veins locally (Fig. 5.5-5(f)), although the folding could also be explained as a result of later pressure solution of the chert beds parallel to a bedding-parallel foliation.

Silicification of volcanic rocks is frequently associated with chert veining, as the veins become more common upwards, towards capping chert horizons. Mutual cross-cutting relationships between carbonaceous and translucent chert veins observed in both the Barberton and Nondweni greenstone belts indicate that these chert types formed contemporaneously. These observations also suggest that most, if not all, of the silicification of the extrusive lava sequences and the overlying sedimentary horizons took place contemporaneously, associated with seafloor hydrothermal activity that led to the formation of chert veins.

5.5-6.4. Silicification and Seafloor Alteration

Submarine alteration of basaltic rocks is a characteristic feature of modern oceanic crust. Two types of alteration are generally distinguished: low-temperature ($< \sim 150\,°C$) basalt–seawater interaction, which preferentially affects the uppermost few hundred metres of oceanic crust, and high-temperature ($> \sim 150\,°C$) hydrothermal alteration related to hydrothermal systems near areas of magmatic activity, such as along mid-ocean ridges (Staudigel, 2003; German and von Damm, 2003). Alkali metals, such as K, are released from basalts at temperatures of $150\,°C$ and above (Seyfried, 1987), whereas the solubility of most other metals increases significantly only at temperatures above $350\,°C$ (Seewald and Seyfried, 1990). Most metals are leached from basalt during high-temperature hydrothermal alteration and interaction with up to ~400 °C fluids, whereas Mg and some-times Na are added. In contrast, low-temperature alteration generally results in a gain of alkali metals. However, element fluxes can be variable at different depths in crustal sections (Staudigel et al., 1996).

In the Nondweni greenstone belt, MgO has been lost from the alteration zones rather than added, while K has been enriched. The alteration of basaltic glass to palagonite in-volves uptake of K, Rb, and Cs and results in depletion of Na, Ca, and Mg (Staudigel and Hart, 1983). Fe is relatively immobile during sea floor alteration (Staudigel, 2003), although it is strongly enriched in seafloor hydrothermal fluids (German and von Damm, 2003). The large degree of depletion of Fe in the Nondweni alteration zones point to the more reducing conditions of Archaean seawater (cf. Van Kranendonk et al., 2003), which may have been able to leach Fe efficiently, even under relatively low-temperature condi-tions.

The loss of most elements with the addition of, and dilution by, SiO_2 is not surprising, but it means that large volumes of fluids must have passed through the sequence to account for this substantial change in bulk chemistry. There are a number of possible sources for the silica in the alteration zones, including mineral breakdown reactions in the subsurface. However, much of the silica was probably derived from seawater, as Archaean seawater had a much higher level of dissolved silica, because of the absence of silica-secreting organisms at that time (Siever, 1992). REE patterns of highly silicified komatiites from Barberton are consistent with seawater as the source of the silica (Hofmann, 2005a), which is also in agreement with the submarine depositional setting of the greenstone succession. The silica may have been precipitated as a result of a decrease in silica solubility, when hydrothermal fluids cooled during ascent through the crust or mixed with much cooler ocean waters near the seafloor, probably coupled with sudden changes in Eh and pH conditions.

The enrichment of As in the alteration zones also points to an hydrothermal environ-ment, as this element is commonly enriched in hydrothermal fluids. Arsenic contents are generally in the order of 1–10 ppm in common igneous and sedimentary rocks (Price and Pichler, 2006), whereas the highest value found in Nondweni samples was 61 ppm in a green chert, indicating significant enrichment (Fig. 5.5-7(a,c)). Arsenic was probably leached from rocks in the subsurface during hydrothermal circulation and then reprecipi-tated together with the silica near the seafloor. Arsenic is a biologically toxic element, and its presence in hydrothermal fluids of the Archaean seafloor may have represented a lim-iting factor for early microbial life forms. However, some modern microbes have evolved various detoxification strategies, while some chemoautotrophs that obtain energy through oxidation or reduction of As have been described in several hydrothermal systems (Plant et al., 2003). Similar chemoautotrophs may well have inhabited the hydrothermal systems that are preserved in the Nondweni greenstone belt.

5.5-6.5. Significance of Sea Floor Alteration for the Early Earth

The type of silicification described for the Nondweni greenstone belt seems to be char-acteristic of the pre-Mesoarchaean rock record. Silicification is pervasive in the Barberton

Greenstone Belt (Hofmann, 2005a), and shows identical distributions in greenstone belts of the Pilbara Craton (Kitajima et al., 2001; Van Kranendonk and Pirajno, 2004; Van Kranendonk, 2006), to an extent that the relationship between chert and underlying silica alteration zones can be used as a reliable stratigraphic way-up indicator (Versfeld, 1988). Silicification of volcanic rocks is also common feature in the Eoarchaean of Greenland (Fedo and Whitehouse, 2002). Neoarchaean greenstone belts do not show such pervasive silicification. This clearly suggests secular changes in the processes that led to silicification.

In Neoarchaean greenstone belts, silicification is commonly associated with base metal deposits, such as in the Noranda area in Canada (Gibson et al., 1983). In this area, strong silicification of andesitic lava flows is confined to the upper part of a submarine, extrusive volcanic sequence and is overlain by massive sulphide deposits, suggesting relatively high-temperature, black smoker-type submarine hydrothermal activity. Silicification is also common in modern submarine hot springs, such as the Trans-Atlantic Geotraverse (TAG; Humphries and Tivey, 2000) and the Sea Cliff (Zierenberg et al., 1995) hydrothermal fields, and develops either in the relatively low temperature ($<100\,^{\circ}$C) part of the hydrothermal system, or during a low-temperature stage in its history.

Hydrothermal processes and the circulation of seawater through the uppermost part of the crust took place during deposition of the bedded chert precursor and resulted in silicification of sedimentary and volcanic material immediately below the sea floor, in a way very similar to what has been described from Barberton (Hofmann, 2005a; Hofmann and Bolhar, 2007) and also the Pilbara (Van Kranendonk, 2006). Fluid convection was probably halted shortly after burial by a new series of volcanic lava flows. This is because the cherts would have acted as a barrier for cold, descending seawater, but in time new hydrothermal convection cells would have developed again in the overlying volcanic rocks, leading to a repeated cycle.

The Archaean hydrothermal fluid was heated seawater, which was enriched in elements leached from the underlying volcanic rocks. The element depletion–enrichment patterns of the alteration zones indicate relatively low-temperature hydrothermal activity. Oxygen isotope data of silica alteration zones in Barberton indicate temperatures in the order of 100–150 $^{\circ}$C (Hofmann et al., 2006). Such relatively low temperatures are in line with the absence of focused hydrothermal upflow zones and the lack of massive sulphides associated with these zones throughout the mafic volcanic sequence of the Nondweni greenstone belt and Barberton, suggesting diffuse upflow of hydrothermal fluids through the volcanic rocks. There is also no evidence that the volcanic rocks of the Nondweni and Onverwacht Groups formed close to an Archaean mid-oceanic ridge. No sheeted dykes have been observed in any Archaean greenstone sequence suggestive of ocean floor spreading. The alteration zones are laterally extensive and tabular, bearing no resemblance to the heavily faulted spreading centres along mid-ocean ridges. Focused discharge zones in the form of epidosites and black smoker-type sulphide deposits are absent. Instead, the Nondweni and Onverwacht Groups are characterized by laterally extensive, submarine lava flows more akin to large submarine shield volcanoes or oceanic plateaus, as also inferred for the Pilbara Craton (Van Kranendonk and Pirajno, 2004; Smithies et al., 2005b).

Each volcanic sequence that is capped by chert is altered at the top. This relationship indicates that the alteration took place during a time interval of relative volcanic quiescence, when there was enough time for the (probably rather slow) deposition of sediments and pervasive silicification before the onset of the next volcanic episode. A difference between Nondweni and Barberton Greenstone Belts is the thickness of the chert-bounded lava packages. While the Onverwacht Group does contain ~15 chert horizons, each bounding a volcanic package in the order of tens to hundreds of metres in thickness, chert horizons are much more abundant in the Nondweni Group and bound lava sequences that are generally only tens of metres thick, which suggests the Nondweni Group to represent a more distal volcanic pile relative to the lava source. The most likely heat source for the hydrothermal processes include heat retained in the probably rapidly extruded lava sequences, and heat conducted into the crust from the lithospheric mantle, i.e., the regional heat flow. Heat from subvolcanic intrusions (cf. Van Kranendonk, 2006) can be discounted as a likely heat source, at least for most alteration zones, as they cannot be directly linked to such intrusions. Because many Neoarchaean greenstone sequences contain equally thick submarine lava plain sequences that are locally covered by thin sedimentary units (e.g., Belingwe greenstone belt; Hofmann et al., 2003), but generally lack the alteration zones described from Nondweni, Barberton and the Pilbara, the regional heat flow must have been much higher during the formation of these Palaeoarchaean sequences (cf. de Wit and Hart, 1993).

5.5-7. CONCLUSIONS

Rocks of Palaeo- to Mesoarchaean age in South Africa are exposed from the Barberton greenstone to the southern margin of the Kaapval Craton. The Barberton Greenstone Belt has emerged as one of the classic Archaean geological provinces worldwide in which to study the nature and processes of early Earth history. Much less research has been done on the smaller greenstone belts south of Barberton. These belts show lithostratigraphic successions and ages similar to the rocks in the Barberton Greenstone Belt, and their study provides an opportunity to investigate Archaean geological processes at a craton scale.

The Witkop Formation of the Nondweni greenstone belt consists of a several kilometres thick pile of submarine ultramafic to mafic lava flows. The volcanic rocks are interbedded with chert horizons, typically a few metres thick, which represent silicified sediments that accumulated on the seafloor during times of volcanic quiescence. Each chert horizon is underlain by a 20–50 metres thick zone of silicified volcanic rocks, where SiO_2 contents increase upsection, in many cases up to 90 wt% SiO_2. Silicification is associated with a depletion of many elements. K_2O (+Rb, Ba) and As are enriched in these zones. The element depletion–enrichment patterns of the alteration zones indicate low-temperature ($<150\,°C$) hydrothermal processes for the origin of the alteration zones. Alteration took place immediately beneath the Archaean seafloor by heated seawater that moved in shallow subseafloor convection cells driven most likely by regional high heat flow. Rapid cooling of the fluid during mixing with colder seawater resulted in a decreased silica solubility and related silicification of volcanic rocks and overlying sea floor sediments, forming the silica alteration zones and bedded cherts. Sediment silicification gave rise to the formation of impermeable caps that were fractured by subsequent pulses of overpressured fluids, which resulted in the formation of chert-filled hydrothermal fractures within and below the silicified sediment horizons.

As the silica alteration zones are characteristic of pre-Mesoarchaean volcano-sedimentary successions – they are always present beneath interflow sedimentary units of mafic to ultramafic volcanic lava piles – they can provide important information on the geological processes acting on the early Earth. For example, the depletion of Fe from the alteration zones is unlike modern day seafloor alteration (see Hanor and Duchac (1990) for the Barberton Greenstone Belt). This indicates that the Archaean ocean floor was a major source for dissolved iron which was removed by Archaean seawater. A more detailed study of the silica alteration zones, coupled with information on the distribution of other types of alteration, such as carbonate alteration, may provide data on element fluxes to the Archaean hydrosphere during basalt-seawater interaction and, thus, on the chemistry of the Archaean oceans.

It should also be noted that the silica alteration zones developed in the Barberton Greenstone Belt and elsewhere are direct evidence for the emanation of warm to hot fluids onto the Archaean seafloor. Hydrothermal activity in the Archaean was not restricted to small areas near mid-oceanic ridges (which have not been identified in the Archaean rock record), but was distributed quite widely on the seafloor. This would have been a suitable environment for hyperthermophiles during the Palaeoarchaean and possibly the birthplace of life in earlier times.

ACKNOWLEDGEMENTS

This work was supported by a competitive research grant from the University of KwaZulu-Natal and a focus area grant from the National Research Foundation of South Africa. Reviews by Angela Riganti and Martin Van Kranendonk are gratefully acknowledged.

Earth's Oldest Rocks
Edited by Martin J. Van Kranendonk, R. Hugh Smithies and Vickie C. Bennett
Developments in Precambrian Geology, Vol. 15 (K.C. Condie, Series Editor)
© 2007 Elsevier B.V. All rights reserved.
DOI: 10.1016/S0166-2635(07)15056-8

607

Chapter 5.6

TTG PLUTONS OF THE BARBERTON GRANITOID-GREENSTONE TERRAIN, SOUTH AFRICA

JEAN-FRANÇOIS MOYEN, GARY STEVENS, ALEXANDER F.M. KISTERS AND RICHARD W. BELCHER[*]

Department of Geology, Geography and Environmental Science, University of Stellenbosch, Private Bag X 01, Matieland 7602, South Africa

5.6-1. INTRODUCTION

Plutonic rocks constitute a large part of Archean terranes and occur mostly in the form of variably deformed orthogneisses. The most common plutonic rocks are a suite of sodic and plagioclase-rich igneous rocks made of tonalites, trondhjemites and granodiorites, collectively referred to as the "TTG" suite. A large body of geochemical and experimental data exists for TTGs, and these studies have led to the general conclusion that TTGs are essentially melts generated by partial melting of mafic rocks, mostly amphibolites (as the dominant melting reaction involves hornblende breakdown) within the garnet stability field. However, the geodynamic setting for the origin of TTGs is still debated and contrasting interpretations are proposed, the most common being melting of the down-going slab in a 'hot' subduction zone setting (e.g., Arth and Hanson, 1975; Moorbath, 1975; Barker and Arth, 1976; Barker, 1979; Condie, 1981; Jahn et al., 1981; Condie, 1986; Martin, 1986, 1994, 1999; Rapp et al., 1991; Rapp and Watson, 1995; Foley et al., 2002; Martin et al., 2005), and melting of the lower part of a thick, mafic crust in an intra-plate settings (e.g., Maaløe, 1982; Kay and Kay, 1991; Collins et al., 1998; Zegers and Van Keken, 2001; Van Kranendonk et al., 2004; Bédard, 2006).

In many cases, TTGs are the oldest component of Archean cratons. They generally appear as polyphase deformed gneissic complexes, commonly referred to as "grey gneisses", which display variable degrees of migmatization. In such units, high finite strains and the tectonic transposition of different TTG phases obscuring original igneous contacts, renders the recognition of original protoliths difficult and detailed geochemical studies on individual magmatic intrusions are not possible. However, in the Barberton granitoid-greenstone terrain (BGGT)[1], many of the TTGs are characterized by weak fabrics and low

---

[*]Present address: Council for Geoscience, Limpopo Unit, P.O. Box 620, Polokwane 0700, South Africa

[1]In this paper, we use "Barberton Granitoid-Greenstone Terrain" (BGGT) as an encompassing term to refer to the whole area of Archean outcrops (plutons and supracrustals), as opposed to the "Barberton belt" *stricto sensu*, that refers only to the supracrustal association.

strain intensities, therefore allowing detailed study of their intrusive relationships, original compositions and comprehensive petrogenesis.

TTGs from the BGGT range in age from ca. 3.55 to 3.21 Ga and the relationship between the greenstone belt and the surrounding TTG "plutons" is complex. The apparent domal pattern of TTG gneisses in tectonic contact with the overlying supracrustal greenstone belt is actually an oversimplification. In fact, each of the "plutons" has its own, distinct emplacement and deformational history (summarized in Table 5.6-1), with some of the "plutons" corresponding to relatively simple magmatic intrusive bodies, whereas others are composite units with complex and protracted emplacement and structural histories and are not really "plutons" in the classical sense. Likewise, the TTGs also have distinct petrological and geochemical natures, and while they all broadly belong to the "TTG" group, are actually petrologically and geochemically complex. Such a diversity points to different petrogenetic histories related to different geodynamic settings.

The TTGs of the BGGT can be divided in to at least two "sub-series": (i) a "low-Sr", commonly tonalitic sub-series; and (ii) a "high-Sr", commonly trondhjemitic sub-series. In most Archean cratons, tonalites and trondhjemites are typically associated in highly strained grey gneiss complexes, which are tectonically interleaved on a mm- to dm-scale, to such a degree that it gives the impression that both lithologies reflect only minor differences in terms of petrogenetic processes. In contrast, in the BGGT tonalites and trondhjemites occur as distinct intrusive bodies with well-defined margins and intrusive contact relationships. This allows their petrogenetic evolution to be studied independently from one another.

In this paper, we demonstrate that the tonalitic and trondhjemitic bodies reflect two fundamentally different magma types, with different origins and evolutions. We propose that the two distinct TTG "sub-series" of the BGGT could reflect the results of two geodynamic environments important in the formation of Archean TTG's, namely formation at the base of a thickened crust, and derivation from a subducting slab.

5.6-2. GEOLOGICAL SETTING

The BGGT formed between ca. 3.51 and 3.11 Ga.[2] Although supracrustal rocks (lavas and sediments) from the belt itself yield a relatively continuous spread of ages from 3559 ± 27 Ma (Byerly et al., 1996; Poujol et al., 2003; Poujol, this volume) to 3164 ± 12 Ma (Armstrong et al., 1990; Poujol et al., 2003), the BGGT predominantly assembled during three or four discrete tectono-magmatic events (Poujol et al., 2003) at 3.55–3.49, 3.49–3.42, 3.255–3.225 and 3.105–3.07 Ga (see also Lowe and Byerly, this volume).

The first two events (3.49–3.55 and 3.42–3.49 Ga) are well represented in the Ancient Gneiss Complex to the east (Kröner, this volume). However, in the BGGT proper, >3.42 Ga rocks are restricted to the high-grade "Stolzburg domain" (Fig. 5.6-1) (Kisters et

---

[2] Ages indicated in millions of years (Ma) correspond to actual, measured ages with reference and error, while dates given in billions of years (Ga) refer to generalized time intervals.

Table 5.6-1. Main field characteristic and ages of Barberton TTG plutons

Pluton	Surface (km^2)	Age (Ma)	Characteristics and emplacement features
1) ca. 3.5 Ga generation			
Steynsdorp pluton	15	3553 ± 4 to 3490 ± 4 (Kröner et al., 1996)	Foliated plutons, transposed contact with greenstones. Two gneissic units (tonalite and granodiorite)
Vlaakplats granodiorite (intrusive in the Steynsdorp pluton)	<1	3540 ± 3 (Kröner et al., 1996)[a]	Fine grained gradioriorite intrusive into Steynsdorp
Elements of the Ngwane gneisses (in the ACG of Swaziland)	~2500	3683 ± 10 (Kröner et al., 1996) to 3213 ± 10 (Kröner et al., 1993)[b]	Tonalitic to trondhjemitic orthogneisses, interlayered with metasediments
2) ca. 3.45 Ga generation			
Stolzburg pluton	~80	3445 ± 3 (Kröner et al., 1991); 3431 ± 11 (Dziggel et al., 2002); 3460 ± 5 (Kamo and Davis, 1994)	Leucocratic trondhjemite, medium to coarse grained. Intrusive (with intrusive breccias and dyke swarm) into the greenstone belt, contact deformed during the 3.2 Ga events
Theespruit pluton	42	3443 ± 4 (Kamo and Davis, 1994); 3440 ± 5 (Kröner et al., 1991); 3437 ± 6 (Armstrong et al., 1990)	The same
Small southern plutons (Theeboom, Eerstehoek, …)	>90	*No published age – probably similar to Theespruit and Stolzburg*	The same
Doornhoek	3.6	3448 ± 4 (Kamo and Davis, 1994)	
Elements of the Ngwane gneisses (in the ACG of Swaziland)	~2500	3683 ± 10 (Kröner et al., 1996) to 3213 ± 10 (Kröner et al., 1993)	Tonalitic to trondhjemitic orthogneisses, interlayered with metasediments

Table 5.6-1. (*Continued*)

Pluton	Surface (km²)	Age (Ma)	Characteristics and emplacement features
Tsawela gneisses (in the ACG)	~150[c]	3455 ± 3 (York et al., 1989) to 3436 ± 6 (Kröner et al., 1993)	Tonalitic orthogneisses, forming a mappable, relatively homogeneous unit in the Ngwane gneisses
3) 3.23–3.21 Ga generation			
Kaap Valley pluton	780	3229 ± 5 (Tegtmeyer and Kröner, 1987); 3227 ± (Kamo and Davis, 1994); 3223 ± 4 and 3226 ± 5 (Layer et al., 1992); 3226 ± 14 (Armstrong et al., 1990)	Dark, coarse-grained, amphibole bearing tonalite. Probably emplaced as a sub-concordant laccolith into the greenstone belt
Nelshoogte pluton	~320	3236 ± 1 (de Ronde and Kamo, 2000); 3212 ± 2 (York et al., 1989)	Composite pluton, dominated by coarse grained, leucocratic trondhjemite Syn- to post tectonically emplaced as a laccolith into the greenstone belt
Badplaas gneisses	~160	3290–3240 Ma (Kisters et al., 2006); Poujol (pers. comm.)	Polyphased gneiss domain, emplaced during a ca. 50 Ma period, made of a variety of mutually intrusive, diversely deformed phases

ACG = Ancient Gneiss Complex. See Kröner et al. (this volume). Surfaces are derived using GIS from the map of Anhaeusser (1981).

[a]With inherited zircons dated at 3702 ± 1 Ma (Kröner et al., 1996).

[b]Gneissic unit, intruded by younger plutons. Probably not only made of orthogneisses, contains some metasedimentary components (Hunter et al., 1978). See Kröner (this volume).

[c]Interlayered with the Ngwane gneisses. Exact extension poorly known.

al., 2003; Moyen et al., 2006; Stevens and Moyen, this volume), which corresponds to the high-grade, "lower" portions of both the "Steynsdorp and Songimvelo terranes[3]" (Lowe, 1994, 1999; Lowe and Byerly, 1999).

5.6-2.1. >3.42 Ga Accretion of the BGGT

The >3.5 Ga event is represented by the mafic and felsic volcanics of the Theespruit Formation (Lowe and Byerly, 1999, this volume, and references therein), which are coeval with the emplacement of the ca. 3.55–3.50 Ga Steynsdorp pluton (Kröner et al., 1996). Little information is available regarding the geological context of their formation.

The 3.42–3.49 Ga event corresponds to the formation of the Komati, Hooggenoeg and Kromberg Formations of the Onverwacht Group (Lowe, 1999b; Lowe and Byerly, 1999, this volume, and references therein), which are mostly located in the lower-grade (upper plate of Kisters et al., 2003) portions of the Songimvelo and Steynsdorp terranes. These three formations are dominantly mafic to ultramafic lavas, with subordinate felsic volcanic rocks and cherts. At the contact between the Hooggenoeg and Kromberg Formations, the ca. 3.44–3.45 Ga "H6" unit (Kröner and Todt, 1988; Armstrong et al., 1990; Kröner et al., 1991a; Byerly et al., 1996) is nearly synchronous with the intrusion of TTG plutons in the Stolzburg domain (Theespruit, Stolzburg, and the minor plutons to the South defined by Anhaeusser et al., 1981). The H6 unit is a thin (few tens of meters) unit of dacitic lava flows and shallow intrusives (geochemically regarded as the extrusive equivalents of the TTG plutons; de Wit et al., 1987) and clastic sediments and conglomerates. This suggests that some topography existed at that stage. The first, well constrained deformation event affecting the belt (D_1) (Lowe et al., 1999) also occurred at about the same time and is interpreted to represent the development of an active margin (oceanic arc) (Lowe, 1999b; de Ronde and Kamo, 2000; Lowe and Byerly, this volume, and references therein) at ca. 3.45 Ga.

Following the D_1 event, the Mendon Formation was deposited in the Stolzburg domain (Songimvelo and Steynsdorp blocks) in the east (Lowe, 1999b), and the Weltvreden Formation in the western terranes, from ca. 3.42 to 3.25 Ga. Based on studies of the volcanic and sedimentary units, a period of quiescence (rift/intracontinental setting) is suggested (Lowe, 1999).

5.6-2.2. Main Orogenic Stage at 3.25–3.21 Ga

The main, "collision" stage (D_{2-5}), occurred between 3.25 and 3.21 Ga. Evidence for an accretionary orogen is presented elsewhere (Lowe and Byerly, this volume; Stevens and Moyen, this volume), and is thus only briefly summarized here. D_2 corresponds to the amalgamation of the various sub-terranes that make up the belt, with the major suture

---

[3]"Terrane" (or "block") is used in this paper to describe a "fault-bounded geological entity with distinct tectonostratigraphic, structural, geochronological and/or metamorphic characteristics from its neighbors (in the sense of Coney et al., 1980)" (Van Kranendonk et al., 1993), as opposed to "terrain", which simply refers to a geographical region or area with no particular tectonic or genetic meaning.

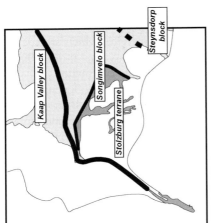

Phanerozoic (Transvaal) cover

ca. 3.1 Ga potassic plutons
(granites & syenites)

BMK: Boesmanskop pluton (syenite)

Dalmein pluton (3215 Ma)

Barberton Greenstone Belt

Greenschist facies

Amphibolite facies

TTGs

ca. 3.2 Ga plutons

ca. 3.45 Ga plutons

**Uit.: Uitgevonden pluton
Hon.: Honingklip pluton
Eers.: Eerstehoek pluton**

ca. 3.55 Ga gneisses

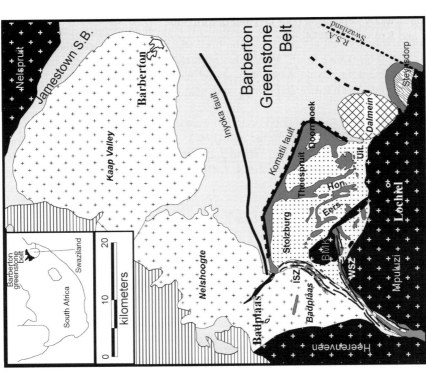

ISZ: Inyoni shear Zone (3.2 Ga - D2)
WSZ: Welverdiend Shear Zone (3.1 Ga - D4)

Fig. 5.6-1. (*Previous page.*) Geological map of the southwestern part of the Barberton Greenstone Belt and surrounding TTG plutons (BGGT). Left: map modified after Anhaeusser et al. (1981). See text and Table 5.6-1 for comments and references. Top right: location map. Bottom right: Structural sketch indicating the position of the main terranes and structures. While the "Songimvelo block" of Lowe (1994) includes part of the Barberton Greenstone Belt, and the adjacent ca. 3.45 Ga plutons in the south, the latter are separated from the former by the Komatii fault, leading to the identification of a distinct "Stolzburg terrane" (Kisters et al., 2003; Kisters et al., 2004; Diener et al., 2005; Diener et al., 2006; Moyen et al., 2006) corresponding to the amphibolite-facies portion of the Songimvelo terrane. The main structure is the Inyoni–Inyoka fault system, separating the western (Kaap Valley block) from the eastern domain (Steynsdorp and Songimvelo blocks, including Stolzburg terrane). Note that the "Onverwacht Group" on both sides of the Inyoka fault actually corresponds to rocks with different stratigraphy and of contrasting ages: 3.3–3.25 Ga to the west, and 3.55–3.3 Ga in the east. Furthermore, the details of the stratigraphic sequences on both sides cannot be correlated, suggesting that the two parts of the belt evolved independently, prior to the accretion along the Inyoka fault (Viljoen and Viljoen, 1969a; Anhaeusser et al., 1981,1983; de Wit et al., 1992; de Ronde and de Wit, 1994; Lowe, 1994; Lowe and Byerly, 1999; Lowe et al., 1999; de Ronde and Kamo, 2000).

---

zone corresponding to the Inyoni–Inyoka fault system (Fig. 5.6-1). Despite the apparently continuous stratigraphy across the fault, the sequences on both sides cannot be correlated (Lowe, 1994, 1999; Lowe et al., 1999; Stevens and Moyen, this volume). The D_2 event is shortly followed by deposition (syn D_3) and deformation (D_4 and D_5) of the <3.22 Ga (Tegtmeyer and Kröner, 1987) Moodies Group conglomerates and sandstones.

The most likely sequence of events for this stage is:

- From ca. 3.25 to 3.23 Ga, syn-tectonic (D_{2a}) deposition of the felsic volcanics and clastic sediments of the Fig Tree Group, probably resulting in the development of a volcanic arc in what is now the terrane west of the Inyoni–Inyoka fault system (Lowe, 1999b; de Ronde and Kamo, 2000; Kisters et al., 2006). The Badplaas gneisses were also emplaced into the western terrane during this period.

- Accretion of the two terranes along the Inyoni–Inyoka fault system at ca. 3.23 Ga (D_{2b}). This was accompanied by high-pressure, low- to medium-temperature metamorphism of the eastern, Stolzburg domain (Dziggel et al., 2002; Diener et al., 2005; Moyen et al., 2006), especially along the fault system, interpreted as a suture zone (Stevens and Moyen, this volume).

- Collision was immediately followed at ca. 3.22–3.21 Ga by extensional collapse of the orogenic pile (Kisters et al., 2003), leading to nearly isothermal exhumation of the high-pressure rocks of the Stolzburg domain along detachment faults (Diener et al., 2005; Moyen et al., 2006) and the emplacement of TTG plutons (Nelshoogte and Kaap Valley plutons). The extension collapse roughly corresponds to the D_3 event of Lowe (1999b) and was synchronous with deposition of (at least part of) the Moodies Group in small, discontinuous, fault-bounded basins (Heubeck and Lowe, 1994a, 1994b). This was immediately followed by diapiric exhumation of the lower crust, and steepening of the fabrics.

– Late, ongoing deformation (D_4–D_5) resulted in strike-slip faulting and folding of the whole sequence (including the Moodies Group). Some late to post-tectonic plutons (e.g., 3215 ± 2 Ma Dalmein pluton; Kamo and Davis, 1994), crosscut all ca. 3.23–3.21 Ga structures.

5.6-2.3. Later Events at ca. 3.1 Ga

At ca. 3.1 Ga, a final orogenic event (not numbered in Lowe's (1999) terminology) resulted in intraplate compression (Belcher and Kisters, 2006a, 2006b) and widespread melting at different crustal levels. This led to the emplacement of voluminous, sheeted potassic batholiths and the development of a network of syn-magmatic shear zones that affected the older "basement" (Westraat et al., 2004). The volumetrically dominant intrusions in the BGGT (Fig. 5.6-1) were emplaced at this time (Maphalala and Kröner, 1993; Kamo and Davis, 1994) and are represented by the Piggs' Peak batholith (east of the BGGT and in Swaziland), Nelspruit batholith (in the north), and the Mpuluzi/Lochiel and Heerenveen batholiths (in the south). Collectively, these rocks are mostly leucogranites, granites and granodiorites, associated with minor monzonites and syenites, and commonly referred to as the "GMS" (granites/granodiorites, monzonites and syenites/syenogranites) suite (Yearron, 2003). Although the GMS suite formed, at least in part, from partial melting of rocks compositionally similar to the 3.5–3.2 Ga rocks of the BGGT (Belcher et al., submitted), the TTG "basement" observed in outcrop across the terrain was unaffected by this melting event.

5.6-3. TTG PLUTONS OF THE BGGT

5.6-3.1. Geology and Field Relationships of TTG Plutons

TTGs of the BGGT belong to three main generations, corresponding to the three geological events outlined above (Table 5.6-1).
– The ca. 3.55–3.50 Ga TTGs, represented by the Steynsdorp pluton (Robb and Anhaeusser, 1983; Kröner et al., 1996), contain a pervasive solid-state gneissosity and occurs mostly as banded gneisses. The protolith of these gneisses is tonalitic (Kisters and Anhaeusser, 1995b; Kröner et al., 1996), although a granodioritic component, possibly related to the remelting of older tonalites or trondhjemites (see below), is also recorded. The Steynsdorp pluton outcrops in a domal antiform (Kisters and Anhaeusser, 1995b), and the contact with the enveloping supracrustals of the Theespruit Formation is tectonic.
– The ca. 3.45 Ga (syn-D_1) TTGs are represented by a number of intrusive bodies in the Stolzburg terrane, located to the south of the main part of the Barberton Greenstone Belt (Viljoen and Viljoen, 1969d; Anhaeusser and Robb, 1980; Robb and Anhaeusser, 1983; Kisters et al., 2003; Moyen et al., 2006). The two most prominent and better defined

intrusions are the Stolzburg and Theespruit plutons. Together with the smaller Doorn-hoek pluton, these plutons intruded the supracrustal rocks of the belt. Further south, several smaller plutons or domains are recognized and form a complex pattern of TTG gneisses and greenstone remnants, partially transposed and dismembered by ca. 3.1 Ga shear zones. These are the Theeboom, Eerstehoek, Honingklip, Weergevonden "cells" and "plutons" of Anhaeusser et al. (1981), see also Robb and Anhaeusser (1983). To the west, the Stolzburg terrane is bounded by the Inyoni shear zone, which is the south-ern extent of the Inyoni–Inyoka fault system. Rocks predominantly from the Stolzburg pluton are foliated and transposed in this shear zone, in a ~500 m wide area. To the north, it is truncated by the extensional detachment corresponding to the Komati Fault (Kisters et al., 2003). Within the terrane, the plutons preserve clear intrusive relation-ships with the surrounding greenstones (Fig. 5.6-2(a)) (Kisters and Anhaeusser, 1995a; Kisters et al., 2003), although the terrane as a whole (granitoids and country rocks) were deformed during D_3 exhumation (Kisters et al., 2003; Diener et al., 2005, 2006; Stevens and Moyen, this volume). The nature of the preserved contacts, which clearly cut across amphibolite-facies foliations (Fig. 5.6-2(a)), the presence of a network of surrounding dykes, and the existence of simultaneous, cogenetic extrusive rocks all suggest that the Stolzburg pluton (and the other plutons of the terrane/domain) intruded under brittle con-ditions, at relatively shallow crustal levels (Kisters and Anhaeusser, 1995a). All the ca. 3.45 Ga plutons are composed predominantly of medium- and/or coarse-grained leuco-trondhjemites (Robb and Anhaeusser, 1983; Kisters and Anhaeusser, 1995a; Yearron, 2003). Minor dioritic dykes are also observed (Yearron, 2003), especially in the margins of the plutons, and in the complex inter-pluton areas.

- The 3.29–3.21 Ga group (syn-D_2 and D_3) of plutons is more composite, and occurs along the northern and southwestern margins of the Barberton Greenstone Belt (Viljoen and Viljoen, 1969d; Anhaeusser and Robb, 1980; Robb and Anhaeusser, 1983).
  - In the south, the 3.29–3.24 Ga (Kisters et al., 2006, Poujol, pers. comm.) Bad-plaas gneisses (and probably the apparently similar Rooihoogte gneisses, west of the 3.1 Ga Heerenveen batholith) are composed of two main suites: an older, coarse-grained leuco-trondhjemitic component that underwent solid-state deformation; and a younger, multiphase intrusive component, made up a variety of typically finer-grained trondhjemites. In proximity to the Inyoni shear zone, most of these intrusions are syn-tectonic: the Batavia pluton, of coarse-grained, leucocratic, porphyritic trondhjemite, is syntectonic in the central part of the Inyoni Shear Zone (Anhaeusser et al., 1981). Further away from the shear zone, the trondhjemites form either irregularly shaped, discontinuous, stockwork-like breccias or small (100 m – 5 km) plugs and intrusions. The long-lived emplacement of the Badplaas pluton, and its composite nature, makes it unique in the BGGT.
  - Further north, the composite 3.23–3.21 Ga Nelshoogte pluton (Anhaeusser et al., 1981, 1983; Robb and Anhaeusser, 1983; Belcher et al., 2005) is dominated by coarse-grained leuco-trondhjemites that are intruded by amphibole-tonalites, particularly

Fig. 5.6-2. Field appearance of the various type of TTG rocks around Barberton Greenstone Belt. (a) Lit-par-lit and cross-cutting intrusive relations between the 3.45 Ga Stolzburg pluton and amphibolites of the Theespruit Formation. (b) Brecciation of Onverwacht Group amphibolites by the 3.23–3.21 Ga Nelshoogte pluton.

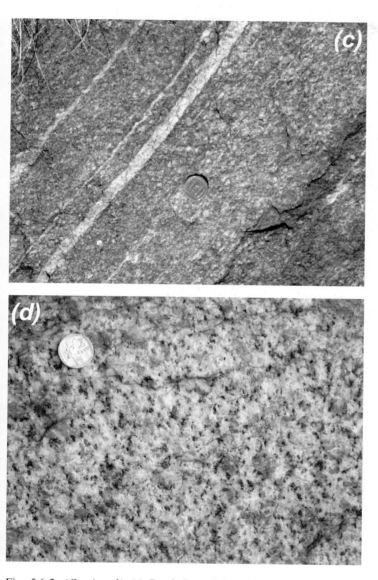

Fig. 5.6-2. (*Continued.*) (c) Banded tonalitic gneisses of the 3.55–3.50 Ga Steynsdorp pluton. (d) Leucocratic coarse-grained trondhjemites from the 3.45 Ga Stolzburg pluton. The Stolzburg pluton shows a pronounced vertical rodding not seen in this image, which is taken on a plane orthogonal to the stretching lineation. Coin for scale in photos (c)–(f).

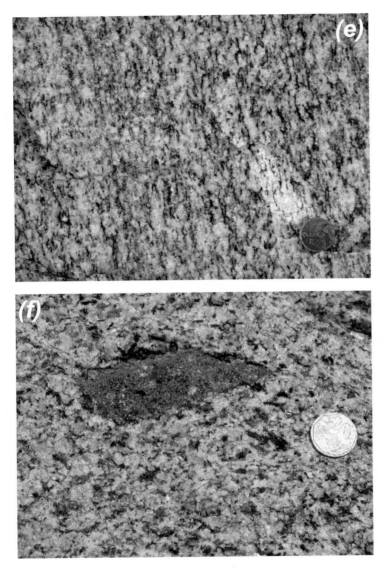

Fig. 5.6-2. (*Continued.*) (e) Trondhjemitic orthogneisses in the 3.23–3.21 Ga Nelsghoogte pluton. (f) Hornblende-bearing tonalites of the 3.23–3.22 Kaap Valley pluton. Microgranular mafic enclaves (Didier and Barbarin, 1991), as seen in this photo, are common. Coin for scale in photos (c)–(f).

along the northern and northeastern margin of the pluton. The pluton was intruded during regional folding, probably as a laccolith, and lit-par-lit intrusive relationships, as well as smaller-scale brecciation with the surrounding greenstone wallrocks, are preserved (Belcher et al., 2005) (Fig. 5.6-2(b)). This is again suggestive of relatively shallow level of emplacement. The domal map pattern reflects late stage folding and steepening of the syn-emplacement, initially flat fabrics.

- The large, 3.23–3.22 Ga Kaap Valley pluton along the northern margin of the Barberton Greenstone Belt is, for the most part, made up of coarse-grained, biotite-amphibole tonalite (Robb et al., 1986), with minor occurrences of amphibole-tonalite (biotite free).

5.6-3.2. The Pristine Character of Barberton TTGs

A rather unique feature of Barberton TTGs is that they represent a group of well-defined, distinct intrusions. Apart from the Badplaas gneisses, they do not constitute a heterogeneous complex of orthogneisses (grey gneisses) like many other TTG complexes that often are polyphase, high strain, transposed and often migmatitic, or even poly-migmatitic, orthogneisses. Although the TTGs around Barberton are all technically gneisses, in the sense that they underwent solid-state deformation after their emplacement (most likely related to the 3.2 Ga D_2–D_3 event: Kisters and Anhaeusser, 1995a; Kisters et al., 2003; Belcher et al., 2005), they still commonly contain original magmatic and emplacement features and textures. In the 3.45 Ga plutons, for instance, deformation occurred 150–200 My after their emplacement and is marked by strong, subvertical rodding (D_3), corresponding to pure coaxial stretching. However, strain intensities are low enough to allow magmatic-looking textures to be preserved, at least in planes perpendicular to the lineation. Likewise, emplacement-related features and intrusive contacts are also occasionally preserved (Kisters and Anhaeusser, 1995a) and deformation did not result in transposition and development of a gneissic fabric, but rather limited textural overprinting along the margin of the plutons and the immediately surrounding wallrocks. Consequently, unlike many grey gneisses terrains in the world, their composition has not been altered by partial melting. The Barberton TTGs thus preserve their true magmatic compositions and present a very good example to study the origin and evolution of TTG magmas (s.s.), as opposed to the geochemistry of grey gneisses complexes (even though these are dominated by TTGs).

Some banded grey gneisses are known in the BGGT. However, they represent part of well-constrained high strain zones, corresponding to 3.23–3.21 Ga (e.g., the Inyoni Shear Zone: Kisters et al., 2004) or 3.1 Ga (e.g., the Weltverdiend Shear Zone: Westraat et al., 2004) tectonic events. Within these zones, the complex orthogneisses are very similar to any other grey gneiss complex in the world, being characterized by transposed, high strain fabrics, amphibolite enclaves, etc. However, the field relationships of the components are obvious, and they clearly formed by deformation of the plutonic rocks and supracrustal remnants.

A second, equally important point is that Barberton TTGs are relatively high-level, intrusive bodies. Although no quantitative data is available, the emplacement mode of all plutons (except, perhaps, part of the Badplaas gneisses) are suggestive of emplacement in the middle or upper crust, under brittle conditions (Kisters and Anhaeusser, 1995a, 1995b; Kisters et al., 2003, 2004; Belcher et al., 2005). Barberton TTG plutons are not migmatitic domes, with liquids and solids still intermingled, nor are they lower-crustal diatexitic bubbles rising diapiricaly. Rather, they are "clean" (purely or mostly magmatic liquids), high-level plutons emplaced sometimes as syn-tectonic magmas and sometimes deformed during subsequent events. As the TTG melts were probably generated at depths greater than 10–12 kbar (see below), this implies that they were emplaced far (at least 15–20 km above) from their source. The present outcrop level is entirely disconnected from the melting domain.

5.6-3.3. Petrology and Mineralogy

Although few Archean geologists would refer to them in this way, TTGs are I-type granites (Chappell and White, 1974) and belong to a calc-alkaline series (Le Bas et al., 1986; Le Maître, 2002). However, they do show significant differences with typical, modern calc-alkaline lavas or arc-related I-type granitoids.

Two main rock-types are represented in the TTG rocks of the BGGT (Fig. 5.6-2(c–e)):

5.6-3.3.1. Leucocratic biotite trondhjemite
Several types of trondhjemites are observed in Barberton TTG plutons (Yearron, 2003). They range from fine- to coarse-grained rocks, with occasional porphyritic varieties; all have similar mineralogy, dominated by plagioclase (oligoclase to andesine; 55–65%), quartz (15–20%), biotite (5–15%) and microcline (~10%). Accessory minerals are apatite, allanite and (magmatic) epidote, with secondary chlorite, sericite and saussurite. It is worth noting that the name "trondhjemite" is synonymous with "leuco-tonalite" and should be used only for rocks with less than 10% mafic minerals, less than 10% alkali feldspar, and more than 20% quartz (Le Maître, 2002). Obviously, some samples of this rock type do not strictly fit the definition, and are "tonalites", "granodiorites", or even "(leuco-) quartz monzonites"; however, the name "trondhjemites" fits most of the samples and is retained for convenience.

5.6-3.3.2. Hornblende tonalite
Hornblende tonalites are found in the Kaap Valley pluton and the northern margin of the Nelshoogte pluton (Robb et al., 1986; Yearron, 2003; Belcher et al., 2005). Smaller, plug-like and isolated tonalitic intrusions also occur in the southern TTG-gneiss terrain around the Schapenburg schist belt (Anhaeusser et al., 1983; Stevens et al., 2002) and along the western margin of the large Mpuluzi batholith (Westraat et al., 2004). They are dominated by plagioclase (oligoclase to andesine; ~60%), interstitial quartz (10–20%), and subhedral hornblende (~15%) with minor biotite and microcline and accessory allanite and ilmenite. Secondary chlorite and epidote develop at the expense of hornblende. In places, more mafic dioritic enclaves are common, displaying the same mineral assemblage as the hornblende tonalites, but in different proportions. With less than 20% quartz, some of the "tonalites" are technically leuco-quartz-diorites (in IUGS terms) (Le Maître, 2002).

In contrast to the trondhjemites, tonalites are absent from the ca. 3.45 Ga group. They are found in parts of the Steynsdorp pluton, and represent the latest (syn- to post-tectonic) stages of the ca. 3.29–3.21 Ga group.

5.6-3.3.3. Minor components

– Mafic dykes are observed as a minor component of many of the plutons, most commonly in the ca. 3.2 Ga plutons. Some dioritic dykes also occur in the ca. 3.45 Ga TTGs, especially along the margins of the individual plutons. The diorites have a mineralogy similar to the "wall rock" trondhjemite or tonalites (Yearron, 2003), but with different mineral proportions (60% plagioclase, 15% quartz, 10% each biotite and amphibole, some microcline – Yearron (2003)). Gabbroic dykes are also reported, but not described (Yearron, 2003) in the Nelshoogte and Kaap Valley plutons.
– Felsic dykes range from leucocratic versions of the TTGs, to plagiogranites, to porphyries and aplites, or pegmatites. All point to some degree of in-situ differentiation, probably fluid assisted, or they are related to the later, ca. 3.1 Ga, event. Collectively however, their volume is too small to represent more than local processes.

Clearly, in the typical "grey gneiss" terrains of most Archean provinces, these diverse components would be interleaved and transposed with the dominant trondhjemites or tonalites, resulting in some difficulty in explaining the scatter of compositions of these gneissic units. This is not the case in the relatively low strain BGGT.

5.6-3.4. Summary

The TTGs of the BGGT are spatially and temporally distinct from one another. Geographically, the ca. 3.45 Ga TTGs are in the east, and younger 3.2 Ga old rocks are in the west, separated from one another by the Inyoni shear zone, which is interpreted to represent a suture zone during ca. 3.25–3.21 Ga orogenesis. This temporal and spatial distinction is also recorded in the compositions of these rocks. The 3.45 Ga generation is only trondhjemitic, whereas the 3.29–3.21 Ga plutons are both trondhjemitic and tonalitic; in this latter group, the tonalites always represent the youngest phases, either as the slightly younger Kaap Valley pluton, or as late intrusive phases in composite plutons. The switch from trondhjemitic (3.29–3.22 Ga) to tonalitic (3.22–3.21 Ga) compositions at ca. 3.22 Ga appears to coincide with a change in geological regime from collision tectonics to orogenic collapse.

5.6-4. GEOCHEMISTRY

Numerous analyses of Barberton TTGs have been published (Anhaeusser and Robb, 1980, 1983; Anhaeusser et al., 1981; Robb et al., 1986; Kleinhanns et al., 2003; Yearron, 2003). Unfortunately, many are either relatively old and were not obtained with modern mass spectrometry techniques, or samples were crushed with carbide tungsten mills, such that the existing database, while extensive, is not particularly consistent and lacks reliable determination for some important elements (Ni, Cr, Ta, Pb). The following discussion is based on 314 analyses from published (see references above) and unpublished data (Table 5.6-2). Unpublished analyses were obtained in 2004–2005 at Stellenbosch University and University of Capetown. Major elements and some traces were analyzed by XRF, whereas most traces including REE have been analysed in-situ by LA-ICP-MS on glass beads.

Table 5.6-2. Representative analysis of Barberton TTGs, for the different plutons studied

	3.55–3.50 Ga			ca. 3.45 Ga			
	Steynsdorp			Stolzburg		Theespruit	
Sample	STY2A	STY1	STY4B	STZ10	STZ11	THE4A	THE6B
Pluton	Steynsdorp	Steynsdorp	Steynsdorp	Stolzburg	Stolzburg	Theespruit	Theespruit
Type	Tonallite	Tonallite	High-K	Trondhjemite	Trondhjemite	Trondhjemite	Trondhjemite
	Low-Sr	Low-Sr	(Low-Sr)	High-Sr	High-Sr	High-Sr	High-Sr
Source	Yearron 2003	Yearron 2003	Yearron 2003	Yearron 2003	Yearron 2003	Yearron 2003	Yearron 2003
Major elements (wt%)							
SiO_2	70.52	65.67	75.11	73.41	70.99	70.34	73.15
TiO_2	0.37	0.46	0.14	0.22	0.26	0.25	0.24
Al_2O_3	15.93	17.70	14.04	14.12	15.25	16.04	15.46
FeOt	2.75	3.75	1.26	1.62	1.73	2.01	1.36
MnO	0.04	0.06	0.02	0.04	0.05	0.03	0.02
MgO	1.25	1.82	0.18	0.46	0.73	1.10	0.46
CaO	2.95	3.89	1.13	1.43	1.61	2.85	2.37
Na_2O	4.60	4.43	3.56	6.45	7.37	5.74	5.32
K_2O	1.40	1.99	4.54	2.20	1.95	1.56	1.55
P_2O_5	0.19	0.24	0.04	0.07	0.06	0.08	0.07
LOI							
Selected ratios							
A/CNK	1.10	1.07	1.09	0.91	0.89	0.98	1.05
K_2O/Na_2O	0.30	0.45	1.28	0.34	0.26	0.27	0.29
CaO/Al_2O_3	0.19	0.22	0.08	0.10	0.11	0.18	0.15
CaO/Na_2O	0.64	0.88	0.32	0.22	0.22	0.50	0.45

Table 5.6-2. (*Continued*)

	3.55–3.50 Ga			ca. 3.45 Ga			
	Steynsdorp			Stolzburg		Theespruit	
Sample	STY2A	STY1	STY4B	STZ10	STZ11	THE4A	THE6B
Pluton	Steynsdorp	Steynsdorp	Steynsdorp	Stolzburg	Stolzburg	Theespruit	Theespruit
Type	Tonallite	Tonallite	High-K	Trondhjemite	Trondhjemite	Trondhjemite	Trondhjemite
	Low-Sr	Low-Sr	(Low-Sr)	High-Sr	High-Sr	High-Sr	High-Sr
Source	Yearron 2003	Yearron 2003	Yearron 2003	Yearron 2003	Yearron 2003	Yearron 2003	Yearron 2003
Trace elements (ppm)							
Sc							
V	38.1	48.3	6.9	15.0	15.0	22.1	13.7
Cr	32.0	73.0	3.3				
Ni	20.4	40.5					
Cu	20.8	6.9		2.0	2.0		
Zn	65.0	97.2	34.0	32.0	44.0	47.6	36.5
Ga							
Ge							
As							
Rb	66.4	100.5	111.7	44.0	54.0	51.3	45.7
Sr	495.5	586.6	166.2	551.0	586.0	623.8	517.6
Y	11.3	13.3	5.5	5.0	5.0	7.0	18.7
Zr	183.8	205.6	128.4	95.0	103.0	127.1	98.4
Nb	7.2	11.1	7.3			4.6	6.7
Sb							
Ba	247.2	316.5	575.4	480.0	374.0	311.6	211.5
Hf							
Ta							
Pb							
Th							
U							

Table 5.6-2. (Continued)

	3.55–3.50 Ga			ca. 3.45 Ga			
	Steynsdorp			Stolzburg		Theespruit	
Sample	STY2A	STY1	STY4B	STZ10	STZ11	THE4A	THE6B
Pluton	Steynsdorp	Steynsdorp	Steynsdorp	Stolzburg	Stolzburg	Theespruit	Theespruit
Type	Tonallite Low-Sr	Tonallite Low-Sr	High-K (Low-Sr)	Trondhjemite High-Sr	Trondhjemite High-Sr	Trondhjemite High-Sr	Trondhjemite High-Sr
Source	Yearron 2003	Yearron 2003	Yearron 2003	Yearron 2003	Yearron 2003	Yearron 2003	Yearron 2003
Selected ratios							
Sr/Y	44	44	30	110	117	90	28
Nb/Ta							
REE(ppm)							
La	24.1	16.0	21.9	13.7	11.6	11.7	38.4
Ce	49.2	27.8	45.7	25.8	21.6	19.8	70.8
Pr							
Nd	12.0	12.1	15.7	11.4	8.6	10.0	26.8
Sm		2.47	2.88	2.33	1.89	1.60	
Eu		0.84	0.50	0.54	0.62	0.54	
Gd		2.78	2.21	1.08	0.87	1.36	
Tb		0.38	0.33			0.18	
Dy							
Ho							
Er							
Tm							
Yb		1.31	0.46	0.54	0.56	0.48	
Lu		0.18	0.09			0.06	
La$_N$	73.0	48.5	66.3	41.6	35.1	35.4	116.4
Yb$_N$		6.0	2.1	2.5	2.6	2.2	
Eu/Eu*		0.99	0.61	1.05	1.49	1.13	
(La/Yb)$_N$		8.1	31.7	17.0	13.8	16.2	

Table 5.6-2. (Continued)

	3.23–3.21 Ga					Nelshoogle					
	Badplass										
Sample	4-7-05A	4-7-13B	4-7-05B	BDP8C	BTV16A	NLG14C	NLG15	NLG21A	NLG5	SKV20	17-99/128
Pluton	Badplaas	Badplaas	Badplaas	Badplaas	Badplaas	Nelshoogle	Nelshoogle	Nelshoogle	Nelshoogle	Kaap	Kaap
Type	Melt-depleted	Melt-depleted	High-K (low-Sr)	Trondhjemite High-Sr	Trondhjemite Low-Sr	Tonallite Low-Sr	Tonallite Low-Sr	Trondhjemite Low-Sr	Trondhjemite Low-Sr	Tonallite Low-Sr	Tonallite Low-Sr
Source	This work	This work	Yearron 2003	Yearron 2003	Yearron 2003	Yearron 2003	Yearron 2003	Yearron 2003	Yearron 2003	Anhaeusser and Robb 1983	This work
Major elements (wt.%)											
SiO_2	70.67	73.47	75.47	71.21	68.67	63.37	63.13	70.11	73.91	60.51	64.51
TiO_2	0.28	0.19	0.12	0.19	0.46	0.46	0.62	0.26	0.16	0.57	0.51
Al_2O_3	16.53	15.19	14.47	16.56	16.39	15.15	17.33	16.06	14.34	16.31	16.11
FeOt	2.02	1.05	0.39	1.83	3.40	5.00	4.04	2.72	1.18	4.10	3.95
MnO	0.06	0.05	0.05	0.03	0.06	0.10	0.04	0.05	0.05	0.06	0.06
MgO	1.04	0.45	0.34	0.62	1.25	4.48	2.63	1.05	0.51	3.27	2.71
CaO	3.59	1.39	0.64	3.32	3.55	4.78	5.18	3.01	1.91	4.32	4.56
Na_2O	3.71	4.57	3.36	5.09	4.54	4.91	5.56	5.28	6.27	5.31	6.45
K_2O	0.99	1.80	3.76	1.09	1.52	1.61	1.31	1.34	1.61	1.20	1.47
P_2O_5	0.13	0.10	0.08	0.08	0.14	0.14	0.17	0.11	0.06	0.19	0.19
LOI	1.68	1.61	1.66							3.47	
Selected ratios											
A/CNK	1.21	1.27	1.34	1.06	1.05	0.82	0.87	1.03	0.92	0.91	0.79
K_2O/Na_2O	0.27	0.39	1.12	0.21	0.33	0.33	0.24	0.25	0.26	0.23	0.23
CaO/Al_2O_3	0.22	0.09	0.04	0.20	0.22	0.32	0.30	0.19	0.13	0.26	0.28
CaO/Na_2O	0.97	0.30	0.19	0.65	0.78	0.97	0.93	0.57	0.30	0.81	0.71
Trace elements (ppm)											
Sc	3.3	2.4	1.8								8.1
V	14.5	10.5	15.1		58.7	86.9	88.6	39.1	15.0		75.7

Table 5.6-2. (Continued)

	3.23–3.21 Ga					Nelshoogle					
	Badplaas										
Sample	4-7-05A	4-7-13B	4-7-05B	BDP8C	BTV16A	NLG14C	NLG15	NLG21A	NLG5	SKV20	17-99/128
Pluton	Badplaas	Badplaas	Badplaas	Badplaas	Badplaas	Nelsnoogie	Nelsnoogie	Nelsnoogie	Nelsnoogie	Kaap	Kaap
Type	Melt-depleted	Melt-depleted	High-K (low-Sr)	Trondhjemite High-Sr	Trondhjemite Low-Sr	Tonallite Low-Sr	Tonallite Low-Sr	Trondhjemite Low-Sr	Trondhjemite Low-Sr	Tonallite Low-Sr	Tonallite Low-Sr
Source	This work	This work	Yearron 2003	Yearron 2003	Yearron 2003	Yearron 2003	Yearron 2003	Yearron 2003	Yearron 2003	Anhaeusser and Robb 1983	This work
Cr	31.4	9.6									74.6
Ni	7.9	5.6									56.3
Cu	15.3	10.1	22.9		14.4	23.2	27.8	13.7	2.0		14.9
Zn	63.7	35.6	25.1	33.1	61.6	83.0	50.2	41.4	20.0		54.7
Ga	29.8	22.8	24.6								17.3
Ge	1.7	0.2									
As	11.1	1.6	4.7								
Rb	45.0	34.7	63.6	40.0	57.4	53.5	39.4	37.7	50.0	48	39.1
Sr	799.0	576.4	253.6	547.8	312.3	460.9	823.1	555.2	272.0	570	572.0
Y	3.9	1.8	2.1		15.2	14.1	7.2	6.5	6.0		9.0
Zr	141.7	78.9	42.5	113.9	154.6	80.7	177.6	96.4	71.0		103.0
Nb	2.6	2.0	1.0	1.9	5.3	3.5	1.7	3.5			3.8
Sb	1.4	0.9	0.9								
Ba	180.8	216.9	297.2	143.4	132.3	266.9	253.1	269.1	222.0	191	237.0
Hf	3.1	1.8	1.3								2.6
Ta	0.2	0.2	0.1								0.2
Pb	13.2	12.1	25.8								4.7
Th	2.3	1.6	2.5								1.7
U	0.6	0.4	1.3								0.5
Selected ratios											
Sr/Y	205	313	124		21	33	114	85	45		64
Nb/Ta	12.5	9.1	9.2								15.7

Table 5.6-2. (Continued)

	3.23–3.21 Ga					Nelshoogle					
	Badplaas										
Sample	4-7-05A	4-7-13B	4-7-05B	BDP8C	BTV16A	NLG14C	NLG15	NLG21A	NLG5	SKV20	17-99/128
Pluton	Badplaas	Badplaas	Badplaas	Badplaas	Badplaas	Nelsnoogie	Nelsnoogie	Nelsnoogie	Nelsnoogie	Kaap	Kaap
Type	Melt-depleted	Melt-depleted	High-K (low-Sr)	Trondhjemite High-Sr	Trondhjemite Low-Sr	Tonallite Low-Sr	Tonallite Low-Sr	Trondhjemite Low-Sr	Trondhjemite Low-Sr	Tonallite Low-Sr	Tonallite Low-Sr
Source	This work	This work	Yearron 2003	Yearron 2003	Yearron 2003	Yearron 2003	Yearron 2003	Yearron 2003	Yearron 2003	Anhaeusser and Robb 1983	This work
REE (ppm)											
La	15.2	12.4	11.1	15.2	13.7	13.3	6.4	14.5	9.8		14.8
Ce	27.2	20.0	18.3	23.4	24.6	25.7	13.3	33.9	18.5		31.7
Pr	2.8	2.0	1.8								3.9
Nd	9.0	7.1	6.6	8.9	13.0	15.0	7.4	12.6	8.0		15.1
Sm	1.79	1.16	1.74	1.50	2.49	4.01	1.48		3.34		2.99
Eu	0.78	0.50	0.43	0.48	0.65	1.06	0.51		0.50		0.96
Gd	1.04	0.61	1.12	1.15	2.24	2.61	1.48		1.02		2.48
Tb	0.19	0.11	0.14	0.07	0.36		0.24				0.34
Dy	0.84	0.41	0.47								1.76
Ho	0.18	0.07	0.11								0.34
Er	0.60	0.16	0.19								0.89
Tm			0.04								0.13
Yb	0.64	0.30	1.03	0.15	1.03	1.14	0.33		0.60		0.80
Lu	0.15	0.04	0.04	0.03	0.11		0.05				0.11
La$_N$	45.9	37.6	33.5	46.0	41.6	40.4	19.3	43.8	29.6		44.9
Yb$_N$	2.9	1.4	4.7	0.7	4.7	5.2	1.5		2.7		3.6
Eu/Eu*	1.75	1.84	0.96	1.12	0.85	1.01	1.06		0.83		1.08
(La/Yb)$_N$	15.8	27.3	7.2	67.5	8.7	7.8	12.9		10.8		12.3

"Type" refers to the types identified in Section 5.6-4.2. "High-K" and "melt-depleted" are the "non-TTG" facies; "low" and "high Sr" denote the two subseries idetified. Major elements are in wt%, traces in ppm. LOI = loss on ignition. FeOt: total iron recalculated as FeO. A/CNK: molecular ratio Al/(Ca+Na+K). La$_N$, Yb$_N$: normalized REE values (Nakamura, 1974). Eu/Eu* = Eu$_N$/(0.5 × (Sm$_N$ + Gd$_N$)) is a measure of the "depth" of the Eu anomaly (a negative Eu anomaly corresponds to Eu/Eu* < 1).

5.6-4.1. Common Characteristics

5.6-4.1.1. Major elements
The two rock types identified above (tonalites and trondhjemites) display some differences in terms of major elements contents. The tonalites are silica-poorer (typically 62–68 wt%), whereas the trondhjemites are more felsic (typically 70–75 wt%). Accordingly, the tonalites are richer in FeO and MgO and marginally poorer in Na_2O, K_2O and CaO. Both the tonalites and the trondhjemites belong to a sub-alkaline, calc-alkaline series (Fig. 5.6-3(a,b)) with their volcanic equivalents being "soda-rhyolites", dacites and minor andesites (for the tonalites). Most of the samples reviewed belong to a medium-K series (Fig. 5.6-3(c)), but a significant part of the 3.23–3.21 Ga group, especially in the Badplaas unit, belongs to a low-K series. Most of BGGT rocks belong to the high-Al TTG group of Barker and Arth (1976). In a (normative) Ab-An-Or diagram (O'Connor, 1965) (Fig. 5.6-4), the data plots mostly in the trondhjemite field (leucocratic facies), extending into the granite field, or in the tonalite and granodiorite fields (hornblende tonalite). All these characteristics are typical of most TTG rocks (Martin, 1994). In Harker diagrams (Fig. 5.6-5), most elements display a compatible behavior, with all samples plotting along similar trends. However, Al_2O_3, K_2O and Na_2O show a different pattern, with much more scatter and differences between samples or sample groups, which will be used to further subdivide the TTG rocks in several sub-series.

5.6-4.1.2. Trace elements
To some degree, all TTGs of the BGGT present comparable features. Like all TTGs, they have low concentrations of compatible transition elements (Ni, Cr, V), relatively low HFSE contents (Ti, Zr, Hf) and moderately high LILE and fluid-mobile elements contents (Rb, Ba, Th). LILE/HFSE ratios are higher than in modern arc-related magmas (Pearce, 1983). One of the most characteristic features of the TTGs from the BGGT is the high Sr contents (typically 500–1000 ppm) and associated low Y values (average 7.8 ppm), which confers a high Sr/Y ratio (typically around 100). In Harker-type diagrams (Fig. 5.6-6), it is possible to identify several groups with different trends; diagrams such as SiO_2 vs Sr/Y or La/Yb, for instance, clearly show that the tonalites (and some of the trondhjemites) on one hand, and most of the trondhjemites on the other hand, define contrasting sub-horizontal and sub-vertical trends, respectively. To some degree, the same grouping can be observed with most of the other elements, Sr being the most discriminating.

REE patterns display high LREE ($La_N = 40$–60) and low HREE ($Yb_N < 5$) contents, corresponding to rather fractionated REE patterns ($(La/Yb)_N = 10$–25) that lack Eu anomalies. This is lower than most TTGs, which have La_N values of $\sim$100, and $(La/Yb)_N$ of 35–40 (Martin, 1994) (Fig. 5.6-7).

These observations imply the existence of a phase with a high partition coefficient (K_d) for Y and the heavy REE at some stage during TTG petrogenesis. Among the common minerals, only garnet and to a lesser degree amphibole (Rollinson, 1993; Bédard, 2006) have adequate K_d values, implying that either (or both) coexisted as solid phases with the magma at some stage of its evolution, and were not entrained in the plutons as observed now.

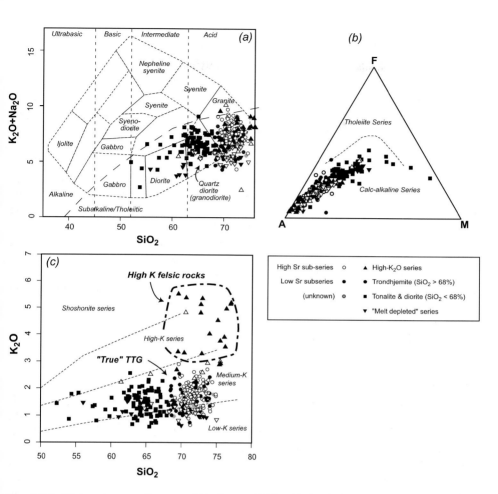

Fig. 5.6-3. Major elements features of Barberton TTGs: (a) Total alkali vs silica (TAS) diagram (Cox et al., 1979); (b) FMA (Fe-Mg-alkali) diagram (Irvine and Baragar, 1971); (c) Silica-potassium diagram (Peccerillo and Taylor, 1976), showing the potassic rocks, probably formed by remelting of earlier TTGs, with a distinctive vertical trend in this diagram. The three diagrams allow us to characterize the TTG rocks as belonging to a sub-alkaline (a), low-to-medium-K (b), and calc-alkaline (c) series. Symbols: analyses are grouped according to their chemistry (Section 5.6-4.2); colours indicate whether the sample belongs to a low- or high-Sr sub-series, whereas the symbol differentiates between "true TTGs" (tonalites or trondhjemites), high-K_2O rocks, and "melt-depleted" samples.

Another significant feature of Archean TTGs, in general, is their variable, but typically low, Nb/Ta ratios (Kambers et al., 2002; Kleinhanns et al., 2003; Moyen and Stevens, 2006). However, the data presented herein, being relatively incomplete, does not document this well.

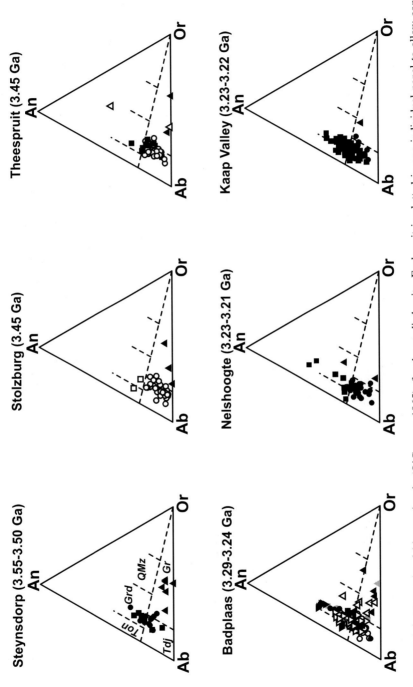

Fig. 5.6-4. Normative feldspar triangle (O'Connor, 1965), for the studied units. Each unit is plotted in one individual panel to allow comparison. Same caption as Fig. 5.6-3. The fields are labeled only in the first panel: Tdj, trondhjemite; Ton, tonalite; Grd, granodiorite; QMz, quartz-monzonite; Gr, granite.

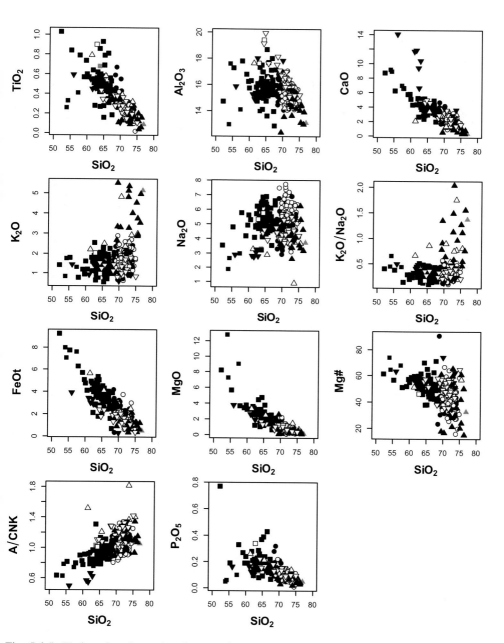

Fig. 5.6-5. Harker plots for major elements for Barberton TTG. FeOt = total iron as FeO; Mg# = molecular ratio 100 Mg/(Mg+Fe); A/CNK = molecular ratio Al/(Ca+Na+K). Symbols as in Fig. 5.6-3.

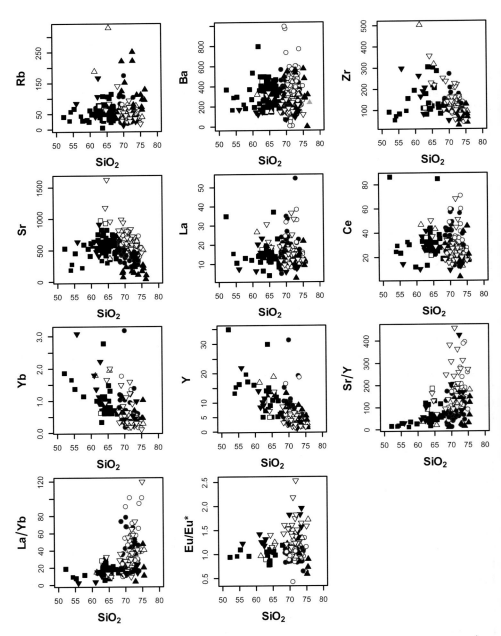

Fig. 5.6-6. Harker plots for selected trace elements and ratios for Barberton TTGs. Eu/Eu* = Eu$_N$/(0.5(Sm$_N$+Gd$_N$)), normalization values after Nakamura (1974). Symbols as in Fig. 5.6-3.

5.6-4.1.3. Discrimination diagrams

In discrimination diagrams (Pearce et al., 1984), TTGs always fall in the "VAG" (volcanic arc granites) field, reflecting their low HFSE, Y and Yb contents. However, no genetic implication should be drawn from this observation. Indeed, geotectonic diagrams like these are based on compilations of analyses of rocks from known tectonic settings. Using these diagrams for Archean rocks (in a genetic sense) implies that Archean magmas formed in similar contexts and via similar processes to modern magmas. In the case of the Archean, this carries the implicit assumptions that: (1) modern-style plate tectonics operated during the Archean; and (2) that its modalities (thermal regimes, rock type presents, etc.) were the same as present-day situations. Both assumptions are far from proven, and therefore geotectonic "discrimination" diagrams should not be used in pre-Phanerozoic times, as pointed out in the original paper by Pearce et al. (1984).

The most classical discrimination diagrams used for the interpretation of TTG petrogenesis, however, reflect their REE, Sr and Y contents (Fig. 5.6-8). In Sr/Y vs Y and La/Yb vs Yb diagrams (Martin, 1986, 1987, 1994), TTGs plot along the Y-axis, distinct from modern, calc-alkaline magmas (and most I-type granites). However, Barberton TTGs tend to cluster in the lower-left corner of both diagrams, close to, or in the overlap area between, the two fields.

5.6-4.2. Distinct Geochemical Types

In addition to the shared characteristics presented above, it is possible to identify several sub-series, with distinct geochemical signatures; these sub-series are most clearly differentiated by their Sr contents and their K_2O/Na_2O nature, defining:
(1) a K_2O poor sub-series, either "low Sr" or "high Sr";
(2) a K_2O rich sub-series;
(3) a high Eu/Eu^*, very low K_2O, "melt-depleted" sub-series.

5.6-4.2.1. Low and high Sr sub-series

Regardless of their petrologic nature (tonalite or trondhjemite), the TTGs of the BGGT can be classified in to sub-series on the basis of their position in a SiO_2-Sr diagram (Fig. 5.6-9(a)).

Although the *absolute* values of Sr abundances are comparable in the two sub-series, the low SiO_2 group defines a lower Sr *trend* for a given SiO_2 value (Fig. 5.6-9(a)). A very similar observation is made by Champion and Smithies (this volume) for TTGs in the Pilbara Craton, although TTGs from the BGGT have collectively higher Sr values than Pilbara rocks (the Barberton low-Sr group has Sr levels comparable to the Pilbara *high*-Sr rocks). In the Pilbara, Champion and Smithies (this volume) observed that Al_2O_3 contents also reflect this difference, with the high-Sr sub-series also being high-Al. This is only partially supported by our data: most of our samples (low and high-Sr together) have the same Al contents as, and plot together with, the Pilbara high-Sr group. On the other hand, a R_1-R_2 diagram (de la Roche et al., 1980), which takes into account most major elements, clearly differentiates between the two sub-series (Fig. 5.6-9(b)). The high-Sr sub-series is

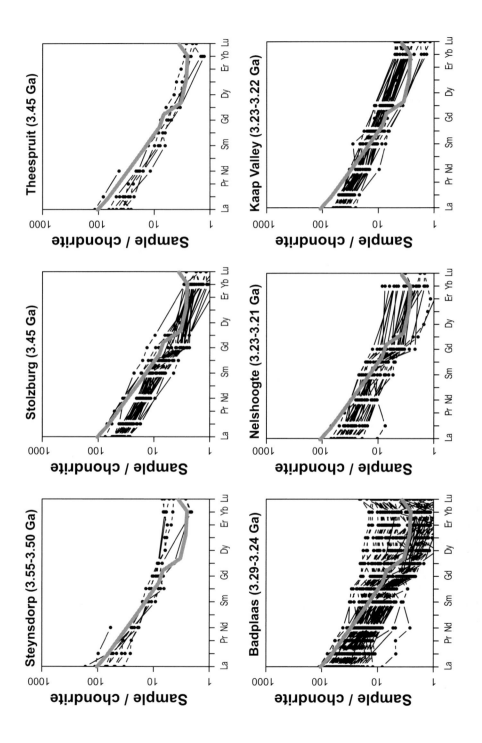

Fig. 5.6-7. (*Previous page.*) REE patterns of Barberton TTGs (normalized to chondrite after Naka-mura (1974). In all diagrams, the thick grey line corresponds to the TTG average of Martin (1994). Note the opposition between the low-Sr plutons (Kaap Valley, Nelshoogte) that mostly plot above the average for HREE, and the high-Sr plutons (Stolzburg, Theespruit) that mostly plot below. Also note the important scatter for the composite Badplaas gneisses.

---

also somewhat more sodium-rich (and with higher Na_2O/CaO) than the low-Sr series. The low-Sr rocks also tend to have higher Y contents (or, rather, a larger range of Y values for a given SiO_2 content), giving them lower Sr/Y ratios.

The high-Sr rocks are mostly trondhjemites, with $68\% < SiO_2 < 75\%$; rare samples have lower SiO_2 contents and are tonalitic. The low-Sr group, in contrast, comprises both tonalites (with $SiO_2 < 68\%$) and trondhjemites ($68\% < SiO_2 < 77\%$). In an O'Connor (1965) normative diagram, the high-Sr group occupies almost exclusively the trondhjemite field, whereas the low-Sr sub-series plot in the tonalite, trondhjemite and granodiorite fields. In the low-Sr group, the tonalites and trondhjemites are clearly differentiated, not only by their SiO_2 contents, but also by the flatter trend of the tonalites in SiO_2-Sr binary diagrams at ca. 600 ppm Sr (Fig. 5.6-9(a)).

The subdivision of TTGs into a low- and high-Al sub-series is not new, and was pro-posed more than 30 years ago (Barker and Arth, 1976). All our samples, however, (like most of the world's TTGs, see for instance Martin (1994)) belong to what would be a high-Al_2O_3 series, according to these definitions; the (relatively subtle) differences we observe reflect subdivisions of Barker's high-Al group.

5.6-4.2.2. High K sub-series

High K_2O felsic rocks form a minor component of, for example, the Steynsdorp and Bad-plaas units. These rocks are not always possible to identify in the field. In some cases, like the granodioritic phases of the Steynsdorp pluton, their K-feldspar rich nature is immedi-ately obvious. Sometimes, however, they are macroscopically indistinguishable from the more common trondhjemites and they can only be identified geochemically. Such "potas-sic" facies have relatively high K_2O/Na_2O (>0.5). They plot in the medium to high-K fields of a SiO_2-K_2O diagram (Fig. 5.6-3(c)) at ca. 70% SiO_2, with no clear trends, and are mostly granites (in an O'Connor diagram) (Fig. 5.6-4). In the R_1-R_2 diagram (Fig. 5.6-9(b)), they clearly plot below (lower R_2 values) the other rock types. They have low Y (mostly <10 ppm), Yb (<1 ppm), and sometimes a slight negative Eu anomaly. Most of the "high K" rocks belong to the low-Sr series (Fig. 5.6-9(a)), with very low Sr contents (<250 ppm). In Sr/Y vs Y or La/Yb vs Yb diagrams, they are virtually indistinguishable from the "normal" (sodic) TTGs, although they tend to plot "below" the field of ordinary TTGs in a Sr/Y vs Y diagram (Fig. 5.6-8), reflecting lower contents of both Sr and Y. The high-K_2O rocks also have high LILE contents (Rb, Ba, U, of course K) and are quite simi-lar to the "enriched TTGs" reported in the Pilbara Craton by Champion and Smithies (this volume).

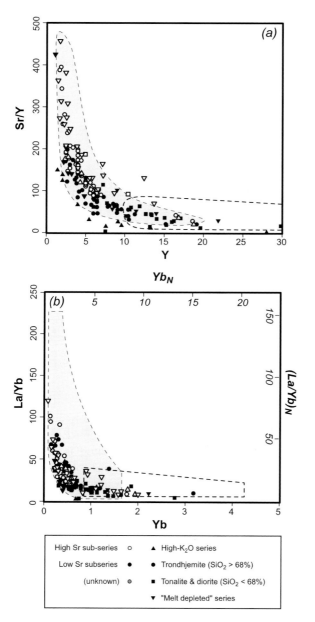

Fig. 5.6-8. Some trace elements characteristics of Barberton TTGs. In both diagrams, the grey field is the TTG field and the stippled field delineates modern calc-alkaline magmas (Martin, 1994). Symbols as in Fig. 5.6-3. The panel (b) has a double scale, both in ppm and in normalized values (Nakamura, 1974).

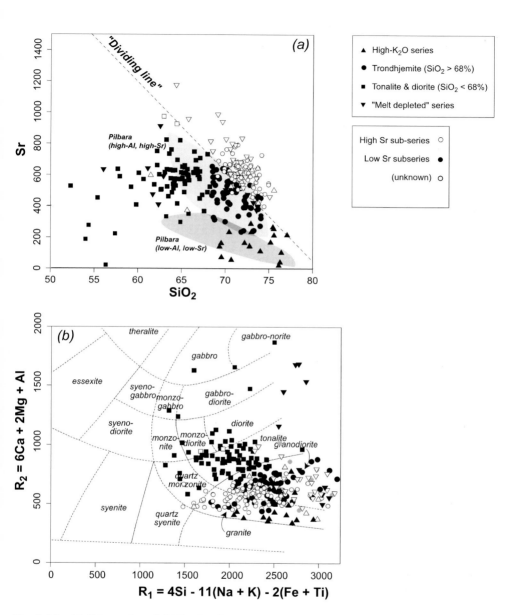

Fig. 5.6-9. (a) SiO_2 vs Sr and (b) R_1–R_2 diagrams for Barberton TTGs. R_1 and R_2 are the cationic parameters of de la Roche (1980). The two sub-series define two distinct trends: a low-Sr trend corresponding to the lower SiO_2 tonalitic facies, but including some of the higher-SiO_2 rocks; and a high-Sr trend mostly corresponding to the high SiO_2 trondhjemites. The low-Sr trend also evolves from higher R_2 values. The high K_2O series has low R_2 values and the "melt-depleted" samples have high R_1. The trend of Pilbara TTGs is from Champion and Smithies (this volume).

5.6-4.2.3. "Melt-depleted" samples

Some rocks with uncommon geochemistry are found in the Badplaas gneisses that belong to a very low K_2O series (around 1% K_2O or less, Fig. 5.6-3(c)), but are not very rich in Na_2O (~4%). They are Al_2O_3-enriched and correspondingly have high, to very high, A/CNK ratios (1.2–1.4), consistent with their chlorite-rich mineralogy (the chlorite is probably a secondary mineral, but reflects an Al-rich composition whatever the primary minerals were; garnet is occasionally observed). Thee rocks have high R_1 values and plot to the right of most other samples in a R_1-R_2 diagram (Fig. 5.6-9(b)). They tend to have high Sr, and high Sr/Y ratios (up to >400), the highest in the whole database (Fig. 5.6-8). Finally, they have small positive Eu anomalies (Fig. 5.6-7). Geochemically, they are therefore the opposite of the high K_2O group.

All these features, combined with the long-lived, multiphase nature of the Badplaas gneisses, suggest that they represent a melt-depleted facies; i.e., restitic rocks out of which some melt has been extracted, represented by the high K_2O rocks in the Badplaas gneisses. Whether the melt extraction reflects the 3.29–3.22 Ga evolution of the Badplaas domain, or rather the later, ca. 3.1 Ga formation of the nearby Heerenveen batholith is uncertain.

5.6-4.3. Summary and Subdivision in the Different Plutons

On a geochemical basis, four groups of rocks were identified; high-K_2O rocks, low-Sr and high-Sr "true TTGs", and "melt-depleted" gneisses of the Badplaas unit. The "melt-depleted" rocks are not, strictly speaking, magmas (although their origin is related to magmatic evolution). The three other types can be distinguished by devising a "ΔSr" vs K_2O/Na_2O diagram. Plotting Sr in a diagram can be misleading, as this parameter is strongly correlated to SiO_2; Sr values alone do not allow differentiation between low and high SiO_2 series, which are distinguished by Sr contents *at a given SiO_2 level*. To overcome this problem, we calculate a new parameter, ΔSr, that represents the distance of an analysis from a reference line in a SiO_2-Sr diagram (Fig. 5.6-9(a)). Here, this line is taken as the dividing line between low and high-Sr sub-series, allowing straight-forward interpretations: low-Sr sub-series rocks have negative ΔSr, while high Sr sub-series samples have positive ΔSr. In our case, the reference line follows the following equation:

$$Sr_{ref} = 4621 - 57.14 \cdot SiO_2,$$

and therefore

$$\Delta Sr = Sr - Sr_{ref}$$

for each individual sample.

K_2O/Na_2O is also correlated to SiO_2, and ideally it would be possible to calculate a "ΔK" parameter in the same way. The benefit would, however, be minimal, since the range of K_2O/Na_2O values between the normal TTGs and the high-K_2O group exceeds the variations within a group.

The K_2O/Na_2O vs ΔSr diagrams presented in Fig. 5.6-10, therefore, allow us to distinguish the main groups.

The ca. 3.55 Ga Steynsdorp pluton is made of two components, a low-Sr tonalitic to trondhjemitic facies, and a high-K_2O unit. Both are now interleaved, but have been identified in the field.

The ca. 3.45 Ga group (Stolzburg and Theespruit plutons) appear as a largely homogeneous population. It is primarily composed of high-Sr trondhjemites, although some examples of low-Sr tonalites are found in the database.

The 3.29–3.21 Ga group is more complex. The older Badplaas gneisses encompass samples belonging to both the high and low-Sr sub-series, together with high-K_2O samples and melt-depleted rocks. The Nelshoogte and Kaap Valley plutons are both made up of low-Sr trondhjemites and low-Sr tonalites; the trondhjemites are dominant in the Nelshoogte pluton, whereas the tonalites form most of the Kaap Valley pluton – a fact somehow obscured in Fig. 5.6-10 by sampling bias. In other words, the 3.29–3.21 Ga group probably records a transition from high-Sr trondhjemites, to low-Sr trondhjemites, to low-Sr tonalites.

5.6-4.4. Isotopes

Whole rock Sr-Nd isotopic data have been published on Barberton TTGs (Barton et al., 1983; Kröner et al., 1996; Yearron, 2003; Sanchez-Garrido, 2006). Unfortunately, only one study (Sanchez-Garrido, 2006) gives combined Sr and Nd data for the studied samples. Collectively, 18 Nd isotopic analyses and 61 Sr data are published, but only 5 combined Sr-Nd analyses. However, combining the (independently) published data allows us to define the probable range of compositions (Fig. 5.6-11).

TTG plutons mostly have isotopic characteristics close to the bulk Earth, with ε_{Nd} values between +4 and −3 and ε_{Sr} between −7 and +5 (I_{Sr} values of 0.6995 to 0.701). This is a commonly observed feature of Archean TTGs (e.g., Bickle et al., 1983; Martin, 1987; Peucat et al., 1996; Whitehouse et al., 1996; Bédard and Ludden, 1997; Berger and Rollinson, 1997; Liu et al., 2002; Whalen et al., 2002; Stevenson et al., 2006; Zhai et al., 2006; Champion and Smithies, this volume). This implies that TTGs are derived from juvenile, or newly extracted sources, either the mantle itself or more probably, basalts recently extracted from the mantle.

In the case of the Barberton TTGs, however, there is a systematic difference between the older (3.45 Ga) and the younger (3.29–3.21 Ga) TTGs, the former having more juvenile characteristics (high ε_{Nd} and low ε_{Sr}) than the latter.

The 3.29–3.21 Ga group was possibly derived from either pre-existing rocks of the Onverwacht Group (Hamilton et al., 1979; Kröner et al., 1996), its high-grade equivalents in the Swaziland Ancient Gneiss Complex (Kröner et al., 1993; Kröner and Tegtmeyer, 1994; Kröner, this volume) or even the Fig Tree Group (Toulkeridis et al., 1999; Sanchez-Garrido, 2006). Alternatively, the relatively enriched signature of the 3.23–3.21 Ga generation could reflect a composite source, including both depleted and enriched (recycled or already emplaced crust?) components.

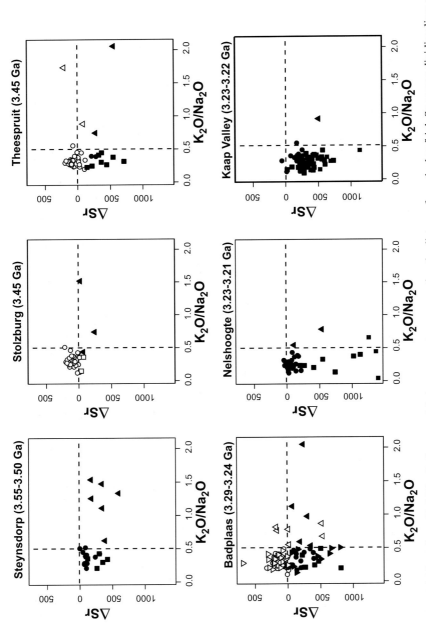

Fig. 5.6-10. ΔSr vs K_2O/Na_2O diagram. ΔSr is a parameter representing the distance from the low/high Sr groups dividing line; see definition in the text. The vertical and horizontal dashed lines correspond, respectively, to the limit between "true TTGs" and "high-K_2O", and between low- and high-Sr groups, effectively defining 4 sub-series (although the high-K_2O, high-Sr is virtually non-represented, such that only three sub-series really exist).

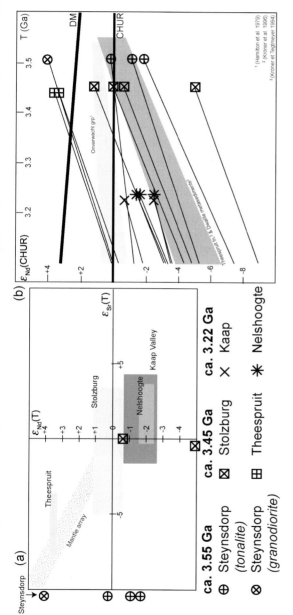

Fig. 5.6-11. Isotopic characteristics of TTG plutons around the Barberton Belt. (a) ε_{Nd} vs ε_{Sr} diagram (Zindler and Hart, 1986). ε_{Nd} values are calculated at the age of formation of these rocks (Table 5.6-1), so that this diagram is therefore not drawn for a specific time. Individual analyses (when both Sr and Nd data are available) are plotted as individual symbols, which correspond here to different plutons. One Sr-Nd analysis showing an aberrant ε_{Sr} value is not plotted. When no coupled analyses are published, the box for each pluton is bounded by the extreme range of Sr isotopic data (in X) and the extreme range of Nd isotopic data (in Y). (b) Nd isotopic evolution diagram, using published data for TTG plutons (individual analyses, Kröner et al., 1996; Yearron, 2003; Sanchez-Garrido, 2006) and for supracrustals (grey fields, Hamilton et al., 1979; Kröner and Tegtmeyer, 1994; Kröner et al., 1996). The light grey band corresponds to Onverwacht Group mafic and ultramafic lavas, the darker grey to Onverwacht metasediments, intermediate and felsic lavas, and to their equivalents in the Ancient Gneiss Complex in Swaziland. Depleted mantle is linearly interpolated from $\varepsilon_{Nd} = 0$ at 4.56 Ga to $\varepsilon_{Nd} = +10$ now (Goldstein et al., 1984). In both cases, note the difference between the ca. 3.45 Ga plutons, with isotopic characteristics intermediate between the depleted mantle and CHUR or the Onverwacht mafics and ultramafics, and the isotopically more evolved ca. 3.23–3.21 Ga plutons, consistent with derivation from an enriched mantle source or from part of the Onverwacht crust. CHUR values are $^{143}Nd/^{144}Nd = 0.512638$; $^{147}Sm/^{144}Nd = 0.1967$; $^{87}Sr/^{86}Sr = 0.7045$; $^{87}Rb/^{86}Sr = 0.0827$ (Goldstein et al., 1984).

On the other hand, a depleted mantle component (or basalts derived from it) must have played at least some role in the origin of the 3.45 Ga generation. This essentially rules out their generation by partial melting of a pre-existing old cratonic crust.

5.6-5. PETROGENESIS OF TTG ROCKS

Different hypothesis (not always mutually exclusive, but separated here for clarity) have been proposed to account for the origin of TTG magmas, in general. The most common are:

(1) Partial melting of mantle, either directly to generate felsic magmas (Moorbath, 1975; Stern and Hanson, 1991; Bédard, 1996), or indirectly to form basaltic or andesitic melts that subsequently fractionate amphibole ± plagioclase ± garnet (Arth et al., 1978; Barker, 1979; Feng and Kerrich, 1992; Kambers et al., 2002; Kleinhanns et al., 2003).

(2) Partial melting of crustal plagioclase + biotite ± quartz-rich lithologies (either metagraywackes or earlier tonalites: Arth and Hanson, 1975; Kröner et al., 1993; Winther, 1996; Bédard, 2006).

(3) Partial melting of mafic lithologies (metabasalts, either as amphibolites or eclogites), either in intraplate conditions in the lower part of a thick oceanic or continental crust (Smithies and Champion, 2000; Whalen et al., 2002; Bédard, 2006; Champion and Smithies, this volume) or in a subducting slab (Arth and Hanson, 1975; Moorbath, 1975; Barker and Arth, 1976; Barker, 1979; Condie, 1981; Jahn et al., 1981; Condie, 1986; Martin, 1986; Rapp et al., 1991; Martin, 1994; Rapp and Watson, 1995; Martin, 1999; Foley et al., 2002; Martin et al., 2005).

These three hypotheses will now be briefly discussed:

5.6-5.1. *TTG as Mantle Melts?*

Felsic magmas can be generated directly from the mantle, assuming very low melt fractions (<5%). Calc-alkaline magmas are generated from wet mantle, typically above active subduction zones. However, both experimental (Mysen and Boettcher, 1975a, 1975b; Green, 1976; Green and Ringwood, 1977; Wyllie, 1977) and theoretical (Jahn et al., 1984; Pearce and Parkinson, 1993; Martin, 1994; Kelemen et al., 2003) approaches show that, in this case, the melts are andesitic (and potassic), rather than tonalites and trondhjemites, as they are formed through the breakdown of potassic hydrous phases (richterite or phlogopite, Millhohlen et al., 1974; Sudo, 1988; Tatsumi, 1989; Tatsumi and Eggins, 1995; Schmidt and Poli, 1998), with no significant amounts of garnet in the residuum.

Alternately, andesitic or basaltic magmas generated in the mantle could fractionate amphibole ± garnet ± plagioclase and evolve towards more felsic, HREE-depleted compositions. This has been shown to be possible both on experimental (Alonso-Perez et al., 2003; Grove et al., 2003) and theoretical (Kambers et al., 2002; Kleinhanns et al., 2003) grounds. However, such a process would require large amounts (up to 75%) of mafic-ultramafic cumulate (Martin, 1994), which are largely missing from the BGGT. Furthermore, as pointed by Bédard (2006), the inferred parental magma – an andesite or andesitic basalt – is an uncommon rock type in the Archean, indeed unknown in Barberton Greenstone Belt both at 3.45 or 3.23 Ga.

5.6-5.2. *TTGs as Melts of Pre-Existing Felsic Lithologies?*

Melting of biotite (or amphibole) – plagioclase – quartz assemblages has been experimentally demonstrated to generate broadly tonalitic to trondhjemitic magmas (Gardien et al., 1995; Patiño-Douce and Beard, 1995; Winther, 1996; Patiño-Douce, 2005). Owing to relatively potassic sources (compared to mafic or ultra-mafic sources), this process results in the formation of magmas with a distinct geochemical signature, characterized by sub-vertical trends in SiO_2-K_2O diagrams, relatively high K_2O/Na_2O values and higher LILE concentrations. In Barberton TTGs, rocks belonging to the high-K_2O group do correspond to this description, and can be safely attributed to the melting of comparatively enriched, relatively potassic (and probably felsic) sources. Again, this corresponds to the interpretation proposed by Champion and Smithies (this volume) for the Pilbara LILE-enriched, "transitional TTGs".

The nature of the felsic source, in the regional context, is uncertain. In the Badplaas unit, the presence of "melt-depleted" gneisses with matching geochemical characteristics suggests that, at least for this unit, the high-K_2O rocks proceed from partial melting of already emplaced TTGs. A similar explanation is likely for the Steynsdorp pluton, where the "potassic" unit represents a sizable volume. On the other hand, in all other studied intrusions, high-K_2O rocks are a minor, very uncommon type, precluding important remelting of the TTGs. High-K_2O rocks could represent late melt mobilization during emplacement; alternately, they could reflect minor source heterogeneities. Indeed, the supracrustal pile of the BGB contains, even in the Onverwacht Group, minor sediment layers or felsic lavas (see above), and is not a perfectly homogeneous pile of basalts. During melting, such heterogeneities would yield potassic, LILE-enriched melts in small volumes. Most of them would be diluted and assimilated into the dominant TTG component, but it is possible that small magma batches are somehow preserved and retain their geochemical characteristics.

At high melt fractions, melting of TTG-like sources would of course generate melts whose composition would be very close to the source, to the point of becoming hardly distinguishable (Bédard, 2006). Bulk recycling of a tonalitic/trondhjemitic crust would, therefore, produce a continuum of compositions, from low melt fraction, high-K_2O liquids, to higher melt fraction, tonalitic to trondhjemitic liquids. Whereas this is more or less observed in the Badplaas unit, such a continuum is lacking from all other plutons, suggesting that bulk recycling of older TTG gneisses typically was not an important process in their generation.

5.6-5.3. *TTGs as Melts of Mafic Lithologies?*

The most common hypothesis for TTG genesis is partial melting of mafic lithologies (metabasalts) dominated by plagioclase and amphibole. This is supported by ample experimental (reviewed in Moyen and Stevens, 2006) and geochemical (reviewed in Martin,

1994) evidence. The major element composition of the TTGs, in general, is explained by fluid-absent melting of plagioclase-amphibole assemblages (Rushmer, 1991; Rapp and Watson, 1995; Vielzeuf and Schmidt, 2001; Moyen and Stevens, 2006); i.e., melting during which water was supplied by the breakdown of hydrous phases (either amphibole, or sometimes epidote). This is an incongruent melting reaction, in which solid products (commonly garnet and/or orthopyroxene) are generated in addition to melt. The dominant melting reactions will be either:

(1) Amphibole + Plagioclase = Melt + Ti-oxides + Orthopyroxene ± Clinopyroxene ±
    Olivine

(Beard and Lofgren, 1991; Rapp et al., 1991; Rushmer, 1991; Patiño-Douce and Beard, 1995; Rapp and Watson, 1995; Zamora, 2000; Vielzeuf and Schmidt, 2001), at pressures below garnet stability (i.e., P < 10–12 kbar); or

(2) Amphibole + Plagioclase = Melt + Garnet + Ti-oxides ± Clinopyroxene

at higher pressures (Rapp et al., 1991; Rapp and Watson, 1995; Zamora, 2000; Vielzeuf and Schmidt, 2001).

While the role of plagioclase accounts for the sodic nature of the melts, the presence of mafic peritectic phases keep them leucocratic, by locking up the Fe and Mg to very high temperatures (>1100 °C). Trace element characteristics are largely due to the presence of garnet in the residuum (either as a preexisting phase, or as a peritectic product), implying melting at pressures above 10–12 kbar. Therefore, there is now a large consensus on the fact that TTGs are the products of partial melting of mafic lithologies in the garnet stability field.

Despite this large consensus, details of the processes involved are debated. Several parameters can affect the melt composition of the melt, and a large part of the debate focuses on "which set of parameters better matches all characteristics of TTGs".

The geodynamic environment of melting is also debated. Indeed, metabasites can reach melting conditions within the garnet stability field, in several conceivable scenarios:

– A commonly proposed model for the generation of Archean TTGs is that they were generated within a subducting slab of oceanic crust. Under presumably hotter Archean conditions, slab melting was probably favored over slab dehydration, resulting in the relatively easy and widespread generation of TTG melts, rather than dehydration of the slab causing mantle wedge fertilization and eventually leading to the formation of andesites (Martin, 1986, 1987, 1994). Such a process is observed in the formation of adakites, which are in many respects modern-day analogues of Archean TTGs (Martin, 1999; Martin et al., 2005). However, this view has been increasingly criticized in the recent years, on several grounds:

  • Modeling of the thermal structure of the slab is inconclusive. There is no definitive proof that slab melting could be a widespread or universal phenomena in the Archean (review in Bédard, 2006, paragraph. 2.3), but there is no definitive proof of the opposite, either. Actually, such models depend critically on too many unconstrained parameters, such as the potential mantle temperature, mantle composition, thicknesses of oceanic and continental lithosphere and crustal thicknesses, to be able to provide better than semi-quantitative answers.

- It has been suggested that the volume of magmas formed by subduction-type processes is not able to generate the large TTG batholiths observed in Archean terranes (Whalen et al., 2002; Bédard, 2006). Such calculations, however, rely on many rather unconstrainable assumptions (thickness of the subducting slab, 3D shape and volume of the TTG intrusions, precise timing of events, etc.). For instance, in the BBGT, many of the younger 3.2 Ga TTG plutons have recently been suggested to represent upfolded, possibly relatively thin laccoliths, rather than voluminous diapiric bodies (Kisters et al., 2003; Belcher and Kisters, 2005). This makes the issue surrounding whether enough magma can be generated or not somewhat less pertinent.
- In the case of slab melting, felsic TTG liquids would form at relatively low melt fractions (Moyen and Stevens, 2006), raising issues surrounding how they are extracted from the source, and their ascent mechanism through a hot mantle wedge. In the case of modern adakites, high Ni, Cr, and Mg contents are ascribed to melt-mantle interactions during ascent (Kelemen, 1995; Smithies, 2000; Martin and Moyen, 2002; Martin et al., 2005). Evidence for similar processes in Barberton TTGs is, however, cryptic (see below).

  Collectively, it seems likely that Archean slab melting could, and did, occur, but was not the universal process as previously assumed.

- Over an active subduction zone, in underplated basalts undergoing subsequent remelting (Gromet and Silver, 1983; Petford and Atherton, 1996). Assuming the overriding plate was thick enough, underplating of basalts would occur at a sufficient depth to be in garnet stability field, and subsequent remelting would indeed generate TTG magmas.
- At the base of a thick crust, either continental or oceanic, either away from any plate boundary (e.g., oceanic plateau: Maaløe, 1982; Kay and Kay, 1991; Collins et al., 1998; Zegers and Van Keken, 2001; Van Kranendonk et al., 2004; Bédard, 2006; Champion and Smithies, this volume) or over tectonically thickened crust (de Wit and Hart, 1993; Dirks and Jelsma, 1998). Many such models involve delamination of the dense lower crust, resulting in heating of the mafic stack and pervasive melting of its base, accompanied by diapiric rise of the melts or partially molten rocks.

The question of the geodynamic site of TTG formation is difficult to answer solely on geochemical grounds; indeed, all environments discussed above allow metabasalts to melt within the garnet stability field and therefore generate sodic felsic melts, similar to TTGs. The differences between these environments will be subtle, at best, and any interpretation in terms of geodynamic environment requires a sound discussion of petrogenetic processes, and a good understanding of the details of the mechanisms affecting TTG melt geochemistry.

5.6-6. PARTIAL MELTING OF AMPHIBOLITES AND CONTROLS ON THE MELT GEOCHEMISTRY

In this section, we focus on the origin of the dominant, "true TTG" lithologies. As demonstrated above, they belong to two main types: a high-Sr, trondhjemitic sub-series, and a low-Sr, tonalitic to trondhjemitic sub-series.

The composition of TTG rocks is a result of several different parameters. Each of them is discussed below, in order to try and assess whether it can account for the difference between the two sub-series.

5.6-6.1. Composition of the Source

Amphibolites (or metabasic rocks in general) actually encompass a diversity of compositions, and experimental studies have used widely different source materials (Moyen and Stevens, 2006), from plagioclase-dominated to amphibole-dominated sources. However, compiling experimental work shows that for major elements, the composition of the source only marginally affects the composition of the melt. This is not surprising, considering the generally eutectic (or at least eutectoid) nature of partial melting of Earth's rock. Whatever the source composition (within reasonable limits), the melting reactions and stoichiometry will be essentially the same, yielding very similar magmas. This, of course, is not true for trace elements, whose content in the melt is strongly tied to the source characteristics.

Interestingly, during melting of biotite and K-feldspar free lithologies (i.e., most Archean crustal lithologies!), potassium behaves as a trace element, as there is no mineral phase that it enters other than by substituting for other ions. The K_2O contents of the melts – and therefore, to some degree, their nature (e.g., granodioritic vs trondhjemitic) - therefore depends largely on the source composition (Sisson et al., 2005). High-K_2O sources will generate high-K_2O melts, which is essentially the conclusion already arrived at for the "high-K_2O" group (Section 5.6-5.2).

The composition of the source (or sources) of Barberton TTG rocks can be at least estimated. For a purely incompatible element (bulk repartition coefficient D = 0), the batch melting equation (Shaw, 1970) can be simplified as $C_1/C_0 = 1/F$ (where C_1: concentration of the melt, C_0: concentration of the source and F: melt fraction). The melt fraction is, of course, unknown. However, in experimental liquids (Moyen and Stevens, 2006), SiO_2 is linearly correlated to F, such that the latter can be at least estimated. Here, we use the following equation:

$$F = \frac{\frac{SiO_2}{52} - 1.525}{-0.011}$$

to estimate the melt fraction.

It is therefore possible to recalculate the (possible) source composition for each sample. The results are plotted in a multi-element, N-MORB normalized diagram (Sun and McDonough, 1989) (Fig. 5.6-12). Importantly, the concentrations predicted are only minimal estimates, as we assumed a D value of 0: if D is higher, the source composition must consequently be higher as well. Obviously, moving to the right of the diagram (towards less incompatible elements), the approximation becomes less correct and the source composition becomes more underestimated.

Two important conclusions can be drawn from these results:
– There are no major differences in terms of probable source compositions between the two groups of TTGs. Both groups can derive from similar sources, suggesting that

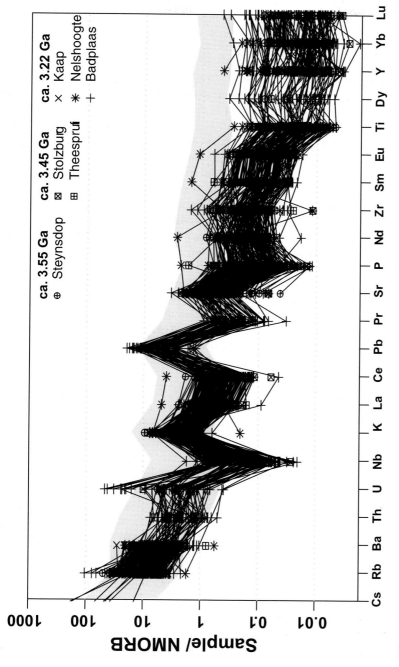

Fig. 5.6-12. MORB-normalized (Sun and McDonough, 1989) multi-elements diagram showing the minimum trace elements concentration of the plausible source of Barberton TTGs (see text). Same caption as Fig. 5.6-9. Grey field: range of compositions for the basalts of the Komatii formation (GEOROC database).

the differences between the low- and high-Sr groups do not reflect diversely enriched sources. Individual plutons show an even bigger homogeneity, except the Badplaas gneisses which display quite a large spread in calculated source composition, consistent with their composite nature in the field.

– The source of all the TTGs was an enriched MORB (>10 times chondritic for the incompatible elements). The apparent negative Nb anomaly that appears is probably an artifact. This calculation predicts minimum estimates for the source concentrations; the more compatible the element, the more underestimated the concentration. Owing to the presence of phases with high affinity for Nb in the residuum (rutile), it is likely that the D value for Nb will be rather high, and much higher than for the neighboring elements. The predicted composition, more enriched than a Phanerozoic MORB, is however, in good agreement with the composition proposed for Archean MORBs (Jahn et al., 1980; Condie, 1981; Jahn, 1994). Regionally, the basalts from the Komatii formation (from the GEOROC database, http://georoc.mpch-mainz.gwdg.de/georoc/Start.asp) also show similar compositions (Fig. 5.6-12), including a small positive Pb anomaly, which is present in the modeled source composition. A similar, slightly enriched source composition is also predicted for (some of the) granitoids in the Pilbara Craton (Champion and Smithies, this volume) and in Finland (Martin, 1994).

5.6-6.2. Conditions (Temperature and Depth) of Melting

To better constrain the melting conditions, we modeled the compositions of primary melts from amphibolites as a function of the P-T conditions. The composition of the final rocks would obviously be modified by further magmatic evolution (e.g., fractional crystallization), as discussed below (Section 5.6-6.3).

5.6-6.2.1. Principle of the model

Based on a parameterization of published experimental data, we proposed a generalized model for vapour-absent partial melting of tholeiitic amphibolites (Moyen and Stevens, 2006). A vapour-absent scenario is favoured for reasons detailed in the cited paper, one of the most compelling being the unlikehood of free water surviving in the crust at 10–20 kbar. Major elements in the melt are interpolated from published melt compositions, with a linear equation of the form $(C_{melt}/C_{source}) = aF + b$, where F is the melt fraction and a and b are two empirically determined coefficients. The a and b coefficients used are slightly modified from Moyen and Stevens (2006), the largest modification affecting the parameters for Na_2O (we now use $a = 0.025$ and $b = 0.60$ in the garnet-amphibolitic domain; $a = 0.060$ and $b = 0.6$ in the eclogitic domain). For high melt fractions ($F > 0.4$), the validity of the approximation becomes doubtful, and we simply calculate the high-F melts as a weighted average of a $F = 0.4$ melt and the source. This approximation is still questionable, but not that important, as $F = 0.4$ corresponds to melts with 62% SiO_2, which is less than most of the rocks studied here.

Trace elements are calculated using an equilibrium melting equation, K_d values from Bédard (2005, 2006), and mineral proportions interpolated from experimental data (Moyen and Stevens, 2006). According to the conclusions above (Section 5.6-6.1 and Fig. 5.6-12), a relatively enriched source composition is used (Sr = 240 ppm and Y = 20 ppm, within the range of the compositions of the non-komatiitic basalts of the Onverwacht Group in GEOROC database).

5.6-6.2.2. *Variations in P-T space*

The single most important parameter controlling the geochemistry of melts from metabasites is the degree of melting: higher degrees of melting (corresponding to higher temperatures) correspond to more mafic melts. Assuming both are primary melts of similar sources, trondhjemites corresponds to melt fractions lower than ca. 20% (Moyen and Stevens, 2006), whereas tonalites reflect melt fractions up to 40–50%. Experimentally, melt fractions sufficiently high to generate a ~65% SiO_2 liquid (equivalent to the tonalites) are attained at ca. 1000 °C, below 15 kbar, but require higher temperatures as pressure goes up (to ca. 1200 °C at 30 kbar) (Moyen and Stevens, 2006). Likewise, CaO/Na_2O values between 0.5 and 1, typical of the tonalitic rocks, correspond to the same P-T range. In contrast, the high silica, low CaO/Na_2O trondhjemites are generated at temperatures below 1000 °C.

The depth of melting controls the nature of the solid phases (residuum) in equilibrium with the TTG melts. There is a potentially major difference between low to medium pressure assemblages (amphibole and plagioclase stable, with garnet present but not abundant, and Ti mostly accommodated in ilmenite), and high pressure (eclogitic) assemblages dominated by clinopyroxene and garnet, with rutile as the main titaniferous phase. To complicate further, even at sub-eclogitic pressures, amphibole and plagioclase are consumed by the melting reactions, such that high melt fractions will coexist with amphibole- and plagioclase-free restites that are mineralogically rutile-free eclogites (Moyen and Stevens, 2006).

Experimentally, both amphibolitic (Winther and Newton, 1991; Sen and Dunn, 1994; Patiño-Douce and Beard, 1995; Rapp and Watson, 1995) and eclogitic (Skjerlie and Patiño-Douce, 2002; Rapp et al., 2003) residuum have been demonstrated to be in equilibrium with TTG liquids. This is unsurprising, since both an eclogitic (clinopyroxene + garnet) and an amphibolitic (amphibole + plagioclase) residuum have similar major elements compositions, except for Na_2O. Sodium is indeed less abundant in eclogitic assemblages, resulting in high-pressure melts that are typically more sodic than their low-pressure counterparts for a given melt fraction (Moyen and Stevens, 2006). But a more important effect is associated with the melt fraction formed. In P-T space, the melt abundance curves are positively sloped, such that at high pressures the same melt fraction is approached only at higher temperatures, as mentioned above.

Combining both parameters allows the identification of low-pressure liquids (relatively high-melt fraction, sodium poor liquids: granodiorites and tonalites) and the high-pressure liquids (lower melt fraction, more leucocratic and more sodic liquids: trondhjemites). A major "dividing line" thus exists, separating tonalites (and granodiorites) from trondhjemites (Fig. 5.6-13). The same division is observed in Barberton TTGs, where the

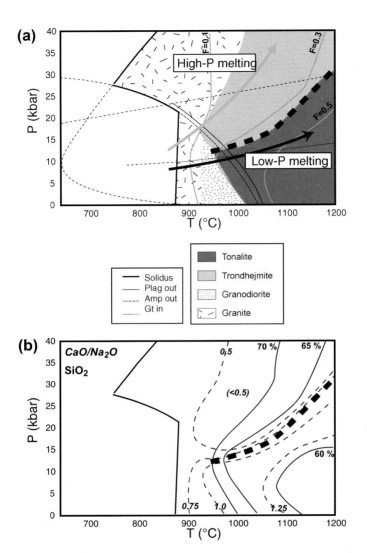

Fig. 5.6-13. Melt composition in PT space, from parameterization of experimental data (Moyen and Stevens, 2006); a "ThB" source (tholeiitic basalt) has been used. (a) Nature of the liquid formed (in O'Connor (1965) systematics) as a function of the P-T conditions of melting. The grey lines represent 10, 30 and 50% melt (*F* value). Fine lines correspond to the solidus and to the mineral stability limits (plag: plagioclase, amp: amphibole, gt: garnet). The two arrows labeled low and high pressure melting graphically display two possible geotherms leading to the formation of trondhjemites and granodiorites to tonalites, respectively. (b) Major element composition of the melts. The lines correspond to iso-values of SiO$_2$ contents and CaO/Na$_2$O ratios of the melts. The thick dashed line is the "tonalite-trondhjemite divide" (panel (a), see text). (c) Sr contents of the melts in P-T space. (d) Sr/Y values of the melts in P-T space.

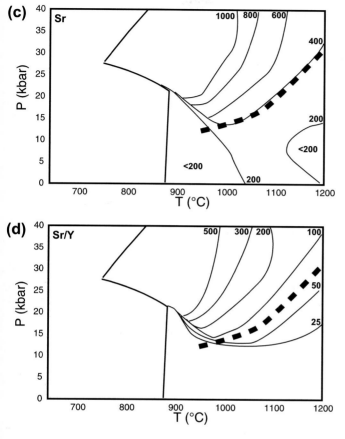

Fig. 5.6-13. (*Continued.*)

"low-Sr" group plots in the tonalite and granodiorite field in O'Connor (1965) diagrams, while the high-Sr rocks are almost exclusively trondhjemitic.

Trace elements provide slightly different information and are far more sensitive to the pressure of melting. Indeed, trace elements will be partitioned in markedly different ways in eclogitic (garnet-clinopyroxene-rutile) and amphibolitic (amphibole-plagioclase-ilmenite) assemblages. In addition, the mode of each mineral also changes with pressure (garnet becomes more abundant at higher pressure). Even within the realm of amphibolitic or eclogitic residues, melt composition vary significantly as a function of depth (Moyen and Stevens, 2006).

For the elements used here (La, Yb, Sr, Y), the main control is exerted by the abundance of high K_d phases; i.e., garnet (for Y and Yb) and plagioclase for Sr. Therefore, trace elements in this case mostly record "pressure" information, with low pressure melts coexisting with plagioclase but not garnet, and having low Sr but high Y and Yb contents, whereas at high pressures, Sr is released because of plagioclase breakdown, but Y and Yb are locked in the garnet. Collectively, low Yb and high Sr/Y melts are produced only at relatively high pressures (>15–20 kbar); below this threshold, higher Yb and lower Sr/Y values are observed. Fig. 5.6-14 summarizes the geochemical trends predicted by both low- and high-pressure melting.

Combining these observations allows the clear discrimination of the two sub-series. High-Sr melts are only trondhjemitic, and they form at high pressure (to the left of the dividing line), plotting in the P-T space from 1000 °C at 15 kbar and below to 1200 °C at 30 kbar. The low-Sr group contains tonalites and granodiorites (in O'Connor's terminology, even if they are trondhjemites on the basis of their field appearance and mineralogy) and forms on the high-temperature side of this divide, at pressures below 15–20 kbar.

It is worth noting that both types denote very contrasting geothermal gradients. High-Sr TTGs formed at relatively low temperatures (probably around 1000 °C), but high pressures (>15 kbar), corresponding to a 15–20 °C/km apparent geotherm. In contrast, the low-Sr group formed at lower pressures (10–15 kbar) and comparable or higher temperatures, corresponding to a distinct geotherm of 30–35 °C/km.

The model used here is dependent on the exact parameters used (position of the mineral stability lines, source composition, etc.). A more detailed treatment of the different cases is presented elsewhere (Moyen and Stevens, 2006). Importantly, however, even if the actual values are dependent on the model parameter, the same logic and the same opposition (low P, low Sr/Y, high F melts vs high P, low F, high Sr/Y melts) remains.

Interestingly, all of the Barberton TTGs are high-Al (Barker and Arth, 1976), and correspond to the Pilbara high-Al group of Champion and Smithies (this volume). These authors proposed that the difference between low Al (and low Sr) and high-Al groups reflects the depth of melting and stability of plagioclase in the residuum. In this model, both sub-series form at pressures above the plagioclase stability field (Fig. 5.6-13), yet the geochemistry of the melts evolves with depth, allowing a distinction between the two sub-series described here.

5.6-6.3. The Role of Fractional Crystallization Following Melting

While fractionation has always been recognized as one possible process affecting TTG composition (e.g., Martin, 1987), it is generally regarded as a minor process that only marginally affects TTG composition. However, it has recently be suggested (Bédard, 2006) that it plays a far bigger role in shaping the trace element composition of Archean TTGs in general (and their high Sr/Y ratio in particular), and that equivalents of Barberton trondhjemites can be generated by fractional crystallization and differentiation of tonalites. The question is actually two-fold: (i) can fractional crystallization turn the low-Sr tonalites into low-Sr trondhjemites; and (ii) can fractional crystallization differentiate (low-Sr) tonalites into high-Sr tonalites?

To investigate the potential effects of fractional crystallization, we modeled the differentiation of a ca. 65% SiO_2 tonalite (Table 5.6-3), using three different mineral assemblages: amphibole + biotite (model 1; Bédard, 2006); plagioclase + amphibole (model 2;

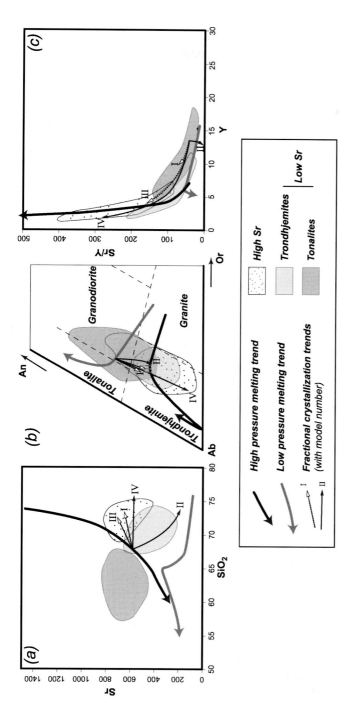

Fig. 5.6-14. Modelled melting and fractionation trends in binary or ternary diagrams. The field of low-Sr (tonalites and trondhjemites), and high-Sr (trondhjemites) are shown for comparison. Heavy arrows: melting trend from the solidus to ca. 1200 °C. Grey: low pressure (13 kbar) melting; black: high-pressure (21 kbar) melting. Thin arrows: fractionation vectors; the length of the arrow corresponds to the biggest possible degree of fractionation (see text and Table 5.6-3). The dotted arrows correspond to models I and III, which do not fit the data. Note how the compositional spread of each individual rock unit is "shaped" by fractionation vectors (model II, hornblende + plagioclase most likely), whereas their position in the diagrams is better explained by the melting trend. (a) Sr vs SiO₂. (b) Part of the Albite (Ab) – Anorthite (An) – Orthoclase (Or) diagram. (c) Sr/Y vs Y.

Table 5.6-3. Modelling fractional crystallization of a tonalite

Mineral compositions and K_D's

Major elements composition

	SiO_2	Al_2O_3	FeO	MgO	CaO	Na_2O	K_D Sr	Y	La	Yb
Amphibole	42.2	13.5	15.5	12.0	12.2	2.0	0.389	2.47	0.319	1.79
Clinopyroxene	52.2	8.7	9.1	9.9	15.7	3.5	0.032	0.603	0.028	0.635
Garnet	38.5	20.9	25.1	5.3	7.6	0.1	0.019	14.1	0.028	23.2
Ilmenite			50.0				0.0022	0.037	0.015	0.13
Plagioclase	60.7	24.6	0.2	0.1	6.0	8.0	6.65	0.138	0.358	0.094
Bioite	36.3	20.2	16.7	12.7	0.0	0.4	0.1	0.07	0.02	0.11
Magnetite			100.0				0.022	0.018	0.015	0.018
Titanite	30.0		0.1		27.2		2.68	5.42	4.73	3.02
Zircon	32.0						20	80	26.6	490
Epidote	36.9	30.9	4.1		23.3	2	4.3		2.05	2.96
Allanite							10.3	9.18	1005	1.59
Apatite					52.0		1.4	17.5	12	13

Source (undifferentiated liquid – low SiO2 tonalite at 64–66% SiO2)

	SiO_2	Al_2O_3	FeO	MgO	CaO	Na_2O	Sr	Y	La	Yb	Sr/Y	La/Yb
Co	65.14	15.57	2.9973	2.49	4.12	5.17	560.0	13.5	15.0	0.8	41.5	18.8

Table 5.6-3. (Continued)

Model 1 (Bedard, 2006 – Amphibole + Biotite)

	SiO₂	Al₂O₃	FeO	MgO	CaO	Na₂O	Sr	Y	Sr/Y	La	Yb	La/Yb
Bulk repartition coefficient D						1.70	0.43	2.36	1.42	2.57		
Cumulate	40.76	14.57	15.78	11.75	10.65							
Fractionated liquids												
5%	66.42	15.62	2.32	2.00	3.78	5.35	576.7	12.6	45.8	14.7	0.7	19.9
10%	67.85	15.68	1.58	1.46	3.39	5.56	594.8	11.7	50.8	14.4	0.7	21.2
15%	69.44	15.75	0.74	0.86	2.97	5.78	614.6	10.8	56.7	14.0	0.6	22.6
20%	71.24	15.82	<0	0.18	2.49	6.04	636.2	10.0	63.8	13.7	0.6	24.3
25%	73.27	15.90	<0	<0	1.94	6.33	660.2	9.1	72.2	13.3	0.5	26.7
30%	75.59	16.00	<0		1.32	6.66	686.7	8.3	82.5	12.9	0.5	28.3
...												
80%	>100	19.58	<0	<0	<0	19.04	1406.0	1.5	923.4	7.7	0.1	120.4

Model 2 (Martin, 1987 – Plagioclase + Amphibole)

	SiO₂	Al₂O₃	FeO	MgO	CaO	Na₂O	Sr	Y	Sr/Y	La	Yb	La/Yb
Bulk repartition coefficient D						5.50	4.09	1.05		0.34	0.76	
Cumulate	52.53	19.88	6.92	4.77	8.36							
Fractionated liquids												
5%	65.80	15.34	2.79	2.37	3.90	5.15	477.8	13.5	35.5	15.5	0.8	19.2
10%	66.54	15.09	2.56	2.24	3.65	5.13	404.3	13.4	30.1	16.1	0.8	19.6
15%	67.36	14.81	2.30	2.09	3.37	5.11	338.8	13.4	25.3	16.7	0.8	20.1
20%	68.29	14.49	2.02	1.92	3.06	5.09	280.8	13.3	21.0	17.4	0.8	20.6
25%	69.34	14.13	1.69	1.73	2.71	5.06	230.0	13.3	17.3	18.1	0.9	21.2
30%	70.54	13.72	1.32	1.51	2.30	5.03	185.8	13.3	14.0	19.0	0.9	21.8
45%	75.46	12.04	<0	0.63	0.65	4.90	88.1	13.1	6.7	22.3	0.9	24.7
...												
80%	>100	<0	<0	<0	<0	3.85	3.9	12.4	0.3	43.6	1.2	37.0

Table 5.6-3. (Continued)

Model 3 (Garnet + Clinopyroxene)

	SiO₂	Al₂O₃	FeO	MgO	CaO	Na₂O	Sr	Y	Sr/Y	La	Yb	La/Yb
Bulk repartition coefficient D							0.03	7.35		0.03	11.92	
Cumulate	45.35	14.80	17.13	7.61	11.62	1.82						
Fractionated liquids												
5%	66.18	15.61	2.25	2.22	3.73	5.35	588.7	9.7	60.4	15.8	0.5	34.5
10%	67.34	15.66	1.43	1.92	3.29	5.54	620.6	6.9	89.8	16.6	0.3	65.6
15%	68.63	15.71	0.50	1.59	2.80	5.76	656.1	4.8	136.4	17.6	0.1	129.5
20%	70.09	15.76	<0	1.21	2.25	6.01	696.0	3.3	212.7	18.6	0.1	266.2
25%	71.74	15.83	<0	0.78	1.62	6.29	741.2	2.2	341.3	19.8	0.0	573.4
30%	73.62	15.90	<0	0.30	0.91	6.61	792.8	1.4	565.8	21.2	0.0	1302.3
...												
80%	>100	18.65	<0	<0	<0	18.57	2687.4	4.9E−04	5.5E+06	71.7	1.9E−08	3.8E+09

Table 5.6-3. (Continued)

Model 4 (Garnet + Epidote)

	SiO$_2$	Al$_2$O$_3$	FeO	MgO	CaO	Na$_2$O	Sr	Y	Sr/Y	La	Yb	La/Yb
Bulk repartition coefficient D							1.01	9.20		1.04	13.08	
Cumulate	37.70	25.90	14.63	2.64	15.46	0.07						
Fractionated liquids												
5%	66.58	15.03	2.39	2.48	3.52	5.44	559.7	8.9	63.1	15.0	0.4	34.8
10%	68.19	14.42	1.70	2.47	2.86	5.74	559.4	5.7	98.3	14.9	0.2	66.7
15%	69.98	13.75	0.94	2.46	2.12	6.07	559.1	3.6	157.0	14.9	0.1	132.7
20%	72.00	12.99	0.09	2.45	1.29	6.45	558.8	2.2	258.0	14.9	0.1	275.4
25%	74.29	12.13	*<0*	2.44	0.34	6.87	*558.5*	*1.3*	*437.7*	*14.8*	*0.0*	*598.9*
30%	76.90	11.14	*<0*	2.43	*<0*	7.36	*558.1*	*0.7*	*770.2*	*14.8*	*0.0*	*1374.6*
...												
80%	*>100*	*<0*	*<0*	*1.89*	*<0*	*25.57*	*551.5*	*2.5E−05*	*2.2E+07*	*14.1*	*2.9E−09*	*4.9E+09*

Major elements composition are calculated using mass balance, and trace elements using Rayleigh's law. The source composition C$_0$ is taken as the average of the low-SiO$_2$ tonalites between 64 and 66% SiO$_2$. Partition coefficients (K$_d$) are taken from Moyen and Stevens (2006). Mineral compositions are either real minerals from TTG gneisses (Martin, 1987), or mineral in equilibrium with melts in experiments (Zamora, 2000). Three models are calculated with different mineral proportions: model 1 (Bédard, 2006): 82% amphibole, 15% biotite, 0.5% magnetite, 0.3% titanite, 0.2% zircon, 1.5% epidote, 0.1% allanite, 0.4% apatite. Model 2 (Martin, 1987): 39.25% amphibole, 1.5% ilmenite, 59.25% plagioclase. Model 3: 50% clinopyroxene, 50% garnet. For each model, the bulk reparation coefficient D and the major elements cumulate composition is given; the major and trace elements composition of the fractionated liquids is given for increasing degrees of fractionation. Impossible values for major elements (<0, meaning that the fractionation cannot process to that stage) are indicated in the left-hand side table; corresponding trace elements values are italicized.

Martin, 1987); garnet + clinopyroxene (model 3) and garnet + epidote (model 4, representing high-pressure fractionation; Schmidt, 1993; Schmidt and Thompson, 1996).

In both sub-series, Al_2O_3 (Fig. 5.6-5) is negatively correlated with SiO_2. This behaviour is not predicted by models 1 and 3; only models 2 (plagioclase + amphibole) and 4 (epidote + garnet) correctly predicts a decrease of Al_2O_3 with differentiation. Sr decreases with increasing SiO_2 (Fig. 5.6-9(a)), as correctly predicted only by model 2. Model 4 also predicts an uncommon behavior for Ni, which, owing to the low K_d of this element in epidote (0.1: Bédard, 2006) and its moderate K_d in garnet ($\sim$1.2), remains at constant concentrations or even increases. This results in the dramatic increase in Ni/Cr ratios predicted during differentiation in model 4. Such behavior is not observed in Barberton TTGs, nor in TTGs elsewhere in the world.

To achieve significant changes in trace elements signatures, high degrees of fractionation are required. This seems difficult to achieve, especially in high viscosity felsic melts. Such a degree of fractionation is also difficult to achieve on geochemical grounds, as fractionation of amphibole+plagioclase (Martin, 1987), for instance, would run out of MgO after about 40% of the crystals are removed from the melt; fractionation of biotite + amphibole (Bédard, 2006) would use up all FeO even faster, after about 20% fractionation (Table 5.6-3). K_2O, and to a lesser degree Na_2O, would likewise be limiting factors. This put an upper boundary on the amount of crystals that can be formed out of the melt in such models and, accordingly, to the effect of fractional crystallization on trace elements.

Starting with a liquid with a Sr/Y of ca. 40, possible fractionation (in terms of major elements) is sufficient to evolve a tonalite (ca. 65% SiO_2) into a trondhjemite (ca. 72% SiO_2), but can not raise the Sr/Y values of the differentiated liquids above 60, 150 and 250 (models 1, 3 and 4, respectively) and Sr/Y actually decreases slightly in model 2. Only the high-pressure fractionation models (3 and 4) have the potential to bring the Sr/Y ratios to the high values featured by the high-Sr trondhjemites.

In summary, only models 2 (plagioclase + amphibole) and 4 (garnet + epidote) can partially fit the data. Model 2 is able to reproduce the trends observed within each rock type, but has only a limited effect and can barely fractionate the tonalites into trondhjemites. It is also unable to change low-Sr rocks into high-Sr rocks and can also not account for the high Sr/Y values in the (high-Sr) trondhjemites, as the fractionation of amphibole + plagioclase has no noticeable effect on Sr/Y values of the melts. Model 4, on the other hand, has a more pronounced effect on the melt compositions, and could result in evolution of low-Sr tonalites into high-Sr trondhjemites. But the fit with the data is poorer (elements such as Sr and Ni are not convincingly modeled). Furthermore, model 4 calls for fractionation of garnet and epidote, a high pressure (>20 kbar: Schmidt, 1993; Schmidt and Thompson, 1996) and high water activity assemblage, regardless of whether the high Sr/Y is related to high pressure melting (as proposed Section 5.6-6.2.2), or to high-pressure fractionation. Nevertheless, it points to evolution at pressures >20 kbar for the high-Sr sub-series, but such pressures are not required for the low-Sr group. Finally, while fractionation of epidote + garnet can change a low-Sr liquid into a high-Sr liquid, the reverse is not true and it appears impossible to fractionate a low-Sr tonalite formed at *shallow* depth (see Section 5.6-6.2.2) under *high*-pressure conditions!

Collectively, it seems that if fractionation played a role in the geochemical evolution of Barberton TTGs, it was only minor. The geochemical trends for at least some of the plutons are shaped by late fractionation (amphibole + plagioclase), probably reflecting liquid-crystal separation during emplacement. It is also possible that the high-Sr (and high Sr/Y) signature of some deeply generated trondhjemites was enhanced by some high-pressure fractionation. But fractionation cannot account for the difference between the low- and high-Sr sub-series: they represent two fundamentally different sub-series, reflecting different conditions of melting (Fig. 5.6-14). In addition, fractionation can barely explain the difference between tonalites and trondhjemites, and in all likelihood this difference also reflects different melting conditions (temperatures).

5.6-6.4. *Possible Interactions with the Mantle*

If melting occurs at great depth (whatever the context, see below), it will most likely occur below a peridotite layer. Therefore, the TTG magma rising to the surface will have to cross a large volume of peridotite and will most likely interact with it, resulting in the formation of "hybrid" TTGs (Rapp et al., 2000; Rapp, 2003; Martin et al., 2005). It has been proposed (Smithies, 2000; Martin and Moyen, 2002) that the secular increase of Mg#, Ni and Cr in TTGs reflects progressively deeper melting, allowing more pronounced interactions with the mantle. At the extreme end of this spectrum of melt-mantle interactions is the formation of "sanukitoids" (Martin et al., 2005). Sanukitoids are characterized by both elevated Mg, Ni and Cr contents and significant LILE and REE enrichments, typically with relatively high K/Na ratios (Moyen et al., 2003). High HFSE levels are also common. This association is not found in any of the Barberton TTG, and we see no evidence for interactions between TTG melts and the mantle in the BGGT.

5.6-7. SUMMARY AND GEODYNAMIC IMPLICATIONS

5.6-7.1. *Petrogenetic Processes for Individual Plutons*

5.6-7.1.1. *The ca. 3.55–3.50 Ga Steynsdorp pluton*
Despite only relatively few analyses being available, the Steynsdorp pluton appears to be made up of two components; low-Sr tonalites and high-K_2O granodiorites. The existing data and the discussion above suggest that the tonalitic component represents relatively low depth, high melt fraction liquids from amphibolites. The granodiorites, interleaved with the tonalites (Kröner et al., 1996), display the characteristics trends and high-K_2O nature of the "secondary" liquids, which formed by remelting of pre-existing TTG, probably equivalents of the associated tonalites.

5.6-7.1.2. *The ca. 3.45 Ga group (Stolzburg and Theespruit plutons)*
Intrusive phases of the Stolzburg and Theespruit plutons are fairly homogeneous. They are leucocratic trondhjemites, mostly belonging to the high-Sr, high-pressure, low-melt fraction group. Isotopically, their source was the most depleted of the studied rocks.

A somewhat surprising feature of the ca. 3.45 Ga high-Sr TTGs, however, is that despite their probable deep origin (>20 kbar; e.g., more than 60 km), there is no clear evidence for interaction with the mantle in the geochemical signature of this group.

5.6-7.1.3. The 3.29–3.24 Ga Badplaas gneisses

The Badplaas gneisses are the most complex and composite unit of the BGGT plutons. They include all 4 rock types identified regionally: high- and low-Sr "true" TTGs, together with high-K_2O rocks and matching "melt-depleted" samples, both probably related to re-melting of the newly emplaced TTGs. The true TTGs belong to the two sub-series, demonstrating that the Badplaas gneisses were formed from sources at different depths. Therefore, it seems that the Badplaas gneisses recorded a long (ca. 50 My) and complex history of melting of a vertically extensive source region, accretion of a "proto Badplaas terrane" and remelting of this terrane, possibly during the ca. 3.22 Ga subduction-collision event. Proper interpretation of the geochemistry of the Badplaas gneisses, however, would require a more detailed, field-constrained study of the different units, which is beyond the scope of the present work.

5.6-7.1.4. The ca. 3.23–3.21 Ga Nelshoogte pluton

The Nelshoogte pluton is a composite intrusion, made up of early trondhjemitic phases belonging to the low-Sr group (although quite close to the boundary with the high-Sr sub-series), intruded by a later set of low-Sr tonalites, clearly cutting across the earlier lithologies. This indicates a succession of melting conditions at moderate depths but with increasing temperatures, consistent with the emplacement of this pluton during orogenic collapse of the BGGT 3.22 Ga "orogen" (Belcher et al., 2005). The relatively enriched isotopic characteristics of the Nelshoogte pluton are consistent with melting of the preexisting Onverwacht (or even Fig Tree) supracrustals, and also support this model.

5.6-7.1.5. The ca. 3.23–3.22 Ga Kaap Valley tonalite

The Kaap Valley pluton is almost exclusively made of phases belonging to the low-Sr group, pointing to shallow, high-melt fraction melting. Isotopic characteristics also suggest a slightly enriched (Onverwacht-like) source, whereas the emplacement history is also consistent with syn-exhumation intrusion. The relatively high REE contents of the Kaap Valley pluton (compared to the other TTGs) has been interpreted as precluding simple derivation by melting of a common source (Robb et al., 1986). Indeed, the isotopic data also points to a slightly different origin for the Kaap Valley tonalite compared to the other TTG plutons. We suggest that these differences mostly reflect melting of the (relatively enriched) Onverwacht supracrustals (mostly mafic and ultramafic lavas, but possibly with incorporation of a minor sedimentary component). The unique nature of the Kaap Valley pluton would, therefore, reflect both a slightly different (more fertile) source and a higher temperature of melting compared to the other TTG plutons, the combination of both parameters resulting in higher melt fractions and the generation of a dominantly tonalitic pluton, unique in the BGGT. It seems, therefore, that the Kaap Valley pluton mostly reflects partial melting of the base of an already formed crust (Onverwacht Group-like).

5.6-7.2. Geodynamic Model

In addition to the geochemical information presented above, the geodynamic evolution of the Barberton Greenstone Belt has been discussed in other papers in this volume. Our geochemical and geodynamical conclusions fit with this model, and allow us to refine it in some aspects. As the geological history of each event is partially erased by subsequent events, it will be presented backwards, starting with the youngest.

5.6-7.2.1. Ca. 3.23–3.21 Ga: main event of terrane accretion

The dominant geological event that shaped the present-day structure of the belt occurred at ca. 3.23 Ga. Structural (de Wit et al., 1992; de Ronde and de Wit, 1994; de Ronde and Kamo, 2000; Kisters et al., 2003) as well as metamorphic (Dziggel et al., 2002; Stevens et al., 2002; Kisters et al., 2003; Diener et al., 2005; Dziggel et al., 2005; Diener et al., 2006; Moyen et al., 2006) studies suggest collision (or arc accretion) between two relatively rigid blocks, separated by the Inyoni–Inyoka tectonic system (Lowe, 1994). The western terrane has largely been overprinted by the ca. 3.25–3.21 Ga rocks (Fig Tree lavas and TTGs), but was probably built on a nucleus of slightly older (3.3–3.25 Ga: de Ronde and de Wit, 1994; Lowe, 1994; Lowe and Byerly, 1999; Lowe et al., 1999; de Ronde and Kamo, 2000) mafic and ultramafic lavas, possibly an oceanic plateau of some sort. The eastern terrane is better preserved and was at this time a composite unit including old lavas and sediments intruded by ca. 3.45 Ga TTGs and overlain by still younger mafic/ultramafic lavas. It is interpreted to represent an oceanic plateau that was modified by a relatively minor subduction event (see below). The accretion itself occurred via under-thrusting (subduction?), and the eastern, high-grade Stolzburg terrane probably represents the lower plate of this event (Fig. 5.6-15).

In this context, the ca. 3.29–3.21 Ga plutons record the transition from pre-collision to post-collision magmatism. The earliest phases formed by deep melting (high Sr parts of the Badplaas gneisses) and their ages correspond to the accretion stage of the BGGT, most probably in a magmatic arc (de Ronde and Kamo, 2000; Kisters et al., 2006). The latest phases (low-Sr rocks in the three units) formed by relatively shallow (10–12 kbar) melting of amphibolites, possibly parts of the Onverwacht Group. The transition from low-Sr trondhjemites (bulk of the Nelshoogte pluton, part of Badplaas gneisses) to low-Sr tonalites (late phases in the Nelshoogte pluton, Kaap Valley pluton) reflect increasing temperatures at the base of the collapsing pile, as commonly observed in post-orogenic collapse (Kisters et al., 2003). Some of the early rocks underwent intracrustal remelting, more or less at the same time (mostly in the Badplaas pluton). Field and structural studies demonstrate that at least some of these plutons formed during orogen parallel extension, all of which is consistent with lower crustal melting of the thickened, dominantly mafic crust during orogenic collapse, and/or possibly during slab breakoff.

From south to north (i.e., from Badplaas to Kaap Valley), there is an overall evolution towards younger and lower silica rocks, reflecting the switch from syn-subduction or

Fig. 5.6-15. Geodynamic model for the evolution of the BGGT, with emphasis on the formation and emplacement of TTG plutons. On the right hand side, a time scale shows the position of the cartoons in the global evolution of the BGGT. Note that, for the cartoons on the left, the time scale is distorted. Also note that the scale is *not* a stratigraphic scale, as the younger stages are at the bottom. Left-hand side cartoons are approximately at the same scale, looking towards the (present-day) northeast; the front section of each block corresponds to a NW-SE cross-section. In each cartoon, the active plutonism is in black, whereas rocks that have already been emplaced are in grey. Symbols denote the melting zone: stars are for melting of amphibolites (grey: deep, generating high-silica trondhjemites, black forming low-silica tonalites); white triangles denote melting of already formed felsic crust. In the top stage (Steynsdorp), two alternatives are proposed: intra-plate accretion of an oceanic plateau, followed by remelting at its base generating low-pressure TTGs (left), or low-pressure melting at the base of a tectonic stack of oceanic crust. The last cartoon shows more or less the relative positions of individual geological elements (that have been only marginally modified by the later, ca. 3.1 Ga events). Plutons: B: Badplaas, N: Nelshoogte, KV: Kaap Valley, S: Stolzburg, Ts: Theespruit. Structures: IF: Inyoka Fault, ISZ: Inyoni Shear Zone. Cartoons are modified from Moyen (2006), the top three are inspired from Lowe (1999).

---

collision to syn-collapse magmatism: the latest, collapse-related magmatic event is better represented in the northern plutons. This could reflect some along-strike differences between the southern segment of the orogen, which involved an already rigid continental nucleus (the already-formed Stolzburg terrane), and the northern segment, where no evidence for rigid crust is documented and which could have been a less consolidated volcanic arc at the time.

5.6-7.2.2. Ca. 3.45 Ga: accretion of the Stolzburg domain
The origin of the continental Stolzburg domain is somewhat obscured by the dominant, ca. 3.23–3.21 Ga collision. The composition, mirroring a deep source, of the 3.45 Ga old Stolzburg and Theespruit plutons suggests that they could have intruded as suprasubduction zone plutons into a small, mafic to ultramafic crustal block. This is consistent with their shallow level of emplacement. Existence of a still older crust (the lower Onverwacht Group and the Steynsdorp pluton) suggests that this subduction occurred along the margin of a pre-existing "proto-continent" (whatever its nature was, the abundance of komatiites suggests that it was probably an oceanic plateau). After the emplacement of the TTG plutons, renewed komatiitic volcanism at ca. 3.45–3.40 Ga has been interpreted as reflecting the rifting of the newly formed crustal nucleus (Lowe et al., 1999).

5.6-7.2.3. Ca. 3.55–3.50 Ga: the early Steynsdorp continental nucleus
The ca. 3.55–3.50 Ga TTGs of the Steynsdorp pluton apparently formed by shallow melting of amphibolite (and quick remelting of the newly formed felsic lithologies). We suggest that this could represent the very start of the cratonization process, through remelting of the lower part of a thick pile of mafic rocks.

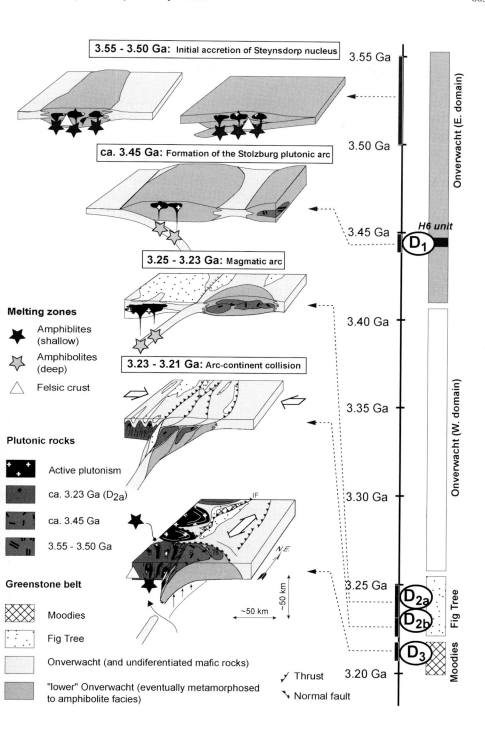

5.6-8. DISCUSSION

5.6-8.1. The Different Sub-Series of TTGs

An important result of this work is the identification of three main types of magmas, all belonging to the wide group collectively referred to as "TTGs", and in fact all being high-Al TTGs (Barker and Arth, 1976; Champion and Smithies, this volume). Firstly, a group of relatively potassic rocks, mostly granites and granodiorites, with some trondhjemites, is derived from melting of relatively felsic, enriched sources such as pre-existing TTGs or felsic components (sediments, felsic lavas) of the supracrustal pile. Secondly, the "true" TTGs are themselves differentiated into high-Sr TTGs that are mostly trondhjemites, and low-Sr TTGs ranging from tonalites to trondhjemites and granodiorites. While fractional crystallization, probably late and emplacement-related, does play a role in shaping the geochemical trends of individual plutons, it cannot explain the first-order differences between low- and high-Sr sub-series, and between tonalites and trondhjemites: the sub-series identified herein correspond to differences between primary melts. The high-Sr sub-series formed by high pressure (>20 kbar) melting of amphibolites, whereas the low-Sr sub-series formed by lower-pressure (and relatively high temperature) melting of amphibolites.

The difference between the three sub-series is important, as each of them corresponds to a significantly different combination of sources and P-T conditions of melting. Any geodynamic reconstitution or tectonic model based on "TTG" magmatism should take into account these differences, as they represent important constraints on our understanding of the crustal evolution of Archean cratons.

5.6-8.2. Comparison with Pilbara Granitoids

5.6-8.2.1. Geochemical observations

TTG granitoids of the same period (3.5–3.2 Ga) are a major lithology in the Pilbara Craton (Champion and Smithies, this volume), such that it is worth drawing some comparisons. In the Pilbara, two main suites of "TTG" (or related) rocks are found; a high-Al, high-Sr group and a low-Al, low-Sr group. In Barberton, two groups of TTGs are also observed (low-Sr tonalites and trondhjemites, and high-Sr trondhjemites), but both of these would fall within the definition of high-Al, high-Sr rocks in the Pilbara. The Pilbara high-Al series ranges from ca. 65–72% SiO_2, broadly corresponding to the range observed in Barberton (low Sr) TTGs. However, the low-SiO_2 series (<68% SiO_2) seem to be less common in the Pilbara than in Barberton (where they form the Kaap Valley pluton). True low-Al series rocks are completely missing from the Barberton rock record. In the Pilbara, the two series are not temporally or spatially distinct, with no clear logic behind the repartition of the types. In contrast, in the BGGT there is a clear repartition of the two rock types; the older plutons (ca. 3.45 Ga) are high-Sr trondhjemites, whereas the younger (3.23–3.21 Ga) plutons are predominantly of the low-Sr series type. Only the complex Badplaas gneisses show, on a smaller scale of a few kilometers, the same degree of internal complexity, both in terms of time and of geochemistry, but again, low-Al rocks are missing from this unit.

Finally, some of the Pilbara TTG (mostly from the low-Al group) are enriched in LILE, including K_2O, which makes them granodioritic rather than trondhjemitic ("transitional TTGs"). They can be compared to the "high K_2O/Na_2O" rocks that we have identified in Barberton TTGs, although we observe a compositional gap between the high-K_2O rocks and the "ordinary" TTGs, rather than the continuous evolution recorded in the Pilbara. In addition, most LILE-enriched TTGs from the Pilbara belong to the low-Al series, which is missing from Barberton. The high K_2O rocks are also rarer in Barberton, where they form minor phases of composite plutons, such as Steynsdorp or Badplaas, and (apart from some dykes) are largely missing from the simple, monogenetic plutons like Stolzburg or even the Nelshoogte and Kaap Valley plutons. However, a large part of what is referred to as the "late GMS suite" (the 3.1 Ga batholiths, clearly distinguished from the older TTG magmatism in our case) in Barberton is actually, geochemically, quite similar to the transitional TTGs of the Pilbara, including relatively low Sr and Al contents (Anhaeusser and Robb, 1983; Yearron, 2003). This suggests that the classical distinction between "TTG gneisses" and "late potassic plutons" (e.g., Moyen et al., 2003) might not be that clear, as the same type of rock can be regarded either as "a LILE-enriched component of the TTG gneisses", or "late potassic plutons", depending on the field relationships. In all cases, the interpretation proposed is quite similar: all groups of rocks are interpreted to reflect the melting of "a LILE-enriched, 'crustal' component" (Champion and Smithies, this volume), "pre-existing felsic lithologies (e.g., tonalites)" (this work), or "the ca. 3.5–3.2 Ga TTG basement" (Anhaeusser and Robb, 1983).

5.6-8.2.2. Petrogenetic models
The petrogenetic models proposed for the Pilbara (Champion and Smithies, this volume) and Barberton (this work) granitoids are quite similar. The "normal" TTGs are regarded as the products of amphibolite melting at different depths, resulting in the distinction between low- and high-Al sub-series. The comparison between Barberton and Pilbara data suggests that the high-Al series can further be subdivided into a low- and high-Sr group. Superimposed on this classification, we observe in Barberton a difference in melt fractions (SiO_2 contents) that leads us to propose different geothermal gradients, as well as different depths of melting, which is apparently not the case in the Pilbara. Fractional crystallization and interactions with a mantle wedge are, in both cases, regarded as minor processes, at best.

"Transitional" (LILE-enriched, high K_2O/Na_2O) facies are regarded as melts of more felsic lithologies. In the Pilbara, Champion and Smithies (this volume) propose that this occurs both at high and low depths, resulting in transitional TTGs belonging both to the low- and high-Al groups. These are interpreted to form at the same time, and in the same regions, as the other TTGs. In Barberton, they mostly belong to the low-Sr group (as in the Pilbara); high-Sr samples are apparently associated with high-Sr "true" TTG plutons (e.g., Stolzburg). We propose that, rather than coeval magmas, they more commonly correspond to later remelting of already emplaced TTGs, at mid-crustal depths.

Our geodynamical inferences differ from these proposed by Champion and Smithies. In the Pilbara, the interlayering of all types of rocks (low and high Al, "normal" and "transitional" TTGs) leads them to propose a model of essentially intracrustal melting of a dominantly basaltic stack, with locally more felsic layers. Progressive differentiation of the crust would lead to increasingly felsic, and increasingly more crustal, sources and account for the relative abundance of transitional TTGs in the later stages. In contrast, in Barberton, the time and space repartition of the different rock types allows them to be fitted into the framework of a "plate tectonics" model (at least at 3.2 Ga, possibly at ca. 3.45 Ga). Here too, the progressive "maturation" of the crust eventually results in the formation of "potassic" magmas (corresponding to the ca. 3.1 Ga GMS suite of the BGGT), forming well defined, younger, clearly distinct batholiths that contrast with the less well-defined "transitional" phase of the Pilbara.

It would then appear that the two cratons followed a somewhat different early evolution. The Pilbara Craton, from 3.45 to 3.3 Ga, apparently evolved essentially in an intra-plate setting (oceanic plateau; see also Van Kranendonk et al. and Smithies et al., this volume), reflected by heterogeneous sources and depth of melting for the granitoids of this time. In contrast, after the initial accretion of "shallow" TTGs (probably through intraplate processes, as well) at ca. 3.55 Ga, the BGGT shows very homogeneous, deeply-originated TTGs at ca. 3.45 Ga. We interpret this to be subduction related. This suggests that some sort of arc fringed the oceanic plateau that was the proto-BGGT at ca. 3.45 Ga, in contrast with the Pilbara nucleus, which is devoid of any such structure.

However, at ca. 3.2 Ga, the evolution of both cratons again becomes similar; e.g., a modern-style arc setting in the Pilbara, based on the geochemistry of ca. 3.12 Ga lavas (Smithies et al., 2007), and a collisional orogenic setting in the Barberton, based on the geochemistry of the ca. 3.2 Ga plutons (Stevens and Moyen, this volume).

5.6-9. CONCLUSIONS

Far from being the homogeneous, monotonous group of rocks that they are commonly assumed to be, TTGs are a complex, composite group encompassing a large family of plutonic rocks showing evidence for a diversity of processes, both in term of emplacement history and geochemistry/petrogenetic history. This suggests that specific attention should be paid to the details of the field relations and geochemistry of the TTG gneisses and to elucidate their intricate histories, as they are more than the simple "basement" to (apparently) more interesting supracrustal lithologies.

The most significant information recorded by the TTG geochemistry is linked to the depth of melting of the source amphibolite. Geochemistry of the high-Al TTGs of the BGGT allows the differentiation between two "sub-series"; a high-pressure and relatively low temperature sub-series of mostly leucocratic trondhjemites ("high Sr sub-series"), and a lower-pressure and higher temperature sub-series of considerably more diverse rocks, ranging from tonalites (and even diorites) to trondhjemites and granodiorites ("low Sr sub-series"). The geothermal gradients of both sub-series record, together with the established tectonic framework of the BGGT, that only the high pressure sub-series (corresponding to the ca. 3.45 Ga plutons and part of the 3.29–3.24 Ga Badplaas gneisses in Barberton) can be regarded as a record of Archean subduction. It seems likely that a similar distinction between low- and high-Sr sub-series may be possible throughout the Archean: at least some recent studies (e.g., Benn and Moyen, in press; Champion and Smithies, this volume) suggest that this distinction applies to other Archean cratons, as well.

The degree of enrichment of the source is also recorded to some degree in the composition of the TTGs, and we can distinguish between "normal TTGs" (melts from amphibolites) and "high-K_2O samples" (melts from more felsic lithologies – either older TTGs, or felsic lavas/sediment components in the source). It is quite possible that further studies will demonstrate further distinctions between more or less enriched sources.

Subordinate factors controlling the composition of TTGs include later fractional crystallization (although reasonable degrees of fractionation do not hugely modify the geochemistry of these rocks) and interaction with mantle rocks (implying some form of lithosphere-scale imbrication of mantle and crust rocks). While minor on a craton scale, these processes can locally be important in the petrogenesis of one specific rock unit, and cannot be *a priori* ignored.

In the BGGT, the evolution from "shallow" tonalites at 3.55–3.50 Ga, to "deep" trondhjemites at 3.45 Ga, to "shallow", complex tonalites and trondhjemites at 3.29–3.24 Ga probably mirrors the formation and evolution of the eastern segment of the Kaapvaal Craton, from the generation of an early crustal nucleus, its subsequent growth via the addition of new material generated along a subduction margin, to its final accretion (and reworking) in a collisional orogen.

ACKNOWLEDGEMENTS

D. Champion and H. Smithies kindly supplied an early draft of their manuscript in this volume that was highly thought-provoking and allowed us to draw fruitful comparisons between our two models. A detailed review by Jean Bédard greatly helped to improve both the content and the form of the manuscript. JFM's post-doctoral fellowship at the University of Stellenbosch was funded by the South African National Research Foundation (grant GUN, 2053698) and a bursary from the Department of Geology, Geography and Environmental Sciences. Running costs were supported by a NRF grant awarded to AFMK (grant no. NRF, 2053186). Access to lands and the hospitality of farmers and residents in and around the town of Badplaas is greatly appreciated.

Earth's Oldest Rocks
Edited by Martin J. Van Kranendonk, R. Hugh Smithies and Vickie C. Bennett
Developments in Precambrian Geology, Vol. 15 (K.C. Condie, Series Editor)
© 2007 Elsevier B.V. All rights reserved.
DOI: 10.1016/S0166-2635(07)15057-X

Chapter 5.7

METAMORPHISM IN THE BARBERTON GRANITE GREENSTONE TERRAIN: A RECORD OF PALEOARCHEAN ACCRETION

GARY STEVENS AND JEAN-FRANCOIS MOYEN

Department of Geology, Stellenbosch University, Matieland, 7130, South Africa

The Barberton Granite Greenstone Terrain (BGGT) has been interpreted to record an accretionary orogeny during which at least two crustal terranes merged along a crustal scale suture zone (de Ronde and de Wit, 1994; Lowe, 1994, 1999; de Ronde and Kamo, 2000). This orogeny has been deemed to be responsible for the main deformation event in the Barberton Greenstone Belt (BGB) (D2), at ca. 3.21 Ga, which is well recorded in the lower parts of the stratigraphy of the belt in the Onverwacht and Fig Tree groups (Viljoen and Viljoen, 1969c; Anhaeusser et al., 1981,1983; Lowe and Byerly, 1999; Lowe et al., 1999). Terrane amalgamation was followed by the deposition of molasses of the Moodies Group, which were themselves subsequently refolded during the late stages of orogeny. In the nearby granitoids, ca. 3.23–3.21 Ga plutons are interpreted as resulting either from arc-type magmatism, or from orogenic collapse (Moyen et al., this volume, and references therein). Relatively high-grade metamorphism in the BGGT is confined to the granitoid domains surrounding the belt and the Theespruit and Sandspruit Formations that form the belt's lower-most stratigraphy. The interior of the belt is typified by lower greenschist facies metamorphism Fig. 5.7-1) (Anhaeusser et al., 1981).

In the modern Earth, accretionary orogens involving collision between oceanic and continental plates are characterized by a particular pattern of regional metamorphic grade distribution. In the lower plate (which is generally linked to a subducted oceanic plate), high pressure and low to medium temperature metamorphism is developed (Chopin, 1984; Bodinier et al., 1988; Ernst, 1988; Chopin et al., 1991; Nicollet et al., 1993; Spear, 1993; Wang and Lindh, 1996), commonly reaching relatively high grades. In the upper plate, lower grade metamorphism develops along typically warmer geotherms (Burg et al., 1984, 1989). This duality of metamorphic types has been recognized as one of the "hallmarks of plate tectonics" and has been proposed as useful in determining the timing of the onset of conventional plate tectonics (Brown, 2007). Thus far, clear evidence for this signature has only been documented from the Proterozoic and Phanerozoic rock record (Brown, 2007).

In contrast, Archean metamorphic conditions are typically interpreted to reflect mostly "hot" and uniform P-T conditions (Percival, 1994; Brown, 2007). Thus, Archean terrains are regarded as lacking metamorphic evidence for collisional orogeny involving oceanic rocks. Furthermore, the typical map pattern of gneissic domes surrounded by narrow, syn-

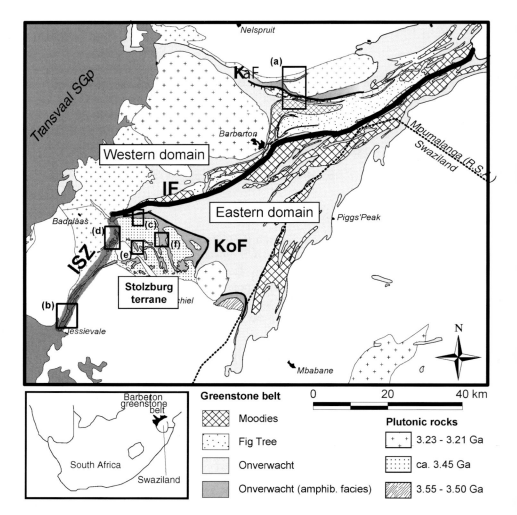

Fig. 5.7-1. Geological map of the Barberton Greenstone Belt (modified after Anhaeusser et al. (1981)). KaF: Kaap River Fault; KoF: Komatii fault; ISZ: Inyoni shear zone; IF: Inyoka–Saddleback fault. The boxes refer to areas were the detailed metamorphic studies reviewed in this paper were conducted. Western domain: (a) Stentor Pluton (Otto et al., 2005; Dziggel et al., 2006), (b) Schapenburg schist belt (Stevens et al., 2002). Eastern domain: (c) Tjakastad schist belt (Diener et al., 2005; Diener et al., 2006), (d) Inyoni shear zone (Dziggel et al., 2002; Moyen et al., 2006), (e) Stolzburg schist belt (Kisters et al., 2003), (f) Central Stolzburg terrane (Dziggel et al., 2002).

formal greenstone belts ("dome and keel patterns") is regarded as contradictory with collision or collision-like processes (Chardon et al., 1996; Choukroune et al., 1997; Chardon et al., 1998; Collins et al., 1998; Hamilton, 1998; Collins and Van Kranendonk, 1999; Van Kranendonk et al., 2004; Bédard, 2006).

Several new studies (summarized in Table 5.7-1) have recently been published on aspects of the metamorphic evolution of the BGGT and, in combination, provide particularly clear insights into the Archean geodynamic processes that shaped the greenstone belt. In this chapter, we review the findings of these studies and show that two fundamentally important aspects emerge. Firstly, that the higher-grade metamorphic margins to the belt are in faulted contact with the lower-grade metamorphic interior, and that these zones are characterized by strong syndeformational isothermal decompression signatures, with peak metamorphic conditions typically reflecting a minimum estimate (particularly for pressure). Secondly, there appear to be two fundamentally different metamorphic signatures in the amphibolite-facies rocks associated with the belt. In the ca. 3.45 Ga and older granitoid-dominated terrane to the south of the belt (Fig. 5.7-1), a relatively low-temperature, high-pressure metamorphic signature is dominant. This contrasts with a significantly higher metamorphic field gradient developed in the amphibolite-facies domains along granite-greenstone contacts on the northern margin of the belt and within greenstone remnants in the far south of the BGGT. The main body of the greenstone belt, although at lower metamorphic grades, also records a signature of relatively high metamorphic field gradient.

In addition to reviewing these metamorphic findings and their significance, this study will propose a model for the development of the dome-and-keel pattern, within the framework of an orogenic process.

5.7-1. EVIDENCE FOR ACCRETIONARY OROGENY IN THE BGGT

5.7-1.1. Stratigraphy

The general stratigraphy of the BGB appears to confirm the importance of tectonic processes in the history of the belt. The stratigraphy of the BGB is subdivided into three main groups, from bottom to top these are the Onverwacht, Fig Tree and Moodies Groups (Viljoen and Viljoen, 1969c; Anhaeusser et al., 1981, 1983; Lowe and Byerly, 1999). The 3.55–3.25 Ga Onverwacht Group predominantly consists of mafic/ultramafic lavas, interstratified with cherts, rare clastic sedimentary rocks and felsic volcanic rocks. The 3.25–2.23 Ga Fig Tree Group is an association of felsic volcaniclastic rocks, together with clastic and chemical [banded iron formation (BIF)] sedimentary rocks. The 3.22–3.21 Ga Moodies Group is made of sandstone and conglomerates.

The Onverwacht, and, to some degree, the Fig Tree, Groups show different stratigraphies in the northwestern and southeastern parts of the BGB (Viljoen and Viljoen, 1969c; Anhaeusser et al., 1981, 1983; de Wit et al., 1992; de Ronde and de Wit, 1994; Lowe, 1994; Lowe and Byerly, 1999; Lowe et al., 1999; de Ronde and Kamo, 2000). In the west, the Onverwacht Group is mostly 3.3–3.25 Ga, whereas it is much older in the eastern part of the belt (3.55-3.3 Ga). Furthermore, the details of the stratigraphic sequences on both sides cannot be correlated, confirming that the two parts of the belt evolved via a similar, yet independent history. The boundary between the two domains is tectonic and corresponds to the Inyonka–Saddleback fault system, described below. This structure spans the length of the belt from the Stolzburg syncline near Badplaas in the south, to the northern extremity at Kaapmuiden.

5.7-1.2. Tectonic History of the BGB

At least five major phases of deformation have been identified in the BGB (de Ronde and de Wit, 1994; Lowe, 1999b; Lowe et al., 1999). Early D_1 (ca. 3.45 Ga old) deformation is occasionally preserved in lower Onverwacht Group rocks. However, the dominant tectonic event recorded in the BGGT occurred between 3.25 and 3.20 Ga. Four (or five) successive deformation phases related to this event are identified. The first (D_{2a}) deformation occurred during the deposition of the sedimentary and felsic volcanic rocks of the Fig Tree Group, at 3.25–3.23 Ga, probably associated with the development of a volcanic arc in what is now the terrane to the west of the Inyoni–Inyoka fault system (discussed below). At ca. 3.23 Ga (D_{2b}), a dominant period of deformation resulted from the accretion of the two terranes along the Inyoni–Inyoka fault system.

The D_2 accretion was immediately followed, at ca. 3.22–3.21 Ga, by the syn-tectonic (D_3) deposition of the sandstone and conglomerates of the Moodies Group in small and discontinuous, fault-bounded basins (Heubeck and Lowe, 1994a, 1994b). The D_3 deformation is at least in part extensional, with normal faulting in the BGB (upper crust) and core complex exhumation followed by diapiric rise of gneissic domes in the lower crust (surrounding granitoids) (Kisters et al., 2003, 2004). This event corresponds to post-collisional collapse. Late, ongoing compression resulted in strike-slip faulting and folding of the whole sequence, including the Moodies Group, during D_4 and D_5 deformation.

5.7-1.3. The Inyoka–Inyoni Fault System

Within the BGB, the main D_2 structure is the "Inyoka–Saddleback fault", which is developed approximately parallel to the northwestern edge of the belt (Lowe, 1994, 1999; Lowe et al., 1999). This fault forms the boundary between the northwestern and southeastern facies of the Onverwacht Group. The fault system also contains several layered mafic/ultramafic complexes (Anhaeusser, 2001), which may correspond to fragments of oceanic crust trapped in a suture zone. On a larger scale, this zone corresponds to a geophysical boundary within the Kaapvaal craton that extends for several hundreds of kilometers along strike and separates two geophysically and geochronologically distinct terranes (Poujol el al., 2003; de Wit et al., 1992; Poujol, this volume). The Inyoka–Saddleback fault consists of a network of subvertical faults that were active during several of the later deformation events described above, leading to a complex history. It is interpreted to be a D_2 thrust, that was steepened during subsequent (D_3–D_5) deformation.

Further south in the granitoid dominated terrane, a ductile north–south trending shear zone runs from the southern termination of the Stolzburg syncline towards the Schapenburg schist belt, some 30 km further south. This zone, called the "Inyoni shear zone" (ISZ: Kisters et al., 2004; Moyen et al., 2006), is a major structure in the granitoid terrane south of the BGB; it separates the ca. 3.2 Ga Badplaas gneisses to the west, from the ca. 3.45 Ga

Stolzburg pluton in the east, mirroring the difference between the relatively young, western "Kaap Valley" block and the older terranes (Songimvelo, etc.; Lowe, 1994) to the east of the Inyoka–Saddleback fault. Thus, the ISZ is possibly a lower crustal equivalent of the Inyoka–Saddleback fault system.

5.7-2. METAMORPHIC HISTORY OF THE EASTERN TERRANE

Amphibolite facies metamorphic domains have been investigated in detail in both the Eastern and Western domains around the BGGT (Fig. 5.7-1). These potentially provide a window into the lower or middle crust of different portions of the orogen.

5.7-2.1. The Stolzburg Terrane

One of the best studied high-grade regions in the BGGT is known as the "Stolzburg terrane" (Kisters et al., 2003, 2004), which crops out to the south of the BGB, and corresponds to a portion of the "Songimvelo block" of Lowe (1994). The Stolzburg terrane is comprised of ca. 3.45 Ga trondhjemitic orthogneisses of the Stolzburg, Theespruit and other plutons. The terrane contains greenstone material in the form of amphibolite-facies greenstone remnants along the pluton margins, as well as amphibolite-facies Theespruit Formation rocks along the southern margin of the BGB (Fig. 5.7-1). The greenstone remnants within the granitoid terrane have been interpreted to be part of the Sandspruit Formation of the Onverwach Group (Anhaeusser et al., 1981, 1983; Dziggel et al., 2002) and consist of metamorphosed mafic and ultramafic metavolcanic sequences, with minor metasedimentary units that comprise thin metachert and metamorphosed BIF. In addition to these typical lower Onverwacht Group lithologies, this area also contains an up to 8 m-thick, metamorphosed clastic sedimentary unit, within which are well-preserved primary sedimentary features, such as trough cross-bedding. A minimum age of sediment deposition is indicated by a 3431 ± 11 Ma age of an intrusive trondhjemite gneiss (Dziggel et al., 2002). The youngest detrital zircons within the metasedimentary rocks are 3521 Ma in age, indicating that the sedimentary protoliths were deposited between ca. 3521 and 3431 Ma (Dziggel et al., 2002), and therefore are not significantly older than the "overlying" Theespruit and Komatii Formations.

The Stolzburg terrane is bounded to the west by the ISZ, which separates it from the 3.23–3.21 Ga Badplaas pluton, which therefore belongs to the Eastern domain. The northern limit of the Stolzburg terrane is the Komati fault, which corresponds to a sharp metamorphic break between the amphibolite-facies Stolzburg terrane and the greenschist-facies rocks of the main part of the BGB (Eastern domain: Kisters et al., 2003; Diener et al., 2004).

Three recent studies are relevant to the metamorphism of this terrane: Dziggel et al. (2002), who studied the metamorphism of rare clastic metasedimentary rocks within greenstone remnants along the southern margin of the Stolzburg pluton; Kisters et al. (2003), who studied the tectonometamorphic history of the northern boundary of the Stolzburg

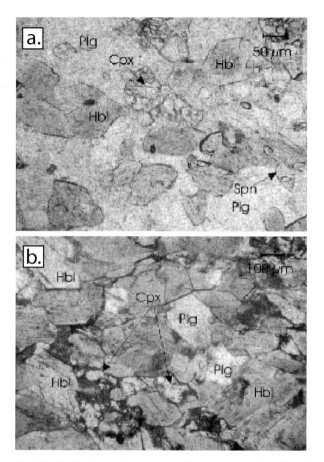

Fig. 5.7-2. Typical peak metamorphic textural relationships (above) and P-T estimates (overleaf) for samples from the central Stolzburg terrane: (a) and (b) represent two examples of the post tectonic peak metamorphic textures. On the P-T diagram; (c) BE1 and BE2 illustrate the peak metamorphic conditions as constrained by two of the samples studied by Dziggel et al. (2002). Schematic andalusite-sillimanite-kyanite phase boundaries are included for reference.

pluton; and Diener et al. (2005), who investigated the tectonometamorphic history of the Tjakastad schist belt (areas c, e, and f on Fig. 5.7-1).

5.7-2.1.1. Peak of metamorphism

Dziggel et al. (2002) documented two types of clastic metasedimentary rocks: a trough cross-bedded, proximal meta-arkose and a planar bedded, possibly more distal, metasedimentary unit of relatively mafic geochemical affinity. The latter are characterized by the peak-metamorphic mineral assemblage diopside + andesine + garnet + quartz. This assemblage (and garnet in particular) is extensively replaced by retrograde epidote. Peak-

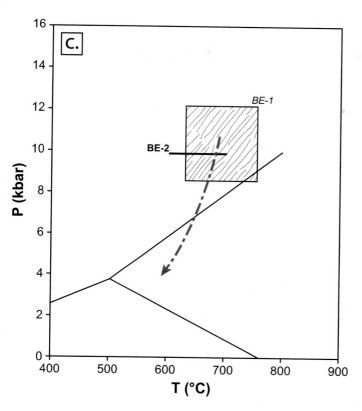

Fig. 5.7-2. (*Continued.*)

metamorphic mineral assemblages of magnesio–hornblende + andesine + quartz, and quartz + ferrosilite + magnetite + grunerite have been recorded from adjacent amphibolites and interlayered BIF units, respectively. In these rocks, retrogression is marked by actinolitic rims around peak metamorphic magnesio–hornblende cores in the metamafic rocks, and by a second generation of grunerite that occurs as fibrous aggregates rimming orthopyroxene in the iron formation. The peak metamorphic textures are typically post tectonic and are texturally mature and well equilibrated. Peak pressure-temperature (PT) estimates, using a variety of geothermometers and barometers, for the peak-metamorphic mineral assemblages in all these rock types vary between 650–700 °C and 8–11 kbar (Fig. 5.7-2). As suggested by the texturally well-equilibrated nature of the assemblages, no evidence of the prograde path is preserved. Dziggel et al. (2002) interpreted the relatively high pressures and low temperatures of peak metamorphism to reflect a tectonic setting comparable to modern continent–continent collisional settings, and suggested that the Stolzburg terrane represents an exhumed mid- to lower-crustal terrane that formed a 'basement' to the BGB at ca. 3230 Ma.

5.7-2.1.2. Contacts with the greenstone belt

The deformed and metamorphosed margins of the Stolzburg terrane in the north, where it abuts the lower grade greenstone belt, have been studied in two separate areas. Kisters et al. (2003) conducted detailed mapping of the contacts between the supracrustal and gneiss domains along the southern margin of the greenstone belt. They documented an approximately 1 km wide deformation zone that corresponds with the position of the heterogeneous and mélange-like rocks of the Theespruit Formation, within which two main strain regimes can be distinguished (Fig. 5.7-3). Amphibolite-facies rocks at, and below, the granite–greenstone contacts are characterized by rodded gneisses and strongly lineated amphibolite-facies mylonites. Kisters et al. (2003) demonstrated that lineations developed in the BGB either side of the Stolzburg syncline are brought into parallelism by unfolding around the inclined fold axis of the syncline, suggesting extension prior to folding; that when rotated into a subhorizontal orientation, the bulk constrictional deformation at these lower structural levels records the originally vertical shortening and horizontal, NE–SW directed stretching of the mid-crustal rocks; that the prolate coaxial fabrics are overprinted by greenschist-facies mylonites at higher structural levels that cut progressively deeper into the underlying high-grade basement rocks; and that these mylonites developed during non-coaxial strain and kinematic indicators consistently point to a top-to-the-NE sense of movement of the greenstone sequence with respect to the lower structural levels. These features suggest bulk coaxial NE–SW stretching of mid-crustal basement rocks and non-coaxial, top-to-the-NE shearing along retrograde mylonites at upper crustal levels is consistent with an extensional orogenic collapse of the belt and the concomitant exhumation of deeper crustal levels.

The dominant peak metamorphic assemblage within preserved amphibolite-facies domains throughout the study area is hornblende + plagioclase + sphene + quartz. Other locally developed assemblages are: garnet + hornblende + plagioclase + sphene + quartz, and garnet + plagioclase + hornblende + calcite + biotite + epidote + quartz in metamafic rocks; and garnet + biotite + muscovite + quartz in a single metapelitic layer. All the garnet-bearing assemblages are confined to specific narrow layers developed parallel to the compositional banding of the rocks (S_0). In all cases, retrogression is associated with the development of later shear fabrics (S_1 in retrograde mylonites) that postdate the peak-metamorphic porphyroblasts.

Kisters et al. (2003) interpreted these features to suggest a primary bulk-compositional control (Fe/Fe+Mg ratios and the presence of carbonate) on the distribution of the garnet-bearing peak-metamorphic assemblages, and that these assemblages are probably metamorphic grade equivalents of the predominant peak assemblage in the amphibolites. Peak P-T conditions were constrained using the assemblages garnet + plagioclase + hornblende + biotite + quartz, and garnet + plagioclase + hornblende + biotite + quartz + epidote + calcite, which yielded P-T estimates of $491 \pm 40\,°C$ and 5.5 ± 0.9 kbar, and $492 \pm 40\,°C$ and 6.3 ± 1.5 kbar, respectively. Retrogression is marked by the development of actinolite + epidote + chlorite + quartz assemblages in the metamafic rocks and muscovite + chlorite + quartz in the metapelitic layer. These conditions are at lower grades than those defined

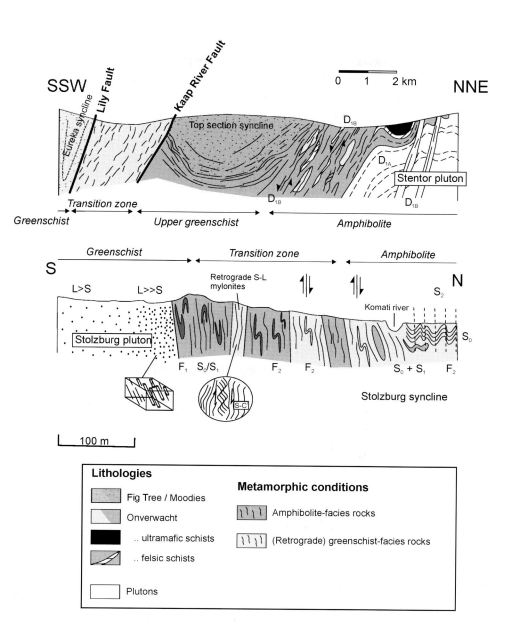

Fig. 5.7-3. Schematic cross-sections across granite–greenstone contacts from the Western and Eastern domains of the BGGT. (a) The low to high grade transition in the Stentor pluton area in the Western domain (after Dziggel et al., 2006). (b) The northern boundary of the Stolzburg terrane against the Eastern domain (after Kisters et al., 2003).

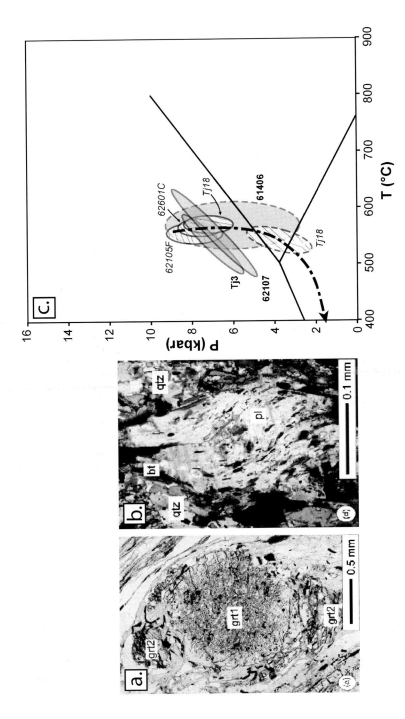

Fig. 5.7-4. Typical peak metamorphic textural relationships (left) and P–T estimates (right) for samples from the Tjakastad area (Diener et al., 2005, 2006): (a) Illustrates two generations of syntectonic garnet development (grt1, grt2); (b) illustrates a deformed plagioclase porphyroblast (pl) in a quartz (gtz) and biotite (bt) matrix; (c) illustrates the P-T conditions of metamorphism calculated using assemblages from the Tjarkastad schist belt. The sample numbers in (c) correspond to those used by Diener et al. (2005).

by Dziggel et al. (2002), but are developed along a similarly low apparent geothermal gradient.

Diener et al. (2004) investigated the tectonometamorphic history of the Tjakastad schist belt (Fig. 5.7-1), which contains remnants of the Theespruit Formation that predominantly includes amphibolites, felsic volcanoclastic rocks, and minor aluminous metasedimentary rocks. The metamafic and metasedimentary rocks record an identical deformational history to the rocks studied by Kisters et al. (2003), some 5 to 10 km to the northwest. Both the peak metamorphic and retrograde assemblages are syntectonic with fabrics developed during exhumation, illustrating the initiation of the detachment at deep crustal levels and elevated temperatures. In contrast with the rocks studied by Kisters et al. (2003), however, the rocks investigated by Diener et al., (2004) provided a better record of the retrograde path. Within the metamafic rocks, more aluminous layers are characterized by the peak metamorphic assemblage garnet + epidote + hornblende + plagioclase + quartz. Within the aluminous metasedimentary unit, an equivalent peak metamorphic assemblage is defined by garnet + staurolite + biotite + chlorite + plagioclase + quartz. These assemblages produce calculated P-T estimates of 7.0 ± 1.2 kbar and 537 ± 45 °C and, 7.7 ± 0.9 kbar and 563 ± 14 °C, respectively (Fig. 5.7-4). In these rocks, the peak metamorphic assemblages are syntectonic, with peak metamorphic porphyroblasts (e.g., staurolite) recrystallised and deformed within the exhumation fabric (Fig. 5.7-4). Within rare low-strain domains in the garnet-bearing amphibolite, retrograde mineral assemblages pseudomorph peak metamorphic garnet. In these sites, a new generation of garnet is developed within the assemblage garnet + chlorite + muscovite + plagioclase + quartz. Calculated P-T estimates from these sites yield conditions of 3.8 ± 1.3 kbar and 543 ± 20 °C, indicating near isothermal decompression (Fig. 5.7-4). This is consistent with the presence of staurolite as part of the peak and retrograde assemblages, with the modeled staurolite stability field in relevant compositions being confined to a narrow temperature range of between 580–650 °C over a pressure range between 10–3 kbar. These calculated P-T conditions are also consistent with the occurrence of sillimanite replacing kyanite within the staurolite-bearing rocks (Diener et al., 2004).

Geochronological constraints, combined with the depths of burial, indicate that exhumation of the high-grade rocks occurred at rates of 2–5 mm/a. This is similar to the exhumation rates of crustal rocks in younger compressional orogenic environments, and when coupled with the low apparent geothermal gradients of ca. 20 °C/km, led Diener et al. (2004) to suggest that the crust was cold and rigid enough to allow tectonic stacking, crustal overthickening and an overall rheological response very similar to that displayed by modern, doubly-thickened continental crust.

5.7-3. METAMORPHISM IN THE WESTERN DOMAIN

The metamorphic history of the Western domain is less well understood than the Stolzberg terrane, as fewer studies have been conducted and these are more widespread, making the relationships between the study areas less obvious. Two studies are relevant to this discussion: the study by Dziggel et al. (2006), who investigated the tectonometa-morphic history of the northern contact of the BGB, where it is in contact with the Stentor pluton [area (a) in Fig. 5.7-1]; the study by Stevens et al. (2002), who investigated the metamorphic history of the Schapenburg schist belt [area (b) in Fig. 5.7-1]. This study area lies along the southern extension of the ISZ, which is believed to anastomose around the Schapenburg schist belt. This belt is included in the Western domain on account of it dis-playing a similar apparent geothermal gradient to that documented by Dziggel et al. (2006). An important difference between the Western and Eastern domains is that the Eastern do-main contains an abundance of granitoid intrusions (Badplaas, Nelshoogte and Kaapvalley) that are essentially syntectonic with the ca 3.23 Ga deformation.

5.7-3.1. Schapenburg Schist Belt

The Schapenburg schist belt is one of several large (approximately 3×12 km) greenstone remnants exposed in the granitoid-dominated terrane to the south of the BGB and is unique in that it contains a well-developed metasedimentary sequence in addition to the typical mafic-ultramafic volcanic rocks (Anhaeusser, 1983). Stevens et al. (2002) conducted an investigation of the metamorphic history of the belt, which is summarized below.

The metasedimentary sequence consists of two distinctly different units. A meta-tuffaceous unit, essentially of granitoid composition, but containing both minor agglom-erate layers and, within low strain domains, well preserved cross-bedding and graded bedding in the southwestern portion of the belt. This unit underlies a rhythmically banded unit of metagreywacke that consists of approximately 10 cm-thick units of formerly clay-rich rock that grade into 1 to 2 cm thick quartz-rich layers. On the basis of both the graded bedding and trough cross-bedding in the underlying meta-tuffaceous unit, the metased-imentary succession can be shown to young to the east. This succession is overlain by Onverwacht Group rocks.

Detrital zircons within the metasedimentary rocks have ages as young as 3240 ± 4 Ma and thus are correlated with the Fig Tree Group in the central portions of the BGB some 60 km to the north, where they are metamorphosed to lower greenschist facies grades.

The Schapenburg schist belt metasedimentary rocks are relatively K_2O-poor and are commonly characterized by the peak metamorphic assemblage garnet + cordierite + gedrite + biotite + quartz ± plagioclase. Other assemblages are garnet + cummingtonite + biotite + quartz, cordierite + biotite + sillimanite + quartz and cordierite + biotite + anthophyllite. In all cases, the post-tectonic peak assemblages are texturally very well equilibrated (Fig. 5.7-5) and the predominantly almandine garnets from all rock types show almost flat zonation patterns for Fe, Mg, Mn and Ca. Consequently, there appears to be no preserved record of the prograde path.

Analysis of peak metamorphic conditions using FeO-MgO-Al_2O_3-SiO_2-H_2O FMASH reaction relations, as well as a variety of geothermometers and barometers, constrained the peak metamorphic pressure-temperature conditions to $640 \pm 40\,^\circ$C and 4.8 ± 1.0 kbar. The maximum age of metamorphism was defined by the 3231 ± 5 Ma age of a syntectonic tonalite intrusion into the central portion of the schist belt. In combination with the age

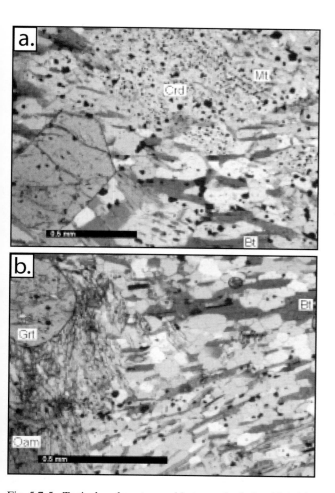

Fig. 5.7-5. Typical peak metamorphic textural relationships (above) and P-T estimates (overleaf) for samples from the Schapenburg schist belt (Stevens et al., 2002): (a) and (b) illustrate the post tectonic character of the peak metamorphic minerals; cordierite (Crd) in (a) and garnet (Grt) and orthoamphibole (Oam) in (b); (c) P-T diagram illustrating the calculated conditions of peak metamorphism. The sample numbers correspond to those used by Stevens et al. (2002).

of the youngest detrital zircons in the metasedimentary rocks, this age demonstrates that sedimentation, burial to mid-crustal depths (~18 km), and equilibration under amphibolite facies conditions were achieved in a time span of between 10–20 Ma.

5.7-3.2. *Stentor Pluton Area*

Dziggel et al. (2006) demonstrated that the granitoid–greenstone contact along the northern margin of the BGB is characterized by a shear zone that separates the generally low-grade,

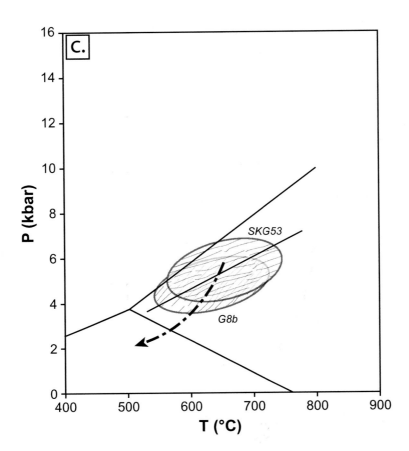

Fig. 5.7-5. (*Continued.*)

greenschist-facies greenstone belt from mid-crustal basement gneisses. In addition, these authors demonstrated that the supracrustal rocks in the hangingwall of this contact are metamorphosed to upper greenschist facies, whereas similar rocks and granitoid gneisses in the footwall are metamorphosed to amphibolite facies. Within the amphibolite facies domain, metamafic rocks are characterized by the assemblages hornblende + plagioclase + quartz; hornblende + plagioclase + clinopyroxene + quartz and hornblende + plagioclase + garnet + clinopyroxene + quartz. Aluminous schists from this domain contain the peak metamorphic assemblage garnet + muscovite + sillimanite + biotite + quartz.

Calculated P-T estimates on these assemblages constrain the peak P-T conditions of metamorphism to between 600 and 700 °C and 5 ± 1 kbar (Fig. 5.7-6). This corresponds to an elevated metamorphic field gradient of ∼30–40 °C/km. The peak metamorphic minerals in this area are syntectonic with fabrics that are interpreted to have formed during exhumation of the high grade rocks at ca 3.23 Ga (Dziggel et al., 2006). Retrograde assemblages that form through the replacement of peak metamorphic clinopyroxene and plagioclase in

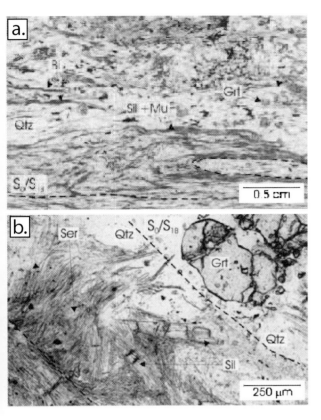

Fig. 5.7-6. Typical peak metamorphic textural relationships (above) and P-T estimates (overleaf) for samples from the Stentor pluton area (from Dziggel et al., 2006). Sample numbers on the P-T diagram are from Dziggel et al. (2006).

the metamafic rocks by coronitic epidote + quartz and actinolite + quartz symplectites yield retrograde P-T conditions of 500–650 °C and 1–3 kbar. Dziggel et al. (2006) interpreted this to indicate that exhumation and decompression commenced under amphibolite facies conditions (as indicated by the synkinematic growth of peak metamorphic minerals during extensional shearing), followed by near-isobaric cooling to temperatures below 500 °C. The last stages of exhumation are characterized by solid state doming of the footwall gneisses and strain localization in contact-parallel greenschist-facies mylonites that overprint the decompressed basement rocks.

The southern margin of the Stentor pluton area is bounded by the Kaap River and Lily faults (Fig. 5.7-1). These correspond to a major metamorphic break, from 6–8 kbar in the amphibolitic domain, to nearly unmetamorphosed supracrustal rocks in the BGB immediately south of the faults (Otto et al., 2005; Dziggel et al., 2006).

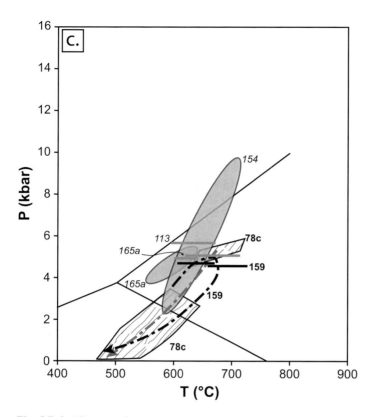

Fig. 5.7-6. (*Continued.*)

5.7-4. INYONI SHEAR ZONE

The Inyoni shear zone (ISZ) is a complex structure extending in a southwesterly direction from the termination of the Stolzburg syncline into the granitoid dominated terrane to the south (area (d) in Fig. 5.7-1). It forms the boundary between the Stolzburg terrane to the east and the Badplaas pluton to the west. Both Dziggel et al. (2002) and Moyen et al. (2006) have investigated the metamorphic history of the ISZ. The shear zone contains a diverse assemblage of greenstone remnants, mostly typical lower Onverwacht Group interlayered metamafic and meta-ultramafic units, with occasional minor BIF horizons, but some clastic metasedimentary rock also occur (Dziggel et al., 2002). The greenstone fragments are enclosed within TTG orthogneisses, components of which were intruded syntectonically during, or close to, the peak of metamorphism.

Structures in the Inyoni shear zone are complex, and result from the interference of:
(1) east–west shortening, resulting in the formation of a predominantly vertical foliation, with symmetrical folds and the development of a crenulation cleavage at all scales (from the map pattern to hand specimen); and

(2) vertical extrusion of the Stolzburg terrane, causing the development of a syn-melt vertical lineation, and folds with vertical axes.

Evidence for earlier structures has also been described, in the form of rootless isoclinal folds in some of the supracrustal remnants.

Both greenschist and amphibolite facies remnants have been described, possibly indicating the imbrication of rocks with diverse metamorphic histories. However, most of the remnants are dominated by metamafic rocks and appear to have been metamorphosed to amphibolite facies grades. The dominant foliation is defined by hornblende in the metamafic rocks, which is cut by syntectonic tonalitic veins with an age of 3229 ± 5 Ma (Dziggel et al., 2006). Metamorphic titanite that formed in association with epidote through retrograde replacement of garnet and plagioclase has an age of 3229 ± 9 Ma (Dziggel et al., 2006).

Dziggel et al. (2002) focused on metasedimentary rocks within the ISZ and produced P-T estimates of amphibolite facies peak metamorphic conditions very similar to those described for the Stolzburg terrane, at 600–700 °C and 8–11 kbar. No information on the prograde history of the rocks could be determined due to the well equilibrated nature of the peak metamorphic assemblages.

In contrast, Moyen et al. (2006) examined the metamorphic record within metamafic samples and produced information on both the prograde and retrograde P-T evolution of this zone. Textural evidence of the prograde metamorphic evolution is recorded in garnet-bearing low-strain domains, such as the cores of certain rootless isoclinal folds, where core-to-rim growth-zoned garnet occurs that have low-temperature mineral inclusions contained within their cores. In these sites, garnet can be shown to have grown simultaneously with albitic plagioclase, as evidenced by euhedral garnets surrounded by plagioclase (Fig. 5.7-7) and albitic inclusions within garnets, sometimes with negative garnet forms. In the same domains, clinopyroxene and quartz are also sometimes intergrown with garnet (Fig. 5.7-7). This assemblage appears to have formed at the expense of a relatively sodic amphibole (Fe-edenite, up to 1.1 sodium atoms per formula unit), partially reequilibrated relicts of which are found within albitic moats around the garnets. These commonly occur as several small, separate relic crystals that are in crystallographic continuity, indicating the original presence of substantially larger crystals. Calculated P-T estimates for this assemblage in a number of sites range from 600–650 °C and 12–15 kbar. Garnet in samples from higher-strain domains generally shows partial replacement by symplectitic coronas of epidote + Fe-tschermakite + quartz symplectite. Calculated P-T estimates from these assemblages produce retrograde conditions of 580–650 °C at 8–10 kbar. The estimated metamorphic conditions constrained by these decompression structures correspond well with peak metamorphic estimates from the nearby clastic sedimentary intercalations within the metavolcanic sequence. Locally, in both the high- and low-strain domains, greenschist-facies chlorite + epidote + actinolite retrogression overprints the amphibolite-facies assemblages.

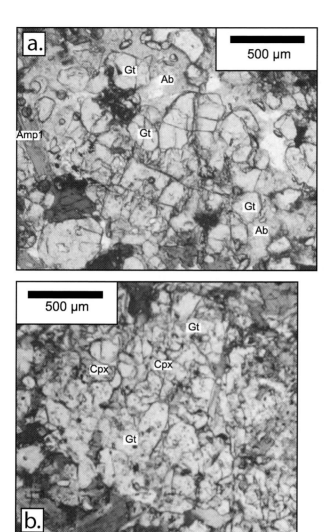

Fig. 5.7-7. Typical metamorphic textural associations (above) and P-T estimates (overleaf) for samples from the Inyoni shear zone (Moyen et al., 2006): (a) Illustrates the small garnet crystals developed in conjunction with albitic plagioclase during the breakdown of sodic hornblende; (b) illustrates an intergrowth of garnet and clinopyroxene; (c) P-T diagram, with sample numbers after Moyen et al. (2006).

5.7-5. DISCUSSION AND CONCLUSIONS

The data presented above support several general observations on the nature of the ca. 3.23 Ga metamorphic event in the BGGT (Figs. 5.7-8 and 5.7-9):

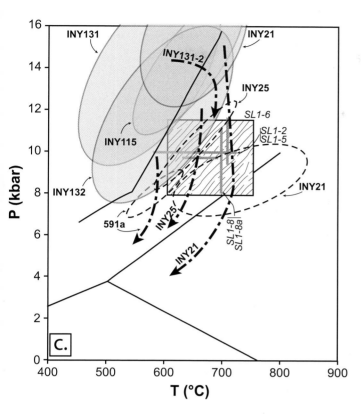

Fig. 5.7-7. (*Continued.*)

(1) The existence of two different thermal regimes in the deep crust of the BGGT. Mid-to lower-crustal rocks from the Western domain generally record apparent geothermal gradients as low as 18–20 °C km^{-1}. Similar rocks from the Eastern domain record apparent geothermal gradients of 30–40 °C km^{-1}.

(2) The granite-greenstone margins in both domains are defined by the presence of amphibolite facies supra-crustal rocks and gneissose granitoids, as part of the deep crustal section. In both cases, a substantial pressure transition of >5 kbar (ca. 15 km) can be documented across the sheared contacts, over just a few kilometers laterally. This transition occurs in a zone of high-strain rocks (up to mylonites) that record a normal sense of movement with the low-grade greenstone belt being down-thrown relative to the surrounding amphibolite-facies gneisses. In essence, these zones define the cuspate granite-greenstone contacts of the "dome and keel" pattern. Peak metamorphism in these areas is syntectonic with the exhumation process, which is continuous as the margin of the uplifted block evolved into greenschist facies conditions.

(3) In those parts of both the Eastern and Western domains, peak metamorphic conditions away from the greenstone belt are post-tectonic. This indicates coherent behavior of

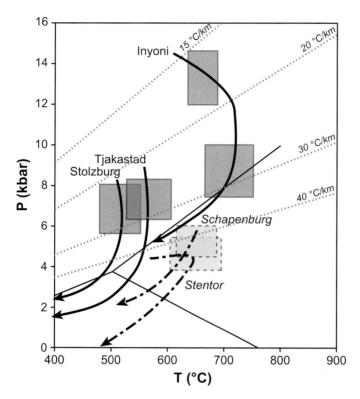

Fig. 5.7-8. Compilation diagram of P-T estimates of the studies discussed in this paper. Strong evidence for decompression exists in the samples from the Inyoni shear zone, the Tjakastad schist belt and the Stentor pluton. The rocks of the Eastern domain (dark shaded boxes) clearly underwent a peak of metamorphism in the kyanite stability field, potentially recording heating during exhumation from greater depths than the recorded pressures indicate. Peak metamorphic conditions in the Western domain (light shaded boxes) were in the sillimanite stability field.

Fig. 5.7-9. (*Next page.*) Proposed geodynamic model for the ca. 3.2 Ga accretionary orogen in the BGGT. All cartoons are at approximately the same scale, looking towards the (present-day) northeast; the front section of each block corresponds to a NW–SE cross-section. In each cartoon, the active plutonism is in black, while the already emplaced rocks are grey. Plutons: B: Badplaas, N: Nelshoogte, KV: Kaapvalley, S: Stolzburg, Ts: Theespruit. Structures: IF: Inyoka–Saddleback fault, ISZ: Inyoni shear zone. Cartoons are modified from (Moyen et al., 2006). Circled letters (A, B, C, D) in the figures correspond to the Theespruit Formation of the Tjakastad schist belt (Diener, 2005), the ISZ samples (Moyen et al., 2006), the Schapenburg schist belt (Stevens, 2002), and the Stentor pluton area (Dziggel et al., 2006), respectively. The P-T evolution of points A, B and C during the assembly and collapse phases of the orogen are illustrated in the P-T diagrams presented below the second and third cartoons.

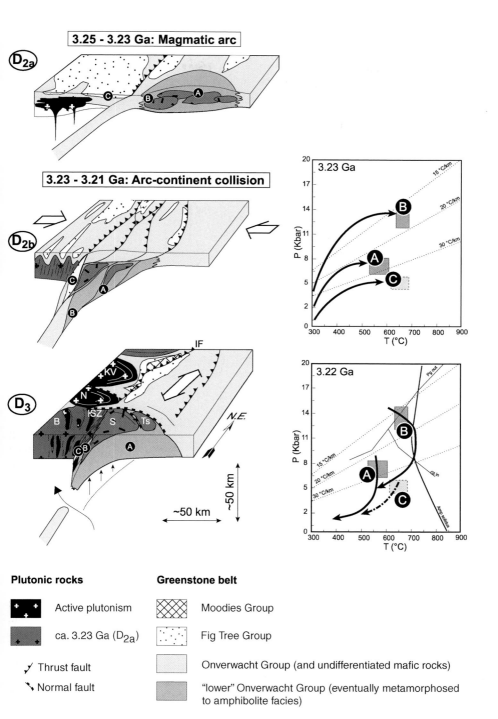

3.25 - 3.23 Ga: Magmatic arc

3.23 - 3.21 Ga: Arc-continent collision

Plutonic rocks

Active plutonism

ca. 3.23 Ga (D$_{2a}$)

Thrust fault

Normal fault

Greenstone belt

Moodies Group

Fig Tree Group

Onverwacht Group (and undifferentiated mafic rocks)

"lower" Onverwacht Group (eventually metamorphosed to amphibolite facies)

Table 5.7-1. A summary of the relevant metamorphic studies

A) Within the BGB proper

Peak assemblage	Retrograde minerals	Sample number	Notes	Method	P (kbars)	S.D. (P, kbars)	T (°C)	S.D. (T, °C)	Ref.
Komatii Formation									
Mafic/ultramafic pillows	Chl + Amp + Pl		peak	Chl, Amp and Pl isopleths + Hbl-Pl	ca. 4		ca. 520		Cloete, 1991, 1999
			retro	Chl, Amp and Pl isopleths + Hbl-Pl	ca. 2.5		ca. 350		"
Onverwacht & Fig Tree Groups, center of the belt									
Diverse, mostly mafic to intermediate lavas	Chl-Trem/Act + Qtz ± Ser ± Cc		Greenschist facies assemblage	Al substitution in chlorite			ca. 320		Xye et al., 1997

B) Amphibolite facies – Western domain

Peak assemblage	Retrograde minerals	Sample number	Notes	Method	P (kbars)	S.D. (P, kbars)	T (°C)	S.D. (T, °C)	Ref.
Stentor pluton									
Metabasites	Hbl + Pg + Qtz ± Cpx ± Grt	Ep + Zo + Act	165a		THERMOCALC (av. PT)	4.6	0.7	595	35	Dziggel et al., 2006
		165a		Grt-Cpx	[5]		670	42	"
		165a		Grt-Hbl	[5]		620	20	"
		154		THERMOCALC (av. P)	6	2.9	[600]		"
		154		THERMOCALC (av. P)	6.5	3.2	[700]		"
		154		Hbl-Pl (ed-tr)	[5]		625	50	"
		154		Hbl-Pl (ed-ri)	[5]		600	40	"

Table 5.7-1. (*Continued*)

	Peak assemblage	Retrograde minerals	Sample number	Notes	Method	P (kbars)	S.D. (P, kbars)	T (°C)	S.D. (T, °C)	Ref
Felsic schist (amphibolite facies)	Grt + Mu + Sili-Bi-Qtz	Ser	113		Hbl-Pl (ed-tr)	[5]		630	40	"
			113		Hbl-Pl (ed-ri)	[5]		675	30	"
			159		THERMOCALC (av. PT)	5.7	1.2	753	190	Dziggel et al., 2006
			159		Grt-Bi (Ganguly and Saxena, 1994)	[5]		640	30	"
			159		Grt-Bi (Hackler and Wood, 1989)	[5]		690	30	"
			159		Pseudosection modelling (THERMOCALC)	3–5.7		575–700		"
Felsic schist (greenschist facies)	Mu + Bi + Qtz + And	Ctd	78c	peak	Pseudosection modelling (THERMOCALC)	5–6		625–725		Dziggel et al., 2006
				retro	Pseudosection modelling (THERMOCALC)	ca. 3.5		475–650		"
Schapenburg Metaturbidite	Grt + Crd + Ged + Bi + Pl + Qtz ± Cumm		SKG53		THERMOCALC (av. PT)	5.4	1.2	654	79	Stevens et al., 2002
					Petrogenetic grid (THERMOCALC)	<5.3 kbar at 650°C				"
			G8b		THERMOCALC (av. PT)	4.8	1.1	633	74	"

Table 5.7-1. (Continued)

C) Amphibolite facies – Eastern domain

Peak assemblage	Retrograde minerals	Sample number	Notes	Method	P (kbars)	S.D. (P, kbars)	T (°C)	S.D. (T, °C)	Ref.
Tjakastad schist belt									
Metasediment $Grt + St + Bi + Chl + Pl + Qtz$	$Grt_{(n+1)} + Chl_{(n+1)} + Pl_{(n+1)} + Mu$	62105F		THERMOCALC (av. PT)	7.9	1.1	556	19	Diener et al., 2005
		62601C		THERMOCALC (av. PT)	7.7	0.9	563	14	"
		Tj18	peak	THERMOCALC (av. PT)	7.2	1	569	18	"
			retro	THERMOCALC (av. PT)	3.8	1.3	543	20	"
Metabasites $Grt + Ep + Pl + Chl + Hbl + Qtz$		62107		THERMOCALC (av. PT)	7	1.6	556	59	Diener et al., 2005
		Tj3		THERMOCALC (av. PT)	7	1.2	537	45	"
		61406		THERMOCALC (av. PT)	6.1	2.7	569	42	"
Stolzburg arm (exhumation)									
Amphibolites $Hbl + Pg + Qtz + Sph \pm Grt$	$Act + Ep + Chl + Qtz$	An1	max, for $XH_2O = 1$	THERMOCALC (av. PT)	5.5	0.9	491	40	Kisters et al., 2003
			min, for $XH_2O = 0.3$						
		PB3	$XH_2O = 0.5$	THERMOCALC (av. PT)	5.1	0.9	458	35	"
				THERMOCALC (av. PT)	8	1.3	557	37	"

Table 5.7-1. (*Continued*)

Peak assemblage	Retrograde minerals	Sample number	Notes	Method	P (kbars)	S.D. (P, kbars)	T (°C)	S.D. (T, °C)	Ref.
Central Stolzburg – Greenstone remnant BE									
Clastic metasediments	Hbl + Pl + Cpx + Qtz	Ser	BE1	Hbl-Pl (ed-ri)	[10]		630–706		Dziggel et al., 2002
				Hbl-Pl (ed-tr)	[10]		668–753		"
				Cpx-Pl-Qtz (Ellis, 1980)	8.2–12.1		[700]		"
Amphibolite	Hbl + Pl + Qtz + Cpx	Ep + Ser + Act	BE2	Hbl-Pl (ed-ri)	[10]		601–652		Dziggel et al., 2002
				Hbl-Pl (ed-tr)	[10]		660–706		"
Inyoni shear zone									
Iron formation	Qtz + Mag + Gru + Opx + Hbl	$Gru_{(n+1)}$ + Act + $Mag_{(n+1)}$	SL1-5	Opx (Witt-Eiskschen and Seek, 1991)	[10]		636–694		Dziggel et al., 2002
Clastic metasediments	Cpx + Pl + Qtz + Grt ± Hbl	Ep ± Ser ± Act	SL1-6	Grt-Cpx (Ellis and Green, 1979)	[10]		639–767		Dziggel et al., 2002
				Grt-Cpx (Ganguly, 1979)	[10]		641–749		"
				Grt-Cpx (Powell, 1985)	[10]		625–756		"
				Grt-Cpx-Pl-Qtz (Powell and Holland, 1988)	8.7–9.9		[700]		"
				Grt-Cpx-Pl-Qtz (Eckert et al., 1991)	9.0–11.0		[700]		"
				Cpx-Pl-Qtz	8.1–11.5		[700]		"

Table 5.7-1. (*Continued*)

Peak assemblage	Retrograde minerals	Sample number	Notes	Method	P (kbars)	S.D. (P, kbars)	T (°C)	S.D. (T, °C)	Ref
		SL1-8		Grt-Cpx-Pl-Qtz (Powell and Holland, 1988)	7.9–10.7		[700]		"
				Grt-Cpx-Pl-Qtz (Eckert et al., 1991)	8.9–11.0		[700]		"
				Cpx-Pl-Qtz	8.0–9.9		[700]		"
		SL1-8a		Grt-Cpx-Pl-Qtz (Powell and Holland, 1988)	8.7–10.4		[700]		"
				Grt-Cpx-Pl-Qtz (Eckert et al., 1991)	9.9–11.9		[700]		"
				Cpx-Pl-Qtz	10.3–11.6		[700]		"
Garnet-amphibolites	Hbl + Ep + Pl + Grt ± Cpx	$Ep_{(n+1)}$ + $Hbl_{(a+1)}$ + $Pl_{(n+1)}$ Chl-Trem/Act	INY131	Grt growth site	THERMOCALC (av. PT)	13.1	4.1	534	92	Moyen et al., 2006
				"	THERMOCALC (av. PT)	14.1	3.8	604	102	"
			INY132	Grt growth site	THERMOCALC (av. PT)	11.7	3.7	599	100	"
			INY115	Grt growth site	THERMOCALC (av. PT)	12.8	2.6	604	61	"
				"	THERMOCALC (av. PT)	14.0	2.2	623	60	"
				"	THERMOCALC (av. PT)	13.0	3.3	695	100	"
				"	THERMOCALC (av. PT)	14.9	2.4	668	66	"
				"	THERMOCALC (av. PT)	16.4	2.7	690	99	"

Table 5.7-1. (*Continued*)

Peak assemblage	Retrograde minerals	Sample number	Notes	Method	P (kbars)	S.D. (P, kbars)	T (°C)	S.D. (T, °C)	Ref.
	(different sites for each sample)		"	THERMOCALC (av. PT)	13.5	2.8	580	74	"
		INY21	Grt core	THERMOCALC (av. PT)	15.7	3.1	646	68	"
			Grt rim	THERMOCALC (av. PT)	9.5	2.2	827	156	"
			"	THERMOCALC (av. PT)	9.4	1.8	758	125	"
			"	THERMOCALC (av. PT)	8.8	1.3	764	95	"
			"	THERMOCALC (av. PT)	6.7	1.4	740	117	"
			"	THERMOCALC (av. PT)	8.7	1.3	830	108	"
			Matrix	THERMOCALC (av. P)	8	1			"
			"	Hbl-Pl (avg of 2 reactions)			725	50	"
			Matrix	THERMOCALC (av. P)	7.5	1			"
			"	Hbl-Pl (avg of 2 reactions)			690	50	"
		591a	Grt breakdown	THERMOCALC (av. PT)	7.9	1.0	585	39	"
		INY25	Grt breakdown	THERMOCALC (av. PT)	10.1	1.7	653	58	"

S.D. = Standard deviation.

the exhumed deep crust, in that the mappable domains discussed here, such as the Stolzburg terrane, represent largely intact deep crustal sections that were exhumed along discrete shear zones along the granite-greenstone contacts. This lack of penetrative post-peak metamorphic deformation internal to the terranes appears inconsistent with the diapiric rise of plastically deforming domes.

(4) The peak P–T estimates for the ISZ, as well as the mélange-like character of the zone (Moyen et al., 2006), confirm this zone as a terrane boundary and the possible trace of the subduction zone that closed to allow crustal collision. The pressures reported for this zone (P $=$ 12 to 15 kbar) are, at present, the highest crustal pressures reported for meso-Archean rocks, and correspond to by far the lowest known apparent geothermal gradients ($12\,^\circ\mathrm{C\,km^{-1}}$) in the Archean rock record. In the modern Earth, the only process capable of producing crustal rock evolution through this P–T domain occurs within subduction zones.

5.7-5.1. The Case for 3.2 Ga Cold Crust and Horizontal Tectonics

The case for cool continental crust in the BGGT prior to 3.23 Ga is convincing. The rocks of the Stolzburg terrane represent an approximately 400 km^2 domain of rocks that were buried to depths of 35–40 km. Internally to this domain, peak metamorphic equilibration occurred, in rocks that were not undergoing deformation, to record an amphibolite facies apparent geothermal gradient no higher than those recorded by younger metamorphic rocks from ocean-continent collision zones. This occurred simultaneously with syntectonic peak metamorphism in the terrane margins, where deformation was driven by the exhumation of the high-grade portions of the thickened crust. The presence of crustal rocks recording pressures of 12–15 kbar and a metamorphic field gradient of $12\,^\circ\mathrm{C\,km^{-1}}$, in the setting of a shear zone containing both metasedimentary and metamafic rocks at variable grades of peak metamorphism suggest that this zone marks the prior existence of a subduction zone.

The abundance of synkinematic trondhjemites in the shear zone is likely to be the result of decompression melting of amphibolites at deeper levels during exhumation. The presence of these melts is possibly important to understanding the documented metamorphic signature. High-strain fabrics confined to synkinematic trondhjemites point to strain localization in the melts, which, in turn, is likely to assist the buoyancy- or extrusion-related exhumation of the rocks. The advective heat transfer associated with the intrusion of these synkinematic magmas also contributes to the syn- to late-collisional heat budget of the collisional belt that acted to partially destroy the evidence for the earlier high-pressure, low-temperature metamorphism. This possibly holds the key to understanding the very high geothermal gradients recorded by the high-grade rocks of the western domain following uplift (50–$60\,^\circ\mathrm{C\,km^{-1}}$), as much of the crust that constitutes the western domain comprises syntectonic magmatic rocks.

5.7-5.2. Are the BGGT Domes Core Complexes?

Sections of the Pilbara Craton in Western Australia and the BGGT show numerous regional-scale similarities. For this discussion, the most notable of these are the typical dome-and-keel geometries between TTG domes and greenstone synforms, and the localized occurrence of high-pressure, low- to medium-temperature metamorphism of the supracrustal sequences (Collins et al., 1998; Van Kranendonk et al., 2002; Van Kranendonk, 2004a). Importantly, despite these similarities, completely different models have arisen for the evolution of the Archean crust in these two areas. Tectonic models proposed to account for the high-grade metamorphism of greenstone sequences in the Pilbara Craton critically hinge on a high rate of heat production in the Mesoarchean crust, that is assumed to be in excess of twice typical modern day rates (e.g., Sandiford and McLaren, 2002) and has produced significant weakening of the crust (e.g., Marshak, 1999). Thus, the high-grade metamorphism, special thermal regime and widespread constrictional-type strains recorded in the Pilbara supracrustal succession are interpreted to indicate the gravitational sinking and burial of denser, mainly mafic and ultramafic greenstone rocks along the flanks of, and between, buoyant and rising TTG diapirs, a model known as partial convective overturn of the crust (e.g., Collins et al., 1998; Van Kranendonk et al., 2002, 2004a).

In contrast, the 3.23 Ga structural evolution of the Stolzburg terrane in the BGGT has been interpreted to be the result of core-complex-like exhumation of the lower crust, probably in a post-collision setting (Kisters et al., 2003). The following points appear to argue strongly against partial convective overturn of the crust in the BGGT of the type proposed for the Pilbara Craton.

(1) The southern contact of the BGB with the Stolzburg terrane is marked by mainly prolate fabrics. These are exhumation fabrics and they are not related to the burial or sinking of the supracrustal sequence.

(2) The highest pressures (8–11 kbar by Dziggel et al. (2002), and 12–15 kbar by Moyen et al. (2006)) are documented from the southern TTG terrain and not in the greenstone sequences. Indeed, the felsic plutonic rocks contain metamorphic assemblages recording significantly higher pressures than the flanking greenstones.

(3) In addition, the low apparent geothermal gradient in the exhumed basement to the south of the BGGT appears inconsistent with an essentially thermally driven process.

Despite these important differences, there are some similarities in the processes proposed for the two areas. After initiation of exhumation of the granitoid domains along the extensional detachment in the southern BGGT, there is a transition from extensional, to buoyancy driven rise, and the final emplacement of the gneissic "domes" may well be aided by the buoyancy contrast between the gneisses and the mafic greenstones. In the later stage, ascent of the coherent basement blocks causes the development of a predominantly linear fabric, with vertical stretching lineations along their margins. Unlike the scenarios proposed in other, similar, dome-and-keel terranes, the ascent here occurs after an initial stage of extensional collapse, and affects not single magma batches (plutons), nor migmatitic complexes, but essentially chunks of solid, composite continental crust made of several well-identified plutons and surrounding volcanosedimentary sequence (within which lithological relationships, including intrusion relationships between the plutons and the supracrustal rocks ca. 200 Ma prior to the orogenic history, are often well-preserved). This is mostly a solid-state process, although syntectonic intrusion into the high-strain margins is common and possibly important in achieving the significant vertical displacement recorded by the magnitude of the metamorphic pressure differences across the margins. This late evolution and steepening of bounding shear zones to close to vertical is not classically known from modern core complexes, but seems to correspond to a unique Archean process that is essentially driven by the buoyancy contracts between the mafic/ultramafic lower sections of the greenstone belt stratigraphy and the granitoid middle and lower crust.

It may be possible that in Archean orogens (at least in the BGGT), crustal thickening followed by orogenic collapse quickly evolves into buoyancy driven, near-vertical emplacement of the lower-crustal domains as a result of the higher density contrast between the heavy upper crust (dominated by mafic/ultramafic rocks) and the felsic lower crust (TTG gneisses), resulting in a density inversion and an unstable density stratification. Such a situation is not commonly attained in modern orogens, where the upper crust is made of lighter gneisses or sediments, and the lower crust of dense eclogites or granulites. It might be tempting to also propose a higher Archean heat production, causing a generally softer lithosphere, and facilitating bulk diapiric rise of the crust. In the case of the BGGT, this does not appear to fit either the relatively low-temperature, high-pressure metamorphic signature of the Western domain, or the strain localization patterns associated with the uplift of this rather rigid crustal block.

Earth's Oldest Rocks
Edited by Martin J. Van Kranendonk, R. Hugh Smithies and Vickie C. Bennett
Developments in Precambrian Geology, Vol. 15 (K.C. Condie, Series Editor)
© 2007 Elsevier B.V. All rights reserved.
DOI: 10.1016/S0166-2635(07)15058-1

Chapter 5.8

TECTONO-METAMORPHIC CONTROLS ON ARCHEAN GOLD MINERALIZATION IN THE BARBERTON GREENSTONE BELT, SOUTH AFRICA: AN EXAMPLE FROM THE NEW CONSORT GOLD MINE

ANNIKA DZIGGEL[a], ALEXANDER OTTO[a], ALEXANDER F.M. KISTERS[b] AND F. MICHAEL MEYER[a]

[a]*Institute of Mineralogy and Economic Geology, RWTH Aachen University, Wüllnerstrasse 2, 52062 Aachen, Germany*
[b]*Department of Geology, University of Stellenbosch, Private Bag X1, Matieland 7602, South Africa*

5.8-1. INTRODUCTION

Orogenic gold deposits are a distinct class of gold deposits, recognized throughout Earth's history, that have been the source of much of the world's gold production (e.g., Groves et al., 1998; Goldfarb et al., 2001). The term orogenic gold deposit refers to structurally controlled gold deposits that occur in deformed and variably metamorphosed host rocks. Orogenic gold deposits form in arc, backarc, or accretionary settings during the late stages of Cordilleran-style tectonics (Kerrich et al., 2000; Goldfarb et al., 2001, Groves et al., 2003), although there are exceptions (Bateman and Hagemann, 2004; Gauthier et al., 2004). Most deposits are situated in second- or third-order structures, or in fault intersections, corresponding to sites of reduced mean stress that result in increased fluid fluxes and fluid localization. Gold is often contained in quartz–(carbonate) veins, and the mineralized structures are spatially closely associated with large-scale compressional and/or transpressional shear zones proximal to craton-scale fault zones (McCuaig and Kerrich, 1998). The mineralized lodes typically formed over a range of P-T conditions of ca. 200–650 °C and 1–5 kbars, and the low salinity, CO_2-rich mineralizing fluids are generally interpreted as having originated from metamorphic devolitilization reactions in the mid and lower crust (e.g., Powell et al., 1991). The mineralization is syn- to late-tectonic and typically postdates the peak of metamorphism in greenschist facies deposits, while mineralization in amphibolite facies gold deposits appears to be synchronous with the peak of metamorphism. High-T gold deposits of this type are termed hypozonal orogenic gold deposits, which form part of a continuum of gold deposits over an extended range of crustal conditions (e.g., Groves et al., 1998).

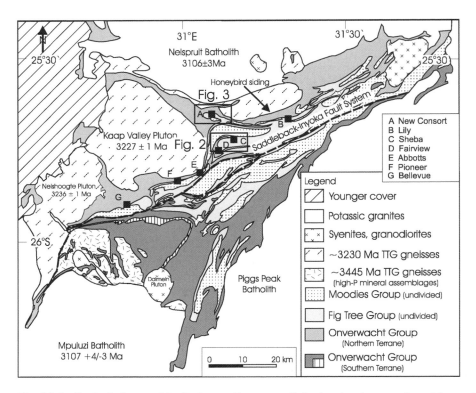

Fig. 5.8-1. Geological map of the Barberton Greenstone Belt, showing the extension of the northern and southern terranes, and the location of the Saddleback–Inyoka Fault System (modified after de Ronde and de Wit (1994)). The occurrence of major gold deposits is restricted to the northern terrane.

The hypozonal New Consort gold mine in the Palaeo- to Mesoarchean Barberton Greenstone Belt, South Africa, forms one of four currently active mines that are clustered along the northern margin of the greenstone belt. The mine is generally interpreted as having formed during the late tectonic evolution of the belt in the Mesoarchean between ca. 3.2 and 3.1 Ga (de Ronde et al., 1991b). As such, it represents one of the oldest known orogenic gold deposits on Earth (Groves et al., 1998; Goldfarb et al., 2001), although older deposits have been documented for the Pilbara Craton (Zegers et al., 2002). The timing of mineralization relative to the metamorphic evolution of the greenstones is a significant problem that relates to this and other Archean gold deposits, as this relationship has profound implications for the fluid source and tectonic setting in which mineralization occurred. In this paper, we review the tectonic and metamorphic setting of this hypozonal gold deposit within the overall tectonic framework of the greenstone belt, in an attempt to characterize the possible tectonic scenarios that were responsible for the formation of these gold deposits.

5.8-2. GEOLOGICAL SETTING

The Barberton Greenstone Belt, South Africa, forms part of the oldest nucleus of the Kaapvaal Craton (e.g., de Wit et al., 1992). It contains a well-preserved ca. 3570–3220 Ma volcano-sedimentary sequence, which is surrounded by various generations of tonalite-trondhjemite-granodiorite (TTG) gneiss domes and sheet-like potassic granites emplaced between ca. 3500–3100 Ma (Fig. 5.8-1: e.g., Armstrong et al., 1990; Kamo and Davis, 1994; de Ronde and de Wit, 1994). The regional structural framework of the belt is dominated by large, upright synforms separated either by thrust faults, or narrow anticlines. A characteristic feature of the belt is that the metamorphic grade is generally low, but increases towards the sheared contacts with surrounding TTG gneisses. This contact has been interpreted as an extensional detachment that separates the greenschist facies greenstone belt from mid-crustal basement gneisses (Kisters et al., 2003; Diener et al., 2005).

The greenstone belt sequence, assigned to the Swaziland Supergroup, has been subdivided into three stratigraphic units. From base to top, these include: (1) the Onverwacht Group, dominated by ultramafic and mafic volcanic rocks; (2) the Fig Tree Group, a meta-turbiditic succession made up of greywackes, shales, and cherts; and (3) the Moodies Group, characterized by coarse-grained clastic sedimentary rocks, mainly including sandstones and conglomerates (Fig. 5.8-1: SACS, 1980). Major differences in age relationships, depositional environments and sediment provenances between rocks to the north and south of the Saddleback-Inyoka Fault System in the centre of the belt (Fig. 5.8-1), suggests that a major suture zone separates the greenstone belt into a northern and southern terrane (e.g., de Ronde and de Wit, 1994; Kamo and Davis, 1994; Lowe and Byerly, 1999, this volume; Lowe, 1994, 1999b). Consequently, the stratigraphy of the two terranes will be described separately.

5.8-2.1. Southern Terrane

Onverwacht Group rocks to the south of the Saddleback–Inyoka Fault range between ca. 3550–3300 Ma. They mainly include mafic to ultramafic volcanic rocks, intercalated felsic volcaniclastic sequences, and minor units of clastic and chemical sedimentary rocks (e.g., Viljoen and Viljoen, 1969a; de Wit et al., 1987b; Kamo and Davis, 1994; Byerly et al., 1996; Dziggel et al., 2006a). At the base, they are intruded by an early generation of TTG granitoid plutons, most of which have been dated at ca. 3445 Ma (e.g., Armstrong et al., 1990). The overlying Southern Facies of the Fig Tree Group comprises shales, sandstones, conglomerates, dacitic to rhyodacitic volcanic and volcaniclastic rocks, and minor chemical sedimentary rocks (Lowe and Byerly, 1999). U-Pb zircon dating of felsic volcaniclastic rocks reveal ages of ca. 3260–3230 Ma (e.g., Kröner et al., 1991a; Byerly et al., 1996). Deposition took place in alluvial, fan-delta, and shallow to moderately deep subaqueous environments (e.g., Lowe and Nocita, 1999). The coarse-clastic sedimentary rocks of the uppermost Moodies Group are dominated by quartz-rich sandstones that were probably deposited in several isolated basins (e.g., Heubeck and Lowe, 1999). The age of the Moodies Group is poorly constrained, however, a minimum age of 3216 +2/−1 Ma (U-Pb zircon dating; Kamo and Davis, 1994) is suggested by the emplacement age of the post-kinematic Dalmein Pluton that crosscuts the Kromberg syncline (Fig. 5.8-1).

5.8-2.2. Northern Terrane

Onverwacht Group rocks to the north of the Saddleback–Inyoka Fault have been assigned to the Weltevreden Formation (e.g., Lowe and Byerly, 1999). In addition to the predominantly mafic and ultramafic volcanic rocks and minor chemical sedimentary rocks, this formation is characterized by abundant layered ultramafic intrusive complexes (Anhaeusser, 2001). A Nd isochron age of 3286 ± 29 Ma on komatiites suggests that the deposition of this unit might correlate time wise with the uppermost Onverwacht Group rocks of the Southern Facies (Lahaye et al., 1995). The base of the Onverwacht Group in this part of the greenstone belt has been intruded by TTG granitoid plutons that yield U-Pb zircon ages of ca. 3230 Ma (Fig. 5.8-1: e.g., Kamo and Davis, 1994). Compared to the 'Southern Facies', the 'Northern Facies' of the Fig Tree Group has a more turbidite-like character, and consists of carbonaceous shales, ferruginous cherts, greywackes, dacitic volcaniclastic rocks, turbiditic sandstones, and minor conglomerates. U-Pb zircon data of 3226 ± 6 Ma and 3225 ± 6 Ma indicate deposition largely coeval with the 'Southern Facies' (Kamo and Davis, 1994; Lowe and Byerly, 1999). The age of the Moodies Group in the Northern Terrane remains uncertain, although Layer (1986) presented evidence that it was overprinted during emplacement of the Kaap Valley Pluton at ca. 3214 ± 4 Ma.

5.8-2.3. Tectonic Evolution and Timing of Gold Mineralization

The greenstone belt and surrounding granitoid–gneiss terrain record a long, complex, and polyphase tectono-magmatic history (e.g., de Ronde and de Wit, 1994; Lowe, 1994, 1999b). The episodic tectonic evolution involved at least three periods of deformation.

The first significant tectonic event (D_1) took place at ca. 3445–3416 Ma, and was restricted to Onverwacht Group rocks in the southern part of the greenstone belt (e.g., de Ronde and de Wit, 1994). The D_1 deformation was closely associated with the intrusion of an early generation of TTG granitoid plutons along the southern margin of the greenstone belt (Fig. 5.8-1; e.g., de Wit et al., 1987a, 1987b; de Wit, 2004).

The D_2 deformation occurred during the late Palaeoarchean (ca. 3227–3230 Ma) and has been ascribed to a short-lived accretionary episode that affected the entire greenstone belt (de Ronde and de Wit, 1994; Kamo and Davis, 1994). The D_2 deformation was associated with the amalgamation of the southern and northern terranes along the Saddleback–Inyoka Fault system and resulted in substantial crustal thickening (e.g., de Ronde and de Wit, 1994; Kamo and Davis, 1994; de Ronde and Kamo, 2000). At this time, early TTG granitoid gneisses and the associated basal sequences of the Onverwacht Group along the southern margin of the greenstone belt, namely the Sandspruit and Theespruit Formations, were buried to mid- to lower-crustal depth, where they experienced high-pressure amphibolite facies metamorphism (Dziggel et al., 2002, 2005; Diener et al., 2005, 2006; Moyen et al., 2006). P-T conditions of up to 650 °C and 15 kbar indicate very low apparent geothermal gradients of ca. 12 °C/km, the lowest apparent geothermal gradient ever been recorded in the Archean rock record (Moyen et al., 2006). Deposition of the Fig Tree Group, as well as the syn-deformational molasse-type sedimentation of the Moodies Group most probably correlate with this event (e.g., de Ronde and de Wit, 1994; Kamo and Davis, 1994; de Ronde and Kamo, 2000). The D_2 deformation is interpreted as a result of crustal convergence and the accretion of the southern terrane during northwards-directed subduction (Diener et al., 2006; Moyen et al., 2006). This interpretation is also supported by the presence of syn (D_2)-tectonic TTG granitoid plutons along the northern margin of the greenstone belt, such as the Nelshoogte and Kaap Valley Plutons (Fig. 5.8-1), which have been dated at 3236 ± 1 Ma and 3227 ± 1 Ma (U-Pb zircon; Kamo and Davis, 1994; de Ronde and Kamo, 2000). Further, high-grade basement gneisses and associated supracrustal rocks along the northern margin of the greenstone belt record distinctly higher geothermal gradients ($\geqslant$30 °C/km) than those in the south. This is consistent with advective heating of upper plate rocks during TTG emplacement (Dziggel et al., 2006b). D_2 accretion was either synchronous with, or immediately followed by, a period of syn-orogenic extension and solid-state doming that eventually resulted in the steepening of fabrics during the orogenic collapse of the belt (e.g., Kisters et al., 2003; Dziggel et al., 2006b).

Currently available data on the timing of gold mineralization suggest that the gold deposits in the Barberton Greenstone Belt formed more than 100 million years after the main accretionary D_2 event (e.g., de Ronde et al., 1991b). Consequently, most workers regard the late-tectonic evolution of the greenstone belt to be of particular importance for the gold mineralization (e.g., de Ronde et al., 1991b, 1992; de Ronde and de Wit, 1994). During D_3, many of the steepened thrusts appear to have been reactivated as strike slip shear zones, when most of the gold appears to have been precipitated (e.g., de Ronde et al., 1992; Robertson et al., 1993). The D_3 deformation was associated with the emplacement of voluminous sheet-like potassic granites to the north and south of the greenstone belt, such as the ca. 3100 Ma Mpuluzi and Nelspruit batholiths (Fig. 5.8-1; e.g., Kamo and Davis, 1994; de Ronde and de Wit, 1994). Time constraints on the shear-zone-hosted gold mineralization are given by a ca. 3126 Ma porphyry dyke that predates shearing at the Fairview Mine (Fig. 5.8-1), and a ca. 3084 Ma age for hydrothermal rutile associated with gold mineralization (U-Pb dating; de Ronde et al., 1991b). This late-tectonic evolution, however, has remained speculative mainly because of the lack of robust geochronological data constraining the age of strike-slip shearing within the greenstone belt, but also due to the lack of knowledge about the influence of the D_3 deformation on the final architecture of the granite–greenstone terrain. Further complicating this picture is that the emplacement of the late potassic granites seems to be related to a craton-wide period of intracratonic magmatism that was driven by the development of a crescent-shaped, juvenile arc along the northern and western margins of the Kaapvaal Craton (e.g., Poujol et al., 2003). Thus, the Mesoarchean magmatic and tectonic accretion at ca. 3.1 Ga appears to have been localized some 200–300 km to the north and west of the Barberton Greenstone Belt, pointing to a rather atypical intracratonic setting for the formation of the orogenic gold deposits.

5.8-3. CHARACTERISTICS OF GREENSCHIST FACIES GOLD DEPOSITS

Although more than 350 gold deposits have been recorded in the Barberton Greenstone Belt, the bulk of the ca. 320 t of gold (more than 85%) has been produced from the Sheba–Fairview, New Consort, and Agnes–Princeton mining complexes in the northern part of the belt (Fig. 5.8-1; e.g., Anhaeusser, 1986; Ward, 1999). The majority of these and associated smaller gold deposits are clustered in a crescent-shaped zone in greenschist facies supracrustal rocks along the eastern margin of the ca. 3230 Ma, tonalitic Kaap Valley Pluton (Fig. 5.8-1). Gold mineralization in these deposits is structurally controlled by east–west trending strike-slip shear zones that are situated in, or in close proximity to, major D_2 structures that have been interpreted to have originated as thrust faults (e.g., de Ronde et al., 1992). The gold deposits within these D_3-related, brittle–ductile shear zones show similar alteration characteristics with gold-bearing quartz–carbonate veins hosted in fuchsite-, sericite-, and carbonate-rich alteration zones. Due to the close spatial association between the gold deposits and the Kaap Valley Pluton, de Ronde et al. (1992) interpreted this tonalite body as a major impermeable barrier for the mineralizing fluids. Fluid inclusion data show that gold was deposited from a H_2O–CO_2–NaCl fluid due to phase separation at conditions of ca. 300 °C and 1 kbar, corresponding to a depth of ca. 3–4 km (de Ronde et al., 1992). Based on the large apparent time gap between regional greenschist facies metamorphism and gold mineralization (>100 million years), as well as the shallow crustal depth of the greenstone belt (ca. 4–8 km; de Beer et al., 1988), the source of the mineralizing fluids was interpreted to be external to the greenstone belt.

In the following sections, we review the characteristics of one of the better documented greenschist facies gold deposits, the Sheba gold mine, in order to emphasize the structural control and relative timing of gold mineralization in the greenschist facies gold deposits of the Barberton Greenstone Belt. This is followed by a more detailed description of the structural and metamorphic evolution of the New Consort gold mine.

5.8-3.1. Sheba Gold Mine

Gold mineralization at the Sheba gold mine is situated in the Sheba Hills, dominated by the arcuate Eureka and Ulundi Synclines of post-Moodies Group age (Figs. 5.8-1 and 5.8-2; e.g., Anhaeusser, 1986). These northerly-verging folds are separated by the Sheba Fault, a regional-scale strike-slip shear zone. Arenaceous Moodies Group rocks dominate the Eureka Sycline to the north of the Sheba Fault, whereas the Ulundi Syncline to the south of the Sheba Fault is made up of Onverwacht and Fig Tree Group rocks (Wagener and Wiegand, 1986; Robertson et al., 1993). Host rocks to the mineralization in this part of the greenstone belt are mainly Fig Tree and Moodies Group sedimentary rocks, including quartzites, shales, sandstones, and greywackes. Within the Ulundi Syncline, Onverwacht Group volcanic rocks (mainly talc–carbonate schists and cherts) outline km-scale, tight- to isoclinal folds that are locally referred to as the Sheba Anticlines (Fig. 5.8-2). The Sheba Fault has been interpreted as an originally low-angle thrust fault that was reactivated during

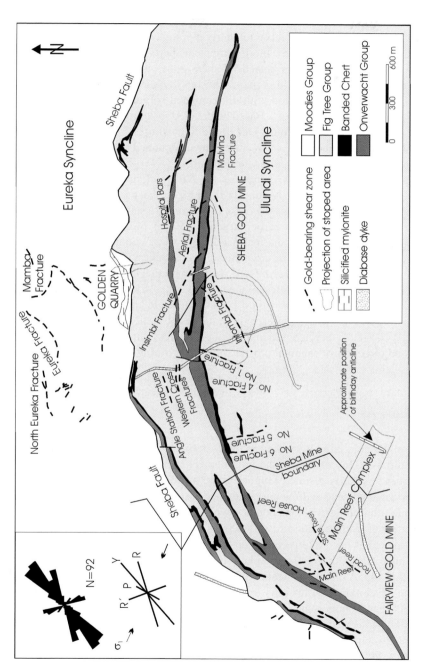

Fig. 5.8-2. Simplified geological map of the area around the Sheba and Fairview mines, including major structures (modified after Wagener and Wiegand (1986)). Inset shows a rose diagram plot illustrating the trend of shear zones in the Main Reef Complex (Robertson et al., 1993), as well as an interpretation of the data.

post-Moodies deformation. During refolding and arcuation of the Eureka and Ulundi Synclines, the Sheba Fault was reactivated as a dextral strike-slip fault (e.g., Robertson et al., 1993). Gold mineralization is predominantly situated in the Ulundi Syncline, where it is spatially closely associated with the steeply southerly dipping contact between greywackes and shales of the Fig Tree Group and the more competent cherts of the Onverwacht Group.

Most of the gold mineralization is hosted by shear zones that dip steeply to the SE and SW (Robertson et al., 1993). Gold is mainly hosted in quartz–carbonate veins along these fractures, or occurs in association with disseminated sulfides (mainly arsenopyrite and pyrite) within the adjacent wall rocks. Based on detailed structural analyses of the Main Reef Complex (MRC) in the westernmost part of the Sheba mine (Fig. 5.8-2), Robertson et al. (1993) interpreted the geometry of fracture zones associated with the gold mineralization as second- and third-order structures formed during WNW–ESE crustal shortening. They distinguished three sets of shear zones: (i) the NE-trending and SE-dipping main shear zones, which are sub-parallel to bedding; (ii) the steep shear zones, which strike E–W and dip steeply to the south, and (iii) the cross shear zones, a set of shallowly southerly dipping shear zones that cross-cut the earlier structures. Zones of high-grade gold mineralization occur at the intersection of the main and cross shear zones, and plunge at low to moderate angles to the NE. This fracture pattern is consistent with the D_3-related, dextral strike slip shearing along the Sheba Fault. In their model, Robertson et al. (1993) interpreted the main shear zones to represent the principal displacement shears (Y-shears), whereas the conjugate shear fractures would correspond to R and P shears (Fig. 5.8-2, inset). The cross shear zones would represent high-angle antithetic (R') shears. The geometry indicates a WNW–ESE directed maximum principal stress during shearing, consistent with the proposed direction of regional crustal shortening during D_3 (e.g., de Ronde and de Wit, 1994; Belcher and Kisters, 2006a).

5.8-4. THE NEW CONSORT GOLD MINE

The New Consort gold mine is situated in the eastern part of the Jamestown schist belt (Figs. 5.8-1 and 5.8-3). In contrast to the gold deposits described above, mineralization at New Consort is hosted by distinctly higher-grade metamorphic rocks. The complexly folded and imbricated volcano-sedimentary sequence is situated in the immediate hanging-wall of the basal granitoid–greenstone contact, which separates the generally greenschist facies greenstone belt from the mid-crustal gneisses of the Stentor Pluton (Dziggel et al., 2006b). This contact has been interpreted as an extensional detachment along which basement gneisses have been exhumed, most likely in the course of the orogenic collapse of the belt at ca. 3230 Ma. A distinctive feature of the gold mineralization at the New Consort gold mine is the development of laterally extensive mineralized horizons in wall-rocks of significantly different metamorphic grade. The gold mineralization is mainly structurally controlled by highly silicified ductile shear zones at, or near, the structural contact between rocks of the Onverwacht and Fig Tree Groups. This contact, locally referred to as the Consort Bar, reaches a thickness of up to several meters, and epigenetic ore shoots may occur in

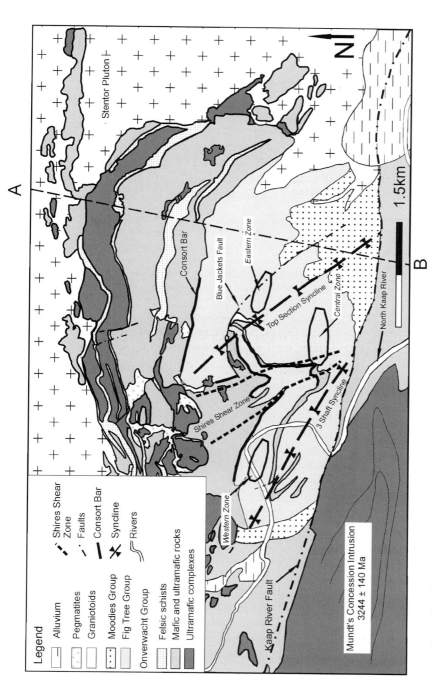

Fig. 5.8-3. Simplified geological map of the New Consort gold mine area in the eastern part of the Jamestown schist belt (modified after Anhaeusser and Viljoen (1965); Anhaeusser (1972)). The Western, Central and Eastern Zones mark the areas investigated by Otto et al. (2007), and are projections from underground. See text for explanation. Section A-B shown in Fig. 5.8-4.

different stratigraphic levels at, or near, this structural break. Stratigraphically, the exposed rocks comprise Onverwacht Group mafic and ultramafic volcanic rocks and intercalated aluminous felsic schists and cherts that are overlain by argillaceous and coarse clastic sedimentary rocks of the Fig Tree and Moodies Groups (Fig. 5.8-3). The volcano-sedimentary sequence has been folded into two major synclines that plunge at low to moderate angles to the SE and E, namely the Top Section and 3 Shaft Synclines (Fig. 5.8-3). The structural and metamorphic setting of the main ore bodies, as well as the characteristics of ore and alteration assemblages associated with the gold mineralization, are outlined below. It should be noted that the sequence of deformation events described in the following refers to the *local* structural evolution based on overprinting structural fabrics developed in the mine workings and surroundings. The nomenclature does not correspond to the regional tectonic events outlined in the chapters on regional geology (e.g. Lowe and Byerly, this volume), although we try to establish a correlation of local and regional fabrics and deformation events.

5.8-4.1. Structural Evolution

The earliest fabric recorded along the granite–greenstone contact is a moderately southerly dipping S_1 foliation. S_1 is parallel to the granite–greenstone contact and to bedding, attributed to the transposition of lithologies (Figs. 5.8-3 and 5.8-4). This D_1 shearing affected both the greenstone sequence, as well as the structurally underlying gneisses of the Stentor Pluton. The orientation of fabrics and kinematic indicators contained within them point to the fact that D_1-related fabrics are associated with the syn- to post-collisional extension of this mid- to lower-crustal segment (Dziggel et al., 2006b). In the area north of the New Consort gold mine, two main structural domains can be distinguished (Fig. 5.8-4): the early S_{1A} fabrics are restricted to the amphibolite facies gneisses of the Stentor Pluton and overlying supracrustal rocks. Associated mineral stretching lineations plunge at low to moderate angles to the W and SW, parallel to the fold axes of open to tight F_{1A} folds. The SW plunge of L_{1A} stretching lineations is also parallel to the general NE–SW trend of constrictional fabrics in the southern parts of the greenstone belt, and thus is in agreement with the postulated direction of orogen-parallel extension (Kisters et al., 2003). In contrast, S_{1B} fabrics are mainly confined to a 0.5–1 km wide high strain belt, where they overprint the earlier fabrics. These fabrics are recorded in both amphibolite and retrograde greenschist facies mylonites (Fig. 5.8-4), pointing to a considerable increase in strain intensity during retrogression. The associated mineral stretching lineations are either downdip, or plunge at low to moderate angles to the SE, parallel to the fold axes of isoclinal intrafolial folds. Kinematic indicators consistently point to a Stentor Pluton-up/greenstone-down sense of movement with a sinistral strike slip component. The S_{1B} fabrics have been linked to the last stages of exhumation that were characterized by solid-state doming of footwall gneisses in response to the increasing buoyancy contrast during extensional shearing (Dziggel et al., 2006b). The folding of the shallowly plunging Top Section and 3 Shaft Synclines has been linked to the D_{1C} deformation (Fig. 5.8-4). Their formation most probably occurred in direct response to the emplacement and solid-state doming of hot, ductile

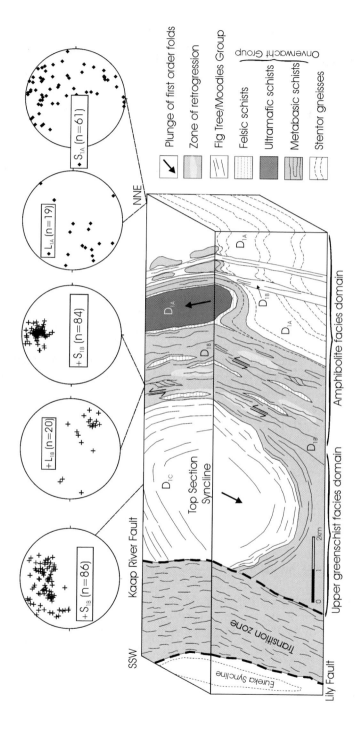

Fig. 5.8-4. Geological cross-section through the granite–greenstone contact, illustrating the structural relations and distribution of metamorphic domains (section A-B in Fig. 5.8-3; modified after Dziggel et al. (2006b)). All stereographic projections are equal area plots and to the lower hemisphere.

basement rocks into shallower crustal levels, resulting in the folding of earlier fabrics and the overlying stratigraphic units. Because this folding also affected the Consort Bar, much of the shearing (but not the mineralization) along this shear zone must predate the regional folding (Fig. 5.8-3).

The gold mineralization has been related to the local D_2 deformation, which has been interpreted to correspond to the regional D_3 tectonic event (Tomkinson and Lombard, 1990; Harris et al., 1995; Otto et al., 2007). In general, the fabrics related to this deformation event remain steep, but are variable in orientation and partitioned into discrete high-strain zones (Otto et al., 2007). In the central parts of the mine, the local D_2 deformation resulted in the formation of a prominent, steeply dipping shear zone system known as the Shires shear zone (Fig. 5.8-3). The Shires shear zone is made up of an anastomozing network of discrete, N–S to NW–SE trending shear zones, and can be up to 350 m wide. Individual shear zones vary from a few cm to 5 m in thickness (Harris et al., 1995). Earlier studies on the kinematics along this shear zone proposed a west-block-up and minor sinistral sense of movement that appears to displace the Consort Bar by 1200 m on surface (Tomkinson and Lombard, 1990; Harris et al., 1995). This suggests NW–SE directed subhorizontal shortening during the D_2 deformation. However, the occurrence of gold in shear zones parallel to, or within, the Consort Bar along the northern, roughly E–W trending, limbs of the Top Section and 3 Shaft Synclines indicates that D_2 shearing was not restricted to the central parts of the mine (Fig. 5.8-3). In these shear zones, the mylonitic S_2 fabrics are essentially coplanar with the regional S_1 foliation, and thus, not easily distinguished. The D_2 deformation is also marked by the intrusion of numerous syn-kinematic pegmatite dykes, in an orientation strongly controlled by D_2 shear zones (Harris et al., 1995). These pegmatites locally crosscut the mineralization, but may also be folded and commonly display mylonitic fabrics along their margins, pointing to a continuous deformation during their emplacement. Rb-Sr and Sm-Nd garnet isotope analyses yield an age of ca. 3040 Ma (Harris et al., 1993). The last deformation event recorded (D_3) is characterized by the development of brittle normal faults, collectively referred to as the Blue Jackets fault system, which crosscut all earlier structures (Fig. 5.8-3; Tomkinson and Lombard, 1990; Harris et al., 1995).

Detailed structural underground mapping by Otto et al. (2007) revealed that the geometry of the three ore bodies currently mined at New Consort varies according to the exact structural position within the mine. The location of the three areas investigated, projected from underground, is shown in Fig. 5.8-3. The Western Zone ore body is located on the shallowly southerly dipping, northern limb of the 3 Shaft Syncline, about 150 m below surface. The mineralization of the Central Zone is located in the Consort Bar within the Shires Shear Zone system and ca. 1400 m below surface. The Eastern Zone ore body is located at the northern limb of the Top Section Syncline, ca. 1400 m below surface and ca. 60 m in the footwall of the Consort Bar, and represents an example of the so-called Footwall Lens mineralization (see Section 5.8-4.3).

In accordance with the structural framework outlined above, Otto et al. (2007) identified 3 phases of deformation (Fig. 5.8-5). In all three localities, the variably dipping S_1 foliation strikes approximately east–west, and is interpreted to correlate with the S_{1B} foliation

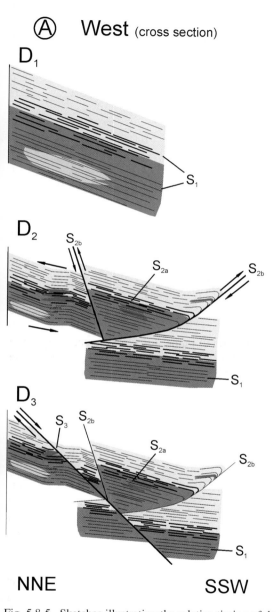

Fig. 5.8-5. Sketches illustrating the relative timing of deformation, mineralization, and magmatism in the (a) Western, (b) Central, and (c) Eastern Zones (modified after Otto et al. (2007)). Note that the intensity of deformation in different lithologies is highly variable, as indicated by the various line spacings.

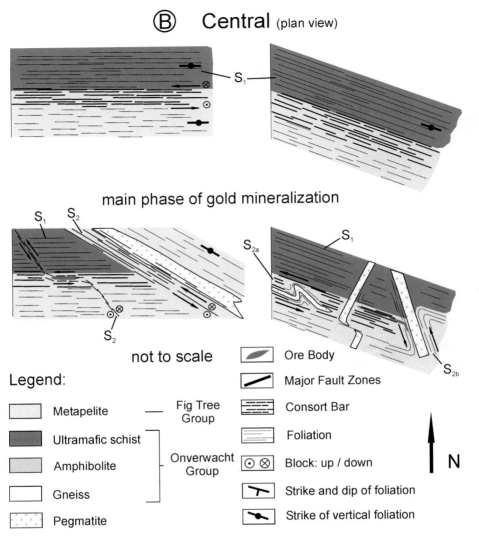

Fig. 5.8-5. (*Continued.*)

of Dziggel et al. (2006b). The D_2 deformation is characterized by: (i) an approximately N–S directed compressional deformation with a top-to-the-north sense of movement along the Consort Bar in the Western Zone (Fig. 5.8-5(a)); (ii) sinistral strike-slip shearing with a minor reverse sense of movement in the Central Zone, resulting in the formation of the NW–SE trending Shires shear zone system that crosscuts the Consort Bar (Fig. 5.8-5(b)); and (iii) contact-parallel sinistral strike-slip shearing in the Eastern Zone, also with a minor reverse sense of movement. Here, shearing was most intense in footwall rocks some 60 m

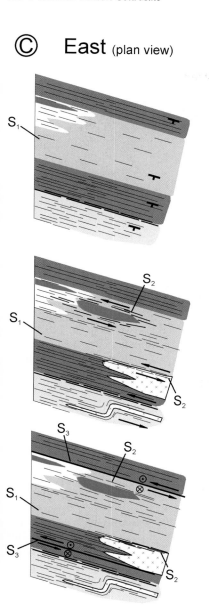

Fig. 5.8-5. (*Continued.*)

below the Consort Bar (Fig. 5.8-5(c)). The ore body in the Western Zone is located within SE plunging dilational zones that are associated with the flat parts of flat-and-ramp structures. The plunge of the ore body is at high angles to the movement direction along the

Consort Bar, and parallel to the SE plunging fold axis of the 3 Shaft Syncline. In contrast, the plunge of the elongate, replacement style ore bodies in the Central Zone is subvertical, and is controlled by the plunge of dilational jogs formed at the intersection of D_1 and D_2 shear zones (Fig. 5.8-5). Similar subvertical plunges of foliation-parallel ore bodies have also been recorded in the Eastern Zone. D_3-related late normal faulting has only been identified in the Western and Eastern Zones.

5.8-4.2. Metamorphism

Regional field mapping of metamorphic isograds along the contact between the greenstone belt and the Stentor Pluton indicates that lower grade rocks systematically overlie higher grade supracrustal rocks and associated basement gneisses (Dziggel et al., 2006b). Four metamorphic domains can be distinguished: (i) an amphibolite facies domain, comprising the Stentor gneisses as well as Onverwacht Group metavolcanic rocks to the south of the granite–greenstone contact (Fig. 5.8-4); (ii) an upper greenschist facies domain, including Fig Tree and Moodies Group rocks of the Top Section and 3 Shaft Synclines; (iii) a mainly greenschist facies transition zone, also made up of Onverwacht Group volcanic rocks; and (iv) a low-grade metamorphic domain that coincides with the predominantly Moodies Group sedimentary rocks of the Eureka Syncline (Fig. 5.8-4). P-T estimates on hornblende + plagioclase + quartz ± garnet ± clinopyroxene and garnet + sillimanite + biotite + muscovite + quartz assemblages within meta-basic and aluminous felsic schists of the amphibolite facies domain, using a variety of conventional geothermometers as well as the average P-T method of the computer program THERMOCALC (Holland and Powell, 1998), indicate peak metamorphic conditions of ca. 600–700 °C and 5 ± 1 kbars (Dziggel et al., 2006b). Combined petrological data and pseudosection modelling reveal a clockwise P-T evolution, comprising the early burial to mid-crustal levels and subsequent exhumation (Fig. 5.8-6). The derived P-T paths point to near isothermal decompression to conditions of ca. 500–600 °C and 1–3 kbars associated with the exhumation of the rocks and extensional shearing, followed by isobaric cooling to temperatures below ca. 500 °C after deformation.

The gold mineralization at New Consort is situated within the amphibolite and upper greenschist facies domains. These are separated by the Consort Bar, suggesting that this shear zone represents a major structural break within the sequence. On the basis of detailed petrology and thermobarometric calculations on unaltered rocks hosting the mineralization, Otto et al. (2007) documented a distinct metamorphic gradient in the mine. Upper greenschist facies conditions prevail in the structurally highest Fig Tree Group rocks in the Western Zone, whereas upper amphibolite facies grades are developed in Onverwacht Group rocks in the structurally deepest part in the mine, the Eastern Zone (Fig. 5.8-7). P-T estimates on Fig Tree Group metapelites in the hangingwall of the Consort Bar range from 500–550 °C in the lowest grade Western Zone to ca. 550–600 °C and 4–5 kbars in the Central and Eastern Zones. This difference in the estimated temperatures within the Fig Tree Group is consistent with increasing P-T conditions with increasing structural depth along a steady state geothermal gradient. In contrast, Onverwacht Group volcanic rocks

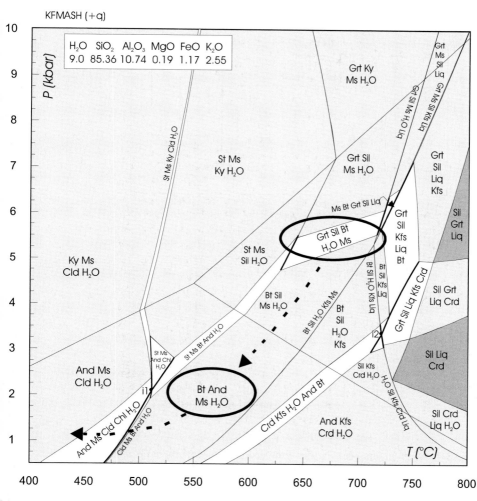

Fig. 5.8-6. KFMASH P-T pseudosection calculated for a garnet-bearing aluminous felsic schist from the Onverwacht Group (modified after Dziggel et al. (2006b)). The bulk composition in mol% is given in the upper left inset.

and intercalated Fig Tree Group sedimentary rocks in the footwall of the Consort Bar record distinctly higher metamorphic grades of ca. 600–700 °C and 6–8 kbars (Fig. 5.8-7), indicating that this shear zone is largely responsible for the observed condensation of metamorphic isograds within the structural sequence. However, despite the similarity in estimated temperatures with those recorded by Dziggel et al. (2006b), the pressures of 6–8 kbar are considerably higher than the 5 ± 1 kbar reported by these workers. Because of the consistency of pressure estimates in both studies, based on different equilibrium assemblages and barometric methods, it seems likely that the pressure difference is real, and not a

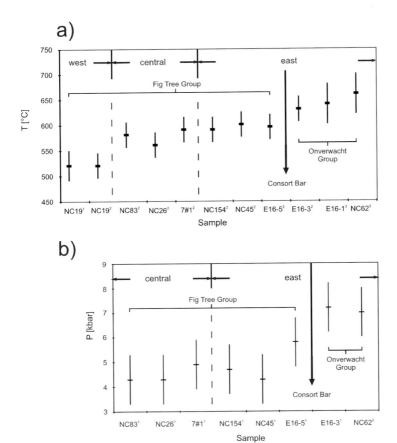

Fig. 5.8-7. Diagram summarizing the estimated peak metamorphic (a) temperature and (b) pressure conditions of host lithologies at the New Consort gold mine (adopted from Otto et al. (2007)). The calibrations used for estimating temperatures include the [1]Grt-Hbl (Ravna, 2000a), the [2]Grt-Bt (Holdaway, 2000) and [3]Hbl-Pl (Holland and Blundy, 1994) thermometers; pressure estimates are based on the [1]Grt-Bt-Pl-Qtz and [2]Hbl-Pl barometers of Wu et al. (2004) and Plyusnina (1982). The error bars represent the errors connected to the respective calibration, and are considerably larger than 1 sigma.

result of errors related to the thermobarometric calculations. The Onverwacht Group rocks containing the higher pressure mineral assemblages documented by Otto et al. (2007) are thus interpreted to represent basement from slightly deeper crustal levels.

5.8-4.3. A Two Stage Gold Mineralization Model

Previous workers distinguished two main types of gold mineralization at the New Consort gold mine (e.g., Tomkinson and Lombard, 1990). The most common mineralization was

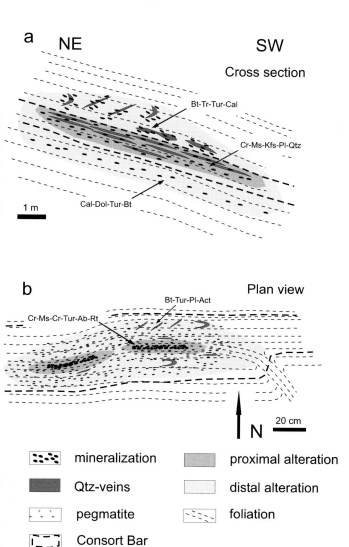

Fig. 5.8-8. Sketches illustrating the characteristics of gold mineralization and alteration in the (a) Western, (b, c) Central, and (d) Eastern Zones. Mineral abbreviations: Act = actinolite, Ap = apatite, Bt = biotite, Cal = calcite, Cr = Cr-rich muscovite, Dol = dolomite, Hbl = hornblende, Kfs = K-feldspar, Ms = muscovite, Pl = plagioclase, Qtz = quartz, Rt = rutile, Tr = tremolite, Ttn = titanite.

termed the Consort Contact mineralization and comprises reefs that are mainly located within the Consort Bar. Ore shoots of this type have been documented to be dominated by arsenopyrite, and examples include the mineralization in the Western and Central Zones. The second type of gold mineralization is associated with the so-called Footwall Lens,

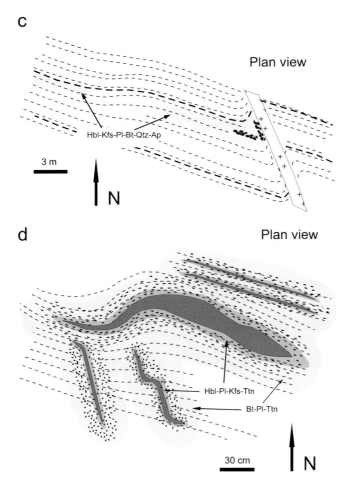

Fig. 5.8-8. (*Continued.*)

a several meter thick competent gneiss body hosted by Onverwacht Group amphibolites below the Consort Bar. This type of mineralization corresponds to the mineralization in the Eastern Zone (Fig. 5.8-5) and is characterized by the presence of loellingite, a characteristic phase of many high-temperature orogenic gold deposits (Neumayer et al., 1993). In general, the style of mineralization within the different ore shoots is variable, and includes massive replacement style ore bodies, and ore bodies in which the mineralization is hosted by quartz veins or occurs disseminated in the wall rocks (Fig. 5.8-8; Otto et al., 2007). The major and trace element characteristics of the mineralized zones throughout the mine indicate an addition of K, Na, Si, As, S, and Au to the altered host rocks, and are consistent with a large hydrothermal system that was operating contemporaneously at different crustal levels (Otto et al., 2006).

Despite the similarity in the major and trace element signatures, the petrology of the sulphide and alteration assemblages changes systematically with metamorphic grade (Table 5.8-1, Fig. 5.8-9), indicating the increasing temperatures to be the key factor controlling the ore and alteration assemblages. Typical sulphide assemblages in the Western and Central Zones include arsenopyrite–pyrrhotite–pyrite, and arsenopyrite–pyrrhotite–chalcopyrite (Fig. 5.8-9(a)). The associated proximal alteration assemblages comprise Cr-muscovite–K-feldspar–plagioclase–quartz in the Western Zone, and Cr-muscovite–tourmaline–plagioclase–rutile in the Central Zone (Fig. 5.8-9(b); Table 5.8-1). In contrast, the footwall lens mineralization in the Eastern Zone is characterized by two distinct mineralization stages (e.g., Tomkinson and Lombard, 1990, Otto et al., 2007). An early, relatively minor high-T mineralization (loellingite–pyrrhotite) is associated with a calc-silicate alteration assemblage comprising garnet–clinopyroxene–amphibole–K-feldspar–quartz–calcite (Fig. 5.8-9(c)). This early assemblage is overprinted by a second stage of mineralization, which is closely associated with cm- to dm-wide quartz veins (Figs. 5.8-4 and 5.8-8). The sulphide assemblage of this late mineralization is arsenopyrite–loellingite–pyrrhotite–chalcopyrite, and the, up to 4 m wide, proximal alteration zone comprises hornblende–plagioclase–K-feldspar–titanite (Fig. 5.8-9(d,e)).

The distinct change in mineralogy of ore bodies and associated alteration zones is further substantiated by temperature estimates on suitable ore and alteration assemblages (Otto et al., 2007). The estimated temperatures of $510 \pm 30\,°C$ and $540 \pm 40\,°C$ in the Western and Central Zones, based on the arsenopyrite and hornblende–plagioclase thermometers of Sharp et al. (1985) and Holland and Blundy (1994), are in agreement with the peak temperatures recorded in Fig Tree Group rocks (Fig. 5.8-10(a)). In contrast, the main (second) phase of gold mineralization in Onverwacht Group rocks of the Eastern Zone clearly post-dates the peak of metamorphism. Temperature estimates of ca. $680\,°C$ for the early mineralization are consistent with the P-T conditions during the peak of metamorphism in these rocks (Fig. 5.8-10(b)) (Otto et al., 2007). This may indicate that the early mineralization in the footwall lens occurred during the local D_1 deformation. The estimated temperatures of $590 \pm 40\,°C$ for the main phase of gold mineralization (Table 5.8-1) are about $100\,°C$ lower than the peak temperatures recorded, and are basically identical to those established for the overlying Fig Tree Group rocks.

5.8-5. CONSTRAINTS ON THE TIMING OF DEFORMATION, METAMORPHISM AND MINERALIZATION

The systematic variations of ore and alteration assemblages and metamorphic conditions at the New Consort gold mine suggest that the timing of gold mineralization was synchronous with the peak of metamorphism in Fig Tree Group rocks, whereas the main phase of mineralization in rocks of the Onverwacht Group clearly postdates the peak of metamorphism there (Fig. 5.8-8). However, while there is a close temporal relation between syn-kinematic gold mineralization and high-temperature metamorphism, there is no clear evidence as to whether this happened during one progressive tectono-metamorphic

Table 5.8-1. Ore and alteration assemblages and temperature estimates of mineralization at the New Consort gold mine

	Ore type	Ore assemblage	Alteration – proximal	Alteration – distal	Geothermometer	Estimated temperature
Western Zone	Consort Bar	Apy–Po–Cp–Ulm ± Stbn ± Sb	Cr–Ms–Kfs–Pl–Qtz ± Bt	Cal–Dol–Tur–Bt	Apy[1]	510 ± 30 °C
	Metapelite	Apy–Py–Po–Cp	Bt–Tur–Tr–Cal			
Central Zone	type I	Apy–Po–Cp ± Py	Cr–Tur–Cr–Ms–Ab ± Rt ± Cr–Spi	Act/Hbl–Tur–Pl–Bt	Hbl–Pl[2]	540 ± 40 °C
	type II	Apy–Lol–Po ± Bi ± Mal	Hbl–Kfs–Pl–Bt–Qtz ± Tur ± Ttn			
Eastern Zone	early	Lol–Po	Cpx–Hbl ± Grt ± Kfs ± Qtz ± Bt ± Cal	Grt–Cpx[5,6]	Grt–Hbl[3,4] 670 ± 25 °C	690 ± 25 °C
	late	Apy–Lol–Po ± Cp	Hbl–Pl–Kfs–Ttn ± Ap ± Cal ± Ep	Bt–Pl–Ttn ± Kfs ± Ap ± Cal ± Ep	Hbl–Pl[2]	590 ± 40 °C

[1]Sharp et al. (1985). [2]Holland and Blundy (1994). [3]Ravna (2000a). [4]Graham and Powell (1984). [5]Ravna (2000b). [6]Sengupta et al. (1989).

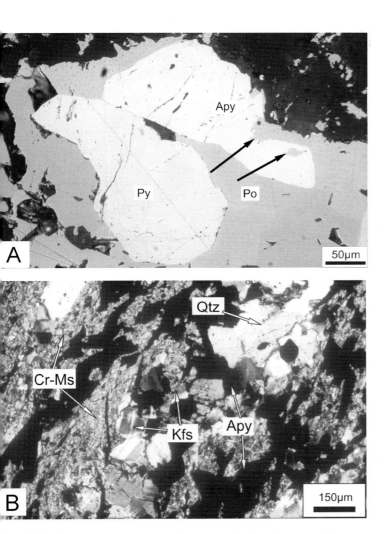

Fig. 5.8-9. Photomicrographs of typical ore and alteration assemblages, showing the close relationship between the ore mineralogy and metamorphic grade: (a) arsenopyrite (Apy) – pyrite (Py) pyrrhotite (Po) assemblage of the Western Zone. Arrows indicate growth of pyrrhotite at the expense of arsenopyrite; (b) associated proximal Cr-muscovite (Cr-Ms) – K-feldspar (Kfs) – quartz (Qtz) alteration assemblage; (c) calc-silicate alteration assemblage of the early mineralization in Eastern Zone, comprising clinopyroxene (Cpx), hornblende (Hbl), garnet (Grt), and quartz; (d) alteration of the late mineralization in the Eastern Zone, comprising a proximal hornblende–plagioclase (Hbl-Pl), and a distal biotite–plagioclase (Bt-Pl) alteration assemblage; (e) relict loellingite (Lol) with gold (Au) inclusions replaced by arsenopyrite (late mineralization, Eastern Zone).

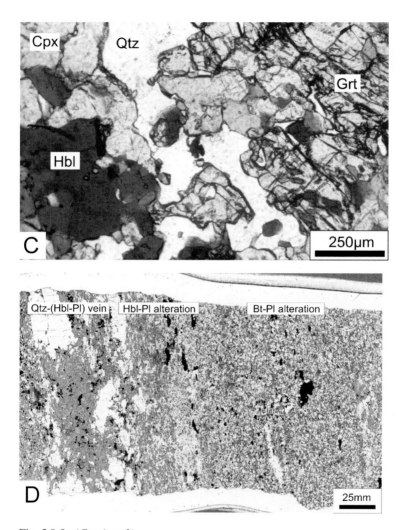

Fig. 5.8-9. *(Continued.)*

event (e.g., during D_3), or whether a two stage metamorphic evolution has to be invoked. In the latter scenario, a second high-temperature event associated with the gold mineralization would essentially mimic the temperature conditions of the country rocks. Thus, the interpretation of the sequence of deformation events can only be speculative and has to rely on the currently available geochronological data.

Dziggel et al. (2006b) interpreted the amphibolite facies metamorphism recorded in Onverwacht Group rocks along the northern margin of the greenstone belt as a result of advective heating due to granitoid plutonism in upper plate rocks during deformation at

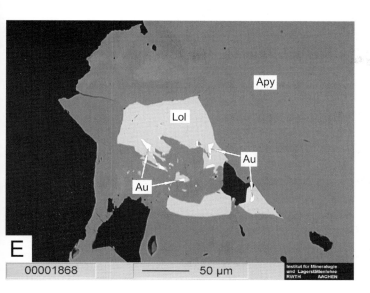

Fig. 5.8-9. (*Continued.*)

ca. 3230 Ma. The proposed timing of regional metamorphism is based on regional tectonic considerations, i.e., the close temporal relationship between high-pressure amphibolite facies metamorphism in basement rocks of the southern terrane, and TTG plutonism in the northern terrane (Fig. 5.8-1), as well as the inferred high geothermal gradient of ca. 30–40 °C/km. A D_2 timing for the peak of metamorphism is further substantiated by U-Pb zircon data from a late-kinematic granite dyke intrusive into and cross-cutting the early S_{1A} fabrics to the east of New Consort (at Honeybird siding, Fig. 5.8-1), which yield an age of 3237 ± 47 Ma (Dziggel et al., unpublished data). As these fabrics are defined by peak metamorphic minerals, this age has to be interpreted as a minimum age for metamorphism and the local D_1 deformation. In addition, $^{40}Ar/^{39}Ar$ laser step heating of magmatic hornblende from the centre of the Kaap Valley Pluton, some 10 km to the west of New Consort, yields a weighted mean date of 3214 ± 4 Ma, suggesting that this tonalite had cooled to temperatures <500 °C shortly after its emplacement. Thus, the peak of metamorphism and development of the S_{1A} foliation in the footwall rocks most likely correlate with the regional D_2 deformation.

The gold mineralization at New Consort, however, postdates the regional D_2 deformation by more than 150 million years, as indicated by the ca. 3040 Ma age on syn-kinematic pegmatites intruding the mineralized shear zones in the deeper parts of the mine (Harris et al., 1993). This age is somewhat younger than the 3084 ± 18 Ma date established for hydrothermal rutile associated with the gold mineralization at the Fairview mine (de Ronde et al., 1991b), and has thus been interpreted to represent a minimum age for the mineralization (Harris et al., 1993). Biotite $^{40}Ar/^{39}Ar$ data from the hydrothermally altered southern margin of the Kaap Valley Pluton give an age of 3035 ± 9 Ma (Layer et al., 1992),

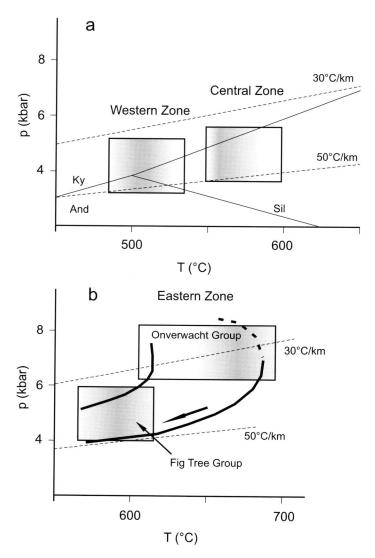

Fig. 5.8-10. P-T diagram showing peak P-T conditions of host lithologies at New Consort (open squares) compared to T estimates of mineralization (grey-shaded areas) in (a) the Western and Central Zones, and (b) the Eastern Zone (modified after Otto et al. (2007)). Errors as in Fig. 5.8-7.

which is similar to the age of the pegmatites. This age is also considerably younger than the 3142 ± 20 Ma established for biotite from the interior of the Kaap Valley Pluton, suggesting that biotite was thermally reset during a later thermal event. On the basis of similar $\delta^{13}C$ values from this alteration zone and from alteration zones in D_3 structures hosting the greenschist facies gold deposits to the south, as well as the elevated gold concentrations observed within the altered tonalites, de Ronde et al. (1992) suggested the age of 3035 ± 9 Ma to represent the age of gold mineralization.

5.8-6. ONSET OF EXTENSIONAL TECTONICS – 3230 MA, 3100 MA, OR MULTISTAGE?

The geochronological evidence presented above indicates that the northern terrane of the Barberton Greenstone Belt was tectonically active for a protracted period of time, and that deformation culminated in two episodes, namely the regional D_2 and D_3 deformation events. While the earliest S_{1A} fabrics correlate with the regional D_2 deformation, the mineralized shear zones formed during the D_3 deformation. The unresolved timing of deformation is central to the debate about whether or not the tectonic processes that resulted in the formation of the gold deposits include the formation of the S_{1B} and S_{1C} fabrics; i.e., the doming stage and subsequent folding of the Top Section and 3 Shaft Synclines (Dziggel et al., 2006b). If these fabrics formed during D_2, this would indicate that the condensed metamorphic gradients are not related to the gold mineralization, and that the latter was superimposed on earlier formed fabrics. A second scenario is that the D_{1B} and D_{1C} fabrics were formed during the D_3 deformation, thus inferring that the exhumation of basement complexes surrounding the Barberton Greenstone Belt records a multistage evolution.

De Ronde and de Wit (1994) interpreted the gold mineralization to be related to a large-scale rapid release of deep crustal fluids during a change in plate motion, i.e. from a dominantly convergent to a transtensional or extensional tectonic setting during ongoing accretionary tectonics at ca. 3080 Ma. The extensional exhumation of the surrounding gneiss domes was regarded to form part of this event, contributing to the steepening of earlier fabrics, and thus, the creation of pathways for the mineralizing fluids. Recent studies, however, have documented the deep burial and subsequent extensional unroofing of the granitoid-dominated terrane to the south of the Barberton Greenstone Belt as being linked to regional D_2 accretion (e.g., Dziggel et al., 2002, 2005; Kisters et al., 2003; Diener et al., 2005, 2006). This D_2-related exhumation of gneiss complexes was interpreted as a result of the orogenic collapse of the belt, based on: (i) U-Pb zircon and titanite ages on exhumation fabrics along the southern margin of the greenstone belt; (ii) the close temporal relationship between these ages and the deposition of coarse-clastic sediments of the syn-tectonic Moodies Group; and (iii) the age of the post-tectonic Dalmein Pluton (3216 ± 1 Ma, Kamo and Davis, 1994), which provides a minimum age on the deformation in the southern terrane (Fig. 5.8-1).

The apparent diachronous nature of late-stage deformation in the greenstone belt may be a direct result of a pronounced localization of the bulk D_3 strain into syntectonic intrusions. This is based on a number of field studies focussing on emplacement features of the D_3-related granite sheets to the south of the greenstone belt, which show that the strain was strongly partitioned into zones of syn-tectonic magma emplacement (e.g., Westraat et al., 2005; Belcher and Kisters, 2006a). The observed temporal and spatial differences between fabrics along the southern and northern granite–greenstone contacts were interpreted as a result of the distinctly different distances to the late granites (Fig. 5.8-1). The close spatial relationship between D_3-related structures and the gold mineralization suggests that this strain partitioning may also be significant for the localization of gold mines along the northern margin of the greenstone belt. Support for this interpretation is given by the consistency of the established NW–SE directed main shortening direction during the emplacement of late granites and the gold mineralization (e.g., de Ronde and de Wit, 1994).

The common occurrence of pegmatite dykes in the mineralized zones at the New Consort gold mine underlines the close temporal and spatial relationships between the formation of D_3 structures and syn-tectonic magmatism. However, a simple hydrothermal event overprinting pre-existing high-temperature fabrics appears to be in conflict with the observed systematic variations of ore and alteration assemblages with increasing peak temperatures in the Fig Tree Group (Fig. 5.8-10). This suggests that the exposed rocks were still hot while they were deformed. On the other hand, the mineral assemblages defining the S_{1A} and S_{1B} fabrics do not indicate any difference in metamorphic grade. This is difficult to reconcile with a two-stage metamorphic evolution separated by more than 150 million years because the footwall rocks should have cooled considerably before the final exhumation, even if they stayed at the same crustal level. Consequently, it was suggested that the S_{1B} and S_{1C} fabrics formed in response to progressive unroofing of basement rocks during regional D_2 deformation (Dziggel et al., 2006b). The progressive nature of the fabric forming events during retrogression is also supported by the development of a compressional crenulation cleavage in rocks of the Onverwacht Group, which correlates with folding of the Top Section and 3 Shaft Synclines (Dziggel et al., 2006b).

As a possible explanation for this dilemma, we propose a two-stage metamorphic evolution, in which the onset of extensional shearing and exhumation correlates with the regional D_2 deformation at 3230 Ma. The early extensional shearing resulted in the formation of the constrictional S_{1A} fabrics, as well as an early doming stage that is indicated by S_{1B} fabrics defined by amphibolite facies mineral assemblages. The second metamorphic event correlates with the regional D_3 deformation between ca. 3100 and 3040 Ma, and is characterized by renewed heating and the reactivation of pre-existing fabrics. The D_3 event comprises the final extensional unroofing of the footwall rocks, as well as the folding of the Top Section and 3 Shaft Synclines under metamorphic conditions corresponding to the upper greenschist/amphibolite facies transition. The formation of retrograde S_{1B} foliations recorded in the footwall of the Consort Bar, as well as the local D_2 shearing, and the gold mineralization, would form part of this event. Thus, in addition to strain localization into syn-tectonic intrusions, we propose that this late-stage exhumation may have played a significant role in focussing the mineralizing fluids into shear zones along the northern margin of the belt.

5.8-7. CONCLUSION

In this paper, we have outlined the structural and metamorphic setting of major gold deposits formed during the late-tectonic evolution of the Barberton Greenstone Belt, at ca.

3080–3040 Ma. We have shown that the gold mineralization postdates regional D_2-related metamorphism at ca. 3230 Ma, and that it was associated with a low-P, high-T metamorphic event during NW–SE directed subhorizontal shortening. While a late timing of gold mineralization within the overall tectonic evolution of an orogen is a characteristic feature of orogenic gold deposits (e.g., McCuaig and Kerrich, 1998), the large time gap of more than 150 million years between accretion and final gold mineralization in the greenstone belt is rather unusual, though not unique. The sum of the geologic, structural and geochronological evidence indicates that the tectonic processes responsible for the gold mineralization were external to the greenstone belt. They were related to a craton-wide period of granite plutonism and associated low-P, high-T metamorphism that occurred during tectonomagmatic activity along the leading edges of the Kaapvaal Craton. The timing relationships also indicate that the source of the mineralizing fluids must have been located in the deeper crust below the greenstone belt, as metamorphic devolatilization reactions within the greenstone sequence can be ruled out (de Ronde et al., 1992). An external fluid source is also consistent with the work by Otto et al. (2007), who concluded metamorphism at deeper crustal levels to be unlikely as a source for the mineralizing fluids because the footwall rocks were already on their retrograde path. Belcher and Kisters (2006a) suggested that by ca. 3100 Ma, the Barberton Greenstone Belt may have corresponded to a contractional continental back arc region, while subduction-related accretion dominated in the area around the Vredefort dome. In line with Moser et al. (2001) conclusion, we suggest that crustal accretion and convergence recorded along the edges of the Kaapvaal Craton may have lead to the delamination of the lithospheric mantle keel, resulting in heat input from asthenospheric upwelling and crustal heating. This would eventually lead to back-arc extension, the onset of which has been suggested to coincide with the deposition of the Dominion Group lavas at ca. 3074 ± 6 Ma (e.g., de Ronde and de Wit, 1994).

ACKNOWLEDGEMENTS

Martin Van Kranendonk is thanked for inviting us to submit this contribution. We are grateful to the management of Barberton Mines Ltd. for support and access to underground workings and sample material. A. Dziggel acknowledges support by the Institute of Mineralogy and Economic Geology, RWTH Aachen, and the German Science Foundation. Work by Alex Otto was funded by the German Academic Exchange Service (DAAD), and student grants of the Society of Economic Geology. Reviews by Martin Van Kranendonk and Steffen Hagemann are greatly appreciated.

PART 6

PALEOARCHEAN GNEISS TERRANES

Earth's Oldest Rocks
Edited by Martin J. Van Kranendonk, R. Hugh Smithies and Vickie C. Bennett
Developments in Precambrian Geology, Vol. 15 (K.C. Condie, Series Editor)
© 2007 Elsevier B.V. All rights reserved.
DOI: 10.1016/S0166-2635(07)15061-1

Chapter 6.1

PALEOARCHEAN GNEISSES IN THE MINNESOTA RIVER VALLEY AND NORTHERN MICHIGAN, USA

MARION E. BICKFORD[a], JOSEPH L. WOODEN[b], ROBERT L. BAUER[c] AND MARK D. SCHMITZ[d]

[a]*Department of Earth Sciences, Heroy Geology Laboratory, Syracuse University, Syracuse, NY 13244-1070, USA*
[b]*U.S. Geological Survey, Stanford – U.S. Geological Survey Ion Microprobe Facility, Stanford University, Stanford, CA 94305-2220, USA*
[c]*Department of Geological Sciences, University of Missouri, Columbia, MO 65211, USA*
[d]*Department of Geosciences, Boise State University, Boise, ID 83725, USA*

6.1-1. INTRODUCTION

Meso-to Paleoarchean gneisses occur along the southern margin of the Neoarchean Superior Craton (Fig. 6.1-1). The most extensive exposure of these rocks is in the Minnesota River Valley (MRV) of southwestern Minnesota (Fig. 6.1-2), but there are also exposures in northern Michigan. Much more is known about the MRV than the northern Michigan occurrences because better exposure in the MRV has elicited more intensive study. Nevertheless, because the terranes share many features of age and rock types, and may have once been continuous in the subsurface, they constitute a major feature on the southern margin of the Superior Craton.

6.1-2. REGIONAL SETTING

Rocks exposed in the narrow erosional window of the MRV (Fig. 6.1-2) have received considerable attention due to their great antiquity (e.g., Catanzaro, 1963; Goldich and Hedge, 1974). However, understanding of their regional setting (Fig. 6.1-1), as the southernmost exposures of the Superior Craton of the Canadian Shield, is primarily based on interpretations of geophysical and sparse drill-hole data. Morey and Sims (1976) recognized the boundary between the northern Minnesota granite–greenstone terrane (Wawa Subprovince) and the much older Archean rocks of the MRV to the south, on the basis of contrasts in their respective aeromagnetic and gravity anomalies. Sims et al. (1980) and Sims (1980) extended this boundary from Minnesota into Wisconsin and the upper peninsula of Michigan and named it the Great Lakes Tectonic Zone (GLTZ; Fig. 6.1-1).

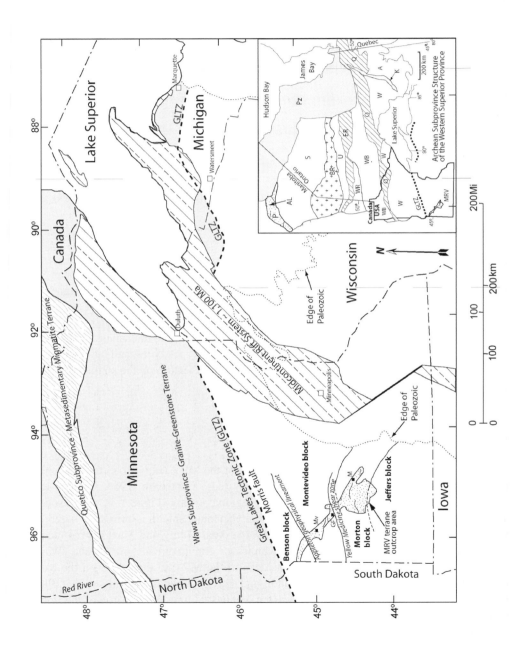

Fig. 6.1-1. (*Previous page.*) Maps showing the regional Archean geologic context for rocks exposed in the Minnesota River Valley terrane. Areas of Proterozoic outcrops other than rocks of the Midcontinent rift system (MCR) are not shown (MCR beneath Paleozoic cover is based on the midcontinent gravity anomaly). Town locations shown in the Minnesota River Valley terrane are: MV – Montevideo, GF – Granite Falls, M – Morton. For details of occurrences near Watersmeet in northern Michigan, readers are referred to Fig. 5 in Sims (1996a). Subprovince locations shown on the inset map of the western Superior province are, from south to north: MRV – Minnesota River Valley, W – Wawa, K – Kapuskasing structural zone, A – Abitibi, Q – Quetico, WB – Wabigoon, WR – Winnipeg River, ER – English River, U – Uchi, BR – Berons River, S – Sachigo, P – Pikwitonei. AL indicates the location of the Assean Lake area. Pz indicates Paleozoic cover west and southwest of Hudson Bay and James Bay. Map sources include: Sims (1991), Southwick and Chandler (1996), and Ontario Geological Survey (1992). After Bickford et al. (2006).

The GLTZ in Minnesota is referred to as the Morris fault segment and marks the northern boundary of the MRV. Seismic reflection data across the Morris fault segment by the Consortium for Continental Reflection Profiling (COCORP) indicate that the fault dips 30–40° toward the north and verges southward (Gibbs et al., 1984; Smithson et al., 1985). In northern Michigan, however, outcrop data indicate that the GLTZ is a north-verging feature (Sims, 1991, 1996a; Sims and Day, 1992). South of Marquette, Michigan, the GLTZ is marked by an approximately 2.2 km wide zone of mylonite (Sims, 1996a).

6.1-3. GEOLOGIC SETTING

6.1-3.1. *Minnesota River Valley*

Aeromagnetic mapping of southwestern Minnesota (Chandler, 1987, 1989, 1991), and detailed gravity and magnetic modeling within the MRV (Schaap, 1989; Southwick and Chandler, 1996), have delineated four crustal blocks in the MRV that are bounded by three east-northeast-trending geophysical anomalies that roughly parallel the Morris fault segment (Fig. 6.1-1). Most of the previous work in the MRV has concentrated on exposures in the Montevideo block and the Morton block, which are separated by the Yellow Medicine shear zone (Fig. 6.1-2). Southwick and Chandler (1996) recognized a belt of moderately-to weakly-metamorphosed mafic and ultramafic rocks known as the Taunton belt, which is parallel to, and partly within, the Yellow Medicine shear zone. They suggested that the Taunton belt may represent either allochthonous greenstone belt rocks that were infolded with older cratonic gneisses, or a remnant of late Archean oceanic crust along a suture between the Montevideo and Morton blocks.

Rocks exposed between Ortonville, near the South Dakota border, and New Ulm, Minnesota (Fig. 6.1-2, inset), include a complex of granitic migmatites, schistose to gneissic amphibolite, metagabbro, and paragneisses. The best known units are the Morton Gneiss (Morton block; Fig. 6.1-2), consisting of tonalitic to granodioritic migmatite with

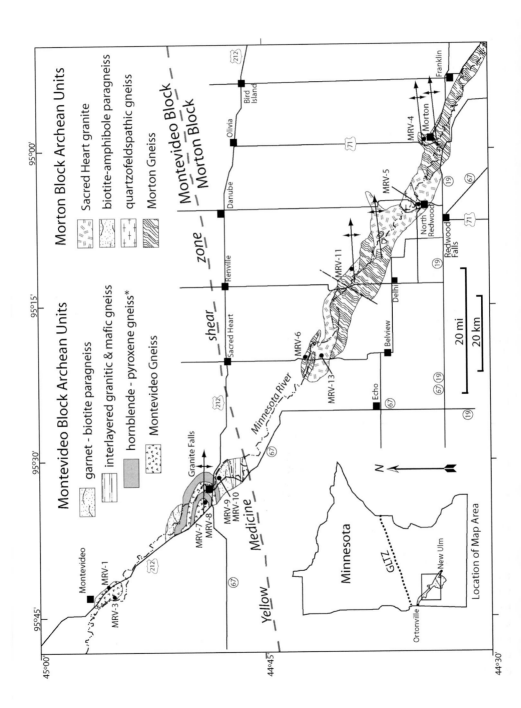

Fig. 6.1-2. (*Previous page.*) Simplified geologic map showing the distribution of major rocks units exposed in the Minnesota River Valley and the locations of the samples collected for this study. The distribution of rocks units is simplified from Himmelberg (1968), Grant (1972), and Southwick (2002). *The hornblende-pyroxene gneiss unit includes the metagabbro unit referred to in the text along the southeastern margin of Granite Falls. After Bickford et al. (2006).

---

abundant amphibolitic enclaves (Fig. 6.1-3), and the Montevideo Gneiss (Montevideo block; Fig. 6.1-2), consisting of banded granitic gneiss, coarse-grained granitic gneiss, and medium-grained granitic gneiss (Fig. 6.1-4). Paragneiss, derived mostly from a greywacke protolith, occurs southeast of Granite Falls and northeast of Delhi. Mafic gneisses, including metagabbro, hornblende–pyroxene gneiss, and amphibolite, occur in the Granite Falls – Montevideo region (Fig. 6.1-2). The Morton Gneiss is intruded by a major, younger, weakly-deformed granite body, the ca. 2600 Ma Sacred Heart Granite (Fig. 6.1-5; Doe and Delevaux, 1980; Bickford et al., 2006; Schmitz et al., 2006). This unit does not intrude the Montevideo Gneiss, nor has any equivalent intrusive body been recognized in the Montevideo block. Rocks in the Ortonville–Odessa area (Benson block; Fig. 6.1-1) include the Ortonville Granite, little-studied granitic gneiss, and paragneiss that have recently been dated by Schmitz et al. (2006).

Field mapping and structural analysis by Lund (1956), Himmelberg (1968), Grant (1972), and Bauer (1980) identified and characterized several periods of ductile deformation within both the Montevideo and Morton blocks. The earliest deformation, D1, produced a well-defined gneissic banding and banding-parallel foliation in all of the granitic gneisses. This banding was subsequently folded by a series of upright F2 folds that plunge gently to the east-northeast (Fig. 6.1-2). The steeply-dipping, easterly-striking axial planes of these folds are consistent with north-northwest directed shortening during Neoarchean accretion of the MRV to the southern margin of the Superior Craton. Bauer (1980) recognized two younger folding events in the Granite Falls and Montevideo areas, and numerous ductile shear zones, up to two meters wide, that deformed the Montevideo Gneiss in both the Montevideo and Granite Falls areas (Himmelberg, 1968; Bauer, 1980; Young, 1987). These zones locally deform tholeiitic diabase dikes with K-Ar hornblende ages of 2080 Ma, but the shear zones are cut by hornblende andesite dikes with K-Ar hornblende ages of 1670 to 1730 Ma (Hanson and Himmelberg, 1967). Although the Morton Gneiss also contains evidence of local zones of ductile shear, detailed structural analysis is lacking.

The most extensive studies of metamorphism in the MRV examined the rocks in the Granite Falls and Montevideo areas where granulite-facies assemblages, including garnet and orthopyroxene, occur in both the mafic and granitic gneisses and a biotite-garnet paragneiss (Himmelberg and Phinney, 1967; Himmelberg, 1968). Outcrops near Morton, Redwood Falls, and Delhi include upper amphibolite-facies assemblages (Grant, 1972; Grant and Weiblen, 1971). Moecher et al. (1986) did a comparative geothermobarometric study of rocks from both the Granite Falls and Morton areas and found temperature ranges of 650 to 750 °C and pressures of 4.5 to 7.5 kbar, using a variety of thermobarometers. They attributed the observed temperature variations to varying degrees of re-equilibration of the

Fig. 6.1-3. Outcrop photograph of Morton Gneiss showing highly deformed, migmatized rock with amphibolite enclaves. Coin is 2.5 cm in diameter.

Fig. 6.1-4. Field photograph of Montevideo Gneiss. Green Quarry, Granite Falls.

mineral assemblages during cooling, but allowed for possible real temperature variations between the granulite and amphibolite facies regions. Himmelberg and Phinney (1967) and Himmelberg (1968) also recognized retrograde metamorphic recrystallization in rocks of the Granite Falls area. They attributed some of the retrograde reactions to re-equilibration during cooling following granulite-facies metamorphism. However, they also attributed some of the reactions to the low-grade thermal metamorphic event that produced ca. 1800 Ma K-Ar ages in the Granite Falls area (Goldich et al., 1961; Hanson and Himmelberg, 1967).

6.1-3.2. Northern Michigan

Ancient gneisses, similar to those of the MRV, are exposed in the Marenisco–Watersmeet area in the western part of the upper peninsula of northern Michigan (Fig. 6.1-1), somewhat to the south of the extension of the Great Lakes Tectonic Zone into Michigan. The only exposures of rocks of confirmed Paleoarchean age are in a domal feature near the town of Watersmeet, where ca. 3.6 Ga tonalitic biotite augen gneiss occurs (Peterman et al., 1980, 1986). The Neoarchean (ca. 2650 Ma) Puritan quartz monzonite occurs in the western part of the Watersmeet area, where it is overlain by Proterozoic metasedimentary rocks. Small outcrops of granite and granitic gneiss, termed the Granite near Thayer (Fritts, 1969), occur between the Paleoarchean tonalitic gneiss and the Puritan quartz monzonite.

Fig. 6.1-5. Field photograph of Sacred Heart Granite.

Undated granite gneiss, assumed to be Paleoarchean, also occurs in domal structures in the Marquette Iron Range at the eastern end of the Upper Peninsula of Michigan (Fig. 6.1-1). As shown in Fig. 6.1-1, these structures occur south of the GLTZ and expose gneisses that were considered parts of the Minnesota River Valley Subprovince by Sims (1996b). Schneider et al. (2002), in a geochronological study of ca. 1873 Ma rhyolite from the Hemlock Formation that overlies Archean basement, found abundant xenocrystic zircons, three yielding ^{207}Pb-^{206}Pb ages in excess of 3.6 Ga, suggesting that, although undated, at least some of the basement gneisses are Paleoarchean.

6.1-4. GEOCHRONOLOGY

6.1-4.1. Minnesota River Valley

The geochemistry and geochronology of Archean rocks exposed in the MRV (Fig. 6.1-2) was reported by Catanzaro (1963), Hanson and Himmelberg (1967), Himmelberg (1968), Goldich et al. (1970), Goldich and Hedge (1974), Farhat and Wetherill (1975), Wilson and Murthy (1976), Goldich et al. (1980), Goldich and Wooden (1980), and Wooden et al. (1980). Geochronological study of these rocks began with the zircon dating of Catanzaro (1963), but the most extensive studies were those of Goldich and Hedge (1974), who reported ages as old as 3.7 Ga for portions of the felsic gneisses. However, dating was complicated by difficulties in isolating older and younger parts of the complex rocks, and by a long and protracted igneous and metamorphic history that includes high-grade metamorphism at ca. 3.05 and 2.6 Ga, as well as a low-grade thermal event at ca. 1.8 Ga. Goldich et al. (1980) presented both Rb-Sr and U-Pb (zircon) data for rocks from the Granite Falls area. The Rb-Sr data are complex, reflecting the prolonged igneous and metamorphic history. Goldich et al. (1980) interpreted data from one locality as indicating a maximum crystallization age of 2.91 Ga, but there is considerable scatter of the analytical data. A second locality yielded similar scattered data, from which Goldich et al. (1980) interpreted a maximum crystallization age of 3.725 Ga. Zircons from a sample of adamellite and granitic gneiss yielded an array of data that was difficult to interpret. Goldich et al. (1980) thought that a minimum age for the older granitic gneiss was 3.23 Ga, but also argued that if the data were interpreted as resulting from Pb loss at ca. 1.80 Ga, a maximum crystallization age of about 3.50 Ga could be extracted.

Goldich and Wooden (1980) reported both Rb-Sr and U-Pb zircon ages from the Morton Gneiss. The Rb-Sr data yielded an isochron age of 3475 ± 110 Ma for tonalitic and related gneisses of the paleosome, but Goldich and Wooden (1980) saw clear evidence for redistribution of Rb and Sr. Neosomal granitic gneisses and aplite dikes yielded a Rb-Sr age of ca. 2700–2600 Ma, but zircon U-Pb data indicated an age greater than 3000 Ma. U-Pb analysis of zircons from a tonalitic gneiss, including previously published data of Catanzaro (1963) and Goldich et al. (1970), yielded data that could be interpreted as indicating a primary age of 3487 ± 123 Ma, with a lower intercept at ca. 1730 Ma. Exclusion of some analyses yielded an upper intercept age of 3590 Ma, with a lower intercept of ca. 1900 Ma. Goldich and Wooden (1980) quoted preliminary Sm-Nd ages by M.T. McCulloch and G.J. Wasserburg for rocks in the Morton area as ranging from 3600 to 3200 Ma. McCulloch and Wasserburg (1978) reported a model age of 3580 ± 30 Ma for two samples of residual clay developed on the Morton Gneiss.

Two recent geochronologic studies have largely resolved the difficulties in dating these complex rocks. Bickford et al. (2006) analyzed zircons with the SHRIMP–RG instrument at Stanford University in a study that focused on the oldest components of the suite of gneisses, whereas Schmitz et al. (2006) used high-precision TIMS methods to study zircons and monazites, mostly from younger components of the rock suite. The results of these studies make clear why earlier investigators had a difficult time unraveling the magmatic, thermal, and metamorphic history of these rocks, for the zircons are complex indeed, many displaying cores with multiple overgrowths (Figs. 6.1-6, 6.1-7, 6.1-8), and the magmatic and thermal history spans 1000 Ma!

Ages of important rock units may be summarized as follows and are compiled in Table 6.1-1. Unless otherwise noted, the ages are derived from U-Pb analyses of zircon and are those of Bickford et al. (2006).

6.1-4.1.1. Montevideo block

1. A sample of Montevideo Gneiss near the village of Montevideo (MRV-1; Fig. 6.1-2) yielded complex zircons with cores and distinct overgrowths (Fig. 6.1-6). Cores yielded

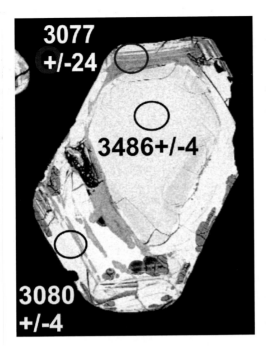

Fig. 6.1-6. Back-scattered scanning electron microscope image of zircon from Montevideo Gneiss sample MRV-1 of Bickford et al. (2006), showing ca. 3080 Ma overgrowth around ca. 3500 Ma core.

a regressed age of 3485 ± 10 Ma, but distinct overgrowths yielded ages of 3388 ± 8 and 3080 ± 4 Ma.

2. A second sample of Montevideo Gneiss, from an abandoned quarry near the village of Montevideo (MRV-3; Fig. 6.1-2), yielded an age of 3385 ± 8 Ma, but one grain, presumably a xenocryst, yielded an age of 3500 ± 3 Ma.

3. A sample of Montevideo Gneiss from the classic 'Green Quarry' site in the town of Granite Falls (MRV-7; Fig. 6.1-2) yielded a zircon age of 3496 ± 9, but well-developed rims on these zircons yielded a regressed age of 2606 ± 4 Ma (Fig. 6.1-7).

4. A second, somewhat more mafic, sample from the Green Quarry site (MRV-8) was thought to represent an early phase of the Montevideo Gneiss. However, it yielded a tight cluster of data points giving a weighted mean average age of 3141 ± 2 Ma, indicating that the sample is of a later, thoroughly disrupted, intrusion.

5. Two samples of garnet-biotite gneiss near Granite Falls yielded a regressed age of 2619 ± 20 Ma, which Bickford et al. (2006) interpreted as the time of high-grade metamorphism. Schmitz et al. (2006) analyzed two monazite crystals from a similar sample, obtaining [207]Pb-[206]Pb ages of 2609 ± 1 Ma and 2595 ± 1 Ma, again interpreting the ages as that of prograde to peak metamorphism.

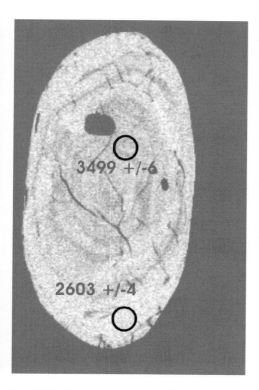

Fig. 6.1-7. Back-scattered scanning electron microscope image of zircon from Montevideo Gneiss sample MRV-7 of Bickford et al. (2006), showing ca. 3500 Ma core overgrown by ca. 2600 Ma rim.

6.1-4.1.2. Morton block

6. A sample near the village of Morton (MRV-4; Fig. 6.1-2) yielded one group of grains with a regressed age of 3516±17 Ma and a second group of grains yielding a regressed age of 3360 ± 9 Ma. Both groups of grains are euhedral and appear primary. A rim around a ca. 3500 Ma core yielded a concordant ^{207}Pb-^{206}Pb age of 3145 ± 2 Ma (Fig. 6.1-8), evidently reflecting zircon growth during the magmatic event that formed the mafic intrusion represented by sample MRV-8 from the Green Quarry site.

7. A second sample of typical Morton Gneiss, collected on Minnesota Highway 9, just north of the Minnesota River (MRV-6; Fig. 6.1-2), yielded results similar to MRV-4. One group of grains has a regressed age of 3529 ± 3 Ma, whereas a second group of grains has a regressed age of 3356±10 Ma. One grain (Fig. 6.1-9) displayed a 3521 ± 4 Ma core, overgrown by a 3370 ± 3 Ma rim, in turn overgrown by a 3178 ± 18 Ma rim.

8. Schmitz et al. (2006) dated zircons from a sample of Morton Gneiss collected near Delhi (their sample MRV96-25; Fig. 6.1-10) by TIMS methods following extensive abrasion to remove younger rims. They obtained a regressed age of 3422 ± 1 Ma.

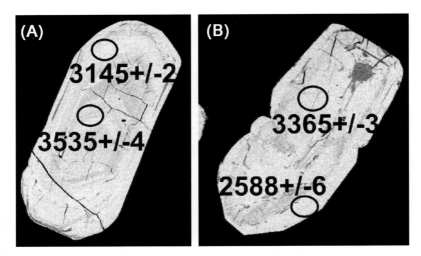

Fig. 6.1-8. Back-scattered scanning electron microscope of zircons from Morton Gneiss sample MRV-4 of Bickford et al. (2006). (A) Ca. 3500 Ma core overgrown by ca. 3140 Ma rim; (B) 3365 Ma core overgrown by ca. 2600 Ma rim.

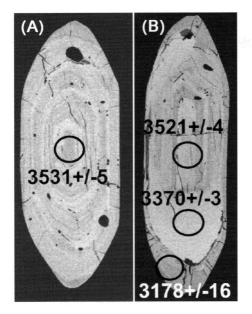

Fig. 6.1-9. Back-scattered scanning electron microscope image of zircons from Morton Gneiss sample MRV-6 of Bickford et al. (2006). (A) 3531 Ma grain with well-developed igneous oscillatory growth zoning; (B) similar grain showing 3370 Ma overgrowth in turn overgrown by ca. 3180 Ma rim.

Table 6.1-1. Summary of age results

Age (Ma)	Montevideo Block	Morton Block	Benson Block
3500 ± 15	Primary crystallization of Montevideo Gneiss	Primary Crystallization of Morton Gneiss	
3440–3420	Zircon-forming event	Primary crystallization of tonalite in Morton Gneiss	
3385 ± 10	Granodiorite intrusion and growth of zircon rims in Montevideo Gneiss	Local intrusion of granodiorite into Morton Gneiss	
3140 ± 5	Mafic Intrusion in Montevideo Gneiss in Granite falls	Growth of rims on older zircons in Morton Gneiss	
3080 ± 10	Growth of zircon rims in Montevideo Gneiss		
2610 ± 10	Granite Falls area: High-grade metamorphism of garnet-biotite gneiss; growth of rims on older zircons in Montevideo Gneiss	Variable growth of rims on older zircons in Morton Gneiss	Metamorphism of granulitic paragneiss
2603 ± 1		Synkinematic emplacement of Sacred Heart granite	Synkinematic emplacement of Ortonville granite
2590 ± 2		Post-kinematic emplacement of syenogranite and aplite	Post-kinematic emplacement of syenogranite and aplite

Interestingly, Bickford et al. (2006) found a core with an age of 3422 ± 9 Ma in their sample MRV-8, in which most zircons are 3141 Ma, and I.S. Williams (2005, personal comm.) obtained a SHRIMP age of 3441 ± 5 Ma for another sample of Morton Gneiss. Further, there is a ca. 3480 Ma peak in the age distribution of detrital zircons in metasedimentary sample MRV-11 of Bickford et al. (2006; see below).

9. A foliated granodiorite to tonalite, collected near the village of North Redwood (MRV-5; Fig. 6.1-2) yielded a regressed age of 3377 ± 19 Ma.
10. Both Bickford et al. (2006) and Schmitz et al. (2006) dated samples of the Sacred Heart Granite (Fig. 6.1-5), a slightly deformed pluton that intrudes older rocks in the Morton block (Fig. 6.1-2). The SHRIMP data of Bickford et al. (2006) yielded a weighted mean average age of 2604 ± 5 Ma for their sample MRV-13 (Fig. 6.1-2), whereas Schmitz et al. (2006) dated the main phase of the granite at 2603 ± 1 Ma (MRV99-4; Fig. 6.1-7) and a syenitic border phase at 2592 ± 1 Ma (MRV96-7; Fig. 6.2-11).

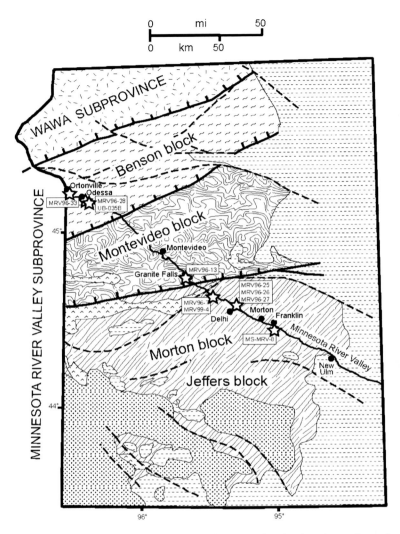

Fig. 6.1-10. Map of Minnesota River Valley area showing locations of samples studied by Schmitz et al. (2006).

Schmitz et al. (2006) also reported a zircon age of 2600 ± 0.4 Ma for a monzogranite intrusion into the Morton Gneiss and a monazite age of 2590 ± 1 Ma for an aplitic dike that intrudes the Morton Gneiss.

11. A feldspar-biotite schist (MRV-11; Fig. 6.1-2) was collected from a site commonly known as 'the cordierite pasture' (cf. Grant and Weiblin, 1971) to study populations of detrital zircons. Zircon grains exhibiting all of the ages determined from the other sam-

Paleoproterozoic (2.5 - 1.6 Ga)

Sioux Quartzite

Paleoproterozoic rocks, undifferentiated

Tear or scissors faults

Archean (3.5 to 2.6 Ga)

Wawa subprovince

Neoarchean greenstone belt lithic association and granitoid intrusions

Minnesota River Valley Subprovince

Neoarchean greenstone belt lithic association, low to moderate metamorphic grade

Mesoarchean to Neoarchean quartzofeldspathic gneiss and granitoid intrusions

Block-bounding shear zones, hatchures indicate dip direction

MS-MRV-8 Sampling site ☆ & numbers

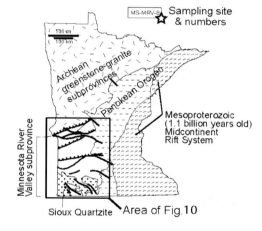

Fig. 6.1-10. (*Continued.*)

ples in this study are present in the detrital population, with well-defined age peaks at about 3520, 3480, 3380, 3140, and 2600 Ma. Ages of 2600 Ma were determined only from rims on the older grains and a few featureless grains, interpreted as indicating formation during high-grade metamorphism of the protolith. These data suggest that the sedimentary protolith of this rock accumulated prior to 2600 Ma but after 3140 Ma.

12. Schmitz et al. (2006) obtained zircon U-Pb data for a tholeiitic diabase dike near Franklin (Fig. 6.1-10) that indicated an age of 2067 ± 1 Ma.

6.1-4.1.3. Benson block

13. Schmitz et al. (2006) obtained monazite ages of 2603 ± 2 Ma for the main monzogranite phase of the Ortonville Granite, 2591 ± 7 Ma for a syenogranite border phase of the Ortonville Granite, and 2594 ± 1 Ma for a granulitic paragneiss in the Ortonville area (Fig. 6.1-10).

6.1-4.2. Northern Michigan

Although clearly disturbed, zircon U-Pb data from tonalitic gneiss exposed in the Watersmeet Dome of northern Michigan (Peterman et al., 1980) indicate that this rock is at least 3410 Ma. In a subsequent study by Peterman et al. (1986), it was suggested that the gneiss could be at least 3562 Ma on the basis of a linear array of discordant points. They also showed that the complex discordant pattern of the U-Pb zircon systems was the result of multiple events at about 2.6–2.8 Ga, 1.7–1.8 Ga, and a young lead-loss event. McCulloch and Wasserburg (1980) reported a Nd model age (T_{CHUR}) of 3.60 ± 0.4 Ga for the tonalitic gneiss, and Barovich et al. (1991) reported Sm-Nd isotopic data for five samples that yielded model ages (T_{CHUR}) ranging from 3.69 to 3.64 Ga. These data suggest that the tonalitic gneiss is probably close to 3.6 Ga. Significantly, the tonalitic gneiss in the Watersmeet Dome is intruded by granite that has a zircon age of 2.75 Ga, and by leucogranitic dikes dated at 2.60 Ga (Peterman et al., 1980). Thus, rocks in this area are approximately the same primary age as those in the MRV and appear to have had a similar subsequent history.

Schneider et al. (2002) studied zircons from metarhyolite of the Hemlock Volcanics in a part of the Marquette Iron Range region of northern Michigan that lies south of the GLTZ (Fig. 6.1-1). Here, Paleoproterozoic rocks were deposited on a basement of older, presumably Archean, rocks, now exposed in domes. Schneider et al. (2002) found, in addition to an igneous population yielding an age of ca. 1870 Ma, a number of ca. 2.8–2.6 Ga Archean xenocrystic zircons and one grain with a ^{207}Pb-^{206}Pb age of 3760 Ma. Thus, it is likely that some of the basement rocks are Neoarchean and perhaps some are Paleoarchean.

6.1-5. DISCUSSION

6.1-5.1. Ages of Migmatitic Gneiss Components in the MRV

It is important to observe that the ancient and complex gneisses of the study area are almost certainly, in their current form, aggregates of rocks of several ages that have been mixed tectonically after multiple intrusive events. This is certainly true for the Morton Gneiss, as indicated by the fact that samples MRV-4 and MRV-6 of Bickford et al. (2006) contain well-developed euhedral populations of both ca. 3500 Ma and 3380 Ma zircons, both of which have been overgrown by 2600 Ma rims. Conversely, sample MRV-5, which cuts banded gneiss more typical of the Morton Gneiss, contains only 3380 Ma zircons. These relationships suggest that samples MRV-4 and MRV-6 are aggregates of older, ca. 3500 Ma, rocks that have been mixed with ca. 3380 Ma rocks. Other data indicate a ca. 3420–3440 Ma component in both the Morton and Montevideo Gneisses. Samples of the Montevideo Gneiss studied by Bickford et al. (2006) contain mostly ca. 3500 Ma zircons that are overgrown by zircon formed at 3380 and 3080 Ma. However, their sample MRV-3, which they interpreted as a granodiorite intrusion into the Montevideo Gneiss, yielded only 3380 Ma zircons and is evidently another example of the 3380 Ma igneous activity displayed by sample MRV-5 in the Morton block.

6.1-5.2. *Thermotectonic History of the Benson, Montevideo, and Morton Blocks*

The data of Bickford et al. (2006) indicate that major common tectonothermal events, recognized in the zircon geochronology, occurred in both the Morton and the Montevideo blocks. For example, crystallization of the earliest igneous protoliths of the Montevideo Gneiss, the granite gneiss at Granite Falls, and the Morton Gneiss occurred at ca. 3500 Ma. The 3380 Ma granodioritic intrusions into the Montevideo Gneiss (MRV-3) and into the Morton Gneiss (MRV-5), the presence of abundant 3380 Ma euhedral zircons in the Morton Gneiss in MRV-6, and the 3380 Ma overgrowths on older cores (e.g., Fig. 6.1-9(c)), indicate that there was a major igneous event at this time in both the Morton and the Montevideo blocks. These 3380 Ma intrusive units contain a foliation, but they locally cut the earlier gneissic banding in both the Morton and Montevideo Gneisses. Although less well documented, there appears to have also been a common, ca. 3420–3440 Ma, igneous event. A mafic intrusion in the granite gneiss at Granite Falls has a well constrained age of 3140 Ma, indicating another intrusive magmatic event at that time; zircon overgrowths of this age are also seen in zircons from the Morton Gneiss (e.g., Fig. 6.1-8(b)).

As discussed by Bickford et al. (2006), the only thermochronologic distinctions between the Morton and the Montevideo blocks are the 3080 Ma event and the variable representation of the 2600 Ma event in the zircons. Only zircons in the Montevideo Gneiss (MRV-1) show overgrowths with ages of ca. 3080 Ma. This event may be associated with intrusion of pegmatite and a red massive granitic phase from this outcrop (cf. Bauer, 1980), which Goldich et al. (1980) referred to as adamellite and for which they reported an estimated age of ~3100 Ma.

Late-kinematic granite bodies are features of the Morton block, into which the Sacred Heart Granite (Fig. 6.1-5) and a monzogranite body were emplaced at ca. 2600 Ma, and the Benson block, into which the Ortonville Granite was emplaced, but late granites of this age have not been found within the Montevideo block. A major metamorphic event at ca. 2600 Ma is particularly well developed in rocks in the Granite Falls area (Bickford et al., 2006; Schmitz et al., 2006), and in granulitic paragneiss in the Benson Block near Ortonville (Schmitz et al., 2006), but it was not detected in rocks in the Montevideo area and is variably represented in samples studied from the Morton block. The granulite-facies mineral assemblages in the Granite Falls area have generally been attributed to this event (e.g., Goldich et al., 1980), and are in contrast to the upper amphibolite-facies assemblages produced by the same event in the Morton block. Bickford et al. (2006) showed that zircons from Morton Gneiss sample MRV-4 (Fig. 6.1-8(c)), and detrital zircons from cordierite-bearing schist, both show well-developed 2600 Ma zircon rims. However, neither the 3380 Ma zircons from their sample MRV-5, nor the complexly-zoned zircons from Morton Gneiss sample MRV-6, contain 2600 Ma rims, despite their proximity to the ca. 2600 Ma Sacred Heart Granite (Fig. 6.1-2). Although further study will be required to fully understand the distribution of the 2600 Ma metamorphic event, the simplest explanation of the emplacement of the Sacred Heart Granite within the Morton block is that it was related to crustal melting during ca. 2600 Ma regional metamorphism associated with crustal thickening. These events are summarized in Table 6.1-1 for both the Morton and the Montevideo blocks.

6.1-5.3. When Were the Benson, Montevideo, and Morton Blocks Assembled?

As noted earlier, Southwick and Chandler (1996) recognized that within the Yellow Medicine shear zone that marks the boundary between the Montevideo and Morton blocks, there is a belt of moderately to weakly metamorphosed mafic and ultramafic rocks, which they named the Taunton belt. They suggested that the Taunton belt may represent either allochthonous greenstone belt rocks that were infolded with older cratonic gneisses, or a remnant of late Archean oceanic crust lodged along a suture between the Montevideo and Morton blocks. A suture origin for the Taunton belt would allow – but not require – a distinct early history between the Montevideo and the Morton blocks. Although the age of the Taunton belt and timing of deformation within the Yellow Medicine shear zone remain unclear, Southwick and Chandler (1996) suggested that their orientation, which is parallel to the structural grain of the rest of the Superior Craton, is consistent with Neoarchean accretion along the southern margin of the Superior Craton. Schmitz et al. (2006) have argued that the abundance and evident synchroneity of late-kinematic granite in the MRV are also consistent with this hypothesis. Such accretion could have been associated with suturing of the Benson, Montevideo, and Morton blocks. However, it is also possible that the several blocks within the MRV were sutured much earlier than 2600 Ma and that displacement along the block-bounding shear zones was a response to stresses generated when the MRV collided with the southern margin of the Superior Craton along a north-dipping suture zone (the Morris fault segment of the GLTZ) but was incapable of being subducted (Southwick and Chandler, 1996).

The data of Bickford et al. (2006) are more consistent with the latter hypothesis, which was also favored by Southwick and Chandler (1996). The similar ages of the rock units and thermal events between the two blocks suggest that the Montevideo and Morton blocks were linked by at least 3380 Ma and possibly as early as 3500 Ma. Bounding shear zones, such as the Yellow Medicine shear zone, and the gently-plunging F2 folds that occur in both the Montevideo and Morton blocks, were probably a result of partitioned shortening and shear strain adjustments during collision between the MRV and the southern margin of the greenstone-granite terranes along the southern margin of the Superior Craton.

6.1-5.4. How Are the MRV and Northern Michigan Terranes Related to the Superior Craton?

It is striking that the ancient gneisses of the MRV and northern Michigan lie along the southern margin of the mostly Neoarchean Superior Craton, and that similarly ancient terranes, such as the Assean Lake terrane (Böhm et al., 2000; 2003) and the Porpoise Cove terrane (David et al., 2003), lie along its northern margin (Fig. 6.1-1). Indeed, other Meso- to Paleoarchean crustal fragments appear to lie mostly on the margins of Neoarchean cratonal blocks, rather than within their interiors. Other examples include the Beartooth Mountains of southwestern Montana, on the southern margin of the Neoarchean Hearne Province, where a 3500 Ma trondhjemitic gneiss has been dated (Mueller et al., 1996) and where detrital zircons from orthoquartzites indicate provenance from rocks as old as 3500–4000 Ma (Mueller et al., 1998); and the recent discovery of evidence for ca. 3700 Ma rocks in the Uranium City area of northern Saskatchewan, on the northern margin of the Hearne Province (Hartlaub et al., 2004a, 2006). These occurrences suggest a relationship between the ancient terranes and the formation of the Neoarchean cratons. If so, the MRV-Northern Michigan terrane is likely related to the formation of the Superior Craton, whose origin has been debated for decades. Much of the debate has centered on whether its elongate greenstone belts and intervening terranes, composed mostly of granitic rocks, were formed by convergent plate tectonic mechanisms (e.g., Card, 1990; Condie, 1994; Thurston, 1994; Percival and Helmstaedt, 2004; Percival et al., 2004b) or whether, during the Neoarchean, different tectonic styles dominated by vertical mechanisms were operative (e.g., Kröner, 1981; Hamilton, 1993). The possible models would appear to include: (1) the Paleo-to Mesoarchean terranes are the remnants of crust that was rifted prior to the assembly of the Neoarchean Superior Craton through modern convergent plate tectonic processes that included the formation of island arcs, oceanic plateaux, and continental arcs; (2) the Paleoarchean terranes are remnants of a rifted former crust, but the Neoarchean terranes of the Superior Craton were formed continuously through rift-related tectonic processes; or (3) the Paleoarchean terranes may be earlier, but disparate crustal fragments that have been accreted to the margins of the Superior Craton.

In recent articles dealing with the western Superior Craton, Percival and Helmstaedt (2004) and Percival et al. (2004a, 2004b) have described and discussed the results of Lithoprobe seismic studies coupled with thorough geologic mapping, structural analysis, and extensive geochronology. They have shown that the Superior Craton is composed of diverse rock assemblages representing domains of older continental crust (ca. 3400–3000 Ma) separated by belts of younger (ca. 2800–2700 Ma), juvenile oceanic material, possibly representing oceanic plateaux, but also including calc-alkaline rocks akin to those in island arc edifices (see also Percival, this volume). Additionally, these authors have shown that continental arcs, formed between ca. 2800–2700 Ma, are well-developed in the northeastern Superior Province. These rocks reveal their involvement with older continental crust in enriched isotopic signatures and the ages of inherited zircons. Rift sequences, represented by clastic sedimentary assemblages overlain by komatiite, basalt, and rhyolite volcanic sequences, evidently formed early in the history of the craton, perhaps prior to

2930 Ma. Although a major question has been whether the younger rocks have been deposited upon a basement of older (ca. 3400–3000 Ma) rocks (e.g., Corfu and Davis, 1992), the presence of belts of juvenile rocks separating the older crustal belts suggests that this is not the case (Tomlinson et al., 2004). Rather, the emerging picture is that the Superior Craton has evolved through processes involving collisional tectonics, as well as possible rifting, during a protracted history over ca. 400 My. These results are consistent with either model 1 or model 3, above.

Schmitz et al. (2006) and Southwick and Chandler (1996) have argued that the nature and orientation of structures within the MRV, as well as the abundance of relatively undeformed 2600 Ma granite, suggests that the ancient rocks were probably sutured to the Superior Craton at about 2600 Ma. These arguments would support model 3, but are not inconsistent with model 1. Whether the terranes of much older rocks (ca. 3800–3500 Ma) on both the northern and southern margins of the Superior Craton can be shown to be remnants of a former ancient continental block that was rifted prior to the formation of the Superior Craton must await more detailed geologic, geochronologic, and geochemical studies.

ACKNOWLEDGEMENTS

Bickford acknowledges support from a grant from the College of Arts and Sciences, Syracuse University. Schmitz was supported by U.S. Army Research Office grant DAAH049510560 to S.A. Bowring.

Earth's Oldest Rocks
Edited by Martin J. Van Kranendonk, R. Hugh Smithies and Vickie C. Bennett
Developments in Precambrian Geology, Vol. 15 (K.C. Condie, Series Editor)   751
© 2007 Elsevier B.V.   All rights reserved.
DOI: 10.1016/S0166-2635(07)15062-3

Chapter 6.2

THE ASSEAN LAKE COMPLEX: ANCIENT CRUST AT THE NORTHWESTERN MARGIN OF THE SUPERIOR CRATON, MANITOBA, CANADA

CHRISTIAN O. BÖHM[a], RUSSELL P. HARTLAUB[b] AND LARRY M. HEAMAN[c]

[a]*Manitoba Geological Survey, Manitoba Industry, Economic Development and Mines, 360-1395 Ellice Ave., Winnipeg, MB, Canada, R3G 3P2*
[b]*Department of Mining Technology, British Columbia Institute of Technology, 3700 Willingdon Avenue, Burnaby, BC, Canada, V5G 3H2*
[c]*Department of Earth & Atmospheric Sciences, 4-18 Earth Science Building, University of Alberta, Edmonton, AB, Canada, T6G 2E3*

6.2-1. INTRODUCTION

The Superior Craton, which forms a substantial portion of the ancient core of North America, represents the largest known Archean craton (Fig. 6.2-1). Paleoarchean crustal remnants are scarce in the Superior Craton (e.g., David et al., 2004), possibly due to extensive Neoarchean crustal formation, recycling and polyphase tectono-metamorphism that resulted from proto-plate tectonic processes (e.g., Card and Ciesielski, 1986; Card, 1990; Thurston et al., 1991; Williams et al., 1992; Lin, 2005; Percival et al., 2006; Percival, this volume). Despite these problems, evidence for a Paleoarchean component to the northwest Superior Craton was identified by Böhm et al. (2000a) in the Assean Lake area, north-central Manitoba, Canada. Further studies (Böhm et al., 2003; Hartlaub et al., 2006) have advanced our knowledge of this ancient region and produced a significant database of mineral and whole-rock isotopic information.

An initial tectonic model for the Assean Lake area (Corkery and Lenton, 1990) indicated that a regionally extensive high-strain zone running through the lake marks the suture between Archean high-grade crustal terranes of the Superior Craton to the southeast, and Paleoproterozoic rocks of the Trans-Hudson Orogen to the northwest (Fig. 6.2-2). Detailed geologic re-mapping (Böhm, 1997b, 1998; Böhm and Corkery, 1999), combined with isotopic and geochemical studies (Böhm et al., 2000a, 2003; Hartlaub et al., 2006), led to a re-interpretation of the crust immediately north of the Assean Lake high-strain zone as Mesoarchean.

This paper describes the age and extent of these Mesoarchean rocks, defined herein as the Assean Lake Complex (ALC). In addition, we describe the Paleoarchean components

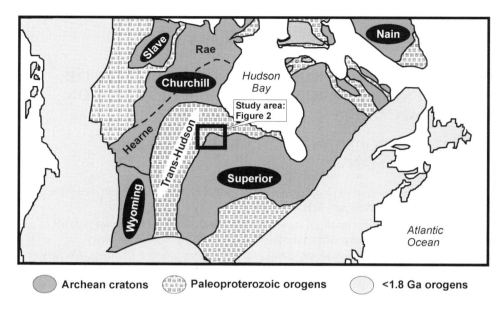

Fig. 6.2-1. Simplified geological map of part of North America highlighting the major Archean cratons.

of the ALC and examine the relationship of this complex to surrounding crustal terranes. This description is of particular interest since the ALC may have been exotic with respect to neighboring high-grade terranes prior to the 2.68–2.70 Ga assembly of the northwest Superior Craton (Davis et al., 1988; Davis and Amelin, 2000; Percival et al., 2006).

6.2-2. PRINCIPAL GEOLOGICAL ELEMENTS OF THE NORTHWESTERN SUPERIOR CRATON MARGIN

The study area straddles the boundary between the Archean Superior Craton and the ca. 1.90–1.84 Ga arc and marginal basin rocks of the Trans-Hudson Orogen, which represents the remains of ca. 1.83–1.76 Ga ocean closure and orogeny (Corrigan et al., 2005; Ansdell, 2005). Within the northwestern part of the Superior Craton (Fig. 6.2-2), the Pikwitonei Granulite Domain and Split Lake Block (Böhm et al., 1999) are separated by the Aiken River deformation zone, but comprise similar, variably retrogressed, granulite-grade rocks. To the north and west of these domains is the Superior Boundary Zone, which is composed of complexly interleaved Archean rocks of the Superior Craton, Paleoproterozoic rocks related to the Trans-Hudson Orogen, and Mesoarchean rocks of the ALC that are bounded

Fig. 6.2-2. Tectonic map of the Superior Boundary Zone region in north-central Manitoba.

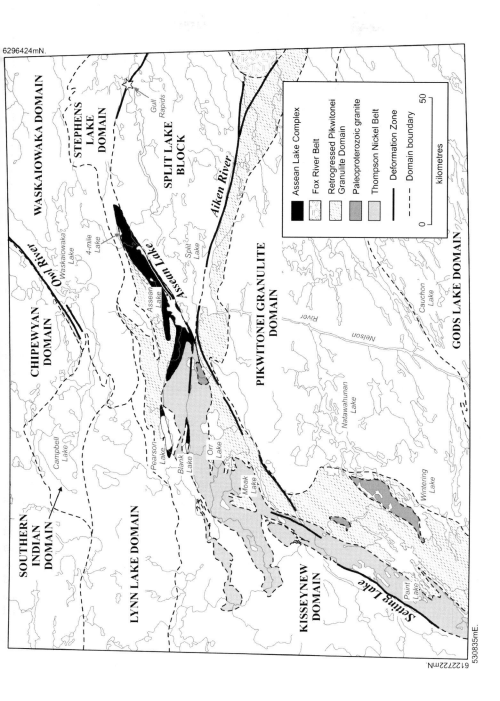

by major deformation zones. The region has been geologically subdivided based on differ- ences in structural trend, aeromagnetic signature, metamorphic grade, lithological nature and age (e.g., Böhm et al., 2000a; Zwanzig and Böhm, 2004). An economically important component of the Superior Boundary Zone is the Thompson Nickel Belt (Peredery et al., 1982; Bleeker, 1990; Machado et al., 1990), one of the most important magmatic nickel- copper sulphide districts in the world. The boundary between the ALC and the Split Lake Block is a major deformation zone, although it is not clear if this represents some sort of suture. To the north of the ALC, another question is the extent to which Meso- and/or Neoarchean rocks underlie Paleoproterozoic rocks of the Trans-Hudson Orogen.

6.2-2.1. Pikwitonei Granulite Domain

The Pikwitonei Granulite Domain is interpreted to represent the middle to deep crustal lev- els of an Archean granite-greenstone terrane (Green et al., 1985). Vestiges of supracrustal belts (Weber, 1978, 1983; Böhm, 1998) remain, but a TTG suite of orthopyroxene-bearing tonalite and granodiorite dominate. Some tonalite gneisses may have 3.0 Ga or older crys- tallization ages, but most were emplaced around 2.7 Ga (Heaman et al., 1986; Böhm et al., 1998, 1999 and unpublished data). The area around Orr Lake (Fig. 6.2-2), formerly referred to as the Orr Lake Block (Lenton and Corkery, 1981; Böhm et al., 2000a), repre- sents a structural and lithological complex hosting a number of terrane fragments. These fragments may include the northeast extension of the Thompson Nickel Belt, variably retrogressed rocks of the Pikwitonei Granulite Domain, Paleoproterozoic intrusive and sed- imentary rocks of the Trans-Hudson Orogen, and possibly fragments of the ancient ALC (Zwanzig and Böhm, 2002).

   Based on field relationships, petrography, and U-Pb geochronology, there is an indi- cation of two, and possibly three, high-grade Archean deformational and metamorphic episodes in the Pikwitonei Granulite Domain (Weber and Scoates, 1978; Hubregtse, 1980; Heaman et al., 1986). Geochronological studies indicate, however, that these events may be diachronous across the region (Heaman et al., 1986; Mezger et al., 1990). A 2695 ± 2 Ma orthopyroxene-bearing granitic dike is the first indication of localized granulite condi- tions. Complex metamorphic zircon populations from felsic and mafic granulites suggest amphibolite-grade metamorphism at ca. 2705–2692 Ma, followed by granulite-grade meta- morphism from 2683–2665 Ma, and possibly also at ca. 2657 Ma, followed by localized amphibolite-facies retrogression at ca. 2636 Ma (Heaman et al., 1986; Böhm et al., 1999 and unpublished data). Estimates of peak pressure and temperature conditions during granulite-facies metamorphism are approximately 6.7–7.3 kbar and 730–770 °C in the southeast and approximately 7.0–7.8 kbar and 780–840 °C in the northwest Pikwitonei Granulite Domain (Mezger et al., 1990). U-Pb zircon ages of ca. 2629 and 2598 Ma from post-granulite pegmatite in the Cauchon Lake area (K. Mezger, unpublished data, 1990) are additional evidence that metamorphic conditions reached amphibolite grade shortly after granulite facies. The presence of orthopyroxene-sillimanite- and sapphirine-bearing rocks at Sipiwesk Lake southwest of the Pikwitonei Granulite Domain indicates that high-grade metamorphism reached maximum intensity at this location (Arima and Barnett, 1984; Macek, 1989).

6.2-2.2. *Split Lake Block*

The Split Lake Block is a tectonic lens of variably retrogressed and reworked Superior Craton margin rocks bounded by the Assean Lake and Aiken River deformation zones (Fig. 6.2-2: Corkery, 1985; Böhm et al., 1999). These deformation zones have been interpreted to represent Neoarchean structures reactivated by Paleoproterozoic tectonism (Böhm et al., 2000a, 2003 and unpublished data; Kuiper et al., 2004a, 2004b). The Split Lake Block is dominated by medium- to coarse-grained granoblastic gneisses which contain hypersthene, diopside and their retrograde equivalents. Although the metamorphic and lithological character of this domain is similar to the Pikwitonei Granulite Domain, the Split Lake Block has been retrogressed and hydrated to a greater degree. Field and petrographic studies (Haugh, 1969; Corkery, 1985; Hartlaub et al., 2004) indicate that the gneisses of the Split Lake Block consist primarily of meta-igneous protoliths of gabbroic to granitic composition. Tonalite and granodiorite are the most volumetrically dominant, but an anorthosite dome is also present in the northeast. Böhm et al. (1999) report three periods of Archean magmatism in the Split Lake Block:

(1) pre-2.9 Ga granodiorite to tonalite magmatism, which is considered to be part of the basement;
(2) a possible period of 2841 ± 2 Ma tonalite magmatism; and
(3) granite intrusion at 2708 ± 3 Ma.

Similarly, granodiorite rocks at the northeast edge of the Split Lake Block at Gull Rapids (Fig. 6.2-2) are ca. 3.16 and 2.86 Ga and form the basement of a ca. 2.70 Ga mafic volcano-sedimentary sequence that contains 2.71 to 3.35 Ga zircon detritus (Bowerman et al., 2004).

Similar to the Pikwitonei Granulite Domain, three high-grade metamorphic events are recognized in the Split Lake Block (Corkery, 1985). Two of these events occurred within a short time span of about 10 My (Böhm et al., 1999). Based on the age of metamorphic zircon overgrowth from enderbite, the older event resulted in hornblende granulite-grade metamorphism at 2705 ± 2 Ma, closely linked to granite magmatism at 2708 ± 3 Ma. A younger granulite-grade peak metamorphic event is constrained at $2695 +4/-1$ Ma based on the age of orthopyroxene-bearing leucosome isolated from mafic gneiss. The youngest significant metamorphic event is localized ca. 2620 Ma amphibolite-grade retrogression (Corkery, 1985; Böhm et al., 1999).

6.2-2.3. *Thompson Nickel Belt*

The Thompson Nickel Belt, which is mainly exposed southwest of Moak Lake (Fig. 6.2-2), contains significant nickel-copper mineralization which has resulted in intense exploration and a wealth of geological and geophysical information that is summarized in Macek et al. (2006). Nickel-copper deposits in the belt are hosted by the Ospwagan Group supracrustal succession (Macek and McGregor, 1998), and are generally associated with ultramafic bodies in contact with sulphide-bearing metasedimentary units (Bleeker, 1990). The Thompson Nickel Belt includes variably reworked, ca. 2.7 Ga (Machado et al., 1987) Archean basement gneiss, Ospwagan Group rocks of probable 2.1–1.89 Ga age (Zwanzig, 2005), and ca. 1.88 Ga ultramafic bodies (Hulbert et al., 2005). The Ospwagan Group is interpreted as platform to marginal basin siliciclastic and chemical sedimentary sequence overlain by mafic volcanic rocks and intruded by felsic to ultramafic Paleoproterozoic bodies.

6.2-2.4. The Trans-Hudson Orogen Northwest of the Assean Lake Complex

The crust northwest of the ALC (Fig. 6.2-2) preserves prograde amphibolite metamorphic assemblages. Unlike the northwest Superior Craton, there is no indication that rocks northwest of the ALC ever attained granulite-grade conditions except in strongly reworked slivers of the Thompson Nickel Belt. Metasedimentary rocks from this area can be correlated with Burntwood Group greywacke and Sickle Group arkose of the Kisseynew domain (Zwanzig, 1990) of the Trans-Hudson Orogen. Northwest of Assean Lake, a belt-like pattern that is continuous with Trans-Hudson Orogen subdivisions to the west, was identified by Lenton and Corkery (1981). The belts are defined by alternating, east- and southeast-trending belts dominated by plutonic (e.g., Chipewyan, Waskaiowaka) and supracrustal (Kisseynew, Lynn Lake, Southern Indian) domains (Fig. 6.2-2). The presence of abundant ca. 2.45 Ga detrital zircons in a metagreywacke at Campbell Lake (Hartlaub et al., 2004), ca. 50 km northwest of Assean Lake (Fig. 6.2-2), is consistent with derivation from the Sask Craton (Ashton et al., 1999), an Archean microcontinent that may underlie much of the Trans-Hudson Orogen northwest of the ALC in Manitoba and Saskatchewan.

6.2-2.5. Stephens Lake Domain of the Trans-Hudson Orogen

North and east of the Split Lake Block are Paleoproterozoic rocks of the Stephens Lake Domain that mark the northeastern edge of the ALC (Fig. 6.2-2). Greywacke- and arkose-derived paragneiss of middle to upper amphibolite grade form the principal rock types in this domain (Haugh and Elphick, 1968; Corkery, 1985). The paragneisses contain minor amounts of amphibolite and quartzite, and the entire package is intruded by tonalite to granite and derived migmatitic gneiss. Metagreywacke and layered granodiorite gneiss from the east end of Stephens Lake have Nd model ages of 1.95 and 2.1 Ga, respectively (Böhm et al., unpublished data, 2000). The fact that both para- and orthogneiss in the area are derived from Paleoproterozoic material is consistent with the interpretation that the Stephens Lake Domain represents the far eastern extension of the Kisseynew Domain of the Trans-Hudson Orogen.

6.2-3. GEOLOGY OF THE ASSEAN LAKE COMPLEX

Shoreline exposures at Assean Lake (Fig. 6.2-2) were first recognized as the 'Assean Lake series' of sedimentary and volcanic rocks by Dawson (1941). The name was subsequently modified to 'Assean Lake Group' (Mulligan, 1957), and expanded to include sedimentary and volcanic rocks of the Ospwagan Group of the Thompson Nickel Belt. Haugh (1969) re-mapped the Assean Lake area and defined three subareas divided by extensive zones of cataclasis. Lithologies such as grey biotite gneiss, lit-par-lit gneiss, pelitic schist, amphibolite, gneissic granite and gabbro were interpreted to be continuous, identical and age-equivalent extensions of rocks in the Ospwagan Lake area southwest of Thompson (Haugh, 1969). Despite the similar composition and metamorphic grade, the Ospwagan Group supracrustal rocks are unrelated to the supracrustal rocks of the ALC (Böhm et al., 2003).

Regional studies at the northwestern Superior Craton margin helped to place the Assean Lake area into a regional context (Corkery, 1985; Corkery and Lenton, 1990), but a pre-3.0 Ga origin for the ALC was not suspected until more recent mapping (Böhm, 1997a, 1997b) and radiogenic isotope studies were conducted (Böhm et al., 2000a). Subsequent studies (Böhm et al., 2003; Hartlaub et al., 2005, 2006) provided more than a dozen Archean U-Pb ages and a significant database of Nd isotope data (Table 6.2-1). This database, combined with numerous complimentary studies in the northwest Superior Craton (e.g., Bowerman et al., 2004; Hartlaub et al., 2004; Kuiper et al., 2004a, 2004b; Zwanzig and Böhm, 2004), provide the basis for the more complete tectonic analysis described herein.

The exposed ALC is an assembly of 090–110° trending crustal segments that have been overprinted by the 060° trending Assean Lake deformation zones (Fig. 6.2-3). Sub-vertical tectonic fabrics are moderately to well developed in all lithological units. The high-strain deformation zone is more than three kilometres wide and grades from a northern cataclastic zone that passes through the northeast arm of Assean Lake (Lindal Bay), into a mylonitic zone in the south (Böhm, 1997a). Rocks can only locally be traced into lower strain equivalents, making protolith determination in the high-strain zones difficult. Although kinematics are complex and polyphase, a main dextral transpressive component has been recognized (Böhm, 1997a; Kuiper et al., 2004a). Supracrustal rocks in the ALC are subdivided into the Clay River assemblage of migmatitic metasedimentary rocks in the southwest and a northeast volcano-sedimentary package termed the Lindal Bay assemblage (Fig. 6.2-3). The Clay River and Lindal Bay assemblages are separated and intruded by abundant orthogneiss ranging in composition from tonalite to granite (central orthogneiss domain).

6.2-3.1. Clay River Assemblage

At the western end of Assean Lake, outcrops are dominated by paragneiss with subordinate orthogneiss (Fig. 6.2-3). A greywacke protolith is well established for upper-amphibolite

Table 6.2-1. Summary of Nd isotopic and U-Pb geochronological data of rocks of the Assean Lake complex

Sample	Lithology[1]	Locality	UTM East NAD83 Z14	UTM North NAD83 Z14
CB96-17	quartzo-feldspatic gn	central Assean Lake	656886	6234446
CB96-22b	granodiorite gn	north Assean Lake	653136	6234546
CB96-42	biotite granodiorite gn	west Assean Lake	648746	6231346
CB96-48a	garnet biotite (ortho)gn	northwest Assean Lake	652026	6232816
CB96-73a	tonalite gn	Assean Lake	652826	6233726
CB97-12	metagreywacke	northeast Assean Lake	659016	6236476
CB97-55a	biotite granodiorite gn	northwest Assean Lake	649146	6232596
CB98-14	layered tonalite gn	Blank Lake	603765	6226126
CB98-21	leucogranite gn	4-Mile Lake	682316	6253476
CB98-24	quartzo-feldspathic granite gn	4-Mile Lake	680246	6253626
CB98-83	pelitic greywacke migmatite	west Assean Lake	647286	6230076
CB98-84	granite augen gn	northeast Assean Lake	659986	6236416
CB00-56	metagreywacke	northeast Assean Lake	659016	6236476
CB00-62a	granite augen gn	northeast Assean Lake	659986	6236416
CB00-71	granodiorite gn	northeast Assean Lake	659386	6236456
CB00-83	pelitic metagreywacke	east Assean Lake	658966	6235826
CB00-102	quartz arenite gn	west Assean Lake	647516	6228926
12-01-217	granodiorite gneiss	Pearson Lake	604071	6233709
97-04-8172	metagreywacke	north Assean Lake	653204	6235247

(continued on next page)

Abbrev.

[1] gn = gneiss.

grade garnet ± sillimanite ± cordierite gneisses (Fig. 6.2-4(a)), but arkose and arenite protoliths have been heavily obscured by recrystallization, mobilization and melt injection. Minor garnetiferous pegmatite, amphibolite, white weathering feldspathic biotite-gneiss, and silicate facies iron formation are locally present. Metasandstone is gneissic and highly variable in quartz content (Fig. 6.2-4(b)). Compositional layering in these rocks, and interlayering with amphibolite, is interpreted as primary sedimentary and volcanic layering that has been enhanced by the development of in-situ and injection mobilizate.

6.2-3.2. Central Felsic Intrusive Rocks (Orthogneiss Domain)

The area between the Clay River and Lindal Bay assemblages is dominated by a sequence of tonalite to granodiorite intrusives and derived gneisses (Fig. 6.2-3). These felsic rocks are variably layered due to injection by later pegmatites (Fig. 6.2-4(c)) along the meta-

Table 6.2-1. (*Continued*)

Sample	Analytical method	U-Pb age (Ma)	Error 2σ abs.	Mineral[2]	Nd model age (Ma)[3]
CB96-17	ID-TIMS			wr	3510
CB96-22b	ID-TIMS			wr	3730
CB96-42	ID-TIMS	3191	5	zc	
	ID-TIMS			wr	4150
CB96-48a	LA-MC-ICPMS	3169	10	zc	
	ID-TIMS			wr	3720
CB96-73a	SHRIMP	3180	6	zc	
	SHRIMP	2680	5	zc	
	ID-TIMS			wr	3630
CB97-12	SHRIMP	3278	19	zc	
	ID-TIMS			wr	3850
	ID-TIMS	2636	10	mz	
CB97-55a	ID-TIMS			wr	3550
CB98-14	ID-TIMS			wr	3590
CB98-21	LA-MC-ICPMS	3206	4	zc	
	ID-TIMS			wr	3530
CB98-24	LA-MC-ICPMS	~3100		zc	
	ID-TIMS			wr	3720
CB98-83	SHRIMP	2607	17	zc	
	ID-TIMS	2444	2	mz	
	SHRIMP	3203	5	zc	
	ID-TIMS			wr	3760
CB98-84	ID-TIMS			wr	n/a
	ID-TIMS	~2620		zc+mz	
CB00-56	LA-MC-ICPMS	3165	27	zc	
CB00-62a	ID-TIMS			wr	3580
CB00-71	ID-TIMS			wr	3520
CB00-83	ID-TIMS			wr	3750
CB00-102	ID-TIMS			wr	3500
12-01-217	ID-TIMS	3185	7	zc	
97-04-8172	LA-MC-ICPMS	~ 3180	19	zc	

(*continued on next page*)

Abbrev.

[2]mz = monazite, wr = whole rock, zc = zircon.

[3]Crustal residence Nd model ages after Goldstein et al. (1984).

Table 6.2-1. (*Continued*)

Sample	$\frac{^{147}Sm}{^{144}Nd}$	$\frac{^{143}Nd}{^{144}Nd}$	Error 2σ abs.	Age interpret.	Reference
CB96-17	0.1276	0.511162	0.000005	model	Böhm et al. (2003)
CB96-22b	0.1242	0.510951	0.000010	model	Böhm et al. (2000a)
CB96-42				igneous?	Böhm et al. (2003)
	0.1283	0.510811	0.000009	model	Böhm et al. (2003)
CB96-48a				igneous	Hartlaub et al. (2005)
	0.1203	0.510861	0.000008	model	Böhm et al. (2003)
CB96-73a				igneous	Böhm et al. (2003)
				metamorphic	Böhm et al. (2003)
	0.1076	0.510615	0.000010	model	Böhm et al. (2000a)
CB97-12				min. detrital	Böhm et al. (2003)
	0.1151	0.51065	0.000010	model	Böhm et al. (2000a)
				metamorphic	Böhm et al. (2003)
CB97-55a	0.0898	0.510251	0.000007	model	Böhm et al. (2000a)
CB98-14	0.1058	0.510598	0.000011	model	Böhm et al. (2000a)
CB98-21				igneous	Hartlaub et al. (2005)
	0.0967	0.510433	0.000008	model	Böhm et al. (2000)
CB98-24				igneous	Hartlaub et al. (2005)
	0.1223	0.510914	0.000006	model	Böhm et al. (2000a)
CB98-83				metamorphic	Böhm et al. (2003)
				metamorphic	Böhm et al. (2003)
				min. detrital	Böhm et al (2003)
	0.1219	0.510881	0.000006	model	Böhm et al. (2000a)
CB98-84	0.1467	0.511315	0.000007	model	Böhm et al. (2000a)
				metamorphic	Böhm, unpublished
CB00-56				metamorphic?	Hartlaub et al. (2006)
CB00-62a	0.1003	0.510478	0.000007	model	Böhm et al. (2003)
CB00-71	0.1206	0.510993	0.000010	model	Böhm et al. (2003)
CB00-83	0.1194	0.510824	0.000009	model	Böhm et al. (2003)
CB00-102	0.1175	0.510939	0.000009	model	Böhm et al. (2003)
12-01-217				igneous?	Zwanzig & Böhm (2002)
97-04-8172				min. detrital	Hartlaub et al. (2005)

morphic fabric. The felsic intrusive rocks are predominantly in structural conformity with most supracrustal units, but in rare cases intrusive contacts crosscut the principal layering in metasedimentary and mafic volcanic rocks. Lenses of paragneiss and amphibolite in orthogneiss are common and provide further evidence that the central orthogneisses intruded the Clay River and Lindal Bay supracrustal rocks.

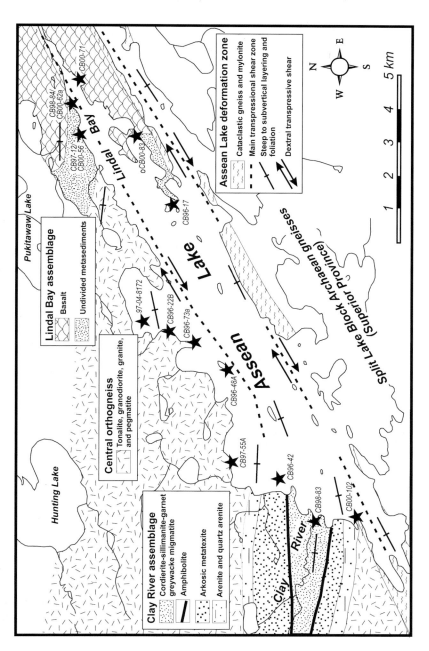

Fig. 6.2-3. Schematic geological map of the Assean Lake area, showing locations of Nd isotope and U-Pb geochronology samples listed in Table 6.2-1.

Fig. 6.2-4.

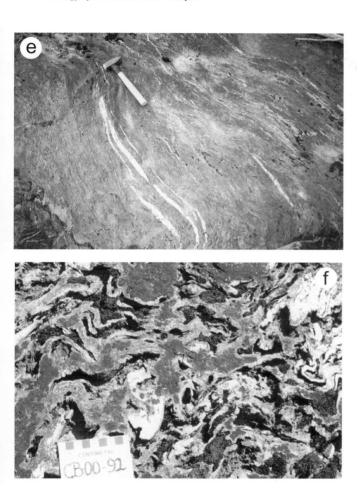

Fig. 6.2-4. (*Continued.*) Outcrop photographs of Mesoarchean rocks at Assean Lake. Clay River metasedimentary rocks: (a) cordierite-sillimanite-garnet-bearing greywacke migmatite; (b) arenite and quartz arenite. Central orthogneisses; (c) hornblende-biotite granodiorite gneiss with pegmatitic melt injection. Lindal Bay supracrustal rocks; (d) highly-strained pillowed basalt; (e) pelitic meta-greywacke; and (f) silicate-facies iron-formation.

6.2-3.3. Lindal Bay Assemblage

The Lindal Bay area of northeast Assean Lake (Fig. 6.2-3) is dominated by ca. 060° trending mafic to intermediate metavolcanic and semipelitic metasedimentary rocks. As in the Clay River assemblage, lithological units can be defined but stratigraphic relationships are difficult to determine due to the complex structural overprinting and lack of continuous outcrop. On the south shore of Lindal Bay, metasedimentary rocks predominate, but interlayered iron formation and amphibolite are also present. On the north shore of the bay, 090–110° trending amphibolite-grade mafic to intermediate volcanic rock (Fig. 6.2-4(d)) and subordinate greywacke is exposed (Fig. 6.2-4(e)). The mafic volcanic rocks vary from fine- to medium-grained, and range from massive to layered at a centimetre-scale. In several locations the layering can be recognized as flattened remnants of volcanic pillows (Fig. 6.2-4(d)). The intermediate and mafic volcanic rocks are geochemically similar to modern ocean-floor basalts, but they have slightly enriched rare-earth elements and Th, and are depleted in Nb. Together this may indicate an enriched mantle component or a volcanic arc setting (Böhm et al., 2003). The mafic volcanic rocks are predominantly basaltic, with subordinate andesitic to dacitic and ultramafic compositions (Böhm et al., 2003). The ultramafic rocks occur as rare, isolated outcrops with locally well-preserved intrusive textures; their age relationship to the paragneiss sequence is uncertain. North to northeast-trending mafic dikes may be coeval with, and appear to be feeders to, the mafic volcanic rocks.

Metasedimentary rocks of the Lindal Bay assemblage are generally highly strained, thinly bedded semipelitic gneisses (Fig. 6.2-4(e)) interlayered with quartz arenite, silicate-facies iron formation (Fig. 6.2-4(f)) and mafic lithic and psammitic greywacke. Staurolite in semipelitic gneiss indicates middle amphibolite-grade peak metamorphic conditions. Both sequences, however, seem to be intruded by tonalite-granodiorite orthogneiss, granite and pegmatite associated with the central orthogneiss domain. No basement to the supracrustal rocks has been located at Assean Lake.

6.2-3.4. Structural Domains of the Assean Lake Area

Based on the studies by Kuiper et al. (2004a) the Assean Lake area can be subdivided into the following structural domains.

(1) The main Assean Lake deformation zone, which contains mylonite, protomylonite and some ultramylonite, dips steeply to the southeast. Fold axes and lineations within the plane of the shear zone indicate southeast-side-up, dextral movement.

(2) Northwest Assean Lake is structurally dominated by east-plunging, open to isoclinal folds (Kuiper et al., 2004a). Genetically related folds may also occur in the Split Lake Block (Kuiper et al., 2004b). If true, the ALC was juxtaposed with the Split Lake Block of the Superior Craton prior to this folding event.

(3) Dextral, southeast-side-up structures predominate on both shores of Lindal Bay. A fault or shear zone may be present between the shores, which would explain the difference in rock types and foliation orientation across the bay. Alternatively, the variation in rock types could be explained by folding. The southern Lindal Bay area lacks the development of mylonite, but rocks have been locally affected by dextral, southeast-side-up shear. This area may represent a window of relatively less deformed crust at the margin of the main Assean Lake deformation zone.

6.2-3.5. *Sm-Nd and Lu-Hf Isotopic Constraints on the Antiquity of the Assean Lake Complex*

Rocks of the Assean Lake area were initially assigned to the Paleoproterozoic Kisseynew Domain, based on paragneiss composition and metamorphic grade (Fig. 6.2-5: e.g., Corkery and Lenton, 1990). In order to test this assumption, a Sm-Nd tracer isotope study was commenced on felsic igneous and metasedimentary gneiss samples at Assean Lake. Crustal residence Nd model ages (T_{CR}; Goldstein et al., 1984) in the Assean Lake area were found to range from ~3.5 to over 4.1 Ga (Table 6.2-1) (Böhm et al., 2003). Results from orthogneiss samples primarily range between 3.5 and 3.7 Ga, but a few samples have Nd model ages $\geqslant 4.0$ Ga. The fact that these samples have Neo- and Mesoarchean U-Pb crystallization ages (Table 6.2-1), yet Eo- to Paleoarchean Nd isotope signatures, indicates significant recycling of ancient crust. Metasedimentary rocks yielded Nd model ages between 3.5 Ga (meta-arenite) and 3.9 Ga (metagreywacke).

The uniformly ancient Nd model ages of the ALC provided the strong isotopic evidence for Paleo- to Mesoarchean crustal material at Assean Lake, and meant that the crustal suture between the Trans-Hudson Orogen and the Superior Craton likely exists northwest of the ALC (Böhm et al., 2000a). In a regional context, the Nd model ages from the ALC are much older than the 2.9–3.3 Ga Nd model ages that dominate the Pikwitonei Granulite Domain and the Split Lake Block. Nd model ages as old as 3.6 Ga do occur in the northwest Superior Craton, but they are extremely rare (Böhm et al., 2000a). The Thompson Nickel Belt has a broader range of Nd model ages, from ~3.5–2.5 Ga, that most likely represent a mix of Archean and Proterozoic crustal components (Böhm et al., 2000b; Zwanzig and Böhm, 2002). North of the ALC, a mix of younger Nd model ages, from 3.4 to 2.1 Ga, likely indicate a transition zone of variably reworked Archean crust and contaminated juvenile Paleoproterozoic sedimentary and intrusive rocks (Böhm et al., 2000a, 2000b).

In order to fully explore the nature of crustal material in the area, laser ablation analyses for in-situ Hf isotope compositions were obtained from Paleoarchean detrital zircons from rocks at Assean Lake (Hartlaub et al., 2006). Ablation runs utilized spot sizes of 40 to 60 μm and focused on the ancient cores where effects of zircon overgrowth and discordance were minimized. The majority of ca. 3.6 to 3.9 Ga zircon grains from Assean Lake rocks have negative ε_{Hf} values between -2 and -10 when compared to the CHUR values of Blichert-Toft and Albarède (1997). These ε_{Hf} values indicate that Assean Lake detrital zircons are derived from evolved, reworked crust. When combined with results from Jack Hills, Australia and the Acasta gneisses, Canada (Amelin et al., 1999, 2000), the Hf isotope data from Assean Lake suggests that there was significant early (>4.0 Ga) crust formation (see also Cavosie et al., this volume, and Kamber, this volume).

6.2-3.6. *U-Pb Age Constraints of the Assean Lake Complex*

U-Pb age data for TTG-type orthogneisses from the Assean Lake Complex (Böhm et al., 2000a, 2003) provide evidence for a major magmatic event at ca. 3.2–3.1 Ga, with minor, ca. 3.5 Ga, crustal inheritance. Several analyzed felsic intrusive samples yielded well-constrained zircon ages around 3.17–3.18 Ga (Böhm et al., 2003), including a U-Pb zircon

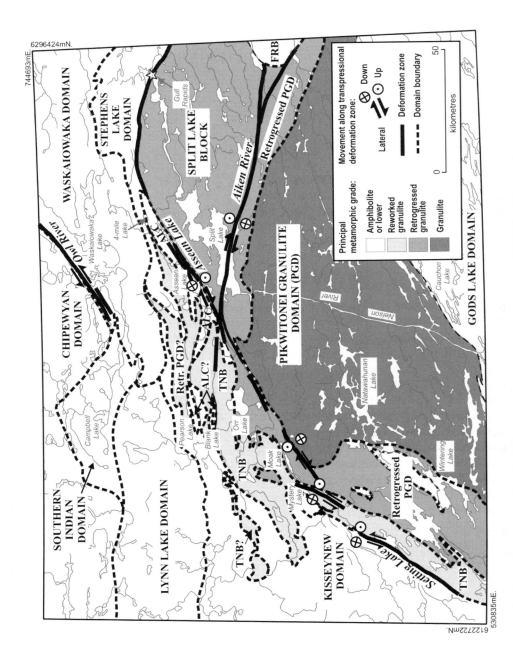

Fig. 6.2-5. (*Previous page.*) Schematic tectonic and metamorphic map of the Superior Boundary Zone region in north-central Manitoba with domain boundaries as in Fig. 6.2-2 and principal movement along main deformation zones as in Kuiper et al. (2005). ALC = Assean Lake Complex, FRB = Fox River Belt, PGD = Pikwitonei Granulite Domain, TNB = Thompson Nickel Belt.

---

age of 3169 ± 10 Ma for a medium grained biotite-rich tonalite gneiss from northwest Assean Lake (Hartlaub et al., 2005). Mesoarchean ages for felsic intrusive rocks are less common in the adjacent high-grade terranes of the northwest Superior Craton. The Split Lake Block, for example, is dominated by Neoarchean igneous rocks. One exception is the Gull Rapids area at the northeast margin of the Split Lake Block (Fig. 6.2-2), ca. 100 km to the east of the ALC in a tectonically similar area along the margin of the Superior Craton (Fig. 6.2-5). In this area, the prevalent orthogneiss was dated in two locations to be ca. 3.16 Ga (Bowerman et al., 2004). At Gull Rapids, a ca. 3 km thick exposed supracrustal package structurally lies on top of the Mesoarchean orthogneiss. Paragneiss in this package is dominated by Neo- and Mesoarchean zircon detritus, mostly younger than 2.9 Ga but with rare zircons as old as 3.35 Ga. The supracrustal rocks at Gull Rapids were likely deposited at ca. 2.71 Ga based on the youngest detrital zircons and ca. 2.68 Ga crosscutting leucogranite dikes (Bowerman et al., 2004). Metasedimentary rocks at Assean Lake, in comparison, have zircon detritus exclusively older than 3.18 Ga (Fig. 6.2-6: Böhm et al., 2003).

In the central Split Lake Block ca. 30 km west of Gull Rapids, garnet-sillimanite-biotite paragneiss contains detrital zircons that range in age from ca. 2.7 to 3.8 Ga (Hartlaub et al., 2005). This unit underwent Neoarchean high-grade metamorphism that, combined with the detrital ages, indicates sedimentation between ca. 2.70–2.68 Ga. The abundance of Meso- and Paleoarchean detrital zircon in the sample suggests that the source terrane likely included the ALC and its cryptic basement. In comparison, more than half of the zircon detritus in the Assean Lake paragneiss samples yielded Paleoarchean ages of between 3.5 and 3.9 Ga (Fig. 6.2-6: Böhm et al., 2003; Hartlaub et al., 2006), confirming the ancient provenance of these paragneisses as indicated by the ca. 3.5–3.9 Ga Nd model ages. No exposures of Paleoarchean basement rocks seem to be preserved or exposed at Assean Lake. Internal and external morphologies of these zircons are described in Böhm et al. (2003), where it is pointed out that inner zircon areas of both detrital and inherited Paleoarchean crystals are typically structureless or show weak, patchy zoning, whereas Mesoarchean igneous zircon growth areas and Neoarchean metamorphic zircon rims typically display oscillatory zoning. Although the timing of metamorphism at Assean Lake is not yet fully understood, 3.14, 2.68 and 2.61 Ga metamorphic zircon overgrowth ages have been reported (Böhm et al., 2003; Hartlaub et al., 2005). In addition, an even older metamorphic event around 3.5 Ga is recorded in metamorphic or altered domains of some Paleoarchean zircons from a metagreywacke at northeast Assean Lake (Fig. 9d in Böhm et al., 2003). 3.5 Ga also is the age of xenocrystic zircon in 3.18 Ga tonalite-granodiorite gneiss from north Assean Lake (sample CB96-73a in Böhm et al., 2003). Ca. 2.68 Ga amphibolite grade metamorphic overprint of the Assean Lake orthogneisses occurred contemporane-

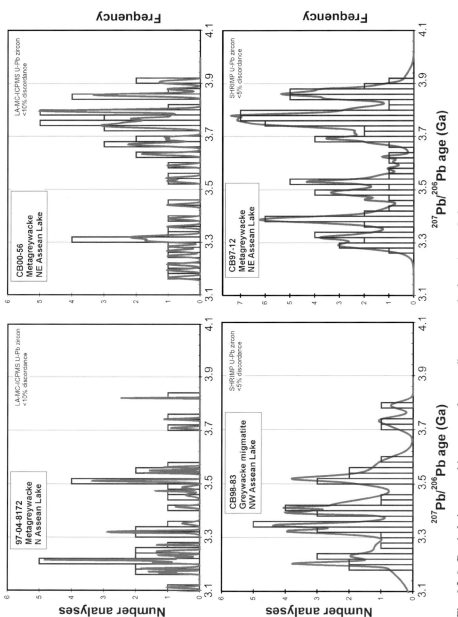

Fig. 6.2-6. Detrital zircon age histograms of metasedimentary rocks from Assean Lake.

ously with peak metamorphic conditions, related partial melting, and voluminous injection of leucocratic granitic magma in the nearby Split Lake Block (Heaman et al., 1986; Böhm et al., 1999; Hartlaub et al., 2004; Bowerman et al., 2004), which represents the earliest temporal link of periods of metamorphism between the two terranes across the Assean Lake deformation zone.

In addition to the zircon age data, U-Pb ages of monazite extend the record of metamorphic mineral growth in the ALC. Several Neoarchean and Paleoproterozoic ages of monazite growth have been identified (Fig. 10 in Böhm et al., 2003), including concordant ages at ca. 1810, 2444 and 2630 Ma. Whereas the 1810 and 2630 Ma monazite ages can be directly correlated with regional peak metamorphism during the Hudsonian and late stage Kenoran (Böhm et al., 1999) orogenies, respectively, the 2444 Ma monazite age from a metagreywacke migmatite at northwest Assean Lake (sample CB98-83 in Böhm et al., 2003) can only be correlated with ca. 2.45 Ga peak metamorphic and igneous activities in the Sask Craton (Ashton et al., 1999).

6.2-4. EXTENT OF THE MESOARCHEAN ASSEAN LAKE COMPLEX

Mapping, isotopic and age data combined with recently acquired high-resolution aeromagnetic data (Coyle et al, 2004) indicate that the Mesoarchean ALC is a crustal slice up to 10 km wide, and has a strike length of at least 50 km (Fig. 6.2-5). The ALC is centered on Assean Lake, and runs along the northern side of the suture with the Split Lake Block. In detail, the margins of the ALC are irregular, transected by shears and fault zones, and reworked by polyphase fold interference patterns in the generally subvertical lithologies (Kuiper et al., 2004a).

6.2-4.1. Eastern Extent of the Assean Lake Complex

Apetowachakamasik Lake (4-mile Lake; Fig. 6.2-7), which is located north of the Split Lake Block and approximately 30 km east-northeast of Assean Lake (Fig. 6.2-2), was recently discovered to contain upper amphibolite grade igneous rocks of Mesoarchean age (Hartlaub et al., 2005). The first hint for potentially Mesoarchean rocks exposed at 4-mile Lake was a ca. 3.7 Ga Nd model age for a quartzo-feldspathic gneiss interpreted as granitic (sample 98-24; Böhm et al., 2000a). Follow-up mapping and isotopic work by Hartlaub et al. (2005) outlined that the boundary between Paleoproterozoic sedimentary rocks of the Trans-Hudson Orogen and Mesoarchean rocks that are most likely associated with the ALC, runs approximately east–west through the center of 4-mile Lake (Fig. 6.2-7). Moreover, aeromagnetic data of the Assean Lake area (Coyle et al., 2004) displays a continuous east-northeast extension of weak, narrow linear magnetic anomalies from Assean Lake into the 4-mile Lake area.

The north shore of 4-mile Lake is predominantly comprised of greenschist facies conglomerate and sandstone, whereas a mixed suite of upper-amphibolite facies Archean

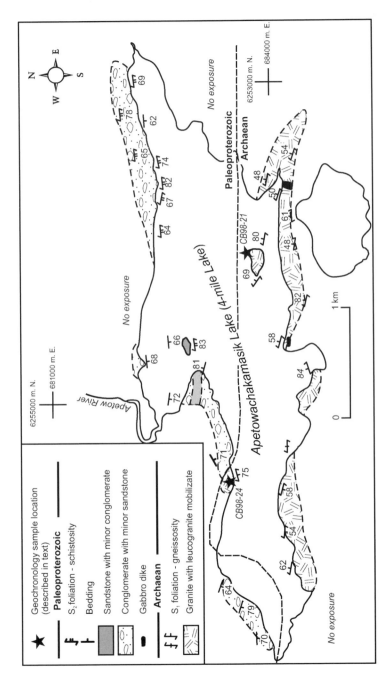

Fig. 6.2-7. Simplified geology of Apetowachakamasik Lake (4-Mile Lake).

granite gneiss with injected leucogranite is exposed along the south shore (Fig. 6.2-7; Hartlaub et al., 2005). The granite gneiss is composed of granite to granodiorite paleosome and leucogranite neosome. Mafic, biotite-rich enclaves that may represent xenoliths of older rock or disrupted dikes are locally abundant. A sample of pink, medium to coarse grained granite gneiss is dominated by 3206 ± 4 Ma zircon prisms interpreted to record the crystallization age of the granite (sample CB98-21 in Table 6.2-1; Hartlaub et al., 2005). A single ca. 2.7 Ga metamorphic zircon and a single ca. 3.6 Ga xenocrystic zircon were also identified in this sample. A sample of migmatitic, quartzo-feldspathic leucogranite, which may represent a large segregation of neosome from the gneiss, has a complex zircon population dominated by ca. 3.1 Ga zircons that are interpreted to record granite crystallization (sample CB98-24 in Table 6.2-1; Hartlaub et al. (2005)). Older zircons with ^{207}Pb/^{206}Pb ages between 3.2 and 3.8 Ga are interpreted to be xenocrystic and may have been derived, at least in part, from the mafic enclaves that are present in the sampled outcrop.

6.2-4.2. Western Extent of the Assean Lake Complex

Exposures along the south shore of Blank Lake, located approximately 50 km west of Assean Lake (Fig. 6.2-2), are dominated by migmatitic, layered tonalite-granodiorite gneiss including various amounts of amphibolite (metagabbro) lenses. A sample of hornblende-biotite granodiorite gneiss yielded a Nd model age of ca. 3.6 Ga (sample 98-14; Böhm et al., 2000a), interpreted to indicate a Meso- or Paleoarchean age for the granodiorite gneiss precursor, and possibly an indication for ALC-type Mesoarchean orthogneiss at Blank Lake.

Ca. 10 km north of Blank Lake, Pearson Lake (Fig. 6.2-5) lies along strike of the intersection of the northeast-trending Superior Boundary Zone and the Owl River deformation zone, the latter presumably separating mixed Archean and Proterozoic crust from dominantly Proterozoic crust (Böhm et al., 2000a). Granodiorite gneiss that is the dominant unit at Pearson Lake (Zwanzig et al., 2001) yielded a U-Pb zircon age of 3185 ± 7 Ma (sample 12-01-217 in Table 6.2-1). This age is interpreted as the time of granodiorite crystallization (Zwanzig and Böhm, 2002), coeval with felsic magmatism at Assean Lake. The Mesoarchean orthogneiss at Pearson Lake is interleaved with Paleoproterozoic- to Neoarchean-derived paragneiss in the north, and is in fault contact with predominantly Paleoproterozoic-derived greywacke-migmatite of the Kisseynew Domain to the south (Fig. 6.2-2; Zwanzig and Böhm, 2002).

6.2-4.3. Northern Extent of the Assean Lake Complex

A single exposure of highly-strained, feldspar-augen granodiorite-tonalite gneiss at Pukitawaw Lake occurs about one kilometer north of Assean Lake (Fig. 6.2-3). Nd isotopic analysis of a sample of this exposure yielded a model age of ca. 3.4 Ga, slightly younger than orthogneiss Nd model ages at Assean Lake (Böhm et al., 2000a, 2003). ID-TIMS U-Pb age dating of four abraded single-grain zircons resulted in slightly (<2%) discordant analyses with a mean ^{207}Pb/^{206}Pb age and concordia upper intercept age of ca. 2.70 Ga

(Böhm, unpublished data, 2001). Due to the consistent ^{207}Pb/^{206}Pb ages, we interpret the 2.70 Ga age to represent the timing of felsic magmatism at Pukitawaw Lake. This age is similar to that of prevalent Neoarchean magmatism in the Split Lake Block and the Pikwitonei Granulite Domain. Consequently, Pukitawaw Lake may represent a sliver of Split Lake Block-related crust along the northern margin of the ALC (Fig. 6.2-2). This sliver could also represent an allochthonous block that was thrust over the ALC, or the ALC was thrust part ways onto the Split Lake Block. Regardless; the exposed ALC likely extends less than a kilometer to the north of Assean Lake.

6.2-4.4. *The Assean Lake Complex–Split Lake Block Connection*

6.2-4.4.1. *Supracrustal rocks*
Unlike the ALC, supracrustal rocks are extremely rare in the Split Lake Block, primarily occurring as mafic granulite with thin interlayered horizons of pelite (Bowerman et al., 2004; Hartlaub et al., 2005). Mafic granulite in the Split Lake Block is fine to medium grained, and locally displays compositional layering. Orthopyroxene $\pm$ garnet melt segregations comprise up to 5% of this unit. The fine to medium grained nature and local compositional layering of the mafic granulite may indicate that its protolith was a mafic volcanic rock. Local garnet rich horizons may represent iron rich interflow sediments. Whole-rock geochemistry of the mafic granulite from Split Lake indicates that the samples have a primitive MORB like signature (Hartlaub et al., 2004). Like the ALC, pelite in the Split Lake Block consists of well layered quartz, feldspar, biotite, garnet and sillimanite with trace sulphides $\pm$ graphite. Detrital zircons have been analyzed from pelite collected from both the Split Lake Block (Hartlaub et al., 2005) and the ALC (Böhm et al., 2003; Hartlaub et al., 2006; Table 6.2-1). Although detritus from both areas contain ancient (>3.6 Ga) grains, the youngest detrital grains in the pelites in the Split Lake Block are ca. 2.70 Ga, whereas the youngest detrital zircons in the ALC are ca. 3.18 Ga. Volcanic rocks have not been directly dated in either the ALC or the Split Lake Block, but the youngest detritus places a maximum age of ca. 3.18 Ga on supracrustal rocks of the ALC and ca. 2.70 Ga for the Split Lake Block. The presence of diverse detrital zircon age populations in sediments indicates that sediment in both the ALC and the Split Lake Block were derived by erosion of continental-type crust.

6.2-4.4.2. *Felsic plutonism*
Although sediments in the ALC and Split Lake Block were deposited at different times, both regions share ca. 3.16–3.20 Ga Mesoarchean plutonism (Böhm et al., 2003; Hartlaub et al., 2004; Bowerman et al., 2004). In the ALC, this plutonism is considered intrusive into the supracrustal package, whereas these plutons are considered basement to the supracrustal rocks of the Split Lake Block. This relationship suggests that subduction was initiated beneath the ALC at circa 3.2 Ga. The >3.5 Ga Nd model ages (Böhm et al., 2000a) of the 3.16–3.18 Ga plutons in the ALC indicates that the subduction derived melts mixed with ancient crustal material, some of which may have been detritus in sedimentary rocks. This period of magmatism built the ALC into a Mesoarchean protocontinent that

may have continued to grow by collision of oceanic arc material. The inclusion of MORB-type mafic rocks in the ALC (Böhm et al., 2003) indicates that oceanic material was caught up in the formation of the ALC protocontinent. Mesoarchean crust of the Split Lake Block may have formed contemporaneously to the ALC as a separate protocontinent, and became later stitched together. Alternatively, the Mesoarchean basement of the northwest Superior Craton may be an extension of the ALC. Regardless, Neoarchean sediments in the Split Lake Block contain some ancient detritus and therefore the ALC and Split Lake Block must have been related and may have amalgamated prior to 2.70 Ga.

6.2-5. CONCLUSIONS

The ALC represents the largest and best studied fragment of ancient crustal material related to the Superior Craton. A local derivation for ancient detritus at Assean Lake appears consistent with the occurrences of ⩾4.0 Ga Nd model ages in igneous rocks of the ALC. Although not exposed at surface, ⩾4.0 Ga crust is considered to be a significant component of the ALC protocontinent. The Paleoarchean component of the ALC likely acted as a protocontinent nucleus that underwent significant crustal growth in the Mesoarchean. Mesoarchean crust of the Split Lake Block may have formed contemporaneously with the ALC as a separate protocontinent or, alternatively, the Mesoarchean basement of the northwest Superior Craton may be an extension of the ALC.

Ancient rocks occur elsewhere around the margins of the Superior Craton (see Bickford et al., this volume and O'Neil et al., this volume). These include a sliver of ca. 3825 Ma volcanic and sedimentary rocks in younger tonalite (ca. 3650 Ma) in northeast Quebec, Canada (David et al., 2004; O'Neil et al., this volume). At the southwest margin of the Superior Craton, the Minnesota River Valley terrane contains a suite of poorly exposed, but complex gneisses that formed at ca. 3.5 Ga and were metamorphosed and injected by tonalite at ca. 3.3–3.4 Ga. (Goldich and Hedge, 1974; Bickford et al., this volume). Neither of these areas has yet been found to contain crustal material ⩾3.9 Ga. Thus the ancient Paleoarchean zircon detritus with ⩾4.0 Ga Hf signatures in the ALC is unlikely to be derived from either of these sources. The only known location with exposed ⩾4.0 Ga crust is the Acasta Gneiss Complex of the Slave Craton (Fig. 6.2-1: Bowring and Williams, 1999; Iizuka et al., this volume). However, a large proportion of ALC zircon detritus is ca. 3.7 to 3.86 Ga (Fig. 6.2-6), an age period missing in the Acasta Gneiss.

ACKNOWLEDGEMENTS

Financial support for this project was provided by Lithoprobe and NSERC grants to L.M. Heaman. Field work was funded by the Manitoba Geological Survey. The author thanks T. Corkery and H. Zwanzig for many helpful discussions and B. and P. Lenton for support with drafting of the figures.

Earth's Oldest Rocks
Edited by Martin J. Van Kranendonk, R. Hugh Smithies and Vickie C. Bennett
Developments in Precambrian Geology, Vol. 15 (K.C. Condie, Series Editor)
© 2007 Elsevier B.V. All rights reserved.
DOI: 10.1016/S0166-2635(07)15063-5

Chapter 6.3

OLDEST ROCKS OF THE WYOMING CRATON

KEVIN R. CHAMBERLAIN[a] AND PAUL A. MUELLER[b]

[a]*Dept. of Geology and Geophysics, 1000 E. University, Dept. 3006, University of Wyoming, Laramie, WY 82071, USA*
[b]*Department of Geological Sciences, Box 112120, University of Florida, Gainesville, FL 32611, USA*

6.3-1. INTRODUCTION

The Wyoming Craton has long been considered one of the oldest cratons, largely on the basis of Nd and Pb isotopic compositions of widespread Meso- to Neoarchean plutonic and metasedimentary rocks, which have been interpreted to reflect Paleoarchean (3.2 to 3.6 Ga) and Eoarchean (3.6 to 4.0 Ga) sources (Wooden and Mueller, 1988; Frost and Frost, 1993; Frost, 1993; Frost et al., 1998; Mueller and Frost, 2006). Paleoarchean and Eoarchean rocks and minerals exist, but they are scarce and only locally preserved (Aleinikoff et al., 1989; Mueller et al., 1992, 1996; Langstaff, 1995). Several relatively large exposures of ca. 3.4–3.2 Ga crust occur in the northwestern Wyoming Craton (e.g., Mogk et al., 1992; Mueller et al., 1996, 2004) and in the Granite Mountain region of central Wyoming (Fisher and Stacey, 1986; Langstaff, 1995; Kruckenburg et al., 2001; Grace et al., 2006). A migmatite from the Wind River Range has also been interpreted to be 3.2 Ga (Aleinikoff et al., 1989) and xenocrystic cores of this age have been found in 2.95 to 2.85 Ga plutons from the Bighorn Mountains (Frost and Fanning, 2006). This chapter will argue that a relatively large Wyoming craton existed by at least 3.3 Ga, and possibly earlier. We have included absolute dates when possible; however, many of the major events in Wyoming province history have ages that overlap Era boundaries.

Archean basement rocks of the Wyoming Craton are exposed primarily in the cores of Cretaceous (Laramide) uplifts in Wyoming and southwestern Montana (Fig. 6.3-1). The older core of the craton abuts Proterozoic orogens in the northwest (Great Falls tectonic zone: O'Neill and Lopez, 1985; Mueller et al., 2002, 2005), southeast (Cheyenne belt: Karlstrom and Houston, 1984; Chamberlain, 1998), and east (Dakota segment of the Trans-Hudson Orogen: Hoffman, 1990; Redden et al., 1990; Dahl et al., 1999; Chamberlain et al., 2002; McCombs et al., 2004). An extensive Neoarchean orogen (ca. 2.67 Ga) comprises the western and southwestern part of the craton in the Teton and Wind River Ranges (Frost, B.R. et al., 2006) and terminates in a poorly understood region of mixed Archean and Proterozoic rocks along the western margin (Foster et al., 2006). Two ad-

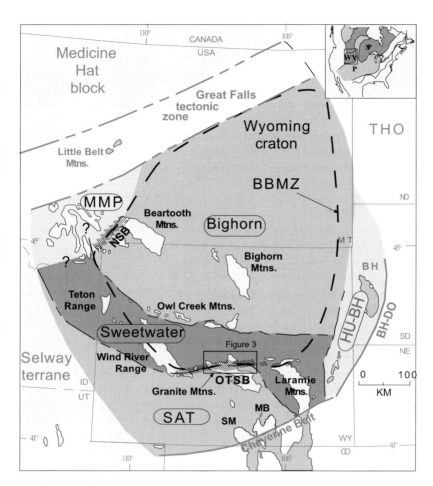

Fig. 6.3-1. Sketch map of Archean Wyoming craton exposed principally in Laramide (Cretaceous) uplifts shown in white, with tectonic subprovinces after Mogk et al. (1992), Chamberlain et al. (2003) and Frost et al. (2006b). Montana metasedimentary province (MMP) exposes some of the largest tracts of Paleoarchaean crust and was restitched to the rest of the craton along the Neoarchean North Snowy block deformation zone (NSB). Bighorn and Sweetwater subprovinces preserve evidence for plume-related and plate tectonics style growth mechanisms, respectively. Southern accreted terranes (SAT) were added to the craton along the 2.65–2.63 Ga Oregon Trail structural belt (OTSB). Hartville Uplift-Black Hills block (HU-BH) may have been exotic to the Wyoming craton until the Proterozoic. The Selway terrane has recently been recognized as separate from the Wyoming province (Foster et al., 2006). The Beartooth-Bighorn magmatic zone (BBMZ) is characterized by nearly coeval TTG and high-K, Meso- to Neoarchean plutonism. Evidence for crust greater than 3.2 Ga in age is restricted to the MMP, BBMZ, and portions of the Sweetwater subprovince. BH-DO, Black Hills and Dakotan orogenies; THO, Trans-Hudson orogen; SM, Sierra Madre; MB, Medicine Bow Mountains. Inset map: WY, Wyoming province, SP, Superior province, P, Proterozoic accreted terranes.

ditional Neoarchean collision zones have been recognized: (1) the North Snowy Block in southwestern Montana, interpreted as an intracratonic deformation zone restitching the Montana metasedimentary province (MMP) to the Wyoming Craton at ca. 2.55 Ga (Mogk et al., 1992; Mueller et al., 1996); and (2) the 2.65–2.63 Ga Oregon Trail structural belt in south-central Wyoming, interpreted as an accretionary front between the older core and relatively juvenile arcs and microcontinents to the south (Chamberlain et al., 2003; Frost, C.D. et al., 2006b; Grace et al., 2006). The Neoarchean collision zones along the western, southwestern, and south-central portions of the craton are within the Sweetwater tectonic subprovince (Chamberlain et al., 2003), a magmatically and tectonically reworked portion of the Paleoarchean craton with hallmarks of plate tectonic growth and processes (Frost et al., 1998; Chamberlain et al., 2003; Grace et al., 2006). The occurrences of rocks and minerals greater than 3.2 Ga are restricted to the Sweetwater subprovince and the cratonic core to the north and east, comprising the Beartooth-Bighorn magmatic zone and MMP (Chamberlain et al., 2003; Mogk et al., 1992; Mueller et al., 1993, 1994, 1996, 2004). Geophysical traces of the Great Falls tectonic zone and Dakota segment of the Trans-Hudson Orogen extend into southern Canada (Boerner et al., 1998; Lemieux et al., 2000), so it is possible that the ancient core underlies much of central and eastern Montana (Peterman, 1981; Carlson et al., 2004). Very little is known about this region, although crustal xenoliths from K-T aged intrusions and are being studied (Carlson et al., 2004; Downes et al., 2004; Facer and Downes, 2005).

6.3-2. KNOWN OCCURRENCES OF EO- AND PALEOARCHEAN ROCKS AND MINERALS

6.3-2.1. *Eoarchean Detrital Zircons and Cores*

There are only seven known occurrences of Eoarchean zircons in the Wyoming Craton (Fig. 6.3-2(a)). Three are from detrital grains in ca. 3.3 Ga to 2.7 Ga quartzites of the eastern Beartooth Mountains, western Beartooth Mountains, and Tobacco Root Mountains (Mueller et al., 1992, 1998), one as detrital grains in ca. 2.86 Ga quartzites of the Granite Mountains (Langstaff 1995), and three as xenocrysts in the Madison Range (Mueller et al., 1995), the Granite Mountains (Kruckenberg et al., 2001), and in the Wind River Range (Aleinikoff et al., 1989).

The Hellroaring plateau area of the eastern Beartooth Mountains has the highest abundance of Eoarchean detrital zircons identified so far in the Wyoming Craton. Although the dominant detrital zircon ages are between 3.2 to 3.4 Ga, six of eight quartzites studied contain Eoarchean grains, including two samples in which approximately 25% of the grains are >3.6 Ga (Mueller et al., 1998). Clusters at 3.7 Ga and 3.9–4.0 Ga document survival of some of the oldest detrital zircons in North America and permit possible linkages between the Wyoming Craton and known potential sources for these ages, such as the North Atlantic and Slave Cratons, respectively (Mueller et al., 1992, 1998). The dominance of

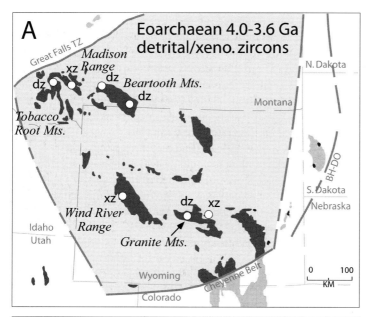

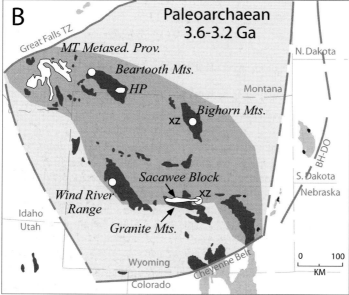

Fig. 6.3-2. (A) Locations of Eoarchean detrital zircons (dz) and xenocrystic zircon cores (xz) in the Wyoming province. (B) Known occurrences of Paleoarchean rocks and minerals. Relatively large tracts occur in the Montana metasedimentary province, Hellroaring plateau (HP) and Sacawee blocks. Darker shading is region where Meso- to Neoarchean plutons have Nd and Pb isotopic evidence for involvement of Paleoarchean crust.

ca. 3.3–3.4 Ga grains in these samples (over 50%) argues for a relatively proximal source of this age at the time of deposition and is used to argue for depositional ages of 2.7–3.2 Ga (Mueller et al., 1992, 1998). Lu-Hf chondritic model ages of 3.3–3.4 Ga from bulk detrital zircons in the quartzites (Stevenson and Patchett, 1990) suggest a major, relatively primitive magmatic pulse produced much of the 3.2–3.4 crust from which these grains were eroded.

The second occurrence of Eoarchean detrital zircons comes from the MMP, in the northwesternmost portion of the craton, where extensive occurrences of metapelite-quartzite-metacarbonate-amphibolite assemblages are structurally interleaved with 3.3–3.1 Ga quartzofeldspathic gneisses (Mogk et al., 1992, 2004; Mueller et al., 1993). Two samples of quartzite have been analyzed for detrital zircon ages. The dominant populations are between 3.2–3.4 Ga, similar to those from the eastern Beartooth Mountains (Mueller et al., 1998). One sample from the Tobacco Root Mountains also contains a single grain dated as 3.93 Ga. A single 3.7 Ga grain has been reported from a 3.3 Ga trondhjemite from the northern Madison Range (Mueller et al., 1995).

In the central part of the Wyoming Craton, Eoarchean detrital zircons have been identified from the basal quartzite in the ca. 2.86 Ga Barlow Gap Group, preserved in the Granite Mountains (Langstaff, 1995; Kruckenberg et al., 2001; Fruchey, 2002). The basal quartzite is in depositional contact (Sutherland and Hausel, 2002) with the ca. 3.26 Ga Sacawee orthogneiss (Langstaff, 1995; Fruchey, 2002). The combined detrital zircon spectra from three samples of this quartzite are dominated by 3.3 Ga grains (61 of 70 analyzed), with a few 3.2–3.1 Ga grains (Langstaff, 1995; Grace et al., 2006), consistent with the local distribution of gneissic and plutonic rocks (Grace et al., 2006). One of these samples had two grains that are 3.44 to 3.48 Ga in age and a single grain that is 3.83 Ga (Langstaff, 1995). The 3 grains greater than 3.3 Ga have no known local sources, although ca. 3.5 Ga trondhjemitic gneiss is preserved in the northwestern Beartooth Mountains and Madison Range, approximately 400 km to the north and west (Mueller et al., 1995, 1996).

The UC Ranch tonalitic orthogneiss, which crops out twenty kilometers east of the Granite Mountain detrital zircon locations, contains Eoarchean xenocrystic zircon cores. The zircon U-Pb systematics of this rock are complex, with the dominant population yielding ages of about 3.3 Ga, but individual grains yielded nearly concordant ID-TIMS dates of 3.42, 3.48, 3.59 and 3.80 Ga (Kruckenberg et al., 2001; Kruckenberg, unpubl.). The magmatic age is interpreted to be 3.3 Ga, based on the dominance of 3.3 Ga zircons, and the older grains are interpreted to be xenocrystic.

Xenocrystic Eoarchean zircon cores also exist in a ca. 3.2 Ga granulite-facies migmatite from the southwestern Wind River Range (Aleinikoff et al., 1989). SHRIMP analyses of the cores of complexly zoned zircons revealed 3.8 and 3.65 Ga domains. The 3.8 Ga core has extremely low U concentrations (34 and 25 ppm in two spots) and high Th/U ratios (0.70 and 0.74) and may have formed in a mafic igneous rock or high-grade metamorphic rock (Aleinikoff et al., 1989).

6.3-2.2. Paleoarchean Rocks of Southwestern Montana; The Montana Metasedimentary Province (MMP)

The mountain ranges of southwestern Montana (e.g., Highland, Ruby, Blacktail, Madison, and Gallatin Ranges and the Tobacco Root Mountains) provide exposures of one of the largest occurrences of Paleoarchean (3.6–3.2 Ga) rocks in the Wyoming Craton (Figs. 6.3-1, 6.3-2(b)). Collectively, these exposures comprise the Montana metasedimentary province (MMP: Mogk et al., 1992; Mueller et al., 1996), which is composed largely of quartzofeldspathic gneiss with distinctive metasupracrustal assemblages containing metapelitic rocks and substantial thicknesses of marbles with lesser quantities of metamorphosed iron formation, and quartzites (Wooden et al., 1988a, 1988b; Mogk and Henry, 1988; Mogk et al., 1992, 2004; D'Arcy and Mueller, 1992; Mueller et al., 1993). SHRIMP U-Pb dates of individual zircons from selected quartzites (e.g., Mueller et al., 1998) indicate principally Paleoarchean rocks in the provenance, but with ages up to 3.93 Ga. Tonalitic to trondhjemitic compositions dominate the oldest gneisses, which are interlayered with metasedimentary rocks and metabasites (Mogk et al., 1992). Direct igneous dates on the older gneisses are in the range 3.3–3.5 Ga, but have inherited grains up to 3.7 Ga (Mueller et al., 1993, 1995, 2004; Krogh et al., 1997). They were metamorphosed to granulite facies and intruded by high-Al trondhjemites at 3.1–3.25 Ga (Mogk et al., 1992).

The MMP gneisses share the distinctive Pb-isotopic characteristic of the rest of the Wyoming Craton; specifically, a high $^{207}Pb/^{204}Pb$ for given $^{206}Pb/^{204}Pb$ (Wooden et al., 1988a, 1988b; Wooden and Mueller, 1988), and have been interpreted to be part of the craton even though they are separated by a ca. 2.55 Ga deformation zone in the North Snowy block of the western Beartooth Mountains (Mogk et al., 1992; Mueller et al., 1996). The deformation zone in the North Snowy block is interpreted as an intracratonic feature (Mogk et al., 1992) and appears to have been a significant lithospheric structure on account that it delineates the northwestern limit of Meso- to Neoarchean magmatism associated with the Beartooth-Bighorn magmatic zone (Mueller et al., 1996). The deformation zone may have formed as early as 3.5 Ga and certainly by 3.25 Ga, as the geologic histories on either side diverge significantly after this time (Mogk et al., 1992; Mueller et al., 1996).

6.3-2.3. Paleoarchean Rocks of the Beartooth Mountains

The crystalline basement rocks of the Beartooth Mountains are dominated by Meso- to Neoarchean metaplutonic rocks of the Beartooth-Bighorn magmatic zone (BBMZ), but several enclaves of Paleoarchean and limited Eoarchean rocks are preserved (Mogk et al., 1992; Mueller et al., 1996, 2006). The best documented ancient gneiss in the Wyoming Craton is a 3.50 Ga trondhjemitic unit exposed in the North Snowy block of the northwestern Beartooth Mountains (Mueller et al., 1996). This rock is interlayered with amphibolite at a meter scale and structurally sandwiched between a greenschist-facies schist unit and an overlying amphibolite-facies nappe. SHRIMP U-Pb data from the gneiss define three main age clusters, with magmatic growth interpreted at 3.50 Ga and two pulses of metamorphic growth at 3.25 and 2.55 Ga (Mueller et al., 1996). The Nd model age of the gneiss is 3.87 Ga and has been interpreted to indicate that the trondhjemitic melt was derived from Eoarchean or Hadean mafic crust (Mueller et al., 1996). Zircons from an amphibolite in the overlying nappe have simple U-Pb systematics and yield an age of 3244 ± 12 Ma (Mueller et al., 1996). Titanite from this rock yield an age of 2.55 Ga, strengthening the interpretation of contraction and metamorphism in the North Snowy block at that time (Mogk et al., 1988, 1992).

Paleoarchean and Eoarchean enclaves are dispersed throughout the rest of the Beartooth Mountains. The best studied and exposed examples occur in the Hellroaring Plateau and Quad Creek regions of the easternmost Beartooth Mountains (Fig. 6.3-2(b): Henry et al., 1982; Mueller et al., 1985, 1992, 1998, 2006; Wooden et al., 1988a, 1988b; Mogk and Henry, 1988). These rocks are primarily metasupracrustal assemblages of pelitic schists, quartzites (including those described above with Eoarchean detrital zircons), metamorphosed iron formation, mafic granulites, meta-ultramafic rocks, and granitic, tonalitic and trondhjemitic gneisses (Mogk et al., 1992). There are no direct U-Pb dates on the metasupracrustal rocks, but detrital zircon data establish a maximum age of deposition of 3.3 Ga (Mueller et al., 1992, 1998). Granulite-facies metamorphism is constrained at 3.25–3.10 Ga, based on Rb-Sr and Pb-Pb isotopic data (Henry et al., 1982; Wooden et al., 1988a, 1988b; Wooden and Mueller, 1988; Mogk et al., 1992).

6.3-2.4. *Paleoarchean Rocks of the Granite Mountains, the Sacawee Block*

The Granite Mountains of south-central Wyoming expose a large tract of 3.1 to 3.3 Ga granitic and tonalitic orthogneiss. The Paleoarchean age of the gneiss was first determined by Peterman and Hildreth (1978). Subsequent mapping, geochronology, and isotopic and chemical analyses have established that the Sacawee orthogneiss, a distinctive, 3.26 Ga, Kspar-megacrystic granitic orthogneiss crops out over a 700 km^2 area (Langstaff, 1995; Kruckenberg et al., 2001; Fruchey, 2002; Grace, 2004; Meredith, 2005; Grace et al., 2006). Several older gneisses have been identified including the UC Ranch orthogneiss (Langstaff, 1995; Kruckenberg et al., 2001; Fruchey, 2002), and all of these Paleoarchean units have been cut by felsic and mafic dykes dated between 2.96 and 3.10 Ga (Kruckenberg et al., 2001; Grace et al., 2006). The Sacawee orthogneiss is locally an augen gneiss and is strongly deformed over most of its exposure. Discrete east-west striking, high-strain zones occur near its margins (Fig. 6.3-3). These shear zones are parts of the Oregon Trail structural belt, and the timing of deformation has been directly dated to 2.63–2.65 Ga in several locations (Grace et al., 2006; Frost, C.D. et al., 2006b). The shear zones are interpreted to be either foreland structures related to accretion of juvenile arc terranes to the south (Grace et al., 2006; Frost, C.D. et al., 2006b), or Neoarchean sutures between the Sacawee block and the rest of the Wyoming Craton to the north (Grace, 2004; Meredith, 2005; Grace et al., 2006).

The Sacawee block has Nd isotopic characteristics that could link it to the Paleoarchean Wyoming Craton, and minor felsic and mafic intrusions coeval with the Beartooth-Bighorn

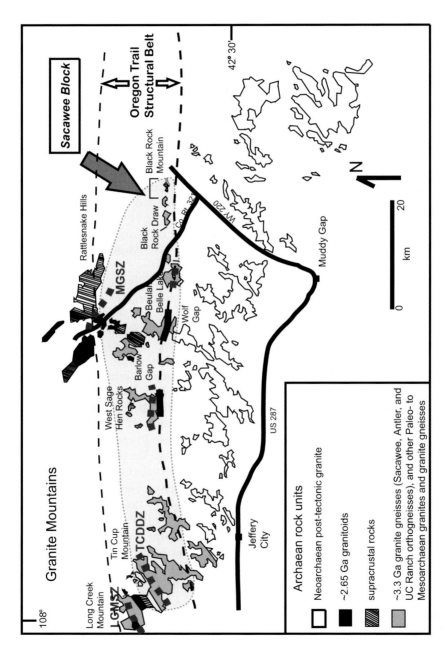

Fig. 6.3-3. Detailed map of Sacawee block from Grace et al. (2006). Long Creek Mountain shear zone (LCMSZ), Tin Cup ductile deformation zone (TCDDZ; Meredith, 2005) and McDougal Gulch shear zone (MGSZ; Langstaff, 1995; Fruchey, 2002) are discrete shear zones within the 2.65–2.63 Ga Oregon Trail structural belt (Chamberlain et al., 2003; Grace et al., 2006).

magmatic zone, which argue for a common history since the Mesoarchean at least, but an allochthonous origin can not be entirely ruled out (Grace et al., 2006).

There do not appear to be any Paleoarchean metasedimentary rocks preserved in the Sacawee block. Various quartzite units are infolded with the Sacawee orthogneiss, but these have been interpreted as basal Barlow Gap Group equivalent strata (ca. 2.86 Ga), based on similar field relations, detrital zircon age spectra, and Nd isotopic compositions (Grace et al., 2006).

The Sacawee orthogneiss is a biotite-bearing, medium to coarse-grained, magnesian, calcic to calc-alkalic granitic gneiss with anastomosing foliation and local augen texture (Langstaff, 1995; Fruchey, 2002; Grace et al., 2006). Alkali feldspar porphyroclasts are up to 3 cm in length and typical mineralogy is 5–15% quartz, 40–50% alkali feldspar, 30–40% plagioclase feldspar, and ~5% biotite. The Sacawee gneiss is high in silica (74%), peraluminous, with less than 4% Na, and LREE-enriched, with a negative Eu anomaly (Fruchey, 2002). It comprises approximately 50–60% of the basement gneiss exposed in the Sacawee block (Langstaff, 1995; Sutherland and Hausel, 2002; Fruchey, 2002). ID-TIMS U-Pb data of single air-abraded zircons yielded an age of 3258 ± 8 Ma, with slight inheritance indicated by one grain dated at ca. 3294 Ma (Fruchey, 2002). Contrary to the lack of evidence for inherited zircons, the Nd isotopic data of the Sacawee orthogneiss are extremely nonradiogenic and require that the crustal source of the Nd was 3.8 Ga, or older (Fruchey, 2002).

In the southeastern Sacawee block, the UC Ranch biotite tonalite crops out south of a chlorite-grade shear zone (Langstaff 1995; Kruckenberg et al., 2001). The UC Ranch orthogneiss is 70% plagioclase, 5–10% biotite with high-Na (4.5–5.5%), 64–70% silica, and a LREE-enriched pattern with no Eu anomaly (Kruckenberg et al., 2001), typical of Archean TTG (Martin, 1994; Grace, 2004). ID-TIMS U-Pb data of single air-abraded zircons yielded $^{207}Pb/^{206}Pb$ dates from 3.27–3.80 Ga and are interpreted to indicate magmatism ca. 3.3 Ga, with a variety of inherited grains (Kruckenberg et al., 2001). Similar tonalitic gneisses crop out south of a ca. 2.65 Ga shear zone 15 km east of the UC Ranch outcrops (Grace, 2004). These undated tonalites also have REE patterns typical of Archean TTG (Grace, 2004).

The rest of the basement gneiss in the Sacawee block is a complex of tonalitic to granitic gneisses and intrusions with several distinct lithologies that have not been mapped in detail (Langstaff, 1995). Contacts between these units are variable, with both intrusive and tectonic relationships locally (Fruchey, 2002). The Sacawee orthogneiss intrudes some of the units and is cut by others (Langstaff, 1995). SHRIMP U-Pb data from a tonalitic gneiss that is cut by the Sacawee orthogneiss yielded an age of 3371 ± 6 Ma (Langstaff, 1995). No inherited cores were detected in the 16 grains analyzed from this sample.

There does not appear to be any Paleoarchean metamorphism preserved in the Sacawee block. Peak metamorphic conditions recorded in the ca. 2.86 Ga Barlow Gap Group are upper amphibolite and are constrained to 2.83 Ga, based on metamorphic zircon dates (Fruchey, 2002; Kruckenberg et al., 2001). Most of the deformation of the Sacawee orthogneiss appears to be associated with the 2.65–2.63 Ga Oregon Trail structural belt, although there may be an earlier deformation at about 3.1 Ga (Grace et al., 2006).

6.3-3. ISOTOPIC EVIDENCE OF ANCIENT CRUST IN MESO- TO NEOARCHEAN
        PLUTONS AND SEDIMENTARY ROCKS

In addition to local occurrences of Eo- to Paleoarchean rocks and minerals, the Nd and
Pb isotopic compositions of widespread Meso- to Neoarchean TTG to high-K granodioritic rocks of the BBMZ (Figs. 6.3-4 and 6.3-5) have been interpreted to indicate an early
crustal history (Wooden and Mueller, 1988; Frost et al., 1998; Frost, C.D. et al., 2006a).
Plutonic rocks of the BBMZ dominate the basement exposures in the Beartooth, Bighorn
and western Owl Creek Mountains (Mogk et al., 1992; Mueller et al., 1996; Frost, C.D.
et al., 2006a). Similar aged rocks also crop out in the northern Wind River Range (Frost,
C.D. et al., 2006a), the Granite Mountains (Kruckenberg et al., 2001; Grace et al., 2006),
and possibly the northern Laramie Mountains (Johnson and Hills, 1976). The BBMZ rocks
have been dated to 2.8–3.0 Ga (Wooden et al., 1988; Kirkwood, 2000; Frost and Fanning,
2006; Frost, C.D. et al., 2006a). Nd model ages from the BBMZ rocks range from juvenile
(2.95–2.85 Ga) to as old as 3.4 Ga (Wooden and Mueller, 1988; Frost et al., 1998; Frost,
C.D. et al., 2006a). Sacawee block and MMP gneisses have less radiogenic Nd isotopic
compositions than these rocks at the time of BBMZ magmatism (Fig. 6.3-4) and could be
related to their crustal sources. The Sacawee block and MMP rocks have Nd model ages up
to 4.0 Ga (Grace et al., 2006) that are much older than their igneous ages and imply an earlier source for the Nd at 3.3 Ga (Frost, C.D. et al., 2006a; Grace et al., 2006). Nd isotopic
compositions of Neoarchean metasedimentary rocks have similarly non-radiogenic values

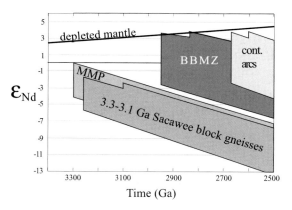

Fig. 6.3-4. Summary figure of Nd isotopic evolution for the Wyoming Craton, data sources are
Mueller et al. (1993), Wooden and Mueller (1988), Frost et al. (1998, 2006a), Grace et al. (2006).
The Montana metasedimentary province (MMP) and Sacawee block gneisses have unradiogenic
Nd compositions that require incorporation of sialic crustal components at least as old as 4.0 Ga.
Beartooth-Bighorn magmatic zone rocks (BBMZ) reflect mixtures of depleted mantle and sialic crust
similar to the older gneisses (Frost et al., 2006a). Neoarchean, calc-alkaline magmatism in the Wind
River Range that have been interpreted as continental arc plutons, also reflect mixtures of depleted
mantle and older crust (Frost et al., 1998).

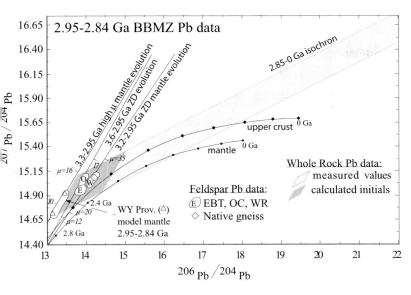

Fig. 6.3-5. Summary figure of Pb isotopic compositions for 2.95–2.84 Ga Beartooth-Bighorn magmatic zone rocks (BBMZ) adapted from Frost et al. (2006a), data from Wooden and Mueller (1988), Frost et al. (2006a). Zartman and Doe (1981) evolution lines (ZD) are for crustal reservoirs that separated from a Zartman and Doe model mantle at 3.6 and 3.2 Ga, respectively. Relatively high μs (15 to 30) are required for these crustal reservoirs to account for the measured Pb isotopic values of BBMZ rocks. The 'high-μ evolution' line tracks a crustal reservoir separated at 3.3 Ga from a high-μ Wyoming province model mantle. WY province model mantle values were calculated after Luais and Hawkesworth (2002) by single-stage growth from Canyon Diablo Pb isotopic values using two samples with depleted mantle Nd isotopic compositions (Table 6.3-1). Estimated μs for this mantle reservoir range from 8.50 to 8.66. The full range of measured and initial Pb isotopic compositions of BBMZ rocks can be accounted for by mixtures of mantle material from a Wyoming province high-μ reservoir at 2.95 to 2.84 Ga and crustal material separated from that mantle reservoir at 3.3 Ga. Reasonable μs of only 8 to 15 are required for the 3.3 Ga crustal reservoirs in this case.

and have been interpreted as evidence for early crust formation in the Wyoming Craton Frost, 1993; Grace et al., 2006).

One of the most distinctive isotopic characteristics of the Wyoming Craton is elevated $^{207}Pb/^{204}Pb$ for a given $^{206}Pb/^{204}Pb$. These values place these rocks consistently above the Zartman and Doe (1981) upper crustal Pb evolution curve (Fig. 6.3-5) and require separation from a Zartman and Doe (1981) model mantle by at least 4.0 to 3.6 Ga (Wooden and Mueller, 1988; Frost et al., 1998). The geographical extent of rocks that share this characteristic (Fig. 6.3-2(b)) include the MMP (Wooden et al., 1988), BBMZ (Wooden and Mueller, 1988; Frost, C.D. et al., 2006a) and Neoarchean rocks of the Laramie Mountains (Chamberlain and Bowring, 2001) and the Teton Range (Frost, B.R. et al., 2006). The Neoarchean rocks of the Sierra Madre and Medicine Bow Mountains have distinctly less radiogenic Pb isotopic compositions than the rest of the craton (Harper, 1997; Chamberlain

Table 6.3-1. Pb and Nd isotopic data from select BBMZ rocks

Sample	206Pb/204Pb	207Pb/204Pb	208Pb/204Pb	U, ppm	Pb, ppm	^{238}U/^{204}Pb	initial 206Pb/204Pb	initial 207Pb/204Pb	initial εNd
Quartzofeldspathic gneisses of central and southern Bighorn Mountains (initials calculated at 2.95 Ga)									
95LH3 wr	16.185	15.377	41.289	1.16	17.0	4.36	13.657	14.832	3.7
Model mantle[a]						8.50	13.145	14.721	
96LH3 wr	16.464	15.414	43.176	0.81	10.0	5.33	13.373	14.746	1.9
Granitic rocks of Bighorn batholith (initials calculated at 2.85 Ga)									
96BH1 wr	17.080	15.556	46.027	0.84	9.0	6.40	13.524	14.834	−1.2
96BH8 wr	14.647	15.046	41.678	0.23	6.3	2.27	13.383	14.790	−1.9
Tonalite from Western Owl Creek Mtns. (2.84 Ga in age)									
98OC4 plag.	13.884	15.009	33.517				13.447	14.921	3.7
Model mantle[a]						8.66			

Notes: Data from Frost et al. (2006). Abbreviations: wr, whole rock; plag., plagioclase.
[a]Wyoming province model mantle reservoir values calculated after Luais and Hawkesworth (2002) using whole rock Pb isotopic values from samples with near mantle Nd values (95LH3 and 98OC4) and single stage growth from Canyon Diablo meteorite values at 4.57 Ga to crystallization age of sample.

et al., 2003), which is one of the reasons the Medicine Bow Mountains and Sierra Madre basement were interpreted as parts of an exotic block accreted in the Neoarchean (Chamberlain et al., 2003). This model has been strengthened by subsequent Nd analyses and has led to the concept of the Southern Accreted Terrane (Souders and Frost, 2006; Frost, C.D. et al., 2006b).

At face value, the elevated $^{207}Pb/^{204}Pb$ values in rocks of the northern and central Wyoming Craton support the incorporation of Eoarchean crust in Meso- to Neoarchean plutons based on interpretations of Nd data. However, many of the BBMZ and Teton Range rocks with mantle-like, juvenile Nd isotopic compositions also display elevated $^{207}Pb/^{204}Pb$ (Table 6.3-1, Fig. 6.3-5: Frost, C.D. et al., 2006a; Frost, B.R. et al., 2006), so at least some of the Pb isotopic characteristics reflect an evolved mantle composition (e.g., Mueller and Wooden, 1988; Frost, C.D. et al., 2006a) that may indicate an episode of early metasomatism as the means of enrichment (Mueller and Wooden, 1988; Wooden and Mueller, 1988). The Wyoming Craton mantle could have acquired its unique composition through incorporation of Hadean to Eoarchean crustal components via subduction and metasomatism in an Archean mantle wedge analog (e.g., Mueller and Wooden, 1988; Wooden and Mueller, 1988).

Mueller and Wooden (1988) also recognized that the Wyoming Pb isotopic signature was distinctly enriched compared to other Archean cratons, and proposed a model that attributed these differences to bulk composition and metamorphic history, with the most depleted signatures related to cratons with early LILE-depleting metamorphic histories (e.g., West Greenland). The presence of the distinctive ^{207}Pb-rich Wyoming signature in Neoarchean rocks was attributed to lack of an early LILE-depleting metamorphic history. More recently, Luais and Hawkesworth (2002) proposed that an early mantle-altering event, such as the separation of the Moon, produced heterogeneous mantle reservoirs with different U/Pb concentrations (but little changed Sm/Nd) in the Hadean, which led to subsequent high- and low-μ ($^{238}U/^{204}Pb$) cratons (see Kamber, this volume). The Slave, Wyoming, Yilgarn, Pilbara, Zimbabwe, Kaapvaal, and northeastern India are high-μ cratons; Greenland, Superior, and southwestern India are low-μ cratons (see also Kamber et al., 2003). For many of these cratons, the characteristic high- or low-μ reservoirs were tapped repeatedly in later times and appear to have accompanied the cratons as mantle keels (Luais and Hawkesworth, 2002).

Regardless of location and formation mechanism, the high-μ reservoir must have formed in the Hadean to account for the low $^{206}Pb/^{204}Pb$ of the measured isotopic compositions. For example, assuming Canyon Diablo meteorite Pb isotopic compositions for initial ratios, evolution of Pb from 4.57 Ga to 2.95 Ga in a reservoir with a μ of 8.5 is required to produce the initial whole rock Pb isotopic compositions of 2.95 Ga BBMZ plutons from the Bighorn Mountains (Table 6.3-1), including one with mantle-like Nd compositions (Frost, C.D. et al., 2006a). Although it is possible that the U-Pb system has been slightly open in these whole rocks and that the data are over-corrected for growth, even the feldspar compositions from the western Owl Creek and eastern Beartooth Mountains require higher μ values than a Zartman and Doe (1981) model mantle by 3.6 Ga (Fig. 6.3-5). Even slight *in situ* growth of Pb in the feldspars, from a μ of less than 0.7, pushes the divergence from a bulk Earth mantle reservoir back to the Hadean. The Wyoming Craton mantle field shown in Fig. 6.3-5 is calculated from two samples with ages between 2.95 to 2.84 Ga and juvenile Nd compositions (Table 6.3-1: see also Frost, C.D. et al., 2006a) and reflects the evolution of a high-μ Wyoming Craton mantle during this period. Variation in measured Pb isotopic compositions with ^{207}Pb/^{204}Pb above these mantle values could be explained by addition of crustal components that are 3.3 Ga and later, which evolved in reservoirs with reasonable μ values of 8–15 (Fig. 6.3-5). In the high-μ mantle model for Pb isotopic evolution, there is no need for significant crustal material older than 3.3 Ga to explain the Pb isotopic values of the Wyoming Craton, although >3.3 Ga crust has been documented. If mantle enrichment did occur independent of crust formation, it would represent a unique aspect of Earth's early chemo-dynamics.

The Nd isotopic compositions of Meso- to Neoarchean metasedimentary rocks have been interpreted to reflect Paleo- to Mesoarchean sources (Frost, 1993), because many of these rocks have Nd model ages of 3.3–3.4 Ga and some as high as 3.8 Ga. Although Sm-Nd isotopic compositions of sediments give a first-order approximation of the crustal residence ages of the Nd sources (e.g., O'Nions et al., 1983; Goldstein et al., 1984), they do not necessarily reflect the average age of the detritus or the age of exposed crust at the time of deposition. Ca. 2.86 Ga quartzites from the Sacawee block yielded Nd model ages of 3.8 Ga, yet detrital zircon dates of these same samples are dominated by 3.3 Ga grains (32 out of 38 analyzed), with the rest between 3.2 to 3.1 Ga (Grace et al., 2006). There was no evidence for older cores within the analyzed zircons. It is likely that the detrital zircons are derived from the underlying ca. 3.3 Ga Sacawee orthogneiss, which also has a Sm-Nd model age of 3.8 Ga, and that the Nd budget of the quartzite is largely inherited from erosional products of the Sacawee block (Grace et al., 2006). In this case, the Nd and detrital zircon data reflect a complex history with an early stage of erosion of 3.8 Ga sialic crust liberating unradiogenic Nd, followed by a magmatic stage at 3.3 Ga that incorporated this Nd into the Sacawee orthogneiss, before erosion of the Sacawee block and deposition in the analyzed quartzite. An alternative path for the Nd, namely assimilation of 3.8 Ga felsic rocks in the Sacawee source region, seems unlikely given the lack of evidence for any xenocrystic 3.8 Ga zircon cores. 3.8 Ga mafic crust would have a mantle-like Nd evolution, due to limited Sm/Nd fractionation, and would not produce the unradiogenic values required in the Sacawee source at 3.3 Ga. With a minimum of 400 My between the two periods of erosion ($\geq$3.3 Ga and 2.86 Ga), there could be a great deal of distance between the 3.8 Ga Nd source and the Wyoming Craton at 2.86 Ga, during deposition.

6.3-4. DISCUSSION

Based on the abundant, geographically widespread evidence for 3.3 Ga crust in the Wyoming Craton, we postulate that a major craton (possibly 200,000 km^2) existed at 3.3 Ga, stretching from the Oregon Trail structural belt in south-central Wyoming to the Great Falls Tectonic zone and Dakotan segment of the Trans-Hudson Orogen (Fig. 6.3-6). Even though the two major exposed terranes of that age – the MMP and Sacawee block –

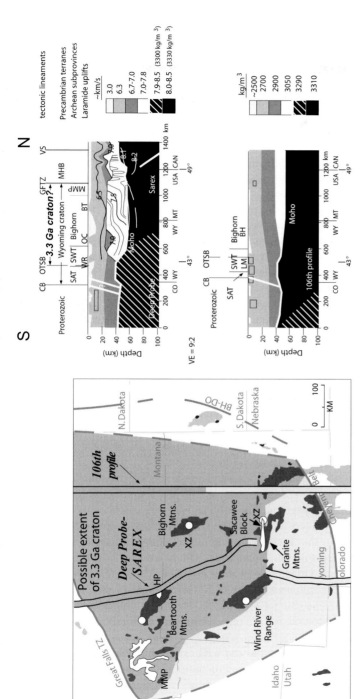

Fig. 6.3-6. Left map figure depicts proposed extent of 3.3 Ga craton in dark shade in the northern Wyoming Craton along with known occurrences of rocks and minerals of that age. Dark shading in the traces of Deep Probe-SAREX seismic transect and 106th longitude gravity profile denotes the occurrence of fast lower crust at 40 km depth. Right figures display the results of geophysical studies after Snelson et al. (1998), Gorman et al. (2002), Chamberlain et al. (2003) and demonstrate possible correlation between fast lower crustal layer and the extent of ca. 3.3 Ga crust.

are bound to the rest of the province by Neoarchean shear zones, both have isotopic characteristics that permit Wyoming Craton ancestry, as do intracratonic interpretations for the deformation (Mogk et al., 1992; Grace et al., 2006). Furthermore, there is considerable evidence for 3.3 Ga crust within the central ancient core, including xenocrystic 3.25 Ga zircon cores in the Bighorn Mountains (Frost and Fanning, 2006) and ca. 3.2 Ga granulite-facies metamorphism and migmatization in the Wind River Range (Aleinikoff et al., 1989) and eastern Beartooth Mountains (Henry et al., 1982; Wooden et al., 1988a, 1988b; Wooden and Mueller, 1988; Mogk et al., 1992). The prevalence of 3.3–3.4 Ga detrital zircons in the Hellroaring plateau quartzites of the eastern Beartooth Mountains argues for relatively proximal sources of that age at the time of deposition at ca. 2.7–3.3 Ga (Mueller et al., 1998), although the extent of sedimentary recycling cannot be sufficiently constrained to make this a certainty.

Although Nd isotopic evidence requires incorporation of 3.8–3.6 Ga sialic crustal components in the source regions of later magmatic rocks, it is unclear whether this implies Paleoarchean rocks in the magma source regions, Paleoarchean detritus incorporated via subduction, or ancient enriched mantle that was not directly (spatially) associated with continental crust. Incorporation of Paleoarchean crustal components could also explain the radiogenic Pb isotopic signatures of most of the Wyoming Craton plutonic rocks, except for those samples with mantle-like Nd at 2.95–2.85 Ga that still carry elevated $^{207}Pb/^{204}Pb$. These samples are better explained by derivation from a Hadean-aged mantle reservoir with high μ. Derivation from a high-μ, Hadean-aged, basaltic proto-crust (e.g., Kamber et al., 2003; Kamber, this volume) could explain the Pb isotopic characteristics and juvenile Nd, because the Sm/Nd ratio would not be expected to fractionate as significantly in basalt formation, but this source would not explain the wide variation in Nd compositions and old Nd model ages of Mesoarchean plutons. The combination of wide variation in Nd isotopic systematics with consistently elevated $^{207}Pb/^{204}Pb$ and presence of Eoarchean detrital zircons argues for at least two sources for the BBMZ plutons; Eoarchean sialic crust and high-μ mantle.

Consequently, we postulate that the Wyoming Craton may have formed largely at about 3.3 Ga on the flanks of an Eoarchean craton, such as the Slave Craton, which was also derived from a high-μ mantle reservoir. The modern Wyoming Craton received Eoarchean detrital zircons from its predecessor nucleus, or neighboring craton, which shared the high-μ mantle reservoir. In this scenario, a significant portion of the Nd budget would need to be dominated by detritus from Eoarchean crust, whereas much of the rock forming material was Paleoarchean and mantle-derived, principally from 3.3 Ga sources. Subducted detritus has exerted a strong influence on the Nd characteristics of plutons in modern arc systems, such as the Aleutians (Singer et al., 1992), and a similar process may have operated for the Wyoming Craton ca. 3.3 Ga (e.g., Wooden and Mueller, 1988; Mueller and Wooden, 1988).

By 3.0–2.85 Ga, the Wyoming Craton was well established, with the widespread plutonism of the BBMZ either stitching older microcontinents, or modifying an older 3.3 Ga craton. Crustal growth at this time is interpreted as Archean-style, with broad plutonic blooms rather than linear belts more typical of later arc systems (Frost et al., 2006a;

Mueller et al., 2006). Potential models include flat-slab subduction of thick mafic crust (e.g., Smithies et al., 2003), or plume-related differentiation such as the Bédard (2006) model that invokes catalyzed delamination and the coupled genesis of Archean crust and sub-continental mantle lithosphere. The presence of nearly coeval TTG and high-K granitic plutonism at 3.3 and 3.0–2.86 Ga suggests multiple pulses of magmatic growth by similar processes, such as predicted by the delamination model of Bédard (2006), which involved pre-existing crust and lithosphere. For the Wyoming Craton, these pulses may have occurred at 3.5, 3.3 and 3.0–2.86 Ga, based on age distributions of preserved rocks.

There also may be geophysical expression of the ancient, 3.3 Ga craton in the form of a 15–25 km thick, seismically-fast, lower crustal layer imaged beneath the Bighorn subprovince by the Deep Probe experiment (Fig. 6.3-6: Snelson et al., 1998; Gorman et al., 2002; Chamberlain et al., 2003). This is arguably a Mesoarchean feature, based on the apparent geographical correlation with BBMZ plutonism (Chamberlain et al., 2003; Frost, C.D. et al., 2006a), but it may also represent the distribution of 3.3–3.1 Ga and earlier crust that was modified during the Neoarchean. The fast lower crustal layer has been imaged all the way north to the Great Falls tectonic zone and gravity models permit it to extend to at least eastern Wyoming and Montana (Fig. 6.3-6: Snelson et al., 1998). The distribution of the fast, lower crustal layer strengthens our interpretation that the ancient craton may exist beneath much of central and eastern Montana.

6.3-5. CONCLUSIONS

Eoarchean and Paleoarchean rocks and minerals are relatively scarce in the Wyoming Craton. However, isotopic data for Meso- and Neoarchean rocks and geophysical evidence are interpreted to indicate that a significant craton (at least 200,000 km^2) existed by ca. 3.3 Ga. In particular, enriched Pb isotopic signatures for widespread Mesoarchean and Neoarchean rocks distinguish the Wyoming Craton from many other Archean cratons and suggest that these rocks formed to a significant degree from an enriched, high-μ mantle reservoir that initially developed in the Hadean. Crustal growth mechanisms in the Paleoarchean and Mesoarchean appear to be characterized by episodic, broad magmatic blooms with coeval TTG and high-K plutonism at 3.5, 3.3 and 3.0–2.84 Ga. Structures and compositions characteristic of lateral accretion and arc magmatism typical of modern plate tectonic processes are not clearly evident until 2.67–2.63 Ga and are confined to the southern part of the craton.

Earth's Oldest Rocks
Edited by Martin J. Van Kranendonk, R. Hugh Smithies and Vickie C. Bennett
Developments in Precambrian Geology, Vol. 15 (K.C. Condie, Series Editor)
© 2007 Elsevier B.V. All rights reserved.
DOI: 10.1016/S0166-2635(07)15064-7

Chapter 6.4

THE OLDEST ROCK ASSEMBLAGES OF THE SIBERIAN CRATON

OLEG M. ROSEN[a] AND O.M. TURKINA[b]

[a]*Geological Institute of the Russian Academy of Sciences (RAS), Pyzhevsky per. 7, Moscow, 019017, Russia*
[b]*Institute of Geology and Mineralogy, United Institute of Geology, Geophysics and Mineralogy, Siberian Branch of RAS, UIGGM, Koptyug Avenue 3, Novosibirsk, 630090, Russia*

6.4-1. INTRODUCTION

The Siberian Craton (Fig. 6.4-1), which occupies an area of approximately 4×10^6 km^2, is overlain largely (70%) by Mesoproterozoic (Riphean) to Phanerozoic sedimentary rocks, whose thickness ranges between 1 and 8 km, averaging approximately 4 km thick. The Siberian Craton consists of large geologic elements (tectonic provinces, composite terranes or superterranes), including component, primordial microcontinents that accreted at ∼2.6 and 1.8 Ga to form a single craton. The oldest rock assemblages in the craton are geographically located in the Aldan and Anabar shields and in the Sharyzhalgay uplift in the far southwestern part of the craton (Fig. 6.4-1(a)), within the Tungus, Aldan and Anabar geological provinces (Fig. 6.4-1(b)). The following provinces are distinguished within the craton: the Tungus granite–greenstone province in the west; the chiefly granulite-facies Aldan and Stanovoi provinces in the southeast; the Anabar province, which is composed chiefly of granulite–gneiss terranes; and the Olenek granite–greenstone province in the northeast (Fig. 6.4-1(b)). The craton is bounded to the north and south by what are essentially Phanerozoic sedimentary fold belts (Taimyr and Verkhoyansk fold belts), whereas volcanic orogenic belts, such as the Paleozoic Central Asian belt and the Mesozoic Mongolia–Okhotsk belt, bound the craton to the south-west and south, respectively.

The antiquity of the rocks within the terranes has been identified through lithological mapping and geochronological studies. U-Pb zircon dates frequently conform to Sm-Nd residence time estimations, and both methods were used to date the rock assemblages and the events that affected them soon after. As outlined in more detail below, the geochronological data yield information that is sufficient to develop the following sequence of geological events:

(1) the first appearance of crust;
(2) later reworking, accompanied by intrusions; and
(3) erosion, marked by deposition of sedimentary rocks.

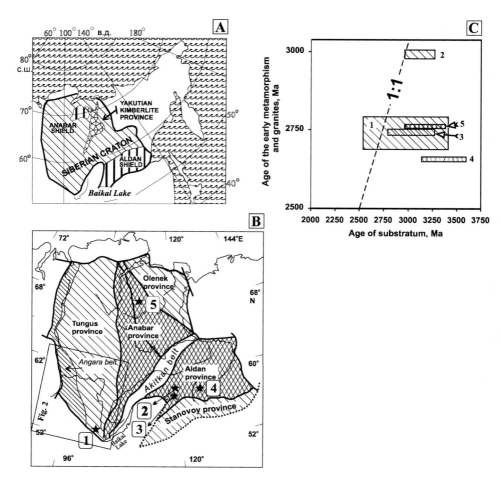

Fig. 6.4-1. The oldest rock assemblages in the Siberian Craton: (A) location of the craton in NE Asia; (B) component elements of the Siberian Craton (Rosen et al., 1994; Rosen, 2003). Rectangle indicates area of Fig. 6.4-2; (C) Age relationships between the age of the oldest rocks and their first metamorphism (mainly granulite facies, accompanied by granite (charnockite) anatexis) in different parts of the Siberian Craton: 1 – Sharyzhalgay uplift polymetamorphic assemblage and greenstone belts, basement of the Tungus granite–greenstone terrane; 2–3 – Olekma granite–greenstone terrane (2 – Tungurcha greenstone belt; 3 – Olondo greenstone belt); 4 – Aldan granulite–gneiss terrane; 5 – Daldyn granulite–gneiss terrane.

Numerous schematic tectonic maps of the Siberian Craton are based on an integrated analysis of predominantly geological and/or geophysical data (see an early review by Yanshin and Borukaev, 1988, and references therein). The concepts of early Precambrian ophiolites (oceanic crust) (Moralev, 1986) and so called "nuclears" (Glukhovsky, 1996;

Moralev and Glukhovsky, 2000) were advanced, and Proterozoic accretion of Archean crustal blocks was presumed (Zonenshain et al., 1989). Isotope age determinations helped to derive historic-geological models of the craton's evolution (see review in Rosen et al., 1994; Rosen, 2002). Following Hoffman (1988), the idea that heterogeneous terranes of various ages represent the principal elements of the craton's structure was advanced; these terranes initially originated and subsequently evolved as independent sialic masses (microcontinents), which later formed a single craton in the process of collision. It was known previously that the basement of the Siberian Craton consists of two principal types of lithologic assemblages, namely, granulite–gneiss and granite–greenstone domains (Petrov et al., 1985). The latest data (Rosen et al., 1994) have shown that these domains have various ages and represent terranes of corresponding compositions up to several tens to hundreds of kilometers across.

Granulite–gneiss terranes consist mainly of orthopyroxene-plagiogneiss (enderbite), two-pyroxene mafic granulites (presumably metamorphosed island-arc volcanics), and metamorphosed carbonate rocks and orthopyroxene-bearing quartzites. All of these rocks are the products of granulite-facies metamorphism. They are deformed into narrow (a few kilometers wide) isoclinal folds that often lie within large antiforms and synforms. These folds are distinctly traceable under the sedimentary cover owing to their high positive linear magnetic fields (ΔTa) and elevated gravity values, which were noted, for instance in Genshaft (1996).

Granite–greenstone terranes include broad, isometric, amoeboid granitoid domains between which linear greenstone belts are squeezed. These belts, which occupy 10–20% of the area for instance in the Aldan Shield (Popov et al., 1990), consist mainly of mafic volcanics and graywackes (greenstone belts) that have been metamorphosed under greenschist and/or amphibolite facies conditions. The granitoid domains are rather distinctly identified from weak negative, nonlinear, and mosaic-like magnetic fields and by low gravity fields, which indicate large buried granite plutons (Khoreva, 1987). In contrast, the linear, variously-oriented greenstone belts and/or mafic and ultramafic intrusions are distinguished by their elevated positive magnetic fields (Gafarov et al., 1978). In the typical areas, a comparison between these genotypes reveals distinct differences between the former and the latter varieties of terrane. It should be noted that the granulite–gneiss terranes (predominantly, granulites) originated in the lower crust, whereas the granite–greenstone terranes are upper-crustal rocks consisting mostly of granitoids. Although theoretically both may crop out together on the erosion surface of a single terrane, the conventional binary subdivision is viewed as a necessary simplification and a useful research tool until more geological and geochronological data are obtained.

Collision zones (suture or clash zones) are the faults that stitch together these terranes and show evidence of tectonic compression and thrusting. Another indicator of their collisional origin (Rosen et al., 2000) and the synchroneity of local metamorphism and granite formation in the collision zones versus the regional nature of metamorphism in the adjacent terranes. This synchroneity probably occurs only as a result of spontaneous heating (thermal relaxation) within thickened crust in a collision zone, based on geological observations (Rosen and Fedorovsky, 2001), and consistent with the modeling of England and

Thompson (1984). The present-day structure of the craton is viewed as the result of the collision and accretion (amalgamation, adhesion) of primordial microcontinents of various ages that were transformed into heterogeneous tectonic blocks, or terranes, in the process of collision (Rosen et al., 1994; Rosen, 2003; Smelov et al., 1998). The collision zones consist of various blastomylonites and tectonites, migmatites, and autochthonous granites, and enclose large allochthonous inliers of adjacent terranes, as well as anorthosite blocks (Lutts and Oksman, 1880; Rosen et al., 1994). These zones, which are a few hundred meters to 30 km wide, are distinctly traceable under the sedimentary cover over distances exceeding 1000 km due to their high, alternating-sign, linear magnetic fields.

The ages outlined in Rosen (2002) show that the age of magma supply to the crust from a mantle source is highly variable between the different terranes. Four groups of magma extraction ages are recognized, at ~3.5, 3.3, 3.0, and 2.5 Ga. Early granulite-facies metamorphism and granite formation in the craton occurred at about 2780–2660 Ma. Although these processes are observed sporadically, it is quite possible that they resulted from collision of continental masses to form a supercontinent (Condie, 1998), namely Pangea-0 (Khain, 2001). After the breakup of this supercontinent at about 2.1 Ga, separate microcontinents formed. Those later participated in accretion of the Siberian Craton in the late Paleoproterozoic. Island arcs formed in the Paleoproterozoic ocean at 1.9 Ga, and were deformed within orogens, including the Akitkan orogenic belt with an age of 2.2–1.8 Ga (Zonenshain et al., 1989; Neymark, 1998).

6.4-2. SHARYZHALGAY UPLIFT

6.4-2.1. Geology and Geochronology

The Sharyzhalgay basement uplift is the southernmost exposed part of the Tungus province of the Siberian Craton (Fig. 6.4-2). This uplift extends for 350 km from the Oka River in the west to the southern part of Lake Baikal. The southwestern boundary of the Sharyzhalgay uplift is the Main Sayan fault (shear zone), and in the northeast uplift is covered with Neoproterozoic-Phanerozoic platform sediments. From NW to SE, several NW- or north-trending suture zones and faults divide the Sharyzhalgay uplift into the Bulun (Erma), Onot, Kitoy and Irkut domains (Fig. 6.4-3). The suture zones are marked by slices and small tectonic blocks of high-pressure ultramafic-mafic and felsic upper mantle and lower crustal rocks (Gladkochub et al., 2001; Sklyarov et al., 2001; Turkina, 2001; Ota et al., 2004). The modern structure of the Sharyzhalgay uplift resulted from Paleoproterozoic collisional events that are constrained by dating of metamorphic zircons and their overgrowth rims in metamorphic rocks, as well as magmatic zircons from collisional granitoids within all domains. The available geochronological data are summarized in Table 6.4-1.

The Irkut and Kitoy domains are dominated by Archean high-grade gneisses, mafic and felsic granulites and amphibolites (Petrova and Levitsky, 1984). Their protoliths are basalt, andesite, and dacite volcanic rocks (Nozhkin and Turkina, 1993; Nozhkin et al.,

Table 6.4-1. Representative age estimations on the Sharyzhalgay uplift

Age, Ma	Method	Zircon type	Rock, complex	Locality*	Reference
Kitoy domain					
3347 ± 370	U-Pb, SHRIMP	Zircon, dark and bright cores	Orthopyroxene-biotite felsic granulite	1	Poller et al., 2005
2623 ± 32	U-Pb, SHRIMP	Metamorphic zircon	Orthopyroxene-biotite felsic granulite	1	Poller et al., 2005
2570 ± 30	U-Pb, SHRIMP	Zircon, old cores	Orthopyroxene-biotite intermediate granulite	1	Poller et al., 2005
2532 ± 12	U-Pb, conv.	Magmatic zircon	Kitoy granite	2	Gladochub et al., 2005
1880 ± 27	U-Pb, SHRIMP	Zircon, young rims	Orthopyroxene-biotite intermediate granulite	1	Poller et al., 2005
1855 ± 9	U-Pb, SHRIMP		Two-pyroxene mafic granulite	1	Poller et al., 2005
Irkut domain					
3390 ± 35	U-Pb, SHRIMP	Zircon, cores	Two-pyroxene intermediate granulite	3	Poller et al., 2005
2649 ± 6	U-Pb, conv.	Metamorphic zircon	Metagabbro	4	Salnikova et al., 2003
2562 ± 20	U-Pb, conv.	Magmatic zircon	Post-granulite granite	4	Salnikova et al., 2003
2557 ± 28	U-Pb, conv.	Magmatic zircon	Pegmatoid granite	5	Salnikova et al., 2003
1876 ± 47	U-Pb, SHRIMP	Zircon, rims	Two-pyroxene intermediate granulite	3	Poller et al., 2005
1866 ± 10	U-Pb, conv.	Granulite zircon	Enderbite	6	Salnikova et al., 2003
1855 ± 5	U-Pb, conv.	Magmatic zircon	Syenite	6	Salnikova et al., 2003
1853 ± 1	U-Pb, conv.	Magmatic zircon	Pegmatoid granite	6	Salnikova et al., 2003
Onot domain					
3386 ± 14	U-Pb, SHRIMP	Magmatic zircon	Trondhjemite gneiss	7	Bibikova et al., 2006
3415 ± 6	U-Pb, SHRIMP	Zircon, cores	Trondhjemite gneiss	7	Bibikova et al., 2006
3351 ± 84	U-Pb, conv.	Magmatic zircon	Trondhjemite gneiss	7	Bibikova et al., 2006
3287 ± 10	U-Pb, conv.	Magmatic zircon	Intrusive(?) tonalite	8	Bibikova et al., 2001
~ 2476	U-Pb, conv. ($^{207}Pb/^{206}Pb$)	Magmatic zircon	Metarhyolite	9	O.M. Turkina, unpubl. data

Table 6.4-1. (*Continued*)

Age, Ma	Method	Zircon type	Rock, complex	Locality*	Reference
1880 ± 17	Ar-Ar	Amphibole	Metarhyolite	9	O.M. Turkina, unpubl. data
1861 ± 1	U-Pb, conv.	Zircon	Shumikha granite	10	Donskaya et al., 2002

*Localities Nos. 1–6 are on Fig. 6.4-3, Nos. 7–10 on Fig. 6.4-4.

2001). These rocks incorporate beds of marbles, metacarbonate-silicate rocks, and garnet-bearing, high-Al gneisses that are linked with a primary sedimentary carbonate-greywacke-pelite association. These metasedimentary-volcanic rocks are called Sharyzhalgay Group. The metavolcanic rocks have major and trace element composition similar to those of subduction-related rock assemblages.

The formation of part of the protoliths of the high-grade rocks began as early as the Paleoarchean. Zircon cores from the intermediate and felsic granulites yielded ages of ca. 3.3–3.4 Ga (Poller et al., 2005), which record the emplacement age of the magmatic precursor to the granulites (Table 6.4-1). Despite magmatic zoning in zircons, the nature of the granulites is still controversial. The magmatic and sedimentary protoliths underwent two stages of granulite-facies metamorphism in the Neoarchean and Paleoproterozoic. Neoarchean ages of ca. 2.6 Ga have been recorded in metamorphic zircon rims from the Kitoy granulites, together with older zircon cores in zircons obtained from metagabbro of the Irkut domain. Neoarchean metamorphism was followed by the emplacement of the Kitoy granite at ca. 2.53 Ga (Gladkochub et al., 2005) and granite veins and pegmatoids (ca. 2.56 Ga) within the Irkut domain (Salnikova et al., 2003). This suggests that juxtaposition of the Kitoy and Irkut domains occurred in the Neoarchean. Several samples of granulites of both domains, as well as the Kitoy granites, have yielded Nd crustal residence times ranging from 3.1–3.4 Ga (Poller et al., 2004).

In contrast, the Onot and Bulun granite–greenstone domains are dominated by low-grade and variably deformed supracrustal metasedimentary-volcanic greenstone sequences and a subordinate grey gneiss basement complex that is preserved as slices and thrust-bounded sheets. In the Onot domain, the lower visible part of the greenstone sequence (the Onot Group) consists of thin-banded gneisses with thin layers of amphibolites, whose protoliths correspond to bimodal felsic and basaltic volcanic rocks (Nozhkin et al., 2001) (Fig. 6.4-4). The upper part of the greenstone section consists of metavolcanic amphibolite, amphibole gneiss intercalated with banded iron formation (BIF), magnesite, dolomite and metasedimentary garnet-cordierite gneisses.

The age of the greenstone sequence in these domains is poorly constrained. Metamorphic zircons from a metarhyolite from the bottom of the greenstone section are 1830 ± 300 Ma. The age of high temperature metamorphism of the greenstones has been more accurately determined by Ar-Ar dating of amphiboles (1880 ± 17 Ma) from the same sample. This age accords well with the U-Pb age of 1861 ± 1 Ma (Donskaya et al., 2002) of

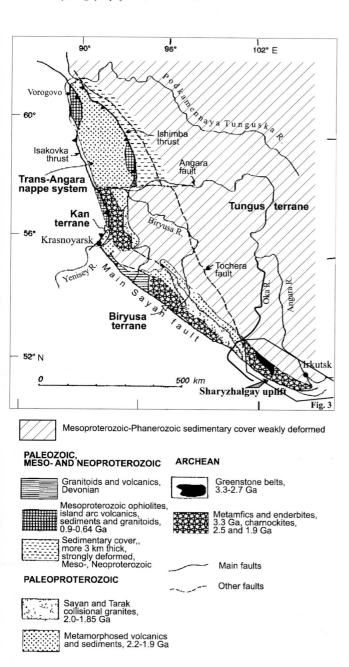

Fig. 6.4-2. Structure of the southern part of the Tungus province (after Nozhkin, 1986, 1999; Postel-nikov and Museibov, 1992; Rosen, 2003; Vernikovsky et al., 2003). Polygon in bottom right indicates area of Fig. 6.4-3.

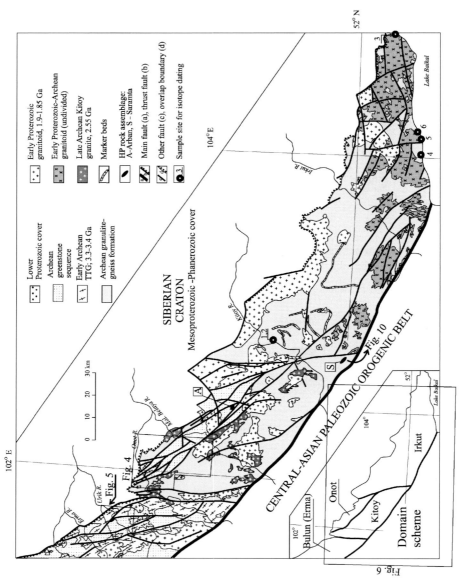

Fig. 6.4-3. Geological map of the Sharyzalgay uplift (according to M.P. Khrenov and others). Areas of Figs. 6.4-4, 6.4-5, and 6.4-10 indicated.

zircons from post-orogenic granites intruding the greenstone sequence. Magmatic zircons from this metarhyolite have strongly discordant Pb isotopic ratios, and the ^{207}Pb/^{206}Pb ratio of the zircons gives a minimum age for the greenstone volcanic rocks at ~2.5 Ga.

The grey gneiss basement complex comprises linear slices and thrust-bounded sheets (up to 10–12 km^2) that are tectonically juxtaposed against elongated blocks of the Onot greenstone sequence (Fig. 6.4-4). The complex is dominated by gneissic to massive biotite tonalites and trondhjemites, whereas biotite and rare amphibole-biotite gneisses are of subordinate occurrence (Turkina, 2004). The biotite gneisses are locally migmatized, and the field relationships suggest that they are probably older than the primary intrusive gneissic to massive trondhjemites. Occasionally, the gneisses contain amphibolite layer-like enclaves of up to tens meters maximal thickness. The gneissic trondhjemites also contain irregular, or lens-shaped, amphibolite bodies. The amphibolite enclaves are interpreted as the remnants of an earlier mafic crust that predated the emplacement of the grey gneiss complex. The formation of gneisses and weakly deformed to foliated trondhjemites is suggested to belong to two distinct magmatic events. U-Pb SHRIMP dating of magmatic zircon grains from trondhjemite gneiss yielded and age of 3386 ± 14 Ma, and one zircon core is dated at 3415 ± 6 Ma (Bibikova et al., 2006). U-Pb conventional multigrain zircon analysis of a weakly foliated trondhjemite yielded an age of 3287 ± 10 Ma.

The structural relations of supracrustal greenstone sequence and grey gneiss basement complex are similar in the Onot and Bulun domains. In the southern part of the Bulun domain, the greenstone sequence is named the Urik Group and includes two rock assemblages (Fig. 6.4-5). The first assemblage comprises amphibolite and amphibole gneiss resembling oceanic floor tholeiite basalts. Similar precursors have been found for high-pressure garnet amphibolites and mafic granulites (Gladkochub et al., 2001) that, together with ultra-mafic rocks and lenses of kyanite schist, make up slices in suture zones. Variable gneisses, ranging in composition from andesitic to dacite metavolcanic rocks, as well as garnet-bearing metasedimentary schists, comprise the second assemblage. The intermediate to felsic gneisses are enriched in incompatible elements and share geochemical features with subduction-related volcanic rocks. The garnet-bearing schists correspond to tuffaceous rocks and graywackes. The grey gneiss basement complex is dominated by foliated to massive porphyry plagiogranite and subordinate plagiogneiss that form sheet-like bodies and slices. Both the gneiss and granitoids correspond to the tonalite-trondhjemite-granodiorite (TTG) series. The basement grey gneisses yield Nd model ages of ca. 3.3–3.5 Ga, close to those of the intermediate to felsic greenstone metavolcanic rocks (T(Nd)$_{DM}$ = 3.3 Ga; see Table 6.4-2 and Fig. 6.4-6). The Early Proterozoic Sayan granites (ca. 1.9 Ga) have intruded the greenstone sequence.

Fig. 6.4-4. (*Next page.*) Geological map of the northwestern Onot granite–greenstone domain (after Nozhkin et al., 2001). Circles mark sampling points for isotope dating, with site numbers corresponding to those in Table 6.4-1.

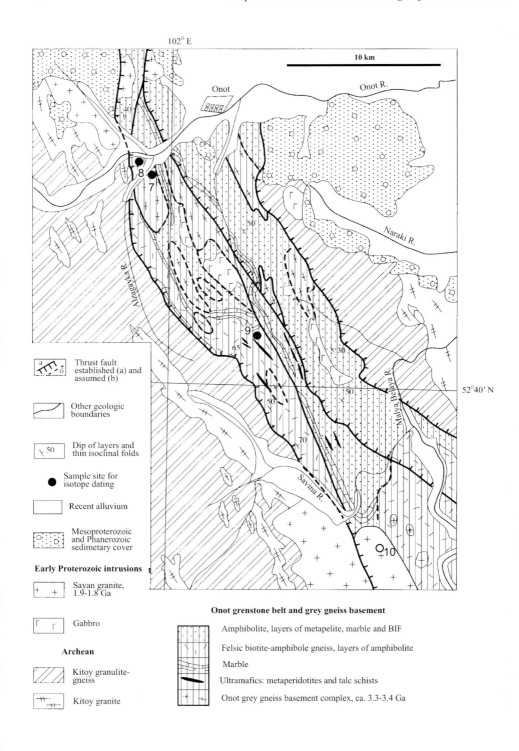

Thrust fault established (a) and assumed (b)

Other geologic boundaries

Dip of layers and thin isoclinal folds

Sample site for isotope dating

Recent alluvium

Mesoproterozoic and Phanerozoic sedimetary cover

Early Proterozoic intrusions

Sayan granite, 1.9-1.8 Ga

Gabbro

Archean

Kitoy granulite-gneiss

Kitoy granite

Onot grenstone belt and grey gneiss basement

Amphibolite, layers of metapelite, marble and BIF

Felsic biotite-amphibole gneiss, layers of amphibolite

Marble

Ultramafics: metaperidotites and talc schists

Onot grey gneiss basement complex, ca. 3.3-3.4 Ga

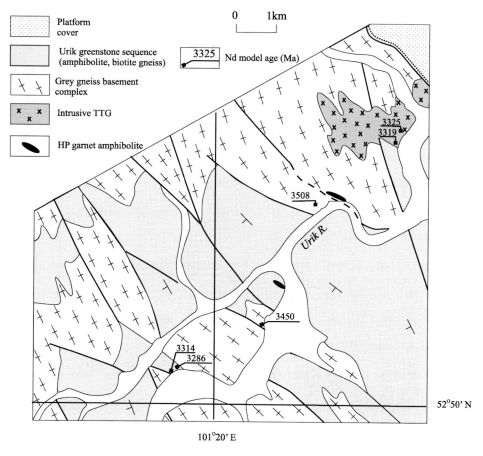

Fig. 6.4-5. Nd isotopic data for the Sharyzhalgay uplift (after Bibikova et al. (2006) and unpublished data of O.M. Turkina).

6.4-2.2. Geochemistry

6.4-2.2.1. The grey gneiss basement complex

Table 6.4-3 shows 19 representative major and trace element analyses of the grey gneiss basement complex. Grey gneisses and granitoids have 63–71 wt% SiO_2, 14.9–16.4 wt% Al_2O_3 and low K_2O/Na_2O ratios (0.2–0.6). In terms of major element composition, they correspond to high-Al tonalites and trondhjemites. Both the gneisses and intrusive trond-hjemites have low Mg# (23–50), similar to other pre-3.0 Ga TTG (Mg# < 50) (Smithies, 2000). The gneisses share all features of typical Archean TTG series (Martin, 1994), such as low HREE and Y abundances, strongly fractionated REE patterns (La/Yb_n = 20–55) and high Sr/Y ratios (23–66) (Turkina, 2004). The medium to coarse-grained intrusive trond-

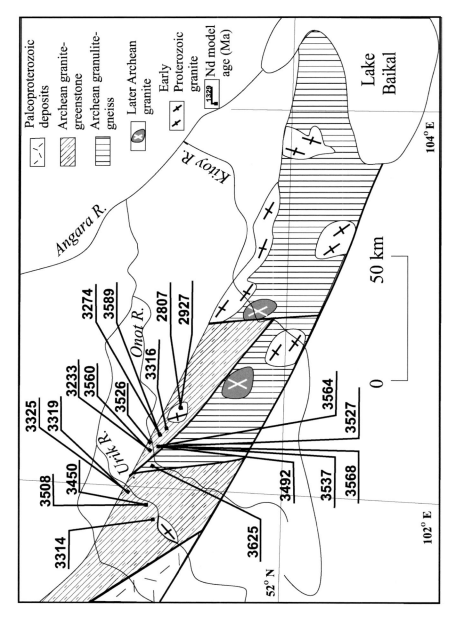

Fig. 6.4-6. Nd isotopic data for the Sharyzhalgay uplift (after Bibikova et al. (2006) and unpublished data of O.M. Turkina).

Table 6.4-2. Sm-Nd isotopic data of Archean rocks from the granite–greenstone domains in the Sharyzhalgay uplift

Sample	Lithology	Nd, ppm	Sm, ppm	$^{147}Sm/^{144}Nd$	$^{143}Nd/^{144}Nd$	T, Ma	T(DM), Ma	Initial ε_{Nd}
Onot domain								
1 63-95	plagiogneiss	18.56	3.08	0.100214	0.510476 ± 10	3400	3564	0.2
2 48-95	plagiogneiss	9.97	1.29	0.078295	0.509955 ± 15	3400	3568	−0.4
3 48-03	plagiogneiss	10.29	1.85	0.108692	0.510695 ± 27	3400	3537	0.7
4 40-03	plagiogneiss	11.72	1.85	0.09542	0.510396 ± 9	3400	3527	0.7
5 49-03	plagiogneiss	15.0	2.40	0.096548	0.510446 ± 25	3400	3492	1.2
6 66-95	trondhjemite	15.91	2.75	0.104269	0.510599 ± 27	3300	3526	−0.4
7 173-95	trondhjemite	19.98	3.19	0.096430	0.510390 ± 4	3300	3560	−1.2
8 152-95	trondhjemite	10.63	1.67	0.095080	0.510617 ± 18	3300	3233	3.9
9 51-03	amphibolite	5.61	1.82	0.195697	0.512746 ± 17	3400*	–	2.6
10 157-95	amphibolite	9.99	3.06	0.184999	0.512495 ± 14	3400*	–	2.4
11 57-03	amphibolite	2.61	0.87	0.200667	0.512785 ± 20	3400*	–	1.1
12 38-03	amphibolite	10.71	3.30	0.186565	0.512392 ± 18	3400*	–	−0.4
13 64-95	amphibolite	8.60	2.78	0.195390	0.512586 ± 11	3400*	–	−0.4
14 2-03	metarhyolite	68.83	13.1	0.115136	0.511019 ± 16	1900*	3274	−11.7
15 6-03	metarhyolite	36.24	7.95	0.132637	0.511227 ± 25	1900*	3589	−12.0
16 19-95	metarhyolite	33.02	7.33	0.134151	0.511408 ± 20	1900*	3316	−8.8
17 20-95	metabasalt	5.26	1.63	0.187182	0.512670 ± 19	1900*	–	3.0
18 80-95	metabasalt	6.85	2.13	0.87490	0.512715 ± 16	1900*	–	3.8
Bulun (Erma) domain								
19 35-04	plagiogneiss	4.55	27.0	0.10180	0.510599 ± 15	–	3450	–
20 60-04	plagiogneiss	6.97	42.4	0.09930	0.510497 ± 19	–	3508	–
21 67-04	tonalite	5.11	34.9	0.08860	0.510402 ± 4	–	3325	–
22 69-04	trondhjemite	3.03	23.5	0.07780	0.510170 ± 5	–	3319	–
23 23-04	metaandesite	13.3	96.1	0.08340	0.510297 ± 10	–	3314	–
24 26-04	metarhyolite	6.64	47.6	0.08430	0.510341 ± 9	–	3290	–

Initial epsilon Nd values were calculated assuming $^{147}Sm/^{144}Nd = 0.1967$ and $^{143}Nd/^{144}Nd = 0.512638$ for CHUR (Jacobsen, Wasserburg, 1984). 3400* – age of country rocks. 1900* – age of metamorphism. Model age values were calculated using depleting mantle parameters $^{147}Sm/^{144}Nd = 0.21365$ and $^{143}Nd/^{144}Nd = 0.513151$ (Goldstein, Jacobsen, 1988).

1–18. Onot domain: 1–13 – grey gneiss basement complex including amphibolite (9–13); 14–18 – metavolcanics. 19–24. Bulun (Erma) domain: 19–22 – grey gneiss basement complex, 23–24 – metavolcanics of the Urik greenstone belt.

hjemites are weakly enriched in HREE (La/Yb$_n$ = 16–29). The primitive mantle (PM)-normalized trace element patterns for the Onot TTG are strongly fractionated, displaying LILE and LREE enrichment and prominent negative Nb and Ti anomalies (Fig. 6.4-7(a)). A narrow range of ε_{Nd} values for grey gneisses, from +1.2 to −0.4, and the ca. 3.5–3.6 Ga Nd model ages (Table 6.4-2) of these rocks reflect a contribution of an ancient crustal source, consistent with the age of ancient zircon cores dated at 3415 ± 6 Ma (Bibikova

Table 6.4-3. Major and trace element analyses of Early Archean grey gneiss basement complex from the Onot and Bulun (Erma) domains

1. Onot domain

	51-03[*]	57-03[*]	157-95[*]	64-95[*]	38-03[*]	26-03	48-03[*]	32-03	49-03[*]	39-03
	1	2	3	4	5	6	7	8	9	10
(wt%)										
SiO_2	49.2	49.13	49.05	49.49	51.2	52.59	65.53	66.64	67.07	67.11
TiO_2	0.95	0.97	1.39	1.30	1.41	0.99	0.45	0.402	0.47	0.44
Al_2O_3	13.69	13.73	13.99	14.28	13.48	13.91	14.88	16.77	14.98	15.84
Fe_2O_3t	13.27	14.47	14.32	15.61	18.15	12.54	6.02	5.43	5.58	6.37
MnO	0.211	0.241	0.22	0.247	0.244	0.179	0.097	0.085	0.079	0.123
MgO	7.62	6.8	7.31	6.32	6.16	5.55	1.6	1.59	1.54	1.79
CaO	11.88	11.66	9.97	9.38	5.4	9.74	3.71	3.41	3.14	1.37
Na_2O	2.72	2.22	2.44	1.15	1.63	3.17	6.52	3.51	4.57	4.9
K_2O	0.26	0.38	0.65	1.05	0.71	0.52	1.15	1.72	1.77	1.87
P_2O_5	0.075	0.078	0.12	0.088	0.142	0.12	0.105	0.073	0.115	0.105
LOI	0.3	0.56	0.84	1.04	0.77	0.26	0.02	0.14	0.16	0.02
Total	100.2	100.2	100.3	99.9	99.3	99.6	100.1	99.81	99.5	99.9
(ppm)										
Th	0.34	0.29	0.73	0.84	1.06	0.74	4.19	2.58	5.56	4.63
Rb	3.7	8.0	16.0	28.0	9.4	11.5	54	105	99	129
Ba	26.5	64.7	79.0	61.0	80.2	28.8	226	228	355	257
Sr	153	124	111	116	117.1	95	320	364	258	177
La	3.38	3.06	5.5	4.3	6.94	5.09	19.37	8.82	23.12	23.73
Ce	7.85	7.71	13	11	15.98	11.45	38.11	17.19	43.95	42.70
Pr	1.27	1.21	2.3	1.9	2.34	1.68	3.95	2.40	5.29	4.89
Nd	6.01	6.23	10	9	10.68	7.91	11.91	9.14	16.41	14.36
Sm	1.85	2.10	3.1	2.7	3.33	2.40	2.30	2.36	3.02	2.17
Eu	0.76	0.84	1.0	1.1	1.12	0.97	0.59	0.64	0.71	0.66
Gd	2.78	2.81	4.5	4.3	4.25	3.29	1.36	1.92	1.90	1.69
Tb	0.44	0.47	0.78	0.77	0.72	0.54	0.17	0.26	0.24	0.26
Dy	3.17	3.33	5.2	4.9	5.15	3.74	1.32	1.66	1.52	1.22
Ho	0.73	0.74	1.1	1.1	1.08	0.82	0.20	0.26	0.27	0.20
Er	2.22	2.31	3.3	3.2	3.48	2.57	0.49	0.79	0.77	0.63
Tm	0.34	0.37	0.47	0.54	0.51	0.41	0.06	0.10	0.12	0.08
Yb	2.13	2.24	3.4	3.3	3.33	2.63	0.37	0.62	0.69	0.43
Lu	0.37	0.36	0.47	0.46	0.55	0.37	0.06	0.09	0.08	0.07
Zr	20.5	18.2	95.0	90.0	49.8	19.5	229	119	288	144
Hf	0.80	0.69	2.4	2.2	1.41	0.76	4.41	2.61	4.74	2.95
Ta	0.17	0.17	0.39	0.29	0.27	0.22	0.02	0.18	0.22	0.01
Nb	2.73	2.87	5.2	4.1	4.63	4.41	3.09	4.29	5.45	4.78
Y	21.7	22.5	31.0	30.0	35.1	25.4	6.21	8.99	9.13	7.67
$(La/Yb)_n$	1.1	0.9	1.1	0.9	1.4	1.3	35.2	9.7	22.6	37.1
Sr/Y							51.6	40.6	28.3	23.1

Table 6.4-3. (*Continued*)

1. Onot domain

	60-95	63-95[*]	44-03	48-95[*]	40-03[*]	173-95[*]	66-95[*]	152-95[*]	153-95
	11	12	13	14	15	16	17	18	19
(wt%)									
SiO_2	67.25	68.82	69.16	70.31	71.31	68.25	68.3	71.68	70.49
TiO_2	0.426	0.46	0.40	0.32	0.37	0.48	0.485	0.313	0.279
Al_2O_3	16.29	14.99	15.82	15.84	15.61	15.34	15.41	14.22	15.73
Fe_2O_3[*]	4.54	4.24	4.45	2.87	2.03	4.8	4.59	3.48	3.26
MnO	0.077	0.073	0.058	0.044	0.09	0.078	0.072	0.062	0.059
MgO	1.96	1.97	1.23	1.2	1.23	1.87	1.93	1.11	0.84
CaO	2.27	2.21	1.84	2.25	1.66	2.81	1.94	1.84	2.37
Na_2O	4.0	3.62	5.79	4.63	5.36	3.76	4.94	4.33	4.85
K_2O	1.95	2.1	1.35	1.43	1.35	2.09	2.02	2.32	2.03
P_2O_5	0.07	0.124	0.081	0.06	0.03	0.106	0.095	0.071	0.034
LOI	0.84	1.37	0.22	0.79	0.7	0.78	0.44	0.54	0.46
Total	99.67	99.9	100.1	99.74	99.7	100.3	100.2	99.97	100.4
(ppm)									
Th	3.0	4.7	5.2	3.1	5.33	4.5	5.7	4.6	3.1
Rb	80	89	65	45	64	94	52	77	76
Ba	250	168	105	245	234	252	206	600	696
Sr	339	219	372	330	257	266	110	219	300
La	17	26	21.30	19.8	24.06	26	28	15.3	15
Ce	32	47	38.67	30	41.77	46	55	31	29
Pr		5.8	4.52	–	4.84	–	–	–	–
Nd	14	19	12.74	11	14.37	18	24	12	10
Sm	2.4	3.1	2.06	1.8	2.52	3.1	4.0	2.0	1.31
Eu	0.73	0.79	0.46	0.58	0.65	0.7	0.98	0.6	0.46
Gd	2.2	2.7	1.12	1.2	1.74	2.2	3.1	1.7	0.9
Tb	0.31	0.33	0.14	0.15	0.22	0.3	0.42	0.25	0.12
Dy		1.8	0.60	–	1.32	–	–	–	–
Ho		0.33	0.07	–	0.23	–	–	–	–
Er		0.91	0.36	–	0.96	–	–	–	–
Tm		0.14	0.05	–	0.12	–	–	–	–
Yb	0.56	0.85	0.26	0.31	0.61	0.73	0.88	0.65	0.35
Lu	0.08	0.16	0.03	0.037	0.10	0.11	0.11	0.09	0.05
Zr	97	211	290	110	136	–	–	–	–
Hf	3.7	4.8	4.71	3	3.19	4.8	4.1	3.8	1.8
Ta	0.41	0.82	0.01	0.29	1.07	0.62	0.73	0.49	0.26
Nb	9.6	6.2	2.62	6.3	12.02	–	–	–	–
Y	8.7	10	2.77	5	7.56	–	10	7	4
$(La/Yb)_n$	20.5	20.6	55.0	43.1	26.7	24.0	21.5	15.9	28.9
Sr/Y	39.0	21.9	134.2	66.0	34.1	–	11.0	31.3	75.0

1–6 – amphibolite inclusions in grey gneiss complex, 7–15 – gneisses, 16–19 – intrusive trondhjemites. Samples analyzed on Sm-Nd isotopes are marked by asterisk.

Table 6.4-3. (*Continued*)

2. Bulun (Erma) domain

	35-04[*]	60-04[*]	62-04	27-04	67-04[*]	69-04[*]
	1	2	3	4	5	6
(wt%)						
SiO_2	65.36	68.91	70.32	70.82	67.88	74.09
TiO_2	0.742	0.458	0.354	0.285	0.426	0.255
Al_2O_3	17.66	16.36	14.87	15.16	16.5	14.35
Fe_2O_3t	5.65	4.55	4.33	2.96	4.39	2.53
MnO	0.106	0.068	0.061	0.053	0.05	0.035
MgO	1.24	1.12	0.93	0.71	0.91	0.56
CaO	3.06	3.29	3.28	2.87	3.08	2.16
Na_2O	3.82	3.81	3.63	5.91	4.33	3.68
K_2O	1.69	1.06	1.25	1.01	1.77	1.77
P_2O_5	0.218	0.116	0.108	0.077	0.147	0.067
LOI	0.91	0.35	0.59	0.41	0.46	0.38
Total	100.4	100.1	99.72	100.3	99.94	99.88
(ppm)						
Th	2.7	5	3.6	3.1	4.0	5.2
Rb	88	66.6	56.81	116	56.59	55.2
Ba	367	178.4	208.4	271.1	425.3	336.9
Sr	346	511.5	439.4	847.7	971.1	554.5
La	45	33.71	25.08	20.03	39.36	33.26
Ce	79	64.3	48.05	36.29	72.17	59.74
Pr	9.1	6.52	4.87	3.55	7.36	5.54
Nd	29	22.7	17.92	12.93	25.93	18.92
Sm	4.7	3.56	2.39	1.95	4.17	2.92
Eu	1.2	0.82	0.77	0.56	1.03	0.6
Gd	4.4	3.05	1.7	1.53	3.33	1.58
Tb	0.56	0.4	0.18	0.17	0.36	0.14
Dy	2.8	2.29	1	0.9	1.96	0.92
Ho	0.48	0.4	0.17	0.15	0.36	0.22
Er	1.4	1.19	0.49	0.4	0.93	0.56
Tm	0.24	0.15	0.07	0.06	0.17	0.06
Yb	1.4	0.95	0.51	0.38	0.67	0.34
Lu	0.24	0.19	0.1	0.08	0.12	0.08
Zr	346.0	362.7	299.2	208	246.4	180.3
Hf	8.7	6.09	4.53	3.37	4.94	4.1
Ta	0.87	0.36	0.14	0.6	0.26	0.2
Nb	10	7.08	5.18	5.22	3.08	2.28
Y	13	14.58	8.67	6.73	8.32	4.1
$(La/Yb)_n$	22.6	23.9	33.2	35.5	39.6	66.0
Sr/Y	26.6	35.1	50.7	126.0	116.7	135.2

1–4 – grey gneiss basement complex, 5–6 – intrusive trondhjemites. Samples analyzed on Sm-Nd isotopes are marked by asterisk. t – All are analyzed as ferric iron.

Table 6.4-4. Major and trace element analyses of Onot and Urik greenstone belt metavolcanics

	2-03[*]	6-03[*]	19-95[*]	271/2	26-03	20-95[*]	21-95	4-95	80-95[*]	113-95	172-95	102-95
	1	2	3	4	5	6	7	8	9	10	11	12
(wt%)												
SiO_2	71.76	70.47	74.68	73.95	75.53	49.9	50	50.09	50.16	46.96	48.59	48.5
TiO_2	0.354	0.549	0.33	0.35	0.35	0.702	0.7	1.25	0.985	0.565	0.853	0.9
Al_2O_3	11.89	11.99	11.18	11.48	11.59	14.33	13	14.17	14.54	16.69	14.74	14.79
Fe_2O_3t	6.42	7.52	4.71	4.72	4.21	11.58	11	13.83	11.96	9.77	12.63	11.57
MnO	0.129	0.109	0.074	0.104	0.077	0.205	0.2	0.236	0.234	0.175	0.191	0.226
MgO	0.43	0.52	0.1	0.1	0.19	8.49	8.2	7.34	7.47	10.45	8.93	7.73
CaO	1.11	1.57	0.94	0.98	0.93	12.42	12	10.92	11.75	10.7	9.62	11.21
Na_2O	3.03	2.68	3.48	2.89	2.78	1.28	1.7	1.9	1.65	1.41	1.72	2.61
K_2O	3.72	4.14	4.24	4.26	4.08	0.34	0.3	0.33	0.33	1.02	0.61	0.69
P_2O_5	0.032	0.109	0.03	0.042	0.03	0.031	0	0.096	0.054	0.032	0.048	0.05
LOI	0.69	0.3	0.12	0.66	0.5	0.72	1.2	0.06	0.67	2.04	1.57	1.2
Total	99.86	99.48	99.9	99.5	99.59	100.0	99.3	100.2	99.8	99.8	99.5	99.5
(ppm)												
Th	12.2	11.3	11.5	18	7.4	0.6	1.0	0.8	0.91	0.27	0.44	0.5
Rb	198.5		152	147	168	12	19	7	7.6	30	9.3	21
Ba	712		833	920	839	56	106		73	120	52	100
Sr	130		84	95	91	117	127	144	113	87	108	112
La	73.22		35	76	36	3.4	4.2	7.7	4.7	2.1	3.2	3.6
Ce	114.9		106	130	71	8.5	9.4	17	11	5.1	8.1	8.7
Pr99.86	15.6						1.5		1.7	0.82	1.4	
Nd	58.86		38	75	34	5.7	6.4	10	7.9	4.0	6.7	6.5
Sm	10.68		9.3	18.2	7.6	1.8	2.0	3	2.4	1.2	2.0	2
Eu	2.06		1.79	2	1.5	0.66	0.73	1.08	0.88	0.47	0.86	0.76
Gd	11.26		10	13.5	7.4	2.3	2.8	3.7	3.06	1.63	2.92	2.6
Tb	1.82		1.6	2.6	1.3	0.42	0.49	0.68	0.54	0.30	0.53	0.48
Dy	11.32						3.3		3.7	2.1	3.5	
Ho	2.28						0.73		0.83	0.44	0.81	
Er	7.09						2.2		2.4	1.3	2.2	
Tm	0.99						0.33		0.36	0.21	0.35	
Yb	6.75		6.1	8	5.3	1.6	2.1	2.6	2.5	1.4	2.4	1.68
Lu	1.12		0.85	1.2	0.88	0.24	0.33	0.38	0.38	0.20	0.37	0.26
Zr	606.8		330	371	349	45	43.0	76	56	32	54	78
Hf	10.6		11	10.5	11.9	0.9	1.50	2.2	1.9	1.2	1.8	2
Ta	1.46		1.3	1.6	1.31	0.09	0.2	0.28	0.17	0.13	0.16	0.12
Nb	25.6		15	15	17	2.6	2.8	5	2.5	1.9	2.4	2.2
Y	89.4		55	78	40	14	19	23	24	13	21	16
Cr						146	167	251	127	628	210	212
Ni						80	81		80	225	132	93
Co						44	53	46	55	53	54	40
V						294	269		250	175	237	238
$(La/Yb)_n$	7.3		3.9	6.4	4.6	1.4	1.3	2.0	1.3	1.0	0.9	1.4

Table 6.4-4. (*Continued*)

	89-220	89-224	92-95	79-95	114-95	82-95	48-03[*]	32-03	49-03[*]	39-03
	13	14	15	16	17	18	19	20	21	22
(wt%)										
SiO_2	48.49	48.7	52.31	52.43	56.49	55.06	65.53	66.64	67.07	67.11
TiO_2	0.45	0.88	0.866	0.907	1.187	1.285	0.45	0.402	0.47	0.44
Al_2O_3	15.4	12.9	13.65	14.96	14.65	16.9	14.88	16.77	14.98	15.84
Fe_2O_3t	10.1	13.81	11.79	11.93	11.27	12.35	6.02	5.43	5.58	6.37
MnO	0.15	0.18	0.236	0.213	0.169	0.25	0.097	0.085	0.079	0.123
MgO	10.3	8.7	7.36	5.4	4.4	2.1	1.6	1.59	1.54	1.79
CaO	11.2	9.6	10.09	8.7	6.4	6.7	3.71	3.41	3.14	1.37
Na_2O	1.59	2.42	1.86	2.1	3.3	4.1	6.52	3.51	4.57	4.9
K_2O	0.81	0.58	0.79	1.5	1.2	0.8	1.15	1.72	1.77	1.87
P_2O_5	0.06	0.06	0.053	0.1	0.1	0.4	0.105	0.073	0.115	0.105
LOI	1.52	2.6	0.96	1.98	0.84	0.6	0.02	0.14	0.16	0.02
Total	100.1	100.4	100.0	100.2	100.5	100.6	100.1	99.81	99.5	99.9
(ppm)										
Thm	0.4	0.6	0.3	3.4	3.6	2.3	4.19	2.58	5.56	4.63
Rb	16	12	29	46	31	17	54	105	99	129
Ba	104	120	133	356	426	178	226	228	355	257
Sr	86	127	131	200	350	230	320	364	258	177
La	1.55	2.6	3.6	19	16	16	19.37	8.82	23.12	23.73
Ce	4.8	7.1	8.6	40	36	40	38.11	17.19	43.95	42.70
Pr99.86							3.95	2.40	5.29	4.89
Nd	3.7	6	5.7	20	18	28	11.91	9.14	16.41	14.36
Sm	1.25	2	1.74	4.3	4.4	8.4	2.30	2.36	3.02	2.17
Eu	0.38	0.76	0.67	1.16	1.25	2.89	0.59	0.64	0.71	0.66
Gd	1.5	2.3	2.1	4.4	4.6	9.6	1.36	1.92	1.90	1.69
Tb	0.27	0.41	0.38	0.74	0.73	1.7	0.17	0.26	0.24	0.26
Dy							1.32	1.66	1.52	1.22
Ho							0.20	0.26	0.27	0.20
Er							0.49	0.79	0.77	0.63
Tm							0.06	0.10	0.12	0.08
Yb	1	1.9	1.48	2.1	1.74	6.2	0.37	0.62	0.69	0.43
Lu	0.15	0.3	0.23	0.3	0.23	0.95	0.06	0.09	0.08	0.07
Zr	22	38	49	99		357	229	119	288	144
Hf	0.5	1.2	1.4	2.6	3	6.8	4.41	2.61	4.74	2.95
Ta	0.06	0.115	0.12	0.43	0.42	0.48	0.02	0.18	0.22	0.01
Nb	1.3	1.5	2.5	5.4		2	3.09	4.29	5.45	4.78
Y	9	17	11	19		61	6.21	8.99	9.13	7.67
Cr	800	480	132	322	56	102				
Ni	197	180	68	130	76	5				
Co	46	50	48	31	53	22				
V	210	240	288	162	227	50				

Table 6.4-4. (*Continued*)

	89-220	89-224	92-95	79-95	114-95	82-95	48-03[*]	32-03	49-03[*]	39-03
	13	14	15	16	17	18	19	20	21	22
$(La/Yb)_n$	1.1	0.9	1.6	6.1	6.2	1.7	35.2	9.7	22.6	37.1
Sr/Y							51.6	40.6	28.3	23.1

1–18 – Onot Group. 1–8 – lower unit: 1–5 – metarhyolites, 6–8 – metabasalts; 9–18 – upper unit: 9–15 – metabasalts, 16–18 – metatuffaceous shales. 19–22 – Urik Group, meta-andesites. Samples analyzed on Sm-Nd isotopes are marked by asterisk.

et al., 2006). The intrusive trondhjemites have more variable ε_{Nd} values (+3.9 to −1.2) and $T(Nd)_{DM}$ ages (3.2–3.6 Ga), suggesting an input from ancient crustal and juvenile mantle-derived sources.

The amphibolites that comprise the enclaves in the grey gneiss complex correspond to low-Mg tholeiitic basalts with nearly flat rare earth (La/Yb_n = 0.9–1.4) and multi-element PM-normalized patterns (Fig. 6.4-7(b)). Based on trace-element compositions, the amphibolites can be subdivided into two types (Turkina, 2004). The first type are relatively depleted in Ti, REE, Th, HFSE and have trace element ratios close to PM values (Nb/La_{PM} = 0.8–0.9, Nb/Th_{PM} = 1–1.2, $Th/La \sim 0.1$) (Fig. 6.4-8). These features are similar to those of Phanerozoic oceanic plateau basalts. The ε_{Nd} (T) values of the first type of amphibolites ranges from 2.6 to 2.4, suggesting an origin from a depleted mantle source (Table 6.4-2). The second type of amphibolites are characterized by a weak negative Nb anomaly (Nb/La_{PM} = 0.7–0.9, Nb/Th_{PM} = 0.5–0.9) and higher Th/La (0.13–0.2). The decreasing Nb/La_{PM} and Nb/Th_{PM} ratios are accompanied by the decrease of ε_{Nd} to −0.4, suggesting contamination of a primary basaltic melt by earlier continental crust or recycled crustal material in the mantle source region due to subduction.

The extreme depletion of TTG rocks in HREE could have resulted from the melting of a garnet-bearing source at a depth of $\geqslant 15$ kbar. This, together with low Mg#, suggests their relation to a shallow-deep subduction process, while the ε_{Nd} values of +1.2 to −0.4 strongly support the involvement of crustal input into TTG melt genesis. In this case, the TTG are interpreted as likely to have resulted from melting of over-thickened crust, consisting of a mix of mafic and sialic components (see also Champion and Smithies, this volume; Moyen et al., this volume). The geochemical affinity of the enclave amphibolites to the oceanic plateau basalts suggests the accretion of an oceanic plateau with an incorporated fragment of ancient sialic crust as the most plausible mechanism for thick crustal growth, accompanied by remelting and TTG formation.

Similarly to the Onot domain, rocks of the Bulun grey gneiss complex correspond to high-Al trondhjemites and tonalities. The grey gneisses are characterized by increased Mg# (30–48), moderately fractionated REE patterns (La/Yb_n = 24–36) and high Sr/Y (27–50) (Fig. 6.4-9(a)). Compared to the grey gneiss, the intrusive trondhjemites and tonalites have lower Mg# (28–38), higher La/Yb_n (40–66) and Sr/Y (117–135). Both rock types have weakly different Nd model ages: ca. 3.3 Ga for trondhjemites versus ca. 3.4–3.5 Ga for

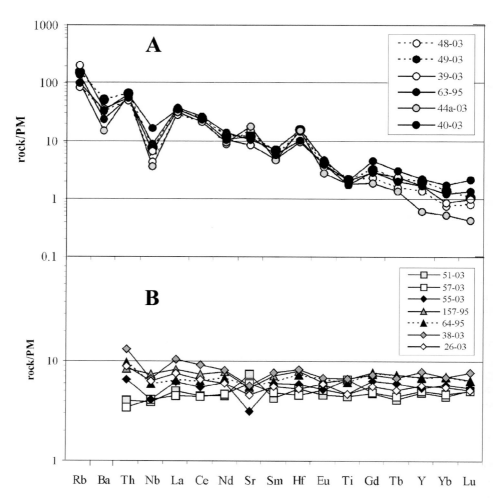

Fig. 6.4-7. Primitive mantle-normalized multi-element patterns for the Onot grey gneisses basement complex in the Sharyzhalgay uplift: (A) gneisses; (B) amphibolites.

gneisses. The close Nd model ages of both the Bulun and Onot grey gneiss basement rocks suggest that these rock complexes were related to the same crust-forming event.

6.4-2.2.2. Onot Group

The lower part of the greenstone sequence consists of a bimodal association dominated by felsic metavolcanic rocks with minor mafic rocks (Table 6.4-4). The metarhyolites are characterized by high FeO/FeO+MgO (>0.9) and enriched REE and HFSE (Fig. 6.4-9(b)). They are geochemically close to intra-plate felsic volcanic rocks, or A-type granites (Nozhkin et al., 2001). Based on T(Nd)$_{DM}$ values of 3.3–3.6 Ga (Table 6.4-2) and low Y/Nb

Table 6.4-5. Major and trace element analyses of the Arban granulites

	38-95	32-98	29-95	36b-95	25-98	36-95	36a-95	35-95	34-95	31-95
	1	2	3	4	5	6	7	8		
(wt%)										
SiO_2	47.2	46.5	49.3	49.35	46.2	45.0	46.82	44.2	67.55	65.48
TiO_2	1.04	0.97	1.12	1.19	1.14	1.7	1.56	2.26	0.39	0.46
Al_2O_3	11.4	14.4	14.4	13.2	13.0	13.2	13.42	12.4	16.15	15.42
Fe_2O_3t	15.8	14.1	14.6	15.65	17.9	21.9	20.78	22.9	4.22	5.55
MnO	0.27	0.26	0.26	0.27	0.29	0.38	0.375	0.37	0.06	0.09
MgO	10.2	8.98	7.24	7.5	7.59	6.64	5.96	6.38	1.16	2.46
CaO	12.5	11.8	10.9	12	11.7	11.1	10.64	11.1	4.14	4.73
Na_2O	0.91	0.81	1.93	0.7	1.04	1.12	0.4	0.4	5.24	4.9
K_2O	0.14	0.55	0.23	0.04	0.13	0.21	0.21	0.03	1.3	0.81
P_2O_5	0.08	0.07	0.09	0.078	0.09	0.15	0.135	0.15	0.1	0.07
LOI	0.28	0.93	0.16	0.18	0.38	0.56	0.38	0.1	0.02	0.22
Total	99.82	99.37	100.2	100.16	99.46	102	100.68	100.3	100.3	100.19
(ppm)										
Th	0.6	–	0.4	0.4	–	0.2	0.2	–	0.8	0.4
Rb	1	5	1.2	0.5	2.9	1.1	0.8	0.5	14	5
Ba	64	–	91	45	–	28	32	45	511	527
Sr	75	46	96	65	78	56	50	28	500	336
La	4.5	3.5	4.7	5.4	3.0	3.8	3.2	3.1	17	17
Ce	11.7	10.5	12.4	13.9	9.5	11.0	11.0	9.0	31	27
Nd	8	7.5	9.1	10	8.5	10.0	10.0	8.7	12	11.4
Sm	2.6	2.4	3.0	3.3	2.8	3.6	3.9	3.8	2.3	2.2
Eu	0.93	1.16	0.98	1.19	1.0	1.38	1.31	1.4	0.95	0.77
Gd	3.2	3.2	4.0	4.4	3.2	5.2	5.2	5.9	2.0	1.8
Tb	0.57	0.75	0.76	0.78	0.83	1.0	0.96	1.1	0.27	0.25
Yb	2.1	2.5	3.3	3.5	3.5	5.7	4.3	6.3	0.5	0.68
Lu	0.3	0.33	0.52	0.54	0.56	0.68	0.67	0.95	0.07	0.09
Zr	36	38	37	33	43	39	37	39	83	67
Hf	1.5	1.6	1.7	1.0	1.3	2.0	1.4	2.4	2.7	2.3
Ta	0.19	0.15	0.2	0.1	0.13	0.2	0.2	0.55	0.13	0.15
Nb	4.4	1.1	4.9	2.7	2.3	5.0	5.0	7.0	3.5	2.7
Y	17.6	17.5	23.3	28.8	27.4	42.4	35.4	32	6.1	4.7
$(La/Yb)_n$	1.4	0.9	1.0	1.0	0.6	0.4	0.5	0.3	22.9	16.9

ratios (2–5), the primary felsic melt can be interpreted as derived from a source with a long crustal residence time. Such a source was probably the grey gneiss basement complex because the studied metarhyolites are very similar to melts produced experimentally from a tonalite source at 10 kbar (Skjerlie and Johnston, 1993). The associated amphibolites correspond to tholeiitic basalts (Mg# = 50–60) that are weakly enriched in LREE (La/Sm_n

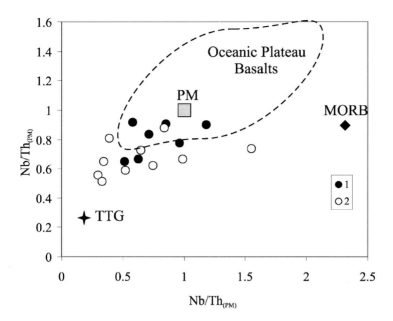

Fig. 6.4-8. Nb/La$_{PM}$ vs Nb/Th$_{PM}$ variation diagram for amphibolitic enclaves in the Onot grey gneiss complex (1) and greenstone amphibolites (2), after Puchtel et al. (1997). PM (primitive mantle) and MORB (mid-ocean ridge basalt) from Sun and McDonough (1989); TTG = average composition of Onot TTG.

= 1.2–1.6) with Nb/La$_{PM}$ = 0.6–0.7 and Nb/Th$_{PM}$ = 0.3–0.8 (Figs. 6.4-8, 6.4-9(b)). The negative Nb anomalies in PM-normalized multi-element diagrams suggest contamination by sialic crust or input of crustal material into the mantle source region. However, the amphibolites from the upper unit are different: they are compositionally close to high-Mg tholeiites (Mg# = 55–68), with lower La/Sm$_n$ (0.8–1.2) and weak to no depletion in Nb (Nb/La$_{PM}$ = 0.6–0.9, Nb/Th$_{PM}$ = 0.3–1.0) (Figs. 6.4-8, 6.4-9(c)) that reflects less contribution of crustal material in their genesis.

6.4-2.2.3. Urik Group

Metabasalts of the first assemblage correspond to low Mg (Mg# = 30–58) and high-Ti (1.4–2.0%) tholeiites, characterized by depleted to enriched LREE patterns (La/Sm$_n$ = 0.6–1.4) and nearly flat PM-normalized multi-element patterns (Fig. 6.4-9(d)) displaying weak negative to positive Nb anomalies (Nb/La$_{PM}$ = 0.8–1.7, Nb/Th$_{PM}$ = 0.9–1.2) that excludes any crustal contribution in their genesis. The Nb depletion is typical only for intercalated leucobasalts (Nb/La$_{PM}$ = 0.4–05, Nb/Th$_{PM}$ ∼ 0.4) and resulted from their enrichment in Th and La but not Nb. These geochemical features imply an intra-oceanic setting of basalt formation.

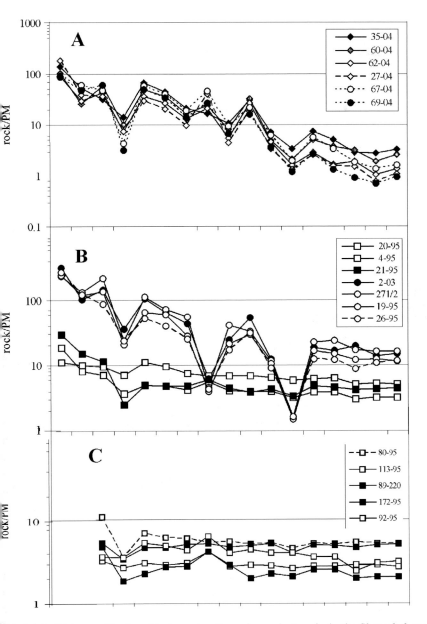

Fig. 6.4-9. PM-normalized multi-element patterns for ancient rocks in the Sharyzhalgay uplift: (A) Bulun grey gneisses of the basement complex; (B) Metarhyolites and metabasalts of the lower part of the Onot greenstone belt; (C) Metabasalts of the upper part of the Onot greenstone belt; (D) Metabasalts and meta-andesites and -dacites of the Urik greenstone belt; (E) Mafic and felsic granulites of the Arban massif.

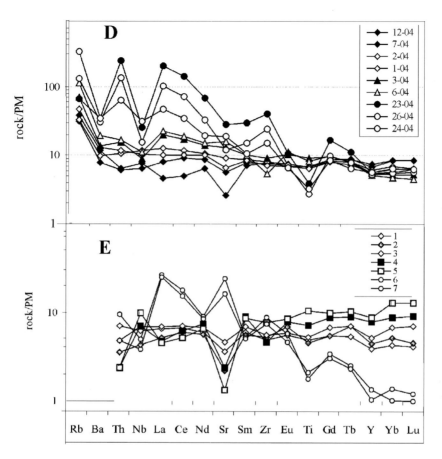

Fig. 6.4-9. (*Continued.*)

In contrast, the second assemblage of intermediate and felsic metavolcanic rocks are strongly enriched in LILE and LREE (La/Yb$_n$ = 7–32) and their multi-element patterns display profound negative Nb and Ti anomalies (Fig. 6.4-9(d)), suggesting a subduction-related origin. These metavolcanic rocks could have been derived from a mostly ancient crustal source, as indicated by Nd model ages of ca. 3.3 Ga.

6.4-2.3. *Lateral Continuity of Early Archean Crust*

The isotopic and geochemical features of the Onot and Bulun domain intermediate and felsic greenstones are indicative of the lateral continuation of early Archean grey gneiss complexes, which are not limited to the exposed sheet- or slice-like gneiss fragments. The evidence for an early Archean crustal contribution to the formation of the intermediate and felsic greenstones of the Bulun domain is indicated by their ancient Nd model ages of

a. 3.3 Ga, and by incompatible trace element enrichments. The geochemical features of the greenstones allow us to suggest a subduction-related genesis in an active continental margin setting that was accompanied by major crustal source melting.

Within the Onot domain is a bimodal greenstone assemblage. Metarhyolites could have been derived from a source with a long crustal residence time. Their compositional affinity to intra-plate volcanic rocks suggests that they formed due to melting of a TTG-like source. Based on experimental data on A-type granite origins (Skjerlie and Johnston, 1993), the source region melted at about ca. 10 kbar; i.e., the crust was at least 35 km thick and thereby implying a significant vertical thickness of the early Archean grey gneiss crust. Moreover, this thick continental crust could be a contaminant for the basaltic magmas, as evidenced by the Nb depletion of amphibolites from the bottom, and to a lesser degree, the middle part of the greenstone sequence. Thus the thick, rigid and stable craton-like continental crust could have been formed by the Neoarchean.

4-2.4. Lower Crustal Surroundings: High-Pressure Granulite and Ultramafic Assemblages

4-2.4.1. The lower crustal assemblage of the Arban granulite massif

The Arban granulite massif comprises a package of gently dipping slices and is located at the southernmost termination of the Onot domain, near the intersection of two regional fault zones dividing the Onot and Kitoy domains (Fig. 6.4-3: Sklyarov et al., 2001). The slices consist of mafic and felsic granulites alternating with thin lenses of garnet-kyanite metasedimentary schists. The felsic granulites consist of plagioclase-quartz, two pyroxenes, biotite ± microcline. The mafic granulites consist of garnet-clinopyroxene-plagioclase-amphibole-orthopyroxene and are strongly deformed and foliated as a result of retrograde metamorphism that formed brown amphibole, followed by actinolite and later chlorite.

There are two granulite varieties; garnet-enriched and leucocratic, with ⩽5–10% quartz and up to 20% quartz and plagioclase, respectively. The magmatic relict zircons from garnet-kyanite schists yielded ages ranging from 2.67 to 2.52 Ga (Table 6.4-1), which marks the time of protolith formation (Sklyarov et al., 2001). Prismatic zircons from felsic granulite yielded older ages of ca. 2.7–2.8 Ga (Turkina, 2001). The magmatic and sedimentary protoliths underwent two stages of metamorphism: Neoarchean to Paleoproterozoic high-pressure (HP) metamorphism (ca. 2.4–2.5 Ga) and Paleoproterozoic granulite-facies metamorphism (ca. 1.85–2.0 Ga). P-T estimates on HP minerals of mafic and felsic granulites are 8.0–9.5 kbar and 690–830 °C and do not accord with the occurrence of kyanite-bearing schist because the boundary of the kyanite stability field is at 10 kbar and 800 °C (Turkina, 2001). Therefore, we suggest there was a retrograde stage during the Paleoproterozoic granulite-facies metamorphism at ca. 1.9 Ga.

The microcline-free felsic granulites correspond to tonalities and trondhjemites (Table 6.4-5). They are characterized by strongly fractionated REE patterns (La/Yb$_n$ = 7–44), enrichment in LILE and LREE, and prominent negative Nb and Ti anomalies (Fig. 6.4-9(e)), suggesting that the precursors melted from a mafic source at 10–15 kbar

(Turkina, 2001). The leucocratic mafic granulites correspond to low-Ti tholeiitic basalts with nearly flat REE patterns (La/Yb_n = 0.9–1.4) and PM-normalized multi-element patterns. Compared to the leucocratic granulites, the garnet-rich granulites are enriched in FeO, TiO_2, HREE and Nb, but depleted in Na_2O, Sr, LREE, Zr. They show LREE depleted patterns (La/Yb_n = 0.3–0.6) and prominent positive Nb (Nb/La_{PM} = 1.3–2.2) and negative Sr and Zr anomalies on PM-normalized multi-element diagrams. The compositions of the garnet-rich and incompatible element depleted mafic granulites are complimentary to those of the TTG-like felsic granulites. The depletion of the garnet-rich mafic granulites in major and incompatible trace elements, which concentrate in tonalitic melts and the felsic granulites, suggests a residual origin for the mafic granulites. Thus, based on mineralogical and geochemical data, the Arban granulites could have resulted from melting of a lower crustal source region, followed by the segregation of melt and residue as a precursor of the felsic and garnet-rich mafic granulites, respectively (Turkina, 2001). The formation of the felsic granulite recorded a Neoarchean stage of TTG magmatism in the Onot domain, at lower crust depths.

6.4-2.4.2. Upper mantle and lower crust ultramafic-mafic rock assemblages

HP ultramafic-mafic rock assemblages occur within the Kitoy and Bulun domains (Sklyarov et al., 1998; Gladkochub et al., 2001; Ota et al., 2004). The Saramta massif is situated at the southern margin of the Kitoy domain, near the Main Sayan Fault (Fig. 6.4-10). Spinel peridotites and garnet websterites are the dominant lithologies. Based on parageneses of minerals and their chemistry, Ota et al. (2004) determined the minimal P-T parameters of the pre-peak (0.9–1.5 GPa, 640–780 °C) and peak (2.3–3.0 GPa, 920–1030 °C) stages of metamorphism. Peak metamorphism was followed by retrograde episodes to conditions of 750–830 °C and 0.5–0.9 GPa. The P-T history of the Saramta garnet websterites implies subduction of these rocks to a depth of ca. 100 km and subsequent exhumation to mid-crustal depths during collision of the Kitoy and Irkut domains, followed by rapid cooling. Like the Arban granulite-facies metamorphism, the eclogite-facies metamorphism of the Saramta massif could have occurred in the early Paleoproterozoic (Ota et al., 2004).

Within the Bulun domain, HP ultramafic-mafic metamorphic rocks in association with metapelites and quartzites compose layers and lenses in low-grade greenstone belts, occasionally confined to thrust and fault zones (Fig. 6.4-5). These ultramafic-mafic rocks record an early phase of HP metamorphism at 690–720 °C and ~1.4 GPa, followed by a low-pressure amphibolite-facies episode at 600 °C and 0.5–0.6 GPa (Sklyarov et al., 1998; Ota et al., 2004). Gladkochub et al. (2001) reported an Sm-Nd isochron age of 1880 ± 90 Ma from a metabasite, suggesting a later LP metamorphic event in younger Paleoproterozoic times, possibly related to collisional events accompanied by migmatization and granite emplacement. Compositionally, the precursors of the ultramafic rocks correspond to residual Fe-rich peridotites similar to those of mid-ocean ridges and back-arc spreading zones. The metagabbro and metapyroxenites possess geochemical affinities to NMORB or TMORB (Gladkochub et al., 2001). Gladkochub et al. (2001) proposed that the ultramafic-mafic rock assemblages represent a fragment of the early Paleoproterozoic oceanic lithosphere, which was buried during plate subduction and later exhumed and

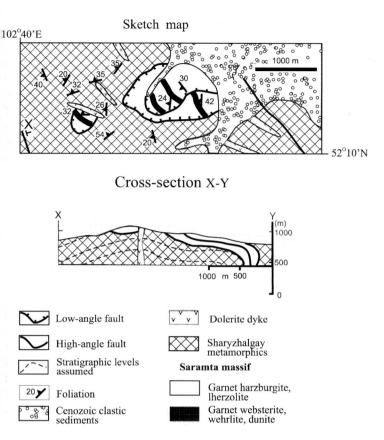

Fig. 6.4-10. Geologic sketch map and cross-section around the Saramta massif in the Kitoy block of the Sharyzhalgay complex (see Fig. 6.4-3 for location) (after Ota et al., 2004).

juxtaposed with LP metamorphic greenstones and grey gneiss complexes during Paleoproterozoic collisional events.

6.4-3. ALDAN SHIELD

6.4-3.1. Geology and Geochronology

The Aldan Shield, in the southeastern part of the Siberian Craton (Fig. 6.4-1(a)), extends almost 1200 km from east to west and 300–400 km across. It consists of the Aldan province in the north and the Stanovoy province in the south, sutured by the Kalar shear zone (Fig. 6.4-11). The Aldan province incorporates the Olekma (in the west) and the Batomga (in the east) granite–greenstone terranes, separated by the Aldan (West Aldan, Nimnyr) and Uchur (East Aldan) granulite–gneiss terranes. Metamorphism at $P \leqslant 6.0$ kbar,

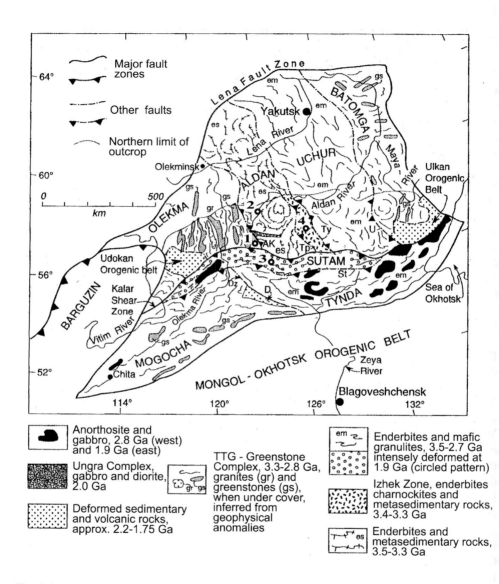

Fig. 6.4-11. Geological outline of the Aldan Shield, including the Aldan province (Olekma, Aldan, Uchur, and Batomga terranes) and the Stanovoy province (Mogocha, Tynda, and Sutam terranes and the Dzheltulak foldbelt); after Rosen et al. (1994) and Rosen (2003). D is the Dzheltulak foldbelt. Principal faults: AK = Aldan–Kilier (Amga zone); St = Stanovoi; Tp = Timpton; Ty = Tyrkanda; U = Ulkan; Dz = Dzheltulak. Circles with numbers 1–4 show localities sampled for isotopic dating.

T = 850–1000 °C conditions was widespread in the Aldan province, but conditions as high as P ⩽ 8.5 kbar, T = 950–1000 °C have been documented for the southern part (Tomilenko, 2006). The observed dip orientations of the schistosity and metamorphic banding suggest that the Aldan province was thrust over the Olekma terrane along the Aldan–Kilier fault (Amga zone). U-Pb zircon dating of collisional granites along this fault yields ages of 1966–1925 Ma. The Uchur terrane was thrust westward over the Aldan terrane along the Tyrkanda–Timpton fault system, where collisional melting took place at 1993–1925 Ma. The Batomga terrane was thrust under the Uchur terrane. Amalgamation of the Aldan province (Superterrane) proceeded in an E to W direction (in the present-day structure) (Kitsul and Dook, 1985; Rosen et al., 1994; Kotov et al., 2005).

The characteristic structural feature of the Aldan and Uchur terranes are large, oval-shaped antiforms up to 100 km across, composed mainly of enderbite and referred to as domes (Glukhovsky and Moralev, 2001; Popov et al., 1989). Both are probably a system of recumbent folds formed under collisional thrusting and later deformed into large-scale domes. The Stanovoy province (Superterrane) was thrust from the south over the Aldan province (superterrane) along the Kalar collision zone that overlies all of the Aldan province terranes. The deep-seated, high-pressure (P ⩽ 11.0 kbar, T = 1000–1100 °C Chogar complex; Tomilenko, 2006) granulite complexes of the Sutam terrane were uplifted along this collision zone. The southern terranes, Tynda and Mogocha, were accreted onto the Sutam terrane along the Stanovoy fault (Rosen et al., 1994; Rosen, 2003, and references therein).

The short outline below relies in large part on studies of geology and geochronology by B.M. Jahn et al. (1998) that summarized new results with most of the previously available data.

The *Olekma granite–greenstone terrane* is dominated by gneisses and granitoids of the TTG series (Fig. 6.4-12). Part of the gneiss protoliths formed as early as the Paleoarchean, as evidenced by SHRIMP U/Pb zircon dates of 3212 ± 8 Ma (Nutman et al., 1990, 1992a). Whereas the majority of gneisses have ages of ca. 3.0 Ga and are practically contemporaneous with granitoids of the multi-phase Amnunnacta, Oldongso and Tungurchakan plutons that are composed of tonalite, trondhjemite and rare gabbro (Table 6.4-6). The grey gneiss domains are separated by four longitudinal greenstone belts that extend for 300 km in length and are up to 30 km wide. These belts are composed of supracrustal sequences and are bounded by blastomylonitic contacts with TTG grey gneiss basement.

Metavolcanic and metasedimentary rocks with carbonate and terrigenous clastic sedimentary units make up the tectonic fragments of the Temulyakite and Tungurcha belts. Compositionally, the metavolcanic rocks range from rare komatiite and dominant tholeiite basalt to intermediate and felsic volcanics (Puchtel et al., 1993). The metakomatiite-basalt, andesite-dacite and terrigenous units dominate the Tokko-Khani greenstone belt

---

Fig. 6.4-12. (*Next page.*) The oldest rock assemblages in the Olekma terrane, Aldan province (Smelov et al., 2001; Kotov et al., 2005; Rosen, 2002). Circles with numbers 1–4 show localities sampled for isotopic dating, shown in Fig. 6.4-11.

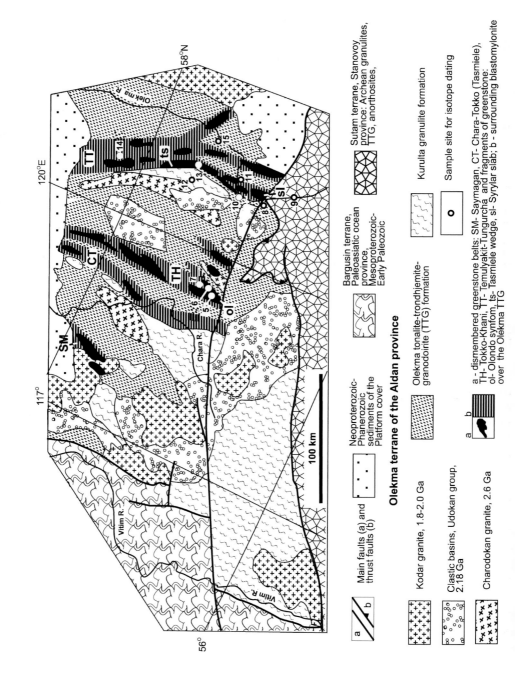

Table 6.4-6. Representative older U-Pb zircon age estimations on the Aldan shield terranes

Age, Ma	Method	Zircon type	Rock, complex	Locality[a]	Reference
Olekma granite–greenstone terrane					
3212 ± 8	U-Pb, SHRIMP	Magmatic zircon	TTG gneiss, Olekma formation	10	Nutman et al., 1992
3018 ± 10	U-Pb, SHRIMP	Magmatic zircon	Tonalite gneiss, Gabbro-diorite-tonalite pluton, Olondo greenstone belt	7	Baadsgaard et al., 1990
3016 ± 8	U-Pb, SHRIMP	Magmatic zircon	Trondhjemite gneiss, syntectonic veins, Temulyakit greenstone belt	8	Nutman et al., 1992
3006 ± 11	U-Pb, SHRIMP	Magmatic zircon	Felsic metavolcanics, Olondo greenstone	6	Nutman et al., 1992
3001 ± 3	U-Pb, conv.	Magmatic zircon	TTG gneiss, Tungurchakan pluton	11	Salnikova et al., 2003
2999 ± 51	U-Pb,	Magmatic zircon	Trondhjemite gneiss, Oldongso pluton	15	Salnikova et al., 2003
2998 ± 9	U-Pb, SHRIMP	Magmatic zircon	Felsic metavolcanics, Olondo greenstone	5	Nutman et al., 1992
2984 ± 22	U-Pb,	Magmatic zircon	Metagabbro, metadiorite, plagiogranite, Amnunnakta pluton	14	Neymark et al., 1993
2983 ± 9	U-Pb, SHRIMP	Magmatic zircon	Felsic metavolcanics, Olondo greenstone	6	Nutman et al., 1992
~2980	U-Pb, conv	Magmatic zircon	Metadacite, Tasmiele greenstone belt	13	Nemchin, 1990
2960 ± 70	U-Pb, conv	Magmatic zircon	Metaandesite-dacite, Olondo greenstone belt	6	Bibikova et al., 1984
2862 ± 7	U-Pb	Magmatic zircon	Tonalite gneiss, Olekma formation	12	Baadsgaard et al., 1990
Aldan granulite–gneiss terrane					
⩾3350, 2700	U-Pb, SHRIMP	The same	TTG gneiss	2	Jahn et al., 1998
3335 ± 3	U-Pb, SHRIMP	Magmatic zircon	TTG gneiss	2	Nutman et al., 1992
⩾3200, 2800	U-Pb,	Zircon	TTG gneiss	2	Jahn et al., 1998
3005 ± 4	U-Pb, conv	Magmatic zircon	Tonalite gneiss	2	Salnikova et al., 1997
Sutam granulite–gneiss terrane					
3131 ± 74	U-Pb, conv	Magmatic zircon	Charonckite	3	Shemyakin et al., 1998

[a]Localities Nos. 1–4 are on Fig. 6.4-16, the others – on Fig. 6.4-17.

(Puchtel and Zhuravlev, 1989; Zhuravlev et al., 1989). The greenstone sequences have been metamorphosed to epidote-amphibolite facies. Kotov (2003) subdivided several ages of greenstones. Earlier greenstones include the Syrylyr and Olondo tectonic fragments. Within the Syrylyr fragment, metasedimentary (meta-sandstone to pelitic) rocks have Nd model ages of 3.2–2.8 Ga and are cut by tonalite sheets dated at 3016 ± 8 Ma (Nutman et al., 1992a). Andesites and dacites of the Olondo fragment yielded magmatic zircon U-Pb ages and the Nd model ages of ca. 3.0 Ga (Tables 6.4-6, 6.4-7). The younger green-stone sequences (Temulyakite, Tungurchakan and Subgan tectonic fragments) consist of metavolcanic rocks with Nd model ages of 3.0–3.1 Ga. Associated metasedimentary rocks vary in T(Nd)$_{DM}$ values from 2.8 to 3.6 Ga (Anisimova et al., 2005). The Subgan green-stone deposits are intruded by gabbro dated at 2910 ± 50 Ma (Nemchin, 1990).

The *Aldan granulite–gneiss terrane* consists of orthogneisses, granitoids and enderbite-gneiss domes separated by metasedimentary deposits. SHRIMP U-Pb dating of tonalitic grey gneisses from the Aldan river yielded a complex ^{207}Pb/^{206}Pb age pattern in the range 2650 to 3350 Ma, whereas the oldest concordant age is 3335 ± 3 Ma (Nutman et al., 1992a). Jahn et al. (1998) reported SHRIMP zircon ^{207}Pb/^{206}Pb ages of $\geqslant$3.2, ~2.8 and 2.2 Ga, emphasizing that the zircon U-Pb systematics have been severely disturbed by later thermal effects. A Mesoarchean age of 3005 ± 4 Ma has been recorded in prismatic zircons from amphibole-biotite tonalitic gneiss, whereas the emplacement of most of granitoids and charnockites occurred in the Paleoproterozoic (1.96–2.4 Ga: Salnikova et al., 1993, 1997). Thus, the Archean crust of the Aldan terrane had been extensively remobilized during Proterozoic tectonothermal events related to collision and amalgamation of crustal terranes (Frost et al., 1998; Nutman et al., 1992a).

Granulites from the *Sutam granulite–gneiss* terrane show U-Pb zircon ages of 3.1 Ga (Shemyakin et al., 1998), Sm-Nd ages of 3.0–3.2 (Kovach et al., 1995a; Jahn et al., 1998), and a tonalite gneiss age of 3.0–3.1 Ga (Jahn et al., 1998; Shemyakin et al., 1998).

6.4-3.2. Sm-Nd Isotopic Data and Important Crust-Forming Events

Based on T(Nd)$_{DM}$ data of 3.4–3.7 Ga (Table 6.4-7, symbols for minerals are according to Kretz (1983)), part of the Olekma grey gneisses have a long crustal prehistory, supporting the contribution of an ancient crustal source to the geological history of this area. Most of grey gneisses have Nd model ages of ca. 3.2–3.3 Ga, which only slightly differ from their emplacement U-Pb ages of ca. 3.0–3.2 Ga. The intrusive TTG rocks (including the Amnunakta, Oldongso and Tungurchakan plutons) and meta-andesite and -dacites have model Nd ages of ca. 3.0 Ga, similar to their U-Pb zircon ages. Within the Olekma terrane, the juvenile character of most of Archean TTG rocks suggests that major crust-forming events occurred at ~3.0 Ga (Jahn et al., 1998).

Across the west margin of the Aldan granulite–gneiss terrane, the T(Nd)$_{DM}$ ages of orthogneisses fall into two groups: one at 3.5–3.6 Ga and another at 3.7–3.8 Ga (see review in Kotov, 2003). This data suggests that ca. 3.0–3.2 gneisses formed through recycling of older crustal material. In this area, most Paleoproterozoic (1.9–2.4 Ga) intrusive granitoids and charnockites also yield Archean Nd model ages, up to 3.3 Ga (Kotov et al., 1993, 1995;

Table 6.4-7. Representative older Sm-Nd age estimates of Aldan shield terranes

Age, Ma	Method	Material	Rock, complex	Locality[a]	Reference
Olekma granite–greenstone terrane					
3687	T(Nd)DM	WR[c]	Grt-Bt-plagiogneiss, Amedichi group	1	Kovach et al., 1995a
3.53[b]	T(Nd)DM	WR	TTG gneiss	10	Jahn et al., 1998
3488	T(Nd)DM	WR	Grt-Bt-gneiss, Kurulta formation, Olomokit block	12	Kovach et al., 1995b
3292	T(Nd)DM	WR	Grt- Bt- plagiogneiss, Amedichi group	1	Kovach et al., 1995a
3232 $\pm$ 199	Sm-Nd isochron	WR	Ultramafic metavolcanics, Tungurcha group	10	Puchtel et al., 1993
3.10[b]	T(Nd)DM	WR	Trondhjemite gneiss	5	Jahn et al., 1998
3094 $\pm$ 430	Sm-Nd isochron	WR	Metagabbro, metadiorite and plagiogranite, Amnuuakta pluton	14	Neymark et al., 1993
3044 $\pm$ 95	Sm-Nd isochron	WR	Tonalite, Olekma formation, Tungurchakan pluton	11	Salnikova et al., 1993
3.03[b]	T(Nd)DM	WR	TTG gneiss	10	Jahn et al., 1998
3.02[b]	T(Nd)DM	WR	TTG gneiss	10	Jahn et al., 1998
3.00[b]	T(Nd)DM	WR	Dioritic gneiss	5	Jahn et al., 1998
2966 $\pm$ 16	Sm-Nd isochron	WR	Mafic and ultramafic metavolcanics, Olondo group	5	Zhuravlev et al., 1989
Aldan granulite–gneiss terrane					
3.64[b]	T(Nd)DM	WR	Tonalitic gneiss	2	Jahn et al., 1998
3.55[b]	T(Nd)DM	WR	Tonalitic gneiss	2	Jahn et al., 1998
3.54[b]	T(Nd)DM	WR	Amphibolite	2	Jahn et al., 1998
3456	T(Nd)DM	WR	Crd-Grt-Bt-gneiss Kurumkan group	2	Kovach et al., 1996
3.28[b]	T(Nd)DM	WR	Granite	2	Jahn et al., 1998
3013	T(Nd)DM	WR	Grt-Sil-Crd-gneiss, Kurumkan group	2	Kovach et al., 1996
3.08[b]	T(Nd)DM	WR	Qtz-dioritic gneiss	2	Jahn et al., 1998
2.95[b]	T(Nd)DM	WR	Qtz-dioritic gneiss	2	Jahn et al., 1998
2.94[b]	T(Nd)DM	WR	Basic granulite	2	Jahn et al., 1998

(*continued on next page*)

Table 6.4-7. (*Continued*)

Age, Ma	Method	Material	Rock, complex	Locality[a]	Reference
Sutam granulite–gneiss terrane					
3.22[b]	T(Nd)DM	WR	Basic granulite, Larba block	3	Jahn et al., 1998
3130 ± 180	Sm-Nd isochron	WR	Grt-Opx-Cpx-mafic granulite, Kurulta–Khani block	9	Jahn et al, 1990
3.07[b]	T(Nd)DM	WR	Tonalitic gneiss	9	Jahn et al., 1998
3038	T(Nd)DM	WR	Opx-GRt-Bt-plagiogneiss, Kurulta–Khani block	9	Kovach et al., 1995a
3000	T(Nd)DM	WR	Tonalite gneiss	3	Shemyakin et al., 1998

Symbols for minerals are from Kretz (1983).
[a]Localities Nos 1–3 are on Fig. 6.4-16, the others – on Fig. 6.4-17.
[b]Data of Jahn et al. (1998) present in Ga as given by authors, included are samples with $^{147}Sm/^{144}Nd < 0.16$.
[c]WR means "whole rock".

Frost et al., 1998). This is used as evidence of granitoid formation by partial melting of earlier Archean sources, with variable input either of mantle or Proterozoic crustal material. The latter could be represented by the metasedimentary rocks with $T(Nd)_{DM}$ ages of 2.3 to 3.5 Ga (Kurumkan Group) (Kovach et al., 1996). Thus, the continental crust of the Aldan terrane started to form in the Paleoarchean and was extensively remobilized in the Paleoproterozoic.

Mafic granulites of the Sutam terrane yield $T(Nd)_{DM}$ ages of 3.0–3.2 Ga, and tonalite gneisses yield ages of ca. 3.0 Ga, similar to the Aldan terrane. The whole-rock Sm-Nd isochron age of 3.13 Ga (Jahn et al., 1998) is interpreted as the age of the main magmatic event in this area.

According to Kotov (2003), there are several periods of crustal-forming processes: ca. 3.5–3.8, 3.2–3.3 and 2.9–3.0 Ga. Paleoarchean ($\geqslant 3.2$ Ga) continental crust extends through the eastern part of the Olekma terrane, the western margin of Aldan terrane, and their junction zone. Juvenile Mesoarchean crust (ca. 2.9–3.0 Ga) underlies most of the Olekma terrane. The oldest (3.0–3.3 Ga) metasedimentary greenstone deposits of the Olekma terrane formed through erosion of Paleoarchean crustal domains, as evidenced from Nd model ages ranging from 3.2 to 3.8 Ga (Anisimova et al., 2005).

Paleoarchean ages do not appear in the other parts of the Aldan Shield. Several Sm-Nd data from the Uchur and Batomga terranes are in the range <2.5–1.9 Ga. These ages are interpreted as younger periods of crust formation in these areas (Kotov, 2003), but confirmation by U-Pb zircon or Sm-Nd-mineral isochron dating is absent and remains to be tested. On the whole, we concur with Jahn et al. (1998) that crustal evolution of the Aldan shield started at ca. 3.5 Ga and culminated at ~3.0 Ga.

6.4-4. ANABAR SHIELD

The Anabar Shield exposes the north-eastern part of the Siberian Craton, which in this area is composed of the Anabar province, including the Magan and Daldyn terranes, and the Olenek province (Birekte terrane) (Rosen et al., 1994, 2000, and references therein; Rosen, 2003) (Figs. 6.4-1 and 6.4-13). High quality age determinations are summarized in Table 6.4-8.

6.4-4.1. Geology and Geochronology

The Magan terrane consists mainly of orthopyroxene-plagioclase gneisses (plagiogneiss), enderbites and charnockites, with rare two-pyroxene metamafic rocks, all of which are named as a whole as the Upper Anabar Group. The northeastern, outcropping part of this terrane includes the Vyurbyur fold belt, composed of meta-carbonates (marbles and calc-silicate rocks), metagraywackes (garnet gneisses), metamafics, and plagiogneisses. This terrane formed at 2.4 Ga (zircon dating) as an active continental margin flanking a basement not older than 3.09 Ga (Sm-Nd model age dating of metagraywackes).

The Daldyn terrane predominantly consists of two-pyroxene metamafic rocks, plagiogneisses (including enderbites), with rare, small charnockite bodies and layers of metacarbonates and orthopyroxene quartzites that comprise the Daldyn Group. An age of ca. 3.3 Ga (SHRIMP U-Pb zircon dating; Rosen et al., 1991) and 3.35 Ga (U-Pb zircon conventional multi-grain dating) appears to be the primary age of the basement, whereas 3.0–3.2 Ga (Sm-Nd whole rock age data) reflects the effects of later contamination. Granulite-facies metamorphism and anatexis occurred at 2.8 and 1.8–2.0 Ga.

In both terranes, metapyroxenite and olivine-bearing meta-ultramafic granulites occur sporadically, as does charnockite. Metamorphic conditions correspond to granulite facies (Fig. 6.4-14). Differences between those two terranes are evident from the proportion of metamafics to plagiogneisses, equating to 1:7 for the Magan and 1:3 for the Daldyn terranes (Rosen et al., 2002).

The rocks are deformed into isoclinal folds with limbs spanning a few kilometers. They are fragmented by faults along strike and are steeply inclined E-NE as a result of thrusting of terranes during collisional orogeny. The layered structure of the observed sequences and the interstratification of plagiogneisses and metamafic rocks with metasedimentary rocks, combined with geochemical data, are used to infer an island arc assemblage.

The Birekte granite–greenstone terrane is covered by the Khapchan fold belt on the west and the Aekit belt on the east, and by platform sedimentary cover rocks. The Khapchan fold belt comprises granulite-facies metagraywackes (garnet gneisses) and metacarbonates. Sm-Nd model ages of metagraywackes are 2.3–2.4 Ga and show that eroded basement (the Birekte terrane) was no older than the Archean-Proterozoic boundary, whereas Sm-Nd mineral isochron dates indicate that granulite-facies metamorphism of these metasedimentary rocks occurred at 1.9 Ga. The Aekit belt is composed of felsic acid volcanic rocks and terrigenous and carbonaceous sedimentary rocks metamorphosed under greenschist-facies conditions and intruded by granitoids at 1.8 Ga.

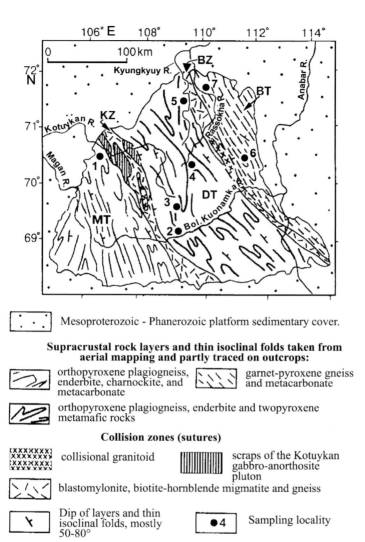

Mesoproterozoic - Phanerozoic platform sedimentary cover.

**Supracrustal rock layers and thin isoclinal folds taken from
aerial mapping and partly traced on outcrops:**

orthopyroxene plagiogneiss, enderbite, charnockite, and metacarbonate

garnet-pyroxene gneiss and metacarbonate

orthopyroxene plagiogneiss, enderbite and twopyroxene metamafic rocks

Collision zones (sutures)

collisional granitoid

scraps of the Kotuykan gabbro-anorthosite pluton

blastomylonite, biotite-hornblende migmatite and gneiss

Dip of layers and thin isoclinal folds, mostly 50-80°

Sampling locality

Fig. 6.4-13. Generalized geology of the Anabar Shield, showing structure and sampling localities for isotopic studies. Sampling localities: 1. Magan terrane supracrustal metamorphics, middle Kotykan River and its left tributary, Vyurbyur and anorthosites sampled 40 km eastward; 2–5 are from the Daldyn terrane. 2. Middle Bolshaya Kuonamka River and the inflow of its left tributary Daldyn. 3. Headwaters of the Daldyn River, from the confluence of the Ulakhan-Daldyn and Aschagyi-Daldyn streams, headwaters of Kotuykan River and the Khatyryk River with its right tributary, the Saryga stream. 4. Headwaters of Rassokha River and its left tributary, the Schenelekh stream. 5. Middle Kyungkyuy River; 6–7 are from the Birekte terrane. 6. Lower reach of the Khaptasynnakh River. 7. Right-hand bank of the Rassokha River. MT, DT, BT = Magan, Daldyn, and Birekte terranes, respectively; KZ, BZ = Kotuykan and Billyakh collision zones.

Table 6.4-8. Representative age estimations on the Anabar shield

Age, Ga	Method	Material	Rock, terrane	Locality[a]	Reference
Magan terrane					
2.85	Sm-Nd isochron	WR[b]	marble, calc-silicate rock, Grt-gneiss	1	Zlobin et al., 1999
2.84–3.09	T(Nd)DM	WR	metagraywackes and metacarbonates	1	Rosen et al., 2000
2.80–3.01	T(Nd)DM	WR	metavolcanic rocks	1	Rosen et al., 2000
2.42 ± 0.02	U-Pb, conv.	Magmatic zircon	twopyroxene-plagioclase gneiss	1	Bibikova et al., 1988
Daldyn terrane					
3.35 ± 0.1	U-Pb, conv.	Zircon	metamafic rock	2	Stepanyuk et al., 1993
3.32 ± 0.1	U-Pb, SHRIMP	Zircon	plagiogneiss-metaandesite	2	Bibikova et al., 1988
3.16 ± 0.2	T(Nd)DM	WR	enderbite	3	Rosen et al., 2000
3.1 ± 0.08	Sm-Nd isochron	WR	plagiogneisses and metamafic rocks	3	Spiridonov et al., 1993
3.0 ± 0.02	U-Pb, conv.	Zircon	plagiogneiss-metaandesite	2	Bibikova et al., 1988
3.00–3.19	T(Nd)DM	WR	quartzites, metagraywackes, and metacarbonates.	3	Zhuravlev and Rosen, 1991
2.76 ± 0.02	U-Pb, conv.	Metamorphic zircon	plagiogneiss (metadacite)	2	Bibikova et al., 1988
2.18 ± 0.03	T(Nd)DM	WR	enderbite	4	Rosen et al., 2000
1.97 ± 0.2	U-Pb, conv.	Metamorphic zircon	Grt-gneiss (metagraywacke)	6	Bibikova et al., 1988
1.94 ± 0.03	Sm-Nd isochron	Minerals	Grt-Pl-gneiss	3	Rosen et al., 2000
1.90 ± 0.07	Sm-Nd isochron	Minerals	enderbite	4	Rosen et al., 2000
1.8 ± 0.02	Sm-Nd isochron	Minerals	2Px-Pl-mafic granulite	2	Stepanyuk et al., 1993
Birekte terrane					
2.32–2.44	T(Nd)DM	WR	garnet gneisses, metagraywackes and metacarbonates	7	Zhuravlev and Rosen, 1991
1.92 ± 0.003	Sm-Nd isochron	Minerals	Grt-gneisses, metagraywackes	7	Rosen et al., 2000
1.91 ±0.014	Sm-Nd isochron	Minerals	Grt-gneisses, metagraywackes	7	Rosen et al., 2000

Table 6.4-8. (*Continued*)

Age, Ga	Method	Material	Rock, terrane	Locality[a]	Reference
Kotuykan collision fault zone					
1.92 ± 0.1	U-Pb, conv.	Magmatic monazite.	syntectonic migmatites	1	Stepanov, 1974
1.87–1.84	U-Pb, conv.	Magmatic zircon	veined Bt-granite	1	Stepanov, 1974
1.84 ± 0.02	U-Pb, conv.	Magmatic zircon	veined Mc-granite	2	Stepanyuk, 1991
Billyakh collision fault zone					
1.97 ± 0.02	U-Pb, conv.	Metamorphic zircons. concordant	Grt-gneisses (metagraywacke)	7	Bibikova et al., 1988
Anorthosites					
2.55 ± 0.05	Sm-Nd isochron	WR	Px-anorthosites, gabbro-norites, and pyroxenites	1	Rosen et al., 2000
2.01	T(Nd)DM	WR	dolerite dyke	1	Rosen et al., 2000

[a]Geographical position shown as point numbers on Fig. 6.4-13.
[b]WR means "whole rock".

Sm-Nd isotope systematics of the Anabar Shield indicate that the granulite–gneiss terranes were separated from a depleted mantle at different times and thus probably represent independent sialic crustal blocks, possibly microcontinents. The average values of ^{147}Sm/^{144}Nd $= 0.11$–0.12 show that the terranes may be categorized as mature continental crust (Fig. 6.4-15).

Collision-related fault zones are interpreted to represent the deeper crustal levels of relict sutures. These structures are steeply inclined (50–80°) décollements along which the terranes were thrust from the ENE to the WSW (in present-day structure) and are represented by zones of between 1–30 km in width, consisting of blastomylonite and melange. The biotite ± hornblende gneisses in the matrix of these zones formed after mylonitization and amphibolite-facies metamorphism. The matrix contains concordant and cross-cutting, vein-like migmatites and bodies of autochthonous granites. Kilometre-scale allochthonous inliers of anorthosite and granulites from adjacent terranes are present. Metamorphosed island arc, calc-alkaline metavolcanics, metacarbonates, quartzites and granodiorite occur within the gneisses, as relics of inferred oceanic rock-forming processes between terranes (Smelov et al., 2002). Syntectonic migmatite and granite from the blastomylonite zones yield 1.8–1.9 Ga metamorphic ages.

Anorthosites and accompanying rocks cover an area of about 1200 km². They form a series of massifs in the northern parts of the Kotuykan zone and also in the Billyakh zone and Magan terrane (Sukhanov et al., 1990). Most of the massifs are composed of

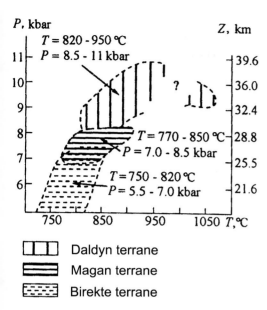

Fig. 6.4-14. Metamorphic P-T parameters in the terranes of the Anabar Shield (Vishnevskiy, 1978; Lutts, 1985; Lutts and Oksman, 1990; Rosen, 1995). Representative mineral assemblages are: Plagiogneisses = Grt_{65-78} + Opx_{44-55} + Pl_{30-46} + Qtz + Mgt ± Cpx; Metamafics = Zpx_{41} + Cpx_{36} + Hbl_{43} + Grt_{66} + Pl_{60} + Mgt; Meta-ultramafics = Opx_{19} + Cpx_{14} + Hbl_{17} + Spl; Metapelites = Opx_{36} + Grt_{43} + Spr_{26} + Crd_{41} + Sil. Abreviations according to Kretz (1993). Subscript number indicates the Fe/(Fe+Mg) ration for mafic minerals, and the anorthite component of plagioclase.

anorthosites. Gabbros, gabbronorites, pyroxenites, and jotunites (monzodiorites) occupy about 20% of the exposed area. All of these rocks lie on a Sm-Nd whole-rock isochron of 2.55 Ga, which may be related to the possible collisional event. Cross-cutting dykes and sills of dolerite, up to 100 m thick and up to 1 km long, also exist in the anorthosite massifs, and yield Sm-Nd whole-rock model ages of ca. 2.01 Ga. These intrusions are contemporaneous with late metamorphism resulting from collision.

On the whole, the laminated plagiogneiss-metamafic assemblage of the Daldyn terrane presents the oldest rocks in the Anabar Shield, at, or older than, 3.2 Ga. On first appearances, the primary assemblage of sediments and volcanic rocks may be interpreted as island arc assemblage.

6.4-4.2. Geochemistry

The first systematic sampling for petrogenic element analyses was completed for an international conference (Rosen, 1990; Bridgwater et al., 1991), followed by trace element analyses using the INAA method. Altogether, 49 samples were analyzed from the Anabar

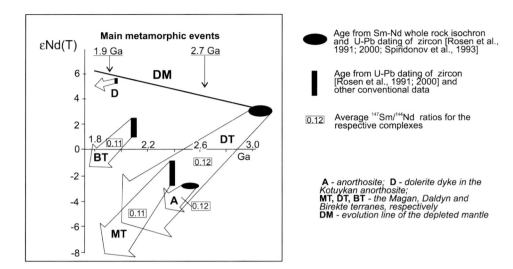

Fig. 6.4-15. Diagram of the Sm-Nd isotope system evolution in the Anabar Shield (Rosen et al., 2000).

Shield. The results allowed for separation of the most widespread varieties, of which representative samples were analyzed by the ICP-MS method, as shown in Table 6.4-9.

The Daldyn terrane is dominated by plagiogneisses (including enderbites), which are characterized by variable SiO_2 content (54–72 wt%) and could have formed from andesitic to rhyodacitic precursors. Concentrations of TiO_2, FeO^*, MgO and CaO decrease as SiO_2 increases, suggesting magmatic fractionation. The gneisses are compositionally similar to low- and intermediate potassium calc-alkaline volcanic rocks ($K_2O/Na_2O = 0.1$–0.3) and they have high contents of CaO and Sr (300–500 ppm) and are thus closely comparable with the Archean TTG series. In comparison with typical Archean TTG, however, the gneisses display a smaller depletion in HREE and Y, have moderately fractionated REE patterns, and have La/Yb_n ratios of 35–47 (Fig. 6.4-16(a)). Primitive mantle-normalized trace element patterns display LILE and LREE enrichment, prominent negative Nb and Ti anomalies, and weak positive Sr and Zr anomalies, which are comparable with those of TTG series rocks. All these geochemical features support a subduction-related genesis, as previously suggested by Rosen et al. (1988, 1991). Plagiogneisses have similar Nd model and U-Pb zircon ages that preclude any crustal contribution in their genesis. The primary intermediate to felsic melts could have been generated either through fractional crystallization of a more mafic magma, or by partial melting of a mafic source. In the latter case, the melt was more likely derived in equilibrium with amphibole, rather than garnet; i.e., at a depth of <30 km, since the gneisses are not extremely depleted in HREE and Y.

Most of the gneisses are extremely depleted in Rb (3–30 ppm) and Th ($\leqslant$2 ppm), similar to the Lewisian granulite-facies gneisses from northwest Scotland (Weaver and Tarney, 1981). These features could be a result of the removal of fluid-mobile elements by means of

Table 6.4-9. Major and trace element analyses of the Daldyn terrane, the Anabar shield

	3[a]				4[a]			
	123-90	128a-90	2-88	131-90	158-90	148-90	155-90	157-90
(wt%)								
SiO_2	61.51	63.2	48.97	70.3	49.46	53.81	72.58	57.05
TiO_2	0.761	0.53	0.982	0.352	1.135	0.937	0.25	0.282
Al_2O_3	16.24	17.85	14.1	16.02	14.03	19.22	15.13	19.36
Fe_2O_3t	6.99	4.7	14.73	2.68	12.64	7.98	1.92	6.05
MnO	0.113	0.061	0.254	0.031	0.219	0.109	0.03	0.11
MgO	3.17	1.72	7.42	0.89	7.39	3.48	0.7	3.9
CaO	5.45	4.92	11.44	3.8	11.27	7.55	3.65	6.34
Na_2O	4.12	5.02	1.91	4.5	2.83	4.43	4.31	5.16
K_2O	0.98	1.06	0.28	0.98	0.6	1.41	0.66	0.88
P_2O_5	0.159	0.266	0.06	0.072	0.097	0.537	0.03	0.03
LOI	0.2	0.58	0.04	0.36	0.72	0.54	0.5	0.71
Total	99.69	99.91	100.19	99.99	100.39	100.00	99.76	99.87
(ppm)								
Th	1.23	1.53	0.074		0.81	0.96		
U	0.36	0.2	0.067		0.13	0.34		
Rb	16.7	8	1.59		5.6	21		
Ba	381	578	38		247	1110		
Sr	319	425	126		235	1837		
La	20	30	2.3		12.4	66		
Ce	39	55	6.7		26	141		
Pr	5.1	6.6	1.2		3.8	20		
Nd	19.2	23	6.7		15.4	79		
Sm	3.6	3.7	2.5		3.6	13.9		
Eu	0.92	0.89	0.91		1.15	3		
Gd	3.5	2.9	3.3		4.1	11.1		
Tb	0.52	0.33	0.63		0.66	1.36		
Dy	2.9	1.6	4.5		4.2	6.8		
Ho	0.53	0.28	0.93		0.82	1.18		
Er	1.54	0.69	2.7		2.2	3.5		
Tm	0.26	0.1	0.43		0.37	0.47		
Yb	1.67	0.58	2.6		2.4	2.9		
Lu	0.22	0.079	0.36		0.33	0.39		
Zr	129	172	42		79	321		
Hf	3	4.4	1.44		2.2	7.2		
Ta	0.69	0.48	0.18		0.59	0.49		
Nb	9.4	4.8	1.93		7.3	8		
Y	17.5	8.3	26		24	39		

[a]Geographical position shown as point numbers on Fig. 6.4-13.

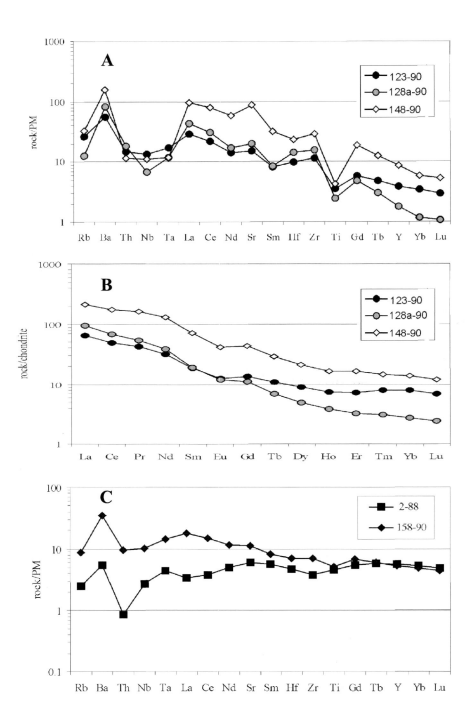

Fig. 6.4-16. Primitive mantle (PM)-normalized multi-element and chondrite-normalized REE patterns for ancient rocks in the Anabar Shield (Daldyn Group): (A) and (B) are data for plagiogneisses and enderbites, respectively; (C) Metamafites.

dehydration under granulite facies metamorphic conditions. On the other hand, Rollinson and Tarney (2005) suggested that such LILE depletion can be a primary feature of the gneisses derived from magmatic precursors depleted in Th and Rb because they formed from a mafic source depleted in fluid-mobile trace elements prior to melting.

Unlike the gneisses, the low-potassium enderbites are enriched in Al_2O_3, Ba, Sr, Zr, Y, REE and show weak negative Eu anomalies. These geochemical features provide evidence for their formation through partial melting of a gneiss or mafic source.

The two-pyroxene metamafic rocks (hereinafter referred as to metamafites) are associated with gneisses and correspond to low-Mg tholeiitic basalts with nearly flat or weakly fractionated REE patterns ($La/Yb_n = 0.6$–3.5). The PM-normalized trace element patterns for metamafites display weak enrichment in Ba and negligible to moderately negative Nb anomalies (Fig. 6.4-16(c)). All metamafites have 1.0 to 1.6 wt% TiO_2 and no correlated negative Nb and TiO_2 anomalies. The metamafites have Nb/La_{PM} ratios of from 0.8 to 0.5, suggesting either minor input of subducted crustal material into the mantle, or contamination of a primary basaltic melt by continental crust during eruption. It should be emphasized that Ce/Nb ratios (2.8–5) for Daldyn metamafites are distinctly lower than those of island arc (Ce/Nb > 6) and back-arc basalts (Ce/Nb > 3) (Saunders et al., 1988). In the Zr/Y vs Nb/Y diagram of Fitton et al. (1997), the metamafites plot within the "Iceland Array", near the PM value; i.e., they are similar to oceanic or mantle plume basalts (Fig. 6.4-17). Many of the Archean non-arc greenstone basalts plot in the same area, near the primitive mantle composition (Condie, 2005).

6.4-5. DISCUSSION

The Siberian Craton incorporates superterranes (provinces) and component terranes. Terranes are subdivided into granite–greenstone and granulite–gneiss varieties, according to the predominant rock types. These different types of terranes are interpreted to represent Archean microcontinents that collided during later assembly at 2.6–2.8 and 2.0–1.7 Ga. During both of these events, granitic melts were generated from older crust within inferred sutures, while regional high-grade metamorphism resulted in geochemical and isotopic contamination of the older crustal material within terranes. This resulted in the fact that U-Pb zircon dates frequently conform to the Sm-Nd residence time estimates.

The southernmost part of the Tungus province in the Sharyzhalgay uplift contains the granite–greenstone Onot and Bulun belts. The Sharyzhalgay granulite–gneiss group aggregating the Irkut and Kitoy granulite–gneiss domains but having unclear outer boundaries surrounds those. The granite–greenstone and granulite–gneiss domains mainly contain

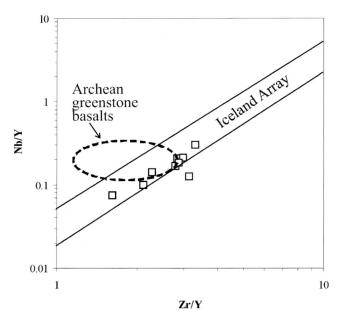

Fig. 6.4-17. Distribution of Daldyn metamafites (open squares) on a Zr/Y–Nb/Y diagram. The "Iceland Array" is from Fitton et al. (1997); the field for Archean greenstone basalts is from Condie (2005).

TTG formation. That is presented as intrusions of tonalite, trondhjemite and granodiorite, as well as gneisses of the same composition and both usually free named as grey gneisses. In the Irkut and Kitoy granulite–gneiss domains, they probably melt out in a shallow sub-duction process from garnet bearing source at ~55 km depth (P = 15 kbar). Included in TTG, amphibolitic enclaves probably originate as oceanic plateau basalts from a mantle plume. The most plausible mechanism for thick crustal growth accompanied by remelt-ing and TTG formation is through the accretion of oceanic plateaux with an incorporated fragment of ancient sialic crust.

The Onot and Bulun granite–greenstone domains yield Sm-Nd WR model ages of TTG, $T(Nd)_{DM}$, of 3.57 Ga and 3.51 Ga, respectively. This seems to be more reliable as melt ages whereas U-Pb zircon ages of 3.41–3.28 Ga mark the events of reworking and contamination of TTG. The greenstone belts in both domains are composed of tholeiite metabasalt of intra-oceanic setting, with basalt having undergone contamination by sialic crust or input of crustal material into the mantle source region. Accompanying felsic volcanic rocks were sourced from the grey gneiss basement complex at ~35 km depth (P ~ 10 kbar). The Onot TTG formed at $T(Nd)_{DM} = 3.59$ Ga and the Bulun TTG at 3.31 Ga. The Onot and Bulun intermediate and felsic greenstones are indicative of the lateral extent of Paleoarchean grey gneiss complexes, which are not limited to the exposed sheet- or slice-like fragments.

The Irkut and Kitoy granulite–gneiss domains contain metavolcanic rocks that originated in a subduction-related environment. Zircon cores from intermediate and felsic granulites yield ages of ca. 3.3–3.4 Ga, whereas the Nd crustal residence time ranges from 3.1 to 3.4 Ga. We suppose that this was the time when the island arc assemblage of the Sharyzhalgay group surrounded the earlier blocks of sialic crust represented by the Onot and Bulun domains.

Thus the thick, rigid and stabile craton-like continental crust of the southernmost Tungus terrane has probably formed in the Mesoarchean. The later collision processes were rather disastrous. The crustal tonalites descended to the crust-mantle boundary (~55 km) and then uplifted (Arban granulite massif, Onot domain). HP-metamorphism at 2.4–2.5 Ga changed to granulitic metamorphism ca. 1.85–2.0 Ga.

Mafic-ultramafic fragments of the crust-mantle sequences (Saramta massif, Kitoy domain) went down to ~100 km depth and were later exhumed and juxtaposed with LP metamorphic greenstone and grey gneiss complexes. HP ultramafic-mafic rocks possibly recorded subduction processes, followed by collision of the Kitoy and Irkut domains. The Sm-Nd isochron age of the last process is 1.88 Ga.

The Aldan Shield, as a large area where the Precambrian crops out, occupies the Aldan and Stanovoy highlands. In the Aldan province, the Olekma granite–greenstone terrane and gneisses and granitoids of the TTG series predominate. TTG gneisses yield $T(Nd)_{DM}$ ages = 3.69–3.49 Ga, whereas tonalite, trondhjemite, metadiorite and metagabbro mainly intruded at 3.1–3.0 Ga. Mafic and ultramafic volcanics in the greenstone belts mainly formed at 3.0–3.2 Ga, as well as felsic volcanics that possibly resulted from extension of the early TTG crust. Associated metasedimentary rocks vary in $T(Nd)_{DM}$ values from 2.8 to 3.6 Ga and suggest deep erosion of the surrounding TTG basement.

In the Aldan granulite–gneiss terrane, tonalitic grey gneisses prevail with $T(Nd)_{DM}$ ages = 3.64–3.46 Ga. The other such gneisses, dioritic gneisses, and granites yield ages of 3.0–3.2 Ga, suggesting formation through recycling of the oldest crustal material. The Kurumkan granulite-facies clastic metasedimentary rocks show two sources, with ages of 3.46 and 3.01 Ga. In the Stanovoy province, 3.22–3.0 Ga tonalite and mafic granulites occupy the Sutam granulite–gneiss terrane. Paleoarchean ages have not been found in the other parts of the Aldan Shield. In the east are Neoarchean and Paleoproterozoic ages, reflecting the presence of widespread younger crust.

The Anabar Shield on the NE side of the Siberian Craton includes only one ancient area – the Daldyn granulite–gneiss terrane. Here, plagiogneisses are the dominant rock type and are compositionally similar to calc-alkaline volcanic rocks. They probably present subduction-related TTG series rocks and are ca. 3.32 Ga. Enderbites appeared at 3.16 Ga, derived through partial melting of a gneissic or mafic source. Metamafites formed at 3.35 Ga and are similar to oceanic or mantle plume basalts. The other plagiogneisses and metamafites, dated as 3.1–3.0 Ga, are possibly reworked varieties of the above rocks. All these magmatic rocks were exposed soon after their origin and were incorporated into sedimentary rocks now having $T(Nd)_{DM}$ ages = 3.00–3.19 Ga.

On the whole, the geochronological data above are sufficient to summarize the following sequence of events: (1) generation of the first crust; (2) later reworking of this crust, accompanied by intrusions; and (3) uplift and erosion, marked by the deposition of sedimentary rocks.

The oldest sialic crust in the Onot and Bulun granite–greenstone domains formed at 3.57–3.51 Ga, and displays evidence of reworking, recycling, and magma intrusion at 3.41–3.28 Ga. In the Olekma granite–greenstone terrane, crust was formed at 3.69–3.49 Ga, and at 3.1–3.0 Ga, and sediments received clastic material from 3.6–2.8 Ga rocks. The Aldan granulite-gneiss terrane first formed at 3.64–3.46 Ga, and was intruded at 3.2–3.0 Ga. In the Daldyn granulite–gneiss terrane, 3.35–3.32 Ga crust formed enderbites that underwent melting at 3.16 Ga, and sedimentary clastic material was sourced from 3.19–3.0 Ga rocks. In the Sutam granulite–gneiss terrane, crust formed at 3.22–3.0 Ga, whereas crust in the Sharyzhalgay granulite–gneiss domain formed at 3.4–3.3 Ga. This data shows clear overlap of some crust-forming processes between these domains, but it is evident that there are two main episodes of crust formation in the origin of the oldest rock assemblages in the Siberian Craton. The first one developed at 3.69–3.46 Ga, followed by crustal reworking at 3.2–3.0 Ga. The second episode occurred at 3.4–3.0 Ga, with reworking at 3.1–3.0 Ga. The oldest sialic crust (3.69–3.46 and 3.4–3.0 Ga) consists mainly of TTG-series rocks interpreted to have formed in a shallow subduction setting, whereas greenstones are similar to oceanic plateau basalts, derived from ahe mantle plume source.

ACKNOWLEDGEMENTS

This study has been supported by Russian Foundation for Basic Research grants 06-05-64332 and 06-05-64572.

PART 7

LIFE ON EARLY EARTH

Earth's Oldest Rocks
Edited by Martin J. Van Kranendonk, R. Hugh Smithies and Vickie C. Bennett
Developments in Precambrian Geology, Vol. 15 (K.C. Condie, Series Editor)
© 2007 Elsevier B.V. All rights reserved.
DOI: 10.1016/S0166-2635(07)15071-4

Chapter 7.1

SEARCHING FOR EARTH'S EARLIEST LIFE IN SOUTHERN WEST GREENLAND – HISTORY, CURRENT STATUS, AND FUTURE PROSPECTS

MARTIN J. WHITEHOUSE[a] AND CHRISTOPHER M. FEDO[b]

[a]*Swedish Museum of Natural History, Box 50007, SE-104 05 Stockholm, Sweden*
[b]*Department of Earth and Planetary Sciences, University of Tennessee, Knoxville, TN 37996, USA*

7.1-1. INTRODUCTION

As the most extensive outcrop of Eoarchean rocks on Earth, the 3.6–3.9 Ga gneiss terrain of southern west (SW) Greenland (Fig. 7.1-1(A)), with its enclaves of supracrustal rocks, represents an obvious target for investigations aimed at finding Earth's earliest life. Dominant tonalite-trondhjemite-granodiorite (TTG) Amîtsoq orthogneisses preserve remnants of supracrustal sequences – the so-called Akilia Association (McGregor and Mason, 1973), including the Isua Greenstone Belt (Appel et al., 1998) – that contain a wide variety of metamorphosed volcanic and sedimentary rocks with the potential to reveal critical information about the suitability of Earth's surface to host life at this time, as well as possible evidence for life itself. In this critical appraisal of the search for Earth's earliest life in SW Greenland, we apply three criteria to evaluate claims: (1) age of the proposed host, (2) suitability of the host as a repository for life, and (3) nature of the evidence for life itself.

The general geology of SW Greenland has been described in detail in numerous publications during the past decade (Nutman et al., 1996, 1999, this volume; Myers and Crowley, 2000) and need not be repeated in detail here. Of specific interest in the search for early life are two localities within the gneiss complex; the Isua Greenstone Belt, located ca. 150 km NE of Nuuk, and the island of Akilia, located ca. 20 km south of Nuuk (Fig. 7.1-1(A)). In terms of the criteria outlined above, these locations differ significantly and are therefore dealt with separately. Following discussion of these case studies, we will consider the possibility that the very presence of banded iron formation (BIF) might itself be a biomarker.

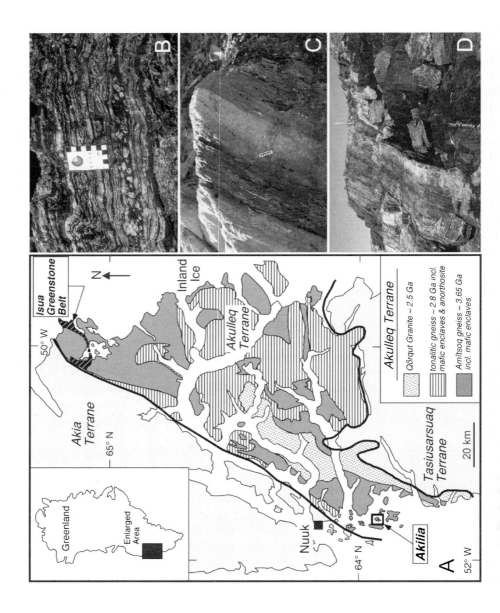

Fig. 7.1-1. (*Previous page.*) (A) Geological map of the Eoarchean of southern West Greenland (modified after Myers and Crowley (2000)) showing the locations of the Isua Greenstone Belt and Akilia. (B) Quartz magnetite BIF, northeast sector, Isua Greenstone Belt. (C) Turbidite locality of Rosing (1999), western sector, Isua Greenstone Belt. (D) Granulite facies quartz-pyroxene rocks, Akilia, outer Gothåbsfjord.

7.1-2. CLAIMS FOR EARLY TERRESTRIAL LIFE

7.1-2.1. Isua Greenstone Belt

Rocks of the Isua Greenstone Belt[1] (IGB) have long represented a fruitful target for early life investigations since the recognition that much of the extensive supracrustal sequence is at least 3.7 billion years old (Moorbath et al., 1973). Despite its promise, however, it must be appreciated that the IGB was metamorphosed at amphibolite facies at least once (Rollinson, 2002), deformed at moderate- to high-strain (Fedo, 2000; Myers, 2001), and in some cases, extensively metasomatised (Rosing et al., 1996) so that the protolith of some rocks remains uncertain.

7.1-2.1.1. Body "fossils" in the IGB Isua
The highly strained nature of much of the IGB clearly complicates the preservation potential of any unicellular body fossil. Despite this, a remarkable claim was made by Pflug and Jaeschke-Boyer (1979), who recognised purported body fossils in a quartzite layer that they interpreted as recrystallised chert that forms part of the extensive BIF in the area (Fig. 7.1-1(b)). The proposed fossils comprise individual single cells and colonies of subspherical objects that were cited as containing organic compounds. Based on the multiple shapes observed, the "organisms" were even given a systematic taxonomic assignment, *Isuasphaera isua* sp.

In terms of the criteria erected for evaluation of early life claims (Table 7.1-1), while the time of deposition of the host rock at 3.7 Ga (Frei et al., 1999) is clearly appropriate, significantly younger intense deformation with concomitant recrystallization of the original chert means that the observable textures are younger in age, possibly Neoarchean, because Paleoarchean Ameralik dykes (White et al., 2000) in the northeast segment of Akilia have been strongly deformed. Detailed thin section analysis of the host rocks for this claim (Bridgwater et al., 1981) demonstrated that the original "cells" and "colonies" actually represent variously shaped limonite-stained fluid inclusions that formed during coarse recrystallization of the fine-grained depositional chert, and so are entirely inorganic in nature. A similar conclusion was reached by Roedder (1981), who assigned the objects in part to the result of dissolution of carbonate, leaving cavities. While clearly discredited 25 years ago, the claim for *Isuasphaera isua* was reiterated by Pflug (2001), who included no

---

[1] The rock association of the Isua supracrustal belt (Allaart, 1976) was recognised as containing all the key elements of a greenstone belt by Appel et al. (1998).

Table 7.1-1. Summary of evidence for early terrestrial life in W. Greenland

Biomarker	Claimed age	Primary ref.	Assessment criteria			Reliability
			Age of host	Suitability of host	Nature of evidence	
Isua Greenstone Belt						
Relict microfossils	3.8 Ga	Pflug and Jaeschke-Boyer, 1979	>3.7 Ga	Extreme deformation precludes preservation of spherical body fossils (1)	Apparently fluid inclusions (2) or cavities (3). Organic matter present probably recent endolithic, cf. (4)	discredited
Bulk rock carbon isotopes	3.7 Ga	Schidlowski et al., 1979	Metacarbonate age unknown (<3.7 Ga)	IGB metacarbonates not original sediments (5, 6)	Range of bulk δ^{13}C not replicated (7)	discredited
Microscale (SIMS) carbon isotopes	>3.7 Ga	Mojzsis et al., 1996	Metacarbonate age unknown (< 3.7 Ga)	IGB metacarbonates not original sediments (5, 6)	Graphite in apatite association not unique to supracrustals (8)	discredited
Bulk rock carbon isotopes	3.78 Ga	Rosing, 1999	>3.7 Ga	Clastic metasediments; suitable host	Bulk δ^{13}C $\approx -19‰$ could be consistent with a biogenic origin	viable
Microscale (SIMS) carbon isotopes	3.8 Ga	Ueno et al., 2002	Cherts >3.7 Ga; metacarbonate age unknown (<3.7 Ga)	IGB metacarbonates not original sediments (5, 6); cherts more suitable	Less extreme fractionation (most δ^{13}C values $> -15‰$) could be abiogenic (authors' assessment)	low

Table 7.1-1. (*Continued*)

Biomarker	Claimed age	Primary ref.	Assessment criteria			Reliability
			Age of host	Suitability of host	Nature of evidence	
Step-release carbon and nitrogen isotopes	3.8 Ga	Nishizawa et al., 2005	>3 7 Ga	Quartz magnetite BIF; suitable host	Bulk $\delta^{13}C \approx -30\%_0$ from high T release inferred to be in diagenetic C_{org} in mt. $\delta^{15}N$ values inconclusive, cf. (9)	viable
Akilia, Gothåbsfjord						
Microscale carbon isotopes	>3.85 Ga	Mojzsis et al., 1996	≥3.65 Ga (10, 11); debatable older age (10, 12), cf. (13–15)	Controversial; chemical sediment (10, 16, 17) or metamorphosed igneous rocks (11, 18–20)?	Graphite not found by subsequent studies (21, 22)	discredited

References: (1) Appel et al., 2003; (2) Bridgwater et al., 1981; (3) Roedder, 1981; (4) Westall and Folk, 2003; (5) Rose et al., 1996; (6) Rosing et al., 1996; (7) van Zuilen et al., 2002; (8) Lepland et al., 2002; (9) van Zuilen et al., 2005; (10) Manning et al., 2006; (11) Fedo et al., 2006; (12) Nutman et al., 1997; (13) Myers and Crowley, 2000; (14) Whitehouse and Fedo, 2003; (15) Whitehouse and Kamber, 2005; (16) Dauphas et al., 2004; (17) Mojzsis et al., 2003; (18) Bolhar et al., 2004; (19) Fedo and Whitehouse, 2002; (20) Whitehouse et al., 2005; (21) Lepland et al., 2005; (22) Nutman and Friend, 2006.

discussion of, or reference to, the earlier criticism, prompting a robust comment by Appel et al. (2003), who pointed out that the presence of organic molecules may readily be explained by recent (post-glacial) ingress of organic contamination, similar to that proposed by Westall and Folk (2003).

7.1-2.1.2. Chemofossils – the significance of isotopically light carbon
The failure of claims for body fossils has given way to the search for biomarkers at Isua, in particular that represented by isotopically light carbon, a proposal first made by Schidlowski et al. (1979), based on the principle that organisms preferentially sequester ^{12}C over ^{13}C during metabolic processes. From Isua, Schidlowski et al. (1979) reported $\delta^{13}C_{PDB}$[2] values between $-6‰$ and $-25‰$ in metacarbonates and as low as $-25‰$ in reduced carbon in a variety of rock types, including BIF and graphitic schist, which they used to suggest that life had been established on Earth as early as 3.8–3.7 Ga. Subsequent high spatial resolution investigations of C-isotopes using secondary ion mass spectrometry (SIMS; Mojzsis et al., 1996; Ueno et al., 2002) also found isotopically light C, as graphite, in IGB rocks, which was assigned to a biogenic origin using the same reasoning as Schidlowski et al. (1979). The metacarbonate and graphite samples analysed by Schidlowski et al. (1979) and Mojzsis et al. (1996), however, have been demonstrated to be the product of metasomatic alteration (van Zuilen et al., 2002) and thus do not represent a suitable supracrustal host for micro-organisms. Furthermore, the origin of the graphite itself was shown to be a secondary result of metasomatic and/or metamorphic processes involving the breakdown of carbonates to reduced carbon (van Zuilen et al., 2002; Lepland et al., 2002), or serpentinisation of ultramafic rocks (Naroaka et al., 1996). Consequently, none of the studies asserting a biologic origin for graphite has survived subsequent scrutiny with regard to our criteria of suitability of host and robustness of evidence (Table 7.1-1).

At the same time that individual claims for early biological activity based on isotopically light carbon have come under intense scrutiny, its role as a unique biomarker has been called into question by several studies that demonstrate abiogenic carbon isotope fractionation pathways that lead to ^{12}C enrichment. For example, a series of inorganic catalytic reactions (Fischer–Tropsch synthesis) have been invoked to generate light carbon at mid-ocean ridges (Holm and Charlou, 2001), while siderite decomposition (McCollom, 2003), deep crustal alkane formation (Sherwood-Lollar et al., 2002) and other experimental studies (e.g., Horita and Berndt, 1999; McCollom and Seewald, 2006) have further demonstrated that isotopically light carbon simply is not a unique biosignature.

Despite the failure of several of the Isua studies and the potential ambiguity of carbon isotopes as a biomarker, two investigations may still be considered as possible documentation of Earth's earliest biology. In the first of these, Rosing (1999) reported bulk rock $\delta^{13}C$ values averaging $-19‰$ from an outcrop interpreted to represent clastic metasedimentary rocks intercalated with pillow basalts and BIF in the western segment of the IGB (Fig. 7.1-1(C)). The host rocks were interpreted as sea floor deposits (specifically, a turbidite sequence), which makes them appropriate as hosts for life, and although the western

---

[2]Carbon isotope variations are normalised to the Pee Dee belemnite standard and usually expressed in parts per thousand (‰).

IGB has been metamorphosed at amphibolite facies in the Neoarchean (the turbidite itself yielded a Pb/Pb regression age of 2.77 Ga; Rosing and Frei, 2004), Sm-Nd data from the same samples record a likely protolith age of 3.78 Ga (Rosing, 1999). In a subsequent investigation, van Zuilen et al. (2002) replicated Rosing's $\delta^{13}C$ values from a turbidite sample from the same outcrop and reported the absence of both siderite and magnetite, arguing against abiogenic generation of the graphite by siderite dissociation. The claim for a biological origin for graphite in the turbidite has not, to date, been questioned and, on the basis of Pb-isotopic evidence for the redox behaviour of U and Th, Rosing and Frei (2004) proposed that the graphite might represent the product of an oxidative photosyn-thetic metabolism (cyanobacteria?) at >3.7 Ga. This latter conclusion has, however, been criticised by Fedo et al. (2006) on the basis that the Pb-isotope interpretation is non-unique. Thus, the light carbon values in the IGB turbidite still stand as a potential indicator of early biology, but the nature of this biology remains unclear.

In the second study, Nishizawa et al. (2005) document the carbon isotope composition of quartz-magnetite BIF from the northeast sector of the IGB, performed by a stepwise combustion experiment. In terms of our criteria, these rocks are undisputedly >3.7 Ga in age (Frei et al., 1999) and of sedimentary origin, thus representing suitable targets for early life studies. In common with the earlier study of van Zuilen et al. (2002) on metacarbonates, isotopically light carbon ($\delta^{13}C \approx -24‰$) released at low temperature (200–400 °C) was assigned by Nishizawa et al. (2005) to contamination and is thus irrelevant to the debate over early life. Carbon released at high temperature (>1000 °C) also yielded light values (two aliquots gave $\delta^{13}C$ values of $-19‰$ and $-30‰$), which was interpreted to represent isotopically light, possibly kerogenous, carbon trapped in magnetite during diagenesis. In support of the kerogenous interpretation, Nishizawa et al. (2005) also reported that negative $\delta^{15}N$ values ($-3‰$), together with C/N ratio from the high temperature release, also fall within the range of Archean kerogens, although van Zuilen et al. (2005) pointed out that such isotope compositions are also similar to those found in metamorphosed basalts and so should be treated with caution as a supporting potential biomarker.

We reiterate that while both of these studies pass the criteria we have erected for viable claims for early biogenic activity (Table 7.1-1), this assessment is also strongly dependent on the reliability of isotopically light C as a biomarker. Furthermore, the metamorphic grade of the IGB rocks needs to be considered as a potential complicating factor in all such studies.

7.1-2.2. Akilia Island

Located 150 km southwest of Isua (Fig. 7.1-1(A)), a small outcrop of rocks on the island of Akilia in the outer Gothåbsfjord region has been the focus of a vigorous debate centered on a claim that these rocks host Earth's earliest (>3.85 Ga) life in the form of isotopically light graphite particles encased in apatite (Mojzsis et al., 1996). In comparison to the Isua claims discussed earlier, all aspects of the Akilia claim have been called into question (Table 7.1-1): the precise age of the host and in particular whether it can unambiguously be demonstrated as older than Isua; the proposed sedimentary protolith of the host rocks; and the very existence of the isotopically light graphite itself.

A detailed summary of the Akilia debate has recently been presented by Fedo et al. (2006) and here we will only summarise the main arguments pertaining to age, nature of the host rock, and nature of the evidence (Table 7.1-1).

7.1-2.2.1. Age of the Akilia claim

Among the many contentious issues surrounding the Akilia claim, the question of the exact age of the host rock remains unresolved. The quartz-pyroxene rocks that host the reported light carbon are not particularly conducive to dating methods that might reveal their protolith age. Nutman et al. (1997) reported the presence of only 2.7 Ga metamorphic zircon from the key sample that hosted the reported light graphite, and similar Neoarchean zircon has been reported from an analogous sample by Fedo et al. (2006). In contrast to these studies, Manning et al. (2006) have reported a substantial number of Eoarchean (ca. 3.65 Ga) metamorphic zircon from a quartz-pyroxene lithology which, together with whole-rock Pb isotope (Whitehouse et al., 2005) and Sm-Nd model age (Fedo et al., 2006) data, clearly point to an Eoarchean protolith. The key question remains, however: can the Akilia rocks on the whole, or any of their individual mineral constituents, be demonstrated unambiguously to be older than any of the claims from the IGB, where a >3.7 Ga age is well-accepted?

If the quartz-pyroxene rocks form part of a stratigraphic succession with their adjacent mafic/ultramafic hosts (but see protolith discussion below), then dating the latter accurately would be appropriate, although dating of the protolith to such rocks, particularly in a polyphase granulite-facies terrain is difficult, given the likelihood of isotopic resetting. Concordant, ca. 3.7 Ga ages have been obtained from Sm-Nd isochron (Moorbath et al., 1997) and feldspar Pb isotope model age (Kamber and Moorbath, 1998) methods. Although the Sm-Nd age might be explained as the result of granulite-facies resetting (e.g., Whitehouse, 1988), such an explanation cannot readily be applied to the Pb isotope data from feldspar in the amphibolites, since it would require wholesale expulsion of any older Pb-isotope signature. Again, the host rocks cannot be used to suggest an age for the Akilia claim in excess of that obtained from any IGB rocks.

The initial claim that the purported biologic relics exceed 3.85 Ga was based on the age of the oldest zircon in a gneiss sheet that was reported to cross-cut adjacent mafic and ultramafic lithologies that were assumed to form part of a stratigraphic sequence with the hosting quartz-pyroxene rocks (Nutman et al., 1997). However, both the cross-cutting relationship and the zircon geochronology have come under intense scrutiny. Myers and Crowley (2000) remapped the cross-cutting outcrop and showed that the claimed discordance did not exist. Whitehouse and Fedo (2003), who studied the same outcrop, documented an isoclinally folded felsic vein along the contact that clearly renders it tectonic in nature. Both of these studies pointed out that the ">3.85 Ga" gneiss sheet (Nutman et al., 1997) appears to be the youngest rock at this particular locality and subsequent detailed U-Pb zircon studies (Whitehouse et al., 1999; Whitehouse and Kamber, 2005) suggest that the oldest zircon in the rock is a relict and that its actual intrusive age may not exceed

3.62 Ga or even 2.7 Ga. Claims that inheritance of older zircon into tonalitic melts is unlikely and would preclude such an interpretation (Mojzsis and Harrison, 2002; Manning et al., 2006) *a priori* assume that the rock crystallised directly from a high temperature liquid rather than representing a mobilised partially melted, low temperature, water-rich crystal-melt mush (Kamber and Moorbath, 1998; Whitehouse and Kamber, 2005). The latter interpretation is supported by the observation that the claimed cross-cutting gneiss sheet appears to cut a polyphase gneiss containing 3.65 Ga magmatic zircon (Whitehouse and Fedo, 2003).

Despite the obvious ambiguities involved in dating using supposedly discordant gneiss sheets with complex zircon populations, this approach has been applied recently to another location on Akilia (Manning et al., 2006). In this case, similar >3.83 Ga zircon cores are interpreted to date the intrusion of a gneiss sheet that cannot be seen to cross-cut mafic/ultramafic rocks of the Akilia enclave, the field relationship instead being inferred to occur beneath covering glacial deposits on the basis of outcrop map interpretation. Given the complexities of the earlier claim, the robustness of this new interpretation, which is inconsistent both with an earlier map (Friend et al., 2002b) and the overall large scale structure of southwest Akilia (Myers and Crowley, 2000), clearly requires further detailed scrutiny.

7.1-2.2.2. Suitability of the host rock

Arguably even more debated than its age, the protolith of the quartz-pyroxene rock also remains a contentious issue. The original description of the outcrop (Fig. 7.1-1(D); Mojzsis et al., 1996) described it as a 5 m wide BIF, finely laminated and so well-preserved that it had clearly escaped the effects of the contemporaneous (the >3.85 Ga age assumed) late heavy meteorite bombardment on Earth. Similarity to quartz-magnetite BIF from the IGB was also claimed, despite the fact that the Akilia rocks are uniformly poor in magnetite, with much of the outcrop possessing only trace quantities. An alternative, non-sedimentary, protolith was proposed by Fedo and Whitehouse (2002), who documented the lithological variation within the 5 m wide zone, as well as evidence for extreme ductile deformation that is the primary cause of the layering as it is observed today. Fedo and Whitehouse (2002) argued on geochemical grounds that many of the rocks are related to each other genetically and can be explained by polyphase metasomatic and metamorphic processes acting on an original mafic/ultramafic igneous precursor. Since then, a number of studies have sought to test a possible sedimentary precursor for part, or all, of the 5 m wide quartz-pyroxene outcrop; a summary of these investigations has been presented by Fedo et al. (2006) and only brief details are presented here.

Trace element geochemistry has been used to compare Akilia quartz-pyroxene rocks with genuine BIF from the IGB. In PAAS[3]-normalised plots of rare earth element data, hydrogenous sediments are expected to display positive anomalies of La, Gd and Y as a result of complexing behaviour in seawater that depends on electron configuration for these elements (e.g., Bau and Moeller, 1993; Kamber and Webb, 2001). Bolhar et al. (2004) and

---

[3]PAAS is Post Archean Average Shale.

Fedo et al. (2006) showed that for Akilia samples, La anomalies are absent and Y anomalies are much smaller than those observed for genuine BIF from the IGB, arguing against a similar protolith.

In their recent assessment of the geochemical data presented by Fedo and Whitehouse (2002), Manning et al. (2006) investigate whether the trace element characteristics of the quartz-pyroxene rocks can be generated by a simple, single-stage, two-component mixing of an ultramafic component and an introduced quartz-rich component. These authors note the failure of this model to explain the full range of variation observed, proposing instead that the trace element patterns can be explained by variable addition of ultramafic detritus into a chemical sediment. As pointed out by Fedo and Whitehouse (2002), the behaviour of certain immobile trace elements that are strongly partitioned into specific mineral phases (e.g., Cr in chromite) also needs to be treated with caution in highly tectonised rocks that are, in some cases, up to 95% SiO_2. It remains possible, however, that complex, polyphase metamorphic and metasomatic processes acting on a heterogeneous igneous precursor might explain the observed trace element variations.

Several stable isotope systems have also been applied to the problem of protolith, with conflicting results. Silicon isotope data presented by André et al. (2006) reveal that negative $\delta^{30}Si$ values typical of Precambrian BIF (including Isua) are not present in a typical sample of the Akilia quartz-pyroxene rock. These authors conclude instead that the quartz in the Akilia rocks, which has a similar Si isotope composition to metamorphic quartz, was probably derived from metamorphic processes and is not part of an original sedimentary rock. In contrast to this conclusion, however, Manning et al. (2006) argue that a sharp transition between oxygen isotope composition of flanking mafic/ultramafic rocks ($\delta^{18}O < 9\permil$) and the quartz-pyroxene rock itself ($\delta^{18}O \approx 13\permil$) is incompatible with large-scale metasomatic processes. This conclusion however, *a priori* assumes that the boundaries of the quartz-pyroxene rocks and their flanking mafic/ultramafic rocks are original contacts and not, like some other contacts between highly strained lithologies in SW Akilia, tectonic in nature (e.g., Whitehouse and Fedo, 2003).

Mojzsis et al. (2003) reported sulphur isotope data from two large chalcopyrite-pyrrhotite aggregate grains in a quartz-pyroxene rock from Akilia, which showed an excess of the ^{33}S isotope ($\Delta^{33}S > 0\permil$), consistent with the presence of sulphur that had undergone mass independent fractionation processes.[4] These data are difficult to interpret, however, because the MIF signature is derived from only a subset of the data that show an unexplained, extremely large, range in mass dependent sulphur isotope fractionation. In a subsequent study of a large number of sulphides from a sample of quartz-pyroxene rock (the very same sample from which the original life claim was made) and a pyroxenite boudin within the 5 m wide zone, Whitehouse et al. (2005) were unable to reproduce either the positive $\Delta^{33}S$ (all data averaged $\Delta^{33}S = 0\permil$) or the range in mass dependent fractionation reported by Mojzsis et al. (2003), and concluded that sulphur isotopes on these secondary minerals provide no convincing evidence for a sedimentary precursor.

---

[4]Mass independent fractionation of sulphur shown by the ^{33}S anomaly is assumed only to be generated by photo-catalysed reactions in a low pO_2 atmosphere (Farquhar et al., 2000) and is therefore a taken as an indicator in the Archean of sulphur derived from a surface reservoir.

The possibility that Fe isotopes might indicate a chemical sedimentary origin (specifically BIF) was proposed by Dauphas et al. (2004), who examined three samples of quartz-pyroxene rocks from Akilia together with a sample of quartz-magnetite BIF from the IGB. These authors reported $\delta^{56}Fe$ values up to $+1‰$ that are consistent with the growing (but still relatively small) empirical and experimental dataset suggesting that fractionated Fe is a feature unique to chemically precipitated sediments, possibly with biological mediation (Johnson et al., 2003). Dauphas et al. (2004) uncritically accepted the >3.83 Ga age for the unit and relied on the initial report (subsequently questioned – see above) of the presence of mass independently fractionated sulphur in support of their claim. By analogy with the observed behaviour of hydrothermally altered basalts, they further argued that generation of isotopically heavy Fe by kinetic fractionation (loss of isotopically light Fe) during metasomatic processes is inconsistent with the observed high Fe/Ti ratios, which instead resemble those typical of BIF. Whether *in situ*, hydrothermally altered, basalts represent reasonable analogues for the high-grade, polyphase metamorphosed rocks of Akilia is, however, open to question. Furthermore, Fedo and Whitehouse (2002) have pointed out that high Fe/Ti ratios are also found in ultramafic layers on Akilia that clearly do not have a BIF precursor. Clearly, further work is needed to determine whether Fe isotopes are indeed a unique indicator of a BIF and/or chemical sedimentary protolith and whether such a signature, when found in such multiply deformed and metamorphosed rocks as those on Akilia, can truly be assigned to *in situ* protolith rather than recycling of already fractionated Fe from an external reservoir (there are rocks of undisputed BIF origin in the Gothåbsfjord), or during development of hydrothermal systems (e.g., Graham et al., 2004; Horn et al., 2006). It is certain that pyroxene did not crystallize on the sea floor as a sedimentary mineral, so its isotopic composition cannot simply be the product of primary processes. Furthermore, both magnetite and pyroxene occur in veins that cross-cut the intense foliation in the rocks, indicating remobilization of Fe-bearing phases long after any inferred time of deposition. Despite these reservations, the Fe isotopes provide the only remaining indicator that the quartz-pyroxene rocks on Akilia might host a component of chemical sediment.

The current polarised state of the debate may be summed up by the contrasting views presented by Fedo et al. (2006) and Manning et al. (2006) from which it is clear there remains little agreement either on protolith, or indeed, the interpretation of geochemical parameters from the same rocks (and even the same dataset).

7.1-2.2.3. Nature of the evidence

Lastly to the carbon isotope evidence itself. From the outset, the unusually large and highly variable uncertainties in SIMS $\delta^{13}C$ measurements (an average $\delta^{13}C$ value of $-37‰$ that is unrepresentative of the bimodal distribution of data, which has two distinct peaks at $-27‰$ and $-44‰$), and the absence of documentation of the analysed particles (first presented by Mojzsis et al. (1996)), were largely overlooked by critical studies that concentrated on the age relationships and nature of the host rocks. One early concern indicated that the apatite host for the purported graphite particles was only 1.5 Ga (Sano et al., 1999b), leaving the age of the graphite as that age, or older. In the IGB, Lepland et al. (2002) demonstrated that the association of graphite with apatite was entirely a product of metasomatism, an abiotic hydrothermal process that occurred at a younger time. Lacking sufficient reason to exclude such an analogous process on Akilia, any graphite that might be found within apatite would be much younger than 3.85 Ga, which is consistent with the known apatite geochronology; there is no known geochronology that can establish an age for the graphite older than known times of deposition in the region.

More serious questions have emerged recently, however, in the form of two studies that have both highlighted the absence of graphite particles in apatite from the very same rock on Akilia from which the original claim was made, as well as several other related samples (Lepland et al., 2005; Nutman and Friend, 2006). It remains unclear exactly what was analysed for C isotopes in the original study (Mojzsis et al., 1996) and until, or unless, clarification emerges – irrespective of debates over age and suitability of the host – the claim for life from Akilia must, for now, be categorised as discredited. Even if graphite is eventually discovered in these rocks as an inclusion within apatite, the lessons learned from Isua (e.g., van Zuilen et al., 2002; Lepland et al., 2002) and experimental studies (McCollom and Seewald, 2006) suggest that neither its presence, nor its carbon isotopic composition, will unambiguously establish a biologic origin.

7.1-3. SIGNIFICANCE OF BANDED IRON FORMATION (BIF)

To date, the search for earliest terrestrial life in SW Greenland has focussed on finding a specific biomarker within target rocks. Arguably such a biomarker might be represented by the specific BIF lithology itself, preserved as a >3.7 Ga, 2-billion-ton deposit in the NE sector of the IGB, as well as in numerous small enclaves in the Gothåbsfjord region, and to the south of the IGB. Despite extensive research, the origin of Precambrian BIF remains enigmatic and there is no general agreement on the possible role of biology in its formation, although it is clear that some better-preserved chert-BIF associations and their commonly spatially-related stromatolites do contain biological remnants in the Neoarchean (e.g., Summons et al., 1999), and even as early as ca. 3.4 Ga (e.g., Van Kranendonk et al., 2003; Tice and Lowe, 2004; Allwood et al., 2006a). Konhauser et al. (2002) quantitatively estimated the amount of bacteria required to generate BIF, demonstrating that microbial oxidation of Fe^{2+} to Fe^{3+}, whether in the form of oxidative photosynthesisers (e.g., Cloud, 1973) or ferroautotrophic species operating in a free-oxygen limited ocean (e.g., Holm, 1989), is clearly a viable mechanism of BIF formation. The alternative of inorganic precipitation, perhaps via photochemical oxidation (Cairns-Smith et al., 1978), remains a viable possibility although, as noted by Konhauser et al. (2002), these early experiments were not conducted in appropriate solutions analogous to likely Archean seawater.

If BIF can be considered as a biomarker, then all occurrences of BIF within the Eoarchean of SW Greenland should be evaluated in terms of their depositional age. The 3.7 Ga age for BIF of the IGB is undisputed. Older examples might occur to the south of the IGB, in a terrane dominated by 3.82 Ga tonalitic gneisses (Nutman et al., 1999), and elsewhere in the region.

7.1-4. PROSPECTS FOR THE DISCOVERY OF EARLY LIFE SIGNATURES

The search for early life on Earth must continue as a two-pronged assault: (1) as a continued search for new, and/or testing of existing, claims for biomarkers in appropriate, unambiguously dated rocks; (2) continued investigation of the origin of BIF and, in particular, whether these can be unambiguously assigned a biogenic origin.

The long-running Akilia debate has shown that high grade metamorphic and strongly deformed rocks are inappropriate targets for such studies. Even at Isua, modification by amphibolite-facies metamorphism has rendered some claims for biological activity invalid, yet the proven antiquity and evident sedimentary precursor of many of the IGB rocks makes this a viable place to continue the search, both to verify existing claims (Table 7.1-1) and to identify and test new prospects. The tools developed in this search will prove invaluable, if lower grade successions of similar age are identified elsewhere.

Finally a comment on scale of evidence is in order. Increasing sophistication of analytical methods, particularly those for *in situ*, high spatial resolution analysis, are increasingly perceived as one of the best ways to identify early life remnants in ancient rocks. While this is probably true, the as-yet unproven possibility that the BIF lithology itself might represent a biomarker could mean that there are 2 billion tons of >3.7 Ga life evident at Isua.

Earth's Oldest Rocks
Edited by Martin J. Van Kranendonk, R. Hugh Smithies and Vickie C. Bennett
Developments in Precambrian Geology, Vol. 15 (K.C. Condie, Series Editor)   855
DOI: 10.1016/S0166-2635(07)15072-6

Chapter 7.2

A REVIEW OF THE EVIDENCE FOR PUTATIVE PALEOARCHEAN LIFE IN THE PILBARA CRATON, WESTERN AUSTRALIA

MARTIN J. VAN KRANENDONK

Geological Survey of Western Australia, 100 Plain St., East Perth, Western Australia 6004, Australia

7.2-1. INTRODUCTION

Evidence for the existence of Paleoarchean life based on evidence in the Pilbara Craton has been proposed since Lambert et al. (1978) described stromatolitic edgewise conglomerates from sedimentary rocks of the Warrawoona Group, which was already known to be of great antiquity at that time (Pidgeon, 1978). Since then, many more claims for early life have been made, based on a variety of types of geological evidence from rocks at a variety of stratigraphic levels in the Pilbara Supergroup of the Pilbara Craton (Figs. 7.2-1, 7.2-2). This evidence can be divided into the following distinct types:

(1) interpreted macroscopic stromatolites[1] in four distinct stratigraphic horizons, based on morphology;

(2) interpreted microfossils in cherty metasedimentary rocks, volcaniogenic massive sulphide deposits, and two levels of syn-depositional, siliceous hydrothermal veins (Awramik et al., 1983; Schopf, 1992, 1993; Rasmussen, 2000; Ueno et al., 2001a, 2001b);

(3) highly negative $\delta^{13}C$ values of carbonaceous material (widely assumed to be kerogen) and ^{13}C-depleted methane in primary fluid inclusions in cherty metasedimentary rocks and siliceous hydrothermal veins, resulting from biological fractionation (Ueno et al., 2001a, 2001b, 2004, 2006);

(4) highly fractionated $\delta^{34}S$ values for microscopic pyrite crystals in the growth zones of coarse barite crystals from sedimentary/hydrothermal rocks (Shen et al., 2001);

(5) Raman spectrographic signatures and organic geochemical characteristics of carbonaceous material from the Strelley Pool Chert (Allwood et al., 2006b; Marshall et al., 2007, this volume);

---

[1]In this paper, stromatolite is used to describe domical, coniform, or laminated sedimentary rock structures of inferred biological origin; "possible stromatolite" is used to describe domical or laminated sedimentary structures of uncertain biological origin; pseudo-stromatolite is used to describe domical or laminated structures in sedimentary rocks that have superficial similarity with stromatolites, but are of (probable) nonbiological origin.

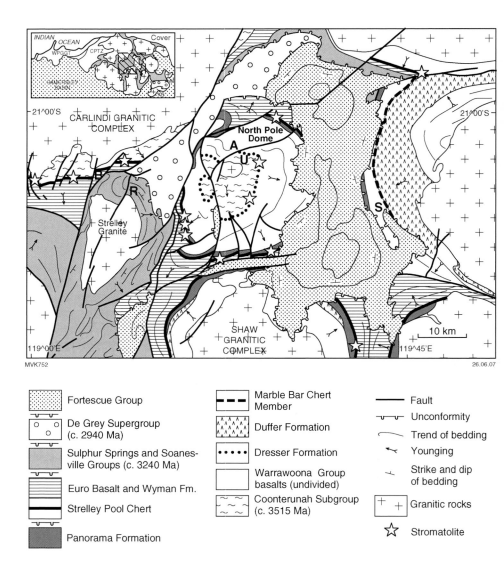

Fig. 7.2-1. Simplified geological map of part of the East Pilbara Terrane, showing locations where putative claims for early life have been made. Bold letters indicate microfossil locations by: A = Awramik et al. (1983); B = Banerjee et al. (2007); R = Rasmussen (2000); S = Schopf (1992, 1993); U = Ueno et al. (2001a, 2001b). Inset shows figure location in Pilbara Craton: WPGGT = West Pilbara granite–greenstone terrane; CPTZ = Central Pilbara Tectonic zone.

(6) ichnofossils in the chilled rinds of weakly metamorphosed pillow basalts of the ca. 3.35 Ga Euro Basalt (Banerjee et al., 2007; see also Furnes et al., 2004).

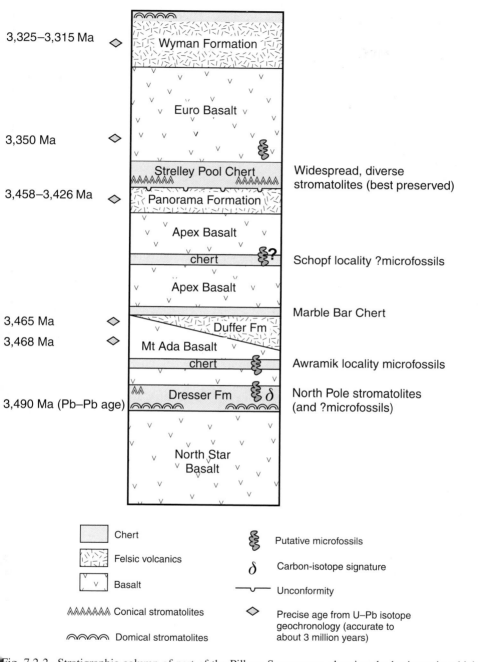

Fig. 7.2-2. Stratigraphic column of part of the Pilbara Supergroup, showing the horizons in which claims for early life have been made.

Yet, except for the most recent of these claims (5 and 6), all other claims have been challenged on the basis of either alternative interpretations of the geological data presented in the original claims, and/or new data gained from more recent research. These include:

(1) a non-biological interpretation of the macroscopic geometry of stromatolites, wherein previously widely accepted morphological criteria have been re-interpreted as structural and/or sedimentological artifacts (e.g., Buick et al., 1981; Lowe, 1994; Lindsay et al., 2003, 2005);

(2) a re-interpretation of microfossils as either more recent contaminants (Neoarchean: Buick, 1984, 1988, 1990), or nonbiological features resulting from mineral crystallization processes within hydrothermal silica veins (Brasier et al., 2002; Garcia-Ruiz et al., 2003);

(3) an interpretation of fractionated $\delta^{13}C$ carbonaceous material in hydrothermal silica veins as a result of inorganic processes, such as Fischer–Tropsch synthesis (cf. Brasier et al., 2002; McCollum, 2003; Lindsay et al., 2005);

(4) a nonbiological origin of highly fractionated $\delta^{34}S$ in microscopic pyrite, based on a hydrothermal interpretation of the hostrock system (Van Kranendonk, 2006; Mojzsis et al., this volume).

The last two claims have been only very recently introduced into the literature and have yet to undergo the type if intense scrutiny that the older claims have undergone. Certainly, the textures and geochemical evidence presented by Banerjee et al. (2007) in support of a biogenic origin for microtubular and microgranular structures in metamorphosed pillow basalts of the ca. 3.35 Ga Euro Basalt (claim 5 above) are compellingly similar to bioalteration structures in modern oceanic basalts (Furnes et al., in press). Similarly, the wide array of geological and analytical evidence presented in support of a biogenic origin for clots of kerogen in clastic sedimentary rocks of the ca. 3.4 Ga Strelley Pool Chert by Allwood et al. (2006b) and Marshall et al. (2007, this volume: claim 6 above) is compelling, although not yet uniquely so.

The arguments for and against a biological interpretation of highly fractionated $\delta^{13}C$ values of carbonaceous material (claim 3 above) are reviewed by Ueno (this volume) and Marshall (this volume), while Mojzsis (this volume) provides a detailed review of the evidence for life from the Pilbara Craton and other terranes based on multiple Sulfur isotope studies (claim 4 above).

In this contribution, I focus on the geological setting and morphological evidence in support of a biological origin for macroscopic stromatolites in the ca. 3.49 Ga Dresser Formation and ca. 3.40 Ga Strelley Pool Chert (claim 1 above), and discuss the geological setting and possible origins of carbonaceous material within hydrothermal silica veins associated with the Dresser Formation and Apex chert (claim 2 above).

7.2-2. GEOLOGICAL SETTING

Claims for early life in the Pilbara Craton reviewed herein are all based on geological evidence from the 3.53–3.16 Ga East Pilbara Terrane, one of five terranes that comprise the northern part of the Pilbara Craton (see Fig. 4.1-1; Van Kranendonk et al., 2007a, this volume). Claims for the presence of ancient life from slightly younger rocks (3.02 Ga) of the Pilbara Craton are presented elsewhere (Kiyokawa et al., 2006).

The East Pilbara Terrane contains a 20 km thick succession of dominantly volcanic rocks, at generally low metamorphic grade, known as the Pilbara Supergroup (see Fig. 4.1-3 in Van Kranendonk et al., this volume). These rocks were deposited in four autochthonous groups from 3.53–3.165 Ga and are unconformably overlain by the ca. 3.02–2.93 Ga De Grey Supergroup and 2.78–2.63 Ga Fortescue Group of the Mount Bruce Supergroup (Van Kranendonk et al., 2006a, 2007a). The groups comprising the Pilbara Supergroup include, from base to top: the 3.53–3.43 Ga Warrawoona Group, the 3.42–3.31 Ga Kelly Group, the 3.27–3.24 Ga Sulphur Springs Group, and the ca. 3.19 Ga Soanesville Group (Rasmussen et al., 2007). Whereas the base of the Warrawoona Group is everywhere an intrusive or sheared intrusive contact with granitic rocks, the lower contact of the younger three groups is marked by an angular unconformity or disconformity (Van Kranendonk, 2000; Van Kranendonk et al., 2006a). The lower three groups are composed predominantly of ultramafic, mafic and felsic volcanic rocks, with generally thin beds of chemical sedimentary rocks and less common clastic sedimentary rocks. Sedimentary formations in these groups are widely accompanied by swarms of syn- to post-depositional seafloor hydrothermal silica $\pm$ barite veins (Van Kranendonk, 2006), a feature also commonly observed in contemporaneous rocks of the Kaapvaal Craton (e.g., Hofmann and Wilson, this volume). The youngest Soanesville Group is composed largely of clastic sedimentary rocks, including sandstone, shale and banded iron-formation, but also includes a thick unit of basalt.

The Pilbara Supergroup was intruded by several generations of granitic rocks at ca. 3.45, 3.42, 3.31, 3.24, 3.17, 3.05, 2.94 and 2.85 Ga. The first four of theses intrusive events were coeval with felsic volcanic formations in the Pilbara Supergroup. These granitic intrusions caused tilting and contact metamorphism of the Pilbara Supergroup that resulted in the dome-and-keel architecture of the terrane (Van Kranendonk et al., 2002, 2004). A more detailed account of the geology of the East Pilbara Terrane is provided by Van Kranendonk et al. (this volume).

7.2-3. USE OF MORPHOLOGY AS AN INDICATOR OF BIOGENICITY

Through the 1980s, claims for early life in the Pilbara Craton were based on the geological setting and morphology of structures developed in finely laminated, low-grade metasedimentary rocks of the Warrawoona Group (Walter et al., 1980; Lowe, 1980, 1983; Walter, 1983). Morphological criteria of stromatolites were widely accepted as evidence of biogenicity, although Buick et al. (1981) expressed caution regarding the carte blanche acceptance of unusual domical features in laminated metasedimentary rocks and erected a rigid set of criteria with which to assess the biogenicity of stromatolites. However, this set of criteria proved too rigid and precludes a biological origin of most stromatolites in Neoarchean to Proterozoic carbonates, which is clearly unacceptable. Lowe (1994) showed, through critical assessment of ancient stromatolites (including those described previously by himself!), that the question of their biogenicity was ambiguous, or could even be disproven. More general observations regarding the difficulty of using morphology of laminated sedimentary materials as an indicator of biogenicity (e.g., Buick et al., 1981; Grotzinger and Rothman, 1996; Pope and Grotzinger, 2000) have led to distrust of morphology as an indicator of biogenicity, despite the fact that this is the only available and widely used criteria to establish the biogenicity of most stromatolites throughout Earth history (e.g., Batchelor et al., 2004; Allwood et al., 2006a); indeed, morphology is the principal criteria for the recognition of most fossils of any age and type. In this paper, I advocate for the use of morphology as a reliable indicator of biogenicity, when used in combination with detailed analysis of geological environment.

7.2-4. DRESSER FORMATION STROMATOLITES

A variety of form of macroscopic stromatolites have been described from the 3.49 Ga Dresser Formation in the North Pole Dome (see Fig. 4.4-2 in Huston et al., this volume), including wrinkly laminated mats, broad domes, columnar forms, and coniform varieties (see Fig. 10 in Van Kranendonk, 2006: Walter et al., 1980; Buick et al., 1981; Walter, 1983; Van Kranendonk et al., 2007c). For >20 years these stromatolites have been accepted as some of the best evidence of early life on Earth, although Lowe (1994) suggested that the domical morphology of the best known of the Dresser Formation stromatolites (cover of Schopf, 1983) was the result of faulting, although he did not address the nature of the wrinkly laminated material that was supposedly deformed into the domical shape.

Previously, the depositional environment of the host rocks was interpreted to be an evaporative, shallow marine basin with restricted water circulation (Groves et al., 1981; Buick and Dunlop, 1990), analogous with present-day Shark Bay, Western Australia, where some of the best examples of living stromatolites are found (Playford and Cockbain, 1976). Buick and Dunlop (1990) interpreted sets of coarse barite crystals oriented parallel to bedding as representing originally evaporative gypsum crystals that were replaced by barite at some younger time. More recent work has shown that, whereas components of this previous model may hold true for some periods of deposition of the Dresser Formation, sediment accumulation occurred in a much more tectonically active environment within the caldera of a felsic volcano, and was accompanied by growth faulting and by multiple pulses of vigorous hydrothermal circulation (Nijman et al., 1998; Van Kranendonk, 2006; Van Kranendonk et al., in press). In these models, barite is regarded as a product of intense acid-sulphate hydrothermal alteration of footwall basalts, according to the model shown in Fig. 7.2-3 (Van Kranendonk, 2006; see also Fig. 7.3-9 in Ueno, this volume).

Previous workers have described the widespread, excellent preservation of sedimentary features and the low grade of metamorphism of sedimentary rocks in the Dresser Formation, despite the great age and tectonic history of this formation (Lambert et al., 1978; Groves et al., 1981; Buick and Dunlop, 1990; Van Kranendonk, 2006). Most of

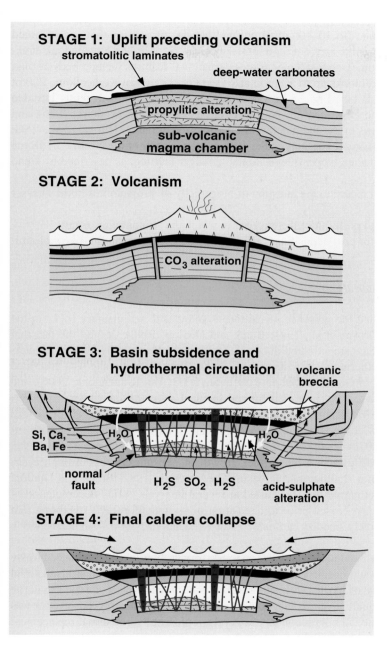

Fig. 7.2-3. Depositional model for the Dresser Formation, showing stromatolite growth under shallow-water conditions during surface uplift that resulted from inflation of a subvolcanic laccolith.

the formation is preserved much as it was deposited and affected only by gentle to moderate tilting of the strata (to 10–40° dips) during episodes of doming that accompanied the emplacement of granitic rocks at 3.46, 3.31, and 3.24 Ga (see Van Kranendonk et al., 2004, 2007a, this volume). Syn-sedimentary deformational features include beds of olistostrome breccia, with blocks up to 10 m in size (see Fig. 13 in Van Kranendonk, 2006), large-scale slump folds (see Fig. 15 in Van Kranendonk, 2006), growth faults intruded by hydrothermal barite and silica (cf. Nijman et al., 1998; Van Kranendonk, 2006), and changes in the thickness of the sedimentary fill across growth faults (Nijman et al., 1998; Van Kranendonk et al., in press). Only in the northwestern corner of the North Pole Dome is the unit strongly affected by post-depositional, younger faulting, as described by Ueno et al. (2001b).

The key features in regard to the question of biogenicity of stromatolites in the Dresser Formation are:

(1) what is the original protolith of the material in which the stromatolites occur;
(2) is there evidence for post-depositional effects giving rise to stromatolite-like morphology?

7.2-4.1. Protolith Materials

Early workers identified the presence of carbonate and clastic sedimentary rocks in the Dresser Formation (Groves et al., 1981; Buick and Dunlop, 1990), despite the fact that most rocks at the surface are silicified due to the combined effects of hydrothermal alteration and Recent weathering (Van Kranendonk, 2006). Primary carbonate was described from one surface outcrop locality (Garcia-Ruiz et al., 2003; Van Kranendonk, 2006), and found to have a characteristic trace element geochemical signature indicative of precipitation from seawater (Van Kranendonk et al., 2003). Drilling below the effects of surface weathering revealed widespread preservation of primary, micritic carbonate (Van Kranendonk et al., 2006c, 2007c). The presence of crystal rosettes at one stratigraphic level (Van Kranendonk et al., 2007c) was used to infer growth of gypsum crystals through evaporation of seawater in a shallow marine basin (Groves et al., 1981; Buick and Dunlop, 1990). Influx of hydrothermal fluids occurred at several intervals during accumulation of the sedimentary pile, but ceased prior to deposition of an unconformably overlying unit of coarse sand and finely bedded carbonate (Van Kranendonk, 2006; Van Kranendonk et al., 2007c).

The finely laminated material that makes up most stromatolites weathers a black to rusty red colour on surface outcrops. In unweathered drill core, it was observed that this material is composed primarily of pyrite, with subordinate, and commonly secondary, sphalerite and barite and macroquartz (Van Kranendonk et al., 2007c). Significantly, however, it was observed that the pyrite, and the secondary minerals, represent hydrothermal replacement minerals of primary carbonate sediment, based on observations of carbonate preserved in gaps between pyrite mineral grains (*op cit.*). Indeed, pyrite was observed by these authors to be a common replacement mineral throughout the formation, indicative of reducing conditions.

7.2-4.2. *Post-Depositional Effects*

Stromatolites in the Dresser Formation occur at several places along and across strike (Nijman et al., 1998; Van Kranendonk et al., 2007c). However, the best developed and most widespread occurrence of stromatolites are at the top of member 1 and in member 2 of Van Kranendonk et al. (2007c). This interval consists of a fining-up sequence, typically 40–120 cm thick, of the following materials, from base to top: centimetre-layered crystalline carbonates with local gypsum crystal rosettes (Fig. 7.2-4(a)); rippled carbonate sediment, also with overprinting gypsum crystal rosettes (Fig. 7.2-4(b,c)); an interval, from 10–40 cm thick, of finely laminated material, locally interbedded with rippled carbonate sediment (Fig. 7.2-2(d)) and local lenses of imbricate flat pebble conglomerate (Fig. 7.2-2(e)). It is this upper interval that contains the most widespread and diverse array of stromatolites. This stratigraphic control of stromatolite growth at the level of demonstrably shallow water, to intermittently exposed, conditions is a significant factor in support of a biogenic origin of Dresser Formation stromatolites (cf. Groves et al., 1981; Buick and Dunlop, 1990; Van Kranendonk, 2006; Van Kranendonk et al., 2007c).

In cross-sectional view, Dresser Formation stromatolites vary from wrinkly laminated mats, to broadly domical forms (Fig. 7.2-4(f)), to roughly coniform varieties, and well-developed coniform types (Fig. 7.2-5). Elsewhere throughout the formation are a wide variety of wrinkly laminated mat and domical forms, but these are commonly developed within finely-laminated material intergrown with coarsely crystalline barite, making identification of primary and secondary morphological effects more difficult.

Several of the textural features presented in Figs. 7.2-4 and 7.2-5 clearly indicate that the morphological traits of stromatolites are *primary* features that formed *during* sediment accumulation (see also Van Kranendonk, 2006). Most important of these is the fact that irregularly laminated, wrinkly mat material is interbedded with cross-laminated clastic sediment (Fig. 7.2-4(d)), providing evidence that deformation was not responsible for the development of the finely laminated material or of its wrinkly texture; this observation counters the model proposed by Lowe (1994). Additional evidence for the undeformed nature of the finely laminated material is given by the primary depositional relationship between finely laminated stromatolite and underlying flat pebble conglomerate shown in Fig. 7.2-4(e).

The other important observational indicator of the primary nature of the stromatolite morphology is the relationship between coniform varieties, sedimentary ripples and gypsum crystal rosettes. The images presented as Figs. 7.2-4 and 7.2-5 are all from within an area of only 10 m^2. They show, beyond any reasonable doubt, the excellent state of preservation of primary sedimentary structures in the rocks (note especially the lack of any indication of penetrative strain in Fig. 7.2-4(c and d)), such that we can be 100% confident that the morphological features of stromatolites in this area are primary features of the sedimentary environment. It is also important to note that apart from the diagenetic gypsum crystal rosettes, this area is free from the effects of later hydrothermal fluid alteration and barite crystal growth, so that the stromatolite morphologies can not be ascribed to these secondary effects.

Fig. 7.2-4. Features of the ca. 3.49 Ga Dresser Formation. (a) Cross-sectional outcrop view of weakly silicified carbonates with gypsum crystal rosettes and broad ripples. (b) Oblique outcrop view of bedding plane with curved ripple marks. (c) Plan outcrop view of bedding plane with gypsum crystal rosettes: note the lack of any penetrative strain. (d) Cross-sectional outcrop view of weakly silicified carbonates with wrinkly laminated stromatolite below and overlying rippled carbonate sediment. Similar ripples from the Strelley Pool Chert were interpreted as indicative of microbial binding by Van Kranendonk et al. (2003). Lower arrow points to small channel; upper arrow points to higher-amplitude ripple, both of which are used to discount structural deformation in the formation of the wrinkly laminations below. (e) Cross-sectional outcrop view of mottled carbonate overlain by fine carbonate laminates and a lens of imbricated flat pebble conglomerate. Coniform stromatolites in Fig. 7.2-5 overlie these conglomerates. (f) Oblique outcrop view of bedding plane with broad domical stromatolite: note the wrinkly laminations on dome flanks.

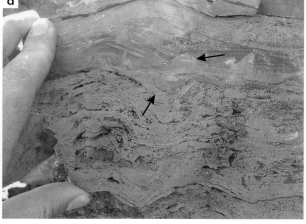

Fig. 7.2-4. (*Continued.*)

7.2-5. STRELLEY POOL CHERT STROMATOLITES

The ca. 3.4 Ga Strelley Pool Chert (Lowe, 1980; Van Kranendonk et al., 2006a) is a widespread unit across the East Pilbara Terrane. It represents the basal formation of the Kelly Group and lies unconformably on rocks of the Warrawoona Group across a sub-aerial erosional unconformity (Buick et al., 1995; Van Kranendonk et al., 2006a, 2007a). The formation consists of three primary members: a basal coarse clastic member; a middle unit of laminated carbonates; an upper unit of coarse clastic sedimentary rocks. Whereas most researchers agree that this unit was deposited under fluviatile to shoreline and shal-

Fig. 7.2-4. (*Continued.*)

low marine environments (Lowe, 1980, 1983; Buick et al., 1995; Van Kranendonk, 2000, 2006; Van Kranendonk et al., 2001b, 2003; Allwood et al., 2006a), Lindsay et al. (2003, 2005) suggested that the middle carbonate unit was deposited from hydrothermal solutions sourced from veins preserved locally in the footwall, based on ideas first presented in Van Kranendonk et al. (2001b).

Coniform stromatolites are widespread in the middle carbonate member of the Strelley Pool Chert, across most of the 220 km diameter of the East Pilbara Terrane (e.g., Fig. 7.2-1: unpublished map data), and have been described in detail by several authors who support a biogenic interpretation (Lowe, 1980; Hofmann et al., 1999; Van Kranendonk, 2000; Van Kranendonk et al., 2003; Allwood et al., 2006a). Alternatively, Lowe (1994) and Lind-

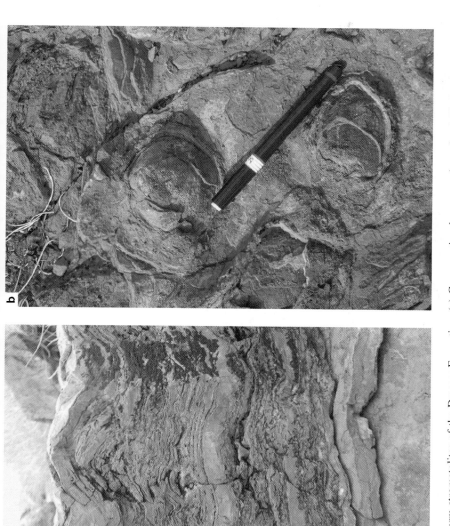

Fig. 7.2-5. Coniform stromatolites of the Dresser Formation. (a) Cross-sectional outcrop view of wrinkly laminated carbonate and coniform stromatolite. (b) Plan outcrop view of bedding plane with coniform stromatolites.

say et al. (2003, 2005) have suggested that coniform structures in the Strelley Pool Chert are abiogenic. Lowe (1994) suggested formation through evaporitic precipitation, whereas Lindsay et al. (2003) concluded that, because "Laminations within the stromatolitic structures are conspicuously isopachous and comparable with abiotic structures formed by direct precipitation of carbonate ... stromatolite structures described from the Strelley Pool Chert are abiotic features deposited by direct precipitation from hydrothermal solutions that were modified by ocean floor currents."

Two aspects pertain to resolution of this controversy: (1) detailed examination of stromatolite morphology, to ascertain whether or not laminations are isopachous; (2) geological setting of the carbonates as either marine or hydrothermal precipitates.

7.2-5.1. Stromatolite Morphology

Stromatolites in the middle carbonate member of the Strelley Pool Chert are most commonly coniform and do locally display isopachous lamination (Fig. 7.2-6(a,b)). However, there are also much larger, more complex forms, including large coniform stromatolites that display incipient branching, and small branching columnar forms that are difficult, if not impossible, to explain as hydrothermal precipitates (Fig. 7.2-6(c,d): see also Allwood et al., 2006a). Detailed mapping show that stromatolite forms vary along strike over distances of 1–1000 m, and form locally onlapping biostromes (Fig. 7.2-7; see also Allwood et al., 2006a).

Closer inspection of laminations shows that most stromatolite forms do not have isopachous layering, but show a variety of layering relationships, including onlap of sediment along the sides of coniform stromatolites (Fig. 7.2-8(a): Van Kranendonk et al., 2003), and growth of coniform stromatolites within, and contemporaneously with, sedimentary rocks deposited from ocean currents (Fig. 7.2-8(b)). These features show that the rocks are not abiogenic precipitates "modified by ocean currents", as suggested by Lindsay et al. (2003), but rather are structures that formed *during* accumulation of carbonate sedimentary material affected by ocean currents.

This relationship is most dramatically displayed in Fig. 7.2-9, which shows a steep-sided coniform stromatolite that displays continuous growth upwards from fine to moderately laminated carbonates at the base (1), thorough an 8-cm-thick bed of flat pebble conglomerate (2), into finely laminated carbonates (3). This unequivocally demonstrates that coniform stromatolites are not precipitates from solutions of either evaporitic or hydrothermal origin and must have formed as a result of living microbial activity.

The same conclusion may be affirmed for small, domical-columnar stromatolites higher up in the middle carbonate unit, where the rocks are more highly silicified and bedding is developed at a finer-scale. These rocks preserve well-developed ripples and contain several horizons with centimeter-high domical-columnar stromatolites defined by rusty red and yellow weathering laminates (Fig. 7.2-10(a–c)). Although these domical-columnar forms display broadly isopachous laminations at outcrop scale (Fig. 7.2-10(b)), thin section observations show this is clearly not the case and that they have many of the characteristic features of more recent stromatolites, including steep-sided growth walls that truncate

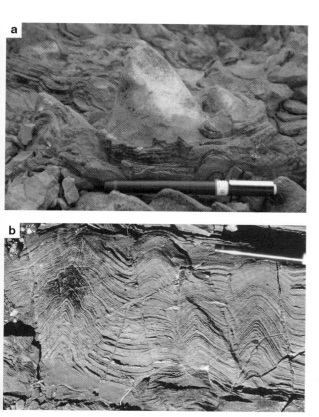

Fig. 7.2-6. Stromatolites of the Strelley Pool Chert. (a) Bedding-parallel outcrop view of large coniform stromatolite. (b) Cross-sectional outcrop view of coniform stromatolites, showing isopachous lamination. Note the incipient branching of the form on the left, the asymmetric limbs of forms on the right, and the way in which the upper form initiates from a planar bedding horizon. (c) Cross-sectional outcrop view of large, complex coniform stromatolite and smaller, branching columnar form (top right; see part d for detail). (d) Cross-sectional outcrop view of small, branching columnar stromatolite: note the steep-sided growth walls.

laminae, internal discordancies of laminae, and growth termination by influx of clastic sediment (Fig. 7.2-10(d)). These morphological features demonstrate dynamic growth within a changing environment and support a biogenic origin for these structures.

7.2-5.2. Geological Setting

Three geological settings have been posed for the carbonates of the Strelley Pool Chert: a shallow marine, evaporative basin (Lowe, 1980, 1983, 1994); a shallow, open marine

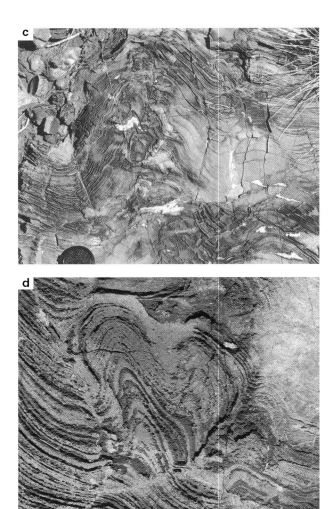

Fig. 7.2-6. (*Continued.*)

environment (Hofmann et al., 1999; Van Kranendonk et al., 2003; Allwood et al., 2006a), or a hydrothermal setting (Lindsay et al., 2003, 2005).

A characteristic feature of the Strelley Pool Chert is that it is recognizable as a distinct formation over a 220 km wide area of the East Pilbara Terrane, with a very regular and predictable internal stratigraphy as described above, and elsewhere (see Fig. 4 in Van Kranendonk, 2006). The formation can be traced in good exposure for 10s of kilometers along strike within greenstone belts, is recognized in all greenstone belts across the terrane, and contains between 5–10 m of finely laminated carbonate in most of the belts. The only real changes within this unit across strike is the thickness of the lower clastic member, which

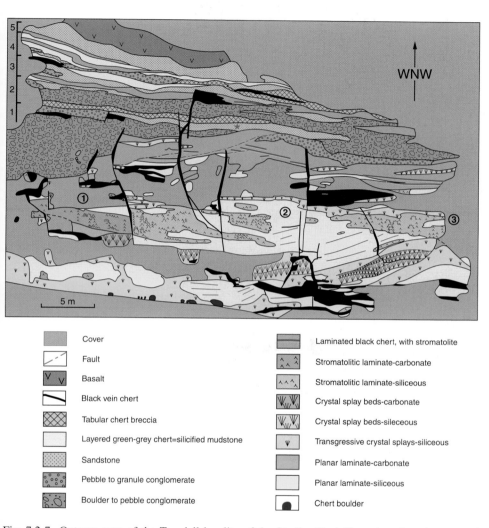

	Cover		Laminated black chert, with stromatolite
	Fault		Stromatolitic laminate-carbonate
	Basalt		Stromatolitic laminate-siliceous
	Black vein chert		Crystal splay beds-carbonate
	Tabular chert breccia		Crystal splay beds-sileceous
	Layered green-grey chert=silicified mudstone		Transgressive crystal splays-siliceous
	Sandstone		Planar laminate-carbonate
	Pebble to granule conglomerate		Planar laminate-siliceous
	Boulder to pebble conglomerate		Chert boulder

Fig. 7.2-7. Outcrop map of the Trendall locality of the Strelley Pool Chert, based on 1 m spaced grid mapping. Circled numbers indicate positions of onlapping coniform stromatolite biostromes (purple areas), which are capped by a layer of crystal splays – possibly originally aragonite, but now medium-grained dolomite – above which stromatolites change morphology to domical-columnar forms (heavy red lines). Note the basal unit of isolated boulders of layered black and red chert, which are encrusted by the overlying carbonate. Star denotes position of sandstone with black, kerogen-rich laminates. Bar in top left indicates position of fining-up successions, interpreted as deposits from a series of receding alluvial fans.

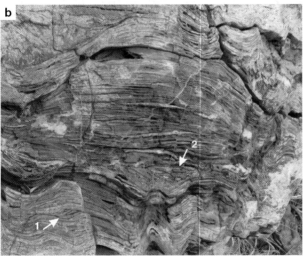

Fig. 7.2-8. Bedding features of (silicified) carbonates of the Strelley Pool Chert at the Trendall locality. (a) Detailed, cross-sectional outcrop view of the flank area of a coniform stromatolite, showing
onlap of carbonate beds against stromatolite flank (arrows) (from Van Kranendonk et al., 2003).
(b) Cross-sectional outcrop view of silicified carbonates, showing coniform stromatolite (bottom
middle) growing within rippled carbonate sediment (arrow 1) and overlain by flat pebble conglomerate (arrow 2) that terminated stromatolite growth. Overlying laminated sediment displays high
amplitude symmetrical swale, whose geometry suggests microbial binding of clastic sediment under
high-energy conditions.

varies from 1 m of boulder and pebble conglomerate to 1000 m of quartz-rich sandstone,
reflecting local changes in depositional environments for this member from fluviatile to
shoreline and shallow marine (Lowe, 1983; Buick et al., 1995; Van Kranendonk, 2000,
2004b; Allwood et al., 2006a).

Fig. 7.2-9. Steep-sided coniform stromatolite of the Strelley Pool Chert, which passes up from finely bedded carbonate at the base, through a layer of flat pebble conglomerate, into finely bedded carbonate at the top: arrows point to coniform peak of the stromatolite in the uppermost carbonate.

Widespread development of weakly upward-radiating crystal splays (e.g., Fig. 7.2-7), together with the presence of desiccation cracks and wind-blown ripples (Fig. 7.2-10(a)) indicate periods of exposure during accumulation of the carbonate member (Van Kranendonk et al., 2003; Allwood et al., 2006a). Significantly, stromatolite abundance is directly linked to distribution of crystal fans, with the latter forming a cap on the large biostromes, above which stromatolite growth is much less abundant and of different morphology (small, domical forms above vs. large coniform varieties below: Fig. 7.2-7). These data suggest that marine precipitation of carbonate rocks in the Strelley Pool Chert occurred under (at least) periodically evaporative conditions.

The widespread and consistent distribution of the carbonate lithofacies argues strongly against a hydrothermal setting of deposition, as these occur within fault-bounded, volcanic terrains and are characteristically highly variable in terms of thickness and composition along strike lengths of 1–10 km. Further evidence against a hydrothermal setting was provided by Van Kranendonk et al. (2003) in the form of trace element geochemical data on the carbonates, which demonstrated that they were precipitated from normal seawater, and not from hydrothermal solutions.

Indeed, the sequence of lithofacies preserved within the Strelley Pool Chert and overlying Euro Basalt – from development of a subaerial erosional unconformity, to shoreline

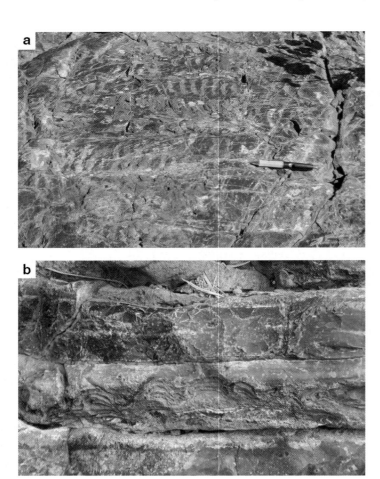

Fig. 7.2-10. Features of the upper part of the carbonate member of the Strelley Pool Chert. (a) Small, crescentic windblown ripples in silicified carbonate. (b) Cross-sectional outcrop view of centimetre-high, domical-columnar stromatolites in silicified carbonate (represented as red lines on Fig. 7.2-7). (c) Plan surface outcrop view of small domical-columnar stromatolites in silicified carbonate. (d) Cross-sectional thin section view of centimetre-high, domical-columnar stromatolites, showing steep-sided growth walls that truncate growth laminae (arrow 1), internal discordancies of laminae (arrow 2), and growth termination by influx of clastic sediment (arrow 3).

deposition of conglomerates and cross-bedded, quartz-rich sandstones, to shallow marine carbonates, to regressive alluvial fans, and finally thick (8 km) pillow basalts – indicates an overall basin-deepening event. Geochemical and geochronological data from the Euro Basalt indicates that this was related to the arrival (uplift and erosion) and eruption (basin deepening and pillow basalt) of a mantle plume (Van Kranendonk and Pirajno, 2004; Smithies et al. 2005b; Van Kranendonk et al., this volume).

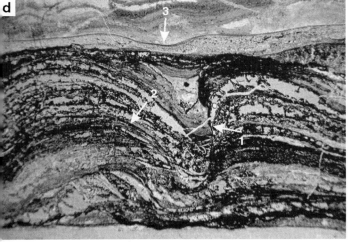

Fig. 7.2-10. (*Continued.*)

7.2-6. ORIGIN OF CARBONACEOUS MATERIAL/MICROFOSSILS IN HYDROTHERMAL SILICA VEINS

Possible microfossils and/or kerogen have been descried from similar geological settings at three different stratigraphic levels in the Pilbara Supergroup (Figs. 7.2-1, 7.2-2). The oldest of these is in the North Pole Dome, where a dense swarm of hydrothermal silica ± barite veins immediately underlie the sedimentary rocks of the ca. 3.49 Ga Dresser Formation. In these veins, Ueno et al. (2001a, 2001b, 2004, 2006) have described thread-like

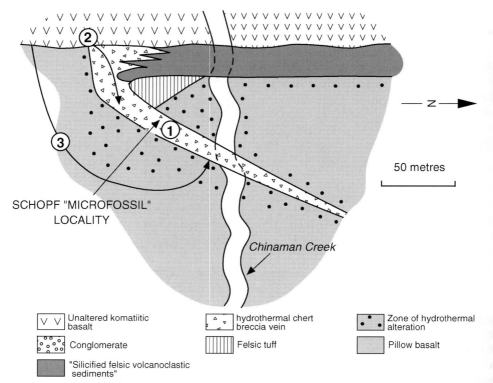

Fig. 7.2-11. Schematic geological cross-section of the Schopf microfossil locality, showing three possible ways that organic material (carbonaceous material/microfossils) could have been entrained into the hydrothermal silica vein. (1) Microbial communities inhabit fissures during cool periods of hydrothermal fluid flow. (2) Microbial communities inhabit the surface margins of the fissure and fall down into the fissures, or be entrained down into the fissures by hydrothermal fluid circulation. (3) Microbes and/or their degraded remnants (kerogen) that inhabit surface seawater or the porosity in the flanking wallrocks could be transported into the veins by downward circulating fluids that form part of the hydrothermal circulation system.

putative microfossils with highly depleted $\delta^{13}C$ values of kerogen and methane-bearing fluid inclusions.

The second locality is the highly controversial Schopf (1992, 1993) microfossil locality, which is located in a hydrothermal breccia vein in the subsurface beneath bedded felsic volcaniclastic rocks (Brasier et al., 2002, 2005; Van Kranendonk, 2006). Controversy rages about the biogenicity of putative microfossils in this vein (Brasier et al., 2002, 2005; Schopf et al., 2002; Schopf, 2004).

The third locality where putative microfossils have been described from a hydrothermal setting is at the top of the Sulphur Springs Group, where thread-like putative microfosils have been described within sulphides from a volcanogenic massive sulphide deposit located directly above a black silica hydrothermal vein system (Rasmussen, 2000).

While it is beyond the scope of this paper to speculate on a biogenic or abiogenic origin of carbonaceous material/microfossils in hydrothermal silica veins in the Pilbara Supergroup (see Marshall, this volume), it is worth discussing how organic material may have been entrained within the veins. I envisage three scenarios, as outlined in Fig. 7.2-11:

1) Microbial communities could directly inhabit the veins/fissures during periods of relative cool fluid circulation;
2) Microbial communities could inhabit the surface margins of the fissures and fall down into the fissures, or be entrained down into the fissures by hydrothermal fluid circulation;
3) Microbes and/or their degraded remnants (kerogen) that inhabit surface seawater or the porosity in the flanking wallrocks could be transported into the veins by downward circulating fluids that form part of the hydrothermal circulation systems.

7.2-7. CONCLUSIONS

Several lines of evidence have been presented since the 1970s in support of the presence of early life during deposition of the Pilbara Supergroup. Although no single piece of evidence is yet accepted by the geological community as *proof* of Paleoarchean life, the collective wealth of data and the diverse nature of the evidence indicates a strong probability that early Earth was colonised by microbial life at 3.5 Ga. Diverse stromatolite morphology in both the Dresser Foramtion and Strelley Pool Chert may indicate that life at this time was already diverse (Van Kranendonk, 2006).

At this point in time, there are three strong pieces of evidence in support of early life: 1) the gross morphology of stromatolites in rocks of proven marine carbonate origin in the Strelley Pool Chert (most convincing) and Dresser Formation; (2) clasts of laminated kerogenous material in clastic sedimentary rocks of the Dresser Formation and Strelley Pool Chert; (3) ichnofossils in the chilled margins of pillow basalts (Furnes et al., 2004; Banerjee et al., 2007).

ACKNOWLEDGEMENTS

Arthur Hickman is gratefully acknowledged for his support of this work. Kath Grey is thanked for discussions regarding biogenicity. Malcolm Walter is thanked for his encouragement and support. Suzanne Dowsett and Allan Blake helped with figures. This paper is published with permission of the Executive Director, Geological Survey of Western Australia.

Earth's Oldest Rocks
Edited by Martin J. Van Kranendonk, R. Hugh Smithies and Vickie C. Bennett
Developments in Precambrian Geology, Vol. 15 (K.C. Condie, Series Editor)   879
© 2007 Elsevier B.V.   All rights reserved.
DOI: 10.1016/S0166-2635(07)15073-8

Chapter 7.3

STABLE CARBON AND SULFUR ISOTOPE GEOCHEMISTRY OF THE CA. 3490 MA DRESSER FORMATION HYDROTHERMAL DEPOSIT, PILBARA CRATON, WESTERN AUSTRALIA

YUICHIRO UENO

Research Center for the Evolving Earth and Planet, Department of Environmental Science and Technology, Tokyo Institute of Technology, Midori-ku, Yokohama 226-8503, Japan

7.3-1. INTRODUCTION

The North Pole area in the Pilbara Craton of Western Australia is well known for the occurrence of the oldest ($\sim$3.5 Ga) putative microfossils (Awramik et al., 1983; Ueno et al., 2001a) and stromatolites (Walter 1980; Hofmann et al., 1999; Van Kranendonk 2006; Allwood et al., 2006a), particularly in the chert-barite dominated sedimentary rocks of the Dresser Formation. However, a biological origin of these putative fossils has been debated (Buick et al., 1981; Buick 1990; Schopf et al., 2002; Garcia Ruiz et al., 2003; Brasier et al., 2002, 2006). Moreover, morphology-based study of prokaryotic fossils provides only limited information for the physiology of possible ancient life.

On the other hand, carbon and sulfur isotope geochemistry is useful to trace specific metabolic activities in the past, including autotrophic carbon fixation, sulfate reduction, and methanogenesis. Shen et al. (2001) reported ^{34}S-depletion of pyrite in bedded and vein barite ($\delta^{34}S_{pyrite} - \delta^{34}S_{barite} = -5$ to -21‰) in the Dresser Formation, which is interpreted as the oldest evidence for microbial sulfate reduction. The recent discovery of ^{13}C-depleted methane ($\delta^{13}C_{CH4} < -56\text{‰}$) in primary fluid inclusions of the Dresser Formation indicates that microbial methanogenesis could have been active in the Paleoarchaean (Ueno et al., 2006).

The depositional environment of the Dresser Formation has long been considered as an evaporitic, shallow marine basin (e.g., Buick, 1990; Lowe, 1983). Hence, most of the palaeontological and geochemical evidence for the putative early life forms have been interpreted from the viewpoint of a photosynthesis-based shallow marine ecosystem, which includes cyanobacteria-like microfossils (Awramik et al., 1983), stromatolites build by photosynthesizers (Walter, 1983), and sulfate-reduction in a locally sulfate-rich evaporitic basin (Shen and Buick, 2004). However, recent geological and geochemical investigations suggest that hydrothermal activity played an important role in the deposition of the chert-barite rocks of the Dresser Formation (Isozaki et al., 1997; Nijman et al., 1999; Runnegar

2001; Ueno et al., 2001a, 2004; Van Kranendonk et al., 2001b; Van Kranendonk and Pirajno, 2004; Van Kranendonk, 2006). The most striking evidence for the hydrothermal activity comes from the occurrence of a prominent swarm of silica ± barite feeder veins, which penetrated along syn-sedimentary growth faults in the footwall to bedded chert-barite rocks (Nijman et al., 1999; Ueno et al., 2001a; Van Kranendonk et al., 2001b; Van Kranendonk and Pirajno, 2004). Hence, it is plausible that the Dresser Formation represents a hydrothermally influenced volcano-sedimentary sequence deposited in a submarine volcanic caldera (Van Kranendonk, 2006, this volume).

With this new depositional setting in mind, we can now re-evaluate the link between hydrothermal activity and the palaeontological and geochemical evidence for life in the Dresser Formation. Furthermore, recent petrological studies confirm that black silica veins, which are abundant in the footwall to the Dresser Formation, commonly contain significant amounts of organic matter (e.g., Ueno et al., 2004), as reported from other Paleo- and Mesoarchaean greenstone terranes (e.g., de Wit et al., 1982; Kiyokawa et al., 2006). The origin of the organic matter in the veins is controversial, but particularly important, because the organic matter has been produced either by biological carbon fixation (Ueno et al., 2004, 2006; Brasier et al., 2006), or by pre-biotic organic synthesis (Brasier et al., 2002; Lindsay et al., 2005; McCollom and Seewald, 2006). In this paper, the mode of occurrence of the Dresser hydrothermal deposits is first briefly summarized. Then, the carbon and sulfur isotope geochemistry of the Dresser deposits is reviewed in relation to the hydrothermal activity.

7.3-2. HYDROTHERMAL DEPOSITS IN THE NORTH POLE AREA

7.3-2.1. The Dresser Formation

The Dresser Formation is exposed in the North Pole Dome of the East Pilbara Terrane, Pilbara Craton (Fig. 7.3-1). The Dresser Formation (Van Kranendonk and Morant, 1998) consists of ca. 2-km thick basaltic greenstones intercalated with three horizons of bedded chert (Fig. 7.3-1). The lowermost chert unit is 1 to 70 m thick and is intercalated with several barite horizons of 0.1 to 5 m thickness. This barite-bearing chert unit corresponds to the "chert-barite unit" previously described by Buick and Dunlop (1990). The other two chert units are thinner (1–13 m) than the chert-barite unit, and contain only rare barite.

The precise age of the Dresser Formation has not been directly determined. Zircon U-Pb dating yields an age of 3458 ± 2 Ma for felsic volcanic rocks of the stratigraphically overlying Panorama Formation in the North Pole area (Thorpe et al., 1992a), indicating an age for the Dresser Formation of >3460 Ma. A model lead age of 3490 Ma (Thorpe et al.,

Fig. 7.3-1. Geological map of the Dresser Formation in the North Pole area, modified from Kitajima et al. (2001).

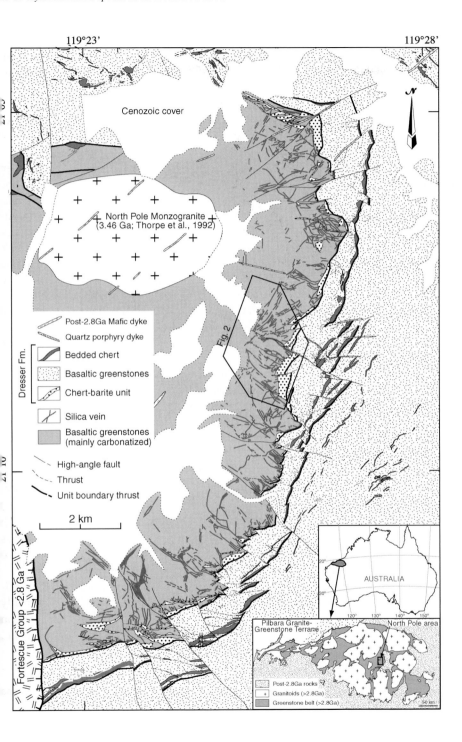

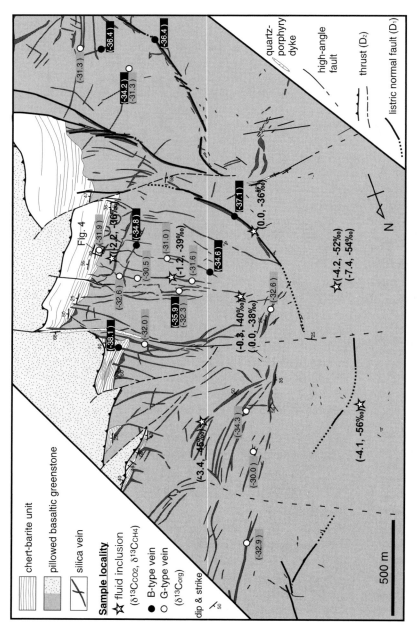

Fig. 7.3-2. Detailed geological map of part of the Dresser Formation, showing carbon isotopic values of analysed organic matter (circles), and of CO_2 and CH_4 in fluid inclusions (stars). Note that the contact between the chert-barite unit and the underlying pillowed basalt represents the sea-floor surface at the time of chert deposition, and therefore that the silica veins developed in the uppermost 1000 m of the oceanic crust. The box indicates the location of Fig. 7.3-4.

1992b) was obtained for galena from barite in the Dresser Formation, which is considered to represent the actual depositional age of the Dresser Formation.

7.3-2.2. *Hydrothermal Silica Vein Swarm*

In the North Pole area, numerous (>2000 identified) silica veins characteristically intruded along normal faults, which developed in pillowed basaltic volcanic rocks below the Dresser Formation (Figs. 7.3-1–7.3-4). They are 0.3–20 m wide and generally >100 m long, with the longest one over 1 km. The veins are massive and are composed mainly of fine-grained silica (<10 μm). Some silica veins show a symmetrical zonation pattern across the dike axis, and sometimes have agate at the center (Fig. 7.3-3(b)), in which the silica shows fan-shape structures grown from the margins toward the center of the veins. This suggests the precipitation of silica from a hydrothermal fluid (cf. Van Kranendonk and Pirajno, 2004).

The silica veins were previously called T-chert, which means "tectonic" chert (Hickman, 1973), chert vein (Nijman et al., 1999), chert dyke (Lindsay et al., 2005), silica dyke (Ueno et al., 2004, 2006), or silica vein (Van Kranendonk, 2006). The term "silica vein" is used in this paper, because "chert" is a term used for a chemical sedimentary rock with a distinct grain size and texture, and is thus not suitable for an intrusive rock, which has a grain size of generally microquartz. Also, the silica veins are clearly distinguished from quartz veins, which consist of white, coarse-grained quartz.

The silica veins terminate into chert beds of the Dresser Formation, but do not cut through the entire chert unit, nor into the overlying pillow basalt (Figs. 7.3-1, 7.3-2, and 7.3-4: Ueno et al., 2004; Van Kranendonk, 2006). The tops of the veins locally show a gradual transition into chert beds, forming a clear T-junction. In the field, repeated cycles of normal faulting, vein formation, and subsequent deposition of chert beds can be recognized (Fig. 7.3-4). These relationships suggest the silica veins were formed intermittently during the deposition of chert beds (Isozaki et al., 1997; Nijman et al., 1999; Ueno et al., 2001a; Van Kranendonk et al., 2001b; Van Kranendonk, 2006).

In addition, 0.1 to 2 m wide barite veins also intruded the footwall basaltic rocks, and some vein barite occurs in silica veins, as growth zones, or as discrete veins (cf. Van Kranendonk, 2006). Similar to the relationship between silica veins and bedded chert, the vein and "bedded" barite show feeder vein-deposit relationships (Nijman et al., 1999). The distribution of vein barite is laterally discontinuous and is generally restricted to the uppermost ~100 m of the pillow lava unit below the Dresser Formation, whereas silica veins occur up to about 1000 m below the formation. This distribution pattern indicates that the hydrothermal fluid circulated at least 1000 m below seafloor at the time of deposition of the Dresser Formation, but that circulation of sulfate-rich fluids was restricted to the shallower part of the system.

7.3-2.3. *Organic Matter in the Silica Veins*

The silica veins contain considerable amounts of organic matter (kerogen), which give the veins their black to gray color (Fig. 7.3-3). Thorough geochemical and petrological

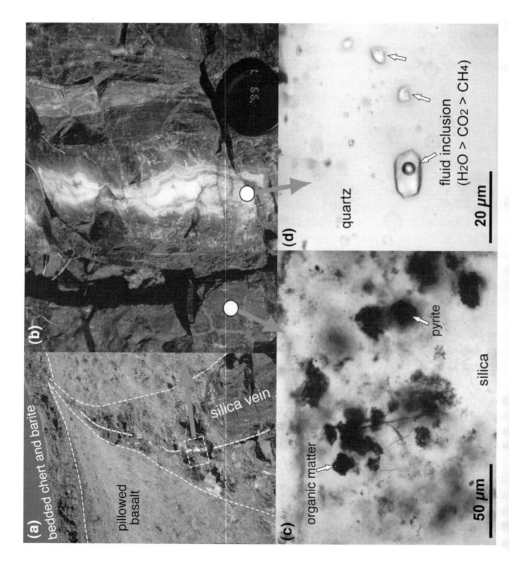

(a) bedded chert and barite
pillowed basalt
silica vein

(b)

(c) organic matter
pyrite
silica
50 μm

(d) quartz
fluid inclusion
($H_2O > CO_2 > CH_4$)
20 μm

Fig. 7.3-3. (*Previous page.*) (a) Photograph of a 1 m wide hydrothermal silica vein leading up to bedded chert and barite. (b) Close-up view of the central part of the silica vein in (a), showing agate-textured quartz core. Width of the photo is about 30 cm. (c) Photomicrograph of a black part of the silica vein, showing irregular-shaped clots of organic matter and pyrite enclosed in a fine-grained silica matrix. (d) Photomicrograph of coarse-grained quartz developed in the central part of the vein, showing H_2O–CO_2 fluid inclusions therein.

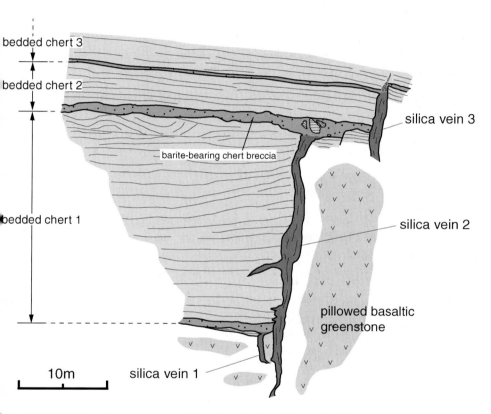

Fig. 7.3-4. Mode of occurrence of the chert-barite beds and silica veins intruding along syn-depositional growth faults (modified from Isozaki et al. (1997)). At least three stages of penetration of the silica vein and subsequent deposition of the bedded chert can be recognized in this outcrop.

investigation of more than 600 hand specimens of the silica veins has revealed that the concentration of organic carbon in the veins would have been originally >1 mg C/g (Ueno et al., 2004). Most veins have suffered post-depositional metasomatism (oxidation) and have lost about 90% of the original organic matter. However, some black silica veins possess an original mineral suite that includes pyrite, Fe-poor sphalerite, and Fe-dolomite (Ueno et al., 2004). These mineral compositions, as well as its silica- and barite-dominated mineral assemblage, suggest that the veins were deposited from relatively low temperature (100–200 °C), reducing hydrothermal fluids (Ueno et al., 2004). This temperature estimate is consistent with homogenization temperatures of fluid inclusions in vein quartz ($\sim$150 °C; Kitajima et al., 2001) and in cavity-filling quartz of pillowed basalt (90–170 °C; Foriel et al., 2004).

Consequently, this suggests that organic carbon was widely distributed in fissures that developed in the upper 1000 m of the seafloor basalt. These fissures acted as conduits for circulating hydrothermal fluids (Van Kranendonk, 2006). The estimated temperature of the fluid is low enough to permit biological activity in the sub-seafloor environment, and is suitable for thermophilic or hyperthermophilic organisms.

7.3-3. CARBON AND SULFUR ISOTOPE GEOCHEMISTRY OF THE DRESSER FORMATION

7.3-3.1. Carbon Isotopic Composition of Organic Matter in the Hydrothermal Veins

The $\delta^{13}C$ values of organic matter in the silica veins and bedded chert in the Dresser Formation are −38 to −30‰ and −31 to −29‰, respectively (Figs. 7.3-2 and 7.3-5). The isotopic variation of the organic matter does not correlate with depth below seafloor surface at that time, but rather with organic carbon concentrations. This indicates that post-depositional alteration would have reduced a considerable amount of the organic carbon in the silica veins and significantly modified its carbon isotopic composition. Hence, we have to first estimate the original isotopic composition of the organic matter before discussing its origin.

Among the silica veins, sulfide-bearing black silica veins contain higher organic carbon concentrations and show more ^{13}C-depleted isotopic compositions than gray sulfide-free silica veins, which formed by alteration of primary sulfide-bearing veins (Ueno et al., 2004). This variation would be produced by the Rayleigh distillation process with a carbon isotopic fractionation factor of 0.9985 (Fig. 7.3-5; Ueno et al., 2004). According to this model, the original organic carbon in the silica veins would have been isotopically hetero-geneous ($\sim$5‰) at around −35‰, and at least some material had initial $\delta^{13}C$ values of $\leqslant$−38‰.

The inferred ^{13}C-depletion of the initial organic carbon can be explained by biological carbon fixation. The isotopic composition of carbonate carbon in the veins is −2‰ (Ueno et al., 2004). This indicates that the equilibrated dissolved CO_2 had $\delta^{13}C$ values of about −4‰ at 100 °C (Mook et al., 1974). This estimate is roughly consistent with $\delta^{13}C$ values of CO_2 entrapped in the primary fluid inclusions (−3 to −7‰; Ueno et al., 2006). Thus, the fractionation between initial organic carbon and dissolved CO_2 (i.e., δ) would have been over 34‰, for at least some of the material.

If the organic carbon was produced by autotrophic organisms, then this fractionation is too large to be associated with Rubisco ($\delta \leqslant 30$‰; e.g., House et al., 2003), the enzyme

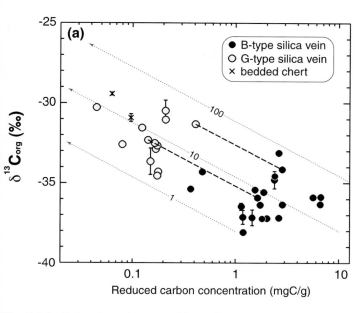

Fig. 7.3-5. Carbon isotopic compositions of organic matter, CO_2, and CH_4 in the silica veins. (a) Relationship between organic carbon concentration and carbon isotopic composition of the organic matter in the silica veins and bedded chert. Dashed lines tie the values from the same rock samples. Doted lines represent Rayleigh-fractionation trajectories with fractionation factor (α) of 0.9985. The three lines labeled 1, 10, and 100 started with an initial $\delta^{13}C$ value of $-38\permil$, and with initial concentrations of organic carbon of 1, 10, and 100 mg C/g, respectively. (b) Relationship between $\delta^{13}C_{CH4}$ and $\delta^{13}C_{CO2}$ values of the extracted volatiles from the fluid inclusions as compared with that of fluids vented from present-day seafloor hydrothermal systems. ^{13}C-depleted CH_4 occurs preferentially in samples rich in primary fluid inclusions (Ueno et al., 2006). (c) The ranges of carbon isotopic compositions of organic matter and carbonate in the silica vein and bedded chert.

utilized by aerobic photoautotrophs. This is consistent with the newly recognized depositional model, in which phototrophs are unlikely to have been involved in the sub-seafloor hydrothermal environment. On the other hand, the large organic carbon fractionation could have been produced via the reductive acetyl-CoA pathway ($\delta \leqslant 42\permil$; e.g., House et al., 2003; Londry and Des Marais, 2003), which is utilized by H_2-dependent chemoautotrophs such as methanogen, acetogen, and some sulfate-reducers. This inferred large fractionation is consistent with the reducing conditions of the sub-seafloor hydrothermal system during the Dresser Formation. In fact, such large fractionations (up to 36‰) have been demonstrated by some anaerobic and thermophilic chemoautotrophs, such as *Methanobacterium thermoautotrophicum* (Fuchs et al., 1979).

Although the geological, petrological, and geochemical characteristics of the organic matter in Dresser Formation hydrothermal veins can be fully explained by metabolic activity of autotrophic organisms in a hydrothermal environment, the data do not eliminate a possible abiological origin for the organic matter. Under hydrothermal conditions

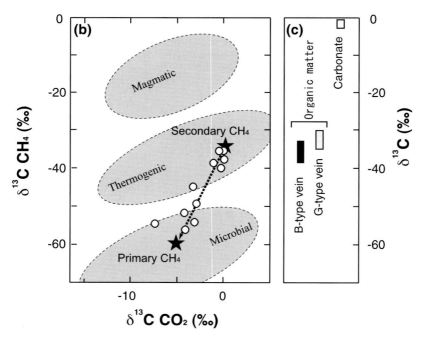

Fig. 7.3-5. (*Continued.*)

of 200–300 °C, it has been shown that abiological Fischer–Tropsch-type (FTT) reactions can produce organic matter with similarly large ^{13}C-depletions (McCollom and Seewald, 2006). However, it is questionable whether FTT synthesis took place under the conditions prevailing during silica precipitation in the Dresser Formation, because likely catalysts for FTT reactions (i.e., native metal and oxide) would not have existed under the reducing conditions exhibited by these veins. Fe-Ni alloy does not occur, despite thorough investigation of more than 300 petrographic thin sections of the silica veins (Ueno et al., 2004). Fe oxides occur only as a secondary mineral in the veins, whereas Fe-Ni sulfide (polydymite) exists as the primary mineral phase in the veins. Significantly, the presence of sulfide is known to poison industrial FTT reactions (Anderson, 1956). The apparent deficiency of an effective catalyst raises doubts as to whether FTT reactions could have produced the organic matter, under the low temperature, sulfidic conditions recorded by Dresser Formation hydrothermal veins. Note that native metal could have existed in the deeper parts of the veins than the site of silica precipitation, such that FTT reactions, if they occurred at all, may have possibly occurred under deeper and much higher temperature conditions ($\gg 200$ °C). If this is the case, however, isotopic fractionation should be smaller than that observed and organic materials should increase with depths in the veins, a feature which does not occur. In the following section, this point is further discussed in relation to the isotopic compositions of methane and carbon dioxide.

7.3-3.2. *Carbon Isotopic Composition of Methane in Fluid Inclusions*

The hydrothermal silica veins and associated quartz veins contain fluid inclusions (Fig. 7.3-3), some of which were entrapped during the original silica precipitation. In-situ laser Raman microspectroscopy confirmed that the fluid mainly consists of H_2O and CO_2, with a minor, but detectable, amount of CH_4 and H_2S (Fig. 7.3-6).

Crushing extraction of the CO_2 and CH_4 in the fluid inclusions coupled with carbon isotope analysis revealed that the $\delta^{13}C$ values of the CO_2 and CH_4 is -7 to $0‰$ and -56 to $-36‰$, respectively (Figs. 7.3-2 and 7.3-5: Ueno et al., 2006). Similar to the isotopic compositions of the organic matter, the $\delta^{13}C_{CH4}$ values do not correlate with the depth below the seafloor surface at that time, but with a mixing ratio of primary and secondary inclusions determined by petrographic observation (Ueno et al., 2006). This relationship indicates that the primary fluid contains ^{13}C-depleted CH_4 and CO_2 ($\delta^{13}C_{CH4} < -56‰$; $\delta^{13}C_{CO2} < -4‰$), and that a secondary fluid was more ^{13}C-enriched ($\delta^{13}C_{CH4} \approx -35‰$; $\delta^{13}C_{CO2} \approx 0‰$).

The large ^{13}C-depletion of the primary CH_4 with respect to CO_2 ($\delta^{13}C_{CO2-CH4} > 52‰$) is comparable to that exhibited by microbial methanogenesis. It is known that methanogenic microbes reduce CO_2 to produce CH_4 with a distinctively large isotope effect ($\delta^{13}C_{CO2-CH4} = 21$ to $69‰$; Conrad, 2005). Although the fractionation may possibly be smaller under a high temperature environment, the large fractionation has also been observed for hyperthermophilic methanogens grown above $80\,°C$ (Botz et al., 1996). Some methanogens can also produce CH_4 by decomposing acetate, which results in fractionations of $7–27‰$ between the acetate and CH_4 (Conrad, 2005). If it is assumed that the potential substrate acetate had a $\delta^{13}C$ value similar to that of organic matter in the silica veins ($\approx -35‰$), then the acetate fermentation process can also explain the observed fractionation of the primary methane and organic matter ($\delta^{13}C_{org-CH4} > 21‰$). Therefore, the carbon isotopic relationship among CH_4, CO_2, and organic matter is consistent with biological methane production by reduction of CO_2 and/or acetate.

The large isotopic fractionation between CO_2 and CH_4 is not to be expected from a magmatic process, which should result in much smaller isotopic fractionation due to high-temperature equilibrium (Fig. 7.3-5). Thermal decomposition of organic matter under this process may have possibly produced $>20‰$ fractionation between primary CH_4 and the organic matter. However, the primary fluid lacks ethane and propane, which should have been co-produced by thermogenesis. Hence, a thermogenic origin of the primary methane is controversial.

Fe-Ni alloy catalyzed FTT reactions at $200–300\,°C$ may produce ^{13}C-depleted methane by $30–50‰$, relative to CO_2 (Horita and Berndt, 1999; McCollom and Seewald, 2006). Even larger isotopic fractionation between CH_4 and CO_2 may be achieved at lower temperature below $200\,°C$ (Horita and Berndt, 1999; McCollom and Seewald, 2006). However, as discussed in the previous section, Fe and Ni would have existed as sulfides at temperatures prevailing during the silica precipitation ($100–200\,°C$). Hence, native metal catalysis cannot be expected to produce the methane in the veins.

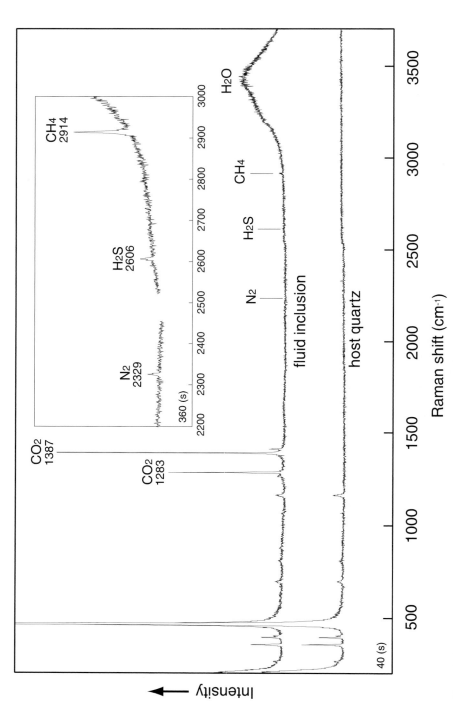

Fig. 7.3-6. Representative Raman spectra of primary fluid inclusions in the silica vein. The fluid mainly consists of H_2O (broad band around 3400 cm^{-1}) and CO_2 (two main peaks at 1283 and 1387 cm^{-1}). Inset shows the result of longer time analysis (360 seconds), indicating the presence of N_2 (2329 cm^{-1}), H_2S (2329 cm^{-1}), and CH_4 (2914 cm^{-1}) in the same fluid.

As already mentioned above, native metal could have existed in deeper parts of the hydrothermal system than the site of silica precipitation (100–200 °C). Thus, methane may have possibly been produced abiotically under much higher temperature conditions. If this was the case, isotopic fractionation between CO_2 and CH_4 should be smaller (less than 30‰ above 300 °C; Horita and Berndt, 1999). Hence deep and high temperature abiotic synthesis does not explain the observed isotopic fractionation of the primary fluid ($\delta^{13}C_{CO2} - \delta^{13}C_{CH4} > 52$‰). Even if abiotic methane in deep and high temperature regimes were introduced into the fluid inclusions, its contribution should be small in the primary fluid preserved in the veins.

Furthermore, the isotopic relationship between the methane and the organic matter in the veins is not expected from FTT reactions (Fig. 7.3-7). Recent experimental studies by McCollom and Seewald (2006) demonstrated that Fe-metal-catalyzed FTT reactions under 250 °C hydrothermal conditions produces ^{13}C-depleted methane ($\delta^{13}C_{CO2} - \delta^{13}C_{CH4} = 36$‰) as well as C_2 to C_{28} hydrocarbons with the same degree of isotopic fractionation ($\delta^{13}C_{CO2} - \delta^{13}C_{hydrocarbons} = 36$‰). They pointed out that the organic matter in the silica veins shows similar degree of fractionation ($\delta^{13}C_{CO2} - \delta^{13}C_{org} \approx 31$‰) and thus may represent higher molecular weight organic compounds polymerized during the FTT synthesis. However, their experiments also demonstrate that significant isotopic fractionation does not occur during the polymerization step of FTT synthesis. Therefore, the observed large ^{13}C-depletion of the primary methane relative to surrounding organic matter in the veins ($\delta^{13}C_{CH4} - \delta^{13}C_{org} < -21$‰) should not result from the Fe-metal-catalyzed FTT reaction (Fig. 7.3-7(b)). In addition, slight isotopic fractionation may possibly occur during the polymerization step of some FTT reactions. It is known that abiotic hydrocarbons in several natural gas reservoirs exhibit progressive ^{13}C-depretion from C_1 to C_4 hydrocarbons (Fig. 7.3-7(a)), which may suggest that isotopically light methane selectively polymerizes into higher hydrocarbons during FTT synthesis (Sherwood-Lollar et al., 2002). This isotopic trend is opposite to that exhibited by the primary methane and organic matter in the Dresser hydrothermal veins. Hence, the inferred FTT processes can not be applied to the formation of organic matter in the Dresser Formation.

In summary, the carbon isotope geochemistry of carbonate, organic matter, methane and carbon dioxide in the Dresser hydrothermal deposits can be fully explained by biological processes including thermophilic chemoautotrophic organisms, especially methanogen. Although the organic matter and methane production by alternative abiological process is not completely dismissed, the observed isotopic relationships are incompatible with our current knowledge of abiotic organic synthesis.

Fig. 7.3-7. (*Next page.*) Relationships between carbon number and its isotopic composition observed in: (a) abiotic hydrocarbons from Kidd Creek (Sherwood-Lollar et al., 2002) and Khibina (Potter et al., 2004) natural gas fields; (b) hydrothermal experiment (McCollom and Seewald, 2006); and (c) the Dresser hydrothermal deposit (Ueno et al., 2004, 2006).

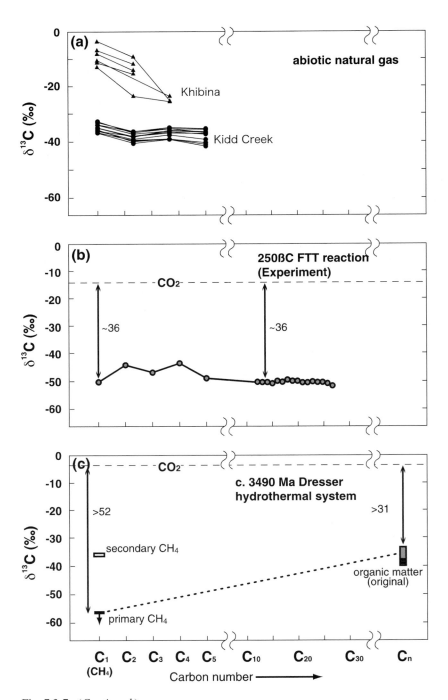

Fig. 7.3-7. (*Continued.*)

7.3-3.3. Sulfur Isotopic Compositions of Pyrite in Barite Deposits

Various forms of sulfur compound occur in the Dresser Formation, and generally show $\delta^{34}S$ values close to 0‰ ($\delta^{34}S = -5$ to $+5$‰; Lambert et al., 1978), which seems to be derived from juvenile magmatic sulfur (Fig. 7.3-8). Exceptionally, bedded and vein barites of the formation contain more ^{34}S-depleted pyrite ($\delta^{34}S = -17$ to $+5$‰; Shen et al., 2001). The observed large isotopic fractionation between the pyrite and the co-existing barite (up to 21‰) is similar to those resulting from the metabolic activity of sulfate-reducing microbes and thus may represent the oldest evidence for microbial sulfate reduction (Shen et al., 2001; Shen and Buick, 2004). Also, this sulfur isotopic fractionation is exceptionally large relative to those reported from other Paleo- and Mesoarchaen deposits, such that microbial sulfate reduction may have been locally developed in sulfate-poor Archaean ocean water (e.g., Canfield, 2005).

The microbial origin of the barite-associated pyrite, however, has been debated, because abiotic reactions under hydrothermal conditions may have produced pyrite with a similar degree of isotopic fractionation (Van Kranendonk, 2006). Although initial models for the Dresser barite suggested that they were deposited as Evaporitic gypsum replaced by barite (Buick and Dunlop, 1990; Lowe 1983), recent geological and petrological investigations have demonstrated a clear hydrothermal origin of the barite (Nijman et al., 1999; Ueno et al., 2001a; Van Kranendonk, 2006) as a non-evaporitic, primary mineral confirmed by X-ray CT analyses (Runnegar, 2001). It is notable that the ^{34}S-depleted pyrite occurs not only in bedded barite, but also in vein barite. Even if the pyrites were produced by microbial sulfate reduction, the depositional environment of the ^{34}S-depleted pyrite should not be a shallow marine setting, but a submarine hydrothermal environment.

Van Kranendonk and Pirajno (2004) proposed that the pyrite and barite were deposited by hydration of magmatic SO_2, which produced both sulfate and sulfide. In this case, equi-

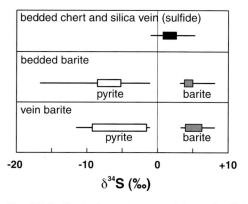

Fig. 7.3-8. Sulfur isotopic compositions of sulfide in chert and veins (black: Lambert et al., 1978; Y. Ueno et al., unpublished data) and disseminated pyrite (white) in bedded and vein barites (gray) (Shen et al., 2001; Y. Ueno et al., unpublished data). The narrow lines indicate the full ranges of observed isotopic compositions, while >70% of the data are within the broad lines.

librium isotopic fractionation between sulfate and sulfide is 22‰ at 250 °C (Ohmoto and Goldhaber, 1997), and is thus comparable to that observed in the barite-pyrite pair described by Shen et al. (2001). Alternatively, Runnegar (2001) and Runnegar et al. (2002) suggested that thermochemical reduction of seawater sulfate could have produced the barite-pyrite association. These abiotic models are roughly compatible with the $\delta^{18}O$ values of the barite ($\delta^{18}O_{SMOW} = 5$ to 10‰; Lambert et al., 1978), which would suggest about 200 °C conditions of precipitation (Runnegar et al., 2002). If this estimate is correct, the temperature is somewhat high to allow microbial activity. However, the oxygen isotope geothermometry of the barite assumes that $\delta^{18}O$ value of Archaean seawater would be the same as that of modern seawater ($\delta^{18}O_{SMOW} = 0$‰). This assumption is still untenable, because the Archaean seawater might have been more ^{18}O-depleted (Shields and Veizer, 2002; Knauth, 2005), giving a much lower temperature estimate for barite deposition. Consequently, both biological and abiological reactions could explain the observed isotopic fractionation between the barite and pyrite.

The origin of the barite-associated pyrite might be further constrained by multiple sulfur isotope analysis. Discovery of non-mass-dependent sulfur isotope fractionation in pre-2.0 Ga sedimentary rocks implies that multiple sulfur isotope ratios ($^{32}S/^{33}S/^{34}S/^{36}S$) could be a useful new tracer for the Archean biogeochemical sulfur cycle (e.g., Farquhar and Wing, 2003). Recent multiple sulfur isotope analysis revealed that the Dresser barite has non-mass dependent isotopic composition, with negative $\Delta^{33}S$ (Ueno et al., 2003; Mojzsis, this volume). This clearly suggests that the barite would have been derived from non-mass-dependently fractionated Archean seawater sulfate, but not from a magmatic source, which should be mass-dependent. The pyrites in both vein and bedded barite also show negative $\Delta^{33}S$ values, again indicating that the pyrite originated from seawater sulfate (Runnegar, 2001; Ueno et al., 2003). This means that reduction of seawater sulfate would be responsible for the ~20‰ ^{34}S-fractionation between the barite and pyrite. However, it is still ambiguous whether the pyrite was produced by thermochemical or microbial sulfate reductions.

7.3-4. CONCLUSIONS

Fig. 7.3-9 summarizes the proposed carbon and sulfur cycles in the Dresser Formation hydrothermal system. The geological and geochemical investigations suggest that the reduced forms of carbon (CH_4 and organic matter) and sulfur (e.g., pyrite) were produced in the hydrothermal system, with significantly large isotopic fractionations. This strongly indicates that low temperature reactions took place in the hydrothermal system, regardless of a biological or abiological interpretation. In principle, large isotopic discrimination could be expected under low temperature conditions, in which chemical reactions are usually prohibited due to kinetic barriers. Hence, the observed large isotopic fractionations imply catalysis that accelerated the low temperature reactions.

For carbon, large isotopic fractionations between CH_4–CO_2 and organic matter-CO_2 are best interpreted as a result of enzymatically catalyzed biological reactions, because effec-

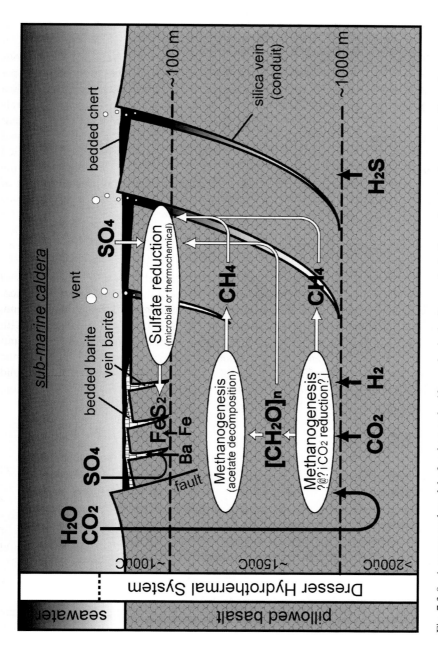

Fig. 7.3-9. A conceptual model of carbon and sulfur cycles in the Dresser hydrothermal system. See text for discussion.

tive catalysts of the FTT reactions (e.g., Fe-Ni alloy) would have been unstable in the low temperature hydrothermal environment. Hence, it is plausible that a microbial ecosystem would have existed in the Dresser hydrothermal system, including a sub-seafloor environment. Methanogens would have gained energy through the conversion of CO_2 and H_2 into CH_4 and acted as a primary producer of organic compounds, which may possibly be utilized by other heterotrophs.

For sulfur, on the other hand, both thermochemical and microbial sulfate reductions could explain the exceptionally large isotopic fractionation between barite and microscopic pyrite crystals embedded therein. Sulfide minerals occur also in bedded chert and in silica veins, but these show no such large isotopic fractionations. Hence, most sulfide minerals in the Dresser Formation seem to be derived from a volcanic source (cf. Van Kranendonk, 2006), but not from seawater sulfate. Although the origin of the barite-associated ^{34}S-depleted pyrite is still ambiguous, sulfate reduction seems to have been restricted to the shallower part of the hydrothermal system (Fig. 7.3-9). This may suggest that the redox gradient beneath the seafloor might have controlled zonation of different types of metabolisms.

ACKNOWLEDGEMENTS

The author would like to express great appreciation and thanks to S. Maruyama and Y. Isozaki for continuous encouragement, support and fruitful discussions over ten years. The author acknowledges the assistance in the field work of M. Terabayashi, Y. Kato, K. Okamoto, T. Ota, T. Kabashima, K. Kitajima, and K. Shimizu. Field collaboration with A. Thorne, K.J. McNamara, and A.H. Hickman was helpful and much appreciated. Constructive comments by M. Van Kranendonk and an anonymous reviewer improved the manuscript. This study was supported by the 21st Century COE Program "How to build habitable planets", Tokyo Institute of Technology, sponsored by the Ministry of Education, Culture, Sports, Technology and Science, Japan.

Earth's Oldest Rocks
Edited by Martin J. Van Kranendonk, R. Hugh Smithies and Vickie C. Bennett
Developments in Precambrian Geology, Vol. 15 (K.C. Condie, Series Editor)
© 2007 Elsevier B.V. All rights reserved.
DOI: 10.1016/S0166-2635(07)15074-X

Chapter 7.4

ORGANIC GEOCHEMISTRY OF ARCHAEAN CARBONACEOUS CHERTS FROM THE PILBARA CRATON, WESTERN AUSTRALIA

CRAIG P. MARSHALL

Vibrational Spectroscopy Facility, School of Chemistry, The University of Sydney, NSW 2006, Australia

7.4-1. INTRODUCTION

One of the major challenges in Earth Sciences is to construct a robust picture of the diversity and timelines of the early forms of life from the fossil record. Biological activity in the early Archaean has been inferred from the occurrence of ^{13}C depleted organic matter (Mojzsis et al., 1996; Rosing, 1999; Schidlowski, 2001; Ueno et al., 2001a, 2001b, 2002, 2004), microfossils (Awramik et al., 1983; Schopf, 1993; Schopf et al., 2002) and stromatolites (Walter et al., 1980; Lowe, 1983; Hofmann et al., 1999; Van Kranendonk et al., 2003; Allwood et al., 2006a) from the Pilbara Craton, Western Australia, and microstructures in volcanic glass from the Barberton Greenstone Belt, South Africa (Kaapvaal Craton) (Furnes et al., 2004, Banerjee et al., 2006). However the biological origin of ca. 3.5 Ga microfossils from the Pilbara Craton, Western Australia, depleted ^{13}C isotopes, and ca. 3.8 Ga graphite from Akilia, western Greenland (North Atlantic Craton) are still contentious (for example, Brasier et al., 2002, 2005; Garcia-Ruiz et al., 2003; Lepland et al., 2005; Lindsay et al., 2005).

Finding evidence for traces of early life on Earth is difficult due to the problems faced in assessing both the syngenicity and the biogenicity of preserved organic matter in Archaean metasedimentary rocks. To date no biomarker molecules have been found in Archaean cherts from the Warrawoona and Kelly Groups, Pilbara Craton, and are probably not retained due to metamorphism. Since the syngeneity of the soluble organic matter fraction is difficult to demonstrate, studies focus on the insoluble organic fraction (kerogen) in order to ascertain biomarkers. The term kerogen is based on an operational definition that refers to disseminated organic matter in sedimentary rocks that is insoluble in non-oxidizing acids, bases, and organic solvents. It is generally accepted that the insoluble organic matter (kerogen) is syngenetic with the host rock and several features of the kerogen confirm its formation simultaneously with the solidification of the siliceous matrix of the cherts. The carbonaceous material in the Warrawoona and Kelly Group cherts are sometimes described as kerogen, or incorrectly described as disordered or amorphous graphite (refer to Sections 7.4-2 and 7.4-5).

The Pilbara Craton is an optimal location to study the early Earth. Rocks contained within the craton have a nearly continuous geological history from >3.5 through to 2.4 Ga, preserved at relatively low metamorphic grade, with large areas of low total strain (Van Kranendonk et al., 2002). Fig. 7.2-2 shows the generalized stratigraphy for the Warrawoona and Kelly Groups, Pilbara Craton, Western Australia. These metavolcanic and metasedimentary rocks contain evidence of hydrothermal systems, stromatolites, and putative microfossils (Dunlop et al., 1978; Lowe, 1980; Walter et al., 1980; Awramik et al., 1983; Schopf, 1993; Hofmann et al., 1999; Van Kranendonk et al., 2003; Ueno et al., 2001a, 2001b, 2004; Van Kranendonk and Pirajno, 2004; Allwood et al., 2006a; Van Kranendonk, 2006).

Much of what we understand about the existence of early life comes from the examination of carbonaceous microstructures or microfossils that have been recovered from Archaean strata. For example, some of the oldest traces of life on Earth have been found in chert horizons occurring in the 3.52–3.43 Ga Warrawoona Group and ca. 3.4–3.31 Ga Kelly Group of the Pilbara Craton, and in the 3.5–3.2 Ga Onverwacht Group, Barberton Greenstone Belt, South Africa. Carbonaceous structures resembling bacteria from the 3.5 Ga Apex chert of the Warrawoona Group have, until recently, been deemed the oldest morphological evidence for life (Schopf, 1993; Schopf et al., 2002). The biological origins of these microstructures were inferred from their carbonaceous (kerogenous) composition, by the degree of regularity of cell shape and dimensions, and by their morphological similarity to extant filamentous prokaryotes. The composition of these carbonaceous microfossils was deduced by Raman spectroscopy, which was proposed as an additional line of evidence for a biological origin of the putative microfossils (Schopf et al., 2002). Similarly, highly negative $\delta^{13}C$ values were obtained from putative microfossils in hydrothermal silica veins in the ca. 3.49 Ga Dresser Formation of the Warrawoona Group (Ueno et al., 2001a, 2001b, 2004). However, there is an ongoing debate about the biogenicity of these putative microfossils, because they occur in hydrothermal silica veins (Brasier et al., 2002; Van Kranendonk, 2006), and that similar structures – in terms of morphology and Raman spectra – can be formed through abiotic reactions to such an extent that these microstructures were even considered as secondary artifacts, formed under hydrothermal conditions (Brasier et al., 2002, 2005; Pasteris and Wopencka, 2002, 2003; Garcia-Ruiz et al., 2003; Lindsay et al., 2005). Additionally, suggestions have also been made that the kerogens may have been introduced into the rocks by later fluid circulation (for example, Buick, 1984) and to have formed as a result of Fischer–Tropsch-type (FTT) synthesis (cf. Brasier et al., 2005; Lindsay et al., 2005).

Therefore, significant interest has been stimulated in elucidating the macromolecular structure of the kerogen in these cherts, in order to discriminate between a biological and non-biological origin. One of the key issues in this debate is to establish the chemical structure of the carbonaceous material – whether kerogenous or graphitic – and whether any genuine and informative molecular or isotopic patterns can be detected from fragmentation products. The following sections will discuss various analytical techniques associated with elucidating the macromolecular structure of organic material in the Warrawoona and Kelly Groups, Pilbara Craton, Western Australia for determining biogenicity. Techniques that are most popular, and furthermore controversial – namely δ^{13}C measurements, Raman spectroscopy, and pyrolysis gas chromatography mass spectrometry (py-GC/MS) – will be discussed in Sections 7.4-3, 7.4-4, and 7.4-5, respectively. Techniques including electron microscopy (TEM and SEM), electron energy loss spectroscopy (EELS), solid state ^{13}C nuclear magnetic resonance (NMR) spectroscopy, and Fourier transform infrared (FTIR) spectroscopy will be discussed together in Section 7.4-6.

7.4-2. PETROGRAPHY OF TYPICAL CARBONACEOUS MATERIAL IN WARRAWOONA AND KELLY GROUP CHERTS

This section describes the typical appearance and occurrence of kerogen in cherts from the Warrawoona and Kelly Groups. These cherts contain considerable amounts of kerogen as well as possible microfossils. The chert dykes contain the carbonaceous microfossil like structures, while the stratiform chert comprises the amorphous structured kerogen (termed clots). Presented here is an example of kerogen in stratiform chert from the Strelley Pool Chert. The kerogen in chert samples consist of clots or clasts of black material finely disseminated through a matrix of polygonal microcrystalline quartz. The samples also contain variable quantities of silicified rock and mineral grains (Fig. 7.4-1). In samples from bedded cherts, the carbonaceous clots and clasts, together with the other grains, define a laminated rock fabric (Fig. 7.4-1) that is concordant with primary bedding and sedimentary fabrics observed in outcrop and across the area. This indicates that the organic matter was deposited as part of the original sediment and is not a younger contaminant. No microfossils, like those described by Schopf et al. (1993), were observed in these samples from the Strelley Pool Chert. The kerogen from all these chert samples are black, indicating the cherts have been subjected to a thermal history exceeding 250 °C (Hunt, 1996).

7.4-3. CARBON ISOTOPE (δ^{13}C) MEASUREMENTS

7.4-3.1. Bulk Carbon Isotope (δ^{13}C) Measurements on Isolated Kerogen

Previous isotopic δ^{13}C compositions conducted from isolated kerogen from the Warrawoona Group typically vary from -28.3 to $-35.8‰$ relative to coexisting carbonate minerals (e.g., Ueno et al., 2001a, 2001b, 2004; Schopf et al., 2002; Brasier et al., 2002, 2005; Lindsay et al., 2005). The kerogens from the Warrawoona Group are depleted in ^{13}C to a degree typically ascribed to biological processes. For example, organic matter in more recent sediments dominated by input from photosynthetic organisms typically exhibits carbon isotope compositions in the range of -25 to $-30‰$ (Hayes et al., 1983). Also modern autotrophic prokaryotes can produce ^{13}C depletions of 0 to $\sim40‰$ during carbon fixation (House et al., 2003). And methaotrophs can fractionate ^{13}C to values of $-60‰$. Thus the carbon isotopic compositions of these ancient kerogens are coincident

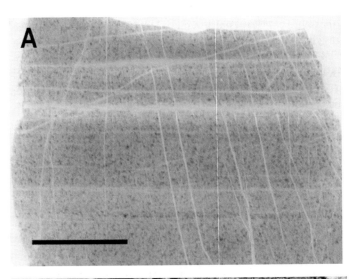

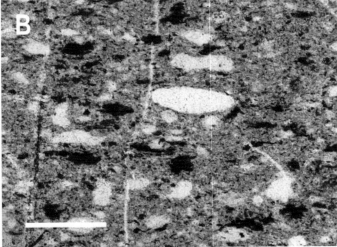

Fig. 7.4-1. Sedimentary fabric of representative sample thin sections. (A) Sample 120803-8: laminated fabric defined by the density and alignment of carbonaceous grains and particles in the chert matrix. Scale = 0.5 cm. (B) 140803-2: carbonaceous material (black) occurring as clasts and clots distributed among rounded silicified grains, such as the large pale grain above centre.

with the range of values resulting from biological fixation, possibly indicating that biological processes could conceivably account for their isotopic composition (cf. Ueno et al., 2001a, 2001b). Recently, however, the use of carbon isotope compositions pertaining to potential biological contributions of kerogen in rock samples from the early Earth, and thereby the implications for dating the emergence of life of Earth, have come under intense scrutiny. This reassessment of depleted ^{13}C values is due to the reconsideration of geological context and new geochemical observations of Archaean localities.

Recent field and laboratory studies (Brasier et al., 2002, 2005; Lindsay et al., 2005) have suggested a non-biological source for kerogen in cherts from the Warrawoona and Kelly Groups. These recent studies are based largely on the discovery that putative kerogenous microfossils from the Apex chert were retrieved from a hydrothermal silica vein rather than a sedimentary chert unit as described by the original author. The more recent studies have suggested that the kerogens were formed by abiotic organic synthesis in a hydrothermal setting by a Fischer–Tropsch-type (FTT) or related process. Industrial Fischer–Tropsch synthesis is superficially a simple process whereby organic compounds are produced abiologically by catalytic hydrogenation and reductive polymerization of CO or CO_2. The reaction involves passing anhydrous CO or CO_2 and hydrogen over a catalytic bed which leads initially to the breaking of the carbon-oxygen bond, which is then replaced by a carbon–carbon or carbon–hydrogen bond. This produces a mixture of short linear chain hydrocarbons or alcohols. Whilst the FTT synthesis is well established in industry, it is not well understood in the geological environment. Nonetheless, it has become commonly accepted in the Earth and planetary sciences that organic compounds are readily synthesized by FTT or related reactions during fluid-rock interactions under hydrothermal conditions. Despite the widespread perception that FTT reactions are continuous and a substantial source of organic matter in submarine hydrothermal systems and elsewhere within the Earth's crust, definitive identification of organic compounds with an abiotic origin from natural systems has been elusive. McCollom and Seewald (2006) have shown that the abiotic synthesis of organic compounds (C_2–C_{28} *n*-alkanes) under putative hydrothermal conditions yield organic products depleted in ^{13}C to a degree typically attributed to biological processes (~ -48 to $-52\permil$). These isotopic compositions of *n*-alkanes experimentally produced where using a starting carbonate of $-15\permil$. Therefore, if applying McCollom and Seewald (2006) results to Warrawoona chert, organic matter should be ~ -32 to $36\permil$ because most Archaean carbonates including Warrawoona are isotopically normal ($\delta^{13}C \sim 0\permil$). These findings confirm what has been known for many years about the fundamentals of C-isotopic fractionation (Urey, 1947). The C-isotopic compositions of bulk samples alone are insufficient to differentiate biotic from abiotic Archaean carbonaceous materials.

7.4-3.2. Ion Microprobe and NanoSIMS

Recent advances in high-mass-resolution ion-microprobe techniques permit in situ analyses of carbon isotopic variations at the scale of $<10\,\mu m$. Application of ion-microprobe was pioneered by House et al. (2000) on Proterozoic microfossils. This approach was shown to be very useful for deducing physiological aspects of morphological microfossils. Ueno et al. (2001a) applied this technique to much older Archaean microfossils from silica veins in the ~ 3.5 Ga chert-barite unit described by Buick and Dunlop (1990) in the North Pole area. The $\delta^{13}C$ values of the carbonaceous filaments and kerogen clots range from -42.4

to $-30.5\text{\textperthousand}$. Subsequently after the pioneering studies of House et al. (2000) and Ueno et al. (2001a), much interest has been expressed in ion-microprobe and NanoSIMS to investigate Archaean microfossils. The recent preliminary work is summarized here. Ion microprobe studies have been used to determine carbon isotope and nitrogen to carbon ratio of individual carbonaceous structures. For example van Zuilen et al. (2006) determined N/C ratio of 0.006 and a $\delta^{13}C$ value around $-35\text{\textperthousand}$ from carbonaceous structures in the Footbridge Chert in the upper part of the Kromberg Formation, Barberton Greenstone Belt. Previous attempts to obtain a carbon image along with measuring the carbon isotope composition has been made by using ion microprobe, as above. However, the maximum spatial resolution (approximately 1 µm) that can be attained with ion microprobe is of the same order of magnitude as the size of fossil cells and larger than that of bacteria. Therefore, the resolution of the optical image remains higher than that of the cells, and thus the direct comparison between the two images is impossible. Recently, this has been overcome by using NanoSIMS, since the spatial resolution of this technique can reach 0.05 µm. NanoSIMS is a well established technique used in materials science. The benefits of this technique are only just being exploited in geology, in particular with regard to Precambrian studies. Wacey et al. (2006) have recently used this technique to investigate a potentially important assemblage of partially mineralized microtubes (5–10 µm in diameter) in $\sim$3.5 Ga siliceous metasedimentary rocks from Western Australia. They used NanoSIMS to geochemically map the minerals within the tubes, linings of the walls of the tubes and zonations within the mineral host, analyzing for biologically important elements (for example, C, N, P, S, K, and Fe), as well as carbon isotope variations between the phases.

7.4-4. RAMAN SPECTROSCOPY

7.4-4.1. Raman Spectroscopy of Carbonaceous Materials

Raman spectroscopy has been used since the early 1970s for the study of carbonaceous materials. For an extensive review on Raman spectroscopy of carbonaceous materials, refer to Dresselhaus and Dresselhaus (1982 and references therein). However, the short summary that follows is intended to highlight some salient features in understanding the Raman spectra acquired from kerogens from the Warrawoona and Kelly Groups.

For an ideal graphitic crystal (space group D_{6h}^4 with unlimited translational symmetry) only one first-order band, the G (graphite) band is exhibited at 1580 cm^{-1} corresponding to an ideal graphitic lattice vibrational mode with E_{2g2} symmetry (Tuinstra and Koenig, 1970). It has been assigned as a C–C stretching in the longitudinal symmetry axis of the graphite plane. The paradigm for the structure of graphite is that of a staggered stacking of flat layers of carbon atoms. Individual layers, sometimes referred to as graphene sheets, are weakly bonded to each other, and are composed of strongly bonded carbon atoms at the vertices of a network of regular hexagons in a honeycomb pattern. When crystallographic defects are introduced into graphite a number of terms can be used to describe the resultant disordered carbon products such as, turbostratic carbon, disordered carbon and the ultimate end product of a disordered carbon network – amorphous carbon. Depending on the source and structure of carbon, a number of types of disordered carbons may be described such as, highly orientated pyrolytic graphite, polycrystalline graphite, pyrolytic carbon, and glassy carbon.

Disordered carbonaceous materials, as shown by the representative curve-fitted spectra of isolated kerogen from the Strelley Pool Chert of the Kelly Group (Fig. 7.4-2), on the other hand, exhibit additional first-order bands (D or Defect bands), which are known to be characteristic of disordered sp^2 carbons and increase in intensity relative to the G band with further disorder introduced into the carbonaceous network. The most intense of the D bands is the D1 band, which appears at $\sim$1350 cm^{-1} and corresponds to a disordered carbon lattice vibration mode with A_{1g} symmetry. This vibrational mode has been suggested to arise from graphene layer carbon atoms in immediate vicinity of a lattice disturbance such as the edge of a graphene layer (Katagiri et al., 1988; Wang et al., 1990), or a heteroatom (Wang et al., 1990). Another first-order band pertaining to structural disorder is the D2 band at $\sim$1620 cm^{-1} which can be observed as a shoulder on the G band. This shoulder becomes further developed in more disordered carbonaceous materials and the G and D2 bands merge, until a single feature is observed around 1600 cm^{-1}, which produces an apparent band broadening and up-shifting of the G band. The D2 band corresponds to a graphitic lattice mode with E_{2g} symmetry (Dresselhaus and Dresselhaus, 1981; Al-Jishi and Dresselhaus, 1982; Cuesta et al., 1994) and is assigned to a lattice vibration involving graphene layers at the surface of a graphite crystal (Dresselhaus and Dresselhaus, 1982). The relative intensities of both the D1 and D2 bands increase with increasing excitation wavelength, which can be attributed to resonance effects (Matthews et al., 1999). The high signal intensity between the G and D1 band maxima can be attributed to another band at $\sim$1550 cm^{-1}, which has been assigned to amorphous carbon fraction of organic molecules, fragments or functional groups (Cuesta et al., 1994; Jawhari et al., 1995) and has been designated D3. In pure amorphous carbon materials this band becomes the dominant broad intense feature of the carbon first-order spectrum. The band at $\sim$1350 cm^{-1}, D1, exhibits a shoulder at $\sim$1200 cm^{-1}, which is denoted as D4 (Fig. 7.4-2). This band has been tentatively attributed to sp^2–sp^3 bonds or C–C and C=C stretching vibrations of polyene-like structures.

The second-order spectrum of carbonaceous materials shows several bands at $\sim$2450, $\sim$2695, $\sim$2735, $\sim$2950 and $\sim$3248 cm^{-1}. These second-order Raman bands are assigned to both overtone scattering ($2 \times 1360 = 2735$ cm^{-1}, the most intense, $2 \times 1620 = 3248$ cm^{-1} a weak but sharp band) and combination scattering ($1620 + 830 = 2450$ cm^{-1}, $1580 + 1355 = 2950$ cm^{-1}). Lespade et al. (1982) attributes the appearance of these second-order bands to three-dimensional structural ordering. In particular, the splitting of the band at 2700 cm^{-1} (S band) into the doublet of the G'_1 and G'_2 at 2695 and 2735 cm^{-1} occurs for well crystallized graphite, which arises when carbonaceous materials acquire triperiodic structure. The stacked spectra shown in Fig. 7.4-3 show pronounced absorptions at 1350 (D band) and 1600 cm^{-1} (G band, which is a combination of the G and D$'$ bands) in the first-order region, while the second-order region contains a non-intense S1

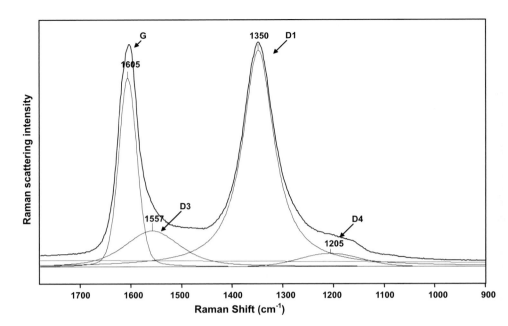

Fig. 7.4-2. A representative Raman first-order curve fitted spectrum showing additional defect bands at 1200 and 1550 cm^{-1} (D″ band) of kerogen isolated from the Strelley Pool Chert.

band at 2700 cm^{-1}. Little to no 3-D structural ordering is present in these samples, which is shown by the low band intensity in the second-order region (Lespade et al., 1982). The isolated kerogen from the Strelley Pool Chert consists of small crystallites with biperiodic structure.

7.4-4.2. Summary of Raman Spectroscopic Studies of Kerogens from the Pilbara Craton

Raman spectroscopy is non-intrusive, non-destructive and particularly sensitive to the distinctive carbon signal of carbonaceous (kerogenous) organic matter, it is an ideal technique for such studies. Schopf et al. (2002) have used this technique to investigate graphitic, geochemically highly altered, dark brown to black carbonaceous filaments that have been inferred to be remnants of ancient microbes, and show that it can provide powerful evidence of biogenicity of even such poorly preserved microstructures. They analysed microbial fossil filaments that had been three dimensionally permineralized in cryptocrystalline cherts from one subgreenschist facies (2.1 Ga Gunflint Formation), and three geochemically more altered (greenschist facies) Precambrian geological units: (1) a cylindrical prokaryotic filament from a domical stromatolite of the 770 Ma Skillogalee Dolomite of South Australia; (2) a cellular trichome from a domical stromatolite of the 2.1 Ga Gunflint Formation of Ontario, Canada; (3) two prokaryotic filaments from flat-laminated microbial mats of the 3.38 Ga Kromberg Formation of South Africa; and (4) the oldest putative microfossils

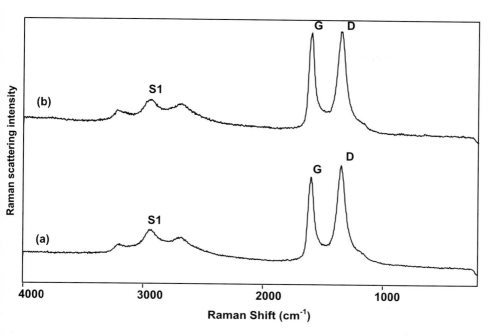

Fig. 7.4-3. Stacked Raman spectra are shown for the carbonaceous material isolated from the Strelley Pool Chert (a) 1904-11 and (b) 120803-8, which shows pronounced absorptions at 1350 (D band) and 1600 cm^{-1} (G band which is a combination of the G and D′ bands) in the first-order region and the second-order region contains a non-intense S1 band at 2700 cm^{-1}. Note the D band intensity is greater for the spectrum acquired from 1904-11.

known – five specimens, representing 4 of 11 described taxa of cellular microbial filaments (Schopf, 1993) that are permineralized in organic-rich clasts of the 3.47 Ga Apex chert, Warrawoona Group, Pilbara Craton.

They noted Raman spectra with vibrational bands at 1350 cm^{-1} and 1600 cm^{-1} which are characteristic of carbonaceous (kerogenous) materials and commonly designated D (disordered) and G (graphitic), respectively, because of their presence in various forms of graphite. The results establish the kerogenous composition of the microscopic structures studied. Raman spectra also show that the filaments are embedded in fine-grained quartz and are devoid of virtually all other mineral phases that are identifiable by Raman analysis. Schopf et al. (2002) led to the conclusion that by correlating directly molecular composition with filament morphology, the Raman images establish unequivocally that the specimens are composed of carbonaceous material – kerogen.

Brasier et al. (2002, 2005) have posed the question: Can high-resolution Raman spectroscopy be used to determine whether the Apex chert structures are composed of abiotic graphite (Brasier et al., 2002) or of biogenic kerogen (Kudryavtsev et al., 2001; Schopf et al., 2002)? Brasier et al. (2002) suggest that both the Apex chert 'microfossils' and carbonaceous groundmass are composed of amorphous carbonaceous matter. Their preliminary examination concluded that there was no strong signature for any other carbon-based material present, other than amorphous graphite. Perhaps the better term to use here would be amorphous carbon. However, inspection of their Raman spectra would indicate that the carbon is not amorphous but rather is disordered sp^2 carbon, which is a better choice of terminology. Furthermore, the nomenclature can be confused in the geological literature by the use of "organic" and "kerogen", implying higher-order carbon compounds, whereas they also resemble graphite. Exhaustive work by Wopenka and Pasteris (1993) indicates that the graphitic signature is a widely observed feature of many rock types. Like Pasteris and Wopenka (2002, 2003), Braiser et al. (2002, 2005) conclude that Raman spectra, when used alone and without control studies (for example, Schopf et al., 2002) should not be taken to imply or exclude biogenicity.

Tice et al. (2004), in their studies of carbonaceous materials from the Onverwacht and Fig Tree Groups of the Barberton Greenstone Belt, and the Marble Bar Chert Member, Pilbara Craton, used Raman spectroscopy as a geothermometer. Many of the metasedimentary rocks are cherts composed largely of silica with trace amounts of carbonaceous material, and they lack metamorphic mineral assemblages that might record their thermal histories. The lack of a conventional high-resolution geothermometer for many Barberton rocks makes interpretation of both the degree of alteration of carbonaceous material and the metamorphic history of the rocks problematic. Because of the abundance of carbonaceous cherts in the Onverwacht Group and Warrawoona Group, and the general lack of diagnostic metamorphic mineral assemblages in many sedimentary units, Raman spectroscopy of carbonaceous materials as a geothermometer could be a useful tool for constraining the thermal history of these, and similar, rocks. Several studies have demonstrated the geothermometer potential of Raman micro-spectroscopy of partially carbonized and graphitized carbonaceous material (Beyssac et al., 2002; Jehlicka et al., 2003; Spötl et al., 1998; Wopenka and Pasteris, 1993; Yui et al., 1996). Once carbonaceous material has reached approximately greenschist-facies metamorphism, it undergoes a characteristic loss of noncarbon atoms (for example, hydrogen, oxygen, and nitrogen) and conversion to increasingly large graphite crystallites, which are reflected by its Raman spectra.

Previous studies have shown that D/G height, width, and area ratios vary systematically with increasing metamorphic grade in metapelites (Beyssac et al., 2002; Jehlicka et al., 2003; Spötl et al., 1998; Wopenka and Pasteris, 1993; Yui et al., 1996). Raman micro-spectroscopy of carbonaceous material in the Onverwacht and Fig Tree Groups in the Barberton Greenstone Belt indicates a relatively low degree of metamorphism. In particular, the stratigraphically continuous Hooggenoeg, Kromberg, and Mendon Formations of the Onverwacht Group, and the overlying Fig Tree Group, have all been subjected to temperatures at most equal to those reflected in chlorite-zone shales, that is, between 300–400 °C (Bucher and Frey, 1994). In addition, samples from the Marble Bar Chert Member of the Pilbara Craton have been heated to the same extent as samples from Barberton. Significantly, Tice et al. (2004) concluded that Raman spectra could be used to test the antiquity of putative microfossils in rocks of comparably heated terranes. Any "microfossil" yielding a Raman spectrum without well-developed D and G bands, or displaying bands corresponding to functional groups of thermally unstable organic compounds, is a recent contaminant (for example, porphyrins). Conversely, a Raman spectrum indicating meta-morphism comparable to that of the surrounding rocks indicates that the carbonaceous material has been in place since the time of maximum heating.

Allwood et al. (2006b) investigated the Raman spectra acquired from carbonaceous materials in the Strelley Pool Chert (Pilbara Craton). The purpose of their study was to determine whether primary structural characteristics of organic molecules may have survived to the present day. They used Raman spectral parameters to identify variations in molecular structure of the carbon and determine whether original characteristics of the carbonaceous materials have been completely thermally overprinted, as would be expected during approximately 3.4 billion years of geologic history. To the contrary, they found that the molecular structure of the carbonaceous materials varies depending on the sedimentary layer from which the sample was collected, and the inferred original palaeoenvironmental setting of that layer, as determined by other geochemical and geological data. Thus, Allwood et al. (2006b) argue that the spectral characteristics of the carbonaceous materials reflect original palaeoenvironments that varied through time from warm hydrothermal settings to cooler marine conditions and a return to hydrothermal conditions. Raman spectroscopy also showed that organic matter is present in trace amounts in association with putative stromatolites in the Strelley Pool Chert, which significantly were previously thought to be devoid of organic remains. Furthermore, the Raman spectra of kerogen associated with stromatolites indicate lower thermal maturity compared to the kerogen in non-stromatolitic hydrothermally altered deposits in overlying and underlying rocks. Significantly, this indicates that the stromatolites are not abiotic hydrothermal precipitates – as proposed by Lindsay et al. (2005) – but support previous studies that suggest they were formed in a cool marine environment that may have been more favorable to life (Hofmann et al., 1999; Van Kranendonk et al., 2003).

Marshall et al. (2007) conducted Raman spectroscopic analysis on isolated kerogen from the Strelley Pool Chert that had undergone addition treatment compared to the standard kerogen isolation procedure. After the acid insoluble residue (kerogen) was isolated it was solvent extracted to remove any potential low molecular weight contaminating hydrocarbon. Further treatment of the isolated kerogens involved extraction with dichloromethane by ultrasonication ($\times 3$), then with n-hexane ($\times 3$). To remove residue trapped bitumen, the carbonaceous material was swelled twice by ultrasonication in pyridine at 80 °C for 2 h. The pyridine was removed by centrifugation and the kerogen was re-extracted with methanol and three times with dichloromethane. This "clean" kerogen was also used for pyrolysis studies detailed in Section 7.4-5. Figs. 7.4-2 and 7.4-3 show examples of the Raman carbon first-order spectra for the isolated kerogens from the Strelley Pool Chert that are typical of spectra obtained from disordered sp^2 carbons with low 2-D ordering (biperiodic structure). The implications of the Raman results show low 2-D ordering throughout the carbonaceous network, indicating the incorrect usage of the term *graphite* in the literature to describe the kerogen or carbonaceous material in Pilbara cherts. In addition, geothermometry data was determined by Raman spectroscopy which revealed that the kerogens have a low degree of 2-D structural organization and can be compared with other spectra of carbonaceous material taken from chlorite up to biotite metamorphic zones (for example, Wopenka and Pasteris, 1993; Jehlicka et al., 2003, and references therein), representing lower to mid greenschist facies (Yui et al., 1996), and indicating peak temperatures between $\sim$300–400 °C through to 400–500 °C (Bucher and Frey, 1994).

7.4-4.3. Implications of Raman Spectroscopy

The carbon first-order spectra for these isolated kerogens from Pilbara Craton cherts are typical spectra obtained from disordered sp^2 carbons, and have a similar line-shape to the spectra acquired by Schopf et al. (2002), Brasier et al. (2002, 2005), Tice et al. (2004), Allwood et al. (2006b), and Marshall et al. (2007). The results and subsequent interpretation clearly show that the organic matter in the Pilbara cherts are not graphitic as previously reported. Geochemical maturation or metamorphism of almost all naturally occurring organic matter, whether biological, abiological or meteoritic in origin, might be expected to give rise to essentially similar assemblages of thermally stable products – interlinked PAHs that have experienced geological conditions that result in carbonization and graphitization. Therefore, Raman spectroscopy of over-mature kerogen cannot provide definitive evidence of biogenicity by itself.

7.4-5. PYROLYSIS GAS CHROMATOGRAPHY MASS SPECTROMETRY (PY-GC/MS)

Currently, little pyrolysis has been performed on isolated kerogen from Pilbara Craton cherts. Skrzypczak et al. (2004, 2005, 2006) and Marshall et al. (2007) are the only studies detailing pyrolysis experiments on isolated kerogens from these rocks. This section outlines their preliminary results and subsequently discusses if any biogenic information can be obtained from the carbonaceous cherts of the Pilbara Craton.

Skrzypczak et al. (2004, 2005, 2006) analyzed isolated kerogen using the standard HCl/HF procedure from the Strelley Pool Chert by solid state ^{13}C NMR spectroscopy, FTIR spectroscopy, and py-GC/MS. Taken together, these techniques revealed the occurrence of unbranched long chains covalently linked to the macromolecular network of the kerogen. They report that low temperature heating reveals that these chains do not correspond to trapped hydrocarbons. Further, they suggest that the lack of associated branched alkanes in the pyrolysate and the occurrence of an even-over-odd carbon number predominance in the C_{10}–C_{18} alkanes points to a biological origin for this macromolecular organic material. In addition, they note the presence of sulfur containing compounds in the pyrolysis products, which they attribute to the co-occurrence of sulfur and carbon in the kerogen, possibly reflecting sulfate reducing bacteria.

Organic contamination of geological samples appears to be a common phenomenon. Insoluble organic matter is composed of a covalently bound cross-linked polymer-like macromolecular network in which potential contaminating solvent soluble/extractable bitumen from a younger geological source may become trapped. Therefore, when studying

Archaean samples, it is desirable to investigate solvent extracted isolated kerogen in order to eliminate false biomarker signals arising from potentially contaminating younger organic matter trapped in the macromolecular structure. Marshall et al. (2007) further treated their kerogens to remove potential contamination from low molecular weight hydrocarbons. They performed catalytic hydropyrolysis (HyPy) on their solvent extracted kerogens instead of standard pyrolysis due to the higher yields of products routinely obtained when using an effective hydrogen donor (Love et al., 1995). HyPy is particularly appropriate for hydrogen-lean carbonaceous materials such as Archaean kerogens. For example, when applied to pre-oil-window kerogens in the presence of a dispersed catalytically-active molybdenum sulfide phase, more than 85% of material is converted to soluble hydrocarbons (Roberts et al., 1995). A continuous flow of high-pressure hydrogen ensures that product rearrangements are minimal, thereby suppressing the recombination of pyrolysis products into a solvent-insoluble char and minimizing alteration to organic structures and stereochemistries (Love et al., 1995). Furthermore, and most importantly with respect to Archaean kerogens, standard pyrolysis techniques do not necessarily discriminate between absorbed and bound organic species. Clearly, methods that remove all adsorbed molecules from the carbonaceous material matrix before release of covalently bonded moieties are suitable for resolving any contamination issues. Hydropyrolysis is a temperature-programmed method which facilities the use of an initial low temperature treatment to drive off residual volatiles (Brocks et al., 2003).

Fig. 7.4-4 displays a representative TIC (Total Ion Chromatogram) of a polyaromatic hydrocarbon (PAH) fraction prepared from the hydropyrolysate of isolated kerogen from the Strelley Pool Chert. An initial HyPy treatment up to 330 °C was performed to remove any residual bitumen prior to this high temperature run (to 520 °C). Over 99 wt% of aromatic hydrocarbons were released in the latter high temperature step. Significantly this adds confidence that, in general, the aromatic compounds reported represent genuine kerogen-bound molecular constituents. The Strelley Pool Chert isolated kerogen hydropyrolysates contain a diverse range of 1 to 7-ring PAH compounds, with phenanthrene (3-ring) or pyrene (4-ring) PAH as the major components. Although the chromatograms are complex, the PAHs do not show a high degree of branched alkylation of aromatic side-chains and C_1- and C_2-substituted PAHs are the dominant alkylated forms. The PAH profiles are fairly similar to those observed by Brocks et al. (2003) from HyPy of 2.5 Ga kerogens from the Hamersley Group (Western Australia).

The principal bound PAH released by HyPy have stable carbon isotopic ($\delta^{13}C$) signatures between -29 to $-36‰$ which when averaged are slightly ^{13}C-enriched in comparison to the bulk kerogen values (Table 7.4-1). This implies that the intractable larger PAH clusters (10–15 ring) comprising the bulk of the kerogen matrix are ^{13}C-depleted by 0–6‰ in comparison with the quantitatively minor 1–4 ring PAH components and this likely results from preferential incorporation of ^{12}C during fusion of aromatic rings into larger structural units during kerogen maturation.

Fig. 7.4-5 is a SIC (Summed Ion Chromatogram) of a Strelley Pool Chert isolated kerogen hydropyrolysate sample that reveals the dominant parent PAH distribution (non-alkylated) in the aromatic fraction of the hydropyrolysate. The hydropyrolysate shows

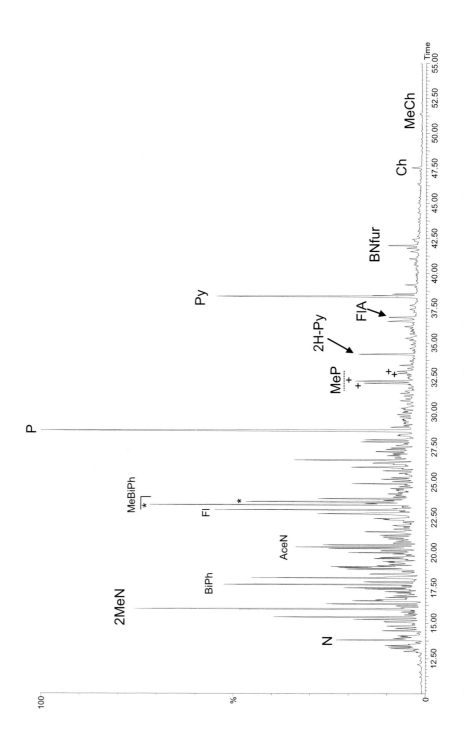

Fig. 7.4-4. (*Previous page.*) Total ion chromatogram (TIC) of the aromatic hydrocarbons generated from HyPy of the insoluble organic matter in SPC 1904-16 (after an initial HyPy pre-treatment to 30 °C to remove residual bitumen). Aromatics amenable to gas chromatography in the SPC hydropyrolysates generally contain from 1-ring to 7-ring polyaromatic clusters, with phenanthrene (3-ring) and pyrene (4-ring) as the principal constituents. N = naphthalene, 2 MeN = 2-methylnaphthalene, BiPh = 1,1'-biphenyl, AceN = acenaphthene, Fl = fluorene, MeBiPh = methylbiphenyls, P = phenanthrene, MeP = methylphenanthrenes, 2H-P = dihydropyrene, FlA = Fluoranthene, Py = pyrene, BNfur = benzonaphthofuran, Ch = chrysene, MeCh = methylchrysene.

The presence of naphthalene, phenanthrene, methylphenanthrene, fluoranthene, pyrene, methylpyrene, chrysene, methylchrysene, benzo[ghi]perylene, benzo[e]pyrene, and up to (7-ring PAH) coronene. For comparison, an unmetamorphosed Mesoproteozoic isolated kerogen from the Urapunga 4 drill core that intersects the 1.4 Ga Velkerri Formation, Roper Group, NT, Australia was analysed. The total PAH profiles for Strelley Pool Chert and Urapunga 4 kerogens have similar features (Fig. 7.4-6), with the less mature Urapunga 4 aromatic compounds not surprisingly exhibiting a greater degree of alkylation. Moreover, these are distinct from those obtained from HyPy treatment of the insoluble carbonaceous material found in Murchison meteorite (Sephton et al., 2005).

Detectable amounts of alkanes, exhibiting a mature distribution (Fig. 7.4-7) were observed in all hydropyrolysates and n-alkanes up to n-C_{23} are evident in TICs. Much higher yields of aliphatic products (alkanes plus alkenes) were released from HyPy treatment of the Mesoproterozoic Urapunga 4 kerogen (138 mg g TOC^{-1} of total aliphatics) and in this case this is largely the result of covalent bond cleavage of bound aliphatic components of the kerogen. It is significant, however, that the proportions of monomethyl-branched to linear alkanes is similar for both Strelley Pool Chert and the Mesoproterozoic kerogen (Fig. 7.4-8) and both contain only trace amounts of pristane and phytane. The quantities of n-C_{14}–n-C_{20} alkanes released from sequential dual temperature HyPy treatment of Strelley Pool Chert kerogens generally constitute less than 0.3 wt% of the kerogen matrix and most were released at high temperature following a preliminary low temperature pretreatment

Table 7.4-1. Compound-specific stable carbon ($\delta^{13}C$) isotopic composition (‰ vs PDB) of selected aromatic compounds released from HyPy of Strelley Pool kerogens 1904-11 and 1904-16 as determined by GC-IRMS analysis

Sample	kerogen	BiPh	MeBiPh*	P	FlA	Py	BNfur
1904-11	−34.0	n.d.	n.d.	−32.9	−31.4	−35.8	n.d.
				(0.04)	(1.31)	(0.12)	
1904-16	−35.0	−29.3	−31.0	−30.5	−29.6	−32.9	−31.1
		(0.21)	(0.47)	(0.55)	(0.85)	(0.16)	(0.32)

BiPh = 1,1'-biphenyl, MeBiPh = methylbiphenyls, P = phenanthrene, Py = pyrene, BNfur = benzonaphthofuran. ( ) Standard deviation from 2 or more analyses.
* Average of two resolvable peaks.

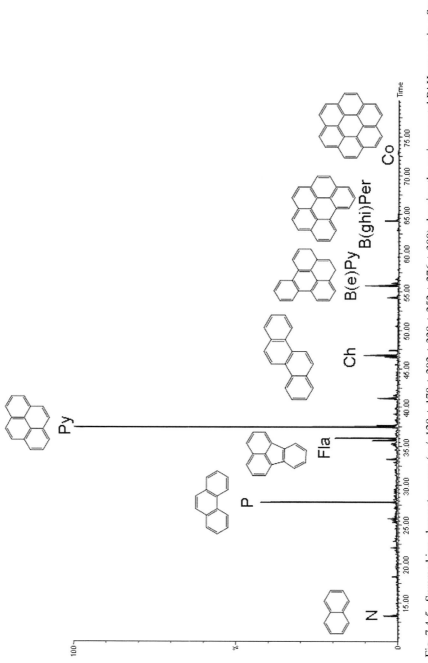

Fig. 7.4-5. Summed ion chromatograms (m/z $128 + 178 + 202 + 228 + 252 + 276 + 300$) showing the main parental PAH present in a Strelley Pool Chert hydropyrolysate for sample 120803-5.

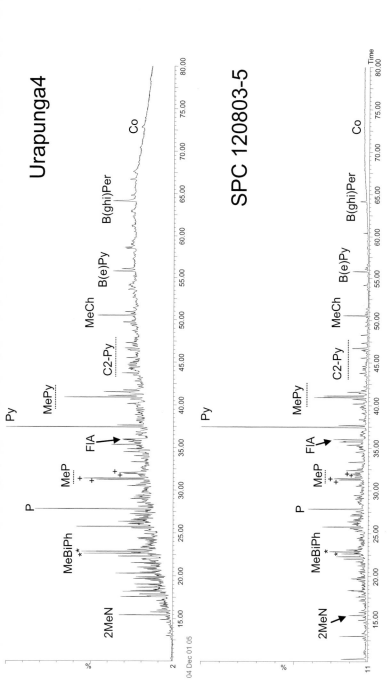

Fig. 7.4-6. Total ion chromatogram (TIC) of the total aromatic compounds generated from HyPy of the insoluble organic matter in Strelley Pool Chert 120803-5 (after an initial HyPy pre-treatment to 330 °C to remove residual bitumen) in comparison with those released from the Mesoproterozoic Urapunga 4 kerogen. 2MeN = 2-methylnaphthalene, MeBiPh = methylbiphenyls, P = phenanthrene, MeP = methylphenanthrenes, FlA = Fluoranthene, Py = pyrene, MePy = methylpyrenes, C2-Py + C2-alakylated pyrenes, MeCh = methylchrysene, B(e)Py = benzo(e)pyrene, B(ghi)Per = benzo(ghi)perylene, Co = coronene.

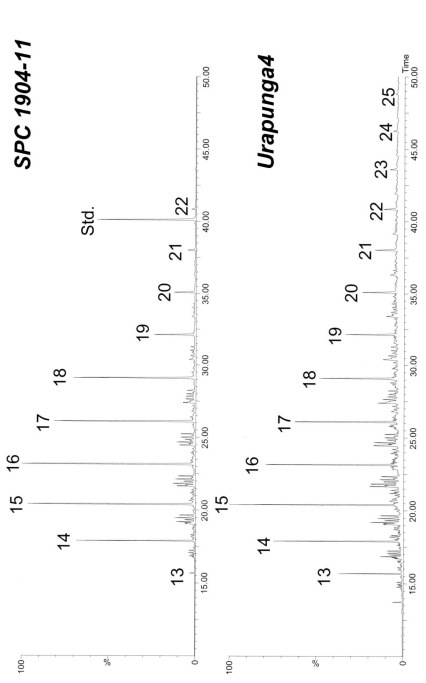

Fig. 7.4-7. m/z 85 ion chromatograms for Strelley Pool Chert 1904-11 and Urapunga 4 (Mesoproterozoic) HyPy products, showing similar distributions of *n*-alkanes and methyl-branched alkanes (MMAs). Numbers (13–25) refer to the carbon chain length of *n*-alkanes. Not surprisingly, the *n*-alkane profiles for the less mature Urapunga 4 sample extend to higher carbon numbers.

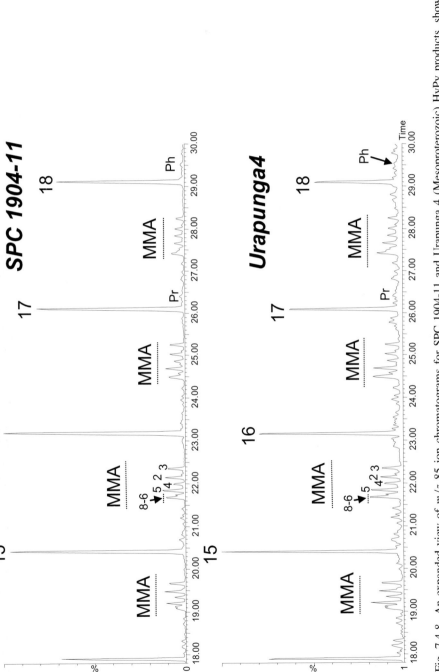

Fig. 7.4-8. An expanded view of m/z 85 ion chromatograms for SPC 1904-11 and Urapunga 4 (Mesoproterozoic) HyPy products, showing the similar distributions of methyl-branched alkanes (MMAs). Numbers associated with MMA clusters (2–8) refer to the position of the methyl-substituent in the linear alkane chain. Other numbers (15–18) refer to the carbon chain length of *n*-alkanes.

(to 330 °C). It was observed that these alkanes were proportionally more abundant relative to aromatic constituents in chromatograms obtained for the most recalcitrant kerogen samples with the lowest atomic H/C ratios. The alkane products were most likely trapped in closed micropores of the kerogen, so not covalently-bound but still not accessible to solvent extraction, and were only released after disruption of the host matrix by cleavage and heating to high temperature during HyPy treatment. The lack of any appreciable alkene content (formed from dehydrogenation reactions that accompany bond cleavage) for Strelley Pool Chert HyPy products supports this interpretation. The alkanes released from Strelley Pool Chert kerogens are unlikely to be contaminants because they are largely (>70 wt%) generated in the high temperature HyPy step and they exhibit an unusual mature alkane distribution unlike typical Phanerozoic petroleum fluids.

7.4-6. OTHER STRUCTURAL ELUCIDATION ANALYTICAL TECHNIQUES APPLIED TO PILBARA KEROGENS

This section discusses various analytical techniques ranging from electron microscopy to spectroscopy that have been applied to kerogens isolated from the Pilbara cherts. Obtaining information about the chemical and physical macromolecular structure of insoluble carbonaceous materials or kerogens is best accomplished with an combination electron microscopic, pyrolytic, and spectroscopic techniques. Sharp and DeGregorio (2003) investigated residual carbonaceous material isolated from the Apex chert of the Warrawoona Group using Electron Energy Loss spectroscopy (EELS). They noticed that the residual carbonaceous material under High Resolution Transmission Electron Microscopy (HRTEM) appears amorphous, is very pure, and is mostly distributed along grain boundaries between quartz crystals. The EELS spectra obtained by Sharp and DeGregorio (2003) are very similar to those of kerogen and amorphous carbon spectra obtained from ca. 2.1 Ga microfossils from the Gunflint Formation that are widely held to be biogenic.

Derenne et al. (2004) and Skrzypczak et al. (2004, 2005) reported that the carbonaceous material isolated from a single chert sample of the "lower chert" (Towers Formation, now referred to as the Dresser Formation: Van Kranendonk et al. (2006a)) at North Pole B deposit barite mine (metamorphic grade of prehnite-pumpellyite to lower greenschist facies: Van Kranendonk and Pirajno (2004)) in the Warrawoona Group is apparently highly aliphatic, and contains aliphatic chains that are branched with abundant C-O and C-N functionalities, using a combination of Electron Paramagnetic Resonance (EPR) spectroscopy, solid state ^{13}C Nuclear Magnetic Resonance (NMR) spectroscopy, Fourier transform infrared (FTIR) spectroscopy, and pyrolysis gas chromatography mass spectrometry (py-GC/MS). In contrast, Westall and Rouzaud (2004), who investigated samples from the Pilbara Craton and Barberton Greenstone Belt, and Rouzaud et al. (2005), who investigated the sample from the Dresser Formation using HRTEM, showed that the carbonaceous material was not amorphous but consisted of stacks of a few, short, nanometric, wrinkled sheets with relatively wide interspacing, indicating a moderate stage of thermal maturity. Marshall et al. (2004a, 2004b, 2007) have shown a polycyclic aromatic macromolecular structure for kerogen isolated from the Strelley Pool Chert by the use of FTIR spectroscopy, solid state ^{13}C NMR spectroscopy, Raman spectroscopy and HyPy-GC-MS.

7.4-7. IMPLICATIONS FOR THE INTERPRETATION OF SIGNALS FROM EARLY LIFE IN THE STRELLEY POOL CHERT

This section is based on new work by Marshall et al. (2007), which is primarily focused on using HyPy-GC-MS as an analytical technique to characterize the macromolecular structure of kerogen contained within the Strelley Pool Chert. Obvious similarities exist between both aliphatic and aromatic product profiles obtained from HyPy of Strelley Pool Chert kerogens in comparison with those generated from one unmetamorphosed Mesoproterozoic kerogen sample (from the 1.4 Ga Velkerri Formation of the Roper Group, Australia). Since this Mesoproterozoic kerogen is biologically-derived, containing detectable bound hopane and other biomarkers, then the match in molecular profiles offers arguably the most compelling evidence to date for a biogenic origin of Strelley Pool Chert organic matter. Sequential dual temperature experiments on two Strelley Pool Chert kerogens suggest that the aromatic hydrocaron products detected were predominantly covalently-bound to the kerogen matrix, whereas the alkanes were most likely trapped in the micropores of the kerogen and only released after disruption of the host matrix by some covalent bond cleavage and heating to high temperature. These molecular products are not structurally representative of the large interlinked PAH clusters comprising the average bulk of the organic matter (which are not GC-amenable in any case) and quantitatively comprise less than 2 wt% of the bulk kerogen. Marshall et al. (2007) has shown that the extent of alkylation of the aromatic hydrocarbon products generated from the HyPy technique can still sensitively assess the relative thermal maturity ordering of Strelley Pool Chert kerogens. With further study, the molecular and isotope patterns of HyPy products (both trapped alkanes and bound aromatics) offers an attractive opportunity for unraveling the origins of this thermally transformed organic matter.

Ultimately, it is of prime importance to use the spectroscopic, molecular and isotopic results generated to discriminate, if possible, a biogenic from abiogenic mechanism for Archaean kerogen formation. Recently, significant enrichment of ^{12}C has been reported for organic compounds synthesized abiotically by Fischer–Tropsch processes under laboratory simulation of hydrothermal conditions (McCollom and Seewald, 2006). Theoretically, this could account for the light stable carbon isotopic signatures of organic matter in the Strelley Pool Chert, but the effects of varying temperature, pressure, the extent of CO/CO$_2$ precursor conversion, and composition of catalytic mineral matrices on the extent of stable carbon isotopic fractionation observed would still need to be investigated thoroughly. It is yet to be demonstrated why the kerogen should be so consistently ^{13}C-depleted (-28 to -35‰) throughout the Strelley Pool Chert unit if an abiotic synthesis pathway is primarily responsible, given that temperature and pressure in hypothesized vent systems, and the degree of conversion of CO/CO$_2$ and at the site(s) of organic synthesis, are likely to vary enormously. A biological origin for the carbonaceous materials also accords well with the

observation of biological stromatolites in the Strelley Pool Chert (Hofmann et al., 199
Van Kranendonk et al., 2003).

In respect to arguments that have been put forward for an abiogenic origin of the Pilba
organic matter (for example, Lindsay et al., 2005), no detailed molecular mechanism ha
been proposed to explain how low amounts of simple apolar organics (predominantly lo
molecular weight n-alkanes up to n-C_{32} and much lower amounts of analogous n-alcohol
produced from a putative Fischer–Tropsch synthesis can be transformed into significar
quantities of highly aromatic kerogen across wide lateral distances in the Pilbara Craton
Nor has any process leading to abiogenic kerogen under such conditions been demonstrate
experimentally. It has been assumed that geochemical maturation or metamorphism of a
most all naturally occurring organic matter, whether biological or abiological in origir
would ultimately give rise to essentially similar thermally stable products – condense
PAHs bound within a macromolecular matrix. No insoluble organic residue formatio
was reported, however, from laboratory simulations of hydrothermal organic synthes
at temperature of 250 °C and 325 bar (McCollom and Seewald, 2006) and this remain
an unexplained aspect of the abiogenic kerogen formation theory. Functionalized lipic
(alcohols, fatty acids) can become covalently bound into kerogen when biomass is art
ficially matured in the laboratory under hydrothermal conditions (Gupta et al., 2004) bι
this appears to require the presence of biopolymers (for example, polysaccharides and pro
teins) which degrade and act as reactive nuclei for polymerization. Problems arise whe
attempting to explain (i) how dissolved apolar organics from hydrothermal fluids can b
concentrated by adsorption or encapsulation in minerals (water being an effective organi
solvent at temperatures 200–300 °C) or (ii) how the saturated hydrocarbons and alcoho
can be efficiently cross-linked together presumably initially via an aliphatic polymeric ma
trix which itself is efficiently fragmented under the same hydrothermal conditions (Lewar
2003).

The most plausible route for forming aromatic-rich kerogen from low molecular weigh
alkanes and alcohols produced from abiotic Fischer–Tropsch synthesis is possibly throug
pyrobitumen formation, where a recalcitrant residue is left behind after thermal crackin
of liquid hydrocarbon constituents at high pressure (ca. 500 bar). However, this is a poorl
understood process, and the molecular and isotopic systematics involved in pyrobitume
formation have not been reported to date in any great detail in the literature. More work i
required to elucidate whether significant quantities of kerogen can be formed from aqueou
processing at high T (200–300 °C) and high P (500 bar) of petroleum condensates an
Fischer–Tropsch synthesis products, and if so, whether the composition of the insolubl
residue is similar or distinct to Strelley Pool Chert kerogens.

Our preferred explanation at this stage is that the kerogen found in the Strelley Poo
Chert was most likely formed from diagenesis and subsequent thermal processing of bio
logical organic matter and that the consistent maturity ordering observed in both Rama
spectroscopic and HyPy molecular products represents differing degrees of subsequer
thermal processing of kerogen due to the hydrothermal activity and burial regime encoun
tered. While persuasive evidence has been obtained in this investigation which suggest
biological origin for the insoluble carbonaceous material, such an interpretation is not de

finitive. Future work will concentrate on analyzing molecular and isotopic patterns from kerogen hydropyrolysates in detail with GC-MS and GC-IRMS but using Raman spectroscopy to screen samples from numerous localities and identify those which are the less thermally altered. Herein, we have identified a wide range of maturities in organic matter in black chert zones in the Strelley Pool Chert. The least altered of these offer the best chance to identify unambiguous molecular biosignatures or isotopic trends in organic carbon that might be diagnostic for the processes that formed it.

7.4-8. CONCLUSIONS

Much of what we understand about the existence of early life comes from the examination of carbonaceous microstructures or microfossils that have been recovered from Archaean strata. Carbonaceous structures resembling bacteria from 3.5 Ga Apex cherts of the Warrawoona Group in Western Australia have, until most recently, been deemed the oldest morphological evidence for life (Schopf et al., 2002; Brasier et al., 2002). The biological origins of these microstructures were inferred from their carbonaceous (kerogenous) composition, by the degree of regularity of cell shape and dimensions, and by their morphological similarity to extant filamentous prokaryotes (Schopf et al., 2002). The composition of these carbonaceous microfossils was deduced by Raman spectroscopy, which was proposed as an additional line of evidence for a biological origin of the putative microfossils (Schopf et al., 2002). In addition, highly negative $\delta^{13}C$ values from similar putative microfossils in hydrothermal silica veins in the ca. 3.49 Ga Dresser Formation were interpreted as biological (Ueno et al., 2004). A re-examination of the Apex chert by Brasier et al. (2002) has, however, called into question the biogenicity of the filamentous carbonaceous microstructures. Instead they suggested that the carbonaceous microstructures are probably secondary artifacts formed from Fischer–Tropsch-type reactions associated with sea-floor hydrothermal systems. Brasier et al. (2002) suggest that similar microstructures in terms of morphology, Raman spectra, and strongly depleted ^{13}C isotopes can be formed through abiotic reactions to such a point that these microstructures were even considered as secondary artefacts, formed under hydrothermal conditions, onto which inorganic carbon had condensed (Brasier et al., 2002; Pasteris and Wopencka, 2002, 2003; Garcia-Ruiz et al., 2003). Therefore, significant interest has been stimulated in elucidating between a biological and nonbiological origin of these controversial microstructures and the insoluble organic matter in these Archaean cherts. One of the key issues in this debate is to establish whether the chemical structure of the macromolecular organic material is consistent with this being either thermally mature kerogen or ordered graphite and whether any genuine and informative molecular or isotopic patterns can be detected from fragmentation products.

Raman spectroscopy has been used to demonstrate a carbonaceous composition for putative microfossils in Early Archaean age rocks. In some cases Raman spectroscopy has been used to infer a biological origin of putative microfossils (Schopf et al., 2002). However, studies have shown that abiological and biological organic matter that has undergone metamorphism displays similar Raman spectral properties (Pasteris and Wopenka, 2003). This is because thermal maturation or metamorphic processes that affect rocks during burial can be expected to produce essentially the same set of thermally stable carbonaceous products (interlinked polyaromatic hydrocarbons, or PAHs) for almost all naturally occurring organic matter, whether biological or abiological in origin. Therefore, Raman spectroscopy of metamorphosed organic matter cannot by itself provide definitive evidence of biogenicity.

The bulk, ion microprobe, and NanoSIMS derived $\delta^{13}C$ values vary from -28.3 to $-42.4‰$, which for modern sediments would be typically ascribed to biological processes. Recently however (Brasier et al., 2002, 2005; Lindsay et al., 2005; McCollom and Seewald, 2006), the use of bulk carbon isotope compositions for unambiguously assessing biological contributions to carbonaceous material preserved in Archaean rocks have come under intense scrutiny. McCollom and Seewald (2006) have recently shown that the abiotic synthesis of organic compounds (C_2–C_{28} *n*-alkanes) under laboratory-simulated hydrothermal conditions can yield organic products depleted in ^{13}C to such a degree usually diagnostic of biological isotopic fractionation ($-36‰$ depletion in organic carbon, in the form of isotopically uniform *n*-alkanes, estimated relative to source carbon dioxide was observed), at least when overall CO_2/CO reductive conversions are low.

So, additional key evidence from molecular and compound-specific isotopic patterns, as well as spectroscopic characterization, is necessary to differentiate biotic from abiotic carbonaceous inputs to Earth's oldest preserved sediments. Hence, there is significant interest for Precambrian palaeobiologists to develop new means/techniques to prove the biogenicity of Archaean microfossils and kerogen. Thus far, it has been shown that the combination of elemental analysis, FTIR spectroscopy, Raman spectroscopy, ^{13}C NMR spectroscopy, and catalytic hydroprolysis combined with GC-MS analyses shows that kerogen isolated from the Warrawoona and Kelly Groups has not reached the graphite stage but consists of a macromolecular network of large polycyclic aromatic units (most probably >15 ring aromatic clusters being most common) containing predominantly short-chain or no aliphatic substituents, covalently crosslinked together. But new work by Marshall et al. (2007) shows obvious similarities between molecular hydrocarbon profiles generated from catalytic hydropyrolysis (HyPy) of 5 Strelley Pool Chert kerogens (3.4 Ga) in comparison with a mature Mesoproterozoic kerogen (ca. 1.4 Ga) isolated from a sediment from the Velkerri Formation of the Roper Group in Western Australia, and with the aromatic hydrocarbon profiles reported previously from HyPy of other mature Mesoproterozoic and late Archaean kerogen (Brocks et al., 2003). This is consistent with a biogenic origin for this early Archaean organic matter, although such an interpretation is not definitive at this stage. Further work is required to test whether consistent molecular and compound-specific isotopic patterns can be generated from a larger set of Archaean kerogens, particularly in comparison with any abiogenic kerogen standards that can be produced in the laboratory from aqueous processing under realistic hydrothermal conditions of temperature and pressure.

This section on organic geochemistry of Pilbara carbonaceous cherts highlights the challenging nature of determining a biological origin of kerogen in Earth's oldest rocks.

Clearly this current work, and any future research, will have substantial implications for interpreting the early rock record with respect to the origin and early evolution of life. Discriminating between true microbial fossils and "pseudo-fossils" is thus a central issue for the study of the appearance and evolution of life on Earth. The simple morphology is not considered as valid criterion for the identification of a microfossil. Ultimately, it is of prime importance to use a combination of morphology, geological context, spectroscopic, molecular, and isotopic results to discriminate, if possible, a biogenic from abiogenic mechanism for Archaean kerogen formation.

ACKNOWLEDGEMENTS

CPM would like to thank financial support from the Australian Research Council. Dr Gordon Love is acknowledged for all his help and assistance in this research. Dr Abigail Allwood is also acknowledged for her help.

Earth's Oldest Rocks
Edited by Martin J. Van Kranendonk, R. Hugh Smithies and Vickie C. Bennett
Developments in Precambrian Geology, Vol. 15 (K.C. Condie, Series Editor)
© 2007 Elsevier B.V. All rights reserved.
DOI: 10.1016/S0166-2635(07)15075-1

Chapter 7.5

SULPHUR ON THE EARLY EARTH

STEPHEN J. MOJZSIS

Department of Geological Sciences, Center for Astrobiology, University of Colorado, 2200 Colorado Avenue, Boulder, CO 80309-0399, USA

7.5-1. INTRODUCTION

On the young Earth, the surface zone of the planet was shaped by physical and chemical interactions between the crust, hydrosphere, atmosphere and the emergent biosphere, on the geochemical cycles of the biologically important elements (H, C, N, O, P, S, Fe and others). An explicit record of the early co-evolution of the atmosphere and biosphere, with the geosphere, has been notoriously difficult to investigate. By its very nature the atmosphere is ephemeral, and the result has been that a direct record of its chemical evolution remained frustratingly out of reach. It is now reasonably well understood that the geochemical behaviour of the multiple sulphur isotopes are the best tool for investigations of the long term evolution of the surface system, and that they provide a uniquely powerful proxy for long-term changes in atmospheric chemistry and other key processes (Thiemens, 1999, 2006). The geochemistry of sulphur is complex. There is a long history of study on the topic so that a single review on the generalities of sulphur biogeochemistry on the early Earth cannot hope to cover all progress in this rapidly changing field (please see Ohmoto and Goldhaber (1997), Canfield and Raiswell (1999), Canfield (2001), Farquhar and Wing (2003) and Seal (2006) for comprehensive reviews). Instead, the goal here will be to outline several key aspects of sulphur that are of particular interest to studies of Earth's oldest rocks (limited to before ca. 3.5 Ga), and to make the case that a comprehensive program of retrospective multiple sulphur isotope analyses is required on samples for which prior $^{34}S/^{32}S$ data have been compiled (Strauss, 2003).

When one attempts to review processes operative on the early Earth, global environmental aspects unique to the young planet need to be acknowledged. In the first billion years, higher crustal heat flow and enhanced mantle outgassing, widespread (and rapid?) crustal recycling and considerable changes in continental volume, as well as the emergence of life and its subsequent take-over of several important aspects of chemistry of the surface zone, were all factors that influenced the various sub-cycles of the global sulphur cycle schematically represented in Fig. 7.5-1. Furthermore, exogenous processes unique to the young planet were also important, even if they are more difficult to account for in physical models. These processes included a higher frequency of asteroid and comet impacts,

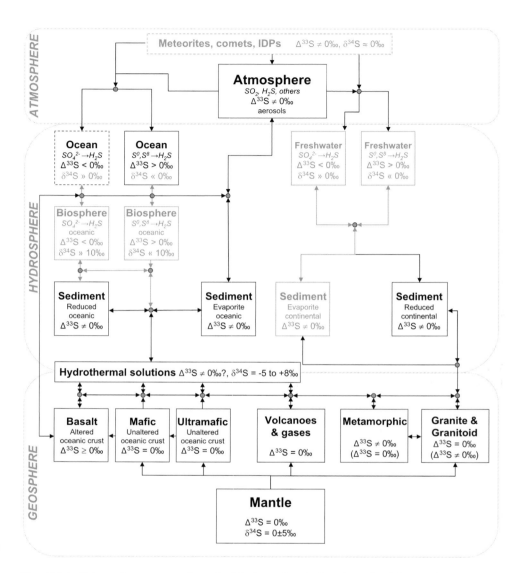

Fig. 7.5-1. Schematic representation of global geochemical cycles of sulphur and the average sulphur isotope compositions of important reservoirs in the Earth system. The interface between the geosphere and hydrosphere is dominated by exchange of sulphur between seawater and rocks by fluids, weathering, dissolution and re-precipitation. Sulphide will precipitate with an adequate supply of dissolved iron (and other metals) and can be abiogenic or biogenic. Sulphur may also be sequestered in evaporates, brine pools or crustal fluids as sulphate, sulphide or elemental sulphur. Chemical weathering is the principal mechanism that returns sulphur from rocks and sediments on land to the hydrosphere. Hydrothermal activity and subduction transports sulphur back to the geosphere. Submarine dissolution of basalt and other rocks also returns S to the hydrosphere.

Fig. 7.5-1. (*Continued.*) Volcanism injects sulphur gases to the hydrosphere and atmosphere. Volcanic gases in the atmosphere can become transformed by light-driven reactions and returned to sediments from aerosol deposition. Subduction will bring all components back to the mantle. Extraterrestrial input is minimal, but could be important in a few cases of large impacts.

an enhanced rate of extraterrestrial dust delivery compared to present, as well as reduced solar luminosity (but higher ultraviolet flux) from the young Sun.

However, there exists a further and yet more fundamental limitation to what can be known about the earliest Earth. A geologic record of the interactions outlined in Fig. 7.5-1 has apparently been lost for the first ca. 750 My due to the effective recycling of all crust except for ca. 4.4–3.8 Ga zircons contained in some younger sediments (see Cavosie et al., this volume). Sedimentary rocks ⩾3.7 Ga in age, which could potentially preserve a direct sample of the ancient sulphur cycle, occur only as rare and deformed enclaves in high-grade Archaean metamorphic granitoid-gneiss terranes. Nevertheless, given recent advances in instrumental techniques, experimental and observational geomicrobiology and molecular biochemistry, chemical oceanography, and access to an increasingly large sample base of ancient rocks, our knowledge of early Earth environments has accumulated to the point where it is now perhaps worthwhile to revisit what we know of the interactions between the geochemical cycles of sulphur at that time. Specifically:

(1) Sulphur comes in four stable isotopes of well-defined relative abundance. What is the origin of sulphur to the Earth? How do changes occur to the abundance ratios of the sulphur isotopes? What is the significance of mass-independent isotope fractionation as opposed to 'normal' mass-dependent effects in the early Earth record?

(2) The infall of meteoritic and cometary debris to the planets was greater in the first billion years than at present. What contributions could extraterrestrial sources have made to the global geochemical sulphur cycle in the first billion years?

(3) Subsequent to primary outgassing, the atmosphere underwent chemical changes. What clues can sulphur provide to the composition and chemical evolution of the atmosphere prior to about 3.5 Ga?

(4) Sulphur compounds are central to many fundamental biochemical processes. What role might sulphur have played in the establishment of the biosphere? Could sulphur isotope fractionations serve as a robust isotopic biomarker in highly metamorphosed rocks?

(5) All terranes older than about 3.5 Ga have been strongly transformed by metamorphism. How well is the sulphur isotope record preserved in the oldest rocks?

(6) Banded iron-formations (BIFs) are the most common sedimentary rock preserved from the early Archaean, but their derivation remains enigmatic. What can sulphur isotopes inform us of the origins of the BIFs?

7.5-2. ORIGIN OF SULPHUR TO THE EARTH

The elemental inventory of the planets was determined by the composition of the nebular gas and dust from which the solar system formed. In stars greater than 8 times solar mass, and at temperatures near 1.5×10^9 K, oxygen fusion begins as a result of the slow (s-process) fusion of neutrons (n), protons (p) and ^{4}He nuclei (α) to a variety of seed nuclei (Cameron, 1979; Woosley and Weaver, 1986). In this scenario, the various sulphur isotopes can be produced from the oxygen-burning nuclear reaction:

$$^{16}O\left(^{16}O, \gamma\right) \rightarrow {}^{32}S; \ {}^{32}S\left(^4He, \gamma\right) \rightarrow {}^{36}S; \ {}^{32}S(n, \gamma) \rightarrow {}^{33}S; \ {}^{33}S(n, \gamma) \rightarrow {}^{34}S$$

Here, it can be seen how the major oxygen-oxygen reaction $^{16}O(^{16}O, \gamma) \rightarrow {}^{32}S$ yields the abundant ^{32}S isotope and energy. Neutrons evolved from side reactions such as $^{16}O(^{16}O, n) \rightarrow {}^{31}S$ can result in further s-processes and the formation of the minor isotopes (Seeger et al., 1965). As a result, there are four stable isotopes of sulphur which have the following relative abundances as defined by the meteorite sulphur standard Cañon Diablo troilite (Hultson and Thode, 1965; see Ding et al., 2001 and Beaudoin et al., 1994):

$$^{32}S = 95.02\%; \ {}^{33}S = 0.75\%; \ {}^{34}S = 4.21\%; \ {}^{36}S = 0.02\%$$

The high relative abundance of ^{32}S can therefore be understood from the fact that it is directly synthesized from O-burning reactions. Sulphur is an exceptionally stable element. It has an even atomic number ($Z = 16$) and the nuclide ^{32}S has both a mass number (A) and a neutron number (N) that are multiples of 4 (the mass of the He nucleus), so that $A = Z + N = 16 + 16 = 32$. Since the minor S isotopes are more closely tied to rates and types of various s-process reactions which also include synthesis from ^{4}He-fusion with the stable isotopes of Si (^{28}Si, ^{29}Si and ^{30}Si), the relative abundances of ^{33}S, ^{34}S and ^{36}S are much lower. This inventory of the different sulphur isotopes was imparted to the solar nebula at some time prior to the accretion of the planets.

7.5-2.1. Primordial Sulphur

The formation time of Earth and other planets, and the elemental inventory of Earth, has been deduced in part on the basis of comparison with primitive meteorites (type 1 carbonaceous chondrites; CI) and the solar photosphere. The average compositions of CI meteorites and the Sun have been compiled by numerous workers (e.g., Anders and Grevesse, 1989), who showed that CI:solar photosphere displays a near 1:1 correspondence in sulphur content normalized to Si, so that bulk meteoritic sulphur is probably a reasonably good approximation for bulk planetary sulphur isotopic compositions. Primitive carbonaceous chondrites contain approximately 5.4 wt% sulphur (Dreibus et al., 1995) and S is the tenth most abundant element in the solar system (Anders and Ebihara, 1982). Sulphur ranks as eighth in abundance for the whole Earth (silicates + core; Morgan and Anders, 1980) and the Bulk Silicate Earth concentration is about 250 ppm (McDonough and Sun, 1995). A summary of the various concentrations of sulphur and sulphur-containing compounds in the terrestrial reservoirs in the atmosphere, hydrosphere, biosphere and geosphere is provided in Table 7.5-1. The sulphur isotopic composition of the bulk Earth is assumed to be close to that of the major S-phases of iron meteorites. Because the record of sulphur is geochemical, a brief review of the stable isotope geochemistry of sulphur is warranted.

7.5-2.2. Sulphur Isotope Fractionations

For low atomic number elements, the mass differences between the various isotopes are large enough for many physical, chemical, and biological processes or reactions to change the relative proportions of the isotopes (reviewed in Hoefs (1997)). In normal physical-chemical interactions on Earth, two isotope effects – equilibrium and kinetics – result in isotope fractionation. Due to these fractionation processes, products often develop unique isotopic compositions (normally expressed as ratios of heavy to light isotopes) that may be indicative of their source, or of the processes that formed them (Faure, 1986, and references therein). Reaction rates depend on the ratios of the masses of the isotopes and their zero-point energies. As a general rule, bonds between the lighter isotopes are several calories less than bonds between heavy isotopes. The lighter isotopes tend to react more readily and become concentrated in the products, and the residual reactants tend to become enriched in the heavy isotopes. The magnitude of the fractionation strongly depends on the reaction pathway utilized and the relative energies of the bonds broken and formed by the reaction. Slower reactions (for instance those proceeding at lower temperatures) tend to show larger isotopic fractionation effects than faster ones.

Equilibrium isotope-exchange reactions involve the redistribution of isotopes among various chemical species or compounds. During equilibrium reactions, the heavier isotopes preferentially accumulate in the compound with the higher energy state. Kinetic isotope fractionations occur when systems are out of isotopic equilibrium, such that forward and back reaction rates are not identical. When reactions are unidirectional, reaction products become physically isolated from the reactants. Biological processes are kinetic isotope reactions that tend to be unidirectional. Both diffusion and metabolic discrimination against the heavier isotopic species occurs because of the differences in zero-point energies of the different isotopes. This can result in significant fractionations between the substrate (isotopically heavier) and the biologically mediated product (isotopically lighter). Many reactions can take place either under purely equilibrium conditions or be affected by an additional kinetic isotope fractionation, and metabolic (oxido-reduction) reactions will preferentially select molecules with lighter isotopes (Schidlowski et al., 1983).

To follow the various isotopic fractionation pathways, partitioning of stable isotopes between two substances A and B is expressed by use of the isotopic fractionation factor (α):

$$\alpha(A - B) = R_A / R_B$$

where R is the ratio of the heavy to light isotope (e.g., $^{34}S/^{32}S$ or $^{33}S/^{32}S$). Values for α tend to be close to 1.00 so that differences in isotopic compositions are usually expressed in parts-per-thousand, or 'per mil' (‰). Kinetic fractionation factors are typically described

Table 7.5-1. Average concentration of sulphur in various reservoirs on Earth (values are in 10^{-6} g/g unless otherwise indicated)

Reservoir	Reduced S H_2S	Elemental S S_2 (S^0, S_8)	Oxidized S SO_2 [SO]	Sulphate SO_4^{2-}	Other OCS	Total S
Atmosphere[a]	3–10×10^{-5}	$\sim 1.5 \times 10^{-5}$ b	2–9×10^{-5}	–	5×10^{-4}	$<1 \times 10^{-3}$
Biosphere	–	–	–	–	–	1000^{c}
Hydrosphere						
Rivers[d]	–	–	–	11.5	–	–
Oceans	–	–	–	2784^{e}	–	$\sim 2800^{f}$
Shale	–	–	–	–	–	2400^{g}
Granite	–	–	–	–	–	260^{h}
Mafic rocks and Ultramafics	–	–	–	–	–	~ 1000–4000^{i}
Volcanic gases[j]	0.057–3.21	3×10^{-4}–1.89	0.006–47.7 [0.03–0.06]	–	2×10^{-5}–0.08	
Comet particles						95480^{k}
IDPs						41664^{l}

Notes:

[a]present troposphere (Warneck et al., 1988); [b]as CS_2; [c]dry weight of tissue (Berner and Berner, 1996 – citing Zinke, 1977); Schlesinger, 1997); [d]Stumm and Morgan (1996); [e]Millero and Sohn (1992); [f]Li (1991); [g]Turekian and Wedepohl (1961); [h]Mason and Moore (1982); [i]Wedepohl and Muramatsu (1979); [j]Symonds et al. (1994) values expressed in mol%; [k]Kissel et al. (1986); [l]Schramm et al. (1988).

in terms of various enrichment or discrimination factors and these are defined in various ways by different researchers (see O'Neil, 1986).

7.5-2.3. Mass-Dependent Fractionation of Sulphur Isotopes

The numerous paths of sulphur between the different reservoirs shown in Fig. 7.5-1 often involve changes in oxidation states that are modulated through both abiotic and biotic equilibrium and kinetic reaction processes, and the different degrees of S-isotope fractionations depend on these modes of isotopic exchange. In order to track these fractionations in nature, sulphur isotope compositions of substances are expressed with the conventional δ notation as:

$$\delta^{34}S_{CDT}(\text{‰}) = \left(R^{34}_{sample}/R^{34}_{CDT} - 1\right) \times 1000$$

$$\delta^{33}S_{CDT}(\text{‰}) = \left(R^{33}_{sample}/R^{33}_{CDT} - 1\right) \times 1000,$$

where $R^{34} = {}^{34}S/{}^{32}S$, $R^{33} = {}^{33}S/{}^{32}S$ and the reference standard is the pyrrhotite end-member phase troilite (FeS) from the Cañon Diablo meteorite (CDT; see Beaudoin et al., 1994). Similarly, this expression holds for $\delta^{36}S$, however the value has not usually been reported because the trace abundance of ${}^{36}S$ makes the measurement difficult with normal mass-spectrometric techniques (e.g., Hulston and Thode, 1965; Hoering, 1989; Gao and Thiemens, 1989). As reviewed elsewhere (Mojzsis et al., 2003; Farquhar and Wing, 2003), most conventional sulphur isotope studies omitted the minor isotope ${}^{33}S$ with the view that there would be no new information conveyed by the measurement of ${}^{33}S/{}^{32}S$ (cf. Hoering, 1989; Lundberg et al., 1994). This is because normal physical, chemical, or biological processes fractionate isotopes with predictable mass-dependent behaviour (Urey, 1947; Bigeleisen and Mayer, 1947). In other words, on sulphur three-isotope plots of $[({}^{33}S/{}^{32}S)/({}^{34}S/{}^{32}S)]$ or $[({}^{36}S/{}^{32}S)/({}^{34}S/{}^{32}S)]$, the relative mass differences are such that variations in $\delta^{33}S$ and $\delta^{36}S$ ought to be correlated with those in $\delta^{34}S$. Mass-dependent fractionation laws arise from the kinetic and equilibrium reactions that fractionate sulphur in approximate proportion to their relative mass difference. Because the mass difference between ${}^{33}S$ and ${}^{32}S$ is about half of that between ${}^{34}S$ and ${}^{32}S$, any mass-dependent isotope effects in $\delta^{33}S$ would be about half those in $\delta^{34}S$. Mass-dependent fractionation (MDF) processes lead to isotopic compositions that can be described by the equation:

$$\left({}^{33}S/{}^{32}S\right)/\left({}^{33}S/{}^{32}S\right)_{CDT} = \left[\left({}^{34}S/{}^{32}S\right)/\left({}^{34}S/{}^{32}S\right)_{CDT}\right]^{\lambda}.$$

In this treatment, linear regression in (x, y) with $y = \ln({}^{33}S/{}^{32}S)$ and $x = \ln[({}^{34}S/{}^{32}S)/({}^{34}S/{}^{32}S)_{CDT}]$ has slope λ. Theoretical analyses of equilibrium MDF laws define slopes of $\lambda \approx 0.515$ (Bigeleisen and Mayer, 1947; Hulston and Thode, 1965). York regressions through large datasets from sulphur standards measurements yield values $\lambda = 0.515 \pm 0.005$ (Farquhar et al., 2000a, 2000b, 2000c; Mojzsis et al., 2003; Farquhar and Wing, 2003; Ono et al., 2003; Hu et al., 2003; Papineau et al., 2005, 2006; Papineau and Mojzsis, 2006; Cates and Mojzsis, 2006; Whitehouse et al., 2005; Ono et al., 2006, 2007; Papineau et al., 2007). Empirical determinations of ${}^{33}S/{}^{32}S$ are obtained by fixing λ at 0.515 with

$(^{34}S/^{32}S)_{V\text{-}CDT} = (22.6436)^{-1}$ (Ding et al., 2001), so that regressions through standards data with fixed λ result in intercepts that equal $\ln[(^{33}S/^{32}S)_{V\text{-}CDT}]$. This yields $(^{33}S/^{32}S)_{CDT}$ values $= (126.3434)^{-1}$, and with the thousands of standards measurements reported thus far, mass-dependent values have been narrowly defined about $[(^{34}S/^{32}S)/(^{33}S/^{32}S)]_{CDT} = 0 \pm 0.1\text{‰}(2\sigma)$.

Different modes of mass-dependent fractionations can arise from different kinetic and equilibrium effects (Bigeleisen and Wolfsberg, 1958). What is little-discussed but germane to sulphur isotope geochemistry is that the scatter often seen in $[(^{34}S/^{32}S)/(^{33}S/^{32}S)]$ data can deviate slightly from orthodox mass-dependency because different slopes in λ may result from these different mass-dependent modes (e.g., Young et al., 2002) captured in the compositions of natural standards. In light of this observation, Papineau et al. (2005) argued that MDF trajectories on a three-isotope plot should be projected as a band that encompasses the different mass-dependent laws, rather than as a simple 'terrestrial (mass-) fractionation line'. Based on this analysis, the width of the York-MDF band is defined as the standard deviations in y about the fixed-slope York regression through all standards in (x, y)-space with a slope $\lambda = 0.515 \pm 0.005$ when transformed to $\delta^{34}S_{CDT}$ vs $\delta^{33}S_{CDT}$ space. This scrutiny of the various modes of mass-dependent sulphur isotope fractionations in the three-isotope system yields external errors in the range of $\pm 0.20\text{‰}$. Other, more detailed studies that explore the quadruple sulphur isotope system at high resolution with laser fluorination methods is reviewed in Ono et al. (2003, 2006, 2007).

Admittedly, there remains considerable room for improvement in these measurements and it is expected that when synthetic sulphur standards become available, the external errors will decrease and finer details of sulphur isotope fractionations will be revealed. Theoretical treatments of the MDF boundaries within the range of possible MDF types show that they plot as shallow curves, and the York MDF-bandwidth (approximately $\pm 0.1\text{‰}$, 2σ in the theoretically best possible case) increases slightly with absolute values of $(\delta^{33}S/\delta^{34}S)_{CDT}$ (Fig. 7.5-2). However, some unusual photolytic atmospheric reactions that occur in the gas-phase, and some nuclear effects in space, produce sulphur that significantly deviates from mass-dependent fractionation laws in what has come to be known as 'mass-independent fractionation'.

7.5-2.4. *Mass-Independent Fractionation of Sulphur Isotopes*

In order to better grasp the significance of mass-independent sulphur isotope effects captured in the geologic record of the early Earth, it is worthwhile to briefly summarise the initial discovery of this phenomenon in the oxygen isotopes. Mass-independent fractionation (MIF) behaviour of stable isotopes was found in the oxygen isotopic compositions (^{16}O, ^{17}O, ^{18}O) of carbonaceous meteorites (Clayton et al., 1973). In extensive work since, Clayton and others have shown that on a three-isotope oxygen plot[1], anhydrous minerals in carbonaceous meteorites lie along a line of slope of $+1$ (Fig. 7.5-3) instead of a

---

[1]$(\delta^{17}O$ vs $\delta^{18}O$; where $\delta^{1x}O = ((^{1x}O/^{16}O)_{sample}/(^{1x}O/^{16}O)_{V\text{-}SMOW} - 1) \times 1000\text{‰}$, x is either 7 or 8 and V-SMOW is the Vienna standard mean ocean water).

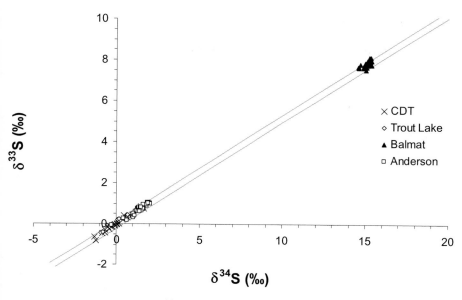

Fig. 7.5-2. Three-isotope plot (δ^{33}S vs δ^{34}S) for various sulphide standards that are used to define the York-MDF bandwidth. For clarity, the error bars for the individual measurements have been omitted (from Papineau et al., 2005).

slope of approximately +0.5 expected to define normal mass-dependent isotope effects on Earth. The initial interpretation that nucleosynthetic processes produced mass-independent isotopic signatures was re-visited by Thiemens and Heidenreich (1983) who showed experimentally that the origin of MIF oxygen isotopes was likely photochemical (Fig. 7.5-4). Photolysis of diatomic oxygen under ultraviolet radiation produces MIF oxygen isotopes in the residual molecular oxygen (O_2) and in the product ozone (O_3). Fig. 7.5-4 defines a theoretical mass fractionation (band), which is the regime wherein all O-isotope MDF processes plot according to the equation:

$$\Delta^{17}O = 1000 \times \left\{ \left(\frac{\delta^{17}O}{1000} + 1 \right) - \left(\frac{\delta^{18}O}{1000} + 1 \right)^{\lambda} \right\}.$$

In this case, $\Delta^{17}O$ is the permil deviation from mass-dependent fractionation and λ represents the slope, which expresses the mass-dependent fractionation relationship between $\delta^{17}O$ and $\delta^{18}O$ (approximately $\delta^{17}O = 0.52 \times \delta^{18}O$). Mass-dependent fractionation processes result in isotopic compositions that have a $\Delta^{17}O$ of 0‰, while mass-independent processes that arise from ultraviolet photolysis will result in $\Delta^{17}O \neq 0$‰, or significant deviation from the mass-dependent fractionation band. In Fig. 7.5-4, product ozone (O_3) from the photolysis of molecular oxygen (O_2) has a positive $\Delta^{17}O$, which means that it is significantly more enriched in ^{17}O than it would if the fractionation process was purely

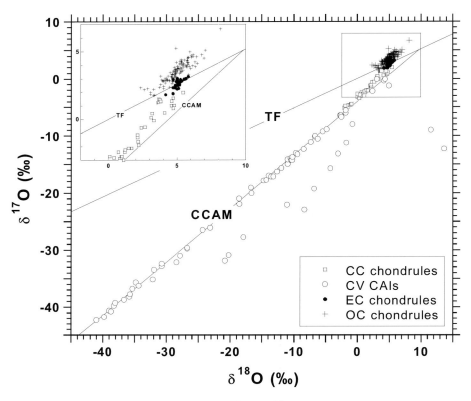

Fig. 7.5-3. Compilation three-isotope plot ($\delta^{17}O$ vs $\delta^{18}O$) of anhydrous minerals in carbonaceous meteorites showing MIF oxygen isotopes (from McKeegan and Leshin, 2001).

mass-dependent. The isotopic composition of residual O_2 also shows MIF in isotopic mass-balance and has a negative $\Delta^{17}O$, which is a depletion of ^{17}O relative to isotopic compositions characteristic of mass-dependent processes. The slope of the data in Fig. 7.5-4 is close to $+1$, while the slope for MDF processes is about $+0.5$. These results convincingly demonstrated that UV photochemical reactions can lead to products and residual reactants that exhibit mass-independent fractionation. Note that the range of $\delta^{18}O$ and $\delta^{17}O$ in Fig. 7.5-4 for the photochemical experiments is different than that observed in carbonaceous meteorites (Fig. 7.5-3). This could mean that photochemical reactions are more efficient in a controlled experiment and/or that isotopic dilution occurred from mixing mass-independently fractionated components within dominantly mass-dependently fractionated oxygen reservoirs. Further progress in our understanding is hampered by the fact that the actual mechanism(s) for mass-independent fractionation have yet to be fully elucidated with quantum-chemical theory (Thiemens, 1999; Mauersberger et al., 1999; Gao and Marcus, 2001; Deines, 2003).

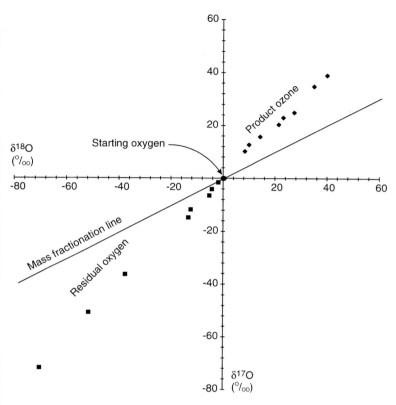

Fig. 7.5-4. Three-isotope plot (δ^{17}O vs δ^{18}O) for oxygen isotopes in product O_3 and residual O_2 after irradiation of O_2 with ultraviolet light from electrical discharge (from Thiemens, 1999).

Sulphur is isoelectronic with oxygen[2] and has also been found to display mass-independent isotope fractionation. Multiple sulphur isotope data are reported using the conventional notation as:

$$\Delta^{33}S = 1000 \times \left\{ \left(\frac{\delta^{33}S}{1000} + 1 \right) - \left(\frac{\delta^{34}S}{1000} + 1 \right)^{\lambda} \right\}$$

$$\Delta^{36}S = 1000 \times \left\{ \left(\frac{\delta^{34}S}{1000} + 1 \right) - \left(\frac{\delta^{36}S}{1000} + 1 \right)^{\lambda} \right\},$$

where $\Delta^{33}S$ and $\Delta^{36}S$ are the permil deviations from mass-dependent isotopic fractionation processes, $\delta^{3x}S = ((^{3x}S/^{32}S)_{sample}/(^{3x}S/^{32}S)_{CDT} - 1) \times 1000\%_0$ where $x = 3, 4$ or 6, and λ represents the mass-dependent fractionation relationship between $\delta^{33}S$ and $\delta^{34}S$ ($\delta^{33}S = 0.52 \times \delta^{34}S$) or between $\delta^{36}S$ and $\delta^{34}S$ ($\delta^{36}S = 1.9 \times \delta^{34}S$). Like oxygen

---

[2]Only mass-dependent fractionations have been reported for Selenium (Johnson and Bullen, 2004). Tellurium does not appear to display MIF, although searches have been conducted (Fehr et al., 2005).

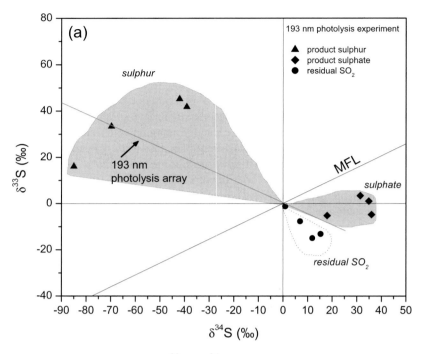

Fig. 7.5-5. Three isotope plots (δ^{33}S vs δ^{34}S) showing (a) experimental results with an ArF excimer laser on SO_2 gas and (b) Precambrian sedimentary sulphide and sulphate multiple sulphur isotope compositions (data from Farquhar et al., 2001).

Δ^{17}O, sulphur Δ^{33}S and Δ^{36}S values $\sim 0\permil$ result from MDF processes such as kinetic, equilibrium and oxido-reduction reactions. Laboratory experiments have shown that mass-independently fractionated sulphur isotopes can arise in gas-phase reactions with starting mixtures of SO_2, H_2S and CS_2 (Zmolek et al., 1999; Farquhar et al., 2000a, 2000b, 2001). Results from photolysis experiments with an ArF excimer UV laser at 193 nm in a SO_2 gas are shown in Fig. 7.5-5(a). These experiments documented MIF sulphur isotopes in product S^0, SO_4^{2-} and residual SO_2. The Δ^{33}S values of these S-phases are positive for product elemental sulphur so that they plot above the mass-dependent fractionation band, and negative for product sulphate (i.e., they plot below the MDF band).

The MIF behaviour of three sulphur isotopes (^{32}S, ^{33}S and ^{34}S) in photochemical experiments resembles multiple sulphur isotopes from Precambrian sedimentary sulphides and sulphates as seen in Figs. 7.5-5(b) and 7.5-6. Similar to the results for oxygen isotopes, the main difference between the plots is that the range of δ^{33}S and δ^{34}S values is larger in the photochemical experiments (tens of permil) and much smaller (a few permil) for natural samples that exhibit mass-independent behaviour such as found in pre-2.45 Ga sedimentary sulphides and sulphates. As for the oxygen case discussed above, this may

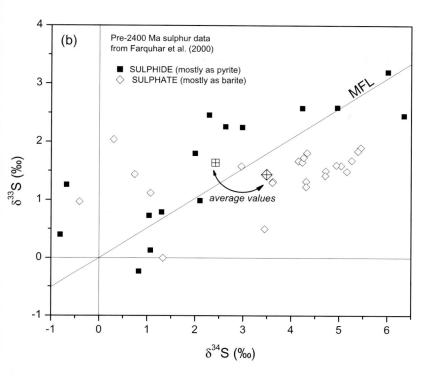

Fig. 7.5-5. (*Continued.*)

arise from more efficient photochemical reactions in controlled experiments and/or from isotopic dilution occurring in natural systems.

Barring further detailed experimental verification from a wider range of UV wavelengths (Lyons, 2006), data from ancient sulphur are consistent with the view that short wavelength solar radiation (i.e., $\lambda < 220$ nm) penetrated deeply in the early terrestrial atmosphere in the absence of an effective ozone screen to lead to significant MIF in the geologic record (Farquhar et al., 2001). It is interesting to note that the higher extreme UV (EUV) output observed for young solar-type stars (Wood et al., 2002) requires that in the early Earth's atmosphere the incident solar EUV flux fell from $\sim$30$\times$ to approximately 4$\times$ present between 4.4 to 3.5 Ga (Zahnle, 2006). The evolution in the incident UV flux has yet to be accounted for in our models for the generation of mass-independent fractionation of sulphur isotopes in the ancient atmosphere. Mass-independently fractionated signatures in early Precambrian sedimentary sulphides and sulphates could be coupled to results of experiments that explore the magnitude of MIF under specific atmospheric regimes and UV intensities at different wavelengths. These data could then be used to trace the chemical evolution of the atmosphere in relation to the long-term interaction of the surface zone with the evolving young Sun.

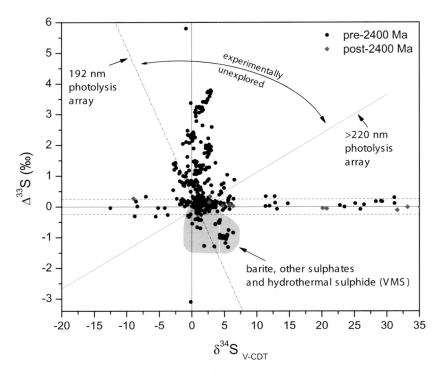

Fig. 7.5-6. Plot of published $\Delta^{33}S$ vs $\delta^{34}S$ for sulphides and sulphates from pre-2.45 Ga samples (triangles) with data from samples younger than ca. 2 Ga (modified from Farquhar and Wing, 2003). The field of isotope compositions for barite and hydrothermal sulphide (volcanogenic massive sulphide 'VMS' deposits) is shown. Some data for pre-2.45 Ga samples tend to plot with an aspect ratio similar to the 193 nm photolysis array of SO_2 and SO (discussed in Farquhar et al., 2001; Pavlov and Kasting, 2002).

It is thought that the early Earth experienced significant mass input from extraterrestrial sources from the sweep-up of remnant debris from solar system formation. Therefore, it makes sense to explore whether extraterrestrial sources of sulphur could have been important to the earliest sulphur cycle.

7.5-3. EXTRATERRESTRIAL INPUTS TO THE SULPHUR CYCLE ON THE EARLY EARTH

Dust delivery has always been important to planets, and interplanetary dust was likely a major contributor to the surface inventory of biogenic elements at the formation and early evolution of terrestrial planets (Maurette et al., 2000; Pavlov et al., 1999; Marty and Yokochi, 2006). Micrometeorites and interplanetary dust particles dominate the contemporary extraterrestrial matter flux to Earth on the order of 10^3–10^4 tons/yr (Love and Brownlee, 1993), or about 2–3 orders of magnitude above the calculated present-day meteorite flux (Bland, 2005).

In the last decade evidence has emerged that many (or most) stars are born with circumstellar disks with abundant dust and rocky debris (Habing et al., 2001; Carpenter et al., 2005). The solar system's original disk became severely depleted by planetesimal accretion and by planet formation within about the first 100 My. Some debris disk systems observed around young (<1 Gyr old) stars by the *Spitzer* telescope look brighter than others of similar age and may represent dust generated from evaporation of comets and/or planetary-scale bombardment events (Kenyon and Bromley, 2005) well after the primary phase of planetary accretion should have ended (Beichman et al. 2005; Rieke et al., 2005; Meyer et al., 2006). If cometary dust production rates from comet Hale-Bopp (Bocklée-Morvan et al., 2004) are any measure of the composition of the early disk dust for our solar system, then the combined sulphur species measured for coma dust (OCS, CS_2, H_2CS) amounts to a mere ~0.65% of the comet's water content. At first glance, it appears that late implantation of extraterrestrial sulphur from dust would have been a relatively minor player in the chemistry of early planetary surfaces and atmospheres. Since some meteorite components show mass-independently fractionated sulphur isotopes, it is worth exploring whether these inputs were at all significant to the post-primary accretion sulphur budget of the early Earth.

7.5-3.1. Sulphur in Extraterrestrial Dust

Interplanetary dust particles compositionally resemble the CM type carbonaceous chondrites (Kurat et al., 1994). Murchison (CM) meteorite contains 3.3% S (Wasson and Kallemeyn, 1988), and has $^{34}S/^{32}S$ and $^{33}S/^{32}S$ values comparable to terrestrial crust and mantle (Gao and Thiemens, 1993a) with average $\Delta^{33}S = 0.013 \pm 0.054‰$. These values are well within the terrestrial mass-dependent field of the sulphur isotopes. Other primitive meteorites such as Orgeuil (CI), considered by some to be a representative of Jupiter family comet compositions (Gounelle et al., 2006), contains upwards of 5% total sulphur. Orgeuil has $^{34}S/^{32}S$ and $^{33}S/^{32}S$ values slightly enriched in the heavier isotopes compared to Murchison, and preserves an average $\Delta^{33}S$ value of $-0.005‰$ (Gao and Thiemens, 1993a). The majority of meteoritic materials that have been analysed so far for the minor sulphur isotopes indicate that inhomogeneities in S isotope compositions possibly related to nucleosynthetic, spallation reactions on Fe, or gas-phase (ion-molecule) photolytic reactions in space, were mixed-in for the most part in the early solar system (Hulston and Thode, 1965; Gao and Thiemens, 1993a, 1993b). Values of $\Delta^{33}S$ not equal to 0‰ have been found in a class of igneous meteorites called the ureilites (Farquhar et al., 2000b), in organic residues from carbonaceous chondrites (Cooper et al., 1997), in the acid-resistant 'phase Q' of the Allende meteorite (Rees and Thode, 1977), and in various phase separates of chondrite (Rai and Thiemens, 2007) as well as achondrite (Rai et al., 2005) meteorites. Greenwood et al., 2000a) raised the possibility that the increased rates of extraterrestrial matter input to the planets of the early solar system from later bombardments could have implanted heterogeneous distributions of the sulphur isotopes to the early surface.

7.5-3.2. Sulphur and the Late Heavy Bombardment

We have no direct measure of the influx of extraterrestrial dust to Earth before about 3.85 Ga during the so-called 'late lunar cataclysm' or 'late heavy bombardment' (Tera et al., 1974; Ryder, 1990). Hartmann et al. (2000) calculated that the total mass of impactors to the Moon at ca. 3.9 Ga was approximately 6×10^{21} g, and that the amount of mass accumulated by the Moon in the 4.4–4.0 Ga timeframe was roughly equivalent to the mass influx that formed the major lunar basins during the late heavy bombardment (4.1–3.85 Ga; reviewed in Mojzsis and Ryder (2001)). Therefore, the total amount of dust accumulated to the Moon during the Hadean could have been on the order of 1.2×10^{22} g. Several studies have attempted to acquire information on the rate of accumulation and intensity of impacts at the tail end of the bombardment epoch from the study of Earth rocks. Koeberl and Sharpton (1988) searched for shock features in quartz from ~3.7–3.8 Ga rocks from Isua (West Greenland). Later, Koeberl et al. (1998, 2000) looked for impact shocked zircons from the same Isua rocks, and of the hundreds of crystals studied, none showed any evidence of optically resolvable shock deformations (Koeberl, 2003). Mojzsis and Ryder (2001) and Anbar et al. (2001) could likewise find no unequivocal chemical evidence from platinum-group elements for the late heavy bombardment in sedimentary rocks with ages at least to ~3.83 Ga (Nutman et al., 1997; Manning et al., 2006). On the other hand, Schoenberg et al. (2002) reported tungsten isotope anomalies in ca. 3.7–3.8 Ga paragneisses from West Greenland and Labrador (Canada) which they interpreted to be remnants of the late bombardment in younger materials. However, follow-up studies with platinoid elements, Cr and other isotopic systems (e.g., Frei and Rosing, 2005) could find no corroborating evidence for cosmic materials in the Isua sediments (Koeberl, 2006). To further compound these difficulties, the various studies cited above explored rocks with formation times (ca. 3.7–3.8 Ga) that likely did not overlap with the main period of late heavy bombardment to the inner solar system (Kring and Cohen, 2002).

Since evidence for the bombardment epoch from terrestrial rocks remains elusive, all estimates of the role of extraterrestrial matter to the early Earth must be treated with caution. Marty and Yokochi (2006) estimated the total range of all extraterrestrial material (including dust and other meteoritic debris) collected on Earth's surface from 4.4–3.8 Ga to have been in the range of 2.4–7.2×10^{23} g. With the average sulphur content of CM and CI meteorites provided above as a guide, this would amount to an accumulation of about 0.8–3.6×10^{22} g of extraterrestrial sulphur to the planet integrated over the first 600 My. Although significant when compared to the cumulative mass delivered to the Moon as provided in Hartmann et al. (2000), this amount is still $\leqslant 1\%$ of the total sulphur inventory for all sediments + seawater presently in the surface zone of the Earth (Holser et al., 1988). In terms of MIF sulphur, Farquhar et al. (2002) argued that bulk extracts of many different meteorite classes indicate that $\Delta^{33}S$ values are homogeneous and range between -0.01 and $+0.04‰$. Earth's mantle did not inherit S-isotope heterogeneities from its own formation. It appears that terrestrial sulphur was dominantly indigenous to the Earth since primary accretion ceased prior to ~4.4 Ga and that extraterrestrial matter was, overall, a relatively minor contributor to the global sulphur cycle to the early Earth. This is important, because mass-independently fractionated sulphur isotopes documented in the oldest Earth rocks and minerals can with reasonable confidence be assigned an origin from terrestrial atmospheric reactions.

7.5-4. SULPHUR AND THE HADEAN EARTH SURFACE STATE

No terrestrial rocks have yet been found that yield ages older than ca. 4.03 Ga (Bowring and Williams, 1999). What we do know of the geochemical evolution of Hadean Earth comes primarily from the study of pre-4.0 Gyr old Hadean detrital zircons shed in younger sediments captured in the Mount Narryer (Froude et al., 1983) and the Jack Hills (Compston and Pidgeon, 1986) sedimentary belts in Western Australia (see Cavosie et al., this volume).

7.5-4.1. Surface Environments in the Hadean

A subset of Hadean zircons, which are up to 4.38 Gyr old (Harrison et al., 2005, 2006) are enriched in heavy oxygen (Wilde et al., 2001; Mojzsis et al., 2001; Peck et al., 2001; Cavosie et al., 2005, 2006; Trail et al., 2007). This observation, coupled with characteristic trace element patterns (Maas and McCulloch, 1991; Maas et al., 1992; Peck et al., 2001; Crowley et al., 2005) has provided mutually consistent evidence that an evolved rock cycle was present within ~150 My of the formation of the Moon (cf. Whitehouse and Kamber, 2002; Nemchin et al., 2006; Coogan and Hinton, 2006). Detailed studies of Hadean zircons from $^{176}Hf/^{177}Hf$ systematics show that large volumes of continental crust were present throughout the Hadean (Amelin et al., 1999; Harrison et al., 2005, 2006; cf. Valley, 2006). Zircon thermometry based on Ti contents of individual pre-4.0 Ga Jack Hills grains also shows that formation temperatures cluster narrowly around 680 °C (Watson and Harrison, 2005). The simplest explanation for this result is that low-temperature wet minimum-melt conditions during granite genesis prevailed in Hadean melts that gave rise to the oldest zircons (Watson and Harrison, 2006). Studies of pre-4 Ga zircons now provide a rather firm foundation to arguments that water and a sustained atmosphere was present within the first 100 My of the formation of the solar system. What data for the earliest surface state might possibly be gleaned from sulphur isotopes?

7.5-4.2. Sulphide in Hadean Zircon

Some Hadean zircons contain inclusions of diverse minerals, but few studies have provided comprehensive results bearing on identified inclusions in these samples (Maas et al., 1992; Cavosie et al., 2004; Crowley et al., 2005). In fact, zircons with noticeable inclusions are often avoided during the selection process as ion probe spot analyses that overlap

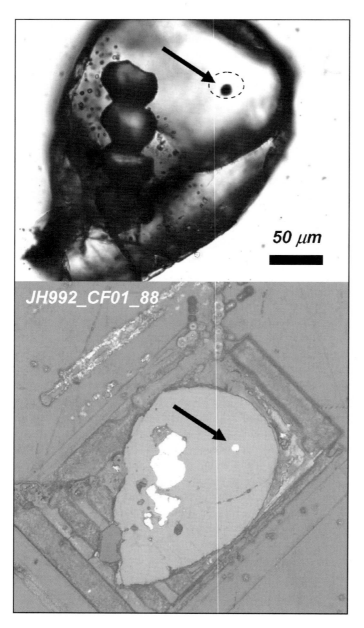

Fig. 7.5-7. Optical micrographs of Hadean zircon sample JH992_CF01_88 from Mojzsis et al. (2001). The location of a Ni-rich pyrite inclusion is shown by the arrow. The multicollector ion microprobe analysis area is indicated by the dashed oval (transmitted light micrograph at top, reflected light below).

inclusions may affect data (e.g., Amelin et al., 1999). Individual Hadean zircons have been shown to contain polyphase domains of albite, Ca-Al silicate, muscovite/quartz/Fe-oxide, abundant quartz, K-feldspar, and phosphates such as monazite, apatite and xenotime (Maas et al., 1992; Cavosie et al., 2004; Crowley et al., 2005). In these studies, it is important to distinguish between primary melt inclusions and secondary/exsolution products of zircon. Mineral inclusions encapsulated inside zircons are only of relevance to host rock characteristics if it is possible to show that they represent patches of melt now solid at normal surface temperatures and specifically of magmatic origin. Such inclusions are often ovoid and chemically incompatible with zircon. For example, in sample JH992_88 from Mojzsis et al. (2001), a small (<10 μm diameter) Ni-rich pyrite inclusion was identified near the centre of a 96% concordant 4.105 Ga zircon with igneous $[Th/U]_{Zr}$ values $= 0.52$ (Mojzsis and Harrison, 2002). This zircon has no visible alteration or cracks leading to or from the inclusion (Fig. 7.5-7). Zircon JH992_88 has $\delta^{18}O_{VSMOW}$ values from two separate analysis spots of $+5.9 \pm 0.7‰$ and $+6.3 \pm 0.4‰$ (reported in Mojzsis et al., 2001), which are slightly enriched in ^{18}O compared to average crust and mantle oxygen ($+5.5‰$; Eiler et al., 1997), but still within the field of I-type magmas (White and Chappell, 1977). The sulphide contained ~ 2 wt% Ni as determined by wavelength dispersive spectroscopy by electron microprobe, which along with oxygen values, is consistent with a mafic- (or ultramafic?) origin for zircon and its inclusion (e.g., Fleet, 1987). Three separate measurements of the multiple sulphur isotope composition of this sulphide were obtained following the techniques described in Mojzsis et al. (2003). The $\delta^{34}S$ values yield a range of -8.1 to $-10.1‰$ (1σ uncertainties were $\pm 0.17‰$); these values are lower in $^{34}S/^{32}S$ than average MORB and mantle (Chaussidon et al., 1987) but comparable to those reported for hydrothermally altered oceanic crust sulphides (Alt, 1995). Two of the three $\Delta^{33}S$ values measured for the inclusion are within 1σ error of the York-MDF band ($+0.5 \pm 0.5‰$ and $-0.9 \pm 1.2‰$), and the third has a slightly ^{33}S-depleted value of $-1.3 \pm 0.7‰$ (Table 7.5-2). The absence of resolvable $\Delta^{33}S$ in 2/3 of the measurements may in part be a consequence of the low S count rates for this minute sulphide inclusion. It is obvious that

Table 7.5-2. Sulphur isotope composition of sulphide inclusion in JH992_CF01_88 by in situ ion microprobe multicollection

spot	$^{34}S/^{32}S$	$\pm 1\sigma$	$^{33}S/^{32}S$	$\pm 1\sigma$		
CF01_88@1	4.4640E–02	7.89E–06	7.9399E–03	5.53E–06		
CF01_88@2	4.4549E–02	7.60E–06	7.9464E–03	3.94E–06		
CF01_88@3	4.4603E–02	7.09E–06	7.9399E–03	9.28E–06		
spot	$\delta^{34}S_{VCDT}$	$\pm 1\sigma$	$\delta^{33}S$	$\pm 1\sigma$	$\Delta^{33}S$	$\pm 1\sigma$
CF01_88@1	−8.11	0.18	−5.53	0.70	−1.35	0.71
CF01_88@2	−10.12	0.17	−4.71	0.50	0.50	0.52
CF01_88@3	−8.92	0.16	−5.53	1.17	−0.93	1.18

Notes: zircon $^{207}Pb/^{206}Pb$ age (1σ) $= 4105 \pm 11$ Ma; $[Th/U]_{Zr} = 0.52$; $\delta^{18}O_{Zr} = +5.9 \pm 0.7‰$ and $+6.3 \pm 0.4‰$ at two separate spots as reported in Mojzsis et al. (2001). Average ^{32}S count rates were: 3.62E7 to 4.23E7 cps.

further analyses of this kind are needed on larger sulphide inclusions in Hadean zircon when they can be found. Other sulphide inclusion studies in resistant mineral hosts such as in diamonds (Chaussidon et al., 1987) have been used to draw conclusions about the nature of sulphur in the mantle as well as crust of the Archaean. For instance, Farquhar et al. (2002) reported the presence of resolvable $\Delta^{33}S$ anomalies in Archaean eclogite-type diamonds from the Orapa kimberlite (Kaapvaal Craton, Botswana) and inferred from these results that atmospheric MIF sulphur was transferred to the mantle. If mass-independently fractionated sulphur isotopes are preserved in Hadean zircons, as hinted at by the single analysis that provided a slightly negative $\Delta^{33}S$ composition, it may be possible in the future with a greatly expanded data set to place some general constraints on the coupling between Hadean mantle, crust and atmosphere.

It is interesting to note that in the conceptual model of Farquhar et al. (2002) for the early Archaean sulphur cycle, sulphur dioxide expelled from volcanoes and transformed by ultraviolet radiation can be deposited to surface reservoirs as aerosols and ultimately recycled into the mantle. In their schema, sulphate in oceanic crust accumulates uniformly

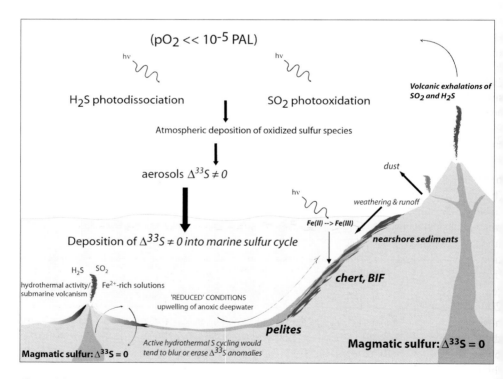

Fig. 7.5-8. A schematic model for sulphur cycles on the early Earth (from Mojzsis et al., 2003). The figure represents a passive margin environment, for a model of the possible contribution of mass-independent sulphur to subduction products on the early Earth see Farquhar et al. (2002).

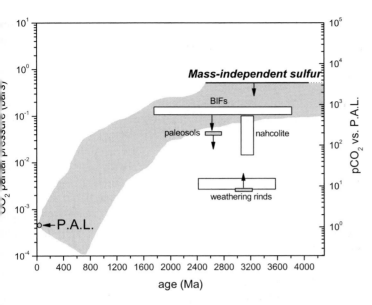

ig. 7.5-9. Changing CO_2 levels in the Earth's atmosphere over time expressed as both partial pres-
ure of CO_2 in bars (left), and against present atmospheric level (PAL; right). The grey shaded region
; the range of concentrations required to compensate for the fainter young Sun (Kasting, 1993).
)ther data points shown are reviewed in (Rollinson, 2007). The multiple sulphur isotopes provide a
trong upper constraint on the limit of CO_2 partial pressure to below 0.5 bar as far back as 3.83 Ga,
nd perhaps before 4.0 Ga (Cates and Mojzsis, 2007b). The same may be said for partial pressures
f methane, which had to be below the threshold necessary for haze formation (Pavlov et al., 2001a,
001b).

egative $\Delta^{33}S$ values (Farquhar and Wing, 2003) which can become reduced to sulfide
biologically or thermo-chemically), subducted, and recycled into arc magmas (Fig. 7.5-8).
   A key feature of these arguments is that the atmosphere must have been transparent to
iltraviolet and that an ozone UV shield was not present for MIF sulphur isotopes to be
mparted to crustal rocks (Farquhar and Wing, 2003). This is because ultraviolet absorp-
ion by the Rayleigh scattering tail of the CO_2 absorption spectrum at partial pressures
hat exceed $\sim$0.5 bar would have blocked UV sufficiently (Shemansky, 1972) to stop the
eactions responsible for mass-independent sulphur isotope fractionations (Farquhar et al.,
:001). Under these conditions, and with most solar models which show that solar lumi-
iosity was $\sim$70% of present day values, other greenhouse gases were required to keep
he Hadean Earth from freezing over (Sagan and Mullen, 1972). Abundant methane is the
•rime candidate for a greenhouse gas that supplemented water vapor and CO_2 throughout
he Hadean-Archaean to have kept the Earth above the freezing point of water (Kasting,
:005). Sagan and Chyba (1997) and Pavlov et al. (2001a, 2001b) suggested that pho-
olysis of methane in the mesosphere by wavelengths shorter than 145 nm could have
;enerated an organic haze that served as an UV shield early in Earth's history to stabilise

ammonia. If, however, short wavelength UV sulphur dioxide photolysis accounts for the Archaean mass-independently fractionated sulphur isotopes, data are inconsistent with the presence of a high-altitude UV shield, whether from ozone, dense CO_2 or an organic haze (Fig. 7.5-9).

It may also be that the atmosphere was mostly transparent to UV for much of the Hadean and the planet was cool (Zahnle, 2006), although the lack of evidence for glaciations older than $\sim$2.9 Ga (Young et al., 1998) apparently weakens this hypothesis.

7.5-5. SULPHUR AND THE EARLY ARCHAEAN ATMOSPHERE

Quantitative constraints on the composition of the Archaean-Proterozoic atmosphere have been attempted for many years based on investigations of redox-sensitive minerals in the geologic record used as input parameters for model calculations. The prevailing view has been that free oxygen (O_2) levels were substantially lower than present values, with estimates of $pO_2 \ll 10^{-2}$ present atmospheric level (PAL) before about 2.3 Ga (e.g., Cloud, 1968, 1972; Walker, 1990; Holland, 1984, 1994; Kasting, 1993; Rye et al., 1997; Rye and Holland, 1998; Farquhar et al., 2000a; Pavlov and Kasting, 2002). Alternatively, some have argued for an O_2-rich surface zone, or at least the condition that the atmosphere was not anoxic, for much or all of the Archaean (Dimroth and Kimberley, 1976; Towe, 1990; Ohmoto, 1997; Watanabe et al., 1997). Regardless of these different arguments for an oxic Archaean, it has not been specified whether these models would also apply to the Hadean Earth. Several reviews of the geologic evidence used to support the contention that the Archaean was anoxic by Holland (1984, 1994) noted detrital uraninite and pyrite grains in pre-2.3 Ga sediments as well as other geochemical data (e.g., Fe(II)/Fe(III) in palaeosols) as indicative of global anoxia at time of deposition (Rye and Holland, 1998; Rasmussen and Buick, 1999).

As outlined in Mojzsis et al. (2003), the extensive production of mass-independently fractionated S isotopes formed on Earth in gas-phase reactions in UV transparent anoxic atmospheres, and preserved in terrestrial rocks and minerals in the absence of enough MDF oceanic sulphate to dilute mass-independently fractionated sulphur products to the surface zone, potentially provides three fundamental constraints on the nature of the early Earth's atmosphere:

(1) Separate mass-dependent arrays of slope $m \approx 0.5$–0.52 within MIF δ^{33}S vs δ^{34}S data from ancient sulphur minerals show that effective oxidation of sulphur in the surface zone and homogenization of reduced and oxidised sulphur reservoirs by biological cycling was either absent or suppressed. In a model study, Pavlov and Kasting (2002) argued that in addition to the palaeosol and detrital mineral data cited above, pO_2 levels had to be below $\sim$$10^{-6}$ PAL throughout the Archaean. A conclusion of this work, and separate experimental studies of microbial sulphate reduction capabilities of organisms at low seawater SO_4^{2-} concentrations (Habicht et al., 2002), is that sulphate was severely limited in the oceans for microbial sulphate reduction before the rise of atmospheric oxygen in the Proterozoic (Papineau et al., 2005, 2007; Papineau

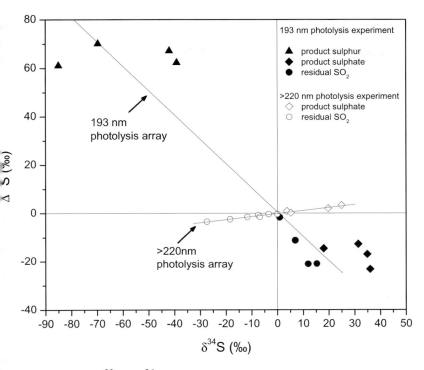

Fig. 7.5-10. Plot of $\Delta^{33}S$ vs $\delta^{34}S$ for closed cell photolysis experiments of SO_2 gas mixtures irradiated at 193 nm (from Farquhar and Wing, 2003; data in Farquhar et al., 2001).

and Mojzsis, 2006). This observation is important because in standard $^{34}S/^{32}S$ isotopic analyses of ca. 3.5 Ga barite deposits from the North Pole (Warrawoona Group, Pilbara Craton, Western Australia), Shen et al. (2001) concluded that a 20‰ range in $\delta^{34}S$ values was evidence for effective microbial sulphate reduction by 3.5 Ga. As noted previously, photolysis reactions that involve $SO_2 + SO$ and other gases (e.g., H_2S, CS_2) in anoxic atmospheres are known to produce large MIF effects (Fig. 7.5-10) with a range in $^{34}S/^{32}S$ of product sulphur that can mimic the entire range of biological fractionations of the sulphur isotopes. This observation underscores the now critical importance of multiple isotope measurements in all subsequent sulphur studies of ancient rocks, and warrants a comprehensive re-visit of pre-2.3 Gyr old terranes previously sampled for sulphur (reviewed in Strauss (1997, 2003)).

2) Experiments to delimit the range of conditions necessary for UV light-driven photolysis reactions with sulphur gases deep in the atmosphere, and to investigate the magnitude of isotopic fractionations in the mass-independently fractionated products (reviewed in Thiemens (1999) and Farquhar et al. (2001)), have succeeded in reproducing the large range of MIF observed in Archaean sulphides. A consensus view has emerged that column abundances of O_2 and O_3 before ~2.3 Ga were very low if, for

example, 190 nm UV and other wavelengths could penetrate deep into the atmosphere to result in widespread MIF. Farquhar et al. (2000a, 2001) showed that this result can be used to place an upper limit on pO_2 in the Archaean of $<10^{-2}$ PAL and an upper limit on pCO_2 of ~ 0.5 bar (1.5×10^3 PAL). Methane may have been present, but did not produce aerosols to attenuate UV (Fig. 7.5-9).

(3) In Fig. 7.5-6 the principal laboratory reaction that has been documented to be most consistent with the data thus far obtained for MIF S in the geologic record involves SO_2 photolysis at 193 nm UV (Farquhar and Wing, 2003). Pavlov and Kasting, 2002 used one-dimensional photochemical models of atmospheres at various O_2 concentrations and found that the transfer of MIF S to the surface by aerosol deposition is only possible in anoxic conditions of the early Earth (e.g., Kasting et al., 1989, 1992). The proposed mechanism is that MIF in sulphur aerosols are preserved from homogenization in oceanic sulphate reservoirs only in atmospheres where $pO_2 \ll 10^{-6}$ PAL. Pavlov and Kasting (2002) concluded that it is not possible to preserve MIF $\delta^{33}S/\delta^{34}S$ in high-O_2 atmospheres; in these models, the demise of MIF S in the geologic record is reached even before the $pO_2 = 10^{-6}$ PAL threshold was overtaken in the Paleoproterozoic (Farquhar et al., 2000a; Mojzsis et al., 2003; Ono et al., 2003; Bekker et al., 2004), sometime between 2.35 and 2.42 Ga during glaciations (Papineau et al., 2006, 2007).

As ever larger data sets become generated for multiple S isotopes in early Archaean rocks, results provide an increasingly robust indication that a chemically reactive, UV transparent, anoxic atmosphere existed on Earth prior to about 2.35 Ga. This anoxic regime stretched as far back as ~ 3.83 Ga, and perhaps well into the Hadean. The diverse $\Delta^{33}S$ values for sulphides reported from Archaean and early Proterozoic rocks means that up to several percent of the marine sulphur cycle was affected by atmospheric deposition of MIF S aerosols following residence in the atmosphere.

A consequence of planetary-scale deposition of sulphur aerosols (Fig. 7.5-8) is that they could have been tapped as a food source for early microbial communities that metabolise elemental sulphur and sulphate.

7.5-6. SULPHUR AND EARLY LIFE

Sulphur has played a central role in biological systems since the origin of life and has followed the genomic and metabolic diversification of organisms since their emergence some 4 b.y. ago (Fraústo da Silva and Williams, 2001; DeDuve, 2005). A landmark in studies of the evolutionary relationships of organisms was established with the comparative sequence analysis of genomes used to infer phylogenetic relationships (Woese and Fox, 1977; Woese, 2000). It is now widely held that the evolutionary history of life is recapitulated by its molecular biochemical inheritance (Pace, 1997, 2001), and that all life can be classified into three Domains within the small subunit ribosomal RNA Phylogenetic Tree: Bacteria, Archaea and Eukarya. In this context, a major objective to analyse the sulphur isotope compositions of Archaean sediments has been to identify and trace the long-term evolution of microbial metabolisms (e.g., Monster et al., 1979; Schidlowski et al., 1983), and lately, to be able to relate the divergence of different sulphur metabolisms to events in the geologic and paleontological record (Canfield, 2004). This rationale is based on several key observations: (i) some of the most deeply branching lineages in the universal phylogenetic tree that reside close to the 'Root' population which gave rise to all extant life, can metabolize sulphur compounds; (ii) significant sulphur isotopic fractionations can result from enzymatic processing of S-containing molecules; and (iii) sulphides preserved in the geological record have a large range of isotopic compositions, and data show that this range has increased in time (reviewed in Canfield and Raiswell (1999)).

7.5-6.1. *Sulphur Chemistry and the Origin of Life*

Reduced sulphur as H_2S, and its organic derivatives in the thiols ($R'-SH$) and the thioesters $(CO-S)$, can form organo-sulphur compounds from the dehydration condensation of carboxylic acid ($R'-COOH$) and thiol in the simple reaction:

$$R'\overset{O}{\underset{OH}{||}} \quad + \quad R'-SH \quad \longrightarrow \quad R'\overset{O}{\underset{SR'}{||}} \quad + \quad OH_2$$

These compounds were likely participants in the most basic proto-metabolic processes in the origin of life. Furthermore, no discussion of sulphur biochemistry is complete without taking into account the rich chemistry of metal-sulphur compounds (Fraústo da Silva and Williams, 2001). In all contemporary organisms, transition metal sulphide clusters play a central catalytic role in biological energy conversion systems (Beinert, 2000) and have been important since the establishment of the earliest metabolic cycles (Rees and Howard, 2003). It is the connection between the essential role of thioesters and the predominance of mineral sulphides in hydrothermal and volcanic vents that has led a number of workers to speculate how life could have emerged in such environments (e.g., Russell and Hall, 1997; Russell and Martin, 2004). Hydrothermal cells form wherever there is hot rock in contact with water, and such environments may have been widespread in the Hadean crust. It is thus highly likely – because of the ubiquity of hydrothermal environments away from a surface zone affected by impacts and intense UV – that the deep biosphere was amongst the earliest on Earth, if not the first (Russell and Arndt, 2005). As a basis for an iron-sulphide theory for the origin of life, Wächtershäuser (1988, 1992) proposed a 'fast origin' by an autotrophic proto-metabolism of low-molecular weight constituents (Fig. 7.5-11) in hydrothermal environments. Along these lines, Cody et al. (2000) presented experimental support for the formation of pyruvic acid ($CH_3-CO-COOH$) from formic acid ($HCOOH$) in the presence of organo-metallic sulphur compounds such as nonylmercaptane ($Fe_2(CH_3S)_2-(CO)_6$), and iron sulphide (FeS_2) at high temperatures and pressures consistent with shallow oceanic crust near volcanic centres (250 °C and 200 MPa). Mixture of these reaction products with lower temperature fluids that emanated further away from vent centres, such as is found at the crust-ocean interface, would have been more

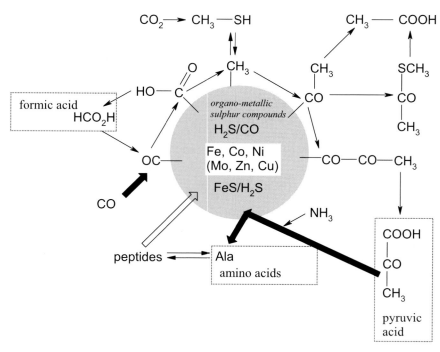

Fig. 7.5-11. Basic proto-biotic reactions that could have taken place in the 'Iron-Sulphur World' (modified after Wächtershäuser (2000)).

conducive to the survivability of other important biomolecules such as amino acids and nucleosides (Orgel, 1998). In a separate analogy with the H_2S/CO system, the formation of FeS_2 in the simple reaction:

$$FeS\,(\text{pyrrhotite}) + H_2S \rightarrow FeS_2\,(\text{pyrite}) + H_2$$

produces molecular hydrogen. The FeS/H_2S couple is a strong reductant that could have played an important supporting role in prebiotic organic synthesis under crustal conditions common to the Hadean Earth.

    Therefore, a self-consistent picture has emerged that hydrothermal zones in shallow Hadean oceanic crust were, if not the birthplace of life on Earth as advocated by Russell and Arndt (2006) and Russell et al. (2005), at least were the crucible wherein early biological forms found a ready source of fluids rich in reduced carbon and more complex organic molecules (Shock and Schulte, 1998). In this scenario, electron acceptors such as S^0 either from recycled sulphur in oceanic crust, from sulphidic hydrothermal fluids, or transferred from the water column to the crustal plumbing system (Fig. 7.5-1) to be catalysed by carbonylated iron-sulphur clusters (Cody et al., 2000), could have provided both a kick-start to, and a rich food source for, the Hadean biosphere.

    A primordial repose for life in deep hydrothermal vent systems that metabolized reduced sulphur compounds is further supported by molecular phylogenetic studies. Bio-

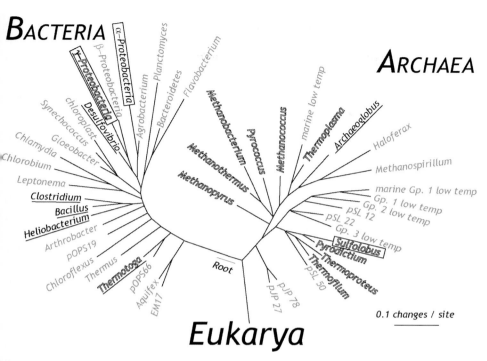

BACTERIA

ARCHAEA

Eukarya

0.1 changes / site

Fig. 7.5-12. Phylogenetic tree based on ssu 16s rRNA comparisons which shows Bacterial and Archaeal lineages of extant organisms capable of elemental sulphur reduction (framed characters), sulphate reduction (underlined lineages) and sulphide oxidation (boxed lineages). This figure is modified from Pace (1997) with information from Klein et al. (2001), Canfield and Raiswell (1999), Stetter and Gaag (1983).

chemical evidence shows that some of the earliest biomes were located in deep hydrothermal environments. A crucial caveat to these studies is that before we actually know how life on Earth began there is no definite requirement that life started in a mild-, medium- or hot-hydrothermal vent field, even if it is a compelling and convenient place for the origin (Nisbet and Sleep, 2001). Sulphate and other oxidised S species figure prominently as electron acceptors for those deeply branching microbes that perform microbial sulphate reduction, yet were probably of limited use in the absence of significant sulphate before the advent of on oxygenated world in the Proterozoic. Instead, the ancient metabolic cycles of life that arose out of the metal-S chemistry described above in all probability metabolized volcanogenic sulphur gases such as H_2S, SO_2 and sulphur compounds including elemental S.

7.5-6.2. *Sulphur Metabolic Styles of Early Life*

Many hyperthermophilic, deep-branching organisms in the domains Bacteria (*Thermotogales, Proteobacter*) and in various Archaea (sub-Domain Crenarcheota and *Pyrococcus*,

Thermoplasma in sub-Domain Euryarcheota) are known to reduce elemental sulphur to H_2S with both H_2 and organic compounds as an electron donor (Fig. 7.5-12). This pathway could have evolved from proto-metabolic metal-sulphur cycles, and presently involves a short electron transport chain controlled by the activity of key enzymes including sulphur reductase, polysulphide reductase and hydrogenase (Laska et al., 2003). The antiquity of this metabolism is further evident in that both deep-branching bacterial and archaeal lineages share this metabolic style (Stetter and Gaag, 1983; Stetter, 1996; Canfield and Raiswell, 1999), an observation that strongly suggests that elemental sulphur reduction (ESR) might be an inherited trait from one of the earliest biomes that resided in a high-temperature environment (Corliss, 1990).

Supposing that ESR is truly ancient, it still remains unknown whether the last common ancestral population at the base of the phylogenetic tree (labelled 'Root' in Fig. 7.5-12) included the genes for this particular metabolic style. Elemental sulphur reduction could very well have evolved later in one of the domains and subsequent lateral gene transfer occurred between domains. The lateral-transfer argument holds for all the discussions on the early rise of metabolisms interpreted from extant phylogenies if they cannot be specifically tied to molecular or isotopic biomarkers in the geologic record. It is anticipated that future phylogenetic comparisons with amino acid sequences of specific enzymes will better constrain the branching sequence of lineages. If a large number of lineages that perform a specific metabolic pathway are known from both the bacterial and archaeal domains, evidence would seem to favour a deeply ancient origin. A critical clue may be that no known Eukaryal organism has this metabolic style. Anaerobic respiration with elemental sulphur is performed by several bacteria and archaea, but has only been investigated in a few organisms in detail (Hedderich et al., 1998). As was noted by Canfield and Raiswell (1999), organisms that do ESR can readily obtain elemental sulphur from rapid cooling of volcanogenic sulphur gases where:

$$SO_2 + 2H_2S \leftrightarrow 3S^0 + 2H_2O.$$

This reaction shifts the equilibrium in favor of S^0 (Grinenko and Thode, 1970). For example, with S^0 present, microbes can perform the reaction:

$$4S^0 + 4H_2O \rightarrow 3H_2S + SO_4^{2-} + 2H^+$$

and in the presence of iron hydroxides, which were abundant in the oceans before 1.9 Ga (see Holland, 1984; Bjerrum and Canfield, 2002; Thamdrup et al., 1993):

$$H_2S + 4H^+ + 2Fe(OH)_3 \rightarrow 2Fe^{2+} + S^0 + 6H_2O$$

As implied by Fig. 7.5-11, the rich chemistry at the interface of metal-sulphur and biochemistry is not restricted to iron, even if that particular branch of inorganic biochemistry has always been dominated by Fe-S (Daniel and Danson, 1995). Many metals which happen to be abundant in most mafic and ultramafic rocks readily interact with sulphur, such as Zn, Mo, Co and Ni. Average MORB contains 79 ppm Zn, 0.46 ppm Mo, 47 ppm Co and 150 ppm Ni (Hertogen et al., 1980; Newsom et al., 1986; Hofmann, 1988). Transition metal-sulphur chemistry figures prominently in the biochemistry of many deeply-branching microbes (reviewed in Konhauser, 2007). Most of the various catalytic roles of metal ion + S clusters other than Fe await experimental verification from analysis of controlled (culture) experiments and such experiments represent an important new avenue for research into the question of a biological origin of sulphur isotope fractionations in the geologic record.

Microbial sulphate reduction (MSR) is an anaerobic metabolic pathway in which SO_4^{2-} is used as an electron acceptor to be reduced to H_2S with electron donors such as H_2 and/or a variety of organic molecules to produce hydrogen sulphide in the general form:

$$SO_4^{2-} + 5H_2 \rightarrow H_2S + 4H_2O$$

Sulphate reduction in bacteria and archaea can proceed via dissimilatory or assimilatory pathways that either excrete H_2S or insert it in organic compounds. A few key enzymes and compounds in MSR are illustrated in Fig. 7.5-13. The timing of evolution of these genes is not accurately known, but qualitative information can be obtained from molecular phylogeny of sulphate reducers.

7.5-6.3. *Biological Sulphur Isotopic Fractionations*

Sulphate-reducers preferentially metabolize $^{32}SO_4^{2-}$ rather than $^{34}SO_4^{2-}$ to produce isotopically light H_2S with a range of $\delta^{34}S$ values between -4 and $-46\permil$ under non-limiting concentrations of SO_4^{2-} (Canfield and Raiswell, 1999, and references therein). Microbially-produced H_2S can react with dissolved Fe^{2+} or other metal cations to precipitate sulphides in the simple reaction:

$$2H_2S + Fe^{2+} \rightarrow FeS_2 + 2H_2$$

Diagenetic sulphides formed from H_2S and FeS (or other metal-sulfide) reactions carry the combined mass-balanced sulphur isotopic composition without significant isotopic fractionation (Butler et al., 2004; Böttcher et al., 1998). These isotope ratios can then become incorporated and preserved in the geological record of authigenic sulphides. The sulphur isotopic composition of sedimentary sulphides can therefore be used to pinpoint the isotopic fractionation between dissolved SO_4^{2-}, and product H_2S as an isotopic biosignature to trace the early evolution of microbial sulphur metabolisms.

It has been experimentally verified that microbial sulphate reduction under limiting SO_4^{2-} concentrations (i.e., lower than 200 µM) has insignificant effect on the $\delta^{34}S$ value of product H_2S (Habicht et al., 2002). If the concentration of SO_4^{2-} in the environment is low, microbial sulphate reduction produces H_2S (and thence sulphide) with approximately the same $\delta^{34}S$ value as the available $\delta^{34}S_{sulphate}$. The $\delta^{34}S$ of sulphides from ancient sediments can also be used to trace the concentration of dissolved SO_4^{2-} in seawater at the time of sedimentation. Sedimentary sulphides with negative $\delta^{34}S$ values (i.e., $<-10\permil$) are consistent with microbial sulphate reduction in high seawater SO_4^{2-} concentrations. Variability in sulphur isotopic fractionations during MSR can also depend strongly on the

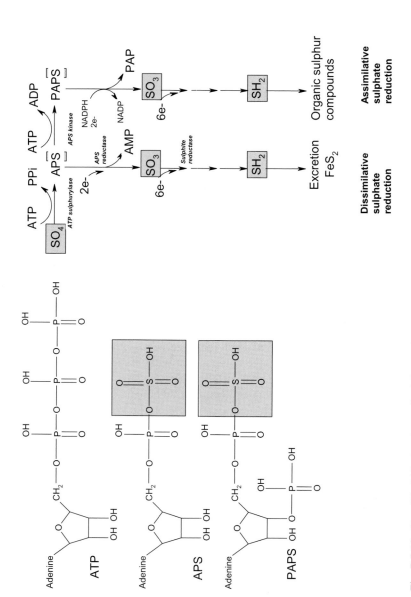

Fig. 7.5-13. Simplified biochemistry of sulphate reduction showing (left) key intermediate compounds produced during sulphate reduction and (at right) schemes of assimilative and dissimilative sulphate reduction (modified from Madigan et al. (2000)).

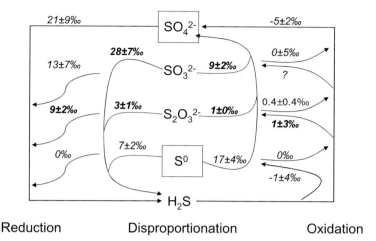

Reduction   Disproportionation   Oxidation

Fig. 7.5-14. Isotopic fractionation in the modern sulfur cycle showing the values for average sulfur fractionations (± standard deviations) associated either with reduction or oxidation. Each disproportionation reaction couples the production of both H_2S and SO_4^{2-} (from Habicht et al., 1998).

sources of organic compounds consumed during growth (Kleikemper et al., 2004) and on the different lineages of sulphate reducers (Detmers et al., 2001). Some organisms can also disproportionate thiosulphate ($S_2O_3^{2-}$), SO_3^{2-} and/or S^0, and the disproportionation of sulphur-containing compounds produces different molar amounts of both H_2S and SO_4^{2-} with isotopic fractionations characterized by enrichments of ^{34}S in SO_4^{2-} and depletions of ^{34}S in H_2S (Böttcher, 2001; Habicht et al., 2002). Natural microbial communities are known to produce sulphide significantly more depleted in ^{34}S than is expected from microbial sulphate reduction alone. The disproportionation of intermediate sulphur-compounds metabolized by sulphate reducers can recycle sulphur through yet greater degrees of isotopic fractionation and this has been used to explain the observed large range of $\delta^{34}S$ in modern sediments (Habicht and Canfield, 2001; Canfield and Thamdrup, 1994). A schematic representation of the different sulphur isotopic fractionations is shown in Fig. 7.5-14. Sulphur oxidation does not significantly fractionate $\delta^{34}S$ values but returns oxidized sulphur compounds to the environment, which in the context of a hierarchical microbial community can then be recycled as electron acceptors for microbial sulphate reduction or disproportionation reactions.

7.5-6.4. Non-Authigenic Sulphur in Sediments

Because sulphur is so widely distributed in the Earth and occurs in oxidized forms (sulphate), neutral state (elemental S^0) and reduced forms (sulphides), it readily travels throughout the rock cycle. Criteria used to distinguish sedimentary from hydrothermal sulphides include the crystal habit, sulphur isotopic composition, and sometimes an unusual abundance of trace metals. Detrital sulphides, which can sometimes be identified from their rounded appearance and occurrence with other characteristic detrital minerals, are older than the depositional age of the sediment. Sulphide formation can postdate the origin of host sedimentary rocks if they formed from the crystallization of remobilized sulphur from post-depositional circulating fluids. Hydrothermal fluids can assimilate sulphur from host sedimentary rocks during water-rock interactions and therefore carry the $\delta^{34}S$ value of the host rock. Pyrite in sediments may have the isotopic signature of microbial sulphate reduction, and subsequent hydrothermal circulation can transport Cu that will react with pyrite to form chalcopyrite with the same $\delta^{34}S$ as the precursor pyrite (Ohmoto and Gold-haber, 1997; Papineau and Mojzsis, 2006). High concentrations of trace metals such as Co, Zn and Ni in sulphides may be used as an indicator of diagenetic conditions or interaction with magmatic or other fluids.

The $\delta^{34}S$ value of sulphur in igneous rocks from the upper mantle is in the range of $-0.7 \pm 5.0\%_0$ (Chaussidon and Lorand, 1990) and more generally is in the range of -3.0 to $+3.0\%_0$ (Ohmoto, 1986). Sulphur isotopes of Archaean magmas (Hattori and Cameron, 1986) have mass-dependently fractionated $\Delta^{33}S$ values between -0.3 and $+0.3\%_0$, unless the magma originated primarily from recycled Archaean sediments with a nonzero $\Delta^{33}S$ and the MIF sulphur isotopes were isolated from dilution by mantle sulphur (e.g., Farquhar et al., 2002; Mojzsis et al., 2003). Metamorphic effects only slightly fractionate sulphur isotopes, and these minor effects are mass-dependent. Metamorphism can potentially modify the $\Delta^{33}S$ value of sedimentary sulphides but it is thought that this can only be achieved by interaction with sulphur-containing metamorphic fluids such that a pre-existing MIF signature is either diluted or remobilized by the fluid. At 900–950 °C, the self-diffusion of S in sulphides (pyrrhotite) is slow, ($\log D = 10^{-16} \, m^2 \, s^{-1}$; Condit et al., 1974), which will tend to preserve mass-independent sulphur isotope signatures even in metamorphic sulphides, but an important caveat to this is that it must otherwise occur in the absence of massive S-rich fluid infiltration and replacement (Papineau and Mojzsis, 2006).

Because of these competing factors, it must be noted that near-zero $\Delta^{33}S$ values for Archaean sedimentary rocks may not necessarily imply high levels of atmospheric O_2 at time of deposition. Mass-independently fractionated sulphur isotopes can be diluted by mass-dependently fractionated sulphur, especially if sulphate production is enhanced in local Archaean environments (Huston et al., 2001a). This point probably explains the observation that some sulphides with nonzero $\Delta^{33}S$ values from Archaean sediments co-exist in samples that also have mass-dependently fractionates sulphur isotopes (Mojzsis et al., 2003; Hu et al, 2003; Farquhar et al., 2000a; Whitehouse et al., 2005). The presence of a mass-independently fractionated sulphur isotopic signature is consistent with sedimentary sulphur that cycled through the atmosphere. However, to propose that mass-independent sulphur isotopes were absent at time of deposition requires other lines of evidence such as $\delta^{34}S$, sulphide composition, morphology and petrogenesis to firmly assign a sedimentary source of sulphur. In sum, without more data to support a sedimentary source, the near-zero $\Delta^{33}S$ values of these samples cannot be used to constrain atmospheric O_2 at time of deposition.

The crystal habit of the analysed sulphide crystals from metasedimentary rocks can vary from anhedral to euhedral and may be related to degree of recrystallization related to meta-

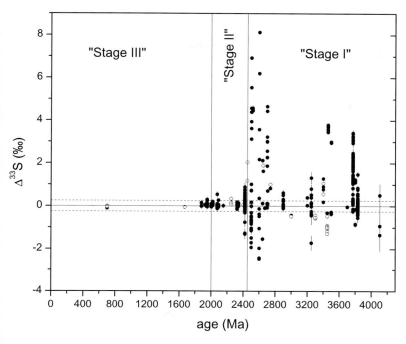

Fig. 7.5-15. Sulphur multiple isotope data (expressed as Δ^{33}S) for sedimentary sulphides and sulphates as a function of age (replotted after Farquhar et al. (2000); see also Seal (2006)). A large range in Δ^{33}S values for samples older than 2.4 Ga most likely reflects a change in atmospheric oxygen concentration past the $pO_2 = 10^{-6}$ PAL threshold (Pavlov and Kasting, 2002). Filled symbols are data from sulphides, open symbols are sulphates.

morphism. All Archaean sedimentary rocks experienced some degree of metamorphism, which can significantly modify the original sulphide morphologies. However, analysis of visible light and electron micrographs of sulphides thus far analysed in situ have failed to show any obvious correlations between sulphur isotopic composition and grain shape and/or size (e.g., Papineau and Mojzsis, 2006). Papineau et al. (2005, 2007) have argued that the habit of a sulphide grain is a weak criterion to assess the sedimentary origin of the sulphur, unless specific features can be identified, such as sulphide nodules. Metamorphic recrystallization is probably responsible for most of the large euhedral crystals observed in samples of various ages. Visible light, as well as electron micrograph images have so far failed to show clear evidence for sulphide overgrowths on pre-existing sulphide cores in the hundreds of analysed sulphides from various Archaean localities worldwide.

7.5-6.5. *Tracking the Antiquity of Sulphur Metabolisms*

A consequence of the model for anomalous Δ^{33}S sulphur aerosol formation is that pyrite, formed via elemental sulphur reduction by organisms to H_2S and reaction with Fe(II),

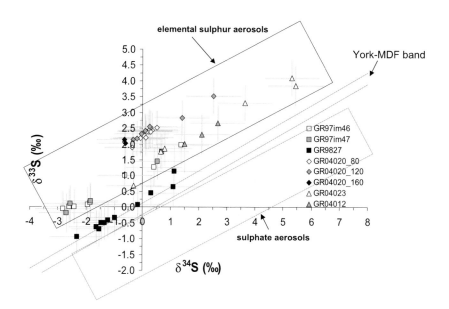

Fig. 7.5-16. Three-isotope plot of sulphur isotope data from sulphides analysed in Isua supracrustal belt metasediments showing the MDF bands (grey lines) reported in Papineau and Mojzsis (2006). The separate MIF arrays in the data shows that effective homogenization of the S reservoirs by mixture with sulphate was suppressed in the anoxic Eoarchean oceans. Mojzsis et al. (2003) noted that the paucity of negative $\Delta^{33}S$ in early Archaean rocks argues against widespread MSR at that time and could provide crucial clues about BIF origins.

would retain positive $\Delta^{33}S$ signatures, while pyrite from sulphate reduction would be expected to preserve negative $\Delta^{33}S$ values. The latter case may be seen in published $\Delta^{33}S$ data for sedimentary rocks $\leqslant 2.5$ Ga where pyrites preserve both negative and positive mass-dependent fractionations (Fig. 7.5-15). In the majority of Archaean samples for which multiple sulphur isotope data are available, most sulphides have strongly positive $\Delta^{33}S$ values. Mojzsis et al. (2003) and Papineau and Mojzsis (2006) interpreted these data to be inconsistent with sulphate reduction and more in line with elemental sulphur reduction. The presence of such positive $\Delta^{33}S$ values provides a strong case that the source of sulphur in these systems was cycled through the atmosphere and introduced to the oceans in the early Archaean, and separate linear mass-dependent arrays in $\delta^{33}S/\delta^{34}S$ space show that the sulphur can be subsequently isolated from re-homogenisation with volcanic and hydrothermal sources (Fig. 7.5-16).

Elemental sulphur (S^0, S_8) aerosols derived from mass-independent isotope fractionations from atmospheric reactions and implanted into the oceans would have provided a ready source of sulphur to a plethora of environments at the global scale from the Hadean onwards. In the absence of oxidative weathering of surface pyrite or other ready sources of sulphate, anoxic waters contain no more than trace sulphate over $HS^- + H_2S$ (Thorstenson,

1970). Although far from a consensus view, the current paradigm holds that the Archaean oceans were generally poor in sulphate relative to present values (Holland, 1984) at least before the pervasive oxygenation of the surface environment (Papineau et al., 2007).

7.5-7. SULPHUR ISOTOPES IN EARLY ARCHAEAN ROCKS

Sulphides are often preserved in ancient marine sediments but sedimentary sulphate minerals have yet to be reported in rocks older than about 3.5 Ga (Huston and Logan, 2004). The oldest recorded evidence of sedimentary sulphate occurs as ~3.4–3.5 Ga barite from Western Australia, South Africa and India (reviewed in Van Kranendonk (2006)). It is because of the paucity of early Archaean multiple sulphur isotope data that the sulphur isotopic composition of seawater sulphate during the 3.5–3.2 Ga interval is not well established, and not established at all for before 3.5 Ga. However, it is not unreasonable to suppose that seawater $\delta^{34}S_{sulphate}$ at the time of deposition of Isua sedimentary protoliths at ca. 3.8 Ga (see below) was not significantly different from the 3.4–3.5 Ga seawater $\delta^{34}S_{sulphate}$ of between +2.7 and +8.7‰ (Strauss, 2003).

As opposed to the conventional $^{34}S/^{32}S$ measurements, an increasingly large body of multiple sulphur isotope data has been collected in the last few years for several of the key metasedimentary units in Western Australia (ca. 3.5 Ga) and West Greenland (ca. 3.8 Ga). These studies have brought to light several important features of the early Archaean surface state.

7.5-7.1. Sulphides and Sulphates from the (ca. 3.49 Ga) Dresser Formation, Warrawoona Group, Western Australia

Microscopic sulphide (pyrite) grains in ca. 3.49 Ga barite beds from the Dresser Formation, Warrawoona Group at North Pole (Western Australia) have $\delta^{34}S$ values up to 24‰ lighter than co-existing sulphate (barite) from the same unit. These data were interpreted by Shen et al. (2001) as evidence of microbial sulphate reduction in the Paleoarchaean, an interpretation that could also be viewed consistent with phylogenetic arguments that suggest an early evolution of microbial sulphate reduction (Fig. 7.5-12). However, to assign an actual time of relative phylogenetic divergence episodes on the RNA phylogenetic tree (Pace, 2001) should be attempted with extreme caution because changes per nucleotide/time have not yet been established with confidence. Smaller ranges of $\delta^{34}S$ values in pyrite from Mesoarchaean marine metasedimentary rocks of between −2.5 and +8.5‰ have been put forth as evidence for sulphate-rich Archaean oceans with concentrations >1 mM that existed under $\gg 10^{-13}$ PAL O_2 (Ohmoto and Felder, 1987; Ohmoto, 1997, 1999; Ohmoto et al., 1993). Such concentrations are about 5 times higher than what is expected from experiments that show similarly small sulphur isotopic fractionations when microbial sulphate reduction occurs at seawater sulphate concentrations of less than 0.2 mM (Habicht et al., 2002). If the Archaean atmosphere was oxygen-rich and stabilised sufficient sulphate in seawater to form the large barite deposits, it requires distinctive interpretations of

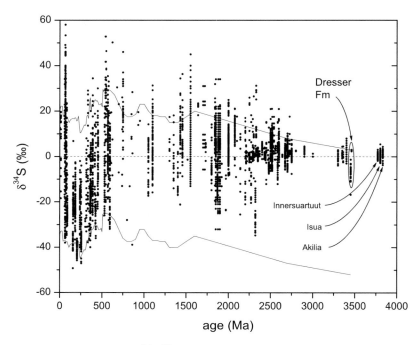

Fig. 7.5-17. Evolution of $^{34}S/^{32}S$ sulphur isotope ratios in the geological record. The two upper lines represent the band of $\delta^{34}S_{sulfate}$ values and the diamonds are compiled published $\delta^{34}S_{sulfide}$ data for sedimentary rocks. Before $\sim$1.7 Ga, constraints on $\delta^{34}S_{sulfate}$ are scarce and the lower line is a 55‰ fractionation from contemporaneous seawater sulphate (from Shen et al., 2001).

geologic data and of thermodynamic arguments for oxidative weathering of siderite and detrital heavy minerals (Ohmoto, 1999; cf. Rasmussen et al., 1999), a different explanation of BIF genesis, and alternative interpretations of $\delta^{13}C_{org}$, $\delta^{13}C_{carb}$ and $\delta^{34}S_{sulphide}$ data for Archaean sedimentary rocks (Ohmoto, 1997). The proposed evolution for seawater $\delta^{34}S_{sulphate}$ is illustrated as a continuous band in Fig. 7.5-17. The observation that the range of $\delta^{34}S_{sulphide}$ values has increased over time – specifically during the Paleoproterozoic and the Neoproterozoic – is instead most easily explained by significant accumulations of atmospheric O_2 after about 2.5 Ga which helped stabilise oceanic sulphate at higher concentrations (Holland, 1984). The generally small ranges of $\delta^{34}S$ values in Archaean metasedimentary rocks appear to indicate that either sulphate-reducing microbes were frustrated by the generally low concentration of seawater sulphate and limited to specific microenvironments where sulphate was present, or they had not yet evolved.

The data of Shen et al. (2001) were collected in cherty barite-containing rocks from the Dresser Formation. The rocks experienced low degrees of metamorphism and slight deformation (Van Kranendonk, 2006), which argues against extensive post-diagenetic alteration of the sulphur isotopes. Groves et al. (1981) interpreted the depositional setting for these units as sedimentary deposits with precipitated gypsum (later replaced by barite) from brine ponds separated from the sea by a sand berm. Buick and Dunlop (1990) postulated the origin of the precipitated sulphate from originally low sulphate sea water that had seeped into briny lagoons and evaporated to concentrate sulphate. These authors also suggested that sulphate may have been locally supplemented by the phototrophic oxidation of volcanogenic sulphide.

In a series of papers, Van Kranendonk et al. (2001), Van Kranendonk and Pirajno (2004) and Van Kranendonk (2006) presented an alternative explanation for the origin of the sulphate with new high-resolution mapping and geochemical analyses of the North Pole rocks. This work convincingly showed that the extensive bedded chert + barite units of the Dresser Formation formed during discrete episodes of volcanogenic hydrothermal circulation during exhalative cooling of a felsic magma chamber. A similar conclusion was reached for the origin of sulphates in the 3.2–3.6 Ga Panorama volcanic-hosted sulphide district, also in Western Australia (Huston et al., 2001a). Under conditions analogous to contemporary S-rich volcanism at Mt. Pinatubo (Philippines), a silicic magma chamber with associated caldera collapse and seawater incursions will evolve highly oxidized (Scaillet et al., 1998) volatile-laden saline fluids (Newton and Manning, 2005). Such conditions stabilise SO_2 and lead to the expulsion of oxidised fluids, which in the hydrolysis reaction (after Hattori and Cameron, 1986):

$$4H_20 + 4SO_2 \leftrightarrow H_2S + 3H^+ + 3HSO_4^-$$

leads to the highly localised enrichment of sulphate, without the intervention of oxygenic photosynthesizers. In the case of the Dresser Formation barites, locally recharged submarine brine-pools with reactive Ba^{2+} leached from basalt would have formed barite (cf. Van Kranendonk, 2006). Sulphate would have been rapidly scavenged by Ba^{2+}; the solubility of barite is low ($\Delta H_r^\circ = 6.35$ kcal mol^{-1}, log $K = -9.97$; Stumm and Morgan (1996)), and Ba^{2+} leached from basalt (average Archaean basalt contains 569 ppm Ba; Condie (1993)) was precipitated as barite syndepositionally with pyrite.

Multiple S-isotope measurements of sulphate-sulphide pairs provide a crucial clue to the origin of the sulphur in the Dresser rocks. Barites from Dresser sample GSWA 169711 (A-I) reported in Farquhar et al. (2000a) have average $\Delta^{33}S$ values $= -0.98 \pm 0.02\%$ and average $\delta^{34}S_{VCDT} = +4.9\%$ that form a well-defined ($r^2 = 0.977$) linear array of MDF slope $\lambda = 0.51 \pm 0.04$ (Fig. 7.5-18(a)). It is thus apparent from these data that the sulphur responsible for this barite resided for some time in the Archaean atmosphere before it was deposited in the water and stabilised as sulphate. Core rock samples of black chert with finely disseminated pyrite, also from the Dresser Formation (core sample GSWA 169729), have average $\Delta^{33}S$ values $= +3.67 \pm 0.03\%$ and average $\delta^{34}S_{VCDT} = +2.70 \pm 0.01\%$ (Table 7.5-3) that form a linear array ($r^2 = 0.893$) of non-MDF slope $\lambda = 0.837 \pm 0.16$ (Fig. 7.5-18(b)). When these data are plotted in $\Delta^{33}S$ vs $\delta^{34}S$ space (Fig. 7.5-19), it is apparent that they are consistent with samples that experienced SO_2 (or SO) photolysis at short UV (e.g., 193 nm) wavelengths (compare with Fig. 7.5-6).

The differences in pyrite and barite multiple sulphur isotope compositions from the Dresser rocks show that deposition of MIF sulphur as pyrite and barite was swift once aerosols reached the water column. Because neither hydrothermal nor biological cycling

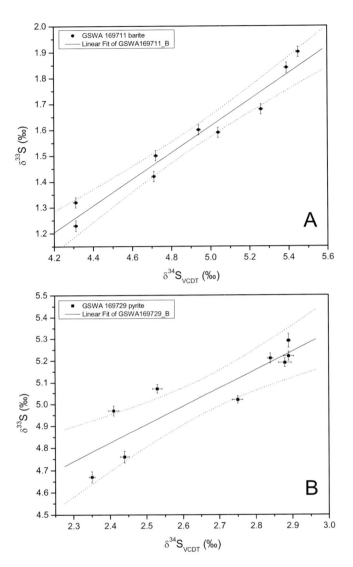

Fig. 7.5-18. A. Plot of multiple S-isotope measurements of barites from Dresser sample GSWA 169711 (A-I) reported in Farquhar et al. (2000a) shows a well-defined ($r^2 = 0.977$) linear array of MDF slope $\lambda = 0.51 \pm 0.04$. B. Core samples of pyrite from black chert from the Dresser Formation sample GSWA 169729 form a less well-defined linear array ($r^2 = 0.893$) of non-MDF slope $\lambda = 0.837 \pm 0.16$.

of sulphate had sufficient time to homogenise the $\Delta^{33}S$ values, it may be the case that neither process was important at time of deposition.

Table 7.5-3. Sulphur isotope composition of sulphides in Dresser Mine core sample 169729-1B by in situ ion microprobe multicollection

spot	$^{34}S/^{32}S$	$\pm 1\sigma$	$^{33}S/^{32}S$	$\pm 1\sigma$	$\delta^{34}S_{VCDT}$	$\pm 1\sigma$	$\delta^{33}S$	$\pm 1\sigma$	$\Delta^{33}S$	$\pm 1\sigma$
G1_s1	4.5135E−02	6.32E−07	8.0006E−03	1.67E−07	2.89	0.01	5.22	0.02	3.73	0.03
G1_s2	4.5114E−02	5.63E−07	7.9969E−03	2.06E−07	2.44	0.01	4.76	0.03	3.50	0.03
G1_s3	4.5110E−02	3.74E−07	7.9962E−03	1.98E−07	2.35	0.01	4.67	0.02	3.45	0.03
G2_s1	4.5113E−02	6.50E−07	7.9986E−03	1.88E−07	2.41	0.01	4.97	0.02	3.72	0.03
G2_s2	4.5134E−02	2.24E−07	8.0011E−03	2.53E−07	2.89	0.00	5.29	0.03	3.79	0.03
G2_s3	4.5119E−02	5.82E−07	7.9993E−03	1.63E−07	2.53	0.01	5.07	0.02	3.75	0.02
G3_s1	4.5134E−02	7.65E−07	8.0003E−03	1.63E−07	2.88	0.02	5.19	0.02	3.69	0.03
G3_s2	4.5132E−02	3.42E−07	8.0005E−03	1.81E−07	2.84	0.01	5.21	0.02	3.74	0.02
G3_s3	4.5128E−02	6.38E−07	7.9990E−03	1.33E−07	2.75	0.01	5.02	0.02	3.59	0.02
G4-s1	4.5140E−02	6.76E−07	8.0014E−03	2.10E−07	3.00	0.01	5.32	0.03	3.76	0.03

Notes: Nomenclature is grain number + ion microprobe spot number. Error demagnification and corrections for instrumental mass bias were applied to the data. Details for these measurement conditions are reported in Papineau et al. (2005).

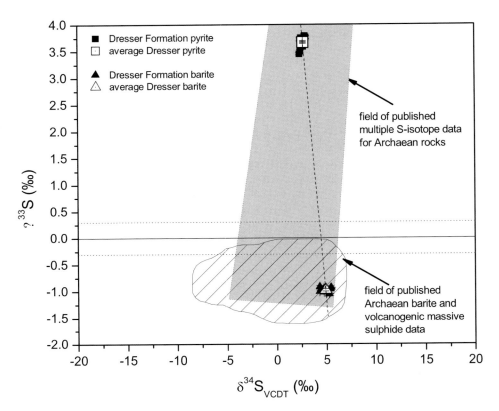

Fig. 7.5-19. Plot of Δ^{33}S vs δ^{34}S for sulphides and sulphates from the 3.46 Ga Dresser Fm. (sulphate data from Farquhar et al. (2000)). The field of isotope compositions for barite and hydrothermal sulphide (volcanogenic massive sulphide 'VMS' deposits) is also labelled. The data for the sulphide-sulphate pair tend to plot with an aspect ratio similar to the 193 nm photolysis array of SO$_2$ and SO (see Fig. 7.5-6).

7.5-7.2. *Sulphides from the (3.8–3.7 Ga) Isua Supracrustal Belt, West Greenland*

Amphibolites of dominantly basaltic composition from the ca. 3.77 Ga Isua supracrustal belt (ISB), 150 km northeast of Nuuk, West Greenland, have δ^{34}S values between -1.0 and $+2.6$‰ (Monster et al, 1979), consistent with a magmatic origin, and an Archaean mantle source with near-zero δ^{34}S values (Hattori and Cameron, 1986). Rocks of sedimentary protolith, such as banded iron-formations (BIF) and garnet-biotite schists of probable ferruginous clay-rich sedimentary origin are recognizable throughout the ISB (reviewed in Nutman et al. (1984a)). Because these rocks are of marine sedimentary origin (Dymek and Klein, 1988) they are important candidate materials to host information relevant to the sulphur geochemistry of the Archaean oceans. In line with these observations, it was proposed some time ago that sulphur isotopes in sulphides from the metamorphic equiva-

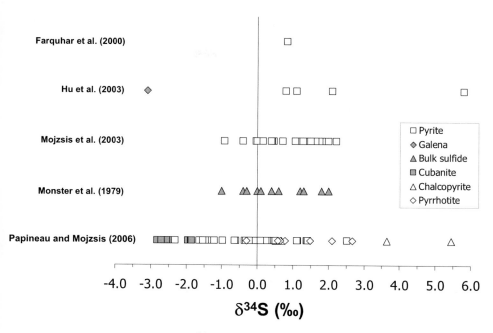

Fig. 7.5-20. Comparison of published $\delta^{34}S$ values for various sulfide phases in Isua supracrustal belt metasedimentary rocks (from Papineau and Mojzsis, 2006).

lents of pelagic sedimentary rocks could preserve isotopic evidence of the early biosphere (Monster et al., 1979; Schidlowski et al., 1983).

Bulk sulphides from Isua banded iron-formations have $\delta^{34}S$ values in the range of -1.0 to $+2.0‰$, comparable to values measured from associated garnet-amphibolite rocks of basaltic composition noted above, which volumetrically dominate the lithotypes of the ISB. Similar $\delta^{34}S$ values for pyrite in other BIF units from Isua range between -1.2 to $+2.5‰$, and were interpreted by Strauss (2003) to reflect magmatic and hydrothermal processes in the sulphur cycle, with no apparent biological fractionation. The total range of $\delta^{34}S$ values measured for metasedimentary rocks of the ISB extends over $\sim9‰$ and is centred around $+1‰$ (Fig. 7.5-20). This $\delta^{34}S$ range is similar to other sulphur isotope ranges that characterize some Mesoarchaean marine sediments (Ohmoto et al., 1993; Ohmoto and Felder, 1987). A possible explanation is that the small variations in $\delta^{34}S$ of sulphides from Isua metasediments reflect mass-dependent fractionation processes during metamorphism, which depend on temperature, sulphide phase, cooling time, etc. Alternatively, such a range may be consistent with biological fractionation at time of deposition, but the range in $\delta^{34}S$ is small and as outlined previously, microbial sulphate reduction was either suppressed or absent at Isua time. More data are required to test the hypothesis that elemental sulphur reduction or some other early metabolic cycling of sulphur was present when the Isua sediments formed, especially if the range of $\delta^{34}S$ for co-existing sulphur phases can be expanded further.

Published sulphur isotope analyses of anhedral pyrites that occur as interstitial blebs in biotite from various quartz-biotite schists ('metapelites' of clay-rich origin, as opposed to banded iron-formations) from the ISB showed positive $\Delta^{33}S$ values between +1.06 and +1.43‰, and a range of $\delta^{34}S$ between -2.82 and +5.44‰ (Mojzsis et al., 2003; Papineau and Mojzsis, 2006). Banded iron-formation samples have also been investigated for their multiple sulphur isotopes from the northeast sector of the ISB. Many of these rocks contain bands of large (<200 µm) pyrite grains within grunerite+magnetite and quartz bands. Published data for subhedral to anhedral sulphide for the ISB can be used to define maximum ranges of $\delta^{34}S$ data for sulphides ($n = 161$) as between -3.1 and +5.8‰, and for reported $\Delta^{33}S$ data ($n = 99$) between -0.87 and +3.41‰. A compendium of published and recent $\delta^{34}S$ data from metasedimentary rocks from the ISB is presented in Fig. 7.5-20.

From the standpoint of environmental chemistry, the multiple sulphur isotope data for the Isua supracrustal belt provides evidence for an anoxic atmosphere at 3.8 Ga and low concentrations of dissolved seawater sulphate. Under these constraints, it is interesting to consider whether the early Archaean oceans were severely limited in many oxidants necessary for various microbial respiration pathways. The dominant biome could have been carbon-fixing autotrophs as suggested by ^{13}C-depleted graphite particles reported from Isua (Ueno et al., 2002; Rosing, 1999; Mojzsis et al., 1996). This conclusion would mean that the low abundance of oxidants in early Earth environments was not conducive to an extensive biosphere nor to metabolically diverse microbial communities before the Paleoproterozoic transition from a globally anoxic world to the beginnings of a oxygenated planet (Papineau and Mojzsis, 2006; Papineau et al., 2007).

7.5-7.3. *Sulphides from the ⩾3.83 Ga Akilia Association, Akilia Island, West Greenland*

As outlined in Manning et al. (2006, and references therein), the supracrustal rocks of the Akilia association at the type locality of Akilia, an island $\sim$30 km S of Nuuk, West Greenland, have undergone a complicated metamorphic history that involved recrystallization under granulite facies conditions (Griffin et al., 1980). The complexity of these rocks, compounded with a long history of 'reconnaissance studies' from this terrane, has been the principal reason for current debates in the interpretation of protoliths for the Akilia enclaves. Detailed analysis has shown that on the south-western peninsula of Akilia are two units of ferruginous quartz-pyroxene rocks sandwiched between amphibolites (Manning et al., 2006). The mineral composition of the units is dominated by quartz, clinopyroxene, magnetite, hornblende, sulphides and minor accessory phases including zircon and graphite (Nutman et al., 1997, McKeegan et al., 2007). The age of the quartz-pyroxene rock has been the subject of debate. One interpretation holds that the units are greater than the ca. 3.83 Ga zircon U-Pb ages for tonalitic orthogneisses, some of which were shown to intersect the supracrustal enclave (Manning et al., 2006). Published U-Th-Pb depth-profiles from Akilia orthogneiss zircons by ion microprobe (Mojzsis and Harrison, 2002a) and Pb-

Pb ages on apatite separates in the quartz-pyroxene rock (Sano et al., 1999b; cf. Mojzsis et al., 1999) indicate a complicated metamorphic history with regional major events at $\sim$3.73 Ga, $\sim$3.65 Ga, $\sim$2.65 Ga and $\sim$1.50 Ga (reviewed in Cates and Mojzsis, 2006). The protolith of the quartz-pyroxene rocks was originally interpreted as banded-iron formation (BIF) by Nutman et al. (1997) and was used by them to earmark these particular samples for biosignatures analyses of C-isotopes (e.g., McKeegan et al., 2007). Although a continued subject of debate, detailed geochemical results published in the last few years provide supporting evidence for a sedimentary for these rocks. These data include major, minor, trace element and rare earth distributions (Mojzsis and Harrison, 2002b; Friend et al., 2002c; Nishizawa et al., 2005; Manning et al., 2006), δ^{56}Fe and δ^{57}Fe values[3] (Dauphas et al., 2004), δ^{18}O values (Manning et al., 2006) and Δ^{33}S values (Mojzsis et al., 2003). In an alternative interpretation, REE patterns and trace elements have been instead used to argue for a metasomatized ultramafic igneous protolith for the Akilia quartz-pyroxene rocks (Fedo and Whitehouse, 2002; Bolhar et al., 2004; cf. Johannesson et al., 2006). Also, Whitehouse et al. (2005) were unable to corroborate previously reported mass-independent sulphur isotopes values for these units (Mojzsis et al., 2003). Lastly, Manning et al. (2006) showed that only an origin as chemical sediment with minor detrital input satisfies all field, petrologic and geochemical tests, and because the Akilia quartz-pyroxene rocks represent a primary lithology that was deposited in the original volcanosedimentary succession, the rocks have the same age as the enclave as a whole or $\geqslant$3825 Ma.

Whole-rock sulphur isotopic analyses of Akila quartz-pyroxene sample GR9707 reported in Farquhar et al., 2000) revealed Δ^{33}S $= +0.12$‰. Subsequent ion microprobe measurements of sulphides in Akilia sample GR9707 reported in Mojzsis et al. (2003) revealed a large positive Δ^{33}S anomaly at $+1.47 \pm 0.14$‰ (2σ). The data also showed a population of positive Δ^{33}S values that ranged between $+0.84$ to $+0.30$‰. A relatively large range ($\sim$12‰) of δ^{34}S was also observed (-6.7 to $+5.2$‰) from this sample. These data are not that different from reported values for various ISB metasedimentary rocks. In a separate study on a different Akilia sample, Whitehouse et al. (2005) measured the multiple sulphur isotope composition of quartz-pyroxene sample G91-26 (Nutman et al., 1997) and found no clear evidence for mass-independently fractionated sulphur. Their Δ^{33}S data ranged from -0.55 to $+0.35$‰ and they describe a range in δ^{34}S of $+0.02$ to $+2.45$‰, substantially smaller than that reported in Mojzsis et al. (2003). Based on their analysis, Whitehouse et al. (2005) came to the conclusion that multiple sulphur isotopes could not be used to infer protolith for highly metamorphosed rocks.

It is likely that the complexity of the Akilia quartz-pyroxene rocks revealed by the various isotopic and trace element data cited above is manifest in heterogeneities in the sulphur isotopes at the outcrop scale. Further detailed work is needed on these rocks, but fortunately, Fe-rich chemical metasedimentary rocks with abundant sulphides turn out to be relatively common in Akilia association enclaves beyond the 'type-locality' on Akilia Island. Indeed, rocks of the 'Akilia type' in mineralogy, bulk composition and antiquity can found throughout scattered throughout the southern limits of the Itsaq Gneiss Complex of

---

[3] The δ^{5x}Fe(‰) $= ((^{5x}$Fe$/^{54}$Fe$)_{\text{sample}}/(^{5x}$Fe$/^{54}$Fe$)_{\text{IRMM-014}} - 1) \times 1000$ and IRMM-014 is the Fe-isotope standard 014 of the Institute of Reference Materials and Measurements.

West Greenland as well as in other early Archaean terranes of the North American Craton (Cates and Mojzsis, 2006, 2007a, 2007b, 2007c; Dauphas et al., 2007).

7.5-7.4. *Sulphides from the ⩾3.77 Ga Akilia Association Enclaves on Innersuartuut Island, West Greenland*

Finely-laminated quartz + magnetite ± garnet banded iron-formations from Innersuartuut island (approximately 30 km south of Akilia at N63°50.227′, W51°41.173′) were sampled for sulphur isotope analysis and reported in Cates and Mojzsis (2006). This rock preserves quartz-magnetite bands that apparently escaped the pervasive strain/recrystallization/silica-mobility recorded by Akilia association units on Akilia Island. In this regard, it most resembles samples 119233 from Ugpik, and 119217 from an unspecified field locality on Innersuartuut, as described in McGregor and Mason (1977).

Multiple sulphur isotope geochemistry of this rock provides a useful and independent (from Akilia Island) benchmark for the preservation of mass-independent sulphur in amphibolite to granulite-grade supracrustal enclaves (Fig. 7.5-21). Sulphides are present as pyrrhotite blebs in the quartz-magnetite matrix with fine intergrowths of SiO_2. The analysed pyrrhotites preserve consistently large $\Delta^{33}S$ values between $+1.31 \pm 0.19\%$ and $+1.49 \pm 0.19\%$ (2σ external errors) and record a small range in $\delta^{34}S$ of $\sim 1.5\%$. Other ferruginous quartz-pyroxene rocks also reported in Cates and Mojzsis (2006) from the Innersuartuut localities are similar in both mineralogy and gross geochemical composition to the samples G91-26 of Mojzsis et al. (1996) and GR9707 of Mojzsis et al. (2003). Like the Akilia Island rocks, these samples contain abundant pyrrhotite and occasional chalcopyrite with subhedral habit in a dominantly quartz + hedenbergite matrix. All analysed pyrrhotite in the Innersuartuut samples preserve MIF S isotopes with large $\Delta^{33}S$ values between $+1.90 \pm 0.25\%$ and $+3.15 \pm 0.25\%$ (2σ external errors) and range in $\delta^{34}S$ by $\sim 2.5\%$ (Cates and Mojzsis, 2006). Although they record a smaller total range in $\delta^{34}S$ compared to sample GR9707 ($\sim 12\%$) from Akilia Island (Mojzsis et al., 2003), the Innersuartuut samples preserve exceptionally large $\Delta^{33}S$ values akin to those reported for ISB banded iron-formation sample 248474 (Baublys et al., 2004; ca. $+3.3\%$) that was reanalysed by Whitehouse et al. (2005). The Innersuartuut rocks are also notable in that they preserve a unit of quartz-biotite-garnet schist what contains abundant pyrrhotite (~ 1 mode%). All analysed sulphides from this particular unit, which Cates and Mojzsis (2006) interpreted to be a paragneiss of probable pelitic origin, contain MIS sulphur isotopes with well-resolved

---

Fig. 7.5-21. Compilation plot of $\Delta^{33}S$ vs $\delta^{34}S$ for sulphides and sulphates from pre-3.7 Ga metasediments in the pre-3.77 Ga Isua supracrustal belt and the pre-3.83 Ga Akilia association rocks from Akilia (island) and Innersuartuut (West Greenland). Data are from Farquhar et al. (2000), Mojzsis et al. (2003), Hu et al. (2003), Whitehouse et al. (2005), Cates and Mojzsis (2006), Papineau and Mojzsis (2006), and unpublished data. Compare the different trends in multiple isotope end-member compositions to the 193-nm results in Fig. 7.5-6.

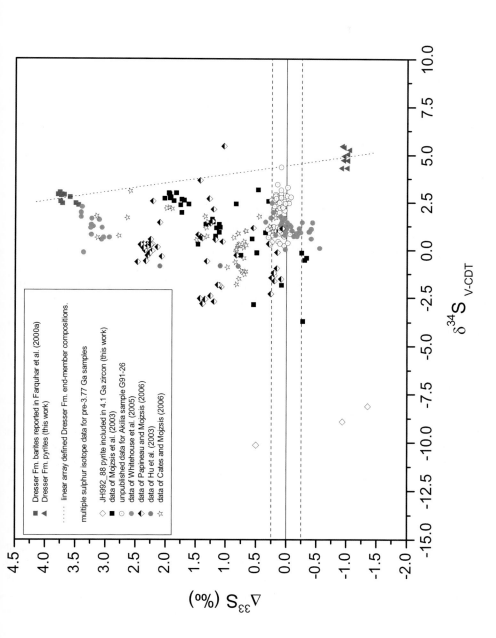

Dresser Fm. barites reported in Farquhar et al. (2000a)
Dresser Fm. pyrites (this work)

...... linear array defined Dresser Fm. end-member compositions.

multiple sulphur isotope data for pre-3.77 Ga samples

◇ JH992_88 pyrite included in 4.1 Ga zircon (this work)
data of Mojzsis et al. (2003)
■ data of Mojzsis et al. (2003)
○ unpublished data for Akilia sample G91-26
● data of Whitehouse et al. (2005)
◆ data of Papineau and Mojzsis (2006)
● data of Hu et al. (2003)
☆ data of Cates and Mojzsis (2006)

Δ^{33}S values, between $+0.67 \pm 0.21\%$ and $+0.98 \pm 0.12\%$ (2σ external errors), and range in δ^{34}S by $\sim 3\%$.

These results for mass-independent sulphur isotopes in amphibolite- to granulite-grade schists from various early Archaean localities in West Greenland underscore the observation that Δ^{33}S values in a rock containing accessory sulphur minerals can be preserved. The simplest explanation for the small spread of δ^{34}S values in these data for the oldest known rocks of marine sedimentary origin is by re-precipitation of sulphides during metamorphic processes. Dilution by invasive fluids (H_2S etc.) which carried exotic sulphur does not appear to be an important process in rocks that contain no more than ~ 1 mode percent sulphur (Ohmoto, 1986). Scatter in Δ^{33}S values is minimal for each these units, which suggests that no significant foreign component of sulphur was added to the system since formation as aqueous sediments in the early Archaean ocean.

7.5-8. CONCLUDING REMARKS: BANDED IRON-FORMATIONS, SULPHUR AND LIFE ON THE EARLY EARTH

The Archaean banded iron-formations are regarded solely as a chemical sediments that have been enriched in metals from volcanic/hydrothermal sources (reviewed in Polat and Frei (2005)) and probably represent part of a relatively deep (> 100 m) marine sedimentary facies derived from either the photo-oxidation of upwelled Fe(II)-rich seawater to ferro-ferri-oxyhydroxides (Braterman et al., 1983) or biological precipitation and/or oxidation of insoluble iron oxides (e.g., Konhauser et al., 2002). In either of these models, the Fe-precipitates subsequently underwent diagenetic transformation to magnetite. Models for biological modulation of iron-formation deposition via anoxygenic photosynthetic Fe(II) oxidation, and the nature of microbially-mediated iron-oxide deposition and organic matter recycling (Kappler et al., 2005), are still in their early stages of development. Cloud (1973) postulated that ferrous iron [Fe(II)$_{aq}$] oxidation by molecular oxygen after cyanobacteria evolved on Earth was responsible for the banded iron-formations. However, it has been argued that the anoxygenic photoautotrophic bacteria are the most ancient type of photosynthetic organisms (Xiong et al., 2000), and these can readily catalyse Fe(II) oxidation under anoxic conditions (Widdel et al., 1993). The Fe isotope compositions of banded iron-formations are consistent with the possibility that anoxygenic photoautotrophs such as purple bacteria played a role in their deposition. Calculations based on experimentally determined Fe(II) oxidation rates by anaerobic photoautotrophic bacteria under low-light regimes representative of ocean depths of a few hundred meters suggest that, even in the presence of cyanobacteria, anoxygenic phototrophs living beneath a wind-mixed surface layer provide the most likely explanation for BIF deposition in a stratified ancient ocean and the absence of Fe in Precambrian surface waters (Kappler et al., 2005). Following Gross (1980), the Archaean BIFs such as those at Isua and in the Akilia association tend to resemble the 'Algoma-type' which are thought to have been deposited in small basins around island arcs (Jacobsen and Pimentel-Close, 1988). Enrichment of seawater in ferrous iron and the transformation to abundant Fe_3O_4 in marine sediments is considered a signature style of sedimentation under anoxic conditions on the early Earth, when Fe(II) concentrations were high in seawater (Holland, 1984; Bjerrum and Canfield, 2002).

Banded iron-formations are strongly enriched in sulphur relative to average Archaean crust and hydrothermal fluids; the average S composition of Isua iron-formation is taken to be close to that of the international banded iron-formation standard IF-G (Govindaraju, 1994) at 700 ppm. In the Archean, BIF sulphur was introduced to the water column from both atmospheric and magmatic exhalative sources at time of deposition (Huston and Logan, 2004), and the major sulphide phases present in the Isua and Akilia rocks are pyrite and pyrrhotite (with minor chalcopyrite and cubanite). As noted by Farquhar and Wing (2003, and references therein), an important consequence of the model for MIF-carrying aerosol formation in the Archaean atmosphere is that sulphide formation via elemental sulphur reduction by microbial life – perhaps deposited as greigite: Fe_3S_4, or mackinawite: $Fe_{1-x}S$ (e.g., Mann et al., 1990) which was later transformed to pyrite: FeS_2 and metamorphosed to pyrrhotite: $\sim Fe_7S_8$ – would be expected to retain positive $\Delta^{33}S$ signatures (Mojzsis et al., 2003) with $> 10\permil$ spread in $\delta^{34}S$ (Canfield and Raiswell, 1999). Sulphide formed from microbial sulphate reduction would be expected to preserve negative $\Delta^{33}S$ values with a potentially large spread in $\delta^{34}S$ values. Sulphur aerosols provided a ready source of sulphur to a plethora of microbial environments at the global scale on the early Earth, at issue of course is whether this oxidative chemistry was actually exploited by life, as perhaps $Fe(II)_{aq}$ was by early microbial communities for an electron donor (Johnson et al., 2003; Dauphas et al., 2004). Could then the low abundance of oxidants in early Archaean environments globally have severely limited the early biosphere before the gradual build-up of oxygen (Papineau and Mojzsis, 2006)? If the Archaean biosphere did in fact modulate BIF deposition, it still remains to be seen whether it left traces of itself in the sulphur isotopes.

Of the principal elements involved in biological activity (C, N, O, H, S, P, Fe etc.), sulphur is probably the most versatile for geochemistry. As summarised by Goldschmidt (1954), chemists have long recognised that sulphur participates in many of the most important processes in inorganic geochemistry, and serves as a bridge between the inorganic world and the biochemistry contained in the central metabolic machinery of all classes of organisms. Because of its chemical reactivity, large mass difference between multiple stable isotopes (^{32}S, ^{33}S, ^{34}S and ^{36}S), diverse redox chemistry through multiple valence states (2− to 6+), and tendency to form both ionic and covalent bonds in a wide variety of minerals, sulphur and its isotopes have been used to trace igneous, sedimentary, metamorphic, hydrothermal, atmospheric and biological processes since the early days of geochemistry (Thode et al., 1949). New inroads in transition metal stable isotopic systems used to trace changes in the surface state of the Earth with time, such as Mo (reviewed in Anbar and Knoll (2002)) and Fe (reviewed in Johnson et al. (2003)), also figure prominently in its phase-space as the various metal-sulphides. How the chemistry of the 'inorganic bridge' relates to the distribution of inorganic components in the cell ('metallomics') and in response to the evolution of the surface zone as traced by the multiple sulphur isotopes, represents an exciting new horizon that is ripe for exploration. Future work will be to unite transition metal isotope systematics with the chemistry of the multiple sulphur isotopes preserved in sedimentary sulphides.

ACKNOWLEDGEMENTS

Discussions and debates on the various topics presented herein with A.D. Anbar, G.L. Arnold, A. Bekker, P.S. Braterman, R. Buick, N.L. Cates, M. Chaussidon, R.N. Clayton, C.D. Coath, N. Dauphas, J. Farquhar, C. Fedo, R. Frei, C.R.L. Friend, J.P. Greenwood, T.M. Harrison, K. Hattori, H.D. Holland, B.S. Kamber, A.J. Kaufman, A. Kappler, J. Karhu, J.L. Kirschvink, M.J. Van Kranendonk, A. Lepland, D. Lowe, J.R. Lyons, C.E. Manning, B. Marty, K.D. McKeegan, S. Moorbath, A.P. Nutman, H. Ohmoto, S. Ono, N.R. Pace, D. Papineau, A.A. Pavlov, M. Rosing, D. Rumble, B. Runnegar, M. Schidlowski, A.K. Schmitt, J.W. Schopf, H. Strauss, M.H. Thiemens, M.J. Whitehouse, B. Wing, Y. Watanabe and M. Van Zuilen have helped to refine and focus this work. N. Pace at the University of Colorado is thanked for access to his library of Archaea and Bacteria molecular phylogenies. Constructive reviews by N.L. Cates, M.J. Van Kranendonk and two anonymous reviewers helped a great deal to improve the manuscript. Our work on the evolution of sulphur isotopes was supported by the National Science Foundation grants EAR9978241 and EAR0228999 to SJM, and since 2000 by the University of Colorado Center for Astrobiology through a cooperative agreement with the NASA Astrobiology Institute. Correspondence and request for materials should be addressed to mojzsis@colorado.edu.

Earth's Oldest Rocks
Edited by Martin J. Van Kranendonk, R. Hugh Smithies and Vickie C. Bennett
Developments in Precambrian Geology, Vol. 15 (K.C. Condie, Series Editor)   971
© 2007 Elsevier B.V.   All rights reserved.
DOI: 10.1016/S0166-2635(07)15076-3

Chapter 7.6

THE MARINE CARBONATE AND CHERT ISOTOPE RECORDS AND THEIR IMPLICATIONS FOR TECTONICS, LIFE AND CLIMATE ON THE EARLY EARTH

GRAHAM A. SHIELDS

Geologisch-Paläontologisches Institut, Westfälische-Wilhelms Universität, Correnstr. 24, 48149 Münster, Germany

7.6-1. INTRODUCTION

Long-term trends in the isotopic composition of seawater are caused by changes to the globally integrated fluxes of that element into the global ocean. In the case of light elements, these are affected additionally by isotopic fractionations that are to differing extents metabolism-, and temperature-dependent. By comparing the isotopic records of ancient marine sedimentary rocks with those of more recent times, we can decipher trends in the surface environment and in biogeochemical cycling through time. Unfortunately, pre-3 Ga isotope records are rare and derive from only three major terrains: the ~3.7–3.8 Ga Isua supracrustal belt of the North Atlantic Craton in SW Greenland, the ~3.5–3.2 Ga Pilbara Supergroup of the Pilbara Craton, Western Australia, and the ~3.5–3.2 Ga Barberton Greenstone Belt of the Kaapvaal Craton, southern Africa. Nevertheless, where isotopic data from Earth's oldest rocks can be shown to be consistent with well established, long-term trends through younger strata, they can provide us with important insight into early Earth System evolution.

Marine carbonate rocks are routinely analysed for their C and O, as well as Sr isotope ratios, while siliceous rocks may be analysed for O, and more recently Si isotope compositions. Seawater $^{87}Sr/^{86}Sr$ is controlled by the relative influence of continental weathering relative to mantle input on ocean composition and so can be used to trace first-order changes in global tectonics, such as crustal evolution and supercontinent amalgamation and reconstruction. The persistence of modern day-like carbon isotope systematics as far back as existing rock archives allow has been used to demonstrate the antiquity of life on Earth, but recent studies call into question the very earliest of these records. Ancient carbonates and cherts are isotopically lighter with respect to oxygen than their more recent counterparts, which has been explained by hot ocean temperatures of at least 70 °C on the early Earth or changing seawater $^{18}O/^{16}O$. In this contribution, I review the evidence for secular changes in the Sr, C and O isotope ratios of sedimentary rocks on Earth and reflect on their significance for crustal evolution, the antiquity of life and climatic evolution on the early Earth.

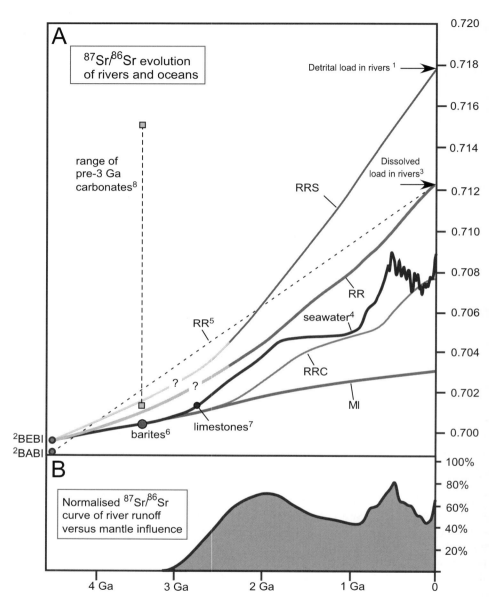

Fig. 7.6-1. Sr isotope evolution of seawater and ocean fluxes (A) and its implications for the relative influence of continental weathering on ocean composition (B). The seawater $^{87}Sr/^{86}Sr$ curve[4] in (A) has been constructed from the lowermost ratios for each time interval, e.g., lowest $^{87}Sr/^{86}Sr$ ratios of ca. 2.7 Ga limestones of southern Africa are 0.7011–3[7]. The river runoff (RR) curve has been determined by assuming a modern-like 9:11 relationship between Sr input from marine carbonate weathering (RRC) and silicate weathering (RRS), respectively. (*Continued on next page.*)

Fig. 7.6-1. (*Continued.*) The RRC curve assumes that carbonate rocks undergoing weathering preserve the isotopic composition of the seawater from which they precipitated and have a skewed age distribution, with an average age of 250 Ma[3], and so lags seawater $^{87}Sr/^{86}Sr$. The RRS curve is an idealistic representation based on predicted crustal evolution (O'Nions et al., 1979); other authors assume much earlier crustal Rb/Sr differentiation with minimal isotopic evolution[5]. Ocean crust alteration provides less radiogenic Sr to the oceans (MI). The curve in (B) assumes that seawater $^{87}Sr/^{86}Sr$ results from simple binary mixing between RR and MI, and shows that the influence of continental weathering was less prior to 2.5 Ga, and negligible prior to 3.0 Ga. Literature sources are: Bickle (1994); [2]McCulloch (1994); [3]Peucker-Ehrenbrink and Miller (2006); [4]Shields and Veizer 2002), Veizer et al. (1999); [5]Kamber and Webb (2001); [6]McCulloch (1994); [7,8]Veizer et al. (1989); Zachariah (1998).

7.6-2. THE SR ISOTOPE COMPOSITION OF THE EARLY OCEAN

The Sr isotope composition of modern seawater ($^{87}Sr/^{86}Sr_{sw}$) is known very precisely at just under 0.7092 (Fig. 7.6-1(A)). Because of the high concentration of Sr ($\sim$8 ppm) and its long residence time ($\sim$4–8 Myr) in seawater (Holland, 1984), the global ocean is isotopically homogeneous with respect to Sr, while secular trends in $^{87}Sr/^{86}Sr_{sw}$ through geological time can be attributed to globally integrated changes in the relative influence of two major Sr sources to the ocean; river runoff (RR) and submarine hydrothermal exchange of Sr (MI) during ocean crust alteration by seawater (Veizer, 1989). Because the $^{87}Sr/^{86}Sr$ ratios of these two fluxes are known for modern Earth – $\sim$0.7124 (Palmer and Edmond, 1989; Peucker-Ehrenbrink and Miller, 2006) and $\sim$0.703 (Hofmann, 1997), respectively – we know that the influence of rivers dominates over that of ocean crust-seawater interactions by a ratio of about 2:1. In order to estimate trends in the relative influence of 'riverine Sr' versus 'mantle Sr' back through time in the same fashion, we need to reconstruct not only the $^{87}Sr/^{86}Sr_{sw}$ of past oceans, but also the long-term isotopic evolution of these two compositional end-members, which is easier said than done. The $^{87}Sr/^{86}Sr_{sw}$ curve is continually being improved (Veizer et al., 1999; Shields and Veizer, 2002), while the isotopic evolution of the upper mantle, for the purpose of the present discussion, is relatively uncontroversial. However, the riverine Sr flux is more difficult to reconstruct due to the incongruent leaching behaviour of Rb versus Sr in carbonate and silicate minerals of the upper crust (Goldstein, 1988), uncertainties regarding the effect of crustal evolution on continental freeboard and on the $^{87}Rb/^{86}Sr$ of silicate minerals, and the likelihood of unsystematic changes in the age, and therefore $^{87}Sr/^{86}Sr$, of rocks undergoing weathering through time.

Because of the tendency of carbonates to incorporate radiogenic ^{87}Sr, e.g., from Rb-rich clay minerals, during post-depositional recrystallisation, lowermost $^{87}Sr/^{86}Sr$ ratios are most likely to best represent the isotopic ratio of the contemporaneous ocean. In most cases, this assumption can be confirmed by the systematic correlation between low $^{87}Sr/^{86}Sr$ ratios and high Sr contents (e.g., Veizer et al., 1989). The Archean $^{87}Sr/^{86}Sr_{sw}$ record (Shields and Veizer, 2002) is based mainly on marine carbonate rocks that were deposited

in greenstone belts, and barites that have ambiguous origins. Marine carbonate rocks de
posited before about 2.0 Ga exhibit lowermost $^{87}Sr/^{86}Sr$ ratios that approach the isotopi
evolutionary curve of the upper mantle (Fig. 7.6-1(A)), while prior to the Neoarchean, th
two are virtually indistinguishable at current resolution. For example, marine carbonate
from ca. 2.7 Ga greenstone belts in Zimbabwe and Canada yield $^{87}Sr/^{86}Sr$ values as lov
as 0.7011–0.7013, respectively (Veizer et al., 1989), which is very similar to $^{87}Sr/^{86}Sr$ es
timates for the contemporaneous, partially depleted upper mantle of close to, or slightl
below, 0.7010 (Machado et al., 1986).

All older carbonate rocks so far analysed, with the exception of ca. 3 Ga marbles fron
India (Zachariah, 1998) (Fig. 7.6-1(A)), yield consistently radiogenic and highly variabl
$^{87}Sr/^{86}Sr$ ratios >0.702, most likely indicative of isotopic alteration, while ca. 3.49 G
barites from the Warrawoona Group of the Pilbara Craton in Australia are interpreted t
provide maximum $^{87}Sr/^{86}Sr$ constraints for the upper mantle, and minimum $^{87}Sr/^{86}Sr$ con
straints for contemporaneous seawater, at 0.7005 (McCulloch, 1994). Despite the scarcit
of firm constraints, it seems likely that $^{87}Sr/^{86}Sr_{sw}$ closely paralleled the isotopic evolu
tion of the upper mantle before about 3 Ga, and that the oceans were "mantle buffered"
(Veizer et al., 1982) throughout the Archean due to vigorous circulation of seawater vi
submarine hydrothermal systems (McCulloch, 1994; Van Kranendonk, 2006). Althoug
protocontinents appear to have existed prior to this time (Campbell, 2003), the $^{87}Sr/^{86}S$
isotopic evolution of Rb-enriched crustal materials appears to have had negligible influenc
on $^{87}Sr/^{86}Sr_{sw}$ until after 2.5 Ga.

With the exponential decline of internal heat dissipation on a cooling Earth, the vigo
of the hydrothermal system also declined, while at the same time the flux of radiogeni
strontium from growing continents brought in by rivers started to assert itself (Veizer et al.
1982). This tectonically controlled transition from "mantle-" to "river-buffered" ocean
across the Archaean/Proterozoic transition (Fig. 7.6-1(B)) is a first order feature of terres
trial evolution, with consequences for other isotope systematics and for the redox state o
the ocean/atmosphere system (Goddéris and Veizer, 2000; Veizer and Mackenzie, 2003).

7.6-3. THE C ISOTOPE COMPOSITION OF EARLY MARINE CARBONATE ROCK!

The two dominant exogenic reservoirs of carbon are carbonate rock, formed either a
a marine precipitate or during alteration of seafloor ocean crust, and organic matter ir
sediments. Carbonate and reduced carbon are linked in the carbon cycle via atmospheri
CO_2 and carbon dioxide species dissolved in the hydrosphere. Carbon deposition is bal
anced by the mantle carbon flux with $\delta^{13}C$ of about $-5‰_{PDB}$ (Hayes and Waldbauer
2006) [light stable isotope ratios are commonly expressed in per mil (‰) deviations rela
tive to international standards such as PeeDee Belemnite (PVB or V-PDB) for carbonate
and standard mean ocean water (SMOW or V-SMOW) for waters and minerals]. In th
absence of autotrophic photosynthesis, this would also be the isotopic composition of sea
water (Broecker, 1970). However, $\delta^{13}C$ for the total dissolved carbon (TDC) in moderi
seawater is $\sim 1 \pm 0.5‰$ (Kroopnick, 1980; Tan, 1988) because of a kinetic isotope effec

1at serves to enrich organic matter in the lighter isotope, ^{12}C. Most of this enrichment erives from the isotope-discriminating properties of RuBP carboxylase, the key enzyme f the Calvin cycle (Schidlowski, 2001). Buried organic matter is consequently enriched 1 ^{12}C by $\sim$28‰ with respect to buried carbonate rock, which results from a combination f a 30‰ fractionation between dissolved inorganic carbon (DIC) in surface seawater and edimentary organic carbon, and a 2‰ fractionation between DIC and precipitated carbonte (Hayes and Waldbauer, 2006). The former isotopic discrimination has varied through me between globally integrated mean values of 25‰ and 50‰ (Des Marais et al., 1992), argely because of changing atmospheric CO_2 levels and the changing relative importance f methanotrophic metabolisms, which are associated with greater fractionations. The later isotopic fractionation of 2‰ between DIC and carbonate minerals seems likely to hold or all kinds of marine carbonate minerals as far back as 3.45 Ga (Hayes and Waldbauer, 006).

Because the source and sink C-isotopic fluxes need to be balanced on geological timecales, the average $\delta^{13}C_{carb}$ of $\sim$0‰ (Schidlowski et al., 1983; Shields and Veizer, 2002) Fig. 7.6-2) and $\delta^{13}C_{org}$ of ~ -28‰ imply that carbon has generally been removed from 1e exogenic system into the crust as an approximately 9:2 mixture of carbonate minerals ɔ organic matter, according to the mass balance:

$$\delta^{13}C_{influx} = f_{org} . \delta^{13}C_{org} + (1 - f_{org}) . \delta^{13}C_{carb} \tag{1}$$

vhere $f_{org} = C_{org}(C_{org} + C_{carb})$. The marine carbonate and kerogen $\delta^{13}C$ records provide, herefore, strong support for the existence of a global microbial ecosystem throughout the narine sedimentary record and perhaps back as far as 3.8 Ga (Schidlowski et al., 1983; chidlowski, 2001). Archean kerogen shows increasing evidence for metamorphic over->rint that, when accounted for by extrapolating back to near primary H/C ratios, indicates hat initial carbonate-kerogen isotopic discrimination was perhaps as high as 50‰ during he Neoarchean (Fig. 7.6-2). Such low $\delta^{13}C$ values for kerogen suggest the influence of nethanogenic bacteria (e.g., Ueno et al., 2006), and further imply that the 9:2 ratio was 1ot constant through time and may even have been as high as 10:1 during parts of the Neoarchean (Des Marais et al., 1992). How far back in Earth history can carbon isotopes ›e used to support the existence of life on Earth?

The oldest, arguably marine, carbonate isotope data derive from the Isua supracrustal belt (ISB), SW Greenland, which contains the oldest rocks on Earth interpreted to be of nitially sedimentary origin. However, the origin of both carbonate and graphite in the ISB s the subject of considerable controversy (Schidlowski, 2001; van Zuilen et al., 2003). 'or a generation, the $\delta^{13}C$ values of carbonates and graphites from Isua have been considered to represent the earliest evidence for life on Earth (Schidlowski et al., 1979). These original interpretations considered that the high and unusually variable $\delta^{13}C$ values of ISB graphite, in particular, were related to its complex amphibolite-grade metamorphic history, while biogenic interpretations were based on the assumption that the graphite-bearing netacarbonate rocks of the ISB were metamorphosed *sediments*. By contrast, other detailed studies, recently exemplified by van Zuilen et al. (2003), argue that "most of the graphite in the ISB occurs in carbonate-rich *metasomatic* rocks (metacarbonates), while

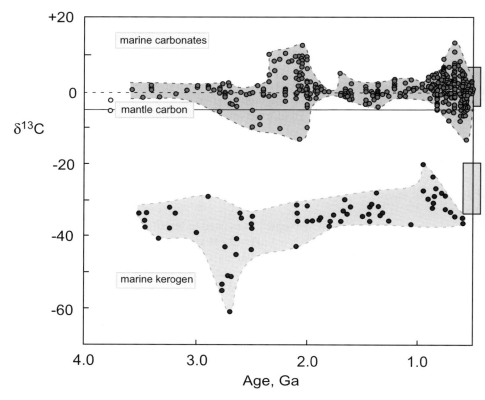

Fig. 7.6-2. Plot of age versus $\delta^{13}C$ of marine carbonates (after Shields and Veizer, 2002) and δ^{13} of kerogen corrected for thermal alteration (after Des Marais, 2001). Boxes on right-hand axis ref to Ediacaran and Phanerozoic ranges. $\delta^{13}C$ carb has averaged about 5‰ higher than mantle carbo $\delta^{13}C$ since ca. 3.4 Ga, indicating the continual existence of a significant autotrophic microbial com munity since that time. Older $\delta^{13}C$ records from Greenland seem likely to be of non-marine orig and are shown as open circles (see text for further explanation).

sedimentary units, including banded iron formations and metacherts, have exceedingly lo graphite concentrations". These observations, together with isotopic arguments, support metasomatic origin for "most, if not all" carbonate minerals (calcites, ferroan dolomite and siderites) from Isua, while all the graphite is now interpreted to derive from the therm decomposition of secondary, metasomatic siderite (van Zuilen et al., 2003). Nevertheless metasedimentary rocks, such as banded iron formations and metacherts, of the ISB als contain small amounts of carbonate minerals (<5%) that are not necessarily of metasc matic origin (van Zuilen et al., 2003). $\delta^{13}C$ compositions of these carbonates are als highly variable, but range up to normal marine values (−9‰ to +1‰), whereas associ ated metacherts exhibit high $\delta^{18}O$ values that are only slightly lower than nonmetamorphi Archean cherts (Perry et al., 1978; Perry and Lefticariu, 2003).

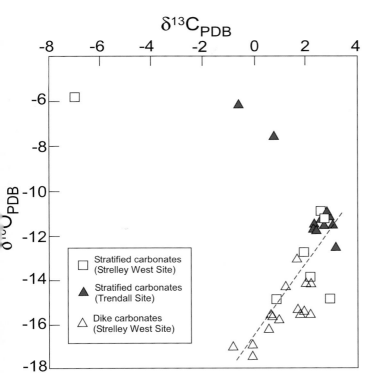

Fig. 7.6-3. Crossplot showing stable carbon and oxygen isotopic compositions in ‰PDB of carbonates from the ca. 3.4 Ga Strelley Pool Chert at two sites in W. Australia (after Lindsay et al., 2005). Three outlying samples are interpreted to have undergone isotopic exchange during replacement; all other samples show positive correlation that could represent a diagenetic trend, implying that primary carbonate values were close to $+2/+3‰$ and $-12/-11‰$ for $\delta^{13}C$ and $\delta^{18}O$, respectively.

The oldest, undisputed marine sediments derive from the ca. 3.4 Ga Strelley Pool Chert of the Kelly Group (Van Kranendonk et al., 2003, 2006a) that exhibit $\delta^{13}C_{carb}$ values between $-8‰$ and $+4‰$ with an average of $+1.0 \pm 2.2‰$ (Lindsay et al., 2005; Veizer et al., 1989). Interestingly, $\delta^{13}C$ data are similar to those from the slightly younger, middle Archean Barberton Greenstone Belt that range between $-4‰$ and $+3‰$, averaging $0.2 \pm 1.6‰$ (Schidlowski et al., 1975; Veizer et al., 1989). The Kelly Group has only been subjected to lower greenschist-grade metamorphism (Van Kranendonk, 2000); however, only few studies have reported detailed mineralogical and petrographic information for samples used in isotope studies. In this regard, data for carbonates in the Strelley Pool Chert stand out as an exception (Lindsay et al., 2005). In that study, a clear correlation in least altered samples between high $\delta^{13}C$ and $\delta^{18}O$ values could be demonstrated (Fig. 7.6-3), which could reflect isotopic exchange during fluid overprinting (Hofmann et al., 1999), possibly at elevated temperatures, or mixing with expelled C-rich fluids derived from the degassing of CO_2 at depth (Lindsay et al., 2005). Whichever scenario is preferred, stratified ferroan dolostones of the c. 3.40 Ga Strelley Pool Chert appear to be systematically better preserved, isotopically speaking, than vein-related carbonates, with least altered, presumably near marine values of +2 to +3‰ for $\delta^{13}C$ and −12 to −11‰ for $\delta^{18}O$. The high $\delta^{13}C$ ratios of carbonates from the Strelley Pool Chert are consistent with all other Archean marine carbonate rocks (Fig. 7.6-2), while their close association with stromatolite reef complexes (Hofmann et al., 1999; Van Kranendonk et al., 2003; Allwood et al., 2006a) provides strong support for the existence of a microbial community of global extent as far back as 3.40 Ga, with additional supporting evidence from the Mesoarchean Barberton Greenstone Belt (Tice and Lowe, 2004). Although available isotopic evidence is consistent with oxygenic photosynthesis, it does not exclude the possibility that only methanogenic and anoxygenic photoautotrophic bacteria contributed to the primary microbial biomass at this time (Schidlowski et al., 1983; Des Marais, 2001).

Existing constraints on primary $\delta^{13}C$ of middle Archean kerogen average ~ -36‰ (see compilations in Hayes et al., 1983; Strauss and Moore, 1992; Hayes and Waldbauer, 2006) and derive mainly from cherts and shales of the Fig Tree and Onverwacht Groups of the Barberton Greenstone Belt, which have been metamorphosed to greenschist facies. This value is similar to the lowest, and presumably least altered, $\delta^{13}C$ values from cherts of the Warrawoona Group (Hayes et al., 1983), although similarly depleted carbonaceous material from cherts of the Pilbara Supergroup are interpreted as abiogenic (Lindsay et al., 2005; Marshall, this volume). Carbonate associated with weathered basalts in the Warrawoona Group exhibit $\delta^{13}C_{carb}$ values of −0.3‰ (Nakamura and Kato, 2004), similar to contemporaneous, sedimentary marine carbonates. Assuming that these isotopic constraints are representative, Eq. (1) tells us that about 14% of the carbon being delivered by outgassing of mantle CO_2 was being buried as reduced, organic carbon by 3.4 Ga (Hayes and Waldbauer, 2006).

7.6-4. THE O ISOTOPE COMPOSITION OF EARLY MARINE SEDIMENTARY ROCKS

Oxygen isotopes, ^{18}O, ^{17}O and ^{16}O, undergo fractionation with respect to each other during many surface processes, such as evaporation, condensation, precipitation and clay mineral formation. Lower temperatures lead to larger isotopic fractionations, thus allowing the $^{18}O/^{16}O$ ratio of minerals to be used as a paleothermometer, as well as a tracer of surface processes within the hydrological cycle. However, water is almost ubiquitous in the surface environment, and so isotopic exchange with fluids after burial is a constant source of concern when interpreting the $\delta^{18}O$ values of ancient rocks. The hydrological cycle favours the retention of ^{16}O in clouds, leading to negative $\delta^{18}O$ values in rainwater and snow compared with seawater. Therefore, isotopic exchange with meteoric groundwater and/or at elevated temperatures will lower the $\delta^{18}O$ values of marine authigenic precipitates after burial. Conversely, precipitation in a restricted marine environment affected by elevated evaporation rates leads to enrichment of ^{18}O in carbonate or chert minerals.

The $\delta^{18}O$ values of marine carbonates and cherts depends on the isotopic composition of porewaters, and the temperature-dependent fractionation during initial precipitation and possible recrystallisation to a thermodynamically more stable mineral phase, i.e., calcite or dolomite in the case of calcium carbonate minerals, or quartz in the case of silica minerals. Other factors such as pH are of only secondary importance, while metabolic or vital effects can be neglected for the Precambrian. If the isotopic composition of seawater is known, then the $\delta^{18}O$ values of well preserved carbonate minerals and chert can be used to determine the approximate temperature of formation. However, if seawater $\delta^{18}O$ is unconstrained and/or the pristine, open marine nature of the mineral phase cannot be guaranteed, then paleotemperature estimates will be ambiguous. This is the basis of the long-standing controversy regarding the ^{18}O-depletion of ancient carbonate and chert minerals (Muehlenbachs, 1998; Veizer et al., 1999). One school of thought maintains that seawater $\delta^{18}O$ has remained fixed throughout Earth history (Gregory, 1991; Muehlenbachs, 1998) and that low $\delta^{18}O$ values must relate to later alteration or higher surface temperatures in the past (Knauth and Lowe, 2003). The opposing viewpoint (Veizer et al., 2000; Kasting et al., 2006) considers that seawater $\delta^{18}O$ can change through time, implying that $\delta^{18}O$-based paleoclimate studies (Knauth and Epstein, 1976; Knauth and Lowe, 1978, 2003; Robert and Chaussidon, 2006) systematically overestimate past ocean temperatures (Wallmann, 2001).

It has long been noted that early Precambrian cherts yield anomalously low $\delta^{18}O_{SMOW}$ values compared with today (Fig. 7.6-4). This has been interpreted in terms of higher ocean temperatures of 75 °C or more during the Archean (Knauth and Epstein, 1976; Knauth and Lowe, 1978, 2003). Most of these data derived from the ca. 3.4 Ga Onverwacht Group of the Barberton Greenstone Belt, South Africa (Fig. 7.6-4), while coverage through the rest of the Precambrian was, and still is, sparse (Perry and Lefticariu, 2003; Robert and Chaussidon, 2006). New data (Knauth and Lowe, 2003) from the Swaziland Supergroup are consistent with previous results (Perry, 1967; Knauth and Lowe, 1978), in that the highest $\delta^{18}O$ value found in these cherts ($+22\%_{SMOW}$) is about 8‰ lower than the lowest values of "representative" shallow marine cherts of the Devonian, while Barberton Greenstone Belt cherts are on average depleted in ^{18}O by about 10‰ compared with recent cherts (Perry and Lefticariu, 2003). By comparison, metacherts of the ca. 3.8 Ga Isua supracrustal belt show even lower maximal $\delta^{18}O$ values, of $+20\%_{SMOW}$ (Perry et al., 1978) and are at higher metamorphic grade than their Barberton (and Pilbara) counterparts. If seawater $\delta^{18}O$ during the early-mid Archean was comparable to today (Muehlenbachs, 1998), and these maximal $\delta^{18}O$ values represent an open marine isotopic signature, then such low $\delta^{18}O$ implies that early oceans were hot, with ambient temperatures of between 55 and 90 °C (Perry and Lefticariu, 2003).

Additional support for hot early oceans has come from a silicon isotope study of marine cherts (Fig. 7.6-4). In their study, Robert and Chaussidon (2006) report a positive correlation between $\delta^{18}O$ and $\delta^{30}Si$ in Precambrian cherts. Unlike $\delta^{18}O$, $\delta^{30}Si$ values in silica do not so much depend on temperature as on the isotopic composition of the medium from which the chert precipitated. The authors argue that seawater $\delta^{30}Si$ is controlled by the difference between the temperature of the oceans and that of hydrothermal fluids. The

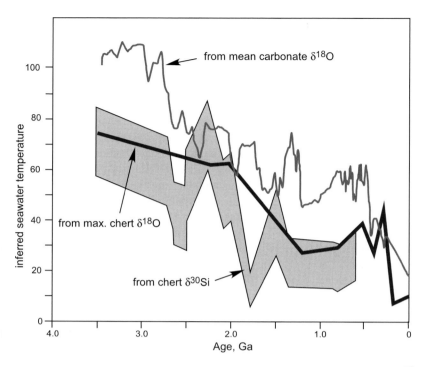

Fig. 7.6-4. Paleotemperature estimates based on the assumption that seawater $\delta^{18}O$ has remained unchanged throughout Earth history. Maximum chert $\delta^{18}O$ line is from Knauth and Lowe (1978) and Robert and Chaussidon (2006); mean carbonate $\delta^{18}O$ line is from data in Shields and Veizer (2002) and Jaffrés (2005); $\delta^{30}Si$ constraints are from Robert and Chaussidon (2006). Mid-Archean marine cherts and carbonates fit the long-term trend of decreasing lower $\delta^{18}O$ with age, and imply implausibly hot climates before 450 Ma (see text for explanation).

observed trend towards higher $\delta^{30}Si$ values between 3.5 and 0.8 Ga could therefore be interpreted as reflecting a progressive decrease in ocean temperature of approximately the same magnitude as that discerned from the $\delta^{18}O$ record. Conversely, this correlation could instead reflect a progressive increase in the globally integrated mean temperature of hydrothermal alteration (Shields and Kasting, 2007), in which case low chert $\delta^{18}O$ values could simply reflect lower seawater $\delta^{18}O$.

The use of cherts to reconstruct past ocean temperature is controversial as the open marine environment of formation of many Precambrian cherts has been questioned (de Wit et al., 1982; Perry and Lefticariu, 2003). In this regard, it may not be meaningful to compare directly the isotopic compositions of (1) greenstone-associated early diagenetic cherts of the early and middle Archean, some of which have been interpreted to be of replacive hydrothermal origin (e.g., Van Kranendonk, 2006), (2) iron formation-associated early diagenetic chert bands formed between 2.5 and 1.8 Ga, or (3) peritidal, evaporite-associated cherts of the Meso- and Neoproterozoic, with predominantly biomediated pelagic cherts

•f the Phanerozoic. Despite such reservations, the general 10‰ depletion of Archean herts is mirrored by an equivalent isotopic depletion in contemporaneous and most later 'recambrian marine carbonates (Shields and Veizer, 2002), suggesting that neither later lteration nor restricted settings was the primary factor controlling the controversial $\delta^{18}O$ rend (Kasting and Ono, 2006).

Like cherts, carbonates commonly show a wide range of $\delta^{18}O$ values (Fig. 7.6-5), which elates to later alteration by meteoric fluids, isotopic exchange at high temperatures, as well s precipitation in ^{18}O-enriched evaporitic settings. However, marine carbonates are less mbiguous than cherts in terms of their depositional environment and possible hydrother- nal overprint, because open marine signatures can be confirmed from Sr and C isotope tudies on the same samples. In addition, some early diagenetic, low-Mg calcite cement ypes, such as molar-tooth structure, can be traced through two billion years back into the Archean (Shields, 2002), thus providing confidence that isotopic trends are meaningful. In everal cases it can be demonstrated that evaporite-related carbonates are systematically nriched in ^{18}O (e.g., Kah, 2000; Bau and Alexander, 2006), indicating that environmental rends are preserved, and that high, outlying $\delta^{18}O_{carb}$ values are not necessarily represen- ative of the global ocean. As with the chert record, ancient marine carbonates are known o be anomalously depleted with respect to their more recent counterparts; however, the ^{18}O trends of these two mineral groups are quite different (Fig. 7.6-4). In the carbon- te record, low $\delta^{18}O$ values are found throughout the period before 450 Ma (Shields and Veizer, 2002), while the chert record shows modern day-like $\delta^{18}O$ and $\delta^{30}Si$ values already luring the Mesoproterozoic (Knauth and Lowe, 1978), and Paleoproterozoic (Robert and Chaussidon, 2006).

Well preserved, low-Mg calcite brachiopod shells (Veizer et al., 1999) and early dia- ;enetic, low-Mg calcite cements (Johnson and Goldstein, 1993) of the early Phanerozoic xhibit the lowest $\delta^{18}O$ values of the Phanerozoic. Highest $\delta^{18}O$ values of these demonstra- •ly well-preserved marine minerals are generally between 0–5‰ ^{18}O-enriched relative to lighest values for early marine cements (molar-tooth structure) of the Archean and Protero- :oic (Fig. 7.6-5). Because mean ocean temperatures could not realistically have exceeded 15 °C since the evolution of vertebrates, low carbonate $\delta^{18}O$ during the early Phanero- :oic has been interpreted as primarily due to increasing seawater $\delta^{18}O$ from $-5‰_{SMOW}$ ince 500 Ma, with temperature a secondary factor only (Veizer et al., 2000). Some well- •reserved Archean limestone units (Abell et al., 1985a, 1985b; Bishop et al., 2006) even xhibit $\delta^{18}O$ values (Fig. 7.6-5) that are mostly within the normal range of well-preserved, arly Ordovician brachiopod shells (Shields et al., 2003), which is inconsistent with any najor decrease in surface temperatures since at least 2.9 Ga. If low seawater $\delta^{18}O$ per- .isted throughout the entire Precambrian, then temperatures are unlikely to have reached ar outside the range of Phanerozoic oceans. An apparent maximum depletion of 5‰ from he Neoarchean to the Cambrian/Ordovician (cf. Shields et al., 2003; Bishop et al., 2006) mplies a maximal temperature difference of $\sim$20 °C. In other words, ocean temperatures ligher than about 55 °C seem to be highly unlikely during the late Archean. But what about he pre-3.0 Ga world?

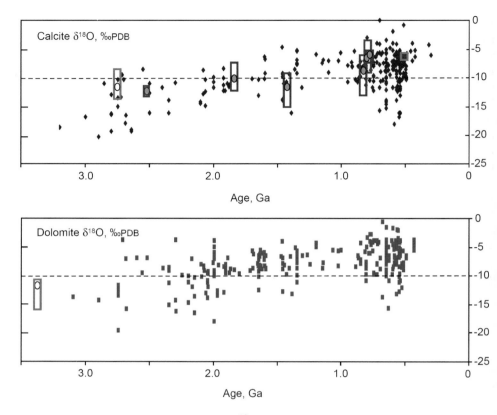

Fig. 7.6-5. Marine limestone and dolostone $\delta^{18}O$ plotted as mean values of data from stratigraphic units with sample size >2 (Shields and Veizer, 2002; Jaffrés, 2005). 98 sample sets characterized as clearly altered or from clearly evaporitic settings have been removed from a total of 785. Boxes with filled circles refer to the range and mean $\delta^{18}O$ values, respectively, of early marine, originally low-Mg-calcite cements (Johnson and Goldstein, 1993; Bishop et al., 2006; Shields, unpubl. data); open circles are from representative isotopic studies of Archean carbonate rocks (Abell et al., 1985a; Lindsay et al., 2005). With the exception of intervals of cool climate, calcites and dolomites of the Precambrian and early Phanerozoic are up to 5–10‰ depleted in ^{18}O with respect to later Phanerozoic and modern equivalents (Veizer et al., 1999; Shields and Veizer, 2002; Shields et al., 2003). The persistence of low $\delta^{18}O$ into the early Ordovician implies however that changing seawater $^{18}O/^{16}O$ and not temperature is likely to have been the major controlling factor over this trend. This implies that surface temperatures were not as extreme on the early Earth as implied in Fig. 7.6-4, and most likely lower than 55 °C since 3.5 Ga.

The existing early-mid Archean $\delta^{18}O$ record is unfortunately sparse and is rendered less useful by the frequent absence of mineralogical and petrographic details in publications. Nevertheless, if we consider the previously discussed constraints on pristine marine dolomite $\delta^{18}O$ from the ca. 3.4 Ga Strelley Pool Chert (Lindsay et al., 2005) of

$\sim -11\%_{PDB}$ (Figs. 7.6-3, 7.6-5), it is clear that, even considering a systematic difference between calcite and dolomite of $\sim 2\%_0$ (Shields and Veizer, 2002), such values are not considerably depleted with respect to Neoarchean calcite. Similar values have also been reported for the Barberton Greenstone Belt of South Africa (Veizer et al., 1989; Abell et al., 1985b). Therefore, the above temperature constraints may also apply to the earlier Archean, too, assuming that seawater $\delta^{18}O$ was at least as low as its early Phanerozoic estimated level of $-5\%_{SMOW}$.

Significant changes of $>2\%_0$ in seawater $\delta^{18}O$ can only be induced through variations in the relative proportion of low- to high-temperature ($>350\,°C$) alteration of crustal silicates either on the continents or beneath the sea floor (Wallmann, 2001). According to Kasting et al. (2006), the higher Archean mantle heat flux would have favoured shallower penetration of hydrothermal fluids into the ocean crust, which would have in turn enhanced the influence of low-temperature alteration. This would imply that the early Phanerozoic $\delta^{18}O_{sw}$ estimate of $-5\%_{SMOW}$ could only have been lower, and not higher during the early Archean. If correct, then Archean ocean temperatures were no hotter, and probably significantly cooler, than $\sim 55\,°C$, which is more consistent with independent temperature constraints from the geological record and atmospheric modelling (Kasting and Ono, 2006).

7.6-5. SUMMARY

Although well preserved marine sedimentary rocks of the early-mid Archean are scarce, existing isotope constraints can still provide us with important insight into early Earth System evolution by comparison with the more recent isotopic record. Sr isotope trends indicate that pre-2.7 Ga oceans were dominated by chemical exchange with juvenile magmatic rocks or mantle input, while post-2.7 Ga oceans were increasingly influenced by river input from a geochemically differentiated continental crust. This implies a relative waning of hydrothermal seafloor circulation relative to continental weathering that appears to be associated with a major crustal building episode around the Archean-Proterozoic boundary. The persistence of modern day-like carbon isotope systematics as far back as existing rock archives allow has been used to demonstrate the antiquity of life on Earth. Although some recent studies call into question the very earliest of these records, ca. 3.2–3.4 Ga carbonates and cherts preserve isotopic signatures consistent with microbial metabolisms and imply that at least 14% of the carbon being delivered by outgassing of mantle CO_2 was buried as reduced, organic carbon already by 3.4 Ga. The early evolution of photosynthesis on Earth is consistent with the widespread occurrence of sedimentary features interpreted to be biogenic in origin, such as stromatolites and wrinkled laminations in Earth's most ancient rocks. Early Earth climate is currently the subject of much debate. The oxygen isotopic compositions of marine minerals have been used to imply extremely high temperatures ($>70\,°C$) on the early Earth. Alternative explanations for low $\delta^{18}O$ in ancient sedimentary carbonates and cherts imply that temperatures typically did not exceed $55\,°C$ and could have remained within modern limits as far back as 3.5 Ga.

PART 8

TECTONICS ON EARLY EARTH

Earth's Oldest Rocks
Edited by Martin J. Van Kranendonk, R. Hugh Smithies and Vickie C. Bennett
Developments in Precambrian Geology, Vol. 15 (K.C. Condie, Series Editor)
© 2007 Elsevier B.V. All rights reserved.
DOI: 10.1016/S0166-2635(07)15081-7

Chapter 8.1

VENUS: A THIN-LITHOSPHERE ANALOG FOR EARLY EARTH?

VICKI L. HANSEN

Department of Geological Sciences, University of Minnesota Duluth, 231 Heller Hall, 1114 Kirby Drive, Duluth, MN 55812, USA

8.1-1. INTRODUCTION

Plate tectonics requires a specific global-scale rheology. The Hadean to Eoarchaean Earth likely lacked this rheology locally, regionally, or perhaps even globally. And yet heat was transferred to the surface, and the planet cooled. It can be difficult to envision regional or global processes other than plate tectonics, in part due to the elegance and comprehensiveness of the plate tectonic model. Yet early Earth was significantly different that modern Earth: bolides must have impacted the surface; magma oceans or seas likely existed locally or globally; the 'lithosphere' may have been weak, marked by a ductile-solid rather than a brittle-solid rheology. All of these factors would influence the tectonic processes, as well as the preserved record of operative processes. Venus – Earth's sister planet – of similar age, density, size, inferred composition, and inferred heat budget as Earth, might be expected to cool through similar, terrestrial plate tectonic processes. However, Venus lacks any evidence of plate tectonics (e.g., Solomon et al., 1991; Phillips and Hansen, 1994; Nimmo and McKenzie, 1998). Thus, Venus might provide a rich arena in which to stretch ones' tectonic imagination with respect to non-plate tectonic processes of heat transfer for an Earth-like planet, providing for means to test geologic histories against multiple hypotheses, aimed at understanding possible early Earth processes (e.g., Gilbert, 1886; Chamberlin, 1897).

Contemporary Earth differs from the Hadean to Eoarchaean Earth; contemporary Venus likely also differs from ancient Venus, and certainly differs from contemporary Earth. Although there is much debate about Venus' evolution, it might be that at least a part of the history recorded on Venus' surface provides clues to processes on very early Earth. Venusian conditions now, and in the past, are perhaps more akin to environmental conditions of the early Earth (Lecuyer et al., 2000). Venus' atmosphere, ~95 bars of supercritical CO_2, forms a strong blanket of insulation with a current surface temperature of ~475 °C and higher temperatures likely in the past (Bullock and Grinspoon, 1996, 2001; Phillips et al., 2001). Venus is homogeneously hot, resulting in an ultra-dry environment with little sediment formation, transport, or deposition, although it may have experienced a wetter past (Donahue and Russell, 1997; Donahue et al., 1997; Donahue,

1999; Lecuyer et al., 2000; Hunten, 2002). Venus' ~970 recognized impact craters represent some of the most pristine impact features in the solar system. The pristine nature of Venus' surface and these craters are testament to a lack of sediment. The current surface shows only minimal amounts of weathering and erosion, with typical weathering rates estimated at $<10^{-3}$ mm/yr (Campbell et al., 1997). With essentially no difference in diurnal, or equatorial-polar temperatures, Venus lacks regionally organized surface winds.

Venus' crust is believed to be homogeneously basaltic (e.g., Grimm and Hess, 1997), due in part to a lack of water, which plays a critical role in the formation of granitic magma. Venus' dense atmosphere also likely affects volcanic, and presumably tectonic processes in that supercritical CO_2 acts more like a conducting layer than a convective layer with regard to heat transport (Snyder, 2002). Venus' atmosphere may have inhibited crystallization and solidification of lava, contributing to low lava viscosity, and perhaps volcanotectonic styles significantly different that of uniformitarian Earth. Venus preserves a record of an ancient era in which the lithosphere (or crust) was globally thin, followed by contemporary Venus with a thick immobile lithosphere. The interplay of atmosphere and lithospheric processes deserves attention in Venus tectonic-volcanic investigations, and may also play an important role in Hadean to Eoarchaean terrestrial processes.

Venus' atmosphere also blocks optical light, completely veiling the surface prior to development of radar technology. The NASA *Magellan* mission, returned incredible views through Venus' clouds providing detailed, and globally comprehensive, images of the surface – and with this astounding data set, glimpses of a whole new world of tectonic processes. These data are digital in form, global in coverage, and accessible via the world-wide-web.

It is likely that Venus and Earth were most similar at birth. Earth's more contemporary plate tectonic processes destroyed much of the surface record of its early history. Venus' lack of plate tectonics means that Venus might preserve a better record of its formative years, and the tectonic processes that shaped it. Thus, although we cannot travel back in time to the Hadean to Archean on Earth, perhaps we can travel through space to consider possible global-scale tectonic processes in an environment possibly akin to that of early Earth.

In this contribution, I briefly discuss four different types of Venusian tectonomagmatic features: (1) radial coronae, (2) Artemis, (3) crustal plateau fabrics, and (4) deformation belts. Each of these types of features likely formed on thin lithosphere, and in some cases weak lithosphere, and none record plate tectonic-related processes (although their modes of formation may be debated, quite enthusiastically in some cases). Radial coronae and Artemis might represent lithospheric signatures of diapirs, though resulting from compositional and thermal buoyancy, respectively. Hypotheses proposed for crustal plateau evolution include mantle downwelling, mantle plumes, and solidification of huge lava ponds. The latter is favored herein, and may provide clues to the evolution of terrestrial magma ocean surfaces. Crustal plateaux (or the lava ponds they represent) may owe their origin to ancient large impact events. Thus Venus provides a reminder that exogenic processes likely play critical roles on all terrestrial planets, particularly in early planet evolution. Deformation belts form linear high strain zones separated by low strain domains, and may record density inversion of the crust, similar to terrestrial granite-greenstone terrains. These belts provide a caution for interpreting orogen-scale linearity as compelling evidence for plate tectonic processes.

8.1-2. VENUS OVERVIEW

Venus and Earth share many similarities, yet they also have profound differences. Venus, 0.72 AU from the Sun, is 95% Earth's size and 81.5% Earth's mass. Solar distance, similar mean density, and cosmo-chemical models for solar system evolution lead to the inference that Venus and Earth share similar bulk composition and heat producing elements (Wetherill, 1990). Data from Soviet Venera and Vega landers indicate surface element abundance consistent with basaltic composition, although the limited data could accommodate other compositions (Grimm and Hess, 1997). Slow retrograde motion makes a Venus day longer than its year (243 and 225 Earth days, respectively), a factor that may contribute to Venus' lack of a magnetic field (Yoder, 1997). Atmospheric composition (96% CO_2, 3.5% N_2 and 0.5% H_2O, H_2SO_4, HCl and HF), surface pressure ($\sim$95 bars) and temperature ($\sim$475 °C) might be similar to Earth's early atmosphere (Lecuyer et al., 2000).

Venus' surface conditions are intimately related to its dense caustic atmosphere, which includes three cloud layers 48–70 km above the surface. The upper atmosphere rotates at a rate of $\sim$300 km/hr, circulating in four Earth days. The clouds reflect visible light and block optical observation. The dense atmosphere results in negligible diurnal temperature variations and an enhanced global greenhouse that makes a terrestrial-style water cycle impossible. Given high surface pressure and temperature, CO_2 exists as a supercritical fluid. Venus lacks obvious evidence of weathering, erosion, and sediment transport and deposition processes, or extensive sedimentary layers clearly deposited by wind or water. Although Venus is presently ultra-dry, the past role of water is unknown. Isotopic data are consistent with, but do not require, extensive reservoirs of water $\geqslant$1 billion years ago (Donahue and Russell, 1997; Donahue et al., 1997; Donahue, 1999; Lecuyer et al., 2000; Hunten, 2002). A lack of water renders Venusian (current) crustal rock orders of magnitude stronger than terrestrial counterparts, even given Venus' elevated surface temperature (Mackwell et al., 1998), a factor critical to topographic support. For example, Maxwell Montes, Venus' highest point (11 km above mean planetary radius, MPR, $\sim$6052 km) could only be 5 m.y. old under Venus' current surface conditions if the rock was typical 'wet' terrestrial basalt (Grimm and Solomon, 1988).

Most workers assume that Venus' mantle is similar in composition and temperature to Earth's mantle. A reasonable working hypothesis is that Venus' effective mantle viscosity is similar to that of Earth, and similarly has a strong temperature-dependent viscosity profile. However, some workers consider Venus' mantle to be stiffer than Earth's due to presumed drier conditions (e.g., 10^{22} Pa s, Kaula (1990); 10^{22-24} Pa s, Turcotte et al. (1999)). Volatiles are of course important in understanding Venus' interior – particularly with regard to viscosity. However, interior volatile values and compositions are currently unconstrained. Lack of volatiles will increase strength and increase the mantle solidus. In contrast, the presence of volatiles will decrease strength and decrease the mantle solidus. Large viscosity contrasts are likely across thermal boundary layers: notably across the lithosphere and core-mantle boundaries. Furthermore, it is possible, and perhaps likely, that mantle viscosity structure has changed through time, and there is no guarantee that contemporary mantle viscosity represents the viscosity structure that accompanied the formation of surface features, and particularly not the surface features discussed herein. The same caution is of course true for the Hadean to Archaean Earth. It does seem clear, however, that, in contrast to Earth, Venus currently lacks a low viscosity asthenosphere, and most likely did so for the duration of its recorded surface history. The lack of an asthenosphere is no doubt critical to Venus' lack of plate tectonics.

Venus' lack of plate tectonics and terrestrial surficial processes (glaciation, erosion, and deposition), results in preservation of a unique surface record of tectonomagmatic processes. Large portions (perhaps all) of the lithosphere have not been completely recycled to the mantle. (In contrast to most workers, Turcotte et al. (1999) propose for episodic lithospheric overturn driven by turbulent mantle flow.) Nor has Venus' surface been extensively dissected, carved or buried as is common on Earth and Mars. Although Venus preserves ~1000 impact craters, its dense atmosphere has shielded its surface from extensive cratering, associated 'gardening', and the development of a thick impact regolith like the Moon. Thus, Venus' surface provides a unique record of non-plate tectonomagmatic processes. Until recently, Venus' atmosphere has veiled this surface from view, but with radar technology the veil has fallen away, allowing us to study this unique surface in incredible global detail. In this contribution, I focus on information gleaned from the surface through the NASA *Magellan* mission.

8.1-2.1. Venus Data

The NASA *Magellan* mission collected four global remote data sets: emissivity, high-resolution gravity, altimetry, and synthetic aperture radar (SAR) images (Ford and Pettengill, 1992; Ford et al., 1993). These data, together with early data from Soviet Venera missions, Pioneer Venus and Arecibo, provide views of Venus' surface. The Soviet Venera Landers provided visible glimpses of the surface and compositional information at a few locations (Barsukov et al., 1986; Surkov et al., 1986). The *Magellan* data, ancillary documentation and software are available through the Planetary Data System [PDS, http://pds.jpl.nasa.gov/]. Emissivity, not discussed herein, is chiefly controlled by dielectric permittivity and surface roughness (Pettengill et al., 1992). Gravity data can resolve features >400 km, and provides clues to subsurface architecture, although interpretations are nonunique. Altimetry data (spatial resolution of ~8 km by ~20 km, along- and across-track; vertical resolution ~50 m) resolves long-wavelength features and morphology, but most topographic features related to primary and secondary structures are only resolvable using SAR data. SAR data (~100 m/pixel), which covers 98% of the surface with local overlap among mapping cycles, allows for geomorphic and geological interpretations, including geologic surface histories.

8.1-2.1.1. SAR data

SAR data (available at http://pdsmaps.wr.usgs.gov/maps.html) was collected in three cycles: left- (cycle 1) and right-look (cycle 2), and stereo (cycle 3). The effective resolution (Zimbelman, 2001) of SAR images depends, in part, on the features of interest; as a general rule, features, other than lineaments, should be >300 m (Fig. 8.1-1). SAR data from cycles 1 and 3 can be combined to provide true stereo (3D) views (Plaut, 1993), although cycle 3 data is limited. A combination of SAR and altimetry data results in synthetic stereo, 3D, views (Kirk et al., 1992), with near global coverage. Cautions for interpretation of geologic features and histories are outlined in a variety of contributions (Wilhelms, 1990; Ford et al., 1993; Tanaka et al., 1994; Hansen, 2000; Zimbelman, 2001). Global geologic mapping is underway as part of the NASA-USGS VMap (1:5,000,000) program (http://astrogeology.usgs.gov/Projects/PlanetaryMapping/PGM_home.html).

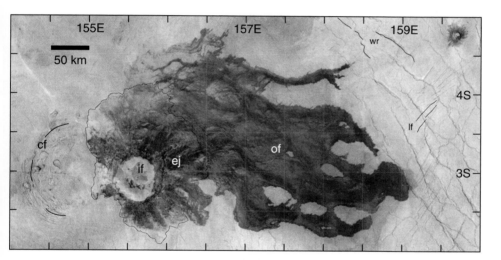

Fig. 8.1-1. Inverted, left-illumination SAR image of Markham Crater, Venus, with extensive outflow material. Flows might be related to the impact event, or they represent subsurface magma that escaped to the surface as a result of bolide impact. Note the central peak, and radar-smooth (bright) flooded interior (if). The crater rim and ejecta (ej) and outflow material (of) appear mostly radar-rough (dark), whereas the basal material is generally radar smooth (bright). Both the ejecta and the outflow material preserve various radar-backscatter facies indicating a range of backscatter properties (likely roughness) within individual geologic material units. The black line marks the limit of the ejecta deposit. Basal material is cut by concentric fractures (cf) to the west of Markham, and by linear (though somewhat sinuous) NW-trending wrinkle ridges (wr, topographic ridges) and NE-trending linear (straight) fractures (lf, narrow troughs) to the east. A small impact crater with a flooded interior and radar-rough ejecta occurs in the NE corner of image.

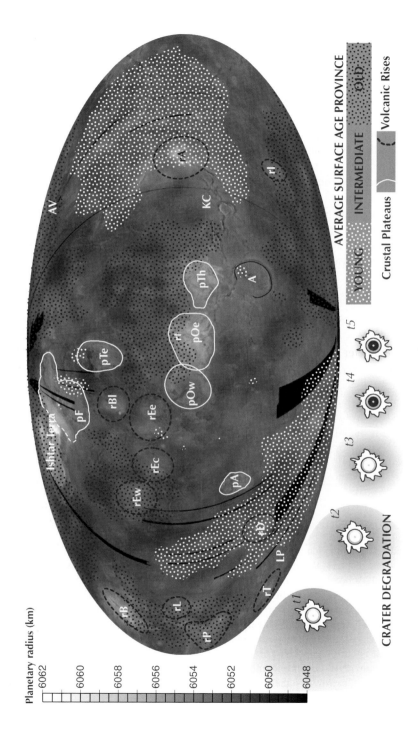

8.1-2.2. Venus Features

Magellan data permit first order characterization of Venus' surface, which is divisible into the lowlands (80%), mesolands (10%), and highlands (10%), based on altimetry. The lowlands, which lie at or below MPR, include relatively smooth low-strain surfaces called plains, or planitiae, and linear deformation belts (Banerdt et al., 1997). The mesolands lie at intermediate elevations and host many coronae, quasi-circular tectonomagmatic features, and chasmata – regional scale linear troughs decorated with tectonic lineaments. Highland regions include volcanic rises, crustal plateaux, and the unique feature Ishtar Terra (Hansen et al., 1997) (Fig. 8.1-2). Volcanic rises are large (1500–2500 km diameter) domical regions, 1–3 km high, marked by local radial volcanic flows. They are widely accepted as contemporary (that is, currently thermally supported) surface expressions of deep mantle plumes on thick lithosphere (e.g., Phillips et al., 1981, 1991; McGill, 1994; Phillips and Hansen, 1994; Smrekar et al., 1997; Nimmo and McKenzie, 1998). Crustal plateaux, similar in planform to rises, but steep sided and flat topped, host unique tectonic fabrics called ribbon-tessera terrain. Plateaux lie 0.5–4 km above their surroundings, the result of shallow isostatic support (Hansen et al., 1997, and references therein). A wide variety of volcanic landforms, preserved at a range of scales, occur across the surface, generally independent of elevation (Head et al., 1992; Crumpler et al., 1997). Hundreds of thousands of volcanic shields, 1–20 km diameter (Guest et al., 1992; Addington, 2001), occur in shield fields (<300 km diameter regions) and as 'shield terrain' (Aubele, 1996; Hansen, 2005) distributed across millions of km^2; lava flows up to hundreds of km long are commonly associated with volcanoes, coronae, and fractures (Crumpler et al., 1997). Volcanic forms are generally consistent with basaltic compositions (e.g., Bridges, 1995, 1997; Stofan et al., 2000). Venus also displays unique narrow channels (1–3 km wide) that trace across the lowlands for tens or hundreds of km (up to the ~6900 km long Baltis) (Baker et al., 1997). Although all scientists agree that the channels are fluid cut, many questions remain

---

Fig. 8.1-2. (*Previous page.*) Mollwiede projection of *Magellan* altimetry with average model surface age (AMSA) provinces (data from Phillips and Izenberg (1995)) and major geologic features including crustal plateaux (Alpha (pA), Fortuna (pF), eastern and western Ovda (pOe, pOw), Phoebe (pP), Tellus (pTe), and Thetis (pTh)) and volcanic rises (Alta (rA), Beta (rB), Bell (rBl), Dione (rD), western, central and Eastern Eistla (rEw, rEc, rEe), Imdr (rI), and Themis (rT)). Phoebe (pP) is transitional between a plateau and a rise (Grimm, 1994; Simons et al., 1997; Hansen and Willis, 1998; Phillips and Hansen, 1998). Crater degradation stages show youngest (t1) to oldest (t5) changes in crater morphology; with time and degradation, an crater loses its halo and its interior fills with lava (Izenberg et al., 1994). Three relative AMSA provinces – old, intermediate and young – are defined based on impact crater density and impact crater degradation stage (Phillips and Izenberg, 1995). Figure locations include: Khabuchi Corona, KC (Fig. 8.1-3); Artemis, A (Fig. 8.1-4), Alpha Region, pA (Fig. 8.1-5), Ovda ribbon-tessera terrain, rt (Fig. 8.1-6), Atlanta-Vinmara deformation belts, AV (Fig. 8.1-7), and Lavinia Planitia deformation belts, LP (Fig. 8.1-8). Modified from Hansen and Young (2007).

debated: Are channels erosional or constructional? Do they represent thermal or mechanical processes? What was the nature of the fluid? What is the substrate? Were channels constructional, down cut, or formed by subsurface stoping? (e.g., Baker et al., 1992, 1997; Komatsu and Baker, 1994; Gregg and Greeley, 1993; Bussey et al., 1995; Williams-Jones et al., 1998; Jones and Pickering, 2003; Lang and Hansen, 2006).

8.1-2.3. Venus' Surface, Time and Cautions

Any discussion of Venus geology is not complete without a brief discussion about time. SAR images provide high-resolution views of the surface, which allow determination of cross cutting relations and relative history. However, global, and even regional, correlation of geologic units, or interpreted events, commonly involve circular reasoning given the fundamental 2D nature of remote sensing data (Hansen, 2000). Furthermore, absolute geologic time cannot currently be constrained on Venus. To date, impact crater density provides the only hope of constraining absolute time on planet surfaces other than Earth. Impact crater 'dating' might be viable on Moon, Mars and Mercury due to the extremely high number of total craters and the wide range in surface crater density. Impact crater dating is ultimately a statistical exercise, and includes several geological challenges (Hartmann, 1998). Venus lacks small craters due to screening by the dense atmosphere. Small craters typically comprise the largest number of carters on a planetary surface, with crater density ages dependent on binning across a range of crater diameters – a technique not possible on Venus (McKinnon et al., 1997). In addition, Venus' craters are distributed in near random fashion (Schaber et al., 1992; Phillips et al., 1992; Hauck et al., 1998). The low number of craters and near random spatial distribution prohibit robust temporal constraints for individual geomorphic features or geologic units (Campbell, 1999). The minimum size area that can be dated statistically by crater density alone is 20×10^6 km^2, or 4.5% of the surface (Phillips et al., 1992). Some workers propose age constraints based on combining morphologically similar features/units into large composite regions for crater density dating (e.g., Namiki and Solomon, 1994; Price and Suppe, 1994; Price et al., 1996). Because these works implicitly assume the combined features formed synchronously, the analyses are circular and lack temporally robust conclusions. Furthermore, dating such large surfaces, even if contiguous, requires assumptions that severely limit the uniqueness of any temporal interpretation (Campbell, 1999). In short, even large surfaces (20×10^6 km^2) are effectively 'undatable'.

Impact crater density analysis of Venus results, at best, in determination of average model surface age (AMSA) provinces – the integrated age of a huge region. Venus records a *global* AMSA (that is, an AMSA for the entire surface) of $\sim 750 + 350/-400$ Ma, based on total impact craters and impactor flux (McKinnon et al., 1997). This global AMSA could be accommodated by a wide range of possible surface histories – conceptually similar to a terrestrial ε_{Nd} average mantle model age (e.g., Farmer and DePaolo, 1983). The global AMSA must be met by any hypothesis, but it provides few unique requirements.

Strom et al. (1994) explored catastrophic (Schaber et al., 1992) versus equilibrium (Phillips, 1993) resurfacing models though Monte Carlo modeling. They varied the areal coverage and iterations of resurfacing from 50%, 25%, 10%, 0.03% and 0.01% of the surface, and considered the final crater distribution (random or not) and the number of embayed craters. The first three experiments yielded low crater embayment, as observed (e.g., Phillips et al., 1992; Schaber et al., 1992; Herrick et al., 1997), but not random crater distribution. In contrast the last two experiments met the random distribution criteria, but predicted high crater embayment. Thus, Strom et al. (1994) called for catastrophic volcanic resurfacing of Venus, ~500 Ma, with ~3 km thick flood lava emplaced globally over a 10–100 m.y. event. The longer the 'catastrophic' event, the more embayed craters, and thus the more at odds with the data. In keeping with catastrophic resurfacing, Basilevsky and Head (1996, 1998, 2002) proposed that Venus displays a coherent global stratigraphy with basal tessera terrain buried by 1–3 km thick flood lava, emplaced quickly and recently.

Although crater density alone cannot delineate statistically distinct, temporally defined, regions, Phillips and Izenberg (1995) subdivided the surface into three AMSA provinces using impact crater density *and* crater morphology (Fig. 8.1-2). Izenberg et al. (1994) recognized a temporal sequence of impact crater degradation allowing the division of craters into relative age groups. Young craters display haloes and radar-rough interiors; old craters lack haloes and show radar-smooth (presumably flooded) interiors. The AMSA provinces – which represent relative rather than absolute age provinces – cannot constrain the age of individual geologic features or units, but rather they represent an average age of an integrated history of these surfaces, reflecting geologic processes that would lead to formation, modification, or destruction of impact craters. Because crater formation is global, and because craters are mostly pristine, the critical factor would seem to be process(es) of crater destruction. Although no individual geologic units or features are robustly temporally constrained, individual features, or groups of features, might show spatial patterns with respect to the three AMSA provinces, and such patterns might provide clues to the relative temporal evolution. But such spatial correlation should never be accepted as a robust age. The presence of three AMSA provinces does, however, provide strong evidence against the hypotheses of global catastrophic volcanic resurfacing of Venus (e.g., Schaber et al., 1992; Strom et al., 1994) and global episodic lithospheric overturn (Turcotte, 1993; Turcotte et al., 1999), which each require a single global AMSA (Phillips and Izenberg, 1995). Hansen and Young (2007) evaluate resurfacing hypotheses with implications for Venus evolution, a topic outside the limits of the current contribution. The important point for the current discussion is that absolute age is unconstrained across Venus, although most workers agree that Venus likely experienced an early Era marked by globally thin lithosphere (<30 km), followed by contemporary Venus marked by thick (100–300 km) lithosphere (e.g., Solomon, 1993; Grimm, 1994; Solomatov and Moresi, 1996; Phillips et al., 1997; Schubert et al., 1997; Hansen and Willis, 1998; Brown and Grimm, 1999; Phillips and Hansen, 1998). The timing of the global transition from thin to thick lithosphere is unconstrained. Venus currently lacks a sharp asthenosphere boundary (Phillips and Hansen, 1994; Phillips et al., 1997; Schubert et al., 1997), perhaps the most important feature of terrestrial plate mechanics.

8.1-3. THIN LITHOSPHERE TECTONOMAGMATIC FEATURES

Terrestrial provinces that preserve views into early Earth typically provide a record of crustal depth, but a regional plan-view record is extremely limited. Venus, on the other hand, provides essentially only a plan-view. Venus' plan-view is continuous and available at an amazing scale of observation, such that geologic histories (and hence temporal dimension – albeit only relative time) might be interpreted, with appropriate cautions.

8.1-3.1. *Radial Coronae: Surface Expression of Compositional Diapirs?*

Coronae (Barsukov et al., 1984), commonly considered unique to Venus, are circular to quasi-circular features typically marked by a raised rim or annulus that displays concentric annular structures (fractures, faults or folds), and variable tectonic and volcanic features, including radial fractures and extensive lava flow deposits (Fig. 8.1-3). Coronae range in size from 60–1050 km diameter (200 km median), and number about 500 (Stofan et al., 1992, 2001). Most coronae occur in chains (68%) or clusters (21%) spatially associated with mesoland chasmata and volcanic rises, respectively; limited coronae (11%) occur as isolated features in the lowlands (Stofan et al. 1992, 1997, 2001; DeLaughter and Jurdy, 1999). Coronae, meaning crown, was initially a descriptive term, but it has evolved into a term which commonly carries genetic connotations. Coronae are widely accepted as representing the surface manifestation of mantle diapirs forming tectonomagmatic 'blisters' in/on the lithosphere (e.g., Stofan et al., 1992, 1997; Squyres et al., 1992a; Janes et al., 1992; Janes and Squyres, 1995; Koch and Manga, 1996; Smrekar and Stofan, 1997). Diapiric models propose evolution characterized by: central doming, radial fracturing, volcanism, eventual reduction of interior topography, production of an annular ring, and possible late subsidence. The wide range of coronae characteristics might represent stages of corona evolution, or they might indicate that features collectively referred to as coronae include genetically unrelated features. For example, some or all coronae could represent volcanic calderas, impact craters, or Rayleigh–Taylor instabilities in a density stratified lithosphere (e.g., Squyres et al., 1992a; Nikolayeva, 1993; Hamilton 1993, 2005; Schultz, 1993; McDaniel and Hansen, 2005; Vita-Finzi et al., 2005; Hoogenboom and Houseman, 2006). The spatial association of corona chains and clusters with chasmata and rises, respectively, favors endogenic (over exogenic) formation for these types of coronae. In short, all coronae might not have formed by the same processes. Discussion here focuses on coronae marked by radial fractures.

Khabuchi Corona (Fig. 8.1-3) represents a typical 'radial corona', within an equatorial coronae chain. Radial coronae fit the predictions of diapiric models with radial fractures and flows that formed broadly synchronously, followed by temporally overlapping concentric fracture formation, with emergence of additional surface flows. Radial coronae seem to lie dominantly within corona-chasmata chains and in clusters associated with volcanic rises, and record rich histories involving broadly contemporaneous fracturing, folding, volcanism, and presumably subsurface magmatism (Hamilton and Stofan, 1996; Stofan et al., 1997; Copp et al., 1998; Chapman, 1999; Hansen and DeShon, 2002). Radial coronae with

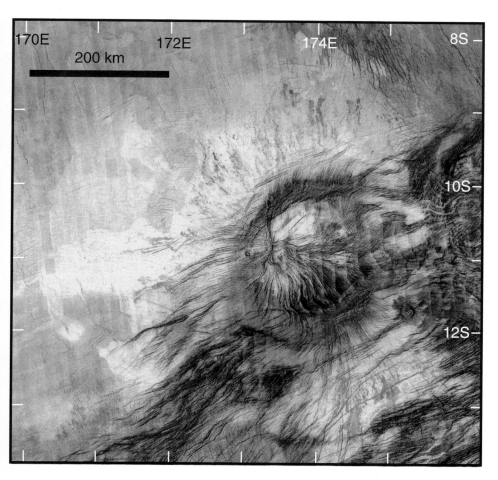

Fig. 8.1-3. Inverted, left-illumination SAR image of Khabuchi Corona, marked by radial and concentric fractures and radial volcanic flows. Interior region sits topographically high relative to the surroundings.

diameters <350 km likely formed on relative thin crust/lithosphere (5–10 km; Janes and Squyres, 1995; Koch and Manga, 1996), and result from diapirs driven by compositional buoyancy (Hansen, 2003). The periodic spacing of chained coronae may reflect the depth (150–250 km) to a layer of mantle instability that served as the source layer (Hamilton and Stofan, 1996).

Radial corona, likely related to diapiric rise and associated deformation of cover material in the form of radial fracturing, concentric shedding of cover (forming concentric folds or fractures), and synchronous volcanism might be broadly analogous to the formation of some terrestrial granite-greenstone belts (e.g., Rey et al., 2003). Indeed, clustered coronae associated with volcanic rises are proposed to result from a deep mantle plume, similar to hypotheses proposed for granitoid doming in the Pilbara Craton (e.g., Pawley et al., 2004; Van Kranendonk et al., 2004; Smithies et al., 2005b) and the northeast Superior Province (e.g., Bédard et al., 2003). Hoogenboom and Houseman (2006) propose that coronae (they do not differentiate corona type) result from lithospheric density inversions (density inversions are discussed further in the section on deformation belts). Presumably, topography would subside with time, being thermally supported, but the tectonomagmatic signature would remain. Similarly, chained coronae – postulated to form above cylindrical mantle upwellings – might be analogous to Archean granite-greenstone terrains, which might represent just a fragment of an originally much more expansive terrain. If coronae existed in the early Earth, they might have played a critical role in early tectonic processes contributing to planet cooling and heat transfer, as well as perhaps crustal differentiation. If coronae formed in a subaqueous environment, they could have harbored early life forms with interaction of aqueous systems (e.g., Van Kranendonk, 2006). Numerous questions with regard to coronae evolution on Venus remain unanswered and controversial.

Radial fracture patterns also form giant radial dike swarms across Venus (e.g., Ernst et al., 2001, 2003). Some overlap exists between features mapped as radial coronae and as giant radial dike swarms; the two types of features could be genetically related, or the overlap may be serendipitous. Radial dike swarms typically have radii that far exceed that of coronae annuli. For example, the radial dike swarm that centers on Heng-O Corona (1010 km diameter annulus) has a radius of >1000 km. Although giant radial dike swarms occur on Venus and Earth, the oldest known terrestrial giant radial dike swarm is Early Proterozic – far younger than Hadean to Eoarchaean. The formation of giant radial dike swarms require huge expanses of strong lithosphere, or plates. Therefore, the occurrence of giant radial dike swarms might place a minimum temporal limit on the existence of global scale plates, and provide robust temporal constraints on lithosphere rheology. Indeed Venus' giant radial dike swarms cross cut, and are therefore younger than, both ribbon tessera terrain and deformation belts (Ernst et al., 2003), discussed below. Although giant radial dike swarms exist on both planets, they likely formed relatively late in planet evolution and therefore are not discussed further herein. On Earth the formation of such features might reflect global conditions ripe for modern plate tectonic processes.

8.1-3.2. *Artemis: Surface Expression of a Large Mantle Plume?*

The formation of Artemis, the largest circular feature on Venus, and perhaps the largest circular feature in the solar system, remains a puzzle. Artemis comprises a huge topographic welt, 2600 km in diameter, that includes a paired circular trough (150–200 km wide; ~1–1.5 km-deep) and outer rise (200 km wide) (Fig. 8.1-4). Artemis defies geomorphic classification: it is similar in size to crustal plateaux and volcanic rises, yet topographically more akin to many coronae. Artemis has been classified as a corona (Stofan et al., 1992), but given its large size, this classification is questionable (Stofan et al., 1997, Hansen, 2002). Herein the feature is simply referred to as 'Artemis', following Hansen (2002). Artemis' trough describes a partial circle that extends clockwise from ~12:00 to 10:30 in

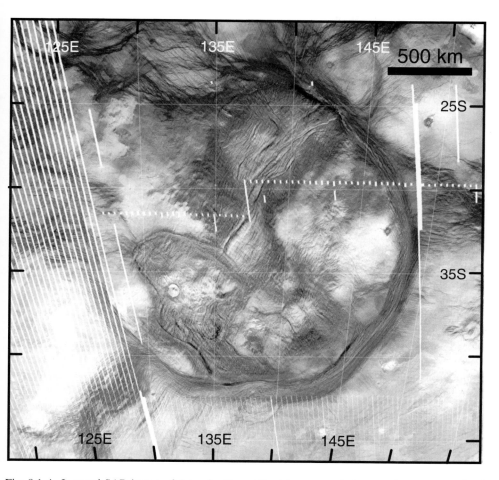

Fig. 8.1-4. Inverted SAR image of Artemis. The white strips, most obvious on the western side of the image, mark gaps in the SAR data. Artemis is defined by the circular feature (dark), which forms a 150–200 km wide topographic trough marked by closely-space (1–2 km) linear structures that parallel the associated portion of the trough. The interior hosts four tectonomagmatic centers marked by radial lineaments and flows, and preserves a penetratively developed linear fabric that generally trends northeast, becoming radial near tectonomagmatic centers.

an analog clock framework, with ends that gradually disappear both topographically and structurally. Short wavelength structures (<1 km) within the trough parallel the trend of the trough. Brown and Grimm (1996) mapped Artemis' trough and rise structures in detail, and Hansen (2002) mapped the interior, trough, and adjacent region in reconnaissance fashion. A 1:5,000,000 scale geologic map (V-48, Artemis) is under review with the U.S. Geological Survey (Bannister and Hansen, 2006). The interior, which sits 2–4 km above the adjacent lowlands, hosts four tectonomagmatic centers – three marked by radial fractures and flows. A fifth possible center, marked by radial fractures but lacking obvious flows, overlaps a portion of the southern trough. A penetratively developed, ~500 m wavelength, fabric occurs across much of Artemis, trending generally northeast, but taking on a radial character near three of the tectonomagmatic centers (Bannister, 2006; Bannister and Hansen, 2006). The trough hosts trough-parallel structures, likely a combination of folds, faults and scarps. Radial extension fractures and trough-concentric wrinkle ridges dominate the rise outboard from the trough, with concentric wrinkle ridges continuing outward for hundreds of kilometers.

Four different hypotheses have been proposed for the formation of Artemis: (a) and (b) related to a subduction interpretation (Brown and Grimm, 1995, 1996; Spencer, 2001); (c) as Venus' largest impact structure (Hamilton, 2005); (d) as the surface expression of a large mantle plume on thin lithosphere (Griffiths and Campbell, 1991; Smrekar and Stofan, 1997; Hansen, 2002). Part of the challenge of understanding Artemis' formation is tied to how Artemis is defined. Is Artemis composed of interior, rim and outer rise that are genetically related; or did each of these regions form separate from one another, with the interior representing a sort of 'captured' real estate?

The subduction hypothesis stems from the topographic asymmetry from the outer high, across the trough, and into the interior, similar in profile to terrestrial subduction zones (McKenzie et al., 1992; Schubert and Sandwell, 1995). Artemis has an apparent depth of compensation of ~200 km, which has been interpreted as evidence of a subducted slab (Schubert et al., 1994; Brown and Grimm, 1995), but it might also represent underplated material or melt residuum within a plume context. Gravity analysis, fraught with assumptions including assumed single depths of compensation, results in non-unique interpretations. Brown and Grimm (1995, 1996) proposed that Artemis Chasma resulted from northwest-directed subduction beneath Artemis' interior; they further suggested that Artemis Chasma includes three distinct trough segments. The trough from: ~2:30 to 6:30 represents a subduction zone marked by ~250 km of under-thrusting of lowlands to the southeast under Artemis' interior; 12:00 to 2:30 represents an associated trough dominated by left-lateral displacement; and 6:30–10:30 represented an older feature, genetically unrelated to the other two segments. Spencer (2001) interpreted a part of Artemis' interior as a region of major crustal extension similar to a terrestrial metamorphic core complex. Although, Spencer (2001) did not place the study within a regional context, he inferred that the proposed extension related to regional subduction.

Compelling arguments against the subduction hypothesis include: (1) the angle of subduction required by the tight curvature of Artemis trough is not geometrically viable on a Venus-sized planet; (2) documented continuity of structures along the entire trough in a trough parallel fashion clockwise from 12:00 to 10:30, and a shared central location of trough topography, trough structures, radial fractures and wrinkle ridges, support the interpretation that the various features of Artemis are genetically related; (3) kinematic arguments would require right-lateral displacement along the southwestern part of the trough to accompany subduction, yet interior flows traverse the southwestern trough margin; and (4) within this same region, interior graben extend across the trough to the exterior, providing further evidence that this portion of the trough did not experience right-lateral displacement. These observations collectively argue for the evolution of Artemis through a single coherent process, rather than serendipitous alignment of two or more unrelated events as required within the context of the subduction hypothesis (Hansen, 2002; Bannister, 2006; Bannister and Hansen, 2006).

Hamilton (2005) asserted that Artemis records the impact of a huge bolide on a cold solid Venus at ~4–3.5 Ga. Unfortunately the 'hypothesis' lacks details, or even clarifying statements or predictions. The impact hypothesis for Artemis formation does not consider many first-order aspects of Artemis, including topography and geologic relations. Artemis' topographic form, with a narrow (100–150 km) circular trough surrounding a raised interior, is opposite to that of large impact basins on Mars and the Moon, with circular rims surrounding interior basins. For example, Mars' Hellas Crater, widely accepted as impact in origin, forms a 2000 km diameter, 6–8 km deep, circular basin surrounded by a greatly modified, but still present, outer rim. Hamilton (2005) infers that early Venus would have been rheologically similar to Mars during the formation of Hellas and therefore, within the context of the impact hypothesis, the two huge impact features should show similar first-order character. Large impact basins also commonly show multiple ring morphology (Hartmann, 1998), features Artemis clearly lacks. Contrary to the assertion by Hamilton (2005), there is no evidence that the northwest margin of Artemis (the arc between 9:30–12:00 in the analogue clock model) is buried beneath other constructs (Brown and Grimm, 1996; Hansen, 2002; Bannister, 2006; Bannister and Hansen, 2006); yet such a large impact basin would be expect to show a complete circular structure. Finally, the impact hypothesis does not address the formation of documented interior tectonomagmatic features, or penetrative fabric, despite the inference that Artemis represents a coherent set of features formed within a geological instant of time.

Currently the most viable hypothesis for Artemis formation seems to be the surface manifestation of a mantle plume on thin lithosphere, consistent with its large size and circular planform. Gravity-topography analysis, though non-unique, is consistent with at least partial dynamical support for Artemis (Simons et al., 1997). As a deep mantle plume rises toward the lithosphere, the lithosphere will be uplifted, and, if the strength of the lithosphere is exceeded, radial fractures could form above the plume head. Alternatively, if the lithosphere were sufficient heated, it might develop a penetrative tectonic fabric. A circular trough could also form, as illustrated in laboratory experiments aimed at modeling the interaction of thermal plumes with the lithosphere (Griffiths and Campbell, 1991). In Griffiths and Campbell's (1991) experiments, as a plume head approached the rigid horizontal boundary, it collapsed and spread laterally. A layer of surrounding 'mantle', squeezed between the plume and the surface, resulted in a gravitationally trapped asymmetric instability and led to the formation of an axisymmetric trough. In addition, the interior squeeze layer might lead to convection on a scale much smaller than that of the original plume. These smaller-scale instabilities could interact with the lithosphere inside the axisymmetric trough and become manifested as interior tectonomagmatic centers. It was on the basis of these experiments that a plume model for Artemis formation was proposed following initial release of Magellan SAR data (Griffiths and Campbell, 1991). Finite-element models of the interaction of a large thermal plume with lithosphere, aimed at modeling corona topography, also show development of an axisymmetric trough above large thermal mantle plumes (Smrekar and Stofan, 1997). In this case, the trough results from lithospheric delamination. Delamination might contribute to a hybrid model that incorporates aspects of plume-lithosphere interactions with signatures that some workers propose might be better addressed through subduction. Fundamentally, it seems that Artemis' formation may have resulted, at some first-order level, from the interaction of a deep mantle plume and relatively thin lithosphere, and as such it may provide valuable clues to the possible formation of Archaean terrestrial plumes and the structures generated therein. It is possible that Artemis could hold clues for processes transitional between plume-dominated and plate-dominated (e.g., Bédard et al., 2003; Bédard, 2006).

8.1-3.3. Crustal Plateaux: Analog for Ancient Magma Ocean Surfaces?

Crustal plateaux (Figs. 8.1-2 and 8.1-5) host distinctive deformation fabrics (Fig. 8.1-6), herein called ribbon-tessera terrain following terminology of Hansen and Willis (1996, 1998). Scientists generally agree that crustal plateaux are isostatically supported in the shallow crust or mantle, as evidenced by small gravity anomalies, low gravity to topography ratios, shallow apparent depths of compensation, and consistent admittance spectra (see citations in Phillips and Hansen (1994) and Hansen et al. (1997)). Spatial correlation of plateau topography and tectonic fabrics strongly suggests that the thickening (uplift) mechanism and surface deformation are genetically related (Bindschadler et al., 1992a, 1992b; Bindschadler, 1995; Hansen et al., 1999; Ghent and Hansen, 1999). Researchers also widely accept that arcuate-shaped inliers of characteristic ribbon-tessera terrain within the lowland represent ancient collapsed crustal plateaux remnants (e.g., Bindschadler et al., 1992b; Phillips and Hansen, 1994; Bindschadler, 1995; Ivanov and Head, 1996; Hansen et al., 1997; Hansen and Willis, 1998; Ghent and Tibuleac, 2000).

   Two basic questions emerge with respect to plateau formation.
1. How were plateau surfaces deformed and concurrently uplifted?
2. How did plateaux collapse?
Initially two end-member hypotheses emerged in response to the first question – the downwelling and plume hypotheses. The downwelling hypothesis involves concurrent crustal thickening and surface deformation due to subsolidus flow and horizontal lithospheric accretion associated with a cold mantle diapir beneath ancient thin lithosphere (e.g., Bindschadler and Parmentier, 1990; Bindschadler et al., 1992a, 1992b; Bindschadler, 1995). The plume hypothesis accommodates thickening and deformation via magmatic underplating and vertical accretion due to interaction of a large deep-rooted mantle plume with ancient thin lithosphere (Hansen et al., 1997; Hansen and Willis, 1998; Phillips and Hansen, 1998; Hansen et al., 1999, 2000). Both hypotheses call for time-transgressive deformation of ancient *thin* lithosphere above individual spatially localized regions, and both embrace the suggestions that a root of thickened crust supports each plateau and that plateau collapse results from lower crustal flow. Recently published finite element modeling illustrates, however, that the range of preserved crustal plateau morphologies and arcuate ribbon-tessera terrain inliers is difficult to achieve through lower crustal flow at

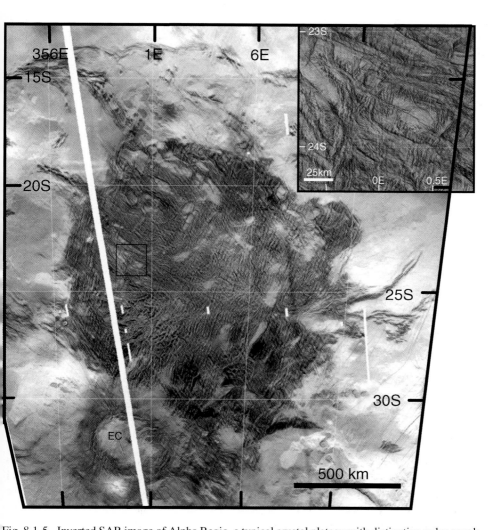

Fig. 8.1-5. Inverted SAR image of Alpha Regio, a typical crustal plateau with distinctive radar-rough (dark) terrain residing in a elevated plateau above the adjacent radar-smooth lowlands (bright); the circular feature that overlaps Alpha along its southwest margin is younger Eve Corona, EC. Inset (box) shows detail of the ribbon-terrain fabric: periodically spaced parallel ridges and troughs, trend north; fold ridges trend west-northwest. Short wavelength folds (~1 km) parallel longer-wavelength folds, but occur below image resolution here. Smooth, light-colored regions represent radar-smooth surfaces, interpreted as areas covered by low viscosity lava flows. White bold lines mark data gaps and indicate the spacecraft track.

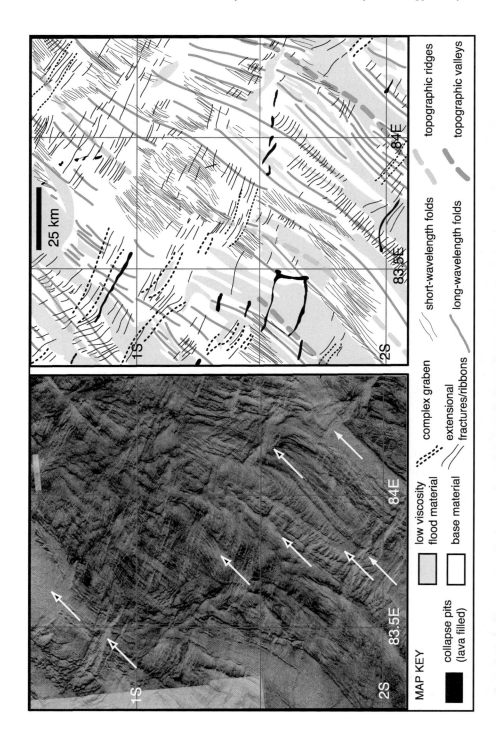

geologically reasonable time scales (Nunes et al., 2004). Thus, neither the range in plateaux elevations, nor the inliers of ribbon tessera-terrain are addressed by either the downwelling or the plume hypotheses.

Additionally, neither of these hypotheses address all characteristics of crustal plateaux, and each carries specific burdens. Challenges for downwelling include: (a) a predicted domical form (Bindschadler and Parmentier, 1990) rather than the observed plateau shape; (b) lower crustal flow called upon for crustal thickening requires 1–4 billion years, well outside reasonable time constraints (Kidder and Phillips, 1996); and (c) formation of documented short-wavelength extensional structures (ribbon fabrics) requires a high geothermal gradient (Hansen and Willis, 1998; Gilmore et al., 1998), which is difficult to justify in a relatively cold downwelling environment (Hansen et al., 1999). The plume hypothesis can accommodate formation of a plateau shape and extensional features. However, both extensive contractional strain, and formation of short-wavelength folds are difficult to accommodate (Ghent at al., 2005). Although the plume hypothesis addresses formation of late long-wavelength folds (or warps), which record very little shortening (<1%), early layer shortening, and/or large amounts of layer shortening, would present a serious challenge for the plume hypothesis. In addition, Gilmore et al. (1998) argue that formation of ribbon fabrics requires a geothermal gradient well above that expect within the environment of a plume-lithosphere interaction.

Despite deep divides within the crustal plateau debate, SAR image mapping on both sides leads to four, mutually agreed upon, observations: (1) plateaux host both contractional structures (folds) and extensional structures (ribbons, extensional troughs, graben), which are generally mutually orthogonal; (2) there are multiple suites of folds, defined by wavelength; (3) there are multiple suites of extensional structures, defined by spacing; and (4) low viscosity fluid, presumably lava, fills local to regional topographic lows. Despite these agreements, controversy exists as to the relative timing of flooding and deformation, and until recently, the amount of shortening has been unconstrained.

Detailed SAR image mapping aimed at addressing the timing of deformation and flooding, and placing limits on shortening strain, yielded new observations and refined geologic histories for plateau surfaces (Fig. 8.1-6), resulting in the proposal of a third hypothesis – the lava-pond hypothesis (Hansen, 2006). Geologic relations call for progressive deformation of an initially very thin layer (10s to 100 m) developed across individual plateaux. The layer shortened, forming ductile folds, and extended in an orthogonal direction along brittle structures (ribbons). With additional shortening, earlier formed short-wavelength structures

---

Fig. 8.1-6. (*Previous page.*) Inverted left-illumination SAR image and interpretive map of a region within crustal plateau eastern Ovda Regio, illustrating the nature of the ribbon-tessera terrain fabric along the crest of a long-wavelength (∼100 km) fold. Note local flooding of medium-wavelength fold troughs preserved in crests, limbs and trough of long-wavelength folds. Also note late collapse pits and associated lava deposits. Arrows with black heads indicate locations where flooding postdated local deformation; arrows with white heads indicate locations where deformation postdated local flooding. See Hansen (2006) for details.

were carried piggyback on younger, progressively longer-wavelength folds. Local flooding accompanied progressive deformation of the increasingly thicker surface layer. Low viscosity flood material leaked from below into local structural lows. Early flooded lows were carried piggyback on younger, longer-wavelength structures (Fig. 8.1-6). Subsurface liquid (magma) formed a sharp decrease in viscosity with depth, required by structural constraints, and served as the source of flood material. Early terrestrial magma oceans may have followed crystallization processes akin to plateaux surface evolution, although testing this might require identifying large tracts of ancient surfaces, rather than subsurface exposures.

The lava-pond hypothesis calls for progressive solidification and deformation of the surface of huge individual lava ponds, each with areal extent marked by individual plateaux. Ribbon-tessera terrain represents lava pond 'scum'. Individual lava ponds resulted from massive partial melting in the shallow mantle caused by large bolide (20–30 km diameter) impact on thin lithosphere (Hansen, 2006). Melt rose to the surface leaving behind a lens of low-density mantle residuum (e.g., Jordon, 1975, 1978). This hypothesis follows the recent suggestion that the terrestrial greater Ontong-Java Plateau formed as a result of large bolide impact on thin lithosphere (i.e., Ingle and Coffin, 2004; Jones et al., 2005), following earlier suggestions (Rogers, 1982; Price, 2001). Isostatic adjustment in the mantle, resulting from the low-density residuum lens, raised a solidified lava pond to plateau stature. Later, local mantle convection patterns could variably strip away the low-density residuum root, resulting in subsidence and/or ultimate collapse of individual plateaux. Remnants of distinctive ribbon-tessera terrain fabrics could survive as a record of an ancient lava pond. Thin surface deposits could partially or completely cover the fabrics, obscuring or erasing, respectively, evidence for individual lava ponds. The lava pond hypothesis addresses the detailed ribbon-tessera history of orthogonal folding and extension at a wide range of wavelengths from 0.1 km to tens of km, as well as the formation and subsequent collapse of ancient crustal plateaux.

Massive partial melting within the shallow mantle could result from: (a) a large bolide impact with ancient thin lithosphere, (b) rise of an extremely hot deep mantle plume beneath ancient thin lithosphere, or (c) a plume spawned by large bolide impact on thin lithosphere. In any case, crustal plateaux require thin lithosphere (as with the downwelling and plume hypotheses), and they owe their topographic stature to a low-density mantle residuum lens, rather than thickened crust. A bolide impact mechanism for melt-generation is favored because the formation of a lava-pond necessitates a large volume of magma at the surface at one time. Balancing formation of massive melt, yet preserving a local lithosphere able to support a large lava pond seems a challenge to address within the context of a plume hypothesis. In contrast, a 20–30 km bolide would simply punch through the lithosphere into the mantle forming a large 'hole', but the lithosphere across a several thousand-km scale could retain its strength – although it might likely be riddled with fractures (Jones et al., 2005). Ivanov and Melosh (2003) state that large bolide impact cannot generate huge volumes of melt, yet others present convincing counter arguments, particularly if a large bolide impacts hot, thin lithosphere (Jones et al., 2005; Elkins-Tanton and Hager, 2005). Clearly such lines of inquiry are in nascent stages of investigation. Hot thin lithosphere, critical to formation of huge melt volumes, might be easily accommodated on ancient Venus, or on early Earth. In addition, a huge body of lava might cool slowly because Venus' dense CO_2 atmosphere acts more like a conductive layer than a convection layer in terms of heat transfer (Snyder, 2002). This brings to mind how early Earth's atmosphere might also affect heat transfer processes, and lava solidification.

The bolide impact and lava pond hypotheses also provide a mechanism to concentrate radiogenic elements in early-formed crust, with possible further differentiation into a subsurface felsic layer beneath a more mafic surface 'scum'. Crustal scale lithologic/density/radiogenic stratification at a map-scale similar to Venusian crustal plateaux is proposed for terrestrial granite-greenstone terrains (e.g., West and Mareschal, 1979; Mareschal and West 1980; Collins et al., 1998; Chardon et al., 2002; Rey et al., 2003; Sandiford et al., 2004), although a lava pond mechanism has not been considered to date. Surely early Earth was bombarded by bolides, which likely affected the early lithosphere. Bolides could have contributed to early mantle differentiation processes, including residuum formation, which could in turn lead to cratonization (e.g., Jordon, 1975, 1978; Bédard, 2006). Large bolide impacts may have contributed to the formation, and preservation of early crust.

8.1-3.4. Deformation Belts

Although circular features dominate Venus' surface, it also preserves large-scale linear features, including: wrinkle ridges, distributed across huge tracts of the surface; extensive fracture belts, thousands of km long and hundreds of km wide; and zones of focused strain, called deformation belts. Deformation belts, first recognized in Venera data (Basilevsky and Head, 1988), rise $\sim$1 km above their surroundings in the lowlands, and host 1-km wide ridges that mark folds or graben (Frank and Head, 1990; Kryuchkov, 1990). Deformation belts commonly occur in groups separated by inter-belt regions. Belts are 100–250 km wide and 100 to >1000 km long; inter-belt regions are $\sim$100–400 km wide and are elongate to equant (Solomon et al., 1992; Squyres et al., 1992b). The periodic nature of deformation belts is particularly apparent in the Atlanta-Vinmera region (Fig. 8.1-7). In early works, the periodicity of deformation belts was proposed as resulting from either harmonic buckling instabilities driven by regional compression, likely the result of the coupling of large-scale mantle convection with the lithosphere (Zuber, 1987, 1990), or widespread contraction of the crust, with deformation belts elevated by focused thrusting (Frank and Head, 1990). Both models addressed constraints derived from low resolution Venera data, and called for region crustal shortening with post-deformational flooding of the inter-belt regions. Analysis of Magellan data revealed several challenges to these early models, including: superposed contraction and extension structures, along strike changes from contraction to extension structures, syntectonic volcanism, orthogonal deformation belts, and evidence for strain localization within the deformation belts, as opposed to evidence for burial of inter-belt deformation (e.g., Squyres et al., 1992b; Phillips and Hansen, 1994; Addington, 2001; Rosenberg and McGill, 2001; Young and Hansen, 2005). The origin of deformation belts remains enigmatic. I briefly review geologic relations within

Fig. 8.1-7. (*Previous page.*) Inverted SAR image of the Atlanta-Vinmara deformation belts; close-up images (i and ii) illustrate local detail; and a sketch of anastomosing deformation belt patterns (upper right corner). Dark regions (radar-rough) characterize the deformation belts, whereas homogeneous gray areas (radar-smooth) represent intervening regions of low strain. White lines indicate exposures of ribbon-tessera terrain. The large image covers ~2500 km in a north-south direction.

---

two deformation belt provinces; the broadly parallel, but anastomosing Atalanta-Vinmara belts of the northern hemisphere, and the orthogonal belts of Lavinia Planitia preserved in the southern hemisphere. I compare these provinces with terrestrial granite-greenstone terrains.

The Atalanta-Vinmara belts form an impressive array of anastomosing deformation (Fig. 8.1-7), generally interpreted as resulting from global-scale shortening normal to their trends. Within the belts, smooth ~1 km wide ridges generally define folds, although fractures and graben occur locally, and even locally dominate. Deformation features grade outward from each belt, whether marked by contractional or extensional structures. Both the belts and inter-belt regions preserve outcrops of ribbon-tessera terrain, which clearly predated belt formation. The belts and the inter-belt regions also preserve evidence of localized volcanic activity in the form of small shields. Volcanism predated, accompanied, and postdated deformation. Collectively, these relationships illustrate that the belts represent high strain zones, or strain localization, compared to low strain inter-belt domains.

Lavinia Planitia preserves generally orthogonal deformation belts ~50–200 km wide, and up to ~800 km long that form topographic highs and record concentrated strain (Fig. 8.1-8). Strain corresponds to belt orientation. NE-trending belts display folds, whereas NW-trending belts exhibit fractures and graben (Fig. 8.1-8). Belts that trend between these orientations host folds and fractures in patterns that reflect plan-view non-coaxial shear. ENE-trending belts record right-lateral shear whereas NNW (to N)-trending belts record left-lateral shear (Koenig and Aydin, 1998; Hansen, 2006, unpublished mapping). Inter-belt regions record relatively low strain, with contractional (wrinkle ridges) and extensional (fractures) structures parallel in trend to their respective counterparts within the belts. A strikingly simple pattern represented by a single regional bulk strain ellipse emerges across Lavinia, suggesting that deformation within and between belts occurred broadly synchronously. As in the case of Atalanta-Vinmara, ribbon-tessera terrain and shields occur in both the belt and inter-belt domains; shields broadly predated, accompanied and post-dated deformation. Coronae-sourced flows locally embay and bury eastern deformation belts.

8.1-4. EARLY EARTH ANALOGUES

The Atalanta-Vinmara and Lavinia regions share first-order characteristics and histories, although the shapes of their low strain regions differ: elongate versus equant, respectively.

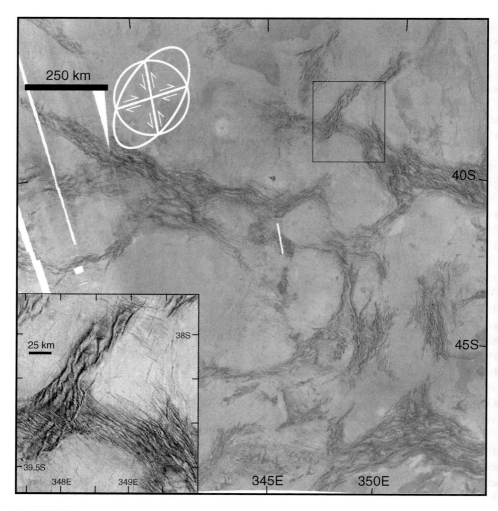

Fig. 8.1-8. Inverted SAR image of Lavinia Planitia deformation belts. Belt-parallel folds characterize NE-trending belts whereas belt-parallel extension fractures characterize NW-trending belts. Inset shows detail of NE-trending belt marked by fold ridges and an orthogonal NW-trending belt marked by extensional structures. Low inter-belt regions display low strain; wrinkle ridges (NE-trending) and extension fractures (NW-trending) parallel deformation belt folds and fractures, respectively. Box indicates location of inset SAR detail. Regional bulk strain ellipse (orientation, not magnitude) matches strain in individual deformation belts as a function of belt orientation shown. See text.

Their map-view stain patterns are similar in plan-view to terrestrial granite-greenstone terrains, which display crustal views. In this analogy, low strain inter-belt domains correspond to granite domes, and deformation belts correspond to greenstone belts. The Atalanta-Vinmara region is more akin to the Yilgarn and Superior Provinces – each displaying elongated granite-greenstone patterns; whereas Lavinia mimics the Pilbara and Dharwar cratons, displaying more equant patterns. The Yilgarn and Superior Provinces have been widely interpreted as evidence for terrestrial Archaean plate tectonic processes, due in large part to the linear structural pattern that extends for thousands of km, and taken to record progressive accretion of distinct arc terranes similar to the modern North American Cordillera (e.g., Card, 1990; Van Kranendonk, 2003c, 2004a, and references therein). However, the linear pattern across millions of km^2 preserved in the Atalanta-Vinmara province did not result from plate tectonic processes.

Terrestrial granite-greenstone terrains are variably interpreted as the result of two end-member models: diapirism and sagduction, versus the accretion of arc terranes within a plate tectonic framework (e.g., Mareschal and West, 1980; Choukroune et al., 1997; Card, 1990; de Wit, 1998; Lin, 2005). This debate has raged for decades and shows little sign of subsiding (e.g., Van Kranendonk, 2004; Van Kranendonk et al., 2004; Cawood et al., 2006). Although terrestrial belts preserve moderately deep crustal views, Venus' deformation belts provide a surface plan-view, and it is more extensive than the view presented by Archaean cratons. In addition, although the role of plate tectonics can be debated in terrestrial cases, it is highly unlikely that plate tectonic processes operated on Venus. Thus the Venusian examples might provide clues for crustal scale density inversion processes. The low strain regions in Venusian deformation belts could mark sites of subsurface diapirism, or low density crust which has moved upward, and the deformation belts could represent sagduction, with local high strain and thickening of a surface cover layer as it sheds off the low strain regions. The Venusian belts, like terrestrial granite-greenstone terrains, show broadly synchronous deformation and volcanism. In the case of the Venusian provinces, synchronous volcanism and contractional deformation has been difficult to explain to date, but both processes might be predicted and addressed within the context of a partial convective overturn model (cf. Collins et al., 1998).

First order similarities of deformation-belt terrains and granite-greenstone terrains beg for future comparative study with fundamental first-order benefits for Venusian and terrestrial studies. Do deformation belts provide evidence of an ancient density-layered crust on Venus? Venus is currently believed to boast a basaltic undifferentiated crust based on Venera composition data and hypsometric relations. A compositionally layered crust with surface basalt and a felsic subsurface could accommodate current constraints, yet could also provide a mechanism for deformation belt formation. Perhaps a density/composition/isotopic-layered crust marks a common stage of terrestrial planet formation. An early-formed felsic (and thus also radiogenic) rich layer could serve to insulate the underlying mantle, leading to partial melting and subsequent formation of mafic melt, which could in turn make its way to the surface where it could form a high-density layer, and perhaps a thermal blanket. Could the higher geothermal gradient expected in early terrestrial planet evolution, together with blanketing and insulation (e.g., Rey et al., 2003; Sandiford et al., 2004) lead to rheological softening of the layered crust, leading in turn to ductile flow and subsequent density inversion? Perhaps a subsurface felsic layer rich in radiogenic elements, could insulate the underlying mantle, leading to partial melting without requiring a plume (e.g., Rey et al., 2003; Bédard, 2006). Crustal density inversion processes would be variably arrested with cooling, or with loss of water – each of which would inhibit crustal ductility. Crustal inversion would not require regional scale horizontal shortening or extension, although such processes could accompany crustal inversion. The Atalanta-Vinmara and Lavinia regions might preserve evidence of a regionally extensive weak crustal rheology – that is, a regionally extensive crust as a ductile solid, as opposed to a brittle-solid crust, following suggestions for a weak terrestrial Archaean crust (e.g., Choukroune et al., 1997; Chardon et al., 2003; Sandiford et al., 2004; Bédard, 2006; Cagnard et al., 2006). Clearly the ideas presented here require further study, but perhaps comparison of Venus' deformation belts and Archaean granite-greenstone terrains could lead to new understanding of both geological provinces, and even early processes of crust formation. Conceivably, granite-greenstone terrains are a natural evolutionary process in the formation of terrestrial planet crust; they may not require mantle plumes, but rather a precursor stratified crust that insulated the underlying mantle and led to partial melting.

8.1-5. SUMMARY

Heat transfer processes drive terrestrial planet evolution. The heat budget, structure and rheology likely change throughout the evolution of terrestrial planets. Similarities and differences between Venus and Earth provide a valuable tectonic experiment that might provide critical clues to understanding the earlier history of our own planet, which may have experienced early tectonic processes quite different than its current trademark plate tectonics. Within the last 15 years, Venus' surface has become visible and accessible to anyone with world-wide-web connection, making Venus 'field work' accessible and inexpensive. Venus' environmental boundary conditions are certainly different than contemporary Earth, and may be more similar in many ways to Earth's Hadean to Archean Era. Any understanding of Venus tectonic processes and planet evolution are surely still in a nascent stage, and yet Venus serves as a rich tectonic playground to stretch one's imagination, and to challenge one to think beyond the elegance of plate tectonics in constructing hypotheses of ancient terrestrial planet processes.

Earth's Oldest Rocks
Edited by Martin J. Van Kranendonk, R. Hugh Smithies and Vickie C. Bennett
Developments in Precambrian Geology, Vol. 15 (K.C. Condie, Series Editor)
© 2007 Elsevier B.V. All rights reserved.
DOI: 10.1016/S0166-2635(07)15082-9

Chapter 8.2

THE EARLIEST SUBCONTINENTAL LITHOSPHERIC MANTLE

W.L. GRIFFIN AND S.Y. O'REILLY

*Key Centre for the Geochemical Evolution and Metallogeny of Continents (GEMOC),
Department of Earth and Planetary Sciences, Macquarie University, NSW 2109, Australia*

8.2-1. INTRODUCTION

Continental crust on the modern Earth is underlain by a subcontinental lithospheric mantle (SCLM), which consists dominantly of variably depleted ultramafic rocks (dunite, harzburgite and lherzolite, all of which may contain garnet ± spinel). The SCLM ranges in thickness from a few tens of kilometres beneath active rift zones, to >250 km beneath some Archean cratons. We know a good deal about the petrography and composition of the SCLM through studies of xenoliths in volcanic rocks, and from exposed massifs thrust into the crust of orogenic belts. These studies show that the composition of the SCLM has changed through time, and that SCLM thick enough to permit the formation of diamonds may only have formed in Archean time.

In this chapter we examine the evidence for the composition and origin of the SCLM, with particular focus on the Archean. We will show that Archean SCLM is both more depleted, and much more widespread, than currently understood, and discuss how its generation led to the formation of buoyant (and hence unsubductable) continental nuclei, which may have influenced both the preservation of early crust and the nature of early plate tectonics.

8.2-2. SCLM COMPOSITION

The composition of the SCLM can be estimated from exposed orogenic massifs and from xenolith suites; each has its advantages and disadvantages. Most massifs represent relatively shallow SCLM sections, and have been strongly deformed during their emplacement, but exceptions may be found in some ultra-high-pressure zones, such as western Norway (Brueckner and Medaris, 1998, 2000). Xenolith suites sample larger vertical sections of the SCLM, but the relationships between different rock types are seldom obvious, and sampling (either by the volcano or by the geologist) may not be representative. Few volcanic rocks carry extensive xenolith suites, but many carry xenocrysts, representing disaggregated mantle wall-rocks. Garnet xenocrysts, in particular, can be used to estimate the composition of the SCLM, and this technique provides a broader basis for mapping the SCLM (Griffin et al., 1998, 1999a).

Table 8.2-1. Comparisons of Gnt-SCLM with median analyses of xenolith suites

	Kaapvaal <90 MA Calc. Gnt. Lherzolite	Kaapvaal Median Lherz. Xen.	Kaapvaal <90 MA Calc. Gnt. Harzburgite	Kaapvaal Median Harz. Xen.	Daldyn Calc. Gnt. Lherzolite	Daldyn Median Lherz. Xen.	Daldyn Calc. Gnt. Harz.	Daldyn Median Harz. Xen.
no. samples	335	79	64	24	390	18	180	3
SiO_2	46.0	46.6	45.7	45.9	45.8	44.3	45.4	42.2
TiO_2	0.07	0.06	0.04	0.05	0.05	0.04	0.02	0.09
Al_2O_3	1.7	1.4	0.9	1.2	1.2	1.0	0.4	0.6
Cr_2O_3	0.40	0.35	0.26	0.27	0.31	0.37	0.18	0.37
FeO	6.8	6.6	6.3	6.4	6.5	7.6	6.1	7.4
MnO	0.12	0.11	0.11	0.09	0.11	0.13	0.11	0.10
MgO	43.5	43.5	45.8	45.2	44.9	45.2	47.2	47.8
CaO	1.0	1.0	0.5	0.5	0.7	1.0	0.2	1.0
Na_2O	0.12	0.10	0.06	0.09	0.08	0.07	0.03	0.07
NiO	0.27	0.28	0.30	0.27	0.29	0.29	0.32	0.31

Table 8.2-1. (*Continued*)

	S. Australia Calc. Gnt. Lherzolite	Mt. Gambier (SA) Median Xenolith	Obnazhennaya Calc. Gnt. Lherzolite	Obnazhennaya Med. Lherz. Xen.	E. China Calc. Gnt. Lherzolite	E. China Med. Lherz. Xen.	Vitim Calc. Gnt. Lherzolite	Vitim Med. Lherz. Xen.
no. samples	365	19	160	19	150	17	30	16
SiO_2	44.4	44.2	44.9	42.6	44.5	45.5	44.5	44.5
TiO_2	0.07	0.04	0.09	0.00	0.15	0.16	0.15	0.16
Al_2O_3	1.9	1.9	2.4	1.8	3.8	3.8	3.7	4.0
Cr_2O_3	0.41	0.44	0.42	0.44	0.40	0.44	0.40	0.37
FeO	7.8	7.6	7.9	8.4	8.0	8.2	8.0	8.0
MnO	0.13	0.13	0.13	0.13	0.13	0.14	0.13	na
MgO	43.2	43.5	41.7	44.7	39.1	38.1	39.3	39.3
CaO	1.6	1.6	2.1	1.4	3.4	3.3	3.3	3.2
Na_2O	0.13	0.05	0.17	0.06	0.27	0.23	0.26	0.32
NiO	0.30	0.29	0.28	0.26	0.25	0.25	0.25	0.25

Notes: Lherzolite and harzburgite garnets classified by Ca-Cr relationships. Al_2O_3 corresponding to each garnet type was calculated from the Cr content of the rock; other elements were calculated from Al_2O_3 using the algorithms of Griffin et al. (1999a), with SiO_2 by difference. See Griffin et al. (1999a) for sources of xenolith data.

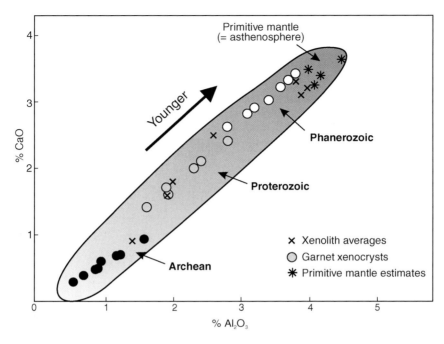

Fig. 8.2-1. Estimates of the mean CaO and Al$_2$O$_3$ contents in SCLM sections of different tectonothermal age, based on the compositions of garnet xenocryst populations (circles) and xenolith populations (crosses) in volcanic rocks.

Table 8.2-1 gives estimates of average SCLM composition for a number of localities worldwide. These localities are classified in terms of the tectonothermal age of the overlying crust, defined as the age of the last major thermal event (Griffin et al., 1999a). *Archons* are areas where the crust has been unaffected since $\geqslant 2.5$ Ga; *Protons* experienced tectonism 2.5–1.0 Ga; *Tectons* have been formed or modified $\leqslant 1.0$ Ga ago. The range in SCLM composition can be illustrated in terms of Ca and Al, two elements that are extracted from the mantle during partial melting events (Fig. 8.2-1). It is apparent that *Archons* are underlain by strongly depleted SCLM (low Ca and Al), while the SCLM beneath *Tectons* is only weakly depleted, and *Proton* SCLM tends to be intermediate between these two extremes.

8.2-3. ARCHEAN SCLM

8.2-3.1. Archean SCLM Is Unique

It is important to recognize that Archean SCLM is not simply *more* depleted than younger SCLM, it is *differently* depleted. This can be illustrated by comparison of depletion trends in peridotite suites representing lithospheric mantle of different tectonothermal age (Fig. 8.2-2). Proton and Tecton SCLM (xenoliths and orogenic massifs), as well as abyssal peridotites, ophiolitic peridotites, and xenoliths from island-arc settings, all have one important feature in common: as Al decreases, Fe contents remain constant within a narrow range. In contrast, Archon peridotites have lower Fe at low Al contents, and even show a weak positive correlation between Fe and Al. A similar situation is seen for Cr: in younger suites, Cr remains constant or (in some island-arc xenolith suites) rises as Al decreases, while in Archon xenolith suites Cr decreases together with Al. This reflects a fundamental genetic difference, which will be discussed below.

It is widely accepted that Archon xenolith suites also show high Si/Mg (high opx/olivine) compared to younger SCLM or oceanic peridotites. However, this effect, which might reflect metasomatism by Si-rich fluids (Boyd, 1989; Bell et al., 2005), really is common only in the suite from the kimberlites in the SW part of the Kaapvaal Craton (Fig. 8.2-2). This well-studied suite is the backbone of our current understanding of Archean SCLM.

Archon SCLM also contains rock types, especially harzburgites with low-Ca, high-Cr garnets, which are essentially absent in younger SCLM, and these rock types commonly show a strong stratification (Fig. 8.2-3) that is rarely seen in younger SCLM.

This high degree of depletion gives Archon SCLM some unique properties. Because it contains highly magnesian olivine and orthopyroxene, and less garnet and clinopyroxene, it has a lower density than more fertile Proton or Tecton SCLM. On a normal cratonic geotherm, a section of Archon SCLM 150–200 km thick is significantly less dense than the underlying asthenosphere, and hence is buoyant (Fig. 8.2-4). This buoyancy makes it nearly impossible for Archean cratonic SCLM to "delaminate" from its overlying crust and sink into the convecting mantle. The high degree of depletion also, at least initially, would produce extreme dehydration of the SCLM, which would result in lower viscosity and greater strength (Lenardic and Moresi, 1999).

Tecton SCLM, which is more fertile, has a higher room-temperature density than Archon SCLM (Table 8.2-2). On an elevated geotherm, as is commonly found under young Tectons (O'Reilly and Griffin, 1985), a typical 60–100 km section of Tecton SCLM will be less dense than the underlying asthenosphere, but on cooling to a lower geotherm it is likely to become negatively buoyant, and hence tectonically unstable. The delamination of Tecton SCLM following an orogeny appears nearly inevitable, and the upwelling of hot asthenospheric mantle to replace the delaminated SCLM may be responsible for large-scale post-tectonic granitic magmatism, as heating drives magmas out of the lower crust (e.g., Zheng et al., 2006a).

Fig. 8.2-2. (*On two following pages.*) Variation of individual oxides with Al_2O_3 in xenolith populations from Archons, Protons, Tectons and island arcs, and in peridotites from ophiolites, orogenic massifs and ocean-floor settings. White area in each plot outlines the field of Tecton peridotites, for comparison. Dashed lines under 'Tectons' show composition of Primitive Mantle, marked by a star.

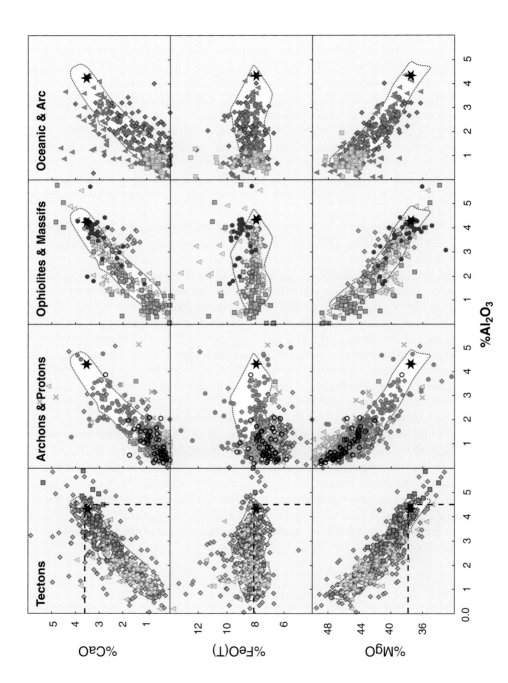

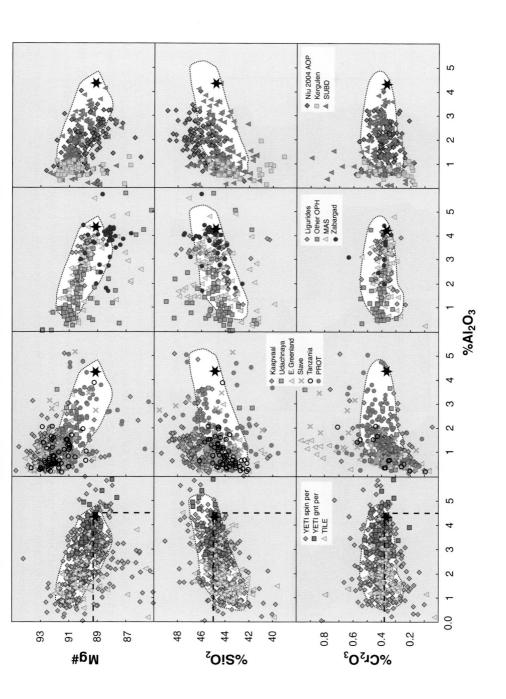

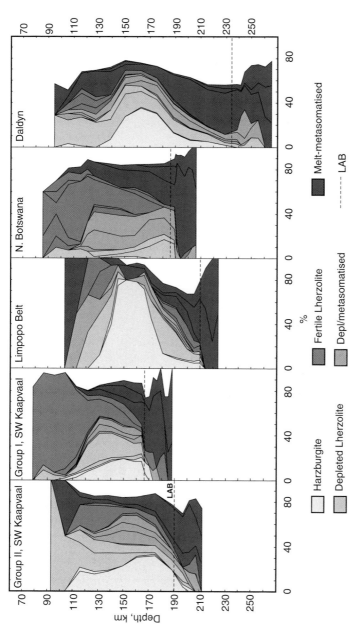

Fig. 8.2-3. Chemical Tomography sections for SCLM beneath selected Archons (SW Kaapvaal Craton, Limpopo Belt, and the Daldyn kimberlite field of Yakutia) and a representative Proton (Northern Botswana). The sections represent the relative proportions of different rock types at each depth, derived from the analysis of garnet xenocrysts in kimberlites. For each grain, a temperature derived from its Ni content is referred to a local geotherm to derive a depth; the major- and trace-element composition of each grain defines its original host-rock type and the nature of metasomatic overprints; see O'Reilly and Griffin (2006) for details of methodology. The 'lithosphere-asthenosphere boundary' (LAB) represents the deepest extent of garnets with <10 ppm Y, representing moderately to highly depleted peridotites. The differences between the sections sampled by Group II (ca. 120 Ma) and Group I (ca. 90 Ma) kimberlites in the SW Kaapvaal Craton reflects metasomatism and heating of the SCLM over this time interval.

Table 8.2-2. Estimates of mean SCLM composition

	Archons				Protons			
	Aver. Archon Gnt. SCLM	Average xenoliths	Kaapvaal aver. high-T lherzolite	Archon "Primitive" estimate	Aver. Proton Gnt. SCLM	Aver. Proton xenoliths	Aver. massif	Proton SCLM (preferred)
SiO_2	45.7	49.4	44.3	42.7	44.7	43.9	45.2	44.6
TiO_2	0.04	0.05	0.17	0.01	0.09	0.04	0.09	0.07
Al_2O_3	0.99	1.14	1.74	0.40	2.1	1.6	2.0	1.9
Cr_2O_3	0.28	0.40	0.30	0.34	0.42	0.40	0.38	0.40
FeO	6.4	6.8	8.1	6.60	7.9	7.9	7.9	7.9
MnO	0.11	0.10	0.12	0.15	0.13	0.12	0.11	0.12
MgO	45.5	45.7	43.3	49.10	42.4	43.9	41.6	42.6
CaO	0.59	0.81	1.27	0.20	1.9	1.3	1.9	1.7
Na_2O	0.07	0.05	0.12	0.10	0.15	0.08	0.13	0.12
NiO	0.30	0.30	0.26	0.30	0.29	0.22	0.28	0.26
Mg#	92.7	92.0	90.5	93.0	90.6	90.8	90.4	90.6
Cr#	0.16	0.19	0.10	0.36	0.12	0.15	0.11	0.12
ol/opx/cpx/gnt(spin)	69/25/2/4		70/18/4/8	84/13.5/2/(0.5)				70/17/6/7
density, g/cc	3.30		3.36	3.29				3.35
Vp, km/s (0.1 Gpa, 20 °C)	8.36		8.31	8.38				8.31
Vp, 5 GPa, 900 °C	8.38		8.34	8.40				8.34
Vs, km/s (0.1 Gpa, 20 °C)	4.87		4.83	4.89				4.83
Vs, 5 GPa, 900 °C	4.70		4.66	4.71				4.66

Table 8.2-2. (*Continued*)

	Tectons				Models	
	Aver. Tecton Gnt. SCLM	Average Tecton Gnt. peridotite	Aver. spin. peridotite McDonough	Average Tecton spinel peridotite	Prim. Mantle McD. & Sun	Prim. Mantle Jagoutz et al.
SiO_2	44.5	45.0	44.0	44.4	45.0	45.2
TiO_2	0.14	0.16	0.09	0.09	0.20	0.22
Al_2O_3	3.5	3.9	2.3	2.6	4.5	4.0
Cr_2O_3	0.40	0.41	0.39	0.40	0.38	0.46
FeO	8.0	8.1	8.4	8.2	8.1	7.8
MnO	0.13	0.07	0.14	0.13	0.14	0.13
MgO	39.8	38.7	41.4	41.1	37.8	38.3
CaO	3.1	3.2	2.2	2.5	3.6	3.5
Na_2O	0.24	0.28	0.24	0.18	0.36	0.33
NiO	0.26	0.24	0.26	0.27	0.25	0.27
Atomic ratios						
Mg#	89.9	89.5	89.8	89.9	89.3	89.7
Cr#	0.07	0.07	0.10	0.09	0.05	0.07
ol/opx/cpx/gnt(spin)	60/17/11/12			66/17/9/(8)	57/13/12/18	
density, g/cc	3.37			3.35	3.39	
Vp, km/s (0.1 Gpa, 20 °C)	8.31			8.33	8.33	
Vp, 5 GPa, 900 °C	8.34			7.99*		
Vs, km/s (0.1 Gpa, 20 °C)	4.81			4.83	4.81	
Vs, 5 GPa, 900 °C	4.65			4.54*		

Notes: Modes calculated using average compositions for minerals from rocks of similar composition. Density, Vp and Vs calculated using modes, mineral compositions and the spreadsheet of Hacker and Abers (2003; H-S averages).

*Values at 1.5 Gpa, 900 °C, on typical Tecton geotherm (O'Reilly and Griffin, 1985).

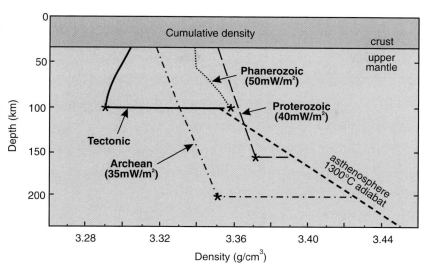

Fig. 8.2-4. Variation of density with depth for typical Archon, Proton and Tecton SCLM, relative to a fertile, convecting asthenospheric mantle with an adiabatic temperature gradient and a surface temperature of 1300 °C. The Archon and Proton SCLM sections are buoyant relative to the asthenosphere; Tecton SCLM is buoyant while warm (Southeast Australia geotherm of O'Reilly and Griffin (1985)), but becomes gravitationally unstable when cooled to a typical Phanerozoic geotherm. After Poudjom Djomani et al. (2001).

The depletion of Archon SCLM, coupled with the low cratonic geotherm of Archon areas, also gives it a distinctive seismic signature. The compressional-wave velocity (Vp) and the shear-wave velocity (Vs) of olivine and orthopyroxene rise as Mg# increases (and density decreases). This means that highly depleted SCLM will have higher seismic velocities at the same depth and temperature than more fertile SCLM (Table 8.2-2). The compositional variations within the SCLM account for ca. 25% of the total range in its seismic velocity worldwide, while temperature variations account for ca. 75% (Deen et al., 2006).

8.2-3.2. What Is Its Bulk Composition?

8.2-3.2.1. Methods of estimating composition

Table 8.2-1 lists some estimates of the average composition of Archean SCLM, based on xenolith and garnet–xenocryst suites. The agreement between the xenolith-based and the xenocryst-based estimates is encouraging in terms of validating the xenocryst-based methods. However, it may also reflect a basic flaw in the ways that SCLM composition is estimated. The garnet–xenocryst methods can only estimate the composition of the garnet-bearing portion of the SCLM, and the agreement with xenolith estimates may simply reflect a long-term bias by many research groups toward the analysis of garnet peridotites. This bias is natural, because pressure-temperature estimates are a key aspect of most xenolith studies, and estimates of pressure are possible only in garnet-bearing samples.

8.2-3.2.2. What do these estimates represent?

In studies of Archon xenolith suites, samples without garnet have received relatively little attention, though they clearly represent an important part of the mantle assemblage. In fertile peridotites, as found beneath many Tectons, garnet is only stable at depths greater than about 60 km, and this transition occurs at higher pressures in more depleted peridotites. In most cratonic SCLM sections, the garnet peridotite–spinel peridotite transition occurs at 80–100 km depth (Fig. 8.2-3). Even at greater depths, garnet will be absent in highly Al-depleted rocks, which will have assemblages of olivine + opx ± spinel (Cr-rich chromite).

Most studies of garnet–peridotite suites have implicitly or explicitly regarded the range of garnet and clinopyroxene contents as reflecting different degrees of melt extraction from a fertile protolith. Cox et al. (1987) recognized that in garnet lherzolite xenoliths, garnet and clinopyroxene commonly are spatially related to one another, and suggested that both phases had been exsolved from high-T Al-rich opx. However, these relationships could also reflect metasomatic introduction of garnet and cpx into a deleted harzburgite, effectively refertilising a depleted residue (e.g., Simon et al., 2003; Bell et al., 2005).

Analysis of SCLM sections using garnet xenocrysts provides further insights into the extent of metasomatic processes. Fig. 8.2-3 compares two sections representing the SCLM of the SW Kaapvaal Craton (South Africa) sampled at different times by Group II (>110 Ma) and Group I (≤90 Ma) kimberlites. The progressive reduction in the proportion of depleted harzburgites and depleted lherzolites is apparent; other important features are the dramatic increase in the level of phlogopite-related metasomatism (at shallow depth) and melt-related metasomatism (at greater depths). This refertilization of the SCLM means that the xenolith suites from the Group I kimberlites, which provide the bulk of the samples included in the estimate of Archon SCLM composition shown in Table 8.2-1, are strongly biased toward highly modified material.

A compelling illustration of this refertilization process has been provided by detailed studies of the occurrence of garnet peridotite within orogenic massifs in western Norway. The Proterozoic continental crust that encloses these massifs was subducted to considerable depths during the Caledonian orogeny (Griffin and Brueckner, 1980). In the peridotite massifs, small volumes of garnet lherzolite, typically interlayered with eclogite and garnet pyroxenites, occur within very large volumes of dunite/harzburgite. Beyer et al. (2004, 2006) used *in situ* and whole-rock Re-Os methods to demonstrate that the dunite/harzburgite is Archean (ca. 3.0 Ga) in age, while the garnet lherzolites represent zones of Proterozoic refertilization. This refertilization process produces chemical trends (Fig. 8.2-5) that mimic, in reverse, the depletion trends illustrated in Figs. 8.2-1 and 8.2-2, and produces mantle compositions that reproduce much of the range seen in garnet lherzolite xenoliths from Archon and Proton settings.

Seismic tomography images over southern Africa also provide chastening evidence that the widely accepted estimates of Archon SCLM composition are strongly biased. In

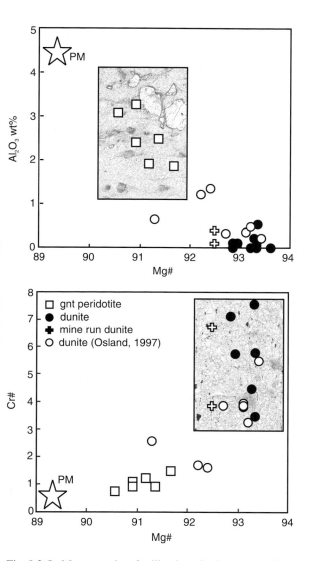

Fig. 8.2-5. Metasomatic refertilization of a depleted peridotite massif, Western Norway. Garnet peridotites (thin section photo in upper box) were produced by Proterozoic infiltration of mafic melts into Archean dunites (thin section photo in lower box). The refertilization trend (next page) mimics the compositional range of the classic Archean xenolith population from the SW Kaapvaal Craton, which usually is interpreted as a depletion trend ('oceanic trend' of Boyd (1999)). After Beyer et al. (2006).

Fig. 8.2-6, white areas show regions of high shear-wave velocity (Vs) in the 100–175 km depth range, corresponding to the depths of most xenolith suites. Petrophysical modelling shows that these white areas represent highly depleted SCLM, with low geotherms. Short-

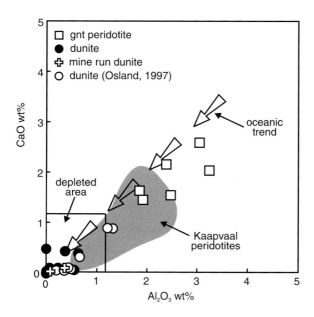

Fig. 8.2-5. (*Continued.*)

wavelength variations in Vs are most likely to reflect compositional differences; metasomatic refertilization will produce rocks with higher density and lower Vs (Griffin et al., 1999a; Deen et al., 2006), represented by the darker red tones. It is apparent that the best-studied, classic xenolith suites from this region are derived from the edges of the high-Vs cratonic cores, representing the more strongly modified part of the SCLM. The more depleted material making up most of the cratonic SCLM has hardly been sampled at all.

One exception may be provided by the 500 Ma kimberlites of the Limpopo Belt (Fig. 8.2-3), which sampled a very depleted, strongly stratified section with abundant garnet harzburgites at 130–170 km depth. After the intrusion of these kimberlites, the SCLM beneath this area was strongly modified by dike swarms related to the Karoo Traps (Griffin et al., 2003b), and today it has a Vs comparable to the Kimberley area (Fig. 8.2-6).

Based on these considerations, it seems probable that most of the cratonic SCLM, and thus most of the original Archon SCLM, probably was made up of highly depleted rocks, similar to the Archean dunite/harzburgite massifs of western Norway. Similar material makes up the bulk of xenolith suites from eastern and western Greenland (Bernstein et al., 1998, 2006). It also is known from the xenolith suites of the Kaapvaal Craton (Boyd, 1989; Bell et al., 2005), the Siberian Craton (Griffin et al., 1999b) and the Slave Craton of Canada (Griffin et al., 1999c), but probably is seriously under-represented in published studies, relative to its volume. A revised estimate for the composition of this 'pristine' Archean SCLM is given in Table 8.2-1.

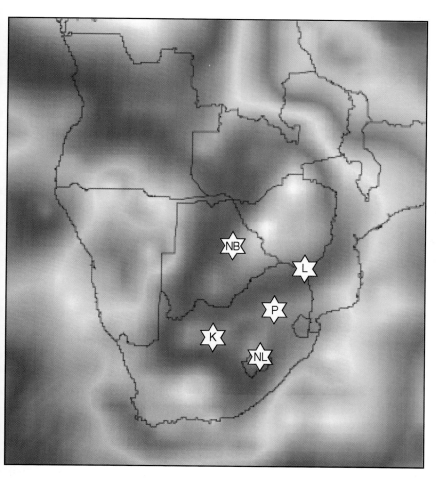

Fig. 8.2-6. Seismic-tomography image (100–175 km) of southern Africa (after Deen et al., 2006; tomography by S. Grand), with location of well-documented xenolith suites (K, Kimberley area; NL, Northern Lesotho; P, Premier mine) and Chemical Tomography sections (Fig. 8.2-3; L, Limpopo Belt; NB, Northern Botswana). White-pink colours indicate high seismic velocities, related to strong depletion and low geotherms; yellow-green colours indicate low velocities related to fertile compositions and higher geotherms.

8.2-3.3. How Old Is It?

The most robust isotopic system available for measuring the ages of SCLM peridotites is provided by the decay of ^{187}Re to ^{187}Os. During partial melting of the mantle, Re is extracted into the fluid phase, while Os remains concentrated in the residue; the removal of the melts from the system tends to freeze in the isotopic composition of the residual Os.

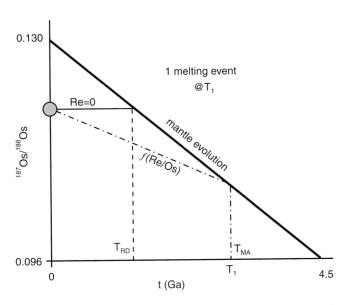

Fig. 8.2-7. Calculation of T_{RD} and T_{MA} model ages in the Re-Os isotopic system, referred to a mantle-evolution curve. T_{RD} ages represent minimum ages for an episode of melt extraction; T_{MA} ages may give a better estimate, but uncertainties arise because the Re/Os ratio is subject to post-melting modification.

The $^{187}Os/^{188}Os$ of the rock, or constituent minerals, can then be referred to a model for the evolution of Earth's Re-Os system to derive a model age that approximates the timing of the melt-extraction event (Fig. 8.2-7). Rhenium-depletion (T_{RD}) model ages assume that any Re in the sample was added recently (and thus is ignored); T_{MA} model ages assume that Re in the system was not extracted completely during partial melting, so that T_{MA} ages typically are older than T_{RD} ages for the same sample. This methodology has been widely applied to dating of SCLM samples (Pearson, 1999; Pearson et al., 2004).

However, there is an important complication to this otherwise elegantly simple method. In most SCLM peridotites, Os is strongly concentrated in trace sulfide phases, and sulfides can be highly mobile under mantle pressures and temperatures. The first detailed study of the Re-Os systematics of these trace phases (Alard et al., 2000, 2002) found that some sulfides, typically monosulfide solid solutions enclosed in primary silicate phases, have high Os contents and distinctly unradiogenic Os (with low $^{187}Os/^{188}Os$, and hence old T_{RD} and T_{MA}), and have been interpreted as residual after partial melting. However, other Ni- or Cu-rich sulfides, typically interstitial to the primary silicates, commonly contain more radiogenic Os, suggesting introduction of sulfide melts at later times.

These observations imply that the model ages derived from the whole-rock samples represent mixtures of >1 generation of sulfides, and hence are unlikely to date any real event. However, it is possible to obtain precise Os model ages on single grains of sulfides, using *in situ* analytical techniques (LAM-MC-ICPMS; Pearson et al., 2002) and thus to isolate

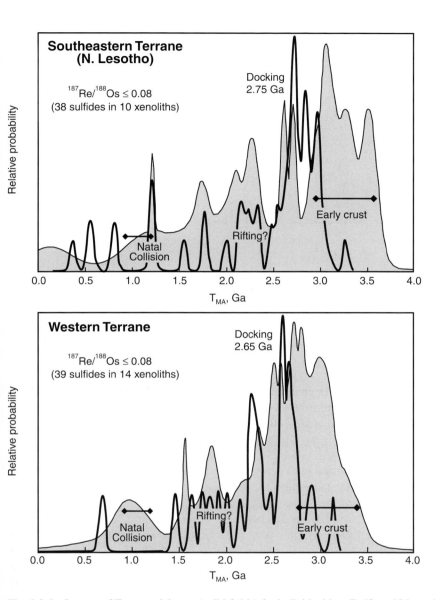

Fig. 8.2-8. Spectra of T_{MA} model ages (solid fields) for individual low-Re/Os sulfide grains in peridotite xenoliths from two terranes within the Kaapvaal Craton (after Griffin et al., 2004). In each case the peaks in sulfide age distribution correspond to major tectonic events in the crust, implying a close crust-mantle linkage. Black lines show spectra of model ages for whole-rock xenolith samples from the literature; most of these are younger than the main sulfide peaks, reflecting the mixing of several generations of sulfide in each rock.

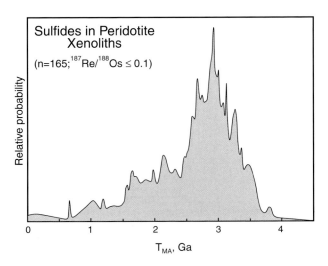

Fig. 8.2-9. Cumulative spectrum of model ages for low-Re/Os sulfides in mantle-derived xenoliths from Archons of southern Africa, Siberia and North America, showing a major peak in Mesoarchean time, and a lack of Hadean ages. Data from references given in text, and authors' unpublished data.

the oldest grains in a given sample. When the two methods are compared on samples from the same area (Fig. 8.2-8), we find sulfides with ages that are significantly older than the oldest whole-rock model ages, and typically correspond to the oldest U-Pb ages recorded in zircon grains from the overlying crust. Groups of younger sulfide model ages appear to correlate with tectonic events in the crust, and may reflect fluid movements through the SCLM at those times (Griffin et al., 2004b). This interpretation assumes that the sulfide-bearing fluids are derived from the convecting mantle; if it is correct, then only the oldest groups of ages may record the timing of SCLM formation events. A steadily growing database (Fig. 8.2-9) shows a major peak centred on ca. 3.0 Ga; suggesting that much of the Archean SCLM formed in Paleoarchean to Mesoarchean time.

Although the *in situ* sulfide dating method has been applied in a number of Archean areas (Alard et al., 2002; Aulbach et al., 2004; Griffin et al., 2002, 2004b, unpubl. data), very few reliable model ages >3.6 Ga have been found; the same is true of whole-rock data (Pearson, 1999). While we do not have xenoliths of the SCLM from beneath areas that contain Hadean crustal rocks, the apparent lack of Hadean model ages beneath the oldest cratons seems to point in one direction: we have no samples of Hadean SCLM. If the SCLM came into being only in Archean time, what does this tell us about the evolution of the early Earth?

8.2-3.4. How Did It Form?

A common model used to explain the formation of Archean SCLM is by the accumulation of subducted slices of oceanic mantle (e.g., Helmstaedt and Gurney, 1995; 'lithospheric stacking'). While this model implies the operation of processes similar to modern plate tectonics, it may be a misleading analogy. Detailed seismic tomographic images of modern subducting slabs (e.g., Replumaz et al., 2004) show that most descend steeply into the Earth, down to at least the 660 km discontinuity, rather than accumulating at shallow depths beneath the continents. In areas where shallow subduction is observed, it rarely extends far under the continents, and does not produce thick SCLM.

Depending on the time scale for lithosphere formation, the subduction-stacking process might be expected to produce a recognizable age pattern, with a decrease in age with depth, and laterally across cratons. No such picture can be recognized in the whole-rock Re-Os data (Pearson, 1999). The more detailed sulfide Re-Os data tend to show an overall decrease in model ages with depth beneath the Kaapvaal craton (Griffin et al., 2004b), but old sulfides and younger ones coexist in single hand specimens, reflecting multiple generations of sulfide addition. The apparent trend to younger ages with depth is therefore best interpreted as the result of repeated metasomatism of the deep lithosphere. In detail, the sulfide data indicate that each crustal terrane within the craton carried its own lithospheric root during the assembly of the craton, rather than the root being built up gradually after craton formation (Griffin et al., 2004b).

More importantly, the subduction-stacking models founder on the marked differences in geochemical trends between Archean SCLM and Phanerozoic examples of depleted mantle, such as abyssal peridotites, island-arc xenolith suites and ophiolites (Fig. 8.2-2). In Archean xenolith suites, the positive correlations between Fe, Cr and Al suggest that no Cr-Al phase (i.e. spinel or garnet) was present on the liquidus during melting. In contrast, highly depleted peridotites produced at shallow levels in modern environments show geochemical patterns that appear to be controlled by the presence of spinel during melting.

The lithospheric stacking models for the Archean SCLM have drawn support from melt-modelling studies, which estimate the conditions of formation for peridotites by comparing their compositions with the products of experimental partial-melting studies (e.g., Walter, 1998; Herzberg, 1999). Several of these studies have concluded that typical Archean garnet lherzolite xenoliths ('typical Archean mantle') could have formed at relatively shallow depths, analogous to abyssal oceanic peridotites, and been subducted to form the SCLM. However, this approach implicitly assumes that the xenoliths are the products of a single stage of melt extraction. It cannot be validly applied to the garnet lherzolites, if they are recognized as the multi-stage products of extreme depletion followed by (often repeated) metasomatic re-fertilization.

A more realistic estimate of the formation conditions for the pristine Archean mantle may be possible by applying the same approach to the most depleted rocks, such as the Norwegian dunite/harzburgites, or the Greenland xenolith suites. These rocks correspond to residues formed at very high degrees of melting, at relatively high pressures (5–6 Gpa; Beyer et al., 2006). We now find these rocks at depths comparable to their original depth of melting, and at shallower depths as well – testimony to their buoyancy relative to the underlying asthenosphere (Poudjom Djomani et al., 2001; Fig. 8.2-4) – and melt extraction may have continued over a large depth range.

The Archean SCLM as defined by xenolith suites is unique; if the 'pristine' Archean SCLM is even more highly depleted, its uniqueness is even more pronounced. The formation of these highly depleted volumes thus appears to be related to specifically Archean processes, which might include megaplumes or massive mantle overturns (Davies, 1995). It is also unlikely to be coincidental that the production of highly depleted Archean SCLM coincides in time with the large-scale production of komatiitic magmas, which probably reflect high degrees of partial melting, requiring rapid decompression (see review by Hoatson et al. (2006)).

Cogent evidence for the contribution of plumes to the SCLM is provided by the occurrence of diamonds with mineral inclusions that are stable only under lower-mantle conditions (ferropericlase, CaSi- and MgSi-perovskites, and others; Harte et al. (1999); Davies et al. (1999)). In the Slave Craton, such diamonds are found in the lower part of a strongly layered SCLM, which may represent the head of a ca. 3.2 Ga plume that underplated a highly depleted SCLM ca. 150 km thick (Griffin et al., 1999c).

The development of Archean lithospheric keels might be related simply to much more intensive plume activity in the hotter Archean Earth. Smithies et al. (2005b) used the detailed stratigraphic succession of basaltic rocks erupted from 3.52–3.24 Ga in the Pilbara Craton (West Australia) to show the presence of two mantle sources. A series of plumes is invoked to provide a continuous source of high-Ti basalts, which show little stratigraphic variation in composition. These are interlayered with low-Ti basalts that reflect a source which became progressively more depleted over the ca. 300 Myr eruptive history. This depleted source is interpreted as the residues of earlier plumes, repeatedly melted by the heat from later plumes, gradually building up a thick keel of depleted harzburgites.

The cratonic keel beneath the Pilbara Craton was sampled by kimberlites that intruded ca. 1800 Ma ago (Wyatt et al., 2002). The Chemical Tomography section derived from the xenocrysts in the Brockman Creek kimberlite (Fig. 8.2-10) shows that it is one of the most depleted Archean mantle samples known. In particular, the proportion of garnets with metasomatic trace-element signatures is exceptionally low, even compared to the 500 Ma Limpopo Belt section (Fig. 8.2-3). It is unlikely to be a coincidence that our oldest mantle section also is the most depleted and the least metasomatized. These data strongly support the suggestion that the Archean SCLM formed through the accumulation of plume-related melting residues, rather than through the 'stacking' of subducted oceanic slabs.

8.2-4. IS THERE MORE ARCHEAN SCLM THAN WE THINK?

The overall secular evolution in SCLM composition (Fig. 8.2-1) has been interpreted in terms of a progressive change, from the Archean to the Phanerozoic, in the processes that generated the SCLM at different stages of Earth's evolution (Griffin et al., 1998, 1999a, 2003). However, we now recognize that refertilization of the pristine depleted Archean SCLM produces a similar trend. Detailed analysis of the tectonic history of the crustal rocks overlying each of the sampling localities shown in Fig. 8.2-1 reveals that in most of the Proton areas, the original crust was Archean, and was reworked during Proterozoic time. This is exemplified by the SCLM section for northern Botswana (Fig. 8.2-3), where

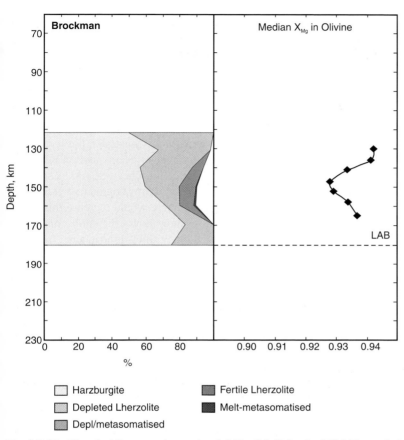

Fig. 8.2-10. Chemical Tomography section (cf. Fig. 8.2-3) for the SCLM beneath the Pilbara Craton, Western Australia, sampled at 1800 Ma by the Brockman Creek kimberlite. This is the oldest, and the most depleted, section presently available.

supracrustal rocks overlying the Archean crust at the craton margin were metamorphosed at ca. 1.9 Ga and intruded by ca. 1.2 Ga granites. The SCLM section (sampled at ca. 90 Ma) is typically 'Proterozoic', with few depleted rocks and an abundance of metasomatized and fertile garnet peridotites. However, Re-Os dating of xenoliths from the Letlhakane kimberlite in this area yields Archean model ages (Carlson et al., 1999).

These observations suggest that the original SCLM beneath many Protons was formed in Archean time, and simply has been progressively refertilized. True Proton SCLM may only be present under some areas with juvenile Proterozoic crust, such as the Colorado Plateau; in such areas, we find SCLM that is only mildly depleted relative to estimates of the Primitive Mantle.

This view of the secular evolution process emphasizes the importance and uniqueness of the Archean SCLM. It also implies that much of the SCLM beneath Protons worldwide, as well as that beneath Archons, originally was formed in Archean time (3.6–3.0 Ga?). The formation of the Archean SCLM was one of the major events in Earth history, and the Paleoarchean period may have been a remarkable interregnum between the Hadean and a more modern Earth.

8.2-5. IMPLICATIONS FOR CRUSTAL EVOLUTION

The lack of Hadean sulfide ages (Fig. 8.2-9) is an ambiguous constraint, but it suggests that all of the Archean SCLM was formed in a small number of events, over a relatively short period in Paleoarchean to Mesoarchean time. 'Typical' Archean SCLM is strongly buoyant relative to the convecting mantle, even when cooled to low conductive geotherms (Fig. 8.2-4), and the higher degree of depletion suggested here would increase this buoyancy. Numerical dynamic modelling has been used to show how efficiently convection processes in the Earth will destroy crust, or even relatively thick lithosphere (Lenardic and Moresi, 1999). However, when a thick, relatively rigid lithospheric mantle, generated by melt extraction, is added to models that incorporate a realistic mantle structure, the longevity and stability of continental nuclei increases dramatically (O'Neill et al., 2006).

It thus appears that there is a direct connection between the generation of SCLM and the survival of the continental crust. In the early Earth, buoyant masses of residual, highly depleted peridotite, produced by large plumes or mantle overturns, may have acted as 'life-rafts' that preserved early proto-continental crust. In this scenario, the scarcity of Hadean crustal relics may be evidence that the processes that generated the SCLM seen beneath Archean cratonic areas had not yet begun to operate. This uniquely Archean tectonic regime may have coexisted with a more 'modern' regime that included a form of plate tectonics (Griffin et al., 2003a). However, the same principles that drive the delamination and subduction of modern plates (Fig. 8.2-4) would tend to destroy evidence of this regime, in the form of crustal rocks. The rare examples of such ancient rocks, such as in West Greenland, may only have survived in areas where this type of crust was obducted onto a "life-raft" of buoyant SCLM.

If, as suggested here, much of the present continental crust is underlain by SCLM produced in Archean time, then Archean crust may also be much more extensive than suggested by current outcrop patterns. Detailed studies of crustal xenoliths in basalts and kimberlites are beginning to show evidence that ancient lower crust can be preserved while younger igneous rocks "resurface" the upper crust (e.g., Zheng et al., 2004a, 2006b). These observations have implications for models of crustal growth and recycling through time, which remain to be explored.

ACKNOWLEDGEMENTS

The ideas presented here have been developed through interactions with many colleagues and PhD students. We thank especially Norman Pearson, Olivier Alard, Craig

O'Neill, Graham Begg and Martin Van Kranendonk for stimulating discussions that contributed directly to this paper. This work has been supported by funding from the Australian Research Council and Macquarie University. This is contribution 485 from the ARC National Key Centre for Geochemical Evolution and Metallogeny of Continents (www.es.mq.edu.au/GEMOC).

Earth's Oldest Rocks
Edited by Martin J. Van Kranendonk, R. Hugh Smithies and Vickie C. Bennett
Developments in Precambrian Geology, Vol. 15 (K.C. Condie, Series Editor)
© 2007 Elsevier B.V. All rights reserved.
DOI: 10.1016/S0166-2635(07)15083-0

Chapter 8.3

ANCIENT TO MODERN EARTH: THE ROLE OF MANTLE PLUMES IN THE MAKING OF CONTINENTAL CRUST

FRANCO PIRAJNO

Geological Survey of Western Australia, 100 Plain St., East Perth, Western Australia 6004, Australia

8.3-1. INTRODUCTION

The fundamental processes that shape planetary surfaces are impact cratering, volcanism, magmatic activity and tectonism. Following the early segregation of a crust, mantle and core, volcanism and magmatic activity produced secondary crusts of predominantly mafic composition. Partial melting of these secondary crusts, in turn, led to their recycling and the formation of primitive continental crust. On Earth, no record of the primary and secondary crusts remain, which were formed sometime between 4.5 and 4.1 Ga (see Taylor, this volume; Kamber, this volume).

Schubert et al. (2001, p. 627) stated that "one of the major products of the Earth's thermal evolution is the formation of the continental crust" and continental growth is the net gain in volume and mass of continental crust during processes that can both add or subtract juvenile or crustal material. Addition processes include subduction zones related volcanic arcs, mantle plumes, and asthenospheric upwellings, whereas subtraction processes would include A-type subduction of one continental slab under another (Bird, 1978; Howell, 1995), and lithospheric and/or lower crustal delamination (Schubert et al., 2001). Continental crust constitutes about 25% of the planet's surface on modern Earth and the source of this crust was the mantle, although the precise mechanisms and timing of extraction are not well constrained (Rogers and Santosh, 2004).

In this contribution, following an introduction on modern continental crust and on mantle convection and plumes, I examine the role of mantle plumes in the making of continental crust and how these affected global geodynamics in the Archaean, Proterozoic and Phanerozoic. For the Archaean, I use two examples of Archaean granite-greenstone terranes: the East Pilbara Terrane of the Pilbara Craton in Western Australia and the Abitibi belt of the Superior Craton in Ontario, Canada, to show that much of the juvenile material in the former was derived from mantle plume activity, whereas the latter was formed through a combination of subduction-related arcs and magmatism related to mantle plume events. For the Proterozoic, I discuss the link between mantle plumes with the assembly and breakup of supercontinents. In the Phanerozoic, the geodynamic regime is dominated by subduction systems, although mantle plume activity still occurs today and accounts for ~20% of crustal growth. The more recent geodynamic events show that lithospheric and lower crustal delamination may have played a major role in parts of eastern Asia, Europe and western North America. These delamination processes, which produced widespread alkaline volcanism, are associated with destruction, at least in eastern Asia, of continental crust.

The Archean and Proterozoic Eons are characterised by a predominance of greenstone belts (Archaean), and the development of epicontinental sedimentation in basins and of continental rifting (Proterozoic), whereas the Phanerozoic Eon is essentially based on the first appearance of exoskeletal invertebrates and the radiation of life. However, these traditional, and indeed convenient divisions, are not entirely coincident with global thermal events. Condie (1997) considered age distributions in juvenile continental crust and related them to time-integrated evolution of thermal regimes in the mantle. It was proposed that these thermal regimes began with layered convection and buoyant subduction (Stage 1), and evolved to catastrophic mantle overturns between 2.8 and 1.3 Ga (Stage 2), and finally to whole mantle convection (Stage 3). Thus, it appears that the crustal history, with its tectonic and magmatic record, is closely linked with mantle dynamics. In this paper, it is suggested that the role of mantle plumes in providing juvenile material to the crust decreased from the Archaean through the Proterozoic to the Phanerozoic, and that mantle dynamics can be considered within the framework of five evolutionary stages: (1) >3.0 to ca. 2.8 Ga; (2) 2.8–2.4 Ga; (3) 2.4–1.0 Ga; (4) 1.0–0.6 Ga and (5) 0.6–present.

8.3-2. THE CONTINENTAL CRUST

Based on the velocity distribution of seismic waves, the present-day continental crust has a layered structure with two main subdivisions (Fig. 8.3-1):
- upper crust (compressional wave velocities, Pv, of 5.9 to 6.3 $km\,s^{-1}$)
- lower crust (Pv of between 6.5 and 7.6 $km\,s^{-1}$).

The boundary between the upper and lower crust is widely displayed as the Conrad discontinuity at a depth of about 20 km (Wedepohl, 1995), although this does not appear to be present everywhere. Recent estimates of the mean composition of the upper crust indicate that it has a granodioritic to quartz dioritic composition. The lower crust is characterised by strong seismic reflections, probably caused by layered mafic and mafic–ultramafic intrusions (Percival et al. 1989). In the western Alps, the Ivrea-Verbano zone is a 10 km thick mafic igneous complex that is considered an exposed slice of lower crust and possibly an exposure of the crust–mantle boundary (Sinigoi et al., 1995). The higher seismic velocities in the lower crust suggest either a more mafic composition, and/or that it has high-pressure mineral assemblages. Following early models of basaltic or gabbroic compositions, there are three possibilities for the composition of the lower crust (see Pirajno, 2000 for overview and references). In one, the lower crust is "dry" (there are only mineral phases that do not contain water, such as pyroxene) and formed by felsic to intermediate rocks (granitic to dioritic) that have been subjected to high pressure modifications to granulite facies. The second is that the lower crust is "wet" (contains water-bearing mineral phases, such as

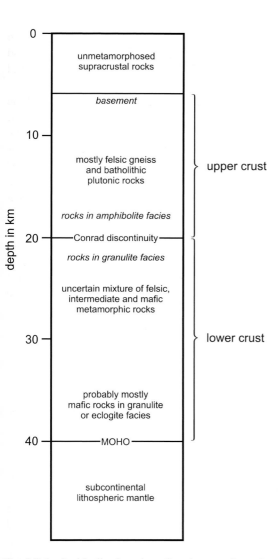

Fig. 8.3-1. An idealised section of modern continental crust. After Rogers and Santosh (2004).

amphibole) and is predominantly made up of amphibolite, with a mineral assemblage consisting of amphibole, plagioclase, epidote and Fe-rich garnet. The third possibility is that the lower crust is composed of gabbroic anorthosite (predominantly feldspar, lesser pyroxene, with minor garnet, quartz and kyanite).

It is probable, however, that the composition of the continental crust is more complex than the above-mentioned possibilities and that in reality it is laterally heterogeneous, with one or the other of the above cases dominating in a given region. Measurements of crustal

Poisson's ratio σ (the ratio of P to S waves velocity) by Zandt and Ammon (1995) revealed a general increase of σ with the age of the crust. This finding supports the presence of a mafic lower crust beneath cratons (Zandt and Ammon, 1995).

The crust is separated from the subcontinental lithospheric mantle by the Mohorovicic discontinuity, commonly called Moho (Fig. 8.3-1). Based on various considerations, the nature of the Moho beneath the continents is likely to be a boundary between the silica-rich rocks of the crust and the ultrabasic rocks of the upper mantle. The nature of the Moho is considered to reflect either a physical discontinuity (phase change, within a single composition) or a chemical change (the lithospheric mantle below the Moho is of ultrabasic composition, i.e. dunite or peridotite) (Wyllie, 1971)

8.3-2.1. The Growth of Continental Crust

As mentioned above, the origin of the continental crust is not completely understood. The general consensus is that much of the crust would have formed in the earliest part of the Earth's geological evolution (Fig. 8.3-2), following initial planetary accretion, and was formed by differentiation of magma from the mantle (the secondary and tertiary crusts of Taylor and McLennan (1985)). The debate as to whether the continental crust grew gradually through geological time or whether it was, and is, being maintained in steady-state processes of addition and destruction is still very much an open question (Hofmann, 1997). Sylvester et al. (1997) used the Nb/U ratios of crust and mantle in an attempt to solve the problem. They compared the Nb/U ratio of 30 in stony meteorites

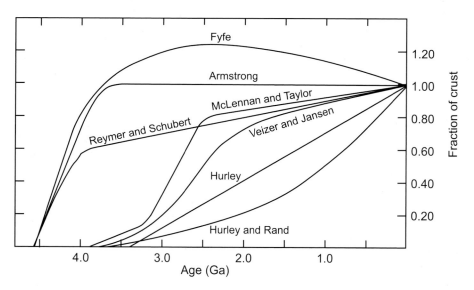

Fig. 8.3-2. Crustal growth models. After Taylor and McLennan (1985 and references cited therein) and Condie (2004). Most of these models suggest rapid rates of crustal growth before 2.5 Ga. See text for details.

to that of the primitive mantle, with the ratio of 47 in present-day "residual" mantle. These authors related this difference to the creation of continental crust, which has a Nb/U ratio of 10. This means that Nb tends to remain in the residual mantle, whereas U is transferred to the crust, via melt transfer of differentiated magmas. Sylvester et al. (1997) found that the Nb/U ratios of 2.7 Ga lavas are close to 47, implying that a similar amount of continental crust had been formed since the Archaean as compared with modern day Earth. As detailed later in this contribution, remnants of the earliest crust can be found in the Archaean Na-rich tonalite-trondhjemite-granodiorite (TTG) suite. Subsequent additions to this original Archaean crust were made during successive igneous and tectonic events that are still occurring today. Recent views propose that Archaean continents grew by the accretion of oceanic plateaux and island arcs, so that a supercontinent became established by about 2.7 Ga (e.g., Percival, this volume; Rogers and Santosh, 2004).

It is generally recognised by most planetary scientists that early thermal instabilities in the mantle, caused by accretionary energy and the decay of radionuclides, resulted in a vigorously convective mantle in the early Earth (cf. Davies, 1999). Models of the geotectonic evolution of planet Earth, from the Hadean to present-day, are based on the thermal regimes of the mantle, its dynamics and the role of convection in the establishment and the making of continental crust. The tectonic regimes of the primitive Earth were substantially different from those of later ages, due to progressive cooling and related changes in the convective patterns of the mantle (Schubert et al., 2001).

The Hadean began with the accretion of the Earth until the oldest known rocks, the ca. 4.03 Ga Acasta Gneiss, and any direct geological evidence for the period between accretion and the oldest rocks is almost completely lost from the geological record, presumably because of intense asteroid bombardments (Martin, 2005: but see Kamber, this volume; Cavosie et al., this volume: Iizuka et al., this volume). Various growth models suggest that from the Hadean to the end of the Archaean, 70–75% of juvenile continental crust was generated and extracted from the mantle (Martin, 2005) (Fig. 8.3-2). The Archaean mantle was hotter, perhaps vigorously convecting, and with abundant mantle plume activity, as recorded by komatiite fields, which are common in greenstone belts. Indeed, some Archaean greenstone belts are considered as a kind of large igneous province (LIPs), related to mantle plume events (Eriksson and Catuneanu, 2004; Van Kranendonk and Pirajno, 2004; Van Kranendonk et al., this volume).

Apart from changes in their shape, estimated crustal growth curves show continuity (Fig. 8.3-2), but there is also evidence that there were major episodes of continental crustal growth at 2.7–2.6 Ga, 1.9–1.8 Ga, 0.5–0.2 Ga, while peak mantle plume events are recorded at 3.0 Ga, 2.8 Ga, 2.5 Ga, 1.8–1.7 Ga, 1.1–1.3 Ga, 0.4 Ga, and 0.25–0.1 Ga (Condie, 2000; 2004; Abbott and Isley, 2002; Groves et al., 2005). A major change is recorded at ca. 2.5 Ga, at the Archaean–Palaeoproterozoic boundary, when a mantle plume-dominated tectonic regime was replaced by a tectonically quieter period (Barley et al., 2005), when mantle plumes were perhaps less numerous but more intense and extensive. Proterozoic mantle plumes impinged onto Neoarchaean supercontinents, leading to cycles of breakup and assembly (Condie, 2004; Rogers and Santosh, 2004). Tectonic regimes changed across the Neoproterozoic–Cambrian boundary, with mantle plumes still active, but with marked changes in the extent and frequency of subduction zones, which tend to dominate this period. The remainder of juvenile continental crust (ca. 30%) was added during Neoproterozoic and Phanerozoic times, some of which was extracted from mantle plumes, some from asthenospheric mantle upwelling during lithospheric delamination, and some extracted in subduction zones.

8.3-3. MANTLE PLUMES

Mantle convection is driven by three fundamental processes: heat loss from the core (about 20%), internal heating from radioactive decay (about 80%) and cooling from above (sinking of lithospheric slabs) (Condie, 2001). Mantle convection is responsible for the movements of lithospheric plates, earthquakes, magmatism, surface volcanic activity and indeed most of the geological and tectonic processes manifested in the crust. Other surface phenomena or physical manifestations that can be related to a convective mantle are geoid, gravity, and heat flow anomalies. Details on the principles and concepts of convection in the Earth's mantle can be found in Davies (1999). Mantle convection is time-dependent and as the rate of heat production decreases, the planet cools and eventually convection slows or stops altogether. Convection is driven by buoyancy anomalies that form at thermal boundary layers (Campbell and Davies, 2006). There are two boundary layers in the Earth's mantle; an upper boundary layer (lithosphere) at ca. 660 km, and the core-mantle boundary (CMB) at ca. 2890 km. Temperature differences at the boundary layers result in convective upwellings, which laboratory experiments show that are likely to be columnar with a tail and a head developing as the column moves through the mantle (Campbell and Davies, 2006). This constitute what is known as a mantle plume. The plume head is cooler than the tail, because it contains entrained material from the surrounding cooler mantle.

Current ideas on mantle plumes posit that they are "jets", "narrow upwelling currents", or "narrow cylindrical conduits" of hot, low-density material originating either from the CMB (or D'' thermal boundary layer: one-layer mantle model), and/or from the 660 km discontinuity (two-layer mantle model) (Davies, 1999; Schubert et al., 2001; Campbell and Davies, 2006) (Fig. 8.3-3). The general consensus is that most large and long-lived plumes originate from the CMB, at the D'' thermal boundary layer, and are caused by heat from the outer core and ensuing thermal instability. Some researchers suggest that this thermal instability is in response to the sinking of cool lithospheric slabs that cascade through the 660 km discontinuity into the lower mantle and accrete onto the CMB (mantle avalanches; see Schubert et al. (2001, pp. 454–456)). Kellogg et al. (1999) suggested that mantle plumes originated as a result of slab avalanches, and may be carrying the geochemical and isotopic signatures of recycled slab material. This is supported by evidence from seismic tomography of slab avalanches to 2700 km depth (Grand et al., 1997).

Courtillot et al. (2003) identified three types of plumes: (1) primary, or deep, plumes, originating from the D'' layer; (2) secondary plumes, originating from the top of large

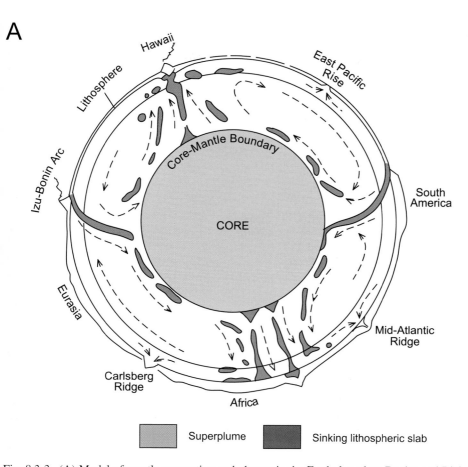

Fig. 8.3-3. (A) Model of mantle convection and plumes in the Earth; based on Davies and Richard (1992) and Jacobs (1992). (B) Types of mantle plumes in the modern Earth, according to Courtillot et al. (2003); (1) deep mantle plumes rise from the core–mantle boundary, forming hotspots such as those of Hawaii and Reunion or large plumes (superplumes), such as those under Africa and the Pacific; (2) secondary plumes may emerge from these large plumes forming discrete igneous provinces; (3) shallow mantle plumes may rise from the upper mantle, as a result of lithospheric delamination, following the collapse of collision orogens, or in back-arc rifts. (*Continued on next page.*)

domes of deep plumes or superplumes; (3) tertiary, or Andersonian, plumes, originating from near the 660 km discontinuity and linked to tensile stresses in the lithosphere (Fig. 8.3-3(b)). The superplumes of Courtillot et al. (2003) are located under Africa and the central Pacific Ocean, where two massive mantle upwellings are evidenced by high crustal elevation (superswells) and by corresponding regions of low shear wave (Vs) velocity anomalies in the mantle (Gurnis et al., 2000).

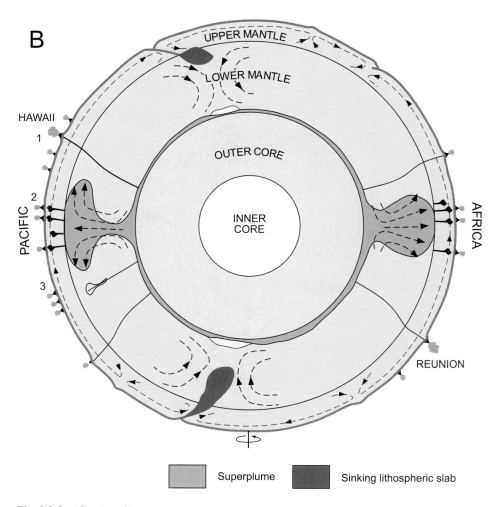

Fig. 8.3-3. (*Continued.*)

The "Andersonian plumes" of Courtillot et al. (2003) represent shallow plumes and may be considered an alternative to the theory of deep mantle plumes. Indeed, a number of researchers do not subscribe to the plume paradigm and maintain that hotspots and flood basalts are not caused by deep plumes (see Foulger et al., 2005, and references therein). Shallow plumes may arise as a result of ductile delamination of the lower lithosphere, which induces upwelling of the asthenosphere, manifested at the surface by uplift, rifting, and the eruption of alkali basalts (Elkins-Tanton, 2005). I return to this mechanism to explain tectono-magmatic events in the Cenozoic and associated destruction of mantle lithosphere and lower crust.

Partial melting in the plume head occurs by adiabatic decompression yielding lower temperature and lower-Mg melts (tholeiitic basalts), whereas melting in the high-temperature tail yields high temperature and high Mg-melts (picrites, komatiites) (Campbell et al., 1989). The surface expression of mantle plumes is typically manifested by doming of the crust – reflected as topographic swells of 1000–2000 km in diameter and 2000–4000 m elevations (above sea level), by rift basins, and by intraplate volcanism (Şengör, 2001; Ernst and Buchan, 2003). Regions of intraplate anorogenic volcanism are commonly called "hotspots", a loose term that essentially refers to the concept of a stationary heat source in the mantle and high heat flow related to magma advection (Wilson, 1963; Schubert et al., 2001). Geodetic data show that several hotspot regions correlate with rises in the gravitational equipotential surface (geoid high), probably reflecting the buoyancy of heated lithosphere (Perfit and Davidson, 2000). Uplift is followed by subsidence due to loss of buoyancy of the plume head, or removal of magma from the top of the plume, thermal decay, or a combination of all three. Subsidence and crustal sagging cause the formation of sedimentary basins, characterized by the deposition of extensive aprons of siliciclastics, carbonates and evaporites, commonly overlain by continental flood basalts and/or transected by related dyke swarms.

In addition to normal plume events, there appear to be major pulses of heat transfer in the evolution of the Earth, in which a number of plumes impinge on to the base of the lithosphere. These plume events, called superplumes (Larson, 1991; Ernst and Buchan, 2002), have important implications in terms of possible links with supercontinent cycles and time–space distribution of metalliferous deposits (Fig. 8.3-5) (Barley and Groves, 1992; Abbott and Isley, 2002). Superplume events have been recognised at various times in the geological record, and correlate with the growth of continental crust.

The eruption and intrusion of great volumes of mafic and ultramafic melts is attributed to the rise and impingement of mantle plumes on continental and oceanic lithospheric plates. These large-scale emplacements of mafic rocks are termed Large Igneous Provinces (LIPs; Coffin and Eldhom, 1992). The eruptions of these mafic melts form vast fields of lava flows and associated igneous complexes, up to 7×10^6 km^2 in areal extent (e.g., Central Atlantic Province; Marzoli et al., 1999; Siberian flood basalt province; Nikishin et al., 2002), or lines of oceanic islands, such as the Emperor-Hawaiian chain, thousands of kilometres long. In the ocean these vast lava fields are known as oceanic plateaux, such as the Ontong Java and Kerguelen plateaux (Taylor, 2006; Wallace et al., 2002). Oceanic plateaux probably constitute the greatest contribution to crustal growth, both ancient and modern. Iceland and Kerguelen, where mantle plumes interact with mid-ocean ridges, provide modern examples of continental crust nucleation processes (Grégoire et al., 1998). In the ancient record, such a continental nucleus is represented by the East Pilbara Terrane, which was built by the emplacement of successive volcanic plateaux, as explained below.

The time-dependence of mantle convection and hence of plume activity appears to be substantiated by isotopic (e.g., Nd and Sr) and trace element systematics (Stein and Hofmann, 1994), and by the distribution of ages of juvenile continental crust (Condie, 1997). Plots of ^{143}Nd/^{144}Nd and ^{87}Sr/^{86}Sr as a function of the age of orogens tend to follow a linear growth pattern, from a primitive mantle (low ^{143}Nd/^{144}Nd and ^{87}Sr/^{86}Sr values), to a highly depleted upper mantle (higher ^{143}Nd/^{144}Nd and ^{87}Sr/^{86}Sr values), with various scatters being accounted for by mixing of older continental material with more juvenile upper mantle material (Stein and Hofmann, 1994). Stein and Hofmann (1994) theorised that mantle flow patterns alternate between two-layer and single-layer (or whole mantle) convection, which give rise to "Wilsonian periods" and "mantle overturn and major orogenies" (MOMO) periods, respectively.

Condie (2004) proposed that there are two types of mantle plumes: long-lived and short-lived. The former, called shielding superplumes (>200 Ma duration), are the result of the shielding effect of a supercontinent, which will cause upwelling of mantle plumes followed by the fragmentation of the supercontinent: these do not appear to be linked to production of large volumes of juvenile crust. The short-lived plume events (<100 Ma duration), called catastrophic superplume events, are associated with accretion of volcanic arcs and addition of juvenile crust.

A decrease in the frequency of mantle plume activity with time is demonstrated by the progressive decrease in the thickness and geotherms of the sub-continental mantle lithosphere (SCLM; Groves et al., 2005; O'Reilly et al., 2001) (Fig. 8.3-4). Thus, an evolution from a mantle plume dominated regime in the Archaean, to a more "buoyant" style tectonics in the Proterozoic, to modern plate tectonics in the Phanerozoic, clearly reflects the trend of a cooling Earth (Groves et al., 2005).

This trend is possibly linked with a change over geological time from whole mantle convection of the early Earth to two-layer mantle convection in the Phanerozoic. The latter is responsible for shallow mantle plumes, which are linked to the upper thermal boundary layer (Fig. 8.3-3(b)) and result in igneous provinces, dominated by alkaline chemistry. As mentioned previously, shallow mantle plumes may form as a result of asthenospheric upwelling due to lithospheric mantle and lower crustal delamination. Tectonic collapse of collisional orogens is inferred to occur as a result of thinning of negatively buoyant mantle lithosphere, with detachment and sinking of orogenic roots or of subduction slabs. This extensional collapse engenders the rise of asthenospheric mantle, during the process of delamination tectonics (Platt and England, 1993).

8.3-4. THE ARCHAEAN

The Archaean record is dominated by granite-greenstone terranes, for whose origin opinions differ from plate-tectonic models dominated by convergent margins and subduction zones to intraplate models that call on vertical tectonics and the activity of mantle plumes. Granite-greenstone terranes vary somewhat in their nature, structure and geodynamic history, as perhaps best exemplified by contrasts between the East Pilbara Terrane and West Pilbara Superterrane in the Pilbara Craton (Hickman, 2004; Smithies et al., 2005b; Van Kranendonk et al., 2007a) and the Yilgarn Craton of Western Australia. Whatever the dominant tectonic regime(s) in the Archaean, there is little doubt that during this time juvenile continental crust was formed.

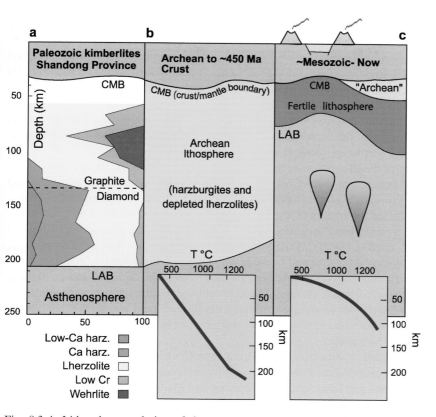

Fig. 8.3-4. Lithosphere evolution of the eastern part of the Sino-Korean Craton. (A) subcontinental mantle derived from garnet data. (B) Ordovician palaeogeotherm derived from xenocrysts in kimberlites, with Archaean mantle thickness. (C) subcontinental lithospheric mantle section and geotherm derived from xenoliths in Tertiary basalts. CMB crust–mantle boundary, LAB lithosphere–asthenosphere boundary. After Griffin et al. (1998b) and O'Reilly et al. (2001). Figure and caption courtesy of S. O'Reilly of Macquarie University, Sydney.

As mentioned in the Introduction, if 70–75% of the continental crust was formed before 2.5 Ga, this must mean that the Archaean must have been a period of considerable magmatic and tectonic activity (Martin, 2005). Neodymium isotope systematics (ε_{Nd}) clearly show that most Archaean rocks have $\varepsilon_{Nd} > 0$, indicating a mantle that underwent extraction of continental crust. Values of $\varepsilon_{Nd} < 0$ appear at about 2.7 Ga, and are indicative of greater, or increasing crustal contributions. Archaean terranes essentially consist of a gneissic basement, greenstone belts (mafic-felsic volcanic and sedimentary rocks), and late granitic intrusions. The gneissic basement rocks represent the earliest juvenile crust and are sodic, commonly referred to as TTG (tonalite-trondhjemite-granodiorite; Martin et al., 2005). TTG typically have initial Sr ratios and $\varepsilon_{Nd} \geqslant 0$, suggesting that their parental magmas were sourced from the mantle and not from a pre-existing continental crust. The TTG suites differ from subduction-related granitic batholiths in their limited range of SiO_2 abundances and increasing Na_2O with SiO_2, and their Sr_i and ε_{Nd} values. There is a vast literature on TTG and their origin, on which the general consensus is that these intrusions were derived from the melting of mantle-derived basaltic magmas that were metamorphosed to amphibolite facies (Martin, 2005; Martin et al., 2005, and references therein). Whether these basaltic magmas were subduction generated, or mantle-plume generated is an important issue in the context of what was the dominant role for the generation of juvenile crust (see Van Kranendonk et al., this volume).

Underplating of basaltic magmas, as shown in seismic profiles, is a major contributor to the growth of continental crust (Condie, 2001). Furthermore, accretion of mantle plume heads onto the lithosphere is a reasonably well documented process, based on Nd-Sr isotopic systematics for the basalts of the 0.9 Ga Arabian-Nubian Shield, which led to the conclusion that a plume head became accreted onto the base of the lithosphere (Stein and Hofmann, 1992, cited in Condie, 2001; Stein, 2003). In Archaean times, oceanic plateaux may have been common features and I concur with Condie (2001) that it is possible, if not likely, that the first continents may have been formed through the accretion and/or stacking of a series of oceanic plateaux, and/or collision of mid-ocean ridges with oceanic plateaux. For example, the Ksotomusha greenstone belt in the Baltic Shield was an oceanic plateau that was tectonically accreted to a continental margin, thereby becoming a new segment of continental crust (Puchtel et al., 1998).

The first postulated supercontinent may have been Ur (Fig. 8.3-5), which stabilised at around 3.0 Ga and included the Pilbara (Western Australia), Kaapvaal (South Africa), and Dharwar, Singhbhum, and Bhandara (India) cratons, and Madagascar (Rogers and Santosh, 2004). According to Condie (2004), frequent collisions and accretions in the Archaean between oceanic plateaux and fragments of continental blocks must have been common between 2.75 and 2.65 Ga in Laurentia, Baltica and Siberia, and between 2.68 and 2.65 Ga in Western Australia and southern Africa. Juvenile crust production reached a maximum at about 2.7 Ga, with the formation of a Late Archaean supercontinent (?Arctica, ?Kenorland) (Fig. 8.3-5). Widespread rifting of Archean crust occurred at about 2.45 Ga, probably associated with a superplume event, as suggested by the presence of large dyke swarms in many cratons (e.g., Matachewan dyke swarm in the Superior Province, Canada and the Widgiemooltha dyke swarm in the Yilgarn Craton, Western Australia). The final breakup of this Late Archaean supercontinent was at around 2.2–2.0 Ga. I return to this topic in the section dealing with the Proterozoic.

8.3-4.1. Pilbara Craton, Western Australia

The Pilbara Craton, together with the Kaapvaal Craton in South Africa, may have formed the Vaalbara continent, which was possibly part of the first supercontinent Ur (Fig. 8.3-5). The Pilbara Craton consists of five terranes (Van Kranendonk et al., 2006a): (1) East Pilbara Terrane; (2) Roeburne Terrane; (3) Sholl Terrane; (4) Regal Terrane and (5) Kurrana Terrane. These terranes exhibit different structural and tectonic styles. The 3.53–3.17 Ga East

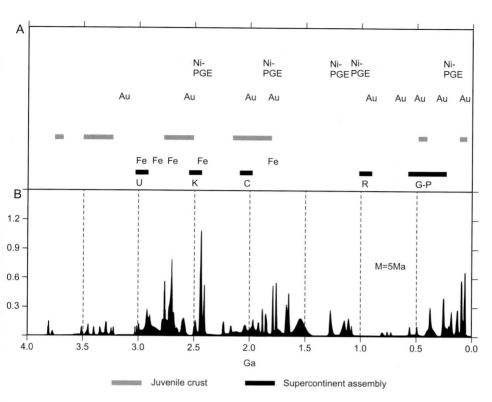

Fig. 8.3-5. Mantle plume time series (mafic–ultramafic intrusions, flood basalts and dyke swarms), weighted M = 5 Ma; U is Ur, K Kenoran, C Columbia, R Rodinia and G-P Gondwana-Pangea supercontinents; approximate age position of major metallogenic events include iron formation (Fe), gold (Au) and nickel–platinum group elements in mafic–ultramafic intrusions (Ni-PGE). Juvenile crust additions are temporally associated with plume events. Based on data from Abbott and Isley (2002), Condie (2000, 2004) and Kerrich et al. (2005).

Pilbara Terrane is characterized by a typical dome-and-keel structural pattern, with dome-shaped granitic complexes and intervening synclinal greenstone belts (Hickman, 2004). By contrast, the 3.27–3.11 Ga West Pilbara Superterrane is characterized by an elongate northeast-trending structural pattern dominated by large-scale upright folds and faults, similar in style to and not unlike that of the Yilgarn Craton (Hickman, 2004; Myers, 1997). These two terranes testify to two different tectonic regimes and geodynamic evolutionary histories (Hickman, 2004; Van Kranendonk et al., 2007a). The structure and horizontal-type tectonism of the West Pilbara Superterrane are consistent with sea floor spreading, subduction zones and arc settings, similar to Phanerozoic plate-tectonic style (e.g., Smith et al., 1998; Smithies et al., 2005a), whereas the East Pilbara Terrane has a "vertical" or gravity-driven style of tectonism (Van Kranendonk et al., 2004). In this section, I focus on the East Pilbara Terrane, leaving out the West Pilbara Superterrane because it has features that are also found in the Abitibi Belt, which is treated in the next section.

8.3-4.1.1. East Pilbara Terrane

The East Pilbara Terrane is stratigraphically represented by the Pilbara Supergroup, which contains a nearly 20 km thick succession of low-strain, low-metamorphic grade rocks spanning a period of ca. 300 million years, from 3.53 to 3.17 Ga (see Van Kranendonk et al., this volume). The Pilbara Supergroup is divided into the Warrawoona and Kelly Groups, followed upward by the Sulphur Spring and Soanesville Groups. The Pilbara Supergroup is overlain by the sedimentary succession of the De Grey Supergroup, which was deposited between 3.02 and 2.93 Ga. Massive outpourings of tholeiitic flood basalts with associated sedimentary and volcaniclastic rocks followed approximately 170 million years later and constitute the 2.775–2.63 Ga Fortescue Group (discussed below).

Van Kranendonk and Pirajno (2004) and Hickman and Van Kranendonk (2004) proposed that the Pilbara Supergroup can be interpreted as a series of submarine volcanic plateaux that were formed one above the other and that subsequent deformation was the result of vertical movements due to the diapiric-style emplacement of granitoids, during episodes of partial convective overturn. The ultramafic–mafic and felsic volcanic cycles of the supergroup are consistent with eight mantle plume events between 3.52 and 3.42 Ga (Hickman and Van Kranendonk, 2004).

Smithies et al. (2005b) recognized that the basaltic rocks of the Pilbara Supergroup can be geochemically divided into high Ti ($TiO_2 > 0.8$ wt%) and low-Ti ($TiO_2 < 0.8$ wt%), with the high-Ti rocks making up about 65% of the entire basaltic package. Smithies et al. (2005b) interpreted the differences between the high- and low-Ti basalts in terms of two distinct sources. The high-Ti basalts derived from a primitive mantle source, most likely a mantle plume, as also documented by the presence of komatiites. There is no significant change of the high-Ti basalts with time. The secular changes of the low-Ti basalts, about 35% of the succession, are explained in terms of successive melting of a source that included increasingly depleted and newly added plume material over a period of some 300 million years (Smithies et al., 2005b, this volume). The source of the low-Ti basalts is considered to have been stationary beneath the Pilbara proto-craton and is likely to represent the residue after the extraction of the earlier high-Ti basalts. Thus, it is argued that the high- and low-Ti basalts reflect separate sources and their temporal relationship suggests that successive mantle plumes provided a direct source of the high-Ti basalts and the heat required for the melting of an increasingly refractory source of the low-Ti basalts (Smithies et al. 2005b).

8.3-4.2. Abitibi-Wawa Greenstone Belt, Superior Craton, Canada

The 2.7 Ga Abitibi-Wawa greenstone belt in the Superior Craton of Canada provide abundant evidence of an interesting mix of mantle plume-derived magmatism (komatiites, tholeiitic basalts) and arc volcanic products at 2.72–2.70 Ga (Mueller et al., 2002; Wyman et al., 2002).

The Abitibi-Wawa greenstone belt is the largest in the world and was accreted with volcanic arcs forming a cratonic nucleus between 2.9 and 2.65 Ga (see Percival et al., this volume). Polat et al. (1998) studied part of the Wawa belt, and interpreted the geodynamic evolution as the result of collision and juxtaposition of subduction-related arc terranes and oceanic plateaux formed by mantle plumes. Wyman et al. (2002) interpreted the lithotectonic association as a series of intraoceanic arcs, consisting of: 2.75–2.7 Ga arc tholeiites and calc-alkaline rocks; a 2.71 Ga, depleted tholeiite–boninite suite; plume-related oceanic plateaux (2.73–2.705 Ga tholeiites and komatiites); adakites or high Mg andesites (slab melting in arc settings); and syn-late tectonic batholithic TTG intrusions, reflecting melting of subducted oceanic crust or plateau material. Shoshonites are also present and are interpreted as late-tectonic, second-stage melting of refractory sub-arc mantle, fertilised by late fluids developed during arc uplift and extensional collapse.

A very different interpretation was favoured by Ayer et al. (2002), using U-Pb geochronological data and focussing on the southern Abitibi greenstone belt. Instead of a collage of terranes, these authors favoured an autochthonous assemblage, which comprises nine supracrustal assemblages, effectively representing nearly continuous volcanism ranging in composition from mantle plume-related komatiitic and tholeiitic basalts to calc-alkaline mafic to felsic subduction-related volcanic rocks. Ayer et al. (2002) advocated a large scale and long-lived interaction between mantle plume magmatism and subduction zone related volcanic arcs, accompanied by extensive mixing of different magmas.

In the Abitibi greenstone belt, komatiite lavas stratigraphically interlayered with Mg- and Fe-rich tholeiitic basalts are attributed to mantle plume activity. The tholeiitic basalts of this association have a geochemical signature that is similar to that of Phanerozoic plateau basalts (e.g., flat REE patterns). The komatiites include both Al-undepleted (Al_2O_3/TiO_2 15–25) and Al-depleted ($Al_2O_3/TiO_2 < 15$) types (e.g., Sproule et al., 2002). The Al-depleted types have positive Nb anomalies, which are akin to high μ ocean island basalts. The boninite-depleted tholeiite association is interlayered with the basalt-komatiite association. This interfingering relationship has been interpreted as evidence of contemporaneous mantle plume and volcanic arc magmatism. The adakitic-high Mg andesite-basalt association is temporally and spatially linked with the calc-alkaline island arc volcanic rocks and appears to be related to intra-arc extension and rifting. It is interesting to note that volcanogenic massive sulphide (VMS) deposits are numerous within the group.

Wyman et al. (2002) found that the Abitibi-Wawa volcanic arc and oceanic plateau lithologies, when plotted on Nb/U vs. La/Sm$_{PM}$ diagrams, form a line that passes through the average Archaean continental crust. These researchers reiterated that this, in conjunction with other trace element systematics and field observations, support a model of tectonically imbricated intraoceanic volcanic arcs and oceanic plateaux formed during a "hybrid" tectonic process that involves both subduction systems and mantle plumes (Fig. 8.3-6).

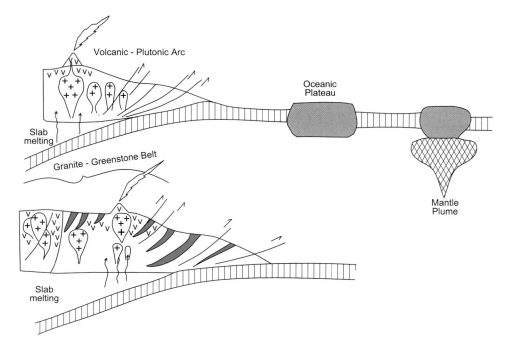

Fig. 8.3-6. Geodynamic evolution of the Schreïbner-Hemlo and White River-Doyohessarah green-stone belts according to Polat et al. 1998. The greenstones were formed by collision and accretion of arc-related volcanic successions and oceanic plateaux, intruded by various generation of granitoids during episodes of slab melting and partial melting of crustal materials.

8.3-4.3. Neoarchaen Continental Flood Volcanism, the Fortescue Group, Western Australia

Proxies of mantle plume events include komatiites, flood basalts, dyke swarms and layered mafic–ultramafic intrusions. Based on these proxies, a number of Precambrian superplume events have been postulated at 2.75–2.7 Ga, 2.45 Ga, and 2.0–1.9 Ga (Fig. 8.3-5) (Condie, 2004). Archaean continental flood basalt-dominated volcanism (CFB) is represented by the ca 3.0 Ga Dominion Group at the base of the Witwatersrand Supergroup in South Africa, and by two ca 2.7 Ga successions; the Ventersdorp Supergroup in South Africa and the Fortescue Group in Western Australia. The Ventersdorp Supergroup is an 8-km thick succession of dominantly sub-aerially erupted tholeiitic basalts, komatiites, andesites and pyroclastics that was emplaced onto the Kaapvaal Craton granitic gneiss basement and the Witwatersrand Supergroup, whereas the Fortescue Group was emplaced onto the Pilbara Craton. The Ventersdorp and Fortescue have similar stratigraphic and compositional features, including the presence of komatiites, and their near contemporaneity has led some authors (e.g., Nelson, 1998a) to suggest that these ancient CFB may have been the result of a single global-scale thermal event, or superplume event (Condie, 2004, and references therein). Both are characterized by a succession of continental komatiitic basalts, tholeiitic basalts and sedimentary rocks that provide good evidence for several episodes of plume magmatism and continental breakup in the 2.7–2.6 Ga period (Fig. 8.3-5: cf. Blake et al., 2004).

In the Pilbara Craton, successive pulses of plume magmatism in the early stages of development were directly responsible for the induced anatexis of the earlier volcanic plateaux, resulting in the emplacement of granitic intrusions and associated felsic volcanism (Van Kranendonk et al., 2002, 2006a, this volume; Smithies et al., 2003). Bédard (2006) suggested that it would be plausible for the base of Archaean oceanic plateaux to partially melt to produce a first generation of TTG magmas. The cycle could have been repeated again, with partial melting of previous-cycle granitic material, as well as oceanic plateau rocks (see Smithies et al., this volume; Champion and Smithies, this volume). These repeated cycles of greenstone melting at depth and remelting of earlier granitic intrusion, eventually led to the building of continental crust, setting the stage for truly continental flood basalt volcanism in the late stages of development. It is after this stage that the next pulse of mantle plume magmatism acquires a distinct continental character, with the eruption of the ca. 2.775 Ga Mt. Roe Basalt of the Fortescue Group.

The concept outlined above is schematically illustrated in Fig. 8.3-7. This model (Pirajno, 2000; Van Kranendonk et al., 2007a), is not dissimilar from the four-step model proposed by Bédard (2006), which he called catalytic delamination tectonomagmatic, and is more detailed in terms of TTG generation in that it envisages three cycles of felsic magmatism. In both models, formation and growth of continental crust is a process that continues for several 100s of million years, providing that both the causative mantle plume(s) and the overlying lithospheric plate remain stationary.

8.3-5. THE PROTEROZOIC

The transition from Archaean tectonic styles (dominated by plume-arc activity) towards a period dominated by subduction of cooler and less buoyant oceanic crust, heralds the Proterozoic Eon. From ca. 2.6–2.5 Ga onward, there was a general change from shallow, or flat, to steep subduction, concomitant with decreased frequency and intensity of plume activity, with the exception of the 1.9 Ga superplume event (Fig. 8.3-5). This geodynamic development resulted in the increase of continental freeboard, which allowed the deposition of thick siliciclastic successions and of iron formations (Kerrich et al., 2005). It is well-established that during the Proterozoic there was development of large sedimentary basins or platforms, supporting the idea of widespread cratonisation and the presence of large continental masses.

The Palaeoproterozoic (ca. 2.5 to 1.8 Ga) is essentially characterised by the assembly and breakup of supercontinents, intracontinental rifting and anorogenic magmatism, attributed to long-lived large scale mantle upwellings (Aspler and Chiarenzelli, 1998). In

A

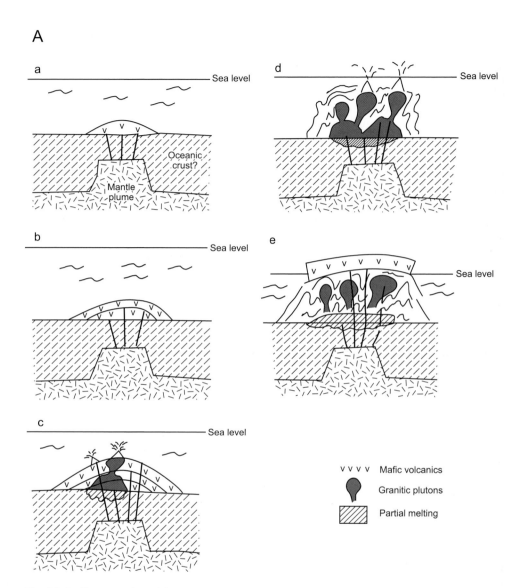

Fig. 8.3-7. Conceptual model (not to scale) of oceanic plateau accretion in the East Pilbara Terrane (Western Australia), in response to a series of mantle plumes or a ?long-lived mantle plume. For simplicity, in this model only three cycles of adiabatic melting are represented. These result in the stacking of oceanic plateaux (a, b, and c). A new mantle plume, or ongoing mantle plume activity induced partial melting of earlier plateau material producing granitic rocks and local eruption of felsic volcanics (c); this is followed by re-melting of both mafic and felsic material from the previous cycles, resulting in gradual build-up of continental crust. (*Continued on next page.*)

Fig. 8.3-7. (*Continued.*) Finally, diapiric uprise of granitic plutons (d) results in the deformation of the oceanic plateaux forming the observed granite-greenstone patterns, as described by Hickman and Van Kranendonk (2004) and Smithies et al. (2005). The newly created crust is stabilised and emerges till the next pulse of plume activity (e), some 200 Ma later produces the first continental flood basalt (Fortescue Group). The Fortescue lavas together with the Ventersdorp Supergroup lavas in South Africa, represent some of the earliest continental flood basalt.

---

the northern hemisphere there was a Neoarchaean supercontinent, Kenorland, whereas in the southern hemisphere the Zimbabwe, Kaapvaal and Pilbara cratons may have formed a Neoarchaean Zimvaalbara continent, possibly including Archaean blocks from the São Francisco Craton (present day South America and Africa) and India (Aspler and Chiaren-zelli, 1998).

The idea that supercontinents have been cyclically forming and fragmenting throughout most of Earth's geological history is not new. One hypothesis (Nance et al., 1988) holds that the aggregation of a supercontinental mass results in the accumulation of heat in the mantle beneath it. This eventually leads to heat dissipation by means of rifts that break the supercontinent into fragments. The fragments then drift and coalesce through collision processes and a new supercontinent is then assembled.

In the context of mantle plumes, fragments of continental crust tend to aggregate above plume downwellings to form supercontinents. These supercontinents, with their large mass, inhibit mantle cooling and promote the build-up of heat at the base of the crust. In this way, convective upwellings form, and result in the uplift of the lithosphere, injection of melts into the crust, and melting of the crust. This is accompanied by continental rupture and dispersal of continental fragments, which then reaggregate over mantle downwellings to start another cycle (Gurnis, 1988). Barley et al. (1998a) argued that the duration of the full cycle is approximately 360 Ma. Condie (1997, 2000) suggested that there are no data to prove the existence of a supercontinent prior to the late Archaean, although Roger and Santosh (2004) proposed that the first supercontinent, Ur, may have been assembled at 3.0 Ga.

Thus, large mafic–ultramafic intrusions and giant dyke swarms are one of the hallmarks of Proterozoic geology, which supports the contention of activity of large and deep mantle plumes for this Eon. Underplating of cratonic areas by Proterozoic mantle plumes probably formed much of the mafic lower crust. This is supported by the abundance and size of mafic–ultramafic intrusions and dyke swarms that range in age between ca. 2.4 and 2.0 Ga. These include the Great Dyke of Zimbabwe (ca. 2.5 Ga), the Jimberlana and Binniringie dykes of Western Australia (ca 2.4), and the Bushveld Igneous Complex (ca. 2.05 and the largest exposed layered intrusion in the world). The large scale, anorogenic magmatism at 1.5–1.3 Ga further added to building of continental crust, leading to the creation of the Rodinia supercontinent and the formation of large orogenic belts, such as the Pan African orogens (Windley, 1995).

8.3-6. THE PHANEROZOIC

At least 80% of Phanerozoic and modern continental crust is primarily extracted from mantle wedges above subduction zones that are enriched in water and volatiles derived from the downgoing oceanic crust slab, whereas about 20% constitutes crustal additions from oceanic plateaux (Martin, 2005). It is pertinent to note that the total length of present-day subduction margins is estimated at about 43,000 km (Leat and Larter, 2003). Volcanic arcs produce continental crust from mafic precursors, especially from intra-oceanic arcs in which thick mafic and ultramafic underplating are revealed by seismic data (e.g., Izu-Bonin arc; Suyehiro et al., 1996).

Following the breakup of the Rodinia supercontinent, the Phanerozoic was characterised by a re-assembly to form Pangea between 400 and 200 Ma. The breakup of Pangea was heralded by the emplacement of continental flood basalts, layered intrusions, and related magmatic Ni and PGE mineralisation (Fig. 8.3-5). Continuing breakup and dispersal of continental fragments led to the inception of subduction systems, magmatic arcs, and metal deposits that are typically formed at convergent margins, such as porphyry and epithermal systems. Barley and Groves (1992) pointed out that the preservation in the geological record of ore systems is dependent on whether these are part of interior or peripheral orogens, as defined by Murphy and Nance (1991). Peripheral orogens are those that form at continental margins, adjacent to an ocean, as for example along the Cordilleran and Andean side of the Americas. Interior orogens, develop during closure of oceanic basins and continental collision, resulting in crustal thickening and uplift, such as the Alpine-Himalayan and central Asian orogenic systems.

The Central Asian Orogenic Belt (CAOB) provides a prime example of interior orogens, Phanerozoic accretionary tectonics, and continental crust growth (Jahn, 2004; Sengör and Natal'in, 1996; Sengör et al., 1993). The CAOB is a complex collage of fragments of ancient microcontinents and arc terranes, fragments of oceanic volcanic islands (e.g., seamounts), some volcanic plateaux, oceanic crust (ophiolites), and successions of passive continental margins. The amalgamation of these terranes occurred at various times in the Palaeozoic and Mesozoic and was accompanied by various phases of magmatism, ranging in age from Ordovician (ca. 450 Ma) to Triassic-Cretaceous (ca. 220–120 Ma) that resulted in the emplacement of large volumes of granitic intrusions (Jahn, 2004) and mafic volcanic rocks, accompanied by lesser volumes of mafic–ultramafic systems (Pirajno et al., in press). Herein I discuss the example of a selected region, NW China (Xinjiang Province), where two major orogenic belts and volcano-sedimentary basins provide evidence of new continental crust.

8.3-6.1. NW China

Three major thermal events affected NW China, between the Carboniferous and the Permian-Triassic periods: Tian Shan, Tarim and Emeishan. These events, briefly discussed below, variably manifested by the eruption of basaltic lavas, mafic–ultramafic intrusions and A-type granitic magmatism, were responsible for juvenile crustal growth in the region.

Xia et al. (2003, 2004) reported on the widespread occurrence of ca. 345–325 Ma Carboniferous volcanic rocks in the Tian Shan orogenic belt. The Carboniferous rift-related volcanic
rocks constitute the Tian Shan large igneous province (LIP) of approximately 210,000 km^2,
which Xia et al. (2004) pointed out is only a minimum estimate, because these volcanic
rocks extend into Kazakhstan, Kyrgyzstan and Mongolia. In the Tian Shan, the volcanic
succession attains thicknesses varying from several hundred metres to over 13 km. Volcanic rocks of the Tian Shan LIP belong to the tholeiitic, alkaline and calc-alkaline series.
The reason for these variations is probably due to heterogeneities of basement lithologies
and the subcontinental lithospheric mantle (Xia et al., 2002). Trace element geochemistry
and Sr-Nd isotopic data ($^{87}Sr/^{86}Sr_{(i)} = 0.703$–$0.705$ and $\varepsilon_{Nd}(T) = +4$ to $+7$) reported by
Xia et al. (2003) indicate that the Tian Shan LIP originated from asthenospheric OIB-type
mantle source, with contributions from crustal and subcontinental lithospheric mantle.

In the Tarim Basin, tholeiitic volcanic rocks, mafic dykes, ultramafic rocks and syenites
form a LIP, estimated at greater than 200,000 km^2 in the western and southwestern parts
of the Basin (Yang et al., 2006; Chen et al., 2006). K-Ar dating yields a range of ages from
277 ± 4 to 288 ± 10 Ma, whereas Ar-Ar dating shows a plateau age of 278.5 ± 1.4 Ma (Yang,
pers. comm., 2006). The Huangshan mafic–ultramafic intrusions dated at ca. 270 Ma and
the Kalatongke intrusions at ca. 284 Ma may be part of this thermal event, as may be
Permian volcanic rocks in the Mongolian orogenic zone, which yielded Rb-Sr isochrons of
ca. 270 Ma (Zhu et al., 2001). Although not yet well defined, this magmatic province may
have a huge extension, from the Tarim block to the northern margin of the North China
Craton, a distance in excess of 3000 km.

The ca. 260–250 Ma Siberian-Emeishan thermal event affected large parts of Central
and Western Asia (Dobretsov, 2005) (Fig. 8.3-8). The Emeishan LIP covers an area of
at least 250,000 km^2 (Chung et al., 1998) in SW China (Yunnan, Sichuan and Guizhou
provinces) and NW Vietnam. The Emeishan LIP consists of a succession of predominantly
tholeiites, locally associated with picritic and rhyolitic lava flows. In addition to lava flows
are mafic–ultramafic layered complexes, dikes and sills, syenite and other alkaline intrusions (Xiao et al., 2004). Precise radiometric ages on mafic and ultramafic intrusions range
from 259 to 262 Ma, and those of lavas from 246 to 254 Ma; the peak magmatism was probably between 251 and 253 Ma (Dobretsov, 2005 and references therein). The Emeishan
LIP was more or less synchronous with the Siberian Traps which together may represent
a Permian superplume event that affected much of the Asian continent (Fig. 8.3-8 and see
Dobretsov, 2005).

8.3-6.2. A-Type Magmatism and Phanerozoic Crustal Growth

Extensive post-orogenic magmatism in the entire region of NW China is characterised by
A-type granitic intrusions and mafic-ultramafic complexes. This includes an $\sim$200 km belt
of A-type peralkaline granites in the Altay Orogen, emplaced into an extensional setting
(Han et al., 1997). These peralkaline granites have Y/Nb ratios from 1.39 to 3.33 and
$\varepsilon_{Nd}(T)$ values ranging from $+5.1$ to $+6.7$, suggesting a mantle source (Han et al., 1997).
In the Sawuer region in the western Junggar, ca. 290 to 297 Ma A-type alkali feldspar gran-

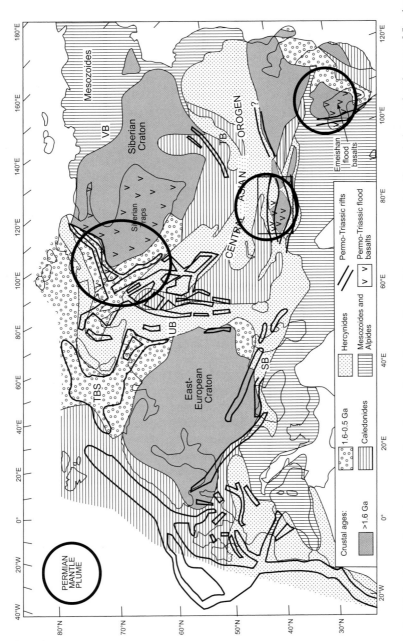

Fig. 8.3-8. Schematic geology of the Eurasian continent showing Permo-Triassic continental rifts and plume-related areas of flood volcanism. Although episodes of flood volcanism are of short duration (approx. 1–2 Myr), mantle plume (or superplume) activity lasted for almost 100 Myr (345–250 Ma), as a result of which juvenile crust was added to the Eurasian terranes (see Fig. 8.3-10). NW China approximately within boxed area. After and slightly modified from Nikishin et al. (2002). See text for details.

ites (Zhou et al., 2006) formed in a post-collisional extensional setting and are associated with Permian volcanic rocks (e.g., Tarim LIP referred to above: Zhou et al., 2006). Chen and Arakawa (2005) suggested that these rocks were derived from a parental magma of mantle origin, and were formed from partial melting of juvenile lower crust and from the differentiation of a basaltic underplate. The magmas thus produced are associated with a post-collisional extensional regime.

This, locally alkaline and peralkaline, magmatism extends for thousands of km from Xinjiang, through southern and inner Mongolia, to North China, where granitic rocks were emplaced in a very large Permian-Triassic rift system of central Asia in two stages: Permian-Triassic (300–250 Ma) and Late Triassic-Cretaceous (210–120 Ma) (Hong et al., 1996, 2004; Vladimirov et al., 1997; Nikishin et al., 2002; Jahn 2004) (Fig. 8.3-8). The formation and emplacement of such large quantities of magma in this rift system are probably associated with extension linked to the impingement of mantle plumes, probably the Siberian-Emeishan plume, or superplume event (Jahn, 2004; Dobretsov, 2005).

As mentioned in the introduction, most models of continental crustal growth in central Asia suggest accretionary and collision processes involving microcontinental fragments, volcanic arcs and oceanic plateaux and crust, resulting in the formation of sutures, some of which contain mafic–ultramafic rocks (ophiolites). Crustal and lithospheric thinning induce thermal disturbances, which cause a sequence of partial melting events first in the mantle, then in the lower crust (Barker et al., 1975). Large mafic magma chambers can rise from mantle plumes and underplate and intrude the lower crust. Seismic transects in the Baltic Shield and Sea provided evidence supporting the model of mafic underplating of the lower crust and the intrusion of mafic sheets (BABEL Working Group, 1993). In this region, high-temperature, dry and fluorine-rich magmas (rapakivi granites) were associated with mantle-derived mafic magmas (Puura and Flodén, 1999). I suggest that the arrival of a mantle plume(s) induced extension and rifting, followed by the emplacement of felsic and mafic–ultramafic magmas, following a scenario schematically illustrated in Fig. 8.3-9.

8.3-6.3. Destruction of Cenozoic Lithosphere and Lower Crust: Neotectonic Eastern China

Mantle lithosphere delamination and upwelling of asthenospheric mantle is a process that destroys continental crust. This process appears to be a prevailing feature in the Cenozoic and could perhaps represent a dominant modern style of crust–mantle interaction. Examples can be found in Western Europe (Wilson and Downes, 1991), North America (Abraham et al., 2005), and northeastern China. The latter is discussed here, because the volcanic provinces (Fig. 8.3-10) provide a nearly continuous sampling of the mantle for the last 180 Ma. Studies of xenoliths recovered from Cenozoic basalts in the Archaean (>2.5 Ga) Sino-Korean Craton in northeastern China have shown that the underlying mantle lithosphere is relatively thin (70–80 km) and characterised by high heat flow (50–100 mW/m^2) (Menzies et al., 1993; Menzies and Xu, 1998; O'Reilly and Griffin, 1996; Griffin et al., 1998b; Xu, 2001; O'Reilly et al., 2001) (Fig. 8.3-4). This is unlike other Archaean cratons, such as Kaapvaal in South Africa and Yilgarn in Western Aus-

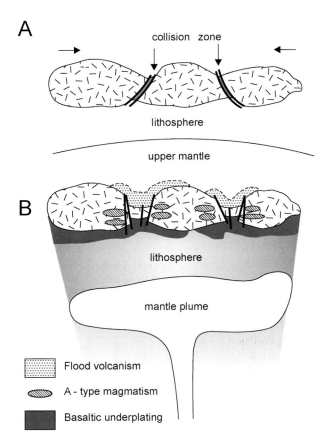

Fig. 8.3-9. Model illustrating juvenile crustal additions and intracontinental tectono-magmatic evolution in NW China in Carbonifeous-Permian times, from (A) Collision and accretion of terranes
of Altay and Tien Shan orogens; (B) mantle plume(s) impinge onto the lithosphere, causing underplating of the lower crust with production of A-type granitoids and continental flood basalts, during
post-orogenic extension. Note that the suture zones in A act as zones of weakness during extensional
stresses resulting in rift basins that are intruded by felsic and mafic–ultramafic plutons and filled with
flood-type volcanic rocks.

tralia, which are underlain by cold ($\sim$40 mW/m^2) and thick ($>$150 km) lithospheric keels.
Geochemical studies of these xenoliths show contrasting compositions of sub-continental
lithospheric mantle (SCLM). For example the Phanerozoic mantle (called Tecton) is more
fertile than the Proterozoic (Proton) and the Archaean (Archon) mantle. Furthermore, the
strongly positive ε_{Nd} values of mineral phases from peridotite xenoliths indicate that the
shallow lithospheric mantle beneath the Sino-Korean Craton is depleted and similar to
ocean island basalt (OIB). As mentioned above, Ordovician kimberlites show that a thick
($\sim$200 km) and cold lithospheric keel beneath the Sino-Korean Craton was present during

the Palaeozoic. In contrast the Cenozoic lithosphere has elevated geotherms and is less than 80 km thick. This requires the removal of about 120 km of cold, refractory Archaean keel from beneath the Sino-Korean Craton, and its replacement by a more fertile OIB-type mantle (Xu, 2001).

The thermo-tectonic evolution of the lithospheric mantle beneath eastern China is envisaged to have resulted from lithospheric extension, delamination and asthenospheric upwelling, with the possibility of deep mantle plumes being ruled out, given the nature of the volcanism in the region (Fig. 8.3-10). The onset of this lithospheric extension can be linked to the Triassic collision between the North China block (part of the Sino-Korean Craton) and the Yangzte Block along the Qinling orogenic belt (Xu, 2001), and the collision of the North China block with the Siberian Craton in the Jurassic-Cretaceous (Wang et al., 2006). Lithospheric thermal weakening and erosion proceed as a result of conductive heating by the upwelling asthenosphere.

Mesozoic-Cenozoic intraplate tectonism and magmatism affected much of the eastern margin of mainland Asia, extending from Far East Russia, through the Baikal and Mongolia regions, to north-eastern and eastern China. These phenomena are broadly linked with a complex series of events, involving subduction, collision, post-collision collapse and rifting. The Mesozoic and Cenozoic extensional tectonics of mainland China and the rest of eastern Asia are the surface expressions of shallow mantle dynamics in the region. These may relate to asthenospheric upwellings linked to lithospheric delamination in response to subduction systems and rifting following collision between Siberia and Mongolia–North China (Jurassic-Cretaceous) and in the western Pacific (Cenozoic) (Barry and Kent, 1998).

8.3-7. CONCLUDING REMARKS

Heat loss from the Earth's interior is achieved by a combination of lithospheric plate motions (plate tectonics) and through the emplacement of large igneous provinces (mantle plumes). This heat loss varied through time in response to the progressive cooling of the planet. A trend from nearly continuous mantle plume activity in the early Archaean to more time-interspersed plume activity during the Proterozoic and Phanerozoic may be a consequence of diminished deep mantle flow in a cooling Earth (Groves et al., 2005). This trend is possibly linked with a change over geological time from whole mantle convection of the early Earth, to a two-layer mantle convection in the Phanerozoic, leading to the modern tectonic and magmatic processes related to asthenospheric upwelling due to delamination.

Major episodes of Archaean crust formation would correlate with the rise of deep mantle plumes. An important aspect is that whole mantle convection facilitates higher rates of juvenile crust production, whereas lower growth rates are associated with layered mantle convection (Condie, 2000).

In Wilsonian periods, the upper and lower mantle are separated by the 660 km phase-change boundary layer, and the upper mantle is depleted in highly incompatible elements. During the Phanerozoic, several plumes originate from the base of the 660 km boundary layer, resulting in cycles of opening and closing of oceans, generation of oceanic

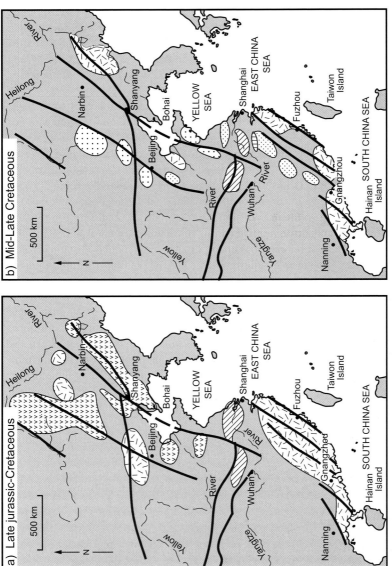

Fig. 8.3-10. Mesozoic-Cenozoic intracontinental volcanism in eastern China. After Pirajno et al. (2006). These volcanic rocks were emplaced in extensional settings (graben style rift basins) and are associated with dyke swarms and A-type granites. Collision of continental plates, North China block with the Yangtze craton in the south and the Siberian craton in the north, followed later by the westward subduction of the west Pacific, all contributed to a succession of post-collisional collapses due to lithospheric delamination and pulses of asthenospheric mantle upwellings. The effects of this type of mantle dynamics may result in partial destruction of lithosphere and lower crust (see Fig. 8.3-4). (*Part (c) and legend on next page.*)

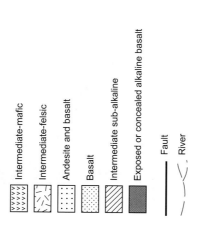

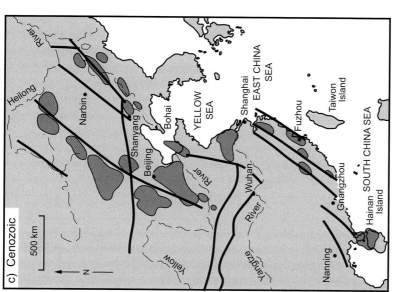

Fig. 8.3-10. (*Continued.*)

crust, subduction, magmatic arc activity, and accretion. The upper mantle becomes progressively depleted in incompatible elements by transfer to the continental and oceanic crusts. Wilsonian periods, however, lead to accumulation of cold lithospheric material at the 660 km boundary layer, which is eventually breached, so that the cold material descends into the lower mantle, down to the CMB (Kellogg et al., 1999; Schubert et al., 2001). This results in the rise of major mantle plumes from the CMB to the surface. Melting of plume heads results in pools of mafic–ultramafic magma that underplate the crust and erupt to form oceanic plateaux and continental flood basalts. This major overturn, or catastrophic downwellings to, and resultant plume upwellings from, the D'' layer, are associated with high rates of continental crust formation or major orogenies (MOMO), accompanied by partial depletion of lower mantle and replenishment of the upper mantle in incompatible elements.The aggregation of continental masses (supercontinents), their rupture, drift and renewed collision, may all be related to heat transfer and dissipation from the interior of the planet to the surface, which are caused by perturbations and convection in the mantle.

It can be concluded that mantle convection through time (e.g., Yokochi and Marty, 2005), may have been: (1) hot, turbulent convection in the Hadean and early Archaean; (2) whole mantle convection in the Late Archaean and Proterozoic; (3) two-layer convection in the Phanerozoic.Thus, based on an analysis of the role of mantle plumes in the building of continental crust, as presented in this contribution, stages of mantle dynamics through time may be considered as follows:

- Stage 1; >3.0 to ca. 2.8 Ga, high intensity and frequency of mantle plumes in a vigorously convecting mantle, stacking of plume-derived oceanic plateaux and limited flat subduction systems, leading to the first build up of continental crust;
- Stage 2; 2.8 Ga to 2.45 Ga, formation and accretion of greenstone belts formed by a combination of subduction-related magmatic arcs, plume-derived oceanic plateaux and microcontinents. First emplacement of continental flood basalts (LIPs);
- Stage 3; 2.45–1.0 Ga, continental rifting and assembly of supercontinents, with significant mantle plume activity, including a superplume event at 1.9 Ga;
- Stage 4; 1.0–0.6 Ga, supercontinent breakup, with common subduction systems;
- Stage 5; 0.6 Ga to present, two-layer mantle convection, full scale Wilsonian cycles; large mantle plumes and LIPs; lithospheric delamination and shallow plumes (upwelling mantle asthenosphere), with alkaline dominated magmatism.

ACKNOWLEDGEMENTS

I thank Martin Van Kranendonk for the invitation to contribute to this book, for the frequent stimulating discussions and the insightful comments on the manuscript. Thanks are also due to Dallas Abbott for her comments. I am grateful to Murray Jones and Michael Prause of the Geological Survey of Western Australia, for drafting the figures, and to Eunice Cheung for library services. S. O'Reilly provided a copy of Fig. 8.3-3. This contribution is published with the permission of the Executive Director of the Geological Survey of Western Australia.

Earth's Oldest Rocks
Edited by Martin J. Van Kranendonk, R. Hugh Smithies and Vickie C. Bennett
Developments in Precambrian Geology, Vol. 15 (K.C. Condie, Series Editor)
© 2007 Elsevier B.V. All rights reserved.
DOI: 10.1016/S0166-2635(07)15084-2

Chapter 8.4

EO- TO MESOARCHEAN TERRANES OF THE SUPERIOR PROVINCE AND THEIR TECTONIC CONTEXT

JOHN A. PERCIVAL

Geological Survey of Canada, 601 Booth St., Ottawa, Ontario, Canada, K1A 0E8

8.4-1. INTRODUCTION

Because of its exceptional exposure and detailed knowledge base, the Superior Province has become a classic example of Neoarchean accretionary tectonics (Langford and Morin, 1976; Card, 1990; Williams et al., 1992; Stott, 1997). A major influence on the tectonic style was the existence of microcontinental terranes that were involved in some of the earliest arc-continent and continent-continent collisions (Percival et al., 2004b; 2006b).

In 1968, Goodwin proposed that the Superior Province had grown through accretion of oceanic material onto a continental nucleus, and although this concept has evolved significantly through four decades of research that includes a new generation of maps, geophysical imaging, and acquisition of almost 2000 high-precision U-Pb ages, it remains a tenable hypothesis. This paper reviews the nature, age and origin of the diverse microcontinental terranes now known within the Superior Province, and describes their role in the Neoarchean tectonic evolution.

Several recent papers have provided overviews of aspects of Superior Province geology, including regional tectonic evolution (e.g., Williams et al., 1992; Stott, 1997; Percival et al., 2006b), metallogeny (Card and Poulsen, 1998; Percival, 2007), geochronology (Skulski and Villeneuve, 1999), crust and mantle geophysics (e.g., Kay et al., 1999a, 1999b; Craven et al., 2001; Kendall et al., 2002; White et al., 2003; Mussachio et al., 2004).

Three microcontinental terranes of the Superior Province – the Northern Superior, Hudson Bay, and Minnesota River Valley – are considered in detail elsewhere in this volume (Böhm et al., this volume; O'Neil et al., this volume; Bickford et al., this volume). The purpose of this contribution is to examine how these and other ancient terranes of the Superior Province correspond in terms of geological history, and to describe their assembly during a series of Neoarchean (2.72–2.68 Ga) orogenies.

8.4-2. TECTONIC FRAMEWORK

The Superior Province forms the Archean core of the Canadian Shield (Fig. 8.4-1, inset map). It has been tectonically stable since ca. 2.6 Ga, having occupied a lower plate setting during most Paleoproterozoic and Mesoproterozoic tectonism that affected its margins.

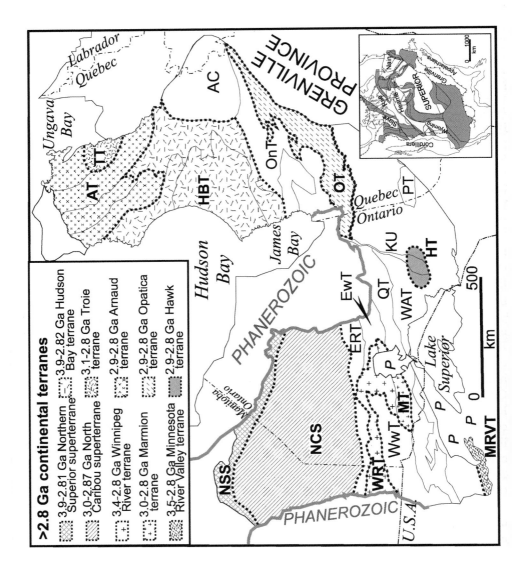

>2.8 Ga continental terranes

3.9-2.81 Ga Northern Superior superterrane	3.9-2.82 Ga Hudson Bay terrane
3.0-2.87 Ga North Caribou superterrane	3.1-2.8 Ga Troie terrane
3.4-2.8 Ga Winnipeg River terrane	2.9-2.8 Ga Arnaud terrane
3.0-2.8 Ga Marmion terrane	2.9-2.8 Ga Opatica terrane
3.5-2.8 Ga Minnesota River Valley terrane	2.9-2.8 Ga Hawk terrane

Fig. 8.4-1. Modified tectonic framework for the Superior Province, showing age range of continental domains (identified in legend), distribution of oceanic domains and metasedimentary belts (unornamented) and Proterozoic cover (*P*). Subdivisions of the Superior Province modified after Card and Ciesielski (1986), Percival et al. (1992), Leclair et al. (2006) and Percival et al. (2006b). Abbreviations: AC: Ashuanipi complex; AT: Arnaud terrane; ERT: English River metasedimentary terrane; EwT: Eastern Wabigoon terrane; HBT: Hudson Bay terrane; HT: Hawk terrane; KU: Kapuskasing uplift; MRVT: Minnesota River Valley terrane; MT: Marmion terrane; NCS: North Caribou superterrane; NSS: Northern Superior superterrane; OnT: Opinaca metasedimentary terrane; OT: Opatica terrane; PT: Pontiac metasedimentary terrane; QT: Quetico metasedimentary terrane; TT: Troie terrane; WAT: Wawa–Abitibi terrane; WRT: Winnipeg River terrane; WwT: Western Wabigoon terrane; Inset: tectonic map of North America (after Hoffman, 1989) showing location of the Superior Province.

A first-order feature of the Superior Province is its linear subprovinces of distinctive lithological and structural character, accentuated by subparallel boundary faults (e.g., Card and Ciesielski, 1986). Trends are generally east west in the south, WNW in the northwest, and NW in the northeastern Superior (Fig. 8.4-1). Recent work based on isotopic and zircon inheritance studies has revealed fundamental age domains across the Superior Province (Fig. 8.4-1). At least eight terranes of Eoarchean to Mesoarchean age are recognized, in spite of pervasive Neoarchean magmatism, metamorphism and deformation. "Terranes" are defined as tectonically bound regions with internal characteristics distinct from those in adjacent regions prior to assembly of the Superior Province. They may comprise several "domains" with distinct lithological characteristics that share a common basement. Terrane boundaries are commonly marked by synorogenic flysch deposits. "Superterranes" carry metamorphic or deformational evidence of at least one amalgamation event before Neoarchean Superior Province assembly. "Subprovince" is a general term describing a region with similar geological and geophysical characteristics.

The oldest crust (up to 3.8 Ga) occurs in the Northern Superior superterrane of the northwestern Superior (NSS; Skulski et al., 2000; Böhm et al., this volume) and Hudson Bay terrane to the northeast (O'Neil et al., this volume). To the south, a large region of ca. 3.0 Ga crust, the North Caribou superterrane (Stott and Corfu, 1991; Stott, 1997), has been interpreted as a continental nucleus during assembly of the Superior Province (cf. Goodwin, 1968; Thurston et al., 1991; Williams et al., 1992; Stott, 1997; Thurston, 2002). Farther south, the Winnipeg River (WR) and Marmion (MM) terranes are relatively small continental fragments dating back to 3.4 and 3.0 Ga, respectively (Beakhouse, 1991; Tomlinson et al., 2004). In the far south, the Minnesota River Valley terrane (MRV) of unknown extent contains remnants of crust as old as ca. 3.6 Ga (Goldich et al., 1984; Bickford et al., 2006, this volume). To the east, the Opatica terrane has ancestry in the 2.8–2.9 Ga range (Davis et al., 1995), as does the Arnaud terrane in the far northeast (Leclair et al., 2006).

Domains of oceanic ancestry, identified by submarine volcanic rocks with juvenile isotopic signatures, lack of inherited zircon and absence of clastic sedimentary input, separate most of the continental fragments (Fig. 8.4-1). These dominantly greenstone-granite terranes generally have long strike lengths and record geodynamic environments including oceanic floor, plateaux, island arc and back-arc settings (e.g.,, Thurston 1994; Kerrich et al., 1999). Examples include parts of the Oxford–Stull and southern La Grande domains in the north, the western Wabigoon terrane in the west, and the Wawa–Abitibi terrane in the southern Superior Province (Fig. 8.4-1).

Still younger features, the English River, Quetico and Pontiac metasedimentary terranes (Breaks, 1991; Williams, 1991), separate some of the continental and oceanic domains. Extending up to 1000 km along strike, these 50–100 km wide belts of metagreywacke, migmatite and derived granite appear to represent thick syn-orogenic wedges (Davis, 1996, 1998, 2002), deposited, deformed and metamorphosed during collisional orogeny.

8.4-3. MICROCONTINENTAL FRAGMENTS

The ancient tectonic building blocks of the Superior Province comprise terranes of continental affinity, with tectonostratigraphic histories independent of those of neighbouring regions prior to Neoarchean amalgamation (Percival et al., 2004a). The Eo- to Mesoarchean terranes, separated by Neoarchean mobile belts, will be described in terms of lithology, history and boundaries, beginning in the northwest, continuing through the south, and terminating in the northeast (Fig. 8.4-1).

8.4-3.1. Northern Superior Superterrane

Dominated by granitic and gneissic rocks, the poorly exposed Northern Superior superterrane at the northern fringe of the Superior Province (Figs. 8.4-1 and 8.4-2) has been recognized on the basis of isotopic evidence (Skulski et al., 1999). In the west, the data include ca. 3.5 Ga orthogneiss from the Assean Lake block (Böhm et al., 2000a) and supracrustal units including iron formation, mafic-to intermediate volcanic rocks, and greywacke with detrital zircon ages up to 3.9 Ga (Böhm et al., 2003). Orthogneiss with >3.5 Ga inherited zircon ages has been recognized to the east (Skulski et al., 2000; Stone et al., 2004).

Ancient rocks throughout the Northern Superior superterrane have been strongly reworked by metamorphism and magmatism. At Assean Lake (Fig. 8.4-2), tonalite-trondhjemite-granodiorite (TTG) magmatism occurred at 3.2–3.1 Ga, and amphibolite-facies metamorphism at 2.68 and 2.61 Ga (Böhm et al., 2003). In the Yelling Lake area of Ontario (Fig. 8.4-2), magmatism at 2.85–2.81 Ga was followed by metamorphism and further magmatism at 2.74–2.71 Ga (Skulski et al., 2000).

Fig. 8.4-2. (*Next page.*) Location of features referred to in the text. Abbreviations: ALB: Assean Lake block; BRPC: Berens River plutonic complex; ELMC: English Lake magmatic complex; GLTZ: Great Lakes tectonic zone; NSGB: North Spirit greenstone belt; NCGB: North Caribou greenstone belt; NKF: North Kenyon fault; SSGB: Savant-Sturgeon greenstone belt; WLGB: Wallace Lake greenstone belt; YL: Yelling Lake. Additional abbreviations as in Fig. 8.4-1.

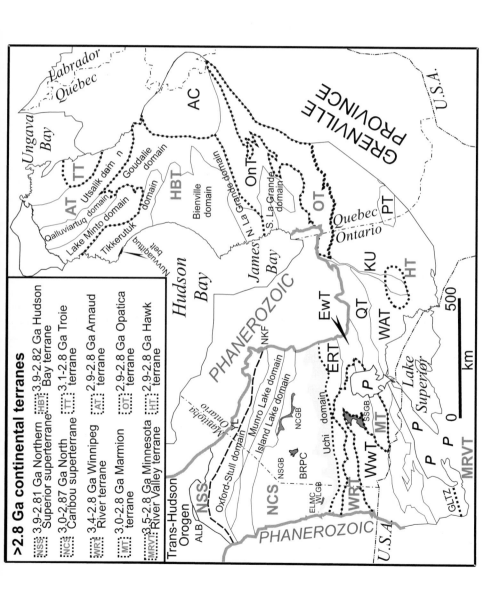

The Northern Superior superterrane is bounded to the south by the North Kenyon fault, which juxtaposes it with the Oxford–Stull domain, a region of 2.84–2.71 Ga juvenile crust (Skulski et al., 2000), and to the northwest by the Paleoproterozoic Trans-Hudson Orogen (Fig. 8.4-1). Its northern and eastern limits are undefined owing to Paleozoic cover and water of Hudson and James bays.

8.4-3.2. Hudson Bay Terrane

The Hudson Bay terrane (Leclair et al., 2006; Fig. 8.4-1) is located in the northeastern Superior Province, a region characterized dominantly by plutonic rocks and north-northwesterly structural trends. Plutonic rocks include widespread charnockitic (pyroxene-bearing) granitoid intrusions (Stern et al., 1994; Skulski et al., 1996; Rabeau, 2003; Bédard, 2003; Boily et al., 2004, 2006b; Stevenson et al., 2006). Supracrustal rocks occur as narrow, linear, lithologically diverse belts (<20 km wide by <120 km long) separated by broad plutonic complexes (Percival and Card, 1992; Percival et al., 1994; Skulski et al., 1996; Lin et al., 1996; Percival and Skulski, 2000; Boily and Dion, 2002; Berclaz et al., 2003, 2004; Leclair et al., 2002a, 2006). The supracrustal belts are generally metamorphosed to amphibolite facies, but also occur at granulite and rarely, greenschist facies.

The Hudson Bay terrane is characterized by ancient Nd model ages, zircon inheritance ages and rare Meso- and Eoarchean rock ages in several sub-regions, including the northern La Grande subprovince, Bienville, Tikkerutuk and Goudalie domains (Fig. 8.4-2). In the northern La Grande subprovince, Mesoarchean basement (3.36–2.79 Ga) is unconformably overlain by clastic rocks (Roscoe and Donaldson, 1988) and 2.75–2.73 Ga volcanic strata (Goutier and Dion, 2004). Igneous rocks include komatiites and related ca. 2.82 Ga sills of probable rift origin (e.g., Skulski et al., 1988). Supracrustal rocks extend to the northeast as isolated volcano-sedimentary belts (Gosselin and Simard, 2001; Thériault and Chevé, 2001; Simard et al., 2002). The Bienville domain is essentially composed of plutonic and gneissic rocks including TTG, granite-granodiorite plutonic suites and their pyroxene-bearing equivalents (ca. 2.74–2.69 Ga) (Ciesielski, 2000; Simard et al., 2004; Roy et al., 2004). They enclose rare supracrustal and older tonalitic units, many of which can be correlated with units to the north. The plutonic rocks contain inherited zircons ranging from 3.20–2.74 Ga and display Nd model ages as old as 3.36 Ga (Skulski et al., 1998; Isnard and Gariépy, 2004; Boily et al., 2004, 2006b; unpublished data from J. David, written comm., 2006). The Tikkerutuk domain in the west is characterized by abundant clinopyroxene-bearing tonalite-diorite, enderbite (2.73–2.69 Ga) and granite (2.72–2.69 Ga) enclosing older tonalitic units (2.84–2.75 Ga), with rare enclaves of tonalitic gneiss (ca. 3.02 Ga) (Percival et al., 2001 and references therein; unpublished data from J. David, written comm., 2006). Several isolated belts of supracrustal rocks (2.76–2.70 Ga) occur mainly in the northern part of the domain (Maurice et al., 2005, 2006). The western Tikkerutuk domain is distinguished by generally older Nd model ages (T_{DM} = 3.2–4.0 Ga) (Nd isotopic data from Stern et al., 1994; Skulski et al., 1996; Rabeau, 2003; Boily et al., 2004, 2006b; Stevenson et al., 2006). It also contains the ca. 3.8–3.6 Ga Nuvvuagittuq belt (Fig. 8.4-2: David et al., 2003, 2004; O'Neil et al., this volume), the oldest volcano-plutonic complex recognized in the Superior Province. The 8 km^2 belt consists mainly of Mg-rich amphibolite, with some ultramafic sills (Francis, 2004), metaconglomerate, and metamorphosed iron formation of metasedimentary origin (Dauphas et al., 2007; O'Neil et al., this volume). Geochemical fingerprints suggest an arc-like setting (Cates and Mojzsis, 2006). A tuffaceous unit returned a U-Pb zircon age of 3823 ± 18 Ma, whereas tonalite gave 3650 ± 5 Ma and Nd model ages between 3.8 and 3.9 Ga (David et al., 2004). Detrital zircons from conglomerate indicate oldest source ages around 3.73 Ga (Cates and Mojzsis, 2006).

The Goudalie domain to the east is characterized by several small (up to 10 × 30 km) supracrustal belts (2.88–2.71 Ga) surrounded by a plutonic complex including older (to 3.01 Ga), coeval and younger (2.72–2.67 Ga) units. The Vizien greenstone belt contains tectonically intercalated 2.725 Ga volcanic rocks of continental arc affinity and 2.79 Ga juvenile oceanic plateau rocks and was postulated to include a Neoarchean suture zone (Skulski et al., 1996; Lin et al., 1996). The Goudalie domain records inherited zircon and Nd model ages in the 2.8–3.3 Ga range (Rabeau, 2003; Boily et al., 2004, 2006b).

8.4-3.3. North Caribou Superterrane

The North Caribou superterrane (Fig. 8.4-1: Thurston et al., 1991) is the largest tract of Mesoarchean rocks in the Superior Province (Stott, 1997). It is characterized by widespread evidence for crust with ca. 3.0 Ga mantle extraction ages (Stevenson, 1995; Stevenson and Patchett, 1990; Corfu et al., 1998; Hollings et al., 1999: Henry et al., 2000), and displays evidence for continental breakup, as well as an amalgamation event prior to 2.87 Ga (Stott et al., 1989; Thurston et al., 1991). Mesoarchean units have been variably reworked by subsequent Archean magmatic and deformational events.

Within the greenstone belts, thin supracrustal packages are preserved sporadically across the North Caribou superterrane (Thurston and Chivers, 1990; Thurston et al., 1991). They consist of a lower, quartz-rich, coarse clastic unit, locally unconformable on basement, overlain by carbonate, iron formation, basaltic and komatiitic volcanic units. In different areas, the sequences have been interpreted as platformal cover strata (Thurston and Chivers, 1990) and as plume-related rift deposits (Hollings and Kerrich, 1999; Hollings, 2002; Percival et al., 2002, 2006a). Evidence for plume-related rifting is based on the presence of quartz arenite and komatiitic rocks with enriched mantle geochemical signatures (Tomlinson et al., 1998, 2001; Hollings and Kerrich, 1999).

Parts of the North Caribou superterrane have been assembled from older fragments (Stott, 1997), although the early history is generally obscured by younger plutonism. Evidence of early tectono-metamorphism comes from the North Caribou greenstone belt (Fig. 8.4-2), where ca. 2.98 and 2.92 Ga volcanic assemblages are intruded by the 2.87 Ga North Caribou pluton, which is interpreted to postdate regional deformation and metamorphism (Stott et al., 1989).

The Oxford–Stull domain (Fig. 8.4-2: Thurston et al., 1991) represents the largely juvenile, 2.88–2.73 Ga continental northern margin of the 3 Ga North Caribou superterrane that was tectonically imbricated with oceanic crustal fragments (Syme et al., 1999; Corkery et al., 2000; Skulski et al., 2000; Stone et al., 2004). Its tectonostratigraphic features (Corkery et al., 2000) include 2.84–2.83 Ga tholeiitic mafic sequences and calc-alkaline arc volcanic rocks with juvenile to locally enriched Nd isotopic composition, which are unconformably overlain by <2.82 Ga sedimentary rocks that contain <2.94 Ga detrital zircons (Skulski et al., 2000). Synvolcanic plutons associated with 2.84–2.72 Ga calc-alkaline volcanism are isotopically juvenile in the Oxford–Stull domain, but have <3 Ga Nd model ages further south, reflecting the influence of thicker North Caribou crust (Skulski et al., 2000). This package was juxtaposed on D_1 faults with submarine, depleted tholeiitic basalts prior to intrusion of 2.78 Ga tonalite (Corkery et al., 2000). Submarine arc volcanic rocks covered the composite basement prior to ca. 2.72 Ga (Lin et al., 2006). Unconformably overlying the shortened continental margin collage is a 2.722–2.705 Ga successor arc of calc-alkaline to shoshonitic volcanic and associated sedimentary rocks (Brooks et al., 1982; Corkery and Skulski, 1998; Corkery et al., 2000; Skulski et al., 2000; Stone et al., 2004; Lin et al., 2006) that record an influx of local and exotic detrital zircons ranging in age from 2.704 to 3.65 Ga (Corkery et al., 1992; Corkery et al., 2000; Lin et al., 2006).

The Munro Lake and Island Lake domains (Fig. 8.4-2) comprise plutonic rocks with several small supracrustal belts in the northern North Caribou superterrane (Stone et al., 2004; Parks et al., 2006). In the Munro Lake domain, rift-related, quartzite locally interbedded with komatiite overlies 2.883–2.865 Ga tonalite (Stone et al., 2004; Corkery et al., in prep.). Tonalite and granodiorite plutons across the Munro and Island Lake domains have U/Pb ages ranging from 2.88–2.70 Ga, and have 3.05–2.71 Ga Nd model ages reflecting variable recycling of North Caribou age crust (Turek et al., 1986; Stevenson and Turek, 1992; Skulski et al., 2000).

To the south, the Island Lake domain includes 2.89, 2.85 and 2.74 Ga volcanic sequences in a series of structural panels (Parks et al., 2006). Diverse clastic sedimentary sequences were deposited synvolcanically at <2.84 >2.744 Ga, and post-volcanically at <2.71 Ga. All sequences have detrital zircon U/Pb ages that range from 2.938–2.711 Ga (Corfu and Lin, 2000), consistent with North Caribou provenance.

The central North Caribou superterrane, which is dominated by 2.745–2.698 Ga calc-alkaline, peraluminous and sanukitoid granitoid plutons of the Berens River plutonic complex (Fig. 8.4-2: Stone, 1998; Corfu and Stone, 1998a), preserves several remnants of ca 3.0 Ga basement crust. Some of the oldest rocks are 3.02 Ga felsic volcanic rocks of the North Spirit assemblage (Fig. 8.4-2: Corfu and Wood, 1986), with juvenile 3.1 and younger Nd model ages (Stevenson, 1995).

8.4-3.3.1. English Lake complex

One of the largest and best preserved remnants of North Caribou crust occurs in the southwestern corner, where Krogh et al. (1974) first recognized 3.0 Ga rocks. Here, rocks of the English Lake complex (Fig. 8.4-2) consist of mantle-derived ultramafic through tonalitic compositions (Whalen et al., 2003). Detailed petrological, geochemical and geochronological studies provide insight into petrogenetic processes active in formation of the English Lake complex.

The complex consists predominantly of weakly metamorphosed tonalite and gneissic equivalents, with diorite, gabbro, anorthosite, hornblendite and metabasite (Fig. 8.4-

Fig. 8.4-3. Photographs of well preserved, Mesoarchean, magmatic and metamorphic textures of the English Lake magmatic complex: (a) multiple cross-cutting relationships among tonalitic, mafic and anorthositic magmatic phases, which have yielded ages between 3007 and 2992 Ma (Whalen et al., 2003); (b) graded magmatic layering in anorthosite and gabbroic anorthosite; note rafted gabbroic fragment; (c) ultramafic breccia showing fragments of pyroxenite (pale) and gabbro (dark) in matrix of igneous hornblendite; (d) garnet-clinopyroxene-hornblende-plagioclase metabasite; metamorphic zircon from this rock yielded a U-Pb age of 2997 Ma (Percival et al., 2006a).

8(a–d); Percival et al., 2006a) that yield U-Pb zircon crystallization ages in the range 3006–2992 Ma (Whalen et al., 2003). Mafic through felsic components share common geochemical attributes, including low trace-element abundances, slight LREE enrichment, and generally prominent negative Th and Nb anomalies (Whalen et al., 2003). These features have been interpreted as the signature of arc magmas produced in a mantle wedge fluxed by fluids derived from serpentinite dehydration (op. cit.). At deep structural levels,

Fig. 8.4-3. (*Continued.*)

mafic layers up to tens of metres thick (Fig. 8.4-3(d)) contain the metamorphic assemblage garnet-clinopyroxene-hornblende-plagioclase-quartz, which provides P-T estimates in the range 12 kbar, 850 °C (Percival et al., 2006a), much greater than the regional background erosion level (ca. 4 kbar; Stone, 2000). Metamorphic zircons yielded a syn-magmatic U-Pb age of 2997 ± 4 Ma (Percival et al., 2006a), suggesting that the crust attained a thickness of at least 36 km during magmatic construction.

A deformation fabric within the complex includes ductile high-strain zones up to several metres wide with dextral transcurrent kinematic indicators (Percival and Whalen, 2000). The shear zones are transected by dykes and plutons of late-tectonic biotite granodiorite, dated at 2941 ± 2 Ma (Percival et al., 2006a). The observations suggest that a Mesoarchean (ca. 2940 Ma) tectonomagmatic event affected the North Caribou superterrane in this area. It may correspond to a >2920 Ma D$_1$ event identified nearby in the Wallace Lake belt,

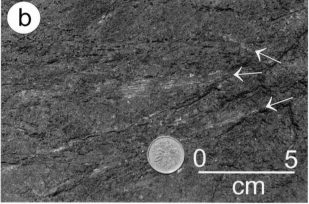

Fig. 8.4-4. Representative rock types of the Mesoarchean Lewis-Storey rift assemblage; (a) quartz arenite with centimetre-scale bedding; (b) spinifex-textured komatiite (white arrows mark acicular olivine crystals).

which is characterized by high-strain zones and tectonic inversion (Sasseville, 2002; Sasseville et al., 2006).

8.4-3.3.2. Lewis-Storey Rift Assemblage

The Lewis-Storey assemblage is a sedimentary-volcanic sequence that sits unconformably on 3.0 Ga tonalitic basement at the southwestern margin of the North Caribou superterrane (Percival et al., 2002, 2006a). Basal arkosic grit (Bailes and Percival, 2005) is overlain by thinly laminated, fine-grained quartz arenite units up to 3 m thick (Fig. 8.4-4(a)), including fuchsitic horizons. Quartz arenite carries a single population of detrital zircons with SHRIMP ages centered around 2991 ± 4 Ma, suggesting local derivation (Percival et al., 2006a). A unit of spinifex-textured komatiite (Fig. 8.4-4(b)) and derived serpentinite structurally overlies the quartz arenite and is succeeded by ca. 5 m of oxide and carbonate facies iron formation. Sills of mafic and rare felsic composition intrude the basement and sedimentary-volcanic sequence. A quartz-porphyritic rhyolite sill cutting basal grit yielded igneous zircons of 2978 ± 3 Ma (Percival et al., 2006a).

The sedimentary-volcanic sequence and associated mafic to felsic sills are interpreted to mark an ancient (ca. 2980 Ma) rifted margin of the North Caribou superterrane. Basal grit may have developed through chemical weathering of tonalite during uplift preceding rifting (cf. Rainbird and Ernst, 2001). The presence of komatiite, generated in anomalously hot mantle, and possible evidence of early uplift and weathering, support a model of plume-driven rifting (e.g., Bleeker et al., 1999). An age gap of 255 million years separates the Lewis-Storey assemblage from juvenile Neoarchean rocks to the southwest, which are juxtaposed along a Neoarchean (ca. 2.70 Ga) D_1 transcurrent shear zone (Percival et al., 2006a). These observations lead to the conclusion that an open ocean existed from ca. 2980 to 2700 Ma, and that a 280 My Wilson cycle is recorded at the southeastern margin of the North Caribou superterrane.

8.4-3.3.3. Uchi domain

The Uchi domain preserves ca. 300 My of tectonostratigraphic evolution along the southern margin of the North Caribou superterrane (Fig. 8.4-2: Stott and Corfu, 1991; Corfu and Stott, 1993a; 1996; Hollings et al., 2000; Sanborn-Barrie et al., 2001; Young et al., 2006). Chronostratigraphic correlations have been established within greenstone belts over a strike length of at least 500 km.

The stratigraphic sequence built on North Caribou basement, records 2.99–2.96 Ga rifting (Tomlinson et al., 1998; Sanborn-Barrie et al., 2004), several episodes of continental arc magmatism (2.94–2.91, 2.89, 2.87, 2.745–2.734 Ga; Henry et al., 2000; Sanborn-Barrie et al., 2001, 2004), intra-arc rifting, several phases of deformation and associated sedimentation. A deformation event and unconformity or disconformity separate Mesoarchean from Neoarchean strata across the Uchi domain. Juvenile volcanic rocks (2.73–2.713 Ga) along the southern margin may have formed on thin North Caribou crust, or may be part of an accreted oceanic terrane (Percival et al., 2006a).

8.4-3.4. Winnipeg River Terrane

The Winnipeg River terrane includes plutonic rocks exposed north and east of the western Wabigoon volcanic domain. It consists of the Winnipeg River subprovince of Beakhouse (1991), a >500 km long terrane composed of Neoarchean plutonic rocks with Meso- to Paleoarchean inheritance; and a Neoarchean plutonic domain to the southeast that contains scattered remnants of Mesoarchean crust and isotopic evidence for recycled 3.4–3.0 Ga material (Tomlinson and Percival, 2000; Tomlinson et al., 2004; Whalen et al., 2002, 2004). With inheritance back to ca. 3.4 Ga (Henry et al., 1998; Tomlinson and Dickin, 2003), the Winnipeg River terrane is distinct from the Northern Superior and North Caribou superterranes to the north and the Marmion terrane to the south (described below). It also carries a long record of magmatic and structural events (Corfu, 1988; Percival et al., 2004b; Melnyk et al., 2006).

The Mesoarchean history of the Winnipeg River terrane has remained cryptic due to extensive overprinting by Neoarchean magmatism and deformation. Tonalitic rocks are the oldest units recognized, and include both gneissic (3.17 Ga; Corfu, 1988; 3.32–3.05 Ga; Davis et al., 1988; Melnyk et al., 2006) and foliated varieties (3.04 Ga; Krogh et al., 1976). Some of these old rocks show still older Nd isotopic signatures, with model ages in excess of 3.4 Ga (Henry et al., 2000; Tomlinson and Dickin, 2003) and zircon inheritance. Similar isotopic signatures characterize younger (2.88, 2.84 and 2.83 Ga) tonalitic rocks, reflecting the antiquity of the basement (Beakhouse and McNutt, 1991; Beakhouse et al., 1988). Mafic volcanic belts older than ca. 3.0 Ga (Davis et al., 1988), and ca. 2.93 and 2.88 Ga intermediate volcanic rocks in the eastern Savant-Sturgeon greenstone belt (Sanborn-Barrie and Skulski, 1999) are also considered part of the Winnipeg River terrane. These rocks, which underlie "rift to drift" continental tholeiites of the 2.88–2.75 Ga Jutten assemblage, represent the earliest evidence for rifting of Winnipeg River basement (Skulski et al., 1998; Sanborn-Barrie and Skulski, 2006).

A complex structural-metamorphic history characterizes the Winnipeg River terrane during the Neoarchean. Metasedimentary rocks contain detrital zircons as young as 2.72 Ga, indicating that rocks now at amphibolite to granulite facies had been deposited after 2.72 Ga (Melnyk et al., 2006). The supracrustal rocks and older gneisses were folded (D_3) between 2.717 and 2.713 Ga, prior to syntectonic injection of 2.713–2.707 Ga tonalite and granodiorite sheets accompanying D_4 horizontal extensional deformation (op. cit.). Upright folding during D_5 deformation took place after 2.705 Ga, and younger upright F_6 folds indicate a period of north–south compression associated with emplacement of 2.695–2.685 Ga granite and granodiorite (op. cit.). Late pegmatites and granites intruded during a dextral transpressive (D_7) regime (op. cit.).

The eastern part of the Winnipeg River terrane represents a broad (200 km wide) transverse corridor of granitoid rocks separating the volcanic-dominated eastern and western Wabigoon domains (Figs. 8.4-1, 8.4-2). Small, east-trending greenstone belts have ages >3.075 to 2.703 Ga (Davis et al., 1988; Tomlinson et al., 2002; 2003). Dated granitoid units have yielded ages in the range 3.075–2.680 Ga (Davis et al., 1988; Whalen et al., 2002) and some of the oldest rocks have ε_{Nd} values of −1 to +1, suggesting derivation from even older crustal sources (Tomlinson et al., 2004). At least five generations of Neoarchean structures (D_1-D_5) have been recognized in complex tonalitic gneisses (Brown, 2002; Percival et al., 2004b).

8.4-3.4.1. Marmion terrane
The Marmion terrane (Figs. 8.4-1, 8.4-2) consists of 3.01–2.999 Ga Marmion tonalite basement (Davis and Jackson, 1988; Tomlinson et al., 2004), upon which several greenstone belts formed between 2.99 and 2.78 Ga (Hollings and Wyman, 1999; Stone et al., 2002; Tomlinson et al., 2003). In contrast to Winnipeg River-type crust with 3.4 Ga ancestry to the north, the Marmion terrane appears to have been juvenile at 3.0 Ga. It may have been tectonically accreted to the Winnipeg River terrane by ca. 2.92 Ga (Tomlinson et al., 2004)

or may have formed by magmatic addition of juvenile crust at the Winnipeg River margin. The Marmion terrane has experienced little, if any, Neoarchean (i.e., 2.745–2.72 Ga) magmatic activity, in contrast to the eastern Winnipeg River terrane to the north, and to the Wabigoon terrane to both the west and east.

8.4-3.5. Hawk Terrane

Mesoarchean strata have been identified within the predominantly juvenile, Neoarchean Wawa–Abitibi terrane, in the form of 2.89–2.88 Ga volcanic rocks of the Hawk assemblage (Turek et al., 1992). Plutonic equivalents (2.92 Ga) are present within a granitoid complex structurally beneath the supracrustal belt (Moser, 1994). Inherited 2.8–2.9 Ga zircons occur in granitoid rocks 100 km (Krogh and Moser, 1994) and 200 km (Ketchum et al., 2006) to the east, suggesting the existence of an extensive sialic substrate in the central Abitibi–Wawa terrane.

8.4-3.6. Minnesota River Valley Terrane

The poorly exposed Minnesota River Valley terrane (Figs. 8.4-1, 8.4-2: MRVT) consists predominantly of granitoid and gneissic rocks with zircon inheritance dating back to ca. 3.5 Ga and mantle extraction ages close to 3.6 Ga (Bickford et al., 2006, this volume). Its complex history includes Mesoarchean magmatic events at 3.48–3.50, 3.36–3.38, and 3.14 Ga, and metamorphism at 3.38 and 3.08 Ga, followed by Neoarchean (2.62–2.60 Ga) magmatism and metamorphism (Bickford et al., 2006), and a low-grade overprint at 1.8 Ga (Goldich et al., 1984; Schmitz et al., 2006).

The Great Lakes tectonic zone (Fig. 8.4-2) is the unexposed boundary between the Minnesota River Valley terrane and Wawa–Abitibi subprovince, identified from aeromagnetic images (Sims and Day, 1993). It is inferred to dip northward based on the presence of isotopic inheritance in plutons of the Vermilion district of the southern Wawa–Abitibi subprovince (Sims et al., 1997).

8.4-3.7. Opatica Belt

Bordering the Abitibi belt on the north, this predominantly metaplutonic belt consists of units ranging from 2.82 Ga tonalite, through 2.77–2.70 Ga tonalite-granodiorite, to 2.68 Ga granite and pegmatite (Benn et al., 1992; Sawyer and Benn, 1993; Davis et al., 1995). Polyphase deformation includes early shear zones, possibly as old as 2.72 Ga (Benn et al., 1992; Davis et al., 1995), overprinted by 2.69–2.68 Ga south-vergent structures. Sawyer and Benn (1993) concluded that the early structures of the Opatica terrane resulted from west-directed thrusting and were overprinted by a southward-propagating fold-and-thrust belt related to overthrusting of the northern Wawa–Abitibi terrane.

8.4-3.8. Arnaud Terrane

The Arnaud terrane (Fig. 8.4-1: Leclair et al., 2006) consists of plutonic and sparse supracrustal rocks with mantle extraction ages in the 2.9–2.8 Ga range (Stern et al., 1994; Stevenson et al., 2006). It comprises several sub-regions including the Lake Minto, Qalluviartuuq and Utsalik domains (Fig. 8.4-2). Lake Minto domain is characterized by abundant paragneiss, migmatite and diatexite of sedimentary origin, with hornblende- and pyroxene-bearing granitic intrusions (2.73–2.69 Ga; Parent et al., 2001; 2003). Supracrustal rocks of the Kogaluc belt include 2.76 Ga volcanic rocks and <2.748 Ga greywacke probably correlative with widespread paragneiss (Skulski et al., 1996). With the exception of a single enclave of tonalite gneiss with an age of 3.125 Ga and inheritance of ca. 3.5 Ga (Percival et al., 1992), the Lake Minto domain has inherited zircon and Nd model ages of 2.8–2.92 Ga (unpublished data from J. David, written comm., 2006; Rabeau, 2003; Boily et al., 2004, 2006b; Stevenson et al., 2006). The Qalluviartuuq domain to the east is characterized by several large (up to 10 × 120 km) supracrustal belts (Percival et al., 1995; Bourassa, 2002; Leclerc, 2004), engulfed in widespread, coeval tonalite-trondhjemite plutonic suites (2.85–2.77 Ga). These rocks were cut by tonalite-trondhjemite-granodiorite (2.76–2.71 Ga), enderbite and granite (2.73–2.67 Ga). The Utsalik domain forms a prominent broad arcuate zone of plutonic rocks including enderbite, granodiorite and granite (2.74–2.69 Ga), with rare supracrustal belts and gneissic tonalites (<2.79 Ga; Percival et al., 2001; Leclair et al., 2002b). Isotopic data for Utsalik units indicate recycling of older sialic crust that does not exceed ca. 2.93 Ga, except at one locality (3.01 Ga).

8.4-3.9. Troie Terrane

The Troie terrane in the far northeastern Superior Province is recognized on the basis of zircon inheritance and Nd model ages in excess of 3.0 Ga (Percival et al., 2001; Rabeau, 2003; Boily et al., 2006b; Leclair et al., 2006; unpublished data from J. David, written comm., 2006). This region consists mainly of a massif of 2.74–2.73 Ga enderbite and granodiorite (Madore et al., 2000, 2002; Madore and Larbi, 2001; Bédard, 2003; Bédard et al., 2003), surrounded by a rim of tonalite (2.88–2.75 Ga) that encloses several small supracrustal belts as well as local younger belts (ca. 2.725 Ga; Leclair et al., 2003). These units are intruded by plutons of granodiorite, granite and monzonite (ca. 2.73–2.69 Ga). The zircon inheritance and Nd isotopic data reflect the presence of older recycled crust (<3.1 Ga). The Mesoarchean units resemble those of the northern La Grande subprovince, with which they have been linked on the basis of age, lithology and isotopic characteristics (Leclair et al., 2006).

8.4-3.10. Eo- to Mesoarchean Terrane Comparison

Ages of igneous and thermotectonic events identified within individual terranes are displayed schematically in Fig. 8.4-5. Bickford et al. (2006, this volume) pointed out similarities between the Minnesota River Valley and Northern Superior terranes in terms of crystallization ages (cf. Böhm et al., 2003). The correlation is less compelling considering

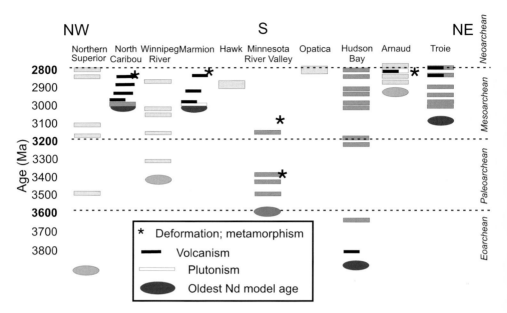

Fig. 8.4-5. Age distribution of magmatic and metamorphic events in microcontinental terranes of the Superior Province (sources listed in text).

differences in crustal residence age and thermotectonic history between the two terranes. Similarities between the Northern Superior and Hudson Bay terranes have been previously noted (e.g., Böhm et al., 2003), a correlation that is supported by detrital zircon ages and the proximity of the two terranes. Other continental fragments with common age features include the juvenile 3.0 Ga North Caribou and Marmion terranes. However, their different rift histories appear to preclude origins in a common parent terrane (Tomlinson et al., 1999). Juvenile terranes with common 2.9–2.8 Ga histories include the Hawk, Opatica and Arnaud terranes, allowing possible common origins.

Possible linkages among some of the younger Mesoarchean terranes present the possibility of common ancestry. Further work is required, including identification and dating of rift sequences, to establish whether correlations are tenable. At present, the older terranes appear to have had independent histories prior to their amalgamation in the Neoarchean. Matches for Superior Province microcontinental fragments may be found in other Archean cratons that emanated from "Superia" (Bleeker, 2003).

8.4-4. HISTORY OF TECTONIC ASSEMBLY

8.4-4.1. Mesoarchean Tectonism

Scattered evidence exists for assembly of Mesoarchean terranes into superterranes prior to Neoarchean collisions that led to final amalgamation of the Superior Province and pervasive thermal overprinting. For example, in the Minnesota River Valley terrane, tectonothermal events are recognized at 3.38 and 3.08 Ga on the basis of foliation development and zircon overgrowths (Bickford et al., 2006), possibly resulting from terrane assembly prior to 2.6 Ga (Bickford et al., this volume). Evidence for ca. 2.94 Ga tectonism in the southwestern North Caribou superterrane is derived from dated shear zones (Percival et al., 2006a) and tectonic inversion of supracrustal rocks prior to 2.92 Ga in the same region (Sasseville et al., 2006). Within the central North Caribou superterrane, the 2.87 Ga North Caribou pluton cuts juxtaposed 2.98 and 2.93 Ga assemblages (Stott et al., 1989; Thurston et al., 1991). In the Winnipeg River terrane, Tomlinson et al. (2004) inferred incorporation of the Marmion domain prior to 2.92 Ga. Two deformation events in northern Winnipeg River terrane gneisses are loosely bracketed between 2.88 and 2.72 Ga (Melnyk et al., 2006). In the Marmion domain to the south, a cryptic Mesoarchean event is recognized on the basis of a titanite age of ca. 2.81 Ga (Davis and Jackson, 1988) in 3.0 Ga tonalite.

In the Arnaud terrane, Percival and Skulski (2000) reported evidence for Mesoarchean amalgamation of juvenile terranes. A shear zone separating dominantly 2.84–2.83 Ga, tholeiitic and calc-alkaline basalt of the Qalluviartuuq belt from calc-alkaline andesite of the Payne Lake sequence is cut by tonalite dated at 2.81 Ga. Kyanite occurrences along the boundary were attributed to Mesoarchean Barrovian-style metamorphism in contrast to the regional Neoarchean andalusite-sillimanite facies series (Percival and Skulski, 2000).

8.4-4.2. Neoarchean Tectonism

Five Neoarchean orogenies are responsible for assembly of ancient Superior Province terranes (Fig. 8.4-6(a)), trapping Neoarchean oceanic terranes (Fig. 8.4-6(b, c)).

The ca. 2.72–2.71 Ga *Northern Superior orogeny* represents collision between the Northern Superior and North Caribou superterranes with inferred southward subduction polarity, based on the presence of arc magmatism in the North Caribou superterrane, and south-over-north shear-zone movement (Fig. 8.4-6(b): Corkery et al., 2000; Skulski et al., 2000; Lin et al., 2006; Parks et al., 2006). Collision and uplift of the Northern Superior superterrane are recorded by the appearance of 3.5 and 3.6 Ga detrital zircon in <2.711 Ga synorogenic sedimentary rocks of the northern North Caribou superterrane (Corkery et al., 2000).

Arc magmatism across the southern North Caribou superterrane was the precursor to the *Uchian orogeny* in which the ca. 3.4 Ga Winnipeg River terrane docked from the south (Stott and Corfu, 1991; Corfu et al., 1995; Stott, 1997). The suture zone is marked by <2.704 Ga synorogenic English River flysch deposits (Fig. 8.4-6(c)). Both the English River and Winnipeg River terrane were rapidly buried and heated by southward overthrusting of the North Caribou superterrane, illustrated prominently on seismic reflection profiles (White et al., 2003).

Parts of the central Superior Province were assembled into a superterrane during the *Central Superior orogeny* just prior to collision with the amalgamated Northern Superior – North Caribou superterrane. The Winnipeg River – Marmion terrane had rifted by ca. 2.88 Ga, leading to formation of Wabigoon oceanic crust to the west. Collision between

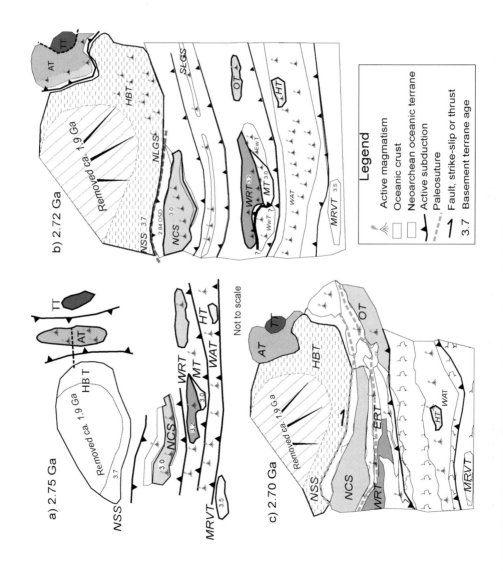

Fig. 8.4-6. Schematic evolutionary model for growth of the Superior Province. Abbreviations as in Fig. 8.4-1. (a) At 2.75 Ga, independent microcontinental fragments are separated by tracts of oceanic crust of unknown dimension. (b) By 2.72 Ga, the Northern Superior superterrane had impinged on the northern margin of the North Caribou superterrane to initiate the composite Superior superterrane. The western Wabigoon terrane begins to impinge on the southwestern Winnipeg River terrane margin. NLGS: northern La Grande subprovince; SLGS: southern La Grande subprovince. (c) Between 2.70 and 2.69 Ga, the Wawa–Abitibi terrane (WAT) docks with the composite Superior superterrane, accompanied by deposition of syn-orogenic Quetico flysch in the intervening trench, its burial and metamorphism. Arc magmatism continues in the oceanic WAT and post-orogenic granitic magmatism is widespread across the composite Superior superterrane to the north.

---

the Neoarchean oceanic terrane and the Winnipeg River – Marmion terrane occurred between 2.71 and 2.70 Ga (Davis and Smith, 1991; Sanborn-Barrie and Skulski, 1999, 2006; Percival et al., 2004b; Melnyk et al., 2006).

The *Shebandowanian orogeny* brought the oceanic Abitibi–Wawa terrane into juxtaposition with the amalgamated northern Superior collage at ca. 2.695 Ga (Fig. 8.4-6(c): Corfu and Stott, 1986, 1998; Stott, 1997). Subduction polarity is inferred to have been to the north based on arc-related magmatism in the Wabigoon – Winnipeg River superterrane (Percival et al., 2006b), and the suture is marked by the thick, synorogenic Quetico turbidite wedge. Seismic profiles across the boundary zone indicate gently northward dipping reflectivity (White et al., 2003).

The Minnesota River Valley terrane docked with the northern Superior collage during the ca. 2.68 Ga *Minnesotan orogeny*. Northward subduction polarity is indicated by reflection geometry (Fig. 8.4-4: White et al., 2003) and by the isotopic signature of old crust beneath the southern Wawa–Abitibi terrane (Sims et al., 1997). The unexposed Great Lakes tectonic zone (Fig. 8.4-3) is the probable suture (Sims and Day, 1993), although in central Quebec, the <2.682 Ga, syorogenic Pontiac flysch marks the boundary (Mortensen and Card, 1993; Davis, 2002). The subduction polarity also appears to be northward, based on north-dipping seismic reflectivity (Calvert and Ludden, 1999) and the presence of peraluminous granite in the southern Abitibi (Feng and Kerrich, 1991, 1992; Chown et al., 2002).

8.4-5. INFLUENCE OF MICROCONTINENTAL TERRANES ON TECTONIC STYLE

Some Superior Province microcontinental terranes were substantially recycled during the Neoarchean, to the extent that few of their primary characteristics are known. In particular, the surficial extent and thickness of the crust and lithosphere would have determined to what extent the sialic fragments behaved as modern continents. Evidence from sparse rift sequences suggests volcanic-dominated margins compatible with the warm mantle – weak crust model of Hynes and Skulski (2005). However, the preservation level is inadequate to define the shape or extent of rift margins.

During convergent tectonism, the internal microcontinents were located in upper plate settings, where they underwent significant magmatic inflation prior to collision. The North

Caribou superterrane appears to have been the upper plate with respect to convergence from both the north (Northern Superior) and south (Winnipeg River). In contrast, the Northern Superior and Minnesota River Valley terranes experienced only minor Neoarchean magmatism, consistent with the lower plate setting inferred on the basis of seismic images (White et al., 2003; Mussachio et al., 2004; Craven et al., 2004; Percival et al., 2006b). In northern Quebec, the Hudson Bay and Arnaud terranes have an extensive record of Neoarchean magmatism (2.78–2.68 Ga) that has been alternatively explained in terms of a suprasubduction zone continental margin setting (Stern et al., 1994; Percival et al., 1994, 2001; Percival and Mortensen, 2002), or plume effects (Bédard, 2006; Boily et al., 2006b). Petrogenetic studies of Neoarchean granitoid suites have generally suggested Andean settings for mixing between mantle-derived arc magmas and older sialic crust (e.g., Stern et al., 1994; Whalen et al., 2002); however, according to Bédard (2006), similar calc-alkaline compositions with evolved isotopic signatures could have been produced in a non-subduction regime through a combination of plume-related magmatism and lithosphere delamination.

One of the consequences of extensive arc magmatism in upper plates prior to collision may have been thermal softening that led to a characteristic thermotectonic style. For example the North Caribou superterrane records widespread calc-alkaline and late sanukitoid magmatism between 2.745 and 2.704 Ga (Corfu and Stone, 1998), followed closely by deposition of English River synorogenic greywacke, polyphase deformation, and peak metamorphism by 2.692 Ga (Corfu et al., 1995). The upper plate may have imposed a hot iron effect on rocks of the lower plate, which had also experienced recent magmatism (Whalen et al., 2002; Percival et al., 2004b), leading to rapid attainment of low-P, high-T metamorphic conditions (Percival et al., 2006b). Thermal softening through widespread plutonism may also explain the modest erosion levels of the upper plate following the collision. The North Caribou upper plate may have been too weak to support much topographic load during collision with the Winnipeg River microcontinental terrane, such that shortening was accommodated partly through ductile orogen-parallel flow in the deep crust (cf. Moser et al., 1996), rather than through mountain building.

Correlation of the Northern Superior superterrane and Hudson Bay terrane across Hudson Bay is supported by their Eo- to Mesoarchean ancestry, although the Neoarchean history differs. In the west, structural trends are parallel to its southwestern margin, whereas in the northeast, the northerly structural grain follows the eastern margin (Fig. 8.4-1). The broad curvature evident across the Superior Province could therefore have been inherited from the original shape of the Northern Superior-Hudson Bay superterrane (Fig. 8.4-6). If valid, this interpretation implies the existence of strong continental lithosphere by 2.7 Ga, providing a mechanical basis for the concept of accretion around an ancient nucleus.

ACKNOWLEDGEMENTS

This summary is the result of projects carried out by the Geological Surveys of Canada, Manitoba, Ontario and Quebec over the past two decades, in collaboration with several universities and the Jack Satterly Geochronology Laboratory. I would particularly like to acknowledge contributions and discussions with coworkers including Al Bailes, Wouter Bleeker, Ken Card, Tim Corkery, Fernando Corfu, Don Davis, Alain Leclair, Vicki McNicoll, Mary Sanborn-Barrie, Tom Skulski, Greg Stott, Phil Thurston and Joe Whalen. Tom Skulski is thanked for a helpful review, and Martin Van Kranendonk for editorial comments.

Earth's Oldest Rocks
Edited by Martin J. Van Kranendonk, R. Hugh Smithies and Vickie C. Bennett
Developments in Precambrian Geology, Vol. 15 (K.C. Condie, Series Editor)   1087
© 2007 Elsevier B.V. All rights reserved.
DOI: 10.1016/S0166-2635(07)15085-4

Chapter 8.5

EARLY ARCHEAN ASTEROID IMPACTS ON EARTH: STRATIGRAPHIC AND ISOTOPIC AGE CORRELATIONS AND POSSIBLE GEODYNAMIC CONSEQUENCES

ANDREW GLIKSON

Department of Earth and Marine Science and Planetary Science Institute, Australian National University, Canberra, ACT 0200, Australia

8.5-1. INTRODUCTION

The heavily cratered surfaces of the terrestrial planets and moons testify to their long-term reshaping by asteroid and comet impacts and related structural and melting processes. Following accretion of Earth from the solar disc at ca. 4.56 Ga (see Taylor, this volume), the Earth–Moon system is believed to have originated by a collision between a Mars-size planet and Earth, followed by episodic bombardment by asteroids and comets, with a peak documented at ca. 3.95–3.85 Ga – the Late Heavy Bombardment (LHB) – recorded on the Moon, but not on Earth due to the paucity and high grade metamorphic state of terrestrial rocks of this age (Wilhelm, 1987; Ryder, 1990, 1991, 1997; see Iizuka et al., this volume). Major impact episodes on Earth are recorded from rocks dated at 3.46, 3.26–3.24, 2.63, 2.56, 2.48, 2.02 Ga (Vredefort), 1.85 Ga (Sudbury), 0.58 Ga (Acraman), 0.368–0.359 Ga (late Devonian), 0.214 Ga (late Triassic), 0.145–0.142 Ga (late Jurassic), 0.065 Ga (K–T boundary), and 0.0357 Ga (late-Eocene) (World Impact List GSC/UNB; Grieve and Shoemaker, 1994; Grieve and Pesonen, 1996; Grieve and Pilkington, 1996; French, 1998; Simonson and Glass, 2004). Some of these impact events coincided with, and are referred to as the direct cause of, mass extinctions and/or the radiation of species (Alvarez, 1986; Keller, 2005; Glikson, 2005a). These impacts represent a small part of the extraterrestrial impact record, as the retention of any impact-generated structure is a direct function of the preservation and/or destruction of crust. By analogy, the continuous re-accretion of sulphur on Io and ice on Europa obliterates their impact records, as does volcanic re-surfacing on Venus and plate tectonics on Earth. Estimates of the impact incidence on Earth is inferred from the lunar cratering record, and the current asteroid and comet flux, which suggests the formation of some 100–200 craters of diameter (D) D $\geqslant$ 100 km, and 5–10 craters of D $\geqslant$ 300 km since ca. 3.8 Ga (Neukum et al., 2001; Ivanov and Melosh, 2003; Glikson et al., 2004). Most of these craters would have impacted on hitherto subducted oceanic crust on Earth. The larger continental impacts are less likely to be destroyed by subduction, and resulted in deeper excavation of the continental crust and thicker ejecta layers, thereby being generally better preserved than their counterparts that impacted oceanic crust. Glikson (2001) estimated that only between 8–17% of the larger impacts (D $\geqslant$ 100 km) is recorded as impact structures and/or ejecta. However, assuming a relatively smaller proportion of large projectiles, over 50% of the larger impacts have been recorded at this stage (Ivanov and Melosh, in Glikson et al., 2004).

Models suggest impact triggering of igneous, tectonic, or plate tectonic events (Green, 1972, 1981; Grieve, 1980; Alt et al., 1988; Jones et al., 2002; Elkins-Tanton et al., 2005; Glikson and Vickers, 2005), but this correlation has not been directly demonstrated on Earth. Ivanov and Melosh (2003) questioned impact-induced volcanic activity, but Glikson et al. (2004) state: 'Indeed, a few giant impact events certainly occurred on Earth in post-LHB history, and a few might have encountered the fortuitous circumstances necessary for them to enhance magma production. This is an important topic for future study.' Lowe et al. (1986, 1989) remarked on the potential tectonic significance of the coincidence between 3.26–3.24 Ga impact horizons recorded in the Barberton Greenstone Belt (Kaapvaal Craton) and the abrupt break between the underlying mafic–ultramafic volcanic crust (3.55–3.26 Ga Onverwacht Group) and the overlying turbidite, felsic volcanic, banded iron formation and conglomerate-dominated succession (3.26–3.225 Ga Fig Tree and Moodies Groups), which includes the earliest observed granite-derived detritus (Fig. 8.5-1). This age period correlates broadly with a similar break in the Pilbara Craton, Western Australia, where the mafic–ultramafic volcanic rocks of the ca. 3.255–3.235 Ga Sulphur Springs Group are unconformably overlain by Soanesville Group clastic sedimentary rocks, including a local unit of olistostrome breccia (Glikson and Vickers, 2005; Van Kranendonk et al., 2006a). Widespread 3.27–3.24 Ga plutonic igneous rocks (dominantly granitic) occur in both the Kaapvaal Craton and Pilbara Craton (see Poujol, this volume, and Van Kranendonk et al., 2007a, this volume).

8.5-2. 3.47 GA IMPACT EVENTS

Microkrystite spherule-bearing lenses in chert and in the matrix of diamictite have been reported from the 3.47 Ga Antarctic Creek Member (ACM) of the Mount Ada Basalt, in the North Pole Dome area of the East Pilbara Terrane, Pilbara Craton, Western Australia (Lowe et al., 1989; Byerly et al., 2002; Van Kranendonk et al., 2006a). The diamictite, defined as ACM-S2, consists of a 0.6–0.8 m thick unit of spherule-bearing, chert-intraclast conglomerate, which is separated from the main ACM-S3 unit by dolerite and felsic volcanics/hypabyssals. Zircon dating from the spherule-bearing sandstone unit yielded an age of 3470 ± 2 Ma (Byerly et al., 2002). The microkrystite spherules are discriminated from angular to subangular detrital volcanic fragments by their high sphericities, inward-radiating fans of sericite, relic quench textures, and Ni–Cr–Co relations. SEM-EDS (scanning electron microscope coupled with electron-dispersive spectrometer) and LA-ICP-MS (laser induced coupled plasma mass spectrometer) analysis indicate high Ni and Cr in sericite-dominated spherules, suggesting a mafic composition of source crust. Ni/Cr and Ni/Co ratios of the spherules are higher than in associated Archaean tholeiitic

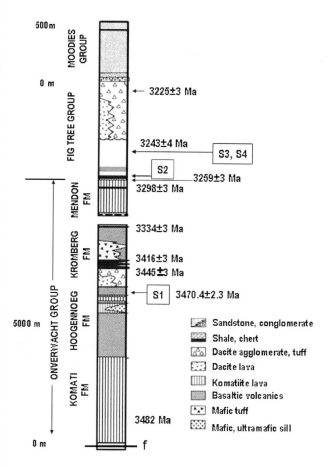

Fig. 8.5-1. Composite columnar section of the Barberton Greenstone Belt stratigraphy, showing the principal groups and formations and U-Pb zircon isotopic age determinations (after Byerly and Lowe, 1994; Byerly et al., 1999; Lowe et al., 2003).

basalts and high-Mg basalts, rendering possible contamination by high Ni/Cr and Ni/Co chondritic components. The presence of multiple bands and lenses of spherules within chert, and scattered spherules in arenite bands within ACM-S3, may signify redeposition of a single impact fallout unit or, alternatively, multiple impacts. Controlling parameters include: (1) spherule atmospheric residence time; (2) precipitation rates of colloidal silica; (3) solidification rates of colloidal silica; (4) arenite and spherule redeposition rates; and (5) arrival of an impact-generated tsunami. The presence of spherule-bearing chert fragments in ACM-S3 may hint at an older spherule-bearing chert (?S1). Only a minor proportion of spherules is broken and the near-perfect sphericities of chert-hosted spherules and arenite-hosted spherules constrain the extent of shallow water winnowing of the originally delicate glass spherules. It is suggested that the spherules were either protected by rapid burial or, alternatively, disturbance was limited to a short term high energy pertur-bation such as may have been affected by a deep-amplitude, impact-triggered tsunami wave.

An impact spherule unit precisely correlated with spherules of the ACM occurs in a 30 cm to 3 m thick chert unit, H4c in the upper part of the Hooggenoeg Formation of the Onverwacht Group, Kaapvaal Craton (Lowe et al., 2003). Zircons in this unit have been dated at 3470 ± 6.3 Ma, identical in age to the Pilbara unit (Byerly et al., 2002). The spherules constitute a bed of medium- to coarse-grained, current-deposited sandstone, 10–35 cm thick and interbedded with fine-grained tuffs and black, or black-and-white, banded cherts. In sections where H4c is ~100 cm thick, the unit generally rests directly on, or just a few centimeters above, altered basaltic to komatiitic volcanic rocks. In the type section, H4c is 2.8 m thick, but varies from ~50 cm to 6 m thick within 500 m along strike. In this section, the unit is underlain by up to 160 cm, and overlain by 90 cm, of pale greenish to light gray sericitic chert and black carbonaceous chert.

8.5-3. THE 3.26–3.24 GA BARBERTON IMPACT CLUSTER

The uppermost, predominantly mafic–ultramafic volcanic sequence of the Onverwacht Group consists of an assemblage of komatiitic volcanics and their hypabyssal and altered equivalents, capped by ferruginous chert. This unit is known as the Mendon Formation (Fig. 8.5-1) and has been dated by U-Pb zircon from a middle chert unit as 3298 ± 3 Ma (Byerly, 1999). Unconformably overlying the Mendon Formation is the Mapepe For-mation, the basal unit of the Fig Tree Group, consisting of a turbidite–felsic volcanic–ferruginous sedimentary rock association dated in the range of 3258 ± 3 Ma to 3225 ± 3 Ma (Lowe et al., 2003) (Fig. 8.5-1). The S_2 impact fallout unit coincides with the base of the clastic Mapepe Formation of the Fig Tree Group, whereas the S_3 and S_4 impact fallout units occur within clastic sedimentary rocks and felsic pyroclastic rocks, 110–120 m above the basal contact. In places, the S3 unit occurs unconformably above deeply incised Mendon Formation komatiites.

Detailed documentation of the S_2, S_3 and S_4 impact fallout units (Fig. 8.5-1) reveals a wide range of diagnostic field, petrographic, mineralogical, geochemical and isotopic cri-teria identifying extraterrestrial components mixed with volcanic and other detritus (Lowe and Byerly, 1986b; Lowe et al., 1989, 2003; Kyte et al., 1992, 2003; Byerly and Lowe, 1994; Shukloyukov et al., 2000; Kyte, 2002; Glikson, 2005b, 2005c; Reimold et al., 2000). Principal criteria include: (1) impact fallout units that can be correlated in below-wave base environments between basins and sub-basins, independently of facies variations of the host sedimentary rocks (Lowe et al., 1989, 2003); (2) microkrystite spherules that have inward-radiating quench textures and centrally offset vesicles (Glass and Burns, 1988) (Fig. 8.5-2), distinct from outward radiating textures of volcanic varioles, which may also contain microphenocrysts of feldspar and quartz (Glikson, 2005b, 2005c); (3) microkrys-tites that contain quench-textured and octahedral Ni-chromites (NiO < 24%), including a

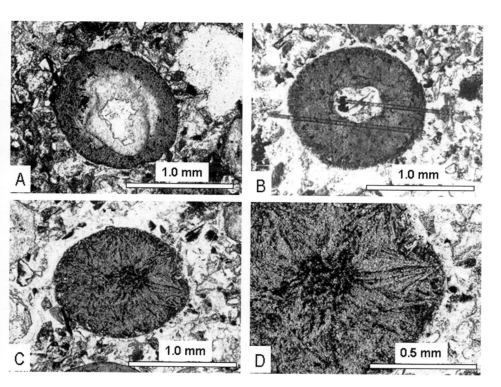

Fig. 8.5-2. Microphotograph of microkrystite spherules in sample SA306-1 (A–D) and SA315-5 (E–H) from the ~3.24 Ga S$_3$ impact fallout unit, Mapepe Formation, Barberton Greenstone Belt. The spherules are dominated by chlorite, displaying inward radiating fans, central to offset vesicles and rims of microcrystalline quartz and are set in a micro-fragmental matrix containing angular clasts of volcanic rock and chert. Plane polarized light. A – spherule with chlorite rim, micro-crystalline mantle and siliceous core; B – chlorite spherule with central siliceous vesicle, showing LA-ICPMS burn traces; C – spherule consisting of inward-radiating chlorite fans, interpreted as replacing quench textures of original pyroxene and olivine; D – detail of C; E – Spherule occupied by radiating palimpsest needles after ?olivine and/or pyroxene representing quench texture; F – spherule with inward-radiating chlorite sheafs and a central quartz-filled vesicle; G – spherule with inward-radiating ferruginous chlorite fans, light-coloured chlorite mantle and siliceous core; H – fragment of spinifex-textured komatiite.

high Co, Zn and V variety unknown in terrestrial chromites (Byerly and Lowe, 1994); and (4) microkrystites that contain nickel and PGE nanonuggets (Kyte et al., 1992; Glikson and Allen, 2004; Glikson, 2005b). The Ni-rich chromites are distinct from terrestrial type Ni-chromites associated with sulphide (NiO < 0.3%) (Czamanske et al., 1976; Grove et al., 1977), and have low Fe$_2$O$_3$/FeO ratios, possibly consequent on condensation under low-oxygen Archaean atmospheric conditions.

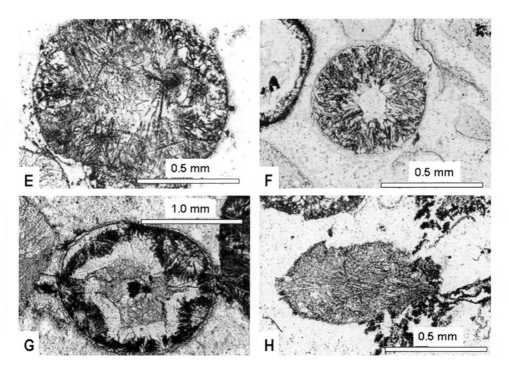

Fig. 8.5-2. (*Continued.*)

Attempts at estimating the magnitude of the impact events and the composition of the target crust are based on inferences derived from spherule radii (O'Keefe and O'Hara, 1982; Melosh and Vickery, 1991) and geochemical mass balance calculations of asteroid size, based on PGE (Kyte et al., 1992; Byerly and Lowe, 1994) and $^{53}Cr/^{52}Cr$ isotope ratios (Shukloyukov et al., 2000; Kyte et al., 2003; Glikson and Allen, 2004). Early estimates based on spherule size distribution (mean 0.85 mm), and the iridium flux (5.3 mg/cm^2), deduced a 20–30 km diameter size of the parental asteroid. Lowe et al. (1989) referred to the maximum size of the microkrystite spherules (<4 mm) to estimate projectile size at 20–50 km in diameter. From the high mean Ir levels (116 ppb) in impact layer S4, Kyte et al. (1992) suggested a FeNi projectile with Ir levels higher than chondrites (~500 ppb), whereas high mean Cr level (1350 ppm) suggest a komatiitic target rock. These authors estimate the Ir-flux for S4 at 2.5 gr/cm^2 – some 30 times greater than that for the K–T boundary Ir flux (0.08 gr/cm^2). Byerly and Lowe (1994) estimate an Ir flux of 5.3 gr/cm^2 for the S$_2$ and S$_3$ impact layers, consistent with a chondritic impactor (Ir 500 ppb) with a diameter ~30 km. A FeNi impactor would be 4–10 times smaller, but inconsistent with the maximum spherule sizes. From a bimodal size distribution of spherules in the SA306-1 sample, with a diameter range of 0.4–2.0 mm, Byerly and Lowe (1994) suggested early flux of small, Ir-rich, Ni spinel-bearing spherules (mean diameter ~0.65 mm) and a late

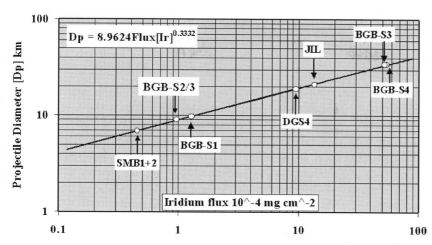

Fig. 8.5-3. Correlation between the flux of iridium (in units of 10^{-4} mg cm^{-2}) and the diameter of chondritic projectile (Dp), based on mass balance calculations assuming mean unit thickness, mean Ir concentration and global distribution of fallout ejecta (after Glikson, 2005). Unit symbols – R_p (projectile radius) = $\sqrt[3]{V_p/[4/3]\pi}$; $V_p = M_p/d_p$; $M_p = F_G/C_p$; $F_G = (A_E/D_S * Ts)$. (A_E – measured element (E) abundance in fallout unit (ppb); D_S – mean specific density of fallout sediments (mg/cm^3) (assumed as 2.65 gr/cm^3); C_E – weight of element E in mg per cm^3 ($C_E = A_E * D_S$); T_S – mean stratigraphic thickness of spherule unit (in mm); F_E – local mean flux of element E in mg per 100 mm^2 (cm^2) surface ($F_E = C_E * T_S$); F_G – inferred global flux of element E in mg per cm^2 surface; $F_G = A_E/D_S * T_S *$ Earth surface area (S_E); C_p – assumed concentration of element E in projectile (ppb) ($C1$ – Ir-450 ppb); M_p – weight of projectile ($M_p = F_G/Cp$); d_p – assumed specific density of the projectile ($C1$ – 3.0 gr/cm^3); V_p – volume of projectile ($V_p = M_p/d_p$). For unit symbols refer to Table 8.5-1.

flux of larger, mostly Ir-poor, spinel-free spherules (mean diameter $\sim$1.25 mm), suggesting early ejection of projectile-rich components and a projectile 24 km in diameter. These authors showed the SA306-1 impact spinels have higher Ni/Fe and lower Fe^{+3}/Fe^{+2} than Phanerozoic impact spinels, hinting at a less oxidizing Archaean atmosphere. From Cr isotope-based estimates of the global thickness and mean Ir concentration of impact units (S$_2$; $\sim$20 cm, 3 ppb: S$_3$; $\sim$20 cm, 300 ppb: S$_4$; $\sim$10 cm, 100 ppb), Kyte et al. (2003) suggested a parental asteroid 50–300 times the mass and 3–7 times the diameter of the $\sim$10 km-large K–T Chicxulub asteroid, consistent with estimate made by Glikson (2005b: Fig. 8.5-3).

Given projectile/crater diameter ratios in the range of 0.05–0.1, these impacts would result in impact basins at least 300–700 km large. The dominantly mafic to ultramafic composition of the impact ejecta requires that these basins were formed in mafic–ultramafic-dominated regions of the Archaean Earth, probably oceanic crust. Impacts of this magnitude can be expected to have had major consequences on the impacted crustal plates, greenstone–granite terrains, and underlying mantle regions.

8.5-4. STRATIGRAPHIC AND ISOTOPIC AGE CORRELATIONS BETWEEN THE 3.26–3.24 GA IMPACTS, UNCONFORMITIES, FAULTING AND IGNEOUS EVENTS

Lowe et al. (1989) observed: 'The transition from the 300-million-year-long On-verwacht stage of predominantly basaltic and komatiitic volcanism to the late orogenic stage of greenstone belt evolution suggests that regional and possibly global tectonic re-organization resulted from these large impacts.' Evidence consistent with impact-induced effects in the Kaapvaal Craton is provided by U-Pb zircon ages of the Nelshoogte trond-hjemite (3236 ± 1 Ma) and the Kaap Valley tonalite (3227 ± 1 Ma), suggesting plutonic emplacement within about $\sim$5–10 My after the S3 and S_4 impacts (3243 ± 4 Ma). The tonalite-trondhjemite geochemistry of the granitoids suggests partial melting of mafic crustal sources.

Close temporal correlations are observed between the Onverwacht Group – Fig Tree Group transition and the transition between the Kelly Group (end of felsic magmatism at ca. 3290 Ma) and the 3255–3235 Ma Sulphur Springs Group in the East Pilbara Terrane of the Pilbara Craton (Fig. 8.5-4). In the Sulphur Springs Group, there is a change from $\sim$3255–3235 Ma ultramafic-mafic-felsic volcanic rocks and silicified epiclastic rocks of the Sulphur Springs Group and deposition of overlying sedimentary rocks (banded iron formation and turbidites) of the Soanesville Group (Fig. 8.5-5). Eruption of the Sulphur Springs Group has been variously interpreted in terms of back arc rifting (Brauhart, 1999), or mantle plume events, the latter including local caldera subsidence above a syn-volcanic laccolith (Van Kranendonk, 2000; Van Kranendonk et al., 2002; Pirajno and Van Kranen-donk, 2005; Smithies et al., 2005b). Whereas in places the boundary above the marker chert (silicified epiclastic rocks) at the top of the Sulphur Springs Group appears conformable, it was clearly lithified prior to deposition of the overlying, local unit of olistostrome brec-cia (Hill, 1997) (Figs. 8.5-5 and 8.5-6) and more widespread turbiditic sandstones of the overlying Soanesville Group (Van Kranendonk et al., 2006a).

The following stratigraphic and age relations are outlined between the Barberton Green-stone Belt (BGB) and greenstone belts of the East Pilbara Terrane (PGB) (Fig. 8.5-4):

(A) BGB – S_2 impact spherule unit directly underlying 3258 ± 3 Ma felsic tuff. PGB – sedimentary rocks of the Leilira Formation (ca. 3255 ± 4 Ma: Buick et al., 2002) at the base of the Sulphur Springs Group, but no impact spherules identified;

(B) BGB – the S_3 and S_4 impact spherule units of the Mapepe Formation, Fig Tree Group, directly overlie 3243 ± 4 Ma felsic tuff. PGB – 3235 ± 3 Ma felsic volcanic rocks in the Kangaroo Caves Formation, Sulphur Springs Group, but no impact spherules identified.

In the West Pilbara Superterrane of the Pilbara Craton, a similar lithological change is seen in the Roebourne Group, where komatiite and basalt of the >3270 Ma Ruth Well For-mation are overlain by 3.27–3.25 Ga sandstone, shale, banded iron-formation, and felsic volcanic rocks of the Nickol River Formation, and accompanied by emplacement of the 3.27–3.26 Ga Karratha Granodiorite. Indeed, major plutonic activity at $\sim$3.275–3.225 Ga

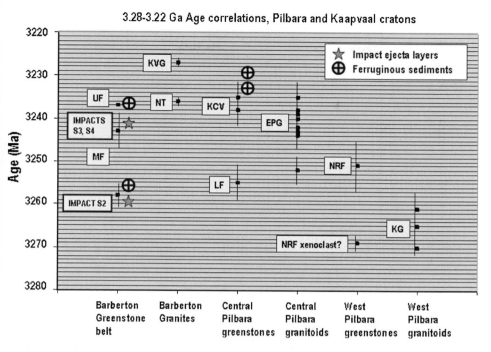

Fig. 8.5-4. Isotopic age correlations between 3.28–3.22 Ga units in the Kaapvaal Craton and Pilbara Craton. Solid squares and error bar lines – U-Pb ages of volcanic and plutonic units; stars – impact ejecta layers; circled crosses – ferruginous sediments; MF – Mapepe Formation; UF – Ulundi Formation; NT – Nelshoogte tonalite; KPG – Kaap Valley granite; LF – Leilira Formation; KCF – Kangaroo Cave Formation; EPG – Eastern Pilbara granites; NRF – Nickol River Formation; KG – Karratha Granodiorite.

(Fig. 8.5-4), defined as the Cleland Supersuite, is documented throughout much of the Pilbara Craton (Van Kranendonk et al., 2006).

The S_2 impact ejecta unit in the south-eastern part of the Barberton Greenstone Belt is overlain directly by 20–30 m thick banded ferruginous chert, the Manzimnyama Jaspilite Member, of the basal Mapepe Formation (Lowe and Nocita, 1999). These widespread sedimentary rocks in the southern part of the Barberton Greenstone Belt include oxide facies BIF, jaspilite, and hematite-rich shale. The S_3 impact ejecta unit in the northern part of the Barberton Greenstone Belt underlies ferruginous sediments of the Ulundi Formation, including jaspilite, ferruginous chert and black chert deposited under quiet, deep-water conditions (Lowe et al., 2003). By contrast, deposition of S_2 and S_3 impact units along the Onverwacht anticline, which formed an antecedent rise located between deeper basins to the northwest and southeast, occurred under high-energy, shallow water conditions (Lowe et al., 2003).

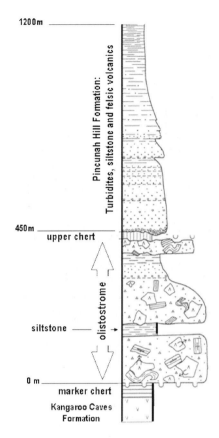

Fig. 8.5-5. Schematic section through the top of the Sulphur Springs Group (Kangaroo Caves Formation – 3235 ± 3 Ma), overlying olistostrome, siltstone, ferruginous siltstone and felsic volcanics of the Pincunah Hill Formation (after Hill, 1997).

In the Pilbara Craton, uppermost stratigraphic levels of the felsic volcanic Kangaroo Caves Formation contain BIF, as rafts in olistostrome and as a more widespread unit, up to 1000 meters thick (Van Kranendonk, 2000). The sequence is disconformably overlain by arenites and turbidites and ferruginous shale and silicified equivalents up to 100 m thick in some areas. The potential significance of a relationship between impacts and BIF is underpinned by similar observations in 3.46, 2.63 and 2.56 Ga impact units (Table 8.5-1; Glikson, 2006).

8.5-5. POSSIBLE GEODYNAMIC CONSEQUENCES

For a present-day mid-ocean geotherm ($25\,^\circ\mathrm{C\,km^{-1}}$), assuming yet higher Archean geothermal gradients and possibly smaller-scale convection cells and plate dimensions

Fig. 8.5-6. Large blocks of felsic volcanics, chert and siltstone in the olistostrome, basal Pincunah Hill Formation, Sulphur Springs, Central Pilbara (after Hill, 1997).

(Lambert, 1983), thin (<5 km) crustal regions underlain by shallow asthenosphere (<50 km) can be expected to have occupied large parts of the Archean oceanic crust. Low angle impacts in deep-water ocean will be partly to largely absorbed by the water column. A high-angle impact by a 20 km diameter projectile on thin oceanic crust underlying <2 km water depth would result in a 300 ± 100 km diameter submarine crater. As originally modeled by Green (1972, 1981), downward compression of the transient crater followed by rebound of asthenosphere originally located at $\sim$40–50 km would result in intersection of the peridotite solidus by rising mantle diapers and in partial melting. Insofar as morphometric estimates after Grieve and Pilkington (1996) may apply to asthenospheric uplift, the amount of rebound may be similar to that of a central uplift $SU = 0.086 Ds^{1.03}$, namely $\sim$30 km. However, Ivanov and Melosh (in Glikson et al., 2004) point out that such large degrees of rebound, deduced from craters formed in high-viscosity crust, may not apply to the mantle.

Elkins-Tanton et al. (2005) modeled the magmatic consequences of very large impacts, suggesting that a 300 km radius crater excavating 75 km thick lithosphere will produce 10^6 km^3 magma through instantaneous in-situ decompression of the mantle and updoming of the asthenosphere, triggering longer-term mantle convection and adiabatic melting, in particular under high geothermal gradients. These volumes are similar to those estimated by Ivanov and Melosh (2003). The short eruption spans of large volumes of some

Table 8.5-1. Archaean and early Proterozoic asteroid impact fallout units and associated ferruginous sediments, Pilbara Craton, Western Australia, and Kaapvaal Craton (Barberton Greenstone Belt – BGB), South Africa

Age/stratigraphy	Impact unit/s	Overlying ferruginous sediments	Reference
3470.1 ± 1.9 Ma: ACM-1, Antarctic Creek Member, Mount Ada Basalt, Warrawoona Group, Pilbara Craton	Silica-sericite spherules in ~1 m thick chert breccia/conglomerate	Overlain by felsic hypabyssal/volcanics	A
3470.1 ± 1.9 Ma: ACM-2, Antarctic Creek Member, Mount Ada Basalt, Warrawoona Group, Pilbara Craton	Silica-sericite spherules within ~14 m thick chert, arenite and jaspilite	~10 m thick jaspilite overlying spherule unit ACM-1	A
3470.4 ± 2.3 Ma: BGB–S1A and BGB–S1B, upper Hooggenoeg Formation, Onverwacht Group, Kaapvaal Craton	Two units of silica-chert spherules within 30–300 cm thick unit of chert and arenite	–	B
3258 ± 3 Ma: BGB–S$_2$, base of the Mapepe Formation, Fig Tree Group, Kaapvaal Craton	<310 cm thick silica–sericite spherules	MJM (Manzimnyama Jaspilite Member): BIF + jaspilite + ferruginous shale (<20 m) and shale common above BGB–S$_2$	B
3243 ± 4 Ma to 3225 ± 3 Ma: BGB–S$_3$ and BGB-4, lower Mapepe Formation, Fig Tree Group, Kaapvaal, Craton	S$_3$ – 10–15 cm-thick to locally 2–3 m thick silica-Cr sericite-chlorite spherules S$_4$ – 15 cm thick arenite with chlorite-rich spherules	BGB–S$_3$ is overlain by ferruginous sediments of the Ulundi Formation in the northern part of the BGB	B
2629 ± 5 Ma: JIL, top Jeerinah Formation, Fortescue Group, Hamersley Basin.	Hesta – 80 cm thick carbonate-chlorite spherules and spherule-bearing breccia; 60 cm thick overlying debris flow	Marra Mamba iron-formation, immediately above ~60 cm thick shale unit overlying JIL	D
?2.63 Ga: Monteville Formation, West Griqualand Basin, west Kaapvaal Craton	5 cm thick spherule layer	Carbonate hosted	E

Table 8.5-1. (*Continued*)

Age/stratigraphy	Impact unit/s	Overlying ferruginous sediments	Reference
?2.56 Ga: Reivilo Formation, West Griqualand Basin, western Kaapvaal Craton	1.8 cm thick spherule unit	Carbonate-hosted	E
2562 ± 6 Ma: SMB-1, top of Bee Gorge Member, upper Wittenoom Formation, Hamersley Group, Hamersley Basin	<5 cm thick K feldspar-carbonate-chlorite spherules in carbonate turbidite	Ferruginous siltstone (Sylvia Formation), banded iron formations (Bruno Member)	D
2562 ± 6 Ma: SMB-2, top of Bee Gorge Member, upper Wittenoom Formation, Hamersley Group, Hamersley Basin	<20 cm thick K feldspar carbonate-chlorite spherules within turbidite	Ferruginous siltstone (Sylvia Formation), banded iron formations (Bruno Member)	D
?2.63 ?2.56–2.54 Ga: Carawine Dolomite, Hamersley Group, Hamersley Basin	K-feldspar-carbonate-chlorite spherules in tsunami-generated carbonate-chert megabreccia	Carbonate megabreccia-hosted microkrystite spherules	E
2481 ± 4 Ma: S4, Shale Macroband, Dales Gorge Member, Brockman Iron-formation, Hamersley Group, Hamersley Basin	10–20 cm K-feldspar-stilpnomelane spherules at top of 2–3 m of ferruginous volcanic tuffs	Located 38 m above the base of the Brockman Iron-formation, Hamersley Basin	D
~2.5–2.4 Ga: lower Kuruman Formation, West Griqualand Basin, west Kaapvaal Craton	1 cm thick spherule unit overlain by 80 cm breccia	Located 37 m above base of banded ironstones	E
1.85–2.13 Ga: Graensco, Vallen, Ketilidean, southwest Greenland	20 cm thick spherule unit	Carbonate-hosted spherules	E

References: A – Byerly et al. (2002), B – Lowe et al. (2003), D – Trendall et al. (2004), E – Simonson and Glass (2004).

plateau basalts (Deccan: 2.10^6 km^3 in 0.3 Ma; Emeishan: 9.10^5 km^3 in 0.3 Ma; Siberian: $3–4.10^6$ km^3 in 0.4 Ma) are consistent with volumes deduced by these authors for excavation of large-radii impact craters. According to Elkins-Tanton et al. (2005), impacts of this magnitude will have world-wide effects for which 'ample evidence could remain in the rock record.'

Ingle and Coffin (2004) suggested a possible impact origin for the short-lived $\sim$120 Ma Ontong Java Plateau (OJP). Jones et al. (2002) conducted hydrocode simulations (Wünnemann and Ivanov, 2003) to test whether impact-triggered volcanism is consistent with the OJP, using dry lherzolite melting parameters, a steep oceanic geotherm and a vertical incidence by a dunite projectile (diameter 30 km: velocity 20 km/s). The thermal and physical state of the target lithosphere is critical for estimates of melt production. Model calculations suggest impact-triggered melting in a sub-horizontal disc of $\sim$600 km diameter down to $>$150 km depth in the upper mantle forms within $\sim$10 minutes of the impact, whereas most of the initial melt forms at depths shallower than $\sim$100 km. The volume of ultramafic melt would reach $\sim$2.5 $\times$ 10^6 km^3, ranging from superheated melts within 100 km of ground zero, to partial melting with depth and distance. The total melt volume would reach $\sim$7.5 $\times$ 10^6 km^3 of basalt if heat were distributed to produce 20–30% partial mantle melting.

Lowe et al. (1989) observed the possible tectonic significance of the S2–S4 Barberton impact layers in view of their juxtaposition with, and immediately above, the boundary between the $\sim$3.55–3.26 Ga, predominantly mafic–ultramafic volcanic rocks of the Onverwacht Group and the overlying Fig Tree Group association of turbidite-felsic volcanic-banded iron formation. The evidence presented here and in earlier papers (Glikson and Vickers, 2005) shows that the 3.26–3.24 Ga impact cluster coincides with termination of the protracted evolution of greenstone–TTG (tonalite-trondhjemite-granodiorite) systems in the BGB and was accompanied by volcanic activity of the Sulphur Springs Group (Pilbara Craton) and by plutonic events in both the Kaapvaal Craton (Nelshoogte and Kaap Valley plutons) and the Pilbara Craton (Cleland Supersuite; Fig. 8.5-4). The full scale of the $\sim$3.26–3.24 Ga bombardment may not have been identified to date, and the three recorded impact events may represent a minimum number. Iridium and chrome isotope-based mass balance calculations, assuming a global distribution of the impact layers – an assumption justified by the tens of cm thickness of Achaean spherule layers as compared to the global $\sim$3–4 mm thick K–T boundary layer (Alvarez, 1986). Spherule size analysis and the ferromagnesian composition of the impact ejecta suggest asteroids diameters of 20–50 km diameter, excavating impact basins 200–1000 km large and located in mafic–ultramafic crustal regions.

8.5-6. LUNAR CORRELATIONS

The Late Heavy Bombardment (LHB) of the Earth and Moon has been interpreted in terms of the tail-end of planetary accretion or, alternatively, a temporally distinct bombardment episode during 3.95–3.85 Ga (Tera and Wasserburg, 1974; Ryder, 1991). Some of the largest lunar mare basins contain low-Ti basalt, which possibly represents impact-triggered volcanic activity, including Mare Imbrium (3.86 ± 0.02 Ga) and associated 3.85 ± 0.03 Ga K, REE, and P-rich-basalts (KREEP; Ryder, 1991). Isotopic Ar-Ar dating of lunar spherules (Culler et al., 2000), when combined with earlier Ar-Ar and Rb-Sr isotopic studies of lunar basalts (BVTP 1981: Fig. 8.5-7), suggest the occurrence of post-LHB impact and volcanic episodes. Possible cause-effect relationships between impact and volcanic activity pertain in Oceanus Procellarum (3.29–3.08 Ga) and the Hadley Apennines (3.37–3.21 Ga). Such relationships gain support from laser ^{40}Ar/^{39}Ar analyses of lunar impact spherules (sample 11199; Apollo 14, Fra Mauro Formation: Fig. 8.5-7), showing a significant age spike at 3.18 Ga, near the boundary of the Late Imbrian (3.9–3.2 Ga) and the Eratosthenian (3.2–1.2 Ga) as defined by the cratering record (Wilhelms, 1987). 34 lunar impact spherules yield a mean age of 3188 ± 198 Ma, whereas seven spherule ages with errors <100 My yield a mean age of 3178 ± 80 Ma. The small size of the sample and the large errors limit the confidence in these data. However, the combined evidence suggests that the period 3.24 ± 0.1 Ga experienced a major impact cataclysm in the Earth–Moon system, resulting in renewed volcanic activity in some of the lunar mare basins approximately at 3.2 Ga, as well as a Rb-Sr age peak about 3.3 Ga. As in the case of lunar impact spherules, large errors on the Ar-Ar and Rb-Sr isotopic ages preclude precise correlations with terrestrial events. Further isotopic age studies of terrestrial and lunar materials are required in order to establish a $\sim$3.2 Ga bombardment in the Earth–Moon system.

8.5-7. SUMMARY

(1) The early Archean rock record from the Kaapvaal and Pilbara Cratons preserves evidence for two major impact clusters on Earth, one at 3.47 Ga, the other at between 3.26–3.24 Ga.

(2) A substantial body of petrological, geochemical, mineralogical and isotopic evidence for an extraterrestrial impact origin of the Barberton 3.26–3.24 Ga spherule units includes inward-radiating quench-textured and offset vesicles of microkrystite spherules, PGE abundance levels, PGE ratios, Ni/Cr, Ni/Co, V/Cr and Sc/V ratios, occurrence of quench-textured and octahedral nickel-rich chromites distinct from terrestrial chromites, and meteoritic ^{53}Cr/^{52}Cr anomalies (Shukloyukov et al., 2000; Kyte et al., 2003).

(3) Mass balance calculations, based on the Ir flux and the ^{53}Cr/^{52}Cr anomalies (Byerly and Lowe, 1994) and on maximum spherule radii (Melosh and Vickery, 1991), suggest asteroid diameters in the order of 20–50 km, implying impact basins 200–1000 km large.

(4) Lowe and Byerly's (1989) perception of potential significance of the 3.26–3.24 Ga impacts in view of their juxtaposition with, and position immediately above, the top of the >12 km thick, predominantly mafic–ultramafic volcanic sequence of the Onverwacht Group is corroborated by precise U-Pb ages.

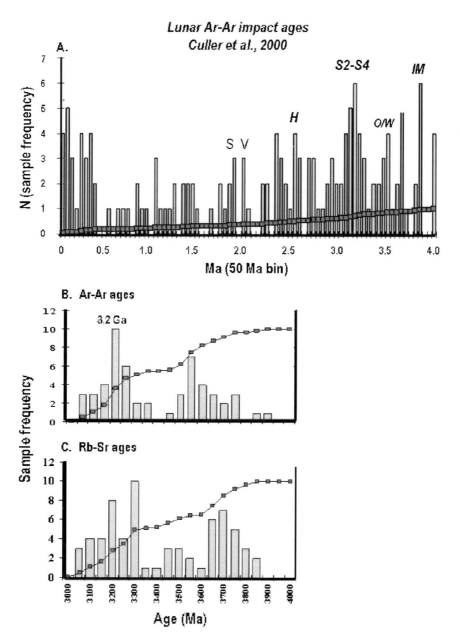

Fig. 8.5-7. (A) Frequency distribution of Rb-Sr isotopic ages of mare basalts; (B) frequency distribution of Ar-Ar isotopic ages of mare basalts (A and B after Basaltic Volcanism of the Terrestrial Planets, 1981); (C) Frequency distribution of Ar-Ar ages of lunar impact spherules based on data by Culler et al. (2000).

(5) The occurrence of banded iron formation, jaspilite and ferruginous siltstone above the 3.26 Ga and 3.24 Ga impact units in the Barberton Greenstone Belt, and at equivalent stratigraphic levels in the Pilbara Craton, may indirectly suggest soluble Fe-enrichment of sea water closely following major impact events, possibly representing erosion of impact-triggered mafic volcanics under the low-oxygen fugacity conditions of the Archaean atmosphere, similar to other impact units in the Pilbara Craton (Glikson, 2006).

Earth's Oldest Rocks
Edited by Martin J. Van Kranendonk, R. Hugh Smithies and Vickie C. Bennett
Developments in Precambrian Geology, Vol. 15 (K.C. Condie, Series Editor)
© 2007 Elsevier B.V. All rights reserved.
DOI: 10.1016/S0166-2635(07)15086-6

1105

Chapter 8.6

TECTONICS OF EARLY EARTH

MARTIN J. VAN KRANENDONK

Geological Survey of Western Australia, 100 Plain St., East Perth, Western Australia 6004, Australia

> "There is nothing permanent except change."
> *Heraclitus*

The Principle of Uniformitarianism has served the geoscience community well over the past two centuries, since Hutton (1785) and Lyell (1830) first recognised that the present is the key to the past. The application of this principle has allowed us to infer the operation of some form of plate tectonic processes back through much of geological time, certainly as far back as the beginning of the Paleoproterozoic, and perhaps back through much of the Archean (see overview in Van Kranendonk (2004a)), to at least 3.2 Ga (Smithies et al., 2005a), or 3.6 Ga (Nutman et al., this volume), or possibly even 4.2 Ga (Cavosie et al., 2005b, this volume).

However, there are many who suggest that plate tectonic processes in the Archean, if they existed at all, were *not* the same as those we are familiar with on modern Earth (e.g., Hamilton, 1998, 2003; McCall, 2003; Stern, 2005). The differences relate to a hotter Archean mantle and resultant differences in the physical characteristics of both the mantle and the crust (see Preface to this volume, but also Park, 1997). Indeed, the papers presented in the first chapter of this book indicate that the Earth changed dramatically though the first billion or so years of its development, as remnant heat from accretion and from the decay of short-lived radionuclides slowly dissipated and the planet evolved from a molten ball at 4.50 Ga to a planet with a dynamic, and differentiated crust (e.g., Kramers, 2007; Taylor, this volume; Kamber, this volume). Thus, when considering the tectonic evolution of this period in Earth history, one need carefully weigh the Principles of Uniformitarianism against those of Secular Change.

What was the nature of Earth's early crust and when, and how, did plate tectonics begin? These are the questions that fascinate researchers of early Earth. A somewhat more overlooked question, but one that deals with a fundamental observation in regards to early Earth studies, and perhaps one of key importance in understanding early Earth tectonics, is: Why are there two fundamentally different types of Archean crust, namely granite-greenstone terrains and high-grade gneiss terrains (e.g., Windley and Bridgwater, 1971)?

In this review is summarised the observations presented in this book on the nature and origin of early Earth tectonics, and a model is presented for the early tectonic evolution of the planet that is based on the geological record.

8.6-1. PATTERNS OF CRUST FORMATION

Several first order observations can be made from the geological data presented in this book, and from other studies (Fig. 8.6-1). The first observation is that felsic crust was formed on Earth from the very early stages of planetary evolution (4.4 Ga: Wilde et al., 2001; Cavosie et al., this volume). There is no obvious time gap in felsic crust production relating to the Late Heavy Meteorite Bombardment (LHMB), but – perhaps significantly – supracrustal rocks are not preserved prior to the end of this event, at 3.825 Ga. What this indicates is that the LHMB, although potentially quite severe, did not interrupt, or unduly influence, the tectonic regime of very early Earth (up to 3.825 Ga), but that tectonic processes on Earth were sufficiently less intense after the end of the LHMB that panels of supracrustal rocks could be preserved.

The second main observation is that all of the most ancient remnants of crust on Earth, from 4.03–3.55 Ga, are high-grade gneiss terrains, consisting of a complex, highly deformed mixture of inclusions of mostly basaltic and ultramafic rocks in seas of sodic granites (TTGs), with only volumetrically minor inclusions of sedimentary rocks.

The third main observation is that after 3.55 Ga, moderate-size tracts of well preserved, low-grade supracrustal rocks (greenstones) are preserved in the Pilbara and Kaapvaal Cratons. These cratons both contain long-lived, nearly continuous records of volcanic eruption and crustal development from 3.55 through to 3.20 Ga, culminating with the formation of thick lithosphere and stable protocontinental nuclei. Significantly, however, high grade gneiss terranes also continued to form throughout the Paleoarchean, to 3.20 Ga (and beyond), indicating that whatever process(es) formed these gneiss terranes did not stop once pieces of more stable protocontinental lithosphere were able to form and be preserved.

After 3.20 Ga, evidence from the Pilbara Craton shows that modern-style plate tectonic processes had commenced (at least locally) on Earth. This evidence includes rifting of the Pilbara protocraton margins at ca. 3.20 Ga, oceanic arc formation during steep subduction at 3.12 Ga, and crustal growth through terrane accretion at 3.07 Ga (Hickman, 2004; Smithies et al., 2005a; Van Kranendonk et al., 2007a). After this, examples of Neoarchean subduction-accretion processes in the formation of continental crust are widespread and compelling (see Van Kranendonk, 2004a, and references therein; Percival, this volume), indicating plate tectonics was the dominant tectonic process on Earth by 2.8 Ga, although with some significant differences from the mode of plate tectonics that has operated since the Neoproterozoic (e.g., Hamilton, 1998; Stern, 2005; Brown, 2006).

8.6-2. TECTONIC EVOLUTION OF EARLY EARTH

8.6-2.1. The Hadean[1] Eon: Initial Properties and First Crust

Nascent Earth incorporated heat during accretion, through gravitational attraction of gasses, rock, and ices, and later, through accretion of larger planetesimals including the impact of a Mars-size body. Additional heating of the planet came from the process of core formation and the decay of short-lived radioactive elements. The end result was ". . . an Earth that was most likely entirely molten (as was the Moon) around 4500 ± 50 Ma." (see Taylor, this volume). Soon after core formation, a relatively short period (200–300 My) of vigorous fluid convection in the mantle (including the crust) transformed into solid-state convection in a mantle that was probably at least 200–300 °C hotter than now and with perhaps 200 times lower viscosity, and convective velocities 30 times greater, than modern Earth (Davies, this volume).

Isotopic data from ancient Earth rocks shows that Hadean Earth probably formed a crust from ca. 4.45 to 4.35 Ga, with a complimentary depleted mantle reservoir (Kamber, this volume). Although this crust was dominantly basaltic, Kamber (this volume) suggests that internal differentiation was inevitable, and from 4.4 to 4.0 Ga, it appears most likely that Earth formed a crust that had at least partially differentiated to felsic compositions.

Some of the felsic crustal components were the result of wet melting under minimum melting conditions, and the oxygen isotopic compositions of Hadean zircons from Jack Hills record the cooling of Earth's surface to the point where liquid water formed the first oceans before ca. 4.20 Ga (Cavosie et al., this volume). These authors argue for a global transformation to cooler surface conditions and stable surface waters that began the aqueous alteration of crustal rocks at low-temperatures. Whereas some argue for large continents and a cool early Earth during this period (e.g., Harrison et al., 2005), only rare zircon crystals in younger sedimentary and granitic rocks (Nelson et al., 2000; Cavosie et al., this volume; Wyche, this volume) and small gneissic remnants in the Acasta Gneiss Complex of the Slave Craton (Iizuka et al., this volume) provide any physical evidence of this crust, such that the true extent of early felsic crust formation is anything but clear.

Pb isotopic characteristics from younger Archean cratons show that some cratons, but not all, formed from a high-μ ($^{238}U/^{204}Pb$) mantle reservoir that developed initially in the Hadean. This mantle reservoir formed either by metasomatic enrichment, or during an early Earth mantle-differentiation event, such as the separation of the Moon, or through recycling of Hadean crust (Chamberlain and Mueller, this volume; Kamber, this volume). Whatever the cause, it is interesting that this feature formed very early in Earth history and survived through much later times.

Without a preserved rock record, it is not possible to say whether Hadean crust deformed through plate tectonics or through the types of processes that continue to deform the single plate crust of our sister planet, Venus (Hansen, this volume, and references therein). Whether or not plate tectonics began in the Hadean, or if life evolved then, we simply are not yet in a position to say.

---

[1] Hadean: Allusion to Hades, Greek mythical underworld of the dead; not a nice place.

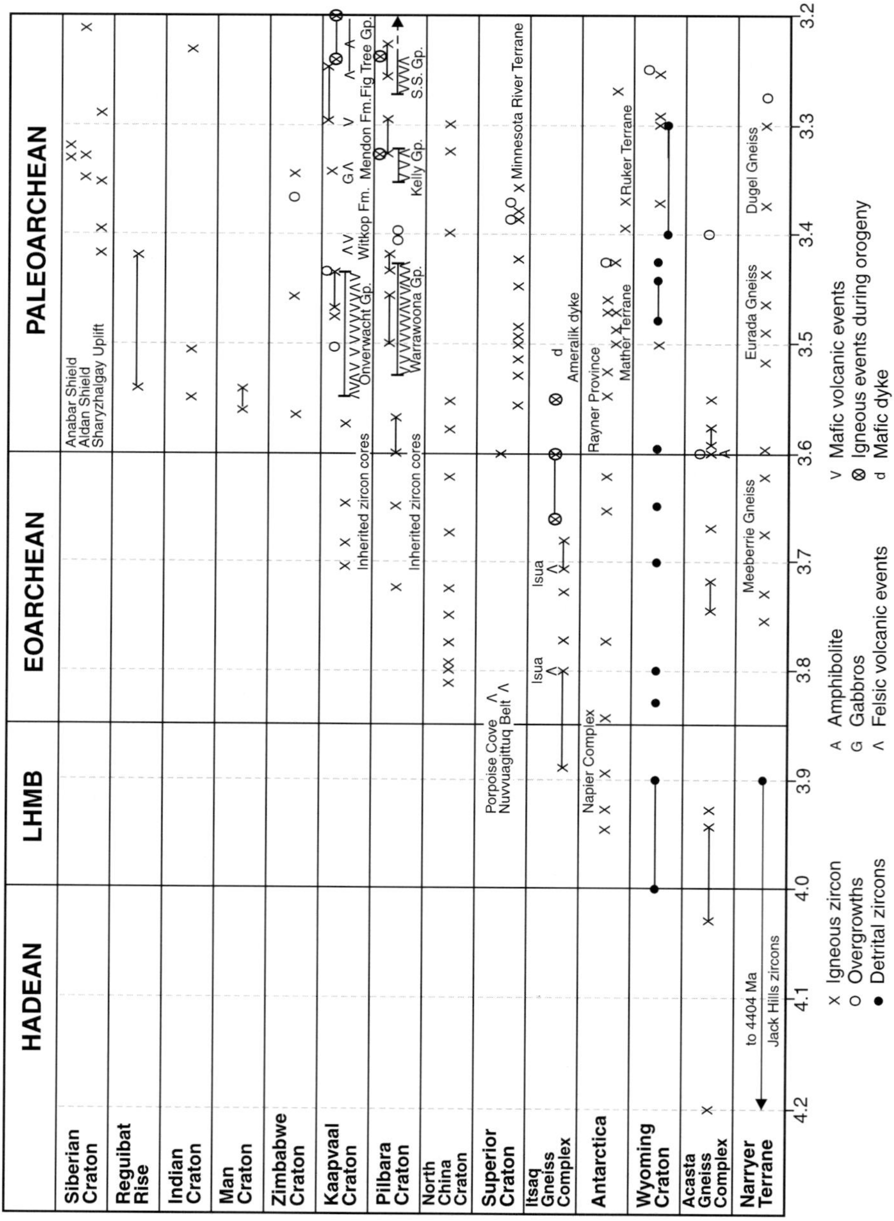

Fig. 8.6-1. Compilation of geochronological data form Earth's oldest rocks, sorted by craton.

8.6-2.2. The Late Heavy Meteorite Bombardment

Pb-isotope data and Ar-Ar dating of lunar impact glasses has shown that there was a significant period of late heavy meteorite bombardment (LHMB) of the Moon at between 4.0 and 3.85 Ga (Tera et al., 1974; Cohen et al., 2000). Since the gravitational force of Earth is larger than that of the Moon, the LHMB event must have also significantly affected Earth. Indeed, the rock record and isotope studies (see Kamber, this volume) seem to indicate that this event almost entirely obliterated the terrestrial protocrust by 3.85 Ga. Analogues with the moon show that the LHMB must have contributed significant new heat to the planet and caused widespread melt extraction from the mantle, possibly resurfacing much of the globe with a new basaltic crust. It is nevertheless clear from the preservation of Hadean to oldest Eoarchean crustal remnants and Hadean zircons that the earliest felsic crust was not completely destroyed by the LHMB. Indeed, the lack of evidence of shock metamorphism in those bits of preserved crust shows that the LHMB was not pervasive on Earth, similar to the way that it was not on the Moon and on Mars. This means that if life had evolved on Earth prior to 3.85 Ga, which seems possible given that life at 3.5 Ga was already diverse (Van Kranendonk, 2006, this volume), it most likely could have survived in refugia through the LHMB and not have had to start all over again from scratch.

8.6-2.3. The Eoarchean Era: Microplume Tectonics

Following the end of the LHMB, at ca. 3.85 Ga, Earth continued to lose heat, following the second law of thermodynamics. This was achieved through a combination of conductive heat loss and heat loss through mantle convection. These processes still operate on modern Earth, but at different rates. On modern Earth, heat flows are highest at mid-ocean ridges (80 mW/m^2) and lowest under Archean continents (41 mW/m^2: data in Meissner, 1986), averaging 70 mW/m^2 (Franck, 1998). The heat is generated from a mantle that is $\sim$1300 °C near the base of the lithosphere (Jackson and Rigden, 1998) and that is convecting in two great hemispherical systems (Wilson, 1988; Pavoni, 1997; Collins, 2003). Values of conductive heat are difficult to estimate for a hotter early Earth, since excess heat near the surface will be accommodated by excess melting, but some estimates are of surface heat flow 2–3 times present day values (Franck, 1998). Estimates of mantle temperature through melting experiments on komatiites suggest that the Paleo- to Neoarchean mantle (ca. 3.46–2.6 Ga) was 200–300 °C hotter than the present Earth (Herzberg, 1992; Nisbett et al., 1993; Arndt et al., 1998). The mantle on Hadean to Eoarchean Earth would have been even hotter (cf. Davies, 1995). Dissipation of this excess mantle heat would be accommodated by increased mantle convection in the form of smaller, more numerous, and more vigorous convection cells and/or a greater number of spreading centres (cf. de Wit and Hart, 1993) and the creation and destruction of greater volumes of crust per unit of time (Bickle, 1978). Spreading rates would have been 2–3 times greater in the Archean, with increased degrees of partial melting at spreading centres generating significantly thicker oceanic crust (ca. 20 km: Bickle, 1978; Sleep and Windley, 1982), but significantly thinner oceanic lithosphere, making it weaker overall. Oceanic crust – because it was young, warm, and thick by the time it reached a downgoing zone – would resist subduction because it was too buoyant (Sleep and Windley, 1982).

Evidence of ductile mylonite-bounded geological terranes in the Itsaq Gneiss Complex of SW Greenland has been used to argue in support of plate tectonic processes in the Eoarchean (Nutman et al., this volume). But does this evidence really indicate the presence of tectonic plates? Terranes and fault-bound slices of different-aged rocks in the Itsaq Gneiss Complex are on the order of 10s of metres to 10s of km wide, and up to 100 km long (see Fig. 3.3-1 in Nutman et al., this volume). Although significant in its own right, this is quite a different scale from terranes accreted onto orogenic margins in modern Earth, which are typically 10s to 100s km wide, and 100s to 1000s of kilometres long (e.g., Irvine et al., 1985; Oldow et al., 1989; Johnston, 2001). Of course the scale of tectonic slices differs in accretionary wedges, but even these are still an order of magnitude larger than those in the Itsaq Gneiss Complex.

An interesting feature of tonalitic rocks from the Itsaq Gneiss Complex is the zircon evidence for an almost continuous series of igneous zircon-growing events over a nearly 100 My period (see Fig. 3.3-5(c–f) in Nutman et al., this volume). Similar long thermal histories have been determined from ca. 4.0 Ga xenocrystic zircons from the Yilgarn Craton of Western Australia (Nelson et al., 2000). If plates at this time were too buoyant to subduct, how were terranes formed and how were they then amalgamated in a subduction zone and subjected to nearly 100 My of continuous magmatism? Such a long duration of subduction-related magmatism in one place only occurs above *very* large, cold, rigid plates that subduct continuously under one position, but it is almost beyond doubt that such large, rigid plates did not exist on the early Earth.

This presents the tectonicist with a conundrum: if we accept that mantle convection cells were smaller and more vigorous on early Earth, and crust too thick and buoyant to subduct, then how was crust recycled back into the mantle?

Vigorous upwelling of mantle would cause voluminous melting to form basaltic and/or komatiitic crust, rapid extension of overlying, newly formed crust, and recycling of this newly formed crust back into the mantle in downwelling zones. It is unlikely that the rapid extension would allow this early crust to develop a depleted and buoyant lithospheric keel that would otherwise protect it from being recycled back into the mantle, as with Paleoarchean to modern crust (see Griffin and O'Reilly, this volume). However, we know from the presence of ancient sedimentary rocks and pillow basalts that liquid water was present on Earth at this time. Evidence form Paleoarchean basaltic crust shows that newly formed basaltic crust would almost certainly have been hydrothermally altered (cf. de Wit et al., 1982; Van Kranendonk, 2006), most likely as a result of the rapid horizontal extension (Nijman and de Vries, 2004; Van Kranendonk et al., 2007a), and downwelling zones could have carried hydrated crust to great depths, facilitating melting and the formation of felsic (TTG) crust, and it seems likely that water was also re-introduced into the upper mantle.

But do the downwelling zones represent subduction zones, at least in the way that we know them today?

Geochemical data from TTG suggests that these volumetrically important Eoarchean rocks are not analogous to modern subduction-related adakite, and were not generated in

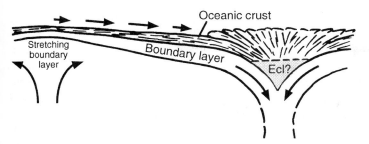

Fig. 8.6-2. Cross-sectional model of early Earth tectonics, showing recycling of oceanic crust in symmetrical downwelling zones in which early felsic crust (high-grade gneiss terranes) may have developed (from Davies, 1995).

steep subduction zones (Smithies, 2000; Martin et al., 2005). Rather, these recent reviews indicate that Eo- to Paleoarchean TTG were either derived from partial melting of basalt at deep crustal levels, or at unusually shallow (low-P) levels on a subducting slab, but did not interact with mantle material. An alternate model for the generation of Eo- to Paleoarchean TTG in flat subduction zones (Smithies et al., 2003) is now no longer favoured, in part because of more recent geochemical data on TTG from the Pilbara and Kaapvaal cratons provide more likely scenarios (see Champion and Smithies, this volume; Smithies et al., this volume), but also because flat subduction zones are essentially amagmatic on modern Earth, since mantle isotherms are depressed in these zones and there is insufficient heat to initiate melting in the downgoing slab.

Early felsic crust (3.85–3.6 Ga) may have been generated in downwelling zones around the margins of upwelling mantle cells, following a geometrical analogy with the Artemis

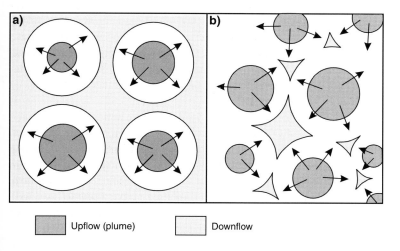

Upflow (plume)   Downflow

Fig. 8.6-3. Plan view model of small cell convection tectonics on early Earth (after Park, 1997).

structure on Venus (see Fig. 8.1-4 in Hansen, this volume), and the previous models of Davies (1992) and Park (1997) (see Figs. 8.6-2 and 8.6-3). In this model, (ultra)mafic crust formed over upwelling zones of convecting mantle, spread rapidly sideways – during which time they were extensively hydrated – and then converged within downwelling zones where they underwent partial melting to form TTG and the first pieces of felsic crust. The highly deformed, migmatitic nature of the early pieces of felsic crust, the juxtaposition of terrane-like slices of crust of different ages within them and the presence of numerous enclaves of mafic and ultramafic material combine to suggest formation of felsic crust within downwelling zones, wherein chaotic mixing of materials and widespread melting of hydrated basaltic crust was taking place. Evidence of long-lived thermal growth histories from zircons in these ancient gneisses (e.g., Nelson et al., 2000), and the juxtaposition of rocks with different geochronological histories in these terranes (cf. Nutman et al., this volume), suggests that the felsic crust may have dipped and bobbed within the downwelling zones over an extended period, probably due to buoyancy of this crust accompanying the generation of felsic melt (cf. Trendall, 2004). I speculate that the early high grade gneiss terranes preserved on Earth today formed only in long-lived zones, where enough material was melted to form significant volumes of felsic crust to prevent it from being entirely recycled back into the mantle. These terranes may have developed an eclogitic root zone, as suggested by Davies (1995), although this would tend to destabilise, rather than help preserve, such pieces of felsic crust.

8.6-2.4. *The Paleoarchean Era: Mesoplume Tectonics*

Widespread preservation of low-grade tracts of upper crustal volcanic and sedimentary rocks commenced after 3.55 Ga, as documented by the Onverwacht Group in the Kaapvaal Craton and the Pilbara Supergroup in the Pilbara Craton. These relatively large pieces of crust (compared with what had been preserved from earlier in Earth history) formed essentially contemporaneously, and continuously, from 3.55–3.20 Ga (Fig. 8.6-1). Crustal growth histories during this time are also recorded from many other Archean cratons. Preservation of upper crust in the Pilbara and Kaapvaal Cratons was made possible by the contemporaneous development and preservation of a strongly depleted, buoyant subcontinental mantle lithospheric keel, which prevented recycling into the mantle (see Smithies et al., 2005a; Griffin and O'Reilly, this volume).

   Formation of a highly depleted lithospheric keel signifies large volumes of melt extraction from a mantle source (Griffin and O'Reilly, this volume). This can be achieved either through subduction, or through melting in the head of one or more mantle plumes. During subduction, however, the mantle wedge is continuously refertilized by influx of new material in the retroflow part above a downgoing slab and it is therefore difficult to envisage how this process could lead to a thick, depleted keel. Rather, melting above a plume head will lead to a depleted, buoyant, residuum (e.g., Griffin and O'Reilly, this volume). Smithies et al. (2005b) showed that depletion of the subcontinental mantle lithosphere beneath the Pilbara Craton evolved progressively from 3.52–3.24 Ga, and that the erupted volcanic rocks had geochemical signatures indicative of derivation from mantle plumes (see also Arndt et al., 2001; Van Kranendonk and Pirajno, 2004). A plume origin for the Onverwacht Group in the Kaapvaal Craton is also probable, based on the presence of interbedded komatiitic and basaltic volcanic rocks, and an upward-younging, autochthonous stratigraphy (Byerly et al., 1996).

Thus, it is suggested that formation and preservation of crust in the Pilbara and Kaapvaal cratons stemmed from voluminous magmatism above a plume head, and coincident development of a complimentary, severely depleted subcontinental lithospheric keel that prevented this crust from being recycled into the mantle in what was still a vigorously convecting early Earth paradigm (see Fig. 4.1-10 in Van Kranendonk et al., this volume). Based principally on the Pilbara example, but possibly also applicable to the Kaapvaal Craton, this type of crust formed over a long-lived upwelling mantle convection cell (or a zone affected by a series of separate plumes) that generated crust with a generally conformable succession of continuously erupted volcanic products spanning >100 My in both cratons (3.53–3.42 Ga in Pilbara; 3.55–3.41 Ga in Kaapvaal). Volcanism was accompanied by the emplacement of contemporaneous granitic rocks (Byerly et al., 1996; Poujol, this volume; Van Kranendonk et al., 2007a, this volume), derived through intracrustal melting and differentiation of mafic magma chambers (Smithies et al., this volume; Champion and Smithies, this volume).

The long duration of volcanism in the early history of these cratons, combined with the evidence for a plume source of the volcanism, indicates a stationary position of the crust over the upwelling plume source. Such a long duration is not unknown for modern mantle plumes (e.g., 130 My duration of the Iceland hotspot; Lawver and Muller, 1994: 85 My duration of the Hawaiian hotspot; Regelous et al., 2003), but the eruptive products of these modern stationary plumes are spread over a wide area as the result of plate motion over the plume source.

In the case of the Paleoarchean Warrawoona and Onverwacht Groups, the evidence for continuous plume volcanism in one place for 100 My may be used to argue *against* significant continental drift and *for* small plates at this early period in Earth history (cf. de Wit and Hart, 1993). Geochemical, stratigraphic, and structural data from these cratons all point to the same conclusion. Long-lived plume volcanism resulted in depletion of the sub-continental mantle lithosphere, resulting in a buoyant, depleted lithospheric keel that helped to preserve these pieces of crust throughout the next 3.5 Gy of Earth history.

Significantly, the formation of high-grade gneiss terrains (composed of seas of TTG with numerous enclaves of mafic and ultramafic rock) continued during this period (e.g., Minnesota River Terrane, Superior Province; Bickford et al., this volume: Siberian Craton; Rosen and Turkina, this volume). This possibly suggests that downwelling zones of the sort suggested for the Eoarchean Era (Fig. 8.6-1) continued to produce felsic crust on Paleoarchean Earth.

Therefore, it appeasrs that three types of crust were being produced during this time period: basaltic and/or komatiitic oceanic crust over zones of upwelling mantle; felsic crust within downwelling zones where oceanic crust was largely consumed and recycled back into the mantle; and thick plateau-like crust over discrete mantle plumes. All of these types of crust were formed without evidence for modern-style, steep, asymmetric subduction and the modern-style plate tectonic regime that it relates to. In the absence of steep, asymmetric subduction and large oceanic plates, it is predicted that back-arcs – which arise through sinking of old, cold, oceanic lithosphere – did not form prior to, or during this period of Earth history.

If modern-style, steep, asymmetric subduction was not yet operating on Paleoarchean Earth, then why the change at 3.55 Ga to production (or at least preservation) of a third type of crust? There is evidence from both the Pilbara and Kaapvaal cratons that the thick supracrustal succession of the lowermost groups in both cratons were deposited on older, high-grade gneissic TTG basement. Perhaps as the planet cooled, mantle convection cells grew larger and areas of felsic crust in downwelling zones grew through lateral accretion. Once some of these areas of felsic crust grew to a size approximating that of the mantle convection cells, then they would be too wide to solely accommodate downwelling. Hot mantle material would rise up beneath these areas of felsic crust and undergo decompression melting to generate the type of hybrid oceanic-continental plateau-type crust inferred for the Pilbara (and Kaapvaal) Craton, with a severely depleted subcontinental mantle lithospheric keel (see Fig. 4.1-10 in Van Kranendonk et al., this volume). Whether the onset of this process represents a stage in the gradual increase in plate size, or the effect of a distinct catastrophic mantle overturn event as suggested by Davies (1995), is unknown.

8.6-2.5. The Mesoarchean Era: Onset of Plate Tectonics

Evidence has been presented elsewhere for the onset of plate tectonics in the Mesoarchean, including rifting of protocontinental margins, steep subduction, and the accretion of exotic terranes (Smithies et al., 2005a; Van Kranendonk et al., 2007a). The evidence for steep subduction is used to infer a jump to even larger tectonic plates on Earth, based on the inference that steep subduction can only occur if subduction is of old, cold oceanic lithosphere. Significantly, the change of tectonic regime identified from the geological record at 3.2 Ga occurs at the same time that modelling shows significant changes in the global mean heat flow (Franck, 1998). This jump in plate size may relate to one of the postulated catastrophic mantle overturn events of Davies (1995) and a change from mesoplumes to whole mantle convection, or a step in the gradual increase in plate size.

But how did plate tectonics start? Analogue studies with Venus show that massive upwelling plumes can spread under the weight of their own magmatic edifice and create arcuate bounding wrinkle ridges and spreading centres that start to look very much like the geometry of Earth plate tectonics (Artemis; Fig. 8.1-4 in Hansen, this volume). Spreading of crust in this fashion across mid-crustal detachments was referred to as continental overflow tectonics by Bailey (1999). This style of lateral spreading of magmatically overthickened crust may have helped to initiate asymmetric subduction on Earth towards the end of the Paleoarchean, leading to classic, steep subduction (at least locally) in the Mesoarchean, and eventually to the plate tectonic paradigm we enjoy today. A change from symmetrical downwelling of crust around the margins of upwelling mantle plumes to asymmetric subduction of oceanic crust may be regarded as a natural progression of events, given that nature rarely acts in a perfectly symmetrical fashion for very long.

8.6-3. TIMESCALE DIVISIONS

The division of Earth's history aids comparison between terranes and understanding of Earth processes and the evolution of life on this planet. Placing definite boundaries on a gradual evolution of the planet is never wholly satisfactory, but given enough data, boundaries can be suggested that fit most of the observations and can be refined as more information becomes available.

In the previous and current divisions of the Precambrian timescale, there is no base to the Archean and the Hadean Eon is loosely characterised (Plumb and James, 1986; Plumb, 1991; Gradstein et al., 2004). Bleeker (2004) suggested that the divisions of the Precambrian timescale were: divorced from the rock record; based on arbitrary, round age numbers; heterogeneous in terms of boundary criteria; and incomplete, with the lower boundary of the Archean undefined. With this in mind, the new compilation of data presented in this book – which is based on the rock record – warrants a re-examination of the divisions of the early history of the Earth.

The geochronological data summarised herein lend gross support to the existing timescale for the early history of the Earth (Fig. 8.1-1). As mentioned above, there is no singular event in the geological record that can be used to place an exact time for the Hadean-Archean boundary: earliest life forms are controversial, and felsic crust formation appears continuous throughout the timespan covered in this volume. Only two notable events are evident from the data. First is the appearance of the oldest supracrustal remnants within high grade gneiss terrains, at ca. 3.825 Ga, a time which closely corresponds to the widely accepted (though formally undefined) Hadean-Archean boundary at 3.85 Ga. I suggest that this existing boundary be formally accepted by the geological community.

The second event is the appearance of well-preserved supracrustal successions in the Pilbara and Kaapvaal cratons, at ca. 3.55 Ga, which is close to the formally accepted base of the Paleoarchean. One could argue that if this criterion is accepted as the basis for the Eo- Paleoarchean boundary, then a more "accurate" age of the boundary could be given as the age of the oldest supracrustal rock in these successions, at 3548 ± 3 Ma (Kröner et al., 1996).

Cavosie et al. (2005b, this volume) suggest that the Hadean Eon apply only from the period of accretion ($T_{zero} = 4.567$ Ga) to 4.2 Ga, when oxygen isotope data from Hadean zircons suggest that magmatic recycling of crust altered at low temperatures in the presence of liquid water oceans occurred. These authors regard this as representing a fundamental change from a truly Hadean Earth. This oxygen isotope data is clearly significant, but was the crustal recycling it infers a sufficiently robust process to produce crust that could withstand the LHMB, and were conditions in Earth then in place that would lead to plate tectonics?

This first, and most tectonically mobile of, eons could be subdivided into three eras: the Paleohadean Era, when the planet was still undergoing the last stages of accretion and was entirely molten (4.567–4.45 Ga); the Mesohadean Era (4.45–4.0 Ga), when the terrestrial protocrust and a depleted mantle reservoir were formed; the Neohadean Era, corresponding to the time of the LHMB (4.0–3.85 Ga), when most of the terrestrial protocrust was destroyed and the planet was largely resurfaced. After this is the Archean Eon, with subdivisions generally as previously determined at 3.6 (or 3.55), 3.2, 2.8 and 2.5 Ga for the end of the Eoarchean, Paleoarchean, Mesoarchean, and Neoarchean, respectively.

ACKNOWLEDGEMENTS

Hugh Smithies is thanked for constructive talks and criticisms on these and many other ideas regarding Archean tectonics over the years. This paper is published with permission of the Executive Director of the Geological Survey of Western Australia.

REFERENCES

Abbott, D.H., Hoffman, S.E., 1984. Archaean plate tectonics revisited 1. Heat flow, spreading rate, and the age of subducting oceanic lithosphere and their effects on the origin and evolution of continents. Tectonics 3, 429–448.

Abbott, D.H., Isley, A.E., 2001. Oceanic upwelling and mantle plume activity: paleomagnetic tests of ideas on the source of the Fe in early Precambrian iron formations. In: Geological Society of America, Special Paper 352, pp. 323–339.

Abbott, D.H., Isley, A.E., 2002. The intensity, occurrence and duration of superplume events and eras over geological time. Journal of Geodynamics 34, 265–307.

Abbott, D., Drury, R., Smith, W.H.F., 1994. Flat to steep transition in subduction style. Geology 22, 937–940.

Abell, P.I., McClory, J., Martin, A., Nisbet, E.G., 1985a. Archaean stromatolites from the Ngesi Group, Belingwe greenstone belt, Zimbabwe; preservation and stable isotopes; preliminary results. Precambrian Research 27 (4), 357–383.

Abell, P.I., McClory, J., Martin, A., Nesbit, E.G., Kyser, K.T., 1985b. Petrography and stable isotope ratios from Archaean stromatolites, Mushandike Formation, Zimbabwe. Precambrian Research 27, 385–398.

Abeysinghe, P.B., Fetherston, J.M., 1997. Barite and fluorite in Western Australia. Western Australia Geological Survey, Mineral Resources Bulletin 17, 97 pp.

Abraham, A-C., Francis, D., Polvé, M., 2005. Origin of recent alkaline lavas by lithospheric thinning beneath the northern Canadian Cordillera. Canadian Journal of Earth Sciences 42, 1073–1095.

Addington, E.A., 2001. A stratigraphic study of small volcano clusters on Venus. Icarus 149, 16–36.

Ahmat, A.L., Ruddock, I., 1990. Windimurra and Narndee layered complexes. In: Geology and Mineral Resources of Western Australia. Western Australia Geological Survey, Memoir 3, pp. 119–126.

Alard, O., Griffin, W.L., Lorand, J.P., Jackson, S.E., O'Reilly, S.Y., 2000. Non-chondritic distribution of the highly siderophile elements in mantle sulfides. Nature 407, 891–894.

Alard, O., Griffin, W.L., Pearson, N.J., Lorand, J.-P., O'Reilly, S.Y., 2002. New insights into the Re-Os systematics of subcontinental lithospheric mantle from in-situ analysis of sulfides. Earth and Planetary Science Letters 203, 651–663.

Albarède, F., Blichert-Toft, J., Vervoort, J.D., Gleason, J.D., Rosing, M., 2000. Hf-Nd isotope evidence for a transient dynamic regime in the early terrestrial mantle. Nature 404, 488–490.

Aleinikoff, J.N., Williams, I.S., Compston, W., Stuckless, J.S., Worl, R.G., 1989. Evidence for an Early Archean component in the Middle to Late Archean gneisses of the Wind

River Range, west-central Wyoming: conventional and ion microprobe U-Pb data. Contributions to Mineralogy and Petrology 101, 198–206.

Al-Jishi, R., Dresselhaus, G., 1982. Lattice-dynamical model for graphite. Physics Review B 26, 4514–4522.

Allaart, J.H., 1976. The pre-3760 m.y. old supracrustal rocks of the Isua area, Central West Greenland, and the associated occurrence of quartz-banded ironstone. In: Windley, B.F. (Ed.), The Early History of the Earth. Wiley, London, pp. 177–189.

Allègre, C.J., 1982. Genesis of Archaean komatiites in a wet subducted plate. In: Arndt, N.T., Nisbet, E.G. (Eds.), Komatiites. Allen and Unwin, London, pp. 495–500.

Allègre, C.J., Birck, J-L., Fourcade, S., Semet, M.P., 1975. Rubidium-87/strontium-87 age of Juvinas basaltic achondrite and early igneous activity in the solar system. Science 187, 436–438.

Allen, M.B., Windley, B.F., Chi, Z., 1992. Palaeozoic collisional tectonics and magmatism of the Chinese Tien Shan, Central Asia. Tectonophysics 220, 89–115.

Allwood, A.C., Walter, M.R., Kamber, B.S., Marshall, C.P., Burch, I.W., 2006a. Stromatolite reef from the Early Archaean era of Australia. Nature 441, 714–717.

Allwood, A.C., Walter, M.R., Marshall, C.P., 2006b. Raman spectroscopy reveals thermal palaeoenvironments of c. 3.5 billion-year-old organic matter. Vibrational Spectroscopy 41, 190–197.

Alonso-Perez, R., Ulmer, P., Müntener, O., Thompson, A.B., 2003. Role of garnet fractionation in H_2O undersaturated andesite liquids at high pressure. Lithos 73 (1–2), S116.

Alt, A.D., Sears, J.W., Hyndman, D.W., 1988. Terrestrial mare: The origins of large basalt plateaus, hotspot tracks and spreading ridges. Journal of Geology 96, 647–662.

Alt, J.C., 1995. Sulfur isotopic profile through the oceanic-crust – sulfur mobility and seawater-crustal sulfur exchange during hydrothermal alteration. Geology 23, 585–588.

Alvarez, W., 1986. Toward a theory of impact crises. Eos 67, 649–658.

Amelin, Y.V., 1998. Geochronology of the Jack Hills detrital zircons by precise U-Pb isotope dilution analysis of crystal fragments. Chemical Geology 146, 25–38.

Amelin, Y., Lee, D.C., Halliday, A.N., Pidgeon, R.T., 1999. Nature of the Earth's earliest crust from hafnium isotopes in single detrital zircons. Nature 399, 252–255.

Amelin, Y., Lee, D.C., Halliday, A.N., 2000. Early-middle Archaean crustal evolution deduced from Lu-Hf and U-Pb isotopic studies of single zircon grains. Geochimica et Cosmochimica Acta 64, 4205–4225.

Amelin, Y., Krot, A.N., Hutcheon, I.D., Ulyanov, A.A., 2002. Lead isotopic ages of chondrules and calcium-aluminium-rich inclusions. Science 297, 1678–1683.

Amelin, Y., Krot, A.N., Twelker, E., 2004. Pb isotopic age of the CB chondrite Gujba and the duration of the chondrule formation interval. Geochimica et Cosmochimica Acta 68, E958.

Amelin, Y., Wadhwa, M., Lugmair, G.W., 2006. Pb-isotopic dating of meteorites using ^{202}Pb-^{205}Pb double-spike: comparison with other high-resolution chronometers. Lunar and Planetary Science XXXVII, abstract 1970.

Anbar, A.D., Knoll, A.H., 2002. Proterozoic ocean chemistry and evolution: A bioinorganic bridge? Science 297, 1137–1142.

Anbar, A.D., Zahnle, K.J., Arnold, G.L., Mojzsis, S.J., 2001. Extraterrestrial iridium, sediment accumulation and the habitability of the early Earth's surface. Journal of Geophysical Research – Planets 106 (E2), 3219–3236.

Anders, E., 1964. Origin, age and composition of meteorites. Space Science Reviews 3, 583–714.

Anders, E., Ebihara, M., 1982. Solar System abundances of the elements. Geochimica et Cosmochimica Acta 53, 2363–2380.

Anders, E., Grevesse, N., 1989. Abundances of the elements: Meteoritic and solar. Geochimica et Cosmochimica Acta 53, 197–214.

Anderson, R.B., 1956. Catalysts for Fischer–Tropsch synthesis. In: Emmett, P.H. (Ed.), Catalysis. Hydrocarbon Synthesis Hydrogeneation and Cyclization. Reinhold, New York, pp. 29–255.

André, L., Cardinal, D., Alleman, L.Y., Moorbath, S., 2006. Silicon isotopes in ~3.8 Ga West Greenland rocks as clues to the Eoarchaean supracrustal Si cycle. Earth and Planetary Science Letters 245, 162–173.

Andreasen, R., Sharma, M., 2006. Solar nebula heterogeneity in p-process samarium and neodymium isotopes. Science 314, 806–809.

Anhaeusser, C.R., 1972. The geology of the Jamestown Hills area of the Barberton Mountain Land, South Africa. Transactions of the Geological Society of South Africa 75, 225–263.

Anhaeusser, C.R., 1973. The evolution of the early Precambrian crust of southern Africa. Philosophical Transactions of the Royal Society of London Ser. A 273, 359–388.

Anhaeusser, C.R., 1976. The geology of the Sheba Hills area of the Barberton Mountain Land, South Africa, with particular reference to the Eureka Syncline. Transactions of the Geological Society of South Africa 79, 253–280.

Anhaeusser, C.R., 1983. The geology of the Schapenburg greenstone remnant and surrounding Archaean granitic terrane south of Badplaas, eastern Transvaal. In: Anhaeusser, C.R. (Ed.), Contributions to the Geology of the Barberton Mountain Land. Geological Society of South Africa, Special Publication 9, pp. 31–44.

Anhaeusser, C.R., 1985. Archean layered ultramafic complexes in the Barberton Mountain Land, South Africa. In: Ayres, L.D., Thurston, P.C., Card, K.D., Weber, W. (Eds.), Evolution of Archean Supracrustal Sequences. Geological Association of Canada, Special Paper 28, pp. 281–302.

Anhaeusser, C.R., 1986. Archaean gold mineralization in the Barberton Mountain Land. In: Anhaeusser, C.R., Maske, S. (Eds.), Mineral Deposits of Southern Africa. Geological Society of South Africa, Special Publications, pp. 113–154.

Anhaeusser, C.R., 2001. The anatomy of an extrusive-intrusive Archaean mafic-ultramafic sequence: the Nelshoogte schist belt and Stolzburg layered ultramafic complex, Barberton greenstone belt, South Africa. South African Journal of Geology 104, 167–204.

Anhaeusser, C.R., 2006. A re-evaluation of Archean intracratonic terrane boundaries on the Kaapvaal Craton, South Africa: Collisional suture zones? In: Reimold, W.U., Gibson, R.L. (Eds.), Processes on the Earth Earth. Geological Society of America, Special Paper 405, pp. 193–210.

Anhaeusser, C.R., Robb, L.J., 1980. Regional and detailed field and geochemical studies of Archean trondhjemitic gneisses, migmatites and greenstone xenoliths in the southern part of the Barberton mountain land, South Africa. Precambrian Research 11, 373–397.

Anhaeusser, C.R., Robb, L.J., 1981. Magmatic cycles and the evolution of the Archaean granitic crust in the eastern Transvaal and Swaziland. In: Glover, J.E., Groves, D.E. (Eds.), Archaean Geology. Geological Society of Australia, Special Publication 7, pp. 457–467.

Anhaeusser, C.R., Robb, L.J., 1983. Chemical analyses of granitoid rocks from the Barbeton Mountain Land. In: Geological Society of South Africa, Special Publication 9, pp. 189–219.

Anhauesser, C.R., Viljoen, M.J., 1965. The base of the Swaziland System in the Barberton–Nordkaap–Louwś Creek area, Barberton Mountain Land. Information Circular, Economic Geology Research Unit, University of the Witwatersrand, Johannesburg, 25, 32 pp.

Anhaeusser, C.R., Mason, R., Viljoen, M.J., Viljoen, R.P., 1969. A reappraisal of some aspects of Precambrian Shield geology. Bulletin of the Geological Society of America 80, 2175–2200.

Anhaeusser, C.R., Robb, L.J., Viljoen, M.J., 1981. Provisional geological map of the Barberton Greenstone Belt and surrounding granitic terrane, eastern Transvaal and Swaziland scale 1:250,000. In: Anhaeusser, C.R. (Ed.), Contributions to the Geology of the Barberton Mountain Land. Geological Society of South Africa, Special Publication 9, pp. 221–223 and insert map.

Anhaeusser, C.R., Robb, L.J., Viljoen, M.J., 1983. Notes on the provisionnal geological map of the Barberton greenstone belt and surrounding granitic terrane, eastern Transvaal and Swaziland (1:250,000 color map). Transactions of the Geological Society of South Africa 9, 221–223.

Anisimova, I.V., Kotov, A.B., Salnikova, E.B., Yakovleva, S.Z., Morozova, I.M., 2005. Age and geodynamic settings of the Archean greenstone belts from western part of Aldan Shield. In: Archean Geology and Geodynamics. First Russian Conference on Problems of Precambrian Geology and Geodynamics, Abstract Volume. Inst. Geol. and Geochron. Precambr., St.-Petersburg, pp. 23–27 (in Russian).

Annen, C., Blundy, J.D., Sparks, R.S.J., 2006. The genesis of intermediate and silicic magmas in deep crustal hot zones. Journal of Petrology 47, 505–539.

Ansdell, K.M., 2005. Tectonic evolution of the Manitoba–Saskatchewan segment of the Paleoproterozoic Trans-Hudson Orogen, Canada. Canadian Journal of Earth Sciences 42, 741–759.

Appel, P.W., 1983. Rare earth elements in the early Archaean Isua iron-formation, West Greenland. Precambrian Research 20, 243–258.

Appel, P.W.U., Fedo, C.M., Moorbath, S., Myers, J.S., 1998. Recognisable primary volcanic and sedimentary features in a low-strain domain of the highly deformed, oldest-known (ca. 3.7–3.8 Gyr) greenstone belt, Isua, Greenland. Terra Nova 10, 57–62.

Appel, P.W.U., Rollinson, H.R., Touret, J.L.R., 2001. Remnants of an Early Archaean (>3.75 Ga) sea-floor, hydrothermal system in the Isua greenstone belt. Precambrian Research 112, 27–49.

Appel, P.W.U., Moorbath, S., Myers, J.S., 2003. *Isuasphaeara isua* (Pflug) revisited. Precambrian Research 126, 309–313.

Arima, M., Barnett, R.L., 1984. Sapphirine-bearing granulites from the Sipiwesk Lake area of the late Archean Pikwitonei granulite terrain, Manitoba, Canada. Contributions to Mineralogy and Petrology 88, 102–112.

Armstrong, R., 1989. 1988 Annual Technical report of the Geological Survey of South Africa. Department of Mineral and Energy Affairs, 27 pp.

Armstrong, R.A., Compston, W., de Wit, M.J., Williams, I.S., 1990. The stratigraphy of 3.5–3.2 Ga Barberton Greenstone Belt revisited: A single zircon microprobe study. Earth and Planetary Science Letters 101, 90–106.

Armstrong, R.A., Compston, W., Retief, E.A., Williams, I.S., Welke, H.J., 1991. Zircon ion microprobe studies bearing on the age and evolution of the Witwatersrand triad. Precambrian Research 53, 243–266.

Armstrong, R.L., 1963. K-Ar dates from West Greenland. Bulletin of the Geological Society of America 74, 1189–1192.

Armstrong, R.L., 1968. A model for Sr and Pb isotope evolution in a dynamic Earth. Reviews of Geophysics 6, 175–199.

Armstrong, R.L., 1981. Radiogenic isotopes: the case for crustal recycling on a near-steady-state no-continental-growth Earth. Philosophical Transactions of the Royal Society of London Ser. A 301, 443–472.

Armstrong, R.L., 1991. The persistent myth of crustal growth. Australian Journal of Earth Sciences 38 (5), 613–630.

Arndt, N.T., Jenner, G.A., 1986. Crustally contaminated komatiites and basalts from Kambalda, Western Australia. Chemical Geology 56, 229–255.

Arndt, N., Albarede, F., Nisbet, E.G., 1997. Mafic and Ultramafic Magmatism. In: de Wit, M.J., Ashwal, L.D. (Eds.), Greenstone Belts. Oxford University Press, Oxford, United Kingdom, pp. 231–254.

Arndt, N., Ginibre, C., Chauval, C., Albarède, F., Cheadle, M., Herzberg, C., Jenner, G., Lahaye, Y., 1998. Were komatiites wet? Geology 26, 739–742.

Arndt, N.T., Nelson, D.R., Compston, W., Trendall, A.F., Thorne, A.M., 1991. The age of the Fortescue Group, Hamersley Basin, Western Australia, from ion microprobe zircon U-Pb results. Australian Journal of Earth Sciences 38, 261–281.

Arndt, N., Bruzak, G., Reischmann, T., 2001. The oldest continental and oceanic plateaus: Geochemistry of basalts and komatiites of the Pilbara Craton, Australia. In: Ernst, R.E., Buchan, K.L. (Eds.), Mantle Plumes: Their Identification Through Time. Geological Society of America, Special Paper 352, pp. 359–387.

Arriens, P.A., 1971. Archaean geochronology of Australia. In: Geological Society of Australia, Special Publication 3, pp. 1–23.

Arth, J.G., Hanson, G.N., 1975. Geochemistry and origin of the Early Precambrian crust of North-eastern Minnesota. Geochimica and Cosmochimica Acta 39, 325–362.

Arth, J.G., Barker, F., Peterman, Z.E., Frideman, I., 1978. Geochemistry of the gabbro-diorite-tonalite-trondhjemite suite of south-west Finland and its implications for the origin of tonalitic and trondhjemitic magmas. Journal of Petrology 19, 289–316.

Ashton, K.E., Heaman, L.M., Lewry, J.F., Hartlaub, R.P., Shi, R., 1999. Age and origin of the Jan Lake Complex: a glimpse at the buried Archean craton of the Trans-Hudson Orogen. Canadian Journal of Earth Sciences 36, 185–208.

Aspler, L.B., Chiarenzelli, J.R., 1998. Two Neoarchean supercontinents? Sedimentary Geology 120, 75–104.

Atherton, M.P., Petford, N., 1993. Generation of sodium-rich magmas from newly underplated basaltic crust. Nature 362, 144–146.

Aubele, J., 1996. Akkruva small shield plains; definition of a significant regional plains unit on Venus. In: Lunar and Planetary Science Conference XXVII, pp. 49–50.

Aulbach, S., Griffin, W.L., Pearson, N.J., O'Reilly, S.Y., Kivi, K., Doyle, B.J., 2004. Mantle formation and evolution, Slave Craton: constraints from HFSE abundances and Re-Os systematics of sulfide inclusions in mantle xenocrysts. Chemical Geology 208, 61–88.

Awramik, S.M., Schopf, J.W., Walter, M.R., 1983. Filamentous fossil bacteria from the Archean of Western Australia. Precambrian Research 20, 357–374.

Ayer, J., Amelin, Y., Corfu, F., Kamo, S., Ketchum, J., Kwok, K., Trowell, N., 2002. Evolution of the southern Abitibi greenstone belt based on U-Pb geochronology: autochthonous volcanic construction followed by plutonism, regional deformation and sedimentation. Precambrian Research 115, 63–95.

Baadsgaard, H., 1973. U-Th-Pb dates on zircons from the early Precambrian Amîtsoq gneisses, Godthaab district, West Greenland. Earth and Planet Science Letters 19, 22–28.

Baadsgaard, H., Nutman, A.P., Bridgwater, D., McGregor, V.R., Rosing, M., Allaart, J.H., 1984. The zircon geochronology of the Akilia association and the Isua supracrustal belt, West Greenland. Earth and Planetary Science Letters 68, 221–228.

Baadsgaard, H., Nutman, A.P., Bridgwater, D., 1986. Geochronology and isotopic variation of the early Archaean Amîtsoq gneisses of the Isukasia area, southern West Greenland. Geochimica et Cosmochimica Acta 50, 2173–2183.

Baadsgaard, H., Nutman, A.P., Samsonov, A.V., 1990. Geochronology of the Olondo greenstone belt. In: Fifth International Conference on Geochronology, Cosmochronology, and Isotope Geology, Abstract Volume. Geological Society of Australia, Publ. 6.

Bagas, L., 2005. Geology of the Nullagine 1:100,000 sheet. Western Australia Geological Survey, 1:100,000 Geological Series Explanatory Notes, 33 pp.

Bagas, L., Smithies, R.H., Champion, D.C., 2003. Geochemistry of the Corunna Downs Granitoid Complex, East Pilbara Granite-Greenstone Terrane, Western Australia. In: Western Australia Geological Survey, Annual Review 2001-02, pp. 61–69.

Bai, J., Huang, X.G., Wang, H.C., Guo, J.J., Yan, Y.Y., Xue, Q.Y., Dai, F.Y., Xu, W.Z., Wang, G.F., 1996. The Precambrian Crustal Evolution of China. Beijing, Geological Publishing House, pp. 1–259 (in Chinese with English abstract).

Bai, J., Li, S.X., He, G.P., Xu, C.L., 1986. The Early Precambrian Geology of Wutaishan. Tianjin, Tianjin Science and Technology Press, pp. 1–475 (in Chinese with English abstract).

Bailes, A.H., Percival, J.A., 2005. Geology of the Black Island area, Lake Winnipeg, Manitoba (parts of NTS 62P1, 7 and 8). Manitoba Geological Survey Geoscientific Report GR2005-2.

Bailey, R.C., 1999. Gravity-driven continental overflow and Archean tectonics. Nature 398, 413–415.

Baker, J., Bizzarro, M., Wittig, N., Connelly, J., Haack, H., 2005. Early planetesimal melting from an age of 4.5662 Gyr for differentiated meteorites. Nature 436, 1127–1131.

Baker, V.R., Komatsu, G., Gulick, V.C., Parker, T.J., 1997. Channels and valleys. In: Bouger, S.W., Hunten, D.M., Phillips, R.J. (Eds.), Venus II. University of Arizona Press, Tucson, AZ, pp. 757–798.

Baker, V.R., Komatsu, G., Parker, T.J., Gulick, V.C., Kargel, J.S., Lewis, J.S., 1992. Channels and Valleys on Venus: Preliminary analysis of Magellan data. Journal of Geophysical Research 97, 13,395–13,420.

Banerdt, W.B., McGill, G.E., Zuber, M.T., 1997. Plains tectonics on Venus. In: Bouger, S.W., Hunten, D.M., Phillips, R.J. (Eds.), Venus II. University of Arizona Press, Tucson, AZ, pp. 901–930.

Banerjee, N.R., Furnes, H., Muehlenbachs, K., Staudigel, H., de Wit, M., 2006. Preservation of ~3.4–3.5 Ga microbial biomarkers in pillow lavas and hyaloclastites from the Barberton Greenstone Belt, South Africa. Earth and Planetary Science Letters 241, 707–722.

Banerjee, N.R., Simonnetti, A., Furnes, H., Staudigel, H., Muehlenbachs, K., Heaman, L., Van Kranendonk, M., 2007. Direct dating of Archean microbial ichnofossils. Geology 35, 487–490.

Bannister, R.A., 2006. Geologic analysis of deformation in the interior region of Artemis (Venus, 34°S 132°E). M.Sc. thesis. University Minnesota Duluth, Duluth, 83 pp.

Bannister, R.A., Hansen, V.L., 2006. Geologic analysis of deformation in the interior region of Artemis (Venus, 34°S 132°E). In: Lunar and Planetary Science Conference XXXVII, 1370.pdf, Lunar and Planetary Institute, Houston, TX.

Barker, F., 1979. Trondhjemite: a definition, environment and hypotheses of origin. In: Barker, F. (Ed.), Trondhjemites, Dacites and Related Rocks. Elsevier, Amsterdam, pp. 1–12.

Barker, F., Arth, J.G., 1976. Generation of trondhjemite-tonalite liquids and Archean bimodal trondhjemite-basalt suites. Geology 4, 596–600.

Barker, F., Wones, D.R., Sharp, W.N., Desborough, G.A., 1975. The Pikes Peak batholith, Colorado Front Range, and a model for the origin of the gabbro-anorthosite-syenit-potassic granite suite. Precambrian Research 2, 97–160.

Barley, M.E., 1982. Porphyry-style mineralization associated with early Archean calc-alkaline igneous activity, eastern Pilbara, Western Australia. Economic Geology 77, 1230–1236.

Barley, M.E., 1986. Incompatible-element enrichment in Archean basalts – A consequence of contamination by older sialic crust rather than mantle heterogeneity. Geology 14, 947–950.

Barley, M.E., 1993. Volcanic, sedimentary and tectonostratigraphic environments of the ~3.46 Ga Warrawoona Megasequence: a review. Precambrian Research 60, 47–67.

Barley, M.E., 1997. The Pilbara Craton. In: de Wit, M.J., Ashwal, L. (Eds.), Greenstone Belts. Oxford University Monographs on Geology and Geophysics, vol. 35. Clarendon Press, Oxford, pp. 657–664.

Barley, M.E., Groves, D.I., 1992. Supercontinent cycles and the distribution of metal deposits through time. Geology 20, 291–294.

Barley, M.E., Pickard, A.L., 1999. An extensive, crustally-derived, 3325 to 3310 Ma silicic volcano-plutonic suite in the eastern Pilbara Craton: evidence from the Kelley Belt, McPhee Dome, and Corunna Downs Batholith. Precambrian Research 96, 41–62.

Barley, M.E., Dunlop, J.S.R., Glover, J.E., Groves, D.I., 1979. Sedimentary evidence for an Archean shallow-water volcanic-sedimentary facies, Eastern Pilbara Block, Western Australia. Earth and Planetary Science Letters 43, 74–84.

Barley, M.E., Sylvester, G.C., Groves, D.I., 1984. Archaean calc-alkaline volcanism in the Pilbara Block, Western Australia. Precambrian Research 24, 285–319.

Barley, M.E., Eisenlohr, B.N., Groves, D.I., Perring, C.S., Vearncombe, J.R., 1989. Late Archaean convergent margin tectonics and gold mineralization: A new look at the Norseman–Wiluna Belt, Western Australia. Geology 17, 826–829.

Barley, M.E., Krapez, B., Groves, D.I., Kerrich, R., 1998a. The Late Archaean bonanza: metallogenic and environmental consequences of the interaction between mantle plumes, lithospheric tectonics and global cyclicity. Precambrian Research 91, 65–90.

Barley, M.E., Loader, S.E., McNaughton, N.J., 1998. 3430 to 3417 Ma calc-alkaline volcanism in the McPhee Dome and Kelley Belt, and growth of the eastern Pilbara Craton. Precambrian Research 88, 3–24.

Barley, M.E., Bekker, A., Krapez, B., 2005. Late Archean to Early Proterozoic global tectonics, environmental change and the rise of atmospheric oxygen. Earth and Planetary Science Letters 238, 156–171.

Barovich, K.M., Patchett, P.J., Peterman, Z.E., Sims, P.K., 1991. Neodymium isotopic evidence for Early Proterozoic units in the Watersmeet gneiss dome, northern Michigan. In: U.S. Geological Survey, Bulletin 1904-G, 7 pp.

Barry, T.L., Kent, R.W., 1998. Cenozoic magmatism in Mongolia and the origin of central and east Asian basalts. In: American Geophysical Union, Geodynamics Series, vol. 27. Washington, DC, pp. 347–364.

Barsukov, V.L., et al., 1984. Geology of Venus from the result of analysis of radar images taken by Venera 15/16 probes: Preliminary data. Geokhimiya 12, 1811–1820.

Barsukov, V.L., Surkov, Y.A., Dmitriev, L.V., Khodakovsky, I.L., 1986. Geochemical study of Venus by landers of Vega-1 and Vega-2 probes. Geokhimiya 14, 275–288.

Barth, M.G., Foley, S.F., Horn, I., 2002. Partial melting in Archean subduction zones: constraints from experimentally determined trace element partition coefficients between eclogitic minerals and tonalitic melts under upper mantle conditions. Precambrian Research 113, 323–340.

Barton, J.M., 1984. Timing of ore emplacement and deformation, Murchison and Sutherland greenstone belts, Kaapvaal Craton. In: Foster, R.P. (Ed.), Gold '82, The Geology, Geochemistry and Genesis of Gold Deposits. Spec. Publ. Geol. Soc. Zimbabwe. A.A. Balkema, Rotterdam, pp. 629–644.

Barton, M.D., 1996. Granitic magmatism and metallogeny of southwestern North America. In: Geological Society of America, Special Paper 315, pp. 261–280.

Barton, J.M., Robb, J.R., Anhaeusser, C.R., Van Nierop, D.A., 1983. Geochronologic and Sr-isotopic studies of certain units in the Barberton granite-greenstone terrane, South Africa. In: Geological Society of South Africa, Special Publication 9, pp. 63–72.

Basilevsky, A.T., Head, J.W., 1988. The geology of Venus. Annual Review of Earth and Planetary Sciences 16, 295–317.

Basilevsky, A.T., Head, J.W., 1996. Evidence for rapid and widespread emplacement of volcanic plains on Venus: Stratigraphic studies in the Baltis Vallis region. Geophysical Research Letters 23, 1497.

Basilevsky, A.T., Head, J.W., 1998. The geologic history of Venus: A stratigraphic view. Journal of Geophysical Research 103, 8531–8544.

Basilevsky, A.T., Head, J.W., 2002. Venus: Timing and rates of geological activity. Geology 30, 1015–1018.

Batchelor, M.T., Burne, R.V., Henry, B.I., Jackson, M.J., 2004. A case for biogenic morphogenesis of coniform stromatolites. Physica A 337, 319–326.

Bateman, R., Hagemann, S.G., 2004. Gold mineralisation throughout about 45 Ma of Archean orogenesis: protracted flux of gold in the Golden Mile, Yilgarn Craton, Western Australia. Mineralium Deposita 39, 536–559.

Bateman, R., Costa, S., Swe, T., Lambert, D., 2001. Archaean mafic magmatism in the Kalgoorlie area of the Yilgarn Craton, Western Australia: a geochemical and Nd isotopic study of the petrogenetic and tectonic evolution of a greenstone belt. Precambrian Research 108, 75–112.

Bau, M., 1996. Controls on the fractionation of isovalent trace elements in magmatic and aqueous systems: evidence from Y/Ho, Zr/Hf, and lanthanide tetrad effect. Contributions to Mineralogy and Petrology 123, 323–333.

Bau, M., Alexander, B., 2006. Preservation of primary REE patterns without Ce anomaly during dolomitization of Mid-Paleoproterozoic limestone and the potential reestablishment of marine anoxia immediately after the "Great Oxidation Event". South African Journal of Geology 109, 81–86.

Bau, M., Moeller, P., 1993. Rare earth element systematics of the chemically precipitated component in Early Precambrian iron formations and the evolution of the terrestrial atmosphere–hydrosphere–lithosphere system. Geochimica et Cosmochimica Acta 57, 2239–2249.

Baublys, K.A., Golding, S.D., Young, E., Kamber, B.S., 2004. Simultaneous determination of $\delta^{33}S_{V\text{-}CDT}$ and $\delta^{34}S_{V\text{-}CDT}$ using masses 48, 49 and 50 in a continuous flow isotope ratio mass spectrometer. Rapid Communications in Mass Spectrometry 18, 2765–2769.

Bauer, R.L., 1980. Multiphase deformation in the Granite Falls – Montevideo area, Minnesota River Valley. In: Geological Society of America, Special Paper 182, pp. 1–18.

Baxter, J.L., Wilde, S.A., Pidgeon, R.T., Fletcher, I.R., 1984. The Jack Hills Metasedimentary Belt: an extension of the early Archaean terrain in the Yilgarn Block, Western Australia. In: Seventh Australian Geological Convention – Geoscience in the Development of Natural Resources. Geological Society of Australia, Abstracts, vol. 12, pp. 56–57.

Baxter, J.L., Wilde, S.A., Pidgeon, R.T., Collins, L.B., 1986. A video presentation on the geological setting of ancient detrital zircons within the Jack Hills Metamorphic Belt, W.A. In: Eighth Australian Geological Convention – Earth Resources in Time and Space. Geological Society of Australia, Abstracts, vol. 15, p. 220.

Beakhouse, G.P., 1991. Winnipeg River subprovince. In: Thurston, P.C., Williams, H.R., Sutcliffe, R.H., Stott, G.M. (Eds.), Geology of Ontario. Ontario Geological Survey, Special Volume 4, Part 1, pp. 279–301.

Beakhouse, G.P., McNutt, R.H., 1991. Contrasting types of late Archean plutonic rocks in northwestern Ontario; implications for crustal evolution in the Superior Province. Precambrian Research 49, 141–165.

Beakhouse, G.P., McNutt, R.H., Krogh, T.E., 1988. Comparative Rb-Sr and U-Pb geochronology of late to post tectonic plutons in the Winnipeg River belt, northwestern Ontario, Canada. Chemical Geology 72, 283–291.

Beard, B.L., Johnson, C.M., Skulan, J.L., Nealson, K.H., Cox, L., Sun, H., 2003. Application of Fe isotopes to tracing the geochemical and biological cycling of Fe. Chemical Geology 196, 43–56.

Beard, J.S., Lofgren, G.E., 1991. Dehydration melting and water-saturated melting of basaltic and andesitic greenstones and amphibolites at 1, 3, and 6.9 kb. Journal of Petrology 32, 365–401.

Beaudoin, G., Taylor, B.E., Rumble, D., Thiemens, M., 1994. Variations in the sulfur isotope composition of troilite from the Cañon-Diable iron meteorite. Geochimica et Cosmochimica Acta 58, 4253–4255.

Beckinsale, R.D., Drury, S.A., Holt, R.W., 1980. 3360 Myr gneisses from the South Indian craton. Nature 283, 469–470.

Bédard, J.H., 2003. Evidence for regional-scale, pluton-driven, high-grade metamorphism in the Archaean Minto Block, northern Superior Province, Canada. Journal of Geology 111, 183–205.

Bédard, J., 2005. Partitioning coefficients between olivine and silicate melts. Lithos 83, 394–419.

Bédard, J., 2006. A catalytic delamination-driven model for coupled genesis of Archaean crust and sub-continental lithospheric mantle. Geochimica et Cosmochimica Acta 70, 1188–1214.

Bédard, J.H., Brouillette, P., Madore, L., Berclaz, A., 2003. Archaean cratonization and deformation in the northern Superior Province, Canada: an evaluation of plate tectonic versus vertical tectonic models. Precambrian Research 127, 61–87.

Bédard, L.P., 1996. Archaean high-Mg monzodiorite plutonic suite: a reevaluation of the parental magma and differentiation. Journal of Geology 104, 713–728.

Bédard, L.P., Ludden, J.N., 1997. Nd-isotope evolution of Archaean plutonic rocks in southeastern Superior province. Canadian Journal of Earth Sciences 34 (3), 286–298.

Beichman, C.A., Bryden, G., Gautier, T.N., Stapelfeldt, K.R., Werner, M.W., Misselt, K., Rieke, G., Stansberry, J., Trilling, D., 2005. An excess due to small grains around the nearby K0V star HD 69830: Asteroid or cometary debris? Astrophysical Journal 626, 1061–1069.

Beinert, H., 2000. A tribute to sulfur. European Journal of Biochemistry 267, 5657–5664.

Bekker, A., Holland, H.D., Wang, P.L., Rumble, D., Stein, H.J., Hannah, J.L., Coetzee, L.L., Beukes, N.J., 2004. Dating the rise of atmospheric oxygen. Nature 427, 117–120.

Belcher, R.W., Kisters, A.F.M., 2006a. Syntectonic emplacement and deformation of the Heerenveen batholith: Conjectures on the structural setting of the 3.1 Ga granite magmatism in the Barberton granite-greenstone terrain, South Africa. In: Reimold, W.U., Gibson, R.L. (Eds.), Processes on the Early Earth. Geological Society of America, Special Publication 405, pp. 211–231.

Belcher, R.W., Kisters, A.F.M., 2006b. Progressive adjustments of ascent and emplacement controls during incremental construction of the 3.1 Ga Heerenveen batholith, South Africa. Journal of Structural Geology 28, 1406–1421.

Belcher, R.W., Kisters, A.F.M., Poujol, M., Stevens, G., 2005. Structural emplacement of the 3.2 Ga Nelshoogte pluton: implications for the origin of dome-and-keel structures in the Barberton granite-greenstone terrain. In: Geocongress, Durban.

Bell, D.R., Gregoire, M., Grove, T.L., Chatterjee, N., Carlson, R.W., Buseck, P.R., 2005. Silica and volatile-element metasomatism of Archean mantle: a xenolith-scale example from the Kaapvaal Craton. Contributions to Mineralogy and Petrology 150, 251–267.

Belousova, E.A., Griffin, W.L., O'Reilly, S.Y., Fisher, N.I., 2002. Igneous zircon: trace element composition as an indicator of source rock type. Contributions to Mineralogy and Petrology 143, 602–622.

Benedix, G.K., McCoy, T.J., Keil, K., Bogard, D.D., Garrison, D.H., 1998. A petrologic and isotopic study of winonaites: evidence for early partial melting, brecciation and metamorphism. Geochimica et Cosmochimica Acta 62, 2535–2553.

Benedix, G.K., McCoy, T.J., Keil, K., Love, S.G., 2000. A petrologic study of the IAB iron meteorites: constraints on the formation of the IAB-winonaite parent body. Meteoritics and Planetary Science 35, 1127–1141.

Benn, K., Moyen, J.-F., in press. The Late Archean Abitibi–Opatica terrane, Superior Province: a modified oceanic plateau. In: Condie, K.C. (Ed.), When Did Plate Tectonics Begin on Planet Earth? Penrose Special Volume. Geological Society of America, Boulder, CO.

Benn, K., Sawyer, E.W., Bouchez, J.L., 1992. Orogen parallel and transverse shearing in the Opatica belt, Quebec: implications for the structure of the Abitibi belt. Canadian Journal of Earth Sciences 29, 2429–2444.

Bennett, V.C., Nutman, A.P., McCulloch, M.T., 1993. Nd isotopic evidence for transient, highly depleted mantle reservoirs in the early history of the Earth. Earth and Planetary Science Letters 119, 299–317.

Benz, W., Slatterly, W.L., Cameron, A.G.W., 1988. Collisional stripping of Mercury's mantle. Icarus 74, 516–528.

Berclaz, A., Godin, L., David, J., Maurice, C., Parent, M., Francis, D., Stevenson, R., Leclair, A., 2003. Geology of the Nuvvuagittuq Belt (3.8 Ga), northeastern Superior Province: towards a multidisciplinary approach [abstract]. In: Québec Exploration 2003, Ministère des Ressources naturelles, Québec, DV 2003-10, 50 pp.

Berclaz, A., Leclair, A., David, J., Maurice, C., 2004. Structural evolution of the Northeastern Superior Province (NESP) over 1 billion years, with emphasis on greenstone belts and their relation with enclosing granitoids [abstract]. Eos, Transactions, CGU/AGU/SEG/EEGS Joint Assembly, Abstract no. V31A-03.

Beresford, S.W., Cas, R.A.F., Lambert, D.D., Stone, W.E., 2000. Vesicles in thick komatiite lava flows, Kambalda, Western Australia. Journal of the Geological Society 157, 11–14.

Beresford, S.W., Cas, R.A.F., Lahaye, Y., Jane, M., 2002. Facies architecture of an Archean komatiite-hosted Ni-sulphide ore deposit, Victor, Kambalda, Western Australia: Implications for komatiite lava emplacement. Journal of Volcanology and Geothermal Research 118, 57–75.

Berger, M., Rollinson, H., 1997. Isotopic and geochemical evidence for crust-mantle interaction during late Archaean crustal growth. Geochimica et Cosmochimica Acta 61 (22), 4809–4829.

Berner, E.K., Berner, R.A., 1996. Global Environment. Water, Air and Geochemical Cycles. Prentice Hall, Upper Saddle River, NJ, 376 pp.

Bernstein, S., Kelemen, P.B., Brooks, C.K., 1998. Depleted spinel harzburgite xenoliths in Tertiary dikes from East Greenland: restites from degree melting. Earth and Planet Science Letters 154, 221–235.

Bernstein, S., Hanghoi, K., Kelemen, P.B., Brooks, C.K., 2006. Ultra-depleted shallow cratonic mantle beneath West Greenland: dunitic xenoliths from Ubekendt Ejland. Contributions to Mineralogy and Petrology 152, 335–347.

Beyer, E.E., Brueckner, H.K., Griffin, W.L., O'Reilly, S.Y., Graham, S., 2004. Archean mantle fragments in Proterozoic crust, Western Gneiss Region, Norway. Geology 32, 609–612.

Beyer, E.E., Griffin, W.L., O'Reilly, S.Y., 2006. Transformation of Archean lithospheric mantle by refertilisation: evidence from exposed peridotites in the Western Gneiss Region, Norway. Journal of Petrology 47, 1611–1636.

Beyssac, O., Rouzaud, J.-N., Goffe, B., Brunet, F., Chopin, C., 2002. Graphitization in a high-pressure, low-temperature metamorphic gradient: A Raman micro-spectroscopy and HRTEM study. Contributions to Mineralogy and Petrology 143, 19–31.

Bhattacharya, S., Kar, R., Misra, S., Teixeira, W., 2001. Early Archean continental crust in the Eastern Ghats granulite belt, India: isotopic evidence from a charnockite suite. Geological Magazine 138, 609–618.

Bibikova, E.V., Drugova, G.M., Kirnozova, T.I., Gracheva, T.V., Makarov, V.A., 1984. Age of volcanism of the Olondo greenstone belt (Eastern Siberia). Doklady Earth Sciences 279 (6), 1424–1428.

Bibikova, E.V., Belov, A.N., Rosen, O.M., 1988. Metamorphic rocks isotope dating in the Anabar shield. In: Markov, M.S. (Ed.), Archean of the Anabar Shield and Problems of the Early Earth's Evolution. Nauka, Moscow, pp. 122–133 (in Russian).

Bibikova, E.V., Morozova, I.M., Gacheva, T.V., Makarov, V.A., 1989. U-Pb age of granulites from the Kurulta complex. In: Rudnik, V.A. (Ed.), The Oldest Rocks of the Aldan–Stanovik Shield, Eastern Siberia, USSR. Excursion Guide, IGCP 280, Lenningrad–Mainz, pp. 89–91.

Bibikova, E.V., Levitsky, V.I., Reznitsky, L.Z., Kirnozova, T.I., Gracheva, T.V., Makarov, V.A., 2001. Archean tonalite-trondhjemite assemblage in the Cis-Sayan salient of the Siberian platform basement: U-Pb, Sm-Nd and Sr isotope data. In: Letnikov, F.A. (Ed.), Geology, Geochemistry and Geophysics. Proceedings of the Russian Basic Research Foundation Conference. Institute of the Earth's Crust, Irkutsk, pp. 175–176 (in Russian).

Bibikova, E.V., Petrova, A., Claesson, S., 2005. The temporal evolution of sanukitoids in the Karelian Craton, Baltic Shield: an ion microprobe U-Th-Pb isotopic study of zircons. Lithos 79, 129–145.

Bibikova, E.V., Turkina, O.M., Kirnozova, T.I., Fugzan, M.M., 2006. Ancient plagiogneisses of the Onot block of the Sharyzhalgay metamorphic massif: isotopic geochronology. Geochemistry International 44 (3), 310–316.

Bickford, M.E., Wooden, J.L., Bauer, R.L., 2004. New SHRIMP U-Pb zircon ages for the Paleoarchean to Mesoarchean rocks of the Minnesota River Valley. In: Geological Society of America Program with Abstracts, Colorado, p. 458.

Bickford, M.E., Wooden, J.L., Bauer, R.L., 2006. SHRIMP study of zircons from Early Archean rocks in the Minnesota River Valley: Implications for the tectonic history of the Superior Province. Bulletin of the Geological Society of America 118, 94–108.

Bickle, M.J., 1978. Heat loss from the Earth: A constraint on Archaean tectonics from the relation between geothermal gradients and the rate of plate production. Earth and Planetary Science Letters 40, 301–315.

Bickle, M.J., 1986. Implications of melting for stabilisation of the lithosphere and heat loss in the Archaean. Earth and Planetary Science Letters 80, 314–324.

Bickle, M.J., 1994. The role of metamorphic decarbonation reactions in returning strontium to the silicate sediment mass. Nature 367, 699–704.

Bickle, M.J., Bettenay, L.F., Boulter, C.A., Groves, D.I., Morant, P., 1980. Horizontal tectonic intercalation of an Archaean gneiss belt and greenstones, Pilbara Block, Western Australia. Geology 8, 525–529.

Bickle, M.J., Bettenay, L.F., Barley, M.E., Chapman, H.J., Groves, D.I., Campbell, I.H., de Laeter, J.R., 1983. A 3500 Ma plutonic and volcanic calc-alkaline province in the Archaean East Pilbara Block. Contributions to Mineralogy and Petrology 84, 25–35.

Bickle, M.J., Morant, P., Bettenay, L.F., Boulter, C.A., Blake, T.S., Groves, D.I., 1985. Archaean tectonics of the Shaw Batholith, Pilbara Block, Western Australia: structural and metamorphic tests of the batholith concept. In: Ayers, L.D., Thurston, P.C., Card, K.D.,

Weber, W. (Eds.), Evolution of Archean Supracrustal Sequences. Geological Association of Canada, Special Paper 28, pp. 325–341.

Bickle, M.J., Bettenay, L.F., Chapman, H.J., Groves, D.I., McNaughton, N.J., Campbell, I.H., de Laeter, J.R., 1989. The age and origin of younger granitic plutons in the Archean of the Pilbara Block, Western Australia. Contributions to Mineralogy and Petrology 101, 361–376.

Bickle, M.J., Bettenay, L.F., Chapman, H.J., Groves, D.I., McNaughton, N.J., Campbell, I.H., de Laeter, J.R., 1993. Origin of the 3500–3300 Ma calc-alkaline rocks in the Pilbara Archaean: isotopic and geochemical constraints from the Shaw Batholith. Precambrian Research 60, 117–149.

Bigeleisen, J., Mayer, M.G., 1947. Calculation of equilibrium constants for isotope exchange reactions. Journal of Chemical Physics 15, 261–267.

Bigeleisen, J., Wolfsberg, M., 1958. Theoretical and experimental aspects of isotope effects in chemical kinetics. Advances in Chemical Physics 1, 15–76.

Bild, R.W., 1977. Silicate inclusions in group IAB irons and a relation to the anomalous stones Winona and Mt. Morris (Wis.). Geochimica et Cosmochimica Acta 41, 1439–1456.

Bindschadler, D.L., 1995. Magellan – a new view of Venus geology and geophysics. Reviews in Geophysics 33, 459–467.

Bindschadler, D.L., Parmentier, E.M., 1990. Mantle flow tectonics: the influence of a ductile lower crust and implications for the formation of topographic uplands on Venus. Journal of Geophysical Research 95, 21,329–21,344.

Bindschadler, D.L., deCharon, A., Beratan, K.K., Head, J.W., 1992a. Magellan observations of Alpha Regio: Implications for formation of complex ridged terrains on Venus. Journal of Geophysical Research 97, 13,563–13,577.

Bindschadler, D.L., Schubert, G., Kaula, W.M., 1992b. Coldspots and hotspots: Global tectonics and mantle dynamic of Venus. Journal of Geophysical Research 97, 13,495–13,532.

Birck, J.-L., Allegre, C.J., 1988. Manganese-chromium isotope systematics and the development of the early solar system. Nature 331, 579–584.

Bird, P., 1978. Initiation of intracontinental subduction in the Himalaya. Journal of Geophysical Research 83, 4975–4987.

Bischoff, A., Geiger, T., Palme, H., Spettel, B., Schultz, L., Scherrer, P., Bland, P., Clayton, R.N., Mayeda, T.K., Herpers, U., Michel, R., Dittrich-Hannen, B., 1994. Acfer 217 – a new member of the Rumuruti chondrite group (R). Meteoritics 29, 264–274.

Bischoff, A., 2000. Mineralogical characterization of primitive type-3 lithologies in Rumuruti chondrites. Meteoritics and Planetary Science 35, 699–706.

Bishop, J.W., Sumner, D.Y., Huerta, N.J., 2006. Molar tooth structures of the Neoarchean Monteville Formation, Transvaal Group, South Africa. II: A wave-induced fluid model. Sedimentology 53, 1069–1082.

Bizzarro, E.A., Baker, J.A., Haack, H., Ulfbeck, D., Rosing, M., 2003. Early history of the Earth's crust-mantle system inferred from hafnium isotopes in chondrites. Nature 421, 931–933.

Bizzarro, M., Baker, J.A., Haack, H., 2004. Mg isotope evidence for contemporaneous formation of chondrules and refractory inclusions. Nature 431, 275–278.

Bjerrum, C.J., Canfield, D.E., 2002. Ocean productivity before about 1.9 Gyr ago limited by phosphorus adsorption onto iron oxides. Nature 417, 159–162.

Black, L.P., James, P.R., 1983. Geological history of the Archaean Napier Complex of Enderby Land. In: Oliver, R.L., James, P.R., Jago, J.B. (Eds.), Antarctic Geoscience. Australian Academy of Sciences, Canberra, pp. 11–15.

Black, L.P., McCulloch, M.T., 1987. Evidence for isotopic re-equilibration of Sm-Nd whole-rock systems in early Archaean crust of Enderby Land, Antarctica. Earth and Planetary Science Letters 82, 15–24.

Black, L.P., Gale, N.H., Moorbath, S., Pankhurst, R.J., McGregor, V.R., 1971. Isotopic dating of very early Precambrian amphibolite facies gneisses from the Godthaab district, West Greenland. Earth and Planetary Science Letters 12, 245–259.

Black, L.P., James, P.R., Harley, S.L., 1983a. The geochronology, structure and metamorphism of early Archaean rocks at Fyfe Hills, Enderby Land, Antarctica. Precambrian Research 21, 197–222.

Black, L.P., James, P.R., Harley, S.L., 1983b. Geochronology and geological evolution of metamorphic rocks in the Field Islands area, East Antarctica. Journal of Metamorphic Geology 1, 277–303.

Black, L.P., Williams, I.S., Compston, W., 1986. Four zircon ages from one rock: the history of a 3930 Ma old granulite from Mount Sones, Enderby Land, Antarctica. Contributions to Mineralogy and Petrology 94, 427–437.

Black, L.P., Harley, S.L., Sun, S.S., McCulloch, M.T., 1987. The Rayner Complex of East Antarctica: complex isotopic systematics within a Proterozoic mobile belt. Journal of Metamorphic Geology 5, 1–26.

Black, L.P., Kinny, P.D., Sheraton, J.W., Delor, C.P., 1991. Rapid production and evolution of late Archaean felsic crust in the Vestfold Block of East Antarctica. Precambrian Research 50, 283–310.

Black, L.P., Sheraton, J.W., Tingey, R.J., McCulloch, M.T., 1992. New U-Pb zircon ages from the Denman Glacier area, East Antarctica, and their significance for Gondwana reconstruction. Antarctic Science 4, 447–460.

Blackburn, C.E., John, G.W., Ayer, J., Davis, D.W., 1991. Wabigoon Subprovince. In: Thurston, P.C., Williams, H.R., Sutcliffe, R.H., Stott, G.M. (Eds.), Geology of Ontario. Ontario Geological Survey, Special Volume 4, Part 1, pp. 303–381.

Blake, T.S., Buick, R., Brown, S.J.A., Barley, M.E., 2004. Geochronology of a Late Archaean flood basalt province in the Pilbara Craton, Australia: constraints on basin evolution, volcanic and sedimentary accumulations, and continental drift rates. Precambrian Research 133, 143–173.

Bland, P.A., 2005. The impact rate on Earth. Philosophical Transactions of the Royal Society A – Mathematical, Physical and Engineering Sciences 363, 2793–2810.

Bleeker, W., 1990. New structural-metamorphic constraints on Early Proterozoic oblique collision along the Thompson Nickel Belt, Manitoba, Canada. In: Lewry, J.F., Stauffer,

M.R. (Eds.), The Early Proterozoic Trans-Hudson Orogen of North America. Geological Association of Canada, Special Paper 37, pp. 57–73.

Bleeker, W., 2003. The late Archaean record: a puzzle in ca. 35 pieces. Lithos 71, 99–134.

Bleeker, W., 2004. Towards a 'natural' timescale for the Precambrian – A proposal. Lethaia 37, 219–222.

Bleeker, W., Davis, W., 1999. The 1991–1996 NATMAP Slave Province Project: Introduction. Canadian Journal of Earth Sciences 36, 1033–1042.

Bleeker, W., Stern, R., 1997. The Acasta gneisses: an imperfect sample of Earth's oldest crust. In: Cook, F., Erdmer, P. (Eds.), Slave-Northern Cordillera Lithospheric Evolution (SNORCLE) Transect and Cordilleran Tectonics Workshop Meeting. Lithosphere Report 56, pp. 32–35.

Bleeker, W., Davis, W.J., Villeneuve, M.E., 1997. The Slave Province: evidence for contrasting crustal domains and complex, multistage tectonic evolution. In: Cook, F., Erdmer, P. (Eds.), Slave-Northern Cordillera Lithospheric Evolution (SNORCLE) Transect and Cordilleran Tectonics Workshop Meeting. Lithosphere Report 56, pp. 36–37.

Bleeker, W., Ketchum, J.W.F., Jackson, V.A., Villeneuve, M.E., 1999. The Central Slave Basement Complex, Part I. Its structural topology and autochthonous cover. Canadian Journal of Earth Sciences 36, 1083–1109.

Blewett, R., 2002. Archaean tectonic processes: a case for horizontal shortening in the North Pilbara granite-greenstone terrane, Western Australia. Precambrian Research 113, 87–120.

Blewett, R.S., Champion, D.C., 2005. Geology of the Wodgina 1:100,000 sheet. Geological Survey of Western Australia, 1:100,000 Geological Series Explanatory Notes.

Blichert-Toft, J., Albarède, F., 1997. The Lu-Hf isotope geochemistry of chondrites and the evolution of the crust-mantle system. Earth and Planetary Science Letters 148, 243–258.

Blichert-Toft, J., Albarede, F., Rosing, M., Frei, R., Bridgwater, D., 1999. The Nd and Hf isotope evolution of the mantle through the Archean; results from the Isua supracrustals, West Greenland, and from the Birimian terranes of West Africa. Geochimica et Cosmochimica Acta 63, 3901–3914.

Bliss, N.W., Stidolph, P.A., 1969. A review of the Rhodesian basement complex. In: Transactions of the Geological Society of South Africa, Special Publication 2, pp. 305–333.

Bocklée-Morvan, D., Crovisier, J., Mumma, M.J., Weaver, H.A., 2004. The composition of cometary volatiles. In: Festou, M., Keller, H.U., Weaver, H. (Eds.), Comets II. University of Arizona Press, Tucson, AZ, pp. 391–423.

Bodinier, J.L., Burg, J.-P., Leyreloup, A., Vidal, H., 1988. Reliques d'un bassin d'arriere arc subducté puis obducté dans la région de Marvejols (Massif Central). Bulletin de la Société Géologique de France 8 (4), 20–34.

Bodorkos, S., Sandiford, M., 2006. Thermal and mechanical controls on the evolution of Archean crustal deformation: examples from Western Australia. In: Benn, K., Mareschal, J.-M., Condie, K. (Eds.), Archean Geodynamics and Environments. Geophysical Monograph Series, vol. 164. American Geophysical Union, pp. 131–147.

Boerner, D.E., Craven, J.A., Kurtz, R.D., Ross, G.M., Jones, F.W., 1998. The Great Falls Tectonic Zone: suture or intracontinental shear zone? Canadian Journal of Earth Sciences 35, 175–183.

Bogard, D.D., 1995. Impact ages of meteorites: a synthesis. Meteoritics 30, 244–268.

Bogard, D.D., Garrison, D.H., 1994. ^{39}Ar-^{40}Ar ages of four ureilites. Lunar and Planetary Science XXV, 137–138.

Bogard, D.D., Garrison, D.H., 1998. ^{39}Ar-^{40}Ar ages and thermal history of mesosiderites. Geochimica et Cosmochimica Acta 62, 1459–1468.

Bogard, D.D., Burnett, D.S., Eberhardt, P., Wasserburg, G.J., 1967. ^{87}Rb-^{87}Sr and ^{40}K-^{40}Ar ages of the Norton County achondrite. Earth and Planetary Science Letters 3, 179–189.

Bogard, D.D., Garrison, D.H., Jordan, J.L., Mittlefehldt, D.W., 1990. ^{39}Ar-^{40}Ar dating of mesosiderites – evidence for major parent body disruption less than 4 Ga ago. Geochimica et Cosmochimica Acta 54, 2549–2564.

Boger, S.D., Miller, J.M., 2004. Terminal suturing of Gondwana and the onset of the Ross-Delamerian Orogeny: the cause and effect of an Early Cambrian reconfiguration of plate motions. Earth and Planet Science Letters 219, 35–48.

Boger, S.D., Wilson, C.J.L., Fanning, C.M., 2001. Early Palaeozoic tectonism within the east Antarctic craton: the final suture between east and west Gondwana? Geology 29, 463–466.

Boger, S.D., Wilson, C.J.L., Fanning, C.M., 2006. An Archaean province in the southern Prince Charles Mountains, East Antarctica: U-Pb zircon evidence for c. 3170 Ma granite plutonism and c. 2780 Ma partial melting and orogenesis. Precambrian Research 145, 207–228.

Böhm, C.O., 1997a. Geology of the Assean Lake area (parts of NTS 64A 1, 2, 3, 4). In: Report of Activities 1997, Manitoba Energy and Mines, Minerals Division, pp. 47–49.

Böhm, C.O., 1997b. Geology of the Assean Lake area. In: Manitoba Energy and Mines, Minerals Division, Preliminary Map 1997 S-3, scale 1:50,000.

Böhm, C.O., 1998. Geology of the Natawahunan Lake area (part of NTS 63P/11). In: Report of Activities 1998, Manitoba Energy and Mines, Geological Services, pp. 56–59.

Böhm, C.O., Heaman, L.M., Corkery, M.T., 1999a. Archean crustal evolution of the northwestern Superior craton margin: U-Pb zircon results from the Split Lake Block. Canadian Journal of Earth Sciences 36, 1973–1987.

Böhm, C.O., Corkery, M.T., 1999b. Geology of the Waskaiowaka Lake area (part of 64A7,8,9,10). In: Manitoba Industry, Trade and Mines, Geological Services, Preliminary Map 1999T-1, scale 1:50,000.

Böhm, C.O., Heaman, L.M., Creaser, R.A., Corkery, M.T., 2000a. Discovery of pre-3.5 Ga exotic crust at the northwestern Suprerior Province margin, Manitoba. Geology 28, 75–78.

Böhm, C.O., Heaman, L.M., Creaser, R.A., Corkery, M.T., 2000b. The northwest Superior Province margin in Manitoba: a single Archean-Proterozoic boundary or heterogeneous transition zone? In: Harrap, R.M., Helmstaedt, H. (Eds.), 2000 Western Superior Transect Sixth Annual Workshop. Lithoprobe Report 77, pp. 13–18.

Böhm, C.O., Heaman, L.M., Stern, R.A., Corkery, M.T., Creaser, R.A., 2003. Nature of Assean Lake ancient crust, Manitoba: a combined SHRIMP-ID-TIMS U-Pb geochronology and Sm-Nd isotope study. Precambrian Research 126, 55–94.

Boily, M., Dion, C., 2002. Geochemistry of boninite-type volcanic rocks in the Frotet-Evans greenstone belt, Opatica Subprovince, Quebec; implications for the evolution of Archaean greenstone belts. Precambrian Research 115, 349–371.

Boily, M., Leclair, A., Maurice, C., Berclaz, A., David, J., 2004. Étude lithogéochimique et isotopique du Nd des assemblages volcaniques et plutoniques de la région sud du Grand-Nord québecois. Ministère des Ressources naturelles, Faune et Parcs du Québec, RP 2004-01, 28 pp.

Boily, M., Leclair, A., Berclaz, A., Labbé, J.-Y., Lacoste, P., Simard, M., Maurice, C., 2006a. Terrane definition in the Northeastern Superior Province. Abstract in Joint Annual Meeting; Geological Association of Canada, Mineralogical Association of Canada and Society of Economic Geologists, Montreal, QC, Canada, vol. 31.

Boily, M., Leclair, A., Maurice, C., Berclaz, A., David, J., 2006b. Étude géochimique et isotopique du Nd des assemblages volcaniques et plutoniques du nord-est de la Province du Supieur (NEPS). Ministère des Ressources naturelles, Québec, Ministère des Ressources naturelles, Québec, GM 62031, 50 pp.

Bolhar, R., Woodhead, J.D., Hergt, J.M., 2002. Comment on: "Growth and recycling of early Archaean continental crust: geochemical evidence from the Coonterunah and Warrawoona Groups, Pilbara Craton, Australia" by Green, M.G. et al. (Tectonophysics 322, 69–88). Tectonophysics 344, 289–292.

Bolhar, R., Kamber, B.S., Moorbath, S., Fedo, C.M., Whitehouse, M.J., 2004. Characterisation of early Archaean chemical sediments by trace element signatures. Earth and Planetary Science Letters 222, 43–60.

Bolhar, R., Kamber, B.S., Moorbath, S., Whitehouse, M.J., Collerson, K.D., 2005. Chemical characterization of Earth's most ancient clastic metasediments from the Isua Greenstone Belt, southern West Greenland. Geochimica et Cosmochimica Acta 69 (6), 1555–1573.

Bonin, B., Ethien, R., Gerbe, M.C., Cottin, J.Y., Feraud, G., Gagnevin, D., Giret, A., Michon, G., Moine, B., 2004. The Neogen to recent Rallier-duBaty nested ring complex, Kerguelen Archipelago (TAAF, Indian Ocean): stratigraphy revisited, implications for cauldron subsidence mechanisms. In: Geological Society of London, Special Publication 234, pp. 125–149.

Bosch, D., Bruguier, O., Pidgeon, R.T., 1996. Evolution of an Archean metamorphic belt: a conventional and SHRIMP U–Pb study of accessory minerals from the Jimperding metamorphic belt, Yilgarn Craton, West Australia. Journal of Geology 104, 695–711.

Boss, A.P., 1990. Solar nebula models: implications for Earth origin. In: Newsome, H.E., Jones, J.H. (Eds.), Origin of the Earth. Oxford University Press, New York, pp. 3–15.

Boss, A.P., 1997. Giant planet formation by gravitational instability. Science 276, 1836–1839.

Boss, A.P., 2003. Rapid formation of outer giant planets by disk instability. Astrophysical Journal 599, 577–581.

Böttcher, M.E., 2001. Sulfur isotope fractionation in the biogeochemical sulfur cycle of marine sediments. Isotopes in Environmental and Health Studies 37, 97–99.

Böttcher, M.E., Smock, A.M., Cypionka, H., 1998. Sulfur isotope fractionation during experimental precipitation of iron (II) and manganese (II) sulfide at room temperature. Chemical Geology 146, 127–134.

Botz, R., Pokojski, H., Schmitt, M., Thomm, M., 1996. Carbon isotope fractionation during bacterial methanogenesis by CO_2 reduction. Organic Geochemistry 25, 255–262.

Boulter, C.A., Bickle, M.J., Gibson, B., Wright, R.K., 1987. Horizontal tectonics pre-dating upper Gorge Creek Group sedimentation, Pilbara Block, Western Australia. Precambrian Research 36, 241–258.

Bourassa, I., 2002. Geology, geochemistry, geochronology and metallogeny of the Cu-Zn-Au-Ag volcanogenic showings of the Archean Duquet Belt, Superior Province, Northern Québec. Thesis. Université du Québec à Montréal, 78 pp.

Bowerman, M.S., Böhm, C.O., Hartlaub, R.P., Heaman, L.M., Creaser, R.A., 2004. Pre-liminary geochemical and isotopic results from the Gull Rapids area of the eastern Split Lake Block, northwestern Superior Province, Manitoba (parts of NTS 54D5 and 6). In: Report of Activities 2004, Manitoba Industry, Economic Development and Mines, Manitoba Geological Survey, pp. 156–170.

Bowring, S.A., Housh, T.B., 1995. The Earth's early evolution. Science 269, 1535–1540.

Bowring, S.A., Van Schmus, W.R., 1984. U-Pb zircon constraints on evolution of Wop-may Orogen, N.W.T. Geological Association of Canada / Mineralogical Association of Canada, Abstract 9, 47 pp.

Bowring, S.A., Williams, I.S., 1999. Priscoan (4.00–4.03 Ga) orthogneisses from NW Canada. Contributions to Mineralogy and Petrology 134, 3–16.

Bowring, S.A., Williams, I.S., Compston, W., 1989a. 3.96 Ga gneisses from the Slave Province, NWT, Canada. Geology 17, 971–975.

Bowring, S.A., King, J.E., Housh, T.B., Isachsen, C.E., Podsek, F.A., 1989b. Neodymium and lead isotope evidence for enriched early Archean crust in North America. Nature 340, 222–225.

Bowring, S.A., Housh, T.B., Isachsen, C.E., 1990. The Acasta Gneisses: remnant of Earth's early crust. In: Newsom, H.E., Jones, J.H. (Eds.), Origin of the Earth. Oxford University Press, New York, pp. 319–343.

Boyd, F.R., 1989. Composition and distinction between oceanic and cratonic lithosphere. Earth and Planetary Science Letters 96, 15–26.

Boyd, F.R., Pokhilenko, N.P., Pearson, D.G., Mertzman, S.A., Sobolev, N.V., Finger, L.W., 1997. Composition of the Siberian cratonic mantle: evidence from Udachnaya peridotite xenoliths. Contributions to Mineralogy and Petrology 128, 228–246.

Boyet, M., Carlson, R.W., 2005. ^{142}Nd isotope evidence for early (>4.53 Ga) global dif-ferentiation of the silicate Earth. Science 309, 576–581.

Boynton, W.V., 1984. Geochemistry of the rare earth elements: meteorite studies. In: Hen-derson, P. (Ed.), Rare Earth Element Geochemistry. Elsevier, Amsterdam, pp. 63–114.

Brandl, G., de Wit, M.J., 1997. The Kaapvaal Craton. In: de Wit, M.J., Ashwal, L.D. (Eds.), Greenstone Belts. Oxford Science Publications, Oxford, pp. 581–607.

Brandl, G., Kröner, A., 1993. Preliminary results of single zircon studies from various Archaean rocks of the Northeastern Transvaal. In: Geological Survey and Mines, 16th Colloquium of African Geology, Mbabane, Swaziland, pp. 54–56.

Brasier, M.D., Green, O.R., Jephcoat, A.P., Kleppe, A.K., Van Kranendonk, M.J., Lindsay, J.F., Steele, A., Grassineau, N., 2002. Questioning the evidence for Earth's oldest fossils. Nature 416, 76–81.

Brasier, M.D., Green, O.R., Lindsay, J.F., McLoughlin, N., Steele, A., Stokes, C., 2005. Critical testing of Earth's oldest putative fossil assemblage from the ~3.5 Ga Apex chert, Chinaman Creek, Western Australia. Precambrian Research 140, 55–102.

Brasier, M.D., McLoughlin, N., Green, O., Wacey, D., 2006. A fresh look at the fossil evidence for early Archaean cellular life. Philosophical Transactions of the Royal Society of London Ser. B 361, 887–902.

Braterman, P.S., Cairns-Smith, A.G., Sloper, R.W., 1983. Photooxidation of hydrated Fe^{2+} – significance for banded iron formations. Nature 303, 163–164.

Brauhart, C., 1999. Regional alteration systems associated with Archean volcanogenic massive sulphide deposits at Panorama, Pilbara, Western Australia. Unpublished Ph.D. thesis. University of Western Australia, 194 pp.

Brauhart, C.W., Morant, P., 2000. Discussion on "Geochemistry and geodynamic setting of volcanic and plutonic rocks associated with early Archean volcanogenic massive sulphide mineralisation, Pilbara Craton" by Vearncombe, S.E., Kerrich, R., 1999. Precambrian Research 98, 243–270. Precambrian Research 104, 95–99.

Brauhart, C.W., Groves, D.I., Morant, P., 1998. Regional alteration systems associated with volcanogenic massive sulfide mineralization at Panorama, Pilbara, Western Australia. Economic Geology 93, 292–302.

Brauhart, C.W., Huston, D.L., Groves, D.I., Mikucki, E.J., Gardoll, S.J., 2001. Geochemical mass transfer patterns as indicators of the architecture of a complete volcanic-hosted massive sulfide hydrothermal alteration system in the Panorama district, Pilbara, Western Australia. Economic Geology 96, 1263–1278.

Brazzle, R.H., Pravdivtseva, O.V., Meshik, A.P., Hohenburg, C.M., 1999. Verification and interpretation of the I-Xe chronometer. Geochimica et Cosmochimica Acta 63, 739–760.

Brearley, A.J., 2003. Nebula versus parent-body processing. In: Davis, A.M. (Ed.), Meteorites, Comets and Planets. Treatise on Geochemistry, vol. 1, Elsevier–Pergamon, Oxford, pp. 247–268.

Brearley, A.J., Hutcheon, I.D., Browning, L., 2001. Compositional zoning and Mn-Cr systematics in carbonates from the Y791198 CM2 carbonaceous chondrite. In: Lunar and Planetary Science XXXII, #1458. Lunar and Planetary Institute, Houston, TX (CD-ROM).

Brearley, A.J., Hutcheon, I.D., 2000. Carbonates in the CM1 chondrite ALH84034: mineral chemistry, zoning and Mn-Cr systematics. In: Lunar and Planetary Science XXXI, #1407. Lunar and Planetary Institute, Houston, TX (CD-ROM).

Brearly, A.J., Jones, R.H., 1998. Chondritic meteorites. In: Papike, J.J. (Ed.), Planetary Materials. Reviews in Mineralogy, vol. 36. Mineralogical Society of America, Washington, DC, pp. 3-1–3-398.

Bridges, N.T., 1995. Submarine analogs to Venusian pancake domes. Geophysical Research Letters 22, 2781.

Bridges, N.T., 1997. Ambient effects on basalt and rhyolite lavas under Venusian, subaerial, and subaqueous conditions. Journal of Geophysical Research 102, 9243.

Bridgwater, D., McGregor, V.R., 1974. Field work on the very early Precambrian rocks of the Isua area, southern West Greenland. Rapport Grønlands Geologiske Undersøgelse 65, 49–54.

Bridgwater, D., Schiotte, L., 1990. The Archean gneiss complex of northern Labrador: a review of current results, ideas and problems. Bulletin of the Geological Society of Denmark 39, 153–166.

Bridgwater, D., McGregor, V.R., Myers, J.S., 1974. A horizontal tectonic regime in the Archaean of Greenland and its implications for early crustal thickening. Precambrian Research 1, 179–197.

Bridgwater, D., Allaart, J.H., Schopf, J.W., Klein, C., Walter, M.R., Barghoorn, E.S., Strother, P., Knoll, A.H., Gorman, B.E., 1981. Microfossil-like objects from the Archaean of Greenland: a cautionary note. Nature 289, 51–53.

Bridgwater, D., Jackson, G., Bostock, H., 1991. The Anabar shield, field conference, Siberia, 1990. Geoscience Canada 18 (1), 28–32.

Brocks, J.J., Love, G.D., Snape, C.E., Logan, G.A., Summons, R.E., Buick, R., 2003. Release of bound aromatic hydrocarbons from late Archean and Mesoproterozoic kerogens via hydropyrolysis. Geochimica et Cosmochimica Acta 67, 1521–1530.

Broecker, W.S., 1970. A boundary condition on the evolution of atmospheric oxygen. Journal of Geophysical Research 75, 3553–3557.

Brown, C.D., Grimm, R.E., 1995. Tectonics of Artemis Chasma: a Venusian "plate" boundary. Icarus 117, 219–249.

Brown, C.D., Grimm, R.E., 1996. Lithospheric rheology and flexure at Artemis Chasma, Venus. Journal of Geophysical Research 101, 12,697–12,708.

Brown, C.D., Grimm, R.E., 1999. Recent tectonic and lithospheric thermal evolution of Venus. Icarus 139, 40–48.

Brown, J.L., 2002. Neoarchean evolution of the western – central Wabigoon boundary zone, Brightsand Forest area, Ontario. Unpublished M.Sc. thesis. University of Ottawa, Ottawa, Ontario.

Brown, M., 2006. Duality of thermal regimes is the distinctive characteristic of plate tectonics since the Neoarchean. Geology 34, 961–964.

Brueckner, H.K., Medaris, L.G., 1998. A tale of two orogens: the contrasting T-P-t history and geochemical evolution of mantle in high- and ultrahigh-pressure metamorphic terranes of the Norwegian Caledonides and the Czech Variscides. Schweizerische Mineralogische und Petrographische Mitteilungen 78, 293–307.

Brueckner, H.K., Medaris, L.G., 2000. A general model for the intrusion and evolution of "mantle" garnet peridotites in high-pressure and ultra-high-pressure metamorphic terranes. Journal of Metamorphic Geology 18, 123–133.

Bruguier, O., Dada, S., Lancelot, J.R., 1994. Early Archean component (>3.5 Ga) within a 3.05-Ga orthogneiss from northern Nigeria, U-Pb zircon evidence. Earth and Planetary Science Letters 125, 89–103.

Bryan, W.B., Thompson, G., Ludden, J.N., 1981. Compositional variation in normal MORB from 22'-25'N: Mid-Atlantic Ridge and Kane Fracture Zone. Journal of Geophysical Research 86, 11,815–11,836.

Bucher, K., Frey, M., 1994. Petrogenesis of Metamorphic Rocks. Springer-Verlag, New York, 318 pp.

Buick, R., 1984. Carbonaceous filaments from North Pole, Western Australia: are they fossil bacteria in Archean stromatolites? Precambrian Research 24, 157–172.

Buick, R., 1988. Carbonaceous filaments from North Pole, Western Australia: are they fossil bacteria in Archean stromatolites? A reply. Precambrian Research 39, 311–317.

Buick, R., 1990. Microfossil recognition in Archean rocks: an appraisal of spheroids and filaments from a 3500 M.Y. old chert-barite unit at North Pole, Western Australia. Palaios 5, 441–491.

Buick, R., Barnes, K.R., 1984. Cherts in the Warrawoona Group: Early Archaean silicified sediments deposited in shallow-water environments. In: Muhling, J.R., Groves, D.I., Blake, T.S. (Eds.), Archaean and Proterozoic Basins of the Pilbara, Western Australia: Evolution and Mineralization Potential. The Geology Department and University Extension, Publication 9. University of Western Australia, pp. 37–53.

Buick, R., Doepel, M.G., 1999. Panorama VMS zinc-copper deposits. In: Ferguson, K.M. (Compiler), Lead, Zinc and Silver Deposits of Western Australia. Western Australia Geological Survey, Mineral Resources Bulletin 15, pp. 80–86.

Buick, R., Dunlop, J., 1990. Evaporitic sediments of early Archaean age from the Warrawoona Group, North Pole, Western Australia. Sedimentology 37, 247–277.

Buick, R., Dunlop, J.S.R., Groves, D.I., 1981. Stromatolite recognition in ancient rocks: An appraisal of irregularly laminated structures in an Early Archaean chert-barite unit from North Pole, Western Australia. Alcheringa 5, 161–181.

Buick, R., Thornett, J.R., McNaughton, N.J., Smith, J.B., Barley, M.E., Savage, M., 1995. Record of emergent continental crust ~3.5 billion years ago in the Pilbara Craton of Australia. Nature 375, 574–577.

Buick, R., Brauhart, C.W., Morant, P., Thornett, J.R., Maniw, J.G., Archibald, N.J., Doepel, M.G., Fletcher, I.R., Pickard, A.L., Smith, J.B., Barley, M.E., McNaughton, N.J., Groves, D.I., 2002. Geochronology and stratigraphic relationships of the Sulphur Springs Group and Strelley Granite: a temporally distinct igneous province in the Archaean Pilbara Craton, Australia. Precambrian Research 114, 87–120.

Bullock, M.A., Grinspoon, G.H., 2001. The recent evolution of climate on Venus. Icarus 150, 19–37.

Bullock, M.A., Grinspoon, D.H., 1996. The stability of climate on Venus. Journal of Geophysical Research 101, 7521–7530.

Bunch, T.E., Keil, K., Olsen, E., 1970. Mineralogy and petrology of silicate inclusions in iron meteorites. Contributions to Mineralogy and Petrology 25, 297–240.

Bureau of Geology and Mineral Resources of Henan Province (BGMRHP), 1989. Regional Geology of Henan Province. People's Republic of China, Ministry of Geology and Mineral Resources, Geological Memoirs, Series 1, vol. 19. Geological Publishing House, Beijing, pp. 1–232 (in Chinese).

Burg, J.-P., Ford, M., 1997. Orogeny through time: an overview. In: Geological Society of London, Special Publication 121, pp. 1–17.

Burg, J.-P., Leyreloup, A., Marchand, J., Matte, P., 1984. Inverted metamorphic zonation and large-scale thrusting in the Variscan Belt: an example in the French Massif Central. In: Geological Society Special Publication 14, pp. 47–61.

Burg, J-P., Delor, C.P., Leyreloup, A., Romney, F., 1989. Inverted metamorphic zonation and Variscan thrust tectonics in the Rouergue area (Massif Central, France): P-T-t record from mineral to regional scale. In: Daly, R.A.C.J.S., Yardley, B.W.D. (Eds.), Evolution of Metamorphic Belts. Geological Society Special Publication 43, pp. 423–439.

Busfield, A., Gilmour, J.D., Whitby, J.A., Simon, S.B., Grossman, L., Tang, C.C., Turner, G., 2001. I-Xe analyses of Tagish Lake magnetite and Monahans halite. Meteoritics and Planetary Science 36 (Supplement), A34–35.

Busfield, A., Gilmour, J.D., Whitby, J.A., Turner, G., 2004. Iodine-Xenon analysis of ordinary chondrite halide: implications for early solar system water. Geochimica et Cosmochimica Acta 68, 195–202.

Bussey, D.B.J., Sorensen, S.-A., Guest, J.E., 1995. Factors influencing the capability of lava to erode its substrate: Application to Venus. Journal of Geophysical Research 100, 6941–6949.

Butler, I.B., Bottcher, M.E., Rickard, D., Oldroyd, A., 2004. Sulfur isotope partitioning during experimental formation of pyrite via the polysulfide and hydrogen sulfide pathways: Implications for the interpretation of sedimentary and hydrothermal pyrite isotope records. Earth and Planetary Science Letters 228, 495–509.

Butler, I.B., Archer, C., Vance, D., Oldroyd, A., Rickard, D., 2005. Fe isotope fractionation on FeS formation in ambient aqueous solution. Earth and Planetary Science Letters 236, 430–442.

BVTP – Basaltic Volcanism of the Terrestrial Planets (BVTP). Basaltic Volcanism Study Project, 1981. Pergamon Press Inc., New York, 1286 pp.

Byerly, G.R., 1999. Komatiites of the Mendon Formation: Late-stage ultramafic volcanism in the Barberton Greenstone Belt. In: Lowe, D.R., Byerly, G.R. (Eds.), Geologic Evolution of the Barberton Greenstone Belt, South Africa. Geological Society of America, Special Paper 329, pp. 189–212.

Byerly, G.R., Lowe, D.R., 1994. Spinels from Archaean impact spherules. Geochimica et Cosmochimica Acta 58, 3469–3486.

Byerly, G.R., Kröner, A., Lowe, D.R., Todt, W., Walsh, M.M., 1996. Prolonged magmatism and time constraints for sediments deposition in the early Archaean Barberton greenstone belt: evidence from the Upper Onverwacht and Fig Tree Groups. Precambrian Research 78, 125–138.

Byerly, G.R., Lowe, D.R., Wooden, J.L., Xie, X., 2002. An Archean impact layer from the Pilbara and Kaapvaal cratons. Science 297, 1325–1327.

Cagnard, F., Brun, J.P., Gapais, D., 2006. Modes of thickening of analogue weak lithospheres. Tectonophysics 421, 145–160.

Cairns-Smith, A.G., 1978. Precambrian solution photochemistry, inverse segregation, and banded iron formations. Nature 276, 807–808.

Calais, E., Ebinger, C., Hartnady, C., Nocquet, J.M., 2006. Kinematics of the East African Rift from GPS and earthquake slip vector data. In: Geological Society of London, Special Publication 259, pp. 9–22.

Calvert, A., Ludden, J.N., 1999. Archean continental assembly in the southeastern Superior Province of Canada. Tectonics 18, 412–429.

Cameron, A.G.W., 1979. Neutron-rich silicon-burning and equilibrium processes of nucleosynthesis. Astrophysical Journal 230, L53-L57.

Cameron, W.E., Nisbet, E.G., Dietrich, V.J., 1979. Boninites, komatiites and ophiolitic basalts. Nature 280 (16), 550–553.

Campbell, B.A., 1999. Surface formation rates and impact crater densities on Venus. Journal of Geophysical Research 104, 21,951–21,955.

Campbell, B.A., Arvidson, R.E., Shepard, M.K., Brackett, R.A., 1997. Remote sensing of surface processes. In: Bouger, S.W., Hunten, D.M., Phillips, R.J. (Eds.), Venus II. University of Arizona Press, pp. 503–526.

Campbell, I.H., 1998. The mantle's chemical structure: insights from the melting products of mantle plumes. In: Jackson, I.N.S. (Ed.), The Earth's Mantle: Composition, Structure and Evolution. Cambridge University Press, Cambridge, pp. 259–310.

Campbell, I.H., 2003. Constraints on continental growth models from Nb/U ratios in the 3.5 Ga Barberton and other Archaean basalt-komatiite suites. American Journal of Science 303, 319–351.

Campbell, I.H., Davies, G.F., 2006. Do mantle plumes exist? Episodes 29, 162–168.

Campbell, I.H., Hill, R.I., 1988. A two-stage model for the formation of the granite-greenstone terrains of the Kalgoorlie–Norseman area, Western Australia. Earth and Planetary Science Letters 90, 11–25.

Campbell, I.H., Griffiths, R.W., Hill, R.I., 1989. Melting in an Archaean mantle plume: heads it's basalts, tails it's komatiites. Nature 339, 697–699.

Candela, P.A., Picolli, P.M., 2005. Magmatic processes in the development of porphyry-type ore systems. Economic Geology, 100th Anniversary Volume, 25–38.

Canfield, D.E., 2001. Biogeochemistry of sulphur isotopes. In: Valley, J.W., Cole, D.R. (Eds.), Stable Isotope Geochemistry. Reviews in Mineralogy and Geochemistry, vol. 43, pp. 607–636.

Canfield, D.E., 2004. The evolution of the Earth surface sulfur reservoir. American Journal of Science 304, 839–861.

Canfield, D.E., 2005. The early history of atmospheric oxygen: Homage to Robert M. Garrels. Annual Review of Earth and Planetary Sciences 33, 1–36.

Canfield, D.E., Raiswell, R., 1999. The evolution of the sulfur cycle. American Journal of Science 299, 697–723.

Canfield, D.E., Thamdrup, B., 1994. The production of S-34-depleted sulfide during bacterial disproportionation of elemental sulfur. Science 266, 1973–1975.

Canup, R.N., 2004. Simulations of a late lunar-forming impact. Icarus 168, 433–456.

Canup, R., Agnor, C.B., 2000. Accretion of the terrestrial planets and the Earth–Moon system. In: Canup, R., Righter, K. (Eds.), Origin of the Earth and Moon. Arizona University Press, Tucson, AZ, pp. 113–129.

Canup, R., Asphaug, E., 2001. Origin of the Moon in a giant impact near the end of the Earth's formation. Nature 412, 708–712.

Cao, G.Q., Wang, Z.P., Zhang, C.J., 1996. Early Precambrian Geology of Western Shandong. Geological Publishing House, Beijing, pp. 1–193 (in Chinese).

Card, K.D., 1990. A review of the Superior Province of the Canadian Shield, a product of Archean accretion. Precambrian Research 48, 99–156.

Card, K.D., Ciesielski, A., 1986. Subdivisions of the Superior Province of the Canadian Shield. Geoscience Canada 13, 5–13.

Card, K.D., Poulsen, K.H., 1998. Geology and mineral deposits of the Superior Province of the Canadian shield. In: Lucas, S. (Coordinator), Geology of the Precambrian Superior and Grenville Provinces and Precambrian Fossils in North America. Geological Survey of Canada, Geology of Canada, vol. 7, pp. 13–194 (Chapter 2).

Carlson, R.W., Hauri, E.H., 2001. Extending the ^{107}Pd-^{107}Ag chronometer to low Pd/Ag meteorites with the MC-ICPMS. Geochimica et Cosmochimica Acta 65, 1839–1848.

Carlson, R.W., Lugmair, G.W., 2000. Timescales of planetesimal formation and differentiation based on extinct and extant radioisotopes. In: Canup, R., Righter, K. (Eds.), Origin of the Earth and Moon. Arizona University Press, Tucson, AZ, pp. 25–44.

Carlson, R.W., Hunter, D.R., Barker, F., 1983. Sm-Nd age and isotopic systematics of the bimodal suite, ancient gneiss complex, Swaziland. Nature 305, 701–704.

Carlson, R.W., Pearson, D.G., Boyd, F.R., Shirey, S.H., Irvine, G., Menzies, A.H., Gurney, J.J., 1999. Re-Os systematics of lithospheric peridotites: Implications for lithosphere formation and preservation. In: Proceedings of the 7th International Kimberlite Conference. Red Roof Design, Cape Town, pp. 99–108.

Carlson, R.W., Irving, A.J., Schulze, D.J., Hearn, B.C., Jr, 2004. Timing of Precambrian melt depletion and Phanerozoic refertilization events in the lithospheric mantle of the Wyoming craton and adjacent Central Plains Orogen. Lithos 77, 453–472.

Carpenter, J.M., Wolf, S., Schreyer, K., Launhardt, R., Henning, T., 2005. Evolution of cold circumstellar dust around solar-type stars. Astronomical Journal 129, 1049–1062.

Carson, C.J., Ague, J.J., Coath, C.D., 2002. U-Pb geochronology from Tonagh Island, East Antarctica: implications for the timing of ultra-high temperature metamorphism of the Napier Complex. Precambrian Research 116, 237–263.

Caro, G., Bourdon, B., Wood, B.J., Corgne, A., 2005. Trace element fractionation in Hadean mantle generated by melt segregation from a magma ocean. Nature 436, 246–249.

Caro, G., Bourdon, B., Birck, J.L., Moorbath, S., 2006. High-precision Nd-142/Nd-144 measurements in terrestrial rocks: Constraints on the early differentiation of the Earth's mantle. Geochimica et Cosmochimica Acta 70 (1), 164–191.

Cas, R., Self, S., Beresford, S., 1999. The behavior of the fronts of komatiite lavas in medial to distal settings. Earth and Planetary Science Letters 172, 127–139.

Casey, J.F., 1997. Comparison of major and trace element geochemistry of abyssal peridotites and mafic plutonic rocks with basalts from the MARK region of the mid-Atlantic Ridge. In: Proceedings of the Ocean Drilling Program, Scientific Results, vol. 153, pp. 181–240.

Cassidy, K.F., Champion, D.C., McNaughton, N.J., Fletcher, I.R., Whitaker, A.J., Bastrakova, I.V., Budd, A.R., 2002. Characterization and metallogenic significance of Archaean granitoids of the Yilgarn Craton, Western Australia. Amira International Limited, AMIRA Project no. P482 / MERIWA Project M281 (unpublished report no. 222).

Cassidy, K.F., Champion, D.C., Krapež, B., Barley, M.E., Brown, S.J.A., Blewett, R.S., Groenewald, P.B., Tyler, I.M., 2006. A revised geological framework for the Yilgarn Craton, Western Australia. Western Australia Geological Survey, Record 2006/8, 8 pp.

Catanzaro, E.J., 1963. Zircon ages in southwestern Minnesota. Journal of Geophysical Research 68, 2045–2048.

Cates, N.L., Mojzsis, S.J., 2006. Chemical and isotopic evidence for widespread Eoarchean ($\geqslant$3750 Ma) metasedimentary enclaves in southern West Greenland. Geochimica et Cosmochimica Acta 70, 4229–4257.

Cates, N.L., Mojzsis, S.J., 2007a. Pre-3750 Ma supracrustal rocks from the Nuvvuagittuq supracrustal belt, northern Québec. Earth and Planetary Science Letters 255, 9–21.

Cates, N.L., Mojzsis, S.J., 2007b. Rare or prevalent Earth? Conditions suitable for life were established rapidly on the young Earth. In: Lunar and Planetary Science Conference XXXVIII, Houston, TX.

Cates, N.L., Mojzsis, S.J., 2007c. Metamorphic history of the pre-3750 Ma Nuvvuagittuq supracrustal belt, Québec In: V.M. Goldschmidt Conference 2007, Cologne, Germany.

Cavosie, A.J., Wilde, S.A., Liu, D., Weiblen, P.W., Valley, J.W., 2004. Internal zoning and U-Th-Pb chemistry of Jack Hills detrital zircons: a mineral record of early Archean to Mesoproterozoic (4348–1576 Ma) magmatism. Precambrian Research 135, 251–279.

Cavosie, A.J., Valley, J.W., Wilde, S.A., 2005a. Magmatic $\delta^{18}O$ in 4400–3900 Ma detrital zircons: A record of the alteration and recycling of crust in the Early Archean. Earth and Planetary Science Letters 235, 663–681.

Cavosie, A.J., Wilde, S.A., Valley, J.W., 2005b. A lower age limit for the Archean based on $\delta^{18}O$ of detrital zircons. Geochimica et Cosmochimica Acta 69, A391.

Cavosie, A.J., Valley, J.W., Wilde, S.A., 2006. Correlated microanalysis of zircon: Trace element, $\delta^{18}O$, and U-Th-Pb isotopic constraints on the igneous origin of complex >3900 Ma detrital grains. Geochimica et Cosmochimica Acta 70, 5601–5616.

Cawood, P.A., 2005. Terra Australis Orogen: Rodinia breakup and the development of the Pacific and Iapetus margins of Gondwana during the Neoproterozoic and Paleozoic. Earth Science Reviews 69, 249–279.

Cawood, P.A., Kröner, A., Pisarevsky, S., 2006. Precambrian plate tectonics: Criteria and evidence. GSA Today 16 (7), 4–11.

Chamberlain, K.R., 1998. Medicine Bow Orogeny: Timing of deformation and model of crustal structure produced during continent-arc collision, ca. 1.78 Ga, southeastern Wyoming. Rocky Mountain Geology 33 (2), 259–278.

Chamberlain, K.R., Bowring, S.A., 2001. Apatite-feldspar U-Pb thermochronometer: a reliable, mid-range (450 °C), diffusion-controlled system. Chemical Geology 172, 173–200.

Chamberlain, K.R., Bauer, R.L., Frost, B.R., Frost, C.D., 2002. Dakotan Orogen: Continuation of Trans-Hudson Orogen or younger, separate suturing of Wyoming and Superior cratons. Geological Association of Canada – Mineralogical Association of Canada Abstracts 27, 18.

Chamberlain, K.R., Frost, C.D., Frost, B.R., 2003. Early Archean to Mesoproterozoic evolution of the Wyoming Province: Archean origins to modern lithospheric architecture. Canadian Journal of Earth Sciences 40, 1357–1374.

Chamberlin, T.C., 1897. The method of multiple working hypotheses. Journal of Geology 5, 837–848.

Champion, D.C., Sheraton, J.W., 1997. Geochemistry and Nd isotope systematics of Archaean granites of the Eastern Goldfields, Yilgarn Craton, Australia: implications for crustal growth processes. Precambrian Research 83, 109–132.

Champion, D.C., Smithies, R.H., 2000. The geochemistry of the Yule Granitoid Complex, East Pilbara Granite-Greenstone Terrane; evidence for early felsic crust. In: Western Australia Geological Survey, Annual Review 1999–2000, pp. 42–48.

Champion, D.C., Smithies, R.H., 2001. Archean granites of the Yilgarn and Pilbara Cratons, Western Australia. In: Cassidy, K.F., Dunphy, J.M., Van Kranendonk, M.J. (Eds.), 4th International Archean Symposium 2001, Extended Abstracts, AGSO – Geoscience Australia, Record 2001/37, pp. 134–136.

Chandler, V.W., 1987. Aeromagnetic map of Minnesota, west-central region, magnetic intensity anomaly. Scale 1:250,000, 2 pls. Map A-6. Minnesota Geological Survey, Minneapolis, MN.

Chandler, V.W., 1989. Aeromagnetic map of Minnesota, southwestern region, magnetic intensity anomaly. Scale 1:250,000, 2 pls. Map A-8. Minnesota Geological Survey, Minneapolis, MN.

Chandler, V.W., 1991. Aeromagnetic map of Minnesota, southeastern region, magnetic intensity anomaly. Scale 1:250,000, 2 pls. Map A-9. Minnesota Geological Survey, Minneapolis, MN.

Chapman, M.G., 1999. Geologic/Geomorphic Map of the Galindo Quadrangle (V–40), Venus Geologic Investigations Series I-2613. U.S. Geological Survey, 1:5,000,000.

Chappell, B.W., White, A.J.R., 1974. Two contrasting granite types. Pacific Geology 8, 173–174.

Chardon, D.H., Choukroune, P., Jayananda, M., 1996. Strain patterns, decollement and incipient sagducted greenstone terrains in the Archaean Dharwar craton (South India). Journal of Structural Geology 18, 991–1004.

Chardon, D.H., Choukroune, P., Jayananda, M., 1998. Sinking of the Dharwar Basin (South India); implications for Archaean tectonics. Precambrian Research 91, 15–39.

Chardon, D., Peucat, J.-J., Jayananda, M., Choukroune, P., Fanning, M., 2002. Archean granite-greenstone tectonics at Kolar (South India): Interplay of diapirism and bulk inhomogeneous contraction during juvenile magmatic accretion. Tectonics 21, 10.1029/2001TC901032.

Chaussidon, M., Lorand, J.P., 1990. Sulfur isotope composition of orogenic spinel lherzolite massifs from Ariege (north-eastern Pyrenees, France) – an ion microprobe study. Geochimica et Cosmochimica Acta 54, 2835–2846.

Chaussidon, M., Albarède, F., Sheppard, S.M.F., 1987. Sulfur isotope heterogeneity in the mantle from ion microprobe measurements of sulfide inclusions in diamonds. Nature 330, 242–244.

Chen, B., Arakawa, Y., 2005. Elemental and Nd-Sr isotopic geochemistry of granitoids from the West Junggar foldbelt (NW China), with implications for Phanerozoic continental growth. Geochimica et Cosmochimica Acta 69 (5), 1307–1320.

Chen, J.H., Wasserburg, G.J., 1981. The isotopic composition of uranium and lead in Allende inclusions and meteoritic phosphates. Earth and Planetary Science Letters 52, 1–15.

Chen, J.H., Wasserburg, G.J., 1990. The isotopic composition of Ag in meteorites and the presence of [107]Pd in protoplanets. Geochimica et Cosmochimica Acta 54, 1729–1743.

Chen, J.H., Papanastassiou, D.A., Wasserburg, G.J., 2002. Re-Os and Pd-Ag systematics in group IIIAB irons and in pallasites. Geochimica et Cosmochimica Acta 66, 3793–3810.

Chen, S.F., Riganti, A., Wyche, S., Greenfield, J.E., Nelson, D.R., 2003. Lithostratigraphy and tectonic evolution of contrasting greenstone successions in the central Yilgarn Craton, Western Australia. Precambrian Research 127, 249–266.

Chen, S.F., Libby, J.W., Wyche, S., Riganti, A., 2004. Kinematic nature and origin of regional-scale ductile shear zones in the central Yilgarn Craton, Western Australia. Tectonophysics 394, 139–153.

Chen, Y.Q., Yang, C.H., Wan, Y.S., Liu, Z.X., Zhang, X.P., Du, L.L., Zhang, S.G., Wu, J.S., Gao, J.F., 2005. Early Precambrian Geological Characters and Anatectic Reconstruction of Crust in North Part of Middle Taihang Mountain. Geological Publishing House, Beijing, pp. 1–191 (in Chinese).

Chen, H.-L., Yang, S.-F., Wang, Q.-H., Luo, J.-C., Jia, C.-Z., Wei, G.-Q., Li, Z.-L., He, G.-Y., Hu, A.-P., 2006. Tarim platform, Early to Middle Permian basalt magmatism and interaction with sedimentary rocks (title translated from Chinese). Geology in China 33, 544–551 (in Chinese).

Chin, R.J., Smith, R.A., 1983. Jackson, W.A. Western Australia Geological Survey, 1:250,000 Geological Series Explanatory Notes, 30 pp.

Choblet, G., Sotin, C., 2001. Early transient cooling of Mars. Geophysical Research Letters 28, 3035–3038.

Choi, S.H., Mukasa, S.B., Andronikov, A.V., Osanai, Y., Harley, S.L., Kelly, N.M., 2006. Lu-Hf systematics of the earliest crust in Antarctica: The ultra-high temperature (UHT) Napier Metamorphic Complex of Enderby Land. Earth and Planetary Science Letters 246, 305–316.

Chopin, C., 1984. Coesite and pure pyrope in high-grade blueschists of western Alps: a first record and some consequences. Contributions to Mineralogy and Petrology 86, 107–118.

Chopin, C., Henry, C., Michard, A., 1991. Geology and petrology of the coesite-bearing terrain, Dora Maira massif, Western Alps. European Journal of Mineralogy 3, 263–291.

Choukroune, P., Bouhallier, H., Arndt, N.T., 1995. Soft lithosphere during periods of Archaean crustal growth or crustal reworking. In: Coward, M.P., Ries, A.C. (Eds.), Early Precambrian Processes. Geological Society of London, Special Publication 95, pp. 67–86.

Choukroune, P., Ludden, J.N., Chardon, D., Calvert, A.J., Bouhallier, H., 1997. Archaean crustal growth and tectonic processes: a comparison of the Superior Province, Canada and the Dharwar Craton, India. In: Burg, J.-P., Ford, M. (Eds.), Orogeny Through Time. Geological Society of London, Special Publication 121, pp. 63–98.

Chown, E.H., Harrap, R., Mouksil, A., 2002. The role of granitic intrusions in the evolution of the Abitibi belt, Canada. Precambrian Research 115, 291–310.

Chung, S.L., Jahn, B.M., Wu, G.Y., Lo, C.H., Bolin, C., 1998. The Emeishan flood basalts in SW China: a mantle plume initiation model and its connection with continental breakup and mass extinction at the Permian–Triassic boundary. In: American Geophysical Union, Geodynamics Series, vol. 27. Washington, DC, pp. 47–58.

Ciesielski, A., 2000. Géologie et lithogéochimie de la partie occidentale de la sous-province de Bienville et des zones adjacentes dans l'est de la Province du Supérieur, Québec. Commission géologique du Canada, Open File 3550, 90 pp.

Class, C., Miller, D.M., Goldstein, S.L., Langmuir, C.H., 2000. Distinguishing melt and fluid subduction components in Umnak Volcanics, Aleutian Arc. Geochemistry Geophysics Geosystems 1, paper number 1999GC000010.

Claoué-Long, J.C., Thirlwall, M.F., Nesbitt, R.W., 1984. Revised Sm-Nd systematics of Kambalda greenstones, Western Australia. Nature 307, 697–701.

Claoué-Long, J.C., Compston, W., Cowden, A., 1988. The age of the Kambalda greenstones resolved by ion-microprobe – implications for Archaean dating methods. Earth and Planetary Science Letters 89, 239–259.

Clayton, R.N., 2003. Oxygen isotopes in meteorites. In: Davis, A.M. (Ed.), Meteorites, Comets and Planets. Treatise on Geochemistry, vol. 1. Elsevier–Pergamon, Oxford, pp. 129–142.

Clayton, R.N., Mayeda, T.K., 1996. Oxygen isotope studies of achondrites. Geochimica et Cosmochimica Acta 60, 1999–2017.

Clayton, R.N., Grossman, L., Mayeda, T.K., 1973. A component of primitive nuclear composition in carbonaceous meteorites. Science 182, 485–488.

Clemens, J.D., Yearron, L.M., Stevens, G., 2006. Barberton (South Africa) TTG magmas: Geochemical and experimental constraints on source-rock petrology, pressure of formation and tectonic setting. Precambrian Research 151, 53–78.

Cloete, M., 1999. Aspects of Volcanism and Metamorphism of the Onverwacht Group Lavas in the Southwestern Portion of the Barberton Greenstone Belt. Memoir of the Geological Society of South Africa 84, 232 pp.

Cloud, P., 1968. Atmospheric and hydrospheric evolution of primitive Earth. Science 160, 729.

Cloud, P., 1972. A working model of the primitive Earth. American Journal of Science 272, 537–548.

Cloud, P., 1973. Paleoecological significance of the banded iron-formation. Economic Geology 68, 1135–1143.

Cody, G.D., Boctor, N.Z., Filley, T.R., Hazen, R.M., Scott, J.H., Sharma, A., Yoder, H.S., 2000. Primordial carbonylated iron-sulfur compounds and the synthesis of pyruvate. Science 289, 1337–1340.

Coffin, M.F., Eldhom, O., 1992. Volcanism and continental breakup: a global compilation of large igneous provinces. In: Geological Society of London, Special Publication 68, pp. 17–30.

Cohen, B.A., Swindle, T.D., King, D.A., 2000. Support for the lunar cataclysm hypothesis from lunar meteorite impact melt ages. Science 290, 1754–1756.

Coleman, R.G., 1989. Continental growth of north-west China. Tectonics 8, 621–635.

Collerson, K.D., Bridgwater, D., 1979. Metamorphic development of early Archaean tonalitic and trondhjemitic gneisses: Saglek area, Labrador. In: Barker, F. (Ed.), Trondhjemites, Dacites and Related Rocks. Elsevier, Amsterdam, pp. 205–273.

Collerson, K.D., Kamber, B.S., 1999. Evolution of the continents and the atmosphere inferred from Th-U-Nb systematics of the depleted mantle. Science 283, 1519–1522.

Collerson, K.D., Campbell, L.M., Weaver, B.L., Palacz, Z.A., 1991. Evidence for extreme fractionation in early Archean ultramafic rocks from northern Labrador. Nature 349, 209–214.

Collins, W.J., 1983. Geological Evolution of an Archean Batholith. Unpublished Ph.D. thesis. La Trobe University, Melbourne.

Collins, W.J., 1989. Polydiapirism of the Archaean Mt. Edgar batholith, Pilbara Block, Western Australia. Precambrian Research 43, 41–62.

Collins, W.J., 1993. Melting of Archaean sialic crust under high a H_2O conditions: genesis of 3300 Ma Na-rich granitoids in the Mount Edgar Batholith, Pilbara Block, Western Australia. Precambrian Research 60, 151–174.

Collins, W.J., 2003. Slab pull, mantle convection, and Pangaean assembly and dispersal. Earth and Planetary Science Letters 205, 225–237.

Collins, W.J., Gray, C.M., 1990. Rb-Sr isotopic systematics of an Archaean granite-gneiss terrain: The Mount Edgar batholith, Pilbara Block, Western Australia. Australian Journal of Earth Sciences 37, 9–22.

Collins, W.J., Van Kranendonk, M.J., 1999. Model for the development of kyanite during partial convective overturn of Archaean granite-greenstone terranes: the Pilbara Craton, Australia. Journal of Metamorphic Geology 17, 145–156.

Collins, W.J., Van Kranendonk, M.J., Teyssier, C., 1998. Partial convective overturn of Archaean crust in the east Pilbara Craton, Western Australia: Driving mechanisms and tectonic implications. Journal of Structural Geology 20, 1405–1424.

Compston, W., Kröner, A., 1988. Multiple zircon growth within early Archean Tonalitic gneiss from the Ancient Gneiss Complex, Swaziland. Earth and Planetary Sciences Letters 87, 13–28.

Compston, W., Pidgeon, R.T., 1986. Jack Hills, evidence of more very old detrital zircons in Western Australia. Nature 321, 766–769.

Compston, W., Lovering, J.F., Vernon, M.J., 1965. Rubidium-strontium age of the Bishopville aubrite and its component enstatite and feldspar. Geochimica et Cosmochimica Acta 29, 1085–1099.

Compston, W., Williams, I.S., Froude, D.O., Kinny, P.D., Ireland, T.R., 1984a. Zircon U-Pb age determinations by ion microprobe on pre-greenstone rocks from the north-west and central Yilgarn Block, West Australia. In: Terra Cognita, Abstracts. European Union of Geosciences, 4 (2), p. 208.

Compston, W., Williams, I.S., Meyer, C., 1984b. U-Pb geochronology of zircons from Lunar breccia 73217 using a sensitive high mass-resolution ion microprobe. Journal of Geophysical Research 89, B525-B534.

Compston, W., Kinny, P.D., Williams, I.S., Foster, J.J., 1986a. The age and lead loss behaviour of zircons from the Isua supracrustal belt as determined by ion microprobe. Earth and Planetary Science Letters 80, 71–81.

Compston, W., Williams, I.S., Campbell, I.H., Gresham, J.J., 1986b. Zircon xenocrysts from the Kambalda volcanics: age constraints and direct evidence for older continental crust below the Kambalda–Norseman greenstones. Earth and Planetary Science Letters 76, 299–301.

Condie, K.C., 1981. Archean Greenstone Belts. Elsevier, Amsterdam, 434 pp.

Condie, K.C., 1986. Origin and early growth rate of continents. Precambrian Research 32, 261–278.

Condie, K.C., 1989. Geochemical changes in basalts and andesites across the Archean-Proterozoic boundary: Identification and significance. Lithos 23, 1–18.

Condie, K., 1993. Chemical composition and evolution of the upper continental crust: Contrasting results from surface samples and shales. Chemical Geology 104, 1–37.

Condie, K.C., 1994. Greenstones through time. In: Condie, K.C. (Ed.), Archean Crustal Evolution. Elsevier, Amsterdam, pp. 85–120.

Condie, K.C., 1997. Plate Tectonics and Crustal Evolution, fourth ed. Butterworth Heinemann, Oxford, 282 pp.

Condie, K.C., 1998. Episodic continental growth and supercontinents: a mantle avalanche connection? Earth and Planetary Science Letters 163, 97–108.

Condie, K.C., 2000. Episodic crustal growth models: afterthoughts and extensions. Tectonophysics 322, 153–162.

Condie, K.C., 2001. Mantle Plumes and Their Record in Earth History. Cambridge University Press, Cambridge, 306 pp.

Condie, K.C., 2004. Supercontinents and superplume events: distinguishing signals in the geologic record. Physics of the Earth and Planetary Interiors 146, 319–332.

Condie, K.C., 2005. Earth as an Evolving Planetary System. Elsevier–Academic Press, Amsterdam, 447 pp.

Condie, K.C., 2005. High field strength element ratios in Archean basalts: a window to evolving sources of mantle plumes? Lithos 79, 491–504.

Condie, K.C., Wronkiewicz, D.J., 1990. The Cr/Th ratio in Precambrian pelites from the Kaapvaal Craton as an index of craton evolution. Earth and Planetary Science Letters 97, 256–267.

Condie, K.C., Macke, J.E., Reimer, T.O., 1970. Petrology and geochemistry of early Precambrian graywackes from the Fig Tree Group, South Africa. Bulletin of the Geological Society of America 81, 2759–2776.

Condie, K.C., Kröner, A., Milisenda, C.C., 1996. Geochemistry and geochronology of the Mkhondo suite, Swaziland: evidence for passive-margin deposition and granulite facies metamorphism in the late Archean of southern Africa. Journal of African Earth Science 21, 483–506.

Condie, K.C., Belousova, E., Griffin, W.L., 2007. Granitic events in space and time: Constraints from igneous and detrital zircon age spectra. Geosphere, in press.

Condit, R.H., Hobbins, R.R., Birchenall, C.E., 1974. Self-diffusion of iron and sulfur in ferrous sulfide. Oxidation of Metals 8, 409–455.

Coney, P.J., Jones, D.L., Monger, J.W.H., 1980. Cordilleran suspect terranes. Nature 288, 329–333.

Conrad, R., 2005. Quantification of methanogenic pathways using stable carbon isotopic signatures: a review and a proposal. Organic Geochemistry 36, 739–752.

Coogan, L., Hinton, R.W., 2006. Do the trace element compositions of detrital zircons reuqire Hadean continental crust? Geology 34, 633–636.

Cooke, D.R., Hollings, P., Walshe, J.L., 2005. Giant porphyry deposits: characteristics, distribution, and tectonic controls. Economic Geology 100, 801–818.

Cooper, G.W., Thiemens, M.H., Jackson, T.L., Chang, S., 1997. Sulfur and hydrogen isotope anomalies in meteorite sulfonic acids. Science 277, 1072–1074.

Copp, D.L., Guest, J.E., Stofan, E.R., 1998. New insights into corona evolution: Mapping on Venus. Journal of Geophysical Research 103, 19,401–19,410.

Corfu, F., 1988. Differential response of U-Pb systems in coexisting accessory minerals, Winnipeg River Subprovince, Canadian Shield: Implications for Archean growth and stabilization. Contributions to Mineralogy and Petrology 98, 312–325.

Corfu, F., 1993. The evolution of the southern Abitibi greenstone belt in light of precise U-Pb geochronology. Economic Geology 88, 1323–1340.

Corfu, F., Ayres, L.D., 1991. Unscrambling the stratigraphy of an Archaean greenstone belt: A U-Pb geochronological study of the Favourable Lake belt, northwestern Ontario. Precambrian Research 50, 201–220.

Corfu, F., Davis, D.W., 1992. A U-Pb geochronological framework for the western Superior Province. In: Thurston, P.C., Williams, H.R., Sutcliffe, R.H., Stott, G.M. (Eds.), Geology of Ontario. Ontario Geological Survey, Special Volume 4, Part 2, pp. 1335–1346.

Corfu, F., Lin, S., 2000. Geology and U-Pb geochronology of the Island Lake greenstone belt, northwestern Superior Province, Manitoba. Canadian Journal of Earth Sciences 37, 1275–1286.

Corfu, F., Noble, S.R., 1992. Genesis of the southern Abitibi greenstone belt, Superior Province, Canada: Evidence from zircon Hf isotope analyses using a single filament technique. Geochimica et Cosmochimica Acta 56, 2081–2097.

Corfu, F., Stone, D., 1998. Age structure and orogenic significance of the Berens River composite batholiths, western Superior Province. Canadian Journal of Earth Sciences 35, 1089–1109.

Corfu, F., Stott, G.M., 1986. U-Pb ages for late magmatism and regional deformation in the Shebandowan belt, Superior Province, Canada. Canadian Journal of Earth Sciences 23, 1075–1082.

Corfu, F., Stott, G.M., 1993a. Age and petrogenesis of two late Archean magmatic suites, northwestern Superior Province, Canada: zircon U-Pb and Lu-Hf isotopic relations. Journal of Petrology 34, 817–838.

Corfu, F., Stott, G.M., 1993b. U-Pb geochronology of the central Uchi Subprovince, Superior Province. Canadian Journal of Earth Sciences 30, 1179–1196.

Corfu, F., Stott, G.M., 1996. Hf isotopic composition and age constraints on the evolution of the Archean central Uchi Subprovince, Ontario, Canada. Precambrian Research 78, 53–63.

Corfu, F., Stott, G.M., 1998. Shebandowan greenstone belt, western Superior Province; U-Pb ages, tectonic implications and correlations. Bulletin of the Geological Society of America 110, 1467–1484.

Corfu, F., Wood, J., 1986. U-Pb zircon ages in supracrustal and plutonic rocks; North Spirit Lake area, northwestern Ontario. Canadian Journal of Earth Sciences 23, 967–977.

Corfu, F., Stott, G.M., Breaks, F.W., 1995. U-Pb geochronology and evolution of the English River subprovince, an Archean low P – high T metasedimentary belt in the Superior Province. Tectonics 14, 1220–1233.

Corfu, F., Davis, D.W., Stone, D., Moore, M., 1998. Chronostratigraphic constraints on the genesis of Archean greenstone belts, northwestern Superior Province, Ontario, Canada. Precambrian Research 92, 277–295.

Corkery, M.T., 1977. Geology of the Waskaiowaka Lake area (part of 64A7,8,9,10). Manitoba Mineral Resources Division, Preliminary Map Series 1977H-4, scale 1:50,000.

Corkery, M.T., 1985. Geology of the Lower Nelson River Project area, Manitoba. Manitoba Energy and Mines, Geological Report, Geological Services/Mines Branch, GR82-1, 66 pp.

Corkery, M.T., Lenton, P.G., 1990. Geology of the Lower Churchill River region. Manitoba Energy and Mines, Geological Services, Geological Report GR85-1 (including Geological Maps GR85-1-1 to GR85-1-9, scale 1:100,000 and 1:250,000).

Corkery, M.T., Böhm, C.O., Heaman, L.M., 1999. Progress report on the northwest Superior Province boundary project. In: Manitoba Industry, Trade and Mines, Geological Services, Report of Activities 1999, pp. 41–43.

Corkery, M.T., Cameron, H.D.M., Lin, S., Skulski, T., Whalen, J.B., Stern, R.A., 2000. Geological investigations in the Knee Lake belt (Parts of NTS 53L). In: Report of Activities 2000, Manitoba Industry, Trade and Mines, Manitoba Geological Survey, pp. 129–136.

Corliss, J.B., 1990. Hot springs and the origin of life. Nature 347, 624.

Corrigan, D, Hajnal, Z., Nemeth, B., Lucas, S.B., 2005. Tectonic framework of a Paleoproterozoic arc-continent to continent-continent collisional zone, Trans-Hudson Orogen,
from geological and seismic reflection studies. Canadian Journal of Earth Sciences 42,
421–434.

Courtillot, V., Davaille, A., Besse, J., Stock, J., 2003. Three distinct types of hotspots in
the Earth's mantle. Earth and Planetary Science Letters 205, 295–308.

Coward, M.P., Lintern, B.C., Wright, L.I., 1976. The pre-cleavage deformation of the sediments and gneisses of the northern part of the Limpopo belt. In: Windley, B.F. (Ed.),
The Early History of the Earth. Wiley, London, pp. 323–330.

Cox, K.G., Smith, M.R., Beswetherick, S., 1987. Textural studies of garnet lherzolites:
evidence of exsolution origin from high-temperature harzburgites. In: Nixon, P.H. (Ed.),
Mantle Xenoliths. Wiley, New York, pp. 537–550.

Cox, K.G., Bell, J.D., Pankhurst, R.J., 1979. The Interpretation of Igneous Rocks. London,
George Allen & Unwin, 450 pp.

Coyle, M., Kiss, F., Oneschuk, D., 2004. First vertical derivative of the magnetic field
and residual total magnetic field, various map sheets, Manitoba. In: Geological Survey of Canada, Open File 4764 to 4785. Manitoba Industry, Economic Development
and Mines, Manitoba Geological Survey, Open File Reports OF2004-3 to -24, scale
1:50,000.

Craven, J.A., Kurtz, R.D., Boerner, D.E., Skulski, T., Spratt, J., Ferguson, I.J., Wu, X.,
Bailey, R.C., 2001. Conductivity of western Superior Province upper mantle in northwestern Ontario. Geological Survey of Canada, Current Research 2001-E6, 6 pp.

Craven, J.A., Skulski, T., White, D.W., 2004. Lateral and vertical growth of cratons: seis-
mic and magnetotelluric evidence from the western Superior transect. In: Lithoprobe
Celebratory Conference. Lithoprobe Report 86, Lithoprobe Secretariat, University of
British Columbia.

Crowley, J.L., 2003. U-Pb geochronology of 3810–3630 Ma granitoid rocks south of the
Isua greenstone belt, southern West Greenland. Precambrian Research 126, 235–257.

Crowley, J.L., Myers, J.S., Dunning, G.R., 2002. Timing and nature of multiple 3700–3600
Ma tectonic events in intrusive rocks north of the Isua greenstone belt, southern West
Greenland. Bulletin of the Geological Society of America 114, 1311–1325.

Crowley, J.L., Myers, J.S., Sylvester, P.J., Cox, R.A., 2005. Detrital zircons from the Jack
Hills and Mount Narryer, Western Australia: Evidence for diverse >4.0 Ga source
rocks. Journal of Geology 113, 239–263.

Crumpler, L.S., Aubele, J.C., Senske, D.A., Keddie, S.T., Magee, K.P., Head, J.W., 1997.
Volcanoes and centers of volcanism on Venus. In: Bouger, S.W., Hunten, D.M., Phillips,
R.J. (Eds.), Venus II, Geology, Geophysics, Atmosphere, and Solar Wind Environment.
The University of Arizona Press, Tucson, AZ, pp. 697–756.

Cuesta, A., Dhamelincourt, P., Laureyns, J., Martinez-Alonso, A., Tascon, J.M.D., 1994.
Raman microprobe studies on carbon materials. Carbon 32, 1523–1532.

Cullers, R.L., Di Marco, M.J., Lowe, D.R., Stone, J., 1993. Geochemistry of a silicified,
felsic volcaniclastic suite from the early Archaean Panorama Formation, Pilbara Block,

Western Australia: an evaluation of depositional and post-depositional processes with special emphasis on the rare-earth elements. Precambrian Research 60, 99–116.

Cumming, G.L., Richards, J.R., 1975. Ore leads in a continuously changing Earth. Earth and Planetary Science Letters 28, 155–171.

Dada, S.S., 1998. Crust-forming ages and Proterozoic crustal evolution in Nigeria: a reappraisal of current interpretations. Precambrian Research 87, 65–74.

Dahl, P.S., Holm, D.K., Gardner, E.T., Hubacher, F.A., Foland, K.A., 1999. New constraints on the timing of Early Proterozoic tectonism in the Black Hills (South Dakota), with implications for docking of the Wyoming province with Laurentia. Bulletin of the Geological Society of America 111 (9), 1335–1349.

Dalziel, I.W.D., 1991. Pacific margins of Laurentia and East Antarctica–Australia as a conjugate rift pair: Evidence and implications for an Eocambrian supercontinent. Geology 19, 598–601.

Daniel, R.M., Danson, M.J., 1995. Did primitive microorganisms use nonheme iron proteins in place of NAD/P? Journal of Molecular Evolution 40, 559–563.

Dann, J.C., 2000. The Komati Formation, Barberton Greenstone Belt, South Africa, Part I: New map and magmatic architecture. South African Journal of Earth Science 103, 47–68.

Dann, J.C., 2001. Vesicular komatiites, ca. 3.5 Ga Komati Formation, Barberton Greenstone Belt, South Africa: Inflation of submarine lavas and origin of spinifex zones. Bulletin of Volcanology 63, 462–481.

Dann, J.C., 2004. The 1.73 Ga Payson Ophiolite, Arizona, USA. In: Kusky, T.M. (Ed.), Precambrian Ophiolites and Related Rocks. Developments in Precambrian Geology, vol. 13. Elsevier, Amsterdam, pp. 73–94.

Dann, J.C., Wilson, A.H., Cloete, M., 1998. Field excursion C1: Komatiites in the Barberton and Nondweni Greenstone Belts. In: IAVCEI – Magmatic Diversity: Volcanoes and Their Roots. Cape Town, p. 61.

Dantas, E.L., Van Schmus, W.R., Hackspacher, P.C., Fetter, A.H., de Brito Neves, B.B., Cordani, U., Nutman, A.P., Williams, I.S., 2004. The 3.4–3.5 Ga Sao do Campestre massif, NE Brazil: remnants of the oldest crust in South America. Precambrian Research 130, 113–138.

D'Arcy, K.A., Mueller, P.A., 1992. Archean metasupracrustal rocks from the western margin of the Wyoming Province. In: 29th International Geological Congress, Abstracts, vol. 29, p. 590.

Dauphas, N., Rouxel, O., 2006. Mass spectrometry and natural variations of iron isotopes. Mass Spectrometry Reviews 25, 515–550.

Dauphas, N., Janney, P.E., Mendybaev, R.A., Wadhwa, M., Richter, F.M., Davis, A.M., VanZuillen, M., Hines, R., Foley, C.N., 2004a. Chromatographic separation and multicollection-ICPMS analysis of Iron: investigating mass-dependent and -independent isotope effects. Analytical Chemistry 76, 5855–5863.

Dauphas, N., van Zuilen, M., Wadhwa, M., Davis, A.M., Martey, B., Janney, P.E., 2004b. Clues from Fe isotope variations on the origin of early Archean BIFs from Greenland. Science 302, 2077–2080.

Dauphas, N., Cates, N.L., Mojzsis, S.J., Busigny, V., 2007. Identification of chemical sedimentary protoliths using iron isotopes in the >3750 Ma Nuvvuagittuq supracrustal belt, Canada. Earth and Planetary Science Letters 254, 358–376.

David, J., Parent, M., Stevenson, R., Nadeau, P., Godin, L., 2003. The Porpoise Cove supracrustal sequence, Inukjuak area: a unique example of Paleoarchean crust (ca. 3.8 Ga) in the Superior Province. Geological Association of Canada Program with Abstracts 28 (http://gac.esd.mun.ca/gac_2003/search_abs/sub_program.asp?sess=98&form=10&abs_no=355).

David, J., Godin, L., Berclaz, A., Maurice, C., Parent, M., Francis, D., Stevenson, R.K., 2004. Geology and Geochronology of the Nuvvuagittuq Supracrustal Sequence: An Example of Paleoarchean Crust (ca. 3.8 Ga) in the Northeastern Superior Province. Eos, Transactions, American Geophysical Union 85 (17), Abstract V23D-03.

David, J., Parent, M., Stevenson, R., Nadeau, P., Godin, L., 2002. La séquence supracrustale de Porpoise Cove, région d'Inukjuak; un exemple unique de croûte paléo-archéenne (ca. 3.8 Ga) dans la Province du Supérieur. Ministère des Ressources naturelles, Québec, DV 2002-10:17.

Davidek, K., Martin, M.W., Bowring, S.A., Williams, I.S., 1997. Conventional U-Pb geochronology of the Acasta gneisses using single crystal fragmentation technique. In: Abstracts of GAC/MAC Annual Meeting, p. A-75.

Davids, C., Wijbrans, J.R., White, S.H., 1997. $^{40}Ar/^{39}Ar$ laserprobe ages of metamorphic hornblendes from the Coongan Belt, Pilbara, Western Australia. Precambrian Research 83, 221–242.

Davies, G.F., 1980. Exploratory models of the Earth's thermal regime during segregation of the core. Journal of Geophysical Research 85, 7108–7114.

Davies, G.F., 1990. Heat and mass transport in the early Earth. In: Newsome, H.E., Jones, J.H. (Eds.), Origin of the Earth. Oxford University Press, pp. 175–194.

Davies, G.F., 1992. On the emergence of plate tectonics. Geology 20, 963–966.

Davies, G.F., 1993. Cooling the core and mantle by plume and plate flows. Geophysical Journal International 115, 132–146.

Davies, G.F., 1995. Punctuated tectonic evolution of the Earth. Earth and Planetary Science Letters 36, 363–380.

Davies, G.F., 1999. Dynamic Earth: Plates, Plumes and Mantle Convection. Cambridge University Press, Cambridge, 460 pp.

Davies, G.F., 2002. Stirring geochemistry in mantle convection models with stiff plates and slabs. Geochimica et Cosmochimica Acta 66, 3125–3142.

Davies, G.F., 2006a. Controls on density stratification in the early Earth. Geochemistry, Geophysics, Geosystems 8 (4), doi: 10.1029/2006GC001414.

Davies, G.F., 2006b. Gravitational depletion of the early Earth's upper mantle and the viability of early plate tectonics. Earth and Planetary Science Letters 243, 376–382.

Davies, R., Griffin, W.L., Pearson, N.J., Andrew, A., Doyle, B.J., O'Reilly, S.Y., 1999. Diamonds from the Deep: Pipe DO-27, Slave Craton, Canada. In: Proceedings of the 7th International Kimberlite Conference. Red Roof Design, Cape Town, pp. 148–155.

Davies, G.F., Richard, M.A., 1992. Mantle convection. Journal of Geology 100, 151–206.

Davis, A.M., Ganapathy, R., Grossman, L., 1977. Pontlyfni: a differentiated meteorite related to the group IAB irons. Earth and Planetary Science Letters 35, 19–24.

Davis, D.W., 1996. Provenance and depositional age constraints on sedimentation in the western Superior transect area from U-Pb ages of zircons. In: Harrap, R.M., Helmstaedt, H. (Eds.), Western Superior Transect Second Annual Workshop. Lithoprobe Report 53, pp. 18–23.

Davis, D.W., 1998. Speculations on the formation and crustal structure of the Superior province from U-Pb geochronology. In: Harrap, R.M., Helmstaedt, H. (Eds.), Western Superior Transect Fourth Annual Workshop. Lithoprobe Report 65, pp. 21–28.

Davis, D.W., 2002. U-Pb geochronology of Archean metasedimentary rocks in the Pontiac and Abitibi subprovinces, Quebec, constraints on timing, provenance and regional tectonics. Precambrian Research 115, 97–117.

Davis, D.W., Amelin, Y., 2000. Constraints on crustal development in the Western Superior Lithoprobe Transect from Hf isotopes in zircons. In: Harrap, R.M., Helmstaedt, H. (Eds.), Western Superior Transect Sixth Annual Workshop. Lithoprobe Report 77, pp. 38–44.

Davis, D.W., Jackson, M., 1988. Geochronology of the Lumby Lake greenstone belt: a 3 Ga complex within the Wabigoon Subprovince, northwest Ontario. Bulletin of the Geological Society of America 100, 818–824.

Davis, D.W., Smith, P.M., 1991. Archean gold mineralization in the Wabigoon Subprovince, a product of crustal accretion: evidence from U-Pb geochronology in the Lake of the Woods area, Superior Province, Canada. Journal of Geology 99, 337–353.

Davis, D.W., Sutcliffe, R.H., Trowell, N.F., 1988. Geochronological constraints on the tectonic evolution of a late Archean greenstone belt, Wabigoon Subprovince, northern Ontario, Canada. Precambrian Research 39, 171–191.

Davis, D.W., Amelin, Y., Nowell, G.M., Parrish, R.R., 2005. Hf isotopes in zircon from the western Superior Province, Canada: Implications for Archean crustal development and evolution of the depleted mantle reservoir. Precambrian Research 140 (3–4), 132–156.

Davis, W.J., Machado, N., Gariépy, C., Sawyer, E.W., Benn, K., 1995. U-Pb geochronology of the Opatica tonalite-gneiss belt and its relationship to the Abitibi greenstone belt, Superior Province, Quebec. Canadian Journal of Earth Sciences 32, 113–127.

Davis, W.J., Gariépy, C., van Breeman, O., 1996. Pb isotopic composition of late Archaean granites and the extent of recycling early Archaean crust in the Slave Province, northwest Canada. Chemical Geology 130, 255–269.

Davy, R., 1988. Geochemical patterns in granitoids of the Corunna Downs Batholith, Western Australia. In: Geological Survey of Western Australia Professional Paper 23, pp. 51–84.

Davy, R., Lewis, J.D., 1986. Geochemistry and petrography of the Mount Edgar Batholith, Pilbara area, Western Australia. Western Australia Geological Survey Report 17.

Dawson, A.S., 1941. Assean-Split Lakes area. Manitoba Mines Branch, Publ. 39-1.

de Beer, J.H., Stettler, E.H., du Plessis, J.G., Blume, J., 1988. The deep structure of the Barberton greenstone belt: a geophysical study. South African Journal of Geology 91, 184–197.

DeDuve, C., 2005. Singularities – Landmarks on the Pathways to Life. Cambridge University Press, Cambridge, UK, 258 pp.

Deen, T., Griffin, W.L., Begg, G., O'Reilly, S.Y., Natapov, L.M., 2006. Thermal and compositional structure of the subcontinental lithospheric mantle: Derivation from shear-wave seismic tomography. Geochemistry, Geophysics and Geosystems, doi: 10.1029/2005GC001164.

Deines, P., 2003. A note on the intra-elemental isotope effects and the interpretation of non-mass-dependent isotope variations. Chemical Geology 199, 179–182.

de Laeter, R.R., Martyn, J.E., 1986. Age of molybdenum-copper mineralization at Coppin Gap, Western Australia. Australian Journal of Earth Sciences 33, 65–72.

de Laeter, J.R., Williams, I.R., Rosman, K.J.R., Libby, W.G., 1981a. A definitive 3350 m.y. age from banded gneiss, Mount Narryer area, Western Gneiss Terrain. In: Western Australia Geological Survey Annual Report 1980, pp. 94–98.

de Laeter, J.R., Fletcher, I.R., Rosman, K.J.R., Williams, I.R., Gee, R.D., Libby, W.G., 1981b. Early Archaean gneisses from the Yilgarn Block, Western Australia. Nature 292, 322–324.

de Laeter, J.R., Fletcher, I.R., Bickle, M.J., Myers, J.S., Libby, W.G., Williams, I.R., 1985. Rb-Sr, Sm-Nd and Pb-Pb geochronology of ancient gneisses from Mount Narryer, Western Australia. Australian Journal of Earth Sciences 32, 349–358.

de la Roche, A., Leterrier, J., Grandclaude, P., Marchal, M., 1980. A classification of volcanic and plutonic rocks using $R_1 - R_2$ diagram and major element analyses – its relationships with current nomenclature. Chemical Geology 29, 183–210.

DeLaughter, J.E., Jurdy, D.M., 1999. Corona classification by evolutionary stage. Icarus 139, 81–95.

Delor, C., Burg, J-P., Clarke, G., 1991. Relations diapirisme-metamorphisme dans la Province du Pilbara (Australie-occidentale): implications pour les régimes thermiques et tectoniques à la Archéen. Comptes Rendus de l'Academie des Sciences Paris 312, 257–263.

DePaolo, D.J., 1981. Neodymium isotopes in the Colorado front range and crust-mantle evolution in the Proterozoic. Nature 291, 193–196.

DePaolo, D.J., Manton, W.I., Grew, E.S., Halpern, M., 1982. Sm-Nd, Rb-Sr and U-Th-Pb systematics of granulite facies rocks from Fyfe Hills, Enderby Land, Antarctica. Nature 298, 614–618.

Derenne, S., Skrzypczak, A., Robert, F., Binet, L., Gourier, D., Rouzaud, J.-N., Clinard, C., 2004. Characterization of the organic matter in an Archaean Chert (Warrawoona, Australia). European Geosciences Union. Geophysical Research Abstracts 6, 03612.

de Ronde, C.E.J., 1991. Structural and geochronological relationships and fluid/rock interaction in the central part of the ~3.2–3.5 Ga Barberton greenstone belt, South Africa. Ph.D. thesis. University of Toronto, 370 pp.

de Ronde, C.E.J., de Wit, M.J., 1994. Tectonic history of the Barberton Greenstone Belt, South Africa: 490 million years of Archean crustal evolution. Tectonics 13, 983–1005.

de Ronde, C.E.J., Kamo, S.L., 2000. An Archaean Arc-Arc collisional event: a short-lived (ca. 3 Myr) episode, Weltevreden area, Barberton greenstone belt, South Africa. Journal of African Earth Sciences 30 (2), 219–248.

de Ronde, C.E.J., Hall, C.M., York, D., Spooner, E.T.C., 1991a. Laser step-heating $^{40}Ar/^{39}Ar$ age spectra from early Archean (~3.5 Ga) Barberton greenstone belt sediments: a technique for detecting cryptic tectono-thermal events. Geochimica et Cosmochimica Acta 55, 1933–1951.

de Ronde, C.E.J., Kamo, S., Davis, D.W., de Wit, M.J., Spooner, E.T.C., 1991b. Field, geochemical and U-Pb isotopic constraints from hypabyssal felsic intrusions within the Barberton greenstone belt, South Africa: implications for tectonics and the timing of gold mineralization. Precambrian Research 49, 261–280.

de Ronde, C.E.J., Spooner, E.T.C., de Wit, M.J., Bray, C.J., 1992. Shear zone-related, Au quartz vein deposits in the Barberton greenstone belt, South Africa: field and petrographic characteristics, fluid properties, and light stable isotope geochemistry. Economic Geology 87, 366–402.

Des Marais, D.J., 2001. Isotopic evolution of the biogeochemical carbon cycle during the Precambrian. In: Valley, J.W., Cole, D.R. (Eds.), Stable Isotope Geochemistry. Reviews in Mineralogy and Geochemistry, vol. 43, pp. 555–578.

Des Marais, D.J., Strauss, H., Summons, R.E., Hayes, J.M., 1992. Carbon isotope evidence for the stepwise oxidation of the Proterozoic environment. Nature 359, 605–609.

Detmers, J., Bruchert, V., Habicht, K.S., Kuever, J., 2001. Diversity of sulfur isotope fractionations by sulfate-reducing prokaryotes. Applied and Environmental Microbiology 67, 888–894.

de Vries, S.T., 2004. Early Archean sedimentary basins: Depositional environment and hydrothermal systems, examples from the Barberton and Coppin Gap greenstone belts. Ph.D. dissertation. University of Utrecht, Netherlands, 160 pp.

de Vries, S.T., Nijman, W., Armstrong, R.A., 2006. Growth-fault structure and stratigraphic architecture of the Buck Ridge volcano-sedimentary complex, upper Hooggenoeg Formation, Barberton Greenstone Belt, South Africa. Precambrian Research 149, 77–98.

de Wit, M.J., 1982. Gliding and overthrust nappe tectonics in the Barberton greenstone belt. Journal of Structural Geology 4, 117–136.

de Wit, M.J., 1983. Notes on a preliminary 1:25,000 geological map of the southern part of the Barberton Greenstone Belt. In: Anhaeusser, C.R. (Ed.), Contributions to the geology of the Barberton Mountain Land. Geological Society of South Africa, Special Publication 9, pp. 185–187.

de Wit, M.J., 1998. On Archean granites, greenstones, cratons and tectonics: does the evidence demand a verdict? Precambrian Research 91, 181–226.

de Wit, M.J., 2004. Archean Greenstone Belts do contain fragments of ophiolites. In: Kusky, T.M. (Ed.), Precambrian Ophiolites and Related Rocks. Developments in Precambrian Geology, vol. 13. Elsevier, Amsterdam, pp. 599–614.

de Wit, M.J., Ashwal, L.D. (Eds.), 1997. Greenstone Belts. Clarendon Press, Oxford, 809 pp.

de Wit, M.J., Ashwall, L.D., 1997. Preface: Convergence towards divergent models of greenstone belts. In: de Wit, M.J., Ashwal, L.D. (Eds.), Greenstone Belts. Oxford University Press, Oxford, UK, pp. i–xvii.

de Wit, M.J., Hart, R.A., 1993. Earth's earliest continental lithosphere, hydrothermal flux and crustal recycling. Lithos 30, 309–335.

de Wit, M.J., Hart, R., Martin, A., Abbott, P., 1982. Archean abiogenic and probable biogenic structures associated with mineralized hydrothermal vent systems and regional metasomatism, with implications for greenstone belt studies. Economic Geology 77, 1783–1802.

de Wit, M.J., Fripp, R.E.P., Stanistreet, I.G., 1983. Tectonic and stratigraphic implications of new field observations along the southern part of the Barberton greenstone belt. In: Anhaeusser, C.R. (Ed.), Contributions to the Geology of the Barberton Mountain Land. Geological Society of South Africa, Special Publication 9, pp. 21–29.

de Wit, M.J., Armstrong, R., Hart, R.J., Wilson, A.H., 1987a. Felsic igneous rocks within the 3.3- to 3.5 Ga Barberton greenstone belt: high crustal level equivalents of the surrounding tonalite-trondhjemite terrain, emplaced during thrusting. Tectonics 6, 529–549.

de Wit, M.J., Hart, R.A., Hart, R.J., 1987b. The Jamestown ophiolite complex, Barberton mountain belt: a section through 3.5 Ga oceanic crust. Journal of African Earth Science 6, 681–730.

de Wit, M.J., Roering, C., Hart, R.J., Armstrong, R.A., de Ronde, C.E.J., Green, R.W.E., Tredoux, M., Peberdy, E., Hart, R.A., 1992. Formation of an Archaean continent. Nature 357, 553–562.

Didier, J., Barbarin, B., 1991. Enclaves and Granite Petrology. Elsevier, Amsterdam.

Diener, J.F.A., Stevens, G., Kisters, A.F.M., Poujol, M., 2005. Metamorphism and exhumation of the basal parts of the Barberton greenstone belt, South Africa: Constraining the rates of Mesoarchaean tectonism. Precambrian Research 143 (1–4), 87–112.

Diener, J.G., Stevens, G., Kisters, A.F.M., 2006. High-pressure, low-temperature metamorphism in the southern Barberton granitoid greenstone terrain, South Africa: a record of overthickening and collapse of Mid-Archean continental crust. In: Benn, K., Mareschal, J.-C., Condie, K.C. (Eds.), Archean Geodynamic Processes. American Geophysical Union, Monograph 164, pp. 239–254.

DiMarco, M.J., Lowe, D.R., 1989a. Shallow-water volcaniclastic deposition in the Early Archean Panorama Formation, Warrawoona Group, eastern Pilbara Block, Western Australia. Sedimentary Geology 64, 43–63.

DiMarco, M.J., Lowe, D.R., 1989b. Stratigraphy and sedimentology of an Early Archean felsic volcanic sequence, eastern Pilbara Block, Western Australia, with special reference to the Duffer Formation and implications for crustal evolution. Precambrian Research 44, 147–169.

Dimroth, E., Kimberley, M.M., 1976. Precambrian atmospheric oxygen – evidence in sedimentary distributions of carbon, sulfur, uranium, and iron. Canadian Journal of Earth Sciences 13, 1161–1185.

Ding, T., Valkiers, S., Kipphardt, H., De Bievre, P., Taylor, P.D.P., Gonfiantini, R., Krouse, R., 2001. Calibrated sulfur isotope abundance ratios of three IAEA sulfur isotope reference materials and V-CDT with a reassessment of the atomic weight of sulfur. Geochimica et Cosmochimica Acta 65, 2433–2437.

Dirks, P., Jelsma, H.A., 1998. Horizontal accretion and the stabilization of the Archaean Zimbabwe craton. Geology 26, 11–14.

Dixon, E.T., Bogard, D.D., Garrison, D.H., 2003. ^{39}Ar-^{40}Ar chronology of R chondrites. Meteoritics and Planetary Science 38, 341–355.

Dobos, S.K., Jacobsen, S.B., Derry, L.A., Goldstein, S.J., 1986. A Nd and Sr isotopic study of metasediments from the Western Gneiss Belt (WGB) of Western Australia. Eos 67, 1265.

Dobretsov, N.L., 2005. 250 Ma large igneous provinces of Asia: Siberian and Emeishan traps (plateau basalts) and associated granitoids. Russian Geology and Geophysics 46 (9), 870–890.

Dodson, M.H., Compston, W., Williams, I.S., Wilson, J.F., 1988. A search for ancient detrital zircons in Zibabwean sediments. Journal of the Geological Society of London 145, 977–983.

Doe, B.R., Delevaux, M.H., 1980. Lead-isotope investigations in the Minnesota River Valley – Late-tectonic and post-tectonic granites. In: Geological Society of America, Special Paper 182, pp. 105–112.

Donahue, T.M., 1999. New analysis of hydrogen and deuterium escape from Venus. Icarus 141–156, 226.

Donahue, T.M., Grinspoon, D.H., Hartle, R.E., Hodges, R.R., 1997. Ion/neutral escape of hydrogen and deuterium: Evolution of water. In: Bouger, S.W., Hunten, D.M., Phillips, R.J. (Eds.), Venus II, Geology, Geophysics, Atmosphere, and Solar Wind Environment. University of Arizona Press, Tucson, AZ, pp. 385–414.

Donahue, T.M., Russell, C.T., 1997. The Venus atmosphere and ionosphere and their interaction with the solar wind: An overview. In: Bouger, S.W., Hunten, D.M., Phillips, R.J. (Eds.), Venus II, Geology, Geophysics, Atmosphere, and Solar Wind Environment. The University of Arizona Press, Tucson, AZ, pp. 3–31.

Donaldson, J.A., De Kemp, E.A., 1998. Archaean quartz arenites in the Canadian Shield: examples from the Superior and Churchill Provinces. Sedimentary Geology 120, 153–176.

Donskaya, T.V., Salnikova, E.B., Sklyarov, E.V., 2002. Early Proterozoic postcollision magmatism at the southern flank of the Siberian Craton: new geochronological data and geodynamic implication. Doklady Earth Sciences 383 (2), 125–128.

Downes, H., MacDonald, R., Upton, B.G.J., Cox, K.G., Bodinier, J.L., Mason, P.R.D., James, D., Hill, P.G., Hearn, B.C., Jr, 2004. Ultramafic xenoliths from the Bearpaw Mountains, Montana, USA; evidence for multiple metasomatic events in the lithospheric mantle beneath the Wyoming Craton. Journal of Petrology 45, 1631–1662.

Drake, M.J., 2001. The eucrite/Vesta story. Meteoritics and Planetary Science 36, 501–513.

Drake, M.J., Righter, K., 2002. Determining the composition of the Earth. Nature 416, 39–43.

Dreibus, G., Palme, H., Spettel, B., Zipfel, J., Wänke, H., 1995. Sulfur and selenium in chondritic meteorites. Meteoritics 30, 439–445.

Drennan, G.R., Robb, L.J., Meyer, F.M., Armstrong, R.A., de Bruiyn, H., 1990. The nature of the Archaean basement in the hinterland of the Witwatersrand Basin: II. A crustal profile west of the Welkom Goldfield and comparisons with the Vredefort crustal profile. South African Journal of Geology 93, 41–53.

Dresselhaus, M.S., Dresselhaus, G., 1981. Intercalation compounds of graphite. Advances in Physics 30, 290–298.

Dresselhaus, M.S., Dresselhaus, G., 1982. In: Cardona, M., Guntherodt, G. (Eds.), Light Scattering in Solids, vol. III. Springer, Berlin, 187 pp.

Drieberg, S.L., 2004. The magmatic-hydrothermal architecture of the Archean volcanic massive sulfide (VMS) system at Panorama, Pilbara, Western Australia. Unpublished Ph.D. thesis. University of Western Australia, 327 pp.

Drummond, B.J., 1988. A review of crust/upper mantle structure in the Precambrian areas of Australia and implications for Precambrian crustal evolution. Precambrian Research 40–41, 101–116.

Drummond, M.S., Defant, M.J., 1990. A model for trondhjemite-tonalite-dacite genesis and crustal growth via slab melting: Archean to modern comparisons. Journal of Geophysical Research 95, 21,503–21,521.

Drummond, M.S., Defant, M.J., Kepezhinskas, P.K., 1996. Petrogenesis of slab-derived trondhjemite-tonalite-dacite/adakite magmas. Transactions of the Royal Society of Edinburgh: Earth Sciences 87, 205–215.

Drury, S.A., Holt, R.W., 1980. The tectonic framework of the South Indian craton: a reconnaissance involving LANDSAT imagery. Tectonophysics 65, 71–75.

Duchac, K.C., Hanor, J.S., 1987. Origin and timing of the metasomatic silicification of an early Archean komatiite sequence, Barberton Mountain Land, South Africa. Precambrian Research 37, 125–146.

Dunlop, J.S.R., Muir, M.D., Milne, V.A., Groves, D.I., 1978. A new microfossil assemblage from the Archaean of Western Australia. Nature 274, 676–678.

Dunn, S.J., Nemchin, A.A., Cawood, P.A., Pidgeon, R.T., 2005. Provenance record of the Jack Hills Metasedimentary Belt: Source of the Earth's oldest zircons. Precambrian Research 138, 235–254.

Dymek, R.F., Klein, C., 1988. Chemistry, petrology and origin of banded iron-formation lithologies from the 3800 Ma Isua supracrustal belt, West Greenland. Precambrian Research 37, 247–302.

Dziggel, A., Stevens, G., Poujol, M., Anhaeusser, C.R., Armstrong, R.A., 2002. Metamorphism of the granite-greenstone terrane south of the Barberton greenstone belt, South Africa: an insight into the tectono-thermal evolution of the 'lower' portions of the Onverwacht Group. Precambrian Research 114, 221–247.

Dziggel, A., Armstrong, R.A., Stevens, G., Nasdala, L., 2005. Growth of zircon and titanite during metamorphism in the granitoid-gneiss terrane south of the Barberton greenstone belt, South Africa. Mineralogical Magazine 69 (6), 1019–1036.

Dziggel, A., Stevens, G., Poujol, M., Armstrong, R.A., 2006a. Contrasting source components of clastic metasedimentary rocks in the lowermost formations of the Barberton greenstone belt. In: Reimold, W.U., Gibson, R.L. (Eds.), Processes on the Earth Earth. Geological Society of America, Special Paper 405, pp. 157–172.

Dziggel, A., Knipfer, S., Kisters, A.F.M., Meyer, F.M., 2006b. P-T and structural evolution of high-T, medium-P basement rocks in the Barberton Mountain Land, South Africa. Journal of Metamorphic Geology 24, 535–551.

Eby, G.N., 1992. Chemical subdivision of the A-type granitoids: petrogenetic and tectonic implications. Geology 20, 641–644.

Eggins, S.M., Woodhead, J.D., Kinsley, L.P.J., Mortimer, G.E., Sylvester, P., McCulloch, M.T., Hergt, J.M., Handler, M.R., 1997. A simple method for the precise determination of >40 trace elements in geological samples by ICPMS using enriched isotope internal standardisation. Chemical Geology 134, 311–326.

Eglington, B.M., Armstrong, R.A., 2004. The Kaapvaal Craton and adjacent orogens, southern Africa: a geochronological database and overview of the geological development of the craton. South African Journal of Geology 107, 13–32.

Eiler, J.M., Farley, K.A., Valley, J.W., Hauri, E., Craig, H., Hart, S.R., Stolper, E.M., 1997. Oxygen isotope variation in ocean island basalt phenocrysts. Geochimica et Cosmochimica Acta 61, 2281–2293.

Elias, M., 1982. Belele, W.A. Western Australia Geological Survey, 1:250,000 Geological Series Explanatory Notes, first ed., 22 pp.

Eldridge, C.S., Barton, P.B., Jr, Ohmoto, H., 1983. Mineral textures and their bearing on formation of the Kuroko orebodies. In: Economic Geology, Monograph 5, pp. 241–281.

Elias, M., 1982. Explanatory notes of the Belele geological sheet. In: Geological Survey of Western Australia, Perth, WA, pp. 1–22.

Elkins-Tanton, L.T., 2005. Continental mechanism caused by lithosphere delamination. In: Geological Society of America, Special Paper 388, pp. 449–461.

Elkins-Tanton, L., Hager, B., 2005. Giant meteoroid impacts can cause volcanism. Earth and Planetary Science Letters 239, 219–232.

Elkins-Tanton, L.T., Hager, B.H., Grove, T.L., 2004. Magmatic effects of the lunar late heavy bombardment. Earth and Planetary Science Letters 222, 17–27.

Elliot, T.T., Plank, T., Zindler, A., White, W., Bourdon, B., 1997. Element transport from slab to volcanic front at the Mariana arc. Journal of Geophysical Research 102, 14,991–15,019.

Endress, M., Bischoff, A., 1993. Mineralogy, degree of brecciation, and aqueous alteration of CI chondrites Orgueil, Ivuna and Alais. Meteoritics 28, 345–346.

Endress, M., Bischoff, A., 1996. Carbonates in CI chondrites: clues to parent body evolution. Geochimica et Cosmochimica Acta 60, 489–507.

Endress, M., Zinner, E., Bischoff, A., 1996. Early aqueous activity on primitive meteorite parent bodies. Nature 379, 701–703.

England, P.C., Thompson, B., 1984. Pressure–temperature–time paths of regional metamorphism. Journal of Petrology 25 (4), 894–955.

Eriksson, K.A., 1978. Alluvial and destructive beach facies from the Archaean Moodies Group, Barberton Mountain Land, South Africa and Swaziland. In: Miall, A.D. (Ed.), Fluvial Sedimentology. Canadian Society of Petroleum Geologists, Memoir 5, pp. 287–311.

Eriksson, K.A., 1979. Marginal marine depositional processes from the Archean Moodies Group, Barberton Mountain Land, South Africa: Evidence and significance. Precambrian Research 8, 153–182.

Eriksson, K.A., 1980. Transitional sedimentation styles in the Moodies and Fig Tree Groups, Barberton Mountain Land, South Africa: evidence favouring an Archaean continental margin. Precambrian Research 12, 141–160.

Eriksson, P.G., Catuneanu, O., 2004. Tectonism and mantle plumes through time. In: Eriksson, P.G., Altermann, W., Nelson, D.R., Mueller, W.U., Catuneanu, O. (Eds.), The Precambrian Earth: Tempos and Events. Elsevier, Amsterdam, pp. 161–163 and pp. 267–270 (Chapter 3).

Ernst, R.E., Buchan, K.L., 2002. Maximum size and distribution in time and space of mantle plumes: evidence from large igneous provinces. Journal of Geodynamics 34, 309–342.

Ernst, R.E., Buchan, K., 2003. Recognising mantle plumes in the geological record. Annual Review Earth and Planetary Sciences 31, 469–523.

Ernst, R.E., Grosfils, E.B., Mége, D., 2001. Giant dyke swarms on Earth, Venus and Mars. Annual Review of Earth and Planetary Sciences 29, 489–534.

Ernst, R.E., Desnoyers, D.W., Head, J.W., Grosfils, E.B., 2003. Graben-fissure systems in Guinevere Planitia and Beta Regio (264 degrees-312 degrees E, 24 degrees-60 degrees N), Venus, and implications for regional stratigraphy and mantle plumes. Icarus 164, 282–316.

Ernst, R.E., Buchan, K.L., Campbell, I.H., 2005. Frontiers in large igneous province research. Lithos 79, 271–297.

Ernst, W.G., 1988. Tectonic history of subduction zones inferred from retrograde blueschist P-T paths. Geology 16, 1081–1084.

Facer, J., Downes, H., 2005. Has the Wyoming craton lost its keel? In: Abstracts of Earthscope in the Northern Rockies Workshop, MSU, Bozeman, MT, Sept. 2005.

Fanning, C.M., Moore, D.H., Bennett, V.C., Daly, S.J., Menot, R.P., Peucat, J.J., Oliver, R.L., 1999. The "Mawson Continent": The East Antarctic shield and Gawler Craton, Australia. In: Programme and Abstracts, 8th International Symposium Antarctic Earth Sciences, Wellington, p. 103 (abstract).

Farhat, J.S., Wetherilll, G.W., 1975. Interpretation of apparent ages in Minnesota. Nature 257, 721–722.

Farmer, G.L., DePaolo, D.J., 1983. Origin of Mesozoic and Tertiary granite in the western United States and implications for the pre-Mesozoic crustal structure 1. Nd and Sr isotopic studies in the geocline of the northern Great Basin. Journal of Geophysical Research 88, 3379–3401.

Farquhar, J., Wing, B.A., 2003. Multiple sulfur isotopes and the evolution of the atmosphere. Earth and Planetary Science Letters 213, 1–13.

Farquhar, J., Bao, H.M., Thiemens, M., 2000a. Atmospheric influence of Earth's earliest sulfur cycle. Science 289, 756–758.

Farquhar, J., Jackson, T.L., Thiemens, M.H., 2000b. A S-33 enrichment in ureilite meteorites: Evidence for a nebular sulfur component. Geochimica et Cosmochimica Acta 64, 1819–1825.

Farquhar, J., Savarino, J., Aireau, S., Thiemens, M.H., 2001. Observation of wavelength-sensitive mass-independent sulfur isotope effects during SO_2 photolysis: Implications for the early atmosphere. Journal of Geophysical Research – Planets 106 (E12), 32,829–32,839.

Farquhar, J., Wing, B.A., McKeegan, K.D., Harris, J.W., Cartigny, P., Thiemens, M.H., 2002. Mass-independent sulfur of inclusions in diamond and sulfur recycling on early Earth. Science 298, 2369–2372.

Farrell, T., 2006. Geology of the Eastern Creek 1:100,000 sheet. Western Australia Geological Survey, 1:100,000 Geological Series Explanatory Notes, 39 pp.

Farrow, D.J., Harmer, R.E., Hunter, D.R., Eglington, B.M., 1990. Rb-Sr and Pb-Pb dating of the Anhalt leucotonalite, northern Natal. South African Journal of Geology 93, 696–701.

Faure, G., 1986. Principles of Isotope Geology. John Wiley and Sons, New York, 589 pp.

Fedo, C.M., 2000. Setting and origin of problematic rocks from the >3.7 Ga Isua greenstone belt, southern West Greenland: Earth's oldest coarse clastic sediments. Precambrian Research 101, 69–78.

Fedo, C.M., Whitehouse, M.J., 2002a. Metasomatic origin of quartz-pyroxene rock, Akilia, Greenland, and implications for Earth's earliest life. Science 296, 1449–1452.

Fedo, C.M., Whitehouse, M.J., 2002b. Origin and significance of Archean quartzose rocks at Akilia, Greenland. Science 298, 917a.

Fedo, C.M., Eriksson, K.A., Krogstad, E.J., 1996. Geochemistry of shales from Archean (~3.0 Ga) Buhwa Greenstone Belt, Zimbabwe: Implications for provenance and source-area weathering. Geochimica et Cosmochimica Acta 60, 1751–1763.

Fedo, C.M., Myers, J.S., Appel, P.W.U., 2001. Depositional setting and paleogeographic implications of Earth's oldest supracrustal rocks, the >3.7 Ga Isua greenstone belt, West Greenland. Sedimentary Geology 141–142, 61–77.

Fedo, C.M., Whitehouse, M.J., Kamber, B.S., 2006. Geological Constraints in Assessing the Earliest Life on Earth: A Perspective from the Early Archaean (>3.7 Ga) of Southwest Greenland. Philosophical Transactions of the Royal Society of London Ser. B 361, 851–867.

Fehr, M.A., Rehkämper, M., Halliday, A.N., Wiechert, U., Hattendorf, B., Gunter, D., Ono, S., Eigenbrode, J.L., Rumble, D., 2005. Tellurium isotopic composition of the early solar system – A search for effects resulting from stellar nucleosynthesis, Sn-126 decay, and mass-independent fractionation. Geochimica et Cosmochimica Acta 69, 5099–5112.

Feng, R., Kerrich, R., 1990. Geochemistry of fine-grained elastic sediments in the Archean Abitibi greenstone belt, Canada: Implications for provenance and tectonic setting. Geochimica et Cosmochimica Acta 54, 1061–1081.

Feng, R., Kerrich, R., 1991. Single zircon age constraints on the tectonic juxtapositon of the Archean Abitibi greenstone belt and Pontiac Subprovince, Quebec, Canada. Geochimica et Cosmochimica Acta 55, 3437–3441.

Feng, R., Kerrich, R., 1992. Geochemical evolution of granitoids from the Archean Abitibi southern volcanic zone and the Pontiac subprovince, Superior Province, Canada: implications for tectonic history and source regions. Chemical Geology 98, 23–70.

Fergusson, K.M., 1999. Lead, Zinc and Silver Deposits of Western Australia. Western Australia Geological Survey, Mineral Resources Bulletin 15, 314 pp.

Fergusson, K.M., Ruddock, I., 2001. Mineral occurrences and exploration potential of the East Pilbara. Geological Survey of Western Australia, Report 81, 114 pp.

Ferry, J.M., Spear, F.S., 1978. Experimental calibration of the partitioning of Fe and Mg between biotite and garnet. Contributions to Mineralogy and Petrology 66, 113–117.

Finucane, K.J., 1936. Marble Bar mining centre, Pilbara goldfield. Aerial Geology and Geophysical Survey of Northern Australia, Western Australia Report 8.

Finucane, K.J., 1937. Halley's Comet centre, Pilbara goldfield. Aerial Geology and Geophysical Survey of Northern Australia, Western Australia Report 10.

Fisher, L.B., Stacey, J.S., 1986. Uranium-lead zircon ages and common lead measurements for the Archean gneisses of the Granite Mountains, Wyoming. In: Peterman, Z.E., Schnable, D.C. (Eds.), Shorter Contributions to Isotopic Research. U.S. Geological Survey, Bulletin 1622, pp. 13–24.

Fisher-Worrell, G.F., 1985. Sedimentology and mineralogy of silicified evaporites in the basal Kromberg Formation, South Africa. Masters thesis. Baton Rouge, Louisiana State University, 152 pp.

Fitton, J.G., Saunders, A.D., Norry, M.J., Hardarson, B.S., Taylor, R.N., 1997. Thermal and chemical structure of the Iceland plume. Earth and Planetary Science Letters 153, 197–208.

Fitzsimons, I.C.W., 2000a. Grenville-age basement provinces in East Antarctica: Evidence for three separate collisional orogens. Geology 28, 879–882.

Fitzsimons, I.C.W., 2000b. A review of tectonic events in the East Antarctic Shield and their implications for Gondwana and earlier supercontinents. Journal of African Earth Science 31, 3–23.

Fitzsimons, I.C.W., 2003. Proterozoic basement provinces of southern and southwestern Australia, and their correlation with Antarctica. In: Yoshida, M., Windley, B.W., Dasgupta, S., Powell, C.McA. (Eds.), Proterozoic East Gondwana: Supercontinent Assembly and Breakup. Geological Society of London, Special Publication 206, pp. 93–130.

Fitzsimons, I.C.W., Kinny, P.D., Harley, S.L., 1997. Two stages of zircon and monazite growth in anatectic leucogneiss: SHRIMP constraints on the duration and intensity of Pan-African metamorphism in Prydz Bay, East Antarctica. Terra Nova 9, 47–51.

Flasar, F.M., Birch, F., 1973. Energetics of core formation: a correction. Journal of Geophysical Research 78, 6101–6103.

Fleet, M.E., 1987. Partition of Ni between olivine and sulfide – the effect of temperature, fO_2 and fS_2. Contributions to Mineralogy and Petrology 95, 336–342.

Fletcher, J.A., 2003. The Geology, Geochemistry and Fluid Inclusion Characteristics of the Wyldsdale Gold-Bearing Pluton, Northwest Swaziland. University of the Witwatersrand, Johannesburg, 215 pp.

Fletcher, I.R., Williams, S.J., Gee, R.D., Rosman, K.J.R., 1983. Sm-Nd model ages across the margins of the Archaean Yilgarn Block, Western Australia – northwest transect into the Proterozoic Gascoyne Province. Journal of the Geological Society of Australia 30, 167–174.

Fletcher, I.R., Rosman, K.J.R., Libby, W.G., 1988. Sm-Nd, Pb-Pb and Rb-Sr geochronology of the Manfred Complex, Western Australia. Precambrian Research 38, 343–354.

Fletcher, I.R., Libby, W.G., Rosman, K.J.R., 1994. Sm-Nd model ages of granitoid rocks in the Yilgarn Craton. In: Western Australia Geological Survey, Report 37, pp. 61–73.

Fletcher, I.R., McNaughton, N.J., Pidgeon, R.T., Rosman, K.J.R., 1997. Sequential closure of K-Ca and Rb-Sr isotopic systems in Archaean micas. Chemical Geology 138, 289–301.

Foley, S.F., Barth, M.G., Jenner, G.A., 2000. Rutile/melt partition coefficients for trace elements and an assessment of the influence of rutile on the trace element characteristics of subduction zone magmas. Geochimica et Cosmochimica Acta 64, 933–938.

Foley, S.F., Tiepolo, M., Vannucci, R., 2002. Growth of early continental crust controlled by melting of amphibolite in subduction zones. Nature 417, 637–640.

Folinsbee, R.E., Baadsgaard, H., Cumming, G.L., Green, D.C., 1968. A very ancient island arc. In: Knopoff, L., Drake, C.L., Hart, P.J. (Eds.), The Crust and Upper Mantle of the Pacific Area. American Geophysical Union, Geophysical Monograph 12, pp. 441–448.

Ford, J.P., Plaut, J.J., Weitz, C.M., Farr, T.G., Senske, D.A., Stofan, E.R., Michaels, G., Parker, T.J., 1993. Guide to Magellan Image Interpretation. NASA-JPL Publication 93-24.

Ford, P.G., Pettengill, G.H., 1992. Venus topography and kilometer-scale slopes. Journal of Geophysical Research 97, 13,103–13,114.

Foriel, J., Philippot, P., Rey, P., Somogyi, A., Banks, D., Menez, B., 2004. Biological control of Cl/Br and low sulfate concentration in a 3.5-Gyr-old seawater from North Pole, Western Australia. Earth and Planetary Science Letters 228, 451–463.

Foster, D., Mueller, P.A., Vogl, J., Mogk, D., Wooden, J.L., Heatherington, A., 2006. Proterozoic evolution of the western margin of the Wyoming craton: implications for the tectonic and magmatic evolution of the northern Rocky Mountains. Canadian Journal of Earth Sciences 43, 1601–1619.

Foulger, G.R., Natalnd, J.H., Presnell, D.C., Anderson, D.L. (Eds.), 2005. Plates, Plumes and Paradigms. Geological Society of America, Special Paper 388, 593 pp.

Franck, S., 1998. Evolution of the global mean heat flow over 4.6 Gyr. Tectonophysics 291, 9–18.

Francis, D., 2003. Mafic cratonic mantle roots, remnants of a more chondritic Archean mantle? Lithos 71, 135–152.

Francis, D., 2004. Mafic magmas of the 3.8 Ga Nuvvuagittuq greenstone belt, Ungava, Québec. American Geophysical Union, Spring Meeting 2004, abstract #V12B-06.

Frank, S.L., Head, J.W., 1990. Ridge belts on Venus: Morphology and origin. Earth, Moon, and Planets 50–51, 421–470.

Fraústo da Silva, J.J.R., Williams, R.J.P., 2001. The Biological Chemistry of the Elements, second ed. Oxford University Press, Oxford, UK, 575 pp.

Frei, R., Polat, A., 2006. Source heterogeneity for the major components of >3.7 Ga Banded Iron Formations (Isua Greenstone Belt, Western Greenland): Tracing the nature of interacting water masses in BIF formation. Earth and Planetary Science Letters 253, 266–281.

Frei, R., Rosing, M.T., 2001. The least radiogenic terrestrial leads: implications for the early Archaean crustal evolution and hydrothermal-metasomatic processes in the Isua Supracrustal Belt (West Greenland). Chemical Geology 181, 47–66.

Frei, R., Rosing, M.T., 2005. Search for traces of the late heavy bombardment on Earth – Results from high precision chromium isotopes. Earth and Planetary Science Letters 236 (1–2), 28–40.

Frei, R., Bridgwater, D., Rosing, M., Stecher, O., 1999. Controversial Pb-Pb and Sm-Nd isotope results in the early Archean Isua (West Greenland) oxide iron formation: Preservation of primary signatures versus secondary disturbances. Geochimica et Cosmochimica Acta 63, 473–488.

Frei, R., Polat, A., Meibom, A., 2004. The Hadean upper mantle conundrum; evidence for source depletion and enrichment from Sm-Nd, Re-Os, and Pb isotopic compositions in 3.71 Gy boninite-like metabasalts from the Isua supracrustal belt, Greenland. Geochimica et Cosmochimica Acta 68, 1645–1660.

French, B.M., 1998. Traces of Catastrophe. Lunar Planetary Science Contribution 954. Lunar and Planetary Institute, Houston, TX, 120 pp.

Friend, C.R.L., Nutman, A.P., 2005a. New pieces to the Archaean terrane jigsaw puzzle in the Nuuk region, southern West Greenland: Steps in transforming a simple insight into a complex regional tectonothermal model. Journal of the Geological Society London 162, 147–163.

Friend, C.R.L., Nutman, A.P., 2005b. Complex 3670–3500 Ma orogenic episodes superimposed on juvenile crust accreted between 3850–3690 Ma, Itsaq Gneiss Complex, southern West Greenland. Journal of Geology 113, 375–398.

Friend, C.R.L., Nutman, A.P., McGregor, V.R., 1987. Late Archaean tectonics in the Faeringehavn-Tre Brodre area, south of Buksefjorden, southern West Greenland. Journal of the Geological Society London 144, 369–376.

Friend, C.R.L., Nutman, A.P., McGregor, V.R., 1988. Late Archean terrane accretion in the Godthab region, southern West Greenland. Nature 335, 535–538.

Friend, C.R.L., Bennett, V.C., Nutman, A.P., 2002a. Abyssal peridotites >3800 Ma from southern West Greenland: field relationships, petrography, geochronology, whole-rock and mineral chemistry of dunite and harzburgite inclusions in the Itsaq Gneiss Complex. Contributions to Mineralogy and Petrology 143, 71–92.

Friend, C.R.L., Nutman, A.P., Bennett, V.C., 2002b. Technical comment on: Origin and significance of Archean quartzose rocks at Akilia, Greenland. Science 298, 917a.

Fritts, C.E., 1969. Bedrock geologic map of the Marenisco-Watersmeet area, Gogebic and Ontonagon Counties, Michigan. U.S. Geological Survey, Miscellaneous Geologic Investigations Map I-576, scale 1:48,000.

Frost, B.R., Avchenko, O.V., Chamberlain, K.R., Frost, C.D., 1998. Evidence for extensive Proterozoic remobilization of the Aldan shield and implications for Proterozoic plate tectonic reconstructions of Siberia and Laurentia. Precambrian Research 89, 1–23.

Frost, B.R., Frost, C.D., Cornia, M.E., Chamberlain, K.R., Kirkwood, R., 2006. The Teton-Wind River domain: a 2.68–2.67 Ga active margin in the western Wyoming Province. Canadian Journal of Earth Sciences 43, 1489–1510.

Frost, C.D., 1993. Nd isotopic evidence for the antiquity of the Wyoming province. Geology 21, 351–354.

Frost, C.D., Fanning, C.M., 2006. Archean geochronological framework of the Bighorn Mountains, Wyoming. Canadian Journal of Earth Sciences 43, 1399–1418.

Frost, C.D., Frost, B.R., 1993. The Archean history of the Wyoming province. In: Snoke, A.W., Steidtmann, J.R., Roberts, S.M. (Eds.), Geology of Wyoming. Geologic Survey of Wyoming, Memoir 5, pp. 58–77.

Frost, C.D., Frost, B.R., Chamberlain, K.R., Hulsebosch, T.P., 1998. The Late Archean history of the Wyoming province as recorded by granitic magmatism in the Wind River Range, Wyoming. Precambrian Research 89, 145–173.

Frost, C.D., Frost, B.R., Kirkwood, R., Chamberlain, K.R., 2006a. The tonalite-trondhjemite-grandiorite (TTG) to granodiorite-granite (GG) transition in the Late Archean plutonic rocks of the central Wyoming province. Canadian Journal of Earth Sciences 43, 1419–1444.

Frost, C.D., Fruchey, B.L., Chamberlain, K.R., Frost, B.R., 2006b. Archean crustal growth by lateral accretion of juvenile supracrustal belts in the southern Wyoming province. Canadian Journal of Earth Sciences 43, 1533–1555.

Froude, C.F., Ireland, T.R., Kinny, P.D., Williams, I.S., Compston, W., Williams, I.R., Myers, J.S., 1983a. Ion-microprobe identification of 4100–4200 Myr old terrestrial zircons. Nature 304, 616–618.

Froude, D.O., Compston, W., Williams, I.S., 1983b. Early Archaean zircon analyses from the central Yilgarn Block. In: Australian National University, Canberra, Research School of Earth Sciences (RSES), Annual Report 1983, pp. 124–126.

Fruchey, B.L., 2002. Archean supracrustal sequences of contrasting origin: the Archean history of the Barlow Gap area, northern Granite Mountains, Wyoming. Unpublished M.Sc. thesis. University of Wyoming, Laramie, 178 pp.

Fryer, B.J., 1977. Rare earth evidence in iron-formations for changing Precambrian oxidation states. Geochimica et Cosmochimica Acta 41, 361–367.

Fryer, B.J., Fyfe, W.S., Kerrich, R., 1979. Archaean volcanogenic oceans. Chemical Geology 24, 25–33.

Fu, B., Page, F.Z., Cavosie, A.J., Clechenko, C.C., Fournelle, J., Kita, N.T., Lackey, J.S., Wilde, S.A., Valley, J.W., in press. Ti-in-zircon thermometry. Contributions to Mineralogy and Petrology.

Fuchs, G., Thauer, R., Ziegler, H., Stichler, W., 1979. Carbon isotope fractionation by *Methanobacterium thermoautotrophicum*. Archives of Microbiology 120, 135–139.

Furnes, H., Banerjee, N.R., Muehlenbachs, K., Staudigal, H., de Wit, M., 2004. Early life recorded in Archean pillow lavas. Science 304, 578–581.

Furnes, H., Banerjee, N.R., Staudigel, H., Muehlenbachs, K., de Wit, M., Van Kranendonk, M., in press. Bioalteration textures in Recent to Paleoarchean pillow lavas: A petrographic signature of subsurface life on Earth. Precambrian Research.

Gafarov, R.A., Leites, A.M., Fedorovsky, V.S., Prozorov, I.P., Savinskaya, M.S., Savinsky, K.A., 1978. Tectonic zones in the basement of the Siberian craton and its continental crust formation stages. Geotectonics 1, 43–58.

Ganapathy, R., Anders, E., 1974. Bulk compositions of the Moon and Earth, estimated from meteorites. In: Proceedings of the 5th Lunar Science Conference, pp. 1181–1206.

Gao, L.Z., Zhao, T., Wan, Y.S., Zhao, X., Ma, Y.S., Yang, S.Z., 2006. Report on 3.4 Ga SHRIMP zircon age from the Yuntaishan Geopark in Jiaozuo, Henan Province. Acta Geologica Sinica 80, 52–57.

Gao, X., Thiemens, M.H., 1989. Multi-isotope sulfur isotope ratios (δ^{33}S, δ^{34}S, δ^{36}S) in meteorites. Meteoritics 24, 269.

Gao, X., Thiemens, M.H., 1993a. Isotopic composition and concentration of sulfur in carbonaceous chondrites. Geochimica et Cosmochimica Acta 57, 3159–3169.

Gao, X., Thiemens, M.H., 1993b. Variations of the isotopic composition of sulfur in enstatite and ordinary chondrites. Geochimica et Cosmochimica Acta 57, 3171–3176.

Gao, Y.Q., Marcus, R.A., 2001. Strange and unconventional isotope effects in ozone formation. Science 293, 259–263.

Garcia-Ruiz, J.M., Hyde, S.T., Carnerup, A.M., Christy, A.G., Van Kranendonk, M.J., Welham, N.J., 2003. Self-assembled silica-carbonate structures and detection of ancient microfossils. Science 302, 1194–1197.

Gardien, V., Thompson, A.B., Grujic, D., Ulmer, P., 1995. Experimental melting of biotite + plagioclase + quartz + or − muscovite assemblages and implications for crustal melting. Journal of Geophysical Research, B, Solid Earth and Planets 100 (8), 15,581–15,591.

Gauthier, L., Hagemann, S., Robert, F., Pickens, G., 2004. New constraints on the architecture and timing of the giant Golden Mile gold deposit, Kalgoorlie, Western Australia. In: Muhling, J., et al. (Eds.), SEG Extended abstracts, Perth, pp. 353–356.

Gay, N.C., 1969. The analysis of strain in the Barberton Mountain Land, eastern Transvaal, using deformed pebbles. Journal of Geology 77, 377–396.

Gee, R.D., 1982. Southern Cross, W.A. Western Australia Geological Survey, 1:250,000 Geological Series Explanatory Notes, 25 pp.

Gee, R.D., Baxter, J.L., Wilde, S.A., Williams, I.R., 1981. Crustal development in the Archaean Yilgarn Block, Western Australia. In: Glover, J.A., Groves, D.I. (Eds.), Archaean Geology. Geological Society of Australia, Special Publication 7, pp. 43–56.

Geng, Y.S., Yang, C.H., Wan, Y.S., 2006. Palaeoproterozoic granitic magmatism in the Luliang area, NCC: constraints from isotopic geochronology. Acta Petrologica Sinica 22, 305–314 (in Chinese with English abstract).

Genshaft, N.S., 1996. Intrinsic factors of the tectonic mobility of cratons. Geotectonics 4, 13–24.

Geological Map of Swaziland, 1:250,000, 1982. Geol. Surv. Mines Dept., Swaziland.

German, C.R., von Damm, K.L., 2003. Hydrothermal processes. Treatise on Geochemistry 6, pp. 181–222.

Ghent, R.R., Hansen, V.L., 1999. Structural and kinematic analysis of eastern Ovda Regio, Venus: Implications for crustal plateau formation. Icarus 139, 116–136.

Ghent, R.R., Phillips, R.J., Hansen, V.L., Nunes, D.C., 2005. Finite element modeling of short-wavelength folding on Venus: Implications for the plume hypothesis for crustal plateau formation. Journal of Geophysical Research 110, doi: 10.1029/2005JE002522.

Ghent, R.R., Tibuleac, I.M., 2000. Ribbon spacing in Venusian tessera: Implications for layer thickness and thermal state. Geophysical Research Letters 29, 994–997.

Gibbs, A.K., Payne, B., Setzer, T., Brown, L.D., Oliver, J.E., Kaufman, S., 1984. Seismic Reflection Study of the Precambrian Crust of Central Minnesota. Bulletin of the Geological Society of America 95, 280–294.

Gibson, H.L., Watkinson, D.H., Comba, C.D.A., 1983. Silicification: Hydrothermal alteration in an Archean geothermal system within the Amulet Rhyolite Formation, Noranda, Quebec. Economic Geology 78, 954–971.

Gilbert, G.K., 1886. Inculcation of the scientific method. American Journal of Science 31, 284–299.

Giles, C.W., Hallberg, J.A., 1982. The genesis of the Archaean Welcome Wells volcanic province, Western Australia. Contributions to Mineralogy and Petrology 80, 307–318.

Gilmour, J.D., Saxton, J.M., 2001. A time-scale for the formation of the first solids. Philosophical Transactions of the Royal Society of London Ser. A 359, 2037–2048.

Gilmour, J.D., Whitby, J.A., Turner, G., Bridges, J.C., Hutchison, R., 2000. The iodine-xenon system in clasts and chondrules from ordinary chondrites: Implications for early solar system chronology. Meteoritics and Planetary Science 35, 445–455.

Gilmore, M.S., Collins, G.C., Ivanov, M.A., Marinangeli, L., Head, J.W., 1998. Style and sequence of extensional structures in tessera terrain, Venus. Journal of Geophysical Research 103, 16,813–16,840.

Gladkochub, D.P., Donskaya, T.V., Mazukabzov, A.M., Salnikova, E.B., Sklyarov, E.V., Yakovleva, S.Z., 2005. The age and geodynamic interpretation of the Kitoi granitoid complex (southern Siberian craton). Russian Geology and Geophysics 46 (11), 1121–1133.

Gladkochub, D.P., Sklyarov, E.V., Menshagin, Yu.V., Mazukabzov, A.M., 2001. Geochemistry of ancient ophiolites of the Sharyzhalgay uplift. Geochemistry International 39 (10), 947–958.

Glass, B.P., Burns, C.A., 1988. Microkrystites: a new term for impact-produced glassy spherules containing primary crystallites. In: Proceedings Lunar and Planetary Science Conference, vol. 18, pp. 455–458.

Glavin, D.P., Kubny, A., Jagoutz, G.W., Lugmair, G.W., 2004. Mn-Cr isotope systematics of the D'Orbigny angrite. Meteoritics and Planetary Science 39, 693–700.

Glikson, A.Y., 1970. Geosynclinal evolution and geochemical affinities of early Precambrian systems. Tectonophysics 9, 397–433.

Glikson, A.Y., 1972. Early Precambrian evidence of a primitive ocean crust and island arc nuclei of sodic granite. Bulletin of the Geological Society of America 83, 3323–3344.

Glikson, A.Y., 1976. Stratigraphy and evolution of primary and secondary greenstones: significance of data from Shields of the southern hemisphere. In: Windley, B.F. (Ed.), The Early History of the Earth. Wiley, London, pp. 257–277.

Glikson, A.Y., 1979. Early Precambrian tonalite-trondhjemite sialic nuclei. Earth Science Reviews 15, 1–73.

Glikson, A.Y., 1980. Uniformitarian assumptions, plate tectonics and the Precambrian Earth. In: Kröner, A. (Ed.), Precambrian Plate Tectonics. Elsevier, Amsterdam, pp. 91–104.

Glikson, A.Y., 2001. The astronomical connection of terrestrial evolution: crustal effects of post-3.8 Ga mega-impact clusters and evidence for major 3.2 ± 0.1 Ga bombardment of the Earth–Moon system. Journal of Geodynamics 32, 205–229.

Glikson, A.Y., 2005a. Asteroid/comet impact clusters, flood basalts and mass extinctions: significance of isotopic age overlaps. Earth and Planetary Science Letters 236, 933–937.

Glikson, A.Y., 2005b. Geochemical and isotopic signatures of Archaean to Palaeoproterozoic extraterrestrial impact ejecta/fallout units. Australian Journal of Earth Science 52 (4–5), 785–798.

Glikson, A.Y., 2005c. Geochemical signatures of Archean to Early Proterozoic mare-scale oceanic impact basins. Geology 133, 125–128.

Glikson, A.Y., 2006. Asteroid impact ejecta units overlain by iron-rich sediments in 3.5–2.4 Ga terrains, Pilbara and Kaapvaal cratons: accidental or cause-effect relationships? Earth and Planetary Science Letters 246, 149–160.

Glikson, A., 2006. Comment on "Zircon thermometer reveals minimum melting conditions on earliest Earth" I. Science, 311 (5762).

Glikson, A.Y., Allen, C., 2004. Iridium anomalies and fractionated siderophile element patterns in impact ejecta, Brockman Iron-formation, Hamersley Basin, Western Australia: evidence for a major asteroid impact in simatic crustal regions of the early Proterozoic Earth. Earth and Planetary Science Letters 220, 247–264.

Glikson, A.Y., Hickman, A.H., 1981. Geochemistry of Archaean volcanic successions, eastern Pilbara Block, Western Australia. Australian Bureau of Mineral Resources, Geology and Geophysics, Record 1981/36, 56 pp.

Glikson, A.Y., Vickers, J., 2005. The 3.26–3.24 Ga Barberton asteroid impact cluster: tests of tectonic and magmatic consequences, Pilbara Craton, Western Australia. Earth and Planetary Science Letters 241, 11–20.

Glikson, A.Y., Davy, R., Hickman, A.H., Pride, C., Jahn, B., 1987. Trace elements geochemistry and petrogenesis of Archaean felsic igneous units, Pilbara Block, Western Australia. Bureau of Mineral Resources, Record 1987/30, 63 pp.

Glikson, A.Y., Ivanov, B., Melosh, H.J., 2004. Impacts do not initiate volcanic eruptions close to the crater: comment and reply. Geology Online Forum, e47–e48.

Glukhovsky, M.Z., Moralev, V.M., 2001. Reconstruction of the tectonic evolution of the Gonam enderbite dome in the Aldan shield. Geotectonics 5, 10–25.

Goddéris, Y., Veizer, J., 2000. Tectonic control of chemical and isotopic composition of ancient oceans: the impact of continental growth. American Journal of Science 300, 434–461.

Goellnicht, N.M., Groves, I.M., Groves, D.I., Ho, S.E., McNaughton, N.J., 1988. A comparison between mesothermal gold deposits of the Yilgarn Block and gold mineralization at Telfer and Miralga Creek, Western Australia: indirect evidence for a non-magmatic origin for greenstone-hosted gold deposits. In: Geology Department and Extension Service, University of Western Australia, Publication No. 12, pp. 23–40.

Goldfarb, R.J., Groves, D.I., Gardoll, S., 2001. Orogenic gold and geologic time: a global synthesis. Ore Geology Reviews 18, 1–75.

Goldfarb, R.J., Baker, T., Dube, B., Groves, D.I., Hart, C.J.R., Gosselin, P., 2005. Distribution, character, and genesis of gold deposits in metamorphic terranes. Economic Geology, 100th Anniversary Volume, 407–450.

Goldich, S.S., Hedge, C.E., 1974. 3,800-Myr granitic gneisses in south-western Minnesota. Nature 252, 467–468.

Goldich, S.S., Wooden, J.L., 1980. Origin of the Morton gneiss, southwestern Minnesota: Part 3. Geochronology. In: Geological Society of America, Special Paper 182, pp. 77–94.

Goldich, S.S., Nier, A.O., Baadsgaard, H., Hoffman, J.H., Krueger, H.W., 1961. The Precambrian Geology and Geochronology of Minnesota. Minnesota Geological Survey, Bulletin 41, 193 pp.

Goldich, S.S., Hedge, C.E., Stern, T.W., 1970. Age of the Morton and Montevideo Gneisses and related rocks, southwestern Minnesota. Bulletin of the Geological Society of America 81, 3671–3696.

Goldich, S.S., Hedge, C.E., Stern, T.W., Wooden, J.L., Bodkin, J.B., North, R.M., 1980. Archean rocks of the Granite Falls area, southwestern Minnesota. In: Geological Society of America, Special Paper 182, pp. 19–43.

Goldich, S.S., Wooden, J.L., Ankenbauer, G.A., Levy, T.M., Suda, R.U., 1980. Origin of the Morton Gneiss, southwestern Minnesota; Part I, Lithology. In: Morey, G.B., Hanson, G.N. (Eds.), Selected Studies of Archean Gneisses and Lower Proterozoic Rocks, Southern Canadian Shield. Geological Society of America, Special Paper 182, pp. 45–50.

Golding, L.Y., Walter, M.R., 1979. Evidence of evaporite minerals in the Archaean Black Flag beds, Kalgoorlie, Western Australia. BMR Journal of Geology and Geophysics 4, 67–71.

Golding, S.D., Young, E., 2005. Multiple sulfur isotope evidence for dual sulfur sources in the 3.24 Ga Sulphur Springs VHMS deposit. Geochimica et Cosmochimica Acta 69 (Suppl.), A449.

Goldschmidt, V.M., 1954. Geochemistry. Clarendon Press, Oxford, UK, 730 pp.

Goldstein, S.J., Jacobsen, S.B., 1988. Nd and Sm isotopic systematics of rivers water suspended material: implications for crustal evolution. Earth and Planetary Science Letters 87, 249–265.

Goldstein, S.L., 1988. Decoupled evolution of Sr and Nd isotopes in the continental crust and mantle. Nature 336, 733–738.

Goldstein, S.L., O'Nions, R.K., Hamilton, P.J., 1984. A Sm-Nd isotopic study of atmospheric dusts and particulates from major river systems. Earth and Planetary Science Letters 70, 221–236.

Goodge, J.W., Fanning, C.M., 1995. 2.5 b.y. of punctuated Earth history as recorded in a single rock. Geology 27, 1007–1010.

Goodge, J.W., Fanning, C.M., Bennett, V.C., 2001. U-Pb evidence of ~1.7 Ga crustal tectonism during the Nimrod Orogeny in the Transantarctic Mountains, Antarctica: implications for Proterozoic plate reconstructions. Precambrian Research 112, 261–288.

Goodrich, C.A., 1992. Ureilites, a critical overview. Meteoritics 27, 227–252.

Goodwin, A.M., 1968. Archean protocontinental growth and early crustal history of the Canadian shield. In: 23rd International Geological Congress, Prague, vol. 1, pp. 69–89.

Goodwin, A.M., 1981. Precambrian perspectives. Science 213, 55–61.

Goodwin, A.M., Ridler, R.H., 1970. The Abitibi orogenic belt. In: Baer, A.J. (Ed.), Symposium on Basins and Geosynclines of the Canadian Shield. Geological Survey of Canada, Paper 70-40, pp. 1–30.

Göpel, C., Manhes, G., Allegre, C.J., 1992. U-Pb study of the Acapulco meteorite. Meteoritics 27, 226.

Göpel, C., Manhes, G., Allegre, C.J., 1994. U-Pb systematics of phosphates from equilibrated ordinary chondrites. Earth and Planetary Science Letters 121, 153–171.

Gorman, A.R., Clowes, R.M., Ellis, R.M., Henstock, T.J., Spence, G.D., Keller, G.R., Levander, A., Snelson, C.M., Burianyk, M.J.A., Kanasewich, E.R., Asuden, I., Hajnal, Z., Miller, K.C., 2002. Deep Probe: imaging the roots of western North America. Canadian Journal of Earth Sciences 39 (3), 375–398.

Gornostayev, S.S., Walker, R.J., Hanski, E.J., Popovchenko, S.E., 2004. Evidence for the emplacement of ca. 3.0 Ga mantle-derived mafic-ultramafic bodies in the Ukrainian Shield. Precambrian Research 132 (4), 349–362.

Gosselin, C., Simard, M., 2001. Geology of the Lac Gayot Area (NTS 23M). Ministère des Ressources naturelles, Québec, RG 2000-03, 28 pp.

Goswami, J.N., Misra, S., Wiedenback, M., Ray, S.L., Saha, A.K., 1995. 3.55 Ga old zircon from Singhbhum-Orissa iron Ore craton, eastern India. Current Science 6, 1008–1011.

Gounelle, M., Zolensky, M.E., 2001. A terrestrial origin for sulfate veins in CI1 chondrites. Meteoritics and Planetary Science 36, 1321–1329.

Gounelle, M., Spurny, P., Bland, P.A., 2006. The orbit and atmospheric trajectory of the Orgueil meteorite from historical records. Meteoritics and Planetary Science 41, 135–150.

Goutier, J., Dion, C., 2004. Géologie et minéralisation de la Sous-province de La Grande, Baie-James. Québec Exploration 2004. Ministère des Ressources naturelles, de la Faune et des Parcs, Québec, DV 2004-06, p. 21.

Govindaraju, K., 1994. 1994 compilation of working values and sample description for 383 geostandards. Geostandards Newsletter 18, 1–158.

Grace, R.L.B., 2004. The oldest fragments of the central Wyoming province: evidence from the Beulah Belle Lake area, northern Granite Mountains. Unpublished M.Sc. thesis. University of Wyoming, Laramie, Wyo., 172 pp.

Grace, R.L.B., Rashmi, L.B., Chamberlain, K.R., Frost, B.R., Frost, C.D., 2006. Tectonic histories of the Paleo- to Mesoarchean Sacawee Block and Neoarchean Oregon Trail structural belt of the south-central Wyoming province. Canadian Journal of Earth Sciences 43, 1445–1466.

Gradstein, F.M., Ogg, J.G., Smith, A.G., Agterberg, F.P., Bleeker, W., Cooper, R.A., Davydov, V., Gibbard, P., Hinnov, L.A., House, M.R., Lourens, L., Luterbacher, H.P., McArthur, J., Melchin, M.J., Robb, L.J., Shergold, J., Villeneuve, M., Wardlaw, B.R., Ali, J., Brinkhuis, H., Hilgen, F.J., Hooker, J., Howarth, R.J., Knoll, A.H., Laskar, J., Monechi, S., Plumb, K.A., Powell, J., Raffi, I., Röhl, U., Sadler, P., Sanfilippo, A., Schmitz, B., Shackleton, N.J., Shields, G.A., Strauss, H., Van Dam, J., van Kolfschoten, T., Veizer, J., Wilson, D., 2004. A Geologic Time Scale 2004. Cambridge University Press, 589 pp.

Graf, J.L., Jr, 1978. Rare earth elements, iron formations and sea water. Geochimica et Cosmochimica Acta 42, 1845–1850.

Graham, C.M., Powell, R., 1984. A garnet-hornblende geothermometer: calibration, testing, and application to the Pelona Schist, Southern California. Journal of Metamorphic Geology 3, 13–21.

Graham, S., Pearson, N., Jackson, S., Griffin, W., O'Reilly, S.Y., 2004. Tracing Cu and Fe from source to porphyry: in situ determination of Cu and Fe isotope ratios in sulfides from the Grasberg Cu–Au deposit. Chemical Geology 207, 147–169.

Grand, S.P., van der Hilst, R.D., Widjyantoro, S., 1997. Global seismic tomography: a snapshot of convection in the Earth. GSA Today 7, 1–7.

Grant, J.A., 1972. Minnesota River Valley, southwestern Minnesota. In: Sims, P.K., Morey, G.B. (Eds.), Geology of Minnesota – A centennial volume. Minnesota Geological Survey, pp. 177–196.

Grant, J.A., 1986. The isocon diagram: a simple solution to Gresen's equation for metasomatic alteration. Economic Geology 81, 1976–1982.

Grant, J.A., 2005. Isocon analysis: A brief review of the method and applications. Physics and Chemistry of the Earth 30, 997–1004.

Grant, J.A., Weiblen, P.W., 1971. Retrograde zoning in garnet near the second sillimanite isograd. American Journal of Science 270, 281–296.

Grantham, G.H., Jackson, C., Moyes, A.B., Harris, P.D., Groenewald, P.B., Ferrar, G., Krynauw, J.R., 1995. P-T evolution of the H.U. Sverdrupfjella and Kirwanveggan, Dronning Maud Land, Antarctica. Precambrian Research 75, 209–229.

Gray, C.M., Compston, W., 1974. Excess ^{26}Mg in the Allende meteorite. Nature 251, 495–497.

Greeley, R., 1987. Planetary Landscapes. Allen & Unwin, London.

Green, A.G., Hajnal, Z., Weber, W., 1985. An evolutionary model of the western Churchill Province and the western margin of the Superior Province in Canada and north-central United States. Tectonophysics 116, 281–322.

Green, D.H., 1972. Archaean greenstone belts may include terrestrial equivalents of lunar mare? Earth and Planetary Science Letters 15, 263–270.

Green, D.H., 1976. Experimental testing of equilibrium partial melting of peridotite under water-saturated, high pressure conditions. Canadian Mineralogist 14, 255–268.

Green, D.H., 1981. Petrogenesis of Archaean ultramafic magmas and implications for Archaean tectonics. In: Kröner, A. (Ed.), Precambrian Plate Tectonics. Elsevier, Amsterdam, pp. 469–489.

Green, M.G., Sylvester, P.J., Buick, R., 2000. Growth and recycling of early Archaean continental crust: geochemical evidence from the Coonterunah and Warrawoona Groups, Pilbara Craton, Australia. Tectonophysics 322, 69–88.

Green, T.H., Pearson, N.J., 1986. Ti-rich accessory phase saturation in hydrous mafic felsic compositions at high P, T. Chemical Geology 54, 185–201.

Green, T.H., Ringwood, A.E., 1977. Genesis of the calc-alkaline igneous rock suite. Contribution to Mineralogy and Petrology 18, 105–162.

Greenwood, J.P., Mojzsis, S.J., Coath, C.D., 2000a. Sulfur isotopic compositions of individual sulfides in Martian meteorites ALH84001 and Nakhla: Implications for crust-regolith exchange on Mars. Earth and Planetary Science Letters 184, 23–35.

Greenwood, J.P., Rubin, A.E., Wasson, J.T., 2000. Oxygen isotopes in R chondrite magnetite and olivine; links between R chondrites and ordinary chondrites. Geochimica et Cosmochimica Acta 64, 3897–3911.

Gregg, T.K.P., Greeley, R., 1993. Formation of Venusian canali: Consideration of lava types and their thermal behaviors. Journal Geophysical Research 98, 10,873–10,882.

Grégoire, M., Cottin, J.Y., Giret, A., Mattielli, N., Weis, D., 1998. The meta-igneous granulite xenoliths from Kerguelen Archipelago: evidence of a continent nucleation in an oceanic setting. Contributions to Mineralogy and Petrology 133, 259–283.

Gregory, R.T., 1991. Oxygen isotope history of seawater revisited: composition of seawater. In: Taylor, H.P., Jr, O'Neil, J.R., Kaplan, I.R. (Eds.), Stable Isotope Geochemistry: A Tribute to Samuel Epstein. Geochemical Society, Special Publication 3. Mineralogical Society of America, Washington, DC, pp. 65–76.

Grew, E.S., 1998. Boron and beryllium minerals in granulite-facies pegmatites and implications of beryllium pegmatites for the origin and evolution of the Archaean Napier Complex of East Antarctica. In: Motoyoshi, Y., Shiraishi, K. (Eds.), Origin and Evolution of Continents. Memoir National Institute of Polar Research, Special Issue 53, pp. 74–92.

Grew, E.S., Manton, W., 1979. Archaean rocks in Antarctica: 2.5 billion year uranium-lead ages of pegmatites in Enderby Land. Science 206, 443–445.

Grieve, R.A.F., 1980. Impact bombardment and its role in proto-continental growth of the early Earth. Precambrian Research 10, 217–248.

Grieve, R.A.F., 1998. Extraterrestrial impacts on Earth – The evidence and the conse-quences. In: Grady, M.M., et al. (Eds.), Meteorites: Flux with Time and Impact Effects. Geological Society of London, Special Publication 140, pp. 105–131.

Grieve, R.A.F., Pesonen, L.J., 1996. Terrestrial impact craters: their spatial and temporal distribution and impacting bodies. Earth, Moon and Planets 72, 357–376.

Grieve, R.A.F., Pilkington, M., 1996. The signature of terrestrial impacts. Australian Geo-logical Survey Organisation Journal, Australian Geology and Geophysics 16, 399–420.

Grieve, R.A.F., Shoemaker, E.M., 1994. The Record of Past Impacts on Earth. The Uni-versity of Arizona Press, Tucson, AZ, pp. 417–462.

Griffin, W.L., Brueckner, H.K., 1980. Caledonian Sm-Nd ages and a crustal origin for Norwegian eclogites. Nature 285, 319–320.

Griffin, W.L., O'Reilly, S.Y., 1987. Is the continental Moho the crust-mantle boundary? Geology 15, 241–244.

Griffin, W.L., McGregor, V.R., Nutman, A.P., Taylor, P.N., Bridgwater, D., 1980. Early Ar-chaean granulite-facies metamorphism south of Ameralik. Earth and Planetary Science Letters 50, 59–74.

Griffin, W.L., O'Reilly, S.Y., Ryan, C.G., Gaul, O., Ionov, D., 1998a. Secular variation in the composition of subcontinental lithospheric mantle. In: Braun, J., Dooley, J.C., Goleby, B.R., van der Hilst, R.D., Klootwijk, C.T. (Eds.), Structure and Evolution of the Australian Continent. American Geophysical Union, Geodynamics Series, vol. 26. Washington, DC, pp. 1–26.

Griffin, W.L., Zhang, A.D., O'Reilly, S.Y., Ryan, G., 1998b. Phanerozoic evolution of the lithosphere beneath the Sino-Korean Craton. In: Flower, M., Chang, S.L., Lo, C.H., Lee, T.Y. (Eds.), Mantle Dynamics and Plate Interactions in East Asia. American Geophysi-cal Union, Geodynamics Series, vol. 27. Washington, DC, pp. 107–126.

Griffin, W.L., O'Reilly, S.Y., Ryan, C.G., 1999a. The composition and origin of sub-continental lithospheric mantle. In: Fei, Y., Bertka, C.M., Mysen, B.O. (Eds.), Mantle Petrology: Field Observations and High Pressure Experimentation: A Tribute to Francis F. (Joe) Boyd. The Geochemical Society, pp. 13–45.

Griffin, W.L., Ryan, C.G., Kaminsky, F.V., O'Reilly, S.Y., Natapov, L.M., Win, T.T., Kinny, P.D., Ilupin, I.P., 1999b. The Siberian lithosphere traverse: mantle terranes and the as-sembly of the Siberian Craton. Tectonophysics 310, 1–35.

Griffin, W.L., Doyle, B.J., Ryan, C.G., Pearson, N.J., O'Reilly, S.Y., Davies, R.M., Kivi, K., van Achterbergh, E., Natapov, L.M., 1999c. Layered mantle lithosphere in the Lac de Gras Area, Slave Craton: Composition, structure and origin. Journal of Petrology 40, 705–727.

Griffin, W.L., O'Reilly, S.Y., Abe, N., Aulbach, S., Davies, R.M., Pearson, N.J., Doyle, B.J., Kivi, K., 2003a. The origin and evolution of Archean lithospheric mantle. Precam-brian Research 127, 19–41.

Griffin, W.L., O'Reilly, S.Y., Natapov, L.M., Ryan, C.G., 2003b. The evolution of lithospheric mantle beneath the Kalahari Craton and its margins. Lithos 71, 215–242.

Griffin, W.L., Belousova, E.A., Shee, S.R., Pearson, N.J., O'Reilly, S.Y.O., 2004a. Archean crustal evolution in the northern Yilgarn Craton: U-Pb and Hf-isotope evidence from detrital zircons. Precambrian Research 131, 231–282.

Griffin, W.L., Graham, S., O'Reilly, S.Y., Pearson, N.J., 2004b. Lithosphere evolution beneath the Kaapvaal Craton. Re-Os systematics of sulfides in mantle-derived peridotites. Chemical Geology 208, 89–118.

Griffiths, R.W., Campbell, I.H., 1990. Stirring and structure in mantle plumes. Earth and Planetary Science Letters 99, 66–78.

Griffiths, R.W., Campbell, I.H., 1991. Interaction of mantle plume heads with the Earth's surface and onset of small-scale convection. Journal of Geophysical Research 96, 18,295–18,310.

Grimm, R.E., 1994. The deep structure of Venusian plateau highlands. Icarus 112, 89–103.

Grimm, R.E., Hess, P.C., 1997. The crust of Venus. In: Bouger, S.W., Hunten, D.M., Phillips, R.J. (Eds.), Venus II, Geology, Geophysics, Atmosphere, and Solar Wind Environment. The University of Arizona Press, Tucson, AZ, pp. 1205–1244.

Grimm, R.E., Solomon, S.C., 1988. Viscous relaxation of impact crater relief on Venus: Constraints on crustal thickness and thermal gradient. Journal of Geophysical Research 93, 11,911–11,929.

Grinenko, V.A., Thode, H.G., 1970. Sulfur isotope effects in volcanic gas mixtures. Canadian Journal of Earth Sciences 7, 1402–1409.

Gromet, L.P., Silver, L.T., 1983. Rare earth element distributions among minerals in a granodiorite and their petrogenetic implications. Geochimica et Cosmochimica Acta 47 (5), 925–939.

Gromet, L.P., Silver, L.T., 1987. REE variations across the Peninsular Ranges Batholith: implications for batholithic petrogenesis and crustal growth in magmatic arcs. Journal of Petrology, 28, 75–125.

Gross, G.A., 1965. Geology of iron deposits in Canada. I. General geology and evaluation of iron deposits. Geological Survey of Canada, Economic Geology Report 22, 181 pp.

Gross, G.A., 1980. A classification of iron formations based on depositional environments. Canadian Mineralogist 18, 215–222.

Gross, G.M., 1983. Ocean sciences review and forecast. Sea Technology 24, 21.

Grotzinger, J.P., Rothman, D.H., 1996. An abiotic model for stromatolite morphogenesis. Nature 383, 423–425.

Grove, D.I., Barrett, F.M., Binns, R.A., McQueen, K.G., 1977. Spinel phases associated with metamorphosed volcanic-type iron-nickel sulphide ores from Western Australia. Economic Geology 72, 1224–1244.

Grove, T.L., de Wit, M.J., Dann, J.C., 1997. Komatiites from the Komati Type Section, South Africa. In: de Wit, M.J., Ashwal, L.D. (Eds.), Greenstone Belts. Oxford University Press, Oxford, UK, pp. 438–456.

Grove, T.L., Elkins, L.T., Parman, S.W., Chatterjee, N., Müntener, O., Gaetani, G.A., 2003. Fractional crystallization and mantle-melting controls on calc-alkaline differentiation trends. Contribution to Mineralogy and Petrology 145, 515–533.

Groves, D.I., Dunlop, J.S.R., Buick, R., 1981. An early habitat of life. Scientific American 245, 64–73.

Groves, D.I., Goldfarb, R.J., Gebre-Mariam, H., Hagemann, S.G., Robert, F., 1998. Orogenic gold deposits: a proposed classification in the context of their crustal distribution and relationship to other ore deposit type. Ore Geology Reviews 13, 7–27.

Groves, D.I., Goldfarb, R.J., Robert, F., Hart, C.J.R., 2003. Gold deposits in metamorphic belts: overview of current understanding, outstanding problems, future research, and exploration significance. Economic Geology 98, 1–29.

Groves, D.I., Vielreicher, R.M., Goldfarb, R.J., Condie, K.C., 2005. Controls on the heterogeneous distribution of mineral deposits through time. In: Geological Society of London, Special Publication 248, pp. 71–101.

Groves, I.M., 1987. Epithermal/porphyry style base- and precious-metal mineralization in the Miralga Creek area, eastern Pilbara Block. Unpublished B.Sc. Honours thesis. University of Western Australia, 82 pp.

Gruau, G., Jahn, B., Glikson, A.Y., Davy, R., Hickman, A.H., Chauvel, C., 1987. Age of the Archaean Talga-Talga Subgroup, Pilbara Block, Western Australia, and early evolution of the mantle: new Sm-Nd evidence. Earth and Planetary Science Letters 85, 105–116.

GSWA, 2006. Compilation of geochronology data, June 2006 update. Geological Survey of Western Australia, on compact disc.

Guan, H., Sun, M., Wilde, S.A., Zhou, X.H., Zhai, M.G., 2002. SHRIMP U-Pb zircon geochronology of the Fuping Complex: Implications for formation and assembly of the NCC. Precambrian Research 113, 1–18.

Guest, J.E., Bulmer, M.H., Aubele, J.C., Beratan, K., Greeley, R., Head, J.W., Micheals, G., Weitz, C., Wiles, C., 1992. Small volcanic edifices and volcanism in the plains on Venus. Journal of Geophysical Research 97, 15,949–15,966.

Guillot, T., Stevenson, D.J., Hubbard, W.B., Saumon, D., 2004. The interior of Jupiter. In: Bagenal, F., Dowling, T.E., McKinnon, W.B. (Eds.), Jupiter. Cambridge University Press, pp. 35–57.

Guimon, R.K., Symes, S.J.K., Sears, D.W.G., Beniot, P.H., 1995. Chemical and Physical studies of type 3 chondrites XII: the metamorphic history of CV chondrites and their components. Meteoritics 30, 704–714.

Guo, J.H., Sun, M., Chen, F.K., Zhai, M.G., 2005. Sm-Nd and SHRIMP zircon geochronology of high-pressure granulites in the Sanggan area. North China Craton: timing of Paleoproterozoic continental collision. Journal of Asian Earth Sciences 24, 629–642.

Gupta, N.S., Briggs, D.E.G., Collinson, M.E., Evershed, R.P., Michels, R., Pancost, R.D., 2004. In situ polymerisation of labile lipids as a source for the aliphatic component of recalcitrant macromolecules in sedimentary materials. Geochimica et Cosmochimica Acta 68, A241–A241 suppl.

Gurnis, M., 1988. Large scale mantle convection and the aggregation and dispersal of supercontinents. Nature 332, 695–699.

Gurnis, M., Mitrovica, J.X., Ritsema, J., van Heijst, H.-J., 2000. Constraining mantle density structure using geological evidence of surface uplift rates: the case of the African Superplume. Geochemistry, Geophysics, Geosystems 1, doi: 10.1029/1999GC000035.

Gutscher, M.A., Maury, R., Eissen, J.-P., Bourdon, R., 2000a. Can slab melting be caused by flat subduction? Geology 28, 535–538.

Gutscher, M.-A., Spakman, W., Bijwaard, H., Engdahl, E.R., 2000b. Geodynamics of flat subduction: seismicity and tomographic constraints from the Andean margin. Tectonics 19, 814–833.

Habicht, K.S., Canfield, D.E., 2001. Isotope fractionation by sulfate-reducing natural populations and the isotopic composition of sulfide in marine sediments. Geology 29, 555–558.

Habicht, K.S., Gade, M., Thamdrup, B., Berg, P., Canfield, D.E., 2002. Calibration of sulfate levels in the Archean ocean. Science 298, 2372–2374.

Habing, H.J., Dominik, C., de Muizon, M.J., Laureijs, R.J., Kessler, M.F., Leech, K., Metcalfe, L., Salama, A., Siebenmorgen, R., Trams, N., Bouchet, P., 2001. Incidence and survival of remnant disks around main-sequence stars. Astronomy and Astrophysics 365, 545–561.

Hacker, B.R., Abers, G.A., 2004. Subduction Factory 3: An Excel worksheet and macro for calculating the densities, seismic waver speeds and H_2O contents of minerals and rocks at pressure and temperature. Geochemistry, Geophysics, Geosystems, doi: 10.1029/2003GC000614.

Haisch, K.E., Lada, E.A., Lada, C.J., 2001. Disk frequencies and lifetimes in young clusters. Astrophysical Journal 533, L131–L136.

Hall, A.L., 1918. The Geology of the Barberton Gold Mining District. Geological Survey of South Africa, Memoir 9, 347 pp.

Halliday, A.N., 2000. Terrestrial accretion rates and the origin of the Moon. Earth and Planetary Science Letters 176, 17–30.

Halliday, A.N., 2003. The origin and earliest history of the Earth. In: Davis, A.M. (Ed.), Meteorites, Comets and Planets. Treatise on Geochemistry, vol. 1. Elsevier–Pergamon, Oxford, pp. 509–557.

Halliday, A.N., Lee, D.-C., 1999. Tungsten isotopes and the early development of the Earth and Moon. Geochimica et Cosmochimica Acta 63, 4157–4179.

Halliday, A.N., Porcelli, D., 2001. In search of lost planets – the palaeocosmochemistry of the inner solar system. Earth and Planetary Science Letters 192, 545–559.

Halliday, A.N., Lee, D.-C., Jacobsen, S.B., 2000. Tungsten isotopes, the timing of metal-silicate fractionation, and the origin of the Earth 1 and Moon. In: Canup, R., Righter, K. (Eds.), Origin of the Earth and Moon. Arizona University Press, Tucson, AZ, pp. 45–62.

Halpin, J.A., Geratikeys, C.L., Clarke, G.L., Belousova, E.A., Griffin, W.L., 2005. In-situ U-Pb geochronology and Hf isotope analyses of the Rayner Complex, east Antarctica. Contributions to Mineralogy and Petrology 148, 689–706.

Hamilton, P.J., Evensen, N.M., O'Nions, R.K., 1979. Sm-Nd dating of Onverwacht Group volcanics, southern Africa. Nature 279, 298–300.

Hamilton, P.J., O'Nions, R.K., Bridgwater, D., Nutman, A., 1983. Sm-Nd studies of Archaean metasediments and metavolcanics from West Greenland and their implications for the Earth's early history. Earth and Planetary Science Letters 62, 263–272.

Hamilton, V.E., Stofan, E.R., 1996. The geomorphology and evolution of Hecate Chasma, Venus. Icarus 121, 171–189.

Hamilton, W.B., 1993. Evolution of Archean mantle and crust. In: Reed, J.C., Jr, et al. (Eds.), Precambrian – Conterminous United States. Geological Society of America, Geology of North America, vol. C-2, pp. 597–614, 630–636.

Hamilton, W.B., 1998. Archean magmatism and deformation were not the products of plate tectonics. Precambrian Research 91, 143–180.

Hamilton, W.B., 2003. An alternative Earth. GSA Today 13, 4–12.

Hamilton, W.B., 2005. Plumeless Venus has ancient impact-accretionary surface. In: Foulger, G.R., Natland, J.H., Presnall, D.C., Anderson, D.L. (Eds.), Plates, Plumes, and Paradigms. Geological Society of America, Special Paper 388, pp. 781–814.

Han, B-F., Wang, S-H., Jahn, B-M., Hong, D-W., Kagami, H., Sun, Y-L., 1997. Depleted-mantle source for the Ulungur River A-type granites from North Xinjiang, China: Geochemistry and Nd-Sr isotopic evidence, and implications for Phanerozoic crustal growth. Chemical Geology 138, 135–159.

Hanmer, S., Greene, D.C., 2002. A modern structural regime in the Paleoarchean (~3.64 Ga); Isua Greenstone Belt, southern West Greenland. Tectonophysics 346, 201–222.

Hanor, J.S., Duchac, K., 1990. Isovolumetric silicification of Early Archean komatiites: geochemical mass balances and constraints on origin. Journal of Geology 98, 863–877.

Hansen, V.L., 2000. Geologic mapping of tectonic planets. Earth and Planetary Science Letters 176, 527–542.

Hansen, V.L., 2002. Artemis: signature of a deep Venusian mantle plume. Bulletin of the Geological Society of America 114, 839–848.

Hansen, V.L., 2003. Venus diapirs: thermal or compositional? Bulletin of the Geological Society of America 115, 1040–1052.

Hansen, V.L., 2005. Venus's shield-terrain. Bulletin of the Geological Society of America 117, 808–822.

Hansen, V.L., 2006. Geologic constraints on crustal plateau surface histories, Venus: The lava pond and bolide impact hypotheses. Journal of Geophysical Research 111, doi: 10.1029/2006JE002714.

Hansen, V.L., DeShon, H.R., 2002. Geologic map of the Diana Chasma quadrangle (V-37). Venus U.S. Geological Survey Geologic Investigations Series I-2752, U.S. Geological Survey, 1:5,000,000.

Hansen, V.L., Willis, J.J., 1996. Structural analysis of a sampling of tesserae: Implications for Venus geodynamics. Icarus 123, 296–312.

Hansen, V.L., Willis, J.J., 1998. Ribbon terrain formation, southwestern Fortuna Tessera, Venus: Implications for lithosphere evolution. Icarus 132, 321–343.

Hansen, V.L., Young, D.A., 2007. Venus's evolution: A synthesis. In: Cloos, M., Carlson, W., Gilbert, M.C., Liou, J.G., Sorenson, S.S. (Eds.), Convergent Margin Terranes and 6 Associated Regions. Geological Society of America, Special Paper 419, 255–273.

Hansen, V.L., Willis, J.J., Banerdt, W.B., 1997. Tectonic overview and synthesis. In: Bouger, S.W., Hunten, D.M., Phillips, R.J. (Eds.), Venus II. University of Arizona Press, Tucson, AZ, pp. 797–844.

Hansen, V.L., Banks, B.K., Ghent, R.R., 1999. Tessera terrain and crustal plateaus, Venus. Geology 27, 1071–1074.

Hansen, V.L., Phillips, R.J., Willis, J.J., Ghent, R.R., 2000. Structures in tessera terrain, Venus: Issues and answers. Journal of Geophysical Research 105, 4135–4152.

Hanson, G.N., Himmelberg, G.R., 1967. Ages of mafic dikes near Granite Falls, Minnesota. Bulletin of the Geological Society of America 78, 1429–1432.

Harley, S.L., 1986. A sapphirine-cordierite-garnet-sillimanite granulite from Enderby Land, Antarctica: implications for FMAS petrogenetic grids in the granulite facies. Contributions to Mineralogy and Petrology 94, 452–460.

Harley, S.L., 1993. Sapphrine granulites from the Vestfold Hills, East Antarctica: geochemical and metamorphic evolution. Antarctic Science 5, 349–402.

Harley, S.L., 1998. On the occurrence and characterisation of ultrahigh temperature crustal metamorphism. In: Treloar, P.J., O'Brien, P. (Eds.), What Drives Metamorphism and Metamorphic Reactions? Geological Society of London, Special Publication 138, pp. 75–101.

Harley, S.L., 2003. Archaean to Pan-African crustal development and assembly of East Antarctica: metamorphic characteristics and tectonic implications. In: Yoshida, M., Windley, B.F. (Eds.), Proterozoic East Gondwana: Supercontinent Assembly and Breakup. Geological Society of London, Special Publication 206, pp. 203–230.

Harley, S.L., Black, L.P., 1997. A revised Archaean chronology for the Napier Complex, Enderby Land, from SHRIMP ion-microprobe studies. Antarctic Science 9, 74–91.

Harley, S.L., Fitzsimons, I.C.W., Buick, I.S., Watt, G., 1992. The significance of reworking, fluids and partial melting in granulite metamorphism, East Prydz Bay, Antarctica. In: Yoshida, Y., Kaminuma, K., Shiraishi, K. (Eds.), Recent Progress in Antarctic Earth Science. Terrapub, Tokyo, pp. 119–127.

Harley, S.L., Snape, I., Black, L.P., 1998. The early evolution of a layered metaigneous complex in the Rauer Group, East Antarctica: evidence for a distinct Archaean terrane. Precambrian Research 89, 175–205.

Harper, C.L., Jacobsen, S.B., 1992. Evidence from coupled ^{147}Sm-^{143}Nd and ^{146}Sm-^{142}Nd systematics for very early (4.5 Ga) differentiation of the Earth's mantle. Nature 360, 728–732.

Harper, K.M., 1997. U-Pb age constraints on the timing and duration of Proterozoic and Archean metamorphism along the southern margin of the Archean Wyoming craton. Unpublished Ph.D. thesis. University of Wyoming, Laramie, 176 pp.

Harris, P.D., Smith, C.B., Hart, R.J., Robb, L.J., 1993. Sm-Nd and Rb-Sr isotope systematics of rare-element pegmatites from the New Consort gold mines, Barberton Mountain Land, South Africa. Information Circular, Economic Geology Research Unit, University of the Witwatersrand, Johannesburg, 265, 20 pp.

Harris, P.D., Robb, L.J., Tomkinson, M.J., 1995. The nature and structural setting of rare-element pegmatites along the northern flank of the Barberton greenstone belt, South Africa. South African Journal of Geology 98, 82–94.

Harrison, T.M., Blichert-Toft, J., Müller, W., Albarede, F., Holden, P., Mojzsis, S.J., 2005. Heterogeneous Hadean hafnium: evidence of continental crust at 4.4 to 4.5 Ga. Science 310, 1947–1950.

Harrison, T.M., Blichert-Toft, J., Müller, W., Albarède, F., Holden, P., Mojzsis, S.J., 2006. Response to Comment on "Heterogeneous Hadean hafnium: Evidence of continental crust at 4.4 to 4.5 Ga". Science 312, doi: 10.1126/science.1123408.

Harrison, T.M., McCulloch, M.T., Blichert-Toft, J., Albarede, F., Holden, P., Mojzsis, S.J., 2006. Further Hf isotope evidence for Hadean continental crust. Geochimica et Cosmochimica Acta 70 (18) Suppl. 1, A234.

Hart, R., Moser, D., Andreoli, M., 1999. Archaean age for the granulite facies metamorphism near the center of the Vredefort structure, South Africa. Geology 27 (12), 1091–1094.

Harte, B., Harris, J.W., Hutchinson, M.T., Watt, G.R., Wilding, M.C., 1999. Lower mantle mineral associations in diamonds from Sao Luiz, Brazil. In: Fei, Y., Bertka, C.M., Mysen, B.O. (Eds.), Mantle Petrology: Field Observations and High Pressure Experimentation: A Tribute to Francis F. (Joe) Boyd. Geochemical Society Special Publication #6. The Geochemical Society, Houston, pp. 125–154.

Hartlaub, R.P., Heaman, L.M., Böhm, C.O., Corkery, M.T., 2003. The Split Lake Block revisited: new geological constraints from the Birthday to Gull rapids corridor of the Lower Nelson River (NTS 54D5 and 6). In: Report of Activities 2003, Manitoba Industry, Trade and Mines, Manitoba Geological Survey, pp. 114–117.

Hartlaub, R.P., Heaman, L.M., Ashton, K.E., Chacko, T., 2004a. The Archean Murmac Bay group: evidence for a giant Archean rift in the Rae Province, Canada. Precambrian Research 131, 345–372.

Hartlaub, R.P., Böhm, C.O., Kuiper, Y.D., Bowerman, M.S., Heaman, L.M., 2004b. Archean and Paleoproterozoic Geology of the northwestern Split Lake Block, Superior Province, Manitoba (parts of NTS54D4, 5, 6 and NTS64A1). In: Report of Activities 2004, Manitoba Industry, Economic Development and Mines, Manitoba Geological Survey, pp. 187–194.

Hartlaub, R.P., Böhm, C.O., Heaman, L.M., Simonetti, A., 2005. Northwestern Superior craton margin, Manitoba: an overview of Archean and Proterozoic episodes of crustal growth, erosion and orogenesis (parts of NTS 54D and 64A). In: Report of Activities 2004, Manitoba Industry, Economic Development and Mines, Manitoba Geological Survey, pp. 54–60.

Hartlaub, R.P., Heaman, L.M., Simonetti, A., Boehm, C.O., 2006. Relics of Earth's earliest crust: U-Pb, Lu-Hf , and morphological characteristics of >3.7 Ga detrital zircon of the western Canadian Shield. In: Reimold, W.U., Gibson, R.L. (Eds.), Processes on the Early Earth. Geological Society of America, Special Paper 405, pp. 75–90.

Hartmann, W.K., 1998. Moons and Planets. Brooks Cole, 528 pp.

Hartmann, W.K., Ryder, G., Dones, L., Grinspoon, D., 2000. The time-dependent intense bombardment of the primordial Earth–Moon system. In: Canup, R.M., Righter, K. (Eds.), Origin of the Earth and Moon. University of Arizona Press, Tucson, AZ, pp. 493–512.

Hattori, K., Cameron, E.M., 1986. Archean magmatic sulphate. Nature 319, 45–47.

Hauck, S.A., Phillips, R.J., Price, M.H., 1998. Venus: Crater distribution and plains resurfacing models. Journal of Geophysical Research 103, 13,635–13,642.

Haugh, I., 1969. Geology of the Split Lake area. In: Manitoba Mines and Natural Resources, Mines Branch, Publication 65-2, p. 87.

Haugh, I., Elphick, S.C., 1968. Kettle Rapids–Moose Lake area. Summary of Geological Fieldwork 1968. In: Manitoba Mines and Natural Resources, Mines Branch, Geological Paper 68-3, pp. 29–37.

Hayes, J.M., Kaplan, I.R., Wedekling, K.W., 1983. Precambrian organic geochemistry, preservation of the record. In: Schopf, J.W. (Ed.), Earth's Earliest Biosphere: Its Origin and Evolution. Princeton University Press, Princeton, pp. 93–134.

Hayes, J.M., Waldbauer, J.R., 2006. The carbon cycles and associated redox processes through time. Philosophical Transactions of the Royal Society of London 361, 931–950.

Head, J.W., Crumpler, L.S., Aubele, J.C., Guest, J.E., Saunders, R.S., 1992. Venus volcanism: Classification of volcanic features and structures, associations, and global distribution from Magellan data. Journal of Geophysical Research 97, 13,153–13,198.

Heaman, L.M., Corkery, M.T., 1996. U-Pb geochronology of the Split Lake Block, Manitoba: preliminary results. In: 1996 Trans-Hudson Orogen, Lithoprobe Report 55, pp. 60–68.

Heaman, L.M., Machado, N., Krogh, T.E., Weber, W., 1986. Preliminary U-Pb zircon results from the Pikwitonei granulite domain, In: Manitoba. Geological Association of Canada – Mineralogical Association of Canada, Joint Annual Meeting, Program and Abstracts, 11, p. 79.

Hedderich, R., Klimmek, O., Kroger, A., Dirmeier, R., Keller, M., Stetter, K.O., 1998. Anaerobic respiration with elemental sulfur and with disulfides. FEMS Microbiology Reviews 22, 353–381.

Heinrichs, T.K., 1980. Lithostratigraphische Untersuchungen in der Fig Tree Gruppe des Barberton Greenstone Belt zwischen Umsoli und Lomati (Sudafrika). Gottinger Arbeiten zur Geologie und Palaontologie 22, 118 pp.

Heinrichs, T.K., 1984. The Umsoli Chert, turbidite testament for a major phreatoplinian event at the Onverwacht/Fig Tree transition (Swaziland Supergroup, Archaean, South Africa). Precambrian Research 24, 237–283.

Heinrichs, T.K., Reimer, T.O., 1977. A sedimentary barite deposit from the Archean Fig Tree Group of the Barberton Mountain Land (South Africa). Economic Geology 72, 1426–1441.

Helmstaedt, H.H., Gurney, J.J., 1995. Geotectonic controls of primary diamond deposits: implications for area selection. Journal of Geochemical Exploration 53, 125–140.

Henry, D.J., Mueller, P.A., Wooden, J.L., Warner, J.L., Lee-Berman, R., 1982. Granulite grade supracrustal assemblages of the Quad Creek area, eastern Beartooth Mountains, Montana. In: Montana Bureau of Mines and Geology, Special Publication 84, pp. 147–159.

Henry, P., Stevenson, R., Gariepy, C., 1998. Late Archean mantle composition and crustal growth in the western Superior Province of Canada: Neodymium and lead isotopic evidence from the Wawa, Quetico, and Wabigoon subprovinces. Geochimica et Cosmochimica Acta 62, 143–157.

Henry, P., Stevenson, R., Larbi, Y., Gariepy, C., 2000. Nd isotopic evidence for Early to Late Archean (3.4–2.7 Ga) crustal growth in the Western Superior Province (Ontario, Canada). Tectonophysics 322, 135–151.

Hensen, B.J., Zhou, B., 1995. A Pan-African granulite facies metamorphic episode in Prydz Bay, Antarctica: evidence from Sm-Nd garnet dating. Australian Journal of Earth Sciences 42, 249–258.

Herd, R.K., 1978. Notes on metamorphism in New Québec. In: Fraserand, J.A., Heywood, W.W. (Eds.), Metamorphism in the Canadian Shield. Geological Survey of Canada, Paper 78-10, pp. 78–83.

Herrick, R.R., Sharpton, V.L., Malin, M.C., Lyons, S.N., Feely, K., 1997. Morphology and morphometry of impact craters. In: Bouger, S.W., Hunten, D.M., Phillips, R.J. (Eds.), Venus II. University of Arizona Press, Tucson, AZ, pp. 1015–1046.

Hertogen, J., Janssens, M.J., Palme, H., 1980. Trace elements in ocean ridge basalt glasses: Implications for fractionations during mantle evolution and petrogenesis. Geochimica et Cosmochimica Acta 44, 2125–2143.

Herzberg, C., 1992. Depth and degree of melting of komatiite. Journal of Geophysical Research 97, 4521–4540.

Herzberg, C., 1999. Phase equilibrium constraints on the formation of cratonic mantle. In: Fei, Y., Bertka, C.M., Mysen, B.O. (Eds.), Mantle Petrology: Field Observations and High-Pressure Axperimentation: A Tribute to Francis R. (Joe) Boyd. Geochemical Society Special Publication #6. The Geochemical Society, Houston, pp. 241–258.

Heubeck, C., Lowe, D.R., 1994a. Depositional and tectonic setting of the Archean Moodies Group, Barberton Greenstone Belt, South Africa. Precambrian Research 68, 257–290.

Heubeck, C., Lowe, D.R., 1994b. Late syndepositional deformation and detachment tectonics in the Barberton Greenstone Belt, South Africa. Tectonics 13, 1514–1536.

Heubeck, C., Lowe, D.R., 1999. Sedimentary petrography and provenance of the Archean Moodies Group, Barberton Greenstone Belt. In: Lowe, D.R., Byerly, G.R. (Eds.), Geologic Evolution of the Barberton Greenstone Belt, South Africa. Geological Society of America, Special Paper 329, pp. 259–286.

Heymann, D., 1967. On the origin of hypersthene chondrites: Ages and shock effects of black meteorites. Icarus 6, 189–221.

Hickman, A.H., 1973. The North Pole barite deposits, Pilbara Goldfield. In: Annual Report of Geological Survey of Western Australia for 1972, pp. 57–60.

Hickman, A.H., 1975. Precambrian structural geology of part of the Pilbara region. In: Annual Report, 1974. Western Australia Geological Survey, pp. 68–73.

Hickman, A.H., 1983. Geology of the Pilbara Block and Its Environs. Western Australia Geological Survey, Bulletin 127, 268 pp.

Hickman, A.H., 1984. Archaean diapirism in the Pilbara Block, Western Australia. In: Kröner, A., Greiling, R. (Eds.), Precambrian Tectonics Illustrated. E. Schweizerbart'sche Verlagsbuchhandlung, Stuttgart, pp. 113–127.

Hickman, A.H., 1997. A revision of the stratigraphy of Archaean greenstone successions in the Roebourne-Whundo area – west Pilbara. In: Western Australia Geological Survey, Annual Review 1996–97, pp. 76–81.

Hickman, A.H., 2004. Two contrasting granite-greenstones terranes in the Pilbara Craton, Australia: evidence for vertical and horizontal tectonic regimes prior to 2900 Ma. Precambrian Research 131, 153–172.

Hickman, A.H., Van Kranendonk, M.J., 2004. Diapiric processes in the formation of Archaean continental crust, East Pilbara Granite-Greenstone Terrane, Australia. In: Eriksson, P.G., Altermann, W., Nelson, D.R., Mueller, W.U., Catuneau, O. (Eds.), The Precambrian Earth: Tempos and Events. Elsevier, Amsterdam, pp. 54–75.

Hildreth, W., Moorbath, S., 1988. Crustal contributions to arc magmatism in the Andes of Central Chile. Contributions to Mineralogy and Petrology 98, 455–489.

Hill, R.E.T., 2001. Komatiite volcanology, volcanological setting and primary geochemical properties of komatiite-associated nickel deposits. Geochemistry: Exploration, Environment, Analysis 1, 365–381.

Hill, R.E.T., Barnes, S.J., Gole, M.J., Dowling, S.E., 1995. The volcanology of komatiites as deduced from field relationships in the Norseman-Wiluna greenstone belt, Western Australia. Lithos 34, 159–188.

Hill, R.I., Campbell, I.H., Compston, W., 1989. Age and origin of granitic rocks in the Kalgoorlie–Norseman region of Western Australia – Implications for the origin of Archaean crust. Geochimica et Cosmochimica Acta 53, 1259–1275.

Hill, R.M., 1997. Stratigraphy, structure and alteration of hanging wall sedimentary rocks at the Sulphur Springs volcanogenic massive sulphide (VMS) prospect, east Pilbara Craton, Western Australia. B.Sc. Hon. thesis. University of Western Australia, 67 pp.

Himmelberg, G.R., 1968. Geology of Precambrian rocks, Granite Falls – Montevideo area, southwestern Minnesota. In: Minnesota Geological Survey, Special Publication Series, SP-5, 33 pp.

Himmelberg, G.R., Phinney, W.C., 1967. Granulite-facies metamorphism, Granite Falls – Montevideo area, Minnesota. Journal of Petrology 8, 325–348.

Hirner, A., 2001. Geology and Gold Mineralization in the Madibe Greenstone Belt, Eastern part of the Kraaipan Terrain, Kaapvaal Craton, South Africa. University of the Witwatersrand, Johannesburg, 220 pp.

Hirose, K., Fei, Y., Ma, Y., Mao, H.-K., 1999. The fate of subducted basaltic crust in the Earth's lower mantle. Nature 397, 53–56.

Hoatson, D.M., Jaireth, S., Jaques, A.L., 2006. Nickel sulfide deposits in Australia: Characteristics, resources and potential. Ore Geology Reviews 29, 177–241.

Hodges, K.V., Bowring, S.A., Coleman, D.S., Hawkins, D.P., Davidek, K.L., 1995. Multistage thermal history of the ca. 4.0 Ga Acasta gneisses. In: American Geophysical Union Fall Meeting, pp. F708.

Hoefs, J., 1997. Stable Isotope Geochemistry. Springer-Verlag, Berlin, 201 pp.

Hoering, T.C., 1989. The isotopic composition of bedded barites from the Archean of southern India. Journal of the Geological Society of India 34, 461–466.

Hofmann, A., 2005a. Silica alteration zones in the Barberton greenstone belt: a window into subseafloor processes 3.5–3.3 Ga ago. In: GEO2005, Abstracts Volume. University of KwaZulu-Natal, Durban, South Africa, pp. 111–112.

Hofmann, A., 2005b. The geochemistry of sedimentary rocks from the Fig Tree Group, Barberton greenstone belt: Implications for tectonic, hydrothermal and surface processes during mid-Archean times. Precambrian Research 143, 23–49.

Hofmann, A.W., 1997. Mantle geochemistry: the message from oceanic volcanism. Nature 385, 219–229.

Hofmann, A.W., 1997. Early evolution of continents. Science 275, 498–499.

Hofmann, A.W., 1988. Chemical differentiation of the Earth: The relationship between mantle, continental crust and oceanic crust. Earth and Planetary Science Letters 90, 297–314.

Hoffman, A.W., 2005. Sampling mantle heterogeneity through oceanic basalts: Isotopes and trace elements. In: Holland, H.D., Turekian, K.K. (Eds.), Treatise on Geochemistry, vol. 2. Elsevier, Amsterdam, pp. 61–101 (Chapter 2.03).

Hoffman, P., 1990. Subdivision of the Churchill province and extent of the Trans-Hudson orogen. In: Lewry, J., Stauffer, M. (Eds.), The Early Proterozoic Trans-Hudson Orogen of North America. Geologic Association of Canada Special Paper 37, pp. 15–39.

Hoffman, P.F., 1988. United plates of America, the birth of a craton: Early Proterozoic Assembly and Growth of Laurentia. Annual Review of Earth and Planetary Sciences 16, 543–603.

Hoffman, P.F., 1989. Precambrian geology and tectonic history of North America, In: Bally, A.W., Palmer, A.R. (Eds.), The Geology of North America – An Overview. The Geology of North America, vol. A. Geological Society of America, pp. 447–512.

Hofmann, A., Bolhar, R., 2007. The origin of carbonaceous cherts in the Barberton greenstone belt and their significance for the study of early life in mid-Archaean rocks. Astrobiology 7, 355–388.

Hoffman, H.J., Grey, K., Hickman, A., Thorpe, R., 1999. Origin of 3.45 Ga coniform stromatolites in Warrawoona Group, Western Australia. Bulletin of the Geological Society of America 111, 1256–1262.

Hofmann, A., Bolhar, R., Dirks, P.H.G.M., Jelsma, H.A., 2003. The geochemistry of Archaean shales derived from a mafic volcanic sequence, Belingwe greenstone belt, Zimbabwe: provenance, source area unroofing and submarine vs subaerial weathering. Geochimica et Cosmochimica Acta 67, 421–440.

Hofmann, A., Bolhar, R., Harris, C., Orberger, B., 2006. Silica alteration zones and cherts as a record of hydrothermal processes on the Archaean seafloor. Geochimica et Cosmochimica Acta 70, Suppl. 1, A257.

Hohenberg, C.M., Brazzle, R.H., Pravdivtseva, O.V., Meshik, A.P., 1998. Iodine-xenon chronometry: The verdict. Meteoritics and Planetary Science 33 (Supplement), A69–70 (abstract).

Hohenberg, C.M., Pravdivtseva, O.V., Meshik, A.P., 2000. Reexamination of anomalous I-Xe ages: Orgueil and Murchison magnetites and Allegan feldspar. Geochimica et Cosmochimica Acta 64, 4257–4262.

Hokada, T., Misawa, K., Shiraishi, K., Suzuki, S., 2003. Mid to late Archaean (3.3–2.5 Ga) tonalitic crustal formation and high-grade metamorphism at Mt Riiser-Larsen, Napier Complex, East Antarctica. Precambrian Research 127, 215–228.

Hokada, T., Harley, S.L., 2004. Zircon growth in UHT leucosome: constraints from zircon-garnet rare earth element (REE) relations in the Napier Complex, Antarctica. Journal of Mineralogical and Petrological Science 99, 180–190.

Holdaway, M.J., 2000. Application of new experimental and garnet margules data to the garnet-biotite geothermometer. American Mineralogist 85, 881–892.

Holland, H.D., 1984. The Chemical Evolution of the Atmospheres and Oceans. Princeton University Press, Princeton, 582 pp.

Holland, H.D., 1994. Early Proterozoic atmosphere change. In: Bengston, S. (Ed.), Nobel Symposium 84, Early Life on Earth. Columbia University Press, New York, pp. 237–244.

Holland, H.D., 2005. Sedimentary mineral deposits and the evolution of Earth's near surface environments. Economic Geology 100, 1489–1500.

Holland, T.J.B., Blundy, J., 1994. Non-ideal interactions in calcic amphiboles and their bearing on amphibole-plagioclase thermometry. Contributions to Mineralogy and Petrology 116, 433–447.

Holland, T.J.B., Powell, R., 1998. An internally consistent thermodynamic dataset for phases of petrological interest. Journal of Metamorphic Geology 16, 309–343.

Hollings, P., 2002. Archean Nb-enriched basalts in the northern Superior Province. Lithos 64, 1–14.

Hollings, P., Kerrich, R., 1999. Trace element systematics of ultramafic and mafic volcanic rocks from the 3 Ga North Caribou greenstone belt, northwestern Superior Province. Precambrian Research 93, 257–279.

Hollings, P., Kerrich, R., 2000. An Archean arc basalt-Nb-enriched basalt-adakite association: the 2.7 Ga Confederation assemblage of the Birch-Uchi greenstone belt, Superior Province. Contributions to Mineralogy and Petrology 139, 208–226.

Hollings, P., Wyman, D.A., 1999. Trace element and Sm-Nd systematics of volcanic and intrusive rocks from the 3 Ga Lumby Lake greenstone belt, Superior Province; evidence for Archean plume-arc interaction. Lithos 46, 189–213.

Hollings, P., Wyman, D.A., Kerrich, R., 1999. Komatiite-basalt-rhyolite associations in northern Superior Province greenstone belts: significance of plume-arc interaction in the generation of the protocontinental Superior Province. Lithos 46, 137–161.

Hollings, P., Stott, G.M., Wyman, D.A., 2000. Trace element geochemistry of the Meen-Dempster greenstone belt, Uchi subprovince, Superior Province, Canada: back-arc development on the margins of an Archean protocontinent. Canadian Journal of Earth Sciences 37, 1021–1038.

Holm, N.G., 1989. The $^{13}C/^{12}C$ ratios of siderite and organic matter of a modern metallifer-ous hydrothermal sediment and their implications for banded iron formations. Chemical Geology 77, 41–45.

Holm, N.G., Charlou, J.L., 2001. Initial indications of abiotic formation of hydrocarbons in the Rainbow ultramafic hydrothermal system, Mid-Atlantic Ridge. Earth and Planetary Science Letters 191, 1–8.

Holser, W.T., Schidlowski, M., Mackenzie, F.T., Maynard, J.B., 1988. Geochemical cycles of carbon and sulfur. In: Gregor, C.B., Garrels, R.M., Mackenzie, F.T., Maynard, J.B. (Eds.), Chemical Cycles in the Evolution of the Earth. John Wiley and Sons, New York, pp. 105–173.

Hon, K., Kauahikaua, J., Denlinger, R., Mackay, K., 1994. Emplacement and inflation of pahoehoe sheet flows: Observations and measurements of active lava flows on Kilauea Volcanoe, Hawaii. Bulletin of the Geological Society of America 106, 351–370.

Hong, D., Wang, S-G., Han, B-F., Jin, M-Y., 1996. Post-orogenic alkaline granites from China and comparisons with anorogenic alkaline granites elsewhere. Journal of South-east Asian Earth Sciences 13, 13–27.

Hong, D., Zhang, J-S., Wang, T., Wang, S-G., Xie, X-L., 2004. Continental crust growth and the supercontinental cycle: evidence from the Central Asian orogenic Belt. Journal of Asian Earth Sciences 23, 799–813.

Hoogenboom, T., Houseman, G.A., 2006. Rayleigh–Taylor instability as a mechanism for corona formation on Venus. Icarus 180, 292–307.

Hopfe, W.D., Goldstein, J.I., 2001. The metallographic cooling rate method revisited: ap-plication to iron meteorites and mesosiderites. Meteoritics and Planetary Science 36, 135–154.

Horan, M.F., Smoliar, M.I., Walker, R.J., 1998. ^{182}W and ^{187}Re-^{187}Os systematics of iron meteorites: chronology for melting differentiation, and crystallization in asteroids. Geochimica et Cosmochimica Acta 62, 545–554.

Horita, J., Berndt, M.E., 1999. Abiogenic methane formation and isotopic fractionation under hydrothermal conditions. Science 285, 1055–1057.

Horn, I., von Blanckenburg, F., Schoenberg, R., Steinhoefel, G., Markl, G., 2006. In situ iron isotope ratio determination using UV-femtosecond laser ablation with application to hydrothermal ore formation processes. Geochimica et Cosmochimica Acta 70, 3677–3688.

Horstwood, M.S.A., Nesbitt, R.W., Noble, S.R., Wilson, J.F., 1999. U-Pb zircon evidence for an extensive early Archean craton in Zimbabwe: A reassessment of the timing of craton formation, stabilization, and growth. Geology 27, 707–710.

Horwitz, R., Pidgeon, R.T., 1993. 3.1 Ga tuff from the Scholl Belt in the west Pilbara: fur-ther evidence for diachronous volcanism in the Pilbara Craton. Precambrian Research 60, 175–183.

Hoskin, P.W.O., Ireland, T.R., 2000. Rare earth element chemistry of zircon and its use as a provenance indicator. Geology 28, 627–630.

Hoskin, P.W.O., Schaltegger, U., 2003. The composition of zircon and igneous and metamorphic petrogenesis. In: Hanchar, J.M., Hoskin, P.W.O. (Eds.), Zircon. Reviews in Mineralogy and Geochemistry, vol. 53, pp. 27–62.

Hou, Z-Q., Ma, H-W., Zaw, K., Zhang, Y-Q., Wang, M-J., Wang, Z., Pan, G-T., Tang, R-L., 2003. The Himalayan Yuolong porphyry copper belt: product of large scale strike-slip faulting in eastern Tibet. Economic Geology 98, 125–145.

Hou, Z.-Q., Qu, X.-M., Wang, S-X., Du, A., Gao, Y-F., Huang, W., 2004. Re-Os age for molybdenite from the Gangdese porphyry copper belt on Tibetan plateau: implication for geodynamic setting and duration of the Cu mineralization. Science in China, Series D 47, 221–231.

House, C.H., Schopf, J.W., McKeegan, K.D., Coath, C.D., Harrison, T.M., Stetter, K., 2000. Carbon isotopic composition of individual Precambrian micro-fossils. Geology 28, 707–710.

House, C.H., Schopf, J.W., Stetter, K.O., 2003. Carbon isotopic fractionation by Archaeans and other thermophilic prokaryotes. Organic Geochemistry 34, 345–356.

Howell, D.G., 1995. Principles of Terrane Analysis – New Applications for Global Tectonics. Chapman & Hall, London, 245 pp.

Hsu, W., Huss, G.R., Wasserburg, G.J., 1997. Mn-Cr systematics of differentiated meteorites. Lunar and Planetary Science XXVIII, 609–610.

Hu, G.X., Rumble, D., Wang, P.L., 2003. An ultraviolet laser microprobe for the in situ analysis of multisulfur isotopes and its use in measuring Archean sulfur isotope mass-independent anomalies. Geochimica et Cosmochimica Acta 67, 3101–3118.

Hubregtse, J.J.M.W., 1980. The Archean Pikwitonei Granulite Domain and its position at the margin of the northwestern Superior Province (central Manitoba). In: Manitoba Department of Energy and Mines, Mineral Resources Division, Geological Paper GP80-3, p. 16.

Hulbert, L.J., Hamilton, M.A., Horan, M.F., Scoates, R.F., 2005. U-Pb zircon and Re-Os isotope geochronology of mineralized ultramafic intrusions and associated nickel ores from the Thompson Nickel Belt, Manitoba, Canada. Economic Geology 100, 29–41.

Hulston, J.R., Thode, H.G., 1965. Variations in the ^{33}S, ^{34}S, and ^{36}S contents of meteorites and their relation to chemical and nuclear effects. Journal of Geophysical Research 70, 3475–3484.

Humphris, S.E., Tivey, M.K., 2000. A synthesis of geological and geochemical investigations of the TAG hydrothermal field: Insights into fluid-flow and mixing processes in a hydrothermal system. In: Dilek, Y., Moores, E.M., Elthon, D., Nicolas, A. (Eds.), Ophiolites and Oceanic Crust: New Insights from Field Studies and the Ocean Drilling Program. Geological Society of America, Special Paper 349, pp. 213–235.

Hunt, J.M., 1996. Petroleum Geochemistry and Geology, second ed. W.H. Freeman and Company, New York, 389 pp.

Hunten, D.M., 2002. Exospheres and planetary escape. In: Mendillo, M., Nagy, A., Waite, J.H. (Eds.), Atmospheres in the Solar System: Comparative Aeronomy. AGU Geophysical Monograph 130, pp. 191–202.

Hunter, D.R., 1979. The role of tonalitic to trondhjemitic rocks in the crustal development of Swaziland and the eastern Trasvaal, South Africa. In: Barker, F. (Ed.), Trondhjemites, Dacites and Related Rocks. Elsevier, Amsterdam, pp. 301–322.

Hunter, D.R., 1991. Crustal processes during Archaean evolution of the southeastern Kaapvaal province. Journal of African Earth Science 13, 13–25.

Hunter, D.R., 1993. The Ancient Gneiss Complex. In: Kröner, A. (Ed.), The Anciet Gneiss Complex: Overview Papers and Guidebook for Excursion. Swaziland Geological Survey Mines Department, Bulletin 11, pp. 1–14.

Hunter, D.R., Wilson, A.H., 1988. A continuous record of Archaean evolution from 3.5 Ga to 2.6 Ga in Swaziland and northern Natal. South African Journal of Geology 91, 57–74.

Hunter, D.R., Barker, F., Millard, H.T., Jr, 1978. The geochemical nature of the Archean Ancient Gneiss Complex and Granodiorite Suite, Swaziland: a preliminary study. Precambrian Research 7, 105–127.

Hunter, D.R., Allen, A.R., Millin, P., 1983. A preliminary note on Archaean supracrustal and granitoid rocks west of Piet Retief. Transactions of the Geological Society of South Africa 86, 301–306.

Hunter, D.R., Barker, F., Millard, H.T., 1984. Geochemical investigation of Archaean bimodal and Dwalile metamorphic suites, Ancient Gneiss Complex, Swaziland. Precambrian Research 24, 131–155.

Hunter, D.R., Smith, R.G., Sleigh, D.W.W., 1992. Geochemical studies of Archaean granitoid rocks in the southeastern Kaapvaal Province: implications for crustal development. Journal of African Earth Sciences 15 (1), 127–151.

Huss, G.R., McPherson, G.J., Wasserburg, G.J., Russell, S.S., Srinivasan, G., 2001. Aluminium-26 in calcium-aluminium-rich inclusions and chondrules from unequilibrated ordinary chondrites. Meteoritics and Planetary Science 36, 975–997.

Huston, D.L., 1999. Stable isotopes and their significance for understanding the genesis of volcanic-hosted massive sulfide deposits: a review. Reviews in Economic Geology 10, 151–180.

Huston, D.L., Large, R.R., 1987. Genetic and exploration significance of the zinc ratio (100Zn/[Zn+Pb]) in massive sulfide systems. Economic Geology 82, 1521–1539.

Huston, D.L., Logan, G.A., 2004. Barite, BIFs and bugs: evidence for the evolution of the Earth's early hydrosphere. Earth and Planetary Science Letters 220, 41–55.

Huston, D.L., Brauhart, C.W., Dreiberg, S.L., Davidson, G.J., Groves, D.I., 2001a. Metal leaching and inorganic sulphate reduction in volcanic-hosted massive sulphide mineral systems: Evidence from the paleo-Archean Panorama district, Western Australia. Geology 29, 687–690.

Huston, D.L., Blewett, R.S., Mernagh, T.P., Sun, S.-S., Kamprad, J., 2001b. Gold deposits of the Pilbara Craton: Results of AGSO Research, 1998–2000. Australian Geological Survey Organisation, Record 2001/10, 86 pp.

Huston, D.L., Sun, S.-S., Blewett, R.S., Hickman, A.H., Van Kranendonk, M., Phillips, D., Baker, D., Brauhart, C.W., 2002. The timing of mineralization in the Archean Pilbara Craton, Western Australia. Economic Geology 97, 733–756.

Huston, D.L., Champion, D.C., Cassidy, K.F., 2005. Tectonic controls on the endowment of Archean cratons in VHMS deposits: Evidence from Pb and Nd isotopes. In: Mao, J.W., Bierlein, F.P. (Eds.), Mineral Deposit Research: Meeting the Global Challenge. Springer, Berlin, pp. 15–18.

Hutcheon, I.D., Olsen, E., 1991. Cr isotopic composition of differentiated meteorites: A search for ^{53}Mn. Lunar and Planetary Science XXII, 605–606.

Hutcheon, I.D., Phinney, D.L., 1996. Radiogenic ^{53}Cr in Orgueil carbonates: chronology of aqueous activity on the CI parent body. Lunar and Planetary Science XXVII, 577–578.

Hutcheon, I.D., Krot, A.N., Keil, K., Phinney, D.L., Scott, E.R.D., 1998. ^{53}Mn-^{53}Cr dating of fayalite formation in the CV3 chondrite, Mokoia: evidence for asteroidal alteration. Science 282, 1865–1867.

Hutcheon, I.D., Weisberg, M.K., Phinney, D.L., Zolensky, M.E., Prinz, M., Ivanov, A.V., 1999. Radiogenic ^{53}Cr in Kaidun carbonates: evidence for very early aqueous activity. Lunar and Planetary Science XXX, abstract 1722.

Hutchison, R., 2004. Meteorites: A Petrologic, Chemical and Isotopic Synthesis. Cambridge University Press, 506 pp.

Hutchison, R., Alexander, C.M.O'D., Barber, D.J., 1987. The Semarkona meteorite: First recorded occurrence of smectite in an ordinary chondrite, and its implications. Geochimica et Cosmochimica Acta 51, 1875–1882.

Hutchison, R., Williams, C.T., Din, V.K., Clayton, R.N., Kirschbaum, C., Paul, R.L., Lipschutz, M.E., 1988. A planetary H-group pebble in the Barwell L6, unshocked ordinary chondrite. Earth and Planetary Science Letters 90, 105–118.

Hutchison, R., Alexander, C.M.O'D., Bridges, J.C., 1998. Elemental redistribution in Tieschitz and the origin of white matrix. Meteoritics and Planetary Science 33, 1169–1179.

Hutton, J., 1785. Theory of the Earth. Abstract from his book published in 1795.

Hynes, A., Skulski, T., 2005. Archean plate tectonics – similarities and differences. Geological Association of Canada Program with Abstracts, 30, p. 92.

Iizuka, T., Hirata, T., 2005. Improvements of precision and accuracy in *in situ* Hf isotope microanalysis of zircon using the laser ablation-MC-ICPMS technique. Chemical Geology 220, 121–137.

Iizuka, T., Horie, K., Komiya, T., Maruyama, S., Hirata, T., Hidaka, H., Windley, B.F., 2006. 4.2 Ga zircon xenocryst in an Acasta gneiss from northwestern Canada: evidence for early continental crust. Geology 34, 245–248.

Iizuka, T., Komiya, T., Ueno, Y., Katayama, I., Uehara, Y., Maruyama, S., Hirata, T., Johnson, S.P., Dunkley, D., 2007. Geology and zircon geochronology of the Acasta Gneiss Complex, northwestern Canada: New constraints on its tectonothermal history. Precambrian Research 153, 179–208.

Iizuka, T., Komiya, T., Maruyama, S., Hirata, T., Johnson, S.P., submitted. In situ Lu-Hf isotopic analysis of zircon from the Acasta Gneiss Complex, NW Canada: Implications for extensive reworking of Hadean continental crust. Chemical Geology.

Ingle, S., Coffin, M.F., 2004. Impact origin for the greater Ontong Java Plateau? Earth and Planetary Science Letters 218, 123–134.

Ingram, P.A.J., 1977. A summary of the geology of a portion of the Pilbara Goldfield, Western Australian. In: McCall, G.J.H. (Ed.), The Archaean, Search for the Beginning. Dowden, Hutchinson and Ross, Stroudsburg, PA, pp. 208–216.

Ireland, T.R., Wlotzka, F., 1992. The oldest zircons in the Solar System. Earth and Planetary Science Letters 109, 1–10.

Irvine, T.N., Baragar, W.R.A., 1971. A guide to the chemical classification of the common volcanic rocks. Canadian Journal of Earth Sciences 8, 523–548.

Irving, E., Woodsworth, G.J., Wynne, P.J., Morrison, A., 1985. Paleomagnetic evidence for displacement from the south of the Coast Plutonic Complex, British Columbia. Canadian Journal of Earth Sciences 22, 584–598.

Isley, A.E., Abbot, D.H., 1999. Plume-related mafic volcanism and the deposition of banded iron formation. Journal of Geophysical Research 104, 15,461–15,477.

Isnard, H., Gariepy, C., 2004. Sm-Nd, Lu-Hf and Pb-Pb signatures of gneisses and granitoids from the La Grande Belt; extent of late Archean crustal recycling in the northeastern Superior Province, Canada. Geochimica et Cosmochimica Acta 68, 1099–1113.

Isozaki, Y., et al., 1997. Early Archean mid-oceanic ridge rocks and early life in the Pilbara Craton, W. Australia. Eos 78, 399.

Ivanov, B.A., Melosh, H.J., 2003. Impacts do not initiate volcanic eruptions. Geology 31, 869–872.

Ivanov, M.A., Head, J.W., 1996. Tessera terrain on Venus: A survey of the global distribution, characteristics, and relation to surrounding units from Magellan data. Journal of Geophysical Research 101, 14,861–14,908.

Izenberg, N.R., Arvidson, R.E., Phillips, R.J., 1994. Impact crater degradation on Venusian plains. Geophysical Research Letters 21, 289–292.

Jackson, I.N.S., Rigden, S.M., 1998. Composition and temperature of the mantle: seismologial models interpreted through experimental studies of mantle minerals. In: Jackson, I.N.S. (Ed.), The Earth's Mantle: Composition, Structure and Evolution. Cambridge University Press, Cambridge, pp. 405–460.

Jackson, M.P.A., 1984. Archaean structural styles in the Ancient Gneiss Complex of Swaziland, southern Africa. In: Kröner, A., Greiling, R. (Eds.), Precambrian Tectonics Illustrated. E. Schweizerbart'sche Verlagsbuchhandlung, Stuttgart, pp. 1–18.

Jackson, M.P.A., Robertson, D.I., 1983. Regional implications of Early-Precambrian strains in the Onverwacht Group adjacent to the Lochiel granite, north-west Swaziland. In: Geological Society of South Africa, Special Publication 9, pp. 45–62.

Jackson, M.P.A., Eriksson, K.A., Harris, C.W., 1987. Early Archean foredeep sedimentation related to crustal shortening: a reinterpretation of the Barberton Sequence, southern Africa. Tectonophysics 136, 197–221.

Jacobs, J., Bauer, W., Spaeth, G., Thomas, R.J., Weber, K., 1996. Lithology and structure of the Grenville-aged (~1.1 Ga) basement of Heimefrontfjella (East Antarctica). Geologische Rundschau 85, 800–821.

Jacobs, J., Fanning, C.M., Henjes-Kunst, F., Olesch, M., Paech, H.-J., 1998. Continuation of the Mozambique Belt into East Antarctica: Grenville-age metamorphism and polyphase Pan-African high-grade events in central Dronning Maud Land. Journal of Geology 106, 385–406.

Jacobs, J., Fanning, C.M., Bauer, W., 2003. Timing of Grenville-age vs. Pan-African medium- to high-grade metamorphism in western Dronning Maud Land (East Antarctica) and significance for correlations in Rodinia and Gondwana. Precambrian Research 125, 1–20.

Jacobsen, S.B., 2005. The Hf-W isotopic system and the origin of the Earth and Moon. Annual Reviews of Earth and Planetary Science 33, 531–570.

Jacobsen, S.B., Pimentel-Close, M., 1988. A Nd isotopic study of the Hamersley and Michipicoten banded iron formations: The source of REE and F in Archean oceans. Earth and Planetary Science Letters 87, 29–44.

Jacobsen, S.B., Wasserburg, G.J., 1984. Sm-Nd evolution of chondrites and achondrites. Earth and Planetary Science Letters 67, 137–150.

Jaffrés, J.B.D., 2005. Development of a model to evaluate seawater oxygen isotope composition over the past 3.4 billion years. Unpublished honours thesis. James Cook University, North Queensland, Australia.

Jagoutz, E., Jotter, R., Kubny, A., Varela, M.E., Zartman, R., Kurat, G., Lugmair, G.W., 2003. Cm?-U-Th-Pb isotopic evolution of the D'Orbigny angrite. Meteoritics and Planetary Science 38 (Supplement), A81 (abstract).

Jahn, B.-M., 1994. Géochimie des granitoïdes archéens et de la croûte primitive. In: Hagemann, R., Jouzel, J., Treuil, M., Turpin, L. (Eds.), La géochimie de la Terre. CEA-Masson.

Jahn, B.-M., 2004. The Central Asian Orogenic Belt and growth of the continental crust in the Phanerozoic. In: Geological Society London, Special Publication 226, pp. 73–100.

Jahn, B.-M., Zhang, Z.Q., 1984. Archaean granulite gneisses from eastern Hebei Province, China; rare earth geochemistry and tectonic implications. Contributions to Mineralogy and Petrology 85, 224–243.

Jahn, B.-M., Auvray, B., Blais, S., Capdevila, R., Cornichet, J., Vidal, F., Hammeurt, J., 1980. Trace elements geochemistry and petrogenesis of Finnish greenstone belts. Journal of Petrology 21, 201–244.

Jahn, B.-M., Glikson, A.Y., Peucat, J.J., Hickman, A.H., 1981. REE geochemistry and isotopic data of Archean silicic volcanics and granitoids from the Pilbara Block, Western Australia: implications for early crustal evolution. Geochimica et Cosmochimica Acta 45, 1633–1652.

Jahn, B., Gruau, G., Glikson, A.Y., 1982. Komatiites of the Onverwacht Group, S. Africa: REE geochemistry, Sm/Nd age and mantle evolution. Contributions to Mineralogy and Petrology 80, 25–40.

Jahn, B.-M., Vidal, P., Kröner, A., 1984. Multi-chronometric ages and origin of Archaean tonalitic gneisses in Finnish Lapland: a case for long crustal residence time. Contribution to Mineralogy and Petrology 86, 398–408.

Jahn, B.-M., Auvray, B., Cornichet, J., Bai, Y.L., Dhen, Q.H., Liu, D.Y., 1987. 3.5 Ga old amphibolites from eastern Hebei Province, China: Field occurrence, petrology, Sm-Nd isochron age and REE geochemistry. Precambrian Research 34, 311–346.

Jahn, B.-M., Auvray, B., Shen, Q.H., Liu, D.Y., Zhang, Z.Q., Dong, Y.J., Ye, X.J., Zhang, Z.Q., Comichet, J., Mace, J., 1988. Archaean crustal evolution in China: The Taishan complex, and evidence for juvenile crustal addition from long-term depleted mantle. Precambrian Research 38, 381–403.

Jahn, B.-M., Gruau, G., Bernard-Griffiths, J., Cornichet, J., Kröner, A., Wendt, I., 1990. The Aldan Shield, Siberia: geochemical characterization, ages, petrogenesis and comparison with the Sino-Korean craton. In: Glover, J.E., Ho, S. (Eds.), Third International Archean Symposium, Extended abstract volume. Geoconferences WA Inc., Perth, Australia, pp. 179–181.

Jahn, B.-M., Gruau, G., Capdevila, R., Cornichet, J., Nemchin, A., Pidgeon, R., Rudnik, V.A., 1998. Archean crustal evolution of the Aldan Shield, Siberia: geochemical and isotopic constraints. Precambrian Research 91, 333–363.

Janes, D.M., Squyres, S.W., 1995. Viscoelastic relaxation of topographic highs on Venus to produce coronae. Journal of Geophysical Research 100, 21,173–21,187.

Janes, D.M., Squyres, S.W., Bindschadler, D.L., Baer, G., Schubert, G., Sharpton, V.L., Stofan, E.R., 1992. Geophysical models for the formation and evolution of coronae on Venus. Journal of Geophysical Research 97, 16,055–16,068.

Javoy, M., 1995. The integral enstatite chondrite model of the Earth. Geophysical Research Letters 22, 2219–2222.

Jawhari, T., Roid, A., Casado, J., 1995. Raman spectroscopic characterization of some commercially available carbon black materials. Carbon 33, 1561–1565.

Jeffery, P.M., Reynolds, J.H., 1961. Origin of excess Xe^{129} in stone meteorites. Journal of Geophysical Research 66, 3582–3583.

Jehlicka, J., Urban, O., Pokorny, J., 2003. Raman spectroscopy of carbon and solid bitumens in sedimentary and metamorphic rocks. Spectrochimica Acta, Part A 59, 2341–2352.

Jenner, F.J., Bennett, V.C., Nutman, A.P., 2006. 3.8 Ga arc-related basalts from Southwest Greenland. Geochimica et Cosmochimica Acta 70, A291.

Jérbak, M., Harnois, L., 1991. Two-stage evolution in an Archean tonalite suite: the Taschereau stock, Abitibi (Quebec, Canada). Canadian Journal of Earth Science 28, 172–183.

Ji, Z.Y., 1993. New data of isotope age of the Proterozoic metamorphic rocks from northern Jiaodong and its geological significance. Geology of Shandong 9, 40–51 (in Chinese with English abstract).

Jian, P., Zhang, Q., Liu, D.Y., Jin, W.J., Jia, X.Q., Qian, Q., 2005. SHRIMP dating and geological significance of Late Archaean high-Mg diorite (sanukite) and hornblende-granite at Guyang of Inner Mongolia. Acta Petrologica Sinica 21, 151–157 (in Chinese with English abstract).

Jin, K., Xu, W.L., Wang, Q.H., Gao, S., Liu, X.C., 2003. Formation time and sources of the Huaiguang migmatitic granodiorite in the Bangbu, Anhui Province: Evidence from SHRIMP zircon U-Pb geochronology. Acta Geologica Sinica 24, 331–336 (in Chinese with English abstract).

Johannesson, K.H., Hawkins, D.L., Cortes, A., 2006. Do Archean chemical sediments record ancient seawater rare earth element patterns? Geochimica et Cosmochimica Acta 70, 871–890.

Johnson, C.M., Beard, B.L., Beukes, N.J., Klein, C., O'Leary, J.M., 2003. Ancient geochemical cycling in the Earth as inferred from Fe isotope studies of banded iron formations from the Transvaal Craton. Contributions to Mineralogy and Petrology 144, 523–547.

Johnson, R.C., Hills, F.A., 1976. Precambrian geochronology and geology of the Box Elder Canyon area, northern Laramie Range, Wyoming. Bulletin of the Geological Society of America 87, 809–817.

Johnston, S.T., 2001. The Great Alaskan Terrane Wreck: reconciliation of paleomagnetic and geological data in the northern Cordillera. Earth and Planetary Science Letters 193, 259–272.

Johnson, T.M., Bullen, T.D., 2004. Mass-dependent fractionation of selenium and chromium isotopes in low-temperatures environments. Reviews in Mineralogy and Geochemistry 55, 289–317.

Johnson, W.J., Goldstein, R.H., 1993. Cambrian seawater preserved as inclusions in marine low-magnesium calcite cement. Nature 362, 335–337.

Johnston, A.D., Wyllie, P.J., 1988. Constraints on the origin of Archean trondhjemites based on phase relationships of Nuuk Gneiss with H_2O at 15 kbar. Contributions to Mineralogy and Petrology 100, 35–46.

Jones, A.P., Pickering, K.T., 2003. Evidence for aqueous fluid-sediment transport and erosional processes on Venus. Journal Geological Society London 160, 319–327.

Jones, A.P., Price, G.D., Price, N.J., DeCarli, P.S., Clegg, R.A., 2002. Impact-induced melting and development of large igneous provinces. Earth and Planetary Science Letters 202, 551–561.

Jones, A.P., Wunemann, K., Price, D., 2005. Impact volcanism as a possible origin for the Ontong Java Plateau (OJP). In: Foulger, G.R., Natland, J.H., Presnall, D.C., Anderson, D.L. (Eds.), Plates, Plumes, and Paradigms. Geological Society of America, Special Paper 388, pp. 711–720.

Jones, C.B., 1990. Coppin Gap copper-molybdenum deposit. In: Australasian Institute of Mining and Metallurgy, Monograph 14, pp. 141–144.

Jordan, T.H., 1975. The continental tectosphere. Geophysics and Space Physics 13, 1–12.

Jordan, T.H., 1978. Composition and development of the continental tectosphere. Nature 274, 544–548.

Kah, L.C., 2000. Depositional $\delta^{18}O$ signatures in Proterozoic dolostones: constraints on seawater chemistry and early diagenesis. In: James, N., Grotzinger, J.P. (Eds.), Carbonate Sedimentation and Diagenesis in the Evolving Precambrian World. SEPM Special Publication, vol. 67, pp. 345–360.

Kaiser, T., Wasserburg, G.J., 1983. The isotopic composition and concentration of Ag in iron meteorites. Geochimica et Cosmochimica Acta 47, 43–58.

Kallemeyn, G.W., Wasson, J.T., 1985. The compositional classification of chondrites: IV. Ungrouped chondritic meteorites and clasts. Geochimica et Cosmochimica Acta 49, 261–270.

Kallemeyn, G.W., Rubin, A.E., Wasson, J.T., 1996. The compositional classification of chondrites: VII. The R chondrite group. Geochimica et Cosmochimica Acta 60, 2243–2256.

Kamber, B.S., Moorbath, S., 1998. Initial Pb of the Amîtsoq gneiss revisited: implication for the timing of early Archaean crustal evolution in West Greenland. Chemical Geology 150, 19–41.

Kamber, B.S., Webb, G.E., 2001. The geochemistry of Late Archaean microbial carbonate: implications for ocean chemistry and continental erosion history. Geochimica et Cosmochimica Acta 65, 2509–2525.

Kamber, B.S., Moorbath, S., Whitehouse, M.J., 2001. The oldest rocks on Earth: time constraints and geological controversies. In: Lewis, C.L.E., Knell, S.J. (Eds.), The Age of the Earth: From 4004 BC to AD 2002. Geological Society of London, Special Publication 190, pp. 177–203.

Kamber, B.S., Ewart, A., Collerson, K.D., Bruce, M.C., McDonald, G.D., 2002. Fluid-mobile trace element constraints on the role of slab melting and implications for Archaean crustal growth models. Contributions to Mineralogy and Petrology 144, 38–56.

Kamber, B.S., Collerson, K.D., Moorbath, S., Whitehouse, M.J., 2003. Inheritance of early Archaean Pb-isotope variability from long-lived Hadean protocrust. Contributions to Mineralogy and Petrology 145, 25–46.

Kamber, B.S., Whitehouse, M.J., Bolhar, R., Moorbath, S., 2005a. Volcanic resurfacing and the early terrestrial crust: Zircon U-Pb and REE constraints from the Isua Greenstone Belt, southern West Greenland. Earth and Planetary Science Letters 240 (2), 276–290.

Kamber, B.S., Greig, A., Collerson, K.D., 2005b. A new estimate for the composition of weathered young upper continental crust from alluvial sediments, Queensland, Australia. Geochimica et Cosmochimica Acta 69, 1041–1058.

Kamo, S.L., Davis, D.W., 1994. Reassessment of Archean crust development in the Barberton Mountain Land, South Africa, based on U-Pb dating. Tectonics 13, 167–192.

Kamo, S.L., Davis, D.W., De Wit, M.J., 1990. U-Pb geochronology of Archean plutonism in the Barberton region, S. Africa: 800 Ma of crustal evolution. In: 7th International Geocongress on Geochemistry, Canberra, p. 53.

Kamo, S.L., Reimold, W.U., Krogh, T.E., Colliston, W.P., 1996. A 2.023 Ga age for the Vredefort impact event and a first report of shock metamorphosed zircons in pseudo-tachylitic breccias and granophyre. Earth and Planetary Sciences Letters 144, 369–387.

Kappler, A., Pasquero, C., Konhauser, K.O., Newman, D.K., 2005. Deposition of banded iron formations by photoautotrophic Fe(II)-oxidizing bacteria. Geology 33, 865–868.

Kareem, K.M., Byerly, G.R., 2002. Observed and predicted crystallization sequence in Barberton komatiite. In: Geological Society of America, Abstracts with Programs, vol. 34, p. 62.

Kareem, K.M., Byerly, G.R., 2003. Petrology and geochemistry of 3.3 Ga komatiites – Weltevreden Formation, Barberton Greenstone Belt. In: Lunar and Planetary Science Conference XXXIV, #2071.

Karlstrom, K.E., Houston, R.S., 1984. The Cheyenne Belt: Analysis of a Proterozoic suture in southern Wyoming. Precambrian Research 25, 415–446.

Kasting, J.F., 1992. Models related to Proterozoic atmospheric and ocean chemistry. In: Schopf, J.W., Klein, C. (Eds.), The Proterozoic Biosphere: A Multidisciplinary Study. Cambridge University Press, pp. 1185–1187.

Kasting, J.F., 1993. Earth's early atmosphere. Science 259, 920–926.

Kasting, J.F., 2005. Methane and climate during the Precambrian era. Precambrian Research 137, 119–129.

Kasting, J.F., Ono, S., 2006. Palaeoclimates: the first two billion years. Philosophical Transactions of the Royal Society of London Ser. B 361, 917–929.

Kasting, J.F., Zahnle, K.J., Pinto, J.P., Young, A.T., 1989. Sulfur, ultraviolet-radiation, and the early evolution of life. Origins of Life and Evolution of the Biosphere 19, 95–108.

Kasting, J.F., Howard, M.T., Wallmann, K., Veizer, J., Shields, G.A., Jaffrés, J., 2006. Paleoclimates, ocean depth, and the oxygen isotopic composition of seawater. Earth and Planetary Science Letters 252, 82–93.

Katagiri, G., Ishida, H., Ishitani, A., 1988. Raman spectra of graphite edge planes. Carbon 26, 565–571.

Kato, Y., Nakamura, K., 2003. Origin and global tectonic significance of Early Archean cherts from the Marble Bar greenstone belt, Pilbara craton, Western Australia. Precambrian Research 125, 191–243.

Kaula, W.M., 1990. Venus: A contrast in evolution to Earth. Science 247, 1191–1196.

Kay, I., Sol, S., Kendall, J.M., Thomson, C., White, D., Asudeh, I., Roberts, B., Francis, D., 1999a. Shear wave splitting observations in the Archean craton of western Superior. Geophysical Research Letters 26, 2669–2672.

Kay, I., Musacchio, G., White, D., Asudeh, I., Roberts, B., Forsyth, D., Hajnal, Z., Koperwhats, B., Farrell, D., 1999b. Imaging the Moho and Vp/Vs ratio in the western Superior Archean craton with wide-angle reflections. Geophysical Research Letters 26, 2585–2588.

Kay, R.W., Kay, S.M., 1991. Creation and destruction of lower continental crust. Geologische Rundschau 80, 259–278.

Keil, K., 2000. Thermal alteration of asteroids: evidence from meteorites. Planetary and Space Science 48, 887–903.

Keil, K., Stöffler, D., Love, S.G., Scott, E.R.D., 1997. Constraints on the role of impact heating and melting in asteroids. Meteoritics and Planetary Science 32, 349–363.

Kelemen, P.B., 1995. Genesis of high Mg# andesites and the continental crust. Contribution to Mineralogy and Petrology 120 (1), 1–19.

Kelemen, P.B., Hanghoj, K., Greene, A.R., 2003. One view of the geochemistry of subduction-related magmatic arcs, with an emphasis on primitive andesite and lower crust. In: Rudnick, R.L. (Ed.), Treatise on Geochemistry, vol. 3: The Crust. Elsevier, Amsterdam, pp. 593–659.

Keller, G., 2005. Impacts, volcanism and mass extinction: random coincidence or cause and effect? Australian Journal of Earth Science 52 (4–5), 725–758.

Kellogg, L.H., Hager, B.H., van der Hilst, R.D., 1999. Compositional stratification in the deep mantle. Science 283, 1881–1884.

Kelly, N.M., Clarke, G.L., Carson, C.J., White, R.W., 2000. Thrusting in the lower crust: evidence from the Oygarden Islands, Kemp Land, East Antarctica. Geological Magazine 137, 219–234.

Kelly, N.M., Clarke, G.L., Fanning, C.M., 2002. A two-stage evolution of the Neoproterozoic Rayner Structural Episode: New U-Pb SHRIMP constraints from the Oygarden Group, Kemp Land, East Antarctica. Precambrian Research 116, 307–330.

Kelly, N.M., Clarke, G.L., Fanning, C.M., 2004. Archaean crust in the Rayner Complex of East Antarctica: Oygarden Group of Islands, Kemp Land. Transactions of the Royal Society of Edinburgh: Earth and Environmental Science 95, 491–510.

Kelly, N.M., Harley, S.L., 2005. An integrated microtextural and chemical approach to zircon geochronology: refining the Archaean history of the Napier Complex, East Antarctica. Contributions to Mineralogy and Petrology 149, 57–84.

Kelly, W.R., Wasserburg, G.J., 1978. Evidence for the existence of ^{107}Pd in the early solar system. Geophysical Research Letters 5, 1079–1082.

Kelsey, D.E., White, R.W., Powell, R., Wilson, C.J.L., Quinn, C.D., 2003. New constraints on metamorphism in the Rauer Group, Prydz Bay, East Antarctica. Journal of Metamorphic Geology 21, 739–759.

Kendall, J.M., Sol, S., Thomson, C.J., White, D.J., Asudeh, I., Snell, C.S., Sutherland, F.H., 2002. Seismic heterogeneity and anisotropy in the western Superior Province, Canada: insights into the evolution of an Archean craton. In: Fowler, C.M.R., Ebinger, C.J., Hawkesworth, C.J. (Eds.), The Early Earth: Physical, Chemical and Biological Development. Geological Society of London, Special Publication 199, pp. 27–44.

Kent, R.W., Hardarson, B.S., Saunders, A.D., Storey, M., 1996. Plateaux ancient and modern: geochemical and sedimentological perspectives on Archaean oceanic magmatism. Lithos 37, 129–142.

Kenyon, S.J., Bromley, B.C., 2005. Prospects for detection of catastrophic collisions in debris disks. Astronomical Journal 130, 269–279.

Kerrich, R., Polat, A., 2006. Archean greenstone-tonalite duality: thermochemical mantle convection models or plate tectonics in the early Earth global dynamics? Tectonophysics 415, 141–165.

Kerrich, R., Wyman, D., Fan, J., Bleeker, W., 1998. Boninite series: Low Ti-tholeiite associations from the 2.7 Ga Abitibi Greenstone Belt. Earth and Planetary Science Letters 164, 303–316.

Kerrich, R., Polat, A., Wyman, D.A., Hollings, P., 1999. Trace element systematics of Mg- to Fe tholeiitic basalt suites of the Superior province: implications for Archean mantle reservoirs and greenstone belt genesis. Lithos 46, 163–187.

Kerrich, R., Goldfarb, R., Groves, D., Garwin, S., 2000. The geodynamics of world-class gold deposits: characteristics, space-time distribution, and origins. In: Hagemann, S.G., Brown, P.E. (Eds.), Gold in 2000. Reviews in Economic Geology, vol. 13, pp. 501–551.

Kerrich, R., Goldfarb, R.J., Richards, J.P., 2005. Metallogenic processes in an evolving geodynamic framework. Economic Geology, 100th Anniversary Volume, 1097–1136.

Kerridge, J.F., Mackay, A.L., Boynton, W.V., 1979. Magnetite in CI carbonaceous chondrites: Origin by aqueous activity on a planetesimal surface. Science 205, 395–397.

Keszthelyi, L., Self, S., 1998. Some physical requirement for the emplacement of long basaltic lava flows. Journal of Geophysical Research 103, B11, 27,447–27,464.

Ketchum, J., Ayer, J., Van Breemen, O., Pearson, N.J., Becker, J.K., submitted. Pericratonic crustal growth of the southwestern Abitibi subprovince, Canada – U-Pb, Hf, and Nd isotopic evidence. Economic Geology.

Khain, V.E., 2001. Tectonics of the Continents and Oceans. Nauchnyi Mir, Moscow, 604 pp. (in Russian).

Khoreva, B.Ya. (Ed.), 1987. Map of the Metamorphic and Granitic Rock Associations of the USSR. Scale 1:10,000,000. Kartfabrika Vsesoyuzn. Geol. Inst., Leningrad (in Russian).

Kidder, J.G., Phillips, R.J., 1996. Convection-driven subsolidus crustal thickening on Venus. Journal Geophysical Research 101, 23,181–23,194.

Kinny, P.D., 1987. An ion microprobe study of uranium-lead and hafnium isotopes in natural zircon. Unpublished Ph.D. thesis. Australian National University, Canberra, ACT, 160 pp.

Kinny, P.D., 1990. Age spectrum of detrital zircons in the Windmill Hill Quartzite. In: Ho, S.E., Glover, J.E., Myers, J.S., Muhling, J.R. (Eds.), Third International Archaean Symposium, Excursion Guidebook. Geology Department and University Extension, University of Western Australia, Publication no. 21, pp. 116–117.

Kinny, P.D., Nutman, A.P., 1996. Zirconology of the Meeberrie gneiss, Yilgarn Craton, Western Australia: an early Archaean migmatite. Precambrian Research 78, 165–178.

Kinny, P.D., Williams, I.S., Froude, D.O., Ireland, T.R., Compston, W., 1988. Early Archaean zircon ages from orthogneisses and anorthosites at Mount Narryer, Western Australia. Precambrian Research 38, 325–341.

Kinny, P.D., Wijbrans, J.R., Froude, D.O., Williams, I.S., Compston, W., 1990. Age constraints on the geological evolution of the Narryer Gneiss Complex, Western Australia. Australian Journal of Earth Sciences 37, 51–69.

Kinny, P.D., Black, L.P., Sheraton, J.W., 1993. Zircon ages and the distribution of Archaean and Proterozoic rocks in the Rauer Islands. Antarctic Science 5, 193–206.

Kirk, R., Soderblom, L., Lee, E., 1992. Enhanced visualization for interpretation of Magellan radar data: Supplement to the Magellan special issue. Journal of Geophysical Research 97, 16,371–16,380.

Kirkwood, R., 2000. Geology, geochronology and economic potential of the Archean rocks in the western Owl Creek Mountains, Wyoming. Unpublished M.Sc. thesis. University of Wyoming, Laramie, 99 pp.

Kissel, J., Brownless, D.E., Buchler, K., Clarke, B.C., Fechtig, H., Grun, E., Hornung, K., Igenbergs, E.B., Jessberger, E.K., Krueger, F.R., Kuczera, H., McDonnell, J.A.M., Morfill, G.M., Rahe, J., Schwehm, G.H., Sekanina, Z., Utterback, N.G., Volk, H.J., Zook,

H.A., 1986. Composition of comet Halley dust particples from Giotto observations. Nature 321, 336–337.

Kisters, A.F.M., Anhaeusser, C.R., 1995a. Emplacement features of Archaean TTG plutons along the southern margin of the Barberton greenstone belt, South Africa. Precambrian Research 75, 1–15.

Kisters, A.F.M., Anhaeusser, C.R., 1995b. The structural significance of the Steynsdorp pluton and anticline within the tectono-magmatic framework of the Barberton Mountain Land. South African Journal of Geology 98, 43–51.

Kisters, A.F.M., Stevens, G., Dziggel, A., Armstrong, R.A., 2003. Extensional detachment faulting and core-complex formation in the southern Barberton granite-greenstone terrain, South Africa: evidence for a 3.2 Ga orogenic collapse. Precambrian Research 127, 355–378.

Kisters, A.F.M., Stevens, G., Van Reenen, D., 2004. Excursion Field Guide: The Kaapvaal Traverse. 17–23 July 2004. Geoscience Africa, University of Witswaterrand, Johannesbourg.

Kisters, A., Belcher, R., Poujol, M., Stevens, G., Moyen, J.F., 2006. A 3.2 Ga magmatic arc preserving 50 Ma of crustal convergence in the Barberton terrain, South Africa. In: AGU Fall Meeting, 11–15 December 2006, San Francisco, USA.

Kitajima, K., Maruyama, S., Utsunomiya, S., Liou, J.G., 2001. Seafloor hydrothermal alteration at an Archaean mid-ocean ridge. Journal of Metamorphic Geology 19, 581–597.

Kitsul, V.I., Dook, V.L., 1985. Endogenic formation regimes and evolution stages of the Early Precambrian lithosphere in the Vitim-Aldan Shield. In: Endogenic Formation Regimes of the Earth's Crust and Ore Deposits in the Early Precambrian. Nauka, Leningrad, pp. 217–235 (in Russian).

Kiyokawa, S., Ito, T., Ikehara, M., Kitajima, F., 2006. Middle Archean volcano-hydrothermal sequence: bacterial microfossil-bearing 3.2 Ga Dixon Island Formation, coastal Pilbara terrane, Australia. Bulletin of the Geological Society of America 118, 3–22.

Kleikemper, J., Schroth, M.H., Bernasconi, S.M., Brunner, B., Zeyer, J., 2004. Sulfur isotope fractionation during growth of sulfate-reducing bacteria on various carbon sources. Geochimica et Cosmochimica Acta 68, 4891–4904.

Klein, M., Stosch, H.-G., Seck, H.A., 1997. Partitioning of high-field strength and rare-earth elements between amphibole and quartz-dioritic to tonalitic melts: an experimental study. Chemical Geology 138, 257–271.

Klein, M., Stosch, H.-G., Seck, H.A., Shimizu, N., 2000. Experimental partitioning of high field strength and rare earth elements between clinopyroxene and garnet in andesitic to tonalitic systems. Geochimica et Cosmochimica Acta 64, 99–115.

Klein, M., Friedrich, M., Roger, A.J., Hugenholtz, P., Fishbain, S., Abicht, H., Blackall, L.L., Stahl, D.A., Wagner, M., 2001. Multiple lateral transfers of dissimilatory sulfite reductase genes between major lineages of sulfate-reducing prokaryotes. Journal of Bacteriology 183, 6028–6035.

Kleine, T., Münker, C., Mezger, K., Palme, H., 2002. Rapid accretion and early core formation on asteroids and the terrestrial planets from Hf-W chronometry. Nature 418, 952–955.

Kleine, T., Mezger, K., Münker, C., Palme, H., Bischoff, A., 2004. ^{182}Hf-^{182}W isotope systematics of chondrites, eucrites and martian meteorites: Chronology of core formation and early mantle differentiation in Vesta and Mars. Geochimica et Cosmochimica Acta 68, 2935–2946.

Kleine, T., Mezger, K., Palme, H., Scherer, E., Munker, C., 2005a. Early core formation in asteroids and late accretion of chondrite parent bodies: Evidence from ^{182}Hf-^{182}W in CAIs, metal rich chondrites and iron meteorites. Geochimica et Cosmochimica Acta 69, 5805–5818.

Kleine, T., Palme, H., Mezger, K., Halliday, A.N., 2005b. Hf-W chronometry of lunar metals and the age and early differentiation of the Moon. Science 370, 1671–1674.

Kleine, T., Halliday, A.N., Palme, H., Mezger, K., Markowski, A., 2006. Hf-W chronometry of the accretion and thermal metamorphism of ordinary chondrite parent bodies. Lunar and Planetary Science XXXVII, abstract 188.

Kleinhanns, I.C., Kramers, J.D., Kamber, B.S., 2003. Importance of water for Archaean granitoid petrology: a comparative study of TTG and potassic granitoids from Barberton Mountain Land, South Africa. Contribution to Mineralogy and Petrology 145, 377–389.

Klemme, S., Blundy, J.D., Wood, B., 2002. Experimental constraints on major and trace element partitioning during partial melting of eclogite. Geochimica et Cosmochimica Acta 66, 3109–3123.

Kloppenburg, A., White, S.H., Zegers, T.E., 2001. Structural evolution of the Warrawoona Greenstone Belt and adjoining granitoid complexes, Pilbara Craton, Australia: implications for Archaean tectonic processes. Precambrian Research 112, 107–147.

Knauth, L.P., 2005. Temperature and salinity history of the Precambrian ocean: implications for the course of microbial evolution. Palaeogeography, Palaeoclimetology, Palaeoecology 219, 53–69.

Knauth, L.P., Epstein, S., 1976. Hydrogen and oxygen isotope ratios in nodular and bedded cherts. Geochimica et Cosmochimica Acta 40, 1095–1108.

Knauth, L.P., Lowe, R.L., 1978. Oxygen isotope geochemistry of cherts from the Onverwacht Group (3.4 billion years) Transvaal, South Africa, with implications for secular variations in the isotopic composition of cherts. Earth and Planetary Science Letters 41, 209–222.

Knauth, L.P., Lowe, D.R., 2003. High Archean climatic temperatures inferred from oxygen isotope geochemistry of cherts in the 3.5 Ga Swaziland Supergroup, South Africa. Bulletin of the Geological Society of America 115, 566–580.

Köber, B., Pidgeon, R.T., Lippolt, H.J., 1989. Single-zircon dating by stepwise Pb-evaporation constrains the Archaean history of detrital zircons from the Jack Hills, Western Australia. Earth and Planetary Science Letters 91, 286–296.

Koch, D.M., Manga, M., 1996. Neutrally buoyant diapirs: A model for Venus coronae. Geophysical Research Letters 23, 225–228.

Koeberl, C., 2003. The late heavy bombardment in the inner solar system: Is there any connection to Kuiper belt objects? Earth, Moon, and Planets 92, 79–87.

Koeberl, C., 2006. Impact processes on the early Earth. Elements 2, 211–216.

Koeberl, C., Sharpton, V.L., 1988. Giant impacts and their influence on the early Earth. In: Papers Presented to the Conference on the Origin of the Earth. Lunar and Planetary Institute, Houston, TX, pp. 47–48.

Koeberl, C., Reimold, W.U., McDonal, I., Rosing, M., 1998. The late heavy bombardment on Earth? A shock petrographic and geochemical survey of some of the world's oldest rocks. In: Origin of the Earth and Moon. LPI Contribution No. 957, Lunar and Planetary Institute, Houston, TX, pp. 19–20.

Koeberl, C., Reimold, W.U., McDonald, I., Rosing, M., 2000. Search for petrographical and geochemical evidence for the late heavy bombardment on Earth in Early Archean rocks from Isua, Greenland. In: Gilmour, I., Koeberl, C. (Eds.), Impacts and the Early Earth. Lecture Notes in Earth Sciences, vol. 91. Springer-Verlag, Heidelberg, pp. 73–97.

Koenig, E., Aydin, A., 1998. Evidence for large-scale strike-slip faulting on Venus. Geology 26, 551–554.

Kohler, E.A., Anhaeusser, C.R., Isachsen, C., 1993. The Bien Venue Formation: a proposed new unit in the northeastern sector of the Barberton greenstone belt. In: 16th International Colloquium on Africa Geology, Ezulwini, Swaziland. Swaziland G.S.M.D., pp. 186–188.

Komatsu, G., Baker, V.R., 1994. Meander properties of Venusian channels. Geology 22, 67–70.

Komiya, T., Maruyama, S., Masuda, T., Nohda, S., Hayashi, M., Okamoto, K., 1999. Plate tectonics at 3.8–3.7 Ga: field evidence from the Isua accretionary complex, southern West Greenland. Journal of Geology 107, 515–554.

Konhauser, K., 2007. Introduction to Geomicrobiology. Blackwell Publishing, Oxford, 425 pp.

Konhauser, K.O., Hamade, T., Morris, R.C., Ferris, F.G., Southam, G., Raiswell, R., Canfield, D., 2002. Could bacteria have formed the Precambrian banded iron formations? Geology 30, 1079–1082.

Konopliv, A.S., Sjorgren, W.L., 1994. Venus spherical harmonic gravity model to degree and order 60. Icarus 112, 42–54.

Kotov, A.B., 2003. Confines on geodynamic models of formation of continental crust of the Aldan shield. Dissertation thesis. Institute of Geology and Geochronology of the Precambrian, St.-Petersburg, 78 pp. (in Russian).

Kotov, A.B., Glebovitsky, V.A., Kazansky, V.I., Salnikova, E.B., Pertsev, N.N., Kovach, V.P., Yakovleva, S.Z., 2005. Age formation confines of the main structures in the central Aldan shield. Doklady Earth Sciences 404 (6), 798–801.

Kotov, A.B., Kovach, V.P., Salnikova, E.B., Glebovitsky, V.A., Yakovleva, S.Z., Berezhnaya, N.G., Myskova, T.A., 1995. Continental crust age and formation stages in the Central Aldan granulite-gneiss terrane – U-Pb and Sm-Nd isotopic data for granitoids. Petrology 3 (1), 87–97.

Kotov, A.B., Morozova, I.M., Salnikova, E.B., Bogomolov, E.S., Belyatsky, B.V., Berezhnaya, N.G., 1993. Early Proterozoic granitoids of NW part of the Aldan granulite-gneiss province: U-Pb and Sm-Nd data. Russian Geology and Geophysics 34 (2), 15–21 (in Russian).

Kouamelan, A.N., Delor, C., Peucat, J.J., 1997. Geochronological evidence for reworking of Archean terrains during the Early Proterozoic (2.1 Ga) in the western Cote d'Ivoire (Man Rise – West African craton). Precambrian Research 86, 177–199.

Kovach, V.P., Kotov, A.B., Salnikova, E.B., Bogomolov, E.S., Belyatsky, 1995a. Age of formation confines of high metamorphic supracrustal complexes in the Aldan shield. In: Earth Crust and Mantle. Abstracts of the Russian Foundation for Basic Research Conference, vol. 2. Institute of the Earth Crust, Irkutsk, pp. 57–58 (in Russian).

Kovach, V.P., Kotov, A.B., Salnikova, E.B., Yakovleva, S.Z., Berezhnaya, N.G., 1995b. Sm-Nd isotope systematic of high metamorphic complexes in the Aldan shield. In: Main Age Confines in the Precambrian Earth Evolution and Their Basis in the Isotope-Geochronology. Conference Abstract Volume. Institute of Geology and Geochronology of the Precambrian, St.-Petersburg, p. 31 (in Russian).

Kovach, V.P., Kotov, A.B., Salnikova, E.B., Berezkin, V.I., Smelov, A.P., 1996. Sm-Nd isotopic systematics of the Kurumkan Subgroup, the Iengra Group of Aldan Shield. Stratigraphy and Geological Correlation 4 (3), 3–10.

Kramers, J.D., 2007. Hierarchical Earth accretion and the Hadean Eon. Journal of the Geological Society London 165, 3–18.

Kramers, J.D., Kreissig, K., Jones, M.Q.W., 2001. Crustal heat production and style of metamorphism: a comparison between two Archaean high grade provinces in the Limpopo Belt, southern Africa. Precambrian Research 112, 149–163.

Kramers, J.D., Tolstikhin, I.N., 1997. Two terrestrial lead isotope paradoxes, forward transport modelling, core formation and the history of the continental crust. Chemical Geology 139, 75–110.

Krapež, B., Brown, S.J.A., Hand, J., Barley, M.E., Cas, R.A.F., 2000. Age constraints on recycled crustal and supracrustal sources of Archaean metasedimentary sequences, Eastern Goldfields Province, Western Australia: evidence from SHRIMP zircon dating. Tectonophysics 322, 89–133.

Kretz, R., 1983. Symbols for rock-forming minerals. American Mineralogist 68, 277–279.

Kring, D.A., Cohen, B.A., 2002. Cataclysmic bombardment throughout the inner solar system 3.9–4.0 Ga. Journal of Geophysical Research – Planets 107, Art. No. 5009.

Krogh, T.E., Moser, D.E., 1994. U-Pb zircon and monazite ages from the Kapuskasing uplift: age constraints on deformation within the Ivanhoe Lake fault zone. Canadian Journal of Earth Sciences 31, 1096–1103.

Krogh, T.E., Ermanovics, I.F., Davis, G.L., 1974. Two episodes of metamorphism and deformation in the Archean rocks of the Canadian shield. In: Carnegie Institution of Washington, Geophysical Laboratory Yearbook, pp. 573–575.

Krogh, T.E., Harris, N.B.W., Davis, G.L., 1976. Archean rocks from the eastern Lac Seul region of the English River gneiss belt, northwestern Ontario. Canadian Journal of Earth Sciences 13, 1212–1215.

Krogh, T.E., Kamo, S., Hess, D.F., 1997. Wyoming province 3300+ gneiss with 2400 Ma metamorphism, northwestern Tobacco Root Mountains, Madison Co., Montana. Geological Society of America, Abstracts with Programs 29 (6), A-408.

Kröner, A.M., 1981. Precambrian plate tectonics. In: Kröner, A. (Ed.), Precambrian Plate Tectonics. Developments in Precambrian Geology, vol. 4. Elsevier, Amsterdam, pp. 56–90.

Kröner, A., 1985. Archean crustal evolution. Annual Reviews of Earth and Planetary Science 13, 264–302.

Kröner, A., Compston, W., 1988. Ion microprobe ages of zircons from early Archaean granite pebbles and greywacke, Barberton greenstone belt, Southern Africa. Precambrian Research 38, 367–380.

Kröner, A., Greiling, R. (Eds.), 1984. Precambrian Tectonics Illustrated. E. Schweizerbart'sche Verlagsbuchhandlung, Stuttgart, 419 pp.

Kröner, A., Tegtmeyer, A., 1994. Gneiss-greenstone relationships in the Ancient Gneiss Complex of southwestern Swaziland, southern Africa, and implications for early crustal evolution. Precambrian Research 67 (1–2), 109–139.

Kröner, A., Todt, W., 1988. Single zircon dating contraining the maximum age of the Barberton greenstone belt, Southern Africa. Journal of Geophysical Research 93, 15,329–15,337.

Kröner, A., Compston, W., Zhang, G.W., Guo, A.L., Todt, W., 1988. Age and tectonic setting of Late Archaean greenstone-gneiss terrain in Henan Province, China, as revealed by single grain zircon dating. Geology 16, 211–215.

Kröner, A., Compston, W., Williams, I.S., 1989. Growth of early Archaean crust in the Ancient Gneiss Complex of Swaziland as revealed by single zircon dating. Tectonophysics 161, 271–298.

Kröner, A., Byerly, G.R., Lowe, D.R., 1991a. Chronology of early Archean granite-greenstone evolution in the Barberton Moutain Land, South Africa, based on precise dating by single grain zircon evaporation. Earth and Planetary Sciences Letters 103, 41–54.

Kröner, A., Wendt, J.I., Tegtmeyer, A.R., Milisenda, C., Compston, W., 1991b. Geochronology of the Ancient Gneiss Complex, Swaziland, and implications for crustal evolution. In: Ashwal, L.D. (Ed.), Two Cratons and an Orogen – Excursion Guidebook and Review Articles for a Field Workshop Through Selected Archaean Terranes of Swaziland, South Africa, and Zimbabwe. IGCP Project 280, Department of Geology, University of the Witwatersrand, Johannesburg, pp. 8–31.

Kröner, A., Hegner, E., Byerly, G.R., Lowe, D.R., 1992. Possible terrane identification in the early Archaean Barberton greenstone belt, South Africa, using single zircon geochronology. Eos, Transactions, American Geophysical Union, Fall Meeting suppl. 73 (43), 616.

Kröner, A., Wendt, J.I., Milisenda, C., Compston, W., Maphalala, R., 1993. Zircon geochronology and Nd isotopic systematics of the Ancient Gneiss Complex, Swaziland, and implications for crustal evolution. In: Kröner, A. (Ed.), The Ancient Gneiss Complex: Overview Papers and Guidebook for Excursion. Swaziland Geological Survey, Mines Department, Bulletin 11, pp. 15–37.

Kröner, A., Hegner, E., Wendt, J.I., Byerly, G.R., 1996. The oldest part of the Barberton granitoid-greenstone terrain, South Africa: evidence for crust formation between 3.5 and 3.7 Ga. Precambrian Research 78, 105–124.

Kröner, A., Cui, W.Y., Wang, S.Q., Nemchin, A.A., 1998. Single zircon ages from high-grade rocks of the Jianping Complex, Liaoning Province, NE China. Journal of Asian Earth Sciences 16, 519–532.

Kröner, A., Jaeckel, P., Brandl, G., 2000. Single zircon ages for felsic to intermediate rocks from the Pietersburg and Giyani greenstone belts and bordering granitoid orthogneisses, northern Kaapvaal Craton, South Africa. Journal of African Earth Sciences 30 (4), 773–793.

Kröner, A., Ekwueme, N., Pidgeon, R.T., 2001. The oldest rocks in West Africa: SHRIMP zircon age for early Archean migmatitic orthogneiss at Kaduna, northern Nigeria. Journal of Geology 109, 399–406.

Kröner, A., Wilde, S.A., Li, J.H., Wang, K.Y., 2005. Age and evolution of a late Archaean to Palaeoproterozoic upper to lower crustal section in the Wutaishan/Hengshan/Fuping terrain of northern China. Journal of Asian Earth Sciences 24, 577–595.

Kröner, A., Wilde, S.A., Zhao, G.C., O'Brien, P.J., Sun, M., Liu, D.Y., Wan, Y.S., Liu, S.W., Guo, J.H., 2006. Zircon geochronology and metamorphic evolution of mafic dykes in the Hengshan Complex of northern China: Evidence for late Palaeoproterozoic extension and subsequent high-pressure metamorphism in the NCC. Precambrian Research 146, 45–67.

Kroopnick, P., 1980. The distribution of ^{13}C in the Atlantic ocean. Earth and Planetary Science Letters 49, 469–484.

Krot, A.N., Keil, K., Goodrich, C.A., Scott, E.R.D., Weisberg, M.K., 2003. Classification of meteorites. In: Davis, A.M. (Ed.), Meteorites, Comets and Planets. Treatise on Geochemistry, vol. 1. Elsevier–Pergamon, Oxford, pp. 83–128.

Kruckenberg, S.C., Chamberlain, K.R., Frost, C.D., Frost, B.R., 2001. One billion years of Archean crustal evolution: Black Rock Mountain, northeastern Granite Mountains, Wyoming. Geological Society of America, Abstracts with Programs 33 (6), A-401.

Krull-Davatzes, A.E., Lowe, D.R., Byerly, G.R., 2006. Compositional grading in an ~3.24 Ga impact-produced spherule bed, Barberton greenstone belt, South Africa: A key to impact plume evolution. South African Journal of Geology 109, 233–244.

Kryuchov, V.P., 1990. Ridge Belts: are they compressional or extensional structures? Earth, Moon, and Planets 50–51, 471–491.

Kudryavtsev, A.B., Schopf, J.W., Agresti, D.G., Wdowiak, T.J., 2001. *In situ* laser-Raman imagery of Precambrian microscopic fossils. Proceedings of the National Academy of Sciences 98, 823–826.

Kuiper, Y.D., Lin, S., Böhm, C.O., Corkery, M.T., 2003. Structural geology of the Assean Lake and Aiken River deformation zones, northern Manitoba (NTS 64A1, 2 and 8). In: Report of Activities 2003, Manitoba Industry, Economic Development and Mines, Manitoba Geological Survey, pp. 105–113.

Kuiper, Y.D., Lin, S., Böhm, C.O., Corkery, M.T., 2004a. Structural geology of Assean Lake, northern Manitoba (NTS 64A1, 2, 8). In: Report of Activities 2004, Manitoba

Industry, Economic Development and Mines, Manitoba Geological Survey, pp. 195–200.

Kuiper, Y.D., Lin, S., Böhm, C.O., Corkery, M.T., 2004b. Structural geology of the Aiken River deformation zone, northern Manitoba (NTS 64A1, 2, 8). In: Report of Activities 2004, Manitoba Industry, Economic Development and Mines, Manitoba Geological Survey, pp. 201–208.

Kump, L.R., Barley, M.E., 2006. Abrupt onset of subaerial volcanism and the rise of atmospheric oxygen. In: Geological Society of America Annual Meeting, Philadelphia, Abstracts with Programs, p. 55.

Kurat, G., Koeberl, C., Presper, T., Brandstatter, F., Maurette, M., 1994. Petrology and geochemistry of Antarctic micrometeorites. Geochimica et Cosmochimica Acta 58, 3879–3904.

Kusky, T.M., 2004. Introduction. In: Kusky, T.M. (Ed.), Precambrian Ophiolites and Related Rocks. Developments in Precambrian Geology, vol. 13. Elsevier, Amsterdam, pp. 1–34.

Kusky, T.M., Polat, A., 1999. Growth of granite-greenstone terranes at convergent margins and stabilization of Archean cratons. Tectonophysics 305, 43–73.

Kusky, T.M., Li, J.H., Tucker, R.T., 2001. The Dongwanzi ophiolite: Complete Archaean ophiolite with extensive sheeted dike complex, North China craton. Science 292, 1142–1145.

Kuznetsov, V.G., Khrenov, P.M. (Eds.), 1982. Geological Map of Irkutsk Region and Contiguous Territories, Scale of 1:500,000. Ministy of Geology of the USSR, Irkutsk.

Kyte, F.T., 2002. Tracers of extraterrestrial components in sediments and inferences for Earth's accretion history. In: Geological Society America, Special Paper 356, pp. 21–38.

Kyte, F.T., Zhou, L., Lowe, D.R., 1992. Noble metal abundances in an early Archaean impact deposit. Geochimica et Cosmochimica Acta 56, 1365–1372.

Kyte, F.T., Shukolyukov, A., Lugmair, G.W., Lowe, D.R., Byerly, G.R., 2003. Early Archean spherule beds: Chromium isotopes confirm origin through multiple impacts of projectiles of carbonaceous chondrite type. Geology 31, 283–286.

Laflèche, M.R., Dupuy, C., Dostal, J., 1992. Tholeiitic volcanic rocks of the late Archean Blake River Group, southern Abitibi greenstone belt: origin and geodynamic implications. Canadian Journal of Earth Sciences 29, 1448–1458.

Lahaye, Y., Arndt, N., Byerly, G.R., Chauvel, C., Fourcade, S., Gruau, G., 1995. The influence of alteration on the trace-element and Nd isotopic compositions of komatiites. Chemical Geology 126, 43–64.

Lamb, S.H., 1984. Structures on the eastern margin of the Archaean Barberton greenstone belt, northwest Swaziland. In: Kröner, A., Greiling, A. (Eds.), Precambrian Tectonics Illustrated. E. Schweizerbart'sche Verlagsbuchhandlung, Stuttgart, pp. 19–39.

Lamb, S.H., Paris, I., 1988. Post-Onverwacht Group stratigraphy in the SE part of Archaean Barberton greenstone belt. Journal of African Earth Sciences 7, 285–306.

Lambert, I., Donnelly, T., Dunlop, J., Groves, D., 1978. Stable isotopic compositions of early Archaean sulphate deposits of probable evaporitic and volcanogenic origins. Nature 276, 808–811.

Lambert, R.St.J., 1983. Metamorphism and thermal gradients in the Proterozoic continental crust. In: Medaris, L.G., et al. (Eds.), Proterozoic Geology: Selected Papers from an International Proterozoic Symposium. Geological Society of America, Memoir 161, pp. 155–166.

Lan, Y.Q., Si, X.M., Li, Z.D., Sun, F.X., 1990. Archaean Metamorphic Geology of Qian'an Area, Eastern Hebei, China. Jilin Science and Technology Press, Jilin, 247 pp. (in Chinese).

Lang, N.P., Hansen, V.L., 2006. Venusian channel formation as a subsurface process. Journal of Geophysical Research 111, doi: 10.1029/2005JE002629.

Langstaff, G.D., 1995. Archean geology of the Granite Mountains, Wyoming. Unpublished Ph.D. dissertation. Colorado School of Mines, Golden, CO, 671 pp.

Langford, F.F., Morin, J.A., 1976. The development of the Superior Province of northwestern Ontario by merging island arcs. American Journal of Science 276, 1023–1034.

Lanier, W.P., Lowe, D.R., 1982. Sedimentology of the Middle Marker (3.4 Ga), Onverwacht Group, Transvaal, South Africa. Precambrian Research 18, 237–260.

Large, R.R., 1992. Australian volcanic-hosted massive sulphide deposits: features, styles and genetic models. Economic Geology 87, 471–510.

Larson, R.L., 1991. Latest pulse of the Earth: evidence for a mid-Cretaceous superplume. Geology 19, 547–550.

Laska, S., Lottspeich, F., Kletzin, A., 2003. Membrane-bound hydrogenase and sulfur reductase of the hyperthermophilic and acidophilic archaeon *Acidianus ambivalens*. Microbiology-SGM 149, 2357–2371.

Lawver, L.A., Müller, R.D., 1994. Iceland hotspot track. Geology 22, 311–314.

Layer, P.W., 1986. Archean palaeomagnetism of southern Africa. Ph.D. thesis. Stanford University, Stanford, CA, 397 pp.

Layer, P.W., Kröner, A., York, D., 1992. Pre-3000 Ma thermal history of the Archean Kaap Valley pluton, South Africa. Geology 20, 717–720.

Leal, L.R.B., Cunha, J.C., Cordani, U.G., Teixeira, W., Nutman, A.P., Leal, A.B.M., Macambira, M.J.B., 2003. SHRIMP U-Pb, [207]Pb/[206]Pb zircon dating, and Nd isotopic signature of the Umburanas greenstone belt, northern São Francisco craton, Brazil. Journal of South American Earth Science 15, 775–785.

Leat, P.T., Larter, R.D., 2003. Intra-oceanic subduction systems: tectonic and magmatic processes. In: Geological Society of London, Special Publication 219, pp. 1–17.

Le Bas, M.J., Le Maître, R.W., Streckeisen, A., Zanettin, B., 1986. A chemical classification of volcanic rocks based on the total alkali-silica diagram. Journal of Petrology 27, 745–750.

LeCheminant, A.N., Heaman, L.M., 1989. Mackenzie igneous events, Canada: Middle Proterozoic hotspot magmatism associated with ocean opening. Earth and Planetary Science Letters 96, 38–48.

Leclair, A.D., 2005. Géologie du nord-est de la Province du Supérieur, Québec. Ministère des Ressources naturelles et de la Faune, DV 2004-04, 19 pp., 1 carte (échelle 1:750,000).

Leclair, A., Berclaz, A., David, J., Percival, J.A., 2002a. Les événements tectonomagmatiques du nord-est de la Province du Supérieur: 300 millions d'années d'évolution archéenne. In: Projet de cartographie du Grand-Nord – Rapport d'atelier. Ministère des Ressources naturelles, Québec, MB 2002-01, pp. 65–67.

Leclair, A., Parent, M., David, J., Dion, D.-J., Sharma, K.N.M., 2002b. Geology of the Lac La Potherie Area (34I). Ministère des Ressources naturelles, Québec, RG 2001-04, 40 pp.

Leclair, A., Berclaz, A., Parent, M., Cadieux, A.M., Sharma, K.N.M., 2003. Géologie 1:250,000, 24L – Lac Dufreboy. Ministère des Ressources naturelles, Québec, carte SI-24L-C2G-03C.

Leclair, A.D., Boily, M., Berclaz, A., Labbé, J.-Y., Lacoste, P., Simard, M., Maurice, C., 2006a. 1.2 billion years of Archean evolution in the northeastern Superior Province. Geological Association of Canada, Abstracts, vol. 31, p. 85.

Leclair, A., Labbé, J.-Y., Berclaz, A., David, J., Gosselin, C., Lacoste, P., Madore, L., Maurice, C., Roy, P., Sharma, K.N.M., Simard, M., 2006. Government geoscience stimulates mineral exploration in the Superior Province, northern Quebec. Geoscience Canada 33, 60–75.

Leclerc, F., 2004. Évolution tectonostratigraphique et métamorphique de la ceinture volcano-sédimentaire de Qalluviartuuq-Payne, nord-est de la Province du Supérieur. M.Sc. thesis. Université du Québec à Montréal, 101 pp.

Lecuyer, C., Simon, L., Guyot, F., 2000. Comparison of carbon, nitrogen and water budgets on Venus and the Earth. Earth and Planetary Science Letters 181, 33–40.

Lee, S.M., 1965. Région d'Inussuaq – Pointe Normand, Nouveau-Québec. Ministère des richesses naturelles, Québec, Rapport géologique 119, 138 pp.

Le Maître, R.W., 2002. Igneous Rocks: A Classification and Glossary of Terms. Cambridge University Press, Cambridge, 236 pp.

Lee, T., Papanastassiou, D.A., 1974. Mg isotopic anomalies in the Allende meteorite and correlation with O and Sr effects. Geophysical Research Letters 1, 225–228.

Lee, T., Papanastassiou, D.A., Wasserburg, G.J., 1976. Demonstration of ^{26}Mg excess in Allende and evidence for ^{26}Al. Geophysical Research Letters 3, 109–112.

Lemieux, S., Ross, G.M., Cook, F.A., 2000. Crustal geometry and tectonic evolution of the Archean crystalline basement beneath the southern Alberta Plains, from new seismic reflection and potential-field studies. Canadian Journal of Earth Sciences 37, 1473–1491.

Lenardic, A., Moresi, L-N., 1999. Some thoughts on the stability of cratonic lithosphere: effects of buoyancy and viscosity. Journal of Geophysical Research 104, 12,747–12,758.

Lenton, P.G., Corkery, M.T., 1981. The Lower Churchill River Project (interim report). Manitoba Department of Energy and Mines, Mineral Resources Division, Open File Report OF81-3, 23 pp.

Lepland, A., Arrhenius, G., Cornell, D., 2002. Apatite in early Archean supracrustal rocks, southern West Greenland: its origin, association with graphite and potential as a biomarker. Precambrian Research 118, 221–241.

Lepland, A., van Zuilen, M.A., Arrhenius, G., Whitehouse, M.J., Fedo, C.M., 2005. Questioning the evidence for Earth's earliest life – Akilia revisited. Geology 33, 77–79.

Lesher, C.M., Arndt, N.T., 1995. REE and Nd isotope geochemistry, petrogenesis, and volcanic evolution of contaminated komatiites in Kambalda, Western Australia. Lithos 34, 127–157.

Leshin, L.A., Rubin, A.E., McKeegan, K.D., 1997. The oxygen isotopic composition of olivine and pyroxene from CI chondrites. Geochimica et Cosmochimica Acta 61, 835–845.

Lespade, P., Al-Jishi, R., Dresselhaus, M.S., 1982. Model for Raman scattering from incompletely graphitized carbons. Carbon 5, 427–431.

Levison, H., Lissauer, J.J., Duncan, M.J., 1998. Modelling the diversity of outer planetary systems. Astronomical Journal 116, 1998–2014.

Lewan, M.D., 2003. Experiments on the role of water in petroleum formation. Geochimica et Cosmochimica Acta 61, 3691–3723.

Li, H.K., Li, H.M., Lu, S.N., 1995. Single grain zircon U-Pb ages for volcanic rocks from Tuanshanzi Formation of Changcheng System and their geological implications. Geochemica 24, 43–48 (in Chinese with English abstract).

Li, J.H., He, W.Y., Qian, X.L., 1997. Genetic mechanism and tectonic setting of a Proterozoic mafic dyke swarm: implications for palaeoplate reconstruction. Geological Journal of China Universities 3, 2–8 (in Chinese with English abstract).

Li, J.H., Qian, X.L., Huang, X.N., 2000. The tectonic framework of the basement of NCC and its implication for early Precambrian cratonization. Acta Petrologica Sinica 16, 1–10 (in Chinese with English abstract).

Li, J.H., Kusky, T.M., Huang, X.N., 2002. Archaean podiform chromitites and mantle tectonites in ophiolitic melange, NCC: A record of early oceanic mantle processes. GSA Today 12, 4–11.

Li, S.G., Hart, S.R., Gou, A.L., Zhang, G.W., 1987. Whole-rock Sm-Nd isotope age of the Dengfeng Group in central Henan and its tectonic significance. Chinese Science Bulletin 22, 1728–1731 (in Chinese).

Li, S.X., Xu, X.C., Liu, X.S., Sun, D.Y., 1994. Early Precambrian geology of Wulashan Region, Inner Mongolia. Geological Publishing House, Beijing, 140 pp. (in Chinese with English abstract).

Li, Y.-H., 1991. Distribution patterns of elements in the ocean: A synthesis. Geochimica et Cosmochimica Acta 55, 3223–3240.

Li, Z.-X., Zhang, L., Powell, C.McA., 1996. Positions of the East Asian cratons in the Neoproterozoic supercontinent Rodinia. Australian Journal of Earth Sciences 43, 593–604.

Libby, W.G., de Laeter, J.R., Armstrong, R.A., 1999. Proterozoic biotite dates in the northwestern part of the Yilgarn Craton, Western Australia. Australian Journal of Earth Sciences 46, 851–860.

Lin, S., 2005. Synchronous vertical and horizontal tectonism in the Neoarchean: Kinematic evidence from a synclinal keel in the northwestern Superior craton, Canada. Precambrian Research 139, 181–194.

Lin, S., Percival, J.A., Skulski, T., 1996. Structural constraints on the tectonic evolution of a late Archean greenstone belt in the northeastern Superior Province, northern Quebec (Canada). Tectonophysics 265, 151–167.

Lin, S., Davis, D.W., Rotenberg, E., Corkery, M.T., Bailes, A.H., 2006. Geological evolution of the northwestern Superior Province: Clues from geology, kinematics and geochronology in the Gods Lake Narrows area, Oxford Lake – Knee Lake – Gods Lake greenstone belt, Manitoba. Canadian Journal of Earth Sciences 43, 749–765.

Lindsay, J.F., Brasier, M.D., McLoughlin, N., Green, O.R., Fogel, M., McNamara, K.M., Steele, A., Mertzman, S.A., 2003. Abiotic Earth – establishing a baseline for earliest life, data from the Archean of Western Australia. Lunar and Planetary Science XXXIV, 1137.

Lindsay, J.F., Brasier, M.D., McLoughlin, N., Green, O.R., Fogel, M., Steele, A., Mertzman, S.A., 2005. The problem of deep carbon – an Archaean paradox. Precambrian Research 143, 1–22.

Lissauer, J.J., 1999. Chaotic motion in the solar system. Reviews of Modern Physics 71, 835–845.

Liu, D.Y., Page, R.W., Compston, W., Wu, J.S., 1985. U-Pb zircon geochronology of late Archaean metamorphic rocks in the Taihangshan–Wutaishan area. Precambrian Research 27, 85–109.

Liu, D.Y., Shen, Q.H., Zhang, Z.Q., Jahn, B.M., Aurvay, B., 1990. Archaean crustal evolution in China: U-Pb geochronology of the Qianxi Complex. Precambrian Research 48, 223–244.

Liu, D.Y., Nutman, A.P., Compston, W., Wu, J.S., Shen, Q.H., 1992. Remnants of >3800 Ma crust in the Chinese part of the Sino-Korean craton. Geology 20, 339–342.

Liu, D.Y., Wilde, S.A., Wan, Y.S., Wu, J.S., Wu, F.Y., Zhou, H.Y., Dong, C.Y., Yin, X.Y., in review. New U-Pb and Hf isotopic data confirm Anshan as the oldest preserved part of the North China Craton. American Journal of Science.

Liu, S.W., Pan, Y.M., Li, J.H., Li, Q.G., Zhang, J., 2002. Geological and isotopic geochemical constraints on the evolution of the Fuping Complex, North China Craton. Precambrian Research 117 (1–2), 41–56.

Lobach-Zhuchenko, S.B., Levchenkov, O.A., Chekulaev, V.P., Krylov, I.N., 1986. Geological evolution of the Karelian granite-greenstone terrain. Precambrian Research 33, 45–65.

Loiselle, M.C., Wones, D.R., 1979. Characteristic and origin of anorogenic granites. In: Geological Society of America, Abstracts with Programs, vol. 11, p. 468.

Londry, K.L., Des Marais, D.J., 2003. Stable carbon isotope fractionation by sulfate-reducing bacteria. Applied and Environmental Microbiology 69, 2942–2949.

Love, G.D., Snape, C.E., Carr, A.D., Houghton, R.C., 1995. Release of covalently-bound alkane biomarkers in high yields from kerogen via catalytic hydropyrolysis. Organic Geochemistry 23, 981–986.

Love, S.G., Brownlee, D.E., 1993. A direct measurement of the terrestrial mass accretion rate of cosmic dust. Science 262, 550–553.

Lovley, D.R., Stolz, J.F., Nord, G.L., Jr., Phillips, E.J.P., 1987. Anaerobic production of magnetite by a dissimilatory iron-reducing micro-organism. Nature 330, 252–254.

Lowe, D.R., 1980a. Archean sedimentation. Annual Review of Earth and Planetary Sciences 8, 145–167.

Lowe, D.R., 1980b. Stromatolites 3,400-Myr old from the Archaean of Western Australia. Nature 284, 441–443.

Lowe, D.R., 1982. Comparative sedimentology of the principal volcanic sequences of Archean greenstone belts in South Africa, Western Australia, and Canada: implications for crustal evolution. Precambrian Research 17, 1–29.

Lowe, D.R., 1983. Restricted shallow-water sedimentation of Early Archean stromatolitic and evaporitic strata of the Strelley Pool Chert, Pilbara Block, Western Australia. Precambrian Research 19, 239–283.

Lowe, D.R., 1992. Major events in the geological development of the Precambrian Earth. In: Schopf, J.W., Klein, C. (Eds.), The Proterozoic Biosphere. Cambridge University Press, Cambridge, pp. 67–76.

Lowe, D.R., 1994a. Abiological origin of described stromatolites older than 3.2 Ga. Geology 22, 387–390.

Lowe, D.R., 1994b. Archean greenstone-related sedimentary rocks. In: Condie, K.C. (Ed.), Archean Crustal Evolution. Elsevier, Amsterdam, pp. 121–170.

Lowe, D.R., 1994c. Accretionary history of the Archean Barberton greenstone belt (3.55–3.22 Ga) southern Africa. Geology 22, 1099–1102.

Lowe, D.R., 1999a. Petrology and sedimentology of cherts and related silicified sedimentary rocks in the Swaziland Supergroup. In: Lowe, D.R., Byerly, G.R. (Eds.), Geologic Evolution of the Barberton Greenstone Belt, South Africa. Geological Society of America, Special Paper 329, pp. 83–114.

Lowe, D.R., 1999b. Shallow-water sedimentation of accretionary lapilli-bearing strata of the Msauli Chert: Evidence of explosive hydromagmatic komatiitic volcanism. In: Lowe, D.R., Byerly, G.R. (Eds.), Geologic Evolution of the Barberton Greenstone Belt, South Africa. Geological Society of America, Special Paper 329, pp. 213–232.

Lowe, D.R., 1999c. Geologic evolution of the Barberton greenstone belt and vicinity. In: Lowe, D.R., Byerly, G.R. (Eds.), Geologic Evolution of the Barberton Greenstone Belt, South Africa. Geological Society of America, Special Paper 329, pp. 287–312.

Lowe, D.R., Byerly, G.R., 1986a. Archean flow-top alteration zones formed initially in a low-temperature sulphate-rich environment. Nature 324, 245–248.

Lowe, D.R., Byerly, G.R., 1986b. Early Archaean silicate spherules of probable impact origin, South Africa and Western Australia. Geology 14, 83–86.

Lowe, D.R., Byerly, G.R., 1999. Stratigraphy of the west-central part of the Barberton Greenstone Belt, South Africa. In: Lowe, D.R., Byerly, G.R. (Eds.), Geologic Evolution of the Barberton Greenstone Belt, South Africa. Geological Society of America, Special Paper 329, pp. 1–36.

Lowe, D.R., Fisher Worrell, G., 1999. Sedimentology, mineralogy, and implications of sili-cified evaporites in the Kromberg Formation, Barberton Greenstone Belt, South Africa. In: Lowe, D.R., Byerly, G.R. (Eds.), Geologic Evolution of the Barberton Greenstone Belt, South Africa. Geological Society of America, Special Paper 329, pp. 167–188.

Lowe, D.R., Knauth, L.P., 1977. Sedimentology of the Onverwacht Group (3.4 billion years), Transvaal, South Africa, and its bearing on the characteristics and evolution of the early Earth. Journal of Geology 85, 699–723.

Lowe, D.R., Knauth, L.P., 1978. The oldest marine carbonate ooids reinterpreted as vol-canic accretionary lapilli: Onverwacht Group, South Africa. Journal of Sedimentary Petrology 48, 709–722.

Lowe, D.R., Nocita, B.W., 1999. Foreland basin sedimentation in the Mapepe Formation, southern-facies Fig Tree Group. In: Lowe, D.R., Byerly, G.R. (Eds.), Geologic Evolu-tion of the Barberton Greenstone Belt, South Africa. Geological Society of America, Special Paper 329, pp. 233–258.

Lowe, D.R., Byerly, G.R., Ransom, B.L., Nocita, B.R., 1985. Stratigraphic and sedimento-logical evidence bearing on structural repetition in Early Archean rocks of the Barberton Greenstone Belt, South Africa. Precambrian Research 27, 165–186.

Lowe, D.R., Byerly, G.R., Asaro, F., Kyte, F., 1989. Geological and geochemical record of 3400-million-year-old terrestrial meteorite impacts. Science 245, 959–962.

Lowe, D.R., Byerly, G.R., Heubeck, C., 1999. Structural divisions and development of the west-central part of the Barberton Greenstone Belt. In: Lowe, D.R., Byerly, G.R. (Eds.), Geologic Evolution of the Barberton Greenstone Belt, South Africa. Geological Society of America, Special Paper 329, pp. 37–82.

Lowe, D.R., Byerly, G.R., Kyte, F.T., Shukolyukov, A., Asaro, F., Krull, A., 2003. Charac-teristics, origin, and implications of Archean impact-produced spherule beds, 3.47–3.22 Ga, in the Barberton Greenstone Belt, South Africa: Keys to the role of large impacts on the evolution of the early Earth. Astrobiology 3, 7–48.

Lu, L.Z., Xu, X.C., Liu, F.L., 1996. Early Precambrian Khondalite Series of North China. Changchun Publishing House, Changchun, 272 pp. (in Chinese).

Lu, S.N., 2002. Preliminary Study of Precambrian Geology in the North Tibet-Qinghai Plateau. Geological Publishing House, Beijing, 125 pp. (in Chinese).

Lu, S.N., Yang, C.L., Li, H.K., Li, H.M., 2002. A group of rifting events in the terminal Palaeoproterozoic in the NCC. Gondwana Research 5, 123–131.

Luais, B., Hawkesworth, C.J., 2002. Pb isotope variations in Archean time and possible links to the sources of certain Mesozoic – Recent basalts. In: Fowler, C.M.R., Ebinger, C.J., Hawkesworth, C.J. (Eds.), The Early Earth: Physical, Chemical and Biologic De-velopment. Geologic Society of London, Special Publication 199, pp. 105–124.

Ludden, J., Hubert, C., Barnes, A., Milkereit, B., Sawyer, E., 1993. A three-dimensional perspective on the evolution of Archaean crust: LITHOPROBE seismic reflection im-ages in the southwestern Superior province. Lithos 30, 357–372.

Ludwig, K.R., 1999. User's Manual for Isoplot/Ex, v2.3, A Geochronological Toolkit for Microsoft Excel. Berkeley Geochronological Center Special Publication No. 1a, 52 pp.

Lugmair, G.W., Galer, S.J.G., 1992. Age and isotopic relationships among the angrites Lewis Cliff 86010 and Angra dos Reis. Geochimica et Cosmochimica Acta 56, 1673–1694.

Lugmair, G.W., Shukolyukov, A., 1997. ^{53}Mn-^{53}Cr isotope systematics of the HED parent body. Lunar and Planetary Science XXVIII, 823–824.

Lugmair, G.W., Shukolyukov, A., 1998. Early solar system timescales according to ^{53}Mn-^{53}Cr systematics. Geochimica et Cosmochimica Acta 62, 2863–2886.

Lugmair, G., Shukolyukov, A., 2001. Early solar system events and timescales. Meteoritics and Planetary Science 36, 1017–1026.

Lund, E.H., 1956. Igneous and metamorphic rocks of the Minnesota River Valley. Bulletin of the Geological Society of America 67, 1475–1490.

Lundberg, L.L., Zinner, E., Crozaz, G., 1994. Search for isotopic anomalies in oldhamite (CAs) from unequilibrated (E3) enstatite chondrites. Meteoritics 29, 384–393.

Lunine, J.I., Coradani, A., Gautier, D., Owen, T.C., Wuchterl, G., 2004. The origin of Jupiter. In: Bagenal, F., Dowling, T.E., McKinnon, W.B. (Eds.), Jupiter. Cambridge University Press, pp. 19–34.

Lutts, B.G., 1985. Magmatism of Mobile Belts of the Early Earth. Moscow, 216 pp.

Lutts, B.G., Oksman, V.K., 1990. Deeply Eroded Fault Zones of the Anabar Shield. Nauka, Moscow, 260 pp.

Lyell, C., 1830. Principles of Geology, Being an Attempt to Explain the Former Changes of the Earth's Surface, by Reference to Causes Now in Operation, vol. 1. John Murray, London, 511 pp.

Lyons, J.R., 2006. Gas-phase mechanisms of sulfur isotope mass-independent fractionation. Eos, Transactions, American Geophysical Union 87, Fall Meeting Supplements, Abstract V11C-0599.

Maaløe, S., 1982. Petrogenesis of Archaean tonalites. Geologische Rundschau 71, 328–346.

Maas, R., McCulloch, M.T., 1991. The provenance of Archean clastic metasediments in the Narryer Gneiss Complex, Western Australia: trace element geochemistry, Nd isotopes, and U-Pb ages for detrital zircons. Geochimica et Cosmochimica Acta 55, 1914–1932.

Maas, R., Kinny, P.D., Williams, I.S., Froude, D.O., Compston, W., 1992. The Earth's oldest known crust: A geochronological and geochemical study of 3900–4200 Ma old detrital zircons from Mt. Narryer and Jack Hills, Western Australia. Geochimica et Cosmochimica Acta 56, 1281–1300.

MacDougall, J.D., Lugmair, G.W., Kerridge, J.F., 1984. Early Solar System aqueous activity: Sr isotope evidence from the Orgueil CI meteorite. Nature 307, 249–251.

Macek, J.J., 1989. Sapphirine coronas from Sipiwesk Lake, Manitoba. Manitoba Energy and Mines, Minerals Division, Geological Paper GP85-1, 42 pp.

Macek, J.J., Bleeker, W., 1989. Thompson Nickel Belt project – Pipe Pit mine, Setting and Ospwagan lakes. In: Report of Field Activities 1989, Manitoba Energy and Mines, Minerals Division, pp. 73–87.

Macek, J.J., McGregor, C.R., 1998. Thompson Nickel Belt project: progress on a new compilation map of the Thompson Nickel Belt. In: Report of Activities, 1998, Manitoba Energy and Mines, Geological Services, pp. 36–38.

Macek, J.J., Zwanzig, H.V., Pacey, J.M., 2006. Thompson Nickel Belt geological compilation map, Manitoba. Manitoba Science, Technology, Energy and Mines, Manitoba Geological Survey, Open File Report, OF2006-33, digital map on CD.

Macgregor, A.M., 1951. Some milestones in the Precambrian of Southern Rhodesia. Proceedings of the Geological Society of South Africa 54, 27–71.

Machado, N., Brooks, C., Hart, S.R., 1986. Determination of initial $^{87}Sr/^{86}Sr$ and $^{143}Nd/^{144}Nd$ in primary minerals from mafic and ultramafic rocks: Experimental procedure and implications for the isotopic characteristics of the Archean mantle under the Abitibi greenstone belt, Canada. Geochimica et Cosmochimica Acta 50, 2335–2348.

Machado, N., Heaman, L.M., Krogh, T.E., Weber, W., 1987. U-Pb geochronology program: Thompson Belt – northern Superior Province. In: Report of Field Activities 1987, Manitoba Energy and Mines, Minerals Division, pp. 145–147.

Machado, N., Krogh, T.E., Weber, W., 1990. U-Pb geochronology of basement gneisses in the Thompson Belt (Manitoba): evidence for pre-Kenoran and Pikwitonei-type crust and early Proterozoic basement reactivation in the western margin of the Archean Superior Province. Canadian Journal of Earth Sciences 27, 794–802.

Mackwell, S.J., Zimmerman, M.E., Kohlstedt, D.L., 1998. High-temperature deformation of dry diabase with application to tectonics on Venus. Journal of Geophysical Research 102, 975–984.

MacPherson, G.J., 2003. Calcium-aluminium-rich inclusions in chondritic meteorites. In: Davis, A.M. (Ed.), Meteorites, Comets and Planets. Treatise on Geochemistry, vol. 1. Elsevier–Pergamon, Oxford, pp. 201–246.

MacPherson, G.J., Davis, A.M., Zinner, E.K., 1995. The distribution of aluminium-26 in the early Solar System – a reappraisal. Meteoritics 30, 365–386.

Madore, L., Larbi, Y., 2001. Geology of the Rivière Arnaud area (25D) and adjacent coastal areas (25C, 25E and 25F). Ministère des Ressources naturelles, Québec, RG 2001-06, 33 pp.

Madore, L., Bandyayera, D., Bédard, J.H., Brouillette, P., Sharma, K.N.M., Beaumier, M., David, J., 2000. Geology of the Lac Peters Area (NTS 24M). Ministère des Ressources naturelles, Québec, RG 99-16, 40 pp.

Madore, L., Larbi, Y., Sharma, K.N.M., Labbé, J-Y., Lacoste, P., David, J., Brousseau, K., Hocq, M., 2002. Geology of the Lac Klotz (35A) and the Cratère du Nouveau-Quebec (southern half of 35H) areas. Ministère des Ressources Naturelles Québec, RG 2002-05.

Mann, S., Sparks, N.H.C., Frankel, R.B., Bazylinski, D.A., Jannasch, H.W., 1990. Biomineralization of ferrimagnetic greigite (Fe_3S_4) and iron pyrite (FeS_2) in a magnetotactic bacterium. Nature 343, 258–261.

Manning, C.E., Mojzsis, S.J., Harrison, T.M., 2006. Geology, age and origin of supracrustal rocks, Akilia, Greenland. American Journal of Science 306, 303–366.

Maphalala, R.M., Kröner, A., 1993. Pb-Pb single zircon ages for the younger Archaean granitoids of Swaziland, Southern Africa. In: Geological Survey and Mines, International Colloquium of African Geology, Mbabane, pp. 201–206.

Mareschal, J.C., West, G.F., 1980. A model for Archean tectonism. 2. Numerical-models of vertical tectonism in greenstone belts. Canadian Journal of Earth Sciences 17, 60–71.

Markowski, A., Quitté, G., Kleine, T., Halliday, A.N., 2005. Tungsten isotopic constraints on the formation and evolution of iron meteorite parent bodies. Lunar and Planetary Science XXXVI, abstract 1308.

Markowski, A., Leya, I., Quitté, G., Ammon, K., Halliday, A.N., Wieler, R., 2006. Correlated helium-3 and tungsten isotopes in iron meteorites: Quantitative cosmogenic corrections and planetesimal formation times. Earth and Planetary Science Letters 250, 104–115.

Marshak, S., 1999. Deformation style way back when: thoughts on the contrasts between Archaean/Paleoproterozoic and contemporary orogens. Journal of Structural Geology 21, 1175–1182.

Marshall, A.E., 2000. Low temperature-low pressure ('epithermal') siliceous vein deposits of the North Pilbara granite-greenstone terrane, Western Australia. Australian Geological Survey Organization, Record 2000/1, 40 pp.

Marshall, C.P., Allwood, A.C., Walter, M.R., Van Kranendonk, M.J., Summons, R.E., 2004a. Spectroscopic and microscopic characterization of the Carbonaceous Material in Archaean Cherts, Pilbara Craton, Western Australia. In: Abstracts of the 21st Annual Meeting of the Society of Organic Petrology, vol. 21. Sydney, New South Wales, Australia, pp. 107–108.

Marshall, C.P., Allwood, A.C., Walter, M.R., Van Kranendonk, M.J., Summons, R.E., 2004b. Characterization of the carbonaceous material in the 3.4 Ga Strelley Pool Chert, Pilbara Craton, Western Australia. Geological Society of America Annual Meeting, Denver, CO. Abstracts with Programs, vol. 36 (5), p. 458.

Marshall, C.P., Love, G.D., Snape, C.E., Hill, A.C., Allwood, A.C., Walter, M.R., Van Kranendonk, M.J., Bowden, S.A., Sylva, S.P., Summons, R.E., 2007. Characterization of kerogen in Archaean cherts, Pilbara Craton, Western Australia. Precambrian Research 155, 1–23.

Marston, J.G., 1979. Copper mineralization in Western Australia. Geological Survey of Western Australia, Mineral Resources Bulletin 13.

Martin, H., 1986. Effect of steeper Archean geothermal gradient on geochemistry of subduction-zone magmas. Geology 14 (9), 753–756.

Martin, H., 1987. Petrogenesis of Archean trondhjemites, tonalites and granodiorites from eastern Finland: major and trace element geochemistry. Journal of Petrology 28, 921–953.

Martin, H., 1993. The mechanisms of petrogenesis of the Archean continental crust – Comparison with modern processes. Lithos 30, 373–388.

Martin, H., 1994. The Archean grey gneiss and the genesis of continental crust. In: Condie, K.C. (Ed.), Archean Crustal Evolution. Elsevier, Amsterdam, pp. 205–259.

Martin, H., 1999. The adakitic magmas: modern analogs of Archean granitoids. Lithos 46, 411–429.

Martin, H., 2005. Genesis and evolution of the primitive Earth continental crust. In: Gargaud, M., Barbier, B., Martin, H., Reisse, J. (Eds.), Lectures in Astrobiology, vol. I. Springer, Berlin, pp. 352–383.

Martin, H., Moyen, J.-F., 2002. Secular changes in tonalite-trondhjemite-granodiorite composition as markers of the progressive cooling of the Earth. Geology 30, 319–322.

Martin, H., Peucat, J.J., Sabate, P., Cunha, J.C., 1997. Crustal evolution in the Early Archean of South America: Example of the Sete Voltas massif, Bahia state, Brazil. Precambrian Research 82, 35–62.

Martin, H., Smithies, R.H., Rapp, R., Moyen, J.-F., Champion, D., 2005. An overview of adakite, tonalite-trondhjemite-granodiorite (TTG) and sanukitoid: relationships and some implications for crustal evolution. Lithos 79, 1–24.

Marty, B., Yokochi, R., 2006. Water in the early Earth. Reviews in Mineralogy and Geochemistry 62, 421–450.

Marzoli, A., Renne, P.R., Piccirillo, E.M., Ernesto, M., Bellieni, G., De Min, A., 1999. Extensive 200-million year old continental flood basalts of the Central Atlantic Magmatic Province. Science 284, 616–618.

Mason, B., Moore, C.B., 1982. Principles of Geochemistry, fourth ed. John Wiley and Sons, New York, 344 pp.

Matthews, M.J., Pimenta, M.A., Dresselhaus, G., Dresselhaus, M.S., Endo, M., 1999. Origin of dispersive effects of the Raman D band in carbon materials. Physics Review B 59, 6585–6588.

Matthews, P.E., Charlesworth, E.G., Eglington, B.M., Harmer, R.E., 1989. A minimum 3.29 Ga age for the Nondweni greenstone complex in the southeastern Kaapvaal Craton. South African Journal of Geology 92, 272–278.

Mauersberger, K., Erbacher, B., Krankowsky, D., Gunther, J., Nickel, R., 1999. Ozone isotope enrichment: Isotopomer-specific rate coefficients. Science 283, 370–372.

Maurette, M., Duprat, J., Engrand, C., Gounelle, M., Kurat, G., Matrajt, G., Toppani, A., 2000. Accretion of neon, organics, CO_2, nitrogen and water from large interplanetary dust particles on the early Earth. Planetary and Space Science 48, 1117–1137.

Maurice, C., Berclaz, A., David, J., Sharma, K.N.M., Lacoste, P., 2005. Geology of the Povungnituk (35C) et de Kovik Bay (35F) areas. Ministère des Ressources naturelles, Québec, RG 2004-05, 41 pp.

Maurice, C., David, J., Leclair, A., Francis, D., 2006. An autochthonous origin for the northeastern Superior Province greenstone belts. In: Geological Association of Canada – Mineralogical Association of Canada, Annual Meeting, Abstracts Volume 31, p. 98.

McCall, J.G.H., 2003. A critique of the analogy between Archaean and Phanerozoic tectonics based on regional mapping of the Mesozoic-Cenozoic plate convergent zone in the Makran, Iran. Precambrian Research 127, 5–18.

McCarthy, T., Rubridge, B. (Eds.), 2005. The Story of Earth & Life – A Southern African Perspective on a 4.6 Billion-Year Journey. Struik Publishers, Cape Town, South Africa, 333 pp.

McCollom, T., 2003. Formation of meteorite hydrocarbons from thermal decomposition of siderite (FeCO$_3$). Geochimica et Cosmochimica Acta 67, 311–317.

McCollom, T.M., Seewald, J.S., 2006. Carbon isotope composition of organic compounds produced by abiotic synthesis under hydrothermal conditions. Earth and Planetary Science Letters 243, 74–84.

McCombs, J.A., Dahl, P.S., Hamilton, M.A., 2004. U-Pb ages of Neoarchean granitoids from the Black Hills, South Dakota, USA: implications for crustal evolution in the Archean Wyoming province. Precambrian Research 130, 161–184.

McCord, T.B., Adams, J.B., Johnson, T.V., 1970. Asteroid Vesta: spectral reflectivity and compositional implications. Science 168, 1445–1447.

McCoy, T., Keil, K., Clayton, R.N., Mayeda, T.K., Bogard, D.D., Garrison, D.H., Huss, G.R., Hutcheon, I.D., Wieler, R., 1996. A petrologic, chemical and isotopic study of Monument Draw and comparison with other acapulcoites: evidence for formation by incipient partial melting. Geochimica et Cosmochimica Acta 60, 2681–2708.

McCoy, T., Keil, K., Clayton, R.N., Mayeda, T.K., Bogard, D.D., Garrison, D.H., Wieler, R., 1997. A petrologic and isotopic study of lodranites: evidence for early formation as partial melt residues from heterogeneous precursors. Geochimica et Cosmochimica Acta 61, 623–637.

McCuaig, T.C., Kerrich, R., 1998. P-T-t-deformation-fluid characteristics of lode gold deposits: evidence from alteration systematics. Ore Geology Reviews 12, 381–453.

McCulloch, M.T., 1987. Sm-Nd isotopic constraints on the evolution of Precambrian crust in the Australian continent. In: Kroner, A. (Ed.), Proterozoic Lithospheric Evolution. American Geophysical Union, Geodynamics Series, vol. 17. Washington, DC, pp. 115–130.

McCulloch, M.T., 1994. Primitive ^{87}Sr/^{86}Sr from and Archean barite and conjecture on the Earth's age and origin. Earth and Planetary Science Letters 126, 1–13.

McCulloch, M.T., Bennett, V.C., 1993. Evolution of the early Earth: Constraints from ^{143}Nd-^{142}Nd isotopic systematics. Lithos 30, 237–255.

McCulloch, M.T., Bennett, V.C., 1994. Progressive growth of the Earth's continental crust and depleted mantle: Geochemical constraints. Geochimica et Cosmochimica Acta 58, 4717–4738.

McCulloch, M.T., Bennett, V.C., 1998. Early Differentiation of the Earth: An Isotopic Perspective. In: Jackson, I. (Ed.), The Earth's Mantle. Cambridge University Press, Cambridge, pp. 127–158.

McCulloch, M.T., Black, L.P., 1984. Sm-Nd isotopic systematics of Enderby Land granulites and evidence for the redistribution of Sm and Nd during metamorphism. Earth and Planetary Science Letters 71, 46–58.

McCulloch, M.T., Wasserburg, G.J., 1978. Sm-Nd and Rb-Sr chronology of continental crust formation. Science 200, 1003–1011.

McCulloch, M.T., Wasserburg, G.J., 1980. Early Archean Sm-Nd model ages from a tonalitic gneiss, northern Michigan. In: Geological Society of America, Special Paper 182, pp. 135–138.

McDaniel, K.M., Hansen, V.L., 2005. Circular lows, a genetically distinct subset of coronae? In: Lunar and Planetary Science Conference XXXVI, 2367.pdf.

McDonough, W.F., Sun, S.-S., 1995. The composition of the Earth. Chemical Geology 120, 223–253.

McGill, G.E., 1994. Hotspot evolution and Venusian tectonic style. Journal of Geophysical Research 99, 23,149–23,161.

McGregor, V.R., 1968. Field evidence of very old Precambrian rocks in the Godthaab area, West Greenland. In: The Geological Survey of Greenland, Report 15, pp. 31–35.

McGregor, V.R., 1973. The early Precambrian gneisses of the Godthaab district, West Greenland. Philosophical Transactions of the Royal Society of London Ser. A 273, 343–358.

McGregor, V.R., Mason, B., 1977. Petrogenesis and geochemistry of metabasaltic and metasedimentary enclaves in the Amîtsoq gneisses, West Greenland. American Mineralogist 62, 887–904.

McGregor, V.R., Friend, C.R.L., Nutman, A.P., 1991. The late Archaean mobile belt through Godthåbsfjord, southern West Greenland: a continent-continent collision zone? Bulletin of the Geological Society Denmark 39, 179–197.

McKeegan, K.D., Davis, A.M., 2003. Early Solar System chronology. In: Davis, A.M. (Ed.), Meteorites, Comets and Planets. Treatise on Geochemistry, vol. 1. Elsevier–Pergamon, Oxford, pp. 431–460.

McKeegan, K.D., Leshin, L.A., 2001. Stable isotope variations in extraterrestrial materials. Reviews in Mineralogy and Geochemistry 73, 279–318.

McKeegan, K.D., Kudryavtsev, A.B., Schopf, J.W., 2007. Raman ion microscope imagery of graphitic inclusions in apatite from older than 3830 Ma Akilia supracrustal rocks, West Greenland. Geology 35, 591–594.

McKenzie, D.P., Bickle, M.J., 1988. The volume and composition of melt generated by extension of the lithosphere. Journal of Petrology 29, 625–679.

McKenzie, D., Ford, P.G., Johnson, C., Parsons, B., Pettengill, G.H., Sandwell, D., Saunders, R.S., Solomon, S.C., 1992. Features on Venus generated by plate boundary processes. Journal of Geophysical Research 97, 13,533–13,544.

McKinnon, W.B., Zahnle, K.J., Ivanov, B.A., Melosh, H.J., 1997. Cratering on Venus: Models and observations. In: Bouger, S.W., Hunten, D.M., Phillips, R.J. (Eds.), Venus II. University of Arizona Press, Tucson, AZ, pp. 969–1014.

McLennan, S.M., Taylor, S.R., Hemming, S.R., 2005. Composition, differentiation and evolution of continental crust: Constraints from sedimentary rocks and heat flow. In: Brown, M., Rushmer, T. (Eds.), Evolution and Differentiation of the Earth's Crust. Cambridge University Press, pp. 93–135.

McPhie, J., Goto, Y., 1996. Lobe and layered structures in dacite sills of the Archean Strelley succession, Western Australia. Eos, Transactions, American Geophysical Union, vol. 77 (Supplement), p. W125.

McSween, H.Y., Jr, Richardson, S.M., 1977. The composition of carbonaceous chondrite matrix. Geochimica et Cosmochimica Acta 41, 1145–1161.

Meissner, R., 1986. The Continental Crust: A Geophysical Approach. International Geophysical Series, vol. 34. Academic Press, 426 pp.

Melnyk, M.J., Cruden, A.R., Davis, D.W., Stern, R.A., 2006. U-Pb ages of magmatism constraining regional deformation in the Winnipeg River subprovince and Lake of the Woods greenstone belt: Evidence for Archean terrane accretion in the western Superior Province. Canadian Journal of Earth Sciences 43, 967–993.

Melosh, H.J., 1990. Giant impacts and the thermal state of the early Earth. In: Newsome, H.E., Jones, J.H. (Eds.), Origin of the Earth. Oxford University Press, New York, pp. 69–83.

Melosh, H.J., Vickery, A.M., 1991. Melt droplet formation in energetic impact events. Nature 350, 494–497.

Menzies, M.A., Xu, Y-G., 1998. Geodynamics of the North China Craton. In: Flower, M., Chng, S.L., Lo, C.H., Lee, T.Y. (Eds.), Mantle Dynamics and Plate Interactions in East Asia. American Geophysical Union, Geodynamics Series, vol. 27. Washington, DC, pp. 155–165.

Menzies, M.A., Fan, W.M., Zhang, M., 1993. Palaeozoic and Cenozoic lithoprobes and the loss of >120 km of Archaean lithosphere, Sino-Korean Craton, China. In: Geological Society of London, Special Publication 76, pp. 71–78.

Meredith, M.T., 2005. A late Archean tectonic boundary exposed at Tin Cup Mountain, Granite Mountain, Wyoming. M.Sc. thesis. University of Wyoming, Laramie, Wyoming, 77 pp.

Meyer, M.R., Backman, D.E., Weinberger, A.J., Wyatt, M.C., 2006. Evolution of circumstellar disks around normal stars: Placing our solar system in context. In: Reipurt, B., Jewitt, D., Keil, K. (Eds.), Protostars and Planets, vol. V. University of Arizona Press, Tucson, AZ, in press.

Mezger, K., Bohlen, S.R., Hanson, G.N., 1990. Metamorphic history of the Archean Pikwitonei Granulite Domain and the Cross Lake Subprovince, Superior Province, Manitoba, Canada. Journal of Petrology 31, 483–517.

Mikhalsky, E.V., Sheraton, J.W., Laiba, A.A., Tingey, R.J., Thost, D.E., Kamenev, E.N., Fedorov, L.V., 2001. Geology of the Prince Charles Mountains, Antarctica. Geoscience Australia Bulletin 247, Australian Geological Survey Organisation, Canberra, 210 pp.

Mikhalsky, E.V., Beliatsky, B.V., Sheraton, J.W., Roland, N.W., 2006a. Two distinct Precambrian terranes in the Southern Prince Charles Mountains, East Antarctica: SHRIMP dating and geochemical constraints. Gondwana Research 9, 291–309.

Mikhalsky, E.V., Laiba, A.A., Beliatsky, B.V., 2006b. Tectonic subdivisions of the Prince Charles Mountains: a review of geologic and isotopic data. In: Futterer, D.K., Damaske, D., Kleinschmidt, G., Miller, H., Tessensohn, F. (Eds.), Antarctica – Contributions to Global Earth Sciences. Springer, Berlin, pp. 69–82.

Miller, D.M., Goldstein, S.L., Langmuir, C.H., 1994. Cerium/lead and lead isotope ratios in arc magmas and the enrichment of lead in the continents. Nature 368, 514–520.

Millero, F., Sohn, M.L., 1992. Chemical Oceanography. CRC Press, Boca Raton, FL, 531 pp.

Millhohlen, G.L., Irving, A.J., Wyllie, P.J., 1974. Melting interval of peridotite with 5.7 per cent water to 30 kbar. Journal of Geology 82, 575–587.

Minister, J.-F., Birck, J.-L., Allegre, C.J., 1982. Absolute age of formation of chondrites studied by the ^{87}Rb-^{87}Sr method. Nature 300, 414–419.

Mittlefehldt, D.W., 2003. Achondrites. In: Davis, A.M. (Ed.), Meteorites, Comets and Planets. Treatise on Geochemistry, vol. 1. Elsevier–Pergamon, Oxford, pp. 291–324.

Mittlefehldt, D.W., Lindstom, M.M., Bogard, D.D., Garrison, D.H., Field, S.W., 1996. Acapulco- and Lodran-like achondrites: Petrology, geochemistry, chronology and origin. Geochimica et Cosmochimica Acta 60, 867–882.

Mittlefehldt, D.W., McCoy, T.J., Goodrich, C.A., Kracher, A., 1998. Non-chondritic meteorites from asteroidal bodies. In: Papike, J.J. (Ed.), Planetary Materials. Reviews in Mineralogy, vol. 36. Mineralogical Society of America, Washington, DC, pp. 4.103–4.130.

Mittlefehldt, D.W., Bogard, D.D., Berkley, J.L., Garrision, D.H., 2003. Brachinites: Igneous rocks from a differentiated asteroid. Meteoritics and Planetary Science 38, 1601–1625.

Moecher, D.P., Perkins, D., III, Leier-Englehardt, P.J., et al., 1986. Metamorphic conditions of late Archean high-grade gneisses, Minnesota River Valley, U.S.A. Canadian Journal of Earth Sciences 23, 633–645.

Mogk, D., Henry, D., 1988. Metamorphic petrology of the northern Archean Wyoming Province, SW Montana: evidence for Archean collisional tectonics. In: Ernst, W. (Ed.), Metamorphism and Crustal Evolution in the Western US. Prentice Hall, Englewood Cliffs, NJ, pp. 363–382.

Mogk, D.W., Mueller, P.A., Wooden, J.L., 1988. Archean tectonics of the North Snowy Block, Beartooth Mountains, Montana. Journal of Geology 96, 125–141.

Mogk, D.W., Mueller, P.A., Wooden, J., 1992. The significance of Archean terrane boundaries: Evidence from the northern Wyoming province. Precambrian Research 55, 155–168.

Mogk, D.W., Mueller, P.A., Wooden, J.L., 2004. Tectonic implications of late Archean-early Proterozoic supracrustal rocks in the Gravelly Range, SW Montana. In: Geological Society of America, 2004 Annual Meeting. Abstracts with Programs, vol. 36 (5), p. 507.

Mojzsis, S.J., Harrison, T.M., 2002a. Establishment of a 3.83-Ga magmatic age for the Akilia tonalite (southern West Greenland). Earth and Planetary Science Letters 202, 563–576.

Mojzsis, S.J., Harrison, T.M., 2002b. Origin and Significance of Archean Quartzose Rocks at Akilia, Greenland. Science 298, 917a.

Mojzsis, S.J., Ryder, G., 2001. Accretion to Earth and Moon ~3.85 Ga. In: Peucker-Ehrenbrink, B., Schmitz, B. (Eds.), Accretion of Extraterrestrial Matter Throughout Earth's History. Kluwer, New York, pp. 423–466.

Mojzsis, S.J., Arrhenius, G., McKeegan, K.D., Harrison, T.M., Nutman, A.P., Friend, C.R.L., 1996. Evidence for life on Earth before 3,800 million years ago. Nature 384, 55–59.

Mojzsis, S.J., Harrison, T.M., Arrhenius, G., McKeegan, K.D., Grove, M., 1999. Origin of life from apatite dating? – Reply. Nature 400, 127–128.

Mojzsis, S.J., Harrison, T.M., Pidgeon, R.T., 2001. Oxygen-isotope evidence from ancient zircons for liquid water at the Earth's surface 4,300 Myr ago. Nature 409, 178–181.

Mojzsis, S.J., Coath, C.D., Greenwood, J.P., McKeegan, K.D., Harrison, T.M., 2003. Mass-independent isotope effects in Archean (2.5 to 3.8 Ga) sedimentary sulphides determined by ion microprobe analysis. Geochimica et Cosmochimica Acta 67, 1635–1658.

Monster, J., Appel, P.W.U., Thode, H.G., Schidlowski, M., Carmichael, C.M., Bridgwater, D., 1979. Sulfur isotope studies in early Archaean sediments from Isua, West Greenland – Implications for the antiquity of bacterial sulfate reduction. Geochimica et Cosmochimica Acta 43, 405–413.

Montelli, R., Nolet, G., Dahlen, F.A., Masters, G., Engdahl, E.R., Hung, S-H., 2004. Finite-frequency tomography reveals a variety of plumes in the mantle. Science 303, 338–343.

Mook, W.G., Bommerson, J.C., Staverman, W.H., 1974. Carbon isotope fractionation between dissolved bicarbonate and gaseous carbon dioxide. Earth and Planetary Science Letters 22, 169–176.

Moorbath, S., 1975. Evolution of Precambrian crust from Strontium isotopic evidence. Nature 254, 395–398.

Moorbath, S., 1977. Ages, isotopes and the evolution of the Precambrian continental crust. Chemical Geology 20, 151–187.

Moorbath, S., 2005. Oldest rocks, earliest life, heaviest impacts, and the Hadean-Archaean transition. Applied Geochemistry 20, 819–824.

Moorbath, S., Taylor, P.N., 1985. Precambrian geochronology and the geological record. In: Snelling, N.J. (Ed.), The Chronology of the Geological Record. Geological Society of London, Memoir 10, pp. 10–28.

Moorbath, S., O'Nions, R.K., Pankhurst, R.J., Gale, N.H., McGregor, V.R., 1972. Further rubidium-strontium age determinations on the very early Precambrian rocks of the Godthåb district: West Greenland. Nature 240, 78–82.

Moorbath, S., O'Nions, R.K., Pankhurst, R.J., 1973. Early Archaean age for the Isua Iron Formation, West Greenland. Nature 245, 138–139.

Moorbath, S., Whitehouse, M.J., Kamber, B.S., 1997. Extreme Nd-isotope heterogeneity in the early Archaean – Fact or fiction? Case histories from northern Canada and West Greenland. Chemical Geology 135, 213–231.

Moores, E.M., 2002. Pre-1 Ga (pre-Rodinian) ophiolites: Their tectonic and environmental indicators. Bulletin of the Geological Society of America 114, 80–95.

Moralev, V.M., 1986. Early Evolution of the Continental Lithosphere. Nauka, Moscow, 119 pp. (in Russian).

Moralev, V.M., Glukhovsky, M.Z., 2000. Diamond-bearing kimberlite fields of the Siberian craton and the early Precambrian metallogeny. Ore Geology Reviews 17, 141–153.

Morant, P., 1998. Panorama Zinc-Copper deposits. In: Berkman, D.A., Mackenzie, D.H. (Eds.), Geology of Australian and Papua New Guinean Mineral Deposits. The Australian Institute of Mining and Metallurgy, Melbourne, Australia, pp. 287–292.

Morbidelli, A., Chambers, J., Lunine, J.I., Petit, J.M., Robert, F., Valsecchi, G.B., Cyr, K.E., 2000. Source regions and time scales for the delivery of water to Earth. Meteoritics and Planetary Science 35, 1309–1320.

Morey, G.B., Sims, P.K., 1976. Boundary between two lower Precambrian terranes in Minnesota and its geological significance. Bulletin of the Geological Society of America 87, 141–152.

Morgan, J.W., Anders, E., 1980. Chemical composition of Earth, Venus and Mercury. Proceedings of the National Academy of Sciences of the United States of America – Physical Sciences 77, 6973–6977.

Morgan, J.W., Walker, R.J., Grossman, J.N., 1992. Rhenium-osmium isotope systematics in meteorites I: Magmatic iron meteorite groups IIAB and IIIAB. Earth and Planetary Science Letters 108, 191–202.

Morgan, J.W., Horn, M.F., Walker, R.J., Grossman, J.N., 1995. Rhenium-osmium concentration and isotope systematics in group IIAB iron meteorites. Geochimica et Cosmochimica Acta 59, 2331–2344.

Mortensen, J.K., Card, K.D., 1993. U-Pb age constraints for the magmatic and tectonic evolution of the Pontiac Subprovince, Quebec. Canadian Journal of Earth Sciences 30, 1970–1980.

Moser, D., 1994. The geology and structure of the mid-crustal Wawa gneiss domain: a key to understanding tectonic variation with depth and time in the late Archean Abitibi-Wawa orogen. Canadian Journal of Earth Sciences 31, 1064–1080.

Moser, D.E., Heaman, L.M., Krogh, T.E., Hanes, J.A., 1996. Intracrustal extension of an Archean orogen revealed using single-grain U-Pb zircon geochronology. Tectonics 15, 1093–1109.

Moser, D.E., Flowers, R.M., Hart, R.J., 2001. Birth of the Kaapvaal tectosphere 3.08 billion years ago. Nature 291, 465–468.

Moyen, J.-F., Stevens, G., 2006. Experimental constraints on TTG petrogenesis: Implications for Archean geodynamics. In: Benn, K., Mareschal, J.-C., Condie, K.C. (Eds.), Archean Geodynamics and Environments. Geophysical Monograph 164. American Geophysical Union, Washington, DC, pp. 149–175.

Moyen, J.-F., Martin, H., Jayananda, M., Auvray, B., 2003. Late Archaean granites: a typology based on the Dharwar Craton (India). Precambrian Research 127 (1–3), 103–123.

Moyen, J.-F., Stevens, G., Kisters, A.F.M., 2006. Record of mid-Archaean subduction from metamorphism in the Barberton terrain, South Africa. Nature 443, 559–562.

Muehlenbachs, K., 1998. The oxygen isotopic composition of the oceans, sediments and the seafloor. Chemical Geology 145, 263–273.

Mueller, P.A., Frost, C.D., 2006. The Wyoming province: a distinctive Archean craton in Laurentian North America. Canadian Journal of Earth Sciences 43, 1391–1397.

Mueller, P.A., Wooden, J.L., 1988. Evidence for Archean subduction and crustal recycling, Wyoming province. Geology 16, 871–874.

Mueller, P.A., Wooden, J.L., Henry, D.J., Bowes, D.R., 1985. Archean crustal evolution of the eastern Beartooth Mountains, Montana–Wyoming. In: Montana Bureau of Mines and Geology, Special Publication 92, pp. 9–20.

Mueller, P.A., Shuster, R.D., Graves, M.A., Wooden, J.L., Bowes, D.R., 1988. Age and composition of a Late Archean magmatic complex, Beartooth Mountains, Montana–Wyoming. In: Montana Bureau of Mines and Geology, Special Publication 96, pp. 6–22.

Mueller, P.A., Wooden, J.L., Nutman, A.P., 1992. 3.96 Ga zircons from an Archean quartzite, Beartooth Mountains, Montana. Geology 20, 327–330.

Mueller, P.A., Shuster, R.D., Wooden, J.L., Erslev, E.A., Bowes, D.R., 1993. Age and composition of Archean crystalline rocks from the southern Madison Range, Montana: implications for crustal evolution in the Wyoming craton. Bulletin of the Geological Society of America 105, 437–446.

Mueller, P.A., Heatherington, A., Weyand, E., Mogk, D., Wooden, J., Nutman, A., 1995. Geochemical evolution of Archean crust in the northern Madison Range, Montana: Evidence from U-Pb and Sm-Nd systematics. Geologic Society of America, Abstracts with Programs 27 (4), 49.

Mueller, P.A., Wooden, J.L., Mogk, D.W., Nutman, A.P., Williams, I.S., 1996. Extended history of a 3.5 Ga trondhjemitic gneiss, Wyoming province, USA: evidence from U-Pb systematics in zircon. Precambrian Research 78, 41–52.

Mueller, P.A., Wooden, J.L., Nutman, A.P., Mogk, D.W., 1998. Early Archean crust in the northern Wyoming province: Evidence from U-Pb ages of detrital zircons. Precambrian Research 91, 295–307.

Mueller, P.A., Heatherington, A.L., Kelly, D.M., Wooden, J.L., Mogk, D.W., 2002. Paleoproterozoic crust within the Great Falls tectonic zone; implications for the assembly of southern Laurentia. Geology 30, 127–130.

Mueller, P., Wooden, J., Heatherington, A., Burger, H., Mogk, D., D'Arcy, K., 2004. Age and evolution of the Precambrian crust of the Tobacco Root Mountains. In: Brady, J.B., Burger, H.R., Cheney, J.T., Harms, T.A. (Eds.), Precambrian Geology of the Tobacco Root Mountains, Montana. Geological Society of America, Special Paper 377, pp. 181–202.

Mueller, P.A., Burger, H.R., Wooden, J.L., Brady, J.B., Cheney, J.T., Harms, T.A., Heatherington, A.L., Mogk, D.W., 2005. Paleoproterozoic metamorphism in the northern Wyoming Province: implications for the assembly of Laurentia. Journal of Geology 113 (2), 169–179.

Mueller, P., Mogk, D., Wooden, J., 2006. Archean crustal evolution in the Northern Wyoming Province: Implications for mantle keels. In: AGU Fall Meeting (V11D-0631).

Mueller, W.U., Marquis, R., Thurston, P. (Eds.), 2002. Evolution of the Archean Abitibi Greenstone Belt and Adjacent Terranes: New Insights from Geochronology, Geochemistry, Structure and Facies Analysis. Precambrian Research 115, 374 pp.

Muhling, J.R., 1990. The Narryer Gneiss Complex of the Yilgarn block, Western Australia: a segment of Archean lower crust uplifted during Proterozoic orogeny. Journal of Metamorphic Geology 8, 47–64.

Mulligan, R., 1957. Split Lake, Manitoba. Geological Survey of Canada, Preliminary Map 10-1956.

Murphy, J.B., Nance, R.D., 1991. Supercontinent model for the contrasting character of late Proterozoic orogenic belts. Geology 91, 469–472.

Musacchio, G., White, D.J., Asudeh, I., Thomson, C.J., 2004. Lithospheric structure and composition of the Archean western Superior Province from seismic refraction/wide-angle reflection and gravity modeling. Journal of Geophysical Research 109, B03304, doi: 10.1029/2003JB002427.

Myers, J.S., 1978. Formation of banded gneisses by deformation of igneous rocks. Precambrian Research 6, 43–64.

Myers, J.S., 1988a. Early Archaean Narryer Gneiss Complex, Yilgarn Craton, Western Australia. Precambrian Research 38, 297–307.

Myers, J.S., 1988b. Oldest known terrestrial anorthosite at Mount Narryer, Western Australia. Precambrian Research 38, 309–323.

Myers, J.S., 1990a. Precambrian tectonic evolution of part of Gondwana, southwestern Australia. Geology 18, 537–540.

Myers, J.S., 1990b. Western Gneiss Terrane. In: Geology and Mineral Resources of Western Australia. Western Australia Geological Survey, Memoir 3, pp. 13–31.

Myers, J.S., 1993. Precambrian history of the West Australian Craton and adjacent orogens. Annual Review of Earth and Planetary Sciences 21, 453–485.

Myers, J.S., 1995. The generation and assembly of an Archaean supercontinent: evidence from the Yilgarn Craton, Western Australia. In: Coward, M.P., Ries, A.C. (Eds.), Early Precambrian Processes. Geological Society of London, Special Publication 95, pp. 143–154.

Myers, J.S., 1997. Archaean geology of the Eastern Goldfields of Western Australia; regional overview. Precambrian Research 83, 1–10.

Myers, J.S., 2001. Protoliths of the 3.8–3.7 Ga Isua greenstone belt, West Greenland. Precambrian Research 105, 129–141.

Myers, J.S., Crowley, J.L., 2000. Vestiges of life in the oldest Greenland rocks? A review of early Archaean geology of the Godthåbsfjord region, and reappraisal of field evidence for >3850 Ma life on Akilia. Precambrian Research 103, 101–124.

Myers, J.S., Hocking, R., 1998. Geological Map of Western Australia, 1:2,500,000, 13th ed. Western Australian Geological Survey.

Myers, J.S., Occhipinti, S.A., 2001. Narryer Terrane. In: Occhipinti, S.A., Sheppard, S., Myers, J.S., Tyler, I.M., Nelson, D.R. (Eds.), Archaean and Palaeoproterozoic Geology of the Narryer Terrane (Yilgarn Craton) and the Southern Gascoyne Complex (Capricorn Orogen), Western Australia – A Field Guide. Western Australia Geological Survey, Record 2001/8, pp. 8–42.

Myers, J.S., Swager, C., 1997. The Yilgarn Craton. In: de Wit, M., Ashwal, L.D. (Eds.), Greenstone Belts. Clarendon Press, Oxford, UK, pp. 640–656.

Myers, J.S., Williams, I.R., 1985. Early Precambrian crustal evolution at Mount Narryer, Western Australia. Precambrian Research 27, 153–163.

Mysen, B., Boettcher, A.L., 1975a. Melting of anhydrous mantle I. Phase relations of natural peridotite at high pressures and temperatures with controlled activity of water, carbon dioxyde and hydrogen. Journal of Petrology 16, 520–548.

Mysen, B., Boettcher, A.L., 1975b. Melting of anhydrous mantle II. Geochemistry of crystals and liquids formed by anatexis of mantle peridotite at high pressures and temperatures as a function of the controlled activities of water, carbon dioxyde and hydrogen. Journal of Petrology 16, 549–593.

Nadeau, P., 2003. Structural investigation of the Porpoise Cove area, Northeastern Superior Province, Northern Quebec. Unpublished thesis. Simon Fraser University, 95 pp.

Nagahara, H., 1992. Yamato 8002: Partial melting residue on the "unique" chondrite parent body. In: Proceedings of the NIPR Symposium on Antarctic Meteorites, vol. 5, pp. 191–223.

Nagao, K., Okazaki, R., Sawada, S., Nakamura, N., 1999. Noble gases and K-Ar ages of five Rumuruti chondrites: Yamato Y-75302, Y-79182, Y-793575, Y-82002, and Asuka-881988. Antarctic Meteorite Research (abstract) 23, 81.

Nägler, T.F., Kramers, J.D., 1998. Nd isotopic evolution of the upper mantle during the Precambrian: Models, data and the uncertainty of both. Precambrian Research 91, 233–252.

Nägler, T.F., Kramers, J.D., Kamber, B.S., Frei, R., Predergast, M.D.A., 1997. Growth of subcontinental lithospheric mantle beneath Zimbabwe started at or before 3.8 Ga: A Re-Os study on chromites. Geology 25, 983–986.

Nakamoto, T., Kita, N.T., Tachibana, S., 2005. Chondrule age distribution and rate of heating events for chondrule formation. Antarctic Meteorite Research 18, 253–272.

Nakamura, K., Kato, Y., 2004. Carbonatization of oceanic crust by the seafloor hydrothermal activity and its significance as a CO_2 sink in the Early Archean. Geochimica et Cosmochimica Acta 68, 4595–4618.

Nakamura, N., 1974. Determination of REE, Ba, Fe, Mg, Na and K in carbonaceous and ordinary chondrites. Geochimica et Cosmochimica Acta 38, 757–775.

Namiki, N., Solomon, S.C., 1994. Impact crater densities on volcanoes and coronae on Venus: implications for volcanic resurfacing. Science 265, 929–933.

Nance, R.D., Worsley, T.R., Moody, J.B., 1988. The supercontinent cycle. Scientific American 259, 44–52.

Naraoka, H., Ohtake, M., Maruyama, S., Ohmoto, H., 1996. Non-biogenic graphite in 3.8-Ga metamorphic rocks from the Isua district, Greenland. Chemical Geology 133, 251–260.

Nelson, D.R., 1996. Compilation of SHRIMP U-Pb Zircon Geochronology Data, 1995. Western Australia Geological Survey, Record 1996/5.

Nelson, D.R., 1997a. Evolution of the Archaean granite-greenstone terranes of the Eastern Goldfields, Western Australia. SHRIMP U-Pb zircon constraints. Precambrian Research 83, 57–81.

Nelson, D.R., 1997b. Compilation of SHRIMP U-Pb Zircon Geochronology Data, 1996. Western Australia Geological Survey, Record 1997/2.

Nelson, D.R., 1998a. Granite-greenstone crust formation on the Archaean Earth: a consequence of two superimposed processes. Earth and Planetary Science Letters 158, 109–119.

Nelson, D.R., 1998b. Compilation of SHRIMP U-Pb Zircon Geochronology Data, 1997. Western Australia Geological Survey, Record 1998/2.

Nelson, D.R., 1999. Compilation of SHRIMP U-Pb Zircon Geochronology Data, 1998. Western Australia Geological Survey, Record 1999/2.

Nelson, D.R., 2000. Compilation of Geochronology Data, 1999. Western Australia Geological Survey, Record 2000/2, pp. 62–65.

Nelson, D.R., 2001. Compilation of SHRIMP U-Pb Zircon Geochronology Data, 2000. Western Australia Geological Survey, Record 2001/2.

Nelson, D.R., 2002a. Compilation of SHRIMP U-Pb Zircon Geochronology Data, 2001. Western Australia Geological Survey, Record 2002/2.

Nelson, D.R., 2002b. Hadean Earth crust: microanalytical investigation of 4.4 to 4.0 Ga zircons from Western Australia. Geochimica et Cosmochimica Acta 66, A549.

Nelson, D.R., 2004. The early Earth, Earth's formation and first billion years. In: Eriksson, P.G., Altermann, W., Nelson, D.R., Mueller, W.U., Catuneau, O. (Eds.), The Precambrian Earth: Tempos and Events. Elsevier, Amsterdam, pp. 3–27.

Nelson, D.R., Trendall, A.F., Altermann, W., 1999. Chronological correlations between the Pilbara and Kaapvaal cratons. Precambrian Research 97, 165–189.

Nelson, D.R., Robinson, B.W., Myers, J.S., 2002. Complex geological histories extending for $\geqslant 4.0$ Ga deciphered from xenocryst zircon microstructures. Earth and Planetary Science Letters 181 (1–2), 89–102.

Nemchin, A.A., 1990. Evolution of the Aldan Shield, Eastern Siberia. In: VII International Conference on Geochronology, Cosmochronology and Isotope Geology, Abstracts volume. Geological Society of Australia, Canberra, p. 70.

Nemchin, A.A., Pidgeon, R.T., 1997. Evolution of the Darling Range batholith, Yilgarn Craton, Western Australia: a SHRIMP zircon study. Journal of Petrology 38, 625–649.

Nemchin, A.A., Pidgeon, R.T., Whitehouse, M.J., 2006. Re-evaluation of the origin and evolution of >4.2 Ga zircons from the Jack Hills metasedimentary rocks. Earth and Planetary Science Letters 244, 218–233.

Neukum, G., Ivanov, B., Hartmann, W.K., 2001. Cratering records in the inner solar system. In: Kallenbach, R., et al. (Eds.), Chronology and Evolution of Mars. Kluwer, Dordrecht, pp. 55–86.

Neumann, E.-R., 1994. The Oslo Rift: *P-T* relations and lithosphere structure. Tectonophysics 240, 159–172.

Neumayr, P., Cabri, L.J., Groves, D.I., Mikucki, E.J., Jackman, J., 1993. The mineralogical distribution of gold and relative timing of gold mineralization in two Archean settings of high metamorphic grades in Australia. Canadian Mineralogist 31, 711–725.

Newsom, H.E., White, W.M., Jochum, K.P., Hofmann, A.W., 1986. Siderophile and chalcophile element abundances in oceanic basalts, Pb isotope evolution and growth of the Earth's core. Earth and Planetary Science Letters 80, 299–313.

Newton, R.C., Manning, C.E., 2005. Solubility of anhydrite, $CaSO_4$, in $NaCl-H_2O$ solution at high pressures and temperatures: Applications to fluid-rock interaction. Journal of Petrology 46, 701–716.

Neymark, L.A., Kovach, V.P., Nemchin, A.A., Morozova, I.M., Kotov, A.B., Vinogradov, D.P., Gorokhovsky, B.M., Ovchinikova, G.V., Bogomolova, L.M., Smelov, A.P., 1993.

Late Archaean intrusive complexes in the Olekma granite-greenstone terrane (Eastern Siberia): geochemical and isotopic study. Precambrian Research 62, 453–472.

Neymark, L.A., Larin, A.M., Nemchin, A.A., 1998. Geochemical, geochronological (U-Pb) and isotopic (Pb, Nd) evidence of anorogenic magmatism in the North Baikal volcano-plutonic belt. Petrology 6 (2), 139–164.

Nichols, R.H., Jr, Hohenberg, C.M., Kehm, K., Kim, Y., Marti, K., 1994. I-Xe studies of the Acapulco meteorite: Absolute I-Xe ages of individual phosphate grains and the Bjurböle standard. Geochimica et Cosmochimica Acta 58, 2553–2561.

Nicolaysen, K., Frey, F.A., Hodges, K.V., Weis, D., Giret, A., 2000. ^{40}Ar/^{39}Ar geochronology of flood basalts from the Kerguelen Archipelago, southern Indian Ocean: implications for cenozoic eruption rates of the Kerguelen plume. Earth and Planetary Science Letters 174, 313–328.

Nicollet, C., Lahlafi, M., Lasnier, B., 1993. Existence d'un épisode métamorphique tardi-hercynien, granulitique de basses pressions, dans le Haut-Allier (Massif Central) : implications géodynamiques. Comptes rendus de l'Académie des Sciences, Série 2 317, 1609–1615.

Niemeyer, S., 1979a. I-Xe dating of silicate and troilite from IAB iron meteorites. Geochimica et Cosmochimica Acta 43, 843–860.

Niemeyer, S., 1979b. ^{40}Ar-^{39}Ar dating of inclusions from IAB iron meteorites. Geochimica et Cosmochimica Acta 43, 1829–1840.

Niemeyer, S., Zaikowski, A., 1980. I-Xe age and trapped Xe components in the Murray (C2) chondrite. Earth and Planetary Science Letters 48, 335–347.

Nieuwland, D.A., Compston, W., 1981. Crustal evolution in the Yilgarn Block near Perth, Western Australia. In: Glover, J.A., Groves, D.I. (Eds.), Archaean Geology. Geological Society of Australia, Special Publication 7, pp. 159–171.

Nijman, W., de Vries, S.T., 2004. Early Archaean crustal collapse structures and sedimentary basin dynamics. In: Eriksson, P.G., Altermann, W., Nelson, D.R., Mueller, W.U., Catuneau, O. (Eds.), The Precambrian Earth: Tempos and Events. Elsevier, Amsterdam, pp. 139–154.

Nijman, W., De Bruin, K., Valkering, M., 1998. Growth fault control of early Archaean cherts, barite mounds, and chert-barite veins, North Pole Dome, Eastern Pilbara, Western Australia. Precambrian Research 88, 25–52.

Nikishin, A.M., Ziegler, P.A., Abbott, D., Brunet, M-F., Cloetingh, S., 2002. Permo-Triassic intraplate magmatism and rifting in Eurasia: Implications for mantle plumes and mantle dynamics. Tectonophysics 351, 3–39.

Nikolayeva, O.V., 1993. Largest impact features on Venus – non-preserved or non-recognizable? In: Lunar and Planetary Science Conference XXIV, Houston, TX, pp. 1083–1084.

Nimmo, F., McKenzie, D., 1998. Volcanism and tectonics on Venus. Annual Reviews of Earth and Planetary Sciences 26, 23–52.

Nisbet, E.G., Sleep, N., 2001. The habitat and nature of early life. Nature 409, 1083–1091.

Nisbett, E.G., Cheadle, M.J., Arndt, N.J., Bickle, M.J., 1993. Constraining the potential temperature of the Archean mantle: A review of the evidence from komatiites. Lithos 30, 291–307.

Nishizawa, M., Takahata, N., Terada, K., Komiya, T., Ueno, Y., Sano, Y., 2005. Rare-earth element, lead, carbon, and nitrogen geochemistry of apatite-bearing metasediments from the ~3.8 Ga Isua Supracrustal Belt, West Greenland. International Geology Review 47, 952–970.

Noldart, A.J., Wyatt, J.D., 1962. The geology of a portion of the Pilbara Goldfield, covering the Marble Bar and Nullagine 4-mile map sheets. Western Australia Geological Survey, Bulletin 115, 199 pp.

Nolet, G., Karato, S-I., Montelli, R., 2006. Plume fluxes from seismic tomography. Earth and Planetary Science Letters 248, 685–699.

Norman, M.D., Borg, L.E., Nyquist, L.E., Bogard, D.D., 2003. Chronology, geochemistry, and petrology of a ferroan noritic anorthosite clast from Descartes breccia 67215: Clues to the age, origin, structure, and impact history of the lunar crust. Meteoritics and Planetary Science 38, 645–661.

Norrish, K., Chappell, B.W., 1977. X-ray fluorescence spectrometry. In: Zussman, J. (Ed.), Physical Methods in Determinative Mineralogy, second ed. Academic Press, London, pp. 201–272.

Norrish, K., Hutton, J.T., 1969. An accurate X-ray spectrographic method for the analysis of a wide range of geological samples. Geochimica et Cosmochimica Acta 33, 431–453.

Nozhkin, A.D., 1986. Main features of composition and structure of the Precambrian complexes of the Yenisey highland. In: Reverdatto, V.V., Khlestov, V.V. (Eds.), Precambrian Crystalline Complexes of the Yenisey Highland. Russian Academy of Sciences Publication, Novosibirsk, pp. 9–18 (in Russian).

Nozhkin, A.D., 1999. Continental margin Paleoproterozoic complexes in the Angara foldbelt and their metallogenic peculiarities. Russian Geology and Geophysics 40 (11), 1524–1544.

Nozhkin, A.D., Turkina, O.M., 1993. Geochemistry of granulites from Kansk and Sharyzhalgay complexes. In: Transactions of the United Institute of Geology, Geophysics and Mineralogy, vol. 817, Novosibirsk, 219 pp. (in Russian).

Nozkin, A.D., Turkina, O.M., Melgunov, M.S., 2001. Geochemistry of metavolcano-sedimentary and granitoid rocks of the Onot greenstone belt. Geochemistry International 39 (1), 27–44.

Nunes, D.C., Phillips, R.J., Brown, C.D., Dombard, A.J., 2004. Relaxation of compensated topography and the evolution of crustal plateaus on Venus. Journal Geophysical Research 109, doi: 10.1029/2003JE002119.

Nutman, A.P., 1984. Early Archaean crustal evolution of the Isukasia area, southern West Greenland. In: Kröner, A., Greiling, R. (Eds.), Precambrian Tectonics Illustrated. E. Schweizerbart'sche Verlagsbuchhandlung, Stuttgart, pp. 79–94.

Nutman, A.P., 2006a. Comment on: Zircon thermometer reveals minimum melting conditions on Earliest Earth. Science 311, 779a.

Nutman, A.P., 2006b. Antiquity of the oceans and continents. Elements 2, 223–227.

Nutman, A.P., Bridgwater, D., 1986. Early Archaean Amîtsoq tonalites and granites from the Isukasia area, southern West Greenland: Development of the oldest-known sial. Contributions to Mineralogy and Petrology 94, 137–148.

Nutman, A.P., Collerson, K.D., 1991. Very early Archean crustal-accretion complexes preserved in the North Atlantic Craton. Geology 19, 791–794.

Nutman, A.P., Cordani, U.G., 1993. SHRIMP U-Pb zircon geochronology of Archean granitoids from the Contendas-Mirante area of the Sao Francisco craton, Bahia, Brazil. Precambrian Research 63, 179–188.

Nutman, A.P., Friend, C.R.L., 2006. Petrography and geochemistry of apatites in banded iron formation, Akilia, W. Greenland: Consequences for oldest life evidence. Precambrian Research 147, 100–106.

Nutman, A.P., Friend, C.R.L., 2007. Adjacent terranes with c. 2715 and 2650 Ma high-pressure metamorphic assemblages in the Nuuk region of the North Atlantic Craton, southern West Greenland: Complexities of Neoarchaean collisional orogeny. Precambrian Research 155, 159–203.

Nutman, A.P., Allaart, J.H., Bridgwater, D., Dimroth, E., Rosing, M.T., 1984a. Stratigraphic and geochemical evidence for the depositional environment of the early Archaean Isua supracrustal belt, southern West Greenland. Precambrian Research 25, 365–396.

Nutman, A.P., Bridgwater, D., Fryer, B., 1984b. The iron-rich suite from the Amîtsoq gneisses of southern West Greenland: Early Archaean plutonic rocks of mixed crustal and mantle origin. Contributions to Mineralogy and Petrology 87, 24–34.

Nutman, A.P., Friend, C.R.L., Baadsgaard, H., McGregor, V.R., 1989. Evolution and assembly of Archean gneiss terranes in the Godthåbsfjord region, southern West Greenland: Structural, metamorphic, and isotopic evidence. Tectonics 8, 573–589.

Nutman, A.P., Chernyshev, I.V., Baadsgaard, H., 1990. The Archean Aldan shield of Siberia, USSR – The search for its oldest rocks and evidence for reworking in the Mid-Proterozoic. In: Glover, J.E., Ho, S. (Eds.), Third International Archean Symposium, Extended Abstract Volume, Geoconferences WA Inc., Perth, Australia, pp. 60–61.

Nutman, A.P., Kinny, P.D., Compston, W., Williams, I.S., 1991. SHRIMP U-Pb zircon geochronology of the Narryer Gneiss Complex, Western Australia. Precambrian Research 52, 275–300.

Nutman, A.P., Chernyshev, I.V., Baadsgaard, H., Smelov, A.P., 1992a. The Aldan shield of Siberia, USSR: the age of its Archean components and evidence for widespread reworking in the mid-Proterozoic. Precambrian Research 54, 195–210.

Nutman, A.P., Chadwick, B., Ramakrishnan, M., Viswanatha, M.N., 1992b. U-Pb ages of detrital zircon in supracrustal rocks older than c. 3000 Ma in southern Peninsular India. Journal of the Geological Society of India 39, 367–374.

Nutman, A.P., Bennett, V., Kinny, P.D., Price, R., 1993a. Large-scale crustal structure of the northwestern Yilgarn Craton, Western Australia: evidence from Nd isotopic data and zircon geochronology. Tectonics 12, 971–981.

Nutman, A.P., Friend, C.R.L., Kinny, P.D., McGregor, V.R., 1993b. Anatomy of an Early Archean gneiss complex: 3900 to 3600 Ma crustal evolution in southern West Greenland. Geology 21, 415–418.

Nutman, A.P., Hagiya, H., Maruyama, S., 1995. SHRIMP U-Pb single zircon geochronology of a Proterozoic mafic dyke, Isukasia, southern West Greenland. Bulletin of the Geological Society Denmark 42, 17–22.

Nutman, A.P., McGregor, V.R., Friend, C.R.L., Bennett, V.C., Kinny, P.D., 1996. The Itsaq gneiss complex of southern West Greenland; the world's most extensive record of early crustal evolution. Precambrian Research 78, 1–39.

Nutman, A.P., Bennett, V.C., Friend, C.R.L., Rosing, M.T., 1997a. ~3710 and ⩾3790 Ma volcanic sequences in the Isua (Greenland) supracrustal belt; structural and Nd isotope implications. Chemical Geology 141, 271–287.

Nutman, A.P., Mojzsis, S.J., Friend, C.R.L., 1997b. Recognition of ⩾3850 Ma waterlain sediments in West Greenland and their significance for the early Archaean Earth. Geochimica et Cosmochimica Acta 61, 2475–2484.

Nutman, A.P., Bennett, V.C., Friend, C.R.L., Norman, M.D., 1999. Meta-igneous (non-gneissic) tonalites and quartz-diorites from an extensive ca. 3800 Ma terrain south of the Isua supracrustal belt, southern West Greenland: constraints on early crust formation. Contributions to Mineralogy and Petrology 137 (4), 364–388.

Nutman, A.P., Friend, C.R.L., Bennett, V.C., McGregor, V.R., 2000. The early Archaean Itsaq Gneiss Complex of southern West Greenland: The importance of field observations in interpreting dates and isotopic data constraining early terrestrial evolution. Geochimica et Cosmochimica Acta 64, 3035–3060.

Nutman, A.P., Friend, C.R.L., Bennett, V.C., 2001. Review of the oldest (4400–3600 Ma) geological and mineralogical record: Glimpses of the beginning. Episodes 24 (2), 93–101.

Nutman, A.P., Friend, C.R.L., Bennett, V.C., 2002a. Evidence for 3650–3600 Ma assembly of the northern end of the Itsaq Gneiss complex, Greenland: implications for early Archean tectonics. Tectonics 21 (1), 10.1029/2000TC001203.

Nutman, A.P., McGregor, V.R., Shiraishi, K., Friend, C.R.L., Bennett, V.C., Kinny, P.D., 2002b. ⩾3850 Ma BIF and mafic inclusions in the early Archaean Itsaq Gneiss Complex around Akilia, southern West Greenland? The difficulties of precise dating of zircon-free protoliths in migmatites. Precambrian Research 117, 185–224.

Nutman, A.P., Friend, C.R.L., Barker, S.S., McGregor, V.R., 2004. Inventory and assessment of Palaeoarchaean gneiss terrains and detrital zircons in southern West Greenland. Precambrian Research 135, 281–314.

Nutman, A.P., Bennett, V.C., Friend, C.R.L., Horie, K., Hidaka, H., 2007. c. 3850 Ma tonalites in the Nuuk region, Greenland: Geochemistry and their reworking within an Eoarchaean gneiss complex. Contributions to Mineralogy and Petrology, doi: 10.1007/S00410-007-0199-3.

Nyquist, L.E., Takeda, H., Bansal, B.M., Shih, C-Y., Wiesmann, H., Wooden, J.L., 1986. Rb-Sr and Sm-Nd internal isochron ages of a subophitic basalt clast and a matrix sample from the Y 75011 eucrite. Journal of Geophysical Research 91, 8137–8150.

Nyquist, L.E., Bansal, B.M., Wiesmann, H., Shih, C-Y., 1994. Neodymium, strontium and chromium isotopic studies of the LEW86010 and Angra dos Reis meteorites and the chronology of the angrite parent body. Meteoritics 29, 872–885.

Nyquist, L.E., Bogard, D.D., Shih, C.Y., Greshake, A., Stoffler, D., Eugster, O., 2001a. Ages and geologic histories of Martian meteorites. Space Science Reviews 96 (1–4), 105–164.

Nyquist, L.E., Reese, Y., Wiesmann, H., Shih, C-Y., Takeda, H., 2001b. Live ^{53}Mn and ^{26}Al in an unique cumulate eucrite with very calcic feldspar (An^{98}). Meteoritics and Planetary Science 36 (Supplement), A151–152.

Nyquist, L.E., Shih, C-Y., Wiesmann, H., Mikouchi, T., 2003. ^{26}Al and ^{53}Mn in D'Orbigny and Sahara 99555 and the timescale for angrite magmatism. In: Lunar and Planetary Science Conference XXXIV, abstract 1388.

Occhipinti, S.A., Reddy, S.M., 2004. Deformation in a complex crustal-scale shear zone: Errabiddy Shear Zone, Western Australia. In: Alsop, G.I., Holdsworth, R.E., McCaffrey, K.J.W., Hand, M. (Eds.), Flow Processes in Faults and Shear Zones. Geological Society of London, Special Publication 224, pp. 229–248.

Occhipinti, S.A., Sheppard, S., Passchier, C., Tyler, I.M., Nelson, D.R., 2004. Palaeoproterozoic crustal accretion and collision in the southern Capricorn Orogen: the Glenburgh Orogeny. Precambrian Research 128, 237–255.

O'Connor, J.T., 1965. A classification for quartz-rich igneous rocks based on feldspar ratios. In: U.S. Geological Survey, Professional Paper 525(B), pp. 79–84.

Ohmoto, H., 1986. Stable isotope geochemistry of ore deposits. Reviews in Mineralogy 16, 491–559.

Ohmoto, H., 1997. When did Earth's atmosphere become oxic? Geochemical News 93, 12–27.

Ohmoto, H., 1999. Redox state of the Archean atmosphere: Evidence from detrital heavy minerals in ca. 3250–2750 Ma sandstones from the Pilbara Craton, Australia: Comment. Geology 27, 1151–1152.

Ohmoto, H., Felder, R.P., 1987. Bacterial activity in the warmer, sulfate-beating Archean oceans. Nature 328, 244–246.

Ohmoto, H., Goldhaber, M.B., 1997. Sulfur and carbon isotopes. In: Barnes, H.L. (Ed.), Geochemistry of Hydrothermal Ore Deposits, third ed. John Wiley & Sons, New York, pp. 517–611.

Ohmoto, H., Kakegawa, T., Lowe, D.R., 1993. 3.4-billion-year-old biogenic pyrites from Barberton, South Africa – sulfur isotope evidence. Science 262, 555–557.

O'Keefe, J.D., Aherns, T.J., 1982. Interaction of the Cretaceous/Tertiary extinction bolide with the atmosphere. In: Geological Society of America, Special Paper 190, pp. 103–120.

Oldow, J.S., Bally, A.W., Ave Lallemant, H.G., Leeman, W.P., 1989. Phanerozoic evolution of the North American Cordillera; United States and Canada. In: Bally, A.W., Palmer, A.R. (Eds.), The Geology of North America – An Overview. The Geology of North America, vol. A. Geological Society of America, Boulder, CO, pp. 139–232.

Olivarez, A.M., Owen, R.M., 1991. The europium anomaly of seawater: implications for fluvial versus hydrothermal REE inputs to the oceans. Chemical Geology 92, 317–328.

Oliver, R.L., James, P.R., Collerson, K.D., Ryan, A.B., 1982. Precambrian geologic relationships in the Vestfold Hills, Antarctica. In: Craddock, C. (Ed.), Antarctic Geoscience. University of Wisconsin Press, Madison, pp. 435–444.

O'Neil, J., Cloquet, C., Stevenson, R.K., Francis, D., 2006. Fe isotope systematics of banded iron formation from the 3.8 Ga Nuvvuagittuq greenstone belt, Northern Superior Pronvince, Canada. Geological Association of Canada, Abstracts, vol. 31.

O'Neil, J.R., 1986. Theoretical and experimental aspects of isotope fractionation. In: Valley, J.W., Taylor, H.P., O'Neil, J.R. (Eds.), Stable Isotope in High Temperature Geological Processes. Reviews in Mineralogy, vol. 16, pp. 1–40.

O'Neill, C.J., Lenardic, A., O'Reilly, S.Y., Griffin, W.L., 2007. Dynamics of cratons in an evolving mantle. Lithos, in press.

O'Neill, J.M., Lopez, D.A., 1985. Character and regional significance of Great Falls Tectonic Zone, east-central Idaho and west-central Montana. American Association of Petroleum Geologists Bulletin 69, 437–447.

O'Nions, R.K., Evenson, N.M., Hamilton, P.J., 1979. Geochemical modelling of mantle differentiation and crustal growth. Journal of Geophysical Research 84, 6091–6101.

O'Nions, R.K., Hamilton, P.J., Hooker, P.J., 1983. A Nd isotopic investigation of sediments related to crustal development in the British Isles. Earth and Planetary Science Letters 63, 229–240.

Ono, S., Eigenbrode, J.L., Pavlov, A.A., Kharecha, P., Rumble, D., Kasting, J.F., Freeman, K.H., 2003. New insights into Archean sulfur cycle from mass-independent sulfur isotope records from the Hamersley Basin, Australia. Earth and Planetary Science Letters 213, 15–30.

Ono, S., Wing, B.A., Johnston, D., Farquhar, J., Rumble, D., 2006. Mass-dependent fractionation of quadruple sulfur isotope system as a new tracer of sulfur biogeochemical cycles. Geochimica et Cosmochimica Acta 70, 2238–2252.

Ono, S., Shanks, W.C., III, Rouxel, O.J., Rumble, D., 2007. S-33 constraints on the seawater sulfate contribution in modern seafloor hydrothermal vent sulfides. Geochimica et Cosmochimica Acta 71, 1170–1182.

Ontario Geological Survey, 1992. Chart A – Archean tectonic assemblages, plutonic suites, and events in Ontario. Ontario Geological Survey, Map 2579.

O'Reilly, S.Y., Griffin, W.L., 1985. A xenolith-derived geotherm for southeastern Australia and its geophysical implications. Tectonophysics 111, 41–63.

O'Reilly, S.Y., Griffin, W.L., 1996. 4-D lithosphere mapping: methodology and examples. Tectonophysics 262, 3–18.

O'Reilly, S.Y., Griffin, W.L., 2006. Imaging chemical and thermal heterogeneity in the subcontinental lithospheric mantle with garnets and xenoliths: Geophysical implications. Tectonophysics 416, 289–309.

O'Reilly, S.Y., Griffin, W.L., Poudjom-Djomani, Y.H., Morgan, P., 2001. Are lithospheres forever? Tracking changes in subcontinental lithopsheric mantle through time. GSA Today 11, 4–10.

Orgel, L., 1998. The origin of life – a review of facts and speculations. Trends in Biochemical Sciences 23, 491–495.

Oslund, R., 1997. Modelling of variations in Norwegian olivine deposits, causes of variation and estimation of key quality factors. Doktor Ingeniør thesis. Norwegian University of Science and Technology, Trondheim, 189 pp.

Ota, T., Gladkochub, D.P., Sklyarov, E.V., Sklyarov, E.V., Mazukabzov, A.M., Watanabe, T., 2004. P-T history of garnet-websterites in the Sharyzhalgay complex, southwestern margin of the Siberian craton: evidence for Paleoproterozoic high-pressure metamorphism. Precambrian Research 132 (4), 327–348.

Otto, A., Dziggel, A., Kisters, A.F.M., Meyer, F.M., 2005. Polyphase gold mineralization in the structurally condensed granitoid-greenstone contact at the New Consort Gold Mine. In: Geological Society of America Abstracts with Programs, Barberton Greenstone Belt, South Africa.

Otto, A., Dziggel, A., Meyer, F.M., 2006. Geochemical signatures of gold mineralization at the New Consort gold mine, South Africa. Beihefte zum European Journal of Mineralogy 18, p. 99.

Otto, A., Dziggel, A., Kisters, A.F.M., Meyer, F.M., 2007. The New Consort gold mine, South Africa: Orogenic gold mineralization in a condensed metamorphic profile. Mineralium Deposita, doi: 10.1007/s00126-007-0135-5.

Oversby, V.M., 1975. Lead isotope systematics and ages of the Archaean acid intrusives in the Kalgoorlie–Norseman area, Western Australia. Geochimica et Cosmochimica Acta 39, 1107–1125.

Pace, N.R., 1997. A molecular view of microbial diversity and the biosphere. Science 276, 734–740.

Pace, N.R., 2001. The universal nature of biochemistry. Proceedings of the National Academy of Sciences of the United States of America 98, 805–808.

Page, F.Z., DeAngelis, M.T., Fu, B., Kita, N.T., Lancaster, P.J., Valley, J.W., 2006. Slow oxygen diffusion in zircon. Geochimica et Cosmochimica Acta 70 (18), Suppl. 1, A467.

Pahlevan, K., Stevenson, D.J., 2005. The oxygen isotope similarity between the Earth and Moon – source region or formation process? Lunar and Planetary Science XXXVI, abstract 2382.

Palme, H., Jones, A., 2003. Solar System abundances of the elements. In: Davis, A.M. (Ed.), Meteorites, Comets and Planets. Treatise on Geochemistry, vol. 1. Elsevier–Pergamon, Oxford, pp. 41–61.

Palmer, M.R., Edmond, J.M., 1989. Strontium isotope budget of the modern ocean. Earth and Planetary Science Letters 92, 11–26.

Papineau, D., Mojzsis, S.J., 2006. Mass-independent fractionation of sulfur isotopes in sulfides from the pre-3770 Ma Isua Supracrustal Belt, West Greenland. Geobiology 4, 227–238.

Papineau, D., Mojzsis, S.J., Coath, C.D., Karhu, J.A., McKeegan, K.D., 2005. Multiple sulfur isotopes of sulfides from sediments in the aftermath of Paleoproterozoic glaciations. Geochimica et Cosmochimica Acta 69, 5033–5060.

Papineau, D., Mojzsis, S.J., Schmitt, A.K., 2007. Multiple sulfur isotopes from Paleopro-terozoic Huronian interglacial sediments and the rise of atmospheric oxygen. Earth and Planetary Science Letters 255, 188–212.

Paris, I.A., 1985. The geology of the farms Josefsdal, Dunbar and part of Diepgezet in the Barberton greenstone belt. Ph.D. thesis. University of the Witwatersrand, Johannesburg, 239 pp.

Paris, I., Stanistreet, I.G., Hughes, M.J., 1985. Cherts of the Barberton greenstone belt interpreted as products of submarine exhalative activity. Journal of Geology 93, 111–129.

Park, J.K., Buchan, K.L., Harlan, S.S., 1995. A proposed giant radiating dyke swarm frag-mented by the separation of Laurentia and Australia based on palaeomagnetism of ca. 780 Ma mafic intrusions in western North America. Earth and Planetary Science Letters 132, 129–139.

Park, R.G., 1997. Early Precambrian plate tectonics. South African Journal of Geology 100, 23–35.

Parks, J., Lin, S., Davis, D.W., Corkery, M.T., 2006. Geochronological constrains on the history of the Island Lake greenstone belt and the relationship of terranes in the north-western Superior Province. Canadian Journal of Earth Sciences 43, 789–803.

Parman, S.W., Grove, T.L., 2004. Petrology and geochemistry of Barberton Komatiites and basaltic komatiites: Evidence of Archean fore-arc magmatism. In: Kusky, T.M. (Ed.), Precambrian Ophiolites and Related Rocks. Developments in Precambrian Geology, vol. 13. Elsevier, Amsterdam, pp. 539–565.

Parman, S., Dann, J.C., Grove, T.L., de Wit, M.J., 1997. Emplacement conditions of ko-matiite magmas from the 3.49 Ga Komati Formation, Barberton Greenstone Belt, South Africa. Earth and Planetary Science Letters 150, 303–323.

Parman, S.W., Grove, T.L., Dann, J.C., 2001. The production of Barberton komatiites in an Archean subduction zone. Geophysical Letter 28, 303–323.

Parman, S.W., Grove, T.L., Dann, J.C., de Wit, M.J., 2004. A Subduction origin for komati-ites and cratonic lithospheric mantle. South African Journal of Geology 107, 107–118.

Pasteris, J.D., Wopencka, B., 2002. Images of Earth's earliest fossils? Nature 420, 476–477.

Pasteris, J.D., Wopencka, B., 2003. Necessary, but not sufficient: Raman identification of disordered carbon as a signature of ancient life. Astrobiology 3, 727–738.

Patchett, P.J., 1983. Importance of the Lu-Hf isotopic system in studies of planetary chronology and chemical evolution. Geochimica et Cosmochimica Acta 47, 81–91.

Patchett, P.J., Kouvo, O., Hedge, C.E., Tatsumoto, M., 1981. Evolution of continental crust and mantle heterogeneity: evidence from Hf isotopes. Contributions to Mineralogy and Petrology 78, 279–297.

Patiño-Douce, A.E., 2005. Vapor-absent melting of tonalite at 15–32 kbar. Journal of Petrology 46 (2), 275–290.

Patiño-Douce, A.E., Beard, J.S., 1995. Dehydration-melting of Bt gneiss and Qtz amphi-bolite from 3 to 15 kB. Journal of Petrology 36, 707–738.

Patterson, C., 1956. Age of meteorites and the Earth. Geochimica et Cosmochimica Acta 10, 230–237.

Pavlov, A.A., Kasting, J.F., 2002. Mass-independent fractionation of sulfur isotopes in Archean sediments: Strong evidence for an anoxic Archean atmosphere. Astrobiology 2, 27–41.

Pavlov, A.A., Pavlov, A.K., Kasting, J.F., 1999. Irradiated interplanetary dust particles as a possible solution for the deuterium/hydrogen paradox of Earth's oceans. Journal of Geophysical Research – Planets 104 (E12), 30,725–30,728.

Pavlov, A.A., Kasting, J.F., Eigenbrode, J.L., Freeman, K.H., 2001a. Organic haze in Earth's early atmosphere: Source of low-^{13}C in Late Archean kerogens? Geology 29, 1003–1006.

Pavlov, A.A., Kasting, J.F., Brown, L.L., 2001b. UV-shielding of NH_3 and O_2 by organic hazes in the Archean atmosphere. Journal of Geophysical Research 106, 23,267–23,287.

Pavoni, N., 1997. Geotectonic bipolarity – evidence of bicellular convection in the Earth's mantle. South African Journal of Geology 100, 291–299.

Pawley, M.J., Collins, W.J., 2002. The development of contrasting structures during the cooling and crystallisation of a syn-kinematic pluton. Journal of Structural Geology 24, 469–483.

Pawley, M.J., Van Kranendonk, M.J., Collins, W.J., 2004. Interplay between magmatism and deformation during the evolution of the domical Archaean Shaw Granitoid Complex, Pilbara Craton, Western Australia. Precambrian Research 131, 213–230.

Pearce, J.A., 1983. The role of sub-continental lithosphere in magma genesis at destructive plate margins. In: Hawkesworth, C.J., Norry, M.J. (Eds.), Continental Basalts and Mantle Xenoliths. Shiva, Nantwich, pp. 230–249.

Pearce, J.A., Parkinson, I.A., 1993. Trace elements models for mantle melting: application to volcanic arc petrogenesis. In: Prichard, H., Alabaster, T., Harris, N.B.W., Neary, C. (Eds.), Magmatic Processes and Plate Tectonics. Geological Society of London, Special Publication 76, pp. 373–403.

Pearce, J.A., Harris, N.B.W., Tindle, A.G., 1984. Trace element discrimination diagrams for the tectonic interpretation of granitic rocks. Journal of Petrology 25, 956–983.

Pearce, J.A., Stern, R.J., Bloomer, S.H., Fryer, P., 2005. Geochemical mapping of the Mariana arc-basin system: Implications for the nature and distribution of subduction components. Geochemistry, Geophysics, Geosystems 6 (7), doi: 10.1029/2004GC000895.

Pearson, D.G., 1999. The age of continental roots. Lithos 48, 171–194.

Pearson, D.G., Canil, D., Shirey, S.B., 2004. Mantle samples included in volcanic rocks: xenoliths and diamonds. In: Holland, H.D., Turekian, K.K. (Eds.), Treatise on Geochemistry, vol. 2. Elsevier, Amsterdam, pp. 171–275.

Pearson, N.J., Alard, O., Griffin, W.L., Jackson, S.E., O'Reilly, S.Y., 2002. In situ measurement of Re-Os isotopes in mantle sulfides by Laser Ablation Multi-Collector Inductively-Coupled Mass Spectrometry: analytical methods and preliminary results. Geochimica et Cosmochimica Acta 66, 1037–1050.

Peccerillo, A., Taylor, S.R., 1976. Geochemistry of Eocene calc-alkaline volcanic rocks from the Kastamonu area, Northern Turkey. Contribution to Mineralogy and Petrology 58, 63–81.

Peck, W.H., Valley, J.W., Wilde, S.A., Graham, C.M., 2001. Oxygen isotope ratios and rare earth elements in 3.3 to 4.4 Ga zircons: Ion microprobe evidence for high $\delta^{18}O$ continental crust and oceans in the Early Archean. Geochimica et Cosmochimica Acta 65, 4215–4229.

Peck, W.H., Valley, J.W., Graham, C.M., 2003. Slow oxygen diffusion rates in igneous zircons from metamorphic rocks. American Mineralogist 88, 1003–1014.

Pellas, P., Fieni, C., Trieloff, M., Jessberger, E.K., 1997. The cooling history of the Acapulco meteorite as recorded by the ^{244}Pu and ^{40}Ar-^{39}Ar chronometers. Geochimica et Cosmochimica Acta 61, 3477–3501.

Peltonen, P., Kontinen, A., 2004. The Jormua Ophiolite: A mafic-ultramafic complex from an ancient ocean-continent transition zone. In: Kusky, T.M. (Ed.), Precambrian Ophiolites and Related Rocks. Developments in Precambrian Geology, vol. 13, pp. 35–72.

Peng, P., Zhai, M.G., Zhang, H.F., Guo, J.H., 2005. Geochronological constraints on the Palaeoproterozoic evolution of the NCC: SHRIMP zircon ages of different types of mafic dikes. International Geological Review 47, 492–508.

Percival, J.A., 1994. Archean high-grade metamorphism. In: Condie, K.C. (Ed.), Archean Crustal Evolution. Developments in Precambrian Geology, vol. 11. Elsevier, Amsterdam, pp. 315–355.

Percival, J.A., 2007. Geology and metallogeny of the Superior Province. In: Goodfellow, W.D. (Ed.), Mineral Resources of Canada: A Synthesis of Major Deposit-Types, District Metallogeny, the Evolution of Geological Provinces, and Exploration Methods. Geological Association of Canada, Special Paper (in press).

Percival, J.A., Card, K.D., 1992. Geology of the Vizien greenstone belt. Geological Survey of Canada, Open file 2495, scale 1:50,000.

Percival, J., Card, K.D., 1994. Geology, Lac Minto – Rivière aux Feuilles. Geological Survey of Canada, Map 1854A (1:500,000).

Percival, J.A., Helmstaedt, H., 2004. Insights on Archean continent-ocean assembly, western Superior Province, from new structural, geochemical, and geochronological observations: introduction and summary. Precambrian Research 132, 209–212.

Percival, J.A., Mortensen, J.K., 2002. Water-deficient calc-alkaline plutonic rocks of Northeastern Superior Province, Canada: significance of charnockitic magmatism. Journal of Petrology 43, 1617–1650.

Percival, J.A., Skulski, T., 2000. Tectonothermal evolution of the northern Minto Block, Superior Province, Québec, Canada. Canadian Mineralogist 38, 345–378.

Percival, J.A., Whalen, J.B., 2000. Observations on the North Caribou terrane – Uchi subprovince interface in western Ontario and eastern Manitoba. In: Geological Survey of Canada, Current Research 2000-C15, 8 pp.

Percival, J.A., Green, A.G., Milkereit, B., Cook, F.A., Geis, W., West, G.F., 1989. Seismic reflection profiles across deep continental crust exposed in the Kapuskasing uplift structure. Nature 342, 416–419.

Percival, J.A., Mortensen, J.K., Stern, R.A., Card, K.D., Bégin, N.J., 1992. Giant granulite terranes of northeastern Superior Province: the Ashuanipi complex and Minto block. Canadian Journal of Earth Sciences 29, 2287–2308.

Percival, J.A., Stern, R.A., Mortensen, J.K., Card, K.D., Bégin, N.J., 1994. Minto block, Superior Province: missing link in deciphering tectonic assembly of the craton at 2.7 Ga. Geology 22, 839–842.

Percival, J.A., Skulski, T., Card, K.D., 1995. Geology, Rivière Kogaluc – Lac Qalluviartuuq region (parts of 34J and 34O). Geological Survey of Canada, Open file 3112, scale 1:250,000.

Percival, J., Skulski, T., Nadeau, L., 1996. Geology, Lac Couture, Québec. Geological Survey of Canada, Open file 3315.

Percival, J., Skulski, T., Nadeau, L., 1997a. Reconnaissance geology of the Pelican–Nantais belt, northeastern Superior province, Québec. Geological Survey of Canada, Open file 3525.

Percival, J., Skulski, T., Nadeau, L., 1997b. Granite-greenstone terranes of the northern Minto Block, northeastern Québec: Pélican–Nantais, Faribault–Leridon and Duquet belts. In: Geological Survey of Canada, Current Research, 1997-C, pp. 211–221.

Percival, J.A., Stern, R.A., Skulski, T., 2001. Crustal growth through successive arc magmatism: Reconnaissance U-Pb SHRIMP data from the northeastern Superior Province, Canada. Precambrian Research 109, 203–238.

Percival, J.A., Bailes, A.H., McNicoll, V., 2002. Mesoarchean breakup, Neoarchean accretion in the western Superior craton, Lake Winnipeg Canada. Geological Association of Canada Field Trip B3 Guidebook, 42 pp.

Percival, J.A., Bleeker, W., Cook, F.A., Rivers, T., Ross, G., van Staal, C., 2004a. PanLITHOPROBE workshop IV: Intraorogen correlations and comparative orogenic anatomy. Geoscience Canada 31, 23–39.

Percival, J.A., McNicoll, V., Brown, J.L., Whalen, J.B., 2004b. Convergent margin tectonics, central Wabigoon subprovince, Superior Province, Canada. Precambrian Research 132, 213–244.

Percival, J.A., McNicoll, V., Bailes, A.H., 2006a. Strike-slip juxtaposition of ca. 2.72 Ga juvenile arc and >2.98 Ga continent margin sequences and its implications for Archean terrane accretion, western Superior Province, Canada. Canadian Journal of Earth Sciences 43, 895–927.

Percival, J.A., Sanborn-Barrie, M., Skulski, T., Stott, G.M., Helmstaedt, H., White, D.J., 2006b. Tectonic evolution of the western Superior Province from NATMAP and Lithoprobe studies. Canadian Journal of Earth Sciences 43, 1085–1117.

Peredery, W.V., Inco Geological Staff, 1982. Geology and nickel sulphide deposits of the Thompson Belt, Manitoba. In: Hutchinson, R.W., Spence, C.D., Franklin, J.M. (Eds.), Precambrian Sulphide Deposits. Geological Association of Canada, Special Paper 25, pp. 165–209.

Perfit, M.R., Davidson, J.P., 2000. Plate tectonics and volcanism. In: Sigurdson, H., et al. (Eds.), Encyclopedia of Volcanoes. Academic Press, San Diego, CA, pp. 89–113.

Pering, C.S., Barnes, S.J., Hill, R.E.T., 1995. The physical volcanology of Archean komatiite sequences from Forrestania, Southern Cross Province, Western Australia. Lithos 34, 189–207.

Perry, E.C., 1967. The oxygen isotope chemistry of ancient cherts. Earth and Planetary Science Letters 3, 62–66.

Perry, E.C., Lefticariu, L., 2003. Formation and geochemistry of Precambrian cherts. In: Mackenzie, F.T. (Ed.), Sediments, Diagenesis, and Sedimentary Rocks. Treatise on Geochemistry, vol. 7. Elsevier, Amsterdam, pp. 99–113.

Perry, E.C., Ahmad, S.N., Swulius, T.M., 1978. The oxygen isotope composition of 3800 m.y. old metamorphosed chert and iron formation from Isukasia, Greenland. Journal of Geology 86, 223–239.

Peterman, Z.E., 1981. Dating of Archean basement in northeastern Wyoming and southern Montana. Bulletin of the Geological Society of America 92 (1), 139–146.

Peterman, Z.E., Hildreth, R.A., 1978. Reconnaissance geology and geochronology of the Precambrian of the Granite Mountains, Wyoming. U.S. Geological Survey, Professional Paper 1055, 22 pp.

Peterman, Z.E., Zartman, R.E., Sims, P.K., 1980. Tonalitic gneiss of early Archean age from northern Michigan. In: Geological Society of America, Special Paper 182, pp. 125–134.

Peterman, Z.E., Zartman, R.E., Sims, P.K., 1986. A protracted Archean history in the Watersmeet gneiss dome. In: U.S. Geological Survey, Bulletin 1655, pp. 51–64.

Petford, N., Atherton, M.P., 1996. Na-rich partial melts from newly underplated basaltic crust: the Cordillera Blanca batholith, Peru. Journal of Petrology 37 (6), 1491–1521.

Petrov, A.F., Gusev, G.S., Tretyakov, F.F., Oksman, V.S., 1985. The Archean (Aldanian) and early Proterosoic (Karelian) megacomplexes. In: Kovalsky, V.V. (Ed.), Structure and Evolution of the Earth's Crust in Yakutia. Nauka, Moscow, pp. 9–39 (in Russian).

Petrova, Z.I., Levitsky, V.I., 1984. Petrology and Geochemistry of Cisbaikal Granulite Complexes. Nauka, Moscow, 201 pp. (in Russian).

Pettengill, G.H., Ford, P.G., Wilt, R.J., 1992. Venus surface radiothermal emission as observed by Magellan. Journal of Geophysical Research 97, 13,091–13,102.

Peucat, J.J., Capdevila, R., Drareni, A., Choukroune, P., Fanning, C.M., Bernard-Griffiths, J., Fourcade, S., 1996. Major and trace element geochemistry and isotope (Sr, Nd, Pb, O) systematics of an Archaean basement involved in a 2.0 Ga, very high-temperature (1000 °C) metamorphic event in Ouzzal Massif, Hoggar, Algeria. Journal of Metamorphic Geology 14 (6), 667–692.

Peucat, J.J., Ménot, R.P., Monnier, O., Fanning, C.M., 1999. The Terre Adélie basement in the East Antarctica Shield: geological and isotopic evidence for a major 1.7 Ga thermal event; comparison with the Gawler Craton in South Australia. Precambrian Research 94, 205–224.

Peucat, J.J., Drareni, A., Latouche, L., Deloule, E., Vidal, P., 2003. U-Pb zircon (TIMS and SIMS) and Sm-Nd whole-rock geochronology of the Gour Oumelalen granulitic basement, Hoggar massif, Tuareg shield, Algeria. Journal of African Earth Science 37, 229–239.

Peucker-Ehrenbrink, B., Miller, M.W., 2006. Marine ^{87}Sr/^{86}Sr record mirrors the evolving upper continental crust. Geochimica et Cosmochimica Acta 70 (18), A487.

Pflug, H.D., 1965. Structured organic remains from the Fig Tree Series of the Barberton Mountain Land. Economic Geology Research Unit, Information Circular 32, University of the Witwatersrand, Johannesburg, 36 pp.

Pflug, H.D., 2001. Earliest organic evolution. Essay to the memory of Bartholomew Nagy. Precambrian Research 106, 79–91.

Pflug, H.D., Jaeschke-Boyer, H., 1979. Combined structural and chemical analysis of 3800-Myr-old microfossils. Nature 280, 483–486.

Phillips, R.J., 1993. The age spectrum of the Venusian surface. Eos 74 (16) (Supplement), 187.

Phillips, R.J., Hansen, V.L., 1994. Tectonic and magmatic evolution of Venus. Annual Reviews of Earth and Planetary Sciences 22, 597–654.

Phillips, R.J., Hansen, V.L., 1998. Geological evolution of Venus: Rises, plains, plumes and plateaus. Science 279, 1492–1497.

Phillips, R.J., Izenberg, N.R., 1995. Ejecta correlations with spatial crater density and Venus resurfacing history. Geophysical Research Letters 22, 1517–1520.

Phillips, R.J., Kaula, W.M., McGill, G.E., Malin, M.C., 1981. Tectonics and evolution of Venus. Science 212, 879–887.

Phillips, R.J., Grimm, R.E., Malin, M.C., 1991. Hot-spot evolution and the global tectonics of Venus. Science 252, 651–658.

Phillips, R.J., Raubertas, R.F., Arvidson, R.E., Sarkar, I.C., Herrick, R.R., Izenberg, N., Grimm, R.E., 1992. Impact crater distribution and the resurfacing history of Venus. Journal of Geophysical Research 97, 15,923–15,948.

Phillips, R.J., Johnson, C.J., Mackwell, S.L., Morgan, P., Sandwell, D.T., Zuber, M.T., 1997. Lithospheric mechanics and dynamics of Venus. In: Bouger, S.W., Hunten, D.M., Phillips, R.J. (Eds.), Venus II. University of Arizona Press, Tucson, AZ, pp. 1163–1204.

Phillips, R.J., Bullock, M.A., Hauck, S.A., II, 2001. Climate and interior coupled evolution on Venus. Geophysical Research Letters 28, 1779–1782.

Pidgeon, R.T., 1978. 3450 M.y. old volcanics in the Archaean layered greenstone succession of the Pilbara Block, Western Australia. Earth and Planetary Science Letters 37, 423–428.

Pidgeon, R.T., 1992. Recrystallisation of oscillatory zoned zircon: some geochronological and petrological implications. Contributions to Mineralogy and Petrology 110, 463–472.

Pidgeon, R.T., Hallberg, J.A., 2000. Age relationships in supracrustal sequences in the northern part of the Murchison Terrane, Archaean Yilgarn Craton, Western Australia: a combined field and zircon U-Pb study. Australian Journal of Earth Sciences 47, 153–165.

Pidgeon, R.T., Nemchin, A.A., 2006. High abundance of early Archaean grains and the age distribution of detrital zircons in a sillimanite-bearing quartzite from Mt Narryer, Western Australia. Precambrian Research 150, 201–220.

Pidgeon, R.T., Wilde, S.A., 1990. The distribution of 3.0 Ga and 2.7 Ga volcanic episodes in the Yilgarn Craton of Western Australia. Precambrian Research 48, 309–325.

Pidgeon, R.T., Wilde, S.A., 1998. The interpretation of complex zircon U-Pb systems in Archaean granitoids and gneisses from the Jack Hills, Narryer Gneiss Terrane, Western Australia. Precambrian Research 91, 309–332.

Pidgeon, R.T., Compston, W., Wilde, S.A., Baxter, J.L., Collins, L.B., 1986. Archaean evolution of the Jack Hills Metsedimentary Belt, Yilgarn Block, Western Australia. Terra Cognita 6, 146.

Pidgeon, R.T., Wilde, S.A., Compston, W., 1990. U-Pb ages of zircons from conglomerate clasts in the Jack Hills Metasedimentary Belt, Yilgarn Craton, Western Australia. In: 7th International ICOG Conference, Abstracts, p. 78.

Pirajno, F., 2000. Ore Deposits and Mantle Plumes. Kluwer Academic Publishers, 507 pp.

Pirajno, F., Van Kranendonk, M.J., 2005. A review of hydrothermal processes and systems on earth and implications for Martian analogues. Australian Journal of Earth Sciences 52, 329–351.

Pirajno, F., Chen, Y.-J., Qi, J.-P., Long, X., 2006. Mesozoic-Cenozoic intraplate tectonics and magmatism in E and NE China. In: IAVCEI 2006. International Conference on Continental Volcanism, Abstracts and Programs, Guangzhou, China, p. 184.

Pirajno, F., Mao, J.-W., Zhang, Z.-C., Zhang, Z.-H., Yang, F.-Q., Chai, F.-M., in press. The association of anorogenic mafic-ultramafic intrusions and A-type magmatism in the Tian Shan and Altay orogens, NW China: implications for geodynamic evolution and potential for the discovery of new ore deposits. Asian Journal of Earth Sciences, Special Issue.

Plank, T., 2005. Constraints from thorium/lanthanum on sediment recycling at subduction zones and the evolution of continents. Journal of Petrology 46, 921–944.

Plant, J.A., Kinniburgh, D.G., Smedley, F.M., Fordyce, F.M., Klinck, B.A., 2003. Arsenic and selenium. In: Treatise on Geochemistry, vol. 9. Elsevier, Amsterdam, pp. 17–66.

Platt, J.P., England, P.C., 1993. Convective removal of lithosphere beneath mountain belts: thermal and mechanical consequences. American Journal of Science 293, 307–336.

Plaut, J.J., 1993. Stereo imaging. In: Ford, J.P., et al. (Eds.), Guide to Magellan Image Interpretation. NASA-JPL Publication 93-24, pp. 33–41.

Playford, P.E., Cockbain, A.E., 1976. Modern algal stromatolites at Hamelin Pool, a hypersaline barred basin in Shark Bay, Western Australia. In: Walter, M.R. (Ed.), Stromatolites. Developments in Sedimentology, vol. 20. Elsevier, Amsterdam, pp. 389–411.

Plumb, K.A., 1991. New Precambrian timescale. Episodes 14, 139–140.

Plumb, K.A., James, H.L., 1986. Subdivision of Precambrian time: Recommendations and suggestions by the commission on Precambrian stratigraphy. Precambrian Research 32, 65–92.

Plyusnina, L.P., 1982. Geothermometry and geobarometry of plagioclase-hornblende bearing assemblages. Contributions to Mineralogy and Petrology 80, 140–146.

Podmore, D.C., 1990. Shay Gap-Sunrise Hill and Nimingarra iron ore deposits. In: Australasian Institute of Mining and Metallurgy, Monograph 14, pp. 137–140.

Podosek, F.A., Cassen, P., 1994. Theoretical, observational, and isotopic estimates of the lifetime of the solar nebula. Meteoritics 29, 6–25.

Poitrasson, F., Halliday, A.N., Lee, D.C., Levasseur, S., Teutsch, N., 2004. Iron isotope differences between Earth, Moon, Mars and Vesta as possible records of contrasted accretion mechanisms. Earth and Planetary Science Letters 223, 253–266.

Polat, A., Frei, R., 2005. The origin of early Archean banded iron formations and of continental crust, Isua, southern West Greenland. Precambrian Research 138, 151–175.

Polat, A., Hofmann, A.W., 2003. Alteration and geochemical patterns in the 3.7–3.8 Ga Isua greenstone belt, West Greenland. Precambrian Research 126, 197–218.

Polat, A., Kerrich, R., 2004. Precambrian arc associations: Boninites, adakites, magnesian andesites, and Nb-enriched basalts. In: Kusky, T.M. (Ed.), Precambrian Ophiolites and Related Rocks. Developments in Precambrian Geology, vol. 13. Elsevier, Amsterdam, pp. 567–598.

Polat, A., Kerrich, R., Wyman, D.A., 1998. The late Archean Schreiber–Hemlo and White River – Dayohessarah greenstone belts, Superior Province: collages of oceanic plateaus and subduction-accretion complexes. Tectonophysics 289, 295–326.

Polat, A., Hofmann, A.W., Rosing, M.T., 2002. Boninite-like volcanic rocks in the 3.7–3.8 Ga Isua greenstone belt, West Greenland: geochemical evidence for intra-oceanic subduction zone processes in the early Earth. Chemical Geology 184, 231–254.

Polat, A., Li, J.H., Fryer, B., Kusky, T., Gagnon, J., Zhang, S., 2006. Geochemical characteristics of the Neoarchean (2800–2700 Ma) Taishan greenstone belt, North China Craton: Evidence for plume–craton interaction. Chemical Geology 230, 60–87.

Poller, U., Gladkochub, D., Donskaya, T., Mazukabzov, A., Sklyarov, E., Todt, W., 2004. Timing of Early Proterozoic magmatism along the southern margin of the Siberian Craton (Kitoy area). Transactions Royal Society Edinburgh: Earth Sciences 136 (95), 353–368.

Poller, U., Gladkochub, D., Donskaya, T., Mazukabzov, A., Sklyarov, E., Todt, W., 2005. Multistage magmatic and metamorphic evolution in the Southern Siberian craton: Archean and Paleoproterozoic zircon ages revealed by SHRIMP and TIMS. Precambrian Research 136, 353–368.

Pope, M.C., Grotzinger, J.P., 2000. Controls on fabric development and morphology of tufa and stromatolites, uppermost Pethei Group (1.8 Ga), Great Slave Lake, northwest Canada. In: Grotzinger, J., James, N. (Eds.), SEPM Special Publication – Carbonate Sedimentation and Diagenesis in the Evolving Precambrian World, pp. 103–122.

Popov, N.V., Smelov, A.P., Dobretsov, H.N., Bogomolova, L., Kartavchenko, V.G., 1990. Olondo Greenstone Belt. Academy of Sciences Publishers, Yakutsk, 170 pp. (in Russian).

Popov, N.V., Zedgenizov, A.N., Berezkin, V.I., 1989. Petrochemistry of the Archean Metavolcanics in the Sunnagin Block of the Aldan Massif. Nauka, Novosibirsk, 78 pp. (in Russian).

Post, N.J., Hensen, B.J., Kinny, P.D., 1997. Two metamorphic episodes during a 1340–1180 Ma convergent tectonic event in the Windmill Islands, East Antarctica. In: Ricci,

A.C. (Ed.), The Antarctic Region: Geological Evolution and Processes. Terra Antartica Publication, Siena, pp. 157–161.

Postelnikov, E.S., Museibov, N.I., 1992. Structure of the Baikal orogeny basement. Geotectonics 6, 37–51.

Potrel, A., Peucat, J.J., Fanning, C.M., Auvray, B., Burg, J.P., Caruba, C., 1996. 3.5 Ga old terranes in the East African craton, Mauritania. Journal of the Geological Society London 153, 507–510.

Potter, J., Rankin, A.H., Treloar, P.J., 2004. Abiogenic Fischer–Tropsch synthesis of hydrocarbons in alkaline igneous rocks; fluid inclusion, textural and isotopic evidence from the Lovozero complex, N.W. Russia. Lithos 75, 311–330.

Poudjom Djomani, Y.H., O'Reilly, S.Y., Griffin, W.L., Morgan, P., 2001. The density structure of subcontinental lithosphere: Constraints on delamination models. Earth and Planetary Science Letters 184, 605–621.

Poujol, M., Anhaeusser, C.R., 2001. The Johannesburg Dome, South Africa: new single zircon U-Pb isotopic evidence for early Archaean granite-greenstone development within the Central Kaapvaal Craton. Precambrian Research 108, 139–157.

Poujol, M., Robb, L.J., 1999. New U-Pb zircon ages on gneisses and pegmatite from South of the Murchison greenstone belt, South Africa. South African Journal of Geology 102 (2), 93–97.

Poujol, M., Robb, L.J., Respaut, J.P., Anhaeusser, C.R., 1996. 3.07–2.97 Ga greenstone belt formation in the northeastern Kaapvaal Craton: Implications for the origin of the Witwatersrand Basin. Economic Geology 91 (8), 1455–1461.

Poujol, M., Robb, L.J., Anhaeusser, C.R., Gericke, B., 2003. A review of the geochronological constraints on the evolution of the Kaapvaal Craton, South Africa. Precambrian Research 127, 181–213.

Poulsen, K.H., Borraidaile, G.H., Kehlenbeck, M.M., 1980. An inverted Archean succession at Rainy Lake, Ontario. Canadian Journal of Earth Sciences 17, 1358–1369.

Powell, R., Will, T.M., Phillips, G.N., 1991. Metamorphism in Archean greenstone belts: Calculated fluid compositions and implications for gold mineralization. Journal of Metamorphic Geology 9, 141–150.

Pravdivtseva, O.A., Hohenberg, C.M., 1999. Observations of mineral specific I-Xe ages in ordinary chondrites. Lunar and Planetary Science XXX, 2047.

Premo, W.R., Tatsumoto, M., Misawa, K., Nakamura, N., Kita, N.I., 1999. Pb-isotopic systematics of lunar highland rocks (>3.9 Ga): Constraints on early lunar evolution. In: Snyder, G.A., Neal, C.R., Ernst, W.G. (Eds.), Planetary Petrology and Geochemistry. GSA, Bellwether Publishing Ltd.

Price, M., Suppe, J., 1994. Mean age of rifting and volcanism on Venus deduced from impact crater densities. Nature 372, 756–759.

Price, M.H., Watson, G., Brankman, C., 1996. Dating volcanism and rifting on Venus using impact crater densities. Journal of Geophysical Research 101, 4637–4671.

Price, N.J., 2001. Major Impacts and Plate Tectonics: A Model for the Phanerozoic Evolution of the Earth's Lithosphere. Routledge, London, 354 pp.

Price, R.E., Pichler, T., 2006. Abundance and mineralogical association of arsenic in the Suwannee Limestone (Florida): Implications for arsenic release during water-rock interaction. Chemical Geology 228, 44–56.

Prinz, M., Nehru, C.E., Delaney, J.S., Weisberg, M., 1983. Silicates in IAB and IIICD irons, winonaites, lodranites and Brachina: a primitive and modified-primitive group. Lunar and Planetary Science XIV, 616–617.

Prinzhofer, A., Papanastassiou, D.A., Wasserberg, G.J., 1992. Samarium-neodymium evolution of meteorites. Geochimica et Cosmochimica Acta 56, 797–815.

Puchtel, I.S., 1992. Mafic-ultramafic rocks and crust-mantle evolution in the Early Precambrian: the Olekma gneiss-greenstone area as an example. Publ. Inst. Petrography, Mineralogy and Ore Deposits of Russia, Russian Academy of Sciences, Moscow, 20 pp. (in Russian).

Puchtel, I.S., Zhuravlev, D.Z., 1989. Petrology and geochemistry of early and later Archean komatiites from Olekma granite-greenstone terrane. In: 28th I.G.C., Abstracts, vol. 2. Washington, DC, pp. 643–644.

Puchtel, I.S., Zhuravlev, D.Z., Samsonov, A.V., Arndt, N., 1993. Petrology and geochemistry of metamorphosed komatiites and basalts from the Tungurcha greenstone belt, Aldan shield. Precambrian Research 62 (4), 399–417.

Puchtel, I.G., Haase, K.M., Hofmann, A.W., Chauvel, C., Kulikov, V.S., Garbe-Schonberg, C.-D., Nemchin, A.A., 1997. Petrology and geochemistry of crustally contaminated komatiitic basalts from the Vetreny Belt, southeastern Baltic Shield: evidence for an early Proterozoic mantle plume beneath rifted Archean continental lithosphere. Geochimica et Cosmochimica Acta 61 (6), 1205–1222.

Puchtel, I.S., Hofmann, A.W., Mezger, K., Jochum, K.P., Shchipansky, A.A., Samsonov, A.V., 1998. Oceanic plateau model for continental crustal growth in the Archaean: a case study from the Kostomuksha greenstone belt, NW Baltic Shield. Earth and Planetary Science Letters 155, 57–74.

Puura, V., Flodén, T., 1999. Rapakivi-granite-anorthosite magmatism – a way of thinning and stabilisation of the Sveofennian crust, Baltic Sea Basin. Tectonophysics 305, 75–92.

Pyke, J., 2000. Minerals laboratory staff develops new ICP-MS preparation method. Australian Geological Survey Organisation Research Newsletter 33, 12–14.

Qian, X.L., Cui, W.Y., Wang, S.Q., Wang, G.Y., 1985. Geology of Precambrian Iron Ores in Eastern Hebei Province, China. Hebei Science and Technology Press, Shijiazhuang, 273 pp. (in Chinese with English abstract).

Rabeau, O., 2003. Étude de l'évolution du néodyme dans la croûte continentale du nord-est de la Province du Supérieur, Nunavik, Québec. Unpublished M.Sc. thesis. Université du Québec à Montréal, 80 pp.

Rai, V.K., Thiemens, M.H., 2007. Mass independently fractionated sulfur components in chondrites. Geochimica et Cosmochimica Acta 71, 1341–1345.

Rai, V.K., Jackson, T.L., Thiemens, M.H., 2005. Photochemical mass-independent sulfur isotopes in achondrite meteorites. Science 309, 1062–1065.

Rainbird, R.H., Ernst, R.E., 2001. The sedimentary record of mantle plume uplift. In: Ernst, R.E., Buchan, K.L. (Eds.), Mantle Plumes: Their Identification Through Time. Geological Society of America, Special Paper 352, pp. 227–245.

Rainbird, R.H., McNicoll, R.J., Theriault, R.J., Heaman, L.M., Abbott, J.G., Long, D.G.F., Thorkelson, D.J., 1997. Pan-continental river system draining Grenville Orogen recorded by U-Pb and Sm-Nd geochronology of Neoproterozoic quartzarenites and mudrocks, northwestern Canada. Journal of Geology 105, 1–17.

Ramberg, H., 1952. The Origin of Metamorphic and Metasomatic Rocks. University of Chicago Press, Chicago, 317 pp.

Ramberg, H., 1967. Gravity Deformation and the Earth's Crust. Academic Press, London, 214 pp.

Ramsay, J.G., 1963. Structural investigations in the Barberton Mountain Land, eastern Transvaal. Transactions of the Geological Society of South Africa 66, 353–401.

Ranen, M.C., Jacobsen, S.B., 2006. Barium isotopes in chondritic meteorites: Implications for planetary reservoir models. Science 314, 809–812.

Ransom, B.L., 1987. The paleoenvironmental, magmatic, and geologic history of the 3,500 Myr Kromberg Formation, west limb of the Onverwacht anticline, Barberton Greenstone Belt, South Africa. Masters thesis. Louisiana State University, Baton Rouge, 103 pp.

Ransom, B., Byerly, G.R., Lowe, D.R., 1999. Subaqueous to subaerial Archean ultramafic phreatomagmatic volcanism, Kromberg Formation, Barberton Greenstone Belt, South Africa. In: Lowe, D.R., Byerly, G.R. (Eds.), Geologic Evolution of the Barberton Greenstone Belt, South Africa. Geological Society of America, Special Paper 329, pp. 151–166.

Rapp, R., 2003. Experimental constraints on the origin of compositional variations in the adakite-TTG-sanukitoid-HMA family of granitoids. In: EGS-AGU-EUG Joint Assembly, Nice, France, 6–11 April 2003.

Rapp, R.P., Watson, E.B., 1995. Dehydration melting of metabasalt at 8–32 kbar: implications for continental growth and crust-mantle recycling. Journal of Petrology 36, 891–931.

Rapp, R.P., Watson, E.B., Miller, C.F., 1991. Partial melting of amphibolite/eclogite and the origin of Archean trondhjemites and tonalites. Precambrian Research 51, 1–25.

Rapp, R.P., Shimizu, N., Norman, M.D., Applegate, G.S., 1999. Reaction between slab derived melts and peridotite in the mantle wedge: experimental constraints at 3.8 GPa. Chemical Geology 160, 335–356.

Rapp, R.P., Shimizu, N., Norman, M.D., 2003. Growth of early continental crust by partial melting of eclogite. Nature 425, 605–609.

Rasmussen, B., 2000. Filamentous microfossils in a 3,235-million-year-old volcanogenic massive sulphide deposit. Nature 405, 676–679.

Rasmussen, B., Buick, R., Holland, H.D., 1999. Redox state of the Archean atmosphere: Evidence from detrital heavy minerals in ca. 3250–2750 Ma sandstones from the Pilbara Craton, Australia: Reply. Geology 27, 1152–1152.

Rasmussen, B., Fletcher, I.R., Muhling, J.R., 2007. In situ U–Pb dating and element mapping of three generations of monazite: Unravelling cryptic tectonothermal events in low-grade terranes. Geochimica et Cosmochimica Acta 71, 670–690.

Ravna, E.K., 2000a. Distribution of Fe^{2+} and Mg between coexisting garnet and hornblende in synthetic and natural systems: an empirical calibration of the garnet-hornblende Fe-Mg geothermometer. Lithos 53, 265–277.

Ravna, E.K., 2000b. The garnet-clinopyroxene Fe^{2+}-Mg geothermometer: an updated calibration. Journal of Metamorphic Geology 18, 211–219.

Rayner, N., Corrigan, D., 2004. Uranium-lead geochronological results from the Churchill River – Southern Indian Lake transect, northern Manitoba. In: Geological Survey of Canada, Current Research 2004-F1, p. 14.

Rayner, N., Stern, R.A., Carr, S.D., 2005. Grain-scale variations in trace element composition of fluid-altered zircon, Acasta Gneiss Complex, northwestern Canada. Contributions to Mineralogy and Petrology 148, 721–734.

Redden, J.A., Peterman, Z.E., Zartman, R.E., DeWitt, E., 1990. U-Th-Pb geochronology and preliminary interpretation of Precambrian tectonic events in the Black Hills, South Dakota. In: Lewry, J.G., Stauffer, M.R. (Eds.), The Early Proterozoic Trans-Hudson Orogen of North America. Geological Association of Canada, Special Paper 37, pp. 229–251.

Redman, B.A., Keays, R.R., 1985. Archaean basic volcanism in the Eastern Goldfields Province, Yilgarn Block, Western Australia. Precambrian Research 30, 113–152.

Rees, D.C., Howard, J.B., 2003. The interface between the biological and inorganic worlds: Iron-sulfur metalloclusters. Science 300, 929–931.

Rees, C.E., Thode, H.G., 1977. S^{33} anomaly in Allende meteorite. Geochimica et Cosmochimica Acta 41, 1679–1682.

Regelous, M., Collerson, M.D., 1996. ^{147}Sm-^{143}Nd, ^{146}Sm-^{142}Nd systematics of early Archaean rocks and implications for crust-mantle evolution. Geochimica et Cosmochimica Acta 60, 3513–3520.

Regelous, M., Hofmann, A.W., Abouchami, W., Galer, S.J.G., 2003. Geochemistry of lavas from the Emperor Seamounts, and the geochemical evolution of Hawaiian magmatism from 85 to 42 Ma. Journal of Petrology 44, 113–140.

Reimold, W.U., Meyer, F.M., Walraven, F., Matthews, P.E., 1993. Geochemistry and chronology of the pre- and post-Pongola granitoids for northeastern Transvaal. In: Geological Survey and Mines, 16th Colloquium of African Geology, Mbabane, Swaziland, pp. 294–296.

Reimold, W.U., Koeberl, C., Johnson, S., McDonald, I., 2000. Early Archaean spherule beds in the Barberton Mountain Land, South Africa: impact or terrestrial origin? In: Koeberl, C., Gilmour, I. (Eds.), Impacts and the Early Earth. Springer-Verlag, Berlin, pp. 100–116.

Ren, J-Y., Tamaki, K., Li, S-T., Zhang, J-X., 2002. Late Mesozoic and Cenozoic rifting and its dynamic setting in Eastern China and adjacent areas. Tectonophysics 344, 175–205.

Replumaz, A., Karason, H., van der Hilst, R.D., Besse, J., Tapponier, P., 2004. 4-D evolution of SE Asia's mantle from geological reconstructions and seismic tomography. Earth and Planetary Science Letters 221, 103–115.

Rey, P.F., Philippot, P., Thebaud, N., 2003. Contribution of mantle plumes, crustal thickening and greenstone blanketing to the 2.75–2.65 Ga global crisis. Precambrian Research 127, 43–60.

Reynolds, D.G., Brook, W.A., Marshall, A.E., Allchurch, P.D., 1975. Volcanogenic copperzinc deposits in the Pilbara and Yilgarn Archaean blocks. In: Australasian Institute of Mining and Metallurgy, Monograph 5, pp. 185–194.

Reynolds, J.H., 1960. I-Xe dating of meteorites. Journal of Geophysical Research 65, 3843–3846.

Richard, P., Shimizu, N., Allègre, J.J., 1976. ^{143}Nd/^{146}Nd, a natural tracer: an application to oceanic basalts. Earth and Planetary Science Letters 31, 269–278.

Richards, J.R., Fletcher, I.R., Blockley, J.G., 1981. Pilbara galenas: precise isotopic assay of the oldest Australian leads; model ages and growth-curve implications. Mineralium Deposita 16, 7–30.

Richardson, W.P., Okal, E.A., Van der Lee, S., 2000. Raylegh-wave tomography of the Ontong-Java Plateau. Physics of the Earth and Planetary Interiors 118, 29–51.

Rieke, G.H., Su, K.Y.L., Stansberry, J.A., Trilling, D., Bryden, G., Muzerolle, J., White, B., Gorlova, N., Young, E.T., Beichman, C.A., Stapelfeldt, K.R., Hines, D.C., 2005. Decay of planetary debris disks. Astrophysical Journal 620, 1010–1026.

Riganti, A., 1996. The northern segment of the early Archaean Nondweni greenstone belt, South Africa: field relations and petrogenetic constraints. Ph.D. thesis. University of Natal, Pietermaritzburg, 384 pp.

Riganti, A., 2003. Geology of the Everett Creek 1:100,000 sheet. Western Australia Geological Survey, 1:100,000 Geological Series Explanatory Notes, 39 pp.

Riganti, A., Wilson, A.H., 1995a. Geochemistry of the mafic/ultramafic volcanic associations of the Nondweni greenstone belt, South Africa, and constraints on their petrogenesis. Lithos 34, 235–252.

Riganti, A., Wilson, A.H., 1995b. Early Archaean fumaroles: evidence in the Nondweni Greenstone Belt, South Africa. In: Extended Abstracts of the Geocongress '95. Geological Society of South Africa, pp. 1177–1180.

Robb, L.J., Anhauesser, C.R., 1983. Chemical and petrogenetic characteristics of Archean tonalite-trondhjemite gneiss plutons in the Barberton Mountain Land. In: Anhauesser, C.R. (Ed.), Contributions to the Geology of the Barberton Mountain Land. Geological Society of South Africa, Special Publication 9, pp. 103–116.

Robb, L.J., Barton, J.M., Jr, Kable, E.J.D., Wallace, R.C., 1986. Geology, geochemistry and isotopic characteristics of the Archaean Kaap Valley pluton, Barberton mountain land, South Africa. Precambrian Research 31, 1–36.

Robert, F., Chaussidon, M., 2006. A palaeotemperature curve for the Precambrian oceans based on silicon isotopes in cherts. Nature 443, 969–972.

Roberts, M.J., Snape, C.E., Mitchell, S.C., 1995. Hydropyrolysis: Fundamentals, two-stage processing and PDU operation. In: Snape, C.E. (Ed.), Composition, Geochemistry and Conversion of Oil Shales. Kluwer, Germany, pp. 277–294.

Robertson, M.J., Charlesworth, E.G., Phillips, G.N., 1993. Gold mineralization during progressive deformation at the Main Reef Complex, Sheba Gold Mine, Barberton Greenstone-Belt, South-Africa. Information Circular, Economic Geology Research Unit, University of the Witwatersrand, Johannesburg, 267, 26 pp.

Roedder, E., 1981. Are the 3,800-Myr-old Isua objects microfossils, limonite-stained fluid inclusions, or neither? Nature 293, 459–462.

Rogers, G.C., 1982. Oceanic plateaus as meteorite impact signatures. Nature 299, 341–342.

Rogers, J.J.M., 1996. A history of the continents in the past three billion years. Journal of Geology 104, 91–107.

Rogers, J.J.W., Santosh, M., 2004. Continents and Supercontinents. Oxford University Press, Oxford, 289 pp.

Rollinson, H.R., 1993. Using Geochemical Data: Evaluation, Presentation, Interpretation. Longman Scientific & Technical, London.

Rollinson, H.R., 2002. The Metamorphic History of the Isua Greenstone Belt, West Greenland. In: Fowler, C.M.R., Ebinger, C.J., Hawkesworth, C.J. (Eds.), The Early Earth: Physical, Chemical and Biological Development. Geological Society of London, Special Publication 199, 329–350.

Rollinson, H., 2007. Early Earth Systems: A Geochemical Approach. Blackwell, Oxford, 285 pp.

Rollinson, H.R., Tarney, J., 2005. Adakites – the key to understanding LILE depletion in granulites. Lithos 79, 61–81.

Roscoe, S.M., Donaldson, J.A., 1988. Uraniferous pyritic quartz pebble conglomerate and layered ultramafic intrusions in a sequence of quartzite, carbonate, iron formation and basalt of probable Archean age at Lac Sakami, Quebec. In: Geological Survey of Canada, Paper 88-1C, pp. 117–121.

Rose, N.M., Rosing, M.T., Bridgwater, D., 1996. The origin of metacarbonate rocks in the Archaean Isua Supracrustal Belt, West Greenland. American Journal of Science 296, 1004–1044.

Rosen, O.M., 1990. Archean geology and geochemistry of the Anabar shield. In: Condie, K.C., Bogdanov, N.A. (Eds.), Guidebook of the International Field Conference for IGCP Projects 217, 247, 280. Institute of Lithosphere, Moscow, 102 pp.

Rosen, O.M., 1995. Metamorphic effects of tectonic movements at the lower crust level: Proterozoic collision zones and terranes of the Anabar Shield. Geotectonics 29 (2), 91–101.

Rosen, O.M., 2002. Siberian craton – a fragment of Paleoporterozoic supercontinent. Russian Journal of Earth Science 4 (2), 103–119.

Rosen, O.M., 2003. The Siberian craton: tectonic zonation and stages of evolution. Geotectonics 37 (3), 175–192.

Rosen, O.M., Fedorovsky, V.S., 2001. Collisional Granitoids and Crustal Sheeting. Nauchny Mir, Moscow, 186 pp. (in Russian).

Rosen, O.M., Andreev, V.P., Belov, A.N., Bibikova, E.V., Zlobin, V.L., Rachkov, V.S., Sonyushkin, V.E., 1988. The Archaean of the Anabar Shield and the Problems of Early Evolution of the Earth. Nauka, Moscow, 253 pp. (in Russian).

Rosen, O.M., Bibikova, E.V., Zhuravlev, D.Z., 1991. The Anabar shield early crust: age and the origin models. In: Ancient Terrestrial Crust: Age and Composition. Nauka, Moscow, pp. 199–224 (in Russian).

Rosen, O.M., Condie, K.C., Natapov, L.M., Nozhkin, A.D., 1994. Archean and Early Proterozoic evolution of the Siberian craton: A preliminary assessment. In: Condie, K.C. (Ed.), Archean Crustal Evolution. Elsevier, Amsterdam, pp. 411–459.

Rosen, O.M., Zhuravlev, D.Z., Sukhanov, M.K., Bibikova, E.V., Zlobin, V.L., 2000. Early Proterozoic terranes, collision zones, and associated anorthosites in the northeast of the Siberian craton: isotope geochemistry and age characteristics. Russian Geology and Geophysics 41 (2), 159–178.

Rosen, O.M., Serenko, V.P., Spetsius, Z.V., Manakov, A.V., Zinchuk, N.N., 2002. Yakutian kimberlite province: position in the structure of the Siberian craton and composition of the upper and lower crust. Russian Geology and Geophysics 43 (1), 1–24.

Rosenberg, E., McGill, G.E., 2001. Geologic map of the Pandrosos Dorsa (V5) quadrangle, Venus. Geologic Investigation Series Map I-2721, U.S. Geological Survey, 1:5,000,000.

Rosing, M.T., 1999. ^{13}C-depleted carbon microparticles in >3700 Ma sea-floor sedimentary rocks from West Greenland. Science 283, 674–676.

Rosing, M.T., Frei, R., 2004. U-rich Archaean sea-floor sediments from Greenland – indications of >3700 Ma oxygenic photosynthesis. Earth and Planetary Science Letters 217, 237–244.

Rosing, M.T., Rose, N.M., Bridgwater, D., Thomsen, H.S., 1996. Earliest part of Earth's stratigraphic record; a reappraisal of the >3.7 Ga Isua (Greenland) supracrustal sequence. Geology 24, 43–46.

Rosing, M.T., Nutman, A.P., Løfqvist, L., 2001. A new fragment of the early Earth crust: the Aasivik terrane of West Greenland. Precambrian Research 105, 115–128.

Rossi, M.J., Gudmundsson, A., 1996. The morphology and formation of flow-lobe tumuli on Icelandic shield volcanoes. Journal of Volcanology and Geothermal Research 72, 291–308.

Rouxel, O., Bekker, A., Edwards, K., 2005. Iron isotope constraints on the Archean and Paleoproterozoic ocean redox state. Science 307, 1088–1091.

Rouzaud, J.-N., Skrzypczak, A., Bonal, L., Derenne, S., Quirico, E., Robert, F., 2005. The high resolution transmission electron microscopy: A powerful tool for studying the organization of terrestrial and extra-terrestrial carbons. Lunar and Planetary Science XXXVI.

Roy, A.B., Kröner, A., 1996. Single zircon evaporation ages constraining the growth of the Archean Aravalli craton, NW Indian shield. Geological Magazine 133, 333–342.

Rubatto, D., 2002. Zircon trace element geochemistry: partitioning with garnet and the link between U-Pb ages and metamorphism. Chemical Geology 184, 123–138.

Rubin, A.E., 1997a. Mineralogy of meteorite groups. Meteoritics and Planetary Science 32, 231–247.

Rubin, A.E., 1997b. Mineralogy of meteorite groups: An update. Meteoritics and Planetary Science 32, 733–734.

Rubin, A.E., 2000. Petrologic, geochemical and experimental constraints on models of chondrule formation. Earth Science Reviews 50, 3–27.

Rubin, A.E., Kalleymeyn, G.W., 1994. Pecora Escarpment 91002: A member of the new Rumuruti (R) chondrite group. Meteoritics 29, 255–264.

Rubin, A.E., Mittlefehldt, D.W., 1992. Mesosiderites – a chronological and petrologic synthesis. Meteoritics 27, 282.

Rundqvist, D.V., Sokolov, Y.M., 1993. Main features of Precambrian metallogeny: In: Rundquist, D.V., Mitrofanov, F.P. (Eds.), Precambrian Geology of the USSR. Elsevier, Amsterdam, pp. 1–6.

Runnegar, B., 2001. Archean sulfates from Western Australia: implications for Earth's early atmosphere and ocean. In: 11th Annual Goldschmidt Conference, Hot Spring, Virginia, p. 3859.

Runnegar, B., Coath, C.D., Lyons, J.R., McKeegan, K.D., 2002. Mass-independent and mass-dependent sulfur processing throughout the Archean. Geochimica et Cosmochimica Acta 66 (15A), A655.

Rushmer, T., 1991. Partial melting of two amphibolites: contrasting experimental results under fluid-absent conditions. Contributions to Mineralogy and Petrology 107, 41–59.

Russell, M.J., Arndt, N.T., 2005. Geodynamic and metabolic cycles in the Hadean. Biogeosciences 2, 97–111.

Russell, M.J., Hall, A.J., 1997. The emergence of life from iron monosulphide bubbles at a submarine hydrothermal redox and pH front. Journal of the Geological Society 154, 377–402.

Russell, M.J., Martin, W., 2004. The rocky roots of the acetyl-CoA pathway. Trends in Biochemical Sciences 29, 358–363.

Russell, M.J., Hall, A.J., Boyce, A.J., Fallick, A.E., 2005. On hydrothermal convection systems and the emergence of life. Economic Geology 100, 419–438.

Russell, S.S., 1998. A survey of calcium-aluminium-rich inclusions from Rumurutiite chondrites: implications for relationships between meteorite groups. Meteoritics and Planetary Science 33, A131–132.

Rutland, R.W.R., 1973. Tectonic evolution of the continental crust of Australia. In: Tarling, D.H., Runcorn, S.K. (Eds.), Implications of Continental Drift to the Earth Sciences. Academic Press, London, pp. 1011–1033.

Ryder, G., 1990. Lunar samples, lunar accretion and th early bombardment of the Moon. Eos, Transactions, American Geophysical Union 70, 322–323.

Ryder, G., 1991. Accretion and bombardment in the Earth–Moon system: the Lunar record. In: Lunar Planetary Institute Contribution 746, pp. 42–43.

Ryder, G., 1997. Coincidence in the time of the Imbrium Basin impact and Apollo 15 Kreep volcanic series: impact-induced melting? In: Lunar Planetary Institute Contribution 790, pp. 61–62.

Rye, R., Holland, H.D., 1998. Paleosols and the evolution of atmospheric oxygen: A critical review. American Journal of Science 298, 621–672.

Rye, R., Kuo, P.H., Holland, H.D., 1997. Atmospheric carbon dioxide concentrations before 2.2 billion years ago. Nature 378, 603–605.

Ryerson, F.J., Watson, E.B., 1987. Rutile saturation in magmas: implications for Ti-Nb-Ta depletion in island-arc basalts. Earth and Planetary Sciences Letters 86, 225–239.

Sackmann, I.-J., Boothroyd, A.I., 2003. Our Sun V. A bright young Sun consistent with helioseismology and warm temperatures on ancient Earth and Mars. Astrophysical Journal 583, 1024–1039.

SACS (South African Committee for Stratigraphy), 1980. Stratigraphy of South Africa: Part 1: Lithostratigraphy of the Republic of South Africa, South West Africa/Namibia and the Republics of Bophuthatswana, Transkei and Venda. Geological Survey of South Africa Handbook, vol. 8, 690 pp.

Sagan, C., Chyba, C.F., 1997. The early faint sun paradox: Organic shielding of ultraviolet-labile greenhouse gases. Science 276, 1217–1221.

Sagan, C., Mullen, G., 1972. Earth and Mars – evolution of atmospheres and surface temperatures. Science 177, 52–56.

Salnikova, E.B., Kotov, A.B., Nemchin, A.A., Yakovleva, S.Z., Morozova, I.M., Bogomolova, L.M., Smelov, A.P., 1993. The age of Tungurchakan Pluton (Olekma granite-greenstone terrane, Aldan shield). Doklady Earth Sciences 331 (3), 356–358.

Salnikova, E.B., Kotov, A.B., Belyatskii, B.V., Yakovleva, S.Z., Morozova, I.M., Berezhnaya, N.G., Zagornaya, N.Yu., 1997. U-Pb age of granitoids in the junction zone between the Olekma and Aldan granulite-gneiss terranes. Stratigraphy and Geological Correlation 5 (2), 101–109.

Salnikova, E.B., Kotov, A.B., Levitskiy, V.I., Morozova, I.M., Berezhnaya, N.G., 2003. Age boundaries of high-temperature metamorphism in crystalline complexes of the Sharyzhalgay uplift of the Siberian Platform: results of U-Pb dating of single zircon grains. In: Proceedings of 2nd Conference on Isotope Geochronology: Isotope Geochronology in the Solution of Problems of Geodynamics and Ore Genesis, St.-Petersburg, pp. 453–455 (in Russian).

Samieyani, A., 1995. Mineralogy and geochemistry of ∼3.5 Ga volcanogenic massive sulfide deposits at the Big Stubby prospect, the Pilbara district, Western Australia. Unpublished M.Sc. thesis. Tohoku University, 90 pp.

Sanborn-Barrie, M., Skulski, T., 1999. 2.7 Ga tectonic assembly of continental margin and oceanic terranes in the Savant Lake – Sturgeon Lake greenstone belt, Ontario. In: Current Research 1999-C, Geological Survey of Canada, pp. 209–220.

Sanborn-Barrie, M., Skulski, T., 2006. Sedimentary and structural evidence for 2.7 Ga continental arc-oceanic arc collision in the Savant–Sturgeon greenstone belt, western Superior Province, Canada. Canadian Journal of Earth Sciences 43, 995–1030.

Sanborn-Barrie, M., Skulski, T., Parker, J.R., 2001. Three hundred million years of tectonic history recorded by the Red Lake greenstone belt, Ontario. In: Current Research 2001-C19, Geological Survey of Canada, 19 pp.

Sanborn-Barrie, M., Rogers, N., Skulski, T., Parker, J.R., McNicoll, V., Devaney, J., 2004. Geology and tectonostratigraphic assemblages, east Uchi, Red Lake and Birch-Uchi belts, Ontario. Geological Survey of Canada, Open File 4256, Ontario Geological Survey, Preliminary Map P.3460, scale 1:250,000.

Sanchez-Garrido, C., 2006. Nature des apports (juvéniles/recyclage) dans la sédimentation orogénique de Schapenburg (3.2 Ga) : approche isotopique Sr et Nd. DEA, Université Blaise-Pascal, Clermont-Ferrand, France, 45 pp.

Sandiford, M., 1985. The metamorphic evolution of granulites at Fyfe Hills: implications for Archaean crustal thickness in Enderby Land, Antarctica. Journal of Metamorphic Geology 3, 155–178.

Sandiford, M., McLaren, S., 2002. Tectonic feedback and the ordering of heat producing elements within the continental lithosphere. Earth and Planetary Science Letters 204, 133–150.

Sandiford, M., Van Kranendonk, M.J., Bodorkos, S., 2004. Conductive incubation and the origin of dome-and-keel structure in Archean granite-greenstone terrains: a model based on the eastern Pilbara Craton, Western Australia. Tectonics 23, TC1009, doi: 10.1029/2002TC001452.

Sangster, D.W., Brook, W.A., 1977. Primitive lead in an Australian Zn-Pb-Ba deposit. Nature 270, 423.

Sano, Y., Terada, K., Hidaka, H., Yokoyama, K., Nutman, A.P., 1999a. Palaeoproterozoic thermal events recorded in the ~4.0 Ga Acasta gneiss, Canada: Evidence from SHRIMP U-Pb dating of apatite and zircon. Geochimica et Cosmochimica Acta 63, 899–905.

Sano, Y., Terada, K., Takahasho, Y., Nutman, A.P., 1999b. Origin of life from apatite dating? Nature 400, 127.

Sarkar, G., Corfu, F., Paul, D.K., McNaughton, N.J., Gupta, S.N., Bishui, P.K., 1993. Early Archean crust in Bastar craton, central India – a geochemical and isotopic study. Precambrian Research 62, 127–137.

Sasaki, S., 1990. The primary solartype atmosphere surrounding the accreting Earth: H_2O-induced high surface temperature. In: Newsome, H.E., Jones, J.H. (Eds.), Origin of the Earth. Oxford University Press, New York, pp. 195–209.

Sasseville, C., 2002. Characteristics of Mesoarchean and Neoarchean supracrustal sequences at the southern margin of North Caribou terrane in the Wallace Lake greenstone belt, Superior Province, Canada. M.Sc. thesis. McGill University, Montreal, Canada.

Sasseville, C., Tomlinson, K.Y., Hynes, A., McNicoll, V., 2006. Stratigraphy, structure, and geochronology of the 3.0–2.7 Ga Wallace Lake greenstone belt, Western Superior Province, SE Manitoba, Canada. Canadian Journal of Earth Sciences 43, 929–945.

Saunders, A.D., Norry, M.J., Tarney, J., 1988. Origin of MORB and chemically-depleted mantle reservoir: trace element constraints. Journal of Petrology, Special lithosphere issue, 415–445.

Sawyer, E.W., Benn, K., 1993. Structure of the high-grade Opatica belt and adjacent low-grade Abitibi subprovince, Canada; an Archean mountain front. Journal of Structural Geology 15, 1443–1458.

Scaillet, B., Clemente, B., Evans, B.W., Pichavant, M., 1998. Redox control of sulfur degassing in silicic magmas. Journal of Geophysical Research – Solid Earth 103 (B10), 23,937–23,949.

Schaap, B.D., 1989. The geology and crustal structure of southwestern Minnesota using gravity and magnetic data. M.Sc. thesis. University of Minnesota, Minneapolis, 78 pp.

Schaber, G.G., Strom, R.G., Moore, H.J., Soderblom, L.A., Kirk, R.L., Chadwick, D.J., Dawson, D.D., Gaddis, L.R., Boyce, J.M., Russell, J., 1992. Geology and Distribution of Impact Craters on Venus: What are they telling us? Journal of Geophysical Research 97, 13,257–13,302.

Schärer, U., Allègre, C.J., 1985. Determination of the age of the Australian continent by single-grain zircon analyses of Mt. Narryer metaquartzite. Nature 315, 52–55.

Scherer, E., Münker, C., Mezger, K., 2001. Calibration of the lutetium-hafnium clock. Science 293, 683–687.

Schidlowski, M., 2001. Carbon isotopes as biogeochemical recorders of life over 3.8 Ga of Earth history: Evolution of a concept. Precambrian Research 106, 117–134.

Schidlowski, M., Eichmann, R., Junge, C.E., 1975. Precambrian sedimentary carbonates: Carbon and oxygen isotope geochemistry and implications for the terrestrial oxygen budget. Precambrian Research 2, 1–69.

Schidlowski, M., Appel, P.W.U., Eichmann, R., Junge, C.E., 1979. Carbon isotope geochemistry of the 3.7×10^9-yr-old Isua sediments, West Greenland: implications for the Archaean carbon and oxygen cycles. Geochimica et Cosmochimica Acta 43, 189–199.

Schidlowski, M., Hayes, J.M., Kaplan, I.R., 1983. Isotopic inference of ancient biochemistries: Carbon, sulfur, hydrogen and nitrogen. In: Schopf, J.W. (Ed.), Earth's Earliest Biosphere – It's Origin and Evolution. Princeton University Press, Princeton, pp. 149–186.

Schiotte, L., Compston, W., Bridgwater, D., 1989. Ion probe U-Th-Pb zircon dating of polymetamorphic orthogneisses from northern Labrador. Canada Journal of Earth Sciences 26, 1533–1556.

Schlesinger, W.H., 1997. Biogeochemistry. Academic Press, San Diego, 588 pp.

Schmidt, M.W., 1993. Phase relations and composition in tonalite as a function of pressure: an experimental study at 650 °C. American Journal of Science 293, 1011–1060.

Schmidt, M.W., Poli, S., 1998. Experimentally based water budgets for dehydrating slabs and consequences for arc magma generation. Earth and Planetary Science Letters 163, 361–379.

Schmidt, M.W., Thompson, A.B., 1996. Epidote in calc-alkaline magmas: an experimental study of stability, phase relationships and the role of epidote in magmatic evolution. American Mineralogist 81, 462–474.

Schmitz, M.D., Bowring, S.A., de Wit, M., Gartz, V., 2004. Subduction and terrane collision stabilize the western Kaapvaal craton tectosphere 2.9 billion years ago. Earth and Planetary Science Letters 222, 363–376.

Schmitz, M.D., Bowring, S.A., Southwick, D.L., Boerboom, T.J., Wirth, K.R., 2006. High-precision U-Pb geochronology in the Minnesota River Valley subprovince and its bearing on the Neoarchean to Paleoproterozoic evolution of the southern Superior Province. Bulletin of the Geological Society of America 118, 82–93.

Schneider, D.A., Bickford, M.E., Cannon, W.F., Schulz, K.J., Hamilton, M.A., 2002. Timing of Marquette Range Supergroup deposition and Penokean collision-related basin formation, southern Lake Superior region: implications for Paleoproterozoic tectonic reconstruction. Canadian Journal of Earth Sciences 39, 999–1012.

Schoenberg, R., Kamber, B.S., Collerson, K.D., Moorbath, S., 2002. Tungsten isotope evidence from ∼3.8-Gyr metamorphosed sediments for early meteorite bombardment of the Earth. Nature 418, 403–405.

Schonwandt, H.K., Petersen, J.S., 1983. Continental rifting and porphyry-molybdenum occurrences in the Oslo region, Norway. Tectonophysics 94, 609–631.

Schopf, J.W. (Ed.), 1983. Earth's Earliest Biosphere: Its Origin and Evolution. Princeton University Press, 543 pp.

Schopf, J.W., 1992. Newly discovered microfossils from the Early Archaean Apex Basalt (Warrawoona Group), Western Australia. In: Schopf, J.W., Klein, C. (Eds.), The Proterozoic Biosphere: A Multi-Disciplinary Study. Cambridge University Press, pp. 25–39.

Schopf, J.W., 1993. Microfossils of the Early Archean Apex Chert: New evidence of the antiquity of life. Science 260, 640–646.

Schopf, J.W., 2004. Earth's earliest biosphere: Status of the hunt. In: Eriksson, P.G., Altermann, W., Nelson, D.R., Mueller, W.U., Catuneanu, O. (Eds.), The Precambrian Earth: Tempos and Events. Developments in Precambrian Geology, vol. 12. Elsevier, Amsterdam, pp. 516–539.

Schopf, J.W., Kudryavtsev, A.B., Agresti, D.G., Wdowiak, T.J., Czaja, A.D., 2002. Laser-Raman imagery of Earth's earliest fossils. Nature 416, 73–76.

Schramm, L.S., Brownlee, D.S., Wheelock, M.M., 1988. The elemental composition of interplanetary dust. In: Lunar and Planetary Science Conference XIX. Houston, TX, pp. 1033–1034.

Schubert , G., Sandwell, D.T., 1995. A global survey of possible subduction sites on Venus. Icarus 117, 173–196.

Schubert, G., Moore, W.B., Sandwell, D.T., 1994. Gravity over coronae and chasmata on Venus. Icarus 112, 130–146.

Schubert, G.S., Solomatov, V.S., Tackely, P.J., Turcotte, D.L., 1997. Mantle convection and thermal evolution of Venus. In: Bouger, S.W., Hunten, D.M., Phillips, R.J. (Eds.), Venus II. University of Arizona Press, Tucson, AZ, pp. 1245–1288.

Schubert, G., Turcotte, D.L., Olson, P., 2001. Mantle Convection in the Earth and Planets. Cambridge University Press, Cambridge, 940 pp.

Schultz, L., Palme, H., Spettel, B., Weber, H.W., Wänke, H., Cristophe Michel-Levy, M., Lorin, J.C., 1982. Allan Hills A77081 – An unusual stony meteorite. Earth and Planetary Science Letters 61, 23–31.

Schultz, P.H., 1993. Searching for ancient Venus. In: Lunar and Planetary Science Conference XXIV, pp. 1255–1255.

Schulze, H., Bischoff, A., Palme, H., Spettel, B., Dreibus, G., Otto, J., 1994. Mineralogy and chemistry of Rumuruti: the first meteorite fall of the new R chondrite group. Meteoritics 29, 275–286.

Scoates, R.F.J., Macek, J.J., 1978. Molson dyke swarm. In: Manitoba Department of Mines, Resources and Environmental Management, Mineral Resources Division, Geological Paper 78-1, p. 53.

Scott, E.R.D., 1972. Chemical fractionation in iron meteorites and its interpretation. Geochimica et Cosmochimica Acta 36, 1205–1236.

Scott, E.R., 1977. Formation of olivine-metal textures in pallasite meteorites. Geochimica et Cosmochimica Acta 41, 693–710.

Scott, E.R.D., Jones, R.H., 1990. Disentangling nebula and asteroidal features of CO_3 carbonaceous chondrite meteorites. Geochimica et Cosmochimica Acta 54, 2485–2502.

Scott, E.R.D., Haack, H., Love, S.G., 2001. Formation of mesosiderites by fragmentation and reaccretion of a large differentiated asteroid. Meteoritics and Planetary Science 36, 869–881.

Seal, R.R., II, 2006. Sulfur isotope geochemistry of sulfide minerals. In: Vaughan, D.J. (Ed.), Sulfide Mineralogy and Geochemistry. Reviews in Mineralogy and Geochemistry, vol. 61, pp. 633–677.

Sears, D.W.G., Kallemeyn, G.W., Wasson, J.T., 1982. The chemical classification of chondrites II: The enstatite chondrites. Geochimica et Cosmochimica Acta 46, 597–608.

Sears, D.W.G., Hasan, F.A., Batchelor, J.D., Lu, J., 1991. Chemical and physical studies of type 3 chondrites – XI: Metamorphism, pairing and brecciation in ordinary chondrites. In: Proceedings of the Lunar and Planetary Science Conference, vol. 21, pp. 493–512.

Seedorff, E., Dilles, J.H., Proffett, J.M., Jr, Einaudi, M.T., Zurcher, L., Stavast, W.J.A., Johnson, D.A., Barton, M.D., 2005. Porphyry deposits: characteristics and origin of hypogene features. Economic Geology, 100th Anniversary Volume, 251–298.

Seeger, P.A., Fowler, W.A., Clayton, D.D., 1965. Nucleosynthesis of heavy elements by neutron capture. Astrophysical Journal 11 (Supplement), 121–166.

Seewald, J.S., Seyfried, W.E., 1990. The effect of temperature on metal mobility in sub-seafloor hydrothermal systems: constraints from basalt alteration experiments. Earth and Planetary Science Letters 101, 388–403.

Self, S., Keszthelyi, L., Thordarson, T., 1998. The importance of pahoehoe. Annual Review of Earth and Planetary Sciences 26, 81–110.

Sen, C., Dunn, T., 1994. Dehydration melting of a basaltic composition amphibolite at 1.5 and 2.0 GPa: implications for the origin of adakites. Contributions to Mineralogy and Petrology 117, 394–409.

Şengör, A.M.C., 2001. Elevation as an indicator of mantle-plume activity. In: Ernst, R.E., Buchan, K.L. (Eds.), Mantle Plumes: Their Identification Through Time. Geological Society of America, Special Paper 352, pp. 183–225.

Şengör, A.M.C., Natal'in, B.A., 1996. Turkic type orogeny and its role in the making of the continental crust. Annual Review Earth and Planetary Sciences 24, 263–337.

Şengör, A.M.C., Natal'in, B.A., 2004. Phanerozoic analogues of Archaean oceanic basement fragments: Altaid ophiolites and ophirags. In: Kusky, T.M. (Ed.), Precambrian

Ophiolites and Related Rocks. Developments in Precambrian Geology, vol. 13. Elsevier, Amsterdam, pp. 675–726.

Şengör, A.M.C., Okurogullari, A.H., 1991. The role of accretionary wedges in the growth of continents: Asiatic examples from Argand to plate tectonics. Eclogae Geologicae Helveticae 84, 535–597.

Şengör, A.M.C., Natal'in, B.A., Burtman, V.S., 1993. Evolution of the Altayd tectonic collage and Paleozoic crustal growth in Eurasia. Nature 364, 299–307.

Sengupta, P., Dasgupta, S., Bhattacharya, P.K., Hariya, Y., 1989. Mixing behavior in quaternary garnet solid-solution and an extended Ellis and Green garnet-clinopyroxene geothermometer. Contributions to Mineralogy and Petrology 103, 223–227.

Sephton, M.A., Love, G.D., Meredith, W., Snape, C.E., Sun, C.-G., Watson, J.S., 2005. Hydropyrolysis: A new technique for the analysis of macromolecular material in meteorites. Planetary and Space Science 53, 1280–1286.

Seyfried, W.E., 1987. Experimental and theoretical constraints on hydrothermal alteration processes at mid-ocean ridges. Annual Reviews Earth and Planetary Sciences 15, 317–335.

Shapiro, L., Brannock, W.W., 1962. Rapid analysis of silicate, carbonate and phosphate rocks. In: U.S. Geological Survey, Bulletin 1144-A.

Sharp, T.G., De Gregorio, B.T., 2003. Determining the biogenicity of residual carbon within the Apex Chert. Geological Society of America, Annual Meeting, Seattle, WA, Paper No. 187-2.

Sharp, Z.D., Essene, E.J., Kelly, W.C., 1985. A re-examination of the arsenopyrite geothermometer – pressure considerations and applications to natural assemblages. Canadian Mineralogist 23, 517–534.

Shaw, D.M., 1970. Trace element fractionation during anatexis. Geochimica et Cosmochimica Acta 34 (2), 237–243.

Shemansky, D.E., 1972. CO_2 extinction coefficient 1700–3000 Å. Journal of Chemical Physics 56, 1582–1587.

Shemyakin, V.M., Glebovitsky, V.A., Berezhnaya, N.G., Yakovleva, S.Z., Morozova, I.M., 1998. On the age of the Sutam block oldest formations. Doklady Earth Sciences 360 (4), 526–529.

Shen, B.F., Luo, H., Li, S.B., Li, J.J., Peng, X.L., Hu, X.D., Mao, D.B., Liang, R.Y., 1994. Geology and Metallization of Archaean Greenstone Belts in North China Platform. Geological Publishing House, Beijing, 202 pp. (in Chinese with English abstract).

Shen, J.J., Papanastassiou, D.A., Wasserburg, G.J., 1996. Precise Re-Os determinations and systematics of iron meteorites. Geochimica et Cosmochimica Acta 60, 2887–2900.

Shen, Q.H., Liu, D.Y., Wang, P., Gao, J.F., Zhang, Y.F., 1987. U-Pb and Rb-Sr isotopic age study of the Jining Group from Nei Mongol of China. Bulletin of the Chinese Academy of Geological Sciences 16, 165–178.

Shen, Q.H., Shen, K., Geng, Y.S., Xu, H.F., 2000. Compositions and Geological Evolution of Yishui Complex, Shandong Province, China. Geological Publishing House, Beijing, 179 pp. (in Chinese).

Shen, Q.H., Song, B., Xu, H.F., Geng, Y.S., Shen, K., 2004. Emplacement and meta-morphism ages of the Caiyu and Dashan igneous bodies, Yishui County, Shandong Province: Zircon SHRIMP chronology. Geological Review 50, 275–284.

Shen, Q.H., Geng, Y.S., Song, B., Wan, Y.S., 2005. New information from surface outcrops and deep crust of Archaean Rocks of the North China and Yangtze Blocks, and Qinling-Dabie Orogenic Belt. Acta Geologica Sinica 79, 616–627 (in Chinese with English abstract).

Shen, Y., Buick, R., 2004. The antiquity of microbial sulfate reduction. Earth Science Reviews 64, 243–272.

Shen, Y., Buick, R., Canfield, D.E., 2001. Isotopic evidence for microbial sulphate reduction in the early Archaean era. Nature 410, 77–81.

Sheppard, S., Occhipinti, S.A., Tyler, I.M., 2003. The relationship between tectonism and composition of granitoid magmas, Yarlarweelor Gneiss Complex, Western Australia. Lithos 66, 133–154.

Sheraton, J.W., Black, L.P., McCulloch, M.T., 1984. Regional geochemical and isotopic characteristics of high-grade metamorphics of the Prydz Bay area: the extent of Protero-zoic reworking of Archaean continental crust in East Antarctica. Precambrian Research 26, 169–198.

Sheraton, J.W., Tingey, R.J., Black, L.P., Offe, L.A., Ellis, D.J., 1987. Geology of Enderby Land and Western Kemp Land, Antarctica. Bulletin Australian Bureau of Mineral Resources 223, 1–51.

Sherwood-Lollar, B., Westgate, T.D., Ward, J.A., Slater, G.F., Lacrampe-Couloume, G., 2002. Abiogenic formation of alkanes in the Earth's crust as a minor source for global hydrocarbon reservoirs. Nature 416, 522–524.

Shields, G.A., 2002. A chemical explanation for the mid-Neoproterozoic disappearance of molar-tooth structure ~750 Ma. Terra Nova 14, 108–113.

Shields, G.A., Kasting, J.F., 2007. No evidence for hot early oceans? Nature 447, doi: 10.1038/nature05830.

Shields, G.A., Veizer, J., 2002. Precambrian marine carbonate isotope database: Version 1.1. Geochemistry, Geophysics, Geosystem 6, 1–12.

Shields, G.A., Carden, G.A.F., Veizer, J., Meidla, T., Rong, J.-Y., Li, R.-Y., 2003. Sr, C and O isotope geochemistry of Ordovician brachiopods: a major event around the Middle-Late Ordovician transition. Geochimica et Cosmochimica Acta 67, 2005–2025.

Shiraishi, K., Ellis, D.J., Hiroi, Y., Fanning, C.M., Motoyoshi, Y., Nakai, Y., 1994. Cam-brian orogenic belt in East Antarctica and Sri Lanka: implications for Gondwana as-sembly. Journal of Geology 102, 47–65.

Shirey, S.B., Walker, R.J., 1998. The Re-Os isotope system in cosmochemistry and high-temperature geochemistry. Annual Review of Earth and Planetary Sciences 26, 423–500.

Shirey, S.B., Wilson, A.H., Carlson, R.W., 1998. Re-Os Isotopic Systematics of the 3300 Ma Nondweni Greenstone Belt, South Africa: Implications for Komatiite Formation at a Craton Edge. In: AGU Spring Meeting, Boston.

Shirey, S.B., Richardson, S.H., Harris, J.W., 2004. Integrated models of diamond formation and craton evolution. Lithos 77, 923–944.

Shock, E.L., Schulte, M.D., 1998. Organic synthesis during fluid mixing in hydrothermal systems. Journal of Geophysical Research – Planets 103 (E12), 28,513–28,527.

Shu, F.H., Shang, H., Gounelle, M., Glassold, A.E., Lee, T., 2001. The origin of chondrules and refractory inclusions in chondritic meteorites. Astrophysics Journal 548, 1029–1050.

Shukolyukov, A., Lugmair, G.W., 1993a. Live iron-60 in the early solar system. Science 259, 1138–1142.

Shukolyukov, A., Lugmair, G.W., 1993b. ^{60}Fe in eucrites. Earth and Planetary Science Letters 119, 159–166.

Shukolyukov, A., Lugmair, G.W., 1997. The ^{53}Mn-^{53}Cr isotope system in the Omolon pallasite and the half-life of ^{187}Re. Lunar and Planetary Science XXVIII, 1315–1316.

Shukolyukov, A., Lugmair, G.W., 2006. Manganese-chromium isotope systematics of carbonaceous chondrites. Earth and Planetary Science Letters 250, 200–213.

Shukolyukov, A., Kyte, F.T., Lugmair, G.W., Lowe, D.R., Byerly, G.R., 2000. The oldest impact deposits on Earth – First confirmation of an extraterrestrial component. In: Gilmour, I., Koeberl, C. (Eds.), Impacts and the Early Earth. Springer-Verlag, Berlin, pp. 99–116.

Siever, R., 1992. The silica cycle in the Precambrian. Geochimica et Cosmochimica Acta 56, 3265–3272.

Sillitoe, R.H., Thompson, J.F.H., 1998. Intrusion-related vein gold deposits: types, tectono-magmatic settings and difficulties of distinction from orogenic gold deposits. Resource Geology 48, 237–250.

Silva, K.E., Cheadle, M.J., Nisbet, E.G., 1997. The origin of B1 zones in komatiite flows. Journal of Petrology 38 (11), 1565–1584.

Simard, M., Gosselin, C., David, J., 2002. Geology of the Maricourt area (24D). Ministère des Ressources naturelles, Québec, RG 2001-07, 41 pp.

Simard, M., Parent, M., David, J., Sharma, K.N.M., 2003. Géologie de la région de la rivière Innuksuac (34K et 34L). Ministère des Ressources naturelles, Québec, RG 2002-10, 46 pp.

Simard, M., Parent, M., David, J., Sharma, K.N.M., 2004. Geology of the Rivière Innuksuac area (34K and 34L). Ministère des Ressources naturelles, Québec, RG 2003-03, 40 pp.

Simon, N.S.C., Irvine, G.J., Davies, G.R., Pearson, D.G., Carlson, R.W., 2003. The origin of garnet and clinopyroxene in "depleted" Kaapvaal peridotites. Lithos 71, 289–322.

Simons, M., Solomon, S.C., Hager, B.H., 1997. Localization of gravity and topography: constraints on the tectonics and mantle dynamics of Venus. Geophysical Journal International 131, 24–44.

Simonson, B.M., Glass, B.P., 2004. Spherule layers – Records of ancient impacts. Annual Review of Earth and Planetary Sciences 32, 329–361.

Sims, J.R., Dirks, P.H.G.M., Carson, C.J., Wilson, C.J.L., 1994. The structural evolution of the Rauer Group, East Antarctica: mafic dykes as passive markers in a composite Proterozoic terrain. Antarctic Science 6, 379–394.

Sims, P.K., 1980. Boundary between Archean greenstone and gneiss terranes in northern Wisconsin and Michigan. In: Geological Society of America, Special Paper 182, pp. 113–124.

Sims, P.K., 1991. Great Lakes Tectonic Zone in Marquette Area, Michigan – Implications for Archean tectonics in north-central United States. In: U.S. Geological Survey, Bulletin 1904-E, pp. E1–E17.

Sims, P.K., 1993. Structure map of Archean rocks, Palmer and Sands 7½ minute quadrangles, Michigan, showing Great Lakes tectonic zone. U.S. Geological Survey, Miscellaneous Investigations Series Map I-2355, scale 1:24,000.

Sims, P.K., 1996a. Great Lakes Tectonics Zone. In: Sims, P.K., Carter, L.M.H. (Eds.), Archean and Proterozoic Geology of the Lake Superior Region, U.S.A., 1993. U.S. Geological Survey, Professional Paper 1556, pp. 24–28.

Sims, P.K., 1996b. Minnesota River Valley Subprovince (Archean Gneiss Terrane). In: Sims, P.K., Carter, L.M.H. (Eds.), Archean and Proterozoic Geology of the Lake Superior Region, U.S.A., 1993. U.S. Geological Survey, Professional Paper 1556, pp. 14–23.

Sims, P.K., Day, W.C., 1992. A regional structural model for gold mineralization in the southern part of the Archean Superior province, U.S.A. U.S. Geological Survey, Bulletin 1904-M, 19 pp.

Sims, P.K., Day, W.C., 1993. The Great Lakes Tectonic Zone – revisited. In: U.S. Geological Survey, Bulletin 1904-S, pp. 1–11.

Sims, P.K., Card, K.D., Morey, G.B., et al., 1980. The Great Lakes tectonic zone; a major crustal structure in central North America. Bulletin of the Geological Society of America 91, 690–698.

Sims, P.K., Kotov, A.B., Neymark, L.A., Peterman, Z.E., 1997. Nd isotopic evidence for middle and early Archean crust in the Wawa subprovince of Superior Province, Michigan, U.S.A. In: Geological Association of Canada, Abstract Volume 23, A137.

Singer, B.S., Myers, J.D., Frost, C.D., 1992. Mid-Pleistocene lavas from the Seguam volcanic center, central Aleutian arc: closed-system fractional crystallization of a basalt to rhyodacite eruptive suite. Contributions to Mineralogy and Petrology 110, 87–112.

Sinigoi, S., Quick, J.E., Mayer, A., Demarchi, G., 1995. Density controlled assimilation of underplated crust, Ivrea–Verbano zone, Italy. Earth and Planetary Science Letters 129, 183–191.

Sircombe, K.N., Bleeker, W., Stern, R.A., 2001. Detrital zircon geochronology and grain-size analysis of a ~2800 Ma Mesoarchean proto-cratonic cover succession, Slave Province, Canada. Earth and Planetary Science Letters 189, 207–220.

Sisson, T.W., Ratajeski, K., Hankins, W.B., Glazner, A.F., 2005. Voluminous granitic magmas from common basaltic sources. Contribution to Mineralogy and Petrology 148, 635–661.

Sjerp, N., 1983. Geological report, Lennons Find Prospect, Pilbara Goldfield, Western Australia. Unpublished Sjerp and Associates report to Centenerary International Pty Ltd.

Skjerlie, K.P., Johnston, A.D., 1993. Fluid-absent melting behavior of an F-rich tonalitic gneiss at mid-crustal pressures: implications for the generation of anorogenic granites. Journal of Petrology 34, 785–815.

Skjerlie, K., Patiño-Douce, A.E., 2002. The fluid-absent partial melting of a zoisite bearing quartz eclogite from 1.0 to 3.2 GPa: implications for melting of a thickened continental crust and for subduction-zone processes. Journal of Petrology 43, 291–314.

Sklyarov, E.V., Gladkochub, D.P., Mazukabzov, A.M., Men'shagin, Yu.V., 1998. Metamorphism of the ancient ophiolites of the Sharyzhalgay massif. Russian Geology and Geophysics 39 (12), 1733–1749.

Sklyarov, E.V., Gladkochub, D.P., Watanabe, T., Fanning, M.K., Mazukabzov, A.M., Menshagin, Yu.V., Ota, T., 2001. Archean supracrustal rocks of the Sharyzhalgay salient and their tectonic implications. Doklady Earth Sciences 377A (3), 278–282.

Skrzypczak, A., Derenne, S., Robert, F., Binet, L., Gourier, D., Rouzaud, J.N., Clinard, C., 2004. Characterization of the organic matter in an Archean chert (Warrawoona, Australia). Geochimica et Cosmochimica Acta (Abstracts volume), A240 (Goldschmidt, Copenhagen).

Skrzypczak, A., Derenne, S., Binet, L., Gourier, D., Robert, F., 2005. Characterization of a 3.5 Billion year old organic matter: Electron paramagnetic resonance and pyrolysis GC-MS, tools to assess syngeneity and biogenicity. Lunar and Planetary Science XXXVI, abstract #1351.

Skrzypczak, A., Derenne, S., Robert, F., 2006. Molecule evidence for life in the 3.46 Ba-old Warrawoona chert. Geophysical Research Abstracts 8, 06203 (EGU06-A-06203).

Skulski, T., Villeneuve, M., 1999. Geochronological compilation of the Superior Province, Manitoba, Ontario, Quebec. Geological Survey of Canada, Open File 3715.

Skulski, T., Hynes, A., Francis, D., 1988. Basic lavas of the Archean La Grande greenstone belt: products of polybaric fractionation and crustal contamination. Contributions to Mineralogy and Petrology 100, 236–245.

Skulski, T., Percival, J.A., Stern, R.A., 1996. Archean crustal evolution in the central Minto block, northern Quebec. In: Radiogenic Age and Isotopic Studies, Report 9: Geological Survey of Canada, Current Research 1995-F, pp. 17–31.

Skulski, T., Sanborn-Barrie, M., Stern, R.A., 1998. Did the Sturgeon Lake belt form near a continental margin? In: Harrap, R.M., Helmstaedt, H. (Eds.), Western Superior Transect Fifth Annual Workshop. Lithoprobe Report 65, Lithoprobe Secretariat, University of British Columbia, pp. 87–89.

Skulski, T., Stern, R.A., Ciesielski, A., 1998. Timing and sources of granitic magmatism, Bienville subprovince, northern Quebec. In: Geological Association of Canada – Mineralogical Association of Canada, Annual Meeting, Abstract Volume 23, p. A-175.

Skulski, T., Percival, J.A., Whalen, J.B., Stern, R.A., 1999. Archean crustal evolution in the northern Superior Province. In: Tectonic and Magmatic Processes in Crustal Growth: A Pan-Lithoprobe Perspective. Lithoprobe Report 75, Lithoprobe Secretariat, University of British Columbia, pp. 128–129.

Skulski, T., Corkery, M.T., Stone, D., Whalen, J.B., Stern, R.A., 2000. Geological and geochronological investigations in the Stull Lake – Edmund Lake greenstone belt and granitoid rocks of the northwestern Superior Province. In: Report of Activities 2000, Manitoba Industry, Trade and Mines, Manitoba Geological Survey, pp. 117–128.

Skulski, T., Hynes, A., Percival, J., Craven, J., Sanborn-Barrie, M., Helmstaedt, H., 2004. Secular Changes in Tectonic Evolution and the Growth of Continental Lithosphere. In: The LITHOPROBE Celebratory Conference: From Parameters to Processes – Revealing the Evolution of a Continent, October 12–15, 2004, Toronto, Ontario, Canada. Lithoprobe Report 86.

Sleep, N.H., Windley, B.F., 1982. Archean plate tectonics: Constraints and inferences. Journal of Geology 90, 363–379.

Smelov, A.P., Zedgenizov, A.N., Parfenov, L.M., Timofeev, V.F., 1998. Precambrian Terranes of the Aldan–Stanovoi shield. In: Metallogeny, Petroleum Potential, and Geodynamics of the North Asian Craton and the Surrounding Orogenic Belts. Santai Publ., Irkutsk, pp. 119–120 (in Russian).

Smelov, A.P., Zedgenizov, V.F., Timofeev, V.F., 2001. Aldano-Stanovoy shield. In: Parfenov, L.M., Kuzmin, M.I. (Eds.), Tectonics, Geodynamics and Metallogeny of the Sakha Republic (Yakutia). Nauka/Interperiodica, Moscow, pp. 81–104 (in Russian).

Smelov, A.P., Bereskin, V.I., Zedgenizov, A.N., 2002. New data on composition, structure and ore deposits of the Kotuykan zone of tectonic melange. Otechestvennaya Geologya 4, 45–49 (in Russian).

Smith, A.J., 1981. The geology of the farms Hooggenoeg 731JT and Avontuur 721JT, southwest Barberton Greenstone Belt. B.Sc. thesis. University of the Witwatersrand, Johannesburg.

Smith, H.S., Erlank, A.J., 1982. Geochemistry and petrogenesis of komatiites from the Barberton greenstone belt, South Africa. In: Arndt, N.T., Nisbet, E.G. (Eds.), Komatiites. Allen and Unwin, London, pp. 347–397.

Smith, J.B., Barley, M.E., Groves, D.I., Krapez, B., McNaughton, N.J., Bickle, M.J., Chapman, H.J., 1998. The Sholl Shear Zone, West Pilbara: evidence for a domain boundary structure from integrated tectonic analyses, SHRIMP U-Pb dating and isotopic and geochemical data of granitoids. Precambrian Research 88, 143–171.

Smith, P.E., Evensen, N.M., York, D., Moorbath, S., 2005. Oldest reliable terrestrial ^{40}Ar-^{39}Ar age from pyrite crystals at Isua, west Greenland. Geophysical Research Letters 32, L21318, 1–4.

Smithies, R.H., 2000. The Archaean tonalite-trondhjemite-granodiorite (TTG) series is not an analogue of Cenozoic adakite. Earth and Planetary Science Letters 182, 115–125.

Smithies, R.H., Champion, D.C., 1999. Late Archaean felsic alkaline igneous rocks in the Eastern Goldfields, Yilgarn Craton, Western Australia: a result of lower crustal delamination. Journal of the Geological Society of London 156, 561–576.

Smithies, R.H., Champion, D.C., 2000. The Archaean high-Mg diorite suite: Links to tonalite-trondhjemite-granodiorite magmatism and implications for early Archaean crustal growth. Journal of Petrology 41, 1653–1671.

Smithies, R.H., Champion, D.C., Cassidy, K.F., 2003. Formation of Earth's early Archaean continental crust. Precambrian Research 127, 89–101.

Smithies, R.H., Champion, D.C., Sun, S.-S., 2004a. Evidence for early LILE-enriched mantle source regions: diverse magmas from the c. 3.0 Ga Mallina Basin, Pilbara Craton, NW Australia. Journal of Petrology 45, 1515–1537.

Smithies, R.H., Champion, D.C., Sun, S.-S., 2004b. The case for Archaean boninites. Contributions to Mineralogy and Petrology 147, 705–721.

Smithies, R.H., Champion, D.C., Van Kranendonk, M.J., Howard, H.M., Hickman, A.H., 2005a. Modern-style subduction processes in the Mesoarchaean: geochemical evidence from the 3.12 Ga Whundo intraoceanic arc. Earth and Planetary Science Letters 231, 221–237.

Smithies, R.H., Van Kranendonk, M.J., Champion, D.C., 2005b. It started with a plume – early Archaean basaltic proto-continental crust. Earth and Planetary Science Letters 238, 284–297.

Smithies, R.H., Van Kranendonk, M.J., Champion, D.C., 2007a. The Mesoarchaean emergence of modern style subduction. Gondwana Research 11, 50–68.

Smithies, R.H., Champion, D.C., Van Kranendonk, M.J., Hickman, A.H., 2007b. Geochemistry of volcanic rocks of the northern Pilbara Craton. Western Australia Geological Survey, Report 104, 47 pp.

Smithson, S.B., Pierson, W.R., Wilson, S.L., et al., 1985. Seismic reflection results from Precambrian crust. In: The Deep Proterozoic Crust in the North Atlantic Provinces. NATO ASI Series, Series C: Mathematical and Physical Sciences, vol. 158. D. Reidel Publishing Company International, pp. 21–37.

Smoliar, M.I., 1993. A survey of Rb-Sr systematics of eucrites. Meteoritics 28, 105–113.

Smoliar, M.I., Walker, R.J., Morgan, J.W., 1996. Re-Os ages of group IIA, IIIA, IVA and IVB iron meteorites. Science 271, 1099–1102.

Smrekar, S.E., Stofan, E.R., 1997. Corona formation and heat loss on Venus by coupled upwelling and delamination. Science 277, 1289–1294.

Smrekar, S.E., Kiefer, W.S., Stofan, E.R., 1997. Large volcanic rises on Venus. In: Bouger, S.W., Hunten, D.M., Phillips, R.J. (Eds.), Venus II. University of Arizona Press, Tucson, AZ, pp. 845–879.

Snape, I.S., Black, L.P., Harley, S.L., 1997. Refinement of the timing of magmatism and high-grade deformation in the Vestfold Hills, East Antarctica, from new Shrimp U-Pb zircon geochronology. In: Ricci, C.A. (Ed.), The Antarctic Region: Geological Evolution and Processes. Terra Antartica Publication, Siena, pp. 139–148.

Snelson, C.M., Henstock, T.J., Keller, G.R., Miller, K.C., Levander, A., 1998. Crustal and uppermost mantle structure along the Deep Probe seismic profile. Rocky Mountain Geology 33 (2), 181–198.

Snyder, D., 2002. Cooling of lava flows on Venus: The coupling of radiative and convective heat transfer. Journal of Geophysical Research 107, 5080–5088.

Sobotovich, E.V., Kamenev, Y.N., Komaristyy, A.A., Rudnik, V.A., 1976. The oldest rocks of Antarctic (Enderby Land). International Geology Review 18, 371–388.

Solomatov, V.S., Moresi, L.N., 1996. Stagnant lid convection on Venus. Journal of Geophysical Research 101, 4737–4753.

Solomon, M., Eastoe, C.J., Walshe, J.L., Green, G.R., 1988. Mineral deposits and sulfur isotope abundances in the Mount Read volcanics between Que River and Mount Darwin, Tasmania. Economic Geology 83, 1307–1328.

Solomon, S.C., 1993. The geophysics of Venus. Physics Today 46, 48–55.

Solomon, S.C., Head, J.W., Kaula, W.M., McKenzie, D., Parsons, B., Phillips, R.J., Schubert, G., Talwani, M., 1991. Venus tectonics: Initial analysis from Magellan. Science 252, 297–312.

Solomon, S.C., Smrekar, S.E., Bindschadler, D.L., Grimm, R.E., Kaula, W.M., McGill, G.E., Phillips, R.J., Saunders, R.S., Schubert, G., Squyres, S.W., Stofan, E.R., 1992. Venus tectonics: An overview of Magellan observations. Journal of Geophysical Research 97, 13,199–13,255.

Solomon, S.C., et al., 2005. New Perspectives on Ancient Mars. Science 30, 1214–1220.

Solvang, M., 1999. An investigation of metavolcanic rocks from the eastern part of the Isua greenstone belt, Western Greenland. Geological Survey of Denmark and Greenland (GEUS) Internal Report, Copenhagen, Denmark, 62 pp.

Song, B., 1992. Isotope geochronology, REE characteristics and genesis of Miyun rapakivi granite. Bulletin of the Chinese Academy of Geological Sciences 25, 137–157 (in Chinese with English abstract).

Song, B., Nutman, A.P., Liu, D., Wu, J., 1996. 3800 to 2500 Ma crustal evolution in the Anshan area of Lianoning Province, NW China. Precambrian Research 78, 79–94.

Souders, A.K., Frost, C.D., 2006. In suspect terrane? Provenance of the Late Archean Phantom Lake Metamorphic Suite, Sierra Madre, Wyoming. Canadian Journal of Earth Sciences 43, 1557–1577.

Southwick, D.L., 2002. Geologic map of Pre-Cretaceous Bedrock in Southwest Minesota. Minnesota Geological Survey, Miscellaneous map Series, Map M-121, scale 1:250,000.

Southwick, D.L., Chandler, V.W., 1996. Block and shear-zone architecture of the Minnesota River Valley Subprovince: implications for late Archean accretionary tectonics. Canadian Journal of Earth Sciences 33, 831–847.

Spaggiari, C.V., 2006. Interpreted bedrock geology of the northern Murchison Domain, Youanmi Terrane, Yilgarn Craton. Western Australia. Geological Survey of Western Australia, Record 2006/10.

Spaggiari, C.V., 2007a. Structural and lithological evolution of the Jack Hills greenstone belt, Narryer Terrane, Yilgarn craton. Western Australia Geological Survey, Record 2007/3, 49 pp.

Spaggiari, C.V., 2007b. The Jack Hills greenstone belt, Western Australia, Part 1: Structural and tectonic evolution over >1.5 Ga. Precambrian Research 155, 204–228.

Spaggiari, C.V., Pidgeon, R.T., Wilde, S.A., 2004. Structural and Tectonic framework of >4.0 Ga detrital zircons from the Jack Hills Belt, Narryer Terrane, Western Australia. Geoscience in a Changing World, GSA Conference Denver 2004. Abstracts with Programs, vol. 36 (5), p. 207 (CD-ROM).

Spaggiari, C.V., Pidgeon, R.T., Wilde, S.A., 2007. The Jack Hills greenstone belt, Western Australia, Part 2: Lithological relationships and implications for the deposition of >4.0Ga detrital zircons. Precambrian Research 115, 261–286.

Spaggiari, C.V., Wartho, J.-A., Wilde, S.A., Bodorkos, S., in review. Proterozoic development of the northern margin of the Archean Yilgarn Craton. Precambrian Research.

Spear, F.S., 1993. Metamorphic Phase Equilibria and Pressure-Temperature-Time Paths. Mineralogical Society of America, Washington.

Spencer, J., 2001. Possible giant metamorphic core complex at the center of Artemis Corona, Venus. Bulletin of the Geological Society of America 113, 333–345.

Spiridonov, V.G., Karpenko, S.F., Lyalikov, A.V., 1993. Sm-Nd age and geochemistry of granulites in the Anabar shield. Geochemistry International 10, 1412–1427.

Spötl, C., Houseknecht, D.W., Jaques, R.C., 1998. Kerogen maturation and incipient graphitization of hydrocarbon source rocks in the Arkoma Basin, Oklahoma and Arkansas: A combined petrographic and Raman spectrometric study. Organic Geochemistry 28, 535–542.

Spray, J.G., 1985. Dynamothermal transition between Archaean greenstone and granitoid gneiss at Lake Dundas, Western Australia. Journal of Structural Geology 7, 187–203.

Sproule, R.A., Lesher, C., Ayer, J.A., Thurston, P.C., Herzberg, C.T., 2002. Spatial and temporal variations in the geochemistry of komatiites and komatiitic basalts in the Abitibi greenstone belt. Precambrian Research 115, 153–186.

Squyres, S.W., Janes, D.M., Baer, G., Bindschadler, D.L., Schubert, G., Sharpton, V.L., Stofan, E.R., 1992a. The morphology and evolution of coronae on Venus. Journal of Geophysical Research 97, 13,611–13,634.

Squyres, S.W., Jankowski, D.G., Simons, M., Solomon, S.C., Hager, B.H., McGill, G.E., 1992b. Plains tectonism on Venus: The deformation belts of Lavinia Planitia. Journal of Geophysical Research 97, 13,579–13,599.

Srinivasan, G., Sahijpal, S., Ulyanov, A.A., Goswami, J.N., 1996. Ion microprobe studies of Efremovka CAIs: potassium isotope composition and ^{41}Ca in the early solar system. Geochimica et Cosmochimica Acta 60, 1823–1835.

Srinivasan, G., Goswami, J.N., Bhandari, N., 1999. ^{26}Al in eucrite Piplia Kalan: plausible heat source and formation chronology. Science 284, 1348–1350.

Stanistreet, I.G., De Wit, M.J., Fripp, R.E.P., 1981. Do graded units of accretionary spheroids in the Barberton greenstone belt indicate Archaean deep water environment? Nature 293, 280–284.

Staudigel, H., 2003. Hydrothermal alteration processes in the oceanic crust. In: Treatise on Geochemistry, vol. 3. Elsevier, Amsterdam, pp. 511–535.

Staudigel, H., Hart, S.R., 1983. Alteration of basaltic glass: mechanisms and significance for the oceanic crust-seawater budget. Geochimica et Cosmochimica Acta 47, 337–350.

Staudigel, H., Plank, T., White, W., Schmincke, H.-U., 1996. Geochemical fluxes during seafloor alteration of the basaltic upper oceanic crust: DSDP sites 417 and 418. In: Bebout, G., Scholl, D.W., Kirby, S.H., Platt, J.P. (Eds.), Subduction: Top to Bottom. Geophysical Monograph, vol. 96, pp. 19–37.

Staudigel, H., Furnes, H., Banerjee, N.R., Dilek, Y., Muehlenbacks, K., 2006. Microbes and volcanoes: A tale from the oceans, ophiolites and greenstone belts. GSA Today 16 (10), 4–10.

Stein, M., 2003. Tracing the plume material in the Arabian-Nubian shield. Precambrian Research 123, 223–234.

Stein, M., Hofmann, A.W., 1992. Fossil plume head beneath the Arabian lithosphere? Earth and Planetary Science Letters 114, 193–209.

Stein, M., Hofmann, A.W., 1994. Mantle plumes and episodic crustal growth. Nature 372, 63–68.

Stepanov, L.L., 1974. Radiogenic age of polymetamorphic rocks in the Anabar shield. In: Early Precambrian Formations in the Central Part of the Arctic and Related Mineral Resources. Arctic Institute Publication, Leningrad, pp. 76–83 (in Russian).

Stepanyuk, L.M., 1991. U-Pb age of the microcline granites in the Anabar shield. Doklady Earth Sciences 10, 127–129.

Stepanyuk, L.M., Ponomarenko, A.N., Yakovlev, B.G., 1993. Genesis and age of zircons in granulites (the Daldyn mafic granulite of the Anabar shield as an example). Mineralogical Journal 15 (2), 40–52.

Stern, R.A., Bleeker, W., 1998. Age of the world's oldest rocks refined using Canada's SHRIMP: The Acasta Gneiss Complex, Northwest Territories. Geoscience Canada 25, 27–31.

Stern, R.A., Hanson, G.N., 1991. Archaean high-Mg granodiorites: a derivative of light rare earth enriched monzodiorite of mantle origin. Journal of Petrology 32, 201–238.

Stern, R.A., Percival, J.A., Mortensen, J.K., 1994. Geochemical evolution of the Minto block: a 2.7 Ga continental magmatic arc built on the Superior proto-craton. Precambrian Research 65, 115–153.

Stern, R.J., 2005. Evidence from ophiolites, blueschists, and ultrahigh-pressure metamorphic terranes that the modern episode of subduction tectonics began in Neoproterozoic time. Geology 33, 557–560.

Stetter, K.O., 1996. Hyperthermophiles in the history of life. In: CIBA Foundation Symposia 202, 1–18.

Stetter, K.O., Gaag, G., 1983. Reduction of molecular sulfur by methanogenic bacteria. Nature 305, 309–311.

Stevens, G., Droop, T.R., Armstrong, R.A., Anhaeusser, C.R., 2002. Amphibolite-facies metamorphism in the Schapenburg schist belt: a record of the mid-crustal response to ~3.23 Ga terrane accretion in the Barberton greenstone belt. South African Journal of Geology 105, 271–284.

Stevenson, D.J., 1988. Planetary evolution – Greenhouses and magma oceans. Nature 335, 587–588.

Stevenson, D.J., 1990. Fluid dynamics of core formation. In: Newsome, H.E., Jones, J.H. (Eds.), Origin of the Earth. Oxford University Press, New York, pp. 231–249.

Stevenson, D.J., Lunine, J.I., 1988. Rapid formation of Jupiter by diffuse redistribution of water vapour in the solar nebula. Icarus 75, 146–155.

Stevenson, I.M., 1968. A Geological Reconnaissance of Leaf River Map-Area, New Quebec and Northwest Territories. Geological Survey of Canada, Memoire 356.

Stevenson, R.K., 1995. Crust and mantle evolution in the late Archean: Evidence from a Sm-Nd isotopic study of the North Spirit Lake greenstone belt, northwestern Ontario. Bulletin of the Geological Society of America 107, 1458–1467.

Stevenson, R.K., Bizzarro, M., 2005. Hf and Nd isotope evolution of litologies from the 3.8 Ga Nuvvuagittuq Sequence, northen Superior Province, Canada. Geochimica et Cosmochimica Acta 69, A391.

Stevenson, R.K., Bizzarro, M., 2006. Petrology and petrogenesis of the Eoarchean Nuvvuagittuq tonalite suite. Geological Association of Canada, Abstracts, vol. 31.

Stevenson, R.K., Patchett, P.J., 1990. Implications for the evolution of continental crust from Hf isotope systematics of Archean detrital zircons. Geochimica et Cosmochimica Acta, 54, 1683–1697.

Stevenson, R.K., Turek, A., 1992. An isotopic study of the Island Lake greenstone belt, Manitoba: crustal evolution and progressive cratonization in the Late Archean. Canadian Journal of Earth Sciences 29, 2200–2210.

Stevenson, R.K., David, J., Parent, M., 2006. Crustal evolution in the western Minto Block, northern Superior Province, Canada. Precambrian Research 145, 229–242.

Stewart, B.W., Papanastassiou, D.W., Wasserburg, G.J., 1994. Sm-Nd chronology and petrogenesis of mesosiderites. Geochimica et Cosmochimica Acta 58, 3487–3509.

St-Onge, M.R., King, J.E., Lalonde, A.E., 1988. Geology, east-central Wopmay Orogen, District of Mackenzie, Northwest Territories. Geological Survey of Canada, Open-File Report 1923, 3 sheets, scale 1:125,000.

Stofan, E.R., Sharpton, V.L., Schubert, G., Baer, G., Bindschadler, D.L., Janes, D.M., Squyres, S.W., 1992. Global distribution and characteristics of coronae and related features on Venus: Implications for origin and relation to mantle processes. Journal of Geophysical Research 97, 13,347–13,378.

Stofan, E.R., Hamilton, V.E., Janes, D.M., Smrekar, S.E., 1997. Coronae on Venus: Morphology and origin. In: Bouger, S.W., Hunten, D.M., Phillips, R.J. (Eds.), Venus II. University of Arizona Press, Tucson, AZ, pp. 931–968.

Stofan, E.R., Anderson, S.W., Crown, D.A., Plaut, J.J., 2000. Emplacement and composition of steep-sided domes on Venus. Journal of Geophysical Research 105, 26,757–26,772.

Stofan, E.R., Tapper, S., Guest, J.E., Smrekar, S.E., 2001. Preliminary analysis of an expanded database of coronae on Venus. Geophysical Research Letters 28, 4267–4270.

Stolz, G.W., Nesbitt, R.W., 1981. The komatiite nickel sulfide association at Scotia: A petrochemical investigation of the ore environment. Economic Geology 76, 1480–1502.

Stone, D., 1998. Precambrian geology of the Berens River area, northwest Ontario. Ontario Geological Survey, Open File Report 5963, 115 pp.

Stone, D., 2000. Temperature and pressure variations in suites of Archean felsic plutonic rocks, Berens River area, northwest Superior Province, Ontario, Canada. Canadian Mineralogist 38, 455–470.

Stone, D., Tomlinson, K.Y., Davis, D.W., Fralick, P., Hallé, J., Percival, J.A., Pufahl, P., 2002. Geology and tectonostratigraphic assemblages, South-central Wabigoon Subprovince. Ontario Geological Survey, Preliminary Map P.3448 or Geological Survey of Canada, Open File 4284, scale: 1:250,000.

Stone, D., Corkery, M.T., Hallé, J., Ketchum, J., Lange, M., Skulski, T., Whalen, J., 2004. Geology and tectonostratigraphic assemblages, eastern Sachigo Subprovince, Ontario and Manitoba. Ontario Geological Survey, Preliminary Map P.3462 or Manitoba Geological Survey Open File OF2003-2 or Geological Survey of Canada, Open File 1582, scale: 1:250,000.

Stott, G.M., 1997. The Superior Province, Canada. In: de Wit, M.J., Ashwal, L.D. (Eds.), Greenstone Belts. Oxford Monograph on Geology and Geophysics, vol. 35. Clarendon, Oxford, pp. 480–507.

Stott, G.M., Corfu, F., 1991. Uchi subprovince. In: Thurston, P.C., Williams, H.R., Sutcliffe, R.H., Stott, G.M. (Eds.), Geology of Ontario. Ontario Geological Survey, Special Volume 4, Part 1, pp. 145–238.

Stott, G.M., Corfu, F., Breaks, F.W., Thurston, P.C., 1989. Multiple orogenesis in northwestern Superior Province. Geological Association of Canada, Abstracts, vol. 14, A56.

Stowe, C.W., 1984. The early Archaean Selukwe nappe, Zimbabwe. In: Kröner, A., Greilings, R. (Eds.), Precambrian Tectonics Illustrated. E. Schweizerbart'sche Verlagsbuchhandlung, Stuttgart, pp. 41–56.

Strauss, H., 1997. The isotopic composition of sedimentary sulfur through time. Palaeogeography, Palaeoclimatology, Palaeoecology 132, 97–118.

Strauss, H., 2003. Sulphur isotopes and the early Archaean sulphur cycle. Precambrian Research 126, 349–361.

Strauss, H., Moore, T.B., 1992. Abundances and isotopic compositions of carbon and sulfur species in whole rock and kerogen samples. In: Schopf, J.W., Klein, C. (Eds.), The Proterozoic Biosphere. Cambridge University Press, pp. 709–798.

Strom, R.G., Schaber, G.G., Dawson, D.D., 1994. The global resurfacing of Venus. Journal of Geophysical Research 99, 10,899–10,926.

Stumm, W., Morgan, J.J., 1996. Aquatic Chemistry. Wiley-Interscience, New York, 1022 pp.

Sudo, A., 1988. Stability of phlogopite in the upper mantle: implications for subduction zone magmatism. M.Sc. thesis. Kyoto University, Kyoto, Japan.

Sukhanov, M.K., Spiridonov, V.G., Karpenko, S.F., 1990. First isochron Sm-Nd dating of the anorthosites in the Anabar shield. Doklady Earth Sciences 310 (2), 448.

Summons, R.E., Jahnke, L.L., Hope, J.M., Logan, G.A., 1999. 2-Methylhopanoids as biomarkers for cyanobacterial oxygenic photosynthesis. Nature 400, 554–557.

Sun, D.Z., Jin, W.S., Bai, J., Wang, W.Y., Jiang, Y.N., Yang, C.L., 1984. The Early Precambrian Geology of Eastern Hebei. Tianjin Science and Technology Press, Tianjin, 273 pp. (in Chinese with English abstract).

Sun, D.Z., Li, H.M., Lin, Y.X., Zhou, H.F., Zhao, F.Q., Tang, M., 1991. Precambrian geochronology, chronotectonic framework and model of chronocrustal structure of the Zhongtiao Mountains. Acta Geologica Sinica 65, 216–231 (in Chinese with English abstract).

Sun, S.-S., Hickman, A.H., 1998. New Nd-isotopic and geochemical data from the west Pilbara – implications for Archaean crustal accretion and shear zone development. Australian Geological Survey Organisation, Research Newsletter, June 1998.

Sun, S.-S., McDonough, W.F., 1989. Chemical and isotopic systematics of oceanic basalts: implications for mantle compositions and processes. In: Saunders, A.D., Norry, M.J. (Eds.), Magmatism in Ocean Basins. Geological Society of London, Special Publication 42, pp. 313–345.

Surkov, Y.A., Moskalyova, L.P., Kharyukova, V.P., Dudin, A.D., Smirnov, G.G., Zaitseva, S.Y., 1986. Venus rock composition at the Vega-2 landing site. Journal of Geophysical Research 91, E215–E218.

Sutherland, W.M., Hausel, W.D., 2002. Preliminary geologic map of the Rattlesnake Hills 1:100,000 Quadrangle, Fremont and Natrona Counties, central Wyoming. Wyoming State Geological Survey Mineral Report 2002-2.

Sutton, J., Watson, J.V., 1951. The pre-Torridonian history of the Loch Torridon and Scourie areas of the north-west Highlands and its bearing on the chronological classification of the Lewisian. Journal of the Geological Society of London 106, 241–296.

Suyehiro, K., Takahashi, N., Arie, Y., Yokoi, Y., Hino, R., Shinohara, M., Kanazawa, T., Hirata, N., Tokuyama, H., Tara, A., 1996. Continental crust, crustal underplating and low-Q upper mantle beneath an ocean island arc. Science, 272, 390–392.

Swager, C.P., Griffin, T.J., Witt, W.K., Wyche, S., Ahmat, A.L., Hunter, W.M., Mc-Goldrick, P.J., 1995. Geology of the Archaean Kalgoorlie Terrane – an explanatory note. Western Australia Geological Survey, Report 48, 26 pp.

Sweetapple, M.T., Collins, P.L.F., 2002. Genetic framework for the classification and distribution of Archean rare metal pegmatites in the North Pilbara Craton, Western Australia. Economic Geology 97, 873–896.

Swindle, T.D., Grossman, J.N., Olinger, C.T., Garrison, D.H., 1991a. Iodine-xenon, chemical, and petrographic studies of Semarkona chondrules: Evidence for the timing of aqueous alteration. Geochimica et Cosmochimica Acta 55, 3723–3734.

Swindle, T.D., Cafee, M.W., Hohenberg, C.M., Lindstrom, M.M., Taylor, G.I., 1991b. Iodine-xenon studies of petrographically and chemically characterized Chainpur chondrules. Geochimica et Cosmochimica Acta 55, 861–880.

Swindle, T.D., Davis, A.M., Hohenberg, C.M., MacPherson, G.J., Nyquist, L.E., 1996. Formation times of chondrules and Ca-Al-rich inclusions: constraints from short-lived radionuclides. In: Hewins, R.H., Jones, R.H., Scott, E.R.D. (Eds.), Chondrules and the Protoplanetary Disk. Cambridge University Press, New York, pp. 77–86.

Swindle, T.D., Kring, D.A., Burkland, M.K., Hill, D.H., Boynton, W.V., 1998. Noble gases, bulk chemistry, and petrography of olivine-rich achondrites Eagle Nest and Lewis Cliff 88763: comparison to brachinites. Meteoritics and Planetary Science 33, 31–48.

Sylvester, P.J., Campbell, I.H., Bowyer, D.A., 1997. Niobium/uranium evidence for early formation of the continental crust. Science 275, 521–523.

Syme, E.C., Corkery, M.T., Bailes, A.H., Lin, S., Skulski, T., Stern, R.A., 1999. Towards a new tectonostratigraphy for the Knee Lake greenstone belt, Sachigo subprovince, Manitoba. In: Harrap, R.M., Helmstaedt, H. (Eds.), Western Superior Transect Fifth

Annual Workshop. Lithoprobe Report 70, Lithoprobe Secretariat, University of British Columbia, pp. 124–131.

Symonds, R.B., Rose, W.I., Bluth, G.J.S., Gerlach, T.M., 1994. Volcanic-gas studies: Methods, results, and applications. In: Carroll, M.R., Halloway, J.R. (Eds.), Volatiles in Magmas. Mineralogical Society of America, Reviews in Mineralogy, vol. 30, pp. 1–66.

Takahashi, K., Masuda, A., 1990. Young ages of two diogenites and their genetic implications. Nature 343, 540–542.

Takeda, H., Bogard, D.D., Mittlefehldt, D.W., Garrison, D.H., 2000. Mineralogy, petrology, chemistry and ^{39}Ar-^{40}Ar and exposure ages of the Caddo County IAB iron: evidence for early partial melt segregation of a gabbro area rich in plagioclase-diopside. Geochimica et Cosmochimica Acta 64, 1311–1327.

Talbot, C.J., 1973. A plate tectonic model for the Archaean crust. Philosophical Transactions of the Royal Society of London Ser. A 273, 413–427.

Tan, F.C., 1988. Stable carbon isotopes in dissolved inorganic carbon in marine and estuarine environments. In: Fritz, P., Fontes, J.C. (Eds.), Handbook of Enviromental Isotope Geochemistry, vol. 3. Elsevier, Amsterdam, p. 171–190.

Tanaka, K.L., Moore, H.J., Schaber, G.G., Chapman, M.G., Stofan, E.R., Campbell, D.B., Davis, P.A., Guest, J.E., McGill, G.E., Rogers, P.G., Saunders, R.S., Zimbelman, J.R., 1994. The Venus Geologic Mappers' Handbook, second ed. Open-File Report, 94-438, USGS, 50 pp.

Tang, Y-J., Zhang, H-F., Ying, J-F., 2006. Asthernosphere-lithospheric mantle interaction in an extensional regime: implications from the geochemistry of Cenzoic basalt from Taihang Mounatins, North China Craton. Chemical Geology 233, 309–327.

Tarney, J., Dalziel, I.W.D., de Wit, M.J., 1976. Marginal basin 'Rocas Verdes' complex from S. Chile: a model for Archaean greenstone belt formation. In: Windley, B.F. (Ed.), The Early History of the Earth. Wiley, London, pp. 131–146.

Tatsumi, Y., 1989. Migration of fluid phases and genesis of basalt magmas in subduction zones. Journal of Geophysical Research 94, 4697–4707.

Tatsumi, Y., Eggins, S., 1995. Subduction Zone Magmatism. Blackwell, Oxford.

Taylor, B., 2006. The single largest oceanic plateau: Ontong-Java-Manihiki-Hikurangi. Earth and Planetary Science Letters 241, 372–380.

Taylor, S.R., 2001. Solar System Evolution: A New Perspective, second ed. Cambridge University Press, 460 pp.

Taylor, S.R., 2004. Why can't planets be like stars? Nature 430, 509.

Taylor, S.R., McLennan, S.M., 1985. The Continental Crust: Its Composition and Evolution. Blackwell Scientific Publications, 312 pp.

Taylor, S.R., Norman, M.D., 1990. Accretion of differentiated planetesimals to the Earth. In: Newsom, H.E., Jones, J.H. (Eds.), Origin of the Earth. Oxford University Press, New York, pp. 29–43.

Taylor, S.R., Taylor, G.J., Taylor, A.L., 2006. The Moon: A Taylor perspective. Geochimica et Cosmochimica Acta 70, 5904–5918.

Tegtmeyer, A., 1989. Geochronologie und Geochemie im Präkambrium des südlichen Afrika. Unpublished doctoral dissertation, University of Mainz, Germany, 270 pp.

Tegtmeyer, A.R., Kröner, A., 1987. U-Pb zircon ages bearing on the nature of early Archean greenstone belt evolution, Barberton Moutainland, Southern Africa. Precambrian Research 36, 1–20.

Tera, F., Wasserburg, G.J., 1974. U-Th-Pb systematics on lunar rocks and inferences about lunar evolution and the age of the Moon. In: Proceedings Lunar Science Conference 5th, pp. 1571–1599.

Tera, F., Papanastassiou, D.A., Wasserburg, G.J., 1974. Isotopic evidence for a terminal lunar cataclysm. Earth and Planetary Science Letters 22, 1–21.

Tera, F., Carlson, R.W., Boctor, N.Z., 1997. Radiometric ages of basaltic achondrites and their relation to the early history of the solar system. Geochimica et Cosmochimica Acta 61, 1713–1731.

Terabayashi, M., Masada, Y., Ozawa, H., 2003. Archean ocean-floor metamorphism in the North Pole area, Pilbara Craton, Western Australia. Precambrian Research 127, 167–180.

Thamdrup, B., Finster, K., Hansen, J.W., Bak, F., 1993. Bacterial disproportionation of elemental sulfur coupled to chemical-reduction of iron or manganese. Applied and Environmental Microbiology 59, 101–108.

Thériault, R., Chevé, S., 2001. Géologie de la région du Lac Hurault (SNRC 23L). Ministère des Ressources naturelles, Québec, RG 2000-11, 49 pp.

Thieblement, D., Delor, C., Cocherie, A., Lafon, J.M., Goujou, J.C., Balde, A., Bah, M., Sane, H., Fanning, C., 2001. A 3.5 Ga granite-gneiss basement in Guinea: further evidence for early Archean accretion within the West African craton. Precambrian Research 108, 179–194.

Thiemens, M.H., 1999. Atmospheric science – mass-independent isotope effects in planetary atmospheres and the early solar system. Science 283, 341–345.

Thode, H.G., McNamara, J., Collins, C.B., 1949. Natural variations in the isotopic content of sulphur and their significance. Canadian Journal of Research 27B, 361.

Thommes, E.W., Duncan, M.J., Levison, H.F., 2002. The formation of Uranus and Neptune among Jupiter and Saturn. Astronomical Journal 121, 2862–2883.

Thordarson, Y., Self, S., 1998. The Roza Member, Columbia River Basalt Group: A gigantic pahoehoe lava flow field formed by endogenous processes? Journal of Geophysical Research 103, 27,411–27,445.

Thorpe, R.A., Hickman, A.H., Davis, D.W., Mortensen, J.K., Trendall, A.F., 1992a. U-Pb zircon geochronology of Archaean felsic units in the Marble Bar region, Pilbara Craton, Western Australia. Precambrian Research 56, 169–189.

Thorpe, R.I., Hickman, A.H., Davis, D.W., Mortensen, J.K., Trendall, A.F., 1992b. Constraints to models for Archaean lead evolution from precise U-Pb geochronology from the Marble Bar region, Pilbara Craton, Western Australia. In: Glover, J.E., Ho, S. (Eds.), The Archaean: Terrains, Processes and Metallogeny. Geology Department and University Extension, Publication 22. The University of Western Australia, pp. 395–408.

Thorstenson, D.C., 1970. Equilibrium distribution of small organic molecules in natural waters. Geochimica et Cosmochimica Acta 34, 745.

Thurston, P.C., 1994. Archean volcanic patterns. In: Condie, K.C. (Ed.), Archean Crustal Evolution. Elsevier, Amsterdam, pp. 45–84.

Thurston, P.C., 2002. Autochthonous development of Superior Province greenstone belts? Precambrian Research 115, 11–36.

Thurston, P.C., Chivers, K.M., 1990. Secular variation in greenstone sequence development emphasising Superior Province, Canada. Precambrian Research 46, 21–58.

Thurston, P.C., Ayres, L.D., Edwards, G.R., Gélinas, L., Ludden, J.N., Verpaelst, P., 1985. Archean bimodal volcanism. In: Ayres, L.D., Card, K.D., Weber, W. (Eds.), Evolution of Archean Supracrustal Sequences. Geological Association of Canada, Special Publication 28, pp. 7–21.

Thurston, P.C., Osmani, I.A., Stone, D., 1991. Northwestern Superior Province: review and terrane analysis. In: Thurston, P.C., Williams, H.R., Sutcliffe, R.H., Stott, G.M. (Eds.), Geology of Ontario. Ontario Geological Survey, Special Volume 4, Part 1, pp. 81–142.

Tian, W., Liu, S.W., Liu, C.H., Yu, S.Q., Liu, Q.G., Wang, Y.R., 2006. Zircon SHRIMP dating and geochemistry of TTG rocks of Shushui Complex in Zhongtiaoshan, Shanxi Province. Progress in Natural Sciences 15, 1476–1484 (in Chinese with English abstract).

Tian, Z-Y., Han, P., Xu, K-D., 1992. The Mesozoic-Cenozoic East China rift system. Tectonophysics 208, 341–363.

Tice, M.M., Lowe, D.R., 2004. Photosynthetic microbial mats in the 3416-Myr-old ocean. Nature 431, 549–552.

Tice, M.M., Lowe, D.R., 2006a. Hydrogen-based carbon fixation in the earliest known photosynthetic organisms. Geology 34, 37–40.

Tice, M.M., Lowe, D.R., 2006b. The origin of carbonaceous matter in pre-3.0 Ga greenstone terrains: A review and new evidence from the 3.42 Ga Buck Reef Chert. Earth Science Reviews 76, 259–300.

Tice, M.M., Bostick, B.C., Lowe, D.R., 2004. Thermal history of the 3.5–3.2 Ga Onverwacht and Fig Tree Groups, Barberton greenstone belt, South Africa, inferred by Raman microspectroscopy of carbonaceous material. Geology 32, 37–40.

Tingey, R.J., 1982. The geologic evolution of the Prince Charles Mountains – an Antarctic Archaean cratonic block. In: Craddock, C. (Ed.), Antarctic Geoscience. University of Wisconsin Press, Madison, pp. 455–464.

Tingey, R.J., 1991. The regional geology of Archaean and Proterozoic rocks in Antarctica. In: Tingey, R.J. (Ed.), The Geology of Antarctica. Oxford University Press, Oxford, pp. 1–73.

Tomilenko, A.A., 2006. Fluid regime of mineral formation at high and middle pressures in the continental lithosphere estimated from fluid inclusion study. Dissertation thesis. Institute of the Earth Crust, Irkutsk, 33 pp. (in Russian).

Tomkinson, M.J., Lombard, A., 1990. Structure, metamorphism, and mineralization in the New Consort Gold Mines, Barberton greenstone belt, South Africa. In: Glover, J.E., Ho,

S.E. (Eds.), Third International Archean Symposium, Perth, 1990. Extended Abstract Volume, Geoconferences (W.A.) Inc., Perth, pp. 377–379.

Tomlinson, K.Y., Dickin, A.P., 2003. Geochemistry and neodymium isotopic character of granitoid rocks in the Lac Seul region of the Winnipeg River subprovince, northwestern Ontario. In: Summary of Field Work and Other Activities, 2003, Ontario Geological Survey, Open File Report 6120, pp. 13-1–13-8.

Tomlinson, K.Y., Percival, J.A., 2000. Geochemistry and Nd isotopes of granitoid rocks in the Shikag–Garden lakes area, Ontario: recycled Mesoarchean crust in the central Wabigoon Subprovince. In: Geological Survey of Canada, Current Research 2000-E12, 11 pp.

Tomlinson, K.Y., Stevenson, R.K., Hughes, D.J., Hall, R.P., Thurston, P.C., Henry, P., 1998. The Red Lake greenstone belt, Superior Province: evidence of plume-related magmatism at 3 Ga and evidence of an older enriched source. Precambrian Research 89, 59–76.

Tomlinson, K.Y., Hughes, D.J., Thurston, P.C., Hall, R.P., 1999. Plume magmatism and crustal growth at 2.9 to 3.0 Ga in the Steep Rock and Lumby Lake area, western Superior Province. Lithos 46, 103–136.

Tomlinson, K.Y., Sasseville, C., McNicoll, V., 2001. New U-Pb geochronology and structural interpretations from the Wallace Lake greenstone belt (North Caribou terrane): implications for new regional correlations. In: Harrap, R.M., Helmstaedt, H. (Eds.), Western Superior Transect Seventh Annual Workshop. Lithoprobe Report 80, Lithoprobe Secretariat, University of British Columbia, pp. 8–9.

Tomlinson, K.Y., Davis, D.W., Percival, J.A., Hughes, D.J., Thurston, P.C., 2002. Mafic to felsic magmatism and crustal recycling in the Obonga Lake greenstone belt, Western Superior Province: evidence from geochemistry, Nd isotopes and U-Pb geochronology. Precambrian Research 114, 295–325.

Tomlinson, K.Y., Davis, D.W., Stone, D., Hart, T., 2003. New U-Pb and Nd isotopic evidence for crustal recycling and Archean terrane development in the south-central Wabigoon Subprovince, Canada. Contributions to Mineralogy and Petrology 144, 684–702.

Tomlinson, K.Y., Stott, G.M., Percival, J.A., Stone, D., 2004. Basement terrane correlations and crustal recycling in the western Superior Province: Nd isotopic character of granitoid and felsic volcanic rocks in the Wabigoon subprovince, N. Ontario, Canada. Precambrian Research 132, 245–274.

Tonks, W.B., Melosh, H.J., 1990. The physics of crystal settling and suspension in a turbulent magma ocean. In: Newsome, H.E., Jones, J.H. (Eds.), Origin of the Earth. Oxford University Press, New York, pp. 151–174.

Torigoye, N., Shima, M., 1993. Evidence for a late thermal event of unequilibrated enstatite chondrites: a Rb-Sr study of Quingzhen and Yamato 6901 (EH3) and Khairpur (EL6). Meteoritics 28, 515–527.

Torigoye-Kita, N., Tatsumoto, M., Meeker, G.P., Yanai, K., 1995. The 4.56 Ga age of the MET 78008 ureilite. Geochimica et Cosmochimica Acta 59, 2319–2329.

Toulkeridis, T., Clauer, N., Kroner, A., Reimer, T., Todt, W., 1999. Characterization, provenance, and tectonic setting of Fig Tree greywackes from the Archaean Barberton Greenstone Belt, South Africa. Sedimentary Geology 124 (1–4), 113–129.

Towe, K.M., 1990. Aerobic respiration in the Archean. Nature 348, 54–56.

Trail, D., Mojzsis, S.J., Harrison, T.M., Schmitt, A.K., Watson, E.B., Young, E.D., 2007. Constraints on Hadean protoliths from oxygen isotopes and Ti-thermometry. Geochemistry, Geophysics, Geosystems, in press.

Trendall, A.F., 2002. The significance of iron-formation in the Precambrian stratigraphic record. In: Altermann, W., Corocron, P.L. (Eds.), Precambrian Sedimentary Environments: A Modern Approach to Ancient Depositonal Systems. Blackwell Science, Oxford, pp. 33–66.

Trendall, A.F., Compston, W., Nelson, D.R., de Laeter, J.R., Bennett, V.C., 2004. SHRIMP zircon ages constraining the depositional history of the Hamersley Group, Western Australia. Australian Journal of Earth Sciences 51, 621–644.

Trieloff, M., Jessberger, E.K., Herrwerth, I., Hoppe, J., Fieni, C., Ghelis, Bourot-Denise, M., Pellas, P., 2003. Structure and thermal history of the H-chondrite asteroid revealed by thermochronometry. Nature 422, 502–506.

Tuinstra, F., Koenig, J.L., 1970. Raman spectra of graphite. Journal of Chemical Physics 53, 1126–1130.

Turcotte, D.L., 1993. An episodic hypothesis for Venusian tectonics. Journal of Geophysical Research 98, 17,061–17,068.

Turcotte, D.L., Morein, G., Malamud, B.D., 1999. Catastrophic resurfacing and episodic subduction on Venus. Icarus 139, 49–57.

Turek, A., Carson, T.M., Smith, P.E., Van Schmus, W.R., Weber, W., 1986. U-Pb zircon ages for rocks from the Island Lake greenstone belt, Manitoba. Canadian Journal of Earth Sciences 23, 92–101.

Turek, A., Sage, R.P., Van Schmus, W.R., 1992. Advances in the U-Pb zircon geochronology of the Michipicoten greenstone belt, Superior Province, Ontario. Canadian Journal of Earth Sciences 29, 1154–1165.

Turekian, K.K., Wedepohl, K.H., 1991. Distribution of the elements in some major units of the Earth's crust. Bulletin of the Geological Society of America 72, 175–192.

Turkina, O.M., 2001. Geochemistry of granulites of the Arban massif (Sharyzhalgay uplift of the Siberian platform). Russian Geology and Geophysics 42 (5), 815–830.

Turkina, O.M., 2004. The amphibolite-plagiogneiss complex of the Onot block, Sharyzhalgay uplift: isotopic-geochemical evidence for the Early Archean evolution of the continental crust. Doklady Earth Sciences 399A (9), 1296–1300.

Turner, G., 1988. Dating of secondary events. In: Kerridge, J.F., Matthews, M.S. (Eds.), Meteoritics and the Early Solar System. University of Arizona Press, Tucson, AZ, pp. 276–288.

Turner, G., Enright, M.C., Cadogan, P.H., 1978. The early history of chondrite parent bodies inferred from ^{40}Ar-^{39}Ar ages. In: Proceedings of the 9th Lunar and Planetary Science Conference, pp. 989–1025.

Turner, G., Knott, S.F., Ash, R.D., Gilmour, J.D., 1997. Ar-Ar chronology of the Martian meteorite ALH84001: Evidence for the timing of the early bombardment of Mars. Geochimica et Cosmochimica Acta 61 (18), 3835–3850.

Turner, G., Harrison, T.M., Holland, G., Mojzsis, S.J., Gilmour, J., 2004. Extinct ^{244}Pu in ancient zircons. Science 306, 89–91.

Tyler, I.M., Fletcher, I.R., de Laeter, J.R., Williams, I.R., Libby, W.G., 1992. Isotope and rare earth element evidence for a late Archaean terrane boundary in the southeastern Pilbara Craton, Western Australia. Precambrian Research 54, 211–229.

Ueno, Y., Isozaki, Y., Yurimoto, H., Maruyama, S., 2001a. Carbon isotopic signatures of individual Archaean Microfossils(?) from Western Australia. International Geology Reviews 43, 196–212.

Ueno, Y., Maruyama, S., Isozaki, Y., Yurimoto, Y., 2001b. Early Archean (ca. 3.5 Ga) microfossils and ^{13}C-depleted carbonaceous matter in the North Pole area, Western Australia: field occurrence and geochemistry. In: Nakashima, S., Maruyama, S., Brack, A., Windley, B.F. (Eds.), Geochemistry and the Origin of Life. Universal Academy Press, Tokyo, pp. 203–236.

Ueno, Y., Yurimoto, H., Yoshioka, H., Komiya, T., Maruyama, S., 2002. Ion microprobe analysis of graphite from ca. 3.8 Ga metasediments, Isua supracrustal belt, West Greenland: Relationship between metamorphism and carbon isotopic composition. Geochimica et Cosmochimica Acta 66, 1257–1268.

Ueno, Y., Rumble, D., Ono, S., Hu, G., Maruyama, S., 2003. Multiple sulfur isotopes of Early Archean hydrothermal deposits from Western Australia. Geochimica et Cosmochimica Acta 67, A500.

Ueno, Y., Yoshioka, H., Maruyama, S., Isozaki, Y., 2004. Carbon isotopes and petrography in ∼3.5 Ga hydrothermal silica dykes in the North Pole area, Western Australia. Geochimica et Cosmochimica Acta 68, 573–589.

Ueno, Y., Yamada, K., Yoshida, N., Maruyama, S., Isozaki, Y., 2006. Evidence from fluid inclusions for microbial methanogenesis in the early Archaean era. Nature 440, 516–519.

Unrug, R., 1997. Rodinia to Gondwana: the geodynamic map of Gondwana supercontinent assembly. GSA Today 7, 1–6.

Urey, H.C., 1947. The thermodynamic properties of isotopic substances. Journal of the Chemical Society (Resumed) 1947, 562–581.

Valley, J.W., 2003. Oxygen isotopes in zircon. In: Hanchar, J.M., Hoskin, P.W.O. (Eds.), Zircon. Reviews in Mineralogy and Geochemistry, vol. 53, pp. 343–386.

Valley, J.W., 2005. A cool early Earth. Scientific American, October, 58–66.

Valley, J.W., 2006. Early Earth. Elements 2, 201–204.

Valley, J.W., Chiarenzelli, J.R., McLelland, J.M., 1994. Oxygen isotope geochemistry of zircon, Earth and Planetary Science Letters 126, 187–206.

Valley, J.W., Peck, J.H., King, E.M., Wilde, S.A., 2002. A cool early Earth. Geology 30, 351–354.

Valley, J.W., Lackey, J.S., Cavosie, A.J., Clechenko, C.C., Spicuzza, M.J., Basei, M.A.S., Bindeman, I.N., Ferreira, V.P., Sial, A.N., King, E.M., Peck, W.H., Sinha, A.K., Wei,

C.S., 2005. 4.4 billion years of crustal maturation: oxygen isotope ratios of magmatic zircon. Contributions to Mineralogy and Petrology 150, 561–580.

Valley, J.W., Cavosie, A.J., Fu, B., Peck, W.H., Wilde, S.A., 2006. Comment on "Heterogeneous hadean hafnium: Evidence of continental crust at 4.4 to 4.5 Ga". Science 312, 5777.

van den Kerkhof, A.M., Kronz, A., Simon, K., Riganti, A., Scherer, T., 2004. Origin and evolution of Archean quartzites from the Nondweni greenstone belt (South Africa): inferences from a multidisciplinary study. South African Journal of Geology 107, 559–576.

van der Hilst, R.D., Kárason, K., 1999. Compositional heterogeneity in the bottom 1000 km of Earth's mantle: towards a hybrid convection model. Science 283, 1885–1888.

van der Laan, S.R., Wyllie, P.J., 1992. Constraints on Archean trondhjemite genesis from hydrous crystallization experiments on Nuk Gneiss at 10–17 kbar. Journal of Geology 100, 57–68.

van Haaften, W.M., White, S.H., 1998. Evidence for multiphase deformation in the Archean basal Warrawoona Group in the Marble Bar area, East Pilbara, Western Australia. Precambrian Research 88, 53–66.

Van Kranendonk, M.J., 1998. Litho-tectonic and structural components of the NORTH SHAW 1:100,000 sheet, Archaean Pilbara Craton. Annual Review 1997–1998. Western Australian Geological Survey, pp. 63–70.

Van Kranendonk, M.J., 2000. Geology of the North Shaw 1:100,000 sheet. Western Australia Geological Survey, 1:100,000 Series Explanatory Notes, 89 pp.

Van Kranendonk, M.J., 2003a. Stratigraphic and tectonic significance of eight local unconformities in the Fortescue Group, Pear Creek Centrocline, Pilbara Craton, Western Australia. In: Annual Review 2001–2002. Western Australia Geological Survey, pp. 70–79.

Van Kranendonk, M.J., 2003b. Geology of the Tambourah 1:100,000 sheet. Western Australia Geological Survey, 1:100,000 Geological Series Explanatory Notes, 59 pp.

Van Kranendonk, M.J., 2004a. Archaean tectonics 2004: A review. Precambrian Research 131, 143–151.

Van Kranendonk, M.J., 2004b. Geology of the Carlindie 1:100,000 sheet. Western Australia Geological Survey, 1:100,000 Geological Series Explanatory Notes, 45 pp.

Van Kranendonk, M.J., 2006. Volcanic degassing, hydrothermal circulation and the flourishing of early life on Earth: new evidence from the Warrawoona Group, Pilbara Craton, Western Australia. Earth Science Reviews 74, 197–240.

Van Kranendonk, M.J., Collins, W.J., 1998. Timing and tectonic significance of Late Archaean, sinistral strike-slip deformation in the Central Pilbara Structural Corridor, Pilbara Craton, Western Australia. Precambrian Research 88, 207–232.

Van Kranendonk, M.J., Morant, P., 1998. Revised Archaean stratigraphy of the North Shaw 1:100,000 sheet, Pilbara Craton. In: Western Australia Geological Survey, Annual Review 1997–1998, pp. 55–62.

Van Kranendonk, M.J., Pirajno, F., 2004. Geological setting and geochemistry of metabasalts and alteration zones associated with hydrothermal chert ± barite deposits in the ca. 3.45 Ga Warrawoona Group, Pilbara Craton, Australia. Geochemistry: Exploration, Environment, Analysis 4, 253–278.

Van Kranendonk, M.J., Saint-Onge, M.R., Henderson, J.R., 1993. Paleoproterozoic tectonic assembly of Northeast Laurentia through multiple indentations. Precambrian Research 63, 325–347.

Van Kranendonk, M.J., Hickman, A.H., Collins, W.J., 2001a. Comment on "Evidence for multiphase deformation in the Archaean basal Warrawoona Group in the Marble Bar area, East Pilbara, Western Australia". Precambrian Research 105, 73–78.

Van Kranendonk, M., Hickman, A.H., Williams, I.R., Nijman, W., 2001b. Archaean Geology of the East Pilbara Granite-Greenstone Terrane Western Australia – A Field Guide. Western Australia Geological Survey, Record 2001/9, 134 pp.

Van Kranendonk, M.J., Hickman, A.H., Smithies, R.H., Nelson, D.R., 2002. Geology and tectonic evolution of the Archean North Pilbara terrain, Pilbara Craton, Western Australia. Economic Geology 97, 695–732.

Van Kranendonk, M.J., Webb, G.E., Kamber, B.S., 2003. Geological and trace element evidence for a marine sedimentary environment of deposition and biogenicity of 3.45 Ga stromatolitic carbonates in the Pilbara Craton, and support for a reducing Archean ocean. Geobiology 1, 91–108.

Van Kranendonk, M.J., Collins, W.J., Hickman, A.H., Pawley, M.J., 2004. Critical tests of vertical vs horizontal tectonic models for the Archaean East Pilbara Granite-Greenstone Terrane, Pilbara Craton, Western Australia. Precambrian Research 131, 173–211.

Van Kranendonk, M.J., Smithies, R.H., Hickman, A.H., Bagas, L., Williams, I.R., Farrell, T.R., 2005. Event stratigraphy applied to 700 m.y. of Archaean crustal evolution, Pilbara Craton, Western Australia. In: Western Australia Geological Survey, Annual Review 2003–2004, pp. 49–61.

Van Kranendonk, M.J., Hickman, A.H., Smithies, R.H., Williams, I.R., Bagas, L., Farrell, T.R., 2006a. Revised lithostratigraphy of Archean supracrustal and intrusive rocks in the northern Pilbara Craton, Western Australia. Western Australia Geological Survey, Record 2006/15, 57 pp.

Van Kranendonk, M.J., Hickman, A.H., Huston, D.L., 2006b. Geology and Mineralization of the East Pilbara – A Field Guide. Western Australia Geological Survey, Record 2006/16, 94 pp.

Van Kranendonk, M.J., Philippot, P., Lepot, K., 2006c. The Pilbara Drilling Project: c. 2.72 Ga Tumbiana Formation and c. 3.49 Ga Dresser Formation, Pilbara Craton, Western Australia. Western Australia Geological Survey, Record 2006/14, 25 pp.

Van Kranendonk, M.J., Smithies, R.H., Hickman, A.H., Champion, D.C., 2007a. Secular tectonic evolution of Archaean continental crust: interplay between horizontal and vertical processes in the formation of the Pilbara Craton, Australia. Terra Nova 19, 1–38.

Van Kranendonk, M.J., Huston, D.L., Hickman, A.H., 2007b. From plumes to accretion: Changing mineralization styles through ~800 million years of crustal evolution in the Pilbara Craton, Western Australia. In: Western Australia Geological Survey, Record 2007/2, pp. 23–25.

Van Kranendonk, M.J., Philippot, P., Lepot, K., 2007c. A felsic volcanic caldera setting for Earth's oldest fossils in the c. 3.5 Ga Dresser Formation of the Warrawoona Group, Pilbara Craton, Western Australia: evidence from outcrops and diamond drillcore. Precambrian Research, in press.

Van Schmus, W.R., Wood, J.A., 1967. A chemical-petrological classification for the chondritic meteorites. Geochimica et Cosmochimica Acta 31, 747–765.

van Zuilen, M., Lepland, A., Arrhenius, G., 2002. Reassessing the evidence for the earliest traces of life. Nature 418, 627–630.

van Zuilen, M.A., Lepland, A., Teranew, J., Finarelli, J., Wahlen, M., Arrhenius, G., 2003. Graphite and carbonates in the 3.8 Ga old Isua Supracrustal Belt, southern West Greenland. Precambrian Research 126, 331–348.

van Zuilen, M.A., Mathew, K., Wopenka, B., Lepland, A., Marti, B., Arrhenius, G., 2005. Nitrogen and argon isotopic signatures in graphite from the 3.8-Ga-old Isua Supracrustal Belt, Southern West Greenland. Geochimica et Cosmochimica Acta 69, 1241–1252.

van Zuilen, M., Chaussidon, M., Rollion-Bard, C., Luais, B., Marty, B., 2006. Carbonaceous cherts of the Barberton Greenstone belt, South Africa. Geophysical Research Abstracts 8, 04975.

Varela, M.E., Kurat, G., Zinner, E., Hoppe, P., Natflos, T., Nazarov, M.A., 2005. The non-igneous genesis of angrites: Support from trace element distribution between phases in D'Orbigny. Meteoritics and Planetary Science 40, 409–430.

Vearncombe, J.R., Barley, M.E., Eisenlohr, B.N., Groves, D.I., Houstoun, S.M., Skwarnecki, M.S., Grigson, M.W., Partington, G.A., 1989. Structural controls on mesothermal gold mineralization: examples from the Archean terranes of southern Africa and Western Australia. In: Economic Geology Monograph, vol. 6, pp. 124–134.

Vearncombe, S.E., 1995. Volcanogenic massive sulphide-sulfate mineralisation at Strelley, Pilbara Craton, Western Australia. Unpublished Ph.D. thesis. University of Western Australia, 152 pp.

Vearncombe, S., Kerrich, R., 1999. Geochemistry and geodynamic setting of volcanic and plutonic rocks associated with Early Archaean volcanogenic massive sulphide mineralization, Pilbara Craton. Precambrian Research 98, 243–270.

Vearncombe, S., Barley, M.E., Groves, D.I., McNaughton, N.J., Mikucki, E.J., Vearncombe, J.R., 1995. 3.26 Ga black smoker-type mineralisation in the Strelley Belt, Pilbara Craton, Western Australia. Geological Society of London Journal 152, 587–590.

Vearncombe, S., Vearncombe, J.R., Barley, M.E., 1998. Fault and stratigraphic controls on volcanogenic massive sulphide deposits in the Strelley Belt, Pilbara Craton, Western Australia. Precambrian Research 88, 67–82.

Veizer, J., Mackenzie, F.T., 2003. Evolution of sedimentary rocks. In: Mackenzie, F.T. (Ed.), Treatise on Geochemistry, vol. 7: Sediments, Diagenesis, and Sedimentary Rocks. Elsevier, Amsterdam, pp. 369–407.

Veizer, J., Compston, W., Hoefs, J., Nielsen, H., 1982. Mantle buffering of the early oceans. Naturwissenschaften 69, 173–180.

Veizer, J., Hoefs, J., Lowe, D.R., Thurton, P.C., 1989. Geochemistry of Precambrian carbonates: II. Archean greenstone belts and Archean sea water. Geochimica et Cosmochimica Acta 53, 859–871.

Veizer, J., et al., 1999. ^{87}Sr/^{86}Sr, δ^{13}C and δ^{18}O evolution of Phanerozoic seawater. Chemical Geology 161, 59–88.

Veizer, J., Goddéris, Y., Francois, L., 2000. Evidence for decoupling of atmospheric CO_2 and global climate during the Phanerozoic eon. Nature 408, 698–701.

Vennemann, T.W., Smith, H.S., 1999. Geochemistry of mafic and ultramafic rocks in the Kromberg Formation in its type section, Barberton Greenstone Belt, South Africa. In: Lowe, D.R., Byerly, G.R. (Eds.), Geologic Evolution of the Barberton Greenstone Belt, South Africa. Geological Society of America, Special Paper 329, pp. 133–150.

Vernikovsky, V.A., Vernikovskaya, A.E., Kotov, A.B., Salnikova, E.B., Kovach, V.P., 2003. Neoproterozoic accretionary and collisional events on the western margin of the Siberian craton: new geological and geochronological evidence from the Yenisey Ridge. Tectonophysics 375 (1–4), 147–168.

Versfeld, J.A., 1988. The geology of the Nondweni greenstone belt, Natal. Ph.D. thesis. University of Natal, Pietermaritzburg, 298 pp.

Versfeld, J.A., Wilson, A.H., 1992a. Nondweni Group. In: Johnson, M.R. (Ed.), Catalogue of South African Stratigraphic Units. South African Commission for Stratigraphy, Council for Geoscience, Pretoria, pp. 19–20.

Versfeld, J.A., Wilson, A.H., 1992b. Toggekry Formation. In: Johnson, M.R. (Ed.), Catalogue of South African Stratigraphic Units. South African Commission for Stratigraphy, Council for Geoscience, Pretoria, pp. 29–30.

Vervoort, J.D., Blichert-Toft, J., 1999. Evolution of the depleted mantle: Hf isotope evidence from juvenile rocks through time. Geochimica et Cosmochimica Acta 63, 533–556.

Vervoort, J.D., Patchett, J.P., 1996. Behavior of hafnium and neodymium isotopes in the crust: constraints from Precambrian crustally derived granites. Geochimica et Cosmochimica Acta 60, 3717–3733.

Vervoort, J.D., Patchett, P.J., Gehrels, G.E., Nutman, A.P., 1996. Constraints on early Earth differentiation from hafnium and neodymium isotopes. Nature 379, 624–627.

Vielzeuf, D., Schmidt, M.W., 2001. Melting reactions in hydrous systems revisited: application to metapelites, metagreywackes and metabasalts. Contribution to Mineralogy and Petrology 141, 251–267.

Viljoen, M.J., Viljoen, R.P., 1969a. The geology and geochemistry of the lower ultramafic unit of the Onverwacht Group and a proposed new class of igneous rocks. In: Geological Society of South Africa, Special Publication 2, pp. 55–86.

Viljoen, R.P., Viljoen, M.J., 1969b. The geological and geochemical significance of the upper formations of the Onverwacht Group. In: Geological Society of South Africa, Special Publication 2, pp. 113–152.

Viljoen, M.J., Viljoen, R.P., 1969c. An introduction to the geology of the Barberton granite-greenstone terrain. In: Geological Society of South Africa, Special Publication 2, pp. 9–28.

Viljoen, M.J., Viljoen, R.P., 1969d. A proposed new classification of the granitoid rocks of the Barberton region. In: Geological Society of South Africa, Special Publication 2, pp. 153–188.

Viljoen, R.P., Saager, R., Viljoen, M.J., 1969. Metallogenesis and ore control in the Steynsdorp Goldfield, Barberton Mountain Land, South Africa. Economic Geology 64, 778–797.

Viljoen, M.J., Viljoen, R.P., Smith, H.S., Erlank, A.J., 1983. Geological, textural, and geochemical features of komatiitic flows from the Komati Formation. In: Geological Society of South Africa, Special Publication 9, pp. 1–20.

Vishnevskiy, A.N., 1978. Metamorphic Complexes of the Anabar Crystalline Shield. Nedra, Leningrad, 214 pp. (in Russian).

Vita-Finzi, C., Howarth, R.J., Tapper, S.W., Robinson, C.A., 2005. Venusian craters, size distribution and the origin of coronae. In: Foulger, G.R., Natland, J.H., Presnall, D.C., Anderson, D.L. (Eds.), Plates, Plumes, and Paradigms. Geological Society of America, Special Paper 388, pp. 815–824.

Vladimirov, A.G., Ponomareva, A.P., Shokalski, S.P., Khalilov, V.A., Kostitsyn, Yu.A., Pnomarchuk, V.A., Rudnev, S.A., Vystavnoi, S.A., Kruk, N.N., Titov, A.V., 1997. Late Paleozoic – Early Mesozoic granitoid magmatism in Altai. Russian Geology and Geophysics 38, 755–770.

Wächtershäuser, G., 1988. Before enzymes and templates – theory of surface metabolism. Microbiological Reviews 52, 452–484.

Wächtershäuser, G., 1992. Groundworks for an evolutionary biochemistry – the iron sulfur world. Progress in Biophysics and Molecular Biology 58, 85–201.

Wächtershäuser, G., 2000. Life as we don't know it. Science 289, 1307–1308.

Wadhwa, M., Shukolyukov, A., Lugmair, G.W., 1998. ^{53}Mn-^{53}Cr systematics in Brachina: A record of one of the earliest phases of igneous activity on an asteroid. In: 29th Lunar and Planetary Science Conference, abstract #1480 (CD-ROM).

Wagener, J.H.F., Wiegand, J., 1986. The Sheba gold mine, Barberton greenstone belt. In: Anhaeusser, C.R., Maske, S. (Eds.), Mineral Deposits of Southern Africa, vol. I. Geological Society of South Africa, pp. 155–161.

Waight, T.E., Maas, R., Nicholls, I.A., 2000. Fingerprinting feldspar phenocrysts using crystal isotopic composition stratigraphy: implications for crystal transfer and magma mingling in S-type granites. Contributions to Mineralogy and Petrology 139, 227–239.

Wakita, K., Metcalfe, I., 2005. Ocean plate stratigraphy in East and Southeast Asia. Journal of Asian Earth Sciences 24, 679–702.

Walker, G.P.L., 1991. Structure, and origin by injection of lava under surface crust, of tumuli, "lava rises", "lava-rise pits", and "lava-inflation clefts" in Hawaii. Bulletin of Volcanology 53, 546–558.

Walker, J.C.G., 1990. Precambrian evolution of the climate system. Global and Planetary Change 82, 261–289.

Wallace, P.J., Frey, F.A., Weis, D., Coffin, M.F., 2002. Origin and evolution of the Kerguelen Plateau, Broken Ridge and Kerguelen archipelago: Editorial. Journal of Petrology 43, 1105–1118.

Wallmann, K., 2001. The geological water cycle and the evolution of marine $\delta^{18}O$. Geochimica et Cosmochimica Acta 65, 2469–2485.

Walsh, M.M., 1992. Microfossils and possible microfossils from the early Archean Onverwacht Group, Barberton Mountain Land, South Africa. Precambrian Research 54, 271–293.

Walsh, M.M., Lowe, D.R., 1999. Modes of accumulation of carbonaceous matter in the Early Archean: A petrographic and geochemical study of the carbonaceous cherts of the Swaziland Supergroup. In: Lowe, D.R., Byerly, G.R. (Eds.), Geological Evolution of the Barberton Greenstone Belt, South Africa. Geological Society of America, Special Paper 329, pp. 115–132.

Walter, M.J., 1998. Melting of garnet peridotite and the origin of komatiite and depleted lithosphere. Journal of Petrology 39, 29–60.

Walter, M.R., 1983. Archean stromatolites: evidence of the Earth's oldest benthos. In: Schopf, J.W. (Ed.), Earth's Earliest Biosphere. Princeton University Press, pp. 187–213.

Walter, M.R., Buick, R., Dunlop, J.S.R., 1980. Stromatolites 3400–3500 Myr old from the North Pole area, Western Australia. Nature 248, 443–445.

Wan, Y.S., 1993. The Formation and Evolution of the Iron-Bearing Rock Series of the Gongchangling Area, Liaoning Province. Beijing Science and Technology Publishing House, Beijing, 108 pp. (in Chinese with English abstract).

Wan, Y.S., Zhang, Z.Q., Wu, J.S., Song, B., Liu, D.Y., 1997. Geochemical and Nd isotopic characteristics of some rocks from the Paleoarchaean Chentaigou supracrustal, Anshan area, NE China. Continental Dynamics 2, 39–46.

Wan, Y.S., Liu, D.Y., Wu, J.S., Zhang, Z.Q., Song, B., 1998. The origin of Mesoarchaean granitic rocks from the Anshan-Benxi area: constraints from geochemistry and Nd isotopes. Acta Petrologica Sinica 14, 278–288 (in Chinese with English abstract).

Wan, Y.S., Song, B., Wu, J.S., Zhang, Z.Q., Liu, D.Y., 1999. Geochemical and Nd and Sr isotopic compositions of 3.8 Ga trondhjemitic rocks from the Anshan area and their significance. Acta Geologica Sinica 73, 25–36 (in Chinese with English abstract).

Wan, Y.S., Song, B., Liu, D.Y., Li, H.M., Yang, C., Zhang, Q.D., Yang, C.H., Geng, Y.S., Shen, Q.H., 2001. Geochronology and geochemistry of a 3.8–2.5 Ga ancient rock belt in the Dongshan scenic park, Anshan area. Acta Geologica Sinica 75, 363–370 (in Chinese with English abstract).

Wan, Y.S., Song, B., Liu, D.Y., 2002. Zircon SHRIMP age of Mesoarchaean meta-argilloarenaceous rock in the Anshan area and its geological significance. Science in China Series B 45, 121–129.

Wan, Y.S., Zhang, Q.D., Song, T.R., 2003a. SHRIMP ages of detrital zircons from the Changcheng System in the Ming Tombs area, Beijing: Constraints on the protolith nature and maximum depositional age of the Mesoproterozoic cover of the NCC. Chinese Science Bulletin 4, 2500–2506.

Wan, Y.S., Song, B., Liu, D.Y., 2003b. Early Precambrian geochronological framework in Eastern China. Geological Report of China Geological Survey, 219 pp. (in Chinese).

Wan, Y.S., Liu, D., Song, B., Wu, J., Yang, C., Zhang, A., Geng, Y., 2005a. Geochemical and Nd isotopic compositions of 3.8 Ga meta-quartz dioritic and trondhjemitic rocks from the Anshan area and their geological significance. Journal of Asian Earth Science 24, 563–575.

Wan, Y.S., Song, B., Yang, C., Liu, D.Y., 2005b. Zircon SHRIMP U-Pb geochronology of Archaean rocks from the Fushun-Qingyuan area, Liaoning Province and its geological significance. Acta Geologica Sinica 79, 78–87 (in Chinese with English abstract).

Wan, Y.S., Song, B., Geng, Y.S., Liu, D.Y., 2005c. Geochemical characteristics of Archaean basement in the Fushun-Qingyuan area, northern Liaoning Province and its geological significance. Geological Review 51, 128–137 (in Chinese with English abstract).

Wan, Y.S., Song, B., Liu, D.Y., Wilde, S.A., Wu, J.S., Shi, Y.R., Yin, X.Y., Zhou, H.Y., 2006. SHRIMP U-Pb zircon geochronology of Palaeoproterozoic metasedimentary rocks in the NCC: evidence for a major Late Palaeoproterozoic tectonothermal event. Precambrian Research 149, 249–271.

Wang, F., Zhou, X.-H., Zhang, L.-C., Ying, J.-F., Zhang, Y.-T., Wu, F.-Y., Zhu, R.-X., 2006. Late Mesozoic volcanism in the Great Xing'an Range (NE China): timing and implications for the dynamic setting of NE Asia. Earth and Planetary Science Letters 251, 179–198.

Wang, L.G., Qiu, Y.M., McNaughton, N.J., Groves, D.I., Luo, Z.K., Huang, J.Z., Miao, L.C., Liu, Y.K., 1998. Constraints on crustal evolution and gold metallogeny in the northwestern Jiaodong, China from SHRIMP U-Pb zircon studies of granitoids. Ore Geology Reviews 13, 275–291.

Wang, X.D., Lindh, A., 1996. Temperature-pressure investigation of the southern part of the southwest Swedish Granulite Region. European Journal of Mineralogy 8, 51–67.

Wang, Z.J., Shen, Q.H., Wan, Y.S., 2004. SHRIMP U-Pb zircon geochronology of the Shipaihe "meta-diorite mass" from Dengfeng County, Henan Province. Acta Geoscientica Sinica 25, 295–298 (in Chinese with English abstract).

Ward, J.H.W., 1999. The metallogeny of the Barberton Greenstone Belt, South Africa and Swaziland. Geological Survey of South Africa, Memoir 86, 116 pp.

Warneck, P., 1988. Chemistry of the Natural Atmosphere. International Geophysical Series, vol. 41. Academic Press, New York, 757 pp.

Wasserburg, G.J., 1985. Short-lived nuclei in the early solar system. In: Black, D.C., Matthews, M.S. (Eds.), Protostars and Planets II. University of Arizona Press, Tucson, AZ, pp. 703–737.

Wasson, J.T., Kallemeyn, G.W., 1988. Compositions of chondrites. Philosophical Transactions of the Royal Society of London Ser. A – Mathematical Physical and Engineering Sciences 325, 535–544.

Watanabe, Y., Naraoka, H., Wronkiewicz, D.J., Condie, K.C., Ohmoto, H., 1997. Carbon, nitrogen and sulfur geochemistry of the Archean and Proterozoic shales of the Kaapvaal Craton, South Africa. Geochimica et Cosmochimica Acta 61, 3441–3459.

Watson, E.B., 1996. Dissolution, growth and survival of zircons during crustal fusion: kinetic principals, geological models and implications for isotopic inheritance. Transactions Royal Society Edinburgh: Earth Sciences 87, 43–56.

Watson, E.B., Cherniak, D.J., 1997. Oxygen diffusion in zircon, Earth and Planetary Science Letters 148, 527–544.

Watson, E.B., Harrison, T.M., 1983. Zircon saturation revisited: temperature and composition effects in a variety of different crustal magma types. Earth and Planetary Science Letters 64, 295–304.

Watson, E.B., Harrison, T.M., 2005. Zircon thermometer reveals minimum melting conditions on earliest Earth. Science 308, 841–844.

Watson, E.B., Harrison, T.M., 2006. Response to comments on "Zircon thermometer reveals minimum melting conditions on earliest Earth". Science 311, doi: 10.1126/science.112.1080.

Weaver, B.L., Tarney, J., 1981. Lewisian gneiss geochemistry and Archaean crustal development models. Earth and Planetary Science Letters 55 (1), 171–180.

Weber, W., 1978. Natawahunan lake. Report of Field Activities 1978, Manitoba Department of Mines, Resources and Environmental Management, Mineral Resources Division, pp. 47–53 (plus Preliminary Map 1978 U-2, scale 1:50,000).

Weber, W., 1983. The Pikwitonei granulite domain: a lower crustal level along the Churchill-Superior boundary in central Manitoba. In: Ashwal, L.D., Card, K.D. (Eds.), A Cross-Section of Archean Crust. Lunar and Planetary Institute, Houston, TX, Technical Report 83-03, pp. 95–97.

Weber, W., Scoates, R.F.J., 1978. Archean and Proterozoic metamorphism in the northwestern Superior Province and along the Churchill-Superior boundary, Manitoba. In: Fraser, J.A., Heywood, W.W. (Eds.), Metamorphism in the Canadian Shield. Geological Survey of Canada, Paper 78-10, pp. 5–16.

Weber, D., Schultz, L., Weber, H.W., Clayton, R.N., Mayeda, T.K., Bischoff, A., 1997. Hammadah al Hamra 119 – A new unbrecciated Saharan Rumuruti chondrite. Lunar and Planetary Science XXVIII, 1511–1512 (abstract).

Weber, H.W., Schultz, L., 1995. Noble gases in Rumuruti-group chondrites. Meteoritics 30, 596.

Wedepohl, K.H., 1995. The composition of the continental crust. Geochimica et Cosmochimica Acta 59, 1217–1232.

Wedepohl, K.H., Muramatsu, Y., 1979. The chemical composition of kimberlites compared with the average composition of three basaltic magma types. Boyd, F.R., Meyer, H.O.A. (Eds.), Kimberlites, Diatremes, and Diamonds: Their Geology, Petrology and Geochemistry. Proceedings of the Second International Kimberlite Conference, vol. 1. American Geophysical Union, Washington, DC, 399 pp.

Weisberg, M.K., Prinz, M., Kojima, H., Yanai, K., Clayton, R.N., Mayeda, T.K., 1991. The Carlisle Lakes-type chondrites: A new grouplet with high $\delta^{17}O$ and evidence for nebula oxidation. Geochimica et Cosmochimica Acta 55, 2657–2669.

Weisberg, M.K., Prinz, M., Clayton, R.N., Mayeda, T.K., Grady, M.M., Franchi, I., Pillinger, C.T., Kallemeyn, G.W., 1996. The K (Kakangari) chondrite grouplet. Geochimica et Cosmochimica Acta 61, 4253–4263.

Wendt, J.I., 1993. Early Archean crustal evolution in Swaziland, southern Africa, as revealed by combined use of zircon geochronology, Pb-Pb and Sm-Nd systematics. Unpublished doctoral dissertation. University of Mainz, Germany, 123 pp.

West, G.F., Mareschal, J.C., 1979. Model for Archean tectonism 1. Thermal conditions. Canadian Journal of Earth Sciences 16, 1942–1950.

Westall, F., Folk, R.L., 2003. Exogenous carbonaceous microstructures in Early Archaean cherts and BIFs from the Isua greenstone belt: Implications for the search for life in ancient rocks. Precambrian Research 126, 313–330.

Westall, F., de Wit, M.J., Dann, J.C., van der Gaast, S., de Ronde, C.E.J., Gerneke, D., 2001. Early Archean fossil bacteria and biofilms in hydrothermally influenced sediments, Barberton Greenstone Belt, South Africa. Precambrian Research 106, 93–116.

Westraat, J.D., Kisters, A.F.M., Poujol, M., Stevens, G., 2005. Transcurrent shearing, granite sheeting and the incremental construction of the tabular 3.1 Ga Mpuluzi batholith, Barberton granite-greenstone terrane, South Africa. Journal of the Geological Society London 162, 373–388.

Wetherill, G.W., 1980. Formation of the terrestrial planets. Annual Review Astronomy, Astrophysics 18, 77–113.

Wetherill, G.W., 1985. Occurrence of giant impacts during the growth of the terrestrial planets. Science 228, 877–879.

Wetherill, G.W., 1990. Formation of the Earth. Annual Review of Earth and Planetary Science 18, 205–256.

Whalen, J.B., Percival, J.A., McNicoll, V.J., Longstaffe, F.J., 2002. A mainly crustal origin for tonalitic granitoid rocks, Superior Province, Canada: Implications for late Archean tectonomagmatic processes. Journal of Petrology 43 (8), 1551–1570.

Whalen, J.B., Percival, J.A., McNicoll, V., Longstaffe, F.J., 2003. Intra-oceanic production of continental crust in a Th-depleted ca. 3.0 Ga arc complex, western Superior Province, Canada. Contributions to Mineralogy and Petrology 156, 78–99.

Whalen, J.B., Percival, J.A., McNicoll, V., Longstaffe, F.J., 2004. Geochemical and isotopic (Nd-O) evidence bearing on the origin of late- to post-orogenic high-K granitoid rocks in the Western Superior Province: Implications for late Archean tectonomagmatic processes. Precambrian Research 132, 303–326.

Whitby, J.A., Ash, R.D., Gilmour, J.D., Prinz, M., Turner, G., 1997. Iodine-xenon dating of chondrules and matrix from the Qingzhen and Kota-Kota EH3 chondrites. Meteoritics and Planetary Science 32 (Supplement), A140.

Whitby, J.A., Burgess, R., Turner, G., Gilmour, J., Bridges, J., 2000. Extinct ^{129}I in halite from a primitive meteorite; evidence for evaporite formation in the early Solar System. Science 288, 1819–1821.

White, A.J.R., Chappell, B.W., 1977. Ultrametamorphism and granitoid genesis. Tectonophysics 43, 7–22.

White, D.J., Lucas, S.B., Bleeker, W., Hajnal, Z., Lewry, J.F., Zwanzig, H.V., 2002. Suture-zone geometry along an irregular Paleoproterozoic margin: The Superior boundary zone, Manitoba, Canada. Geology 30, 735–738.

White, D.J., Musacchio, G., Helmstaedt, H.H., Harrap, R.M., Thurston, P.C., van der Velden, A., Hall, K., 2003. Images of a lower-crustal oceanic slab: Direct evidence for tectonic accretion in the Archean western Superior province. Geology 31, 997–1000.

White, J.C., 2003. Trace-element partitioning between alkali feldspar and peralkalic quartz trachyte to rhyolite magma. Part II: Empirical equations for calculating trace-element coefficients of large-ion lithophile, high field strength, and rare earth elements. American Mineralogist 88, 330–337.

White, R.V., Crowley, J.L., Myers, J.S., 2000. Earth's oldest well-preserved mafic dyke swarms in the vicinity of the Isua greenstone belt, southern West Greenland. Geology of Greenland Survey Bulletin 186, 65–72.

Whitehouse, M.J., 1988. Granulite facies Nd-isotopic homogenisation in the Lewisian Complex of Northwest Scotland. Nature 331, 705–707.

Whitehouse, M.J., Fedo, C.M., 2003. Deformation features and critical field relationships of early Archaean rocks, Akilia, southwest Greenland. Precambrian Research 126, 259–271.

Whitehouse, M.J., Kamber, B.S., 2002. On the overabundance of light rare earth elements in terrestrial zircons and its implications for Earth's earliest magmatic differentiation. Earth and Planetary Science Letters 204, 333–346.

Whitehouse, M.J., Kamber, B., 2005. Assigning dates to thin gneissic veins in high-grade metamorphic terranes: a cautionary tale from Akilia, southwest Greenland. Journal of Petrology 46, 291–318.

Whitehouse, M.J., Fowler, M.B., Friend, C.R.L., 1996. Conflicting mineral and whole-rock isochron ages from the Late-Archaean Lewisian Complex of northwestern Scotland: Implications for geochronology in polymetamorphic high-grade terrains. Geochimica et Cosmochimica Acta 60 (16), 3085–3102.

Whitehouse, M.J., Kamber, B., Moorbath, S., 1999. Age significance of U-Th-Pb zircon data from early Archaean rocks of west Greenland – a reassessment based on combined ion-microprobe and imaging studies. Chemical Geology 160, 201–224.

Whitehouse, M.J., Nagler, T.F., Moorbath, S., Kramers, J.D., Kamber, B.S., Frei, R., 2001. Priscoan (4.00–4.03 Ga) orthogneisses from northwestern Canada – by Samuel A. Bowring and Ian S. Williams: Discussion. Contributions to Mineralogy and Petrology 141, 248–250.

Whitehouse, M.J., Kamber, B.S., Fedo, C.M., Lepland, A., 2005. Integrated Pb- and S-isotope investigation of sulphide minerals from the early Archaean of southwest Greenland. Chemical Geology 222, 112–131.

Widdel, F., Schnell, S., Heising, S., Ehrenreich, A., Assmus, B., Schink, B., 1993. Ferrous iron oxidation by anoxygenic phototrophic bacteria. Nature 362, 834–836.

Wiedenbeck, M., 1990. The duration of tectono-thermal episodes in the late Archaean: constraints from the northwestern Yilgarn Craton, Western Australia. In: Third International Archaean Symposium, Extended Abstracts, Geoconferences, Perth, pp. 477–479.

Wijbrans, J.R., McDougall, I., 1987. On the metamorphic history of an Archaean granitoid greenstone terrane, East Pilbara, Western Australia, using the $^{40}Ar/^{39}Ar$ age spectrum technique. Earth and Planetary Science Letters 84, 226–242.

Wilde, S.A., 1980. The Jimperding Metamorphic Belt in the Toodyay area and the Balingup Metamorphic Belt and associated granitic rocks in the Southwestern Yilgarn Block. Geological Society of Australia, W.A. Division, Excursion Guide, 41 pp.

Wilde, S.A., 2001. Jimperding and Chittering metamorphic belts, southwestern Yilgarn Craton, Western Australia – a field guide. In: Western Australia Geological Survey, Record 2001/12, 24 pp.

Wilde, S.A., Low, G.H., 1978. Perth, W.A. Western Australia Geological Survey, 1:250,000. Geological Series Explanatory Notes, 36 pp.

Wilde, S.A., Pidgeon, R.T., 1990. Geology of the Jack Hills metasedimentary rocks. In: Ho, S.E., Glover, J.E., Myers, J.S., Muhling, J. (Eds.), Third International Archaean Symposium, Perth, Excursion Guidebook. University of Western Australia Publication 21, pp. 82–92.

Wilde, S.A., Middleton, M.F., Evans, B.J., 1996. Terrane accretion in the southwestern Yilgarn Craton: evidence from a deep seismic crustal profile. Precambrian Research 78, 179–196.

Wilde, S.A., Cawood, P.A., Wang, K.Y., 1997. The relationship and timing of granitoid evolution with respect to felsic volcanism in the Wutai Complex, NCC. In: Proceedings of the 30th International Geological Congress, Beijing, vol. 17. VSP International Science Publishers, Amsterdam, pp. 75–87.

Wilde, S.A., Valley, J.W., Peck, W.H., Graham, C.M., 2001. Evidence from detrital zircons for the existence of continental crust and oceans on the Earth 4.4 Gyr ago. Nature 409, 175–178.

Wilde, S.A., Cawood, P.A., Wang, K.Y., Nemchin, A.A., 2005. Granitoid evolution in the Late Archaean Wutai Complex, NCC. Journal of Asian Earth Sciences 24, 597–613.

Wilhelmij, H.R., Dunlop, J.S.R., 1984. A genetic stratigraphic investigation of the Gorge Creek Group in the Pilgangoora syncline. In: Muhling, J.R., Groves, D.I., Blake, T.S. (Eds.), Archaean and Proterozoic Basins of the Pilbara, Western Australia: Evolution and Mineralisation Potential. Geology Department and University Extension, Publication 9. University of Western Australia, pp. 68–88.

Wilhelms, D.E., 1987. The Geological History of the Moon. United States Geological Survey, Professional Paper 1348, 302 pp.

Wilhelms, D.E., 1990. Geologic mapping. In: Greeley, R., Batson, R.M. (Eds.), Planetary Mapping. Cambridge University Press, New York, pp. 208–260.

Williams, D.A.C., Furnell, R.G., 1979. A reassessment of part of the Barberton type area. Precambrian Research 9, 325–347.

Williams, H.R., Stott, G.M., Thurston, P.C., Sutcliffe, R.H., Bennett, G., Easton, R.M., Armstrong, D.K., 1992. Tectonic evolution of Ontario: Summary and synthesis. In: Thurston, P.C., Williams, H.R., Sutcliffe, R.H., Stott, G.M. (Eds.), Geology of Ontario. Ontario Geological Survey, Special Volume 4, Part 2, pp. 1255–1332.

Williams, I.R., 1999. Geology of the Muccan 1:100,000 sheet. Western Australia Geological Survey, 1:100,000 Geological Series Explanatory Notes, 39 pp.

Williams, I.R., 2001. Geology of the Warrawagine 1:100,000 sheet. Western Australia Geological Survey, 1:100,000 Geological Series Explanatory Notes, 33 pp.

Williams, I.R., 2002. Geology of the Cooragoora 1:100,000 sheet. Western Australia Geological Survey, 1:100,000 Geological Series Explanatory Notes, 23 pp.

Williams, I.R., Myers, J.S., 1987. Archaean geology of the Mount Narryer region, Western Australia. Western Australia Geological Survey, Report 22, 32 pp.

Williams, I.R., Walker, I.M., Hocking, R.M., Williams, S.J., 1983. Byro, W.A. Western Australia Geological Survey, 1:250,000 Geological Series Explanatory Notes, first ed., 27 pp.

Williams, I.S., 2001. Response of detrital zircon and monazite, and their U-Pb isotopic systems, to regional metamorphism and host-rock partial melting, Cooma Complex, southeastern Australia. Australian Journal of Earth Sciences 48, 557–580.

Williams, I.S., Collins, W.J., 1990. Granite-greenstone terranes in the Pilbara Block, Australia, as coeval volcano-plutonic complexes; evidence from U-Pb zircon dating of the Mount Edgar batholith. Earth and Planetary Science Letters 97, 41–53.

Williams, I.S., Compston, W., Black, L.P., Ireland, T.R., Foster, J.J., 1984. Unsupported radiogenic Pb in zircon: a cause of anomalously high Pb-Pb, U-Pb and Th-Pb ages. Contributions to Mineralogy and Petrology 88, 322–327.

Williams, S.J., 1986. Geology of the Gascoyne Province, Western Australia. Western Australia Geological Survey, Report 15, 85 pp.

Williams-Jones, G., Williams-Jones, A.E., Stix, J., 1998. The nature and origin of Venusian canali. Journal of Geophysical Research 103, 8545.

Wilson, A.C. (Comp.), 1982. 1:250,000 Geological Map of Swaziland. Geological Survey and Mines Department, Mbabane, Swaziland.

Wilson, A.H., 2003. A new class of silica enriched, highly depleted komatiites in the southern Kaapvaal Craton, South Africa. Precambrian Research 127, 125–141.

Wilson, A.H., Carlson, R.W., 1989. A Sm-Nd and Pb isotope study of Archaean greenstone belts in the southern Kaapvaal Craton, South Africa. Earth and Planetary Science Letters 96 (1–2), 89.

Wilson, A.H., Riganti, A., 1998. A fractionated komatiitic basalt lava lake sequence in the Nondweni greenstone belt. In: International Volcanological Conference. IAVCEI, Cape Town, p. 70.

Wilson, A.H., Versfeld, J.A., 1992a. Magongolozi Formation. In: Johnson, M.R. (Ed.), Catalogue of South African Stratigraphic Units. South African Commission for Stratigraphy, Council for Geoscience, Pretoria, pp. 21–28.

Wilson, A.H., Versfeld, J.A., 1992b. Witkop Formation. In: Johnson, M.R. (Ed.), Catalogue of South African Stratigraphic Units. South African Commission for Stratigraphy, Council for Geoscience, Pretoria, pp. 35–37.

Wilson, A.H., Versfeld, J.A., 1994a. The early Archaean Nondweni greenstone belt, southern Kaapvaal Craton, South Africa, Part I: Stratigraphy, sedimentology, mineralization and depositional environment. Precambrian Research 67, 243–276.

Wilson, A.H., Versfeld, J.A., 1994b. The early Archaean Nondweni greenstone belt, southern Kaapvaal Craton, South Africa, Part II: Characteristics of the volcanic rocks and constraints on magma genesis. Precambrian Research 67, 277–320.

Wilson, A.H., Versfeld, J.A., Hunter, D.R., 1989. Emplacement, crystallization and alteration of spinifex-textured komatiitic basalt flows in the Archaean Nondweni greenstone belt, southern Kaapvaal Craton, South Africa. Contributions to Mineralogy ad Petrology 101, 301–317.

Wilson, A.H., Shirey, S.B., Carlson, R.W., 2003. Archaean ultra-depleted komatiites formed by hydrous melting of cratonic mantle. Nature 423, 858–861.

Wilson, J.F., 1981. The granite-greenstone shield, Zimbabwe. In: Hunter, D.R. (Ed.), Precambrian of the Southern Hemisphere, Elsevier, Amsterdam, pp. 454–488.

Wilson, J.T., 1963. Continental drift. Scientific American 208, 86–100.

Wilson, J.T., 1988. Convection tectonics: some possible effects upon the Earth's surface of flow from the deep mantle. Canadian Journal of Earth Sciences 25, 1199–1208.

Wilson, M., Downes, H., 1991. Tertiary-Quaternary extension-related alkaline magmatism in Western and Central Europe. Journal of Petrology 32, 811–849.

Wilson, W.E., Murthy, V.R., 1976. Rb-Sr geochronology and trace element geochemistry of granulite facies rocks near Granite Falls, in the Minnesota River Valley. In: Proceedings and Abstracts, Institute on Lake Superior Geology, Annual Meeting no. 22, p. 69.

Windley, B.F., 1995. The Evolving Continents, third ed. John Wiley and Sons, Chichester, 526 pp.

Windley, B.F., Bridgwater, D., 1971. The evolution of Archaean low- and high-grade terrains. In: Geological Society of Australia, Special Publication 3, pp. 33–46.

Wingate, M.T.D., Pirajno, F., Morris, P.A., 2004. Warakurna large igneous province: a new Mesoproterozoic large igneous province in west-central Australia. Geology 32, 105–108.

Wingate, M.T.D., Morris, P.A., Pirajno, F., Pidgeon, R.T., 2005. Two large igneous provinces in Late Mesoproterozoic Australia. In: Wingate, M.T.D., Pisarevsky, S.A. (Eds.), Supercontinents and Earth Evolution Symposium. Geological Society of Australia, Abstracts, vol. 81, p. 151.

Winther, T.K., 1996. An experimentally based model for the origin of tonalitic and trondhjemitic melts. Chemical Geology 127, 43–59.

Winther, T.K., Newton, R.C., 1991. Experimental melting of an hydrous low-K tholeiite: evidence on the origin of Archaean cratons. Bulletin of the Geological Society of Denmark 39.

Woese, C.R., 2000. Interpreting the universal phylogenetic tree. Proceedings of the National Academy of Sciences of the United States of America 97, 8392–8396.

Woese, C.R., Fox, G.E., 1977. Phylogenetic structure of Prokaryotic Domain – primary kingdoms. Proceedings of the National Academy of Sciences of the United States of America 74, 5088–5090.

Wolf, R., Anders, E., 1980. Moon and Earth: Compositional differences inferred from siderophiles, volatiles and alkalies in basalts. Geochimica et Cosmochimica Acta 44, 2111–2124.

Wood, B.E., Müller, H.-R., Zank, G.P., Linsky, J.L., 2002. Measured mass-loss rates of solar-like stars as a function of age and activity. Astrophysical Journal 574, 412–425.

Wooden, J.L., Mueller, P.A., 1988. Pb, Sr, and Nd isotopic compositions of a suite of Late Archean igneous rocks, eastern Beartooth Mountains: implications for crust-mantle evolution. Earth and Planetary Science Letters 87, 59–72.

Wooden, J.L., Goldich, S.S., Suhr, N.H., 1980. Origin of the Morton Gneiss, southwestern Minnesota: Part 2. Geochemistry. In: Geological Society of America, Special Paper 182, pp. 57–76.

Wooden, J.L., Mueller, P.A., Mogk, D., 1988a. A review of the geochemistry and geochronology of the Archean rocks of the northern part of the Wyoming Province. In: Ernst, W. (Ed.), Metamorphism and Crustal Evolution in the Western US. Prentice Hall, Englewood Cliffs, NJ, pp. 383–410.

Wooden, J.L., Mueller, P.A., Mogk, D., Bowes, D.R., 1988b. A review of the geochemistry and geochronology of the Archean rocks of the Beartooth Mountains. In: Montana Bureau of Mines and Geology, Special Publication 96, pp. 23–42.

Woosley, S.E., Weaver, T.A., 1986. The physics of supernova explosions. Annual Review of Astronomy and Astrophysics 24, 205–253.

World Impact List. Geological Survey of Canada and University of New Brunswick.

Wronkiewicz, D.J., Condie, K.C., 1987. Geochemistry of Archean shales from the Witwatersrand Supergroup, South Africa: Source-area weathering and provenance. Geochimica et Cosmochimica Acta 51, 2401–2416.

Wu, C.M., Zhang, J., Ren, L.D., 2004. Empirical garnet-biotite-plagioclase-quartz (GBPQ) geobarometry in medium- to high-grade metapelites. Journal of Petrology 45, 1907–1921.

Wu, F-Y., Sun, D-Y., Li, H-M., Jahn, B-M., Wilde, S.A., 2002. A-type granites in northeastern China: Age and geochemical constraints on their petrogenesis. Chemical Geology 187, 143–173.

Wu, F.Y., Zhao, G.C., Wilde, S.A., Sun, D.Y., 2005a. Nd isotopic constraints on crustal formation in the North China Craton. Journal of Asian Earth Sciences 24, 523–545.

Wu, F.Y., Yang, J.H., Liu, X.M., Li, T.S., Xie, L.W., Yang, Y.H., 2005b. Hf isotopes of the 3.8 Ga zircons in eastern Hebei Province, China: Implications for early crustal evolution of the NCC. Chinese Science Bulletin 50, 2473–2480.

Wu, J.S., Geng, Y.S., Shen, Q.H., Liu, D.Y., Li, Z.L., Zhao, D.M., 1991. Important Geological Events in the Early Precambrian North China Platform. Geological Publishing House, Beijing, 115 pp. (in Chinese).

Wu, J.S., Geng, Y.S., Shen, Q.H., Wan, Y.S., Liu, D.Y., Song, B., 1998. Archaean Geology Characteristics and Tectonic Evolution of China–Korea Paleo-Continent. Geological Publishing House, Beijing, 211 pp. (in Chinese).

Wünnemann, K., Ivanov, B.A., 2003. Numerical modelling of impact crater depth-diameter dependence in an acoustically fluidised target. Planetary and Space Science 51, 831–845.

Wuth, M., 1980. The geology and mineralization potential of the Oorschot-Weltevreden schist belt south-west of Barberton – Eastern Transvaal. M.Sc. thesis. University of the Witwatersrand, Johannesburg, 185 pp.

Wyatt, B.A., Mitchell, M., White, B., Shee, S.R., Griffin, W.L., Tomlinson, N., 2002. The Brockman Creek kimberlite, east Pilbara, Australia. In: Extended Abstracts of the 4th International Archean Symposium, AGSO – Geoscience Australia Record 2001/37, pp. 208–211.

Wyche, S., 2003. Menzies, W.A. Western Australia Geological Survey, 1:250,000 Geological Series Explanatory Notes, second ed., 38 pp.

Wyche, S., Nelson, D.R., Riganti, A., 2004. 4350–3130 Ma detrital zircons in the Southern Cross granite-greenstone terrane, Western Australia: implications for the early evolution of the Yilgarn Craton. Australian Journal of Earth Sciences 51, 31–45.

Wyllie, P.J., 1971. The Dynamic Earth. John Wiley & Sons, Inc., New York, 416 pp.

Wyllie, P.J., 1977. Effects of H_2O and CO_2 on magma generation in the crust and the mantle. Journal of the Geological Society London 134, 215–234.

Wyllie, P.J., Wolf, M.B., van der Laan, S.R., 1997. Conditions for formation of tonalites and trondhjemites: magmatic sources and products. In: de Wit, M.J., Ashwahl, L.D. (Eds.), Tectonic Evolution of Greenstone Belts. Oxford University Press, pp. 256–266.

Wyman, D.A., Kerrich, R., 2002. Formation of Archean continental lithospheric roots: the role of mantle plumes. Geology 30, 543–546.

Wyman, D.A., Kerrich, R., Polat, A., 2002. Assembly of Archean cratonic mantle lithosphere and crust: plume-arc interaction in the Abitibi-Wawa subduction-accretion complex. Precambrian Research 115, 37–62.

Xia, L.-Q., Xu, X.-Y., Xia, Z.-C., Li, X.-M., Ma, Z.-P., Wang, L.-S., 2003. Carboniferous post-collisional rift volcanism of the Tianshan mountains, northwestern China. Acta Geologica Sinica 77, 338–360.

Xia, L-Q., Xu, X-Y., Xia, Z-C., Li, X-M., Ma, Z-P., Wang, L-S., 2004. Petrogenesis of Carboniferous rift-related volcanic rocks in the Tianshan, northwestern China. Geological Society of America Bulletin 116, 419–433.

Xiao, L., Xu, Y.G., Mei, H.J., Zheng, Y.F., He, B., Pirajno, F., 2004. Distinct mantle sources of low-Ti and high-Ti basalts from the western Emeishan large igneous province, SW China: implications for plume-lithosphere interaction. Earth and Planetary Science Letters 228, 525–546.

Xie, X., Byerly, G.R., Ferrell, R.E., 1997. IIb trioctahedral chlorite from the Barberton greenstone belt: crystal structure and rock composition constraints with implications for geothermometry. Contributions Mineralogy and Petrology 126, 275–291.

Xiong, J., Fischer, W.M., Inoue, K., Nakahara, M., Bauer, C.E., 2000. Molecular evidence for the early evolution of photosynthesis. Science 289, 1724–1730.

Xiong, X.L., Adam, J., Green, T.H., 2005. Rutile stability and rutile/melt HFSE partitioning during partial melting of hydrous basalt: implications for TTG genesis. Chemical Geology 218, 339–359.

Xu, Y-G., 2001. Thermo-tectonic destruction of the Archaean lithospheric keel beneath the Sino-Korean Craton in China: evidence, timing and mechanism. Physics and Chemistry of the Earth 26, 747–757.

Xu, W.L., Wang, Q.H., Liu, X.C., Wang, D.Y., Guo, J.H., 2004a. Chronology and sources of Mesozoic intrusive complexes in the Xuzhou-Huainan region, central China, constraints from SHRIMP zircon U-Pb dating. Acta Geologica Sinica 78, 96–106.

Xu, W.L., Wang, Q.H., Wang, D.S., Pie, F.P., Gao, S., 2004b. Processes and mechanism of Mesozoic lithospheric thinning in eastern NCC: Evidence from Mesozoic igneous rocks and deep-seated xenoliths. Earth Science Frontiers 11, 309–317 (in Chinese with English abstract).

Yamaguchi, A., Taylor, G.J., Keil, K., 1996. Global crustal metamorphism of the eucrite parent body. Icarus 124, 97–112.

Yamaguchi, A., Taylor, G.J., Keil, K., 1997. Metamorphic history of the eucrite crust of 4 Vesta. Journal of Geophysical Research 102, 13,381–13,386.

Yang, S.-F., Chen, H.-L., Li, Z.-L., Dong, C.-W., Jia, C.-Z., Wei, G.-Q., 2006. Early to Middle Permian magmatism in the Tarim Basin. In: International Conference on Continental Volcanism, IAVCEI 2006, Guangzhou, China, Abstracts & Program, p. 54.

Yanshin, A.L., Borukaev, Ch.B. (Eds.), 1988. Tectonics and Evolution of the Earth's Crust in Siberia. In: Transactions of the Institute of Geology and Geophysics, vol. 173. Nauka, Novosibirsk, 175 pp. (in Russian).

Yashchenko, N.Y., Shekhotikin, V.V., 2000. New data on the tectonomagmatic history of the Ukrainian Shield (Ingul-Ingulets region). Litasfera 12, 76–84.

Yaxley, G.M., Green, D.H., 1998. Reactions between eclogite and peridotite: mantle refertilisation by subduction of oceanic crust. Schweizerische Mineralogische und Petrographische Mitteilungen 78, 243–255.

Yearron, L.M., 2003. Archaean granite petrogenesis and implications for the evolution of the Barberton mountain land, South Africa. Ph.D. thesis. Kingston University, Kingston, UK, 315 pp.

Yin, Q-C., 2005. From dust to planets: The tale told by moderately volatile elements in Chondrites and the Protoplanetary Disk. In: Astronomical Society of the Pacific, Conference Series, vol. 341, pp. 632–644.

Yoder, C.F., 1997. Venusian spin dynamics. In: Bouger, S.W., Hunten, D.M., Phillips, R.J. (Eds.), Venus II. University of Arizona Press, Tucson, AZ, pp. 1087–1124.

Yokochi, R., Marty, B., 2005. Geochemical constraints on mantle dynamics in the Hadean. Earth and Plateary Science Letters 238, 17–30.

York, D., Layer, P.W., Lopez Martinez, M., Kröner, A., 1989. Thermal histories from the Barberton greenstone belt, southern Africa, In: International Geological Congress, Washington, DC, p. 413.

Young, C., 1997. Footwall alteration at the Sulphur Springs Zn-Cu volcanic-hosted massive sulfide (VHMS) prospect, east Pilbara, Western Australia. Unpublished B.Sc. Honours thesis. University of Western Australia, Perth, 135 pp.

Young, D.A., Hansen, V.L., 2005. Poludnista Dorsa, Venus: History and context of a deformation belt. Journal of Geophysical Research 110, doi: 10.1029/2004JE002280.

Young, D.L., 1987. A kinematic interpretation of ductile deformation and shear-induced quartz c-axis fabric development in the Montevideo Gneiss near Granite Falls, S.W. Minnesota. M.Sc. thesis. University of Missouri, Columbia, 201 pp.

Young, D.N., Black, L.P., 1991. U-Pb zircon dating of Proterozoic igneous charnockites from the Mawson coast, east Antarctica. Antarctic Science 3, 205–216.

Young, E.D., Glay, A., Nagahara, H., 2002. Kinetic and equilibrium mass-dependent isotope fractionation laws in nature and their geochemical and cosmochemical significance. Geochimica et Cosmochimica Acta 66, 1095–1104.

Young, E.D., Simon, J.I., Galy, A., Russell, S.S., Tonui, E., Lovera, O., 2005. Supracanonical $^{26}Al/^{27}Al$ and the residence time of CAIs in the solar protoplanetary disk. Science 308, 223–227.

Young, G.M., von Brun, V., Gold, D.J.C., Minter, W.E.L., 1998. Earth's oldest reported glaciation: Physical and chemical evidence from the Archaean Mozzan Group (~2.9 Ga) of South Africa. Journal of Geology 106, 523–538.

Young, M.D., McNicoll, V., Helmstaedt, H., Skulski, T., Percival, J.A., 2006. Pickle Lake revisited: new structural, geochronological and geochemical constraints on greenstone belt assembly, western Superior Province, Canada. Canadian Journal of Earth Sciences 43, 821–847.

Zachariah, J.K., 1998. A 3.1 billion year old marble and the $^{87}Sr/^{86}Sr$ of late Archean seawater. Terra Nova 10, 312–316.

Zahnle, K.J., 2006. Earth's earliest atmosphere. Elements 2, 217–222.

Zamora, D., 2000. Fusion de la croûte océanique subductée : approche expérimentale et géochimique. Ph.D. thesis. Université Blaise-Pascal, Clermont-Ferrand, France, 314 pp.

Zandt, G., Ammon, C.J., 1995. Continental crust composition constrained by measurements of crustal Poisson's ratio. Nature 374, 152–154.

Zartman, R.E., Doe, B.R., 1981. Plumbotectonics – the model. Tectonophysics 75, 135–162.

Zartman, R.E., Jagoutz, E., Bowring, S.A., 2006. Pb-Pb dating of the D'Orbigny and Asuka 881371 angrites and a second absolute time calibration of the Mn-Cr chronometer. Lunar and Planetary Science XXXVII, abstract 1580.

Zeh, A., Klemd, R., Buhlmann, S., Barton, J.M., Jr, 2004. Pro- and retrograde P-T evolution of granulites of the Beit Bridge Complex (Limpopo Belt, South Africa): constraints from quantitative phase diagrams and geotectonic implications. Journal of Metamorphic Geology 22, 79–95.

Zegers, T.E., 1996. Structural, kinematic and metallogenic evolution of selected domains of the Pilbara Granitoid-Greenstone Terrain. Geologica Ultraiectina, Mededelingen van de Faculteit Aardwetenschappen, Universiteit Utrecht, No. 146, 208 pp.

Zegers, T.E., van Keken, P.E., 2001. Middle Archean continent formation by crustal delamination. Geology 29, 1083–1086.

Zegers, T.E., White, S.H., de Keijzer, M., Dirks, P., 1996. Extensional structures during deposition of the 3460 Ma Warrawoona Group in the eastern Pilbara Craton, Western Australia. Precambrian Research 80, 89–105.

Zegers, T.E., de Keijzer, M., Passchier, C.W., White, S.H., 1998a. The Mulgandinnah shear zone; an Archean crustal-scale strike-slip zone, eastern Pilbara, Western Australia. Precambrian Research 88, 233–248.

Zegers, T.E., de Wit, M.J., Dann, J., White, S.H., 1998b. Vaalbara, the Earth's oldest assembled continent? A combined structural, geochronological, and paleomagnetic test. Terra Nova 10, 250–259.

Zegers, T.E., Wijbrans, J.R., White, S.H., 1999. ^{40}Ar/^{39}Ar age constraints on tectonothermal events in the Shaw area of the eastern Pilbara granite-greenstone terrain (W. Australia): 700 Ma of Archaean tectonic evolution. Tectonophysics 311, 45–81.

Zegers, T.E., Nelson, D.R., Wijbrans, J.R., White, S.H., 2001. SHRIMP U-Pb zircon dating of Archean core complex formation and pancratonic strike-slip deformation in the East Pilbara Granite-Greenstone Terrain. Tectonics 20, 883–908.

Zegers, T.E., Barley, M.E., Groves, D.I., McNaughton, N.J., White, S.H., 2002. Oldest gold: deformation and hydrothermal alteration in the early Archean shear-zone hosted Bamboo Creek Deposit, Pilbara, Western Australia. Economic Geology 97, 757–773.

Zhai, M., Kampunzu, A.B., Modisi, M.P., Bagai, Z., 2006. Sr and Nd isotope systematics of Francistown plutonic rocks, Botswana: implications for Neoarchaean crustal evolution of the Zimbabwe craton. International Journal of Earth Sciences 95 (3), 355–369.

Zhang, S.B., Zheng, Y.F., Wu, Y.B., Zhao, Z.F., Gao, S., Wu, F.Y., 2006. Zircon isotope evidence for $\geqslant$3.5 Ga continental crust in the Yangtze craton of China. Precambrian Research 146 (1–2), 16–34.

Zhang, Z-C., Mahoney, J.J., Mao, J-W., Wang, F-S., 2006. Geochemistry of picritic and associated basalt flows of the Western Emeishan Flood Basalt province, China. Journal of Petrology, doi: 10.1093/petrology/eg1034.

Zhao, Z.P., 1993. Precambrian Crustal Evolution of Sino-Korea Platform. Science Press, Beijing, 444 pp.

Zhao, G.C., Cawood, P.A., Wilde, S.A., Sun, M., Lu, L.Z., 2000. Metamorphism of basement rocks in the central zone of the NCC: implications for Palaeoproterozoic tectonic evolution. Precambrian Research 103, 55–88.

Zhao, G.C., Cawood, P.A., Wilde, S.A., Lu, L.Z., 2001a. High-pressure granulites (retrograded eclogites) from the Hengshan complex, NCC: petrology and tectonic implications. Journal of Petrology 42, 1141–1170.

Zhao, G.C., Wilde, S.A., Cawood, P.A., Sun, M., 2001b. Archaean blocks and their boundaries in the NCC: Lithological geochemical, structural and P-T path constraints and tectonic evolution. Precambrian Research 107, 45–73.

Zhao, G.C., Wilde, S.A., Cawood, P.A., Sun, M., 2002. SHRIMP U-Pb zircon ages of the Fuping complex: implications for Late Archaean to Palaeoproterozoic accretion and assembly of the NCC. American Journal of Science 302, 191–226.

Zhao, G.C., Sun, M., Wilde, S.A., 2003. Major tectonic units of the NCC and their Palaeoproterozoic assembly. Science in China Series D 46, 23–38.

Zhao, G.C., Sun, M., Wilde, S.A., Li, S.Z., 2005. Late Archaean to Palaeoproterozoic evolution of the NCC: key issues revisited. Precambrian Research 136, 177–202.

Zhao, G.C., Wilde, S.A., Li, S.Z., Sun, M., Grant, M.L., Li, X.P., 2007. The Dongwanzi ultramafic-mafic intrusion (North China) is not an Archean ophiolite. Earth and Planetary Science Letters 255, 85–93.

Zhao, J-X., McCulloch, M.T., Korsch, R.J., 1994. Characterisation of a plume-related ~800 Ma magmatic event, and its implications for basin formation in central-southern Australia. Earth and Planetary Science Letters 21, 349–367.

Zhao, T.P., Zhai, M.G., Xia, B., Li, H.M., Zhang, Y.X., Wan, Y.S., 2004. SHRIMP dating of zircon from volcanic rocks of the Xiong'er Group: Constraints on the beginning of cover development in the NCC. Chinese Science Bulletin 49, 2342–2349.

Zheng, J.P., Griffin, W.L., O'Reilly, S.Y., Lu, F.X., Wang, C.Y., Zhang, M., Wang, F.Z., Li, H.M., 2004a. 3.6 Ga lower crust in central China: New evidence on the assembly of the NCC. Geology 32, 229–232.

Zheng, J.P., Griffin, W.L., O'Reilly, S.Y., Lu, F.X., Yu, C.M., Zhang, M., Li, H.M., 2004b. U-Pb and Hf-isotope analysis of zircons in mafic xenoliths from Fuxian kimberlites: evolution of the lower crust beneath the NCC. Contributions to Mineralogy and Petrology 148, 79–103.

Zheng, J.P., Griffin, W.L., O'Reilly, S.Y., Zhang, M., Pearson, N.J., Luo, Z., 2006a. The lithospheric mantle beneath the southern Tianshan area, NW China. Contributions to Mineralogy and Petrology 151, 457–479.

Zheng, J.P., Griffin, W.L., O'Reilly, S.Y., Zhang, M., Pan, Y., Pearson, N.J., Lin, G., 2006b. Widespread Archean basement beneath the Yangtze Craton. Geology 34, 417–420.

Zhou, T-F., Yuan, F., Tan, L-G., Fan, Y., Yue, S-C., 2006. Geodynamic significance of the A-type granites in the Sawuer region in west Junggar, Xinjiang: rock geochemistry and SHRIMP zircon age evidence. Science in China Ser. D 49, 113–123.

Zhu, Y-F., Sun, S-H., Gu, L-B., Ogasawara, Y., Jiang, N., Honma, H., 2001. Permian volcanism in the Mongolian orogenic zone, northeast China: geochemistry, magma source and petrogenesis. Geological Magazine 138, 101–115.

Zhuravlev, D.Z., Rosen, O.M., 1991. Sm-Nd age of metasediments among granulites of the Anabar shield. Doklady Earth Sciences 317 (1), 189–193.

Zhuravlev, D.Z., Puchtel, I.S., Samsonov, A.V., 1989. Sm-Nd age and geochemistry of the Olondo greenstone belt volcanics. Izvestiya Academy of Sciences USSR, Geology Ser. (2), 39–49.

Zierenberg, R.A., Schiffman, P., Jonasson, I.R., Tosdal, R., Pickthorn, W., McClain, J., 1995. Alteration of basalt hyaloclastite at the off-axis Sea Cliff hydrothermal field, Gorda Ridge. Chemical Geology 126, 77–99.

Zimbelman, J.R., 2001. Image resolution and evaluation of genetic hypotheses for planetary landscapes. Geomorphology 37, 179–199.

Zindler, A., Hart, R.A., 1986. Chemical geodynamics. Annual Review of Earth and Planetary Sciences 14, 493–571.

Zlobin, V.L., Zhuravlev, D.Z., Rosen, O.M., 1999. Sm-Nd-model age of metacarbonate-gneiss formation of granulite complex in the western Anabar shield, Polar Siberia. Doklady Earth Sciences 368 (1), 95–98.

Zmolek, P., Xu, X.P., Jackson, T., Thiemens, M.H., Trogler, W.C., 1999. Large mass independent sulfur isotope fractionations during the photopolymerization of (CS_2)-^{12}C and (CS_2)-^{13}C. Journal of Physical Chemistry A 103, 2477–2480.

Zolensky, M., Bodnar, R.J., Gibson, E.K., Jr, Nyquist, L.E., Reese, Y., Shih, C-Y., Wiesmann, H., 1999. Asteroidal water within fluid inclusion-bearing halite in an H5 chondrite, Monahans (1998). Science 285, 1377–1379.

Zonenshain, L.P., Kuzmin, V.I., Natapov, L.M., 1989. Geology of the USSR, a Plate Tectonic Synthesis. In: American Geophysical Union, Geodynamics Series, vol. 21. Washington, DC, 242 pp.

Zuber, M.T., 1987. Constraints on the lithospheric structure of Venus from mechanical models and tectonic surface features. Journal of Geophysical Research 92, E541–E551.

Zuber, M.T., 1990. Ridge belts: Evidence for regional- and local-scale deformation on the surface of Venus. Geophysical Research Letters 17, 1369–1372.

Zwanzig, H.V., 1990. Kisseynew gneiss belt in Manitoba: stratigraphy, structure, and tectonic evolution. In: Lewry, J.F., Stauffer, M.R. (Eds.), The Early Proterozoic Trans-Hudson Orogen of North America. Geological Association of Canada, Special Paper 37, pp. 95–120.

Zwanzig, H.V., 2005. Geochemistry, Sm-Nd isotope data and age constraints of the Bah Lake assemblage, Thompson Nickel Belt and Kisseynew Domain margin: relation to Thompson-type ultramafic bodies and a tectonic model. In: Report of Activities 2005, Manitoba Industry, Economic Development and Mines, Manitoba Geological Survey, pp. 40–53.

Zwanzig, H.V., Böhm, C.O., 2002. Tectonostratigraphy, Sm-Nd isotope and U-Pb age data of the Thompson Nickel Belt and Kisseynew north and east margins (NTS 63J, 63P, 63Q, 64A, 64B). In: Report of Activities 2004, Manitoba Industry, Trade and Mines, Manitoba Geological Survey, pp. 102–114.

Zwanzig, H.V., Böhm, C.O., 2004. Northern extension of the Thompson Nickel Belt, Manitoba (NTS 64A3 and 4). In: Report of Activities 2004, Manitoba Industry, Economic Development and Mines, Manitoba Geological Survey, pp. 115–119.

Zwanzig, H.V., Böhm, C.O., Etcheverry, J., 2001. Superior Boundary Zone – Reindeer Zone transition in the Pearson Lake – Odei River – Mystery Lake region (parts of NTS 63P and 64O). In: Report of Activities 2001, Manitoba Industry, Trade and Mines, Manitoba Geological Survey, pp. 51–56.

SUBJECT INDEX